Handbook of Optical Systems
Edited by Herbert Gross

Volume 5:
Metrology of Optical
Components and Systems

Handbook of Optical Systems

Edited by
Herbert Gross

Volume 1: Fundamentals of Technical Optics
Volume 2: Physical Image Formation
Volume 3: Aberration Theory and Correction of Optical Systems
Volume 4: Survey of Optical Instruments
Volume 5: Metrology of Optical Components and Systems
Volume 6: Advanced Physical Optics

WILEY-VCH Verlag GmbH & Co. KGaA

Handbook of Optical Systems

Edited by
Herbert Gross

Volume 5: Metrology of Optical Components and Systems
Bernd Dörband, Henriette Müller, and Herbert Gross

WILEY-VCH Verlag GmbH & Co. KGaA

Bernd Dörband
Carl Zeiss SMT GmbH, Oberkochen, Germany
doerband.mueller@gmx.de

Henriette Müller
Carl Zeiss SMT GmbH, Oberkochen, Germany

Herbert Gross
Carl Zeiss AG, Oberkochen, Germany

Library of Congress Card No.:
applied for

British Library Cataloguing-in-Publication Data
A catalogue record for this book is available from the British Library.

Bibliographic information published by the Deutsche Nationalbibliothek
The Deutsche Nationalbibliothek lists this publication in the Deutsche Nationalbibliografie; detailed bibliographic data is available on the Internet at <http://dnb.d-nb.de>.

Printed in the Federal Republic of Germany.

Printed on acid-free paper.

Cover Design aktivComm GmbH, Weinheim
Composition Kühn & Weyh, Satz und Medien, Freiburg
Printing and Binding betz-druck GmbH, Darmstadt

ISBN 978-3-527-40381-3 (Vol. 5)
ISBN 978-3-527-40382-0 (Set)

Bernd Dörband
Bernd Dörband was born in 1949. He was educated as an optometrist and studied physics at the Technical University in Berlin, Germany. In 1986, he received his Ph.D. from the University of Stuttgart for a thesis on "Analysis of Optical Systems by Means of Optical Metrology and Mathematical Simulation". Since 1986 he has been working with the Carl Zeiss, Oberkochen, Germany, as a specialist for optical metrology. His areas of interest are the design and the development of optical test setups for different kinds of applications with a strong focus on interferometry and surface form measurement techniques. In 1992, he started teaching as a lecturer on optical metrology at the University of Aalen. From 1995 to 2001, Dr. Dörband was head of the metrology department at the Center of Technology of the Carl Zeiss company. In 2001, he accepted a post as a principal scientist in the Lithography Optics Division of Carl Zeiss.

Henriette Müller
Henriette Müller was born in 1952. She studied physics at the University of Jena, Germany, and joined Carl Zeiss, Jena, Germany, in 1976, where she worked in the department of optical metrology. Her main fields of interest were the development of optical test equipment for different applications including infrared sensor systems and microscopic and interferometric devices. From 1994 to 2001, she worked as a scientist and project manager at the Center of Technology of the Carl Zeiss company in Oberkochen, with a focus on interferometry and surface form measurement techniques. In 2001, she started working as a senior scientist in the Lithography Optics Division of Carl Zeiss.

Herbert Gross
Herbert Gross was born in 1955. He studied physics at the University of Stuttgart, Germany, and joined the Carl Zeiss company in 1982, where he has since been working in the department of optical design. His special areas of expertise are the development of simulation methods, optical design software and algorithms, the modeling of laser systems and simulation of problems in physical optics, and the tolerance and the measurement of optical systems. From 1995 to 2009, he was head of the Central Optical Design Department at Zeiss. In 1995, he received his Ph.D. from the University of Stuttgart, Germany, on the modeling of laser beam propagation in the partial coherent region. From 2009 until 2012, he worked as a principal scientist in the Department of Optical Design and Simulation. In 2012, he accepted a professorship at the University of Jena, Germany.

Contents of Volume 1

Handbook of Optical Systems: Vol. 5. Metrology of Optical Components and Systems. First Edition.
Edited by Herbert Gross.

Contents of Volume 2

Contents of Volume 3

Contents of Volume 4

Contents

Handbook of Optical Systems: Vol. 5. Metrology of Optical Components and Systems. First Edition. Edited by Herbert Gross.

Preface

The first two volumes of this handbook series on optical systems covered the basics of technical and physical optics. The third volume deals with aberration theory, performance evaluation and the fundamental layout of systems. Furthermore, the reader was introduced to the techniques used to improve and optimize optical systems and give them the right tolerances for manufacture. In the fourth volume, a summary of the well known optical system types and their classification were reviewed. The goal of this collection is to demonstrate and explain the most important thoughts, principles and properties, which lie behind successful solutions in optical design.

These topics provide the reader with the main framework for understanding the description, design and principles of optical systems.

The important link from the computer results in optical design and simulation to the practical realization is addressed in this fifth volume. If manufacturing aspects are considered, the evaluation of the theoretical predictions needs an accurate characterization and measurement of the individual components and the complete system. It is not the goal of this book series to cover all the manufacturing methods for optical components. But if the performance of a built system does not meet the required quality, it is necessary to analyze the realized data and to find out the reasons for the discrepancy. The possible reasons can be component tolerances or problems in the assembly of the complete system. Therefore this volume is concerned with the testing and metrology of optical components and systems.

Many colleagues and friends have helped to collect, prepare and correct the text and have contributed important material to this volume. We would like to thank them all and apologise if we forget to mention any one of them by name.

Matthias Dreher made substantial corrections and improvements to the chapters on interferometry and component geometry. Hans-Jochen Paul and Volker Weidenhof read the chapter on radiometry carefully and gave important hints for improvement. Frank Höller gave substantial input to the chapter on distance and angle metrology and corrected the related manuscript carefully. Michael Totzeck contributed important suggestions for the chapter on polarimetry. From Frank-Thomas Lentes we received pictures related to striae testing in the chapter on the quality of optical materials. Hannfried Zügge and Günther Seitz carefully corrected the manu-

Handbook of Optical Systems: Vol. 5. Metrology of Optical Components and Systems. First Edition.
Edited by Herbert Gross.

script on texture and imperfections of optical surfaces. Maren Büchner corrected the chapter on the quality of coatings and gave important hints related to practical applications.

At this point, we must thank Ulrike Werner from Wiley VCH for her understanding and never ending patience during the work on this volume. Without her experience, interest and competence, it would have been impossible to finish this book. Due to difficult circumstances, the time required to complete this volume was much longer than initially planned. We apologize for this delay. Also we want to acknowledge Linda Bristow for her fast and perfect language improvement, which has impoved the readability and understandability of the text.

Aalen, Essingen,
January, 2012

Bernd Dörband
Henriette Müller
Herbert Gross

Introduction

In the first volume of this book series, the basics of technical optics are presented. This provides the fundamental knowledge of optics and in particular optical physics, which is necessary to understand optical systems and instruments with their various technological aspects. The second volume deals with the physical principles of image formation from a more theoretical point of view. The topics covered in the second volume provides a deaper understanding of the principles of optical imaging. In the third volume, the special algorithms, methods and techniques are explained, which are necessary to design and optimize optical systems. This content helps in the practical work to develop optical systems efficiently with available tools. In the fourth volume, a collection of systems types is presented, which shows how the basics can be applied to concrete applications. This gives the reader an overview and simultaneously shows how broad the area is and that quite different knowhow is necessary to cover the design of optical systems completely.

The next step in a practical development of systems is the manufacturing of a prototype with a proper tolerance budget of acceptable deviations from the nominal data. The measurement of the overall system quality is an important step in order to characterize the system properties with respect to the specification required. If the performance of the system in reality does not match the theoretical predictions, it is necessary to evaluate the geometrical shape, the roughness and the coating of all the components and the complete system and to analyze the impact of these real conditions and data on the system performance. Therefore the various metrology aspects of the system verification are a consequent next step in the treatment of optical systems. Volume 5 with its theme optical systems deals with these subjects.

The interferometric measurement of optical surfaces is one of most important methods used to characterize optical components. The corresponding techniques of measurement are described in this chapter. The basic principles of interferometers are presented, starting with the basic physics of interference. The next section describes the most important types of interferometric setups, their properties, advantages and drawbacks. After a review of important design principles, the detection of the signals and the various algorithms of data evaluation are described. In the remaining section of this chapter, the techniques of calibration are explained

Handbook of Optical Systems: Vol. 5. Metrology of Optical Components and Systems. First Edition.
Edited by Herbert Gross.

and a comprehensive description about dynamic ranges, accuracy and errors can be found.

Beneath the interferometric measurement of wavefront quality, there are some further quite different possibilities for testing. The most important measurement approaches of non-interferometric wavefront sensing are described in this second chapter. The use of a Hartmann-Shack wavefront sensor or the classical Hartmann setup are discussed and their properties, accuracies and limitations are explained. Some other possible principles are the phase space analyzer, point filtering techniques of special approaches based on filter and Moiré techniques. A further prominent method is the point spread function phase retrieval, which is represented in more detail. Finally some of the major facts and algorithms to calculate Zernike polynomials out of the measurement rough data are discussed in the last section of this chapter.

In the next chapter, the most important techniques and methods about the measurement of radiometric quantities are presented. After a representation of the various definitions of radiometric quantities, the explanation of the basic principles of energy transport follows. The third section contains a comprehensive representation of the various realization possibilities of monochromators, describing absorption filters, Fabry-Perot systems, interference filters, prism and grating monochromators. The measurement of spectral properties including the description of spectrometers is given in the next sections. Finally the discussion of calibrations techniques, accuracy and sources of errors in corresponding measurements are discussed.

The direct analysis of detected images can be used to determine the quality of optical system, if well known object patterns are used for the image formation. The classical star test is one of the prominent methods of this category, which is used in microscopy for a long time. A special discussion on the measurement of distortion can also be found in the fifth section. Further techniques are deflectometry and the projection of patterns or fringes onto test surfaces. These methods, their accuracy and errors are discussed in this chapter.

In the next chapter, the basics of measuring distances and angles are explained. Methods for long range distances and displacements as low-coherence interferometry, linear encoders or frequency combs are discussed. Triangular principles and confocal sensors for shorter values of spatial dimensions are treated. For the quantitative evaluation of tilt and angle metrology the correspondent techniques like angle encoder, autocollimation setups and heterodyne interferometers are described. A special section was added which discusses the surface profile measurement techniques, which are essential for the metrology of optical systems.

In the next chapter, the measurement of polarization properties are presented in a comprehensive way. First the basic principles of polarimetry are given, starting with the Jones matrix formalism and the Stokes-Mueller approach respectively. The major polarizing components and their model descriptions are then presented in detail. In the fourth section, the measurement techniques are explained for the vectors and matrices of the Jones type as well as for Stokes vectors and Mueller matrices. The setup of polarimeters and ellipsometers are described and finally the calibration, accuracy and error sources of the corresponding methods are discussed.

The next chapter deals with the testing of the quality of the materials in optical systems. Here only optical properties of transparent materials are considered commencing with the basic options of specifications where refractive index is considered the most important quantity. In the third section, the transmittance of materials and its measurement is discussed. Special questions such as the testing of inhomogeneity, striae, birefringence, bubbles and inclusions in optical materials are further topics of this chapter.

The geometry of optical components is one of the most important properties that must be tested in optical production. The basic properties of lenses for example are the radii of curvature, the center thickness and the diameter. More sophisticated types of optical surfaces require a closer look at the shape of the surface, which may be non-spherical and therefore can not be described by one radius only. The relative positioning and in particular the centering errors of optical components must be tested too and the corresponding techniques for measurement are represented in this chapter.

The quality of optical surfaces and the measurement of the corresponding properties are the main topics of the next chapter. First the usual specifications are explained and then the description and the metrology of surface texture and imperfections are the subjects examined.

A special aspect of surface quality is the influence of the coating. After an introduction into the basic principles and specifications of coatings, the understanding of the properties of coatings and their simulation models are presented. Special topics such as graded interfaces, the influence of surface roughness and the role of material parameters and dispersion effects are discussed. Finally, the metrology techniques to characterize and measure the properties of coatings are detailed in this chapter.

If a complete system is to be tested, all the individual components and functions need to work together. The functionality of the whole system then must be checked against the specification. There are basic parameters of the system that must be tested, for example focal length, magnification, aperture size and magnification. In addition, the quality of the system must fulfill the requirements of the application. The system performance can be described in quite different quantities, which must be measured. This can be the wavefront quality the point spread function of the modulation transfer function. The principles of testing some further special properties such as transmission, illumination, glare, M^2 beam quality of polarization aberrations are considered in the last section of this chapter. The test procedures of all these properties are described in this chapter, whereas the basic principles are referred to in the previous parts of this book.

46
Interferometry

Handbook of Optical Systems: Vol. 5. Metrology of Optical Components and Systems. First Edition.
Edited by Herbert Gross.

46.1 Introduction

In the fabrication of optical elements the polished surfaces need to be of suitable quality before they are coated, mounted and assembled. The preparation of optical elements includes surface inspections as well as material tests which usually precede the element fabrication procedures. Surface inspection includes checks for surface imperfections and roughness, but the geometrical shape is of the greatest interest. Deviations from a perfect shape are specified by tolerancing certain categories of deviations such as curvature or astigmatic deviation or by tolerancing spatial frequency bands of deviation such as figure, mid-spatial (MSFR) or high-spatial frequencies (HSFR).

The tolerances for optical elements are in the order of μm for simple, low-cost optics and are as low as a few nm for optical elements of lithography systems. Taking the wavelength λ of visible light as the scale, tolerances are specified in the range of a few wavelengths down to a small fraction such as 1/100 or even 1/1000 of a wavelength.

Interference has been proved to be an appropriate tool for measuring any deviations of optical surfaces from a perfect shape.

The performance of optical systems is often tolerated according to the shape of the transmitted or reflected wavefront. When the Strehl Ratio or Maréchal criterion ($RMS < \lambda/14$) are considered then small fractions of the specified wavelength are also a measure of quality [46-135]. Thus interferometry is also a useful tool to check the quality and performance of optical systems.

In most practical applications, two-beam interferometry is used for wavefront inspections. The principle of two-beam interferometry is always the same although the optical arrangements may differ greatly. A light source Q is imaged by an optical object path leading to an image Q_{Obj} and is also imaged by an optical reference path leading to an image Q_{Ref}.

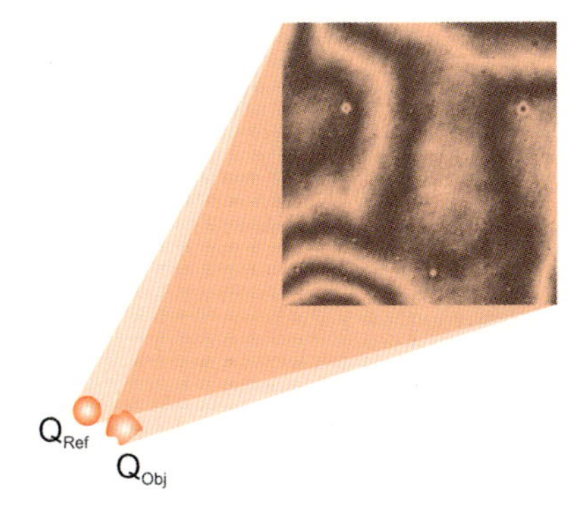

Figure 46-1: Images Q_{Ref} and Q_{Obj} of an extended light source Q acting as two separate coherent light sources. Their light is superimposed and forms an interferogram on a screen.

The object path includes the surface or the system under test. The light emerging from Q_{Ref} is superimposed with that emerging from Q_{Obj} thus leading to an interference fringe pattern if the coherence conditions are fulfilled (see section 46.2.8). The interference fringe pattern is called an interferogram. It contains information about the system or surface errors under test. Since interferometry only delivers phase differences the interferogram can only show how much the surface or system under test deviates from the reference surface or system. Thus a good knowledge of the quality of the reference system is necessary. An interferometric setup can be calibrated to some degree by using measurement results from different positions of the test piece or a reference test piece.

Before the application of modern digital interferometry, interference fringes were judged visually by their "straightness" [46-1]. For a perfect system or surface under test the fringes need to be straight, parallel and equally spaced. Deviations from these conditions can be estimated to the order of a tenth of a fringe. When green light is used the resolution for surface deviations is therefore around 25 nm, due to the fact that the wavefront deviations are twice the surface deviations. Today's digital interferometers are able to resolve wavefronts to much better than 0.1 nm without the need to use shorter wavelengths than the visible spectrum offers. Using sophisticated interferogram analysis algorithms in conjunction with digital cameras and rapid data acquisition techniques, there is a whole range of instruments enabling us to detect wavefront deviations down to 20 pm.

In the following section the conditions and limitations for two-beam interferometry are reported. The basic principles cover the mathematical descriptions of plane and spherical waves as well as their superposition. Since there is no ideal monochromatic point source the necessary coherence conditions for interferometry are explained. An overview on various interferometer principles and optical designs, on detection techniques and algorithms to analyze interferometric fringe patterns will be followed by a description of calibration techniques and accuracy aspects.

46.2 Basic Principles of Interference

In a homogeneous and non-conducting media, the electromagnetic wave equations, derived from Maxwell's equations [46-2], are given by

$$\Delta\vec{E} = \frac{n}{c}\frac{\partial\vec{E}}{\partial t} \tag{46-1a}$$

$$\Delta\vec{B} = \frac{n}{c}\frac{\partial\vec{B}}{\partial t} \tag{46-1b}$$

where

$\vec{E} = \begin{pmatrix} E_x \\ E_y \\ E_z \end{pmatrix}$ is the electrical field strength,

$\vec{B} = \begin{pmatrix} B_x \\ B_y \\ B_z \end{pmatrix}$ is the magnetic induction,

n is the refractive index of the homogeneous, non-conducting medium,
c is the velocity of light in vacuum.
The latter is defined as

$$c = \frac{1}{\sqrt{\varepsilon_0 \mu_0}} = 2.99792458 \times 10^8 \text{ms}^{-1} \tag{46-2}$$

Δ is the Laplace operator symbolizing the operation

$$\Delta = \frac{\partial}{\partial x} + \frac{\partial}{\partial y} + \frac{\partial}{\partial z} \text{ in a cartesian coordinate system or}$$

$$\Delta = \frac{1}{r} \cdot \frac{\partial}{\partial r}\left(r \frac{\partial}{\partial r} \right) + \frac{1}{r \sin\theta} \frac{\partial}{\partial \theta}\left(\sin\theta \frac{\partial}{\partial \theta} \right) \text{ in a polar coordinate system.}$$

ε_0 is the dielectric constant of the vacuum where

$$\varepsilon_0 = 8.85... \times 10^{-10} \frac{As}{Vm}, \tag{46-3}$$

μ_0 is the magnetic permeability of the vacuum where

$$\mu_0 = 4\pi \times 10^{-7} \frac{Vs}{Am}. \tag{46-4}$$

In linear, isotropic and homogeneous media the electrical field strength is coupled to the magnetic field strength. In the following it will therefore be sufficient to restrict the discussions to the electrical field strength [46-2], [46-5], [46-6].

In homogeneous media the vector operation (46-1a) can be separated into three different equations defining the field components in x, y and z. When polarization is of importance all three equations have to be solved. We then speak of vector theory. If polarization does not matter, one equation will be sufficient. We then speak of scalar theory.

46.2.1 Plane Wave

One of the solutions of (46-1a) is the monochromatic harmonic plane wave

$$E(\vec{r}, t) = Ae^{i(\vec{k}\cdot\vec{r}\mp\omega t+\varphi)} \tag{46-5}$$

where
E is the scalar electrical field,,
A is the scalar amplitude of the electrical field,

$$\vec{r} = \begin{pmatrix} x \\ y \\ z \end{pmatrix}$$

is the coordinate vector in the direction from the origin to a point (x,y,z) in space,

$$\vec{k} = \begin{pmatrix} k_x \\ k_y \\ k_z \end{pmatrix}$$

is the propagation vector with magnitude $|k| = 2\pi \cdot n/\lambda$ pointing in the direction of the propagation of the wave, and λ is the wavelength of the harmonic wave.

φ is a phase constant defining the phase of the wave train at the origin for $t = 0$.

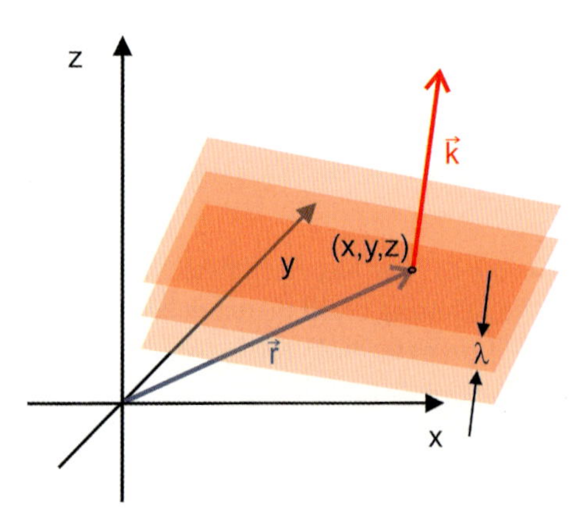

Figure 46-2: A plane wave characterized by its propagation vector $\vec{k}$ and its wavelength λ.

$\mp\omega t$ indicates the wave's direction of travel. We use "–" for travel in the k-direction and "+" for the opposite direction. The direction of travel can be reversed by changing the sign of the propagation vector $\vec{k}$.

If polarization matters, then the amplitude A has to be written in vector form to specify a linearly polarized plane wave in space:

$$\vec{E}(\vec{r},t) = \vec{A}e^{i(\vec{k}\cdot\vec{r}-\omega t+\varphi)} \quad (46\text{-}6)$$

or, in components:

$$\begin{pmatrix} E_x \\ E_y \\ E_z \end{pmatrix} = \begin{pmatrix} A_x \\ A_y \\ A_z \end{pmatrix} e^{i(k_x x+k_y y+k_z z-\omega t+\varphi)}. \quad (46\text{-}7)$$

For an arbitrarily polarized monochromatic wave individual phase constants φ_x, φ_y and φ_z must be considered so that each individual component can be written as in (46-8 a–c).

$$E_x = A_x\, e^{i(k_x x+k_y y+k_z z-\omega t+\varphi_x)} \quad (46\text{-}8a)$$

$$E_y = A_y\, e^{i(k_x x+k_y y+k_z z-\omega t+\varphi_y)} \quad (46\text{-}8b)$$

$$E_z = A_z\, e^{i(k_x x+k_y y+k_z z-\omega t+\varphi_z)}. \quad (46\text{-}8c)$$

Of all three-dimensional waves only the plane wave propagates without changing its shape as long as the medium is free of dispersion. The general form of a three-dimensional plane wave does not need to be harmonic or periodic and can be described by (46-9):

$$\vec{E}(\vec{r},t) = c_1 f\left(\frac{\vec{k}\cdot\vec{r}}{|\vec{k}|} - v\,t\right) + c_2\, g\left(\frac{\vec{k}\cdot\vec{r}}{|\vec{k}|} + v\,t\right) \quad (46\text{-}9)$$

where c_1, c_2 are constants, v is the velocity of the travelling wave and f and g are arbitrary functions that must be twice differentiable.

A surface defined by a constant value of $\vec{k}\cdot\vec{r}$ is called a wavefront. The term $c_1 f\left(\frac{\vec{k}\cdot\vec{r}}{|\vec{k}|} - v\,t\right)$ travels in the direction of the propagation vector with velocity v, whereas the term $c_2\, g\left(\frac{\vec{k}\cdot\vec{r}}{|\vec{k}|} + v\,t\right)$ travels in the opposite direction.

46.2.2 Complex Notation

The complex notation of the monochromatic harmonic plane wave in (46-6) is very convenient for further computation and investigation. However, note that when selecting the complex notation in general only the real part of the expression is of physical meaning. According to Euler's Equation

$$e^{i\phi} = \cos\phi + i\cdot\sin\phi \quad (46\text{-}10)$$

any complex number E can be represented as

$$E = Ae^{i\phi} = A\cos\phi + i\sin\phi = \mathrm{Re}\{E\} + i\,\mathrm{Im}\{E\} \tag{46-11}$$

where
i is the imaginary unit $i = \sqrt{-1}$,
$A = |E| = \sqrt{\mathrm{Re}\{E\} + \mathrm{Im}\{E\}}$ is the magnitude of the complex number and
$\phi = \arctan\dfrac{\mathrm{Im}\{E\}}{\mathrm{Re}\{E\}}$ is the phase or angle of the complex number.

The physical properties of a wave can be associated either with the real part $\mathrm{Re}\{E\} = A\cdot\cos\phi$ or the imaginary part $\mathrm{Im}\{E\} = A\cdot\sin\phi$ of the complex field E.

It is often useful to treat complex numbers as two-dimensional vectors in a complex plane, where the x-axis represents the real part, and the y-axis the imaginary part of the complex number (figure 46-3) [46-3].

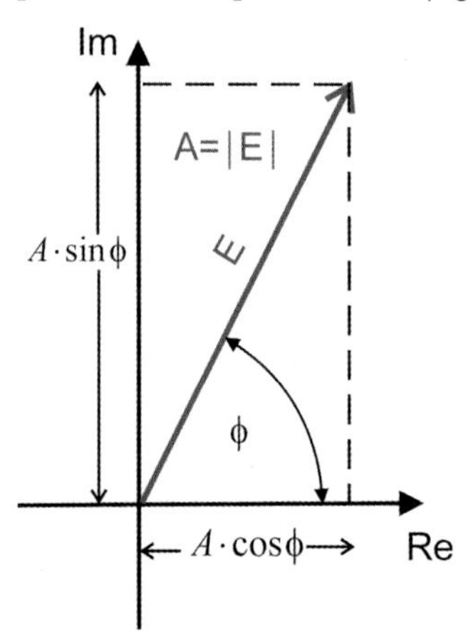

Figure 46-3: Vector representation of a complex number A.

46.2.3
Spherical Wave

A further solutions of (46-1a) is the monochromatic harmonic spherical wave specified by

$$E(r,t) = \frac{A}{r}e^{i(k\cdot r - \omega t + \varphi)} \tag{46-12}$$

where
E is the scalar electrical field: $E = E_x = E_y = E_z$,
A is the scalar amplitude of the electrical field: $A = A_x = A_y = A_z$,
and $r = \sqrt{x+y+z}$ is the magnitude of the coordinate vector $\vec{r}$ in a direction from the point source of the spherical wave to a point (x,y,z) in space. Since the propagation vector $\vec{k}$ and the coordinate vector $\vec{r}$ are always parallel it is sufficient to use the magnitude $k\cdot r$.

For a point source Q at an arbitrary position $\vec{q}$ in space the electrical field can be calculated for any location P in space given by vector $\vec{p}$ according to (46-13).

$$E(\vec{p}, t) = \frac{A}{|\vec{p} - \vec{q}|} e^{i(k \cdot |\vec{p} - \vec{q}| - \omega t + \varphi)} \tag{46-13}$$

Note that (46-13) is valid only for $|\vec{p} - \vec{q}| >> \lambda$.

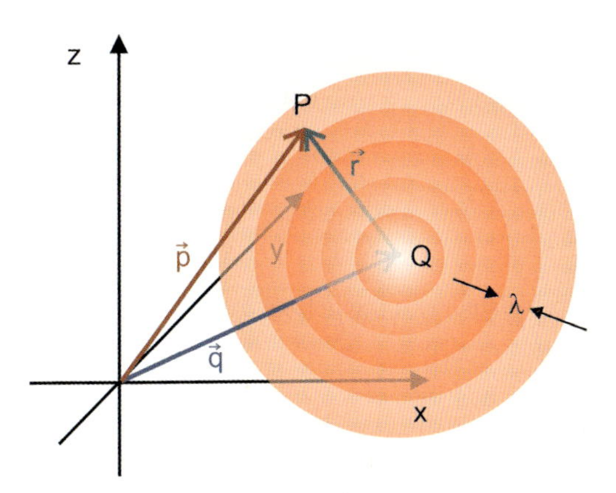

Figure 46-4: Spherical wave with a point source at Q.

46.2.4 Vector Sum of Complex Numbers

The sum of two complex numbers can be calculated with the help of the vector representation in figure 46-5.

$$E = Ae^{i\phi} = E_1 + E_2 = A_1 e^{i\phi_1} + A_2 e^{i\phi_2}. \tag{46-14}$$

The real and imaginary parts of E are now

$$\mathrm{Re}\{E\} = A_1 \cos\phi_1 + A_2 \cos\phi_2 \tag{46-15a}$$

$$\mathrm{Im}\{E\} = A_1 \sin\phi_1 + A_2 \sin\phi_2 \tag{46-15b}$$

thus leading to the amplitude A and phase ϕ of E, respectively

$$A = \sqrt{A_1 + A_2 + 2A_1 A_2 \cos(\phi_2 - \phi_1)} \tag{46-16a}$$

$$\phi = \arctan \frac{A_1 \sin\phi_1 + A_2 \sin\phi_2}{A_1 \cos\phi_1 + A_2 \cos_2} \tag{46-16b}$$

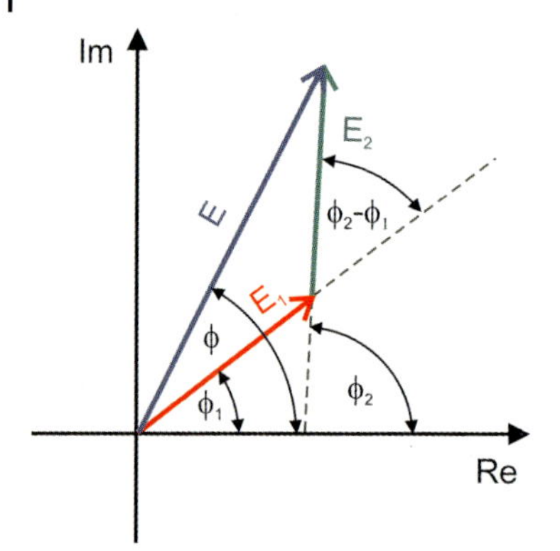

Figure 46-5: The sum of two complex numbers in a complex plane.

The sum of N complex numbers is calculated with the help of the vector representation in figure 46-6.

$$E = Ae^{i\phi} = \sum_{j=1}^{N} E_j = \sum_{j=1}^{N} A_j e^{i\phi_j}. \tag{46-17}$$

Using (46-15a,b) for the real and imaginary part and using addition theorems for trigonometric functions we arrive at a general expression for the amplitude A and phase ϕ of the vector E:

$$\begin{aligned} A &= \sqrt{\left(\sum_{j=1}^{N} A_j \sin\phi_j\right)^2 + \left(\sum_{j=1}^{N} A_j \cos\phi_j\right)^2} \\ &= \sqrt{\sum_{j=1}^{N}\sum_{k=1}^{N} A_j A_k \cos\left(\phi_j - \phi_k\right)} \end{aligned} \tag{46-18a}$$

$$\phi = \arctan \frac{\sum_{j=1}^{N} A_j \sin\phi_j}{\sum_{j=1}^{N} A_j \cos\phi_j}. \tag{46-18b}$$

Figure 46-6 illustrates the sum of five complex numbers in the complex plane by adding the real and imaginary parts.

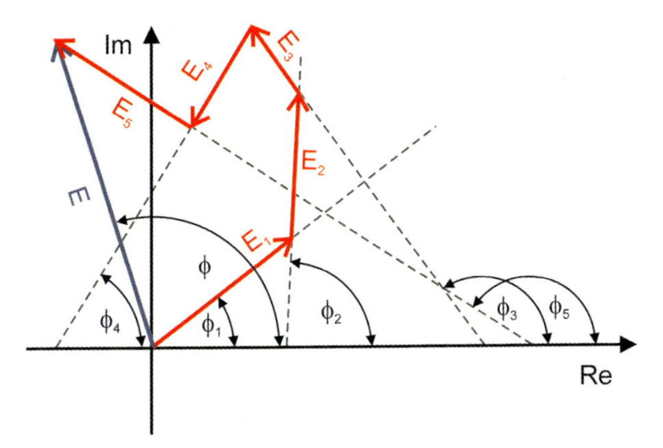

Figure 46-6: The sum of five complex numbers in a complex plane.

Suppose there is a complex function $E(\xi) = A(\xi)e^{i\phi(\xi)}$ that varies with a variable ξ in a known definite way. If we calculate the integral of $E(\xi)$ over ξ (46-19) we will get a resulting complex number E with amplitude A and phase ϕ, which will be calculated according to (46-20a and b).

$$E = Ae^{i\phi} = \int_{\xi=\xi_0}^{\xi_1} E(\xi)\,d\xi = \int_{\xi=\xi_0}^{\xi_1} A(\xi)e^{i\phi(\xi)}d\xi \tag{46-19}$$

$$\begin{aligned} A &= \sqrt{\left(\int_{\xi=\xi_0}^{\xi_1} A(\xi)\,\sin\phi(\xi)\,d\xi\right)^2 + \left(\int_{\xi=\xi_0}^{\xi_1} A(\xi)\,\cos\phi(\xi)\,d\xi\right)^2} \\ &= \sqrt{\int_{\xi=\xi_0}^{\xi_1}\int_{\eta=\xi_0}^{\xi_1} A(\xi)A(\eta)\,\cos\left(\phi(\xi)-\phi(\eta)\right)d\eta d\xi} \end{aligned} \tag{46-20a}$$

$$\phi = \arctan \frac{\int_{\xi=\xi_0}^{\xi_1} A(\xi)\,\sin\phi(\xi)\,d\xi}{\int_{\xi=\xi_0}^{\xi_1} A(\xi)\,\cos\phi(\xi)\,d\xi}. \tag{46-20b}$$

Figure 46-7 shows an example of the integration of a complex function $E(\xi)$.

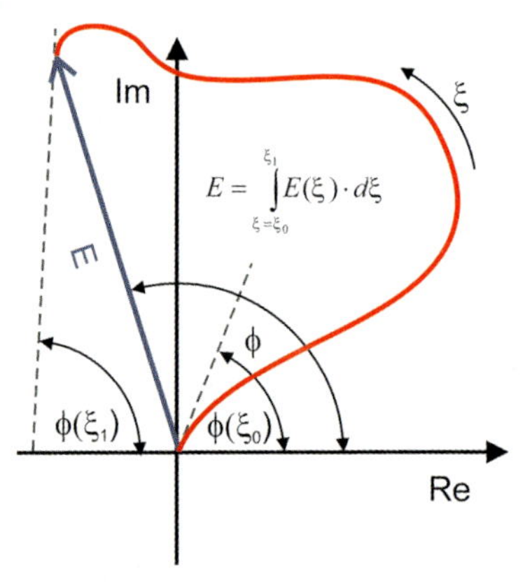

Figure 46-7: Integral over a complex function $E(\zeta)$ in a complex plane.

46.2.5
Interference of Two Plane Waves

We assume two plane waves characterized by their individual propagation directions $\vec{k}_1$ and $\vec{k}_2$ and their amplitudes A_1 and A_2. Their wavelength is λ. We determine the intensity distribution $I(x,y,z)$ in the region of superposition by the square of (46-16a):

$$I(x, y, z) = A_1 + A_2 + 2A_1A_2 \cos \Delta\phi(x, y, z) \tag{46-21}$$

with

$$\Delta\phi(x, y, z) = \phi_2(x, y, z) - \phi_1(x, y, z) = (\vec{k}_2 - \vec{k}_1) \cdot \vec{r} + \Delta\varphi \tag{46-22}$$

and $\Delta\varphi = \varphi_2 - \varphi_1 - \omega\tau$

φ_1 and φ_2 are the initial phases at $t = 0$, whereas τ is a possible time delay between both wave trains due to different optical paths from their source to the area of superposition under consideration. (46-22) does not contain time t any more. The intensity distribution in (46-21) is constant in time, interference patterns or fringes can be observed.

Bright fringes occur at the location where the phase differences $\Delta\phi(x, y, z)$ are equal to a multiple of 2π, dark fringes are located where the phase differences are equal to an odd number of π as in (46-23) specified [46-4 –14].

Bright fringes: $\Delta\phi(x, y, z) = m \cdot 2\pi$ with $m = -\infty, ..., -2, -1, 0, 1, 2, ..., +\infty$
Dark fringes: $\Delta\phi(x, y, z) = (2m + 1) \cdot \pi$.

Surfaces of equal phase differences are equally spaced parallel planes perpendicular to $(\vec{k}_2 - \vec{k}_1)$ with a constant separation of

$$s = \frac{2\pi}{\left|\vec{k}_2 - \vec{k}_1\right|} = \frac{\lambda}{n\left|\vec{e}_2 - \vec{e}_1\right|} \tag{46-23}$$

$\vec{e}_1$ and $\vec{e}_2$ are unit vectors pointing in the direction of $\vec{k}_1$ and $\vec{k}_2$, respectively.

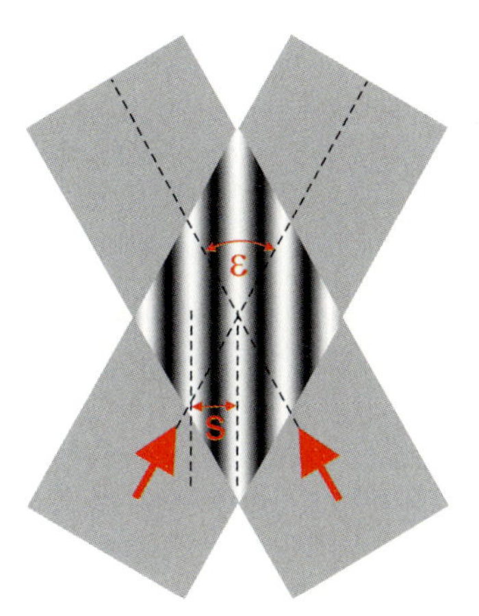

Figure 46-8: The interference of two plane waves in their overlay region. ε is the angle between their two directions of propagation, s is the resulting fringe spacing.

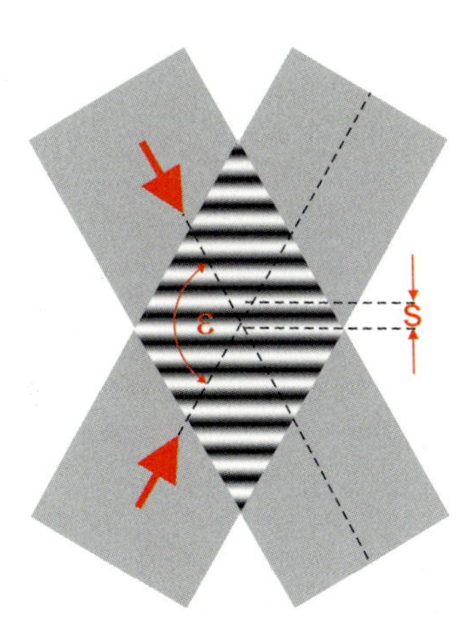

Figure 46-9: The orientation of one plane wave has now changed by 180° relative to figure 46-8. ε is the new angle between the two directions of propagation and s is the new resulting fringe spacing.

In figures 46-8 and 46-9 the angle between the propagation vectors of two plane waves is ε. The fringe spacing is then given by (46-24).

$$s = \frac{\lambda}{2n\,\sin\frac{\varepsilon}{2}}. \tag{46-24}$$

Note that the orientation of propagation is of importance for the orientation and spacing of the fringes. The planes of equal phase (fringe planes) are always perpendicular to $(\vec{k}_2 - \vec{k}_1)$. If one of the waves changes the orientation of travel then its vector $\vec{k}$ changes sign and the fringes change direction by 90°. The fringe spacing is then altered by a factor of $1/\tan\frac{\varepsilon}{2}$.

46.2.6
Interference of Two Spherical Waves

We assume two monochromatic spherical waves of wavelength λ emerging from their origins in $\vec{q}_1$ and $\vec{q}_2$, respectively. We determine the intensity distribution $I(x,y,z)$ in the region of superposition by the square of the sum of two spherical electrical fields according to (46-13) [46-6] :

$$I(x,y,z) = \frac{A_1}{r_1^2} + \frac{A_2}{r_2^2} + 2\frac{A_1}{r_1} \cdot \frac{A_2}{r_2} \cdot \cos\Delta\phi(x,y,z) \tag{46-25}$$

with

$r_1 = |\vec{p} - \vec{q}_1|$ and $r_2 = |\vec{p} - \vec{q}_2|$

$$\Delta\phi(x,y,z) = \phi_2(x,y,z) - \phi_1(x,y,z) = k \cdot (r_2 - r_1) + \Delta\varphi \tag{46-26}$$

$\Delta\varphi = \varphi_2 - \varphi_1 - \omega\tau$ is the relative phase difference of the two waves at their individual origins at $t = 0$, τ is a possible time delay for the start of the waves.

$k = \frac{2\pi \cdot n}{\lambda}$ is the magnitude of the propagation vectors.

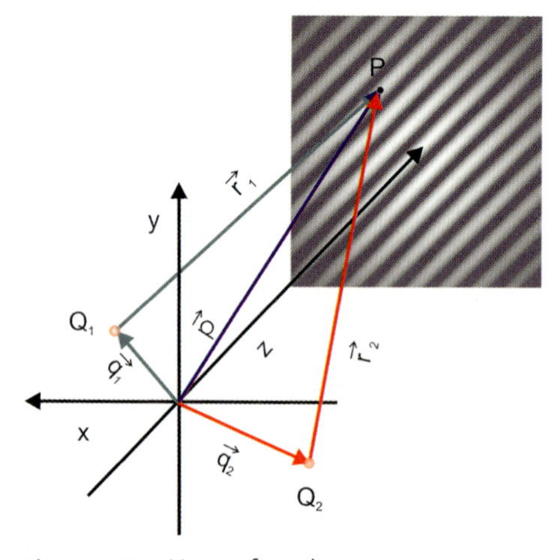

Figure 46-10: Vectors from the origin to two point sources Q_1 and Q_2 and to a point P in space.

Bright fringes occur at the location where the phase differences $\Delta\phi(x, y, z)$ are equal to a multiple of 2π, dark fringes are located where the phase differences are equal to an odd number of π as specified in the following:

Bright fringes: $\Delta\phi(x, y, z) = m \cdot 2\pi$ with $m = -\infty, ..., -2, -1, 0, 1, 2, ..., +\infty$

Dark fringes: $\Delta\phi(x, y, z) = (2m + 1) \cdot \pi$.

Surfaces with equal phase differences are rotational symmetric hyperboloids with their axis going through the source points of the spherical waves.

Let the location of the source points be at

$$\vec{q}_1 = \begin{pmatrix} -q \\ 0 \\ 0 \end{pmatrix} \text{and } \vec{q}_2 = \begin{pmatrix} +q \\ 0 \\ 0 \end{pmatrix}$$

as shown in figure 46-10. According to (46-26) we calculate the phase differences $\Delta\phi(x, y, z)$ by (46-27):

$$\Delta\phi(x, y, z) = k \cdot (\sqrt{(x+q)^2 + y^2 + z^2} - \sqrt{(x-q)^2 + y^2 + z^2}) + \Delta\varphi. \qquad (46\text{-}27)$$

Along the z-axis we have to distinguish between those regions between the source points and those outside the source points.

We have a linear varying phase difference in x

$$\Delta\phi(x, y, z) = 2k \cdot x + \Delta\varphi \text{ for } |x| < q \qquad (46\text{-}28a)$$

and the constant phase difference

$$\Delta\phi(x, y, z) = 2k \cdot q + \Delta\varphi \text{ for } |x| \geq q \qquad (46\text{-}28b)$$

At a distance x_0 the lines of equal phase difference in a y–z plane are concentric circular rings around the x-axis. At a distance z_0 the lines of equal phase difference in an x–y plane are hyperbolas with the y-axis at $x = 0$ as the axis of symmetry.

For $z >> x, y, q$ (46-27) can be approximated by (46-29), which represents straight parallel and equally spaced fringes in the y direction (plane wave approximation).

$$\Delta\phi(x, y, z) \approx k \frac{2qx}{z} + \Delta\varphi. \qquad (46\text{-}29)$$

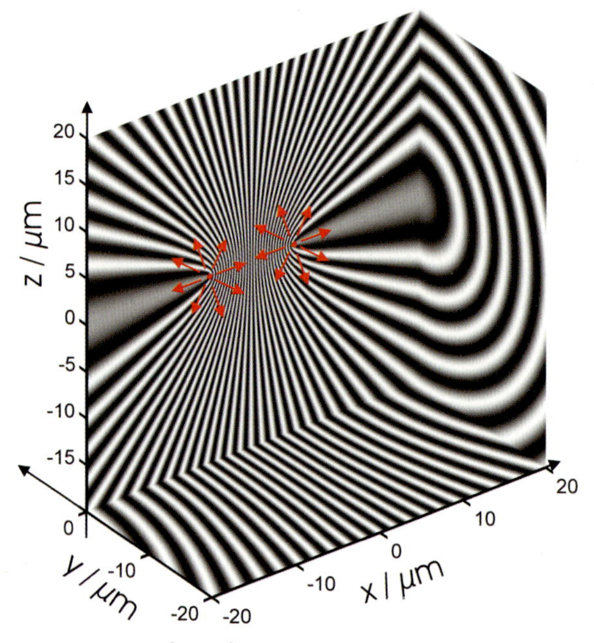

Figure 46-11: Surfaces of equal phase difference for two monochromatic point sources with $\lambda = 0.6328$ µm and $q = 5$ µm (see text). Both waves emerge from their point source and have a relative phase-shift of $\Delta\varphi = 0$.

Note that the amplitude of each wave changes with the distance from its origin. Therefore, the contrast or visibility V of the fringes is not constant in space but varies according to (46-30).

$$V = \frac{I_{max} - I_{min}}{I_{max} + I_{min}} = \frac{2A_1 A_2 r_1 r_2}{A_1 r_2 + A_2 r_1} \tag{46-30}$$

At the source points the fringe contrast is zero, because $r_1 = 0$ or $r_2 = 0$.

The fringe contrast has its maximum $V = 1$ at the locations, where $A_2 r_1 = A_1 r_2$. If there are two sources of equal amplitude $A_1 = A_2$ this is the case when a plane perpendicular to the axis of each point is the same distance from the source points.

If the direction of propagation of one of the spherical waves is inverted, the shape and position of the interference fringes change totally. The surfaces of equal phase difference are now rotationally symmetric ellipsoids with their focal points at the center points of the spherical waves (figure 46-12). The phase differences are calculated according to (46-31). In comparison to (46-26) only the sign of one of the distances has changed.

$$\Delta\phi(x, y, z) = \phi_2(x, y, z) - \phi_1(x, y, z) = k \cdot (r_2 + r_1) + \Delta\varphi. \tag{46-31}$$

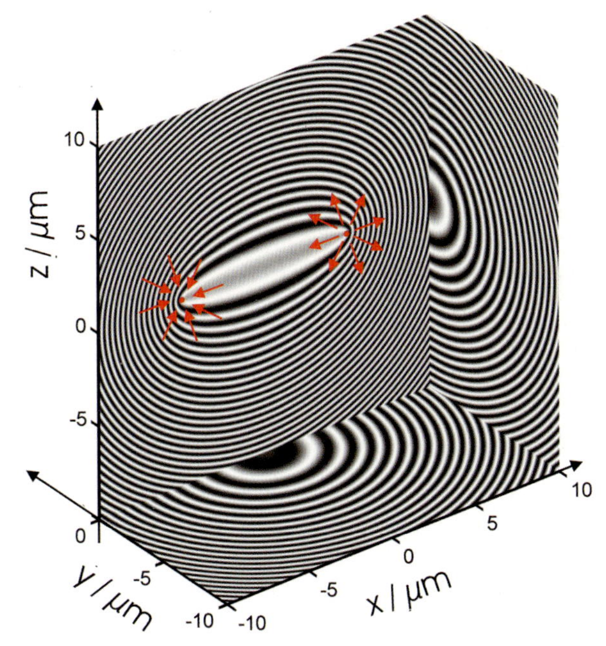

Figure 46-12: Surfaces of equal phase difference for two spherical waves with $\lambda = 0.6328$ µm and $q = 5$ µm propagating in different orientations (see text). One wave emerges from its point source, the other propagates towards its center of curvature. Both waves have a relative phase-shift of $\Delta\varphi = 0$ at $t = 0$.

As in the example above the source point locations will be at

$$\vec{q}_1 = \begin{pmatrix} -q \\ 0 \\ 0 \end{pmatrix} \text{ and } \vec{q}_2 = \begin{pmatrix} +q \\ 0 \\ 0 \end{pmatrix}$$

According to (46-31) we can calculate the phase differences $\Delta\phi(x, y, z)$ as:

$$\Delta\phi(x, y, z) = k \left(\sqrt{(x+q)^2 + y^2 + z^2} + \sqrt{(x-q)^2 + y^2 + z^2}\right) + \Delta\varphi. \tag{46-32}$$

Along the z-axis we again have to distinguish between the regions between the source points and those outside the source points.

We have a linear varying phase difference in z

$$\Delta\phi(x, y, z) = 2k\,x + \Delta\varphi \text{ for } |x| \geq q \tag{46-33a}$$

and a constant phase difference

$$\Delta\phi(x, y, z) = 2k\,q + \Delta\varphi \text{ for } |x| < q. \tag{46-33b}$$

At a distance x_0 the lines of equal phase difference in a y–z plane are circular rings concentric around the x-axis. At a distance z_0 the lines of equal phase difference in an x–y plane are ellipsoids with the y-axis at $x=0$ as the axis of symmetry.

As in the case of two emerging waves the contrast V of the fringes is not constant in space but varies according to (46-30). At the source points the fringe contrast is again zero, the maximum $V=1$ occurs at those points in space, where $A_2\, r_1 = A_1\, r_2$.

46.2.7
Interference of Two Waves with Different Wavelengths

Equation (46-12) shows the general expression for the electrical field of a spherical wave. In the discussions on interferometry we so far have considered only waves of the same wavelength λ. We now assume two interfering spherical waves with different wavelengths, λ_1 and λ_2. The expression for the phase differences then changes to:

$$\Delta\phi(x,y,z) = k_2\, r_2 - k_1\, r_1 - (\omega_2 - \omega_1)\, t + \Delta\varphi \tag{46-34}$$

with $k_1 = \dfrac{2\pi\, n_1}{\lambda_1}$ and $k_2 = \dfrac{2\pi\, n_2}{\lambda_2}$ and
with $\omega_1 = 2\pi\,\dfrac{c}{\lambda_1}$ and $\omega_2 = 2\pi\,\dfrac{c}{\lambda_2}$

The phase difference is no longer stationary but changes in time with a frequency difference of

$$\Delta\omega = 2\pi\, c\,\frac{|\lambda_2 - \lambda_1|}{\lambda_1\,\lambda_2}. \tag{46-35}$$

Consequently, the interference fringes will move in time with a velocity depending on the difference in the wavelength. Since any intensity detector can be integrated over a certain integration time T, the detected intensity signal can be expressed as:

$$I(x,y,z,T) = \frac{A_1}{r_1} + \frac{A_2}{r_2} + \frac{2A_1A_2}{T\, r_1\, r_2} \int\limits_{t=t_0}^{t_0+T} \cos\left(k_2\, r_2 - k_1\, r_1 - \Delta\omega\, t + \Delta\varphi\right) dt. \tag{46-36}$$

The integration of (46-36) leads to:

$$I(x,y,z,T) = \frac{A_1}{r_1} + \frac{A_2}{r_2} + \frac{2A_1A_2}{r_1\, r_2}\,\frac{\sin \Delta\omega\frac{T}{2}}{\Delta\omega\frac{T}{2}}\,\cos\left(k_2\, r_2 - k_1\, r_1 - \Delta\omega\,(t + \frac{T}{2}) + \Delta\varphi\right). \tag{46-37}$$

Integration over a time interval T leads to a decrease in the contrast of the fringes and to a constant phaseshift of $\Delta\omega\frac{T}{2}$. The contrast is modulated by a sinc function, which is zero for $T = m\,\dfrac{\lambda_1\,\lambda_2}{c\,|\lambda_2 - \lambda_1|}$ with $m = 1, 2, 3, \ldots, \infty$

46.2.8 Coherence and Correlation Function

The atoms of a light source do not continuously send out waves. Usually the state of excitation of an atom lasts in the order of 10^{-8} s. During this time a "wave train" is emitted. The length and time of emission of the individual wave trains changes statistically. The resulting frequency distribution is called the "natural linewidth". Since the atoms also move statistically in different directions during emission, the frequency spectrum will be broadened by the Doppler effect. Atoms also collide during their motion and thus broaden the spectrum additionally by interrupting the wave trains (pressure broadening).

The average length of a wave train is called the coherence length l_c. The average duration of emission is called coherence time τ_c. Both are related by (46-38).

$$l_c = c \cdot \tau_c \tag{46-38}$$

To describe the phase correlation of waves in time we define a complex correlation function

$$\Gamma_{12}(r_1, r_2, t, \tau) = \langle E_1(r_1, t+\tau)\, E_2^*(r_2, t)\rangle = \frac{1}{T}\int_{t'=-T/2}^{T/2} E_1(r_1, t'+\tau)\, E_2^*(r_2, t')dt' \tag{46-39}$$

The angle paranthesis denote the time average. T is the integration time of the detector, τ is the initial time delay of the interfering waves.

We call (46-39) the coherence function. In interferometry we shall assume stationary correlations so that Γ_{12} will only depend on τ and the optical paths r_1 and r_2. In a typical interferometric experiment, two identical images of a light source are generated by partial reflection at a beamsplitter. The time delay τ describes the run-time difference between the two interfering waves at the location of observation. Γ_{12} then describes the ability for fringe formation. It is convenient to define a normalized correlation function Γ_{12}, which delivers a quantitative measure of interference contrast or modulation (46-40) [46-2].

$$\gamma_{12}(r_1, r_2, \tau) = \frac{\Gamma_{12}(r_1, r_2, \tau)}{\sqrt{\Gamma_{11}(r_1, 0)\,\Gamma_{22}(r_2, 0)}} = \frac{\Gamma_{12}(r_1, r_2, \tau)}{\sqrt{I_1(r_1)\, I_2(r_2)}}. \tag{46-40}$$

In the case of two spherical waves the normalized correlation function γ_{12} can be written as in (46-41).

$$\gamma_{12}(r_1, r_2, \tau) = \frac{1}{T}\int_{t=0}^{T} e^{i(k(r_1 - r_2) - \omega\tau + \Delta\varphi(t))}\, dt. \tag{46-41}$$

For stationary conditions there is no phase fluctuation and the integral can be omitted.

The intensity distribution can be expressed in terms of the normalized correlation function (46-42).

$$I_{\text{partialcoherent}} = |E_1 + E_2|^2 = I_1 + I_2 + 2\sqrt{I_1 I_2}\,|\gamma_{12}(r_1, r_2, \tau)|\cos\Delta\phi. \tag{46-42}$$

$|\gamma_{12}(r_1, r_2, \tau)|$ determines the *visibility* V of the fringe pattern (46-43).

$$V = \frac{I_{\max} - I_{\min}}{I_{\max} + I_{\min}} = \frac{2\sqrt{I_1 I_2}}{I_1 + I_2}|\gamma_{12}(r_1, r_2, \tau)|. \tag{46-43}$$

V ranges between 0 (for the case of complete incoherence) and 1 (for total coherence). The case $0<V<1$ is called partial coherence.

46.2.9
Interference of Two Monochromatic Waves with Statistically Varying Phase

We assume two monochromatic point sources whose relative phase difference $\Delta\varphi(t)$ changes in time. The interference pattern can then be described by a constant and a time-varying part as in (46-44).

$$\begin{aligned} I &= I_1 + I_2 + 2\sqrt{I_1 I_2}\,\cos\left(k(r_2 - r_1) + \Delta\varphi(t)\right) \\ &= I_1 + I_2 + 2\sqrt{I_1 I_2}\,[\cos k(r_2 - r_1)\,\cos\Delta\varphi(t) - \sin k(r_2 - r_1)\,\sin\Delta\varphi(t)] \end{aligned} \tag{46-44}$$

If we integrate over a time interval T we then arrive at (46-45).

$$\langle I\rangle = \langle I_1 + I_2\rangle + 2\sqrt{I_1 I_2}\,\frac{1}{T}\int_{t=0}^{T}[\cos k(r_2 - r_1)\,\cos\Delta\varphi(t) - \sin k(r_2 - r_1)\,\sin\Delta\varphi(t)]dt \tag{46-45}$$

If $\Delta\varphi(t)$ is evenly distributed between 0 and 2π and the observation interval T is sufficiently long, the integral converges to zero. The correlation function γ_{12} in (46-41), and therefore the visibility of the fringes, becomes zero. We call this the case of complete *incoherence*. No interference fringes are visible and the observed intensity is the sum of the single intensities I_1 and I_2 (46-46).

$$I_{\text{incoherent}} = |E_1|^2 + |E_2|^2 = I_1 + I_2 \tag{46-46}$$

To calculate the intensity distribution of two coherent interfering waves E_1 and E_2 first their sum and then the square of the magnitude is calculated:

$$I_{\text{coherent}} = |E_1 + E_2|^2 = I_1 + I_2 + 2\sqrt{I_1 I_2}\cos\Delta\phi \tag{46-47}$$

46.2.10
Interference of Two Waves which have a Spectrum

Real waves always consist of a variety of wavelengths, which we call a spectrum. Electromagnetic waves of all wavelengths then add up to an electrical field described by (46-48).

$$E = \int_{\omega=\omega_0}^{\omega_1} E(\omega)\, d\omega = \int_{\omega=\omega_0}^{\omega_1} A(\omega) e^{i\phi(\omega)}\, d\omega \tag{46-48}$$

The limits of the spectrum are defined by the lower limit $\omega_0 = 2\pi\nu_0$ and the upper limit $\omega_1 = 2\pi\nu_1$. If the waves are not travelling in a vacuum, we have to consider the dispersion of the medium, such that the index of refraction now becomes a function of the frequency: $n = n(\omega)$.

Using the relation (46-49) we can also use the functional dependence on λ rather than on ω, which sometimes makes the calculations more practical [46-2].

$$\omega = 2\pi\nu = 2\pi \frac{c}{\lambda} \tag{46-49}$$

A plane wave with a spectrum can then be expressed as in (46-50) or (46-51)

$$E(\vec{p}, t) = \frac{1}{\lambda_1 - \lambda_0} \int_{\lambda=\lambda_0}^{\lambda_1} A(\lambda) e^{i\left(\vec{k}(\lambda)\cdot\vec{p} - \frac{2\pi\cdot c}{\lambda} t + \varphi\right)}\, d\lambda \tag{46-50}$$

$$E(\vec{p}, t) = \frac{1}{\nu_1 - \nu_0} \int_{\nu=\nu_0}^{\nu_1} A(\nu) e^{i\left(\vec{k}(\nu)\cdot\vec{p} - 2\pi\nu t + \varphi\right)}\, d\nu \tag{46-51}$$

with $\vec{k}(\lambda) = \frac{2\pi\, n(\lambda)}{\lambda}\, \vec{e}$ or $\vec{k}(\nu) = \frac{2\pi\nu\, n(\nu)}{c}\, \vec{e}$.
where $\vec{e}$ is the unit vector pointing in the direction of propagation. φ is the wave's phase at $\vec{p} = 0$ and $t = 0$.

If we assume a dispersion-free medium with $\vec{k}(\nu) = \vec{k} =$ constant and split off the time-dependent portion in (46-50) and then expand the integral limits to infinity, we arrive at (46-52).

$$E(\vec{p}, t') = e^{i\varphi} \int_{\nu=-\infty}^{\infty} A(\nu) e^{-i2\pi\nu t'}\, d\nu \tag{46-52}$$

with $t' = t - \frac{n}{c}\, \vec{e}\cdot\vec{p}$.

The integral represents the Fourier transform of the spectrum $A(\nu)$ of the light and leads to a time-dependent amplitude at a point $\vec{p}$ in space. If we define an amplitude threshold we can also define a duration for which the wave train shows an amplitude above the threshold. The time between the arrival and the departure of the wave train is then called the *coherence time* τ_c. Since the distance l and time t of a travelling wave are related by the propagation speed c of the wave, a *coherence length* l_c can be found according to (46-38) describing the length of a wave train with a spectrum $A(\nu)$ [46-5], [46-6].

For a spherical wave we can write:

$$E(r,t') = \frac{e^{i\varphi}}{r} \int_{\nu=-\infty}^{\infty} A(\nu) \cdot e^{-i2\pi\nu t'} d\nu. \tag{46-53}$$

Let two waves with an identical spectrum interfere with each other. We use (46-54) to describe the intensity distribution for two plane waves:

$$I(x,y,z) = \frac{1}{\nu_1 - \nu_0} \int_{\nu=\nu_0}^{\nu_1} (A_1(\nu) + A_2(\nu) + 2A_1(\nu)A_2(\nu)\cos(\Delta\phi(x,y,z,\nu)))d\nu \tag{46-54}$$

with

$$\Delta\phi(x,y,z,\nu) = (\vec{k}_2(\nu) - \vec{k}_1(\nu)) \cdot \vec{p} + \Delta\varphi = \frac{2\pi\, n(\nu)\, \nu}{c} (\vec{e}_2 - \vec{e}_1) \cdot \vec{p} + \Delta\varphi \tag{46-55}$$

and we use (46-56) to describe the intensity distribution for two spherical waves:

$$I(x,y,z) = \frac{1}{\nu_1 - \nu_0} \int_{\nu=\nu_0}^{\nu_1} \left(\frac{A_1(\nu)}{r_1} + \frac{A_2(\nu)}{r_2} + 2\frac{A_1(\nu)A_2(\nu)}{r_1 \cdot r_2} \cos(\Delta\phi(x,y,z,\nu)) \right) d\nu \tag{46-56}$$

with

$$\Delta\phi(x,y,z,\nu) = k_2(\nu)\, r_2 - k_1(\nu)\, r_1 + \Delta\varphi = \frac{2\pi\, n(\nu)\, \nu}{c} (r_2 - r_1) + \Delta\varphi. \tag{46-57}$$

To illustrate the resulting effects, we select a dispersion-free material with $n=$ constant and assume constant spectral amplitude distribution A_1 and A_2. For convenience we replace

$$a = \frac{2\pi\, n(\nu)}{c} (\vec{e}_2 - \vec{e}_1) \cdot \vec{p} \tag{46-58}$$

The intensity I is then calculated from (46-54) according to (46-59).

$$I(x,y,z) = A_1 + A_2 + 2A_1A_2 \frac{\sin\frac{a\cdot(\nu_1-\nu_0)}{2}}{\frac{a\cdot(\nu_1-\nu_0)}{2}} \cos\left(\frac{a(\nu_1+\nu_0)+2\Delta\varphi}{2}\right). \tag{46-59}$$

The visibility of the fringes is modulated by a sinc function with a periodicity dominated by $(\nu_1 - \nu_0)$. The first zero crossings are at $a = \pm\frac{2\pi}{\nu_1 - \nu_0}$. Between the first zero crossings we therefore can count m_{fr} fringes given by (46-60).

$$m_{\mathrm{fr}} = \frac{\nu_1+\nu_0}{|\nu_1-\nu_0|} = \frac{\lambda_1+\lambda_0}{|\lambda_1-\lambda_0|} \approx \frac{2\bar{\lambda}}{\Delta\lambda} \tag{46-60}$$

We can consider the number of visible fringes m_{fr} multiplied by the wavelength λ as twice the coherence length l_c. The coherence length l_c is then given by (46-61).

$$l_c = \frac{\lambda_1 \cdot \lambda_0}{|\lambda_1-\lambda_0|} \approx \frac{\bar{\lambda}^2}{\Delta\lambda} \tag{46-61}$$

Figures 46-13 (a) and (b) are illustrations of a fluffing fringe contrast when the spectrum is broadened.

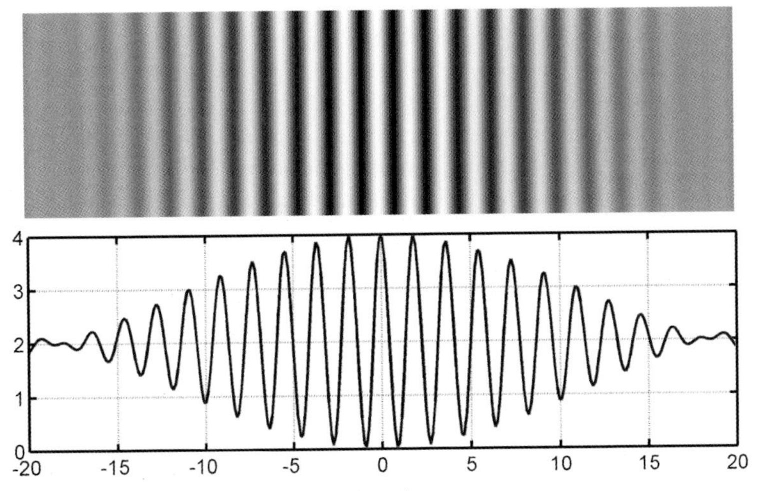

Figure 46-13(a): Interference fringes and irradiance cross-section for two plane waves with $\lambda_1 = 570$ nm and $\lambda_2 = 630$ nm computed by (46-54). According to (46-60) twenty fringes are visible between the first zero crossings of the sinc function.

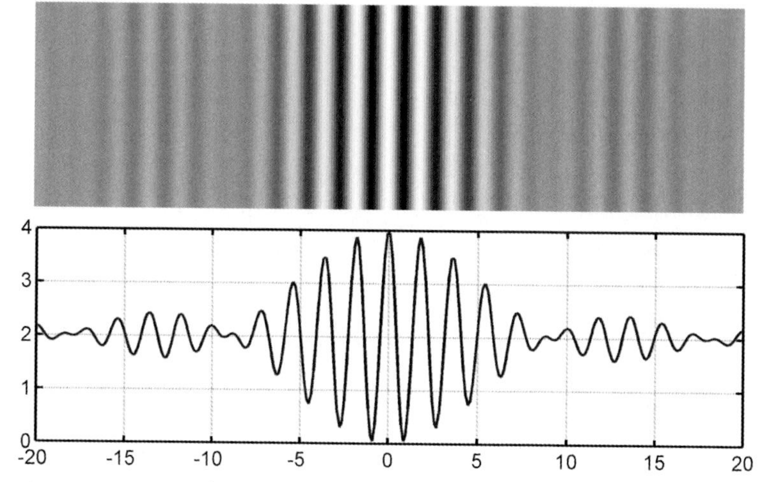

Figure 46-13(b): Interference fringes and irradiance cross-section for two plane waves with $\lambda_1 = 540$ nm and $\lambda_2 = 660$ nm computed by (46-54). According to (46-60) 10 fringes are visible between the first zero crossings of the sinc function. Note that fringes outside the first zero crossings appear shifted by π.

In figure 46-14 we give examples of two identical point sources having a separation $s = 10$ µm. Interference fringes were calculated for a neighborhood of 20 µm × 20 µm for wavelength bands of 400–420 nm, 400–460 nm and 400–500 nm, respectively. In the first row a zero time delay was chosen. In the second row a time delay between the point sources was chosen that corresponds to a delay distance of 5 µm in air. Note that the area of good visibility around the fringe of best visibility becomes smaller when the spectrum broadens. Again the number of fringes between those with zero visibility is given by equation (46-60).

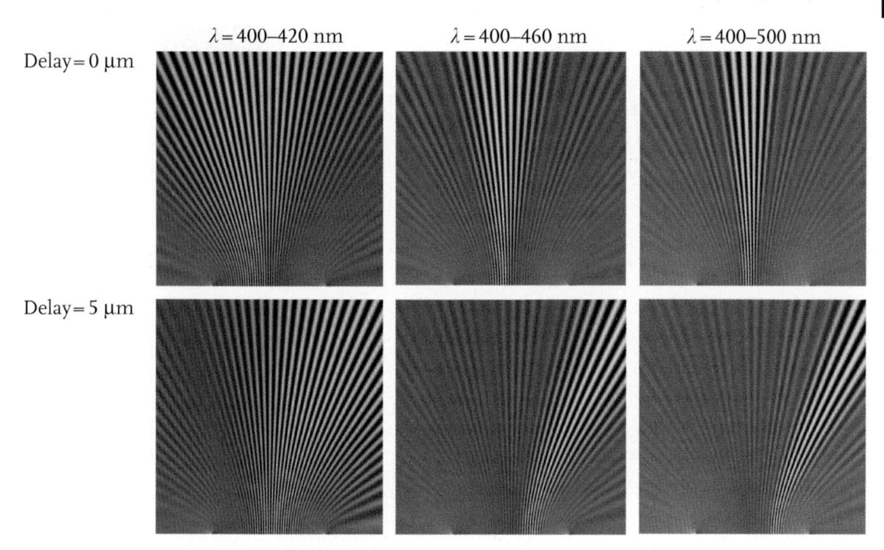

Figure 46-14: Interference fringes for two point sources at the bottom line of the individual drawings having a separation of 10 μm and different spectra. Visibility drops to zero on both sides of the fringe with the best visibility, counting $m_{fr} \approx 2\bar{\lambda}/\Delta\lambda$ fringes in between. For a delay distance of 5 μm between the point sources, the fringe with the best visibility shifts by $5\mu m \bar{\lambda}$.

The coherence length should be measured preferably using a Twyman–Green interferometer (TGI) [46-1]. Figure 46-15 shows a schematic arrangement to test the light of a point source S placed on the optical axis of a collimator C that images S to infinity. A thin beam-splitter BS reflects 50% of the light to the upper plane mirror M_1 and transmits the remaining 50% to the plane mirror M_2. The light hits the mirror surfaces perpendicularly and is reflected back to BS, where 50% are transmitted and 50% reflected. Those portions that are directed to the eyepiece O at the exit of the interferometer are imaged onto the sensor CCD, which detects the intensity distribution of the interfering wavefronts.

If M_1 and M_2 are at equal distances from BS then there is no delay in the two wave trains hitting the mirror surfaces. Therefore the interferogram at CCD will show full contrast with a visibility of $V = 1$.

If we move M_2 along the optical axis by a distance of $\Delta l/2$ the appropriate wave train is delayed by a distance Δl in comparison to the wave train returning from M_1. At the CCD detector a diminished contrast of $0 < V < 1$ will be detected. If Δl is longer than the coherence length l_c of the light, the fringes will not be modulated and the visibility will be zero.

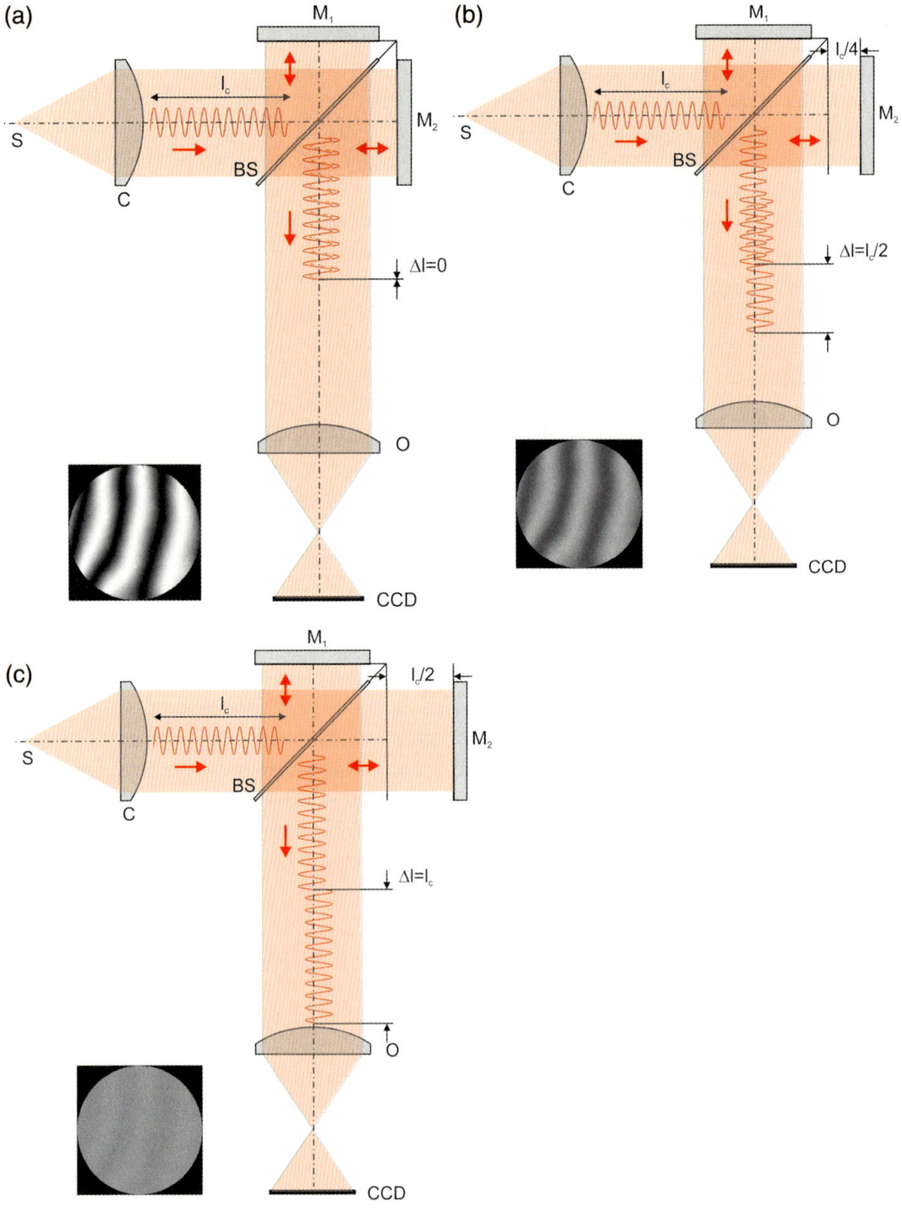

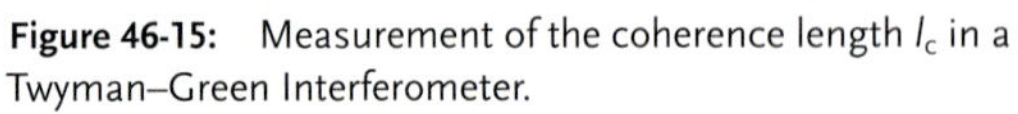

Figure 46-15: Measurement of the coherence length l_c in a Twyman–Green Interferometer.
(a) For $\Delta l = 0$ interference fringes have visibility $V = 1$. (b) For $\Delta l = l_c/2$ interference fringes have visibility $0 < V < 1$. (c) For $\Delta l = l_c$ interference fringes have visibility $V = 0$.
S is the light source under test, C is the collimator, BS is the beam-splitter, M_1 and M_2 are plane mirrors, O is the eyepiece and CCD is the sensor.

46.2.11
Interference with Extended Monochromatic Light Sources

In reality, point sources do not exist in physics. In most cases we assume a real extended light source consisting of an infinite number of adjacent point sources, which are uncorrelated with each other in time. Uncorrelated means that the phase difference of waves emitted by different point sources changes permanently. To calculate the irradiance in the neighborhood of an extended light source we may therefore integrate over the irradiances of individual point sources within the region of the extended light source [46-2].

In the following we analyze the interference pattern caused by two identical images of a monochromatic extended light source. Both images are separated by a distance $2q$ along the x-axis. The resulting interference pattern can be defined by (46-62).

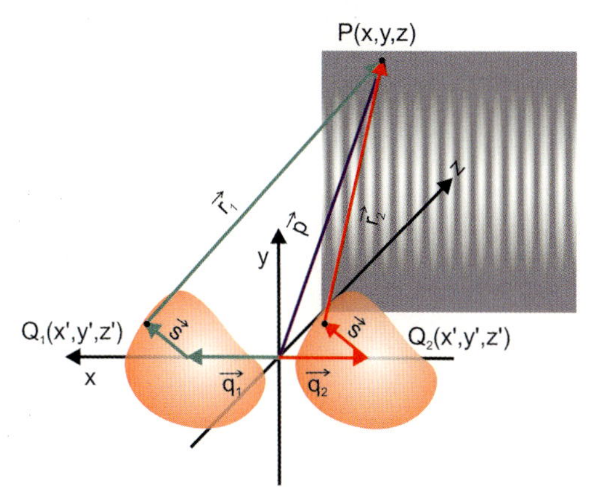

Figure 46-16: Two identical images of an extended light source forming an interference pattern.

$$I(x,y,z) = \frac{1}{S}\iiint_S \frac{A_1}{r_1^2} + \frac{A_2}{r_2^2} + 2\frac{A_1}{r_1}\frac{A_2}{r_2}\cos\Delta\phi(x,y,z)dS \tag{46-62}$$

where S is the region (volume, surface or lineshape) of the extended light source and

$r_1 = |\vec{p} - \vec{q}_1 - \vec{s}|$ and $r_2 = |\vec{p} - \vec{q}_2 - \vec{s}|$ and where

$$\Delta\phi(x,y,z) = \phi_2(x,y,z) - \phi_1(x,y,z) = k(r_2 - r_1) + \Delta\varphi \tag{46-63}$$

is the phase difference between the waves from corresponding point sources Q_1 and Q_2 at point P, $\Delta\varphi$ is their phase difference at Q_1 and Q_2.

As an example we chose two identical line source images of length s oriented along the x-axis and separated by $2q$:

$$\Delta\phi(x,y,z) = k\left(\sqrt{(x-s_x+q)^2+y^2+z^2} - \sqrt{(x-s_x-q)^2+y^2+z^2}\right) + \Delta\varphi \tag{46-64}$$

with $-s/2 \le s_x \le +s/2$ as the coordinate of a point along the line source.

For $z >> x, y, q$ the square-roots can be approximated by neglecting quadratic terms in q.

$$r_1 \approx z + \frac{(x-s_x)^2+y^2}{2z} - \frac{q(x-s_x)}{z} \quad \text{and} \quad r_2 \approx z + \frac{(x-s_x)^2+y^2}{2z} + \frac{q(x-s_x)}{z} \tag{46-65}$$

The resulting phase difference then leads to (46-66).

$$\Delta\phi(x,y,z) = 2k\frac{q(x-s_x)}{z} + \Delta\varphi \tag{46-66}$$

Equation (46-66) represents equally spaced parallel fringes in the y direction whose spatial frequency is determined by k, q and z and whose phase depends on $\Delta\varphi$ and s_x.

The integration then has to be carried out over s_x from $-s/2$ to $+s/2$. Assuming homogeneously radiating sources the final irradiance is given by (46-67).

$$\begin{aligned} I(x,y,z) &= \frac{A_1}{r_1^2} + \frac{A_2}{r_2^2} + 2\frac{A_1}{r_1}\frac{A_2}{r_2}\int_{-s/2}^{s/2}\cos\left(2k\frac{q(x-s_x)}{z} + \Delta\varphi\right)ds_x \\ &= \frac{A_1}{r_1^2} + \frac{A_2}{r_2^2} + 2\frac{A_1}{r_1}\frac{A_2}{r_2}\frac{\sin 2k(qs/z)}{2k(qs/z)}\cos\left(2k\frac{qx}{z} + \Delta\varphi\right) \end{aligned} \tag{46-67}$$

The visibility of the fringes changes according to a sinc-function. The first zero crossing is at $s = \frac{\pi z}{2kq} = \frac{\lambda z}{4q}$ which is equal to the fringe spacing in the observation plane.

Of much more importance is the influence of an extended light source on the phase-shift $\Delta\varphi$ when interferometer arms of unequal length are used. We will denote the arm length difference as cavity length. Figure 46-17 illustrates the effect schematically. An extended light source in an interferometer of unequal arm lengths leads to a low fringe contrast. The contrast goes to zero when the light source reaches a certain dimension. In the following we will discuss this phenomenon in more detail.

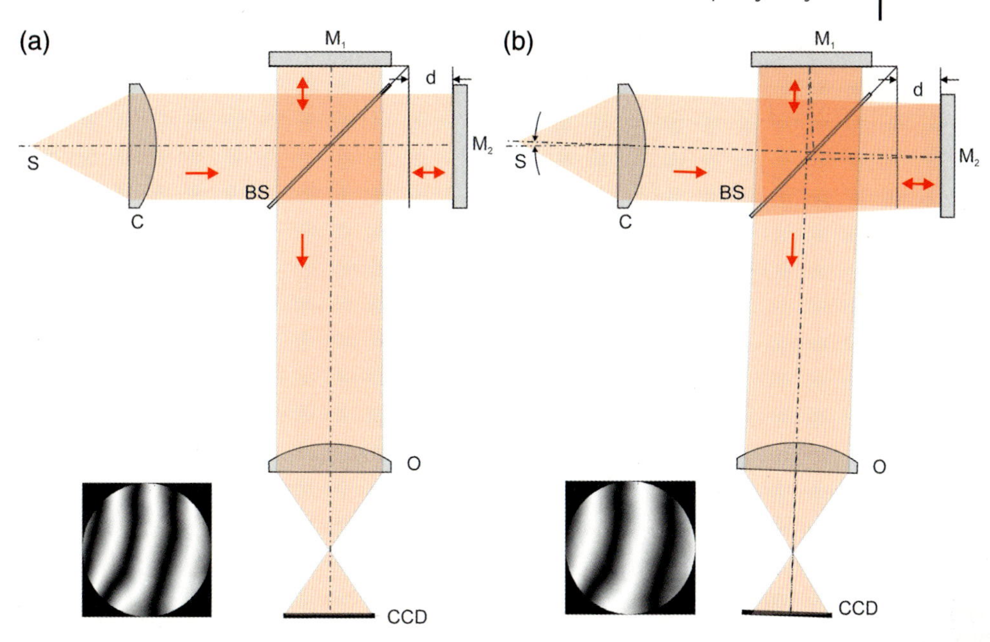

Figure 46-17: Twyman–Green interferometer with cavity length *d*, centered point source (a) and off-axis point source (b) leading to an inclination θ for the test and reference beam. The inclined beams will lead to a smaller optical path difference than in the centered case.

In an interferometer with unequal reference and test paths the detected optical path difference (OPD) between the object and reference beam depends on the inclination θ of the incident beam. When a point source is placed in the front focal point at a distance $s/2$ from the optical axis of a collimator C which has a focal length f, parallel beams will be transmitted by the collimator inclined by an angle θ with

$$\tan\theta = \frac{s}{2f} \tag{46-68}$$

The difference in OPD relative to the centered beam leads to a phase-shift $\Delta\varphi(\theta)$, which can be calculated according to (46-69) and figure 46-18 [46-1].

For a cavity of length d with a refractive index of n_2 the detected OPD will have its maximum when $\theta = 0$. The larger the inclination, the smaller the detected OPD.

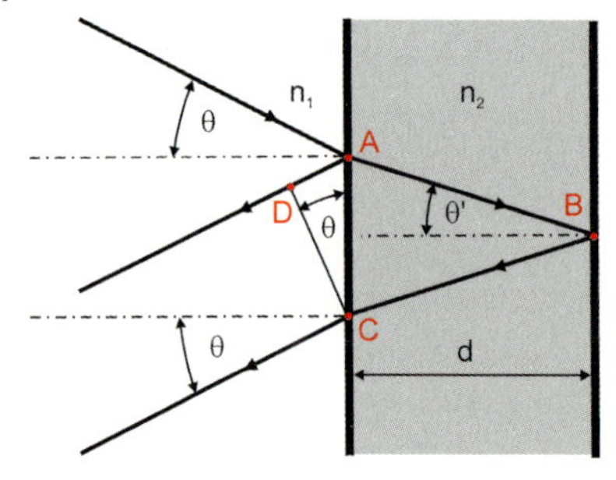

Figure 46-18: A cavity of length d, with refractive indices n_1 in front of and n_2 within the cavity. The inclination angle is θ in front of and θ' within the cavity.

$$\begin{aligned}\Delta\varphi(\theta) &= \frac{2\pi}{\lambda}\left(n_1\bar{A}\bar{D} - n_2(\vec{A}\vec{B} + \vec{B}\vec{C} - 2\mathrm{d})\right) \\ &= \frac{2\pi}{\lambda}\,2\mathrm{dn}_2\left\{1 - \frac{1 - \left(\frac{\mathrm{n}_1}{\mathrm{n}_2}\sin\theta\right)^2}{\cos\left(\arcsin\left(\frac{\mathrm{n}_1}{\mathrm{n}_2}\sin\theta\right)\right)}\right\}\end{aligned} \tag{46-69}$$

Examples are shown in figure 46-19 for a BK7 glass and air cavity of length 1 mm for three different wavelengths. To have sufficient fringe visibility $\Delta\varphi(\theta)$ should be smaller than $\pi/2$. We call the requirement *Rayleigh criterion*. The visibility goes to zero when $\Delta\varphi(\theta_{\max}) = \pi$. Equation (46-70) gives the relation for the maximum inclination angle when an air cavity is used.

$$\theta_{\max,\mathrm{air}} = \arccos\left(1 - \frac{\lambda}{4d}\right) \tag{46-70}$$

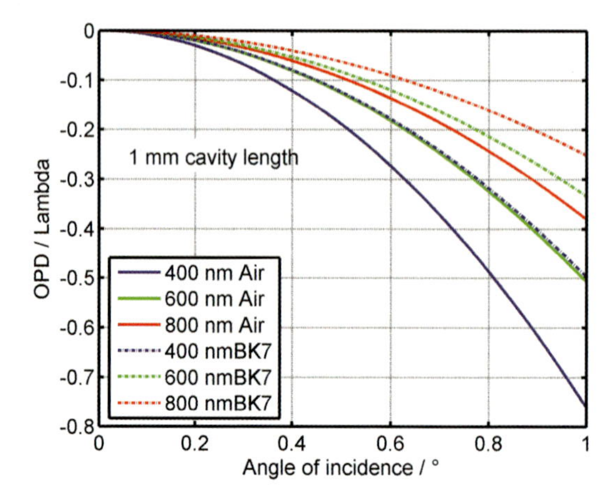

Figure 46-19: The OPD difference to $\theta = 0$ in units of λ for an air and a BK7 cavity of length 1 mm for three different wavelengths. An OPD difference of $\lambda/2$ will lead to zero fringe visibility.

Using (46-68), the maximum diameter $s_{\mathrm{max,air}}$ in conjunction with a collimator of focal length f can be calculated from (46-71).

$$s_{\mathrm{max,air}} = 2f \cdot \tan\left(\arccos\left(1 - \frac{\lambda}{4d}\right)\right) \tag{46-71}$$

Figure 46-20 shows examples of $s_{\mathrm{max,air}}$ (visibility $V = 0$) for a focal length of 1000 mm and wavelengths of 400, 600 and 800 nm in comparison with a maximum s to fulfil the Rayleigh condition.

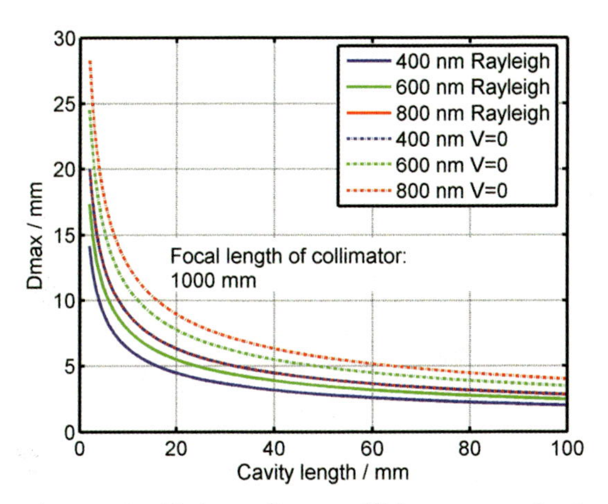

Figure 46-20: Maximum diameter of light source as a function of the air cavity length for a collimator of 1000 mm focal length and three different wavelengths as well as $V = 0$ and the Rayleigh criterion.

46.2.12
Interference of Waves having Different Polarization

When light is reflected at interfaces between different media or transmitted through birefringent media, polarization plays a major role. So far we have assumed that interfering light has the same polarization state. However, interference is disturbed if light from a reference source and an object have different optical paths that alter the state of polarization in different ways.

We consider two linear polarized plane waves traveling in the z direction. Their amplitudes are A_1 and A_2 respectively, their azimuthal orientations are α_1 and α_2 and their phases are ϕ_1 and ϕ_2.

$$E_1 = A_1 \begin{pmatrix} \cos a_1 \\ \sin a_1 \end{pmatrix} e^{i\phi_1} \tag{46-72a}$$

$$E_2 = A_2 \begin{pmatrix} \cos a_2 \\ \sin a_2 \end{pmatrix} e^{i\phi_2} \tag{46-72b}$$

The sum of both waves is calculated from (46-73).

$$E = \begin{pmatrix} A_x e^{i\phi_x} \\ A_y e^{i\phi_y} \end{pmatrix} \tag{46-73}$$

with

$$A_x = \sqrt{A_1^2 \cos^2 a_1 + A_2^2 \cos^2 a_2 + 2A_1 A_2 \cos a_1 \cos a_2 \cos(\phi_2 - \phi_1)} \quad \text{and}$$

$$A_y = \sqrt{A_1^2 \sin^2 a_1 + A_2^2 \sin^2 a_2 + 2A_1 A_2 \sin a_1 \sin a_2 \cos(\phi_2 - \phi_1)} \quad \text{and}$$

$$\phi_1 = \arctan \frac{A_1 \cos a_1 \sin \phi_1 + A_2 \cos a_2 \sin \phi_2}{A_1 \cos a_1 \cos \phi_1 + A_2 \cos a_2 \cos \phi_2} \quad \text{and}$$

$$\phi_2 = \arctan \frac{A_2 \sin a_1 \sin \phi_1 + A_2 \sin a_2 \sin \phi_2}{A_1 \sin a_1 \cos \phi_1 + A_2 \sin a_2 \cos \phi_2}$$

We calculate the irradiance and therefore the interference pattern from (46-74).

$$I = E \cdot E^* = A_x^2 + A_y^2 = A_1^2 + A_2^2 + 2A_1 A_2 \cos(a_1 - a_2) \cos(\phi_2 - \phi_1) \tag{46-74}$$

The contrast of fringes is modulated by a factor $\cos(a_1 - a_2)$. The visibility can be calculated as

$$V = \frac{2A_1 A_2 \cos(a_1 - a_2)}{A_1^2 + A_2^2} \tag{46-75}$$

The fringe visibility vanishes when both linear polarized waves are perpendicular to each other ($a_1 - a_2 = 90^\circ$ or 270°).

Let us assume two waves are linearly polarized with $a_1 = 0°$ and $a_2 = 90°$ such that there are no interference fringes visible when they are superposed. If we add a linear polarizer with an axis orientation θ in the x–y plane the transmitted components of both waves are given by

$$E_1 = A_1 \begin{pmatrix} \cos^2\theta \\ \sin\theta \cos\theta \end{pmatrix} e^{i\phi_1} \tag{46-76a}$$

$$E_2 = A_2 \begin{pmatrix} \sin\theta \cos\theta \\ \sin^2\theta \end{pmatrix} e^{i\phi_2} \tag{46-76b}$$

leading to an interference pattern as in (46-77).

$$I = E \cdot E^* = A_x^2 + A_y^2 = A_1^2 \cos^2\theta + A_2^2 \sin^2\theta + 2A_1A_2 \sin\theta \cos\theta \cos(\phi_2 - \phi_1) \tag{46-77}$$

The visibility of the fringes is then given by

$$V = \frac{2A_1A_2 \sin\theta \cos\theta}{A_1^2 \cos^2\theta + A_2^2 \sin^2\theta} \tag{46-78}$$

For A_1=A_2 and θ =45° the visibility is again V=1 and thus reaches its theoretical optimum (see figure 46-21).

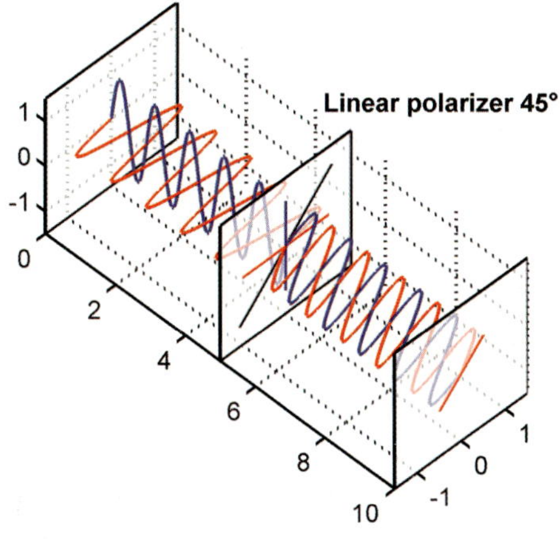

Figure 46-21: Two linearly polarized waves with azimuths of 0° and 90° are filtered by a linear polarizer oriented at 45° to make the appropriate components of both waves interfere.

46.2.13
Interference of Non-spherical Wavefronts

As a wave of arbitrary shape propagates, it continuously changes its shape. Only spherical or plane waves remain spherical or plane during propagation. In this chapter we show that the propagation of a non-spherical wavefront can be approximated by means of a Fourier transform of its deviation from a spherical wavefront.

To calculate the shape of an aspherical wave as it is propagating, we consider a sphere of radius p with its center at the origin of a cartesian coordinate system (figure 46-22).

We consider an infinite number of spherical point sources $P(x',y')$ on the surface of the sphere with wavelength λ, amplitude $A(x',y')$ and an individual phase $\phi(x',y')$.

Huygens' principle now lets us calculate the electrical field at a point Q(x,y,z) by integration over all spherical point sources [46-2]:

$$E(x,y,z) = \frac{1}{S}\iint_S \frac{A(x',y')}{r} e^{i(kr+\varphi(x',y'))} dx'dy' \tag{46-79}$$

where $r = \sqrt{(x-x')^2 + (y-y')^2 + (z-z')^2}$
and S is the area of the radiating portion of the spherical surface. $\varphi(x',y')$is the deviation of the non-sperical wavefront from the sphere.

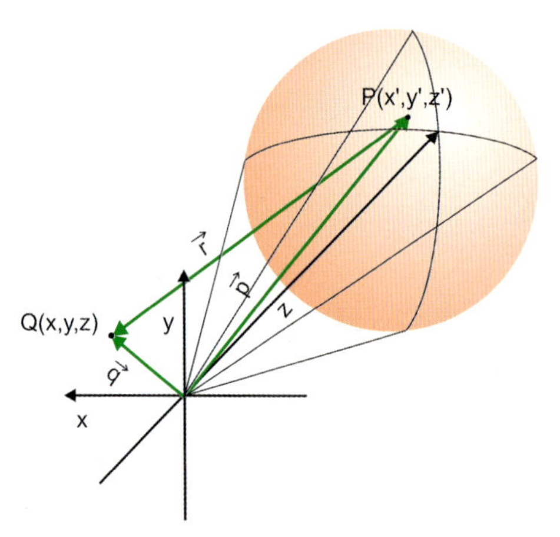

Figure 46-22: An aspherical wavefront is referenced to a sphere with radius p around the origin. A spherical wave at $P(x',y',z')$ with amplitude $A(x',y')$ and phase $\phi(x',y')$ propagates in space. Its amplitude and phase is determined at a point Q(x,y,z).

If we consider a small volume around the origin with dimensions x, y, z small in comparison to the radius p of the spherical surface, the square-root of r can be approximated according to:

$$r \approx p\left(1 - \frac{xx' + yy'}{p^2} - \frac{z}{p}\sqrt{1 - \frac{x'^2 + y'^2}{p^2}}\right) \tag{46-80}$$

Inserting (46-80) into (46-79) and assuming that $\frac{A(x',y')}{r} \approx \frac{A(x',y')}{p}$ we arrive at:

$$E(x,y,z) \approx \frac{e^{ikp}}{p\,S}\iint_S A(x',y')e^{i\varphi(x',y')} e^{-ikz\sqrt{1-\frac{x'^2+y'^2}{p^2}}} e^{-ik\frac{xx'+yy'}{p}} dx'dy' \tag{46-81}$$

We extend the integrals to infinity assuming that $A(x',y')$ describes the limited extension of the radiating area and we introduce new variables ξ and η and functions $f(x,y,z)$ and $F(\xi,\eta,z)$:

$$\xi = \frac{n\,x'}{\lambda\,p} \text{ and } \eta = \frac{n\,y'}{\lambda\,p} \tag{46-82}$$

$$f(x, y, z) = \frac{E(x, y, z) \cdot n^2 \cdot S}{p\,\lambda^2} e^{-ikp} \tag{46-83}$$

$$F(\xi, \eta, z) = A(\xi, \eta) e^{i\varphi(\xi,\eta)} e^{-ikz\sqrt{1-\left(\frac{\lambda\xi}{n}\right)^2-\left(\frac{\lambda\eta}{n}\right)^2}} \tag{46-84}$$

Equation (46-81) now becomes

$$f(x, y, z) = \iint\limits_{\infty} F(\xi, \eta, z) e^{-i2\pi(x\xi+y\eta)} d\xi d\eta \tag{46-85}$$

which is a two-dimensional Fourier transform from which we can calculate the electrical field distribution in a small volume around the origin for an arbitrary amplitude and phase distribution within a sphere of radius p with its center in the origin.

According to the rules of Fourier transformation we calculate the field distribution $F(\xi,\eta,z)$ on a sphere of radius p from a field distribution $f(x,y,z)$ near the origin by the inverse Fourier transform:

$$F(\xi, \eta, z) = \iint\limits_{\infty} f(x, y, z) e^{i2\pi(x\xi+y\eta)} dx dy \tag{46-86}$$

We now can calculate the amplitude and phase of a propagating aspherical wavefront in the neighborhood of $z=0$ by forward Fourier transformation. Selecting a variety of z-values then leads to the amplitude and phase distribution in a volume around the focus of the non-sperical wavefront.

If we want to know the amplitude and phase in the neighborhood of a distance p, we multiply the wavefront E(x,y,0) at $z=0$ with an appropriate quadratic phase factor and then apply an inverse Fourier transformation.

In figure 46-23, four examples are shown to demonstrate the effects of a defocus by 100 μm when an aberrated wavefront associates with a sphere of 50 mm radius of curvature and diameter 40 mm. The examples include a wavefront with 3 λ coma (peak-to-valley), one with 3 λ spherical aberration (peak-to-valley) and one with a phase step of $\lambda/2$, each having a constant amplitude. In the fourth example the round constant amplitude is disturbed by three dust particles which form obscurations; the wavefront being perfectly zero. The lower rows show the amplitude distributions within a region of 200 μm around the focus of the wavefronts. A heavy caustic for coma and spherical aberration is noticed with coma showing the typical "banana"-like shape in the x–z plane. A zero amplitude in a plane containing the optical axis is characteristic for a wavefront with a $\lambda/2$-phase step, whereas the three dust particles do not show a significant characteristic in the region near the focus.

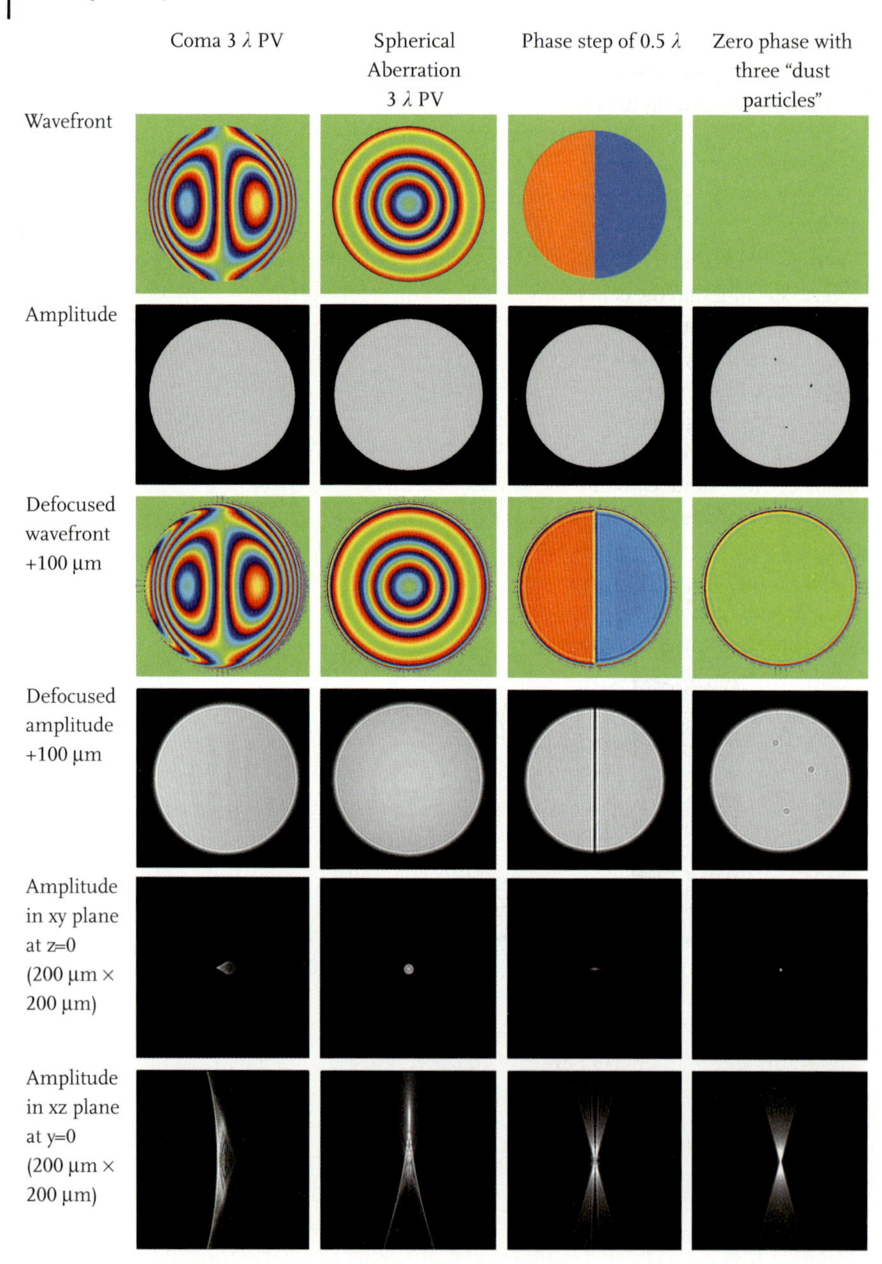

Figure 46-23: Four different non-spherical wavefronts and their propagation states "out of focus" and near the origin (see text).

For all defocused waves, the regions at the edges of the round-shaped areas of amplitude and wavefront are blurred. The neighborhoods of the phase step as well as the obscurations are modulated in amplitude and wavefront.

We now have the tools to calculate the interference effects of waves of non-spherical shape or having obscurations or rims that will alter the phase distribution during propagation according to Huygens' principle.

46.2.14
Interference at Two Plane Parallel Interfaces

Interference occurs in the region of overlay, when light is reflected from two optical surfaces, assuming that the coherence is sufficiently long. For the following considerations we assume three adjacent dielectrical transparent media with refractive indices n_1, n_2 and n_3 separated by two plane parallel optical interfaces of distance d. We call this configuration a Fabry–Perot resonator and the space between the interfaces the "cavity" [46-5], [46-6]. Incident light is reflected back and forth within the cavity and interferes with the directly reflected and transmitted portions. Figure 46-24 shows an incident wave with electrical field E_0 being separated into partial reflections and transmissions, which differ in the number of reflections within the cavity.

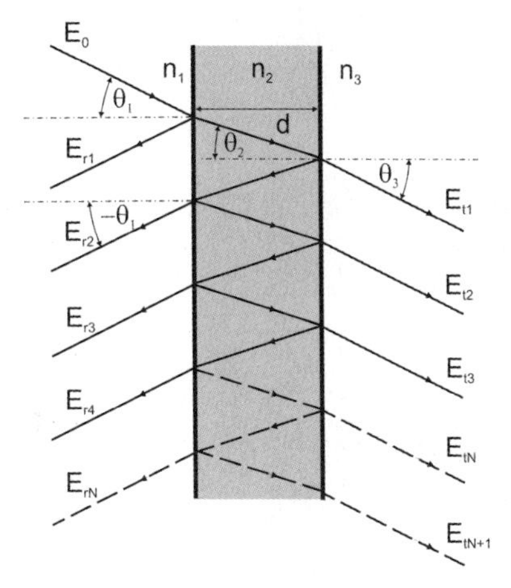

Figure 46-24: Reflection and transmission at a Fabry–Perot resonator consisting of two optical interfaces between three media of different refractive indices.

The Fresnel equations [46-2] for the reflected field components characterize the amplitudes of the reflected portions of an incident wave. Considering two plane parallel interfaces the E-fields perpendicular to the incident plane (TE or s-polarization) are given by (46-87) and (46-88).

$$r_{s1} = \frac{n_1 \cos\theta_1 - n_2 \cos\theta_2}{n_1 \cos\theta_1 + n_2 \cos\theta_2} \tag{46-87}$$

$$r_{s2} = \frac{n_2 \cos\theta_2 - n_3 \cos\theta_3}{n_2 \cos\theta_2 + n_3 \cos\theta_3} \tag{46-88}$$

The E-fields parallel to the incident plane (TM or p-polarization) are given by (46-89) and (46-90).

$$r_{p1} = \frac{n_2 \cos\theta_1 - n_1 \cos\theta_2}{n_2 \cos\theta_1 + n_1 \cos\theta_2} \tag{46-89}$$

$$r_{p2} = \frac{n_3 \cos\theta_2 - n_2 \cos\theta_3}{n_3 \cos\theta_2 + n_2 \cos\theta_3} \tag{46-90}$$

The sum of all individual reflections $E_{r1} \ldots E_{r\infty}$ from an incident complex field E_0 is calculated by (46-91).

$$E_r = E_0 \left(r_1 + \sum_{j=1}^{\infty} (1 - r_1^2)\,(-r_1)^{j-1}\, r_2^j\, (e^{i\phi})^j \right) \tag{46-91}$$

with $\phi = \frac{2\pi}{\lambda}\, 2n_2 d \cos\theta_2$ describing the optical path difference of two consecutive reflections like E_{rm} and $E_{r(m-1)}$ and r_1 and r_2 symbolizing the parallel as well as the perpendicular polarized reflectivity according to (46-87)–(46-90).

Following the rules of infinite series, (46-91) can be simplified to (46-92).

$$E_r = E_0 \left(r_1 + \frac{(1 - r_1^2)\, r_2\, e^{i\phi}}{1 + r_1 r_2\, e^{i\phi}} \right) \tag{46-92}$$

We arrive at the reflected intensity by adding the squares of the real and imaginary parts of (46-92):

$$I_r = I_0 \left(\frac{r_1^2 + r_2^2 + 2r_1 r_2 \cos\phi}{1 + (r_1 r_2)^2 + 2r_1 r_2 \cos\phi} \right) \tag{46-93}$$

Note that the fringes are periodic in ϕ, however their profile does not correspond to an exact cosine. The higher the reflectivities of the two reflecting interfaces, the more the fringe profile deviates from a sinusoidal shape. Only for very small reflectivities can the fringe be called sinusoidal. Figure 46-25 shows examples of transmitted intensities for a cavity length of 1 mm and amplitude reflectivities of $r = r_1 = -r_2 = 0.2, 0.5$ and 0.9. The wavelength is altered from 500 nm to 501 nm to show the varying intensities.

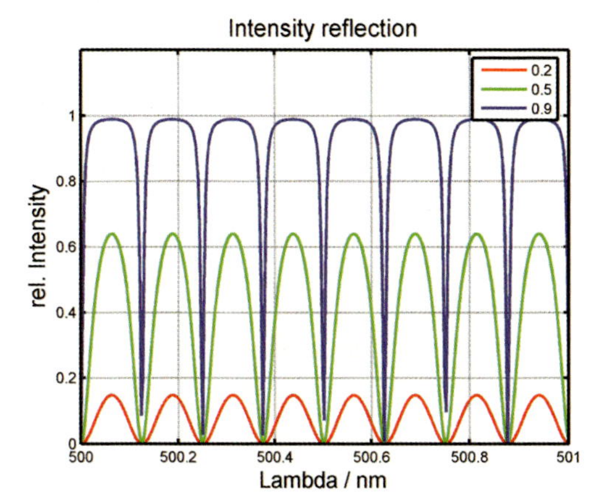

Figure 46-25: Reflected intensities for a cavity length of 1 mm and normal incident light of wavelength 500 nm to 501 nm for amplitude reflectivities $r = 0.2$, 0.5 and 0.9.

The Fresnel equations for the transmitted field components describe the amplitudes of the transmitted portions of an incident wave. For two plane parallel interfaces the E-fields perpendicular to the incident plane (TE or s-polarization) are given by (46-94) and (46-95).

$$t_{s1} = \frac{2n_1 \cos\theta_1}{n_1 \cos\theta_1 + n_2 \cos\theta_2} \tag{46-94}$$

$$t_{s2} = \frac{2n_2 \cos\theta_2}{n_2 \cos\theta_2 + n_3 \cos\theta_3} \tag{46-95}$$

The E-fields parallel to the incident plane (TM or p-polarization) are given by (46-96) and (46-97).

$$t_{p1} = \frac{2n_1 \cos\theta_1}{n_2 \cos\theta_1 + n_1 \cos\theta_2} \tag{46-96}$$

$$t_{p2} = \frac{2n_2 \cos\theta_2}{n_3 \cos\theta_2 + n_2 \cos\theta_3} \tag{46-97}$$

The sum of all individual transmitted fields $E_{t1} \ldots . E_{t\infty}$ from an incident complex field E_0 neglecting a phase factor is calculated by (46-98).

$$E_r = E_0\, t_1 t_2 \sum_{j=0}^{\infty} \left(-r_1 r_2\, e^{i\phi}\right)^j \tag{46-98}$$

where t_1 and t_2 symbolize the parallel as well as the perpendicularly polarized transmittance according to (46-94)–(46-97).

Following the rules of infinite series, (46-98) can be simplified to (46-99).

$$E_t = E_0\left(\frac{t_1 t_2}{1 + r_1 r_2\, e^{i\phi}}\right) \tag{46-99}$$

The transmitted intensity is calculated according to eq. 46-100.

$$I_t = I_0\left(\frac{t_1^2 t_2^2}{1 + (r_1 r_2)^2 + 2 r_1 r_2 \cos\phi}\right) \tag{46-100}$$

When absorption is zero, $I_r + I_t = I_0$. Therefore, the fringes, like the reflected light, do not correspond to an exact cosine. The higher the reflectivities of the two reflecting interfaces, the further the fringe profile of the transmitted light deviates from a sinusoidal shape. Note, however, that the visibility of the transmitted fringes is very poor when the reflectivities are low. It increases with increasing reflectivities to become 1 for $r \to 1$. Figure 46-26 shows examples of transmitted intensities for a cavity length of 1 mm and amplitude reflectivities of $r = r_1 = -\ r_2 = 0.2$, 0.5 and 0.9. The wavelength is altered from 500 nm to 501 nm to show the varying intensities.

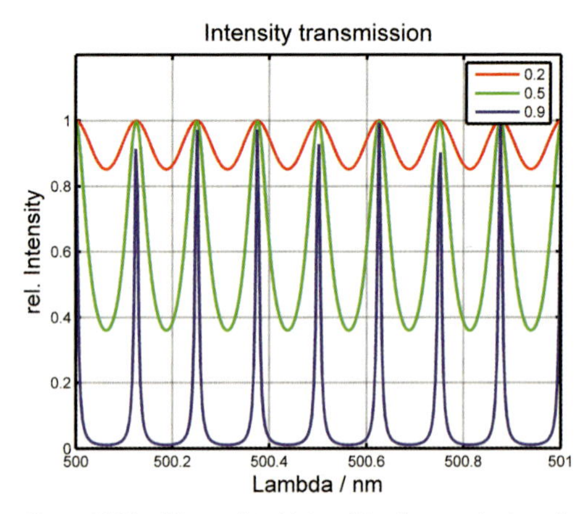

Figure 46-26: Transmitted intensities for a cavity length of 1 mm and normal incident light of wavelength 500 nm to 501 nm for amplitude reflectivities $r = 0.2$, 0.5 and 0.9.

46.2.15
Haidinger Fringes

When light is emitted by a point source at a finite distance from a cavity it enters the two surfaces at different angles θ. Assuming a plane parallel cavity the transmitted and reflected intensities are therefore described by (46-93) and (46-100) keeping in mind that the amplitude reflectivities $r_1(\theta_1,\theta_2)$, $r_2(\theta_2,\theta_3)$ at the two interfaces as well as the optical path difference $\phi(\theta_2)$ are functions of the incidence angle [46-1], [46-2].

We consider a plane thin glass plate and a monochromatic point source E_0 next to it (figure 46-27). Two optical systems of focal length f' are placed in front of and behind the plane plate, respectively . Concentric circular interference rings can be observed on screens in the focal planes of the optical systems. Light reflected or transmitted at an angle θ to the optical axis is focused onto a ring of radius $h = f' \tan\theta$. We call the interference patterns *Haidinger fringes* or *fringes of equal inclination*. Their location is at infinity or, as in our example, in the focal plane of an optical system.

For a glass cavity in air of thickness d and refractive index n the resulting fringe phase ϕ is calculated from (46-101).

$$\phi = \frac{2\pi}{\lambda} 2n\, d \cos\theta' \tag{46-101}$$

Using the law of refraction the internal angle θ' can be replaced by

$$\theta' = \arcsin\left(\frac{\sin\theta}{n}\right) \tag{46-102}$$

to determine the location of the fringes on the two screens.

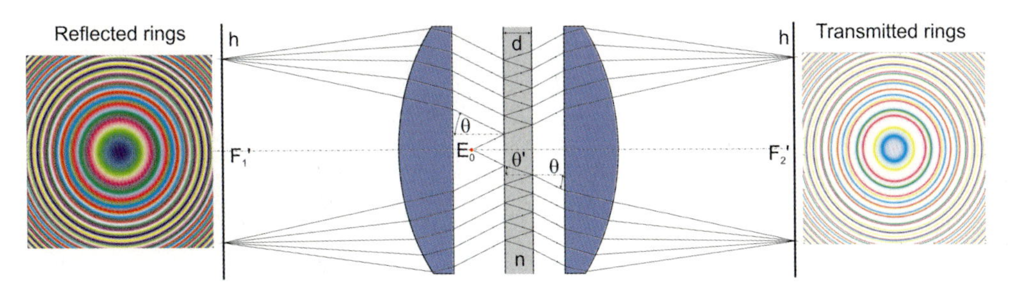

Figure 46-27: Arrangement to observe Haidinger fringes (fringes of equal inclination) in transmission and reflection.

ϕ defines the location of fringe extrema (maxima or minima), however the fringe intensity profiles depend on the amplitude reflectivities according to (46-93) and (46-100). For high reflectivities Haidinger fringes become sharp rings with high visibility in transmission and reflection, whereas for low reflectivities Haidinger fringes show a sinusoidal character with low visibility in transmission. Figure 46-28 shows some examples of transmitted and reflected Haidinger fringes at a plane parallel plate made from BK7 without coating and with a highly reflecting coating.

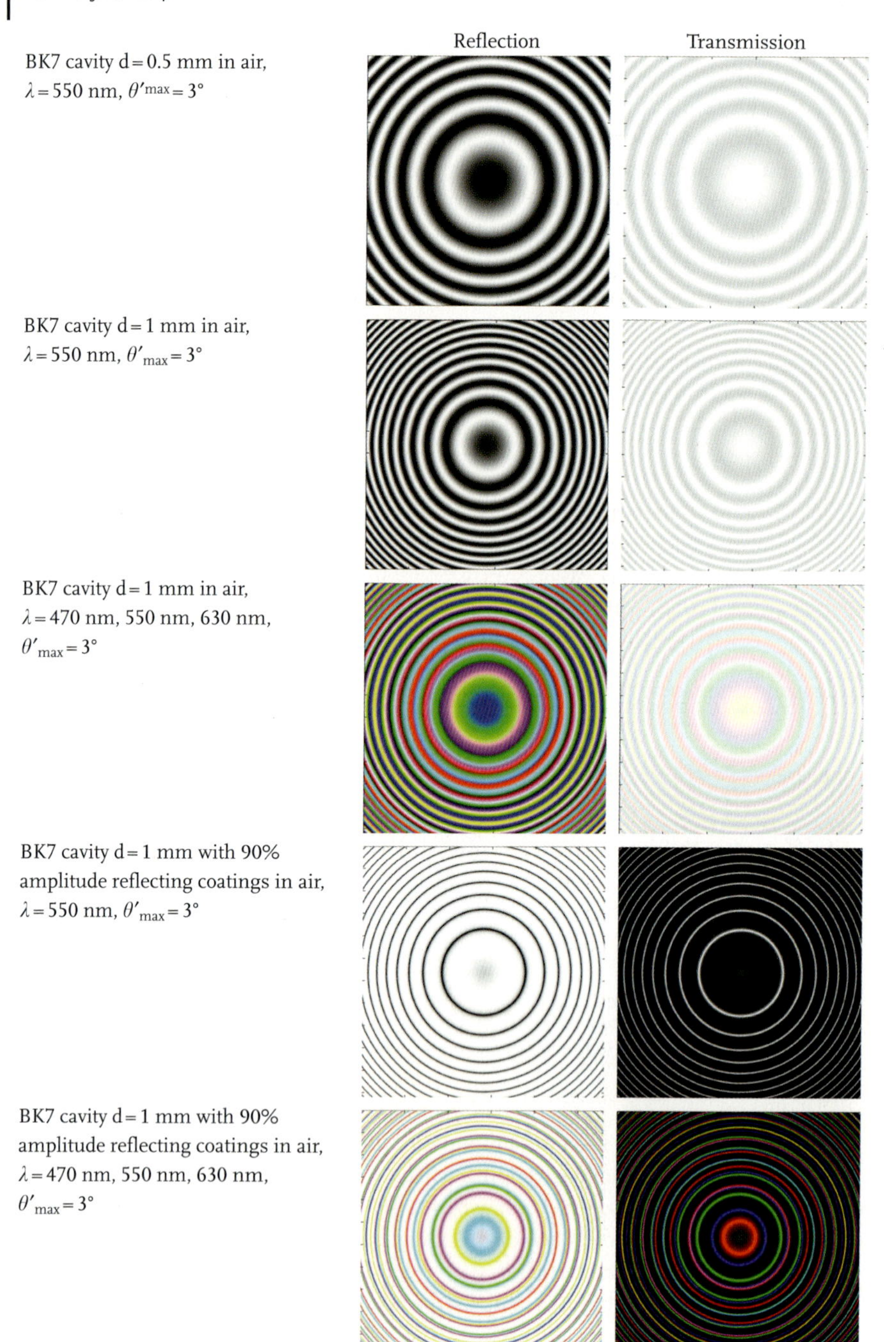

Figure 46-28: Haidinger fringes at a plane parallel plate in air made from BK7 without coating and with a highly reflective coating.

46.2.16 Newton Fringes

When a small point source is placed in the focal point of an optical system (collimator) it can be used to illuminate a cavity at a certain angle θ [46-1]. Assuming an almost plane parallel cavity and an incidence angle of $\theta \approx 0$ the transmitted and reflected intensities can still be calculated from (46-93) and (46-100), but with the expressions for the reflectivities and the optical path differences taking a simple form as shown in (46-103)–(46-104). For normal incidence we therefore have:

$$r_1 = \frac{n_2 - n_1}{n_2 + n_1} \quad (46\text{-}103)$$

$$r_2 = \frac{n_3 - n_2}{n_3 + n_2} \quad (46\text{-}104)$$

We consider a plane thin glass plate and a monochromatic point source at the focal point of a collimator (figure 46-29). The light is directed normal to the glass plate, which is not perfect. We assume that, along the optical axis of the collimator, the plate has a varying thickness of $d(x,y)$ and an inhomogeneous index of refraction $n(x,y)$. We call the interference patterns *Newton fringes* or *fringes of equal thickness*. Their location is near the plane plate surfaces. The fringe phase variations of the reflected and transmitted light are given in (46-105).

$$\phi(x,y) = \frac{2\pi}{\lambda} 2n(x,y)\, d(x,y) \quad (46\text{-}105)$$

In figure 46-29 a beam-splitter is used to observe the Newton fringes for reflection. The fringes show the variation in the optical path differences across the plane plate in terms of λ. In transmission the expected complementary interferogram with low visibility can be observed.

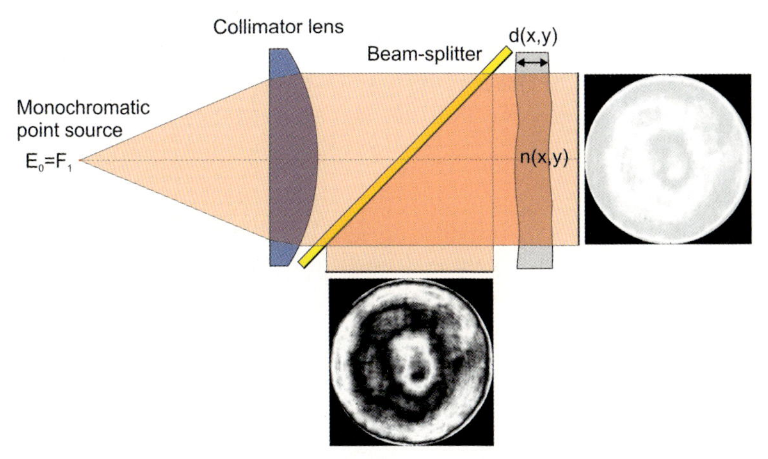

Figure 46-29: Arrangement to observe Newton fringes (fringes of equal thickness) for transmission and reflection.

46.3 Interferometers

Over a period of 300 years many types of interferometers and interferometric arrangements have been invented and used for very different purposes. In this chapter we will concentrate on those that are frequently used for either optical component and system testing, or for distance and angle metrology. The variety is still so great that we need to classify them into groups so that their individual suitability for special applications can be seen more clearly. [46-1].

We will discuss only so-called two-beam interferometers or those that have the primary property of a two-beam interferometer, since they are by far the most frequently used and commercially available interferometers.

In general, we have to distinguish between interferometers using the component or system under test as a substantial part of the interferometer itself and those merely generating an interferogram from an incident wavefront. The latter are called "wavefront sensors" or "wavefront analyzers" because the unknown wavefront, independently of its origin, enters the interferometer to be analyzed according to its wavefront profile and/or amplitude distribution.

A whole variety of classifications for interferometers exist such as:

- common-path
- equal-path
- unequal-path
- single-pass
- double-pass
- multiple-pass
- lateral-shear
- radial-shear
- rotational-shear
- reversal-shear
- two-beam
- multiple-beam.

A further distinction can be made by the way in which the reference and test wavefronts are separated on their way from the light source to the sensor. There are separation methods such as:

- amplitude division
- wavefront partition
- division by scattering
- division by polarization and birefringence
- division by diffraction.

We will refer to these classifications while describing the most important interferometers used in component and system testing.

46.3.1
Newton Interferometer

Isaac Newton (1643–1727) first reported his observation of interference fringes that we now call Newton's Rings while observing a wedge-shaped air gap between the surfaces of two prisms. Joseph Fraunhofer (1787–1826) was the first to use this phenomenon as a measuring tool to test spherical surfaces against a masterpiece. His early death prevented the spread of the technique among the optical community. It was therefore up to August Löber (1830–1912), the senior workshop master at Carl Zeiss, to reinvent what today we call the test glass method or Newton interferometer. It is still the standard inspection procedure in all optical workshops to test spherical surfaces concerning their radius of curvature and deviation from perfect sphericity with resolutions up to $\lambda/10$ [46-1], [46-17], [46-18].

When two polished optical surfaces of concentric shape are placed in close contact with each other, a thin space of air usually remains between them due to the fact that both surfaces deviate slightly in shape. Light passing through these objects, which are in close contact, will be reflected at and transmitted through both of the surfaces. Since the cavity is very small even white light from an extended light source will interfere and demonstrate the well-known Newton fringes (section 46.2.16) in reflection as well as in transmission. However, the transmitted interference fringes show very poor contrast, so that the reflected ones are preferred for surface testing. In an optical workshop, monochromatic light is used to improve fringe contrast and to have a well-defined wavelength as reference. The arrangement for a Newton interferometer is shown in figure 46-30, where a beam-splitter is used to split the illumination and observation path so that an almost perpendicular view onto the test surface is possible.

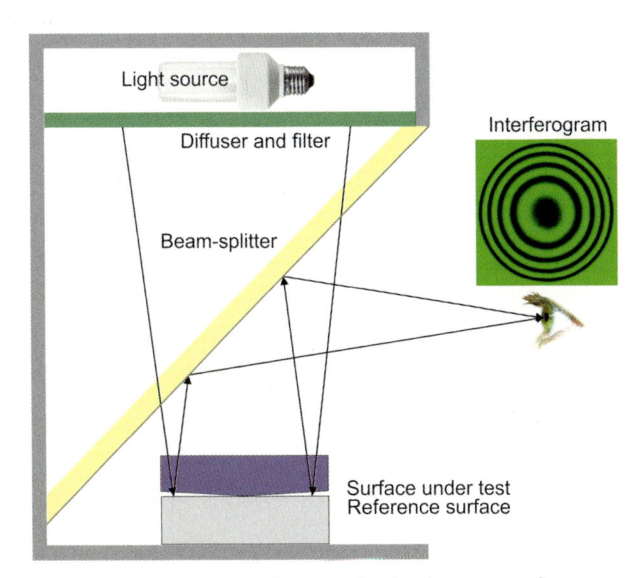

Figure 46-30: Newton interferometer for the observation of interference fringes (Newton rings) between a high-quality reference surface and the surface under test.

Figure 46-31: Some examples of surface deformations viewed in a Newton interferometer using monochromatic or white light.

When a suitably high-quality reference surface is available, the deviations in the surface under test are measured by counting the number of rings or fringes and examining their regularity. The number of rings indicates the difference in radius between the surfaces. This is known as the power, or sometimes as the figure, and is measured in rings. One ring is equivalent to a deviation of half the wavelength. Additionally, the rings may exhibit distortion due to non-uniform shape differences, which we call irregularity. In figure 46-31 an overview of typical deviations is given.

Note that the first example shows straight fringes introduced by leaving a wedge-shaped air space between the surfaces. The deviations from straightness, parallelism and equal spacing of the fringes are then encountered as the error of the surface under test. Note also that, when the air space goes to zero, a dark fringe is visible. Thus, using white light, the "zero" fringe is black and "colorless" whereas the adjacent fringes are more or less colored. This enables the user to identify zones where both surfaces are in direct contact and their deviation is zero. In a white-light arrangement the dark colorless fringe of the contact zone is followed by an almost colorless white fringe. The next dark fringe has a yellow rim on one side of the contact zone and a bluish rim on the opposite side.

Using monochromatic light, contact zones can only be identified when slight pressure is applied to the elements. Fringes then move away from the dark fringe of the contact zone. Figure 46-32 illustrates this effect for concave and convex spherical deviations. In the first case the fringes move towards the center, while in the second case they move away from the center. In figure 46-33 a small wedge-shaped air space is introduced. The fringes move away from the contact zone when slight pressure is applied. The pressure method allows us to determine the "sign", whereas the number of fringes determines the magnitude of the deviation.

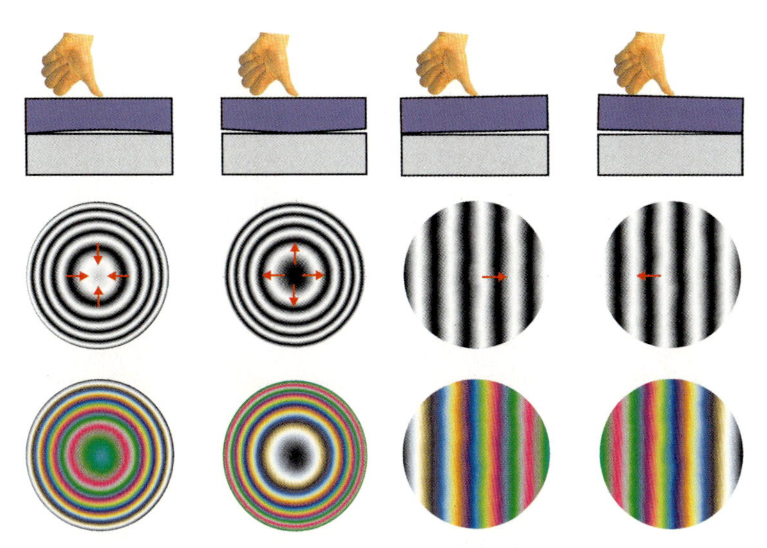

Figures 46-32 and 46-33: When pressure is applied near the center of a test glass arrangement the fringes move away from the contact zone. When white light is used, a dark colorless fringe appears at the contact zone.

The number of fringes to be encountered depends on the angle of observation. Looking through a plane parallel glass plate the number of fringes is given by m according to:

$$m = \frac{2d\cos\theta}{\lambda} \tag{46-106}$$

where d is the thickness of the air gap and θ the angle of observation (figure 46-34). The number of fringes reduces as the observation angle increases.

Fringe deviations can be estimated to a tenth of a fringe-spacing by visual inspection when three to five parallel fringes are introduced. Well-trained specialists are able to detect defects down to $\lambda/20$. When concentric rings are inspected the defects can be estimated to approximately half a ring.

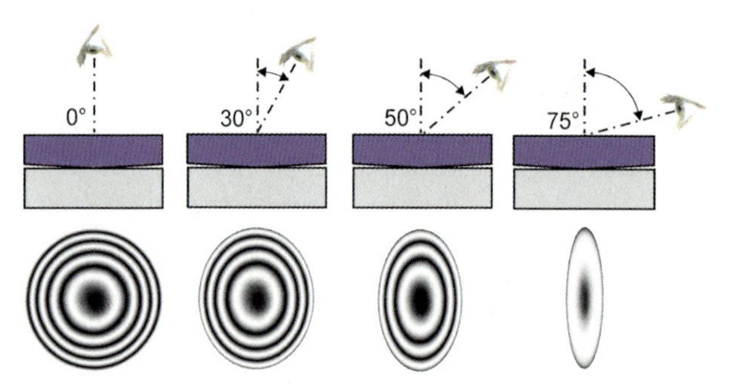

Figure 46-34: The number of visible fringes decreases when the angle of observation is increased.

Test plates or glasses are made flat or spherical to within small fractions of a fringe.

The accuracy of a test glass is only as good as the means used to measure its curvature and residual irregularity. To prevent damage to the surfaces extreme care must be used when placing the reference surface in contact with the actual surface. Dust particles have to be removed from the surfaces by a brush or air stream before contact. When soft glass material is used the risk of scratches is particularly high.

Test glasses are manufactured either as "looking through" types with a spherical back surface designed to help the user to look perpendicularly onto the surface under test from a convenient observation distance or they may be of base-plate type with the flat back-side being placed onto the measurement table (figure 46-35). In the latter case the user is looking through the lens under test, which then has to be polished on both sides.

When test glasses are handled manually they warm up and change their shape in a non-uniform way. For accurate measurements, test glasses and test pieces must have enough time for temperature adaptation. For this reason, test glasses are preferably made from Zerodur, ULE or fused silica, but Pyrex and Duran are also suitable materials.

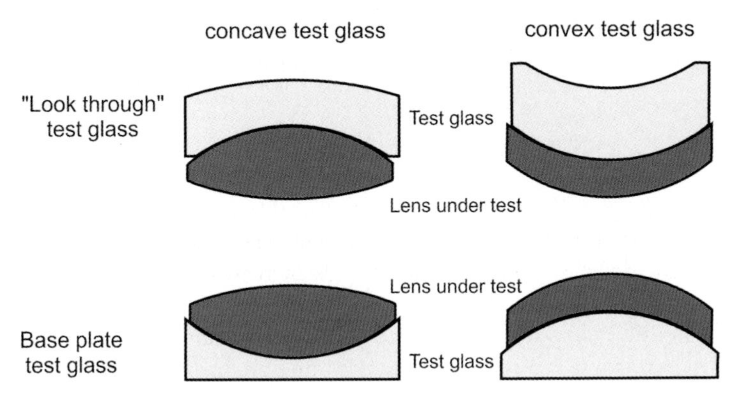

Figure 46-35: Different variations of test glasses used to test convex or concave spherical surfaces.

46.3.2 Fizeau Interferferometer

The main drawback in using test glasses is that, for each curvature, a special test glass must be manufactured. Therefore large storage areas are needed in optical companies to keep the test glasses clean and safe. Test glasses have to be recalibrated from time to time to check their quality and radius of curvature. Contact with lenses under test runs the risk of causing scratches and surface imperfections. Contact with the optical surfaces also limits the resolution of the test method because there is the possibility of dirt between the surfaces and uneven pressure when combining the two elements.

For contactless testing of optical surfaces a Fizeau interferometer can be used which will be described in the following. A Fizeau interferometer basically measures the optical path difference OPD between a reference surface and the surface under test [46-1], [46-19], [46-20].

Figure 46-36 shows a typical Fizeau setup to test spherical surfaces. A small coherent monochromatic light source is placed at the front focal point F_1 of the first collimator. A plane wavefront leaves the collimator and a following beam-splitter divides the wavefront into a transmitted and a reflected portion. The beam-splitter can be a plane parallel plate with a partially reflecting and an anti-reflection coated surface or a plate carrying a small wedge. When testing small optical parts a beam-splitter cube is also useful. The unwanted reflection can then be blocked by the beam-stop as described below. The light reflected from the beam-splitter is not used in the interferometer and must be blocked by an absorbing material. The transmitted light enters a so-called transmission sphere which forms a high-quality spherical wavefront converging toward its back focal point F′. A transmission sphere comprises a final spherical surface concentric to the converging wavefront with its center C_R coinciding with F′. The surface is called the reference surface or Fizeau surface. Since it is not coated it causes a portion of the transmitted light to be reflected. Due to the concentricity of the transmitted wavefront and the reference surface, the returning light mainly follows the same optical path.

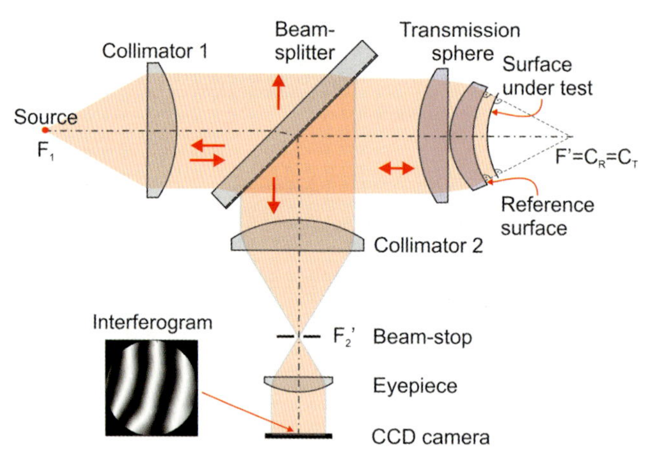

Figure 46-36: Schematic overview of a Fizeau interferometer to test spherical surfaces by means of a transmission sphere.

The light transmitted by the reference surface hits a spherical surface under test, which is also positioned concentric to the reference surface. Its center of curvature C_T coincides with C_R and F′. The reflected wavefront carries twice the deformation of the spherical test surface (figure 46-37a), because the light has to pass the deformed portion before and after reflection. The reflected light interferes with the light reflected by the reference surface as it returns to the beam-splitter. The beam-splitter again divides both waves into a transmitted and reflected portion. A second collimator now focuses the reflected portion to its back focal point F_2', where an opaque beam-stop carrying a small bore is positioned. It lets the reference and test waves pass, but blocks erroneous and unwanted reflections and stray light coming from other surfaces of the setup. The diameter of the bore must be small to block most of the unwanted light, but large enough not to limit the lateral resolution of the system. More details along with a calculation of the optimum size are given in Section 46.4.2.

The front focal point of a small lens system, called the eyepiece, coincides with the back focal point F_2' of the second collimator and forms a Kepler type of telescope. The returning reference and test waves are then plane waves, which will interfere at the plane of a camera sensor. Any deformations in either the reference or test wave result in deformed interference fringes which are captured by the camera.

The benefits of a Fizeau interferometer are as follows.

1. There is no contact between the reference and test surfaces (so no damage or scratches from contact).
2. Each transmission sphere can test a variety of convex or concave spherical surfaces of different radii.

3. The quality of system elements other than the reference surface are of secondary importance because the optical paths of reference and object beams are almost identical.
4. A computer system connected to the camera can evaluate the interferograms by digital means to improve the resolution of surface deviations by a factor of up to 1000 over the test glass method.

The space between the reference surface and the surface under test is called the cavity. All parameters having an effect on the optical path within the cavity are part of the measured result. So it is not only the shape difference between the reference and test surfaces, which is displayed as interference fringe deformations, but also the local and temporal refractive index distribution of the air within the cavity, which affects the result. Since long and short-term drifts in temperature, pressure and humidity continuously cause turbulent or laminar air flows within the cavity, the interference fringes usually move and change their shape. Furthermore, mechanical drifts or vibrations as well as acoustic excitations of the arrangement will have a continuous impact on the results.

This method differs from the test glass method in that the test surface has to be adjusted relative to the reference surface. However, in general, there is no fixation or adjustment help in finding the correct axial position of the test surface. So an absolute measurement of the radius of curvature is not possible in a Fizeau arrangement with only one measurement in one position, without further adjustment. We will see later that a Fizeau interferometer can be used in a special arrangement and by means of special techniques to measure radii of curvature accurately.

In a digital interferometer, where a computer calculates the surface deviations of a spherical test surface, the effects of misalignment such as decenter, tilt and axial misplacement (defocus) are removed by fitting and subtraction of a suitable set of polynomials. In most cases Zernike polynomials [46-1] are selected, especially when round-shaped components are being measured.

The Fizeau arrangement in figure 46-36 uses two collimators of the same size; the first in front of the beam-splitter, the second as an objective lens for the observing telescope at the exit of the interferometer. When convex lenses with large diameters have to be tested, the appropriate transmission sphere, beam-splitter and collimators have to be even larger than the test surface. In this case it makes sense to use only one collimator behind the beam-splitter as shown in figure 46-38. The drawback of this arrangement is the fact that diverging light coming from the light source now crosses the beam-splitter. Astigmatism, spherical aberration and coma will now be introduced to the illuminating beam. However, both the reference and the test beams will carry the same aberrations. If the F-number ($F/\# = f'/D$) of the collimator is sufficient large, for instance > 5, the aberrations caused by the beam-splitter can be neglected. If this is not possible for some reason, the beam-splitter surface next to the light source can be aspherized to create a perfect spherical wavefront.

Deformation of test surface by $\Delta s(x,y)$

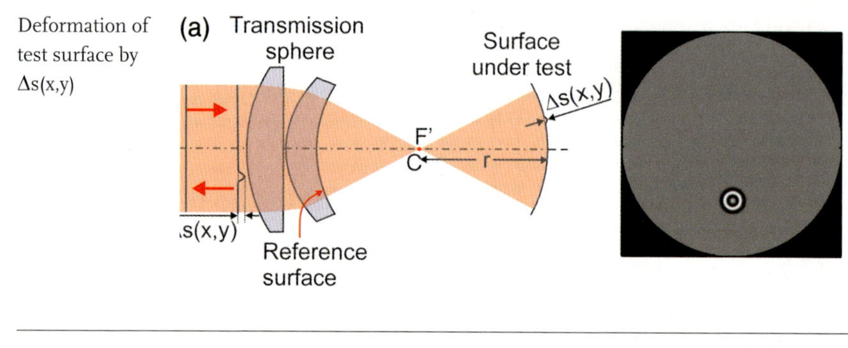

Defocus of test surface by Δz

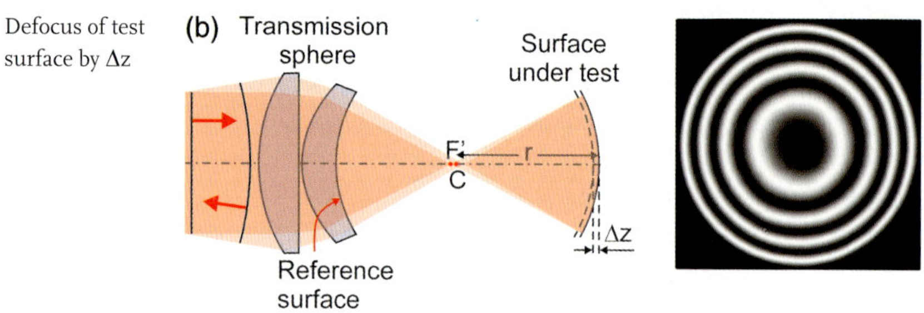

Decenter of test surface by Δy

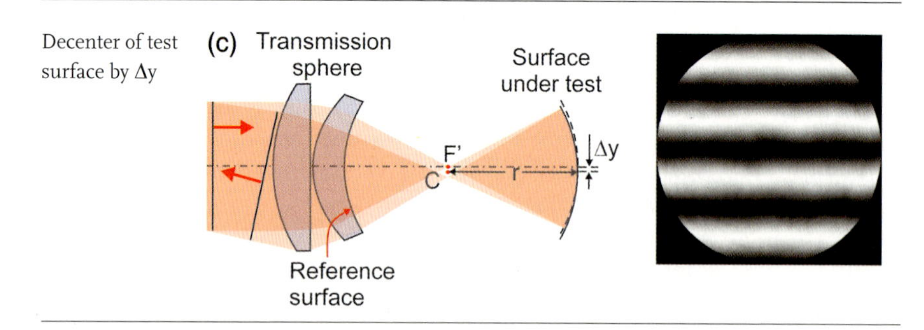

Figure 46-37: Interferograms for deformed (a), defocused (b), and decentered (c) test surfaces in an interferometric arrangement using transmission spheres.

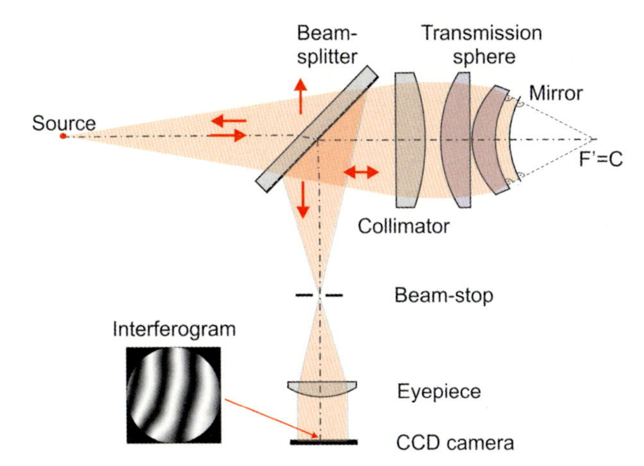

Figure 46-38: The Fizeau interferometer using only one collimator to test spherical surfaces by means of a transmission sphere.

A Fizeau arrangement can also be used to test plane surfaces in reflection (figure 46-39). In this a transmission flat is used instead of a transmission sphere. A transmission flat is a plane parallel plate with an anti-reflection coating on one side and no coating on the side next to the test surface. In most cases the plate carries a small wedge so that possible remaining reflections can be blocked by the beam-stop.

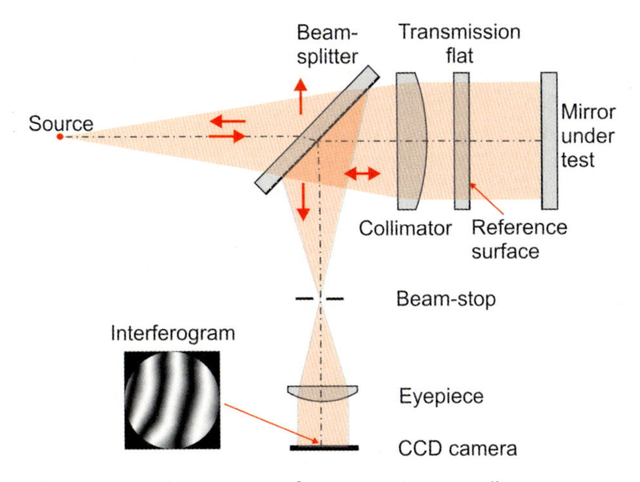

Figure 46-39: The Fizeau interferometer using one collimator to test plane surfaces by means of a transmission flat.

A Fizeau interferometer is also suitable for testing transmissive objects like glass plates or glass blocks carrying plane polished surfaces (figure 46-40). When the reference surface and flat mirror are of good optical quality the optical path difference is mainly influenced by the quality of the surfaces of the transmissive object or the homogeneity of its material. In this way homogeneity testing of raw glass material can be carried out. To avoid the influence of the surfaces of the object sometimes high-quality glass plates, called oil plates, are attached by placing immersion liquid of the same refractive index as the object under test between the object and the oil plates. In this case the object does not need to be polished. The oil plates must carry a wedge to avoid interference between the object's surfaces and the reference flat or flat mirror.

If the transmissive object comprises two parallel polished surfaces, internal reflections will generate an interference pattern that superimposes on the interferogram of the main cavity. The interference pattern cannot usually be interpreted properly. To prevent those effects, the transmissive object has to carry a wedge causing interference fringes that are too fine to be resolved by the CCD camera.

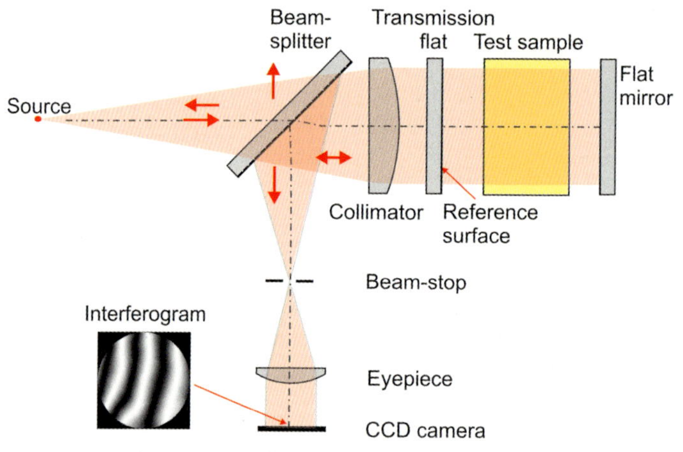

Figure 46-40: The Fizeau interferometer using one collimator to test plano-transmissive objects such as glass blanks by means of a transmission flat and a high-quality flat mirror.

In a Fizeau interferometer the fringe visibility depends on the reflectivities R_1 and R_2 of the reference and test surfaces, respectively. The reflectivity of the beam-splitter does not influence the fringe visibility but determines the irradiance returning to the camera.

Let us assume a non-polarizing beam-splitter having a reflectivity of R_B and a transmissivity of T_B. For an incident irradiance I_0 the interfering beams returning to the camera of a Fizeau interferometer carry the individual irradiances

$$I_{T1} = I_0 \, R_B \, T_B \, R_1 \text{ and } I_{T2} = I_0 \, R_B \, T_B \, (1 - R_1)^2 \, R_2 \tag{46-107}$$

The reflective elements are assumed to be absorption-free and non-diffusing. We also assume that the reflectivity R_1 of the reference surface is small (< 5%) and set $(1 - R_1)^2 \approx 1$.

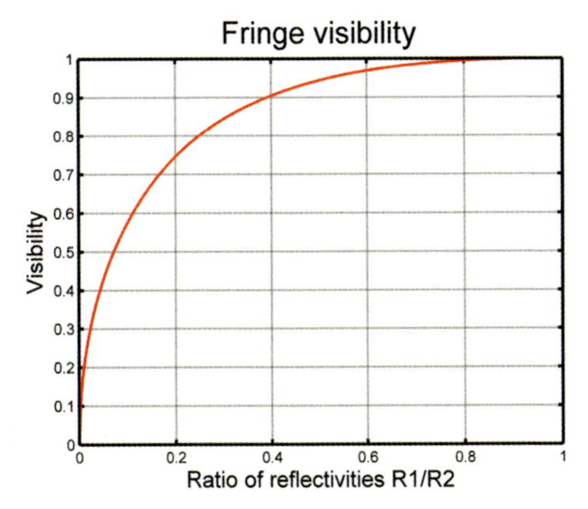

Figure 46-41: Fringe visibility in a Fizeau interferometer for different reflectivity ratios. The reflectivity R_1 of the reference is assumed to be small.

The fringe visibility for a coherent interference signal is then given by

$$V = \frac{2\sqrt{I_{T1} I_{T2}}}{I_{T1} + I_{T2}} \approx \frac{2\sqrt{R_1 R_2}}{R_1 + R_2} = \frac{2\sqrt{k}}{k+1} \tag{46-108}$$

where $k = \dfrac{R_1}{R_2}$ denotes the ratio of reflectivities of the reference and test surfaces. Figure 46-41 shows that fringe visibility is very good for a wide range of ratios but drops down below 50% for ratios smaller than 1:14.

The interfering beams returning to the source are given by

$$I_{S1} = I_0\, T_B^2 R_1 \text{ and } I_{S2} = I_0\, T_B^2 (1 - R_1)^2 R_2 \tag{46-109}$$

The light that is reflected from the beam-splitter when entering the interferometer arm is lost. Its portion together with the lost portion transmitted through the test surface is thus given by

$$I_L = I_0 \cdot R_B + T_B (T_B + R_B)(1 - R_1 - (1 - R_1)^2 R_2) \tag{46-110}$$

Figure 46-42 shows the bias irradiances I_L, $I_S = I_{S1} + I_{S2}$ and $I_T = I_{T1} + I_{T2}$ for reflectivities R_1=0.04 and R_2=0.04. We see that the optimum light efficiency is for a beam-splitter reflectivity of 50%. In this case half of the light is lost on first reflection. Half of the light reflected from the reference and the test surface reaches the camera, the other half returns to the light source.

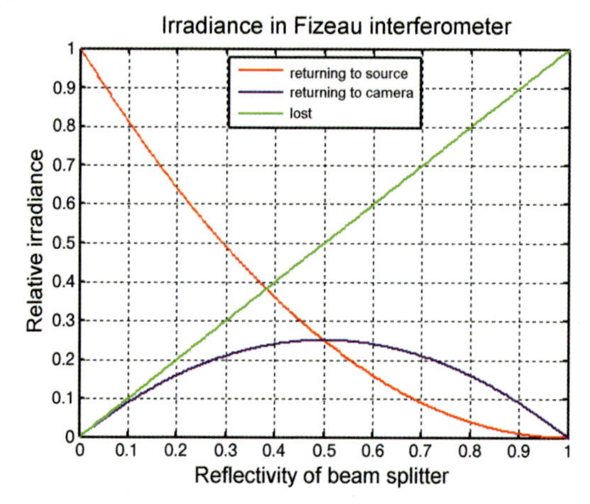

Figure 46-42: Irradiances at different interferometer exits in a Fizeau interferometer with R_1=0.04 and R_2=0.04: I_S (red) returns to the light source and is not used, I_T (blue) is detected at the camera, I_L (green) is lost on first incidence at the beam-splitter and lost on transmission through the test surface.

46.3.3
Twyman–Green Interferometer

In 1881 Albert Michelson introduced an interferometer, which was meant to carry out experiments to prove the existence or non-existence of the so-called world ether, the medium which it was thought was the carrier of electromagnetic waves in the universe. It was later used to provide experimental evidence for special relativity, to discover hyperfine structure in the energy levels of atoms, and to measure the tidal effects of the moon on the earth [46-21].

Like the Fizeau interferometer, the Michelson interferometer operates on the principle of division of amplitude, established by the use of a beam-splitter, which divides the light into two beams of almost equal intensity. Figure 46-43 gives an overview of the basic principle. Illumination is provided by an extended monochromatic or polychromatic source. A beam-splitter of 50% transmittance and 50% reflectance divides and directs the light into the reference and test arms of the interferometer. Both arms end with a plane mirror, which reflects the light back to the beam-splitter. The two beams are subsequently recombined in a common region where interference occurs and fringes can be detected [46-1], [46-21]–[46-23].

A compensator plate, which is identical to the beam-splitter in inclination, thickness and material, is introduced into the interferometer arm adjacent to the reflecting surface of the beam-splitter. The compensator plate is necessary to provide equal optical path lengths in both arms for the total spectrum entering the interferometer. If it is not used then the light in the first interferometer arm would transmit through the beam-splitter three times whereas the light in the second interferometer arm would transmit only once. In the case of broad-band light no fringes would be visible.

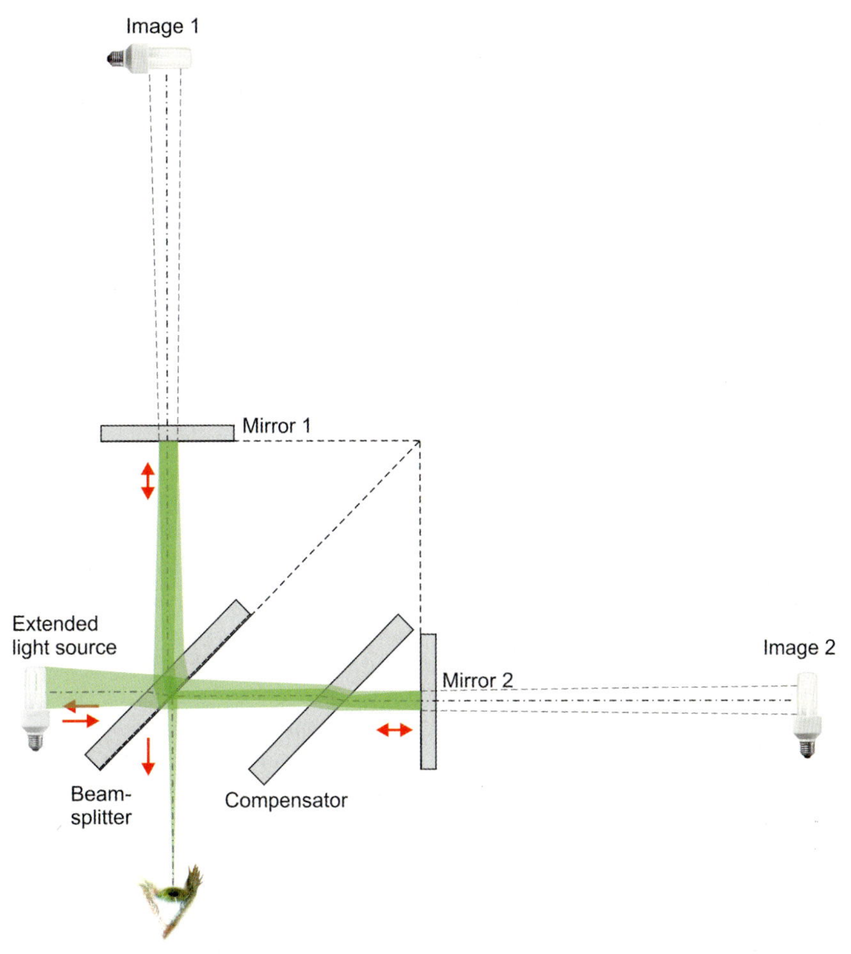

Figure 46-43: A Michelson interferometer in an arrangement to observe fringes of equal inclination.

An observer at the interferometer's exit would see two images of the extended light source behind the mirrors (figure 46-43). If one of the mirrors is moved along the optical axis, the reflected image is moved by twice that distance. The observer would then observe Haidinger fringes of equal inclination.

Although the Michelson interferometer was hardly ever used to test optical surfaces it provided the basic idea which Twyman and Green used to invent an interferometer and then patent it in 1916. It was intended to test prisms and microscope objectives.

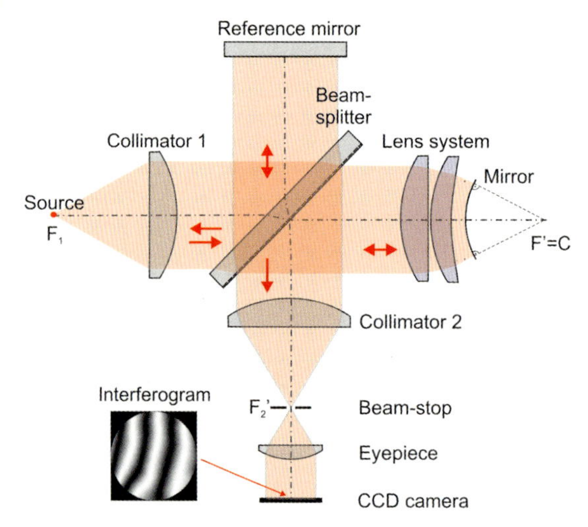

Figure 46-44: Schematic overview of a Twyman–Green interferometer used to test spherical surfaces by means of a lens system.

Figure 46-44 shows a typical Twyman–Green interferometer used to test spherical surfaces. A small coherent monochromatic light source is placed at the front focal point F_1 of the first collimator. A plane wavefront leaves the collimator and a following beam-splitter divides the wavefront into a transmitted and a reflected portion. As in the case of the Fizeau interferometer the beam-splitter can be a plane parallel plate with one partially reflecting and one anti-reflection coated surface or a plate carrying a small wedge. The unwanted reflection can then be blocked by the beam-stop. If the diameter of the interferometer arms is conveniently small a beam-splitter cube can also be used. The light reflected from the beam-splitter travels towards a plane mirror from which it is reflected back to the beam-splitter. The mirror closing the reference arm of the interferometer is called the reference mirror. The light transmitting the beam-splitter enters the test arm of the interferometer. In the case of a spherical test surface it first meets a lens system which forms a high-quality spherical wavefront converging toward its back focal point F'. Note that in the Twyman–Green case this is not what we call a transmission sphere, since it no longer has a final spherical surface which is concentric to the converging wavefront.

The light transmitted by the lens system strikes the spherical test surface whose center of curvature C coincides with F'. The reflected light interferes with the light reflected from the reference mirror after passing the beam-splitter while returning. A second collimator now focuses the reflected portion to its back focal point F_2', where an opaque beam-stop carrying a small bore is positioned. It allows the reference and test waves to pass, but blocks erroneous and unwanted reflections and stray light coming from other surfaces of the setup. The front focal point of the eyepiece coincides with the back focal point F_2' of the second collimator and forms a Kepler-type telescope. The returning reference and test waves are then plane waves which interfere at the plane of a camera sensor. Any deformations in either the reference or test wave result in deformed interference fringes, which are captured by the camera.

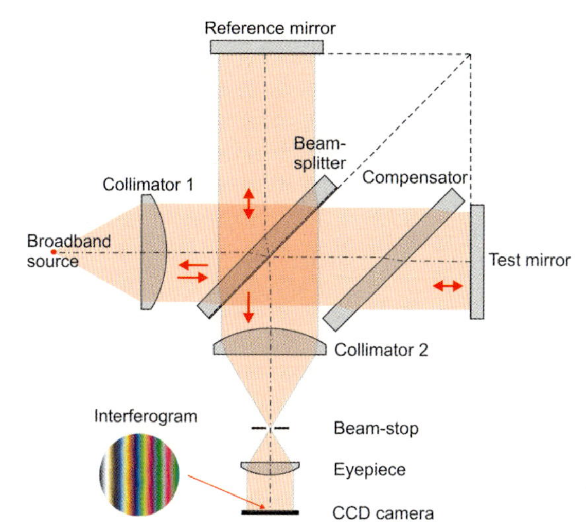

Figure 46-45: Schematic overview of a Twyman–Green interferometer with compensated interferometer arms of equal optical length. An extended broad-band light source can be used, as in the case of a test glass. This then is the same as the Michelson interferometer in an arrangement used to observe Newton fringes of equal thickness.

A Twyman–Green arrangement can also be used to test reflection at plane surfaces. In this case no lens system is necessary to form a spherical test wave. Figure 46-45 shows a schematic overview of a Twyman–Green interferometer used to test a plane mirror. In this example short coherent light is used, which makes it necessary to introduce a compensator plate. If the light source is also extended, the arrangement then becomes the same as the Michelson interferometer in an arrangement used to observe Newton fringes of equal thickness.

The benefits of a Twyman–Green interferometer are as follows.

1. There is no contact between the reference and test surfaces (so no damage or scratches from contact).
2. Each lens system for spherical surface testing can test a variety of convex or concave surfaces.
3. A computer system connected to the camera can evaluate the interferograms by digital means to improve the resolution of surface deviations by a factor of up to 1000 over the test glass method.

In comparison to the Fizeau interferometer there are some differences to be mentioned:

- Since the cavity is the space between the reference and test surfaces it is now no longer restricted to a small region in the neighborhood of the reference surface. Instead, the cavity now includes the spherical lens system, beam-splitter and – on some occasions – the compensator plate as well as the air in between all the components. As explained above, all environmental drifts and parameters have an effect on the optical path within the cavity and are part of the measured result. We no longer have a common path for the reference and test beam as in a Fizeau interferometer except for the interferometer's exit arm.

- When the optical paths in the reference and test arms are adjusted to be of equal length for the total spectrum of the light source, the interferometer acts as a virtual test glass. In this case an extended polychromatic source can be used as in a Newton interferometer (figure 46-30). In a Fizeau arrangement the OPD is always > 0.

As in a Fizeau arrangement an absolute measurement of the radius of curvature is not possible with only one measurement in one position and without further adjusted data points. In a digital Twyman–Green interferometer the effects of misalignment such as decenter, tilt and axial misplacement (defocus) can also be removed by fitting and subtraction of a suitable set of polynomials. The procedure will be described in section 53.4.

The Twyman–Green arrangement in figure 46-44 uses two collimators, which makes the interferometer components very expensive when large convex surfaces have to be tested.

An arrangement also patented by Twyman and Green but sometimes referred to as "Williams interferometer" uses only one collimator behind the beam-splitter as shown in figure 46-46. A concave spherical mirror now replaces the plane mirror in the reference arm. Its center of curvature coincides with the image of the point source. If the interferometer does not have a compensator plate, then the astigmatism, spherical aberration and coma will be different for the reference and test beams. However, if the F-number ($F/\# = f'/D$) of the collimator is sufficiently large, the aberrations caused by the beam-splitter can be neglected. For special cases the reference mirror can be aspherized to match the aberration from the test arm.

In a Twyman–Green interferometer, the fringe visibility only depends on the reflectivities of the reference and test mirrors as given in (46-108). The reflectivity of the beam-splitter does not influence the fringe visibility.

Let us assume a non-polarizing beam-splitter having a reflectivity of R_B and a transmissivity of T_B. For an incident irradiance I_0 the interfering beams at the exit of the interferometer carry the individual irradiances

$$I_{T1} = I_0 \, R_B \, T_B \, R_1 \text{ and } I_{T2} = I_0 \, R_B \, T_B \, R_2 \tag{46-111}$$

The irradiances returning to the source are given by

$$I_{S1} = I_0 \, R_B^2 \, R_1 \text{ and } I_{S2} = I_0 \, T_B^2 \, R_2 \tag{46-112}$$

where R_1 and R_2 denote the reflectivities of the reference and test mirrors, respectively.

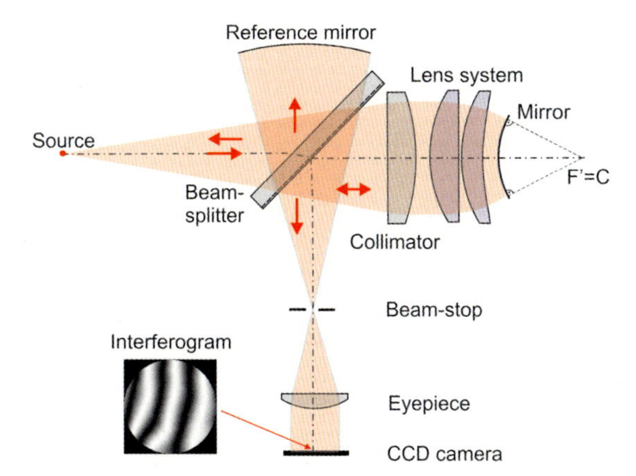

Figure 46-46: A Twyman–Green interferometer using only one collimator to test spherical surfaces by means of a lens system. The reference mirror is spherically concave with its center of curvature coinciding with the image of the light source.

The bias irradiances returning to the source (I_S) and returning to the camera (I_T) can therefore be calculated according to

$$I_T = I_{T1} + I_{T2} = I_0 \, R_B \, T_B \, (R_1 + R_2) \tag{46-113}$$

$$I_S = I_{S1} + I_{S2} = I_0 \, (R_B^2 \, R_1 + T_B^2 \, R_2) \tag{46-114}$$

For $R_1 = R_2 = 0.04$ the irradiances I_T, I_S and the irradiance of the lost light $I_L = I_0 - I_T - I_S$ are shown in figure 46-47. A maximum bias intensity is returned to the camera for R_B=0.5 assuming an absorption-free beam-splitter.

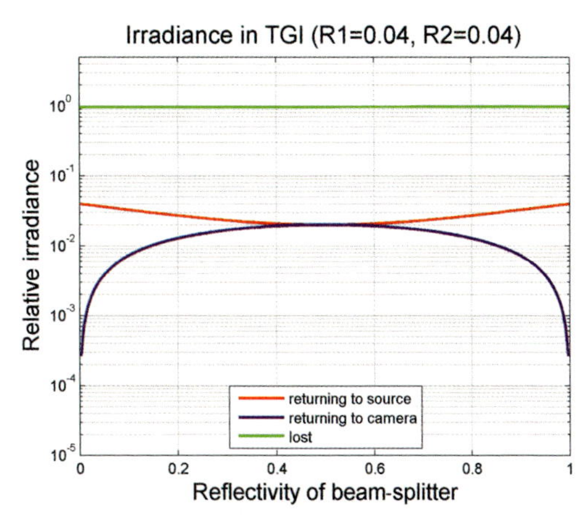

Figure 46-47: Irradiances returning to the source (blue) and returning to the camera (red) in a Twyman–Green interferometer for R_1=0.04 and R_2=0.04 as a function of the beam-splitter reflectivity. I_L (green) is lost on transmission through the reference and the test surface.

46.3.4
Mach–Zehnder Interferometer

The Mach–Zehnder interferometer is named after the physicists Ernst Mach (1838–1916) and Ludwig Zehnder (1854–1949) who developed it to determine the optical path differences caused by a small transparent sample [46-1].

Figure 46-48 shows a Mach–Zehnder interferometer used to test plane transparent objects. A small coherent monochromatic light source is placed at the front focal point F_1 of the first collimator. A plane wavefront leaves the collimator and the following first beam-splitter divides the wavefront into a transmitted and a reflected portion. The beam-splitter can be a plane parallel plate with a partially reflecting and an anti-reflection coated surface or a plate carrying a small wedge. The unwanted reflection can be blocked by the beam-stop in front of the eyepiece at the exit of the interferometer. When small samples are to be tested a beam-splitter cube is also convenient.

The light reflected from the first beam-splitter forms a first interferometer arm and travels towards a first inclined plane mirror (folding mirror 1) from which it is reflected to a second beam-splitter. The light is then again partially transmitted and reflected.

The light transmitted by the first beam-splitter forms a second interferometer arm and travels towards a second inclined plane mirror (folding mirror 2) from which it is reflected to the second beam-splitter. The light is then again partially transmitted and reflected and interferes with the light from the first interferometer arm.

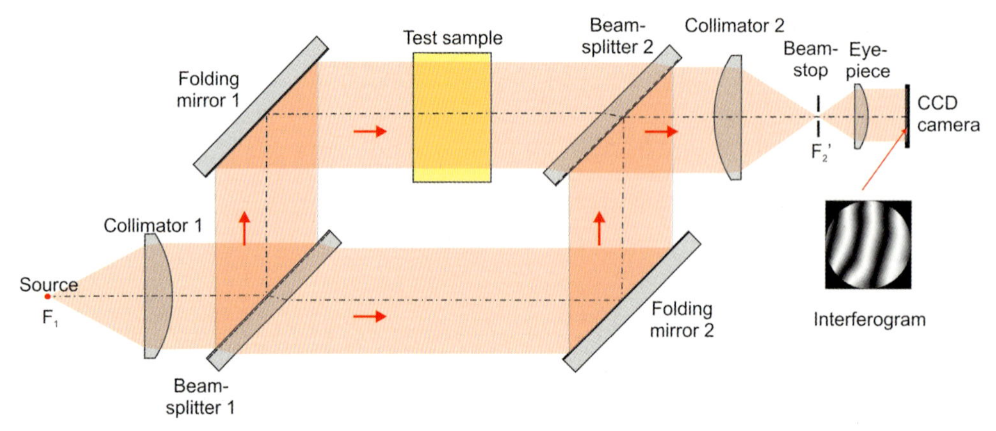

Figure 46-48: Mach–Zehnder interferometer in an arrangement for testing transparent plane objects in transmission.

The Mach–Zehnder interferometer thus has two equivalent exits that can be used to detect the interference signal. The difference between both is a 180° phase-shift of the optical path differences (OPD) to be detected. A second collimator focuses the light at its focal point F_2', where an opaque beam-stop carrying a small bore is positioned. This allows the reference and test waves to pass, but blocks erroneous and unwanted reflections and stray light coming from other surfaces of the setup. The front focal point of the eyepiece coincides with the back focal point F_2' of the second collimator and forms a Kepler-type telescope. The reference and test waves are then plane waves, which interfere at the plane of a camera sensor.

When both folding mirrors and beam-splitters are perfect, the resulting OPD is caused by the refractive index n_T and the thickness d_T of the test sample as well as the refractive index of the surrounding medium n_M. For local variations in the refractive indices and thickness the OPD detected by the interferometer camera can be written as

$$\mathrm{OPD}(x, y) = (n_T(x, y) - n_M(x, y)) \cdot d_T(x, y) \tag{46-115}$$

Note that the test beam passes the test sample only once, which is not the case when testing a transparent object in a Twyman–Green or Fizeau interferometer.

The arrangement shown in figure 46-48 can also be used to test afocal systems such as telescopes, where an incident plane wave mainly remains a plane wave after passing the object under test. Figure 46-49 shows an example of a Kepler-type telescope to be tested in singular pass transmission [46-24]–[46-27].

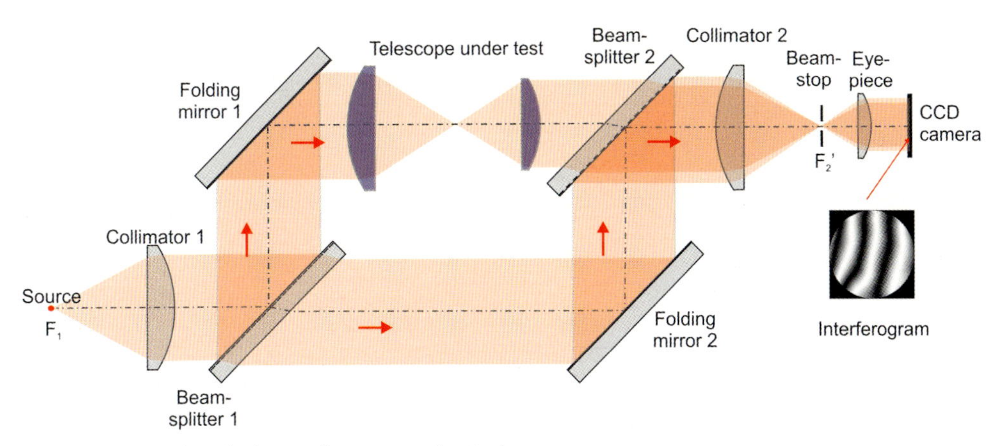

Figure 46-49: Mach–Zehnder interferometer with a Kepler-type telescope in the test arm.

A Mach–Zehnder interferometer can also be used to test plane mirrors in reflection. However, in this case normal incidence cannot be established. In figure 46-50 the two beam-splitters have been combined to a single, large beam-splitter plate. The left half of the plate is coated with a 50% reflective coating on the upside, while the right side has the same coating on the downside. This makes it possible to use a broad-band light source because the optical paths of the test and reference beams are then equal for all wavelengths.

Note that, for oblique incidence, the sensitivity of mirror testing depends on the angle of incidence ε as given in (46-116). *s(x,y)* denotes the surface deviations normal to the x–y plane defining the perfect mirror plane.

$$\mathrm{OPD}(x, y) = 2\, s(x, y) \cdot \cos \varepsilon \qquad (46\text{-}116)$$

Another variation of the Mach–Zehnder interferometer involves combining the beam-splitter and the folding mirror into one single element as shown in figure 46-51 [46-28]. The arrangement is very robust and insensitive to mechanical drifts of the elements because the reference and test arms experience almost the same changes when a separation, decentering or small tilt of the elements is introduced. In this configuration the interferometer is known as the "Jamin interferometer" [46-29].

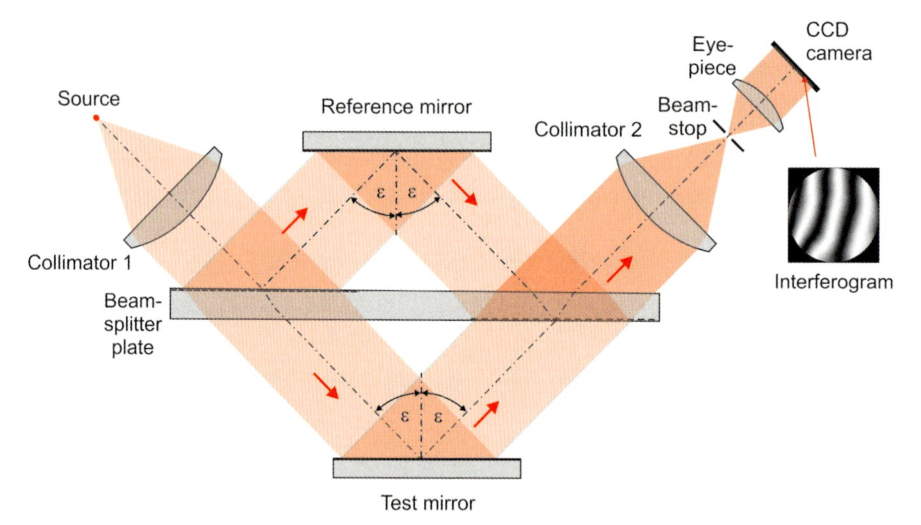

Figure 46-50: The Mach–Zehnder interferometer modified to test plane surfaces in reflection for oblique incidence.

In a Mach–Zehnder interferometer the fringe visibility depends on the reflectivities of the folding mirrors and reflectivities and transmittances of the beam-splitters. In the case of an absorbing test piece its transmittance T also influences the fringe visibility.

Let us assume non-polarizing beam-splitters having reflectivities R_{B1} and R_{B2} and transmittances T_{B1} and T_{B2}. The folding mirrors have reflectivities R_{F1} and R_{F2}. For an incident irradiance I_0 the interfering beams at the exit of the interferometer carry the individual irradiances

$$I_{\mathrm{T1}} = I_0\, R_{\mathrm{B1}}\, R_{\mathrm{F1}}\, T_{\mathrm{B2}}\, T \text{ and } I_{\mathrm{T2}} = I_0\, T_{\mathrm{B1}}\, R_{\mathrm{F2}}\, R_{B2} \qquad (46\text{-}117)$$

The irradiances at the second exit, which is not used, are given by (46-118).

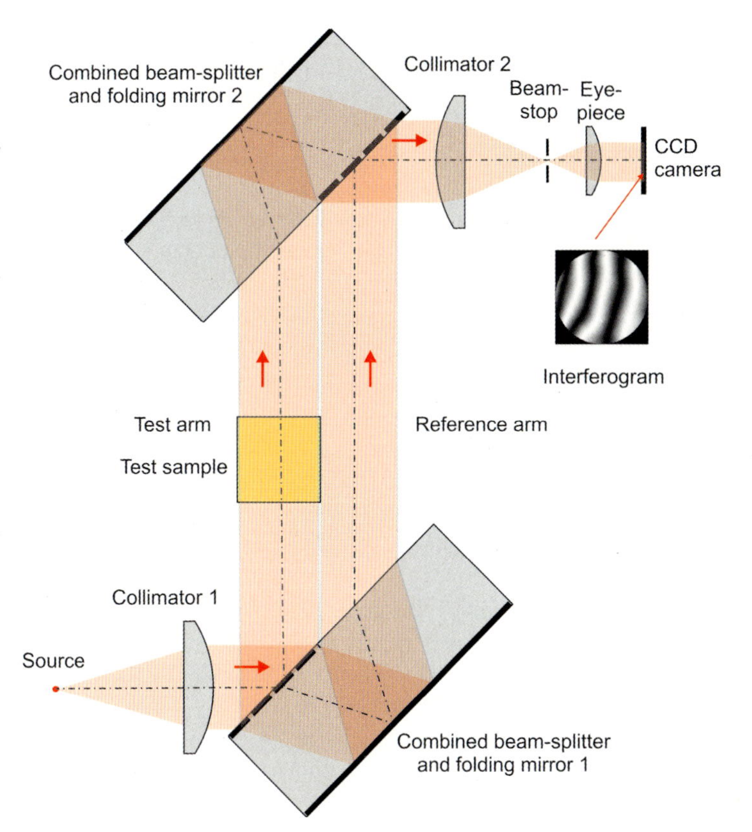

Figure 46-51: The Jamin interferometer as a modification of the Mach–Zehnder interferometer used to test transmission through transparent objects.

$$I_{L1} = I_0 \, R_{B1} \, R_{F1} \, R_{B2} \, T \text{and} I_{L2} = I_0 \, T_{B1} \, R_{F2} \, T_{B2} \tag{46-118}$$

The fringe visibility V_T at the camera is then given by (46-119).

$$V_T = \frac{2\sqrt{R_{B1} \, R_{B2} \, T_{B1} \, T_{B2} \, R_{F1} \, R_{F2} \, T}}{R_{B1} \, R_{F1} \, T_{B2} \, T + T_{B1} \, R_{F2} \, R_{B2}} \tag{46-119}$$

and that at the second exit by (46-120).

$$V_L = \frac{2\sqrt{R_{B1} \, R_{B2} \, T_{B1} \, T_{B2} \, R_{F1} \, R_{F2} \, T}}{R_{B1} \, R_{F1} \, R_{B2} \, T + T_{B1} \, R_{F2} \, T_{B2}} \tag{46-120}$$

Assuming a reflectivity of 1 for the folding mirrors and equal reflectivities of R_B for both non-absorbing beam-splitters, the visibilities are then:

$$V_T = \frac{2\sqrt{T}}{(1+T)} \tag{46-121}$$

$$V_L = \frac{2R_B(1-R_B)\sqrt{T}}{R_B^2\,T + (1-R_B)^2} \tag{46-122}$$

The fringe visibility is constant at the camera exit (first exit) depending only on the transmittance of the test piece. Using the second exit of the interferometer, the fringe visibility also depends on the reflectivity of the beam-splitters. Figure 46-52 shows the visibilities for both exits for transmittances of 1 and 0.1. Note that for $T = 1$ the visibility at the camera will only reach 0.57, whereas it reaches 1 at the second exit for beam-splitter reflectivities of 0.76. For absorbing test samples the second exit therefore leads to better fringe visibilities if the reflectivity of the beam-splitters is selected accordingly.

The bias irradiance has a maximum at the camera exit and a minimum at the second exit for 0.5 beam-splitter reflectivity (figure 46-53). When transmittance of the test piece falls, the maximum for camera irradiance still stays at 0.5 of the beam-splitter reflectivity but with a maximum value of $(1+T)/4$.

The bias irradiance at the second exit reaches T when the beam-splitter reflectivities approximate 1, thus the function becomes asymmetric and reaches its minimum at $R_B = \frac{1-\sqrt{T}}{1-T}$.

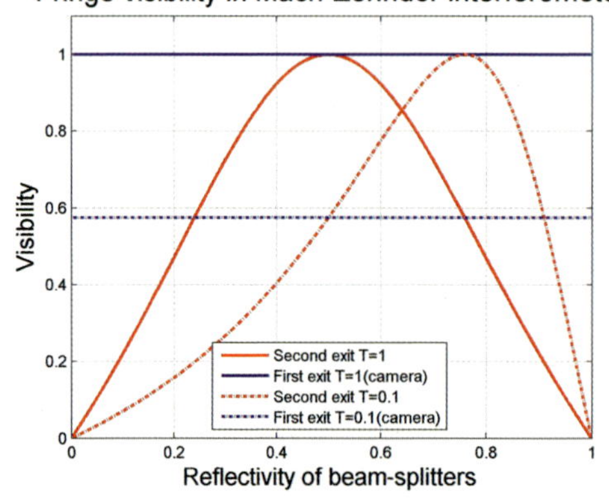

Figure 46-52: Fringe visibility in a Mach–Zehnder interferometer for the regular exit to camera (blue) and the second exit (red) for test-piece transmittances of 1 and 0.1.

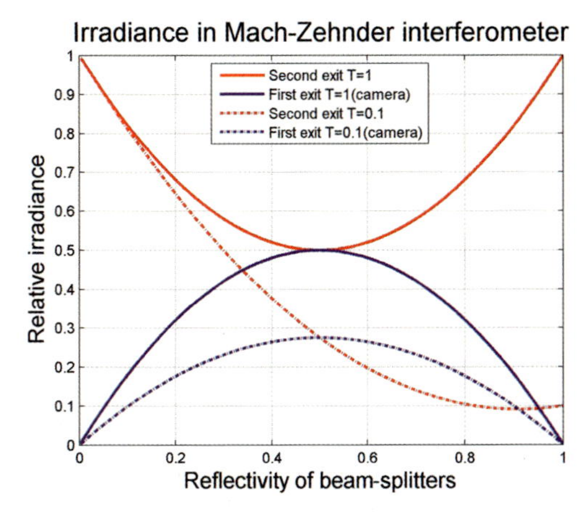

Figure 46-53: Irradiances returning to the camera (blue) and going to the second exit in a Mach–Zehnder interferometer for test-piece transmittances of 1 and 0.1.

46.3.5
Point Diffraction Interferometer

A point diffraction interferometer (PDI) is a two-beam interferometer in which the reference beam is generated by diffraction from a small pinhole or from the end of a fiber. Both beams are coherent and form an interferogram according to their intensities at the location of the detector.

In the arrangement invented by Smartt and Strong (1972) [46-30] a semi-transparent plane plate carrying a small pinhole is placed near the focus of the converging wavefront under test. Diffraction at the pinhole generates a secondary spherical wave which acts as a reference wave and interferes with the undisturbed passing aberrated wave under test.

Figure 46-54 shows a Smartt interferometer arrangement to test transmission of a transparent test sample. A small coherent monochromatic light source is placed at the front focal point F_1 of the first collimator. A plane wavefront leaves the collimator and transmits through a transparent test sample (plane plate, prism or wedge). The aberrated wavefront transmits through a second collimator and converges to the focal point F_2' in which the semi-transparent point diffraction plate carrying a pinhole is placed. F_2' coincides with the front focal point of an eyepiece such that parallel light reaches the following CCD sensor. The perfect spherical wavefront generated by the pinhole acts as the reference wave and meets the CCD sensor as a perfect plane wave, whereas the transmitted aberrated wavefront carries the aberrations from the test sample.

The visibility of the fringes depends very much on the diameter and position of the pinhole within the point source image and it can also be altered by varying the transmittance of the semi-transparent plate. Changing the lateral position of the pinhole introduces tilt to the detected OPD, which means that parallel fringes will be seen in the interferogram. The fringe contrast and the transmitted irradiance become unacceptably low when typically more than five fringes are introduced.

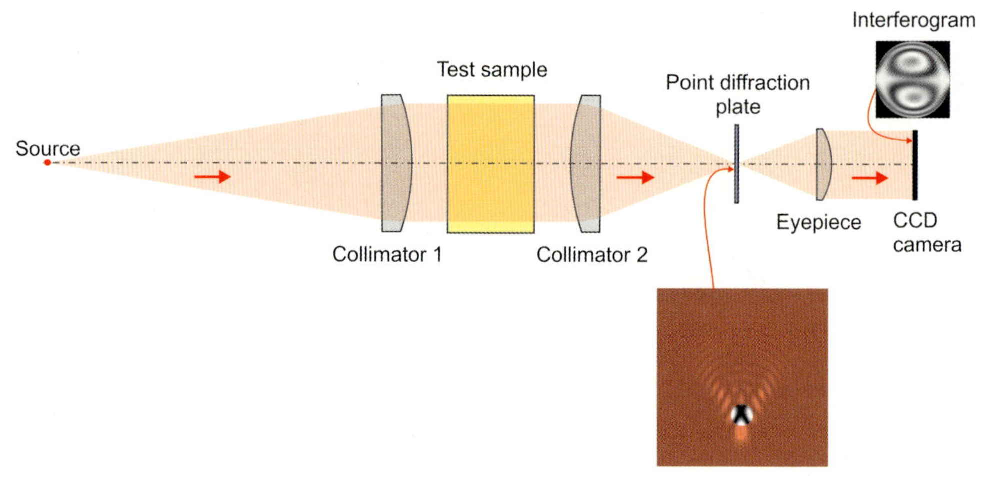

Figure 46-54: The Smartt interferometer in an arrangement to test transmission of a plane test sample. The key element is a semi-transparent point diffraction plate carrying a small pinhole with total transmittance.

Instead of a pinhole it is also possible to use a small circular opaque obscuration on a glass plate carrier to generate the perfect spherical reference wave by diffraction. In either case the quality of the generated spherical wavefront depends on the geometrical quality of obscuration or the pinhole.

As an example a wavefront carrying 3 λ of coma (peak-to-valley) is brought to a focus (figure 46-55a). The irradiance at the location of the point diffraction plate can be calculated by Fourier transformation and is shown in figure 46-55(b). After filtering the electromagnetic field distribution with a constant transparency function of less than 1, leaving a small circular area with a transparency of 1, the interferogram can be calculated by inverse Fourier transformation. Figure 46-56 shows results for three different lateral pinhole positions.

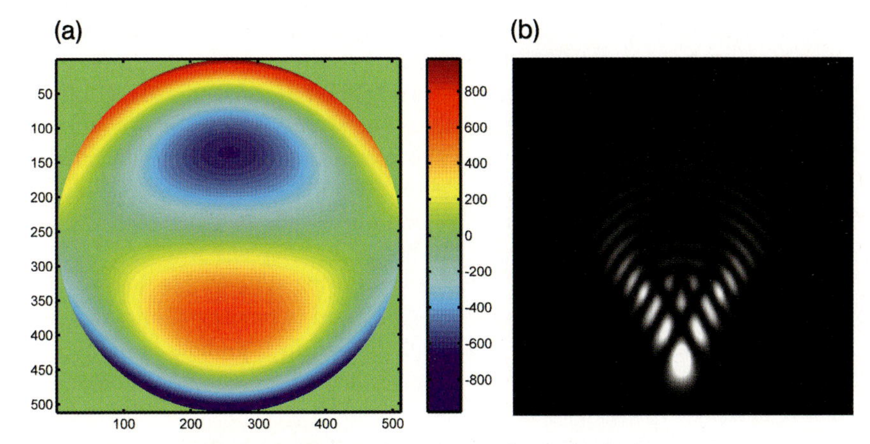

Figure 46-55: An example of an aberrated wavefront with a coma of 3 λ PV: (a) wavefront; (b) irradiance at semi-transparent point diffraction plate.

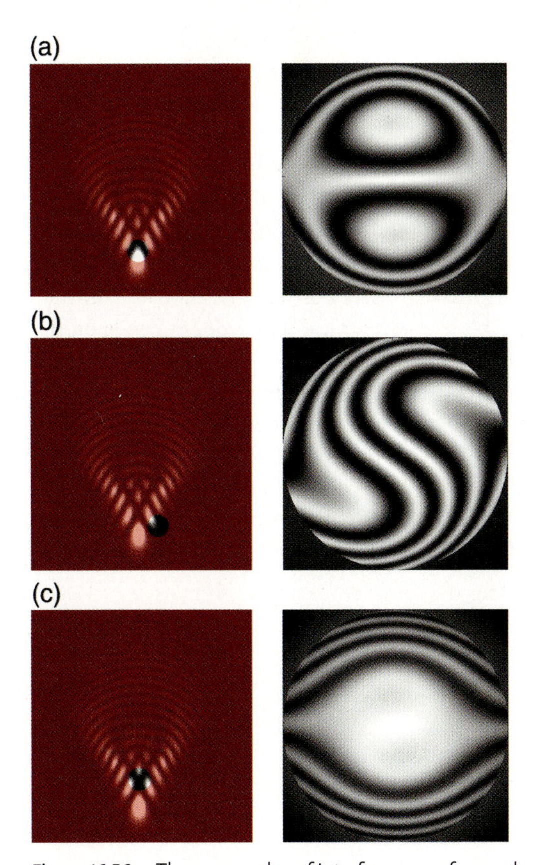

Figure 46-56: Three examples of interferograms for an aberrated wavefront with a coma of 3 λ PV. The pinhole is in three different positions in the point source image plane: (a) pinhole in normal test position; (b) pinhole decentered in x; (c) pinhole decentered in y.

A point diffraction interferometer can be arranged in different ways. It is not always necessary to generate the reference wave by diffraction in a common path arrangement. If the reference and test waves are split off by conventional division of the amplitude as in a Fizeau, Twyman–Green or Mach–Zehnder interferometer (figure 46-57), a setup can be found where the reference wave is additionally filtered by a pinhole. The setup than acts as a wavefront tester, in which the test sample is placed in front of the interferometer. As an example we show a Mach–Zehnder interferometer with a point diffraction modification in the reference arm.

In this setup an aberrated beam is incident upon the system, this then has its amplitude divided into two separate beams. One beam is propagated through one arm unchanged, whereas the other is spatially filtered by a pinhole to produce a clean plane-wave reference beam. The two beams are then superimposed to produce an interference pattern on a CCD camera.

The end of a single-mode optical fiber can also be used to produce an almost perfect spherical wavefront (figure 46-58). In the PDI by G. Sommargren [46-31]–[46-33] the end of an optical fiber is coated with a semi-transparent metallic film. A portion of the spherical wave leaving the fiber hits the concave mirror under test at normal incidence and is reflected back to the end of the fiber. After reflection at the semi-transparent metallic film the light superimposes on the other portion of the spherical wave leaving the fiber such that an interferogram can be displayed onto a CCD camera using an eyepiece with its front focal point at the exit of the fiber. In some arrangements even the eyepiece is avoided to have a lensless interferometer. In this case the image of the test mirror does not coincide with the CCD sensor. A sharp image of sufficient lateral resolution can then be achieved by measuring amplitude and phase of the incident wave and generating a sharp image by wave propagation algorithms.

The critical component in the PDI according to Sommargren is the fiber endface coated with a semi-transparent film. Typical fiber cores have a diameter of three micrometers. The critical extremely well polished region at the end of the fiber has a diameter of about 1 millimeter. Its flatness must be comparable to the desired accuracy of the measurement, but only over a small area around the fiber core. The fiber is usually embedded in a glass substrate and superpolished over the entire assembly. The fiber remains embedded in the substrate during use to ensure stability and ease of mounting.

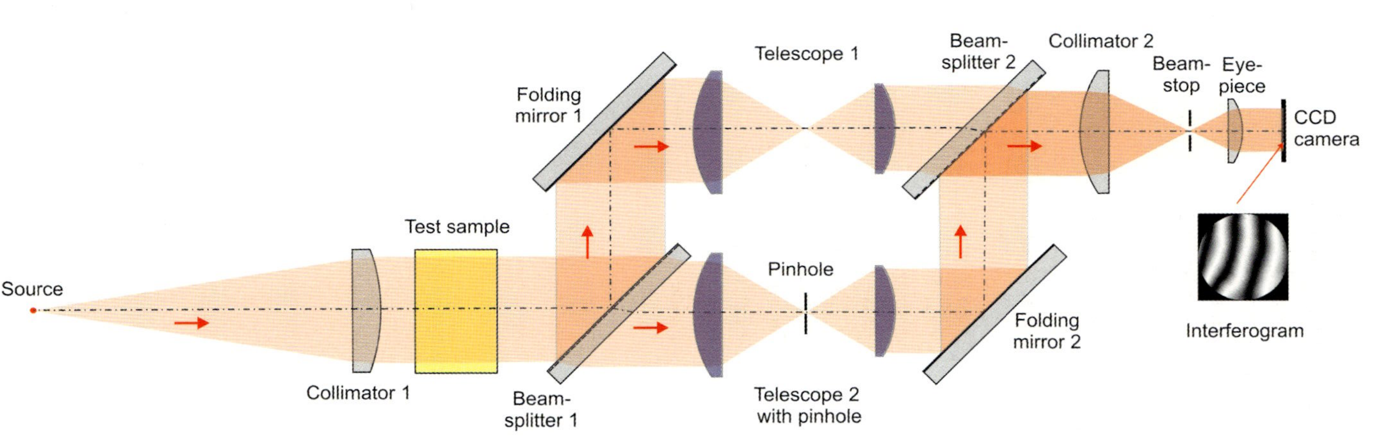

Figure 46-57: Point diffraction interferometer in a Mach–Zehnder arrangement showing the wavefront under test entering the interferometer. The reference and test beams are generated by amplitude division. Two identical telescopes leave the incident aberrated wave mainly unchanged, but in the test arm telescope a pinhole in the focal point generates a perfect spherical wavefront.

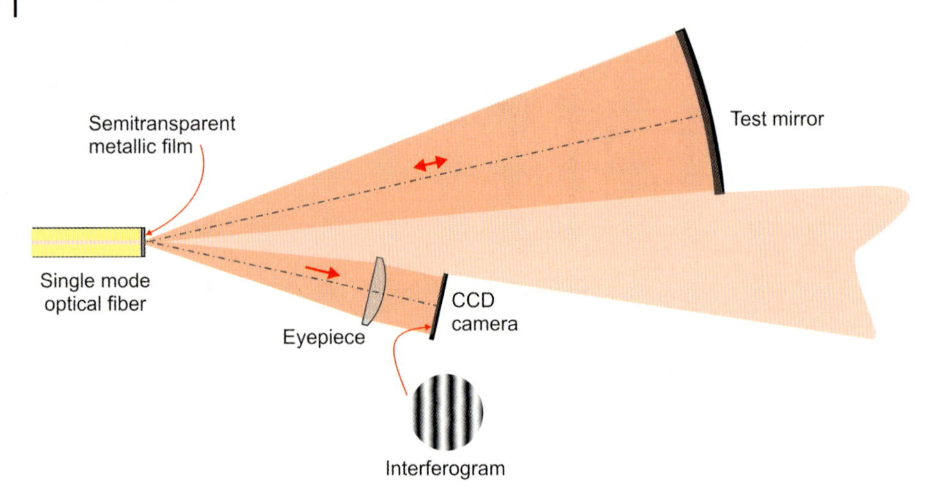

Figure 46-58: Point diffraction interferometer in a Sommargren arrangement with a perfect spherical wavefront leaving the end of a fiber. One portion (the test wave) propagates towards a mirror under test with its focus coinciding with the end of the fiber. The other portion (reference wave) propagates towards an eyepiece that images the test mirror onto a CCD sensor. The light reflected from the test mirror returns to the end of the fiber and is there reflected from a semi-transparent metallic film where it superimposes on the reference wave to form an interferogram, which is then detected by the CCD sensor.

46.3.6 Shearing Interferometer

A shearing interferometer is an instrument in which a copy of a wavefront under test is geometrically deformed or locally displaced (sheared) and superimposed with the original wavefront to form an interferogram that contains information on its shape [46-1], [46-34]–[46-36], [46-40].

With respect to the kind of displacement or deformation of the wavefront copy the following classifications can be made (figure 46-59):

- Lateral shear
  The wavefront copy is laterally displaced along the x- or y-axis with respect to the original wavefront.
- Radial shear
  The wavefront copy is linearly enlarged or reduced relative to the coordinates of the original wavefront, the center coinciding with the center of the original wavefront.
- Rotational shear
  The wavefront copy is rotated around the z-axis going through the center of the original wavefront.
- Reversed shear
  The wavefront copy is flipped relative to the x- or y-axis through the center of the original wavefront.

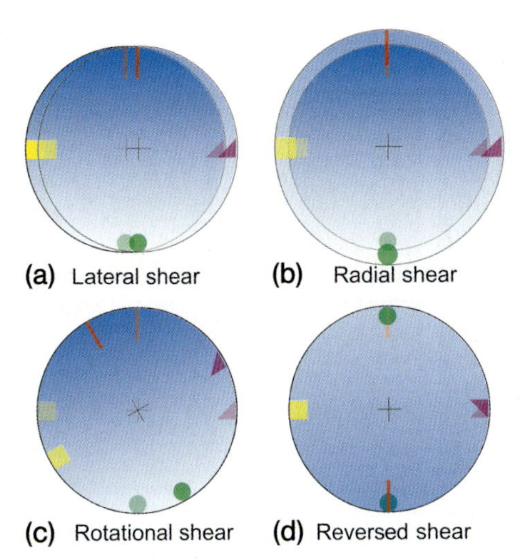

Figure 46-59: Different methods of obtaining shearing wavefronts: (a) lateral shear, (b) radial shear, (c) rotational shear, (d) reversal shear.

When both waves are superimposed, the interferogram is formed by their optical path difference, which is no more the absolute wavefront shape under test, since the reference wave is not known but is itself formed by a variation of the wavefront under test.

Let the phase distribution of the wavefront under test be $\phi(x, y)$. Assuming equal irradiance for the copy of the wavefront the interferogram that is formed can then be described by

$$I(x, y) = I_0(x, y)\,(1 + \cos \Delta\phi(x, y)) \tag{46-123}$$

where $I_0(x, y)$ is the irradiance distribution of the wavefront under test and

$\Delta\phi(x, y) = \phi(x, y) - \phi(x + \Delta x, y + \Delta y)$ for a lateral shear interferogram with Δx and Δy as lateral displacements,

$\Delta\phi(x, y) = \phi(x, y) - \phi(f\,x, f\,y)$ for a radial shear interferogram with f as the magnification factor,

$\Delta\phi(x, y) = \phi(x, y) - \phi(x \cos\varphi + y \sin\varphi, -x \sin\varphi + y \cos\varphi)$ for a rotational shear interferogram with ϕ as the angle of rotation,

$\Delta\phi(x, y) = \phi(x, y) - \phi(\mp x, \pm y)$ for a reversal shear interferogram.

Shearing interferograms can generally be used to determine the original wavefront $\phi(x, y)$ by a more or less complicated integration procedure. It is not possible to fully recover $\phi(x, y)$ from only one shearing interferogram. However, using a

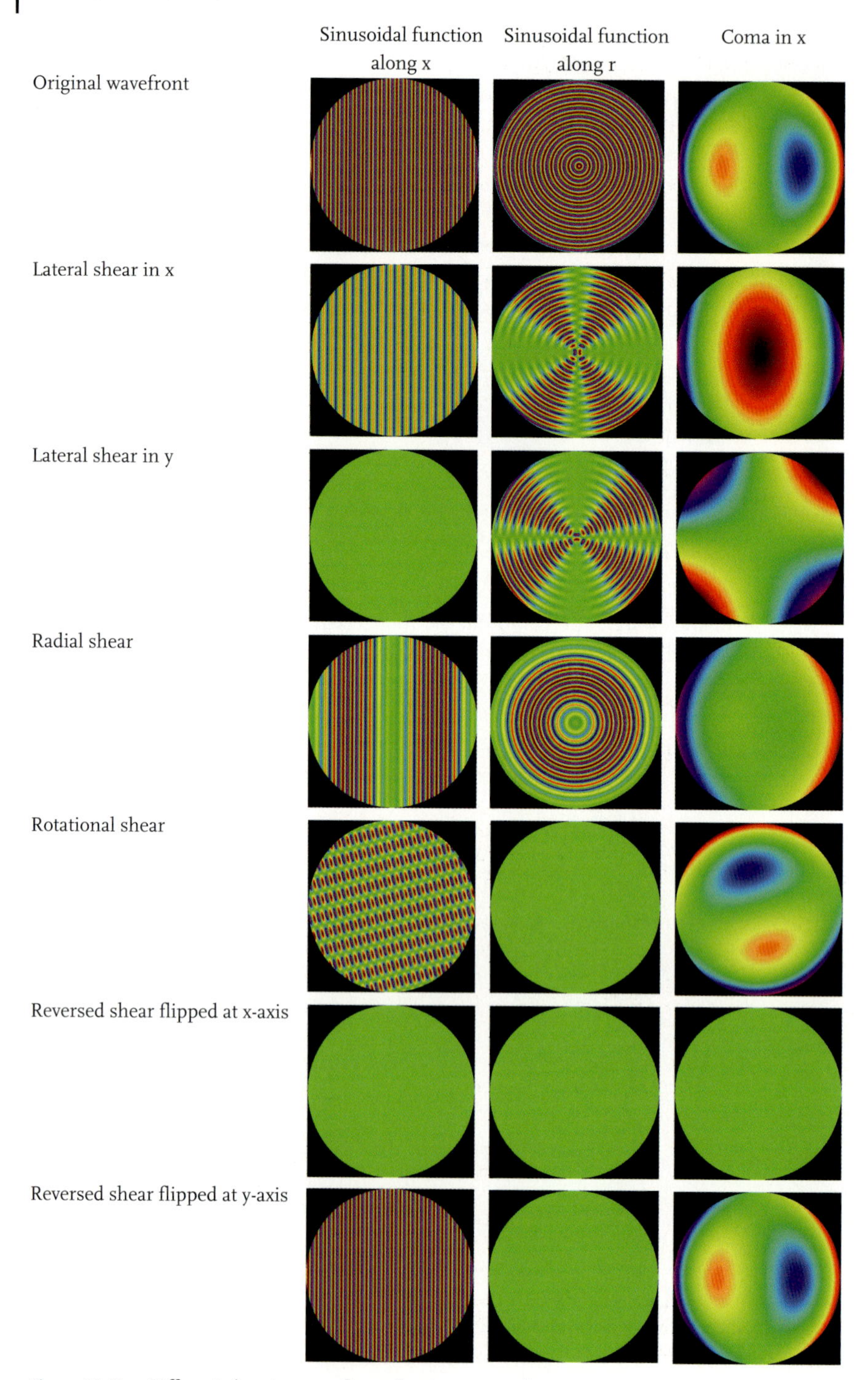

Figure 46-60: Different shearing wavefronts for three examples: sinusoidal phase modulation along *x*, sinusoidal phase modulation along *r*, coma-shaped phase modulation along *x*.

second shearing interferogram in most cases will deliver the missing information. Shearing interferometers are 'blind' to periodic wavefront modulations that correspond to the type and amount of shear. Figure 46-60 gives three examples of wavefronts and the corresponding shearing wavefronts.

The first example is a sinusoidal phase modulation along the *x*-axis. When the lateral shear in *x* is chosen to be a natural number times the sine period, the shear wavefront will be zero. For a shear in *y* the result will always be zero as is the reversed shear has flipped at the x-axis.

The second example is a sinusoidal phase modulation along the radius *r*. For a rotational shear the result will always be zero as will be the reversed shears which have flipped at the *x*- or *y*-axis.

The third example is a coma-shaped phase modulation. Because of its symmetry around the *x*-axis the reversed shear, which has flipped at the *x*-axis, will be zero.

In the following we will concentrate merely on lateral shearing interferometers because they are by far the most frequently used and play an important part as compact, stable and highly resolving wavefront sensors.

Cyclic Interferometer

When equipped with a tilted thick plane parallel plate the cyclic interferometer is a wavefront sensor capable of measuring shears in *x* or *y*. Figure 46-61 gives an overview of the arrangement which resembles a classical Twyman–Green setup.

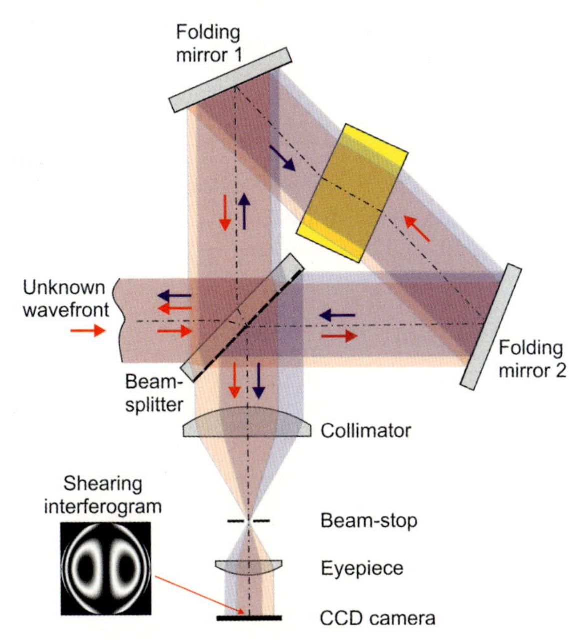

Figure 46-61: Cyclic interferometer equipped with a tilted plane parallel plate which acts as a lateral shear interferometer for variable shears in *x* or *y* when the plane plate is tilted accordingly.

The unknown wavefront enters the interferometer at the beam-splitter, which divides it into a transmitted and a reflected portion. As in the case of the Twyman–Green interferometer the beam-splitter can be a plane parallel plate with a partially reflecting and an anti-reflection coated surface or a plate carrying a small wedge. The unwanted reflection can then be blocked by the beam-stop. If the diameter of the interferometer arms is conveniently small a beam-splitter cube can be used.

The light reflected from the beam-splitter travels towards a plane folding mirror 1 from which it is reflected to the folding mirror 2. The tilted plane plate, which can also carry a small wedge, is placed in between the folding mirrors to laterally shift the incident beam. After reflection from the second folding mirror the beam completes one cycle by reaching the beam-splitter and being reflected towards the interferometer's exit. The light transmitting through the beam-splitter follows the optical path in an anti-cyclic way by first being reflected from folding mirror 2, then passing the tilted plane plate and finally being reflected from folding mirror 1. Both beams converge to the center of a beam-stop with the help of a collimator. An eyepiece with its focal point coinciding with the beam-stop and the back focal point of the collimator directs both parallel beams toward the sensor.

On their cyclic and anti-cyclic path both beams are laterally shifted by an amount ν which depends on the plane plate's tilt angle θ (see(46-124)). The effect in a cyclic interferometer is that one beam is shifted by ν and the other by $-\nu$, such that the resulting lateral shift at the sensor is 2v.

$$\nu = -d \sin\theta \left(1 - \frac{\cos\theta}{\sqrt{\left(\frac{n'}{n}\right)^2 - \sin^2\theta}} \right) \tag{46-124}$$

where d is the thickness of the plane plate, n' is its refractive index and n that of the surrounding medium. Figure 46-62 shows lateral displacements ν for a thickness $d = 30$ mm and three different n' with air as the surrounding medium and for tilt angles from 0° to 90°.

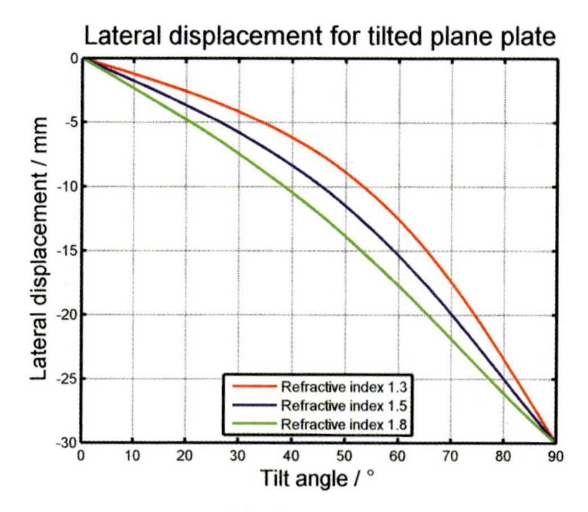

Figure 46-62: Lateral displacement caused by a plane parallel plate of thickness 30 mm and refractive indices of 1.3 (red), 1.5 (blue) and 1.8 (green) as a function of the plane plate's tilt angle.

A cyclic interferometer is always compensated because the optical paths of the clockwise and anti-clockwise traveling beams are equal. However, when broad-band light is used, the dispersion of the plane plate and the beam-splitter introduce lateral shears depending on the wavelength.

Shear Plate

An uncoated thick plane plate is the basis of a very simple lateral shearing interferometer whose shear can also be adjusted by a variable inclination θ of the plane plate. Figure 46-63 shows the principle of a "shear plate" used to split an incident wavefront by amplitude. The wavefront under test is partially reflected at an angle 2θ at the first surface. The transmitted portion is refracted by an angle of θ' and reflected from the second surface. When transmitting through the first surface again its deflection is also 2θ, whereas its lateral displacement is calculated according to (46-125).

$$v = 2d \frac{\frac{n}{n'} \sin\theta \cos\theta}{\sqrt{1 - \left(\frac{n}{n'}\right)^2 \sin^2\theta}} \tag{46-125}$$

where n' is the refractive index of the shear plate, n is the refractive index of the surrounding medium, d is the thickness of the shear plate and θ its tilt angle. Examples are shown in figure 46-64.

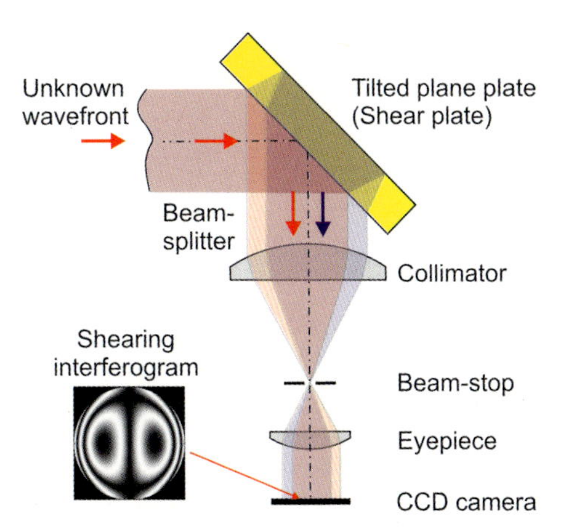

Figure 46-63: Lateral shearing interferometer using a "shear plate" to introduce lateral displacement. Variable shear is introduced by varying the tilt accordingly.

This differs from the cyclic interferometer in that the shear plate is not compensated. Both interfering waves have an optical path difference OPD with a maximum of $2dn$ for $\theta = 0°$. The OPD is given in (46-126) and examples are shown in figure 46-65.

$$OPD = 2d \cdot n' \cdot \sqrt{1 - \left(\frac{n}{n'}\right)^2 \sin^2\theta} \tag{46-126}$$

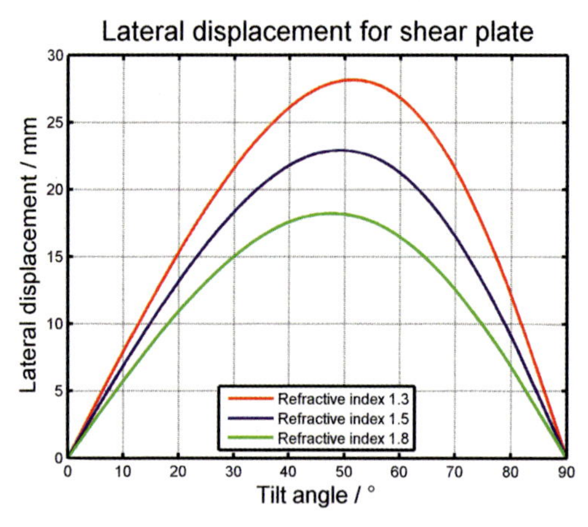

Figure 46-64: Lateral displacement caused by a 30 mm thick plane plate in air tilted by θ for three different refractive indices 1.3, 1.5 and 1.8.

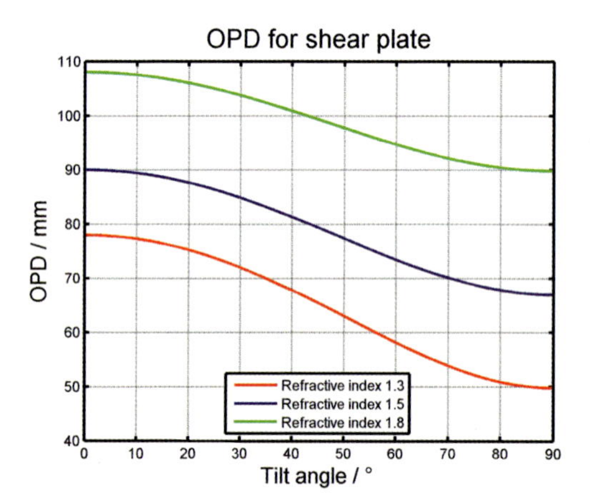

Figure 46-65: Optical path difference between beams reflected from the first and second surface of a 30 mm thick plane plate in air tilted by θ for three different refractive indices 1.3, 1.5 and 1.8.

Grating Angle Shearing Interferometer

For convergent light, lateral shearing can also be introduced by different methods. In this case, the wavefront under test should be copied and displaced in its angular direction of propagation. Figure 46-66 gives an example of an arrangement using a diffraction grating as the shearing element [46-37], [46-38]. The grating generates different diffraction orders in different directions superimposing on a sensor and forming an interferogram containing the information of the unknown wavefront deformation. An eyepiece of focal length f' can be used with its front focal point coinciding with the focus of the wavefront under test to produce nearly plane waves which are laterally shifted by ν according to

$$\nu_m = f' \frac{m\lambda}{\sqrt{p^2 - (m\lambda)^2}} \tag{46-127}$$

where m is the number of the diffraction order, p the grating period and λ the wavelength.

When both first diffraction orders are superimposed their separation is $\nu = \nu_{+1} - \nu_{-1}$. To avoid overlaps of three or more wavefront copies, ν_m should always be greater than half the diameter D of the wavefront under test on the sensor. Figure 46-67 gives an example in which a sinusoidal amplitude modulated grating was used to produce ± first diffraction orders that interfere with the zeroth order and just touch each other. This case is problematic because the subaperture containing the interferogram does not cover the total wavefront aperture and thus is unable to reproduce the total wavefront deformation.

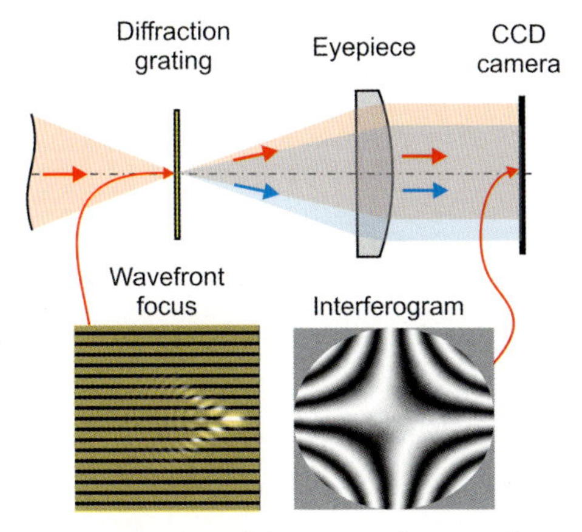

Figure 46-66: Lateral shearing interferometer for converging wavefronts using a diffraction grating to produce angular displacements for different diffraction orders.

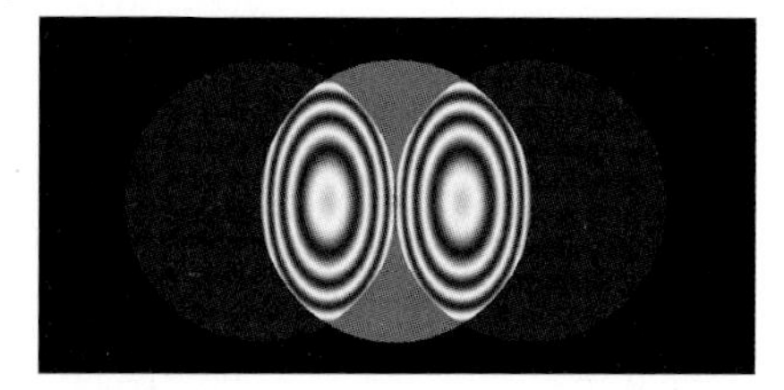

Figure 46-67: Lateral shearing interferograms produced by a sinusoidal amplitude modulated grating. ± first diffraction orders interfere with the zeroth order and touch each other at the center of the zeroth order.

When a double-frequency grating is used, the interfering diffraction orders can be separated more conveniently. The lower frequency grating is selected such that the first and zeroth orders just touch each other. The higher frequency grating causes the lateral displacement needed to produce the lateral shearing interferogram. Equation (46-128) gives the maximum grating period to avoid overlap of the first and zeroth diffraction order,

$$p_{\max} = \lambda\sqrt{\frac{f'}{D}+1} \tag{46-128}$$

where D is the diameter of the wavefront under test projected onto the sensor and f' is the focal length of the eyepiece.

Figure 46-68 shows interferograms produced by a crossed sinusoidal double-frequency grating which simultaneously generates lateral shears in the x and y directions by $D/10$.

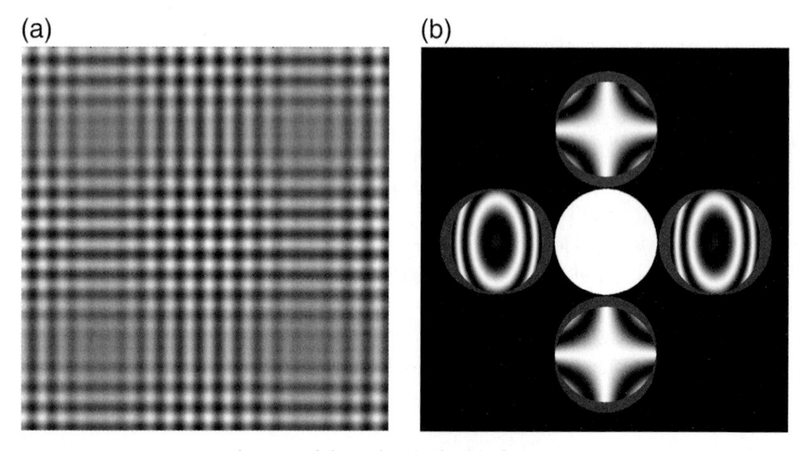

Figure 46-68: (a) Crossed sinusoidal amplitude double-frequency grating. (b) Interferograms produced by (a) in a lateral shearing arrangement according to (46-66). ± first diffraction orders from the lower frequency grating touch the zeroth order and interfere with the ± first orders from the higher frequency grating.

46.4 Interferometer Designs

46.4.1 General Requirements

For the concept of an optical test arrangement the optical designers and engineers have to select the specific optical parts and groups suitable to meet the metrological requirements. They have to consider a broad variety of requirements and constraints that go far beyond the geometric-optical layout of the test setup. The following is a short overview of the aspects that have to be defined in the requirement specifications in the conceptual phase of a test setup.

Optical component test	• Surface figure test: spatial frequency range • Test in transmission or reflection • Spheric, aspheric, plane or free form • Specular or scattering • Coated or uncoated • Metrology mount or original mount • Orientation to gravity • Range and resolution of metrology • Additional tests: radius of curvature, centering, inner centering of aspherical elements, thickness, refractive index, homogeneity, etc.
Optical system test	• Aspects to be tested: Image quality, lateral resolution, magnification and distortion, field curvature, transmission, polarization, etc. • Range and resolution of metrology
Light source	• Wavelength, spectrum • Power • Coherence length • Polarization • Stability
Beam delivery	• Fiber or free space • Beam expansion • Wavefront adaption to NA, field and shape
Sensor	• CCD, CMOS, Infrared or X-ray detector • Number of pixels, pixel size • Sensitivity • Linearity • S/N ratio • Cover glass, microlenses

Optical layout	• Imaging requirements: wavefront range and resolution, lateral resolution • Choice of sensor type • Adaptation to specific test piece • Adjustment tools and facilities
Calibration	• Calibration concept and process • Calibration tools
Coatings of setup components	• Wavelength range • Residual reflectance or transmittance • Polarization
Mechanical mounts for component test pieces	• Isostatic mount, three-point mount or any other • FE correction und budget • Lateral adjustment and fixture • Test-piece loading, adjustment and removal process and tools • Testing in original element mount
Signal processing and evaluation	• Computer and controller units • Algorithms • Implementation, testing, and verification
General mechanical setup	• Vibration isolated table • Housing • Storage areas for test and calibration pieces • Transportation and adjustment eqipment • Accessibility of test volume • Climate stabilization (flow box) • Power supply, pressured air supply
Tolerances of setup components	• Material quality • Surface figures • Roughness • Centering • Imperfections • Adjustment • Intrinsic birefringence
Expected optical artefacts	• Dust particles, scratches, imperfections in components • Speckle effects • Diffraction effects at rims • Stray light and erroneous reflections • Ghost images • Reflections from sensor

Mechanical stability	• Drifts • Vibrations
Environmental influences, drifts	• Temperature • Pressure • Humidity • Gas consistency • Turbulences • Acoustic vibrations • Radiation (light or any other electromagnetic)
Gravitation	• Bending and deformation of components • Introduction of birefringence by tension
Error budget	• Total statistical deviations: repeatabilities, reproducibilities and drifts • Total systematic deviations • Calibration quality and frequency of calibration process • Tolerance budget: production tolerance window, metrology tolerance window (metrology error budget)
Security aspects	• Laser safety • Power safety • Electromagnetic shielding • Protection against cut and squeeze • Maximum weight of test pieces and equipment • Accessibility of test volume • CE qualification etc.
Testing and verification	• Verification test pieces • Calibration • Cross-checks • Verification and validity process (statistical analysis)
Maintenance	• Dust protection, cleanability • Replacement of parts
Productivity and uptime aspects	• Time for standard measurement process • Test-piece preparation and adaptation time • Calibration time and interval • Maintenance and repair time and interval
Costs	• Total cost investment • Total cost for room and energy • Manpower: skills and number of operators • Lifetime of test equipment

In the following we will concentrate only on those aspects that are relevant for the optical layout of a test setup. As an example, for component testing we will choose the Fizeau interferometer since it is by far the most frequently used instrument worldwide.

46.4.2
Definition of Optical Components and Subassemblies

We start with the selection of a suitable light source. If a surface is to be tested in reflection, the wavelength is arbitrary, except for a coated surface. In the latter case a sufficient residual reflectivity is needed, so that a suitable wavelength has to be selected. If an optical component has to be tested in transmission, the operating wavelength usually has to be selected to test the component's final performance.

In any case it needs to be considered that, in an interferometric setup, a wavefront resolution of $\lambda/10$ for visual inspection and up to $\lambda/10\ 000$ for computer analysis can be expected. For component testing, monochromatic light can be used which, in most cases, is highly coherent laser light, so that chromatic corrections will become obsolete.

If a component with quasi-concentric or quasi-parallel surfaces is to be tested, then testing in a Fizeau-type interferometer is not possible, because reflections from both surfaces simultaneously enter the interferometer and cannot be distinguished. In this case a short coherent light source must be used, for example that of a Twyman–Green type interferometer.

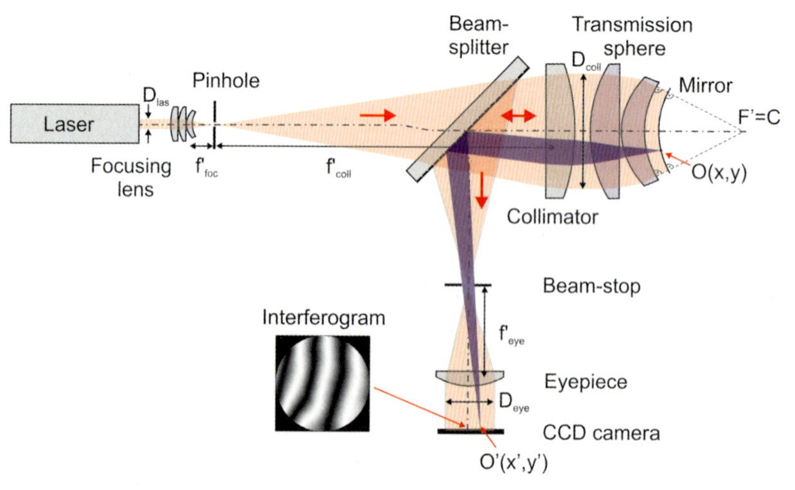

Figure 46-69: Fizeau interferometer to test reflection at a spherical surface.

Light Source, Beam Expansion and Observation

For the general case we select a laser source which delivers a parallel light beam of 0.5 to 1 mm width and a Gaussian irradiance profile. The beam has to be expanded to the diameter of the test piece and at the same time has to be converged or diverged so that it hits the surface in the normal direction. To set up a general purpose instrument that can test a variety of different surfaces such as plane, spherical or aspherical ones with different curvatures, the test beam will usually be expanded to a parallel beam with a maximum diameter of several inches (2″, 4″, 6″, 10″ or 12″ is usual) by a focusing lens and a collimator. To test spherical surfaces an adjacent transmission sphere is added which carries a high-quality spherical reference surface as the last surface which is concentric to the emerging wavefront. An example is shown in figure 46-69. To test plane surfaces an adjacent transmission flat is added, consisting of a plane plate with a slight wedge, carrying the plane reference surface.

Light returning from the test surface is reflected at the beam-splitter, passes a beam-stop and is then imaged onto the sensor by an eyepiece.

The setup therefore basically consists of two telescopes: the first formed by the focusing lens and the collimator and the second by the eyepiece and the collimator.

We define the diameters and focal lengths of the optical elements and laser beam as follows:

D_{las} optically used diameter of laser beam,
D_{coll} optically used diameter of collimator,
D_{sens} optically used diameter of sensor,
f_{foc} focal length of focusing lens,
f_{coll} focal length of collimator,
f_{eye} focal length of eyepiece.

Since all diameters are defined by the selection of the laser, the sensor and the test surface, the remaining degree of freedom is the numerical aperture NA or the so called F-number $F/\#$, which is the ratio of the focal length and diameter of the exit pupil of a lens system used with an infinite object distance. For convenience we select an F-number of > 5 in order to avoid groups with many lens elements requiring correction. We then determine the necessary focal lengths of the optical groups:

$$f'_{foc} = F/\# \cdot D_{las} \tag{46-129a}$$

$$f'_{coll} = F/\# \cdot D_{coll} \tag{46-129b}$$

$$f'_{eye} = F/\# \cdot D_{sens} \tag{46-129c}$$

Note that the components should have some freeboard to avoid diffraction effects at the rims, mounts and stops within the optical groups and to allow the tilted and aberrated wavefronts to pass. Therefore $F/\#$ should be selected to be at least 5 % smaller than is required optically. As a numerical example we choose

$F/\# = 5.0$; $D_{las} = 0.5\text{mm}$; $D_{coll} = 152\text{mm}$ (6″); $D_{sens} = 9.0\text{mm}$.

The focal lengths of the components are then:

$f'_{foc} = 2.5\text{mm}$; $f'_{coll} = 76.0\text{mm}$; $f'_{eye} = 45.0\text{mm}$

Beam-stop

The size of the beam-stop in the back focal point of the collimator now has to be determined. This stop is necessary to block most of the unwanted light. Without it reflections from each optical surface and also stray light and erroneous reflections would fall onto the sensor and disturb the interferogram signal. The minimum diameter of the stop without limiting the spatial bandwidth of the interferogram can be derived from the Nyquist limit for the sensor and from Abbe's law for diffraction-limited resolution.

If the sensor has $N \times N$ pixels, it can resolve N equidistant parallel fringes across the diagonal. Following Nyquist theorem we will use N/2 fringes as maximum fringes resolved along any round diameter. For a wavelength λ the maximum angle between an axial and an oblique beam which are interfering is then given by (46-130).

$$\alpha_{max} = \arcsin \frac{\lambda N}{2 D_{sens}} \approx \frac{\lambda N}{2 D_{sens}} \tag{46-130}$$

The diameter D_{stop} of the stop is then calculated from

$$D_{stop} = 2 f'_{eye} \tan \alpha_{max} \approx 2 f'_{eye} \tan \frac{\lambda N}{2 D_{sens}} \approx \frac{\lambda N f'_{eye}}{D_{sens}} = \lambda\, N\, F/\# \tag{46-131}$$

The appropriate diameter of the beam-stop is defined by the product of the selected F-number of the eyepiece, the number of pixels along one dimension and the wavelength.

Lateral Resolution

We will illustrate the spatial bandwidth limitation of a small beam-stop diameter by the following example.

The test surface in figure 46-69 may have a sinusoidal surface modulation of the following form:

$$S(x, y) = a \sin\left(2\pi \frac{Ny}{c D_{surf}}\right) \tag{46-132}$$

where D_{surf} is the diameter of the surface under test, N the number of pixels along y, a the amplitude of the sinusoidal modulation and c the spatial wavelength in pixel units. In figure 46-70 we show the interferograms for the cases $c = 2$, 4, 8, 16 and irradiance distributions in the plane of the beam-stop. The diameter of the beam-stop is set to allow the first diffraction orders for the case $c = 2$ to pass, which corresponds to the Nyquist limit. The reference surface was tilted slightly in x to introduce six fringes for better visibility of the fringe modulation by the sinusoidal surface structure.

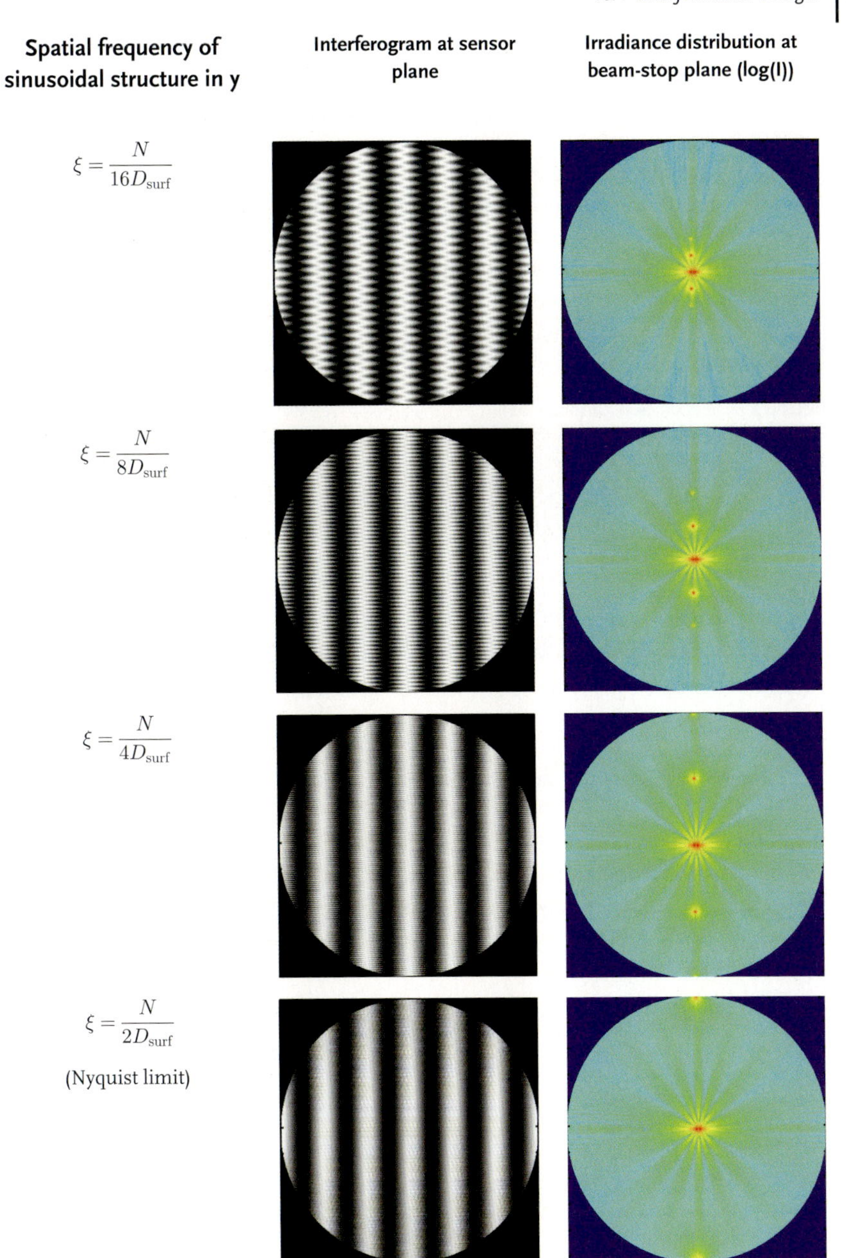

Figure 46-70: Interferograms at the sensor plane and irradiance distribution at the beam-stop plane for various sinusoidal structures of the test surface.

Although coherent light from a point source is being used, there is no infinite depth of focus for an image of the test surface. Every surface structure, whether an amplitude modulation by a dust particle or a phase modulation by a shape deviation or scratch, is the source of a secondary wavefront, which passes through the beam-stop acting as the limiting aperture. The sensor plane therefore has to be focused carefully onto the image plane of the surface under test. Figure 46-71 shows four different interferograms of the same surface with different focus positions of the CCD camera. While the camera was defocused, diffraction effects at the surface rim, around some dust particles and around a scratch on a test surface appeared and broadened. The lateral resolution thus diminishes when the sensor is not in its correct position.

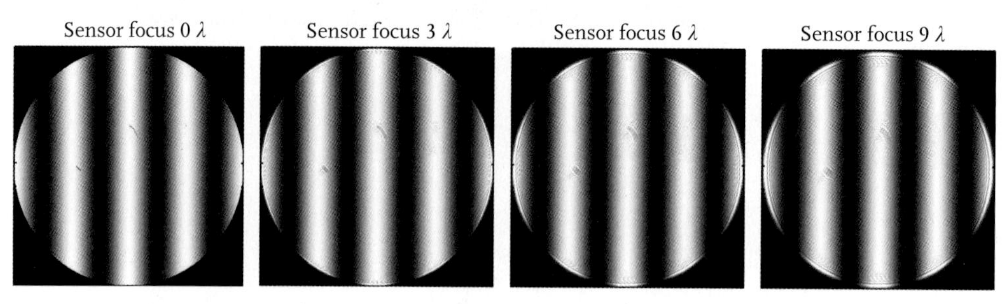

Figure 46-71: Interferograms of a surface with dust particles and a scratch in four different camera focus positions.

In figure 46-72 a sinusoidal surface modulation with varying spatial frequency has been analyzed by applying several different focus positions of the sensor plane. It can be seen that when in exact focus all spatial frequencies up to the Nyqist limit will be detected without loss. A camera defocus corresponding to 0.5 λ damps the higher spatial frequencies, whereas a defocus of 1.0 λ leads to a zero crossing with a following spatial range where we find "false resolution"; a case where "hills" are interpreted as "valleys" and vice versa. Further defocus leads to further zero crossings with alternately correct and false resolutions.

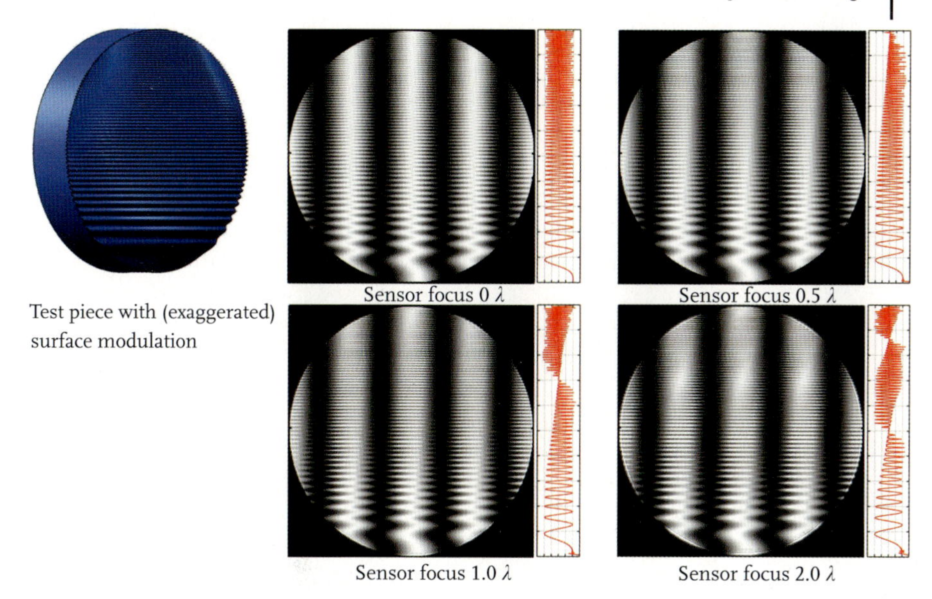

Figure 46-72: Interferograms of a surface with sinusoidal surface modulations of varying spatial frequencies. The corresponding phase profiles detected at the sensor are shown on the right.

Field Curvature

During the design stage of a test setup for an optical system, the field curvature of the system has to be analyzed thoroughly. The aim is to image the total surface under test onto the sensor plane without loss of lateral resolution. However, a residual field curvature, combined with astigmatism, will lead to an image which is partially defocused such that its lateral resolution varies over the sensor surface. Therefore, a first check of spot image diameters will help, when the system is being optimized. The spot images have to be considerably smaller than a pixel to guarantee full lateral resolution (figure 46-73).

Figure 46-73: Fizeau interferometer design for collimator-eyepiece telescope. For a pixel size of 9μm the spot images in the outer zones already limit the lateral resolution.

In the following we will compare three interferometer setups which are identical in their on-axis performance, but which have different collimator and eyepiece arrangements (figure 46-74). Arrangement (a) uses a four-element collimator and an optimized three-element eyepiece. In arrangement (b) a single aspherical lens replaces the collimator leading to a perfect on-axis wavefront. In arrangement (c) the eyepiece has also been replaced by a single aspherical lens. Again the on-axis wavefront is perfect. Note that arrangement (a) has a shorter overall length because the telelens-type collimator has a shorter back focal length.

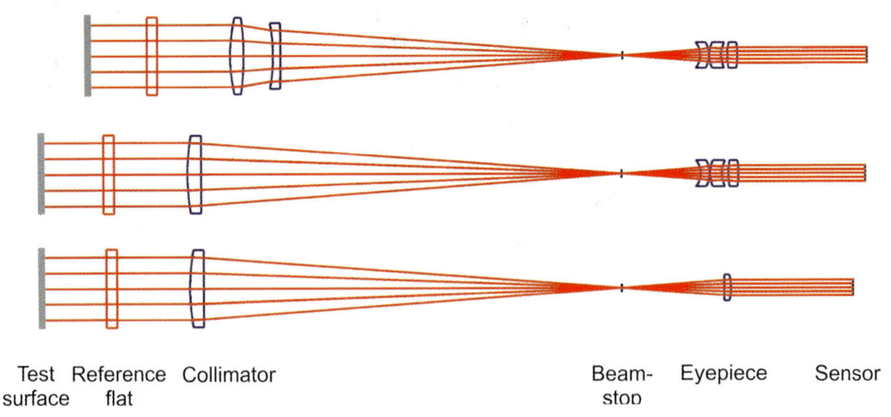

Figure 46-74: Three collimator-eyepiece designs of a Fizeau interferometer telescope: (a) two-element collimator and three-element eyepiece; (b) single aspherical element collimator and three-element eyepiece; (c) single aspheric element collimator and single aspherical element eyepiece.

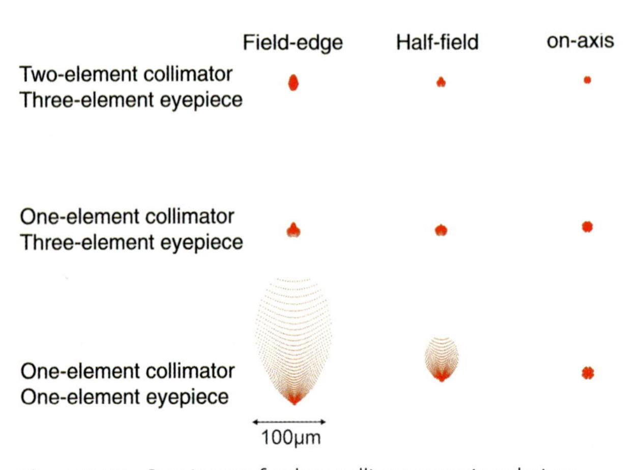

Figure 46-75: Spot images for three collimator-eyepiece designs of a Fizeau interferometer telescope as shown in figure 46-74.

Looking at the spot images of the three configurations on-axis, at half-field and at the edge of the field, we find that a single-element collimator is acceptable in the total field. The resolution does not change considerably over the field and seems to be good enough for a pixel size of around 10 μm unlike the arrangement using the single-element eyepiece. The resolution changes strongly from the center to the edge of the field (figure 46-75).

Distortion

For the design of a proper interferometer telescope the distortion over the field is also important. In the case of a strong distortion the measured surface deviations would have to be re-distorted to exactly match the locations on the test piece for further inspection and correction by computer-controlled polishing machines. A strong distortion would also vary the lateral resolution on the test surface, because some pixel areas are stretched and some are compressed. Note that, in a perfect interferometer, a slightly tilted reference or test surface will lead to straight, parallel and equally spaced fringes. If the interferometer, however, shows strong distortion, the fringes will be bent as if the test surface had a coma-like surface deviation. When the distorted image is corrected by a computer, the fringes will again look perfectly straight.

In figure 46-76 we show the distortion diagrams for the three interferometer arrangements in figure 46-74. The x-axis measures in μm and shows only up to 2 μm for the arrangements (a) and (b). This is much less than a pixel width and will not be noticeable. However, arrangement (c) shows 20 μm deviations which correspond to approximately two pixel widths and this will be of importance for high-precision measurements.

Reduction of Coherent Artefacts

Unfortunately, dust particles and surface imperfections on all optics parts after the pinhole are also the reason for various effects. Scattered light produces speckle effects, and dust particles are most likely to produce so-called "bulls eyes", i.e., interference rings that occur when a secondary spherical wave is diffracted from a dust particle and interferes with the reference and test wave. "Bulls eyes" are related to surfaces not coinciding with the camera plane or its associated images. In an adjusted setup the camera is focused onto the test surface. In normal cases there is no intermediate image between the test surface and camera so that all optical surfaces in between give rise to "bulls eyes". To decrease their impact on the irradiance distribution in the interferogram an extended light source can be used to replace the pinhole (figure 46-77).

For this purpose a rotating diffuser disk (a ground glass or holographic diffuser) can be placed in the front focal plane of the collimator, and the focusing optics is shifted along the optical axis to provide an illuminated circular area on the disk, acting as an extended monochromatic light source. The size of the illuminated area can be adjusted accordingly. Each individual point on the illuminated area is the source of a wavefront traveling through the system, but each wavefront is laterally

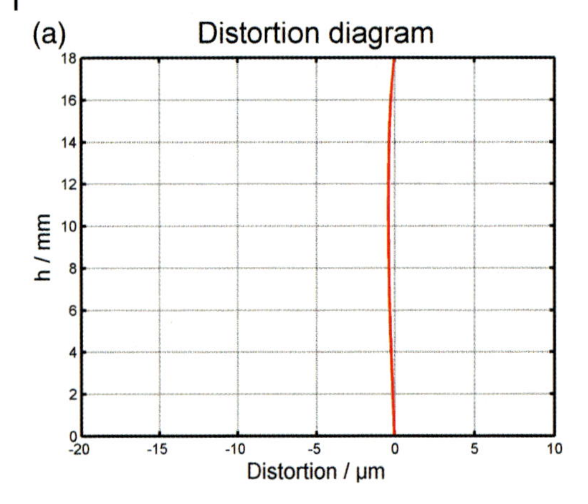

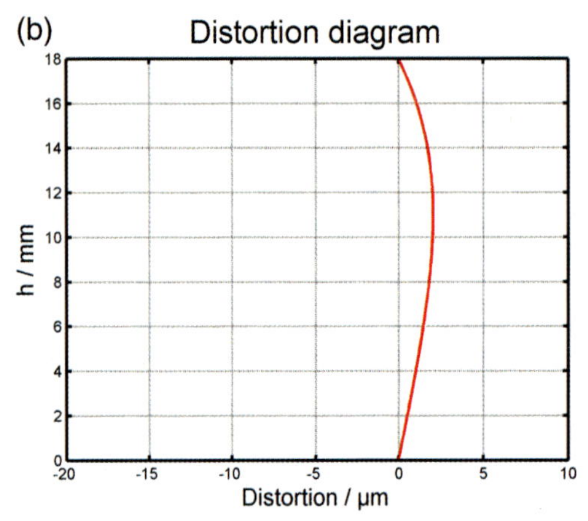

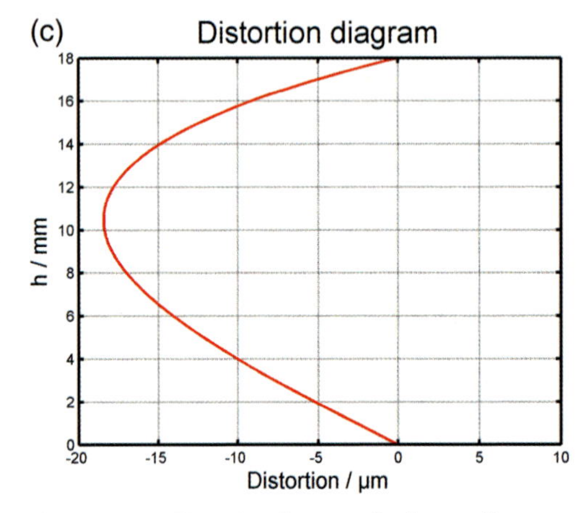

Figure 46-76: Distortion diagrams for three collimator-eyepiece designs of a Fizeau interferometer telescope as shown in figure 4-74.

sheared against its neighbor on all optical surfaces, except for the test surface and the camera. If we consider two separate point sources, their corresponding "bulls eye" irradiances are projected onto the camera and add incoherently. Their relative shear corresponds to the relative shear of the wavefronts at the surface carrying the dust particle. In the case of an extended light source, the irradiance variation at the camera caused by dust particles and imperfections therefore corresponds to the convolution of the projected relative light source distribution $s(x,y)$ in the camera plane with the "bulls eye" irradiances caused by a point source. For disturbances in the collimator the convolution function $s_{col}(x,y)$ is calculated from (46-133) and for disturbances in the eyepiece $s_{eye}(x,y)$ from (46-134).

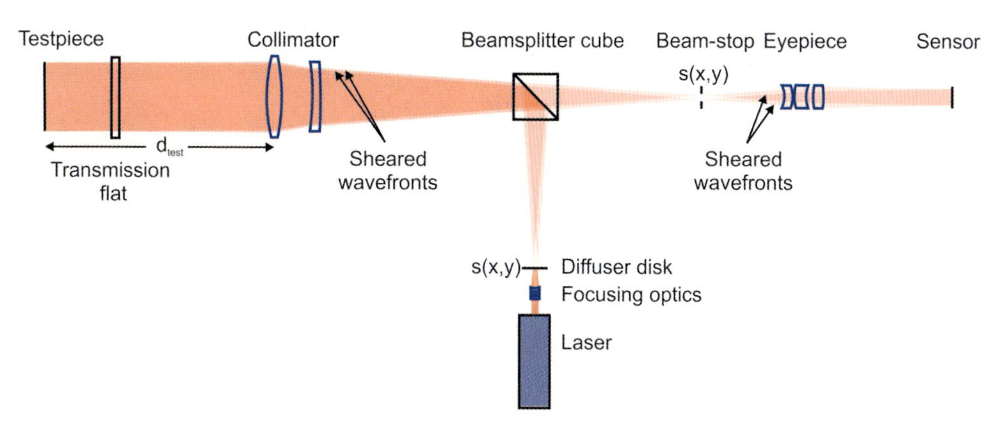

Figure 46-77: An interferometer setup with rotating diffuser disk as the extended light source with the light distribution function $s(x,y)$ causing laterally sheared wavefronts in the collimator and the eyepiece.

$$s_{col}(x,y) = s(x,y)\frac{f'_{eye}d_{test}}{f'_{col}} \tag{46-133}$$

$$s_{eye}(x,y) = s(x,y)\left(1+\frac{f'_{eye}(f'_{col}-d_{test})}{f'_{col}}\right) \tag{46-134}$$

where d_{test} is the distance of the test surface from the collimator (figure 46-77). For $d_{test}=f'_{col}$ the convolution function for collimator disturbances is equal to the light source function $s(x,y)$ divided by the magnification of the Kepler telescope consisting of the collimator and eyepiece. For eyepiece disturbances the convolution function is then equal to $s(x,y)$ (figure 46-78).

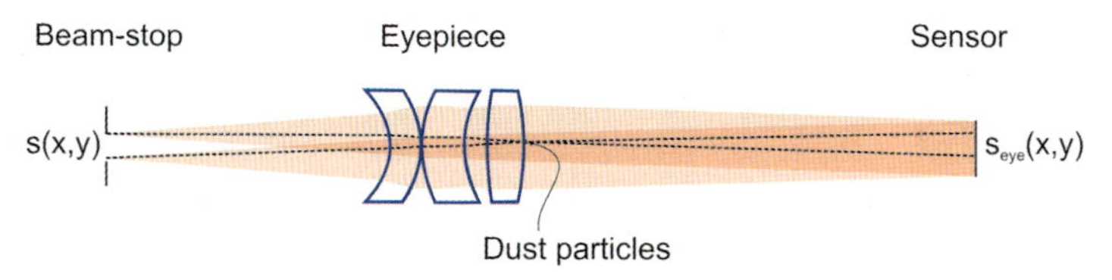

Figure 46-78: Projection of dust particles in the eyepiece onto the sensor.

According to (46-71) the maximum light source diameter is limited to give acceptable fringe visibility. The maximum diameter depends on the cavity length. For a collimator with f'_{coll}= 1000 mm, λ = 632.8 nm and a cavity length of 100 mm, the maximum light source diameter is 3.6mm. If a larger light source diameter is necessary for any reason, the fringe visibility can be regained by introducing a ring-shaped light source [46-42]. For a ring-shaped light source centered round the optical axis, all points on the circle lead to wavefronts that enter the cavity under the same angle. The consequence of this is that interferograms from different points on the circle are equal and add up incoherently to an interferogram without loss of visibility. In this way very long cavity lengths can also be achieved. Figure 46-79 shows a beam expansion unit using a holographic optical element (HOE) to project a circular light distribution onto the rotating diffuser disk centered round the optical axis.

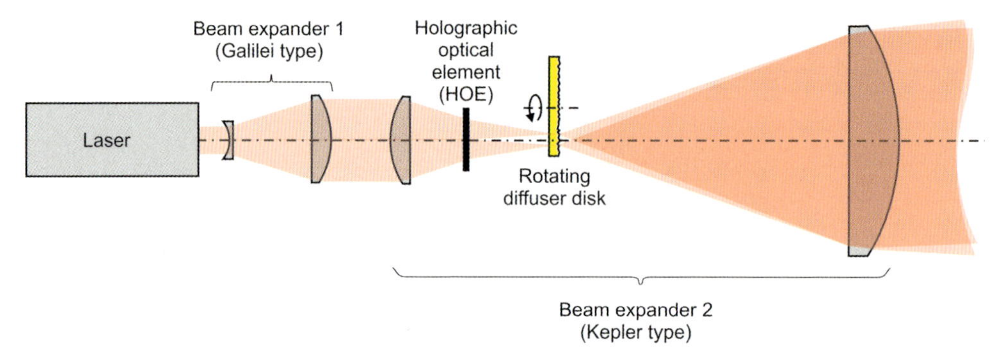

Figure 46-79: A holographic optical element (HOE) projects a circular light distribution onto the rotating diffuser disk to provide an extended light source that can be used in conjunction with very long cavities.

In figure 46-80(a)–(c) examples of the effects from dust particles and imperfections on an optical surface within the interferometer are given. Camera images are shown for: (a) a laser point source; (b) a circular homogeneously radiating extended light source and; (c) a ring-shaped extended light source.

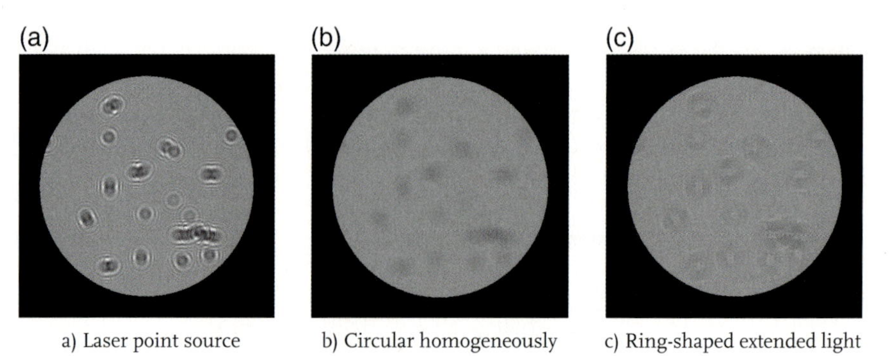

Figure 46-80: Camera images of an interferometer carrying dust particles and imperfections on an optical surface for different types of light sources.

46.5 Detection Techniques and Algorithms

46.5.1 General Considerations

In two-beam interferometry the general task is to determine the phase distribution $\Phi(x,y)$ over a detector pixel area denoted by x,y. Dependent on the interferometric test arrangement the phase distribution holds information, for instance, of a test surface deformation or a refractive index variation across a transmissive test piece.

In this chapter we discuss methods on how to determine $\phi(x,y)$ from irradiance distributions $I(x,y)$ measured with a 2D-detector such as a CCD camera. Many contributions can be found in the literature covering this subject, from which only a subset can be cited [46-43]–[46-62].

The two-beam irradiance distribution is given as [46-1]:

$$I(x,y,t) = I_1(x,y) + I_2(x,y) + 2\sqrt{I_1(x,y)I_2(x,y)}\cos\left(\phi(x,y) + \Delta(x,y,t)\right) \tag{46-135}$$

$I_1(x,y)$ and $I_2(x,y)$ denote the irradiance distribution generated by the individual beams without interference. $\Delta(x,y,t)$ is a well-defined or an unknown phase offset that may vary in time and space. In our further considerations we will use a short form of (46-135) omitting x,y:

$$I = a + b\cos\left(\phi + \Delta\right) \tag{46-136}$$

where $a = I_1(x,y) + I_2(x,y)$ and $b = 2\sqrt{I_1(x,y)I_2(x,y)}$.

Assuming Δ to be a known parameter we have to consider three unknowns: a, b, and ϕ

Therefore at least three interferograms I_1, I_2, I_3 have to be captured to find a solution for ϕ.

$$I_i = a + b\cos\left(\phi + \Delta_i\right)\ i = 1, 2, 3, ... \tag{46-137}$$

with $\Delta_i = \Delta(t_i)$usually taken at different times t_1, t_2, t_3 and $\Delta_1 \neq \Delta_2 \neq \Delta_3$.

Using addition theorems for trigonometric functions the solution for ϕ is then given as

$$\phi = \arctan\frac{I_1(\cos\Delta_3 - \cos\Delta_2) + I_2(\cos\Delta_1 - \cos\Delta_3) + I_3(\cos\Delta_2 - \cos\Delta_1)}{I_1(\sin\Delta_3 - \sin\Delta_2) + I_2(\sin\Delta_1 - \sin\Delta_3) + I_3(\sin\Delta_2 - \sin\Delta_1)} \tag{46-138}$$

46.5.2 Least-squares Phase Detection

A least-squares method can be applied, when the number of interferograms $N > 3$. To derive the linear equations for the least-squares fitting (46-137) has to be transformed to

$$I_i = a + b\cos\phi\cos\Delta_i - b\sin\phi\sin\Delta_i = D_1 + D_2\cos\Delta_i + D_3\sin\Delta_i \tag{46-139}$$

where $D_1 = a$, $D_2 = b \cos D_3 = -b \sin D_2$, D_1, D_2, D_3 are determined by minimizing the variance S according to

$$S = \frac{1}{N}\sum_{i=1}^{N}(D_1 + D_2 \cos\Delta_i + D_3 \sin\Delta_i - I_i)^2 \tag{46-140}$$

The set of simultaneous equations may be written in matrix form as shown.

$$\begin{pmatrix} N & \sum_{i=1}^{N}\cos\Delta_i & \sum_{i=1}^{N}\sin\Delta_i \\ \sum_{i=1}^{N}\cos\Delta_i & \sum_{i=1}^{N}\cos^2\Delta_i & \sum_{i=1}^{N}\cos\Delta_i\sin\Delta_i \\ \sum_{i=1}^{N}\sin\Delta_i & \sum_{i=1}^{N}\cos\Delta_i\sin\Delta_i & \sum_{i=1}^{N}\sin^2\Delta_i \end{pmatrix}\begin{pmatrix} D_1 \\ D_2 \\ D_3 \end{pmatrix} = \begin{pmatrix} \sum_{i=1}^{N} I_i \\ \sum_{i=1}^{N} I_i\cos\Delta_i \\ \sum_{i=1}^{N} I_i\sin\Delta_i \end{pmatrix} \tag{46-141}$$

Denoting the 3×3 matrix in (46-141) as $A = \begin{pmatrix} a_{11} & a_{12} & a_{13} \\ a_{21} & a_{22} & a_{23} \\ a_{31} & a_{32} & a_{33} \end{pmatrix}$, and the vectors as d and b, (46-141) becomes $A\,d = b$. The solution for d is found by inversion: $d = A^{-1}\,b$.

The inverted matrix is given as

$$A^{-1} = \frac{1}{\det(A)}\begin{pmatrix} a_{22}a_{33} - a_{23}a_{32} & a_{13}a_{32} - a_{12}a_{33} & a_{12}a_{23} - a_{13}a_{22} \\ a_{23}a_{31} - a_{21}a_{33} & a_{11}a_{33} - a_{13}a_{31} & a_{13}a_{21} - a_{11}a_{23} \\ a_{21}a_{32} - a_{22}a_{31} & a_{12}a_{31} - a_{11}a_{32} & a_{11}a_{22} - a_{12}a_{21} \end{pmatrix} \tag{46-142}$$

where $\det(A) = a_{11}a_{22}a_{33} + a_{12}a_{23}a_{31} + a_{13}a_{21}a_{32} - a_{13}a_{22}a_{13} - a_{32}a_{23}a_{11} - a_{33}a_{21}a_{12}$

The solution is finally given by

$$\phi = -\arctan\frac{D_3}{D_2} \tag{46-143a}$$

$$I_1 = \frac{1}{2}\left(D_1 - \sqrt{D_1 - D_2 - D_3}\right) \tag{46-143b}$$

$$I_2 = \frac{1}{2}\left(D_1 + \sqrt{D_1 - D_2 - D_3}\right) \tag{46-143c}$$

We can simplify the least-squares procedure by chosing a specific sampling procedure, where measurements are taken at N equally spaced intervals, uniformly spaced in k signal periods defined by

$$\Delta_i = \frac{2\pi(i-1)k}{N} + \Delta_1 \tag{46-144}$$

where $i = 1, 2, ..., N$ and $k \geq 1$ and $N \geq 3$.

Matrix A then becomes a diagonal matrix and matrix inversion is not necessary. The solutions for D_1, D_2, and D_3 are obtained from

$$D_1 = \frac{1}{N} \sum_{i=1}^{N} I_i \tag{46-145a}$$

$$D_2 = \frac{2}{N} \sum_{i=1}^{N} I_i \cos \frac{2\pi k i}{N} \tag{46-145b}$$

$$D_3 = \frac{2}{N} \sum_{i=1}^{N} I_i \sin \frac{2\pi k i}{N} \tag{46-145c}$$

The procedure of equally and uniformly spaced sampling is called synchronous detection.

46.3.3
Error Sources

In practice there are a variety of influences distorting the signal in (46-135). The following is a summary of the most non-linear effects leading to measurement errors [46-62]–[46-72].

1. The light detector has a non-linear response leading to a distorted periodic signal (figure 46-81) that contains not only the base frequency but also higher harmonics.
2. The phase-shifting device has a non-linear response so that sampling is performed at unequally and non-uniformly spaced sampling locations. Miscalibration of the device also leads to detuning error, where sampling occurs at a frequency other than the signal's base frequency.
3. The signal itself contains higher harmonics, as in the case of multiple beam interferometry or the Ronchi test.
4. The base frequency of the signal varies in time when temporal synchronous detection is applied, or the base frequency of the signal varies in space when spatial synchronous detection is applied.

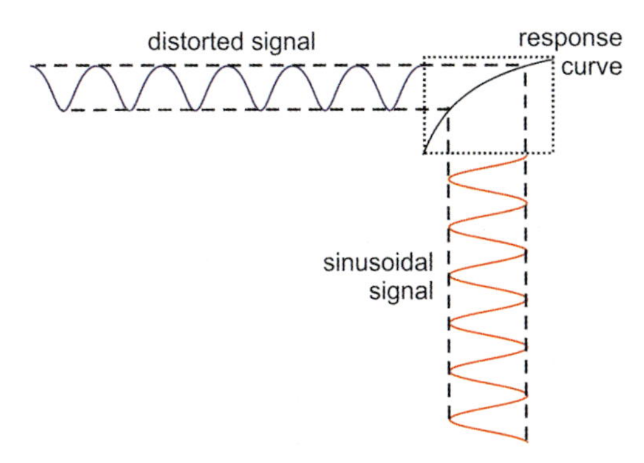

Figure 46-81: Distortion of a sinusoidal signal by the non-linear response of a detector.

The next chapters will show a more detailed description of different phase detection algorithms and will distinguish between temporal and spatial detection techniques used to find the phase information in an interferometric signal.

46.3.4
Phase-shifting Interferometry

In phase-shifting interferometry (PSI) the irradiance signal from (46-135) is varied in time by $\Delta(t)$, which is a well-defined phase-shift, disturbed, however, by statistical environmental effects like vibrations and air turbulences or by systematic phase-shifting errors all described by $\delta(x,y,t)$.

$$I(x,y,t) = I_1(x,y) + I_2(x,y) + 2\sqrt{I_1(x,y)I_2(x,y)}\cos\left(\phi(x,y) + \Delta(t) + \delta(x,y,t)\right) \tag{46-146}$$

The phase-shift $\Delta(t)$ may be applied in a continuous manner, by introducing a permanent frequency shift of $2\pi\nu_0 t$ in the reference beam of the interferometer. This case is called a heterodyne interferometer. The sensor in a heterodyne interferometer integrates the signal over a time interval T, which decreases the visibility of the original signal by a sinc-function given in (46-147). The phase-detection technique is called the integrated bucket technique. Using the short form in (46-136) the detected signal J is calculated from (46-147). The visibility goes to zero for intervals T equal to an integer number of $1/\nu_0$. For T=0 the detected signal is equal to the original function I. Additionally a constant phase-shift proportional to T is introduced, which can be neglected, since it is equal for all pixels.

$$\begin{aligned} J &= \frac{1}{T}\int_{t=0}^{T}\left(a + b\cos\left(\phi + 2\pi\nu_0 t\right)\right)dt \\ &= a + b\sin c\left(2\pi\nu_0\frac{T}{2}\right)\sin\left(\phi + 2\pi\nu_0\frac{T}{2}\right) \end{aligned} \tag{46-147}$$

In the following, the heterodyne technique will be treated as any other technique involving discontinuous phase-shifting (phase-stepping). In this case the signal is measured at several known increments of the phase. The phase has to be stationary for a short time in order to sample the irradiance distribution. The phase is changed rapidly between two consecutive measurements in order to shorten the acquisition process as much as possible. The change in acceleration may cause some vibrations in the system, especially when a mirror is moved and stopped. Figure 46-82 shows four interferograms sampled with phase-shifts of $\pi/2$ between the acquisitions.

Figure 46-82: Interferograms phase-shifted by $\pi/2$ between the individual acquisitions.

We now assume that the signal $I(x,y,t)$ is sampled over a time window and that each sample is multiplied by a weight factor $w(t)$, which we will call the window function [46-73]. Windows will help us to suppress effects from distorted signals and systematic phase-shifting errors. Figure 46-83 shows a periodic signal over three periods and three examples of window functions, a rectangular function being equal to one within the sample time, dropping to zero outside, a so-called Von Hann window described by (46-148) and a Gaussian window described by (46-149).

$$w(t) = \frac{1}{2}\left(1 + \cos\left(2\pi\frac{2t-\tau}{2\tau}\right)\right) \qquad \text{Von Hann window} \tag{46-148}$$

$$w(t) = e^{-\left(\frac{2t-\tau}{\tau}\right)^2 \ln\frac{1}{r}} \qquad \text{Gaussian window} \tag{46-149}$$

where τ is the window width and r is the weight at $t=0$ and $t=\tau$.

If the sampled function $I(x,y,t)w(t)$ is transformed to the frequency domain, the spectrum is denoted by $G(\nu)$

$$G(\nu) = a\,W(\nu) + \frac{b}{2}\left(W(\nu-\nu_0)e^{i\phi} + W(\nu+\nu_0)e^{-i\phi}\right) \tag{46-150}$$

where $W(\nu)$ is the Fourier transform of $w(t)$. $W(\nu)$ is convolved with the original sinusoidal signal such that it is reproduced at the DC term $\nu=0$ and, in the ideal case, at $\nu=\pm\nu_0$. If the signal shows nonlinearities $W(\nu)$ is also reproduced at any higher harmonic $\nu=\pm m\nu_0$ (m=2, 3, 4,...) of the base frequency.

We can determine ϕ from (46-150) by selecting $\nu=\nu_0$ and demanding $W(\nu_0)=0$ and $W(2\nu_0)=0$. In this case the solution for ϕ is given by

$$\phi = \arctan\frac{\mathrm{Im}G(\nu_0)}{\mathrm{Re}G(\nu_0)} \tag{46-151}$$

When $I(x,y,t)$ is sampled at the correct base frequency ν_0 and the integration is carried out over an integer number of periods $W(\nu_0)=0$ and $W(2\nu_0)=0$ are automatically fullfilled for a rectangular window. However, if ν_0 is not properly met (due to a detuning error) or if the sampling occurred under fluctuating conditions (statistical phase variations) the signal peaks are displaced or broadened such that the window filter function now suffers from leakage. The design of the window function $w(t)$ therefore plays a major role.

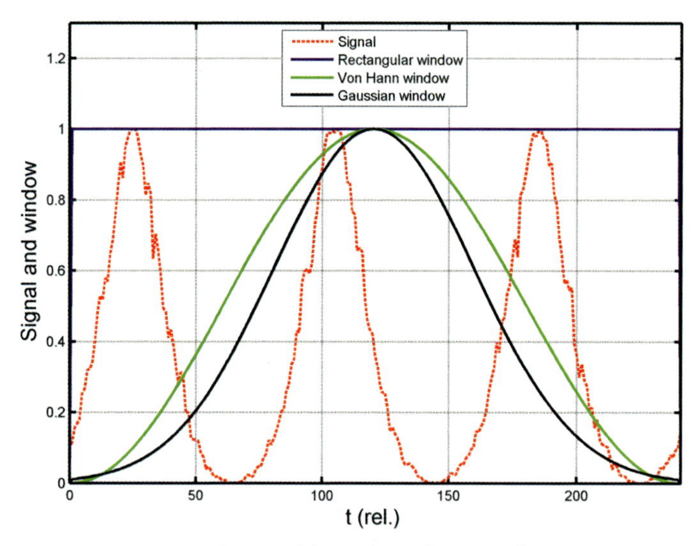

Figure 46-83: Distorted sinusoidal signal over three periods and different window functions $w(t)$: rectangular (blue), Von Hann (green), Gaussian (black).

In figure 46-84 the appropriate presentation in the frequency domain is shown. The signal's first-order peaks are located near the ± third zero crossings of the windows function spectra because three periods have been sampled. Since the gradients of the rectangular spectrum are much higher in the vicinity of the zero crossings than those of the other spectra, the rectangular window will respond much more sensitively concerning detuning, nonlinearities and statistical phase fluctuations.

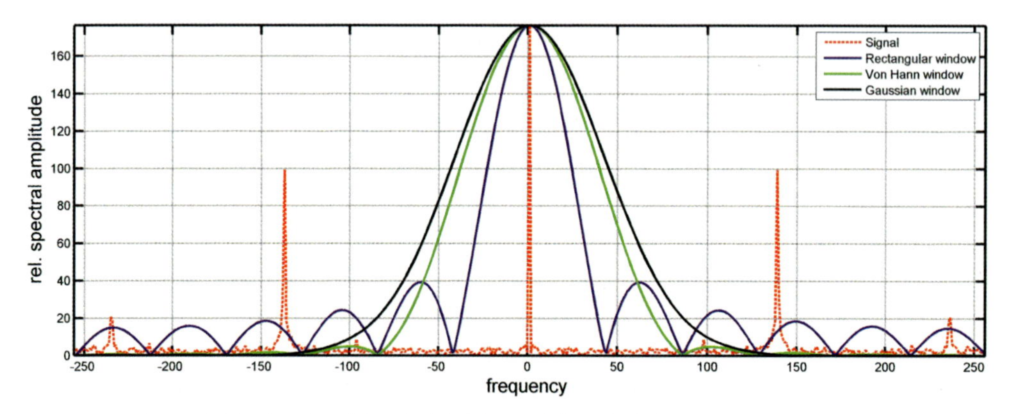

Figure 46-84: Spectra of distorted and detuned sinusoidal signal over three periods and spectra of different window functions $w(t)$: rectangular (blue), Von Hann (green), Gaussian (black).

The special case of a symmetric and real window function leads to the practical important case for N sampled interferograms as shown in (46-152).

$$\phi = -\arctan \frac{\sum_{i=1}^{N} I_i w_i \sin(2\pi(i-1)\Delta)}{\sum_{i=1}^{N} I_i w_i \cos(2\pi(i-1)\Delta)} \tag{46-152}$$

where Δ is the selected phase-shift per step and w_i is the selected window function weight. We simplify (46-152) by setting

$$s_i = w_i \sin 2\pi(i-1)\Delta \text{ and } c_i = w_i \cos 2\pi(i-1)\Delta \tag{46-153}$$

Equation (46-152) is then represented by two discrete convolutions of the interferograms I_i with the convolution kernel elements s_i in the numerator and c_i in the denominator, referred to in table 46-1.

Figure 46-85 gives an example of three different windows used when sampling 16 interferograms in a Fizeau interferometer with a phase-shift of $\pi/2$ between the acquisitions of the individual interferograms. A detuning error of 10% was introduced as well as a statistical error of 0.01 λ RMS for each phase-shift. In the case of a rectangular window, erroneous fringes with twice the frequency of the original fringes in the interferogram are visible. Von Hann and Gaussian windows suppress these effects to a large extent.

There are numerous phase-shift algorithms published in the literature, which can be described by (46-152). Table 46-1 gives an overview of some well-known and thoroughly investigated examples also covering asymmetric kernels. New kernels can be derived, when a phase-shift is added to the cosine and sine function in (46-153).

In general it can be stated that results are of better quality when more samples are taken. For high-quality measurements a set of more than 10 interferograms should be taken. Phase fluctuations can be inhomgeneous due to air turbulences in the cavity so that a larger number of acquisitions must be taken to improve statistical noise in the interferograms. For pure statistical fluctuations the noise in the wavefront result decreases by $1/\sqrt{N}$.

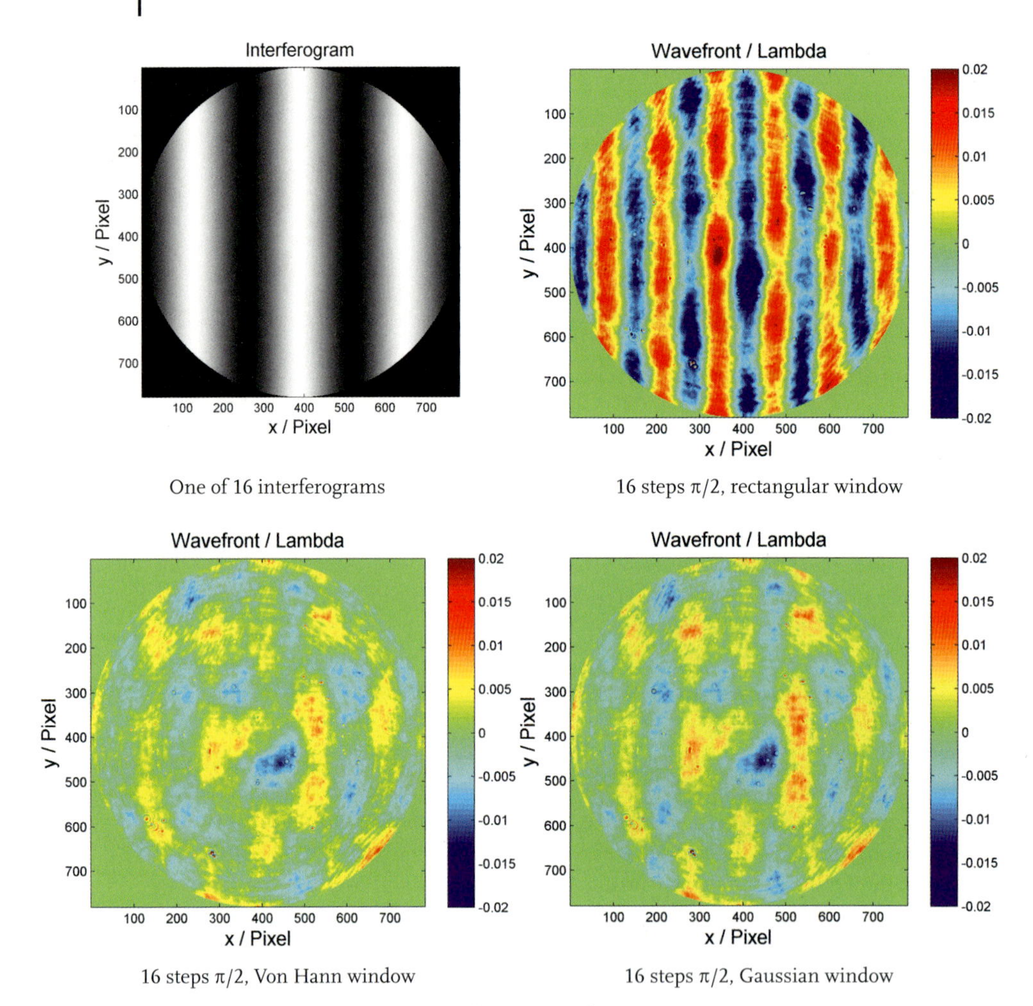

One of 16 interferograms

16 steps $\pi/2$, rectangular window

16 steps $\pi/2$, Von Hann window

16 steps $\pi/2$, Gaussian window

Figure 46-85: Wavefronts calculated from 16 interferograms (one is shown in (a)) with 10% detuning error and a statistical phase fluctuation of 0.01 λ RMS using: (b) a rectangular window; (c) a Von Hann window; (d) a Gaussian window, over 16 frames.

Figure 46-86 shows the numerical results for wavefront deviations when different numbers of steps and different window functions are applied. A systematic detuning error from a $\pi/2$ step of 10% was combined with a 10% deviation from the linearity of the sensor response function. The width of the Von Hann window was selected as the smallest integer multiple of four larger than the number of steps. The Gaussian window was set to $r = 0.01$ according to (46-153). The results show that the RMS wavefront error of a rectangular window reaches a minimum error of 0.01 λ only for N equal to an integer numbers of $2\pi/\Delta$, in this case an integer number of four, since a phase step of $\pi/2$ was selected. Both Gaussian and Von Hann

Algorithm	Δ		1	2	3	4	5	6	7	8	9	10
3 steps 120°	$2\pi/3$	s	1,7321	0	–1,7321							
		c	1	–2	1							
3 steps in inverted T	$\pi/2$	s	–1	2	–1							
		c	1	0	–1							
3 steps in tilted T	$\pi/2$	s	–1	1	0							
		c	0	1	–1							
4 steps in cross	$\pi/2$	s	0	1	0	–1						
		c	1	0	–1	0						
4 steps in X	$\pi/2$	s	–1	–1	1	1						
		c	1	–1	–1	1						
4 steps symmetrical	$2\pi/3$	s	0	1,7321	–1,7321	0						
		c	1	–1	–1	1						
4 steps asymmetrical (Schwider)	$\pi/2$	s	–1	3	–1	–1						
		c	1	1	–3	1						
5 steps	$2\pi/5$	s	–0,5878	–0,9511	0	0,9511	0,5878					
		c	–0,8090	0,3090	1	0,3090	–0,8090					
5 steps (Schwider–Hariharan)	$\pi/2$	s	0	1	0	–1	0					
		c	0,5	0	–1	0	0,5					

Algorithm	Δ		1	2	3	4	5	6	7	8	9	10
5 steps asymmetrical (Schmit–Creath)	$\pi/2$	s	0	3	–3	–1	1					
		c	1	–1	–3	3	0					
6 steps symmetrical	$2\pi/3$	s	0	0,9511	0,5878	–0,5878	–0,9511	0				
		c	0,5	0,3090	–0,8090	–0,8090	0,3090	0,5				
6 steps asymmetrical	$\pi/2$	s	1	3	0	–4	–1	1				
		c	1	–1	–4	0	3	1				
7 steps symmetrical	$2\pi/6$	s	0	1,7321	1,7321	0	–1,7321	–1,7321	0			
		c	1	1	–1	–2	–1	1	1			
8 steps symmetrical (Von Hann τ=8)	$\pi/2$	s	0,0269	0,2183	–0,4889	–0,6802	0,6802	0,4889	–0,2183	–0,0269		
		c	0,0269	–0,2183	–0,4889	0,6802	0,6802	–0,4889	–0,2183	0,0269		
9 steps symmetrical (Von Hann τ=12)	$\pi/2$	s	0	0,5	0	–0,933	0	0,933	0	–0,5	0	
		c	0,25	0	–0,75	0	1	0	–0,75	0	0,25	
9 steps symmetrical (Von Hann τ=12)	$\pi/2$	s	0	0,5	0	–0,933	0	0,933	0	–0,5	0	
		c	0,25	0	–0,75	0	1	0	–0,75	0	0,25	
10 steps symmetrical (Von Hann τ=12)	$\pi/2$	s	–0,1036	0,2620	0,4451	–0,6036	–0,6951	0,6951	0,6036	–0,4451	–0,2620	0,1036
		c	0,1036	0,2620	–0,4451	–0,6036	0,6951	0,6951	–0,6036	–0,4451	0,2620	0,1036

Table 46-1: Some examples of phase-shifting convolution kernels using a phase-shift of Δ, numerator kernels s and denominator kernels c.

windows damp detuning and nonlinearity errors very efficiently when using > 12 steps. The residual wavefront errors will fall below 0.001 λ. When additional statistical phase fluctuations occur, in this example at 0.01 λ RMS, the residual wavefront error is considerably higher (figure 46-87). Classical synchronous detection algorithms are mainly limited by phase fluctuations during the acquisition process. Therefore, high-precision interferometric setups have to be designed with as much mechanical and thermal stability as possible.

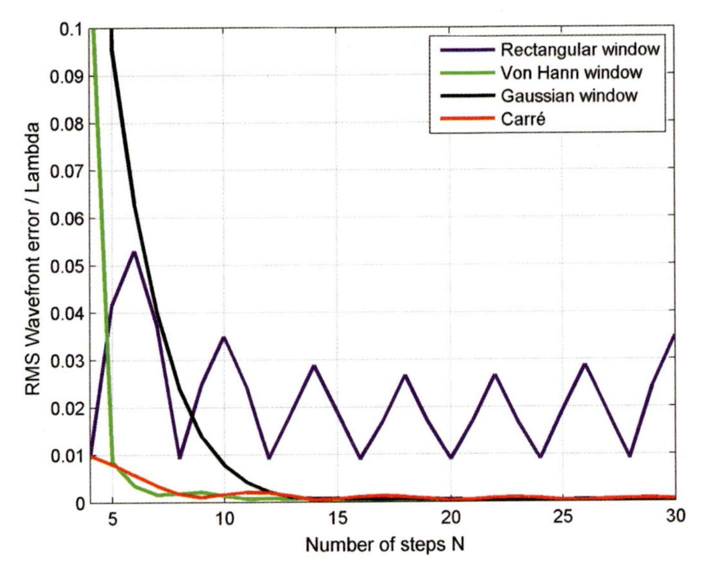

Figure 46-86: RMS wavefront error as a function of number of $\pi/2$ phase steps for 10% detuning error + 10% non-linearity using a rectangular window, a Von Hann window, a Gaussian window and the Carré algorithm. The latter was calculated from the mean of N–3 phase values derived from four adjacent steps.

Synchronous phase-detection algorithms assume an equally spaced and – over an integer number of periods – homogeneously distributed, well known sample rate. If the phase distribution of an interferogram has to be obtained from a sinusoidal irradiance signal with unknown local frequency this is known as asynchronous detection. Δ may then be constant but unknown with respect to the actual wavelength.

The earliest asynchronous algorithm was developed by Carré. Four fringe images are sampled with equally spaced phase-shifts introduced (see (46-154 a–d)). The magnitude of the phase-shift Δ is unknown relative to the signal period.

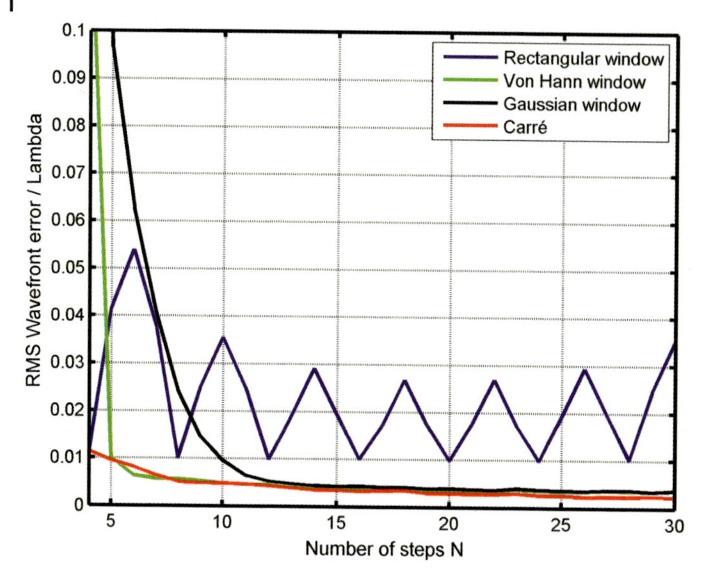

Figure 46-87: RMS wavefront error as a function of number of $\pi/2$ phase steps for 10% detuning error + 10% non-linearity + statistical phase fluctuation of 0.01 λ RMS using a rectangular window, a Von Hann window, a Gaussian window, and the Carré algorithm. The latter was calculated from the mean of N–3 phase values derived from four adjacent steps.

$$I_1 = a + b\cos\left(\phi - \frac{3\Delta}{2}\right) \tag{46-154a}$$

$$I_2 = a + b\cos\left(\phi - \frac{\Delta}{2}\right) \tag{46-154b}$$

$$I_3 = a + b\cos\left(\phi + \frac{\Delta}{2}\right) \tag{46-154c}$$

$$I_4 = a + b\cos\left(\phi + \frac{3\Delta}{2}\right) \tag{46-154d}$$

Using (46-155) the unknown phase-shift Δ is calculated from the four samples, assuming a constant Δ. The resulting phase ϕ is then calculated from (46-156).

$$\Delta = 2\arctan\sqrt{\frac{-I_1 + 3I_2 - 3I_3 + I_4}{I_1 + I_2 - I_3 - I_4}} \tag{46-155}$$

$$\begin{aligned}\phi &= \arctan\left(\frac{I_1 + I_2 - I_3 - I_4}{-I_1 + I_2 + I_3 - I_4}\tan\frac{\Delta}{2}\right)\\ &= \arctan\left(\frac{I_1 + I_2 - I_3 - I_4}{-I_1 + I_2 + I_3 - I_4}\sqrt{\frac{-I_1 + 3I_2 - 3I_3 + I_4}{I_1 + I_2 - I_3 - I_4}}\right)\end{aligned} \quad (46\text{-}156)$$

Equation (46-155) is indeterminate for the simultaneous occurance of $I_1 = I_4$ and $I_2 = I_3$ which is the case for $\phi = 0$. Δ then cannot be determined. There are further indeterminations when Δ is equal to an integer multiple of π, in which case $\tan \Delta/2$ is either 0 or $\pm\infty$.

For slowly varying phase-shifts the Carré algorithm is self-calibrating when a sample of N images is acquired and four adjacent images are selected respectively to derive separate values for ϕ, which are averaged. The algorithm is then defined by

$$\phi = \frac{1}{N-3}\sum_{i=1}^{N-3}\arctan\left(\frac{I_i + I_{i+1} - I_{i+2} - I_{i+3}}{-I_i + I_{i+1} + I_{i+2} - I_{i+3}}\sqrt{\frac{-I_i + 3I_{i+1} - 3I_{i+2} + I_{i+3}}{I_i + I_{i+1} - I_{i+2} - I_{i+3}}}\right) \quad (46\text{-}157)$$

Figure 46-86 shows the results for the Carré algorithm compared to different windowed algorithms described earlier. The results are comparable with those of the Von Hann and Gaussian windows. In the case of a statistically varying phase step as shown in figure 46-87 the dynamically adapted Carré algorithms show an additional small improvement over the other algorithms since an actual average phase step is found for each set of four adjacent samples.

46.3.5
Spatial Carrier Frequency Analysis

Phase-shifting techniques are sensitive towards phase variations during sampling. In an interferometric setup, provisions have to be made to avoid mechanical or acoustic vibrations, air turbulence, or changes in air temperature and pressure. An improvement could be made if the phase were derived from a single interferogram that, of course, had all disturbances "frozen" in a single shot, but many shots could be acquired to form a stable mean phase distribution.

Spatial carrier frequency analysis is able to derive ϕ from a single interferogram [46-76]–[46-78]. When choosing the carrier frequency, two conditions are necessary.

1) The carrier frequency must be selected to be larger than the maximum fringe frequency of the signal,
2) The sign of the carrier frequency has to be selected correctly otherwise the sign of the recovered signal will be flipped over.

If we write (46-136) in a complex form and then define spatial carrier frequencies v_x and v_y in the x and y directions, respectively, we have

$$I(x,y) = a(x,y) + c(x,y)e^{i2\pi(v_x x + v_y y)} + c^*(x,y)e^{-i2\pi(v_x x + v_y y)} \quad (46\text{-}158)$$

with $c(x,y) = \frac{b(x,y)}{2} e^{i\phi(x,y)}$ and $c^*(x,y) = \frac{b(x,y)}{2} e^{-i\phi(x,y)}$

In the same way as (46-150) we will multiply the signal $I(x,y)$ with a spatial window function $w(x,y)$ and will carry out a Fourier transformation leading to (46-159).

$$G(\nu_x, \nu_y) = \left(\tilde{a}(\nu_x, \nu_y) + \tilde{c}(\nu_x - \nu_{0x}, \nu_y - \nu_{0y}) + \tilde{c}^*(\nu_x + \nu_{0x}, \nu_y + \nu_{0y})\right) * W(\nu_x, \nu_y) \tag{46-159}$$

where $\tilde{a}$, $\tilde{c}$, and W denote the Fourier transforms of a, c and w and * denotes the convolution operation.

According to the Takeda algorithm [46-76] one of the side bands of (46-159) is then filtered out and moved to the origin in the frequency spectrum leading to (46-160).

$$F(\nu_x, \nu_y) = \tilde{c}(\nu_x, \nu_y) * W(\nu_x, \nu_y) \tag{46-160}$$

A Fourier back transformation finally leads to the desired result for ϕ as shown in (46-161).

$$\phi(x,y) = \arctan \frac{\mathrm{Im}\{c(x,y)\}}{\mathrm{Re}\{c(x,y)\}} = \arctan \frac{\mathrm{Im}\left\{FT^{-1}(F(\nu_x, \nu_y)\right\}}{\mathrm{Re}\left\{FT^{-1}(F(\nu_x, \nu_y)\right\}} \tag{46-161}$$

Since moving the sideband to the origin is the same as adding a linear phase to the spatial signal, it can also be replaced by subtraction of a best-fitting linear phase from the final result for $\phi\ (x,y)$.

Figure 46-88 shows an example of the main workflow of the Takeda algorithm. The multi-fringe interferogram is multiplied by a window function; in the example a Von Hann window. After Fourier transformation one of the side bands is separated and shifted to the origin. After Fourier back transformation, the phase is calculated according to (46-161).

An alternative technique avoiding Fourier transformation can be introduced by applying well-known temporal phase-shifting techniques to the spatial signal with linear carrier. The main difference is that the sinusoidal signal is now phase-modulated across the data field according to the wavefront under test. The assumption is that, within a small window, the wavefront is considered to be flat and the offset a and modulation b of the signal is assumed to be constant. Equation (46-162a–c) defines the irradiance at three adjacent pixels in the x direction numbered i-1, i, i+1. The carrier frequency in the x direction is defined by $\Delta_i = 2\pi\nu_x i$.

$$I_{i-1} \approx a_i + b_i \cos\left(\phi_i + \Delta_{i-1}\right) = a_i + b_i \cos\left(\phi_i + 2\pi\nu_x i - 2\pi\nu_x\right) \tag{46-162a}$$

$$I_i = a_i + b_i \cos\left(\phi_i + \Delta_i\right) = a_i + b_i \cos\left(\phi_i + 2\pi\nu_x i\right) \tag{46-162b}$$

$$I_{i+1} \approx a_i + b_i \cos\left(\phi_i + \Delta_{i+1}\right) = a_i + b_i \cos\left(\phi_i + 2\pi\nu_x i + 2\pi\nu_x\right) \tag{46-162c}$$

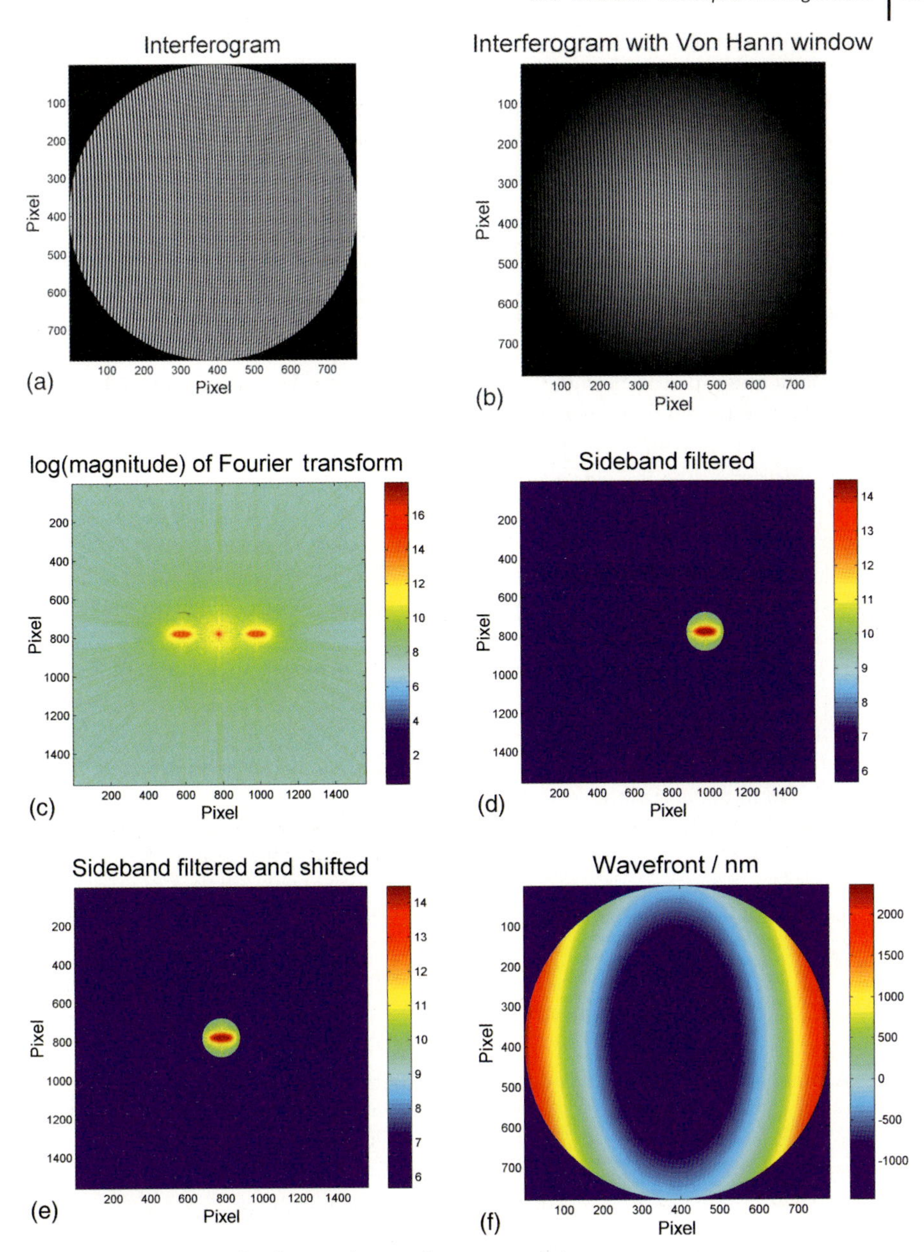

Figure 46-88: An example of a spatial carrier frequency analysis (Takeda algorithm): (a) multi-fringe interferogram; (b) interferogram multiplied by Von Hann window function; (c) Fourier transform (displayed is log(magnitude of Fourier transform)); (d) separated sideband; (e) separated sideband shifted to the origin; (f) wavefront calculated by Fourier back transformation and arctangent of imaginary/real parts.

The approximation is that $a_{i-1} = a_i = a_{i+1}$, $b_{i-1} = b_i = b_{i+1}$, and $\phi_{i-1} = \phi_i = \phi_{i+1}$. The phase ϕ_i at pixel i can then be calculated from (46-163) which is derived from (46-138) defining the evaluation of three interferograms for temporal phase-shifting.

$$\phi_i = \arctan \frac{-I_{i-1} + I_{i+1}}{I_{i-1} - 2I_i + I_{i+1}} \tan \pi v_x - 2\pi v_x i \tag{46-163}$$

If v_x is set to $1/4$ $\tan \pi v_x$ is equal to 1 and the expression becomes quite simple. Equation (46-163) consists mainly of a convolution in the numerator by a kernel $s = [-1\ 0\ 1]$ and a convolution in the denominator by a kernel $c = [1\ -2\ 1]$. The selected carrier frequency determines a factor $\tan \pi v_x$ to be multiplied by the numerator. Note that, after calculation of the arctangent, the linear phase, which was selected to introduce the carrier frequency has to be subtracted.

The analog to temporal phase-shifting of many different algorithms can be applied to determine the phase at the local origin of each interval i. It must be realised that in most of the intervals the fringe frequency is detuned by ϕ_i because of the phase modulation. Therefore a detuning-insensitive algorithm would be advisable. However, in order to establish an effective filter, the convolution kernels should be extended over more than three pixels. In principle all kernels, which are shown in table 46-1, are applicable.

A linear carrier frequency can also be introduced in an arbitrary direction making it necessary to use two-dimensional convolution kernels of size $M \times M$. If a carrier frequency along a 45° diagonal direction is chosen [46-77], [46-78], there are $2M$–1 steps in a direction perpendicular to the fringes and the period is shorter by a factor of $\sqrt{2}$ than in the horizontal or vertical direction.

For an $M \times M$ kernel (M is odd) the phase ϕ_{ij} at pixel i,j is calculated from (46-164).

$$\phi;_{ij} = \arctan \frac{\sum_{k=1}^{M}\sum_{l=1}^{M} s_{kl} I_{i-\frac{M+1}{2}+k,j-\frac{M+1}{2}+l}}{\sum_{k=1}^{M}\sum_{l=1}^{M} c_{kl} I_{i-\frac{M+1}{2}+k,j-\frac{M+1}{2}+l}} - 2\pi v(i+j) \tag{46-164}$$

A suitable 3×3 kernel that is detuning insensitive is given by

$$s = \begin{bmatrix} 1 & -2 & 0 \\ -2 & 0 & 2 \\ 0 & 2 & -1 \end{bmatrix} \qquad c = \begin{bmatrix} 1 & 1 & -1 \\ 1 & -4 & 1 \\ -1 & 1 & 1 \end{bmatrix}$$

The kernel is obtained from a combination of three inverted T algorithms (see table 46-1), shifted by $\pi/2$ ($v = 1/4$) the second with respect to the first and the third with respect to the second, ending with a detuning resistent five-step algorithm. The five elements are distributed over the elements in a diagonal direction and their weights are split according to the numbers of elements in a diagonal.

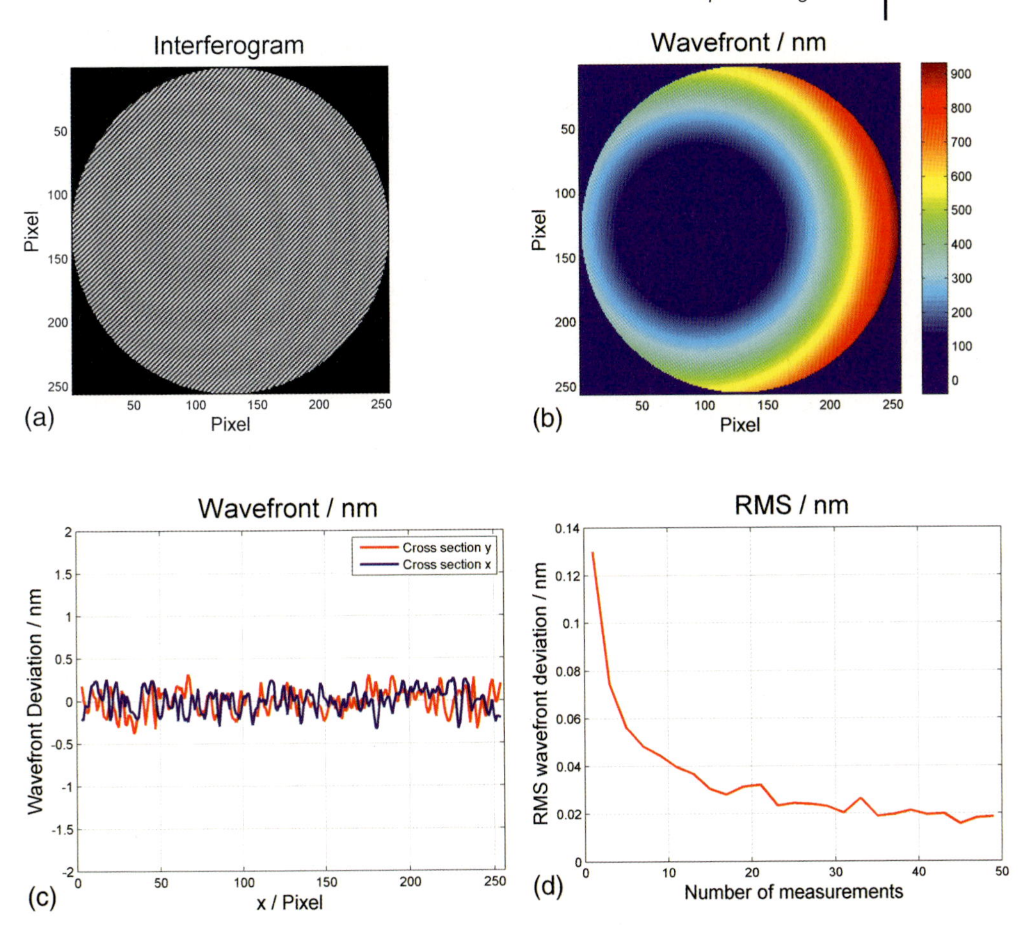

Figure 46-89: Example of a spatial carrier frequency analysis: (a) multi-fringe interferogram; (b) wavefront calculated from a single interferogram; (c) deviations from ideal wavefront due to local detuning of carrier frequency; (d) RMS wavefront error calculated from the average over m phase-shifted individual wavefronts as a function of m.

Figure 46-89 shows an example of an analysis using the kernels above. When a single interferogram is evaluated, detuning errors occur with a wavefront RMS of 0.13 λ as shown in 46-89 (c). The error can be reduced when m wavefronts are averaged where the interferograms have been phase-shifted evenly distributed over an integer number of λ. 46-89 (d) shows the diagram of RMS reduction by averaging over m wavefronts. The curve mainly follows a simple $1/\sqrt{m}$ rule.

Another aspect to be discussed is lateral resolution. Using spatial carrier frequency analysis usually means a reduction in lateral resolution compared to temporal phase-shifting techniques. The latter generates results for each individual pixel whereas in carrier frequency analysis the convolutions use the neighborhood of each pixel to determine the phase value. Figure 46-90 shows the so-called "Height Transfer Function" (HTF) for 3 × 3 kernels as shown above and for temporal phase-shifting techniques [46-79], [46-80].

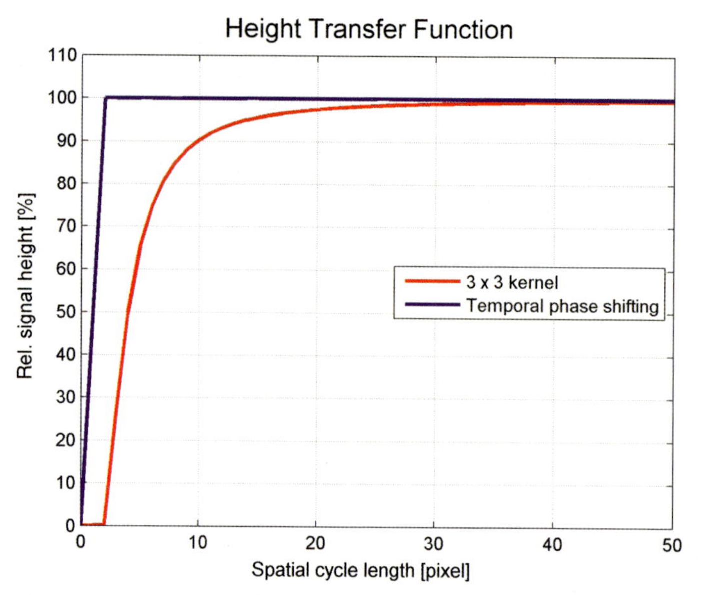

Figure 46-90: Height Transfer Function (HTF) for spatial carrier frequency analysis using 3 × 3 kernels and for temporal phase-shifting techniques.

The HTF shows the relative signal height for sinusoidal wavefronts as a function of different spatial cycle lengths in pixels. For 3 × 3 kernels, a cycle length of 10 pixels is reproduced only with a 90% signal height. For a 100 nm surface waviness with a periodicity of 10 pixels the result would show only 90 nm PV and for a periodicity of five pixels it would show only 60 nm PV. Temporal phase-shifting techniques would always show 100 nm PV down to the Nyquist limit of two pixels per cycle.

Using the Fourier transform technique according to Takeda is also related to a spatial low- pass filter, since isolating the sideband as described above is simply low-pass filtering. The choice of the isolating filter width therefore determines the curve shape and cutoff frequency of the HTF.

46.3.6 Simultaneous phase-shifting interferometry

The advantages of spatial carrier frequency analyses for deriving the wavefront from a single interferogram could also be achieved if interferograms with different phase offsets were captured simultaneously. In this case, the effects of vibrations and air turbulence would be reduced, since the phase-shifts Δ_i would be stable at all times [46-81].

Figure 46-91 shows a Twyman–Green interferometer with polarizing optics in order to obtain four interferograms with phase offsets of 0°, 90°, 180°, and 270° simultaneously. The light that comes from the laser is linearly polarized at 45° with respect to the x and y-axes. It is split off by a dielectric beam-splitter BS 1 into a test and a reference beam. A $\lambda/8$-plate with its axis at 0° is placed in the reference arm. After reflection at the reference mirror and returning to BS 1 both the vertical and horizontal polarization components are shifted by 90°. When they interfere with light returning from the mirror under test, both polarizations generate individual interferograms. They are separated from each other by a polarizing beam-splitter PBS 1 and are captured by two CCD cameras individually.

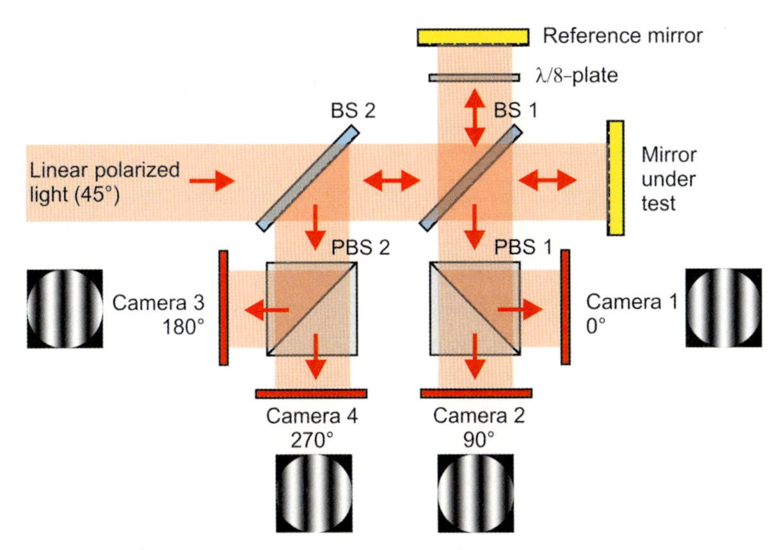

Figure 46-91: Simultaneous phase-shifting interferometer of Twyman-Green type using four CCD-cameras to capture interferograms with phase-shifts of 0°, 90°, 180°, and 270°.

Light that returns to the laser now carries two interferograms of different polarization that are phase-shifted by 180° with respect to those at camera 1 and 2. The beam-splitter BS 2 directs it to a second polarizing beam-splitter PBS 2 that separates the interferograms, which are phase-shifted by 180° and 270° respectively, so that they are captured by two further CCD cameras [46-82].

The use of four different cameras makes the alignment process and the calibration critical when high accuracy is required. An improvement can be made if the four interferograms are projected onto a single CCD camera chip [46-83]. Figure 46-92 shows a Twyman–Green interferometer where the reference and test beams have orthogonal polarization.

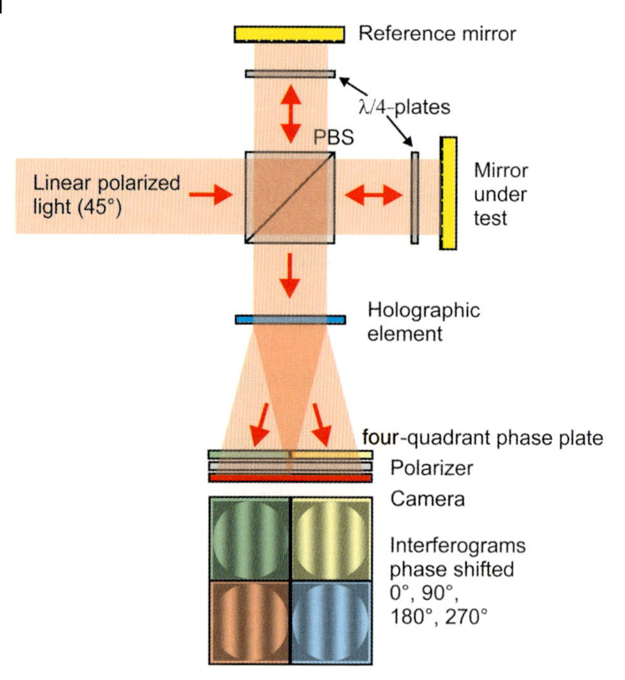

Figure 46-92: A simultaneous phase-shifting interferometer of Twyman–Green type using a holographic element for lateral separation of four interferograms that are phase-shifted by a four-quadrant phase plate by 0°, 90°, 180°, and 270°. A single camera sensor captures all four interferograms simultaneously.

Light from the laser is linearly polarized at 45° with respect to the x and y-axes. The polarizing beam-splitter separates the incident beam into a horizontally and a vertically polarized beam. A $\lambda/4$ phase plate is placed in each arm of the interferometer to reverse the polarization by 90° after reflection at the reference or test mirror and its second passage. After recombination the two beams pass through a holographic element that splits the beam into four separate beams. A birefringent mask with four quadrants of different phase-shifts is placed near the sensor element of a CCD camera. The introduced phase-shifts between the horizontally polarized test and vertically polarized reference beams are 0°, 90°, 180°, and 270°. A polarizer is placed between the sensor and quadrant phase plate with its transmission axis at 45°. In this way all four phase-shifted interferograms are detected simultaneously at the sensor array.

46.3.7 Unwrapping

As shown in the previous chapters, the phase of a wavefront is determined from the irradiance distribution of at least one interferogram by applying the arctangent function. Unfortunately, the arctangent function wraps the phase modulo 2π, such that the resulting phase values are only within the interval $-\pi$ to $+\pi$. We call it the wrapped phase $\phi_w(i,j)$. According to the Nyquist theorem a continuous and differentiable wavefront can be resolved as long as the magnitude of its slope is smaller than 0.5 λ/pixel. Along M pixel a maximum wavefront range of $M\lambda/2$, corresponding to a phase range of $M\pi$, can therefore be detected. In order to reconstruct the complete wavefront range the application of an unwrapping algorithm is necessary to calculate the continuous unwrapped phase $\phi_u(i,j)$.

The relationship between the wrapped phase and the unwrapped phase is given by

$$\phi_u(i,j) = \phi_w(i,j) + 2\pi m(i,j) \tag{46-165}$$

where m is an integer number to be determined by an unwrapping algorithm and i, j are pixel numbers with $1 \leq i \leq M$ and $1 \leq j \leq N$.

The unwrapping algorithm includes the search for $m(i,j)$ for each pixel such that phase differences between all adjacent pixels are smaller than π. For a one-dimensional phase function the phase-unwrapping process is

$$\phi_u(i+1) = \phi_u(i) + V(\phi_w(i+1) - \phi_u(i)) \tag{46-166}$$

where the wrapping function $V(x)$ is that defined in (46-167) and the initial value for $\phi_u(1) = \phi_w(1)$.

$$V(x) = \arctan\left(\frac{\sin x}{\cos x}\right) \tag{46-167}$$

For a two-dimensional function the unwrapping process has to be carried out in the x and y directions. Equation (46-166) is then transformed to

$$\phi_u(i',j') = \phi_u(i,j) + V(\phi_w(i',j') - \phi_u(i,j)) \tag{46-168}$$

where i' and j' denote pixels which are adjacent to (i,j). Adjacent pixels are $(i+1,j)$, $(i-1,j)$, $(i, j+1)$, or $(i, j-1)$.

If the signal to be uwrapped is almost free of noise and the area to be unwrapped consists of a full, unobscured field in which a contiguous, uninterrupted, geometrically simple path can be followed, then phase unwrapping is an almost trivial matter. Figure 46-93 shows an example of a circular-shaped area in which unwrapping is started along the longest row crossing the center. The values of the unwrapped center row are then taken as initial values for half-columns following the path from the center row to the upper and lower rim, respectively [46-84]–[46-86].

If the signal to be unwrapped consists of an arbitrarily shaped contiguous and simply connected region, but interrupted by obscurations and invalid pixel areas such as those shown in figure 46-94(a), a more complex strategy has to be developed for the unwrapping process. An example of a somewhat simple strategy is as follows.

- Define an indicator function $\sigma(i,j)$ and set all values to zero.
- Scan the data field from the upper left corner and set the first occuring pixel $\phi_u(i_0,j_0) = \phi_w(i_0,j_0)$, mark its indicator value $\sigma(i_0, j_0) = 1$.
- Scan the data field from the upper left corner, unwrap all pixels that have valid, unwrapped neighbors and mark them with $\sigma(i, j) = 1$.
- Scan the data field from the lower left corner, unwrap all pixels that have valid, unwrapped neighbors and mark them with σ(i,j) = 1.
- Scan the data field from the upper right corner, unwrap all pixels that have valid, unwrapped neighbors and mark them with $\sigma(i, j) = 1$.
- Scan the data field from the lower right corner, unwrap all pixels that have valid, unwrapped neighbors and mark them with $\sigma(i, j) = 1$.

Figure 46-94 shows the growing areas of unwrapped pixels while the scan direction is altered. With this strategy any shape of contiguous areas can be unwrapped. If areas are strictly separated, no phase connection can be found. In this case each insulated area needs a separate intitial phase value.

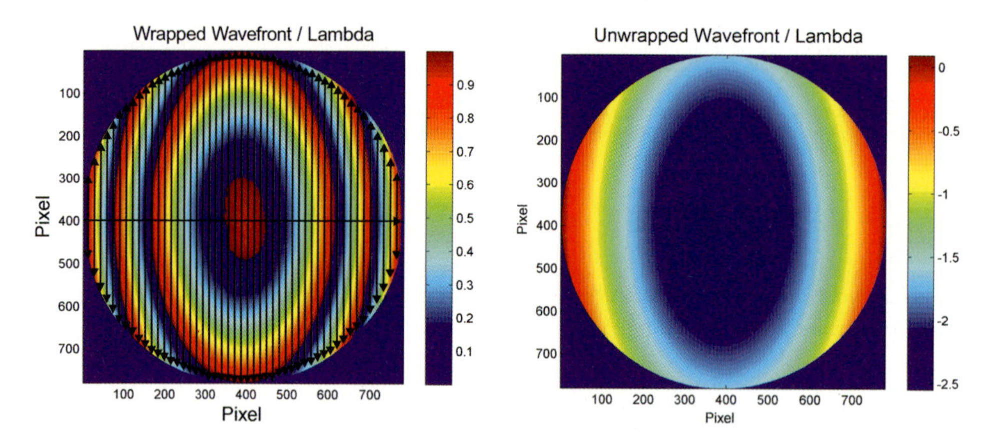

Figure 46-93: A simple phase-unwrapping path used for unobscured, low-noise, simple- shaped wavefronts.

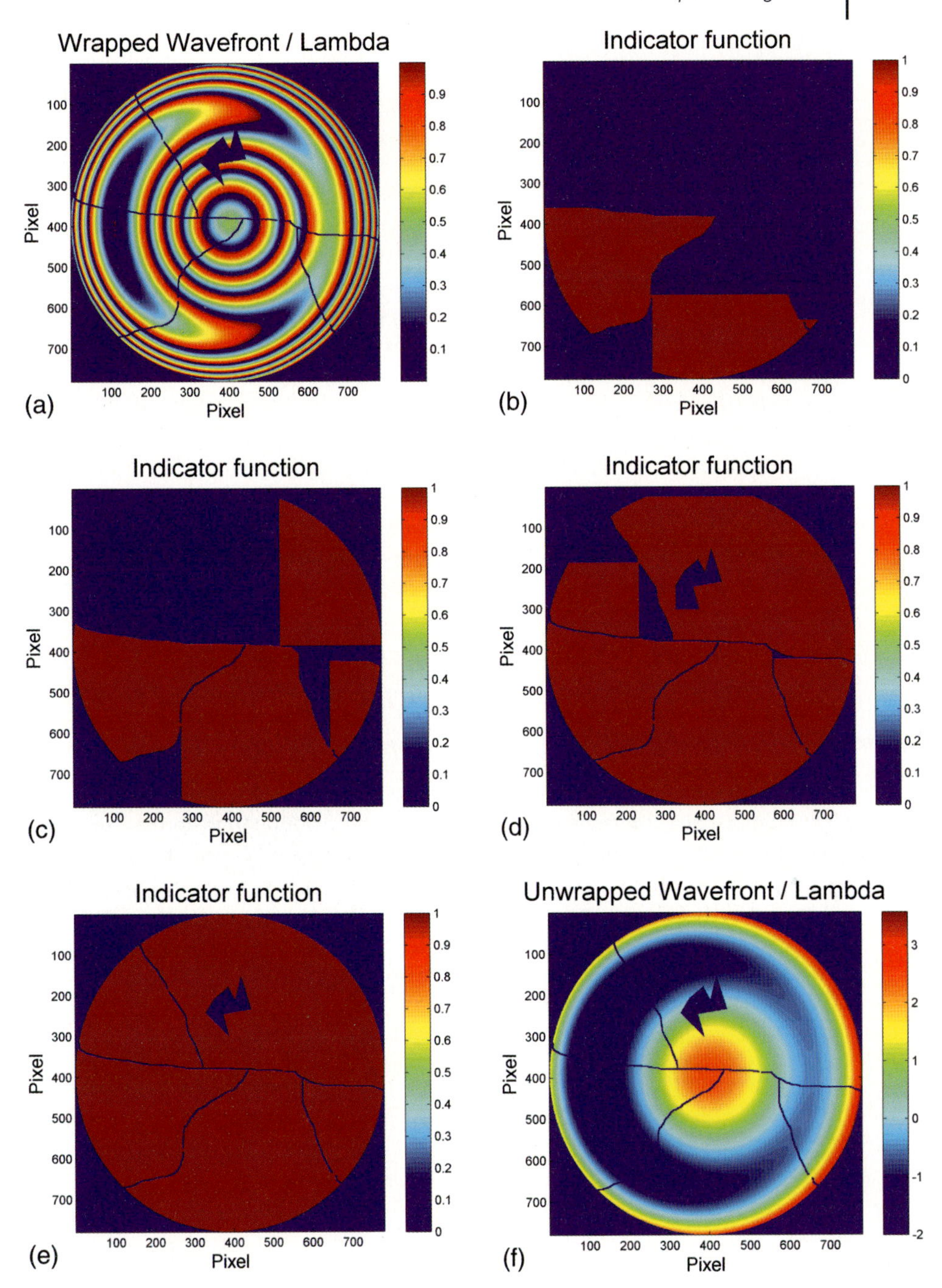

Figure 46-94: Phase-unwrapping strategy for an arbitrarily shaped contiguous, low-noise wavefront showing obscurations and invalid pixel areas: (a) wrapped wavefront; (b)–(e) indicator values after scan from upper left (b); lower left (c); upper right (d); lower right (e); (f) unwrapped wavefront.

In the presence of high-frequency, high-amplitude noise or discontinuous phase jumps, the unwrapping is different. In this case the absolute phase difference between adjacent pixels is greater than π. Before unwrapping those inconsistent phase areas they should be detected and isolated by declaring them as invalid pixels. They can be detected by calculating the sum of the wrapped phase differences in x and y along a square path around each pixel. For valid pixels the sum will be close to zero, while for inconsistent data the sum will deviate considerably from zero. Figure 46-95 shows an example for an interferogram with high- amplitude noise. Unwrapping without the check for inconsistencies leads to unexpected phase jumps in the unwrapped area as seen in figure 46-95(d). Applying the integral over the phase differences around each pixel to detect inconsistent data leads to zero values in the indicator data field to exclude pixels from the unwrapping process. The result in figure 46-95(f) is free of phase jumps, but many unvalid pixels are marked. When the noise increases, the exclusion of invalid pixels is not practical.

Many algorithms have been proposed for phase unwrapping in the presence of noise. Among them are the least-squares integration of wrapped phase differences [46-84] and the regularized phase-tracking unwrapper [46-85].

A rather simple and effective method uses two-dimensional Fourier transformation [46-86] and makes use of basic Fourier calculation rules (see (46-169)).

$$\phi_u(x,y) = \mathrm{Re}\left\{ \frac{1}{2\pi i} F^{-1} \left[\frac{F\left(\frac{\partial \phi_w(x,y)}{\partial x}\right) x + F\left(\frac{\partial \phi_w(x,y)}{\partial y}\right) y}{x^2 + y^2} \right] \right\} \tag{46-169}$$

F denotes the operation of a Fourier transformation, F^{-1} its inverse operation. The wrapped phase data are differentiated in x and y, Fourier transformed and multiplied by x and y, respectively. After normalization, inverse Fourier transformation and division by $2\pi i$, the real part of the expression represents the unwrapped phase data. Figure 46-96 gives an example of a signal carrying a great deal of noise, where conventional unwrapping algorithms will lead to inconsistent phase jumps (figure 46-95(d)). Applying (46-169) will lead to a reasonable, consistent result.

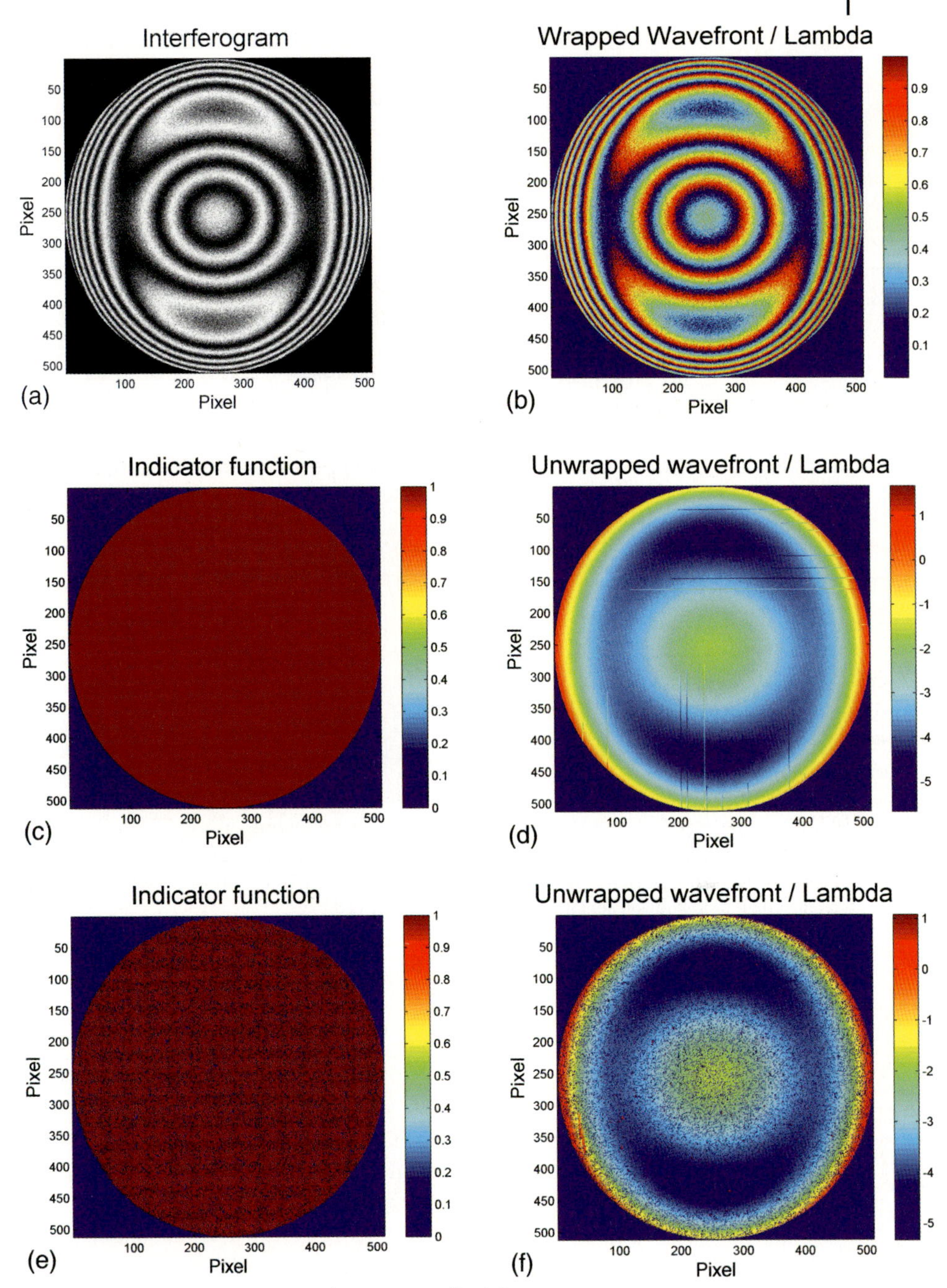

Figure 46-95: Phase unwrapping in the presence of high-frequency, high-amplitude noise: (a) interferogram; (b) wrapped phase; (c) indicator field without check for inconsistencies; (d) unwrapped phase showing unexpected phase jumps; (e) indicator field after check for inconsistencies; (f) unwrapped phase with pixels marked as invalid.

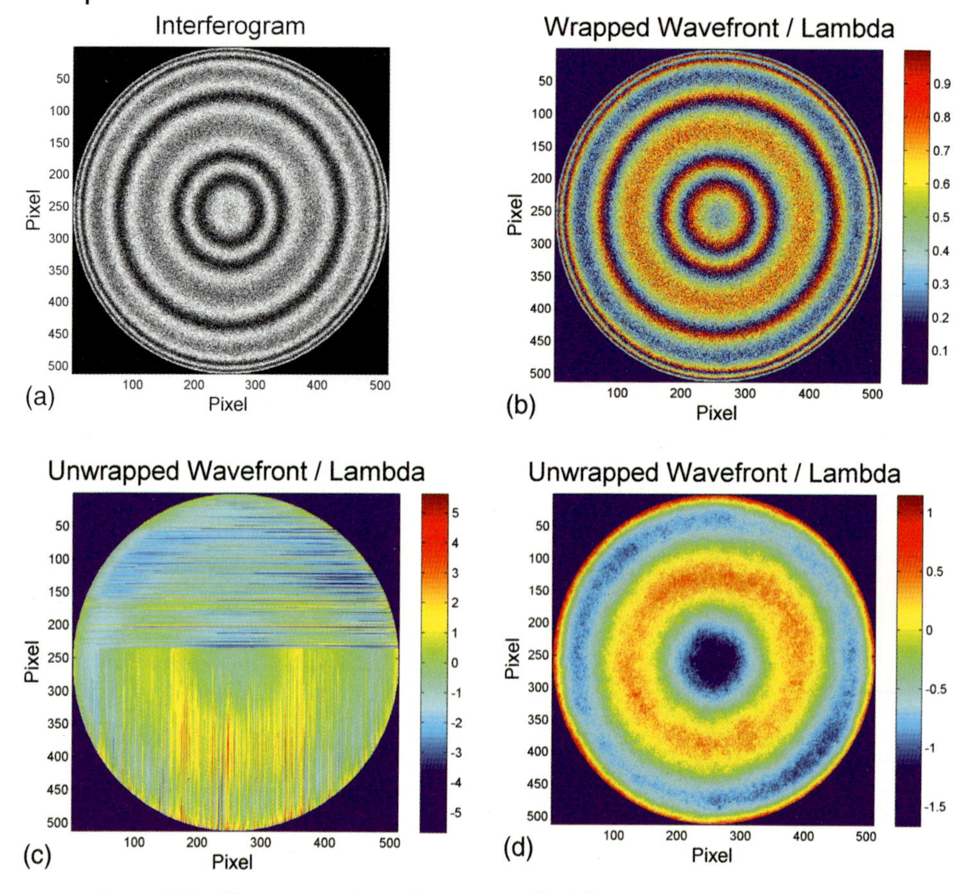

Figure 46-96: Phase unwrapping at the presence of high-frequency, high-amplitude noise: (a) interferogram; (b) wrapped phase; (c) unwrapped phase using conventional unwrapping; (d) unwrapped phase using the algorithm according to (46-169).

46.6 Calibration Techniques

Each metrological instrument must be calibrated in units of a standardized scale and then compared to the target values required, for instance, by the design of the product. Instruments like interferometers carry their own built-in reference surface, whereas in other wavefront- measuring instruments like the Hartmann–Shack the optical component geometry and arrangement acts as an intrinsic reference. All instruments need to be calibrated against a known optical element at least once before being used regularly. Instruments that tend to be unstable need periodic calibrations. Some might even need calibrating before each measurement, while others perhaps only once a year.

For shape-measuring instruments a well-known reference surface would be ideal for calibration purposes. If it is made from a stable material of low thermal expansion like the glass ceramic Zerodur®, together with a well-defined suitable mechanical mount, the reference element can then be used for calibration purposes over a very long time.

The question still remains how to determine the deviation of a reference element from the ideal mathematical target shape. There are generally two ways to determine absolute deviations. This is with the use of:

a) an element whose physical parameters are well-enough known to calculate an error budget that satisfies the instrument's accuracy requirements,
b) a calibration procedure that allows separation of instrument errors from the reference element errors.

46.6.1 Reference Elements with Known Residual Error

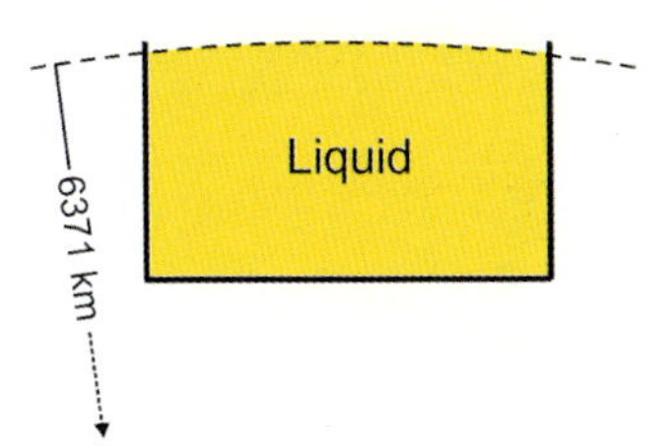

Figure 46-97: Curvature of the surface of a liquid caused by gravity.

An example of a well-known plane surface is the surface of a liquid that is curved mainly by gravitational effects resulting in a radius of curvature, equal to the mean radius of the earth, of 6371 km (figure 46-97). For a surface of diameter D the resulting deviation Δz normal to the surface is given by (46-170).

$$\Delta z = \frac{D^2}{8 \times 6.371 \times 10^9 \text{ mm}} \tag{46-170}$$

For a surface with 1 m diameter the deviation caused by the curvature of the earth is 19.62 nm. Suitable liquids for interferometrical references are mercury, water, silicone oil or cooking oil. However, in practical use, waves on the surface caused by acoustic and mechanical vibrations, turbulence within the liquid caused by thermal and density inhomogeneities, and adhesion effects from the walls of the tank, etc., will substantially influence the results.

When testing aspheres, an absolute calibration is often not possible, because no standard, well-known reference asphere is available. In this case null-systems (compensating systems) are manufactured, where radii, thicknesses, air gaps, and refractive indices are determined as exactly as possible to leave a small error budget, which is assumed to be the most likely systematic deviation from an ideal aspherical shape.

When using computer-generated holograms (CGH) as null-systems their quality substantially defines the error budget. Therefore, the precision of the diffractive structure with respect to position and geometrical shape of the individual lines has to be checked. For ultra-precise measurements, the position and shape of the individual lines have to be measured in order to calculate a corrected wavefront, taken as reference. Bending and surface deformation and the inhomogeneity of the CGH's substrate are the parameters defining the accuracy of the CGH as well as its position relative to the surface under test and also relative to the optics of the instrument.

46.6.2
Calibration Procedures Used to Determine Element Deviations

There are different procedures used to separate errors of the surface under test from the instrument errors [46-87]–[46-103]. Some procedures cover all types of deformation as long as they lie within the spatial spectrum of the instrument. Others can only separate either rotationally symmetric or unsymmetric types of errors. We will begin with the first type.

Three-position Technique

The three-position technique [46-87] is applied to convex or concave spherical surfaces. A test setup is required in which the element under test can be rotated around its optical axis and where the focus of the transmission sphere is accessible. Figure 46-98 shows the arrangement for a concave surface. The first measurement acquires a phase map ϕ_{0° in the usual position of the test surface. For the second measurement the surface is rotated by 180° around the optical axis (ϕ_{180°). For the third measurement the surface is moved with its vertex to the focal point of the transmission sphere (ϕ_{Focus}). We call this position the "cat's eye position". The acquired phase maps can be described mathematically according to (46-171).

$$\phi_{0^\circ} = \phi_{\mathrm{surf}} + \phi_{\mathrm{ref}} + \phi_{\mathrm{trans}} \tag{46-171a}$$

$$\phi_{180^\circ} = \bar{\phi}_{\mathrm{surf}} + \phi_{\mathrm{ref}} + \phi_{\mathrm{trans}} \tag{46-171b}$$

$$\phi_{\mathrm{Focus}} = \phi_{\mathrm{ref}} + \frac{1}{2}(\phi_{\mathrm{trans}} + \bar{\phi}_{\mathrm{trans}}) \tag{46-171c}$$

ϕ_{surf} denotes the phase variation due to the deviation of the surface under test, ϕ_{ref} those of the reference surface (Fizeau surface) of the transmission sphere, while ϕ_{trans} denotes the phase aberration of the transmission sphere in single transmission. The bar over the phase map symbol indicates a 180° (physical or mathematical) rotation of that phase map. The unknown surface deviation can then be calculated from (46-172a), the error due to the transmission sphere including the reference surface can be calculated from (46-172b).

$$\phi_{\text{surf}} = \frac{1}{2}(\phi_{0^\circ} + \bar{\phi}_{180^\circ} - \phi_{\text{Focus}} - \bar{\phi}_{\text{Focus}}) \tag{46-172a}$$

$$\phi_{\text{ref}} + \phi_{\text{trans}} = \frac{1}{2}(\phi_{0^\circ} - \bar{\phi}_{180^\circ} + \phi_{\text{Focus}} + \bar{\phi}_{\text{Focus}}) \tag{46-172b}$$

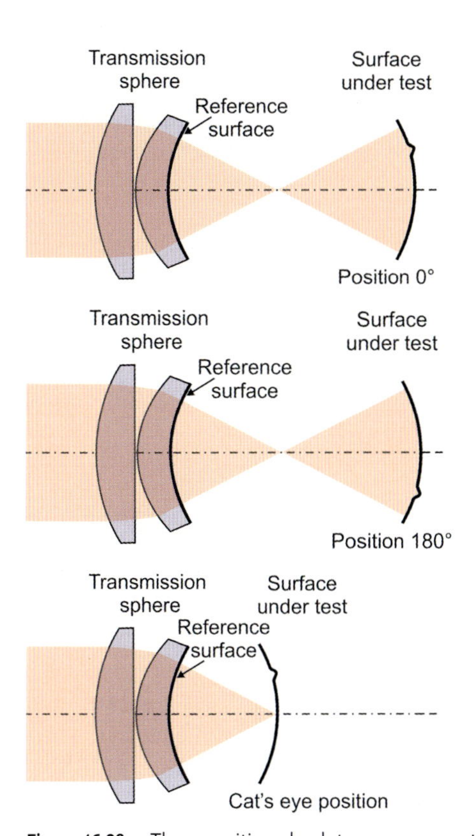

Figure 46-98: Three-position absolute measurement technique used for the separation of surface deviations from the errors of the transmission sphere including its reference surface: (a) regular test position 0°; (b) rotation of test surface by 180°; (c) cat's eye position with focus on the surface's vertex.

The error of the transmission sphere is determined and can be subtracted from future measurements to compensate for the instrument errors. An interferometer is then calibrated and acts as a perfect instrument.

The surface deviations are also determined. The element can now be used as a reference in future calibration procedures in case the instrument or transmission sphere has changed due to thermal or mechanical influences.

There may be test situations where the focus of the transmission sphere is not accessible, for instance, when the back focal length is very long or in the case of a diverging transmission sphere with a virtual focus position. In this case the three-position test is not applicable. However, there are other absolute calibrating procedures that distinguish between the determination of rotationally symmetric and non-rotationally symmetric deviations [46-88], [46-89].

Calibration of Non-rotationally Symmetric Errors

We assume a situation as in figure 46-98(a) and 46-98(b) and simplify (46-171a) by setting $\phi_{\text{int}} = \phi_{\text{ref}} + \phi_{\text{trans}}$, allowing the surface under test to rotate by an angle φ_i in the ith measurement, $i = 1, 2, \ldots N$. Equation (46-173) represents the total phase map as the sum of the interferometer and the surface deviations separated into rotationally symmetric and non-rotationally symmetric terms denoted by the subscripts r and nr, respectively.

$$\phi_{\varphi_i} = \phi_{\text{surf},\varphi_i} + \phi_{\text{int}} = \phi_{\text{surf},\varphi_i,\text{r}} + \phi_{\text{surf},\varphi_i,\text{nr}} + \phi_{\text{int,r}} + \phi_{\text{int,nr}} \tag{46-173}$$

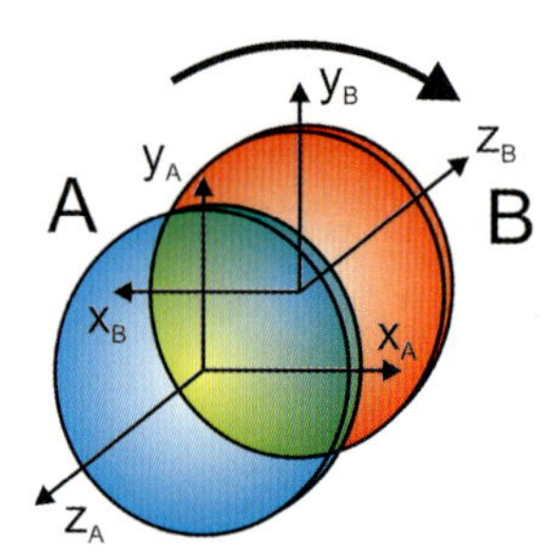

Figure 46-99: Rotation of surface under test (B) relative to the interferometer optics (A) around the common optical axis.

If N rotational positions at an azimuthal spacing of $360°/N$ are measured and averaged out, then all non-rotationally symmetric deviations of the surface under test are averaged out, with the exception of the orders kN ($k = 1,2,\ldots$) $\Delta\phi_{\text{surf,nr}}$, leading to

$$\langle \phi_\varphi \rangle_N = \phi_{\text{surf,r}} + \Delta\phi_{\text{surf,nr},N} + \phi_{\text{int,r}} + \phi_{\text{int,nr}} \tag{46-174}$$

Subtracting (46-174) from (46-171a) for the 0° position leads to

$$\phi_{0°} - \langle \phi_\varphi \rangle_N = \phi_{\text{surf,nr}} - \Delta\phi_{\text{surf,nr},N} \tag{46-175}$$

We thus have a result for the non-rotational terms of the surface under test as long as the residuals $\Delta\phi_{\text{surf,nr}}$ are small enough. If a test with $N = 11$ is assumed, all non-rotationally terms are determined with the exception of the 11th, 22nd, 33rd, ... azimuthal periodicity.

In figure 46-100(a) an example is given of a surface under test generating a wavefront of 53.7 nm RMS. By averaging over 11 equally spaced positions of rotation a residual wavefront of 4.886 nm RMS is achieved as shown in figure 46-100(b).

A much better result can be obtained if the 11 rotations were captured in specifically spaced positions according to the "M+N-method" [46-89]. In addition to N equally spaced rotational positions, we consider a second series of M equally spaced rotational positions. We select M and N as coprime numbers, for instance, 6 and 5. For each series the average is calculated according to (46-174). Their difference leads to

$$\langle \phi_\varphi \rangle_M - \langle \phi_\varphi \rangle_N = \Delta\phi_{\mathrm{surf,nr},M} - \Delta\phi_{\mathrm{surf,nr},N} \tag{46-176}$$

If this phase map is now rotated computationally M times, for example, with the help of a software program, and then averaged out, the n terms drop out leaving only terms with an $M \cdot N$ periodicity:

$$\left\langle \langle \phi_\varphi \rangle_M - \langle \phi_\varphi \rangle_N \right\rangle_M = \Delta\phi_{\mathrm{surf,nr},M} - \Delta\phi_{\mathrm{surf,nr},N\cdot M} \tag{46-177}$$

We combine the result with (46-175) and receive an improved result for the non-rotationally symmetric term of the test surface:

$$\phi_{0^\circ} - \langle \phi_\varphi \rangle_M + \left\langle \langle \phi_\varphi \rangle_M - \langle \phi_\varphi \rangle_N \right\rangle_M = \phi_{\mathrm{surf,nr}} - \Delta\phi_{\mathrm{surf,nr},N\cdot M} \tag{46-178}$$

Figure 46-100(c) shows the result for $M = 6$ and $N = 5$ in comparison to the result of 11 equally spaced measurements in figure 46-100(b). The RMS in the residual wavefront has dropped by a factor of 2.6.

The averaging of non-rotational symmetries can be applied to plane, spherical or rotationally symmetric aspherical surfaces. It fails, of course, when non-symmetrical optical setups, as in the case of free-form surfaces, have to be calibrated.

Shift-rotation Technique

In the case of an inaccessible cat's eye position the three-position technique is not applicable. The non-rotationally symmetric terms can be determined by rotation of the surface under test around the optical axis of the setup. For the separation of the rotationally symmetric terms of the surface under test and the test setup, both have to be shifted laterally relative to each other [46-90], [46-93]. For a spherical test surface this can be done in two ways as shown in figure 46-101(a) and (b).

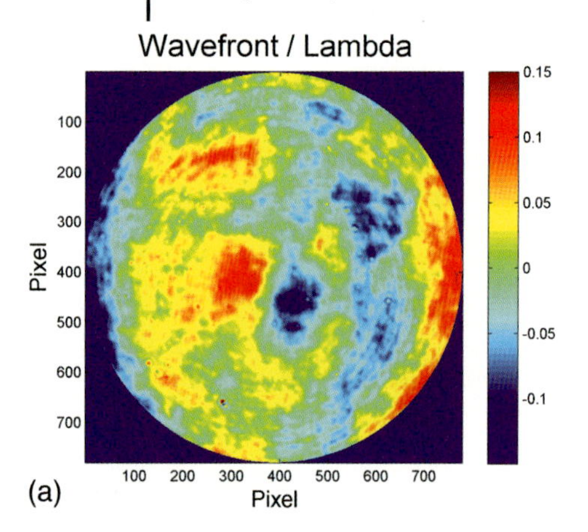

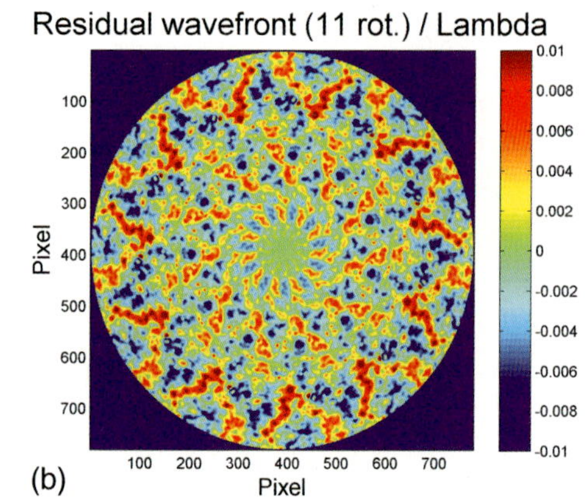

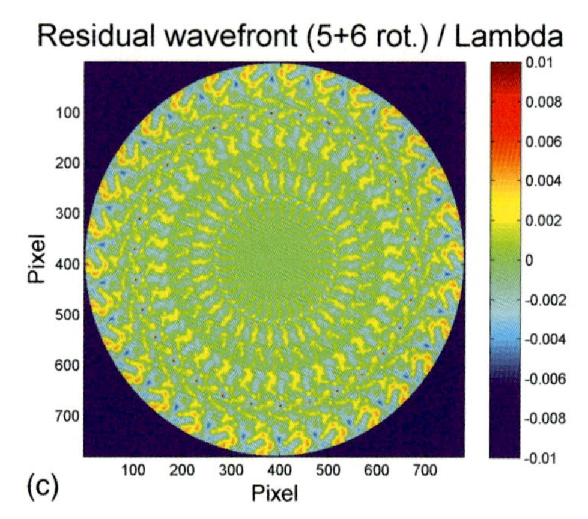

Figure 46-100: Averaging of non-rotationally symmetric errors: (a) original wavefront with non-rotationally symmetric aberrations, RMS = 53.7 nm; (b)average after 11 equally spaced rotations of 32.73°, RMS = 4.886 nm, (c) Average after five and six equally spaced rotations of 72° and 60°, respectively, RMS = 1.892 nm.

In (a) the axis of the rotation table has been tilted along with the test lens so that the center of rotation has now moved laterally relative to the original location. Tilting occurs around the center point of the surface under test by a so that no additional fringes are introduced. In each position the non-rotationally symmetric terms are determined by rotating the test piece. If both results are subtracted from each other the phase deviations caused by the interferometer vanish:

$$\Delta\phi_{\mathrm{surf,r}} = \phi_{\mathrm{surf,r}} + \phi_{\mathrm{int}} - \phi_{\mathrm{surf},a,\mathrm{r}} - \phi_{\mathrm{int}} = \phi_{\mathrm{surf,r}} - \phi_{\mathrm{surf},a,\mathrm{r}} \tag{46-179}$$

$\phi_{\mathrm{surf},r}$ denotes the phase map of the rotationally symmetric terms of the untilted surface and

$\phi_{\mathrm{surf},a,r}$ that of the tilted surface. Figure 46-102 shows an example of appropriate phase maps.

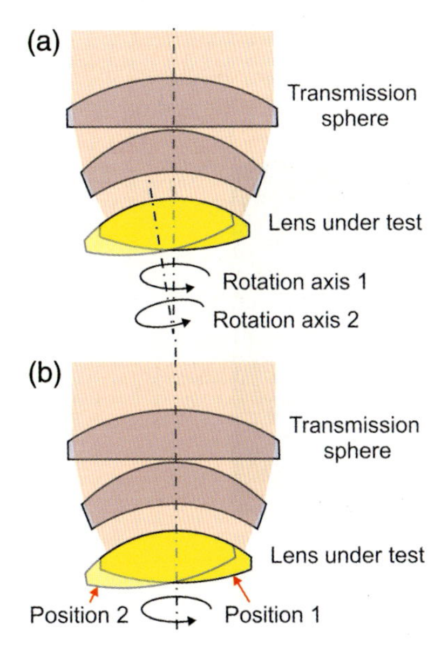

Figure 46-101: Lateral displacement of a surface under test in an interferometric arrangement formed: (a) by tilting the axis of rotation along with the test piece; (b) by tilting the test piece around the center point of the surface under test.

The result for $\phi_{\mathrm{surf,r}}$ is basically achieved by integration of (46-179), however, the procedure is not quite straightforward and includes the following steps:

1. Remove any distortion that might be caused by the transmission sphere or the interferometer optics from the data fields $\phi_{\mathrm{surf,r}}$ and $\phi_{\mathrm{surf},a,\mathrm{r}}$.
2. Transpose the corrected data fields $\phi'_{\mathrm{surf,r}}$ and $\phi'_{\mathrm{surf},a,\mathrm{r}}$ from pixel coordinates to spherical surface coordinates expressed in angle units.
3. Choose an initial value for an arbitrary data point, for instance $\phi'_{\mathrm{surf,r}}(0,0) = 0$.
4. Start a recursive evaluation using the measured and known data field $\Delta\phi'_{\mathrm{surf,r}}$ according to (46-180).

$$\phi'_{\mathrm{surf,r}}(\varphi + (i+1)a, \theta) = \phi'_{\mathrm{surf,r}}(\varphi + ia, \theta) - \Delta\phi'_{\mathrm{surf},a,\mathrm{r}}(\varphi + ia, \theta) \tag{46-180}$$

where φ and θ denote the surface coordinates, a is the angle of tilt of the rotational axis, and $i = 0, 1, 2, \ldots$

After each recursion step a set of values along a radius $r = \sqrt{(\varphi + (i+1)a)^2 + \theta^2}$ is available which serves as input for the next recursion step. In this way a result for each surface data point is generated.

The evaluation of the case as shown in figure 46-101(b) is much more complicated and needs iterative recursion algorithms. Its experimental setup might be less complicated, but while rotating the lens, the border position on the data field constantly changes, which has to be considered in the algorithm.

A simple direct numerical integration replacing step 4 can be applied, if a is chosen to be very small. It is also possible to replace step 4 by a least-squares fitting if the differences of radial polynomials have been precalculated for the appropriate a.

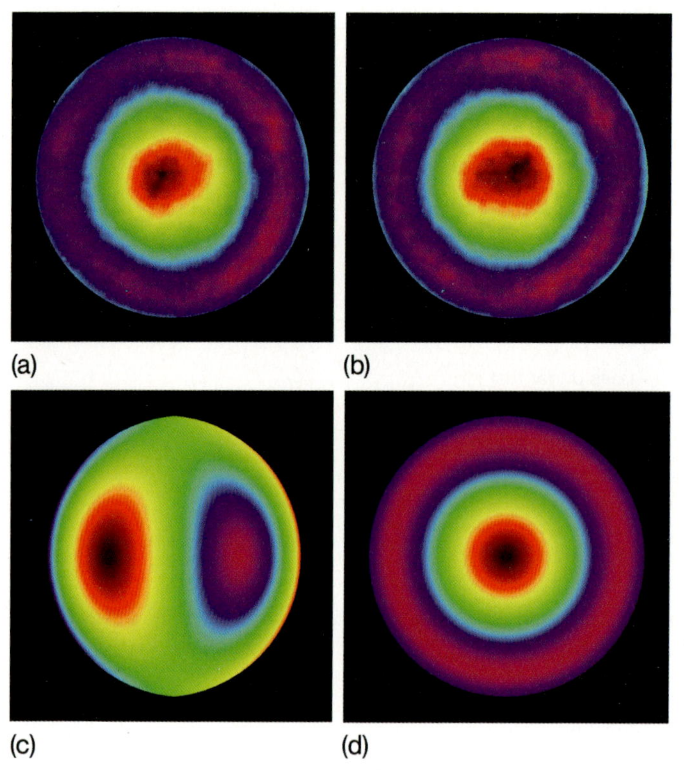

Figure 46-102: Average over different rotational positions of the test surface with: (a) a vertical rotation axis; (b) a tilted rotation axis; (c) the difference between (a) and (b), and ;(d) integration of (c) which is equal to the rotationally symmetric deviations of the test surface.

Calibration of Cross-sections: Three-flat Test

Absolute measurements of the deviation from flatness or sphericity of surfaces can be made by combining relative measurements between three unknown surfaces in a suitable way [46-90]–[46-98], [46-103]. Results are then achieved along a cross-section on each surface rather than over the total surfaces. Figure 46-103 shows an arrangement of three flat surfaces A, B and C combined in a Fizeau arrangement in which A and B have to be transparent to be used as a transmission flat. C can be a non-transparent mirror element. Suppose three wavefronts $W_1(x,y)$, $W_2(x,y)$, $W_3(x,y)$ are acquired from the three arrangements. These are formed by the three surface deviations $z_A(x,y)$, $z_B(x,y)$, $z_C(x,y)$ as shown in (46-181a,b,c).

$$W_1(x, y) = 2(z_A(-x, y) + z_B(x, y)) \quad (46\text{-}181a)$$

$$W_2(x, y) = 2(z_B(-x, y) + z_C(x, y)) \quad (46\text{-}181b)$$

$$W_3(x, y) = 2(z_A(-x, y) + z_C(x, y)) \quad (46\text{-}181c)$$

Along the cross-section in the y direction through the origin the results are calculated from

$$z_A(0, y) = \frac{1}{4}(W_1(0, y) - W_2(0, y) + W_3(0, y)) \quad (46\text{-}182a)$$

$$z_B(0, y) = \frac{1}{4}(W_1(0, y) + W_2(0, y) - W_3(0, y)) \quad (46\text{-}182b)$$

$$z_C(0, y) = \frac{1}{4}(-W_1(0, y) + W_2(0, y) + W_3(0, y)) \quad (46\text{-}182c)$$

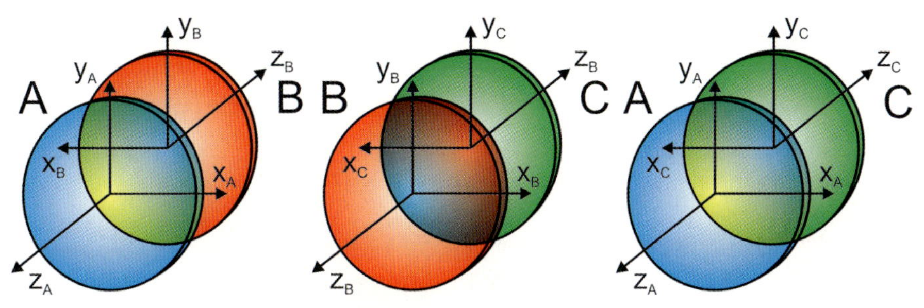

Figure 46-103: A combination of three surfaces A, B and C in a Fizeau arrangement for absolute testing along a cross-section in the y direction through the origin.

The flat surfaces method is known as the three-flat test [46-90]–[46-97]. When combined with averaging by rotation the total surfaces can be determined, because the test delivers rotationally and non-rotationally symmetric terms.

This method can be extended to the combination of spherical surfaces. A and B must be concave surfaces and designed as aplanatic elements to prevent the generation of spherical aberration while the wavefront transmits the element. C can be convex or concave as shown in figure 46-104.

Note that, because of the coordinate inversions caused by the focus crossing the wavefront, the absolute results are achieved along the x-axis perpendicular to the plane of view in figure 46-104. If the radii of curvature for the three surfaces are denoted by r_A, r_B, r_C and the focal length of the transmission sphere by f, the results for the three surfaces are given as

$$z_A\left(\frac{r_A}{f}x, 0\right) = \frac{1}{4}(W_1(x, 0) - W_2(x, 0) + W_3(x, 0)) \quad (46\text{-}183a)$$

$$z_{\mathrm{B}}\left(\frac{r_{\mathrm{B}}}{f}x, 0\right) = \frac{1}{4}(W_1(x, 0) + W_2(x, 0) - W_3(x, 0)) \tag{46-183b}$$

$$z_{\mathrm{C}}\left(\frac{r_{\mathrm{C}}}{f}x, 0\right) = \frac{1}{4}(-W_1(x, 0) + W_2(x, 0) + W_3(x, 0)) \tag{46-183c}$$

Once the absolute deviation along a cross-section is determined, the procedure can be repeated with the surfaces rotated relative to each other. Alternatively, the non-rotationally symmetric terms might be predetermined by the *M*+*N*-method described above. After considering them in the three-plate test, one cross-section is sufficient to determine the unknown rotationally symmetric terms.

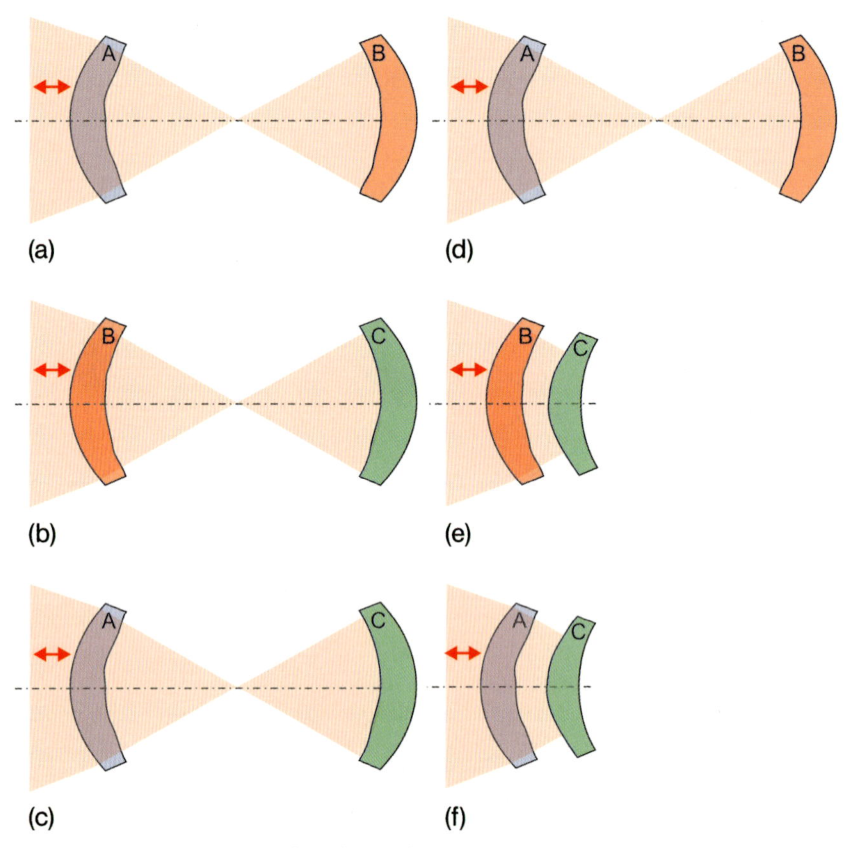

Figure 46-104: Determination of absolute surface deviations along a cross-section in *x* (perpendicular to the plane of view) through the origin for three spherical surfaces. In (a), (b) and (c) all surfaces are concave. In (d), (e) and (f) A and B are concave, C is convex.

Rotating Sphere

In interferometric setups where the focus point of the transmission sphere is accessible, a calibration technique using a rotating sphere can be applied. The method is called "ball averaging" [46-100]–[46-102]. In figure 46-105 the test surface is replaced by a sphere, the center of which coincides with the focus of the transmission sphere. The sphere has to be made of opaque material to prevent reflections from its back surface. The sphere should be rotated in N arbitrary positions denoted by φ_i, θ_i, $i=1$ ... N. Equation (46-184) defines the measured phasemap as the sum of the part caused by the interferometer and that of the sphere in the ith position.

$$\phi_{\varphi_i,\theta_i} = \phi_{\text{sphere},\varphi_i,\theta_i} + \phi_{\text{int}} \tag{46-184}$$

After averaging over N positions, the result is

$$\langle\phi\rangle = \phi_{\text{int}} + \frac{1}{N}\sum_{i=1}^{N} \phi_{\text{sphere},\varphi_i,\theta_i} \tag{46-185}$$

Rotating the sphere at random while making sequential measurements $\phi_{\varphi_i,\theta_i}$ and then averaging over all results amounts to a Monte-Carlo integration of the sphere's shape error over its entire surface. Assuming that the N measurements are uncorrelated samples from a distribution with zero mean, the root-mean-square (RMS) of the mean in (46-185) will be proportional to $1/\sqrt{N}$. If N becomes larger, the RMS of the residual error converges to zero as expressed in (46-186) (figure 46-106).

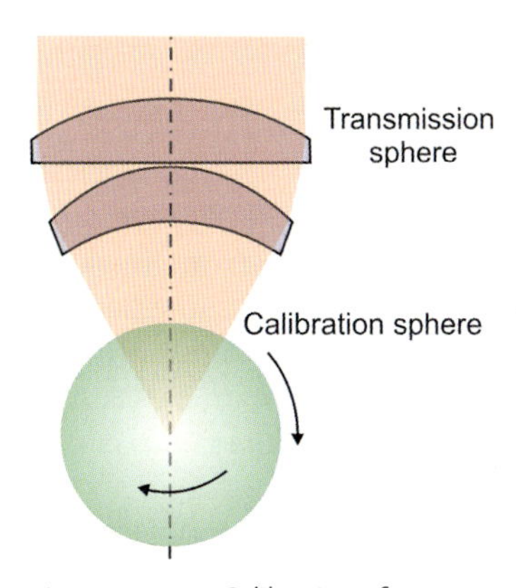

Figure 46-105: Calibration of a transmission sphere using the ball-averaging technique.

$$\lim_{N\to\infty}\left(\frac{1}{N}\sum_{i=1}^{N} \phi_{\text{sphere},\varphi_i,\theta_i}\right) = 0 \tag{46-186}$$

In practical applications several problems arise. The bearing of the sphere and the rotation apparatus must be designed so that no systematic deformation of the sphere will occur which cannot be averaged to zero. An air bearing is preferred if the exhausting air does not disturb the cavity zone. Three smaller balls, each based on ball bearings, arranged in an equilateral triangle might also be adequate. Bearings with considerable friction will generate heat that might partially deform the sphere.

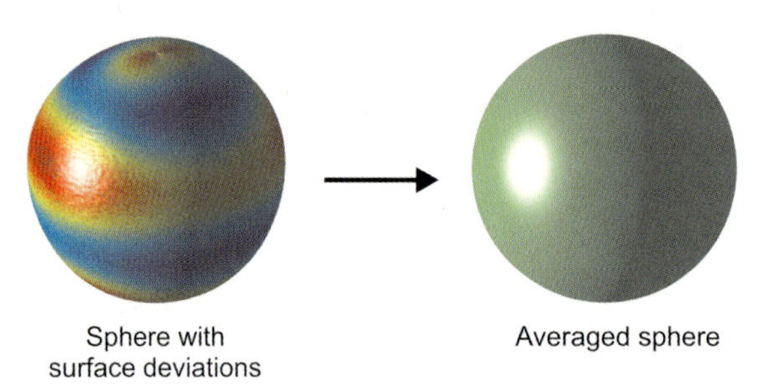

Figure 46-106: A sphere with deformations appears as a perfect sphere when $N \to \infty$ samples are averaged.

The process must be applied statistically in two orthogonal directions to make sure that every point on the sphere passes the sensor area the same number of times per time unit. The process can be accomplished by contacting rotating rubber rolls or, contactless, through air jet nozzles that operate statistically. The latter are applicable if the bearing has very low friction.

Ritchey–Common Test, Hindle Test

When plane mirrors become very large it is often not possible to supply an interferometer with a large aperture covering the entire mirror under test. If, however, a large concave spherical mirror is available it is possible to establish the so-called Ritchey–Common or, as a special case, the Hindle test arrangement. The latter is shown in figure 46-107. The spherical mirror carries a central bore coinciding with the focus point of the transmission sphere through which the test beam passes. The plane mirror under test is located at a distance $d = r/2$ equal to half the radius of curvature r of the concave mirror.

The Ritchey-Common test is the more general case, where the normal to the plane mirror under test is inclined by an angle α relative to the optical axis of the transmission sphere. Figure 46-108 shows an arrangement with $\alpha = 45°$. The sum of the distances d and e is equal to the radius of curvature of the concave spherical mirror: $r = d + e$.

The Hindle and Ritchey–Common tests can be calibrated by an absolute method based on the principle that the topometric sensitivity varies along with the test beam incidence angle over the mirror surface [46-104]–[46-106] as shown in the next section.

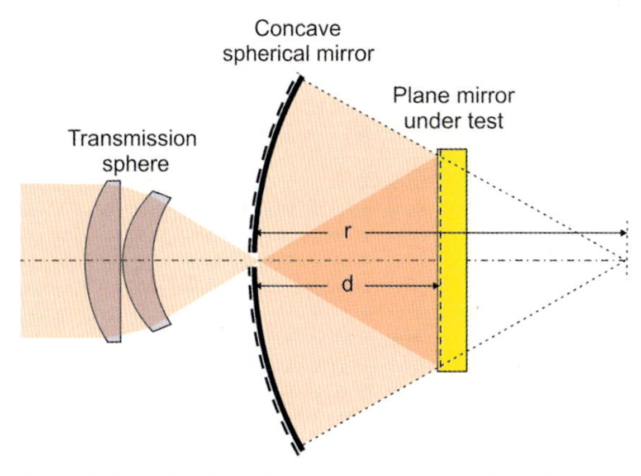

Figure 46-107: Hindle test using a concave spherical mirror with central bore to test a plane mirror.

From figure 46-109, the measured optical path difference OPD for a single-pass test beam incident at an angle ε against the mean surface normal and for a topometric surface modulation h is given by

$$\mathrm{OPD} = \overline{AB} - \overline{AC} = 2h\cos\varepsilon \tag{46-187}$$

The highest sensitivity is therefore achieved at normal incidence, whereas sensitivity goes to zero for $\varepsilon \to 90°$.

Let $z(x,y)$ be the topometric modulation of the mirror under test in mirror coordinates x, y. In a double-pass system a wavefront $W(x',y')$ is measured at the entrance pupil of the transmission sphere with focal length f' in pupil coordinates x', y'. Using (46-187) the relations in equations (46-188a-f) can be derived between $W(x',y')$ and $z(x,y)$ and x',y' and x,y.

$$x = \frac{-dx'}{f'\cos^2 a + x'\sin a} \tag{46-188a}$$

$$y = \frac{-dy'\cos^2 a}{f'\cos^2 a + x'\sin a} \tag{46-188b}$$

$$z(x,y) = \frac{W(x',y')}{g(x',y')} = \frac{\sqrt{x'^2 + (y'^2 + f'^2)\cos^2 a}}{2(f'\cos^2 a + x'\sin a)} W(x',y') \tag{46-188c}$$

$g(x',y')$ is the weight factor denoting the proportionality between $z(x,y)$ and $W(x',y')$.

Back transformation leads to:

$$x' = \frac{-f'x\cos^2 a}{d + x\cdot\sin a} \tag{46-188d}$$

$$y' = \frac{-f'y}{d + x \cdot \sin \alpha} \tag{46-188e}$$

$$W(x', y') = g(x, y)z(x, y) = \frac{2d \cos \alpha}{\sqrt{x^2 + y^2 + 2xd \sin \alpha + d^2}} z(x, y) \tag{46-188f}$$

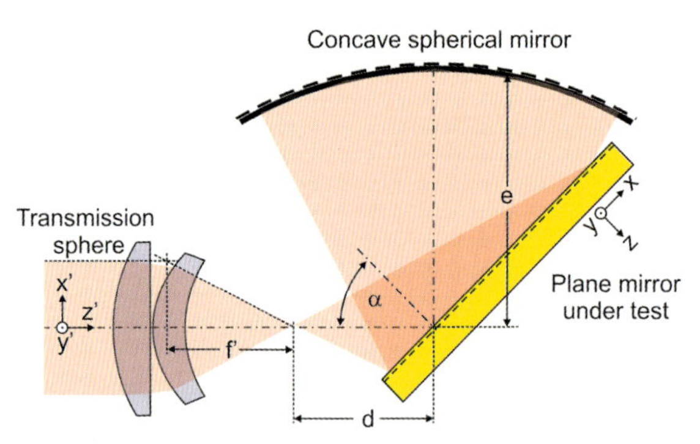

Figure 46-108: Ritchey–Common test using a concave spherical mirror to test a plane mirror tilted by an angle α against the optical axis of the transmission sphere.

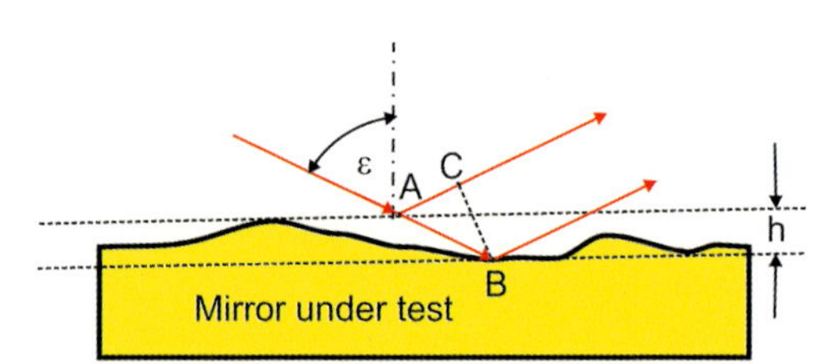

Figure 46-109: Incident beam inclined by ε against the mean normal of the mirror under test with topometric variation of h.

If the mirror under test is circular shaped it can be rotated around its normal as shown in figure 46-110. The wavefront W_i detected in the ith rotational position of the mirror can then be expressed as

$$W_i = g(z_r + z_{nr,i}) + W_{\text{int}} + a_{0i} + a_{1i}x' + a_{2i}y' + a_{3i}(x'^2 + y'^2) \tag{46-189}$$

where g denotes the weight factor as in (46-188(f)), z_r and $z_{nr,i}$ denote the rotationally symmetric and non-rotationally symmetric terms of the mirror deformation, respectively. W_{int} denotes the wavefront caused by the aberrations of the reference surface and the spherical mirror. Misalignment effects in each rotational position are expressed by factors for piston, tilt and focus denoted by a_{0i}, a_{1i}, a_{2i}, and a_{3i}. Their effect on the detected wavefront and on the equivalent mirror deformation is shown in figure 46-111.

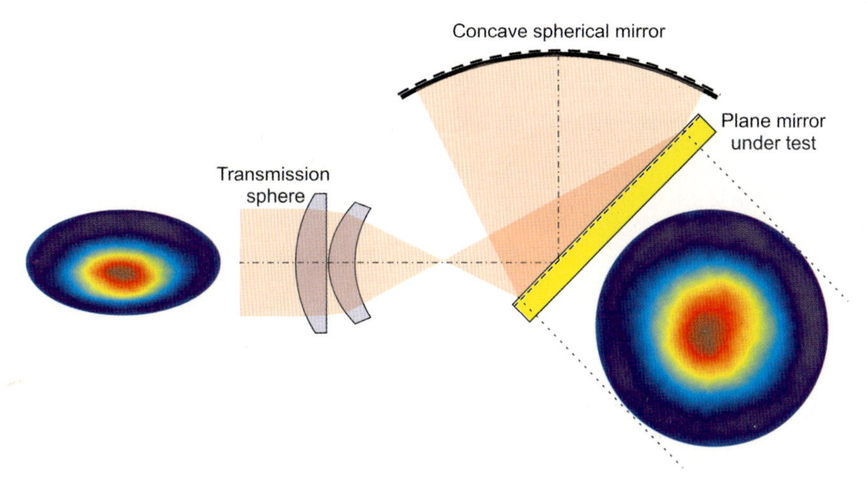

Figure 46-110: Ritchey-Common setup with a circular-shaped mirror under test.

When averaged over N rotational positions the non-rotational terms of the mirror are averaged out as shown in (46-190), neglecting higher order residuals.

$$\langle W\rangle = g\, z_r + W_{\text{int}} + \frac{1}{N}\sum_{j=1}^{N}\left(a_{0j} + a_{1j}x' + a_{2j}y' + a_{3j}(x'^2 + y'^2)\right) \tag{46-190}$$

Subtracting (46-189) from (46-190) leads to the unknown non-rotationally symmetric terms after removing the residual piston, tilt and focus by a least-squares fit.

$$z_{\text{nr},i} = \frac{1}{g}\left(W_i - \langle W\rangle - a_{0i} - a_{1i}x' - a_{2i}y' - a_{3i}(x'^2+y'^2) + \frac{1}{N}\sum_{j=1}^{N} a_{0j} + a_{1j}x' + a_{2j}y' + a_{3j}(x'^2+y'^2)\right) \tag{46-191}$$

Since the removal of misalignment effects might also subtract asymmetric deformations of the mirror, an average over all N measurements has to be calculated after rotating back each data field to the initial rotation position. The non-rotationally symmetric terms of the mirror are finally calculated from

$$z_{\text{nr}} = \frac{1}{N}\sum_{i=1}^{N} R_{\varphi_i}\left[T\left\{\frac{W_i - \langle W\rangle}{g}\right\}\right] - a_{0,res}f_P - a_{1,res}f_X - a_{2,res}f_Y - a_{3,res}f_F \tag{46-192}$$

where T denotes the transformation into mirror coordinates x, y and $R_{\varphi i}$ denotes the back rotation of the data field by $-\varphi_i$. There will be residual terms from misalignment denoted by the coefficients $a_{0,\text{res}}$, $a_{1,\text{ res}}$, $a_{2,\text{res}}$ $a_{3,\text{res}}$ that must be removed by least-squares fitting after transformation to mirror coordinates. The functions f_P, f_X, f_Y, f_F denote the functions of piston, tilt x and y and focus transformed into mirror coordinates as shown in figure 46-111 and equation (46-193a–d).

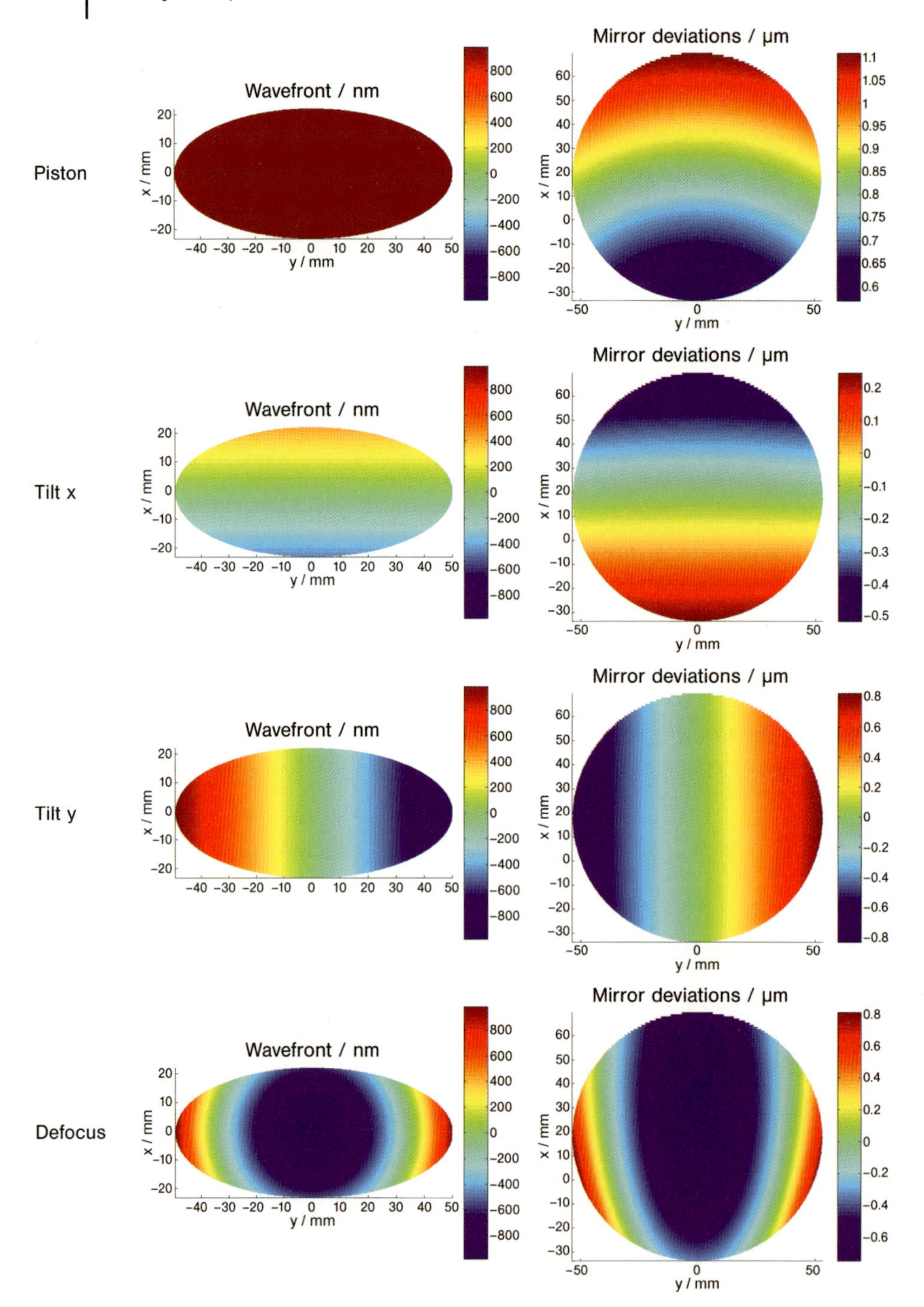

Figure 46-111: Wavefront caused by misalignment in a Ritchey–Common arrangement with a circular shaped mirror under test and the equivalent mirror deformation.

$$f_P(x,y) = \frac{\sqrt{x^2 + y^2 + 2xd\sin\alpha + d^2}}{2d\cos\alpha} \tag{46-193a}$$

$$f_X(x,y) = \frac{-f'x\cos^2\alpha}{d + x\sin\alpha} f_P(x,y) \tag{46-193b}$$

$$f_Y(x,y) = \frac{-f'y}{d + x\sin\alpha} f_P(x,y) \tag{46-193c}$$

$$f_F(x,y) = \frac{f'^2(x^2\cos^4\alpha + y^2)}{(d + x\sin\alpha)^2} f_P(x,y) \tag{46-193d}$$

The rotationally symmetric terms can be determined by laterally displacing the rotation axis of the mirror under test and taking a second set of measurements according to (46-190). The shift-rotation technique can then be applied to determine z_r.

If the mirror under test is not circular shaped the shift-rotation technique cannot be applied for an absolute calibration. For elliptical or other shapes a Ritchey–Common test in different angular positions can be used to derive absolute results. Two different angular positions are sufficient for absolute testing (figure 46-112). Again surface deviations must be distinguished from alignment effects. For simplicity, we assume that the interferometer and the concave spherical mirror have been calibrated before so that they will not be considered further.

Let us assume that two wavefronts $W_1(x',y')$ and $W_2(x',y')$ have been measured corresponding to two angular positions α_1 and α_2. The wavefronts contain unknown contributions of piston, tilt and defocus. After transformation to mirror coordinates x,y there will be two sets of mirror surface data sets, as shown:

$$z_1(x,y) = z_m(x,y) + a_{01}\cdot f_{01}(x,y) + a_{11}\cdot f_{11}(x,y) + a_{21}\cdot f_{21}(x,y) + a_{31}\cdot f_{31}(x,y) \tag{46-194a}$$

$$z_2(x,y) = z_m(x,y) + a_{02}\cdot f_{02}(x,y) + a_{12}\cdot f_{12}(x,y) + a_{22}\cdot f_{22}(x,y) + a_{32}\cdot f_{32}(x,y) \tag{46-194b}$$

$z_m(x,y)$ denotes the true deformation of the mirror under test. The terms a_{01} $f_{01}(x,y)$ to $a_{31}\,f_{31}(x,y)$ and $a_{02}\,f_{02}(x,y)$ to $a_{32}\,f_{32}(x,y)$ denote the equivalent mirror deformations caused by piston, tilt and defocus in the two angular positions, respectively. Figure 46-113 shows an example of equivalent mirror deformations for $\alpha_1 = 35°$ and $\alpha_2 = 55°$. To remove the alignment effects (46-194a) is subtracted from (46-194b) leading to (46-195).

$$\begin{aligned}\Delta z(x,y) &= z_2(x,y) - z_1(x,y)\\ &= a_{02}f_{02}(x,y) + a_{12}f_{12}(x,y) + a_{22}f_{22}(x,y) + a_{32}f_{32}(x,y)\\ &\quad - a_{01}f_{01}(x,y) - a_{11}f_{11}(x,y) - a_{21}f_{21}(x,y) - a_{31}f_{31}(x,y)\end{aligned} \tag{46-195}$$

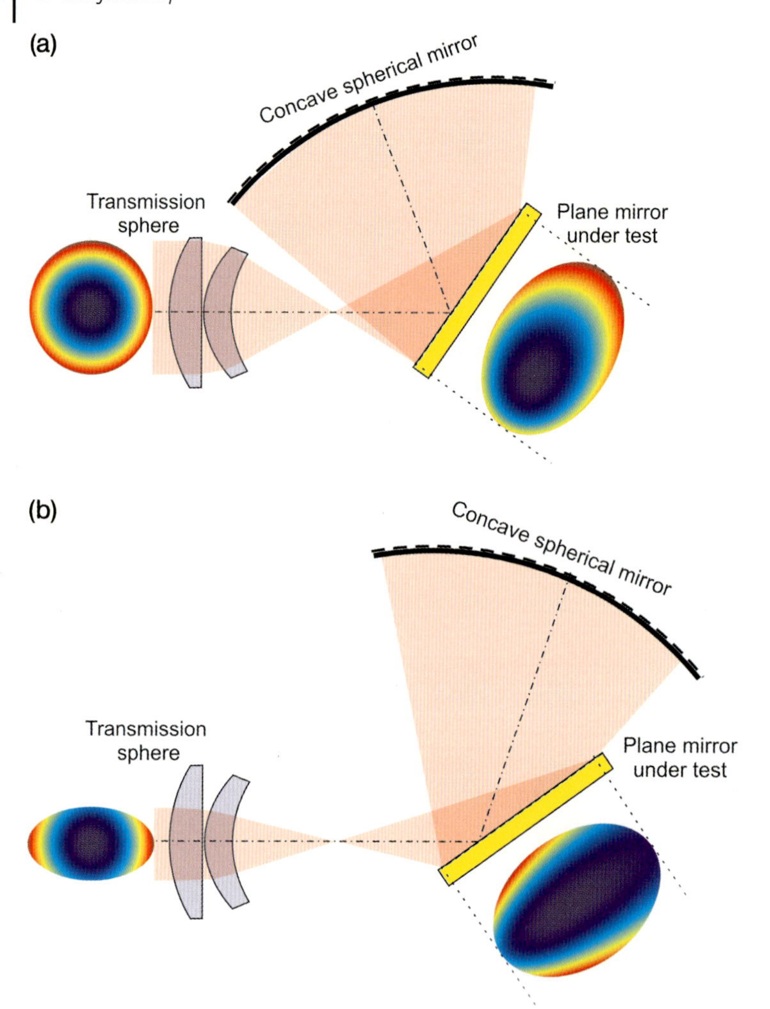

Figure 46-112: Ritchey–Common test in different angular positions of the mirror under test: a) 35°; b) 55° tilt of mirror normal, relative to the optical axis of the transmission sphere.

Since $f_{01}(x,y)$ to $f_{32}(x,y)$ are known from the setup configuration and $\Delta z(x,y)$ was measured, a simple least-squares fit may be applied to determine the unknown coefficients a_{01} to a_{32}. $z_m(x,y)$ can then be determined from (46-194a or b), or – since both results are valid – from an average of both equations leading to

$$z_m(x,y) = \frac{1}{2}\{z_1(x,y) + z_2(x,y) - a_{01}f_{01}(x,y) - a_{11}f_{11}(x,y) - a_{21}f_{21}(x,y) - a_{31}f_{31}(x,y) - a_{02}f_{02}(x,y) - a_{12}f_{12}(x,y) - a_{22}f_{22}(x,y) - a_{32}f_{32}(x,y)\} \tag{46-196}$$

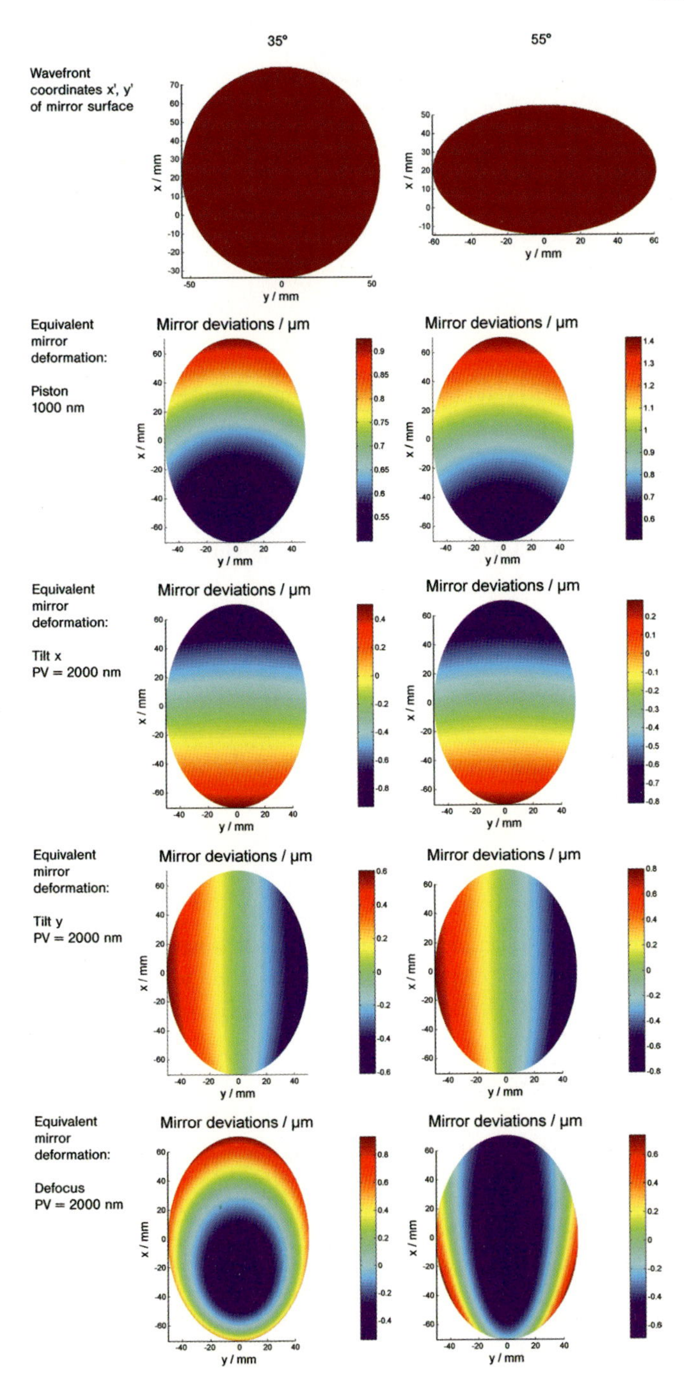

Figure 46-113: Equivalent mirror deformations for piston, tilt and defocus in a Ritchey–Common test for two different angular positions of the test mirror: 35° and 55°.

In practice, the following problem arises: the transformed functions of piston, tilt x and tilt y are very similar (see figure 46-113) for both angles and far from orthogonality. Therefore, piston, tilt x and tilt y for α_2 are excluded from (46-195) to maintain the numerical stability of the algorithm. In practice this is no problem because these parameters are of no interest for the measurement. The removal of the piston introduces a small systematic error which can usually be neglected.

46.7 Dynamic Range

In metrology "dynamic range" (DR) refers to the effective range of values that can be measured by a sensor or metrology instrument. It is defined as the maximum possible signal versus the total noise signal. Usually the dynamic range is limited at one end of the range by saturation of the sensor or by the physical limits of the mechanical motion system of the sensor. The other end of the dynamic range of measurement is limited by sources of random noise or uncertainty in signal levels that are called sensitivity or resolution of the sensor or metrology device. DR is expressed as dimensionless, in decibels (dB) or in bit:

$$\mathrm{DR} = \frac{\text{max signal}}{\text{resolution}} \,[\text{dimensionless}] \tag{46-197a}$$

$$\mathrm{DR} = 20 \cdot \log_{10}\left(\frac{\text{max signal}}{\text{resolution}}\right) [\text{dB}] \tag{46-197b}$$

$$\mathrm{DR} = \log_2\left(\frac{\text{max signal}}{\text{resolution}}\right) [\text{bit}] \tag{46-197c}$$

DR does not refer to the absolute accuracy of a metrology instrument. Non-linearities and systematic deviations are not considered. DR rather refers to the precision of a metrology instrument, which is defined by the reproducibility of the measurement results.

46.7.1 DR in Surface Topometry

We assume that a surface is to be measured over a lateral area of $d_x\, d_y$ with lateral sampling intervals Δx and Δy (figure 46-114). The surface extends in z over a range of $d_z = z_{max} - z_{min}$. A resolution of Δz is required for the metrological problem.

We therefore need a metrology device capable of aquiring $\mathrm{DR}_x\, \mathrm{DR}_y$ sample points each with a dynamic range $\mathrm{DR}_z = z_{max}/\Delta z$. The uncompressed total amount of required information is $Q = \mathrm{DR}_x\, \mathrm{DR}_y \log_2(z_{max}/\Delta z)$ in bit.

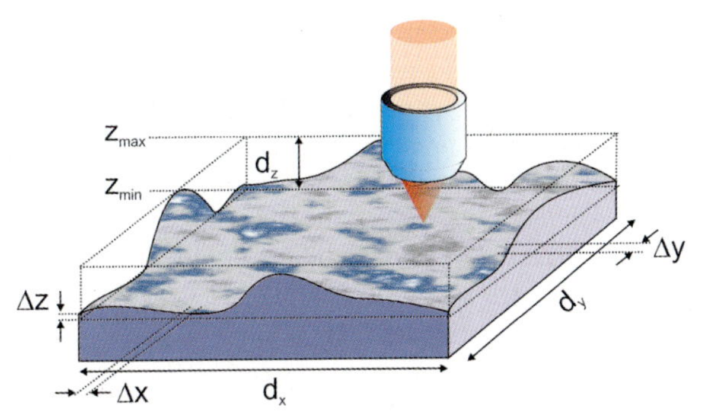

Figure 46-114: The topometric measurement of a surface under test by means of a point-wise measuring sensor. Dynamic ranges are given by $DR_x = d_x/\Delta x$, $DR_y = d_y/\Delta y$, $DR_z = d_z/\Delta z$.

The required dynamic ranges are then given by

$$\mathrm{DR}_x = \frac{d_x}{\Delta x} \tag{46-198a}$$

$$\mathrm{DR}_y = \frac{d_y}{\Delta y} \tag{46-198b}$$

$$\mathrm{DR}_z = \frac{d_z}{\Delta z} \tag{46-198c}$$

If a single point detector is used, either the sensor or the work-piece have to be moved so that all surface points are accessible, each point is acquired with a sufficient resolution in z as well as at a sufficient lateral sampling rate.

As an example, we consider a spherical surface with a radius of curvature of 100 mm and also a diameter of 100 mm. The sagittal height is PV = 13.4 mm. We consider a sample interval of 100 μm and a required resolution $\Delta z = 1$ μm. The required dynamic range DR in z and in x/y and the total amount of information Q are

$\mathrm{DR}_z = 13\,397 = 82.5$ dB ≈ 14 bit
$\mathrm{DR}_{x/y} = 1000 = 60.0$ dB ≈ 10 bit
$Q = 14$ Mbit = 1.75 MB (= Megabyte)

For a good tactile three-coordinate measuring machine (CMM) this is a standard task, but acquiring 785 398 sample points might take several hours.

For high-quality optical surfaces a resolution of 1 nm is necessary for final qualification. The example above then requires a dynamic range and a total amount of information as follows:

$\mathrm{DR}_z = 13\,397\,460 = 142.5$ dB ≈ 24 bit
$\mathrm{DR}_{x/y} = 1000 = 60.0$ dB ≈ 10 bit
$Q = 24$ Mbit = 3 MB

The question arises as to which metrological system will solve the task. A CMM is out of question because the resolution requirement cannot be met.

46.7.2
Dynamic Range of CCD Sensors in Interferometry and Wavefront Sensing

We consider a CCD sensor with M_i M_j pixels. A standard DR for irradiance encoding of a single frame is 10 or 12 bit. In interferometry usually the acquisition of more than one interferogram is necessary in order to determine a phase map. To extend the DR for each pixel the number of acquired interferograms can be increased. For high-quality inspections several hundred or even several thousand frames can be acquired in order to calculate an average phase map of very high DR. Assuming statistical variation of the phase signal the improvement of the RMS-repeatability will be by a factor of $1/\sqrt{N}$ where N is the number of acquired frames.

The maximum fringe signal which a CCD camera can detect is limited by the local fringe density which is given by the Nyquist limit. Assuming a homogeneous fringe density a CCD sensor allows a maximum of $(M_i + M_j)/2$ fringes. Depending on the stability of the test setup and the number of aquired frames 1/100 to 1/10 000 fringes can be resolved. The DR of a fringe acquisition system is therefore given by

$$DR_{\text{fringe}} = \frac{W_{\max} - W_{\min}}{\Delta W} \leq \frac{M_i + M_j}{\Delta W[\text{fringes}]} \tag{46-199}$$

Considering a CCD camera with 1000 x 1000 pixels and a fringe resolution of $\Delta W = 1/10\,000$ fringe leads to

$\text{DR}_{\text{fringe}} \leq 20\,000\,000 = 146.0 \text{ dB} \approx 24 \text{ bit}$

$\text{DR}_{x/y} = 1000 = 60.0 \text{ dB} \approx 10 \text{ bit}$

$Q \leq 24 \text{ Mbit} = 3 \text{ MB}$

In principle, this would be sufficient to test the spherical surface in the example above by means of a plane wave interferometer. However, this is only true if the fringes are spread with uniform density, which they are not in the case of a Fizeau or Twyman–Green interferometer. When a lateral shearing interferometer is used to test spherical surfaces relative to a plane wave a much more uniform fringe density is achieved allowing the use of almost the maximum dynamic range.

In practice several factors will limit the use of the maximum $\text{DR}_{\text{fringe}}$ of a CCD camera-equipped metrology system:

Table 46-2: Limitations and improvement techniques for the dynamic range of fringe- detecting metrology systems.

Limitation	Improvement
Fringes cannot be spread uniformly over the sensor for the general case.	a) Choose more adequate compensating or null system or b) choose a different wavefront-detection technique leading to a more uniform fringe density (i.e., shearing interferometry).
Non-uniform illumination of the CCD sensor due to test-piece, setup or illumination system leads to low signal-to-noise-ratio (SNR).	Improve uniformity of irradiance on CCD sensor.
So-called retrace errors caused by aberrations of the metrology sytem cannot be calibrated for high fringe densities.	Introduce null system or create adequate calibration procedure.
Environmental influences limit the SNR of the sensor.	a) Stabilize environmental conditions, b) acquire more frames for averaging.

Some improvement techniques are mentioned in table 46-2. The combination of results from different techniques measured at different scales is a general solution for the extension of the dynamic range. For instance, combining results from a CMM in combination with those from an adapted interferometric system leads to a very great DR extension.

46.7.3
Lateral DR of CCD Sensors in Interferometry and Wavefront Sensing

So far we have concentrated on the DR_z of a sensor element which usually defines the range and resolution in z or normal to a surface under test. To consider also the lateral $DR_{x/y}$ and resolution we make the following considerations.

Figure 46-115 shows a setup in which a CCD sensor with $M_i M_j$ pixels of size $p_i\, p_j$ is imaged onto a surface under test by means of an optical projection system. The relation between pixel coordinates and lateral coordinates on the test surface are given by

$$x(i,j) = \beta(i,j)\left(p_i(i - i_0)\cos\varphi + p_j(j - j_0)\sin\varphi\right) \tag{46-200a}$$

$$y(i,j) = \beta(i,j)\left(-p_i(i - i_0)\sin\varphi + p_j(j - j_0)\cos\varphi\right) \tag{46-200b}$$

where $\beta(i,j)$ is the magnification and distortion of the projection system, φ is the rotation of the pixel coordinates relative to the x,y coordinates of the test surface and

i_0, j_0 denotes the pixel on the optical axis of the optical system. In the case of rotational symmetry $\beta(i,j)$ is given by

$$\beta(i,j) = \beta_0 + \beta_2 r^2(i,j) + \beta_4 r^4(i,j) + \dots \tag{46-200c}$$

$$r^2(i,j) = (i - i_0)^2 + \left(\frac{p_j}{p_i}(j - j_0)\right)^2 \tag{46-200d}$$

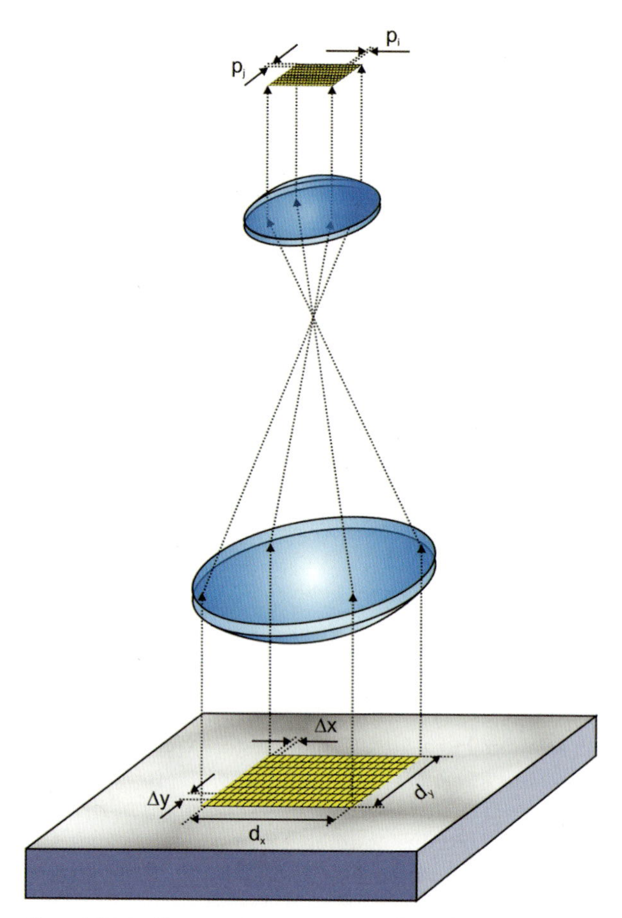

Figure 46-115: Topometric measurement of a surface under test by means of a CCD camera with $M_i \cdot M_j$ pixels of size $p_i \cdot p_j$.

In a setup showing distortion the lateral resolution varies locally in the i and j directions. It is given by

$$\Delta_i(i,j) = \sqrt{\left(\frac{\partial x(i,j)}{\partial i}\right)^2 + \left(\frac{\partial y(i,j)}{\partial i}\right)^2} \tag{46-201a}$$

$$\Delta_j(i,j) = \sqrt{\left(\frac{\partial x(i,j)}{\partial j}\right)^2 + \left(\frac{\partial y(i,j)}{\partial j}\right)^2} \qquad (46\text{-}201b)$$

In a distortion-free system (46-201a,b) lead to $\Delta_i = \beta_0 p_i$ and $\Delta_j = \beta_0 p_j$. The total range of the CCD projected onto the test surface is defined by the four edge points $\{x(1,1), y(1,1)\}$, $\{x(M_i,1), y(M_i,1)\}$, $\{x(1,M_j), y(1,M_j)\}$, $\{x(M_i,M_j), y(M_i,M_j)\}$.

46.7.4
Stitching Technique Used to Extend Lateral DR

In cases where full coverage of the test surface cannot be achieved because no feasible technical solution can be found, the stitching technique is possibly a technique that will greatly extend the lateral dynamic range without loss of lateral resolution or resolution in z [46-107]–[46-112].

When spherical or plane surfaces cannot be totally covered by the aperture of a metrology instrument such as an interferometer, the surface has to be assembled using a set of subapertures. The subapertures overlay their neighboring subapertures by a certain percentage. Each subaperture is captured by an individual measurement. Between the acquisitition of individual subapertures the relative position of the metrology instrument versus the test surface is changed. This can be achieved by adequate movement of the test-piece, while the position of the instrument is static, or vice versa.

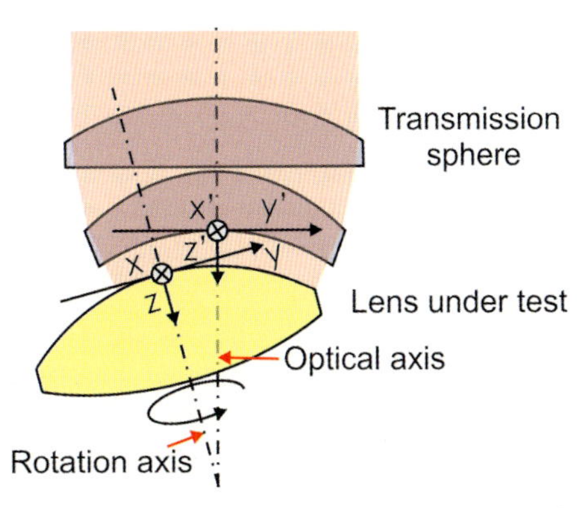

Figure 46-116: Interferometric stitching setup with transmission sphere and lens under test. The lens under test is rotated around an axis, which is tilted relative to the optical axis of the transmission sphere.

Figure 46-116 shows a transmission sphere and a spherical lens under test in an interferometric stitching arrangement. The $F/\#$ of the transmission sphere is larger than that of the test surface therefore no full coverage can be achieved. The lens under test is rotated about an axis which is tilted relative to the optical axis of the transmission sphere while a set of subapertures is acquired. Figure 46-117 shows a set of 12 individual subapertures assembled to form a common data set covering the total surface under test.

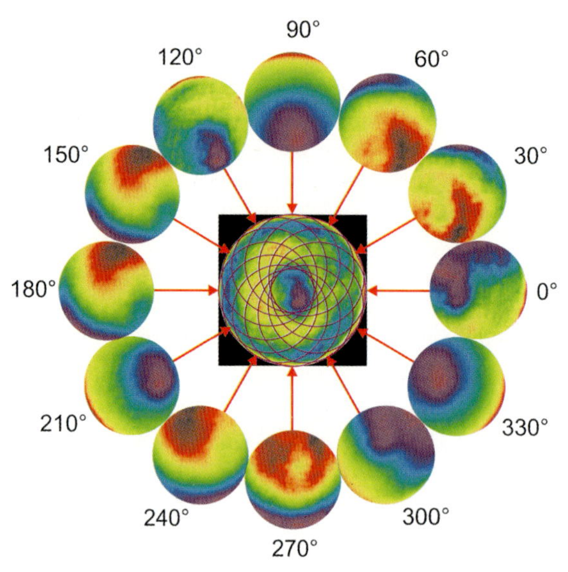

Figure 46-117: Twelve subaperture wavefronts assembled to form a common wavefront covering the total surface under test.

The Stitching procedure produces several problems that need to be resolved in advance.

1. Each subaperture is measured in a coordinate system Σ' coinciding with the coordinate system of the metrology instrument (figure 46-116). To assemble subapertures to a total wavefront they need to be transposed to the coordinate system of the test surface Σ. Any distortion of the instrument optics or the transmission sphere has to be considered.
2. After being transposed each data set has to be interpolated to a common grid that needs to be defined in advance. This is helpful for removal of the misalignment contributions in each individual subaperture wavefront.
3. Each subaperture wavefront still carries terms due to a slight defocus, decenter or tilt of the test subaperture relative to the transmission sphere. These need to be removed in a least-squares sense, minimizing the residuals when all subapertures overlay. Let K be the number of all collected subapertures. The least-squares problem is defined by

$$S = \sum_{i=1}^{M_i} \sum_{j=1}^{M_j} \sum_{k=1}^{K} \sum_{l=1}^{K} \left(\phi_k(i,j) - \phi_{\text{adj},k}(i,j) - \phi_l(i,j) + \phi_{\text{adj},l}(i,j) \right)^2 \qquad \text{(46-202a)}$$

where $\phi_k(i,j)$ denotes the measured phase values of the kth subaperture and $\phi_{\text{adj},k}(i,j)$ denotes the phase terms due to the unknown misalignment of the kth subaperture as described by

$$\phi_{\text{adj},k}(i,j) = a_{0,k}\phi_{\text{def},k}(i,j) + a_{1,k}\phi_{\text{decx},k}(i,j) + a_{2,k}\phi_{\text{decy},k}(i,j) + a_{3,k}\phi_{\text{tiltx},k}(i,j) + a_{4,k}\phi_{\text{tilty},k}(i,j) \quad \text{(46-202b)}$$

$\phi_{\text{def},k}(i,j)$, $\phi_{\text{decx},k}(i,j)$, $\phi_{\text{decy},k}(i,j)$, $\phi_{\text{tiltx},k}(i,j)$, $\phi_{\text{tilty},k}(i,j)$ must be calculated by raytracing for the perfect surface and must be transposed to the coordinate system of the test surface Σ. After defining an initial subaperture $k=0$ by, for instance, removing its individual misalignment terms, (46-202b) can be solved for all $a_{0,k}$, $a_{1,k}$, $a_{2,k}$, $a_{3,k}$, $a_{4,k}$ by a least-squares technique, minimizing the sum S in (46-202a). Note that the matrix to be inverted is of size $(5*(K-1))^2$. Note also that there might be many subaperture combinations k and l that do not share a common pixel area. In those cases their contribution to the sum in (46-202a) is zero.

4. After removal of all misalignment terms in each subaperture the final stitched wavefront results from the mean phase values of all overlaying subapertures.

The stitching technique is not restricted to spherical and plane surfaces but can successfully be applied to rotational symmetric aspherical surfaces as shown in figure 46-118. In this case a CGH generates an off-axis asymmetrical wavefront illuminating a subaperture of the tilted aspherical element rotated around its optical axis analogous to figure 46-117.

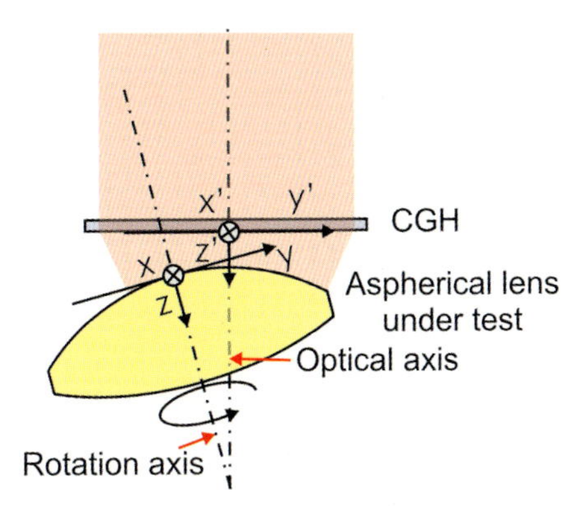

Figure 46-118: An interferometric stitching setup with a computer-generated hologram (CGH) and a lens under test carrying an aspherical surface. The lens under test is rotated around an axis that is tilted relative to the optical axis of the CGH.

46.8 Accuracy and Error Sources

46.8.1 Environmental Limitations

An optical instrument which is dedicated to measure optical beam paths or wavefronts with high resolution always runs the risk of being disturbed by environmental influences such as:

- mechanical vibrations,
- acoustic vibrations,
- radiation from light or heat sources,
- air turbulence and convection,
- changes in temperature, humidity, or pressure of ambient air or,
- heat conduction.

Figure 46-119 illustrates the typical causes affecting a high-resolution optical instrument. Among these are:

- external traffic (highway, railway, airplanes) near the experimental site,
- internal traffic (people walking, transport of goods, running machines) within the building,
- acoustic sources (machines, human voices, radios, computers, etc.),
- radiation sources (people (37°C, 100W/person), heating, illuminations),
- air conditioning, fans, draughts from open doors,
- heat convection through walls and ceilings.

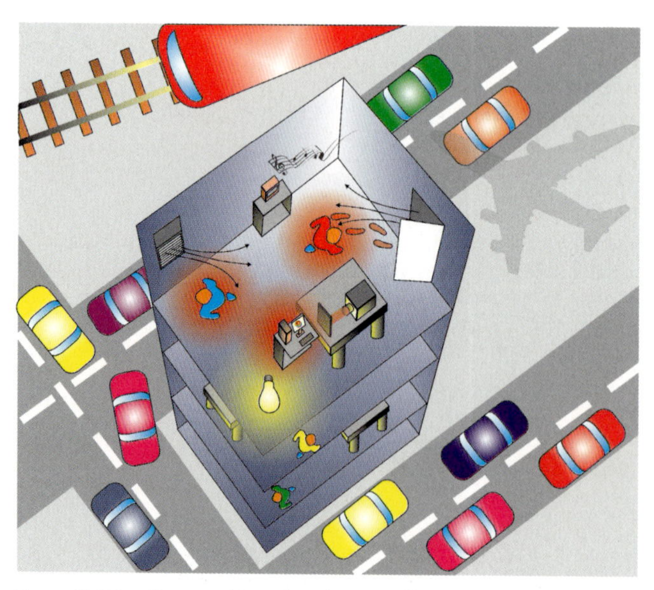

Figure 46-119: Some ambient disturbances affecting high-resolution optical metrology (from [46-114]).

Mechanical Stability

We consider an interference setup designed to measure wavefronts with a resolution of better than $\lambda/100$. We have to make sure that vibrations will not severely disturb the measurement. Taking into account that a series of measurements have to be averaged we nevertheless require a stability of $< \lambda/20$ for each optical component involved. Using visible light this leads to maximum vibration amplitudes of less than 30 nm.

There is a classification of locations where the maximum velocity, rather than displacement or acceleration, is used as a vibration criterion. Table 46-3 shows a generic classification [46-113], where the detail size of the objects to be observed or manipulated is used as a criterion, specifying the maximum RMS velocity of the floor vibration as measured in the vertical and two horizontal directions. The reason for the velocity, rather than the displacement, being selected as the criterion, is related to the exposure mechanism of photographic or CCD cameras, where displacement over time is relevant for the "sharpness" of the image. Considering the requirements for high-resolution interferometers as mentioned above and applying them to an acquisition time of 40 ms/frame will lead us to a maximum velocity of 0.75 μms^{-1}. According to table 46-3 not even the strongest vibration classification VC-E would be sufficient. This makes it essential to use a vibration-isolation system for the table to support the equipment.

To select the appropriate table for the experiment, we first have to measure the floor vibration level, which usually is done with the help of a very sensitive accelerometer and a fast Fourier transform analyzer. The power spectral density (PSD) will mainly determine the type of optical table and vibration-isolation system best suited to minimizing the vibration interference.

We must distinguish between the selection of the tabletop and the isolation system. The latter is primarily related to the severity of the environment, whereas the tabletop relates to the sensivity of the experiment. The ideal mounting system has a rigid tabletop that is mounted to the floor by a combination of springs and dampers. The position of the tabletop is maintained by inertia as the springs compress and relax, whereas the damping mechanism ensures minimum oscillation at resonance and also short settling times.

Vibration-isolation System

Different types of designs are available which have to be selected appropriately, considering the frequency of the floor vibration level [46-114]. A simple form uses air springs with short travel ranges that have a flexible diaphragm on top of a rigid cylinder containing compressed air. During floor vibrations air pressure alternately increases and decreases, while any object placed on top remains stationary.

A more sophisticated design uses a two-chamber air spring, where an upper chamber is connected to the reservoir chamber via a pinhole orifice. When compressed or expanded the air in the system is forced back and forth through the pinholes. The friction caused by this action provides the necessary damping. This technique is also used to damp vertical and horizontal vibrations ranging from 1 to 10 Hz. Combining a vertical isolation chamber with a horizontal vibration-absorbing damper optimizes isolation in both directions.

Table 46-3: Vibration specifications for a variety of applications (from [46-113]).

Vibration class	Max velocity of floor vibration [μinches/s]	Max velocity of floor vibration [μm/s]	Detail Size [μm]	Application
Workshop (ISO)	32 000	812.8	N/A	Distinctly noticeable vibration. Appropriate to workshops and nonsensitive areas.
Office (ISO)	16 000	406.4	N/A	Noticeable vibration. Appropriate to offices and nonsensitive areas.
Residential Day (ISO)	8000	203.2	75	Barely noticeable vibration. Appropriate to sleep areas in most instances. Probably adequate for computer equipment, probe test equipment and low-power (to 50X) microscopes.
Op. Theatre (ISO)	4000	101.6	25	Vibration not noticeable. Suitable for sensitive sleep areas. Suitable in most instances for microscopes to 100X and for other equipment of low sensitivity.
VC-A	2000	50.8	8	Adequate in most instances for optical microscopes to 400X, microbalances, optical balances, proximity and projection aligners, etc.
VC-B	1000	25.4	3	An appropriate standard for optical microscopes to 1000X, inspection and lithography equipment (including steppers) to 3 μm linewidth.
VC-C	500	12.7	1	A good standard for most lithography equipment and inspection equipment (including electron microscopes to 1 μm detail size).
VC-D	250	6.35	0.3	Suitable in most instances for the most demanding equipment including electron microscopes (TEMs and SEMs) and E-Beam systems, operating to the limits of their capability.
VC-E	125	3.175	0.1	A difficult criterion to achieve in most instances. Assumed to be adequate for the most demanding of sensitive systems including long-path, laser-based, small-target systems and other systems requiring extraordinary dynamic stability.

A vibration-isolation system is characterized by its transmissibility, which is defined as the ratio of the amplitude of the transmitted vibration to that of the forcing vibration. Figure 46-120 gives an example.

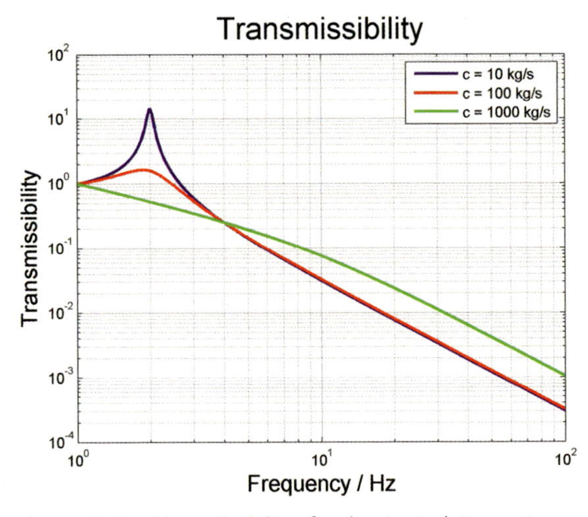

Figure 46-120: Transmissibility of a vibration-isolation system for different dampings.

Tabletop

A good optical tabletop should have:

- good static and dynamic stiffness,
- a first resonance which is higher than 100 Hz,
- good damping properties.

Tabletops have been constructed from granite, concrete, wood, and steel, each with greater or less success. Composite structures have also been used in attempts to improve performance while keeping weight at a realistic level [46-14], [46-15]. The most common of these are so-called honeycomb sandwich tables. Their top and bottom layers are made from steel or aluminum, and the large core is a low-density metal honeycomb, usually steel. The plates provide the stiffness, whereas the honeycomb combines excellent dynamical stiffness with a very low density, while exhibiting fairly good natural damping.

Honeycomb tabletops are available in different sizes and thicknesses. Thicker tabletops are stiffer and have a higher resonant frequency that leads to lower relative tabletop motion. Commercial honeycomb tabletops have mounting holes for attaching equipment, which makes them very flexible when setting up different configurations. Small, rigid flat surfaces, called breadboards are available, which are suitable for laboratory applications that require little space. Most manufacturers can custom-build breadboards from extra-lightweight material using special hole patterns.

A tabletop is characterized by its compliance. The compliance C is the ratio of the displacement x over the applied force amplitude F. Equation (46-203) shows the compliance for a simple system with one degree of freedom, which is equal to the solution of a forced damped vibration.

$$C = \frac{x}{F} = \frac{1}{\sqrt{m^2 \left(f_n^2 - f\right)^2 + (cf)^2}} \quad (46\text{-}203)$$

where
m is the mass being moved or deflected,
c is the damping,
f is the frequency of the applied forced vibration,
f_n is the resonant frequency of the undamped system.

$k = mf_n^2$ is called the stiffness of the system.

The perfect tabletop is an ideal rigid body that does not resonate and therefore has no compliance peaks. When plotted on a log-log scale, an ideal rigid body has a compliance, which is proportional to f^{-2}. Figure 46-121 shows an example of a compliance curve compared with a rigid body. The curve of the real table shows a first resonance maximum at f_n where $A = C(f_n)$. The corresponding value of a rigid body is $B = C_{\text{rigid}}(f_n)$. The ratio $Q = A/B$ is called the maximum amplification factor at resonance, which is a measure of the damping efficiency.

The displacement response δ of a real tabletop can be calculated from (46-204) which expresses the worst-case relative motion between two points on a table at its eigenfrequency f_n.

$$\delta = 2T\sqrt{\frac{1}{32\pi^3}}\, g \sqrt{\frac{Q}{f_n^3}} \sqrt{PSD} \quad (46\text{-}204)$$

where
T is the isolator transmissibility (typically < 0.01 above 10 Hz),
Q is the maximum amplification (dimensionless),
f_n is the table's corresponding resonant frequency measured in Hz,
g is the acceleration due to gravity $= 9.8\ \text{ms}^{-2}$,
PSD is the power spectral density of the floor vibration level measured in $g^2\ \text{Hz}^{-1}$.

Q and f_n are estimated from the compliance curve of the table, T is the specified transmissibility of the isolation system, whereas PSD is estimated from the measured vibration level of the floor.

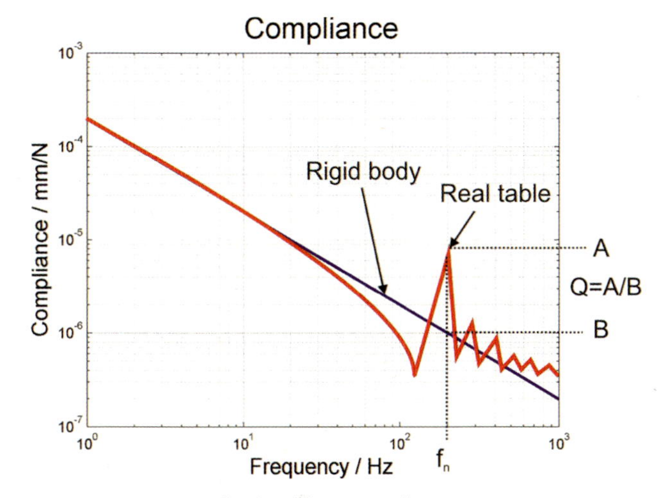

Figure 46-121: Example of a tabletop compliance curve compared with that of a rigid body.

Thermal Stability

Temperature changes in the environment of an optical experiment can severely influence the results. When heated or cooled, the air changes its index of refraction so that the optical path in an optical setup is altered [46-116] –[46-125]. A simplified form of the Edlén formula is given in (46-205). It is useful for hand calculator approximations and is valid only for a wavelength of approximately 633 nm.

$$n_{\mathrm{air},633\,\mathrm{nm}} = 1 + 7.86 \cdot 10^{-4} \frac{p}{273 + T} - 1.5 \cdot 10^{-11} RH \cdot (T^2 + 160) \qquad (46\text{-}205)$$

where p is pressure in kPa, T is temperature in Celsius, and RH is relative humidity in percent. The equation is expected to be accurate within an estimated expanded uncertainty of 1.5×10^{-7}.

(46-205) has been used to generate figure 46-122 showing an optical path variation of $-2.87\ \lambda\ \mathrm{K}^{-1}\mathrm{m}^{-1}$ ($\lambda = 632.8$ nm) per meter of single-path cavity length. When the air pressure changes the optical path is altered as well. During a day, average fluctuations of 0.2 kPa occur as indicated in figure 46-122 by the green and blue curves resulting in an optical path variation of $8.48\ \lambda\ \mathrm{kPa}^{-1} \cdot \mathrm{m}^{-1}$ for $\lambda = 632.8$ nm. Thus, the natural air pressure changes within a day have the same optical effect as temperature changes over 0.6°C.

Temperature changes will also affect the refractive indices and the geometry of the optical components. Table 46-4 shows a list of different optical and non-optical materials and their coefficients of thermal expansion α (CTE), their relative temperature coefficients of refractive index dn_{air}/dT, and their thermal conductivity k.

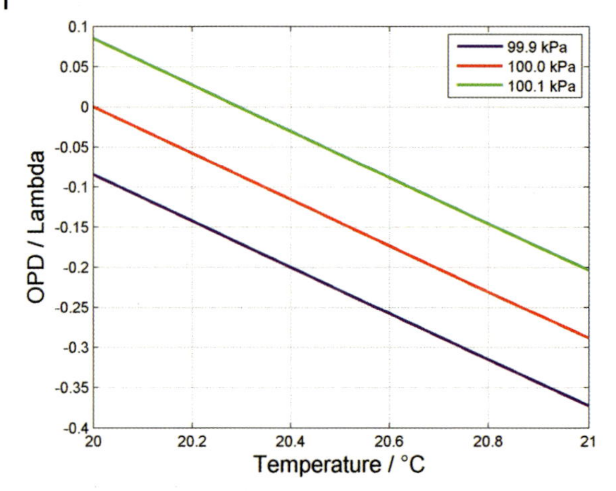

Figure 46-122: Optical path differences ($\lambda = 632.8$ nm) for a 100 mm single-path cavity filled with air, as a function of temperature, for three different air pressures.

Table 46-4: Thermal coefficients of different optical and non-optical materials.

Material	Coefficient of linear thermal expansion (CTE) α [$10^{-6}K^{-1}$ at 20 °C]	Relative temperature coefficients of refractive index dn_{air}/dT at λ = 546.07 nm [10^{-6} K^{-1} at +20°C ...+40°C]	Thermal conductivity k [$Wm \cdot m^{-1} \cdot K^{-1}$]
Polymers	20–500	(not specified)	0.25
Mercury	60	–	8
Lead	29	–	35
Aluminum	22	–	220
Brass	19	–	130
Calcium Fluoride CaF_2	18.9	–10.1	61
Stainless steel	17.3	–	15
Copper	17	–	390
Gold	14	–	316
Nickel	13	–	91
Concrete	12	–	1.7
Iron or steel	11.1	–	55
Carbon steel	10.8	–	33

Material	**Coefficient of linear thermal expansion (CTE) α [$10^{-6}K^{-1}$ at 20 °C]**	**Relative temperature coefficients of refractive index dn_{air}/dT at λ = 546.07 nm [10^{-6} K^{-1} at +20°C ...+40°C]**	**Thermal conductivity k [$Wm \cdot m^{-1} \cdot K^{-1}$]**
Platinum	9	–	72
Granite	8.5	–	3
Optical glass, BK7	8.3	3	1
Optical glass, general	4–16	–6.6 ... 12.5	0.5–1.38
Borosilicate glass, Pyrex	3.3	(not specified)	1
Silicon	3	–	150
Invar	1.2	–	13
Diamond	1.1	7.6	2000
Fused silica, Suprasil	0.51	10.1	1.4
Glass ceramics, Zerodur	0 0.1	14.8	1.5
Air	0.003	–0.9	0.025

When light travels through a plane plate of thickness d and refractive index n, in air, the change in the optical path length ΔOPD due to a change in temperature ΔT will be

$$\Delta \mathrm{OPD} = d\left(\alpha(n-1)+\frac{dn}{dT}\right)\Delta T \tag{46-206}$$

The value in brackets is equal to the specific OPD change $d\mathrm{OPD}/dT$ of a plane plate. Using the specific values in table 46-4 we get the following examples for λ=632.8 nm:

Table 46-5: Specific OPD changes for plane plates made of different optical materials.

Material	Specific OPD change for plane plate [10^{-6} K^{-1}]
Calcium fluoride CaF_2	–1.94
Optical glass, BK7	7.28
Diamond	9.16
Fused silica, Suprasil	10.33
Glass ceramics, Zerodur	14.8

When the temperature of an optical element changes, its focal length is also altered. As an example we consider the focal length of a single lens as it changes with temperature. Equation (46-207) shows the paraxial formula used to calculate the focal length f as a function of temperature T of the lens and the ambient medium.

$$\frac{1}{f(T)} = (n_{\text{lens}}(T) - n_{\text{air}}(T))\left(\frac{1}{r_1(T)} - \frac{1}{r_2(T)}\right) + \frac{d(T)(n_{\text{lens}}(T) - n_{\text{air}}(T))^2}{n_{\text{lens}}(T)r_1(T)r_2(T)} \tag{46-207}$$

where

$$r_{1/2}(T) = r_{1/2}(20^\circ\text{C})(1 + \alpha_{\text{lens}}(T - 20^\circ\text{C}))$$

$$d(T) = d(20^\circ\text{C})(1 + \alpha_{\text{lens}}(T - 20^\circ\text{C}))$$

$$n_{\text{lens/air}}(T) = n_{\text{lens/air}}(20^\circ\text{C}) + \frac{dn_{\text{lens/air}}}{dT} \cdot (T - 20^\circ\text{C})$$

Figure 46-123 shows examples for biconvex lenses of 100 mm focal length made from different materials. Since CaF_2 and fused silica have a strong dn/dT of opposite sign, their focal length changes in opposite directions. In this case the effects are dominated merely by dn/dT, whereas α plays a minor role. Note that, for Zerodur, although the thermal expansion coefficient is zero, the focal length changes rapidly because of the high thermal sensitivity of its refractive index.

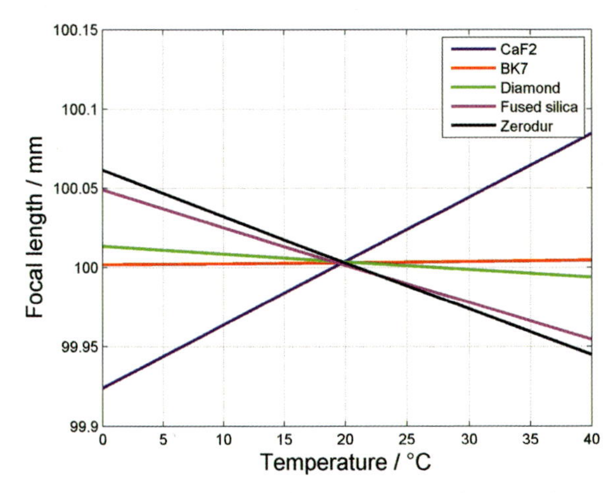

Figure 46-123: Focal length of a single lens as a function of temperature for different materials.

When the ambient temperature changes the lens mounts and individual component support elements are also affected. As a result, air gaps between individual components are altered. The way in which changes are made depends on the mechanical design of the support system and on the materials that have been chosen. If lens elements are glued to support rings, then their thermal expansion coefficients should be selected to be similar to those of the support material to prevent any mechanical tensions being induced as the ambient temperature changes. BK7 and steel can be combined quite well, whereas α for fused silica quite closely matches that of invar.

Heat transportation mechanisms are based on convection, radiation, and conduction. All three play a major role when high-precision optical metrology is set up. In the following some aspects are reported that need to be considered during practical work.

Convection

The process of heat transfer from a solid surface to a moving liquid or a moving gas is called convection. The motion may be natural when driven by the heated or cooled solid, or forced when driven by an external apparatus such as a fan or a pump. The convective process is accomplished in two stages.

1) The heat transfers from the solid to the fluid or gas through a diffusion process adjacent to the surface.
2) The transferred heat is carried away by the motion of the fluid or gas, where the power of the heat transfer is driven by the conductivity of the fluid or gas.

Convection is generated wherever there are solid objects of different temperature than the surrounding air. These objects may be:

- heat radiators,
- light sources,

- power supplies and transformers,
- computers,
- monitors and displays,
- electronic equipment,
- motors and mechanical machines,
- pumps,
- bodies of animals or humans,
- walls, ceilings, floors of different temperature,
- etc.

In natural convection the air is moved by a change in its density. Under gravity the change in density induces the heated air to move up without any externally imposed flow velocity. Natural convection velocities are relatively slow. Optical experiments suffer from convections when “clouds” of heated air pass the metrological beam paths or cavities, changing the refractive index of the air slowly and inhomogeneously.

To prevent disturbances by convection effects optical experiments should be shielded by appropriate housings. Within the housing the temperature should be kept constant by introducing a flow box with an appropriate temperature regulation system. State of the art flow-box systems are able to stabilize T to 50 mK.

In addition, the use of one or several fans should be considered in order to change the velocity of the residual convection clouds within the beam paths. In this way the air flow is partially laminar with turbulent areas, but will have a much greater velocity so that slow drifts are avoided and statistical averaging is made much easier.

Radiation

Thermal or heat radiation is electromagnetic radiation that is emitted from the surface of an object. Whatever their temperature all objects radiate unless their absolute temperature is 0 K. Radiation is generated when heat from the movement of charged particles within atoms is converted to electromagnetic radiation.

When setting up an optical experiment, disturbing sources of heat radiation are the same as those responsible for convection, as mentioned above.

The human body plays a major role since optical materials have to be handled during testing. The radiation power of a human body P_{body} can be estimated using the Stefan–Boltzmann law of radiation [46-134] as expressed in (46-208).

$$P_{\text{body}} = \sigma\, \varepsilon\, A \left(T^4 - T_0^4\right) \tag{46-208}$$

where

T temperature of human body in K,
T_0 ambient temperature in K,
A surface area of human body,
ε emissivity of human body,
σ Stefan–Boltzmann constant $5.67 \times 10^{-8}\ \text{Ws}^{-1}\text{K}^{-4}$.

The emissivity of skin and clothing is near to unity. The surface area of a human body is between 1.5 and 2.0 m^2. Skin temperature is approximately 33°C, but clothing reduces it to about 28°C. In an environment of 20°C the heat radiation of an adult human body is therefore around 90 to 100 W.

Conduction

Heat conduction is the energy transfer from a high-temperature to a low-temperature region due to the process of thermal diffusion. It can take place in solids, liquids and gases. The rate at which heat is conducted through a body per unit cross-sectional area is inversely proportional to the temperature gradient existing in the body. The heat power transferred through a cross-sectional area A is given in (46-209).

$$P_{\text{transfer}} = -k\,A\frac{dT}{dx} \tag{46-209}$$

where k is the thermal conductivity in $W \cdot m^{-1} \cdot K^{-1}$ and dT/dx is the local temperature gradient.

If the materials are touched by a human body they heat up at the point of contact. The imported heat quantity depends mainly on the contact time and the temperature difference of the body and the material. The heat quantity will spread within the material until equilibrium is reached. The necessary time to reach equilibrium depends on the thermal conductivity k of the material. Conventional optical materials have conductivities of between 0.5 and 1.5 $W \cdot m^{-1} \cdot K^{-1}$ which makes them very slow in achieving a homogeneous temperature distribution within the optical element. As a result the refractive index also varies according to its dn/dT.

When a larger lens has been touched even for just for a second it will take half a day until an equilibium within 50 mK is again established. Lenses should therefore be handled by special tools as shown in figure 46-124, to prevent direct contact.

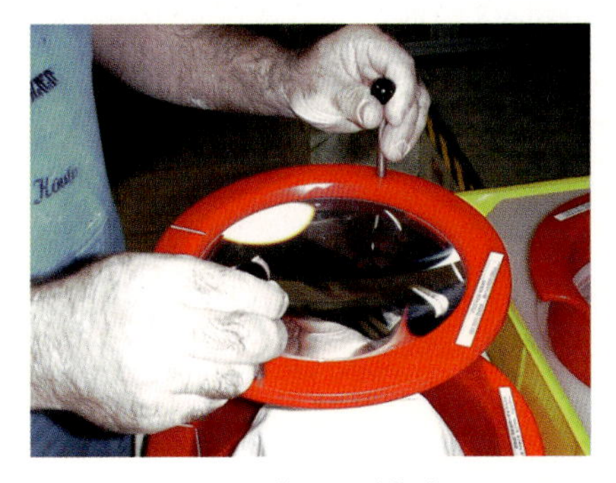

Figure 46-124: Handling tool for lenses to prevent direct contact.

From among the non-optical materials, granite also has low thermal conductivity and therefore tends to warp when heated. Invar is four times, and steel ten times faster in reaching equilibrium, and aluminum is even more than 70 times faster than granite. That is why aluminum is preferred if warping is of major concern for the efficient working of the different parts.

Acoustics

When sound waves propagate through matter they cause local regions of compression and rarefaction. Air is the usual medium supporting sound. Including near-infra and ultrasound, frequencies range from approximately 10 Hz to 20 kHz with wavelengths from 33 m to 16 mm [46-126].

Acoustic vibrations mainly have two different effects on an optical metrology setup.

1. The compressed and rarefied air causes a direct change in the optical path length depending on the sound level of and the distance from, the source.
2. The acoustic vibrations cause mechanical setups to vibrate, the magnitude of excitation depending on the resonance frequency and damping of the mechanical design.

The first effect is in many cases negligible for a short cavity length under normal acoustic conditions. It is only for high sound levels (for instance machines in a close neighborhood) and cavities > 1 m that disturbances have to be considered, when high precision is necessary. Table 46-6 shows a variety of examples and their effect on the optical path change [46-126].

Table 46-6: Objects generating sound and their influence on the optical path in air for a cavity length of 1 m and $\lambda = 632.8$ nm.

Object	Distance from sound source [m]	Relative sound pressure [dB]	Sound pressure p [Pa]	Optical path change per 1 m cavity length for $\lambda = 632.8$ nm [λ]	Optical path change per 1 m cavity length for $\lambda = 632.8$ nm [nm]
Jet aircraft	30	140	200	1.7	1076
Threshold of pain	–	130	63	0.53	335.4
Threshold of discomfort	–	120	20	0.17	107.6
Chainsaw	1	110	6.3	0.053	33.54
Dance floor loudspeaker	1	100	2	0.017	10.76
Diesel motor	10	90	0.63	0.0053	3.35
Highway traffic	5	80	0.2	0.0017	1.08
Vacuum cleaner	1	70	0.063	0.00053	0.34

Object	Distance from sound source [m]	Relative sound pressure [dB]	Sound pressure p [Pa]	Optical path change per 1 m cavity length for $\lambda = 632.8$ nm [λ]	Optical path change per 1 m cavity length for $\lambda = 632.8$ nm [nm]
Conversational speech	1	60	0.02	0.00017	0.11
Average home	–	50	0.0063	0.000053	0.03
Rustling of leaves	100	10	0.000063	0.00000053	0.00
Hearing threshold level		0	0.00002	0.00000017	0.00

The second effect is much more relevant because setups might be excited so that they "ring" if their eigenfrequency coincides with a strong spectral acoustic component. Strong acoustic sources should therefore be avoided in the neighborhood of optical setups. If this is not possible, adequate sound damping prevention might be necessary.

46.8.2 Noise

Noise generally denotes random errors that derive from electrical, optical, mechanical or environmental sources and that behave in a statistical way such that their calibration is not possible.

In an optical measurement device there are usually several noise sources [46-127].

1. Electronic sensor or camera.
2. A light source.
3. An electronic manipulator (for example a phase-shifting device).
4. Speckle effects from rough surfaces.
5. Fast environmental fluctuations.

While the flicker of light sources and the electrical noise in optical manipulators both generally have an effect on the total image, camera noise or optical coherent noise caused by partially rough surfaces within the setup have an uncorrelated effect on individual pixels. In the case of air turbulence within the optical path we might as well refer to noise, because the effects are statistically distributed over the camera target although their spatial bandwidth generally only covers long- and mid-spatial frequencies.

In the following, we will discuss those sources that cause noise effects across a two-dimensional camera target (1., 4. and 5.).

CCD Noise Sources

A CCD camera image generally suffers from three noise sources: photon or shot noise, CCD noise, and readout noise.

Photon noise occurs when the finite number of photons in an optical device is small enough to give rise to detectable statistical fluctuations during a measurement. The average number of photons radiated by a source and collected by a sensor might be constant, but the actual number collected follows a Poisson distribution which, for large numbers of photons, approaches a normal distribution. The standard deviation σ_{photon} of the photon noise is equal to the square-root of the average number of photons $\overline{N}_{\text{photon}}$. Equation (46-210) defines the standard deviation for the number of electrons generated by photon noise. It can also be expressed in camera output units (ADU = Analog-Digital Unit), when the number of electrons per LSB (least significant bit) of the CCD device is known.

$$\sigma_{\text{photon}} = \sqrt{\overline{N}_{\text{photon}}} = \sqrt{\frac{I}{h\nu} t A \eta} \tag{46-210}$$

where

I is the irradiance at the CCD sensor in W m^{-2}
h the Planck constant = $6.62606896 \times 10^{-34}$ Js
ν is the frequency of the photon in Hz
A is the pixel area in m^2
t is the exposure time in s
η is the quantum efficiency.

The quantum efficiency defines the ratio of the number of electron–hole pairs produced per number of photons hitting the photoreactive surface of the CCD camera. The quantum efficiencies of many CCD cameras are between 0.4 and 0.8 for red light around 633 nm wavelength.

CCD noise is due to the CCD image sensor and mainly includes:

- dark current noise caused by thermally induced charge carriers,
- transfer noise due to a shift of charge carriers between the registers of a CCD image sensor,
- fixed-pattern noise causing spatially fixed differences in noise behavior or sensitivity.

Readout noise occurs as the electronic device reads and amplifies the signal. Its components are output amplifier noise, camera noise and clock noise.

Since all noise sources are uncorrelated they add up statistically as shown:

$$\sigma_{\text{noise}} = \sqrt{\frac{I}{h\nu} t A \eta + \sigma_{\text{CCD}}^2 + \sigma_{\text{readout}}^2} \tag{46-211}$$

The signal-to-noise ratio (SNR or S/N) is given by (46-212). It defines the ratio of the signal level to the noise level corrupting the signal. For bright signals it is dominated by the average number of photons collected. An example is shown in figure 46-125.

$$\mathrm{SNR} = \frac{\overline{N}_{\text{photon}}}{\sqrt{\overline{N}_{\text{photon}} + \sigma^2_{\text{CCD}} + \sigma^2_{\text{readout}}}} = \frac{\frac{I}{h\nu} tA\eta}{\sqrt{\frac{I}{h\nu} tA\eta + \sigma^2_{\text{CCD}} + \sigma^2_{\text{readout}}}} \tag{46-212}$$

Figure 46-126(b) shows an example of an interferogram taken with a 10% level of readout and CCD noise (additive) and a 10% level of photon noise (multiplicative).

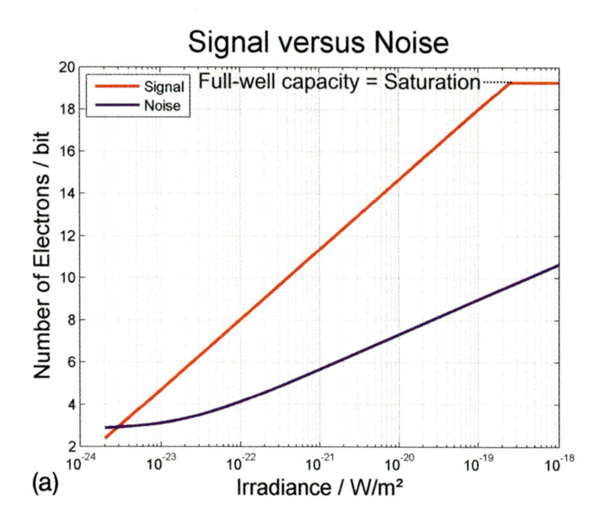

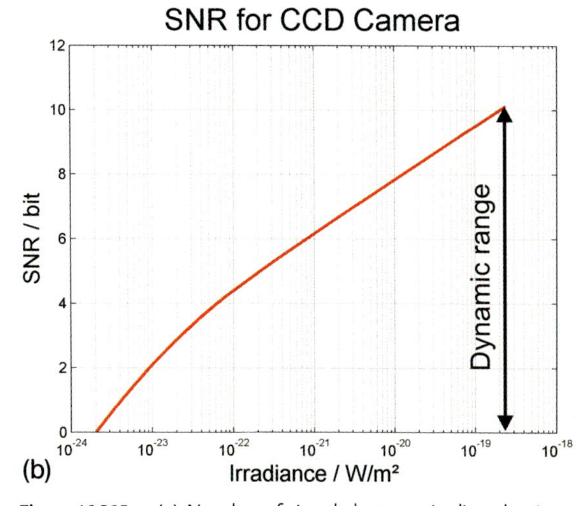

Figure 46-125: (a) Number of signal electrons (red) and noise electrons (blue) in bit, (b) SNR as a function of irradiance at the CCD sensor in bit.

The maximum signal of a CCD sensor is limited by its full-well capacity. The full-well capacity is the largest charge a pixel can hold before saturation. When the charge in a pixel exceeds the saturation level it starts to fill adjacent pixels, a process known as blooming. The full-well capacity of a sensor generally depends on the pixel size. Larger pixels have lower spatial resolution but their greater full-well capacity offers a higher dynamic range. Typical full-well capacities are 45 000 electrons (6.8 μm pixel size) up to 630 000 electrons (24 μm pixel size).

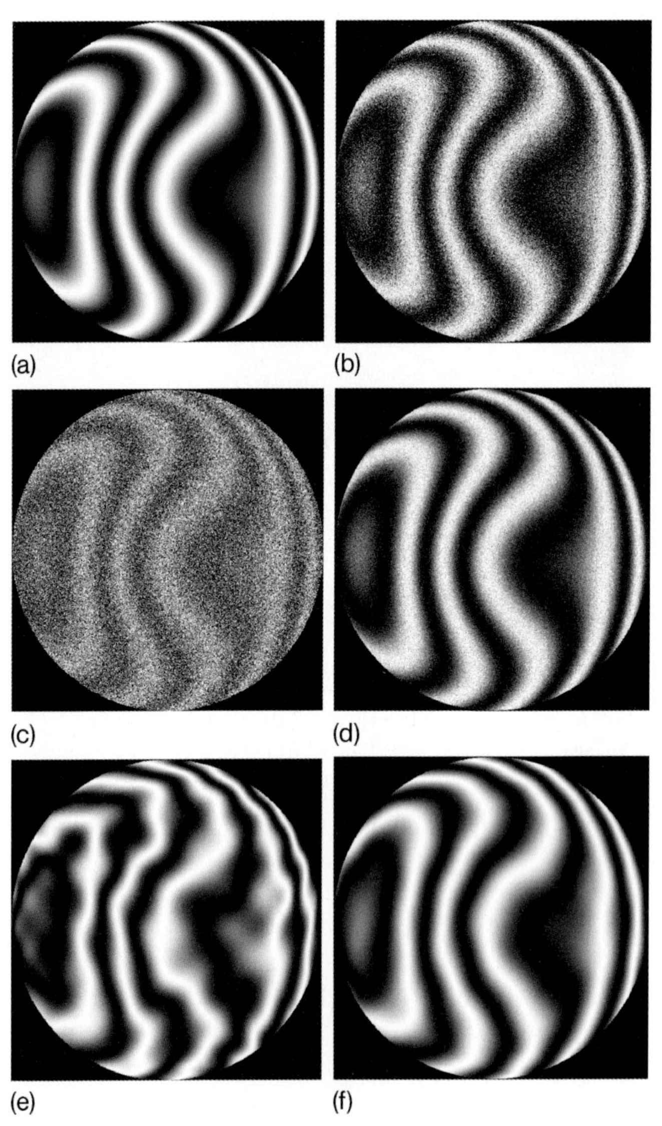

Figure 46-126: Interferogram disturbed by electronic and optical noise: (a) undisturbed; (b) camera noise σ_{photon}=10% + $\sigma_{readout}$=10%; c) additive phase noise $\sigma = \lambda/4$; (d) additive phase noise $\sigma = \lambda/10$; (e) single measurement with air turbulences, PV=0.6 λ; (f) four measurements with air turbulences averaged, PV=0.6 λ.

The dynamic range of a CCD is defined as the full-well capacity divided by the camera noise at the highest irradiance before saturation. The ratio is usually expressed in dB, bit or ADU.

Reducing the noise will extend the dynamic range and this is mainly achieved by:

- increasing the light level to reduce the shot noise,
- using a low-speed readout (slow scan) to reduce readout noise,
- cooling the CCD sensor to reduce dark-current noise (factor 0.5 per 7 – 8°C).

Speckle Effects

A speckle pattern is a spatial high-frequency random intensity pattern produced by the mutual interference of a large number of coherent wavefronts. In an optical instrument it arises when light from a laser is scattered from optical components within the beam paths of the instrument. The reasons for scatter are either surface or bulk defects.

When optical surfaces are ground and polished a certain residual roughness remains. We refer to surface texture meaning the global statistical profile property (roughness) of an optical surface and also refer to small surface irregularities (microdefects), generally less than 1 mm in size. In addition to its texture an optical surface may show localized defects, known as surface imperfections. Examples of surface imperfections are scratches, pits, broken bubbles, sleeks, scuffs and fixture marks [46-128]. The features can range in size from a few angstroms up to several millimeters. In the case of non-random wear processes such as diamond turning, the surface texture may have an inherent symmetrical or periodic pattern.

Transparent material, such as water or glass, usually scatters a very low portion out of the specular beam direction. The size, shape, and distribution of the bulk defects determine the irradiance and the angular distribution of the scattered light.

In the following we will approximate the effects of a disturbing speckle field on a wavefront under test. We assume that the wavefront being tested $E_T = A_T e^{i\phi_T}$ is disturbed by a coherent speckle field $E_S = A_S e^{i\phi_S}$ such that both interfere to form a new wavefront E_{T+S} which is then described as

$$\begin{aligned} E_{T+S} &= A_{T+S} e^{i\phi_{T+S}} \\ &= \sqrt{A_T^2 + A_S^2 + 2A_T A_S \cos(\phi_T - \phi_S)} \cdot e^{i \cdot \arctan \frac{A_T \sin\phi_T + A_S \sin\phi_S}{A_T \cos\phi_T + A_S \cos\phi_S}} \end{aligned} \quad (46\text{-}213)$$

The maximum disturbances of the wavefront under test by the speckle field are then given by

$$\Delta A_{max} = A_{T+S} - A_T = \pm A_S \quad (46\text{-}214a)$$

$$\Delta\phi_{max} = \phi_{T+S} - \phi_T \approx \pm \arctan \frac{A_S}{A_T} \quad (46\text{-}214b)$$

where $A_S << A_T$. The phase disturbance can be further approximated and expressed in λ by

$$\Delta\phi_{\max}[\lambda] \approx \pm\frac{1}{2\pi}\frac{A_S}{A_T} = \pm\frac{1}{2\pi}\sqrt{\frac{I_S}{I_T}} \tag{46-215}$$

where I_S and I_T denote the average irradiances of the wavefront under test and the speckle field, respectively. The phase disturbance by a speckle field of 1% relative irradiance therefore causes a maximum phase error of 0.0159λ. Figure 46-126(c),(d) show interferograms which were disturbed by a normally distributed phase noise of $\sigma = \lambda/10$ and $\sigma = \lambda/4$, respectively.

The scattering of an optical element can be described by the *Bidirectional Scatter Distribution Function* (BSDF), defined as the scattered radiance normalized by the incident irradiance, given by (46-216) [46-129].

$$f(\vartheta_i, \varphi_i, \vartheta_s, \varphi_s) = \frac{\partial\Phi_s}{\partial\omega}\frac{1}{\Phi_i\cos\vartheta_s} \tag{46-216}$$

where $\partial\Phi_s/\partial\omega$ is the scattered power per unit solid angle, Φ_i is the incident power, ϑ_s is the scattering polar angle, φ_s is the scattering azimuthal angle, ϑ_i is the incident polar angle, and φ_i is the incident azimuthal angle, as illustrated in figure 46-127. The BSDF is often used as a polarization-averaged quantity, but it is really a Mueller matrix relating the scattered polarization state to the incident polarization state. Since light may be scattered both in reflection and transmission one may distinguish between the *Bi-directional Reflectance Distribution Function* (BRDF) and the *Bi-directional Transmittance Distribution Function* (BTDF) in defining or measuring scatter (figure 46-128). BSDF can be measured by an array of detectors arranged in an arc or by sweeping a single detector through an arc.

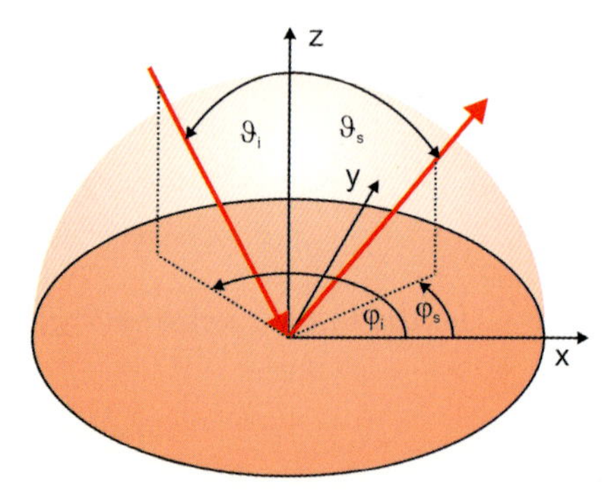

Figure 46-127: Incident and scattered beam at a scattering surface in the *x/y* plane.

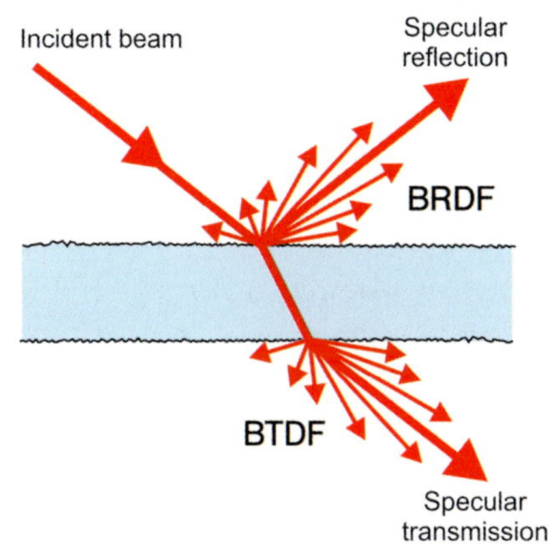

Figure 46-128: Bi-directional reflectance distribution function (BRDF) and the bi-directional transmittance distribution function (BTDF) at a scattering substrate.

Since the scatter of optical components is mainly due to their surface roughness the quality of polishing plays a major role. Measuring the roughness of an element tends to be easier than making scattered light measurements. The roughness of a surface can be specified by a number of different parameters mainly considering the deviations from the mean surface level. See section 54 for roughness metrology techniques.

Speckle effects in optical measurement devices can be reduced by a number of techniques, such as:

- introducing superpolished optical components,
- using an extended light source,
- averaging a number of results while the light source is moved,
- averaging a number of results while the components or the test object rotate.

Environmental Noise

An inhomogeneous temperature distribution within the beam path of an optical measurement device leads to measurement errors that change statistically over time. We can classify them as a form of noise, although on a different lateral and time scale, as in the case of speckle and camera noise. Schlieren (a German word for streaks or striae) are patterns caused by warm air currents from a hot object. Just like any other temperature clouds they drift through the air by convection within the measurement device and cause slight changes in the index of refraction of the air (figure 46-129).

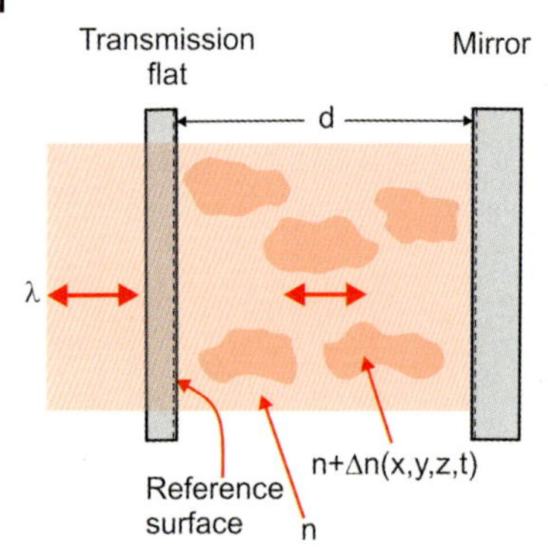

Figure 46-129: Temperature clouds drifting through the cavity of an interferometer.

As a result of this the measured optical path difference OPD(x,y) is disturbed by a current deviation $\Delta n(x,y,z,t)$ from the average index of refraction of the air. In order to approximate the "true" OPD value, a larger number of measurements taken at different times must be averaged (figure 46-126(e),(f)) as shown:

$$\mathrm{OPD}(x,y) = \lim_{N\to\infty} \frac{1}{N} \sum_{i=1}^{N} d(x,y)\,(n + \Delta n_i(x,y,z,t_i)) \qquad (46\text{-}217)$$

In order to lessen the number of measurements which need to be averaged, one or several fans should be introduced to change the velocity of the convection clouds within the beam paths and to homogenize the index of refraction of the air.

46.8.3 Capability of Measurement Systems

Once a measurement system (gauge) is installed it has to be qualified. Before the manufacturing process of any optical part can be controlled or optimized the uncertainty of the measurement system has to be determined. A Measurement System Analysis (MSA) must be carried out. An MSA is a series of specially designed experiments intended to identify the components of the variation in the measurement. It evaluates the entire process of obtaining measurements including a possible dependence on different operators [46-131].

The MSA is defined in a number of published documents including the AIAG's MSA Manual [46-133]. The AIAG (Automotive Industry Action Group) is a non-profit-making association of automotive companies founded in 1982. Because of its general applicability the MSA manual is widely used in different industries.

In order to understand MSA methods several terms need to be defined as explained in table 46-7.

Table 46-7: Some terms used in defining Measurement System Analysis.

Term	Explanation
Accuracy	Variation of a measurement process around the true value. Accuracy includes precision (reproducibility & repeatability) and the systematic offset between the average of measured values and the true value.
Precision	Variation of a measurement process around its average value. Precision can be decomposed further into short- term variation or repeatability, and long-term variation, or reproducibility.
Bias	Difference between the average and the true value.
Linearity	Degree of systematic approximation of the true results by the measured results throughout the range of a measurement system. Ideal linearity is achieved if the measured results versus the true results are represented by a line with slope of 1.0 in an x/y diagram.
Repeatability	Component of measurement precision. Short-term variation that occurs under highly controlled situations (e.g., same metrology instrument, same operator, same setup, same ambient environment).
Reproducibility	Total long and short-term variation when two or more people measure the same unit with the same metrology instrument.
Stability	Variation of the short-term average result over an extended period of time under highly controlled situations (e.g., same metrology instrument, same operator, same setup, same ambient environment).

Accuracy and precision have to be distinguished correctly. While precision describes the variation in the measurement process around its average value, the accuracy includes the variation with respect to the true value as shown in figure 46-130.

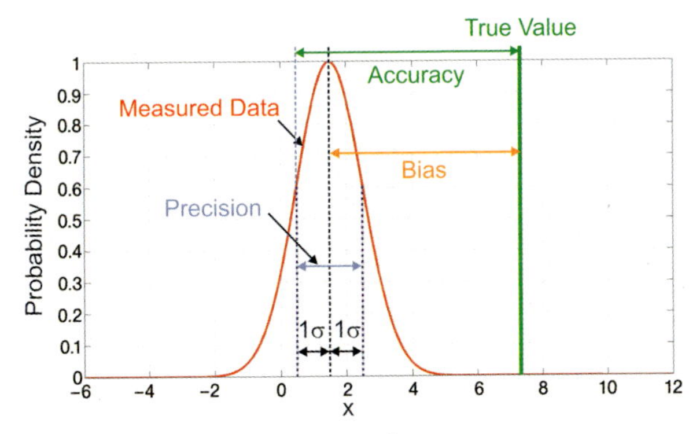

Figure 46-130: Probability density of a measurement process with respect to its average value (precision) and with respect to the true value (accuracy).

The true value of a measurement is, of course, unknown. However, the following methods can be used to estimate the true value with a residual uncertainty.

- Round robin tests of defined samples on different machines at different places and times carried out by different operators. The average is assumed to represent the true value.
- Calibration of the measurement systems by methods and algorithms capable of producing absolute results as in the case of the three-position technique, the shift-rotation technique, or the three-flat test.
- Simulation of the expected measurement deviation by the known or estimated contributions of individual components within the measurement system.

In many cases a lot of effort will be put into the estimation of the true value. Once a master piece has been determined, in an absolute sense, it can be used to calibrate and recalibrate appropriate measurement systems.

The overall variation of a measurement system in a measuring process has to be distinguished from the part-to-part variation (figure 46-131). Therefore, calibration standard elements are also suitable for use in the MSA process.

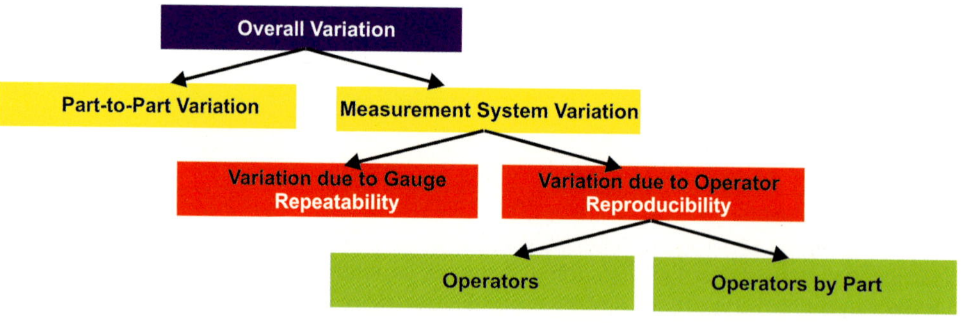

Figure 46-131: Overall variation of a measurement system (gauge) and its components, influenced by the gauge itself and the operators.

There are different MSA techniques that can be applied. A very common technique is the so- called Gauge Repeatability & Reproducibility technique (GR&R or Gauge R&R). If additionally an Analysis of Variance (ANOVA) is applied the technique is called ANOVA GR&R.

46.8.4 Gauge R&R Analysis – Analysis of Repeatability and Reproducibility

The GR&R analysis is a technique used to identify and quantify each component of variation, i.e., variations due to operators and those due to the parts [46-130].

Before the MSA can start the following parameters have to be defined:

y is a quality characteristic to be examined; for example, the coefficient of astigmatism in a measured wavefront,
J is the number of operators,
I is the number of parts,
K is the number of measurements per operator and part.

A result y_{ijk} achieved by operator j on part i at measurement number k can be separated into different components as shown

$$y_{ijk} = \mu + a_i + \beta_j + (a\beta)_{ij} + \varepsilon_{ijk} \quad (46\text{-}218)$$

where
μ is the unknown grand mean,
a_i are random effects of different parts, $i=1, ..., I$
β_j are random effects of different operators, $j=1, ..., J$
$(a\beta)_{ij}$ are random effects of the parts – operator interactions,
ε_{ijk} are random effects due to replication, $k=1, ..., K$.

We assume that the different random components are independent and follow a normal distribution according to

$$f(t) = \frac{1}{\sigma\sqrt{2\pi}} e^{-\frac{(t-\bar{t})^2}{2\sigma^2}} \quad (46\text{-}219)$$

where
$f(t)$ is the probalility density function,
t symbolizes one of the random-effect components,
$\bar{t}$ is the mean value of the random-effect component,
σ is the standard deviation of the random-effect component denoted by

$$\sigma = \sqrt{\frac{1}{N}\sum_{i=1}^{N}(t_i - \bar{t})^2} \quad (46\text{-}220)$$

Figure 46-132 shows one thousand normally distributed samples taken over a certain period of time. 68.3% of the samples lie within the interval 1σ, 95.45% lie within the interval 2σ, 99.73% lie within the interval 3σ.

If one part is selected to qualify GR&R the overall standard deviation is given by

$$\sigma_{\text{overall}} = \sqrt{\sigma_\beta^2 + \sigma_{a\beta}^2 + \sigma_\varepsilon^2} \quad (46\text{-}221)$$

σ_ε is called the repeatability,

$\sqrt{\sigma_\beta^2 + \sigma_{a\beta}^2}$ is called the reproducibility of the measuring system. The repeatability is the uncertainty among replications of k measurements of a given part made by one operator. The reproducibility is the uncertainty among operators for measuring the same part.

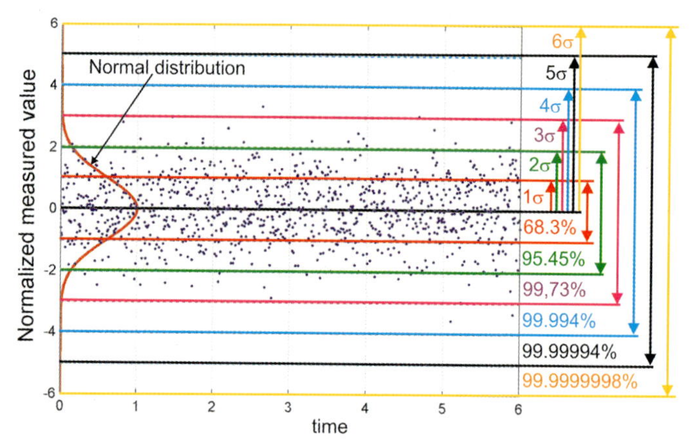

Figure 46-132: One thousand normally distributed samples taken over a certain time.

The analysis of variance method (ANOVA) is the most accurate method of quantifying repeatability and reproducibility. It allows the variability of the interaction between the operators and the parts to be determined [46-132]. The following five sums of squares need to be evaluated:

$$\mathrm{SSA} = \frac{1}{KI}\sum_{j=1}^{J}\left(\sum_{i=1}^{I}\sum_{k=1}^{K} y_{ijk}\right)^2 - \frac{1}{JKI}\left(\sum_{j=1}^{J}\sum_{i=1}^{I}\sum_{k=1}^{K} y_{ijk}\right)^2 \tag{46-222a}$$

$$\mathrm{SSB} = \frac{1}{JK}\sum_{i=1}^{I}\left(\sum_{j=1}^{J}\sum_{k=1}^{K} y_{ijk}\right)^2 - \frac{1}{JKI}\left(\sum_{j=1}^{J}\sum_{i=1}^{I}\sum_{k=1}^{K} y_{ijk}\right)^2 \tag{46-222b}$$

$$\mathrm{SSAB} = \frac{1}{K}\sum_{i=1}^{I}\sum_{j=1}^{J}\left(\sum_{k=1}^{K} y_{ijk}\right)^2 - \frac{1}{JKI}\left(\sum_{j=1}^{J}\sum_{i=1}^{I}\sum_{k=1}^{K} y_{ijk}\right)^2 - \mathrm{SSA} - \mathrm{SSB} \tag{46-222c}$$

$$\mathrm{TSS} = \sum_{i=1}^{I}\sum_{j=1}^{J}\sum_{k=1}^{K} y_{ijk}^2 - \frac{1}{JKI}\left(\sum_{j=1}^{J}\sum_{i=1}^{I}\sum_{k=1}^{K} y_{ijk}\right)^2 \tag{46-222d}$$

$$\mathrm{SSE} = \mathrm{TSS} - \mathrm{SSA} - \mathrm{SSB} - \mathrm{SSAB} \tag{46-222e}$$

From (46-222a–e) numerical values for repeatability, reproducibility, interaction and part variation can be calculated according to

$$\text{Repeatability} = k\sqrt{\text{MSE}} \tag{46-223a}$$

$$\text{Reproducibility} = k\sqrt{\frac{\text{MSA} - \text{MSAB}}{K\,I}} \tag{46-223b}$$

$$\text{Interaction} = k\sqrt{\frac{\text{MSAB} - \text{MSE}}{K}} \tag{46-223c}$$

$$\text{PartVariation} = k\sqrt{\frac{\text{MSB} - \text{MSAB}}{J\,K}} \tag{46-223d}$$

with

$$\text{MSA} = \frac{\text{SSA}}{J-1}$$

$$\text{MSB} = \frac{\text{SSB}}{I-1}$$

$$\text{MSAB} = \frac{\text{SSAB}}{(I-1)(J-1)}$$

$$\text{MSE} = \frac{\text{SSE}}{IJ\,(K-1)}$$

k is a constant and corresponds to the number of selected standard deviations within the variation range of the process. Usually $k = 5.15$ is selected which covers 99% of a normal population ($\sigma = \pm\ 5.15/2$), whereas $k = 6$ contains 99.73% of the measurements.

The precision (R&R) of the measurement system is then given by (46-224) and the total measurement system variation by (46-225).

$$\text{Precision} = \sqrt{\text{Repeatability} + \text{Reproducibility} + \text{Interaction}} \tag{46-224}$$

$$\text{Total System Variation} = \sqrt{\text{Precision} + \text{Part Variation}} \tag{46-225}$$

In the following, an example is given for an ANOVA GR&R evaluation of a measuring instrument. Three operators A, B and C are carrying out two sets of measurements of a certain characteristic on ten different parts. Their results are shown in table 46-8 and figure 46-133. The GR&R results for repeatability, reproducibility, interaction and part variation according to the equations above are graphically interpreted by figure 46-134.

Table 46-8: Measurement results for ten parts taken by operators A, B and C in two different runs.

Operator	Part 1	2	3	4	5	6	7	8	9	10
A	1.1	1.2	1	0.9	1.5	1.2	1.5	1.2	0.8	0.7
A	1.1	1.2	1.1	1	1.4	1.4	1.6	1	0.9	0.6
B	1.2	1.3	0.9	1	1.2	1.1	1.5	1.3	0.9	0.6
B	1.2	1.2	1.1	1.1	1.3	1.4	1.5	1.4	0.9	0.8
C	1.3	1.4	1	1.1	1.3	1.2	1.7	1.4	1	0.8
C	1.2	1.3	1.1	1.2	1.3	1.5	1.8	1.5	1.1	0.9

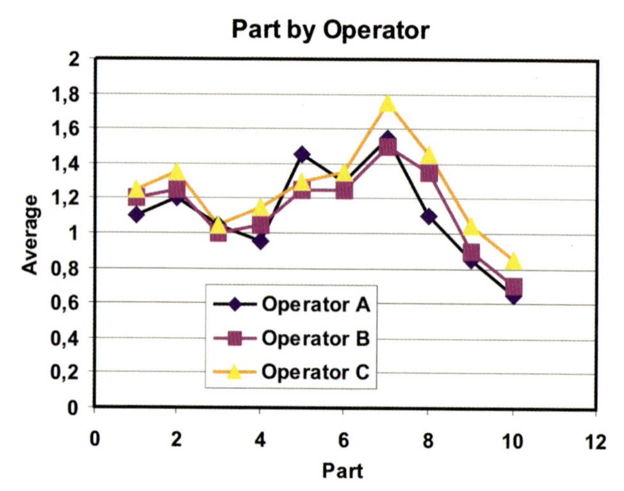

Figure 46-133: Diagram of measurement results from table 46-8 for ten parts taken by operators A, B and C in two different runs.

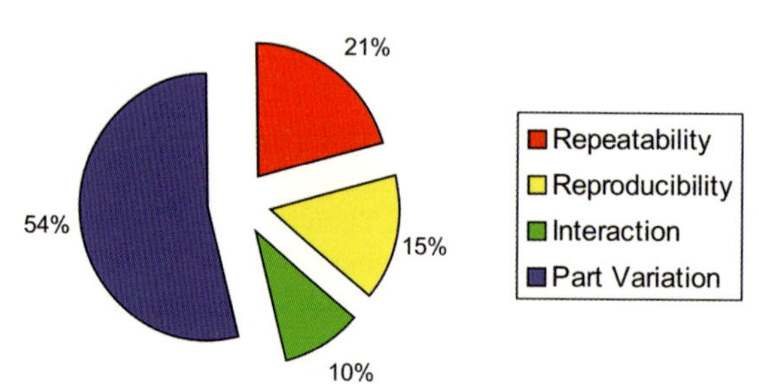

Figure 46-134: GR&R evaluation of the measurement results from table 46-8.

To check whether a measurement system is capable of measuring a given tolerance the *Gauge Capability Ratio* (GCR) or *Precision to Tolerance Ratio* P/T is calculated according to

$$P/T = \frac{k\sigma_{\text{overall}}}{\text{Upper Spec} - \text{Lower Spec}} \qquad (46\text{-}226)$$

The factor k in the nominator is chosen to be 5.15 or 6 covering 99% or 99.73% of the measurements, respectively.

The following rules of thumb can be applied:

$P/T \leq 0.1$ good capability,
$0.1 < P/T \leq 0.3$ acceptable capability,
$P/T > 0.3$ unacceptable capability.

In the latter case the measuring system is currently not capable of measuring the required tolerances. Some adjustment or calibration of the instrument may be necessary or the repeatability of the system is generally not able to resolve the given tolerances.

46.9 Literature

46-1 D. Malacara (Editor), Optical Shop Testing, 3rd Edition (John Wiley & Sons, Inc., Hoboken, New Jersey, 2007).

46-2 M. Born and E. Wolf, Principles of Optics (Pergamon, 1970).

46-3 I. N. Bronstein, K. A. Semendjajew, G. Musiol and H. Mühlig, Taschenbuch der Mathematik (Verlag Harri Deutsch, Frankfurt am Main, 2005).

46-4 M. Francon, Optical Interferometry (Academic Press, New York, 1967).

46-5 E. Hecht. and A. Zajac. Optics. p. 292. (Reading, MA: Addison-Wesley Publishing Co., 1974).

46-6 E. Hecht, Optik (Addison-Wesley (Deutschland), 1994).

46-7 K. Leonhardt, Optische Interferenzen (Wissenschaftliche Verlagsgesellschaft 1981).

46-8 W. H. Steel, Interferometry (Cambridge University Press, 1986).

46-9 P. Hariharan, Optical Interferometry (Academic Press, 2003).

46-10 A. Y. Karasik, Laser Interferometry Principles (CRC Press, 1995).

46-11 P. Hariharan, Basics of Interferometry (Academic Press, 1992).

46-12 K. J. Gasvik, Optical Metrology (John Wiley & Sons, Inc., Hoboken, New Jersey, 2002).

46-13 R. S. Sirohi, Wave Optics and Applications (Orient Longmans, Hyderabad, 1993).

46-14 R. S. Sirohi and F.S. Chau, Optical Methods of Measurement (Marcel Dekker, Inc., New York, Basel, 1999).

46-16 M. Bass (Editor), Handbook of Optics, Vol. II, Devices, Measurements, & Properties, 2nd Edition (McGraw-Hill Inc., New York (1995).

46-17 I. I. Dukhopel, Interference methods and instruments for inspecting optical flats, Soviet Journal of Optical Technology, vol. 38, no. 9, pp. 570-8 (1971)

46-18 I. I. Dukhopel and L. G. Fedina, Interferometric techniques and instruments for checking shapes of spherical surfaces, Soviet Journal of Optical Technology, vol. 40, no. 8, pp. 508–17, (1973).

46-19 L. Dabergerova, Contribution to the problem of construction of an interferometer for contactless testing of spherical optical surfaces,

Jemna Mechanika a Optika, vol. 18, no. 10, pp. 270–72 (1973).
46-20 B. J. Biddles, A non-contacting interferometer for testing steeply curved surfaces, Optica Acta, vol. 16, no. 2, pp. 137–57 (March 1969).
46-21 L. Swenson, The Ethereal Aether – A History of the Michelson Morley Aether Drift Experiment 1880 – 1930, University of Texas Press, Austin (1972).
46-22 F. Twyman, An interferometer for testing camera lenses, Phil. Mag., vol. 42, pp. 777–93 (1921).
46-23 F. Twyman, Prism and Lens Making, 2nd ed. Chapters 11 and 12 (Hilger and Watts Ltd., London, 1952).
46-24 J. Winckler, The Mach–Zehnder interferometer applied to studying an axially symmetric supersonic air jet, The Rev. of Sci. Inst. 19 (5) (1948).
46-25 P. Hariharan, Modified Mach–Zehnder interferometer, Appl. Opt. 8, pp. 1925–26 (1969).
46-26 E. D. Baird, A special interferometer for testing long laser rods, Applied Optics, vol. 9, no. 2, pp. 465–69 (1970).
46-27 B. Gebhard and C. P. Knowles, Design and adjustment of a 20 cm Mach –Zehnder interferometer, The Rev. of Sci. Inst., 37 (1), pp. 12–15 (1966).
46-28 K. Freischlad, Interferometer for optical waviness and figure testing, Proceedings of the SPIE – The International Society for Optical Engineering, vol. 3098, pp. 53–61 (1997).
46-29 V. P. Golubkova and V. P. Kononkevich, Portable interferometer for measuring the refractive index of Measurement Techniques, vol. 14, no. 8, pp. 1162–66 (1971).
46-30 R. N. Smartt and J. Strong, Point diffraction interferometer, Spring meeting of the Optical Society of America, pp. 31, 44 pp. Washington, DC, USA; New York, NY (1972).
46-31 K. A. Goldberg, R. Beguiristain, J. Bokor, H. Medecki, K. Jackson, D. T. Attwood, G. E. Sommargren, J. P. Spallas and R. Hostetler, At-wavelength testing of optics for EUV, Proceedings of the SPIE – The International Society for Optical Engineering, vol. 2437, pp. 347–54 (1995).
46-32 J. P. Spallas, R. E. Hostetler, G. E. Sommargren and D. R. Kania, Fabrication of extreme-ultraviolet point-diffraction interferometer aperture arrays, Appl. Opt. 34, no. 28, pp. 6393–98 (1995).
46-33 E. P. Goodwin and J. C. Wyant, Field Guide to Interferometric Optical Testing, (SPIE Press, 2006).
46-34 D. Malacara, A. Cornejo and M. V. R. K. Murty, A shearing interferometer for convergent or divergent beams, Boletin del Instituto de Tonantzintla, vol. 1, no. 4, pp. 233–39 (1975).
46-35 V. A. Gorshkov and V. G. Lysenko, Study of aspherical wavefronts on a lateral-shearing interferometer, Soviet Journal of Optical Technology, vol. 47, no. 12, pp. 689–91 (1980).
46-36 J. C. Fouere and D. Malacara, Generalized shearing interferometry, Boletin del Instituto de Tonantzintla, vol. 1, no. 4, pp. 227–32 (1975).
46-37 K. Patorski, Grating shearing interferometer with variable shear and fringe orientation, Appl. Opt. 25, no. 22, pp. 4192–98 (1986).
46-38 J. C. Wyant, White-light extended source ac shearing, Annual Meeting of the Optical Society of America, pp. 37, 88 (1972).
46-39 J. Choi, G. M.Perera, M. D. Aggarwal, R. P. Shukla and M. V. Mantravadi "Wedge-plate shearing interferometers for collimation testing: use of a Moiré technique" Appl. Opt. 34 3628–38 (1995).
46-40 J. E. Millerd, N. J. Brock, J. B. Hayes and J. C. Wyant, Instantaneous phase-shift point-diffraction interferometer, Proceedings of the SPIE – The International Society for Optical Engineering, vol. 5531, no. 1, p. 264 (2004).
46-41 K. Freischlad and M. Küchel, Speckle reduction by virtual spatial coherence, Proceedings of the SPIE – The International Society for Optical Engineering, vol. 1755, pp. 38–43 (1993).
46-42 M. Küchel, Reducing coherent artifacts in an interferometer, European Patent EP1390689 (2004).
46-43 D. Malacara, M. Servin and Z. Malcacara, Interferogram Analysis for Optical Testing, 2nd Edition, (CRC Press, 2005).
46-44 D. W. Robinson and G. T. Reid, Interferogram Analysis: Digital Fringe Pattern Measurement Techniques, Institute of Physics Publishing, pp. 94–140 (1993).

46-45 K. Creath, Comparison of phase-measurement algorithms, Proc. Soc. Photo-Opt. Instrum. Eng. 680, pp. 19–28 (1986).

46-46 K. Creath (Ed.) Phase-shifting interferometry techniques, Progress in Optics E.Wolf, Elsevier New York, vol. 26 pp. 357–73 (1988).

46-47 K. Creath and J. Schmit, N-point spatial phase measurement techniques for non-destructive testing, Opt. Las. Eng 24, pp. 365–79 (1996).

46-48 K. Creath and J. Schmit, Interferometry XI: Techniques and Analysis: vol. 4777, SPIE Society of Photo-optical Instrumentation Engineering (2002).

46-49 W. Osten, R. Pryputniewicz, G. T. Reid and H. Rottenkolber (Eds.), Automatic Processing of Fringe Patterns, Proceedings (Academie Verlag, Berlin, 1989).

46-50 W. Jüptner and W. Osten (Eds.), Fringe 1993, The 2nd International Workshop on Automatic Processing of Fringe Patterns, Proceedings, (Adademie Verlag, Berlin, 1993).

46-51 W. Jüptner and W. Osten (Eds.) Fringe 1997, The 3rd International Workshop on Automatic Processing of Fringe Patterns, Proceedings (Academie Verlag, Berlin, 1997).

46-52 W. Jüptner and W. Osten (Ed.), Fringe 2001, The 4th International Workshop on Automatic Processing of Fringe Patterns, Proceedings (Elsevier, Paris, 2001).

46-53 W. Osten (Ed.), Fringe 2005, The 5th International Workshop on Automatic Processing of Fringe Patterns, Proceedings (Springer, 2005).

46-54 G. M. Brown, Interferometry: Techniques and Analysis, Spie Proceedings, vol. 1755 (1993).

46-55 M. Takeda (Ed.), Laser Interferometry IX: Techniques and Analysis, Proceedings, Laser Interferometry IX (1998).

46-56 W. Osten, Interferometry XI: Applications: vol 4778, SPIE Society of Photo-Optical Instrumentation Engineering (2002).

46-57 R. Crane, Interference Phase Measurement, Applied Optics, Vo. 8, No. 3, (1969).

46-58 J. H. Bruning, D. R. Herriott, J. E. Gallagher, D. P. Rosenfeld, A. D. White and D. J. Brangaccio, Digital wavefront measuring interferometer for testing optical surfaces and lenses, Applied Optics, vol. 13, No. 11, (1974).

46-59 R. P. Grosso and R. Crane, Jr., Precise optical evaluation using phase measuring interferometric techniques, in Interferometry, G. W. Hopkins (Ed.) Proc. Soc. Photo-Opt. Instrum. Eng. 192, pp. 65–74 (1979).

46-60 C. Joenathan, Phase-measuring interferometry: new methods and error analysis, Appl. Opt. pp. 4147–55 (1994).

46-61 K. G. Larkin and B. F. Oreb, Design and assessment of symmetrical phase-shifting algorithms, J. Opt. Soc. Am. A 9, pp. 1740–48 (1992).

46-62 I. B. Kong and S. W. Kim, General algorithm of phase-shifting interferometry by iterative least-squares fitting, Opt. Eng. 34, pp. 183–88 (1995).

46-63 K. Kinstaetter, A. W. Lohmann, J. Schwider and N. Streibl, Accuracy of phase-shifting interferometry, Appl. Opt. 27, pp. 5086–89 (1988).

46-64 P. Hariharan, B. F. Oreb, and T. Eiju, Digital phase-shifting interferometry: a simple error-compensating phase calculation algorithm, Appl. Opt. 26, pp. 2504–06 (1987).

46-65 J. Schwider, R. Burow, K.-E. Elssner, J. Grzanna, R. Spolaczyk and K. Merkel, Digital wavefront measuring interferometry: some systematic error sources, Appl. Opt. 22, pp. 3421-32 (1983)

46-66 B. Dörband, Analyse optischer Systeme, Ph.D. Thesis, ITO Stuttgart (1986).

46-67 C. Ai and J. C. Wyant, Effect of piezoelectric transducer nonlinearity on phase-shift interferometry, Appl. Opt. 26, pp. 1116–16 (1987).

46-68 J. van Wingerden, H. J. Frankena and C. Smorenburg, Linear approximation for measurement errors in phase-shifting interferometry, Appl. Opt. 30, pp. 2718–29 (1991).

46-69 K. Creath and P. Hariharan, Phase-shifting errors in interferometric tests with high-numerical aperture reference surfaces, Appl. Opt. 33, pp. 24–25, (1994).

46-70 P. de Groot, Predicting the effects of vibration in phase-shifting interferometry, Optical Fabrication and Testing Workshop, OSA Vol., pp. 189–92 (1994).

46-71 P. de Groot, Vibration in phase-shifting interferometry, J. Opt. Soc. Am. 12, pp. 354–65 (1995).

46-72 P. de Groot, Phase-shift calibration errors in interferometers with spherical Fizeau cavities, Appl. Opt. 34, pp. 2856–63 (1995).

46-73 P. de Groot, Derivation of algorithms for phase-shifting interferometry using the con-

cept of a data-sampling window, Appl. Opt. 34, pp. 4723 –30 (1995).
46-74 K. Freischlad and C. L. Koliopoulos, Fourier description of phase-shifting interferometry, J. Opt. Soc. Am. A7, pp. 546–51 (1990).
46-75 F. J. Harris, On the use of windows for harmonic analysis with the discrete Fourier transform, Proc. IEEE 66, pp. 51–83 (1978).
46-76 M. Takeda, H. Ina and S. Kobayashi, Fourier-transform method of fringe pattern analysis for computer-based topography and interferometry, J. Opt. Soc. Am. 72, pp. 156–60 (1982).
46-77 M. Küchel, The new Zeiss interferometer, Proc SPIE 1332, pp. 655–63 (1990).
46-78 M. Küchel, Interferometer zur Messung von optischen Phasendifferenzen German patent DE 3707331 (1987).
46-79 B. Dörband and J. Hetzler, Characterizing lateral resolution of interferometers: the Height Transfer Function (HTF), Proceedings of the SPIE, Volume 5878, pp. 52–63 (2005).
46-80 P. de Groot and X. Colonna de Lega, Interpreting interferometric height measurements using the instrument transfer function, The 5th International Workshop on Automatic Processing of Fringe Patterns, Proceedings (Springer, 2005).
46-81 R. Smythe and R. Moore, Instantaneous phase measuring interferometry, Optical Eng. vol. 23, No. 4, p. 361 (1984).
46-82 C. L. Koliopoulos, Simultaneous phase-shift interferometer, SPIE Vol. 1531, pp. 119–27 (1991).
46-83 J. E. Millerd and N. J. Brock, Methods and apparatus for splitting imaging and measuring wavefronts in interferometry, US Patent No. 6,304,330 and 6,522,808 (2001).
46-84 D. C. Ghiglia and M. D. Pritt, Two-Dimensional Phase Unwrapping (John Wiley & Sons, Inc., Hoboken, New Jersey, 1998).
46-85 J. M. Huntley, Noise immune phase unwrapping algorithm, Appl. Opt. 28, pp. 3268–70 (1989).
46-86 V. V. Volkov and Y. Zhu, Deterministic phase unwrapping in the presence of noise, Optics Letters vol. 28, No. 22 (2003).
46-87 A. E Jensen, Absolute calibration method for Twyman–Green wavefront testing interferometers, J.O.S.A. 63, 1313A (1973).
46-88 G. Seitz and W. Otto, Method for the interferometric measurement of non-rotationally symmetric wavefront errors, Patent US 7277186 B2 (2000).
46-89 W. Otto, Method for the interferometric measurement of non-rotationally symmetric wavefront errors, Patent US 6839143 B2 (2000).
46-90 J. Grzanna and G. Schulz, Absolute testing of flatness standards at square-grid points, Optics Communications, vol. 77, no. 2–3, pp. 107–12 (1990).
46-91 J. Schwider, Ein Interferenzverfahren zur Absolutprüfung von Planflächennormalen II, Opt. Acta 14, pp. 389–400 (1967).
46-92 C. Ai and J. C. Wyant, Absolute testing of flats by using even and odd functions, Appl. Opt. 32, pp. 4698–705 (1993).
46-93 M. Küchel, A new approach to solve the three flat problem, Optik 112, 381–91 (2001).
46-94 U. Griesmann, Three-flat test solutions based on simple mirror symmetry, Appl. Opt. 45, No. 23 (2006).
46-95 G. Schulz, J. Schwider, C. Hiller and B. Kicker, Establishing an optical flatness, Appl. Opt. 10, no. 4, pp. 929–34 (1971).
46-96 J. Schwider, On the solution of the mirroring problem by interferometric testing of absolute flatness and sphericity, Optics Communications, vol. 5, no. 2, pp. 111–13 (1972).
46-97 G. Schulz and J. Schwider, Interferometric testing of smooth surfaces, Applied Progress in Optics XIII, pp. 93–167 (North Holland, 1976).
46-98 K.-E. Elssner, R. Burow, J. Grzanna and R. Spolaczyk, Absolute sphericity measurement, Appl. Opt. 28, pp. 4649–61 (1989).
46-99 R. E. Parks, C. J. Evans and L. Shao, Calibration of interferometer transmission spheres, Optical Fabrication and Testing Workshop, OSA Technical Digest Series 12, pp. 80–83 (1998).
46-100 E. B. Saff and A. B. J. Kuijlaars, Distributing many points on a sphere, The Mathematical Intelligencer 19, pp. 5–11 (1997).
46-101 C. J. Evans, M. Küchel and C. A. Zanoni, Apparatus and method for calibrating an interferometer using a selectively rotatable sphere, U.S. Patent 6,816,267 (2004).
46-102 U. Griesmann, Q. Wang, J. Soons and R. Carakos, A Simple Ball Averager for Reference Sphere Calibrations, Proc. of SPIE Vol. 5869 (2005).
46-103 R. Freimann and B. Dörband, Method for calibrating an interferometer apparatus, for

qualifying an optical surface, and for manufacturing a substrate having an optical surface, Patent US 7050175 B1 (2003).

46-104 K. L. Shu, Ray-trace analysis and data reduction methods for the Ritchey–Common test, Appl. Opt. 22, No. 12, pp. 1879–86 (1983).

46-105 R. E. Parks, C. Evans and L. Shao, Implementation of the Ritchey–Common Test of 300 mm Wafers, OSA Technical Digest Series, vol. 12, pp. 104–7 (1998).

46-106 Dörband, S. Schulte, F. Schillke and W. Wiedmann, Testing large plane mirrors with the Ritchey–Common test in two angular positions, SPIE Vol. 3739, pp. 330–34 (1999).

46-107 M. Bray, Stitching interferometry for large components, Second Annual International Conference on Solid State Lasers for Application to Inertial Confinement Fusion, Europto series, SPIE, Vol. 3047, Paris (1996).

46-108 M. Bray, Stitching interferometer for large plano optics using a standard interferometer, Optical Manufacturing and Testing II, SPIE, vol. 3134, San Diego (1997).

46-109 M. Bray, Stitching Interferometry: Recent Results and Absolute Calibration, Optical Fabrication, Testing, and Metrology, R. Geyl, D. Rimmer and Lingli Wang (Eds.), Volume 5252, Europto series (SPIE, Saint-Étienne, 2003).

46-110 P. Murphy, G. Forbes, J. Fleig, P. Dumas and M. Tricard, Stitching interferometry: A flexible solution for surface metrology, Optics & Photonics News 14, pp. 38–43 (2003).

46-111 G. Forbes, P. Murphy and J. Fleig, Stitching subaperture data for testing aspheric surfaces, Frontiers in Optics, OSA Technical Digest, paper OTuD5 (2004).

46-112 S. Chen, S. Li, Y. Dai and Z. Zheng, Iterative algorithm for subaperture stitching test with spherical interferometers, J. Opt. Soc. Am. A 23, pp. 1219–26 (2006).

46-113 C. G. Gordon, Generic Criteria for Vibration-Sensitive Equipment, Proceedings of International Society for Optical Engineering (SPIE), Vol. 1619, San Jose, CA, November 4–6, pp. 71–85 (1991).

46-114 A. Patel, Vibration-Free Optical Tabletops Are Critical in the Lab. (Biophotonics International Laurin Publishing Co. Inc., 2000).

46-115 Technical information from CVI Melles Griot, Fundamentals of Vibration Isolation, www.mellesgriot.com (2008).

46-116 M. C. Roggemann and B. Welsh, Imaging Through Turbulance, CRC-Press, 1st edition (1996).

46-117 B. Edlén, The refractive index of air, Metrologia 2, pp. 71–80 (1966).

46-118 B. Edlén, Equation for the refractive index of air, Metrologia 2 (2), pp. 71–80 (1966).

46-119 F. E. Jones, The refractivity of air, J. Res. Natl. Bur. Stand. (U.S.) 86 (1), p. 2730 (1980).

46-120 P. Giacomo, Equations for determination of the density of moist air, Metrologia 18, pp. 33–40 (1982)

46-121 K. P. Birch, F. Reinboth, R. E. Ward, and G. Wilkening, The effect of variations in the refractive index of industrial air upon the uncertainty of precision length measurement, Metrologia 30, pp. 7–14 (1993).

46-122 K. P. Birch and M. J. Downs, An updated Edlen equation for the refractive index of air, Metrologia 30, pp. 155–162 (1993).

46-123 K. P. Birch and M. J. Downs, Correction to the updated Edlén equation for the refractive index of air, Metrologia 31, pp. 315–16 (1994).

46-124 P. E. Ciddor, Refractive index of air: new equations for the visible and near infrared, Appl. Optics 35, pp. 1566–73 (1996).

46-125 G. Bönsch and E. Potulski, Measurement of the refractive index of air and comparison with modified Edlén's formulae, Metrologia 35, pp. 133–139 (1998).

46-126 L. Cremer, Vorlesungen über Technische Akustik (Springer Verlag, Berlin, Heidelberg, New York, 1975).

46-127 G. R. Hopkinson and D. H. Lumb, Noise reduction techniques for CCD image sensors, Journal of Physics E (Scientific Instruments), vol. 15, no. 11, pp. 1214–22 (1982).

46-128 ISO 10110-7, Optics and photonics – Preparation of drawings for optical elements and systems – Part 7: Surface Imperfection Tolerances (2008).

46-129 T. A. Germer and C. C. Asmail, A goniometric optical scatter instrument for bi-directional reflectance distribution function measurements with out-of-plane and polarimetry capabilities, Proceedings SPIE 3141, pp. 220–31 (1997).

46-130 D. W. Hoffa and C. Laux, Gauge R&R: An effective methodology for determining the adequacy of a new measurement system for micron-level metrology, Journal of Industrial Technology, vol. 23, Number 4 (2007).

46-131 R. K. Burdick, C. M. Borror and D. C. Montgomery, A review of methods for measurement systems capability analysis, Journal of Quality Technology, vol. 35 Issue 4, p. 342 (2003).

46-132 D. P. Mood, Explaining ANOVA: A Teaching Hint, Measurement in Physical Education & Exercise Science, Vol. 4 Issue 3 (2000).

46-133 AIAG (Automotive Industry Action Group), Measurement Systems Analysis (MSA), Document No. MSA-3, 3rd edition, 2nd printing, (2003).

46-134 H. Gross, Handbook of Optical Systems, vol. 1, Fundamentals of Technical Optics, Wiley-VCH, Weinheim (2005).

46-135 H. Gross, H. Zügge, M. Peschka, F. Blechinger, Handbook of Optical Systems, vol. 3, Aberration Theory and Correction of Optical Systems, Wiley-VCH, Weinheim (2007).

47
Non-interferometric Wavefront Sensing

Handbook of Optical Systems: Vol. 5. Metrology of Optical Components and Systems. First Edition.
Edited by Herbert Gross.

47.1 Introduction

The interferometric measurement of a wavefront is the most popular and sensitive method of measurement. But several different methods are proposed in the literature, which can be of benefit for tasks with smaller requirements or for special conditions. Some of the most important methods within this class are described in this chapter. In particular, the Hartmann–Shack wavefront sensor and the point-spread function phase retrieval methods are considered in more detail here. Several other techniques are also described shortly. Currently, the Hartmann–Shack sensor is preferred for many measurement tasks due to its ease of use. This method is a modification of the classical Hartmann setup, and therefore it is also represented in considerable detail. Another method, which is not well known in the community is the phase space analyzer. This idea is more fully described in section 47.4 since only a few papers can be found about this interesting method in the literature. It is especially advantageous to use this approach for laser beam quality assessment.

The point spread function retrieval method, has been increasingly developed to produce a very useful technique over the last 20 years. Comprehensive calculations are needed to reconstruct the phase from the caustic data. Therefore, in recent times, large computers are used and the method is of growing interest. One of the main benefits of this method is that the setup is quite simple.

In the last section, we cover the reconstruction of the wavefront from measured gradient data and the calculation of the Zernike aberration coefficients. One point, which, in practice, is not considered sufficiently, is the strong dependence of the Zernike coefficients on the correct normalization radius of the pupil or the diameter of the measured beam.

47.2 Hartmann–Shack Sensor

47.2.1 Principle of the HS Sensor

The Hartmann–Shack wavefront sensor (HS-WFS) can be used to analyze wavefronts [47-1]–[47-4]. It uses a lenslet array and is a further development of the classical Hartmann setup described in section 47.3.

The wave to be inspected is decomposed by a two-dimensional microlens array into a matrix of separate subapertures. This is shown in figure 47-1. The lenses are made refractive or diffractive. Each of the small lenses focuses the corresponding subarea of the wave and the position of the resulting spots is analyzed [47-2]–[47-4]. The geometrical distortion of the grid of focal points seen on a screen provides information about the wave aberrations.

The centroid position of each spot is proportional to the inclination of the average wavefront in the corresponding subaperture and can be evaluated accordingly. The spots will be detected by a two-dimensional sensor, which determines the centroid position in both x- and y-coordinates, and these positions are then analyzed. Figure 47-2 shows a typical spot pattern of a sensor and a three-dimensional picture of the array device. The diameter of the beam is adapted to the size of the sensor, typically a fill factor of 70% – 90% is desired. The shape of the subapertures is quadratic. Circular apertures and hexagonal grid geometries are also possible, but the simplest arrangement is cartesian. One further advantage of this geometry is the disappearance of any dead area on the sensor, when the array is suitably manufactured.

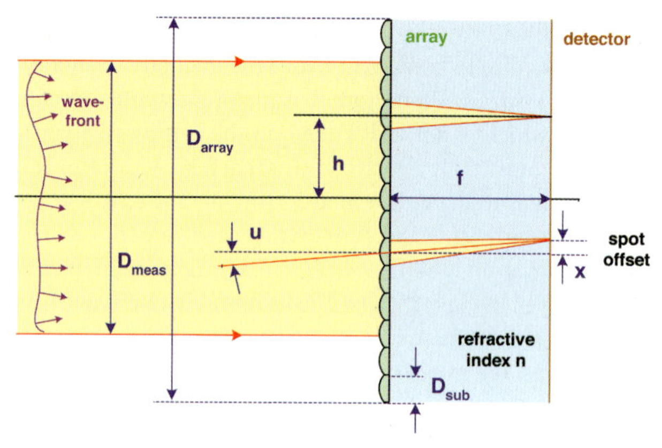

Figure 47-1: Notations and principle of operation of the Hartmann–Shack wavefront sensor.

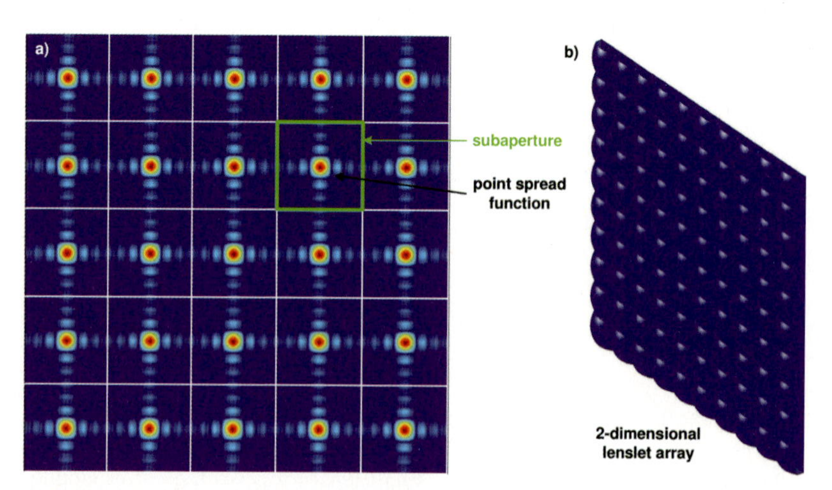

Figure 47-2: Geometry of the Hartmann–Shack wavefront sensor along the optical axis (a). The individual point spread functions of the quadratic subapertures extend to the boundaries of the subapertures in the x- and y-direction, respectively. Part b) shows the array component from a different perspective.

The following quantities are important when describing the system (see figure 47-1):

f: the focal length of the lenslets
N: the number of subapertures in one dimension
D_{meas}: the diameter of the quasi-collimated input beam
D_{sub}: the diameter of a single subaperture
η: the relative illumination factor of the array (filling factor)
P: the pixel size of the sensor

Occasionally a monolithic design is chosen for stability reasons and the microlens array is tightly cemented onto the detector by a glass or a quartz plate. In this case the corresponding index of refraction has also to be taken into account in the focusing distance.

The typical image produced by a wavefront sensor can be seen in Figure 47-3.

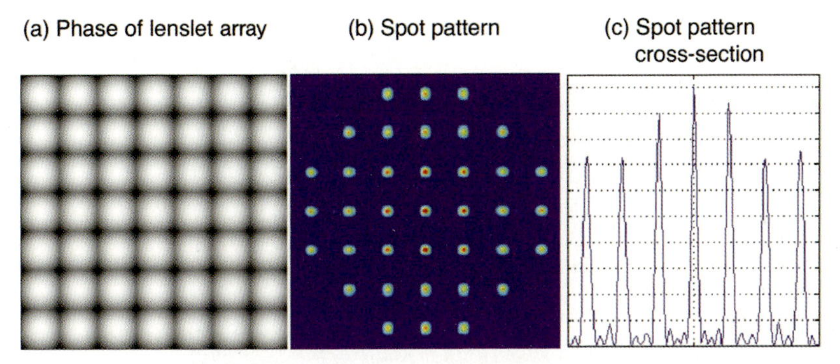

Figure 47-3: Hartmann–Shack wavefront sensor: (a) array phase; (b), spot pattern and (c) cross-section of the spot intensities over the diameter.

If the incoming beam suffers from wave aberrations, there will be a deformation in the grid of the spots in the image plane. In particular, for some simple aberration types, one observes characteristic patterns such as those shown in Figure 47-4. The yellow lines should help to identify the deformation of the grid lines in the spot pattern. The deviations form a rigid cartesian shape describe the centroid offsets due to the aberrations.

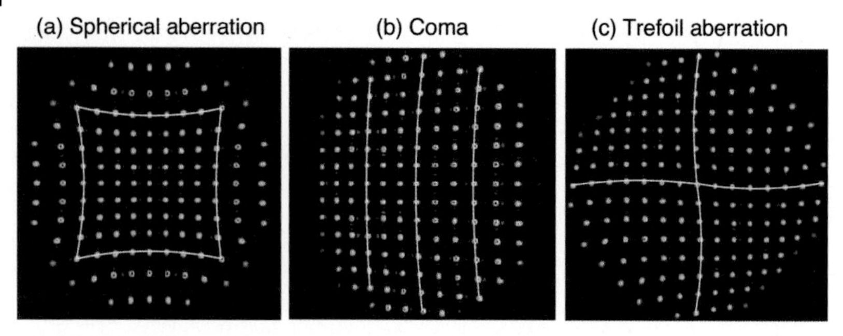

Figure 47-4: Hartmann-Shack wavefront sensor: Spot patterns for selected aberrations are shown. In part a) we consider a spherical aberration; in b) a coma and in c) a trefoil aberration.

The picture of a Hartmann–Shack wavefront sensor signal, taken from measurements, is shown in figure 47-5.

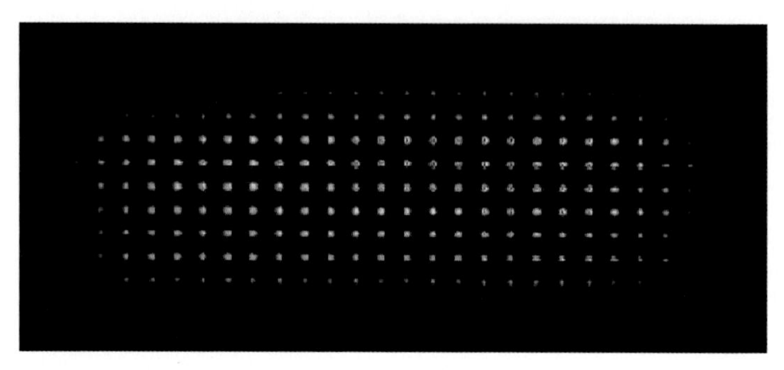

Figure 47-5: Measured picture of a Hartmann–Shack spot pattern.

In reality, the effects of discretization, quantization and noise will change the spot pattern on the camera notably [47-5]. This is illustrated in figure 47-6.

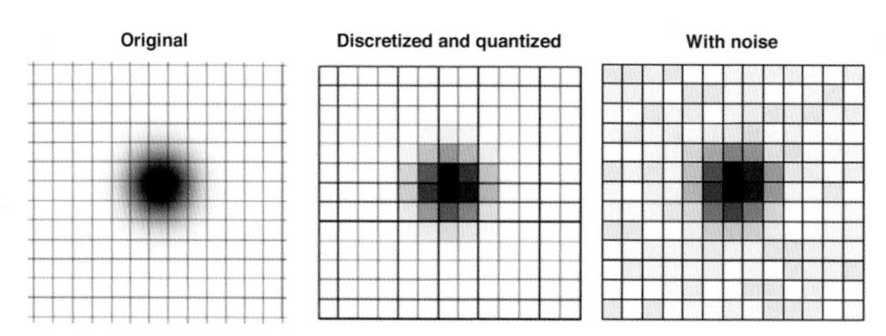

Figure 47-6: Real Hartmann–Shack sensor illumination spot with discretization and noise.

In comparison with other methods for the measurement of wavefront aberrations, the Hartmann–Shack sensor has some special properties, which are listed here:

1. The measurement is almost independent of the degree of coherence. Partially coherent beams can be analyzed as well as coherent beams. In the case of partial coherence, the spot generated by every subaperture is larger and the dynamic range is reduced.
2. The spectral sensitivity is primarily dependent on the properties of the CCD sensor. In principle, all wavelengths, which can be detected with the sensor, can be measured. Depending on the realization of the lenslets, which are mostly refractive, a small dispersion takes place and changes the zero-order parameters. These effects can be controlled by calibration. Furthermore, the material of the array usually has a finite interval of spectral transmission.
3. There is no dependence on the polarization properties of the incoming light.

There are many good contributions in the literature, where the theory of the Hartmann–Shack sensor is described and can be understood from the viewpoint of Fourier optics [47-6]–[47-9]. There are also very special arrangements for Hartmann–Shack sensors which are reported in the literature. A cylindrical array for measuring only in one direction, phased arrays with non-constant array chirp or adaptable focal lengths of the lenslets, obtained by realizing the lenses as liquid crystal devices [47-10] have been proposed.

In addition to these there are different complete evaluation schemes for data processing in a Hartmann–Shack sensor. In [47-9], a special method is published, where the explicit calculation of the centroid coordinates is not necessary and the reintegration of the wavefront from the derivative data is carried out in the Fourier domain.

The accuracy of the Hartmann–Shack sensor is lower, but lies almost in the range of interferometric measurement techniques. There are two different categories of errors, which must be distinguished.

1. The micro accuracy is determined by the precision of the spot centroid evaluation in relation to the pixel size of the detector.
2. The macro accuracy is caused by the discretization of the wavefront into a finite number of subapertures, over which the wave is averaged. The error due to this aspect strongly depends on the shape of the wavefront to be measured

47.2.2
Basic Setup

The sample surface can be tested in autocollimation using a simple setup consisting of a laser or a fiber source for illumination, a beam-splitter, a telescope for matching the beam diameters and an HS-WFS. The almost-ideal common-path portion permits calibration of the sensor to very high accuracy, and the use of a difference signal makes it possible to compensate for residual aberrations introduced by the beam-shaping optics.

It is important for the layout of the setup, as schematically shown in figure 47-7, to achieve a proper image formation of the telescope pupils. The sample surface should be imaged exactly into the camera plane and the array should be located in the pupil plane. The wavefront to be measured must be propagated into the sensor plane with the help of a phase-conserving 4f arrangement [47-11]. For the illumination, an almost point-like source is preferred to avoid the smoothing convolution of the direction spectrum when the source is of finite size.

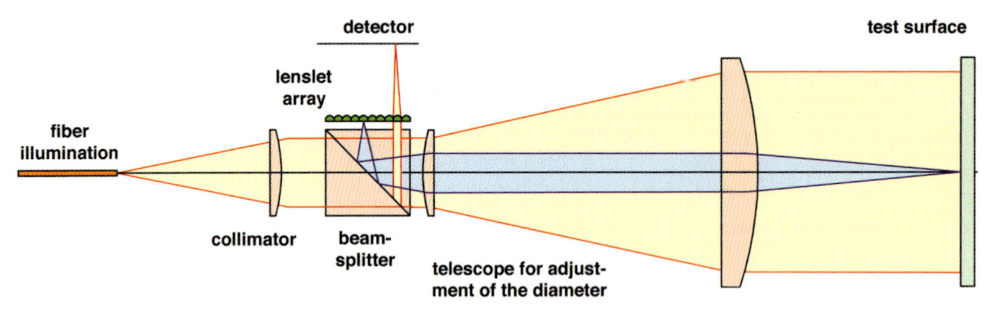

Figure 47-7: Setup for testing of optical components in reflection by a Hartmann-Shack wavefront sensor. The telescope is of the 4f-type to preserve the wavefront shape except for a scaling factor.

Figure 47-8 shows a more general setup for the measurement of a wavefront with a Hartmann–Shack sensor. A stop in the Kepler-type adapting telescope helps to suppress stray light and an additional relay lens between the array and the sensor adapts the size of the sensor for a given array diameter

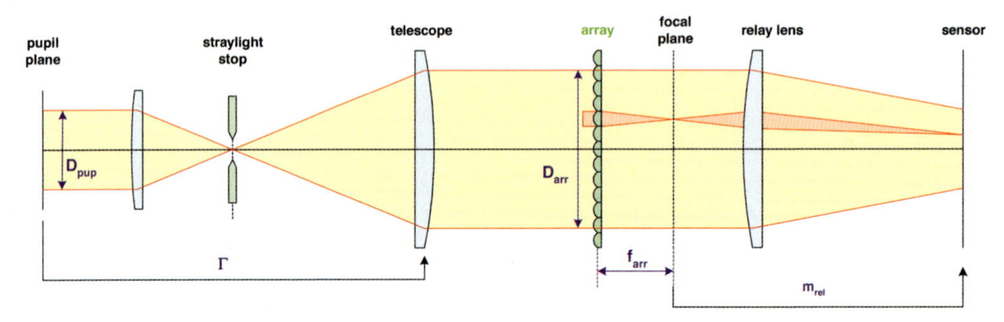

Figure 47-8: Generalized setup for testing the optical components in transmission by a Hartmann–Shack wavefront sensor. The telescope contains a stray light- suppressing stop. Between the array and the sensor, an additional relay lens is located.

47.2.3
Telescope for Diameter Adaptation

In practice, the pupil plane of the lens to be characterized is very often not directly accessible. Furthermore, it is an advantage to use the sensor diameter on a large portion about 80% of the full size. In this case, nearly all the pixel are used and the resolution reaches the estimated size. Therefore, a system is necessary which will adapt the size of the wavefront diameter onto the sensor and which will guarantee that the wavefront is measured in the correct plane. This system is an afocal telescope, which must preserve the shape of the wavefront. For this purpose, either a 4f-system without any change in the diameter or a telescope with focal lengths of f_1 and f_2 is necessary [47-11]. The telescope system is shown in figure 47-9.

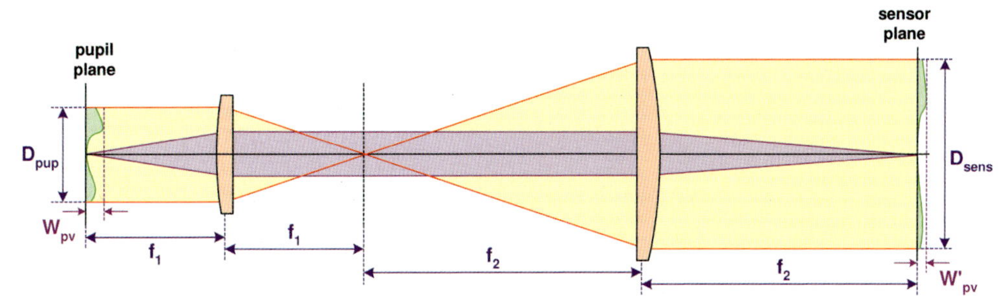

Figure 47-9: Telescope system for the adaptation of the diameters of the pupil and lenslet array.

If the magnification of the telescope is given by Γ, the diameter of the nearly plane wave scales as

$$\Gamma = -\frac{f_1}{f_2} = -\frac{a'}{a} = -\frac{D_{\text{pup}}}{D_{\text{sens}}} \tag{47-1}$$

The size of the phase simultaneously scales inversely. If the diameter is enlarged, the wave aberration decreases by the same factor.

$$W_{\text{sens}} = |\Gamma|\; W_{\text{pup}},\; D_{\text{sens}} = \frac{D_{\text{pup}}}{|\Gamma|} \tag{47-2}$$

In reality, the measured wavefront also contains the residual errors of the imaging telescope. The pupil aberration of the system causes a distortion of the spot pattern and therefore influences the results [47-12]. If the layout of the telescope is telecentric, then in order to achieve quite a good insensitivity for small changes of magnification for slightly defocused pupil locations, the diameter of the system becomes large and the correction of the lens plays a significant role. In this case it is necessary to calibrate these residual aberrations to obtain a good accuracy.

47.2.4
Detector Relay Lens

If an additional relay lens is introduced in the standard setup between the lenslet array and the detector, which images the focal plane of the array with a magnification

$$m_{\mathrm{rel}} = \frac{\Delta x'}{\Delta x} \tag{47-3}$$

then the sensitivity of the sensor system can be changed, because the minimum detectable spot shift is affected and the pixel size remains constant [47-13]. This is shown in figure 47-10.

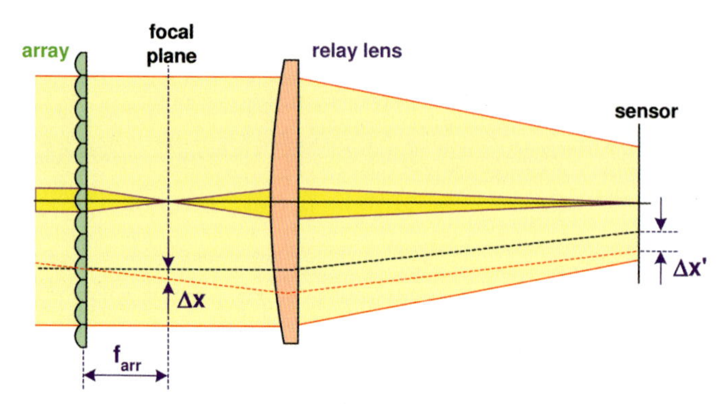

Figure 47-10: Effect of a relay lens between the lenslet array and sensor.

47.2.5
Layout of a Sensor

The following relationships between the phase surface in the exit pupil and the transverse aberration of the spot in the focal plane of the test optics are satisfied [47-13]

$$\Delta x = -\frac{f}{n} \cdot \frac{\partial W}{\partial x} \qquad \Delta y = -\frac{f}{n} \cdot \frac{\partial W}{\partial y} \tag{47-4}$$

It can be seen from these equations, that a refractive index $n > 1$ between the array and the sensor influences the result of the measurement. The Fresnel number of an individual subaperture is defined as

$$N_{\mathrm{F}} = \frac{D_{\mathrm{meas}}^2}{4\lambda f\, N_{\mathrm{sub}}^2 \eta^2} = \frac{D_{\mathrm{sub}}^2}{4\lambda f} \tag{47-5}$$

with the filling factor η according to the equation

$$\eta = \frac{D_{meas}}{D_{array}} \tag{47-6}$$

and the number of subapertures

$$N_{sub} = \frac{D_{array}}{D_{sub}} \tag{47-7}$$

The spot size on the detector is given, assuming uniform illumination of the subapertures and perfect coherence, by

$$D_{spot} = \frac{2\lambda f}{D_{sub}} = \frac{D_{sub}}{2N_F} = \frac{2\lambda f\, N_{sub}\, \eta}{D_{meas}} \tag{47-8}$$

The normalized spot size relative to the subaperture length is then expressed as

$$\frac{D_{spot}}{D_{sub}} = \frac{1}{2\, N_F} \tag{47-9}$$

For a pixel size P and a centroid position determination precision of the order of $k = 1/100$ pixels (this implies a ratio of the spot diameter to the pixel size exceeding approximately 3.5), the angular accuracy of the system is given by

$$\theta_{min} = \frac{k\, P}{f} \frac{m_{rel}}{\Gamma} \tag{47-10}$$

Here m_{rel} is the magnification of the relay lens and Γ is the angle magnification of the preceding adapting telescope. If D_{array} is the diameter of the complete array component, then for the measurable wavefront error one obtains

$$W_{min} = \theta_{min}\, D_{array}\, N_{sub} \tag{47-11}$$

Typically an accuracy of λ / 350 can be achieved. The maximum angle allowed is given by the limitation that the spots should not spread beyond the area assigned to each of them (the region of interest, ROI)

$$\theta_{max} = h \frac{D_{sub}}{2f\,\Gamma^2} \tag{47-12}$$

in this equation, h is a factor with the usual value $h = 1$ for the conventional setup. If special procedures are introduced to enhance the dynamic range by appropriate software algorithms, the value is $h > 1$.

In practice, the only real parameters for the layout of a Hartmann–Shack sensor are the focal length f of the lenslets and the number of subapertures N_{sub}. The goal in an optimization is usually high accuracy, a good spatial resolution and a large dynamic range. If the accuracy is considered, two different definitions must be distinguished: a very precise value of the wave aberration slope at the center of one subaperture, or alternatively a high precision matrix of values across the whole sensor area. Figure 47-11 shows the qualitative behavior of the spot diameter as a function of focal length and lenslet array number [47-14], [47-15]. The curves correspond to lines of constant spot diameter. If the focal length is enlarged, this increases the sensitivity of the result. If the number of subapertures increases, this also increases the spatial resolution. Along the diagonal direction of the chart the accuracy of the measurement increases, while in the opposite direction the dynamic range is enlarged.

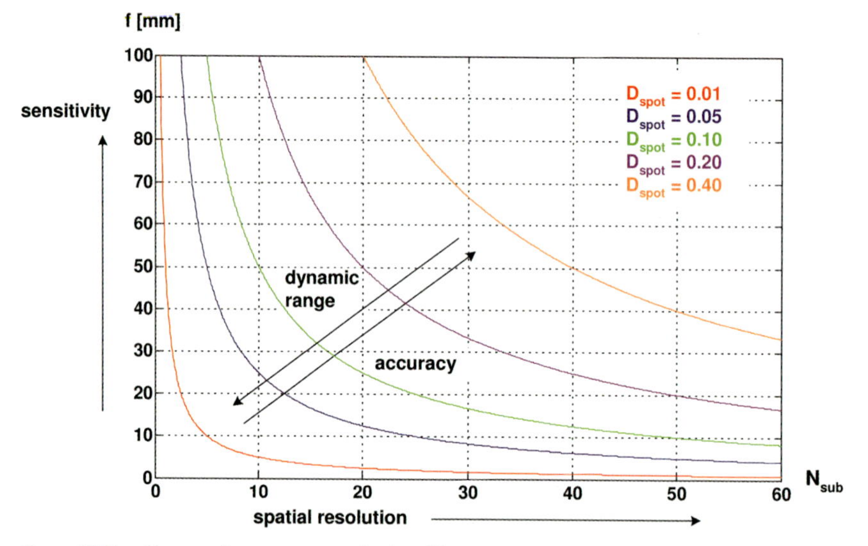

Figure 47-11: Curves of constant spot size in a Hartmann–Shack sensor as a function of the focal length and the number of subapertures. D_{spot} is given in mm.

The ratio between the largest possible and the smallest detectable angle is the dynamic range

$$R = \frac{\theta_{max}}{\theta_{min}} = \frac{h\, D_{sub}}{2kP\, m_{rel}\, \Gamma} \tag{47-13}$$

As can be seen, this is a purely geometrical consideration and in reality is modified by the finite-sized point spread function.

In a similar diagram, the principal degrees of freedom for the layout of a sensor system are very illustrative. Figure 47-12 again shows the f-N_{sub} parameter range. There is only a small band-shaped area for parameter combinations of the sensor which make sense. The lower limit is given by the minimum spot size to get the correct spatial resolution, the upper limit is given by the maximum spot size due to cross-talk problems. Both curves can be deduced from the above formulas. A minimum spatial resolution of the wavefront sampling is assumed to be attained for 10 subapertures and defines a left boundary in the diagram [47-16].

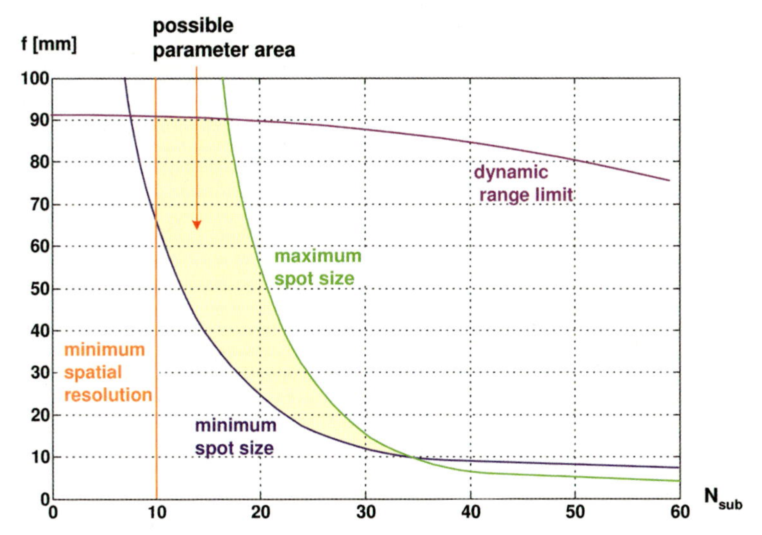

Figure 47-12: Free ranges for the layout parameters f and N_{sub} of a Hartmann–Shack sensor.

47.2.6
Signal Processing

One of the major tasks in the evaluation of measurement data is an accurate determination of the centroids of the individual spots. If the spot size is very small, the pixel size of the sensor limits the precision of the centroid calculation. If the Fresnel number of the lenses is made small so that the spot size increases, the calculation is then much more accurate, but the dynamic upper limit shrinks due to a large extension of the spot inside the subaperture area without allowing large shifts.

Usually, the calculation of the centroids follows the well known simple relationships [47-17], [47-18]:

$$
\begin{aligned}
x_c &= \langle x \rangle = \frac{\iint x\, I(x,y)dxdy}{\iint I(x,y)dxdy} = \frac{1}{P_0} \iint x\, I(x,y)dxdy = \frac{\sum\limits_{mn} x_{mn} I_{mn}}{\sum\limits_{mn} I_{mn}} \\
y_c &= \langle y \rangle = \frac{\iint y\, I(x,y)dxdy}{\iint I(x,y)dxdy} = \frac{1}{P_0} \iint y\, I(x,y)dxdy = \frac{\sum\limits_{mn} y_{mn} I_{mn}}{\sum\limits_{mn} I_{mn}}
\end{aligned}
\tag{47-14}
$$

where P_0 is the power of the beam inside the subaperture. In principle, if a high spatial resolution is obtained in the sensor plane, it is also possible to evaluate the mixed moment $<xy>$ as well as the second-order moments $<x^2>$ and $<y^2>$ of the spot. This allows a more detailed analysis of the spots. For example, the ellipticity of the spot and a broadening of the diameters delivers information on the local curvature of the wavefront and allows an estimation of the residual coherence of the beam inside a subaperture. If the lateral coherence length L_c is not very large in comparison with the subaperture diameter, a partial coherent focusing behavior will be found for a larger spot size.

In reality, the best results for the centroid calculation are obtained when a threshold value for the intensity is defined and the evaluation is done only for pixels above this value [47-8]. There are also quite different algorithms used to determine the centroid position [47-19] and [47-20]. One possibility, which gives very accurate results, is to calculate the correlation of the actual spot with a gaussian-shaped peak function of the same width. Iterative algorithms have been published very recently and appear extremely advantageous [47-21].

If at least 50 or more detector pixels are available for the evaluation of the centroid, an accuracy of typically 1/100 of the pixel size can be achieved for the calculation of the spot position [47-22].

47.2.7 Dynamic Range

The ratio between the largest and the smallest detectable wavefront aberration is defined as the dynamic range of the sensor. The smallest size is given by the resolution of the spot centroid resolution on the sensor and is limited by the pixel size and the underlying algorithm used to localize the centroid as described above. The largest detectable wavefront is given by the extreme elongation of the point spread function, which leaves the subaperture area and then cannot be further distinguished from the neighboring spot. Figure 47-13 shows this case. If the finite size of the point spread function is taken into account, the rough estimation of (47-12) must be refined to the following expression:

$$\theta_{\max} = \frac{D_{\text{sub}} - D_{\text{spot}}}{2f\,\Gamma^2} h \tag{47-15}$$

Hence, together with (47-10) one obtains for the dynamic range of the sensor

$$R = \frac{\theta_{\max}}{\theta_{\min}} = \frac{D_{\text{sub}} - D_{\text{spot}}}{2k\,P\,m_{\text{rel}}} h = \frac{D_{\text{sub}}\,h}{k\,P\,\Gamma\,m_{\text{rel}}} \left(N_f - \frac{1}{2} \right) \leq 4000 \tag{47-16}$$

It should be noted that, in this simple consideration we have assumed that the spot has a compact shape and its size can be defined by a simple number. In reality, we have a diffraction pattern with a complicated intensity distribution and no clearly defined outer margin.

In principle, there are at least two important reasons for enlarged spot sizes in comparison with the above simple consideration.

1. If the light source is non-ideal, the spot is broadened by partial coherence.
2. If the inclination angle of the wave is of a considerable size, the spot suffers from coma and astigmatism and is therefore broadened in comparison with the ideal case.

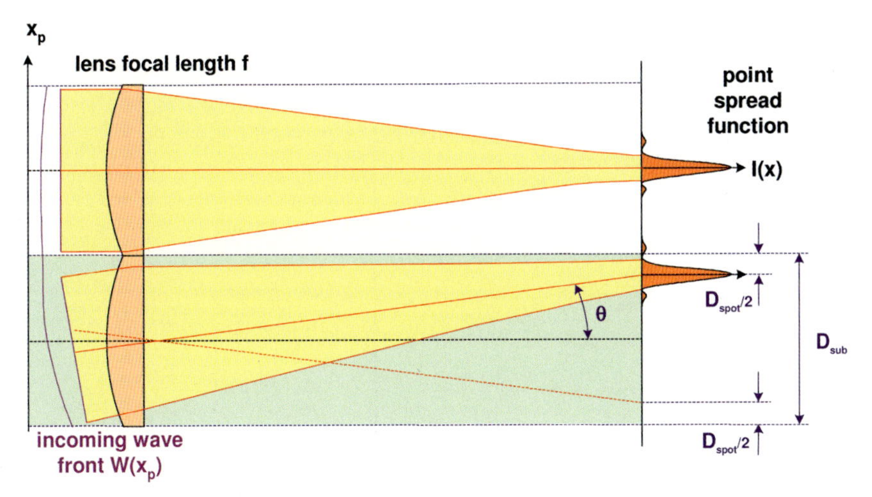

Figure 47-13: The limit of the dynamic range of the Hartmann–Shack wavefront sensor is attained when the point spread functions leaves the corresponding area of a subaperture.

47.2.8 Subaperture Effects

Cross-talk

The size of a spot is given by

$$D_{\text{spot}} = \frac{D_{\text{sub}}}{2N_{\text{F}}} \tag{47-17}$$

The larger the Fresnel number N_F, the smaller the point spread function of the spot due to diffraction at the boundary of the subaperture. If, on the other hand, the Fresnel number is small, the spot has a larger extension. Since the spot is usually diffraction limited, the shape of the spot intensity profile is given by a $(\sin x/x)^2$ behavior. This distribution shows quite a badly converging oscillating side lobe behavior. Therefore, the correct computation of the spot centroid may be a problem. One one hand, it is not easy to determine the region of interest, inside which the centroid is evaluated. On the other hand, it is possible that a part of the point spread function has contributed to a neighboring subaperture and caused an error in the calculation of the spot centroid for both subapertures. This effect increases when the spots are strongly elongated, that is, near the limit of the largest possible wavefront aberration. This corresponds to a decreasing accuracy with increasing spot movement.

Figure 47-14 shows the result of a simulation example for a wavefront sensor. Only one transverse dimension is taken into account. There are two different correction types considered. In the first case, only the nearest-neighbor effect is considered, while in the second case, an intensity cross-talk from all possible subapertures is taken into account.

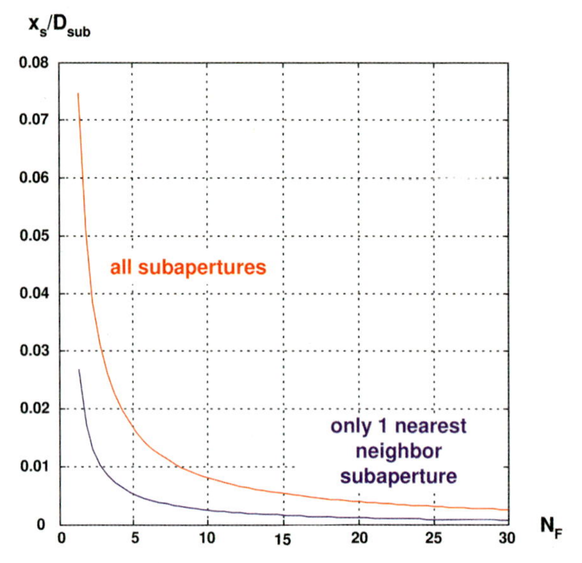

Figure 47-14: Influence of the intensity cross-talk from neighboring subapertures onto the centroid calculation of a wavefront sensor as a function of the Fresnel number.

If the local radius of curvature of the detected wavefront is equal to the focal length of the lenses of the array, then in a simplified geometrical consideration, the spots meet one another at the boundary between two subapertures. This is the largest curvature which can be resolved by the sensor. In a more accurate picture, the finite size of the point spread function decreases the value of this radius of curvature. The condition for this, more exact, case is

$$R_{\min} = f \frac{D_{\text{sub}}}{D_{\text{sub}} - D_{\text{spot}}} = \frac{f}{1 - \frac{1}{2N_F}} \tag{47-18}$$

Figure 47-15 shows this case, together with the purely geometrical consideration, for illustration.

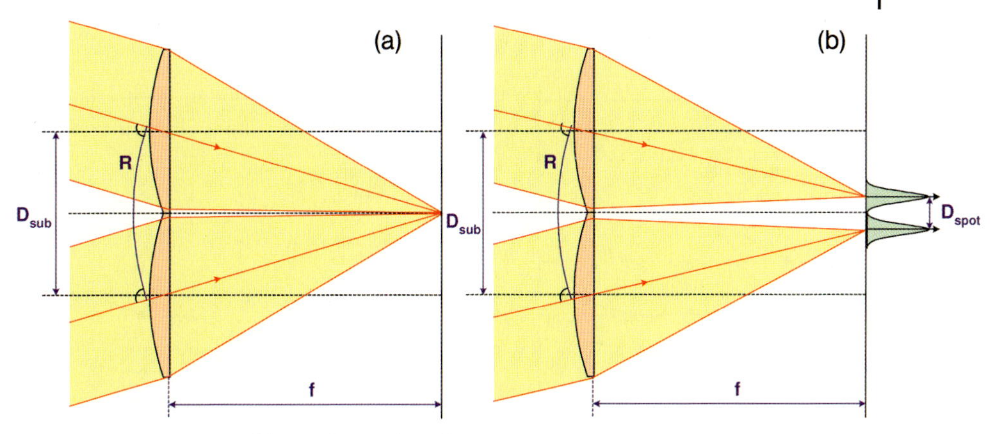

Figure 47-15: Estimation of the smallest radius of curvature, which is detectable with a Hartmann–Shack sensor. Part a) shows the simplified geometrical consideration for a vanishing size of point spread function. In b), the finite size is taken into account.

Subaperture Averaging

The Hartmann–Shack sensor uses a discrete mesh grid to analyze the incoming wavefront. The spots and their centroids belong to a subaperture of finite size and all changes inside this subaperture are averaged. If high frequent wavefront errors are to be detected, this causes a low pass filtering effect. Therefore, this averaging effect causes some systematic errors, which increase with the order of the Zernike coefficients. In figure 47-16, some results are compiled which demonstrate this effect. A system is considered, which has a wavefront of exactly one Zernike polynomial of purely radial type and order n. $n=2$ corresponds to the defocus term, $n=4$ to the primary spherical aberration and so on. The wave is analyzed with a Hartmann–Shack wavefront sensor with a variable number of subapertures N_{sub}. In the figure, the absolute error of the Zernike coefficient, the relative error of the Zernike coefficient, the rms value of the wavefront and the *pv*-value of the wavefront are shown in parts a) – d). It can be seen that the error of the measurement decreases with increasing number of subapertures, which corresponds to smaller areas of the subaperture. On the other hand, the errors are larger for higher orders of the Zernike polynomials because they contain higher spatial frequencies, which are smoothed by the low pass filtering effect. It should be noted, that N_{sub} is the one-dimensional number of subapertures, the total number is N_{sub}^2.

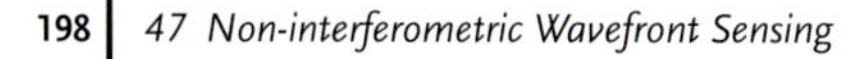

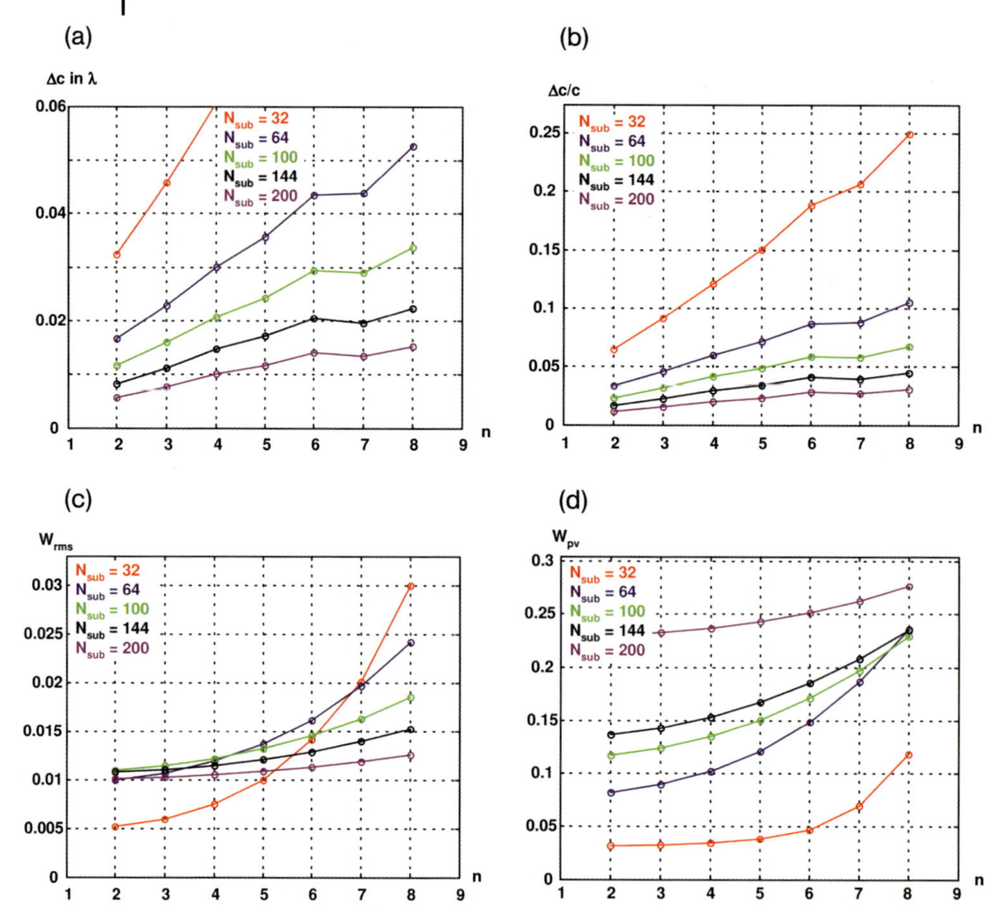

Figure 47-16: Effect of averaging over the subapertures. The result of a calculation for different numbers N_{sub} of subapertures and the effect on the error of the Zernike coefficients in λ is shown in (a); the effect on the relative error of the Zernike coefficients is shown in (b); that on the rms value of the wave aberration in (c) and the peak-valley value of the wave aberration is shown in (d). The abscissa n is the radial order of the Zernike polynomial.

Figure 47-17 shows a very similar result, here only three spherical Zernikes are considered, the error of the coefficients is shown as a function of N_{sub}. It can be seen in a more quantitative way, that a required residual error allows one to determine the minimum necessary number of subapertures.

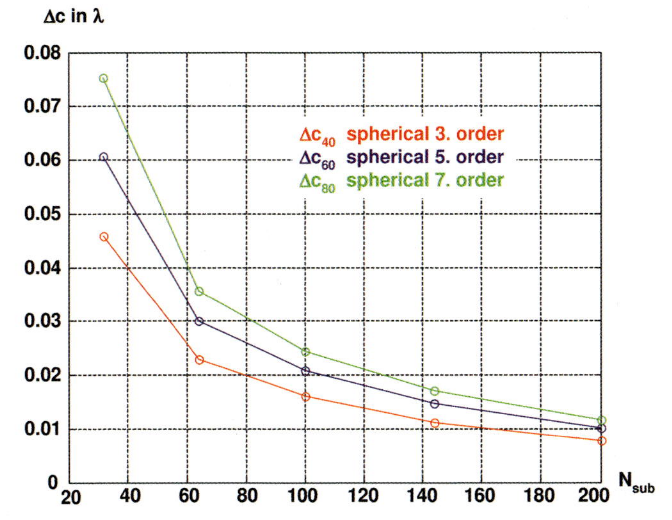

Figure 47-17: Effect of averaging over the subapertures. Shown is the result of a calculation of the error of the lower order spherical Zernike coefficients in λ as a function of the number of subapertures N_{sub}.

Partly Illuminated Subapertures

One special problem occurs at the margin of the illuminated area of the wavefront sensor array. Usually, it is desirable to take a range of 70–90% of the effective area for the signal beam. Otherwise, many subapertures are not used and the spatial resolution is reduced. At the margin of the illuminated region, there are subapertures which are not fully illuminated but which take some energy. This effect is illustrated in figure 47-18.

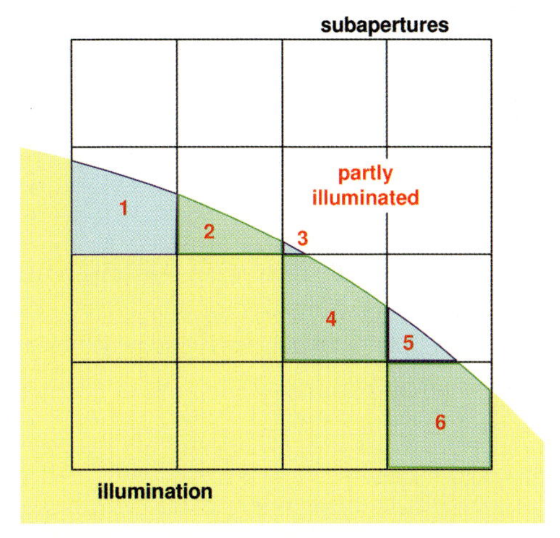

Figure 47-18: Schematic drawing of a marginal region of a wavefront sensor with partly illuminated subapertures (numbers 1-6).

Figure 47-19 shows the result when determining the centroid of a partly illuminated subaperture. The green ray represents the centroid line, which does not pass through the center of the lens. Due to the imaging properties of a typical lens, in the focal plane the lateral offset of the centroid vanishes and produces an exact centroid position. Therefore, the signal from the marginal subapertures can also be used to analyze the wavefront. However as can be seen in the figure, this condition is perturbed if the detector lies in a defocused plane at a distance distance $z \neq f$.

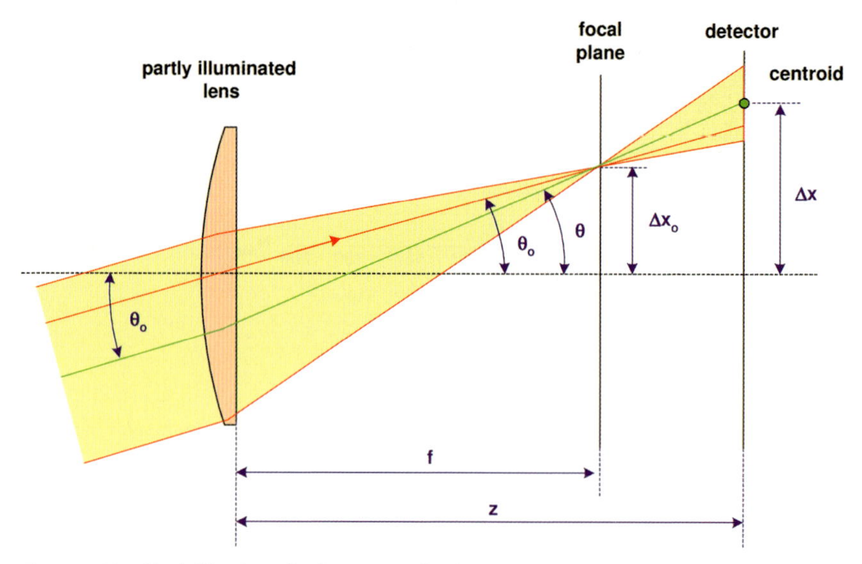

Figure 47-19: Partly illuminated subaperture of a Hartmann–Shack wavefront sensor. In the focal plane, the centroid position of the reduced area is correct.

Figure 47-20 shows three examples of the point spread function of a partly illuminated single subaperture for different truncation geometries. In part a), only 1/4 of the area is illuminated in the y-direction, the spot is correspondingly broadened. In case b), half of the area is illuminated, the effect of a) is reduced. In the third case c), an illumination of 50% is performed along the diagonal direction. In all three geometries, the centroid of the point spread function is conserved.

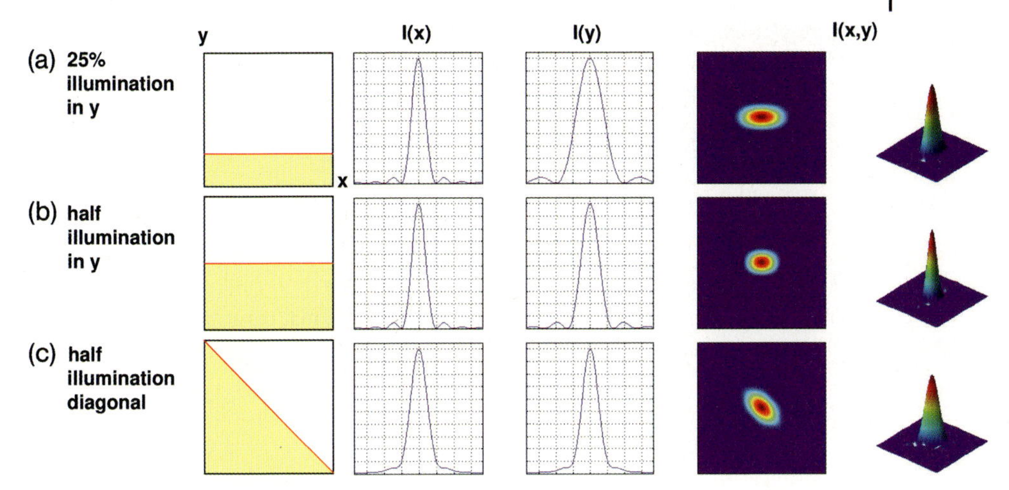

Figure 47-20: Partly illuminated subaperture of a Hartmann–Shack wavefront sensor. In the focal plane, the centroid of the reduced area is correct but the shape of the spot changes.

A similar question arises when a beam is measured that shows an apodization with arbitrary illumination distribution. The sensor also works well in this case, but it should be noted that the diffraction spreading of the small partly illuminated subapertures increases, the dynamical range may be reduced and cross-talk effects can take place.

47.2.9
Accuracy of the HS Sensor

Sources of Error

There are several sources of possible error in the wavefront analysis when using a Hartmann–Shack sensor. The most important aspects are:

1. Centering error of the setup lenslet-array sensor. Here an azimuthal rotation can take place or a tilt in the x- or y-direction can occur.
2. Statistical distributions in the geometry of the array lenses. Differences in the focal lengths of the lenses will influence the accuracy of the measurement. Lateral geometrical errors with the lenslet centers must also be considered.
3. Averaging the wavefront inside the area of a subaperture. This is an inherent principal error of this method.
4. For larger local wavefront tilt angles, the lenslets are used under oblique conditions. This causes coma and influences the centroid determination of the spot.
5. Noise, which is generated by small photon fluences, or which arises in the electronics of the sensor.
6. Errors arising in the reconstruction of the wavefront from the derivative components.

Centering Errors

In the following, the centering tolerances are considered in more detail. This representation follows [47-23]. There are four different types of centering errors in the system of the array and the detector. These are

1. Lateral offset: δx , δy
2. Axial displacement: δz
3. Rotation around the axes x, y: α, β
4. Rotation around the optical axis: γ

The errors of the first kind are interpreted as additional contributions to the centroid offset and cause a tilt error in the wavefront of the size

$$\Delta W_{\delta x, \delta y} = \frac{x \delta x + y \delta y}{f} \tag{47-19}$$

They can be calibrated and therefore are not a severe problem in practice. If an axial shift occurs, we get a scaled sensor distance by a constant factor of the size

$$\Delta W_{\delta z} = \left(1 + \frac{\delta z}{f}\right) W \tag{47-20}$$

which can also be removed by calibration. In practice, the location of the detector can be outside the exact focal plane of the array lenses. This happens when different wavelengths are used for the detection and the chromatic axial aberration of the lenses causes some deviations in the z-direction. Figure 47-21 shows this setup, the sensor is located in a plane at a distance $z > f$ from the principal plane of the lenses. As can be seen in the figure, the determination of the centroid coordinates can still be used for the calculation of the wavefront tilt in the subaperture. However, due to the change in the distance, the relation between the angle and the offset coordinates must be recalibrated.

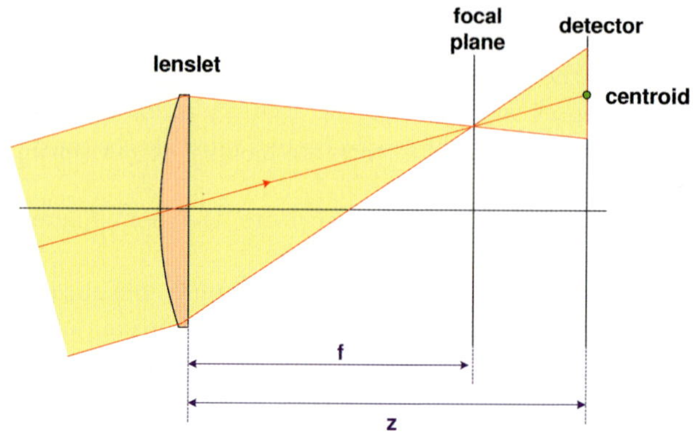

Figure 47-21: Hartmann–Shack sensor with a detector in a defocused plane.

In practice, typically Fresnel numbers in the range 2 –10 are used to get a suitable compromise between accuracy and the largest measurable wavefront aberration. In this range, the low-NA case is realized. Under these conditions, the smallest size of the focused beam is no longer found in the back focal plane of the lens. Therefore, it is a critical task to adjust the position of the sensor relative to the array. In addition, in practice it is usual to assemble the two components in a monolithic structure by filling the space between them with a glass plate and cementing all three parts together. In this case, the change of the back focal length has also to be taken into account.

If there is an azimuthal rotation angle γ between sensor and array, the conditions of integrability for the wave field gradient components

$$W_x = \frac{\partial W}{\partial x}, \; W_y = \frac{\partial W}{\partial y} \tag{47-21}$$

read as

$$\frac{\partial W_y}{\partial x} = \frac{\partial W_x}{\partial y} \tag{47-22}$$

and will no longer be fulfilled. This causes problems in the reconstruction algorithms, which assume the validity of this equation.

If a rotation around the y-axis with angle β takes place, for small angles we obtain a correction term

$$W_{x,\beta} = \frac{x}{f}\left(\frac{1}{\cos\beta} - 1\right) + \frac{1}{\cos\beta}\frac{\partial W}{\partial x} \qquad W_{y,\beta} = \frac{\partial W}{\partial y} \tag{47-23}$$

If P denotes the subpixel accuracy of the centroid determination, a tolerance can be deduced from this equation for the largest allowable tilt angle β

$$\beta_{\max} = \arccos\frac{D_{\text{arr}}}{D_{\text{arr}} + 2P} \tag{47-24}$$

The accuracy of the determination of the spot centroids is one of the most important aspects for the final accuracy of the method. For this calculation, the following points are important:

1. The discrete pixel structure of the sensor.
2. The quantization of the power levels of the sensor.
3. The Poisson statistics of photon noise.
   This is an effect which is especially important when the number of subapertures is large and, correspondingly, the area of one subaperture is small. If N_{phot} denotes the photon count, a standard deviation of the centroid calculation caused by this can be estimated according to (47-24)

$$\sigma_{xc}^{(\text{square})} = \frac{0.74 \cdot \pi}{D_{\text{sub}} \sqrt{N_{\text{phot}}}} \tag{47-25}$$

4. Specification of the region of interest, from which the centroid calculation is performed. In practice, this includes a truncation effect. According to the representation in [47-24], the optimal size of the evaluation regime is given by

$$D_{\text{trunc}} = \frac{3 D_{\text{sub}}}{2\lambda} \tag{47-26}$$

In [47-25], a simple model with a spot of gaussian shape is considered. The error of the calculated centroid position depends on the absolute size of the spot and the number of detector pixels per subaperture. The corresponding behavior is shown in figure 47-22. It can be seen that the error grows significantly when the spot size is smaller than one pixel. Then a range of considerably small errors occurs. This range of operation is broader for larger values N of pixels per subaperture D_{sub}. The error then increases again due to the cross-coupling of the pattern into the neighboring subapertures.

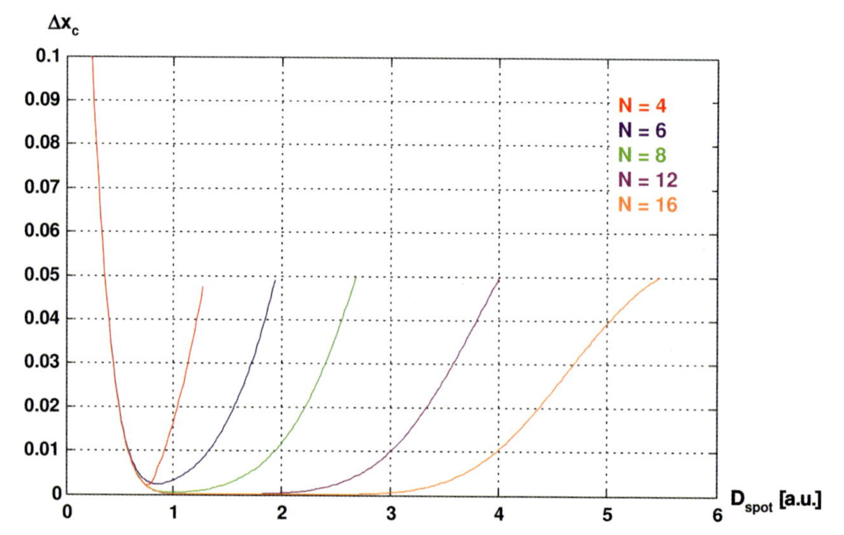

Figure 47-22: Error for the calculation of the centroid of a gaussian spot Δx_c profile (in pixels) as a function of the absolute spot size (in pixels) for different values of the underlying number N of pixels per subaperture.

Calibration

If a point source generates a nearly perfect spherical wave and the distance between the sensor and source is changed, a calibration of the sensor can be performed. The result of such a measurement is shown in figure 47-23. The reconstructed radius of curvature forms a straight line as a function of the distance z [47-26], [47-27]. Using this procedure, a sensor accuracy of approximately $\lambda/60$ peak-valley or $\lambda/500$ rms can be achieved.

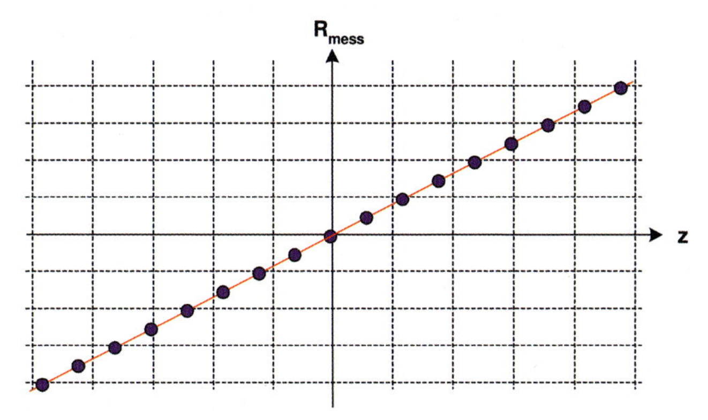

Figure 47-23: Calibration of a Hartmann–Shack sensor by measuring a spherical wave with increasing distance.

Another possible way to calibrate the Hartmann–Shack sensor is by the use of a well-known wedge [47-8].

47.2.10 Modified Setups and Algorithmic Extensions of the Sensor

There are several possible extensions of the simple Hartmann–Shack sensor proposed in the literature to extend the dynamic range. The largest detectable wavefront aberration can be increased by the following special treatments.

1. Extended algorithm for the spot identification [47-28].
   If a more sophisticated algorithm is used, a type of unwrapping can be carried out and the ambiguous occurrence of several spots in the area range of one subaperture can be easily affiliated. This method extends the dynamic range of the Hartmann–Shack sensor extensively. It remains the limitation of a large curvature with a local radius of curvature in the range of the focal length as discussed in section 47.2.8
2. Sequential measuring of subapertures [47-29].
   If a spatial light modulator with adapted lateral resolution is placed in front of the micro lens array, it is possible to switch only one subaperture at a time on and to assign the spot to the transmissive subaperture independently of its location. A better process speed is obtained if special patterns of M subapertures are opened simultaneously. This corresponds to a kind of superaperture with a size of M^2 subapertures. In total, the M^2 measurement must be made in sequence.

A special simple use of this technique masks the neighboring subapertures around one measuring subaperture [47-30]. This increases the dynamic range limitation due to overlap by a factor of 2. If a special masking device is used, which cannot be placed in close contact with the sensor array, a real optic is necessary to generate an additional conjugate plane for the mask.

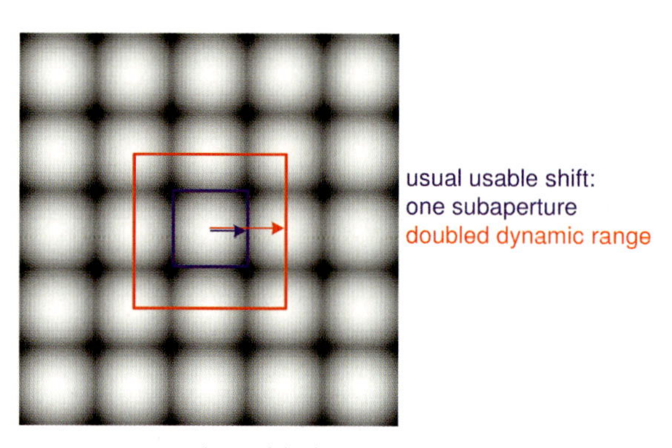

Figure 47-24: Masking of the lens subapertures increases the dynamic range.

3. Astigmatic lens shapes in the array [47-31], [47-32].
   In this special case, the shape of the individual lenslets is made rather differently. A small astigmatic power is added to the spherical part. The azimuthal angle of the principal axes of the astigmatic spot is changed a certain amount from lens to lens. With this kind of signature, the spots of neighboring subapertures can be distinguished, if the mixed moment $\langle xy \rangle$ is evaluated in addition to the first-order moments of the centroid.
4. Measuring the spots in an intermediate plane [47-33].
   If the spots are detected also in an intermediate plane between the sensor and lenslet array, a shift of the spots can be registered unambiguously and the assignment of the spots can be performed easily.
5. Use of cylindrical arrays [47-34].
   If two cylindrical lens arrays are used instead of a two-dimensional array, two separated sets of patterns for the x- and the y-deviation are obtained. The transverse aberrations are coded in deformed lines, which allow for a simple unwrapping by following the lines in the algorithmic evaluation. The cost of this advantage is a division of the energy and additional components in the setup.

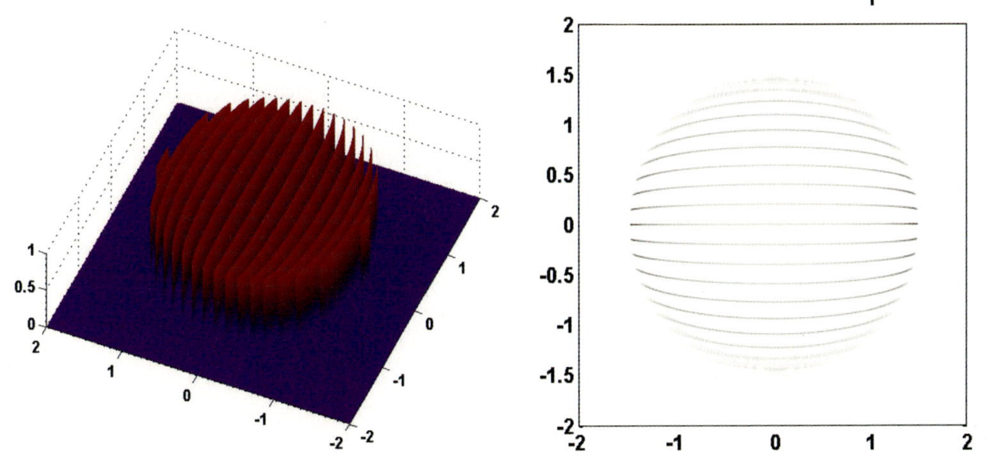

Figure 47-25: Using a cylindrical lens array for one of the transverse aberration components, a set of lines is obtained, from which the unwrapping can be easily calculated.

5. Array with super-resolution filters [47-35].
   In a special modification of the classical Hartmann–Shack wavefront sensor, an additional array with a super-resolving mask for every subaperture is placed close to the sensor array. If the lateral gain of the mask is given by G_T and the Strehl ratio is S, a gain in the centroid variance is given by

   $$R = S^2 \, G_T^4 \tag{47-27}$$

   A simple realization of the mask is a central obscuration with factor 0.75. The remaining ring aperture delivers an increase in sensitivity of approximately 2.1

6. Integrated solution without lenses [47-36].
   If a spot pattern of very small pinholes (approximately 6 μm for visible applications) is used and the sensor distance is also quite small (20 μm), a special compact integrated chip solution of the Hartmann–Shack sensor without lenses can be used. In the near field of the small pinholes, the lateral displacement is obtained and can be detected. This kind of sensor also resembles the classical Hartmann setup and works in the same way as a photon sieve.

47.2.11 Comparison with Interferometer Setup

In table 47-1, a short comparison is given between some of the major properties of a Hartmann–Shack sensor and an interferometer of the Fizeau type with a phase-shifting principle. Corresponding representations can also found in the literature which are very similar [47-37]. The table gives only a very rough qualitative comparison. It can be seen that there are pros and cons for both alternatives. Depending on the precise conditions and requirements of the measurement, either a Hartmann–Shack or an interferometric setup will be the better choice.

Table 47-1: Comparison of a Hartmann–Shack wavefront sensor, with a phase-shifting Fizeau-type interferometer.

	Feature	HS-WFS	PSI-interf.
1	Complexity of the setup	+	
2	Cost of the equipment, additional high-quality components	+	
3	Spatial resolution		+
4	Robust measurement under environmental conditions	+	
5	Noise due to image processing and pixelated sensor		+
6	Perturbation by coherent scattering and straylight	+	
7	Algorithm for reconstructing the wavefront		+
8	Test systems with central obscuration		+
9	Speed of measurement	+	
10	Absolute accuracy		+
11	Measurement possible independently of wavelength, polarization and coherence	+	

47.3 Hartmann Sensor

47.3.1 Introduction

The Hartmann test to measure wavefront aberration is quite an old technique used to qualify optical components and systems [47-38]. The method is quite simple: a diaphragm with a large number of holes is located at two different z-positions after the illumination light passes the component under test. The small pinholes behave almost like rays and the transverse ray deviations can be used to obtain information about aberrations. The classical Hartmann setup is described comprehensively in the literature [47-3], [47-4], [47-39], [47-40]. If the measurement is performed very carefully, the accuracy of the method is quite high and comparable to interferometry [47-41].

The major difference between the classical Hartmann test and the Hartmann–Shack sensor is that the aperture is structured by a pinhole mask. The spot pattern produced by the resulting beam array is recorded in two measuring planes (plane 1 and plane 2). A hole centered at a height y transmits a narrow beam whose heights in the two measuring planes are y_1 and y_2, respectively. The locations of the measuring planes, which should lie in front of and behind the focal position, are denoted by s'_1 and s'_2, respectively. The axial intercept distance corresponding to the incidence height y is then given by (see figure 47-26)

$$s'_y = s'_1 + (s'_2 - s'_1)\frac{y_1}{y_1 + y_2} \tag{47-28}$$

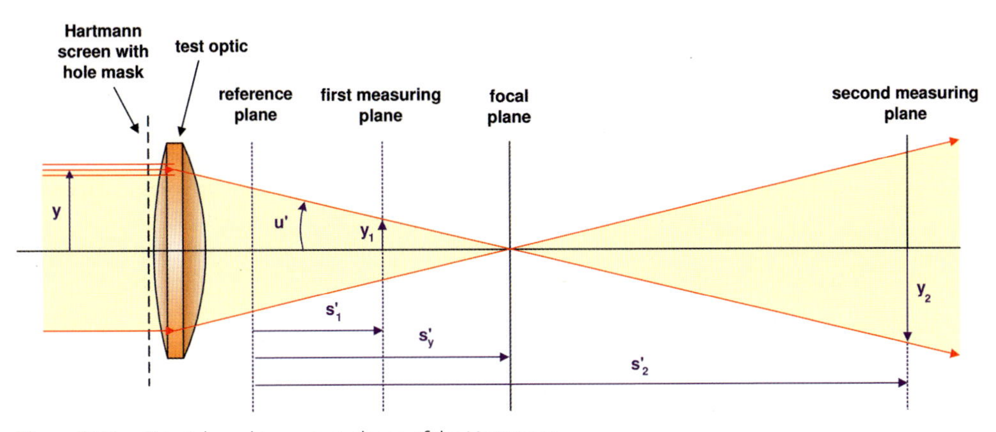

Figure 47-26: Principle and important planes of the Hartmann measuring setup.

On axis the dependence $s'_y(y)$ yields the longitudinal spherical aberration. The transverse aberrations can be obtained from the deviations of $\Delta s'_y$ from a reference image plane assigned to the aberration-free case and integration will then yield the wave aberrations. In the case of rotational symmetry, the equation

$$W'(y) = \int_{u'=0}^{u'_{max}} \Delta s'_y(y) \sin u' \, du' \tag{47-29}$$

holds. In the general case, the goal of the Hartmann test is to determine lateral deviations Δx and Δy in a pre-selected adjustable plane. From the values corresponding to the two measuring planes we get

$$\Delta y = y_1 + (y_2 - y_1)\frac{s'_y - s'_1}{s'_1 + s'_2} \tag{47-30}$$

and

$$\Delta x = x_1 + (x_2 - x_1)\frac{s'_x - s'_1}{s'_1 + s'_2} \tag{47-31}$$

It is reasonable to locate the two measuring planes symmetrically with respect to the image plane since in this case the effect of the errors will be minimized. However this is not a necessary condition.

In reality, the pinholes constituting the Hartmann aperture have a finite size, which leads to macroscopic spots of light in the measuring planes. These spots are photometrically evaluated in order to determine their centroids. The deviation of a centroid from its ideal nominal position gives the lateral deviation, which is assigned to the central ray within the corresponding pinhole. This means that the aberration is averaged across the single pinhole area and the error contribution of each finite-sized hole depends on its relative size and the gradients of the wave to be measured. In principle these are systematic errors.

Alternatively, the Hartmann aperture mask can be located on the image side directly behind the system and used under uniform illumination as a plane for centroid (hole center) determination instead of one of the measuring planes. In an extreme case, the Hartmann mask itself can be used as the first measuring plane. An assumption for this setup is an accurate knowledge of the pinhole positions inside the diaphragm. If the geometrical lateral deviations are plotted as vectors with origins in the subaperture centroids, a diagram, as shown in figure 47-27, is obtained.

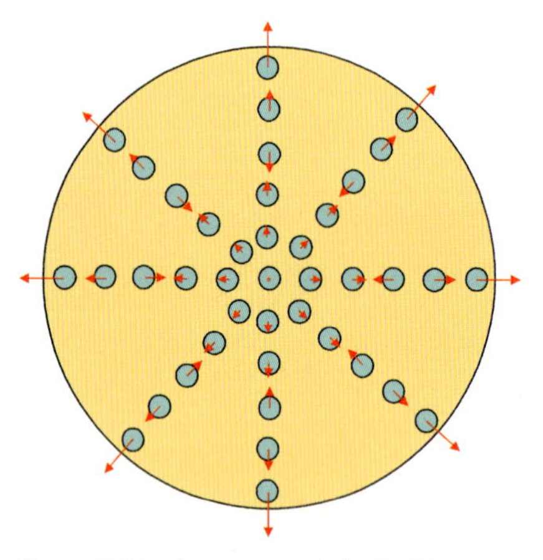

Figure 47-27: Aperture mask for the Hartmann test. The shifts in the centroids, due to aberrations, for every pinhole, are indicated by red arrows.

The separations of the measuring planes from the image plane are chosen on the basis of the spot size in the aperture and the spatial resolution of the detector. The detector should have sufficient resolution allowing the determination of the spot diameter D and the corresponding centroid. This diameter increases with separation from the image plane. It also increases in the vicinity of the image plane when the pinhole diameter is decreased. When the distance chosen between the measuring planes is too large, as a consequence of diffraction, the single spots are strongly broadened and overlap, and cannot be separated and analyzed.

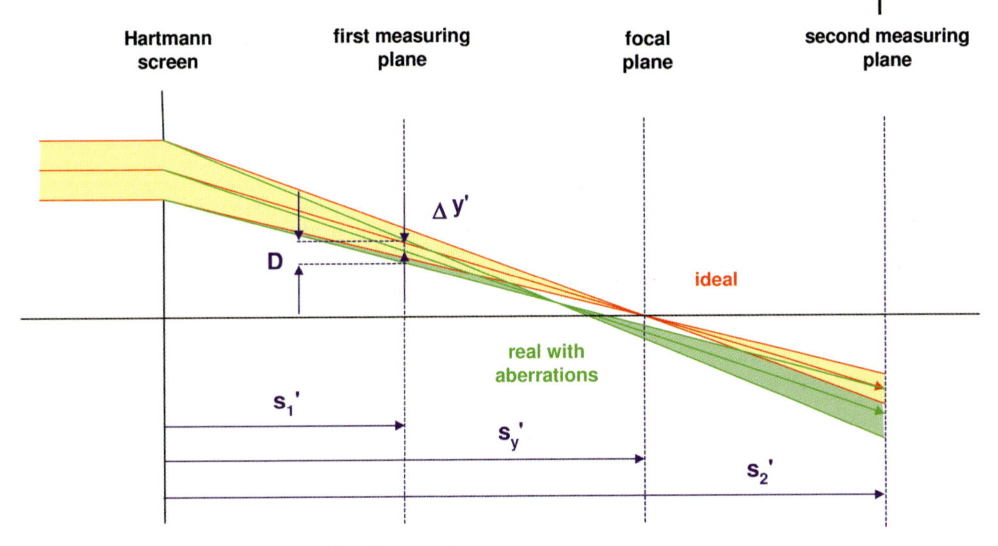

Figure 47-28: Separations between the planes in the Hartmann test and spreading of the beam transmitted through the pinhole.

Figure 47-29 shows a picture of a real Hartmann measurement. A hexagonal pattern is chosen, but leaving out one pinhole near the center and the central pinhole. This guarantees a clear determination of the azimuthal orientation. Since for well-corrected systems the wave aberration near to the optical axis is almost perfect, these two pinholes carry no relevant information. The illumination of the system shows a strong gaussian-like apodization in this case. This is indicated in the section view of figure c) as a dashed blue line. For the calculation of the spot centroids, a minimum threshold above the noise level is defined and also an intermediate threshold between this base level and the peak height. Inside the corresponding areas, the centroid is evaluated. This leads to a region of interest for the centroid calculation with width L as indicated in figure 47-29 part c).

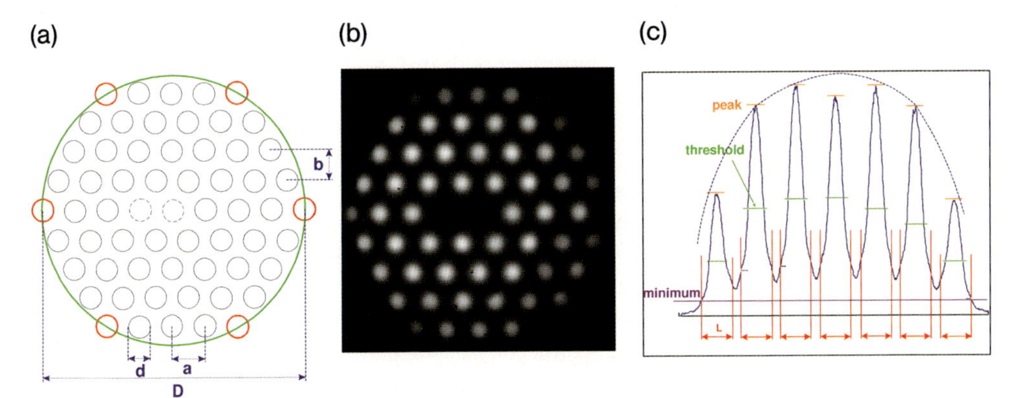

Figure 47-29: Real geometry of a Hartmann test screen (a). The corresponding intensity pattern just behind the screen (b); a section through the spots (c). Two pinholes are omitted to allow a clear determination of the azimuthal position.

The main parameters of the Hartmann setup are as follows.

1. Number of pinholes.
2. Geometry of the pinhole grid.
3. Diameter of the pinholes.
4. Locations of the measuring planes.

These four aspects are discussed in the following in more detail.

Figure 47-30 shows some possible pinhole geometries for the Hartmann screen. In a), a polar arrangement is chosen. This is not really very suitable, because the various data are not isoenergetic in their representative weighting for the pupil area. Part b) shows a classical cartesian grid. The disadvantage of this choice is the bad adaptation of the boundary line to the aperture. The hexagonal geometry in c) and Albrechts grid [47-42] in d) are the most promising ones. They allow for a quite accurate determination of the Zernike coefficients with a minimal number of pinholes. There are also other possibilities known, for example; a spiral curve or a polar grid with adapted ring widths to ensure isoenergetic patches for the pinholes or statistical uniform distributions. In [47-43] it is shown how sensitively the accuracy of the result depends on the geometry of the pinhole grid.

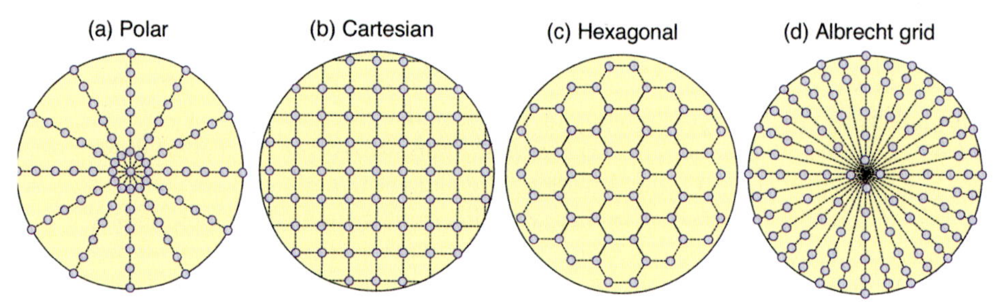

Figure 47-30: Different geometries of spot patterns in a Hartmann test. The cases c) and d) are the most advantageous and are found most frequently in practice.

If the pinholes are chosen to be very small, one has the advantage of a very accurate local definition of the wavefront derivative in the center point of the subaperture. The problem with too-small pinholes is the diffraction effect, which causes a large spreading in the more distant measurement planes. Another problem is the total energy and therefore the signal to noise ratio, which becomes worse when the pinhole diameters are small.

If the individual pinhole beams are no longer separated in the back measuring plane, the calculation of the centroid is impossible or at least becomes erroneous due to a crossover of energy between the channels. If the pinholes are quite large in diameter, the wavefront aberrations across this finite size cause an additional broadening. Figure 47-31 shows some diffraction patterns of a hexagonal Hartmann screen at various distances z for four different sizes of the pinholes. The range left of the red line indicates the regimes, where the spots can be separated and analyzed without difficulty.

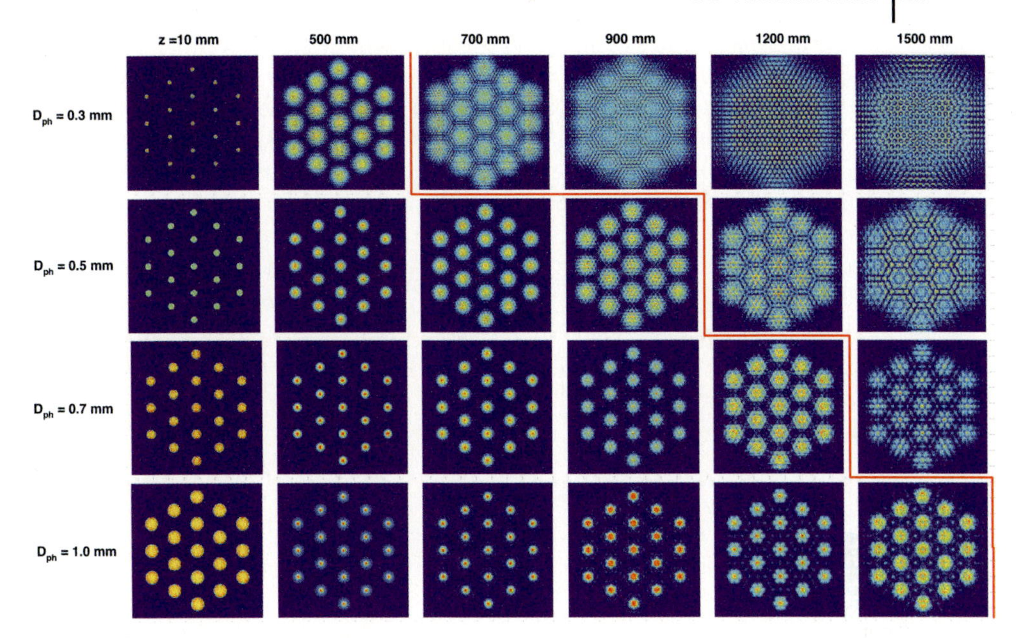

Figure 47-31: Spot pattern of a hexagonal Hartmann test in several z-planes with different diameters of the pinhole. The red line indicates the limit up to which the individual spots can be separated and there is no problem evaluating the aberrations. The diameter of the measured system aperture is $D = 10$ mm.

Figure 47-32 shows the result of a corresponding calculation. In this representation, the lateral coordinates are equally scaled in every plane. The choice of the z-planes is symmetrical around the image plane. It can be seen how the diffraction of the pinholes broadens the spots with increasing distance z from the plane of the diaphragm. Outside the image region, the geometrical spreading allows the spots to be separated again.

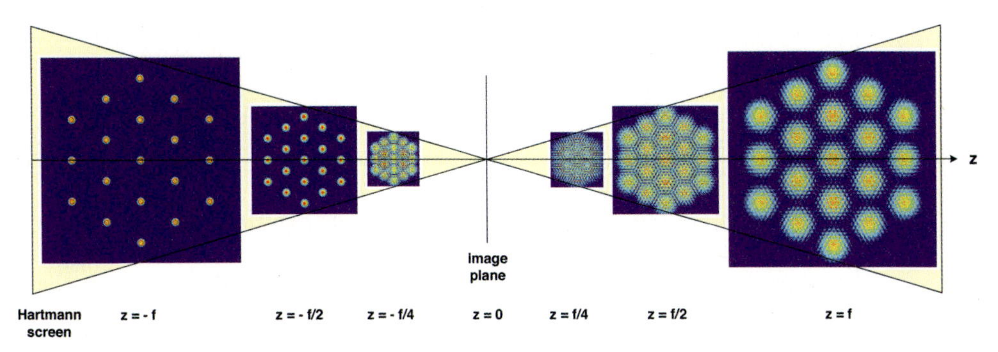

Figure 47-32: Spot pattern of a hexagonal Hartmann test in several z-planes in exactly scaled lateral coordinates.

The number and size of the pinholes determines the energetic throughput of the diaphragm and influences the result via the signal to noise ratio. In figure 47-33, five different sizes of circular pinholes inside a screen with hexagonal geometry containing various numbers of pinholes are considered. The diameter of the pupil for this case is 10 mm, shown is the power transmission of the diaphragm screen. For a typical case of 100 pinholes of 0.2 mm in diameter, one gets a total transmission of $T = 4\%$.

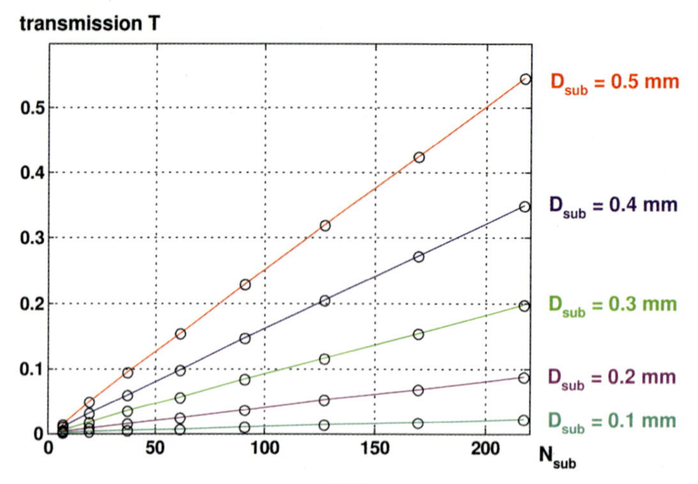

Figure 47-33: Power transmission of different numbers of pinholes with different sizes for a hexagonal geometry of the pinhole pattern.

47.3.2 Accuracy of the Hartmann Method

The evaluation of the Hartmann test uses the well-known relationship between the wave aberration W and the transverse aberrations $\Delta x'$, $\Delta y'$. If R is the radius of the reference sphere, the equations

$$\Delta y' = -R \frac{\partial W(x,y)}{\partial y} \qquad \Delta x' = -R \frac{\partial W(x,y)}{\partial x} \tag{47-32}$$

connect the measured deviations to the wavefront in air. These two equations are given in matrix form, because the centroid offsets are known from the measurement for every pinhole center coordinate. The integration of these equations can be done locally according to

$$W(x,y) = -\frac{1}{R}\int_0^x \Delta x' \, dx - \frac{1}{R}\int_0^y \Delta y' \, dy \tag{47-33}$$

The local integration method is quite sensitive to small perturbations and therefore the results strongly depend on the exact integration path used. It is recommended in this case to use several different integration paths and to calculate the average over these different methods of solution.

Other possibilities are a polynomial fit of the gradient fields, followed by an analytical integration. For example, it is possible to use the Zernike polynomials and to express the coefficients as a weighted sum over all pinhole contributions in the following equation

$$c_{nm} = \frac{\sin u'}{\lambda} \sum_j g_{nm,j}^{(x)} \Delta x_j' + g_{nm,j}^{(y)} \Delta y_j' \tag{47-34}$$

The weighting factors $g_{nm,j}$ depend on the polynomial and the relative position of the pinhole in the diaphragm of the Hartmann screen. If the uncertainty of the pinhole centroids is estimated by Δ, the expected error in a Zernike coefficient can be calculated from the equation

$$\Delta c_{nm} = \Delta \frac{\sin u'}{\lambda} \sqrt{\sum_j \left[\left(g_{nm,j}^{(x)} \right)^2 + \left(g_{nm,j}^{(y)} \right)^2 \right]} \tag{47-35}$$

One special problem in the evaluation of the Hartmann test and the accurate calculation of the Zernike coefficients is the precise knowledge of the margin of the pupil and the pupil radius a. If the illumination is homogeneous and the system pupil has a circular shape, an estimation can be made by assuming the conservation of energy.

$$\frac{P_{\text{ges}}}{\pi a^2} = \frac{P_{\text{spot}}}{D_{\text{sub}}^2} \tag{47-36}$$

from which we get the relation

$$a = D_{\text{sub}} \sqrt{\frac{P_{\text{ges}}}{\pi P_{\text{spot}}}} \tag{47-37}$$

In analogy to the description of the Hartmann-Shack sensor of section 47.2.9, the main sources of error in the Hartmann method are as follows.

1. Averaging the aberrations across the area of the finite size of the pinhole.
2. Noise encountered as Poisson photon noise due to a small photon flux through the small areas of the pinholes.
3. Geometrical errors in the location of the pinhole centers due to manufacturing imperfections. The influence of this error can be removed by calibration.
4. Errors in the exact knowledge of the z-values of the measuring planes. In principle, the major term is the difference between the z-locations of the planes.
5. Errors due to problems in the evaluation algorithms.

47.3.3
Partial Coherent Illumination and Apodization

In general, the Hartmann setup allows the measurement of any state of polarization and coherence. In comparison with the Hartmann–Shack wavefront sensor there is no wavelength dependence. If the illuminating light source is of finite size, additional problems are caused in the broadening of the spot behind the pinhole diaphragm. The corresponding arrangement is shown in figure 47-34. The following terms are used in the theoretical consideration below.

h	distance of the pinholes in the plane of the diaphragm
h'	distance of the pinholes in the camera plane
d_s	pinhole diameter
d'_s	geometrical diameter of the pinholes in the camera plane
d_1	distance between the diaphragm plane and the image plane
d_2	distance between the image plane and the camera plane
f	focal length of the system
D_{obj}	diameter of the field of view due to the finite size of the light source
w	angle of the field of view
u	aperture angle of the pinholes

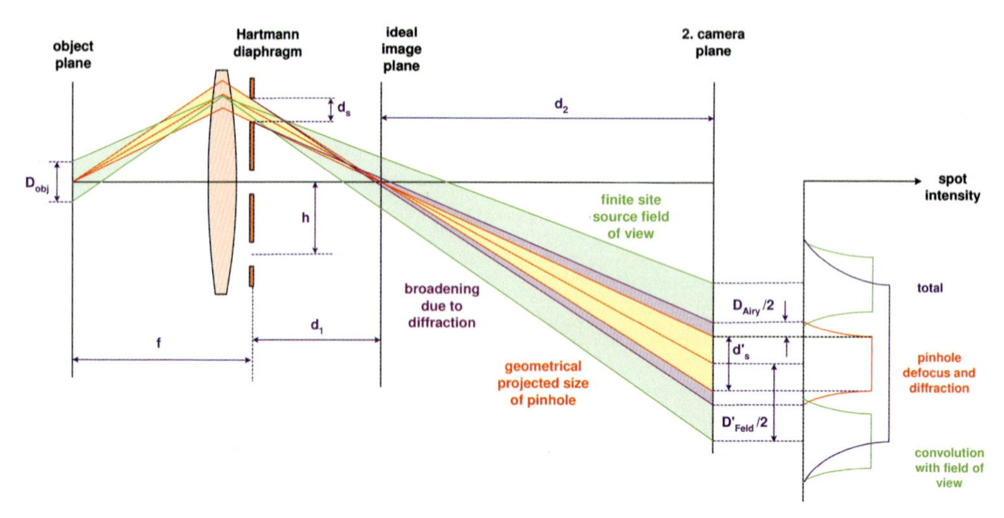

Figure 47-34: Schematic setup of a Hartmann measurement considering the spot size for partial coherent illumination.

The size of the pinhole in the camera plane is given in the geometrical approach by

$$d'_s = \frac{d_2}{d_1} d_s \tag{47-38}$$

The distance between the pinholes in the camera plane is

$$h' = \frac{d_2}{d_1} h \tag{47-39}$$

The diffraction due to the finite pinhole size is given by the Airy diameter (the medium is assumed to be air)

$$D_{\text{Airy}} = \frac{1.22\,\lambda}{\sin u} = \frac{2.44\,\lambda\,(d_1 + d_2)}{d_s} \tag{47-40}$$

With a field of view

$$\sin w = \frac{D_{\text{obj}}}{2f} \tag{47-41}$$

and the magnification of the pinhole

$$D'_{\text{field}} = (d_1 + d_2)\frac{D_{\text{obj}}}{f} \tag{47-42}$$

the total size of the pinhole image is given by

$$d'_s(\text{total}) = \frac{d_2}{d_1} d_s + (d_1 + d_2)\frac{D_{\text{obj}}}{f} + (d_1 + d_2)\frac{2.44\,\lambda}{d_s} \tag{47-43}$$

The spots are separable and can be used to calculate the centroid, if the condition

$$h' > d'_s(\text{total}) \tag{47-44}$$

is valid. According to the van Cittert–Zernike theorem, the coherence function for incoherent illumination is

$$\Gamma = \frac{\pi}{4} D_{\text{obj}}^2 \frac{2\,J_1\left(\frac{\pi\,h\,D_{\text{obj}}}{\lambda f}\right)}{\left(\frac{\pi\,h\,D_{\text{obj}}}{\lambda f}\right)} \tag{47-45}$$

The first zero of this distribution determines the vanishing phase coupling between two neighboring pinholes in the form

$$D_{\text{obj}} > \frac{1.22\,\lambda f}{h} \tag{47-46}$$

If this condition is not fulfilled, there are interference effects between the contributions of the pinholes.

As a rule of thumb [47-40], for a spherical wavefront with radius R and a focal distance f, the optimal size of the pinholes is given by

$$D_{\text{sub}} = \frac{1.22\,\lambda\,R^2}{f} \tag{47-47}$$

47.3.4
Hartmann Measurement of an Apodized Profile

If the illumination light has an intensity distribution over the cross-section of the pupil, the ray representing the centroid of the energy in the pinhole does not pass through the geometric center of the pinhole. This setup is illustrated in figure 47-35. But as can be seen in the picture, the centroid ray intersects the geometric center ray in the image plane. Therefore, if the two measuring planes are used to interpolate the transverse ray aberration in the image plane, an apodization doe not disturb the evaluation of the method in first approximation. Therefore, the behavior is very similar to the Hartmann-Shack sensor setup.

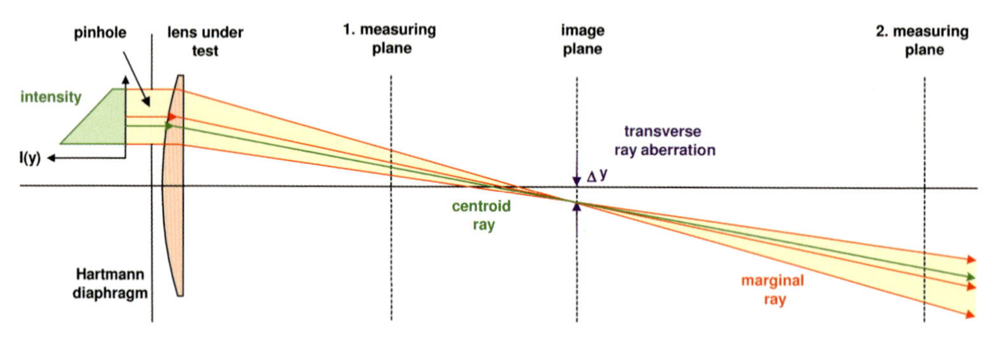

Figure 47-35: Behavior of the Hartmann test in the case of an apodized illumination. In the image plane, the transverse ray aberration is correct, as in the case of homogeneous illumination.

47.3.5
Modified Hartmann Methods

There are some possible variations on the Hartmann setup proposed in the literature. One special idea uses only a single hole in the diaphragm screen, which is scanned at an arbitrary number of positions across the beam cross-section area [47-44]. This gives a larger flexibility for adapting to the required accuracy and there are no longer any problems with cross-talk between the light beams of the holes.

Other setups use every four neighboring pinholes from a larger number of pinholes to extract information about the local curvature and astigmatic angle in this range of the pupil [47-45].

Some suggestions can be found in the literature, which try to protect the measuring method from the cross-talk problems by using a special checkerboard-shaped phase mask [47-46]. With the help of this mask, a Talbot effect takes place and a better light confinement behind the Hartmann pinhole mask is obtained. But this use of the Talbot effect requires a regular cartesian grid geometry for the pinhole mask.

One possible way to increase the number of pinholes in the Hartmann diaphragm is to use a scanning disc with pinholes similar to those of the Nipkow microscope. This enlargement of holes gives a better accuracy [47-47]. The integration of the transverse aberrations to get the wave aberration is a particular source of error when the separation of the individual holes is too large. If, for example, a spiral-shaped pinhole sequence is used, the number of measurable positions can be considerably increased.

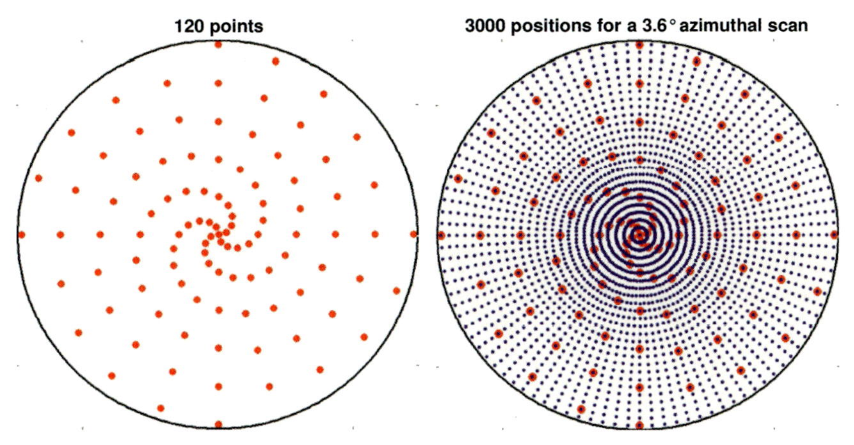

Figure 47-36: Modified pinhole grids for scanning Hartmann pinhole diaphragms.

47.4 Phase Space Analyzer

47.4.1 Introduction

A phase space analyzer is a special optical system with a non-axisymmetrical component, which allows the analysis of the angle or wavefront errors of a beam in a simple way. Figure 47-37 shows the optical system of the phase space analyzer [47-48], [47-49], where the coherent field of a slit is propagated through a special system. A lens and a toroidal lens, which is rotated through 45° around the optical axis, are located at appropriate distances. The lens is a distance a from the slit, the toroidal lens follows at a distance d. The detector plane is at a distance z from the toroidal lens and a distance $b = d + z$ from the spherical lens. In the image plane of the slit, wave aberrations in the slit plane are transformed by the non-orthogonal system into transverse deviations in the spatial domain. By analyzing the deformation of the slit, the wave aberration can be evaluated by geometrical means.

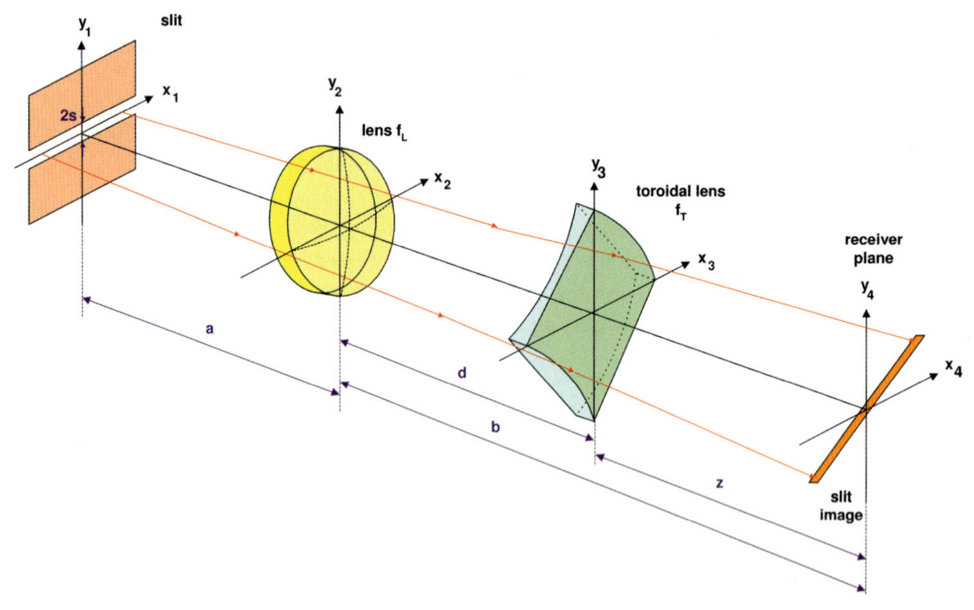

Figure 47-37: Propagation of the coherent field of a slit through the paraxial non-orthogonal ABCD system of the phase space analyzer.

The most interesting component in the phase space analyzer setup is the toroidal lens with the azimuthal rotation angle of 45°. Figure 47-38 shows the effect of this lens on an angle error u in the x-direction. In the lens, this deviation is transformed into an angle v in the y-direction. In part b) of the figure, the vector summation of the focusing effect along the –45° direction and the diverging effect in the +45° direction is illustrated. As a result, a deviation u is transformed into an angle v.

The mixing effect of the rotated toroidal lens can be used to analyze wave or angle aberrations of the incoming beam of the exit pupil for an optical system. Since the illumination is performed in the form of a thin slit, only a radial section of the pupil of the system is observed. The wave aberrations are located along the slit and they are transformed in a perpendicular direction. If a proper imaging of the slit is realized by the measuring setup, the straight slit is deformed and curved, if wave aberrations exist. Figure 47-39 shows some examples of low-order wave aberrations and their impact on the shape of the imaged slit. Figure 47-40 shows a real observed pattern from a metrology tool, which uses three different wavelengths for the measurement of the curve with a lateral offset for the separation of the signals. The inclination of the slit images indicate a defocus, the s-shaped parts show a residual spherical aberration. Different inclination angles for the three colors, show that there is a chromatic axial error in the system.

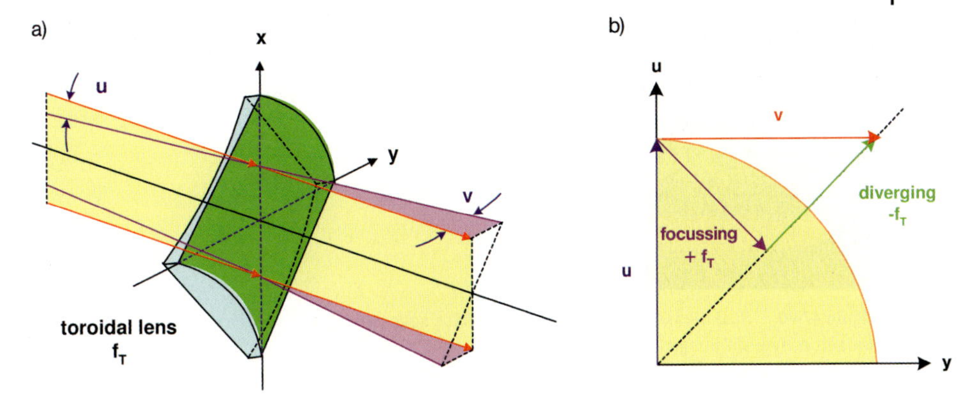

Figure 47-38: Explanation of the x-y-mixing effect of the rotated toroidal lens. Part a) shows the lens and its impact on an angle deviation *u*. Part b) explains this behavior with the focusing and diverging effect of the toroidal lens below 45° in a vector diagram.

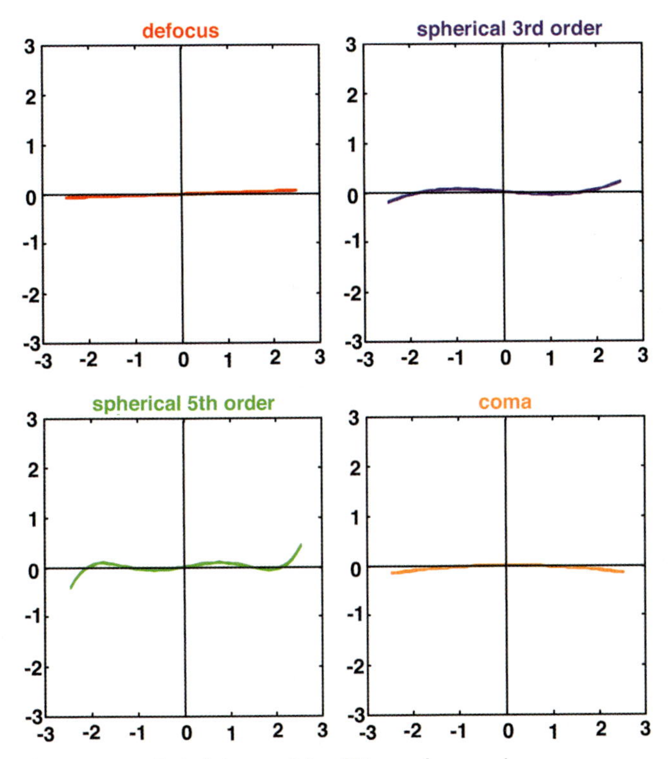

Figure 47-39: Typical shapes of the slit image for some low-order aberrations in the phase space analyzer.

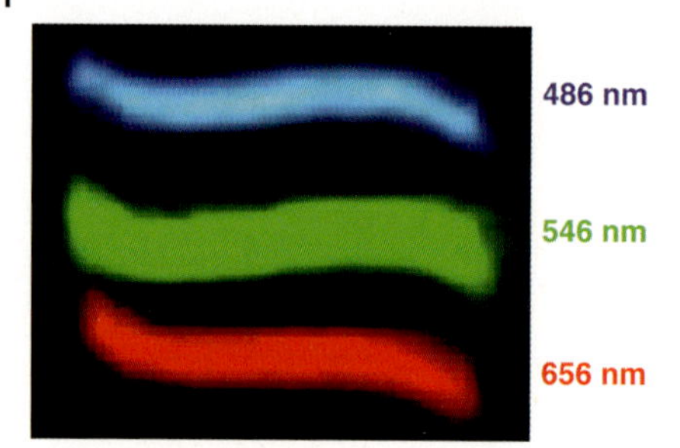

Figure 47-40: Real measured slit images of a phase space analyzer setup. The slit images for three different colours are separated by a lateral shift.

Figure 47-41 shows the deformation of the slit at various distances z for a spherical aberration of $\lambda/2$. The changed shape of the slit image and the effect of defocusing can be seen.

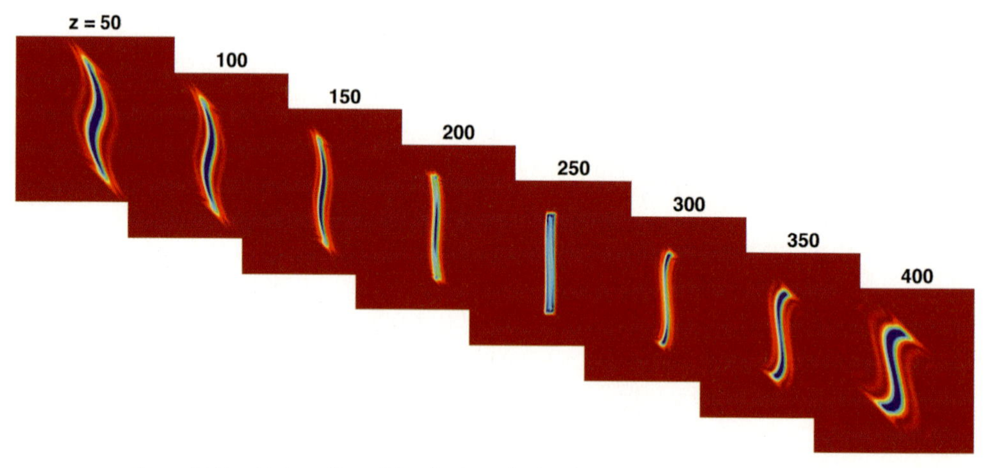

Figure 47-41: Deformation of the slit shape in the phase space analyzer as a function of the distance z in mm for a spherical aberration $\lambda/2$. The correct imaging condition is fulfilled at the location $z = 210$ mm.

47.4.2
Layout Versions

There are several different possibilities for the realization of the phase space analyzer optics. In the first version according to Weber [47-48], [47-49], which is called the type I layout in the following, the slit is exactly imaged in the detector plane

$$\frac{1}{a}+\frac{1}{b}=\frac{1}{f_L} \tag{47-48}$$

Furthermore, the toroidal lens should be a purely Fourier lens, therefore it must be located at a distance

$$d=f_L \tag{47-49}$$

from the spherical lens. Figure 47-42 shows the layout of this special realization of the phase space analyzer in the two major cross-sections.

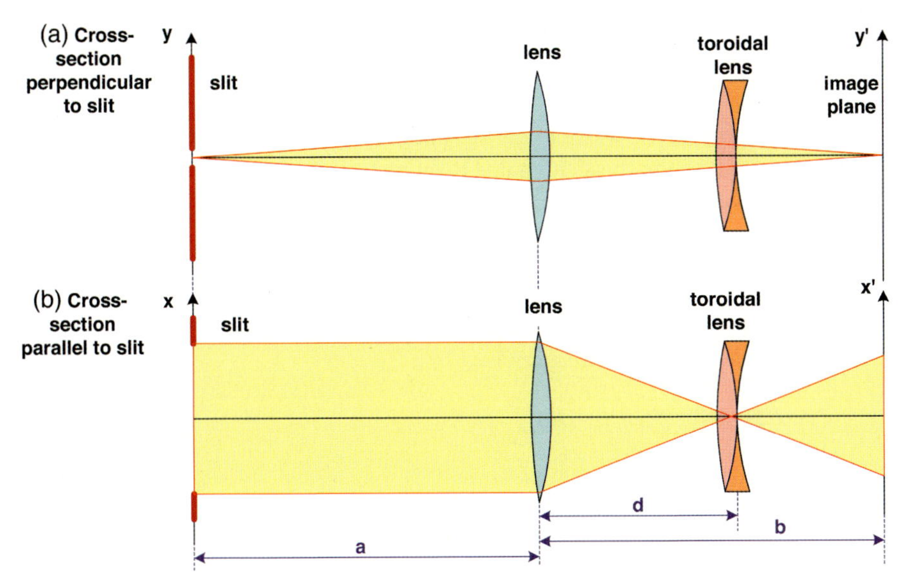

Figure 47-42: Layout of the phase space analyzer of type I in the x and y cross-section.

If these conditions are fulfilled, the transverse deviations in the detector plane are given by

$$x'=\left[1-\frac{b}{f_L}\right]x+\left[\frac{f_L}{f_T}(b-f_L)\right]v \tag{47-50}$$

$$y'=\left[1-\frac{b}{f_L}\right]y+\left[\frac{f_L}{f_T}(b-f_L)\right]u \tag{47-51}$$

As can be seen from these equations, the orientation between an incoming angle u or v is completely switched: u affects y' and v has an impact on x'. The length of the slit in the detector plane is given by

$$x' = \left[1 - \frac{b}{f_L}\right] x \tag{47-52}$$

and the transverse aberration detected in the image plane as determined by the initial angle error u can be calculated by the equation

$$y' = \left[\frac{f_L}{f_T}(b - f_L)\right] u \tag{47-53}$$

The deviations of the slit can be detected effectively, if the slit image itself is quite good and the width is small. Therefore, one can define an efficiency parameter ε, which describes the ratio of the deviation relative to the image width of the slit

$$\varepsilon = \Delta y' / s' \tag{47-54}$$

In this equation, the width of the slit image s' is determined by the geometrical magnification and the broadening due to diffraction. A special layout configuration is the type I case which has a unit magnification $m=-1$ for the lens imaging of the slit.

A second type of layout is of special interest in practice. This so-called type II phase space analyzer is characterized by the condition $a=d=0$ and therefore has the advantage of a very short total track. By geometrical optical means, the spatial deviations in the detector plane are given here by

$$\begin{aligned} x' &= x(1 - b/f_L) - y\,b/f_T + b\,u \\ y' &= y(1 - b/f_L) - x\,b/f_T + b\,v \end{aligned} \tag{47-55}$$

If the width of the slit can be neglected $y=0$ and $v=0$, because the angle errors are only propagated in the direction of the slit, so the equations simplify to

$$\begin{aligned} x' &= x(1 - b/f_L) + u\,b \\ y' &= -x\,b/f_T \end{aligned} \tag{47-56}$$

The inclination of the slit image in the detector plane is, in the ideal case of vanishing angle aberrations $u=0$, given by

$$\tan\varphi = \frac{y'}{x'} = \frac{b f_L}{f_T(f_L - b)} \tag{47-57}$$

This equation shows that, in the type II phase space analyzer, the image of the slit rotates with the distance b of the detector. Usually, the detector is located in the image plane of the spherical lens $b=f_L$. In this special case the deviations are given by

$$\begin{aligned} x' &= b\,u \\ y' &= -\frac{f_L}{f_T} x \end{aligned} \tag{47-58}$$

In table 47-2, the most important conditions and equations for the above discussed versions of the phase space analyzer are shown with one additional measure also defined. The ratio between the spatial deviation $\Delta y'$ and the angle aberration u is called the sensitivity and can be considered as a variable, which shows the sensitivity of the desired signal. It reads

$$\eta = \Delta y' / \Delta u \tag{47-59}$$

Table 47-2: Conditions and equations for the layout of several special versions of the phase space analyzer.

	Type I general	Type I m = –1	Type II general	Type II focused $b = f_L$
Special parameter choice	$\frac{1}{a} + \frac{1}{b} = \frac{1}{f_L}$ $d = f_L$		$a = 0$ $d = 0$	
Magnification m	$m = 1 - \frac{b}{f_L}$	$m = -1$	$m = \frac{b}{f_T}$	$m = \frac{f_L}{f_T}$
Length of the slit	$L' = L\left(1 - \frac{b}{f_L}\right)$	$L' = L$	$L' = L\frac{b}{f_T}$	$L' = L\frac{f_L}{f_T}$
Geometrical width of the slit	$s'_G = s\left(1 - \frac{b}{f_L}\right)$	$s'_G = s$	$s'_G = s\frac{b}{f_T}$	$s'_G = s\frac{f_L}{f_T}$
Broadening of the slit due to diffraction s'_B	$s'_B = \frac{f_L}{f_T}(b - f_L)\frac{\lambda}{s}$	$s'_B = b\frac{\lambda}{s}$	$s'_B = b\frac{\lambda}{s}$	$s'_B = f_L\frac{\lambda}{s}$
Sensitivity $\eta = \Delta y' / \Delta u$	$\eta = (b - f_L) \cdot \frac{f_L}{f_T}$	$\eta = \frac{f_L^2}{f_T}$	$\eta = b$	$\eta = f_L$
Effectiveness $\varepsilon = \Delta y' / s'$	$\varepsilon = \frac{s f_L^2}{s^2 f_T + \lambda f_L^2} u$		$\varepsilon = \frac{s f_T}{s^2 + \lambda f_T} u$	

47.4.3 Evaluation of the Data

If the phase space analyzer is used to quantify the wave aberrations of an objective lens, the image of the slit must be analyzed according to the formulas above. The transition from the wave optical Zernike polynomials to the geometrical transverse angle aberration is quite easy. In the case of a wave aberration $W(\bar{x})$ in the one-dimensional normalized pupil coordinate $\bar{x}_p = x_p/(L/2)$ it follows

$$\Delta x' = -\frac{R}{n}\frac{\partial W(x_p)}{\partial x_p} = -\frac{2R}{nL}\frac{\partial W(\bar{x}_p)}{\partial \bar{x}_p} \tag{47-60}$$

the angle is given by

$$u = \frac{\Delta x'}{R} = -\frac{2}{nL}\frac{\partial W(\bar{x}_p)}{\partial \bar{x}_p} \tag{47-61}$$

If the wavefront is described in fringe Zernike polynomials c_j with scaled values of 1 λ, it follows that

$$W(\bar{x}_p) = c_j \, \lambda \, Z_j(\bar{x}_p)$$
$$u = \frac{\Delta x'}{R} = -\frac{2c_j \, \lambda}{nL}\frac{\partial Z(\bar{x}_p)}{\partial \bar{x}_p} \tag{47-62}$$

The primary slit image can now be analyzed, if the centroid line is determined and the corresponding curve $y_c(x)$ is expanded into a polynomial. Figure 47-43 shows an example of a slit image for spherical aberration. The red line gives the centroid of the intensity distribution perpendicular to the slit extension. The bending of the curve can be analyzed and corresponds to a third-order polynomial in equation (47-62) above.

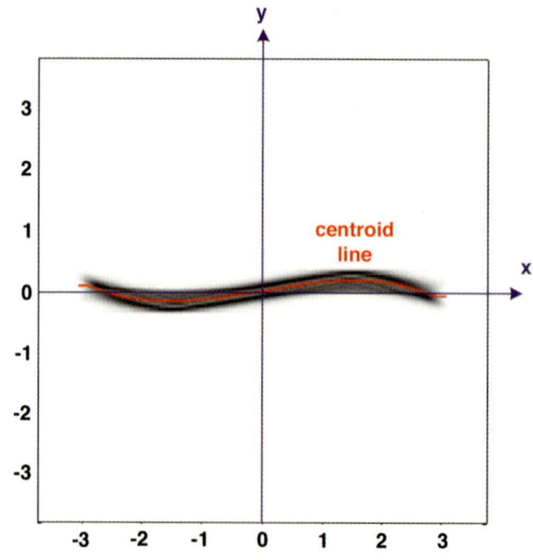

Figure 47-43: Image of the slit together with the centroid line (red) in the case of spherical aberration.

47.4.4
Wave Optical Description

The major function of the phase space analyzer can be understood by geometrical means according to a general ABCD-matrix description. But in reality, a very small slit width causes diffraction effects. The evaluation of the resulting deformed slit image is very effective when there are no perturbations due to diffraction broadening. A special problem in practice is the behavior of the end parts of the slit image. If the aberrations are large, the exact end of the slit image degenerates and is no longer clearly defined. This causes a problem in finding the correct normalization of the pupil size. Therefore, a wave optical calculation of the system has the advantage of producing an optimal layout of all the parameters of the setup and seeing the effects in advance.

For the special case of the type I phase space analyzer with unit magnification, the width of the slit in the image plane is given by a geometrical and also a diffraction contribution. The estimation of the above formulas gives the sum of both effects as

$$s' = s + \frac{b\,\lambda}{s} \tag{47-63}$$

Figure 47-44 shows the result of a numerical calculation of the image slit width for the data $\lambda = 0.5$ µm and $b = 200$ mm. It can be seen that the estimation is quite accurate. The theoretical curve is not exact, since numerical effects play a role and the width of the slit must be defined in the numerical case by an intensity threshold. It should particularly be noticed that there is an optimal slit width, which delivers the smallest image width. This case allows for the most accurate results in finding the transverse deviations of the slit.

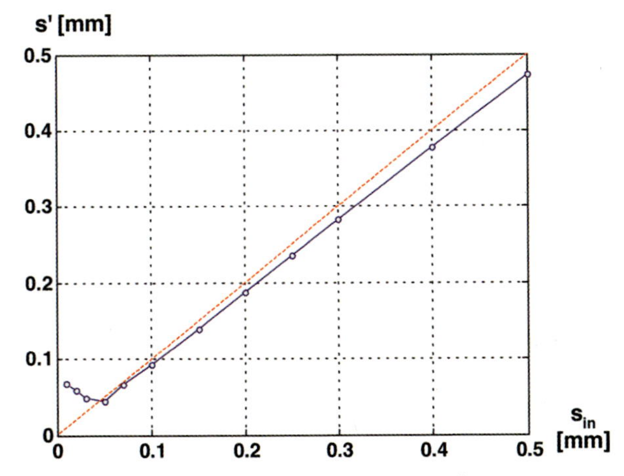

Figure 47-44: Slit width of the image in a type I phase space analyzer as a function of the input slit width.

In the figure 47-45, a series of wave optical simulations is shown. For the type I system in the upper group, the slit width varies from $s = 0.3$ mm to $s = 0.05$ mm for a constant length of 4 mm. The focal lengths are $f_L = 100$ mm and $f_T = 100$ mm. The first three rows represent this phase space analyzer without aberrations and with a spherical aberration of third order with $c_{40} = 1\ \lambda$ and $c_{40} = 2\ \lambda$ respectively. The lower two rows show similar results for a type II phase space analyzer with the focal lengths $f_L = 100$ mm and $f_T = 500$ mm. It can be seen that the definition of the slit image is much better for the type I system. In particular, a very thin slit significantly deforms the image in the case of the type II system. The slit length changes with the aberrations.

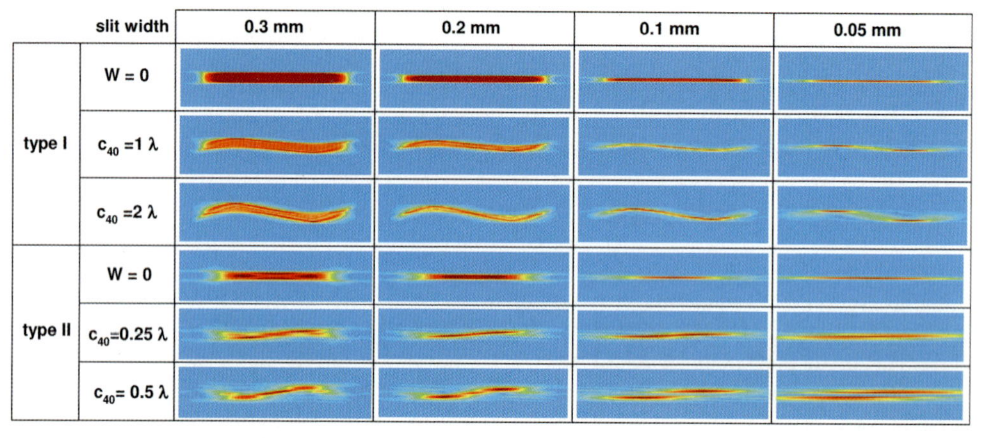

Figure 47-45: Slit images of two different phase space analyzer layouts without (type I) and with (type ii) aberrations for different widths of the slit.

In figure 47-46, an additional example for the type II phase space analyzer is shown. The focal length of the spherical lens is $f_L = 100$ mm, the toroidal lens has a focal length of $f_T = 750$ mm. The slit length is 5 mm, the width of the slit is 0.5 mm. Different values of the wave aberration corresponding to the legend are used. It can be seen that the margin of the slit image and the homogeneity of the slit is perturbed. This effect makes it difficult to obtain an accurate definition of the end of the slit image.

In figure 47-47, an example for the type I phase space analyzer is shown. The focal length of the spherical lens is $f_L = 100$ mm, the toroidal lens has a focal length of $f_T = 100$ mm. The slit length is 4 mm, the width of the slit is 0.25 mm. The magnification is adjusted to be $m = 4$ for distances $a = 125$ mm and $b = 500$ mm. Different values of the wave aberration corresponding to the legend are used. It can be seen that the margin of the slit image and the homogeneity of the slit are only slightly perturbed and are much better than in the previous case. This is a better condition for the evaluation of the centroid line of the slit image. On the other hand, the absolute sensitivity is lower. Notice that the absolute values of the aberration are significantly higher in this example.

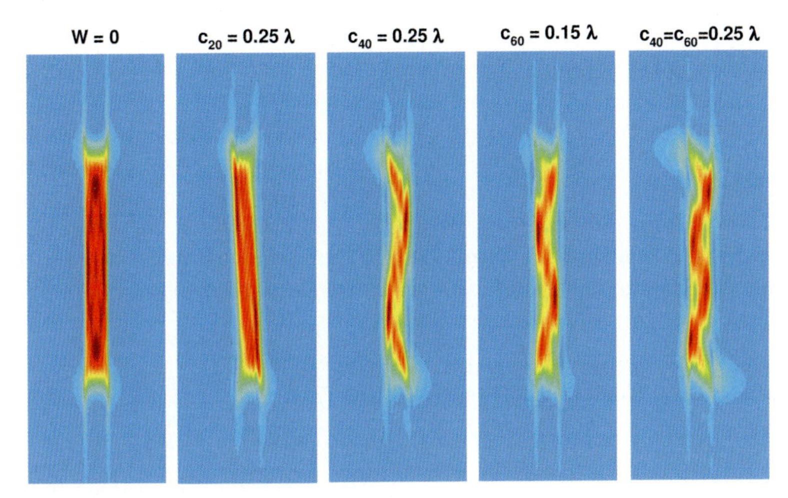

Figure 47-46: Slit images of a phase space analyzer of type II with different values of the wave aberration.

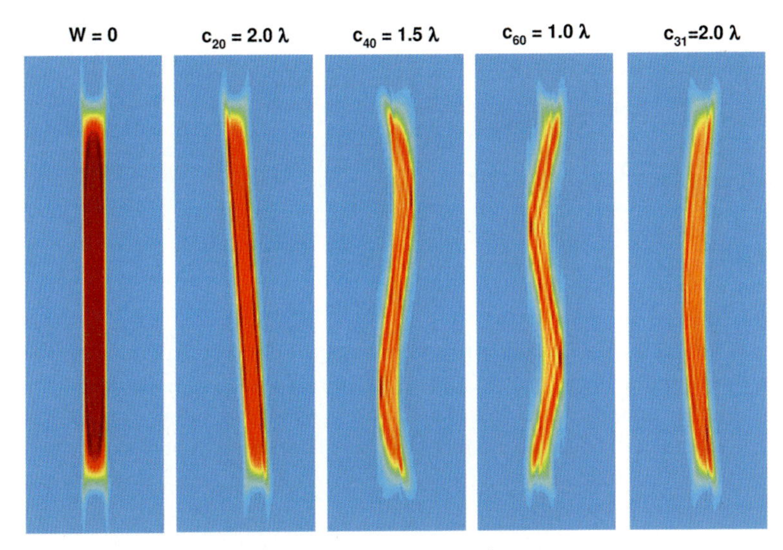

Figure 47-47: Slit images of a phase space analyzer of type I with different values of the wave aberration.

47.5
Point Image Filtering Techniques

There are different types of wavefront sensors, which apparently can be considered as non-interferometric. However, they follow the same rules of diffraction and interference as the classical interferometric test methods, so that there is no real need to distinguish between the different types of wavefront sensor. Some of the test methods described in the following can be simulated by pure geometrical ray optics, particularly when aberrations are greater than the wavelength. Among them are the following.

- The Foucault test (Knife-edge test) (1858), see [47-50].
- The Toepler schlieren method (1866), see [47-51].
- The Ronchi test (1923), see [47-52].
- The Wire test.

In the following, the principle of the point image filtering technique is described (fig. 47-48).

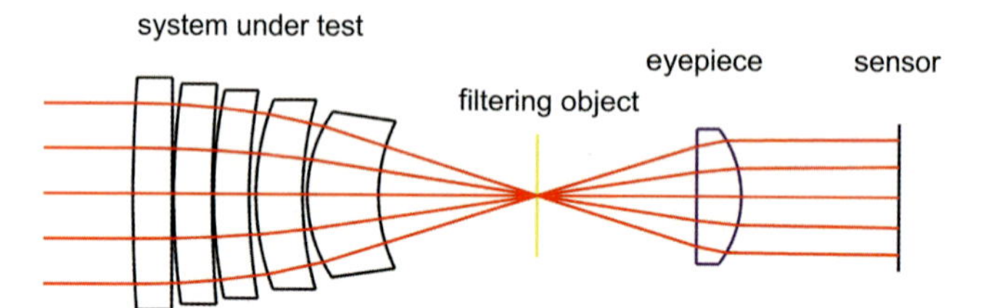

Figure 47-48: Point image filtering setup used for testing an optical system. The eyepiece images the pupil of the system onto the sensor.

A point source illuminates the system or component under test. In the neighborhood of the point source image a filtering object is introduced perpendicular to the optical axis of the arrangement. The wavefront passing the filtering object is adjusted in amplitude or phase in a characteristic way depending on its type of aberration. A variation in irradiance results when the wavefront hits the following sensor. The aberrations can be quantified from the irradiance distribution in the detector plane when the geometry of the setup is known. When an eyepiece is introduced between the filtering object and the sensor, the pupil of the system under test or any of its inner elements can be sharply imaged onto the sensor.

Let us assume the exit pupil of the system to be a sphere with radius of curvature r_0. As indicated in figure 46-49 the optical axis is denoted as z and x and y are the coordinates normal to the optical axis. The aberrated wavefront of the system is described by $W(x,y)$ relative to the exit pupil sphere.

For an aberrated ray emerging from the point $S(x,y)$ of the exit pupil, the crossing point $Q(x,y)$ in the filter plane can be described by (47-64).

$$x_q = x - (r - z_w(x,y))\frac{\partial z_w(x,y)}{\partial x} \tag{47-64a}$$

$$y_q = y - (r - z_w(x,y))\frac{\partial z_w(x,y)}{\partial y} \tag{47-64b}$$

where x and y are the ray coordinates in the exit pupil, r is the distance of the filter plane from the vertex of the exit pupil. The sagittal height $z_w(x,y)$ for the emerging wavefront at x,y is defined by (47-65).

$$z_w(x,y) = \frac{x^2+y^2}{r_0\left(1+\sqrt{1-\frac{x^2+y^2}{r_0^2}}\right)} + W(x,y) \tag{47-65}$$

Equation (47-64) can then be redefined by (47-66).

$$x_q = x - \left(r - r_0 + \sqrt{r_0^2 - x^2 - y^2} + W(x,y)\right) \times \left(\frac{x}{\sqrt{r_0^2 - x^2 - y^2}} + \frac{\partial W(x,y)}{\partial x}\right) \tag{47-66a}$$

$$y_q = y - \left(r - r_0 + \sqrt{r_0^2 - x^2 - y^2} + W(x,y)\right) \times \left(\frac{y}{\sqrt{r_0^2 - x^2 - y^2}} + \frac{\partial W(x,y)}{\partial y}\right) \tag{47-66b}$$

x_q and y_q are called the transverse aberrations of the system. They are zero for each ray when the aberrations $W(x,y)$ of the system are zero and the filter plane (reference plane) is placed at the center point of the exit pupil sphere ($r = r_0$).

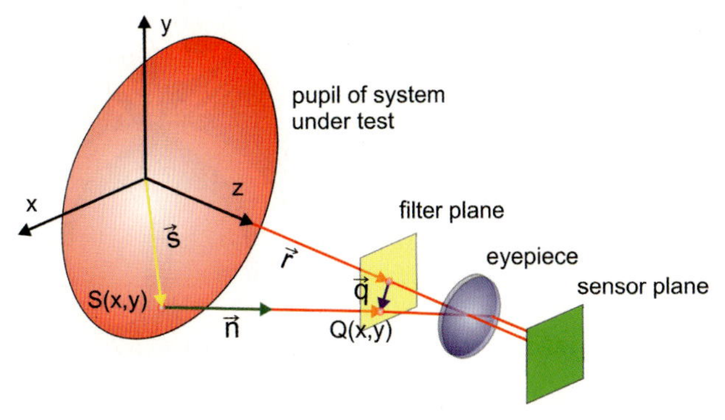

Figure 47-49: Vectors defining a ray emerging from the system under test and crossing the filter plane in a point image filtering setup.

Let us now define a complex filter function $F(x,y)$ according to (47-67), which will be applied in the filter plane.

$$F(x, y) = T(x, y)\, e^{i\psi(x,y)} \tag{47-67}$$

In the case of simple geometric investigations the phase contribution $\psi(x, y)$ of the filter is set to zero so that only the transmitting part is used. We can then define a variety of filters which, in the past, have been used for optical testing and which – except for the wire test – are named after their inventors.

	Filter function	Filter transmittance	Filtered pupil irradiance
Foucault knife edge test	$T(x,y) = \begin{cases} 0 & x < 0 \\ 1 & x \geq 0 \end{cases}$		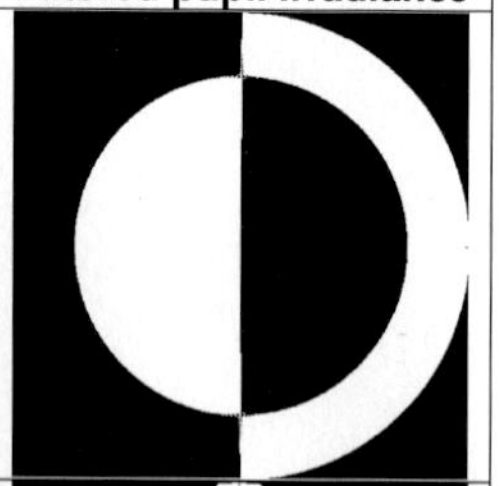
Toepler schlieren test	$T(x,y) = \begin{cases} 0 & \lvert x \rvert > \frac{b}{2} \\ 1 & \lvert x \rvert \leq \frac{b}{2} \end{cases}$		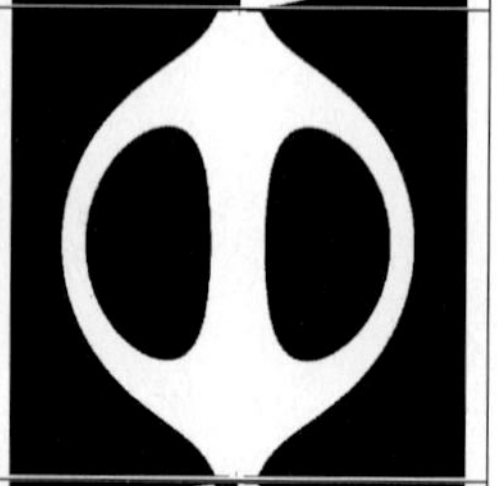
Ronchi test	$T(x,y) = rect(2\pi b)$	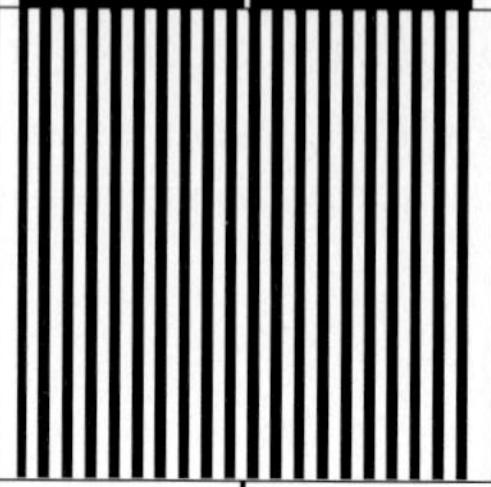	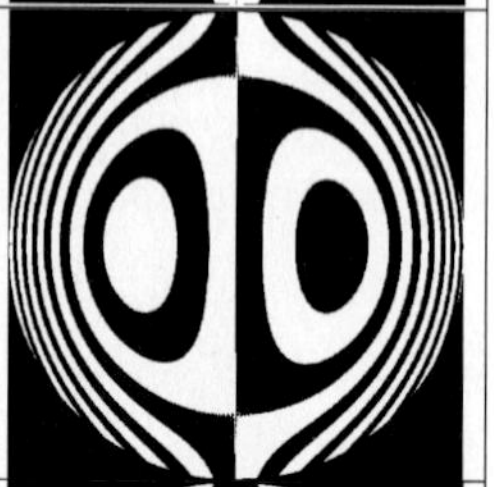
Wire test	$T(x,y) = \begin{cases} 0 & \lvert x \rvert \leq \frac{b}{2} \\ 1 & \lvert x \rvert > \frac{b}{2} \end{cases}$		

Figure 47-50: Filter functions and calculated irradiances in the image of the exit pupil for four different filters: a) Foucault knife-edge test, b) Toepler Schlieren test, c) Ronchi test, d) Wire test. Calculations have been made according to (47-69).

In the case of a diffraction optical calculation the complex electromagnetic field $a(x_q, y_q)$ in the filter plane can be calculated by Fourier transforming the complex pupil function $P(x,y)$. Equation (47-68) defines the relation accordingly. Aberrations are considered small enough for the irradiance in the filter plane mainly too cover a paraxial plane.

$$a(x_q, y_q) = \int_{y=-\infty}^{\infty} \int_{x=-\infty}^{\infty} P(x, y)\, e^{-i2\pi\frac{x_q x + y_q y}{\lambda \cdot r}} dx dy \tag{47-68}$$

with

$$P(x, y) = A(x, y)\, e^{\frac{i2\pi}{\lambda} W(x,y)} \tag{47-69}$$

as the complex pupil function. $A(x,y)$ defines the amplitude transmittance of the system, whereas $W(x,y)$ represents the unknown wavefront aberration of the system under test.

Applying the complex filter function from (47-67) by multiplication with $a(x_q, y_q)$ and back Fourier transformation to the pupil coordinates, we arrive at the complex field in the image of the exit pupil (47-70).

$$A'(x, y) = \int_{y_q=-\infty}^{\infty} \int_{x_q=-\infty}^{\infty} F(x_q, y_q)\, a(x_q, y_q)\, e^{i2\pi\frac{x_q x + y_q y}{\lambda \cdot r}} dx_q dy_q \tag{47-70}$$

In a simulation, the irradiance at the sensor can now be calculated for amplitude-blocking filters as in the case of the Foucault, Ronchi, Toepler and wire test, or for phase-manipulating filters. Test methods using phase filters are as follows.

1. The Smartt interferometer (see section 46.3.5).
2. The Ronchi test using phase gratings (lateral shearing interferometer, see section 46.3.6).
3. The Zernike test (the Smartt interferometer with a $\lambda/4$ phase disk).
4. The Lyot test (the wire test with a $\lambda/4$ phase wire).
5. The Wolter test (the knife edge test with a $\lambda/2$ phase edge).

In figures 47-52 and 47-53 we give an example of all the previously mentioned amplitude and phase-filtering techniques using the phase map shown in figure 47-51. The measured optical surface shows characteristic high-frequency ring-shaped structures from the polishing process with a peak-to-valley magnitude of 170 nm. The example has been chosen because of its considerable wavefront gradient. All examples – except for the Smartt interferometer – are primarily gradient detecting techniques, i.e., according to (47-70) measured irradiances are related to the gradient of the wavefront, which have to be integrated to obtain information about the wavefront.

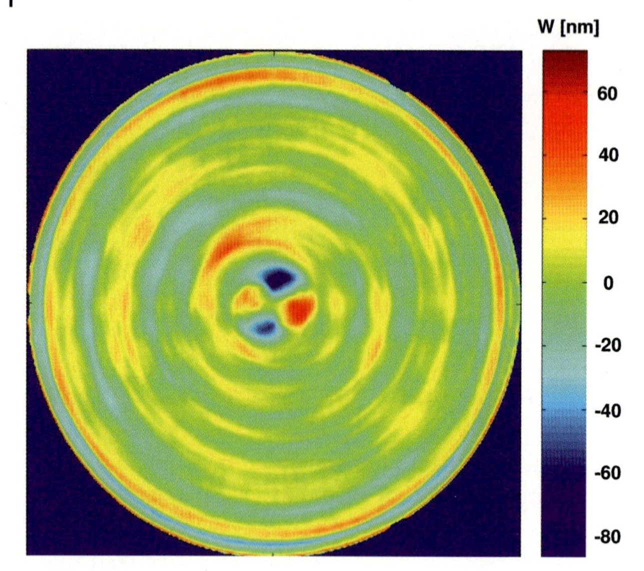

Figure 47-51: Phase map of an optical surface showing primarily ring-shaped structures from the polishing process. *PV* is approximately 170 nm.

It should be noted that most of the techniques are used for visual inspection of optical surfaces or systems, where the aim is to improve the visibility of the intensity-coded surface deviations. Most of the techniques are not very well suited for a quantitatively exact measurement of the surface deviations. It should also be mentioned that – although most of the techniques are very simple to set up and are not related to interferometry – they do show surface defects which are highly resolved in comparison with interferometric techniques, especially when the systems or components have long focal lengths or radii of curvature, respectively.

Knife edge test	$T(x,y)=\begin{cases}0 & x<0\\ 1 & x\geq 0\end{cases}$ $\psi(x,y)=0$	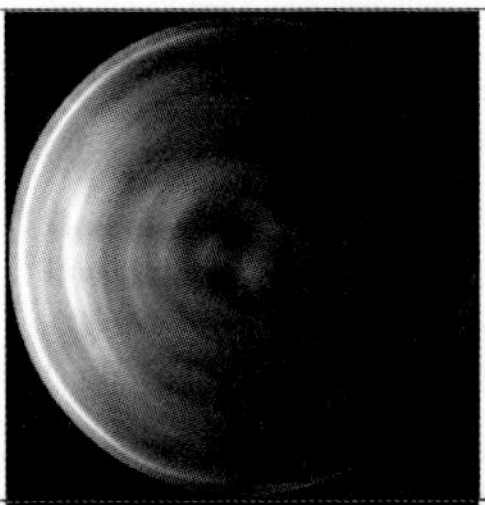
Wolter test (λ/2 knife phase edge)	$T(x,y)=1$ $\psi(x,y)=\begin{cases}0 & x<0\\ i\pi & x\geq 0\end{cases}$	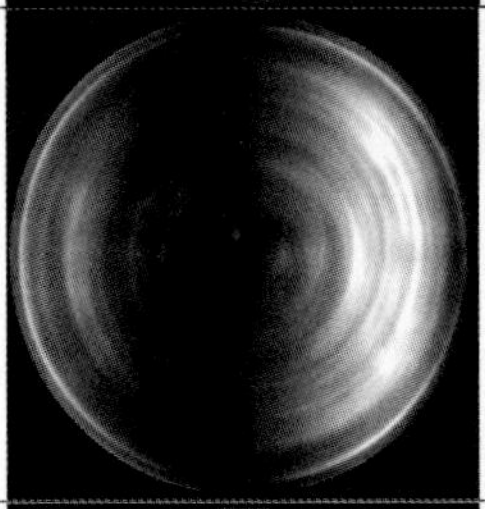
Wire test	$T(x,y)=\begin{cases}0 & \lvert x\rvert\leq\frac{b}{2}\\ 1 & \lvert x\rvert>\frac{b}{2}\end{cases}$ $\psi(x,y)=0$	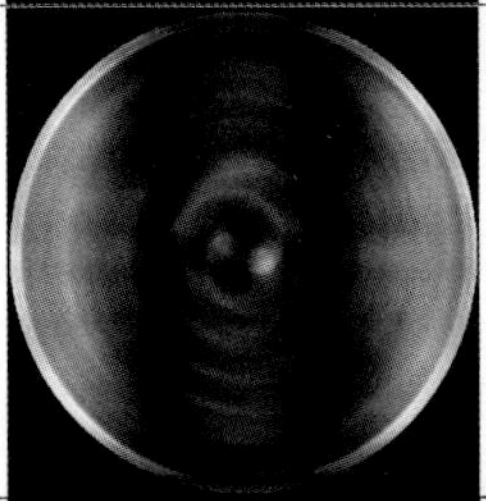
Lyot test (λ/4 phase wire)	$T(x,y)=1$ $\psi(x,y)=\begin{cases}0 & \lvert x\rvert\leq\frac{b}{2}\\ i\frac{\pi}{2} & \lvert x\rvert>\frac{b}{2}\end{cases}$	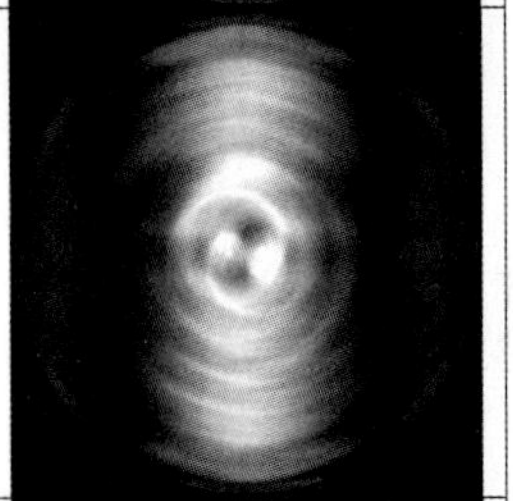
Smartt interferometer	$T(x,y)=\begin{cases}1 & \sqrt{x^2+y^2}\leq\frac{b}{2}\\ <1 & \sqrt{x^2+y^2}>\frac{b}{2}\end{cases}$ $\psi(x,y)=0$	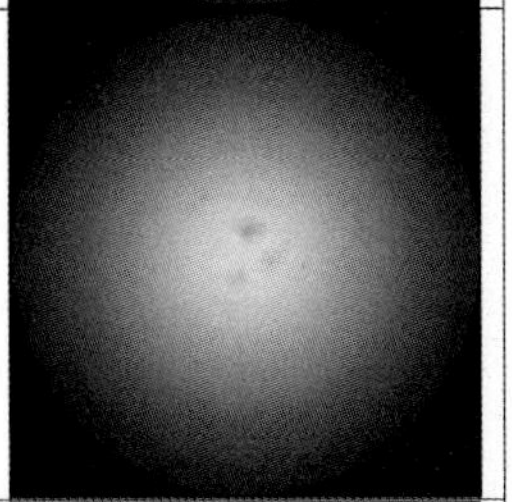
Zernike test (Smartt interferometer with a λ/4 phase disk)	$T(x,y)=1$ $\psi(x,y)=\begin{cases}i\frac{\pi}{2} & \sqrt{x^2+y^2}\leq\frac{b}{2}\\ 0 & \sqrt{x^2+y^2}>\frac{b}{2}\end{cases}$	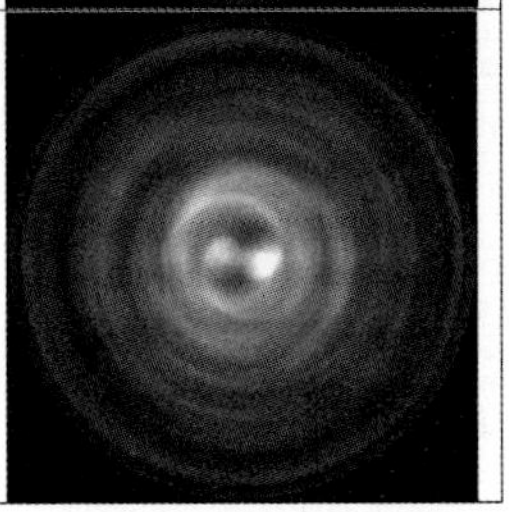

Figure 47-52: Point image amplitude and phase filtering techniques applied to the wavefront shown in figure 47-51.

Ronchi test (rectangular amplitude grid)	$T(x,y)=\frac{1+sign(\cos(2\pi bx))}{2}$ $\psi(x,y)=0$	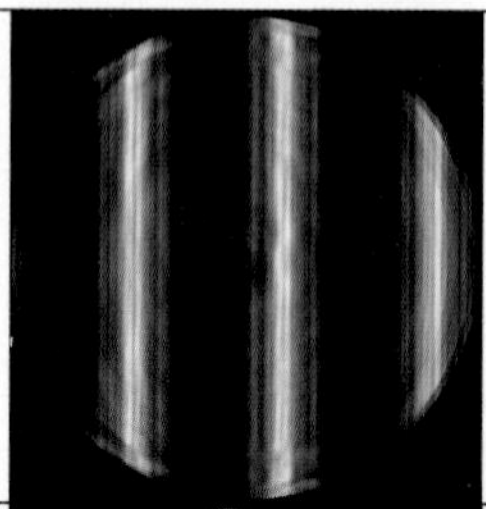
Ronchi test (sinusoidal amplitude grid)	$T(x,y)=\frac{1+\cos(2\pi bx)}{2}$ $\psi(x,y)=0$	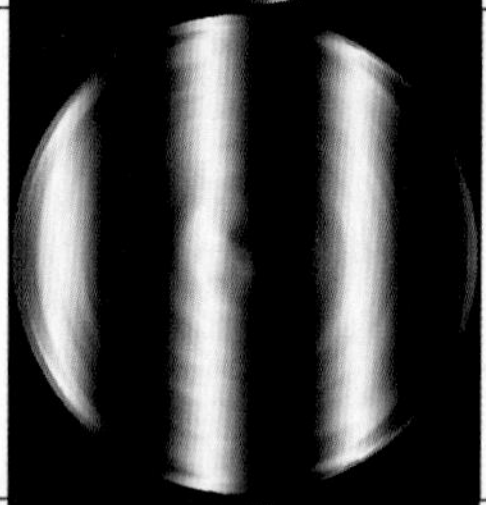
Ronchi test (rectangular λ/2 phase grid)	$T(x,y)=0$ $\psi(x,y)=i\pi\frac{1+sign(\cos(2\pi bx))}{2}$	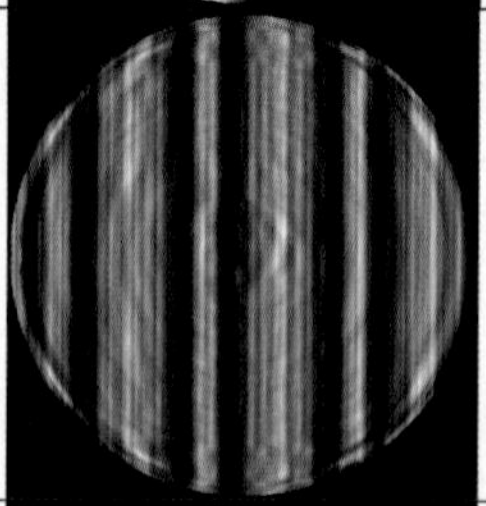
Ronchi test (sinusoidal λ/2 phase grid)	$T(x,y)=0$ $\psi(x,y)=i\pi\frac{1+\cos(2\pi bx)}{2}$	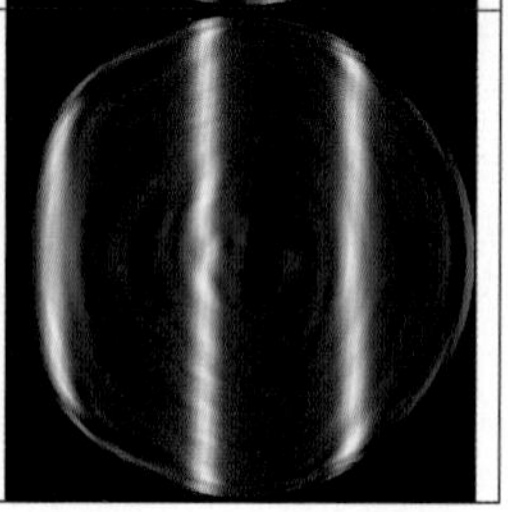

Figure 47-53: Point image amplitude and phase filtering techniques of the Ronchi type applied to the wavefront shown in figure 47-51.

Some of the methods are suitable for use as digital wavefront sensors, where the camera is connected to a computer and the filter is moved by computer control to collect a set of irradiance distributions from which transverse aberrations of the system can be calculated. The Ronchi test and the knife edge test are the preferred methods, since they deliver transverse aberrations in a very direct way, from which wavefront aberrations can be calculated by integration techniques.

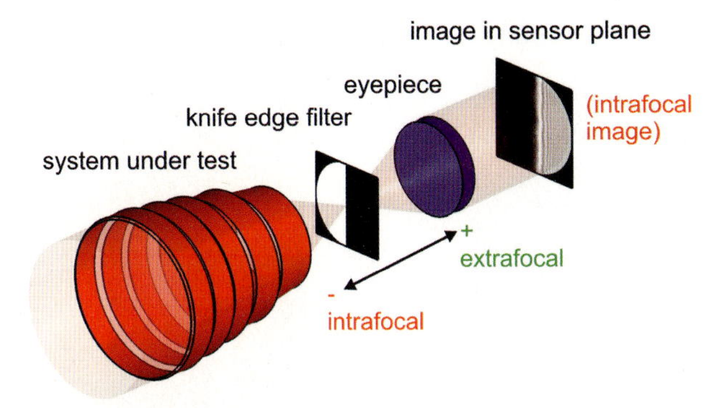

Figure 47-54: Knife edge test with the knife edge filter in the intrafocal position. The shaded side in the knife edge image is flipped when the position of the filter is closer to the system under test than its focal point. In the extrafocal position the shaded sides in the image are equal to the shaded side in the filter plane.

The irradiance distribution in the sensor plane changes when the position of the filter is altered. If information is to be supplied that should enable a computer controlled measurement of transverse aberrations or wavefront maps, then the filter has to be moved in a known way either laterally or axially. We call a position of the filter "intrafocal" when it is closer to the exit pupil than its focal point, or "extrafocal" when it is farther from the exit pupil (figure 47-54). In figure 47-55 we show a set of different intra- and extrafocal positions of a knife edge and the related knife edge images on the sensor plane. Since in the usual setup the camera is focused onto the exit pupil of the camera, the knife edge cannot also be imaged sharply onto the sensor. In intra- and extrafocal positions we therefore see diffraction fringes parallel to the edge image, especially when monochromatic light is used. With the filter at the focal point of the system the irradiance in the sensor plane resembles a landscape with valleys and hills illuminated from the side.

Position of knife edge	Irradiance in knife edge plane	Image in sensor plane (knife edge image)
-10000 nm defocus		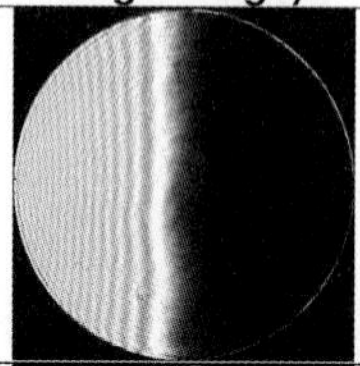
-5000 nm defocus		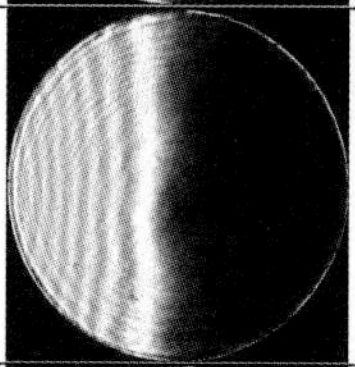
-2000 nm defocus		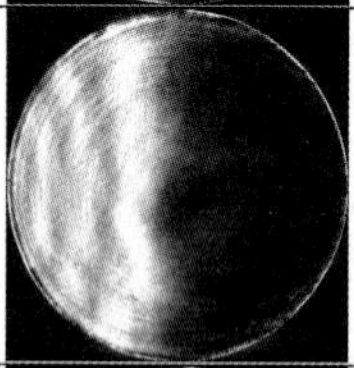
0 nm defocus		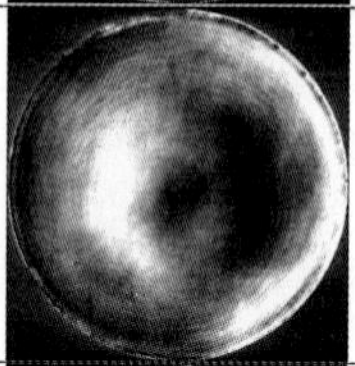
+2000 nm defocus	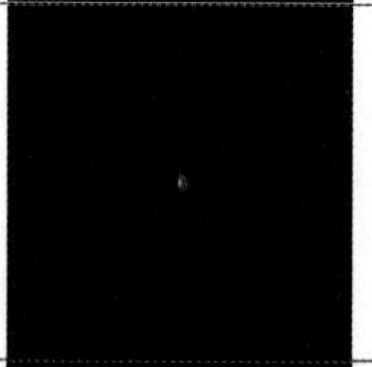	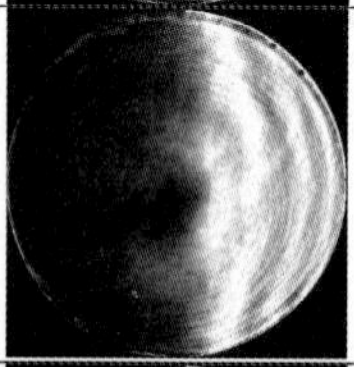
+5000 nm defocus	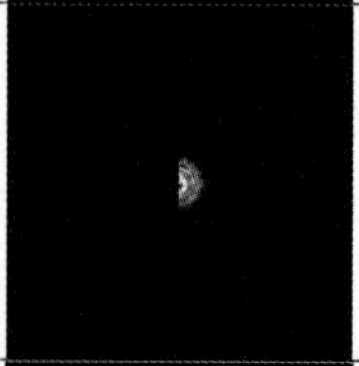	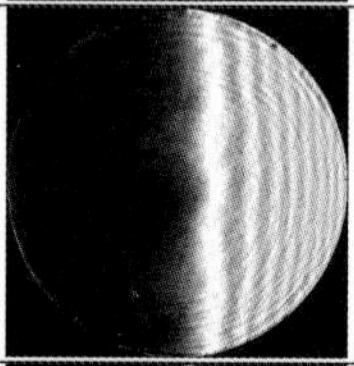
+10000 nm defocus	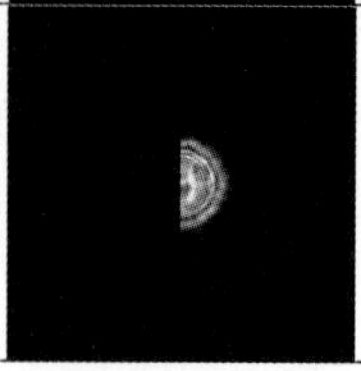	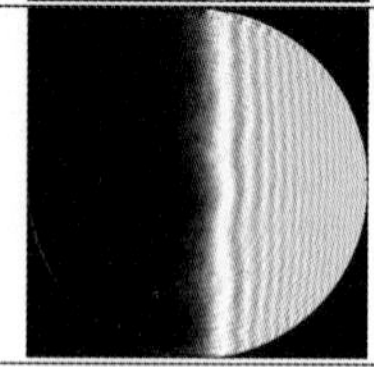

Figure 46-55: Knife edge test with the knife edge filter in different positions. Negative denoted positions are called intrafocal (the knife edge is closer to the system under test than its focal point), positive positions are called extrafocal (the knife edge is further from the system under test than its focal point).

47.6 Other Wavefront Sensor Concepts

47.6.1 Pyramid Curvature Sensor

A pyramid wavefront sensor is a simple sensor used to measure the slope and the curvature of a wave in an easy and quick way. Therefore, it is particularly useful for fast adaptive systems in telescope applications [47-53], [47-54]. The main component of this sensor concept is a pyramid prism as shown in figure 47-56. The pupil is imaged onto a sensor plane by a relay lens. The prism is located near the imaging lens and divides the pupil into four separated parts (1–4) on a four-quadrant sensor, which generates the signals E_1 to E_4. A wavefront error in the system pupil with the corresponding angle, changes the spatial distribution on the prism and so the power contributions on the four quadrants are redistributed. According to the well known evaluation of a four-quadrant detector, the slope of the signals can be obtained by the equations

$$S_x = \frac{E_1 + E_2 - E_3 - E_4}{E_1 + E_2 + E_3 + E_4} \tag{47-71}$$

and

$$S_y = \frac{E_1 - E_2 - E_3 + E_4}{E_1 + E_2 + E_3 + E_4} \tag{47-72}$$

and the mean curvature is calculated from

$$S_c = \frac{E_1 - E_2 + E_3 - E_4}{E_1 + E_2 + E_3 + E_4} \tag{47-73}$$

Due to these simple formulas, the very fast signal processing gives this sensor a high speed. The main advantage of this approach is the simple setup and the lack of moving parts. To obtain a more robust behavior, the prism can be oscillated in a dynamic mode of the sensor. The modulation radius then determines the compromise between accuracy and dynamic range. A more detailed analysis of the sensor and its properties can be found in the literature.

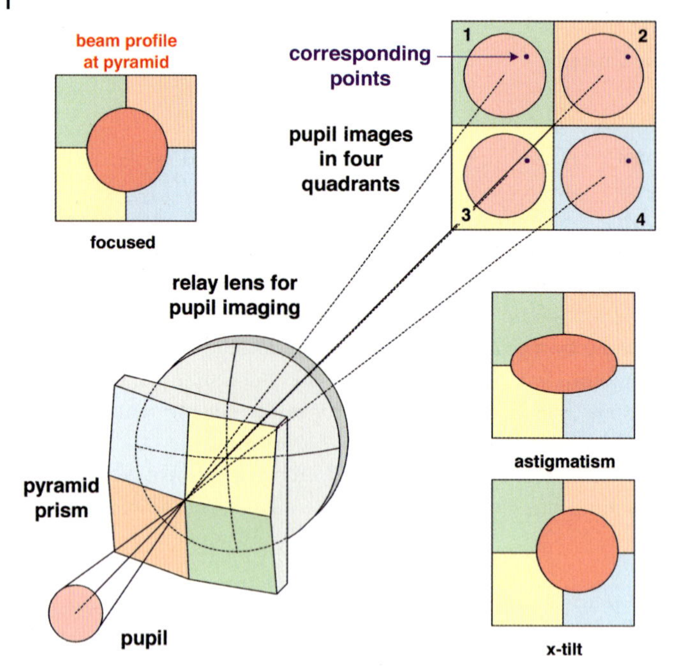

Figure 47-56: Principle of a pyramidal wavefront sensor. Three cases with different spot deviations in the prism plane are shown.

47.6.2
Hartmann–Moire Wavefront Sensor

A special concept, which uses two Hartmann screens at a certain distance from one another, is the Hartmann–Moire sensor [47-55]. It has the advantage of a quite large dynamic range. The second screen is azimuthally rotated against the first one and therefore produces a Moire pattern. It is mainly used for
ophthalmic measurements of the human eye and allows for a measurement in the range of –20D to +18D with an accuracy of 0.1D for spherical errors.

47.6.3
Holographic Modal Wavefront Sensor

A holographic modal sensor contains a special coded hologram [47-56]–[47-58]. The incident beam is divided into multiple focused beams, one pair of beams for each of the Zernike modes. The ratio of the beam power in each pair is a direct measurement of the coefficient of the underlying Zernike function.

The hologram is recorded with a reference beam with known aberrations. The modal weight is quite insensitive on apodization differences or obscurations.

One of the advantages of this sensor concept is its very fast performance because of the analog processing. Defocus aberrations can be measured in the range of –1.0 to +1.0 λ with an accuracy of $\lambda/100$.

47.6.4
Convolution Solvable Pinhole Mask

With the help of a special sampling array it is possible to measure the phase of a given coherent wave or to image a phase object [47-59]. It is necessary to have an array of measuring pinholes inside the mask and also one reference pinhole. The method is called a convolution solvable sampling array (CSSA). An advantageous and typical array pattern is hexagonal. In this case, the reference pinhole lies inside an equilateral triangle of three measuring points. Figure 47-57 shows the geometry of the pinhole grid and figure 47-58 illustrates the principle of the method.

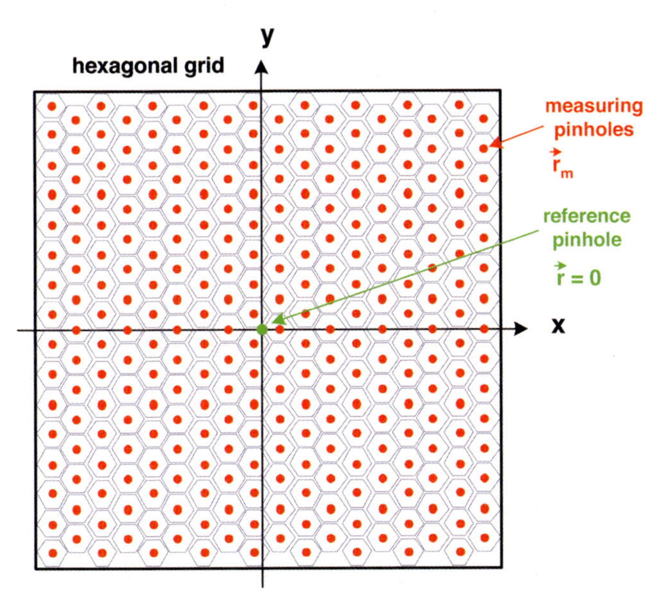

Figure 47-57: Geometry of the pinhole mask with measuring pinholes in a hexagonal array with a reference pinhole in the center.

In the ideal case, the size of the pinholes must be very small, so they can be considered as of negligible extent. In reality, this causes the problem of poor signal energy behind the array screen. In the mathematical description, the pinholes are modelled as ideal δ-functions at the positions r_m. If the reference pinhole is located at the origin of the coordinate system, the measuring mask can be described by the equation

$$P(\vec{r}) = \sum_{m=1}^{M} \delta(\vec{r} - \vec{r}_m) \tag{47-74}$$

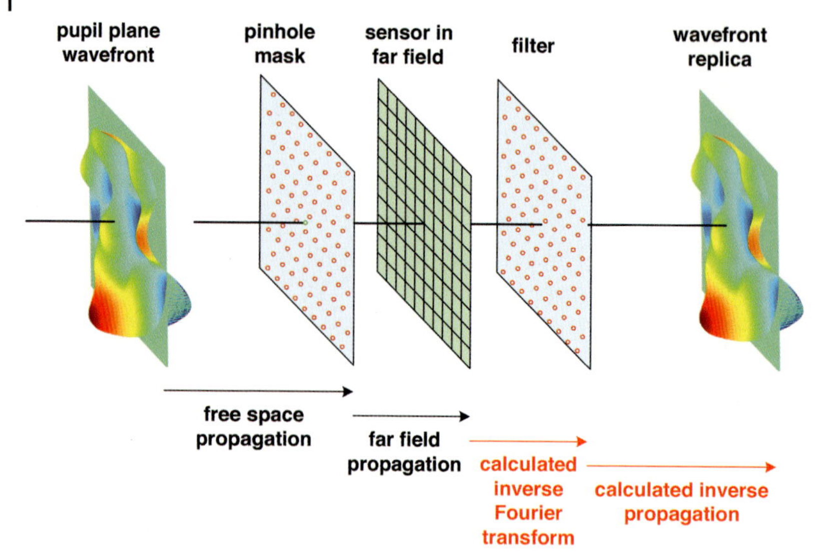

Figure 47-58: Principle of the CSSA method with the propagation part (on the left side) and the computational part (on the right side).

If the incoming field distribution is written as

$$E_0(\vec{r}) = A_0(\vec{r})e^{i\cdot\Phi(\vec{r})} \tag{47-75}$$

the far field pattern of the transmitted light field intensity is given by the following equation, where the reference pinhole is added in the complete mask function

$$I_{\text{far}}(\vec{\rho}) = I_0\left[\hat{F}\left\{E(\vec{r})\left[\delta(\vec{r}) + \sum_{m=1}^{M}\delta(\vec{r}-\vec{r}_m)\right]\right\}\right]^2 \tag{47-76}$$

with the frequency-related coordinates

$$\vec{\rho} = \vec{r}/(z\lambda) \tag{47-77}$$

If now an inverse Fourier transform is performed on this far field and an identical mask function P is applied, we get the expression

$$\begin{aligned} S(\vec{r}) &= P(\vec{r})\hat{F}^{-1}[I_{\text{far}}(\vec{\rho})] \\ &= S_0\sum_{m=1}^{M}\delta(\vec{r}-\vec{r}_m)A_{\text{ref}}A_m\, e^{i\cdot(\Phi_m-\Phi_{\text{ref}})} \\ &= E^*(0)\, E(\vec{r})\, P(\vec{r}) \end{aligned} \tag{47-78}$$

This simplified condition is obtained only if the pinhole pattern satisfies the condition

$$\vec{r}_m - \vec{r}_n \neq \vec{r}_k, \vec{r}_m - \vec{r}_n \neq 0 \tag{47-79}$$

From (47-78) it can be seen that the object wave can be extracted at the positions of the measuring pinholes form the signal S in the spatial domain by

$$E(\vec{r}_m) = \frac{1}{E^*(0)} P(\vec{r}_m) \hat{F}^{-1}[I_{\text{far}}(\vec{\rho})] \tag{47-80}$$

If the far field pattern is recorded by a sensor, the second inverse Fourier transform can be performed by digital signal processing inside the computer. It should be noted that the complete complex function of the object wave is reconstructed by this method.

If the Zernike coefficients of the phase surface are of interest, a phase unwrapping seems to be necessary after the inverse Fourier transform. The distance z between the object and the mask should have a size of at least

$$z \geq \frac{a\,L}{\lambda} \tag{47-81}$$

with the pattern period a and the object size L to avoid aliasing the obtained pattern. The diameter d of the pinholes limits the spatial resolution of the method. One of the drawbacks of this method is a quite weak signal strength when the pinhole size is fairly small.

In the following, a simple numerical example shows typical results for the measurement of a wavefront using the method described here. A mask with 217 pinholes of diameter 20 μm are considered inside a circular pupil of diameter 1.8 mm. Figure 47-59 shows the mask and the far field pattern, as it is obtained on the sensor. Figure 47-60 shows the results of the reconstruction of the wavefront. The computation is performed on a 512 × 512 point grid. The error in the resulting wavefront is $W_{rms} = 0.0011\ \lambda$, the recalculated Zernike coefficients are accurate to better than $\lambda/1000$. In the test calculations small aberrations and no measuring noise are assumed, unwrapping is not necessary in this case.

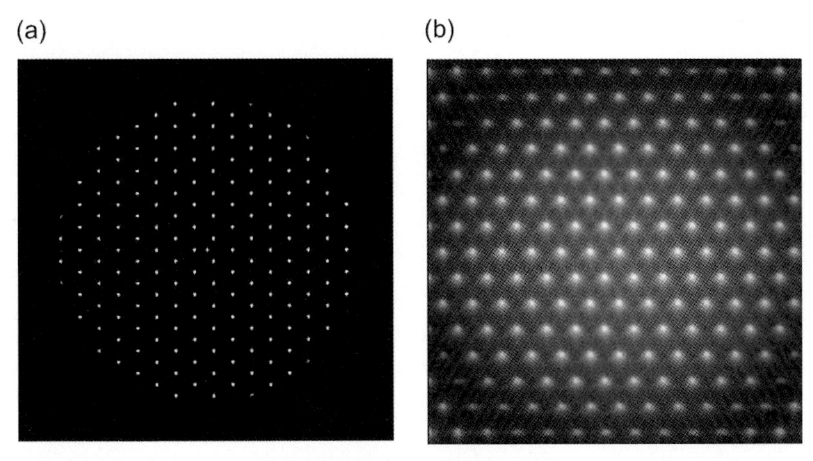

Figure 47-59: Pinhole mask (a) and far field pattern (b) on the sensor. The intensity is shown on a logarithmic scale.

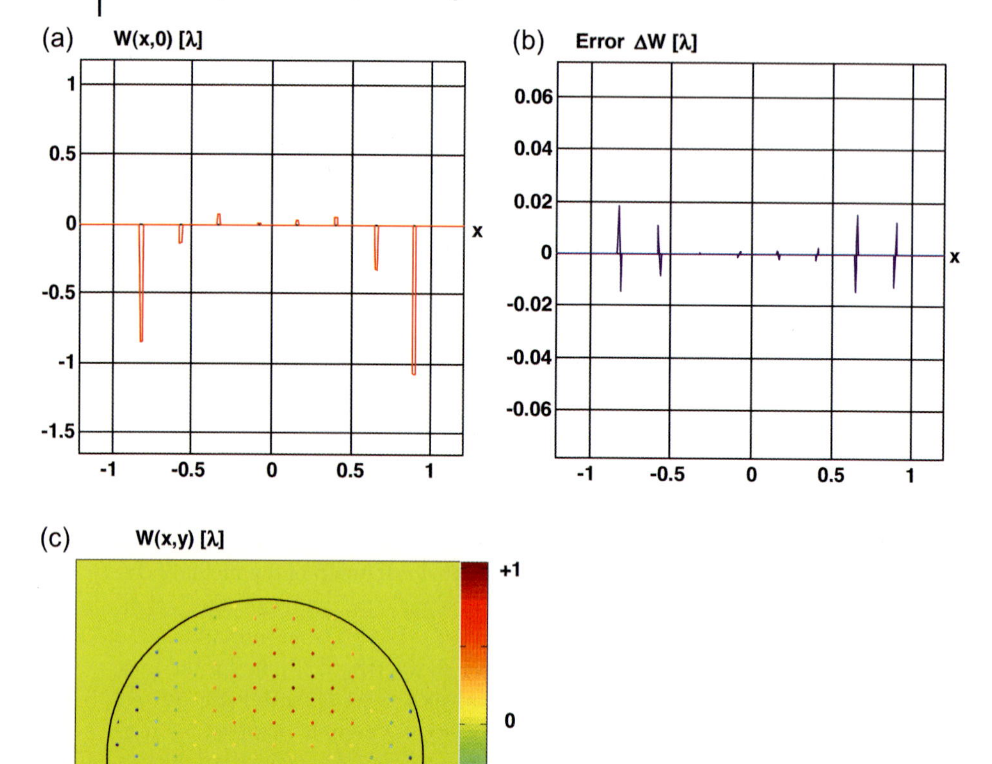

Figure 47-60: Result of the reconstructed wavefront as an x-section (a), the corresponding errors in the wavefront (b) and the two-dimensional pattern of the phase as a pinhole transmission.

47.6.5 Talbot–Moire Interferometer

A Talbot–Moire interferometer is a special setup, which allows the accurate measurement of a wavefront for extreme wavelengths in the x-ray regime as well as for non-coherent illumination conditions [47-60], [47-61]. The basic component of this setup is a phase Ronchi grating G_1 and a second amplitude grating G_2 with half of the grating constant.

If the Ronchi grating has a phase grating with phase step θ, the intensities of the orders with index m are given by [47-62]

$$I_m = \begin{cases} \cos^2(\theta/2) & \text{for} \quad m = 0 \\ 0 & \text{for} \quad m \text{ even} \\ \dfrac{4}{m^2\pi^2}\sin^2(\theta/2) & \text{for} \quad m \text{ odd} \end{cases} \tag{47-82}$$

Therefore the even orders vanish. The zeroth order vanishes if the phase step is equal to $\theta = \pi$. In this case, the low orders have the relative intensities

$$I_0 = 0, I_{\pm 1} = \frac{4}{\pi^2} = 0.405, I_{\pm 2} = 0, I_{\pm 3} = \frac{4}{9\pi^2} = 0.045, I_{\pm 4} = 0.... \tag{47-83}$$

Figure 47-61 shows the typical spectrum of a corresponding Ronchi grating with considerable intensities only in the odd diffraction orders.

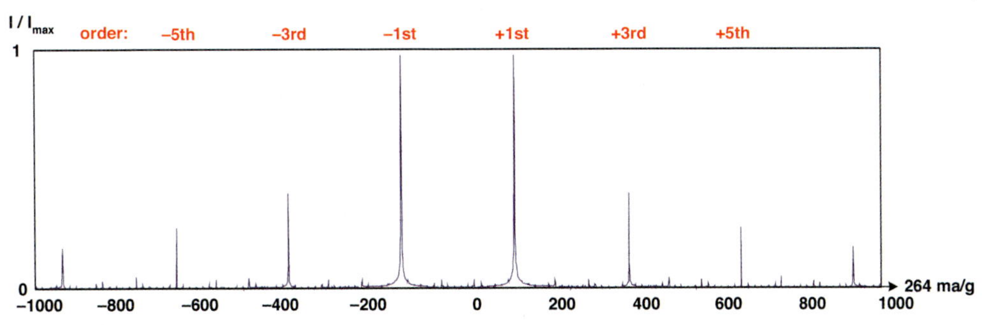

Figure 47-61: Typical spectrum of a Ronchi grating with suppressed even orders.

The intensity distribution $I(x,y,z)$ behind a phase Ronchi phase grating with phase step π as a function of the normalized distance has a periodic contrast as a function of z. This is the well-known Talbot effect [47-60], [47-62]. It can be seen that the contrast is maximal at distances

$$z_m = \frac{g^2}{2\lambda} \cdot \left(\frac{1}{4} + \frac{m}{2} \right) \tag{47-84}$$

with integer m. It should be noticed that in reality the contrast is less significant for larger distances due to diffraction effects and a finite coherence of the illumination source. This periodic behavior with the Talbot length $z_T = g^2/\lambda$ is shown in figure 47-62. It should also be noticed that the phase and intensity profiles are reproduced in alternating order and the sign of both also changes. The spatial frequency of the intensity modulation is twice that of the phase.

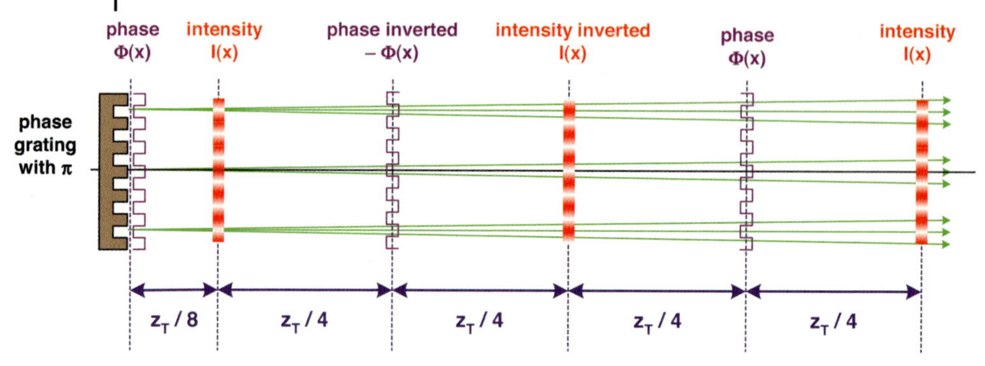

Figure 47-62: Principle of the Talbot effect for a phase grating with periodical axial refocused intensity profiles. The Talbot length $z_T = g^2/\lambda$ scales the period.

The idea of the Talbot–Ronchi interferometer is to use the two first-order diffraction maximal of a Ronchi grating to split the wavefront propagation directions as shown in figure 47-63. In a typical shearing interferometer with grating-based beam separation [47-63], the interference of both partial waves are superposed again after correcting the divergence angle with a second grating of the same type. In the Talbot–Ronchi interferometer an amplitude grating with half of the grating constant $g_2 = g_1/2$ is used as a second grating.

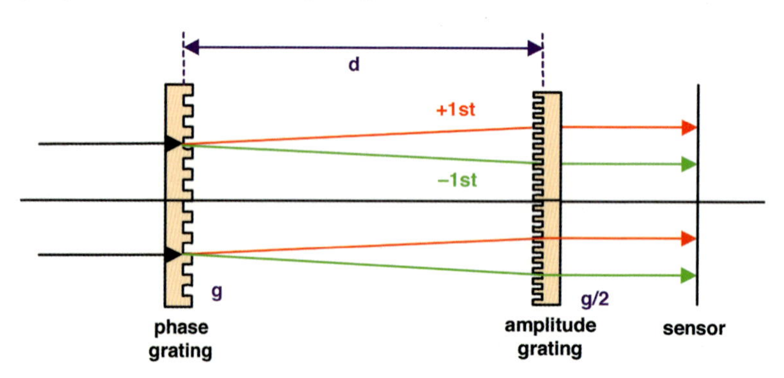

Figure 47-63: Basic setup of a Talbot–Moire interferometer.

The location of the second grating is chosen appropriately in a Talbot-related plane, where the intensity profile is reproduced and therefore the transmission at the second grating is a very sensible function of the transverse beam shift and therefore also of the inclination angle of the incoming wave. Figure 47-64 illustrates this effect. The grating G_2 therefore acts as an analyzer and the effect of an inclination angle φ is a reduced power transmission, which has to be measured as spatially resolved.

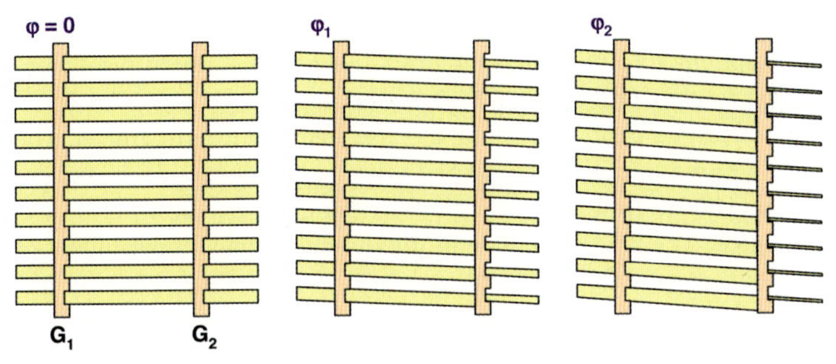

Figure 47-64: Throughput of a Talbot–Moire interferometer used to increase the angle of incidence φ.

In order to measure a wavefront, the pupil or a conjugated replica of it is located at the first grating and the camera sensor is located directly behind the second grating. A phase error in the incoming wavefront causes a lateral displacement of the interference fringes at the position of the analyzing amplitude grating of size

$$\Delta x(x, y) = \frac{\lambda\, d}{2\pi} \frac{\partial W(x, y)}{\partial x} \tag{47-85}$$

From this equation one can see that the sensitivity increases with the distance d.

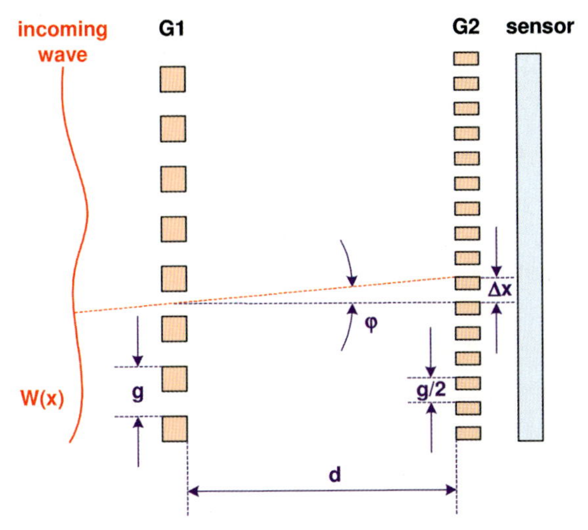

Figure 47-65: Measurement of a wave aberration *W(x)* with the help of a Talbot–Moire interferometer.

If the incoming wave has a spherical curvature R, the Talbot distance between the two gratings has to be modified [47-62] to

$$d = d_0 \frac{R}{r + d_0} \tag{47-86}$$

According to the spatial frequency and the location of the special Talbot planes, there are different options for optimizing the overall length, the spatial resolution, the accuracy and the contrast of the Talbot–Moire interferometer [47-63], [47-64].

The measurement can be performed using a phase-stepping interferometer. The signal processing required to obtain the desired wavefront corresponds to conventional PSI-evaluation [47-65]. In this case the contrast changes periodically if the second grating is scanned. The relative change in these oscillations is independent of the intensity or apodization of the illuminating beam. Therefore, the phase- sensitive image mode is quasi-achromatic. Neither the position nor the lateral position depend on the wavelength.

The transverse spatial resolution of the setup is equal to $\Delta x = g_2 = g_1/2$. Typically, for practical reasons, a transverse displacement of $g_2/10 = g_1/20$ can be detected [47-66]. This and the finite size of the camera pixel defines the accuracy.

According to the well-known principle of the Moire technique, the sensitivity of the system can be improved by a tilt θ of the second grating around the azimuth. This corresponds to an effective grating constant [47-60], [47-67], [47-68]

$$g_{\mathrm{eff}} = \frac{g}{2\sin(\theta/2)} \tag{47-87}$$

This principle is illustrated in figure 47-66.

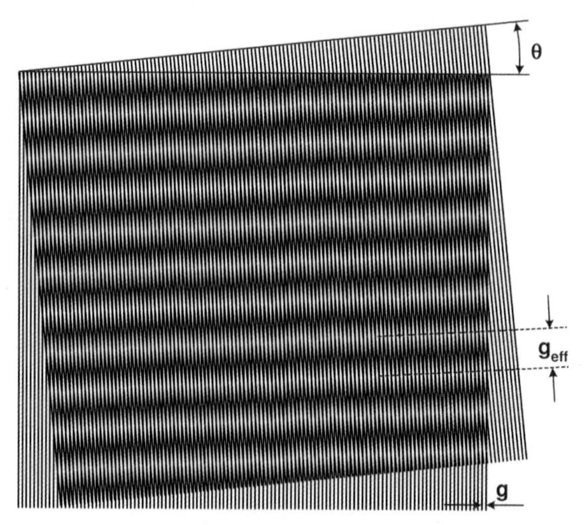

Figure 47-66: Improved sensitivity of the Talbot–Moire interferometer by an azimuthal rotation θ of the second grating.

For a simultaneous measurement of the wave aberration in the *x*- and *y*-directions, several options are possible in order to obtain the data. One variant is to rotate the grating by 90° and to determine the aberration derivatives in the x- and y-direction in a sequential manner. However, it is more appealing to gather the data simultaneously by using a crossed grating or a checkerboard geometry.

Although the technique described is an interferometric method, partial coherent illumination of the setup is possible. In this case, the generation of the Talbot grating at the second grating G_2 is convolved by the finite size of the source. This gives a reduced contrast and the distance d of the grating should be decreased to reduce these broadening effects.

47.7 Point Spread Function Retrieval

47.7.1 Introduction

The recovering of phase information from intensity measurements through focus image stacks has long been an interesting topic [47-69]–[47-72]. Phase retrieval avoids expensive and environmentally sensitive direct phase measurements. The method is applicable to the microscopic imaging of phase objects, to the layout and optimization of phase structures for desired illumination distributions in beam forming setups and also for the metrology of different systems.

In the case of aberration measurements, the concept is of increasing interest, because of some significant advantages in comparison with conventional techniques such as interferometry, Hartmann or Hartmann–Shack sensors [47-73]–[47-79]. The measurements can be done easily employing a movable camera, which should only measure intensities. Furthermore, the investigation of off-axis field points can be carried out without additional effort.

The accuracy of phase retrieval is almost comparable to conventional interferometry, if the parameters and conditions for measurement, the underlying physical model and the analyzing algorithms are suitably chosen.

There have been many publications about phase retrieval over the last three decades and, in particular, the reconstruction algorithms have significantly improved during the last ten years. These retrieval procedures are inverse iterative and numerically ill-conditioned problems. Therefore, the quality of the gathered data and the performance of the algorithms is essential for highly accurate results.

Some recent publications describe phase retrieval without an iterative computation using the so-called extended Zernike approach [47-80]–[47-83]. This method is very fast, but offers only limited possibilities of extending the physical model of the setup.

47.7.2 Transport of Intensity Equation

If the complex field strength in the paraxial wave equation is expressed by the amplitude A and the phase W in the equation

$$E(x, y) = A(x, y)\ e^{ikW(x,y)} \tag{47-88}$$

a system of two coupled differential equations is obtained [47-84]

$$\frac{\partial A}{\partial z} = \frac{A}{2}\left(\frac{\partial^2 W}{\partial x^2} + \frac{\partial^2 W}{\partial y^2}\right) + \frac{\partial A}{\partial x}\frac{\partial W}{\partial x} + \frac{\partial A}{\partial y}\frac{\partial W}{\partial y} \tag{47-89}$$

$$\frac{\partial W}{\partial z} = -\frac{1}{2Ak^2}\left(\frac{\partial^2 A}{\partial x^2} + \frac{\partial^2 A}{\partial y^2}\right) + \frac{1}{2}\left[\left(\frac{\partial W}{\partial x}\right)^2 + \left(\frac{\partial W}{\partial y}\right)^2\right] \tag{47-90}$$

From the first equation for the propagation of the amplitude with z one can calculate the equation for the change in the intensity $I = A^2$ by calculating

$$k\frac{\partial I(x,y)}{\partial z} = -\nabla[I(x,y)\nabla W(x,y)] \tag{47-91}$$

or in a modified form

$$k\frac{\partial I}{\partial z} = -\nabla I \nabla W - I\nabla^2 W \tag{47-92}$$

This is the so-called transport of intensity equation (TIE). It can be seen from this equation that the phase W of the field influences the change in the intensity $I(x,y)$ of the propagation along the z-axis direction. Figure 47-67 illustrates this statement.

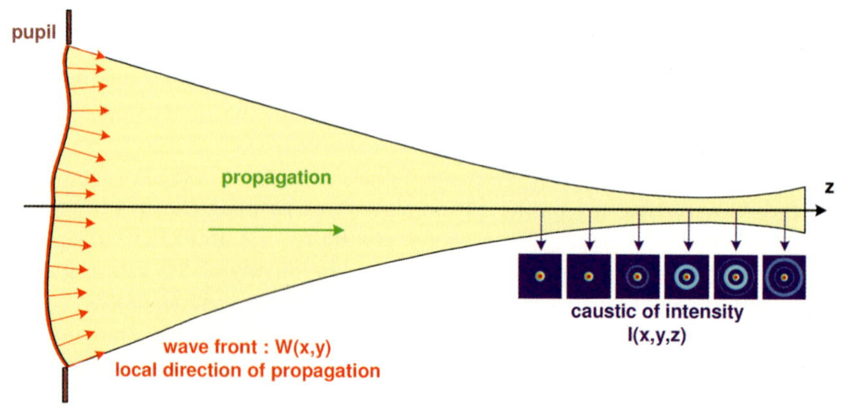

Figure 47-67: Sketch of the fundamental principle of intensity propagation for a wave with phase *W*.

If information is known about the change in intensity $I(x,y,z)$ in three dimensions, the transport of the intensity equation can be solved for W to get the phase of the initial field. As indicated in figure 47-67 by the arrows, (47-92) shows mathematically that the gradient and the curvature of the phase are responsible for the change in the intensity profile with z.

This knowledge enables us to recover the phase information from a measured known variation of the intensity [47-85]–[47-88]. This information may be obtained by measuring the lateral intensity $I(x,y)$ in several z-planes. This can be a z-stack of defocused images behind an optical system. However, notice that this procedure is an ill-conditioned inverse problem. Therefore the algorithm of the calculation is critical and it is desirable to gather redundant information in order to control problems of uncertainty and noise in the basic data.

47.7.3 Principle of Phase Retrieval

The basic principle of classical phase retrieval is sketched in figure 47-68. The wavefront in the exit pupil of an optical system determines the intensity distribution in the focal region of the image space. In particular, the residual aberrations cause significant deviations from the case of a perfect wavefront. Therefore, the three-dimensional intensity distribution contains information about the desired pupil function.

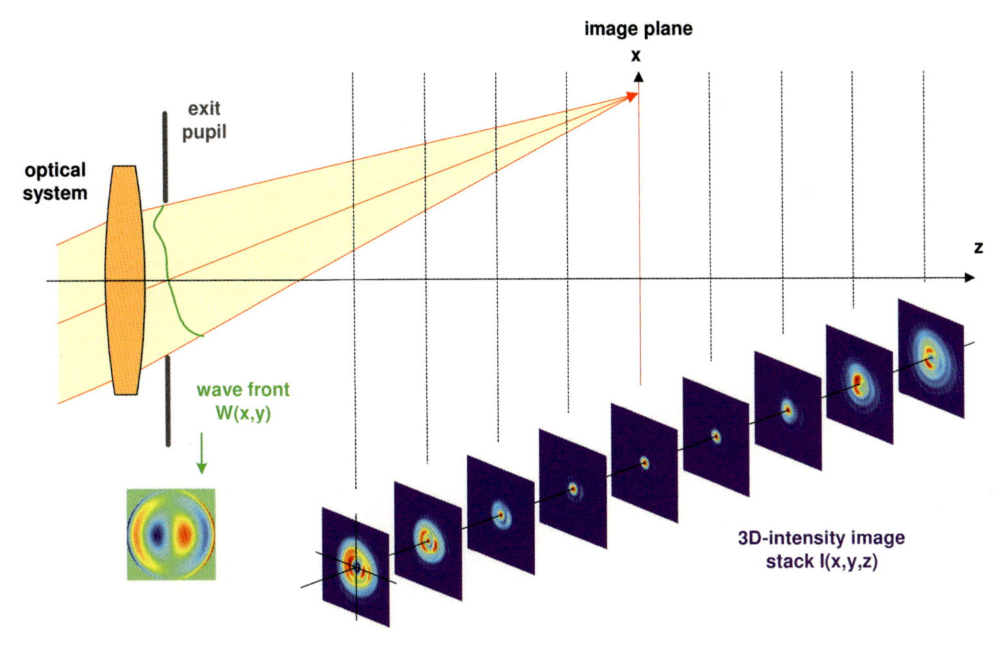

Figure 47-68: The fundamental principle of the phase retrieval method.

If the transport intensity equation is applied to the transfer of light in the optical system, it is possible to obtain the phase $W(x,y)$ of the light in the exit pupil from its propagation in z. As the equation cannot be solved for the phase W explicitly, an iterative solution is necessary. For this purpose, the gradient of the intensity along the optical axis is approximated by detecting several z-planes in the image of a single object point. The modification of the lateral intensity distribution in this z-stack contains the phase information of the light.

In this simple consideration, a point source is assumed, which generates a coherent wavefront. In reality, this condition is often not fulfilled and to get enough energy on the detector, a pinhole of finite size is used in the object space. If this more general setup is considered, a convolution of the known object structure with the unknown transfer functions forms the image. Figure 47-69 shows this setup schematically. In addition to this extension of the underlying model, a uniform illumination of the pupil is often not realized in practice. In this case, the apodization effects of the pupil must be taken into account.

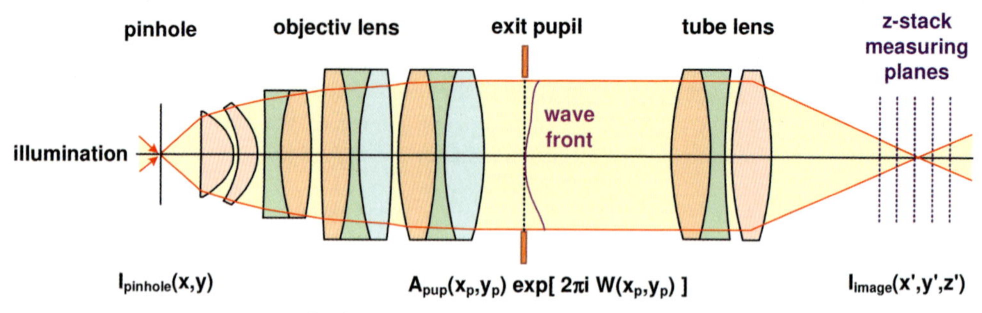

Figure 47-69: Setup for classical phase retrieval and example of measured intensity data as a z-stack in the image space for the case of a microscope objective with tube lens.

If we consider phase retrieval in the phase space, which describes the light in space and angle coordinates, we get the following interpretation [47-89]. The image plane corresponds to the space coordinate and the pupil plane represents the angle. This is shown in figure 47-70 for illustration. The transfer of the light from the pupil to the image corresponds to a rotation of 90° in the phase space diagram. The defocused planes represent smaller or greater rotation angles. Since the field obeys the wave equation and propagates deterministically, from the complex field in one plane or one angle orientation, all other planes can be reconstructed. However, since the reconstruction algorithm is a tomographic method, a high number of defocused planes and a broad spread in the scanned z-interval will improve the accuracy of the method. Furthermore, from this image it can be seen that the accuracy is strongly influenced by the detector pixel size and a suitably chosen magnification. One has to take care that in the focal plane there are enough pixels across the diameter of the small peak of the point spread function in the image plane. On the other hand, in the planes of largest defocus, a truncation of the intensity profiles must be avoided.

From a more general point of view, the use of the defocused planes is one way of using a diversity of states to gather the hidden phase information. This can also be done with other changes in the system behaviour; for example, by switching a filter with different configurations in the pupil [47-90], [47-91] or an added rotatable cylindrical lens [47-92]. The defocusing diversity is the easiest way to observe the system response for different input–output constellations [47-93]. But note that every phase diversity method has its own spatial frequency profile and therefore the recovering of the phase depends on the frequency properties.

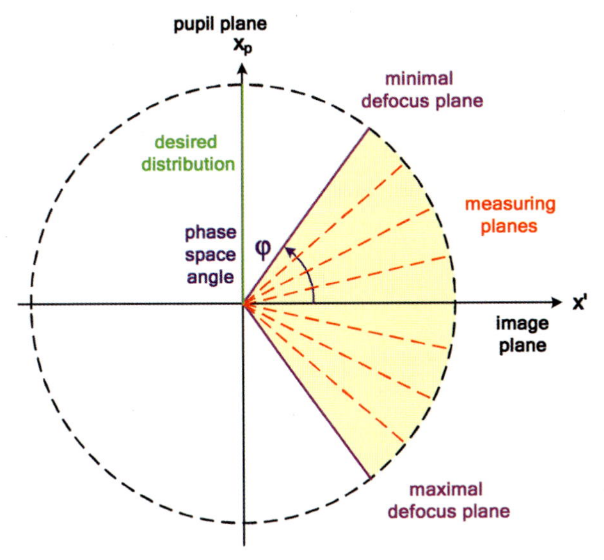

Figure 47-70: Illustration of the phase retrieval problem in the phase space.

The defocusing is particularly insensitive for small frequency components, since the defocus corresponds to a quadratic phase diversity function, which has a zero on the optical axis. Therefore, the low- frequency components, which are located near the pupil axis, are not changed very effectively by defocus and cannot be easily recovered.

In the easiest case of a system with very small aberrations, the effect of the lower order Zernike coefficients for the description of the residual errors can be considered to influence the point spread function independently and linearly. Then it is sufficient to take two symmetrical defocused images and calculate simple differences [47-94], [47-95].

47.7.4
Experimental Settings

One of the critical points in the PSF phase retrieval is the illumination and generation of a quasi point-like object. The dominant problem is to get a sufficiently small pinhole, which is critical in the case of high numerical aperture systems. There are different possible ways of achieving the illumination. Some options are as follows.

1. Physical pinhole in a chromium mask of a size in the range of the Airy diameter or smaller. The pinhole is trans-illuminated by a special system.

2. Epi-illumination and a plane mirror in the object plane. The objective lens is passed twice in this case and the asymmetrical aberrations such as coma are not detected in the first approximation.
3. Use of an identical calibrated or at least an equivalent well-known objective lens. The point object is generated in this case as an image in air. In the case of high numerical aperture this avoids the preparation and manufacturing of a sub-micron sized pinhole.
4. Use of a self-luminous small source in the object plane such as a SNOM cantilever, an excited gold bead or a quantum dot.

The first three possibilities are shown in figure 47-71.

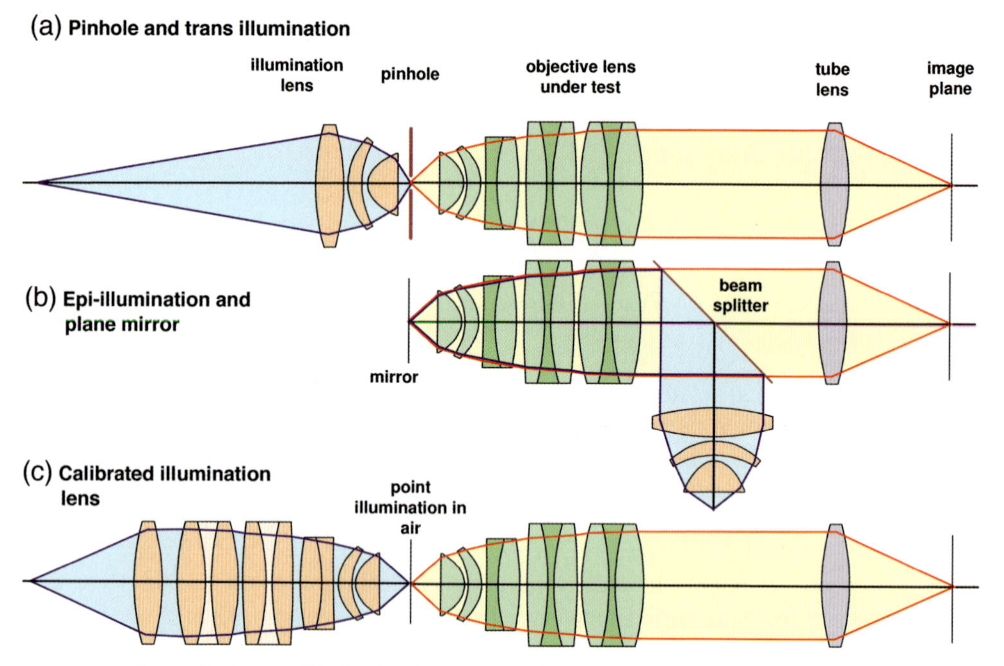

Figure 47-71: Possible illumination setups for the measurement of the point spread function.

In practice, to obtain accurate results it is an important question to set up the experimental conditions appropriately. The critical parameters are as follows.

1. Choice of optical magnification to get a good resolved image stack.
   For a good lateral resolution the size of the point spread function in the image plane is critical. It is recommended to obtain 10–20 camera pixels over the Airy diameter of the intensity profile in the image plane.
2. Choice of the z-planes and the defocusing interval.
   If the number of z-planes is too small, there is insufficient information for the solution of the inversion problem. According to the phase space consideration above, it is always an advantage to have measurements in strongly

defocused planes. One problem with large defocusing is a possible truncation of the signal at the boundary of the camera. A second problem is the decrease in the signal strength with z^2, which means that, in the large defocus range, the signal to noise ratio gets worse.
Typical values are 7–15 z-planes in a range from –3 R_u...+3 R_u with the normalized Rayleigh unit $R_u = n\lambda/NA^2$.
Figure 47-72 shows the typical decrease in the peak intensity for a gaussian apodization and the Airy distribution with uniform pupil illumination for illustration. In the case of a gaussian apodization, the peak is located on the axis; for the Airy distribution, the peak height is always taken at the highest intensity value, which may be on a sidelobe ring for corresponding distances. If z_T is a measure for the axial depth of focus, which is equivalent to R_u but scaled by a factor of size near to 1, the general behavior is approximately a law of Lorentz type [47-81]

$$I_{max}(z) = \frac{I_{peak}}{1 + (z/z_T)^2} \tag{47-93}$$

It can be seen from this formula or figure 47-72 that, in the range of a defocus with 3 R_u, the signal strength defocus falls into a range of 3%–10%, depending on the exact apodization and aberration conditions.

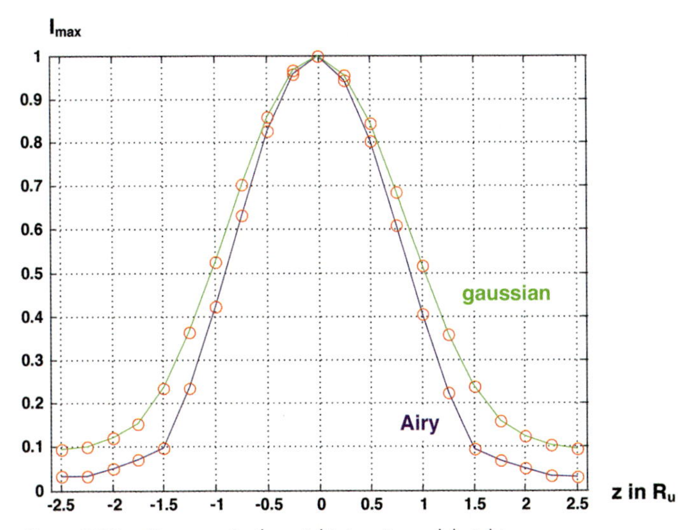

Figure 47-72: Decrease in the axial intensity peak height as a function of defocus, scaled in normalized Rayleigh units for gaussian apodization and Airy distribution, respectively.

3. Illumination setting.
   If an illuminated physical pinhole is used, the size of the pinhole is an essential parameter. If the diameter is in the range [47-78], [47-74]

$$D_{ph} \leq 0.3\ D_{airy} \tag{47-94}$$

the source can be considered to be quasi point-like and the data processing can be done without a deconvolution operation, which is desirable. Figure 47-73 shows the result of a simulation on the influence of the finite size of the pinhole on the image sided point spread function. Two different criteria are used; the intensity at 10% and 50 % of the peak intensity are compared, respectively. It can be estimated from this graph that the condition of (47-94) guarantees an error in the width of the point spread function of less than 8%, which is sufficient for most cases.

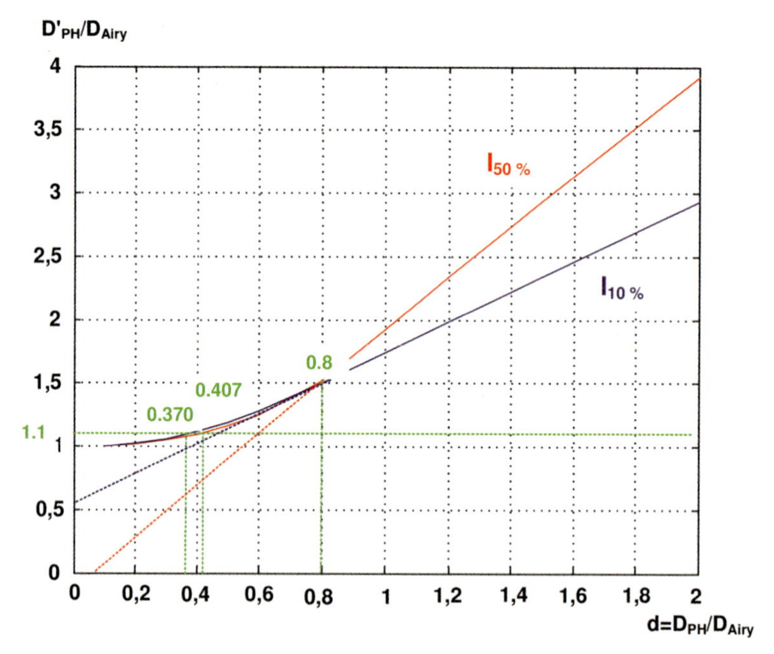

Figure 47-73: Change in the diameter of the point spread function at 10% and 50% of the peak intensity, respectively, due to a normalized pinhole of finite size *d*. The dotted lines indicate the asymptotic linear behavior for large values of the pinhole diameter.

4. Choice of the camera
   The camera used to obtain the image stack pictures must be large enough in diameter to guarantee that, in the largest defocusing setting, there is no relevant truncation of the signal by the detector size.
   Furthermore, to resolve the critical signal heights in the largest defocus position, it is recommended to use at least a 12-bit quantization for the camera dynamic [47-75].

In the measuring process, it is necessary to move the camera in the image space along the optical axis. The accuracy of the corresponding z-settings and the usefulness of the complete stack can be estimated using quite a simple approach. The focal caustic of a perfect diffraction-limited beam without phase perturbations follows a quadratic function of the second intensity moment with increasing distance from the image plane [47-96]. The second moment is defined as

$$M_2(z) = \frac{16}{P} \iint (x^2 + y^2)\, I(x, y, z)\, dx\, dy \tag{47-95}$$

where P is the power of the beam. In the ideal case of an aberration-free wave, the z-dependence can be written as

$$M_2(z) = A_2(z - z_F)^2 + A_1(z - z_F) + A_0 \tag{47-96}$$

where A_0, A_1 and A_2 are the constants of the parabola and z_F is the z-position of the focus. Figure 47-74 shows a practical example of the moments of the measured data and a parabola fitted to this set of points. In this type of representation, it is easy to see that the z-planes are located symmetrically around the image plane and which z-plane best represents the image plane. In practice, it is not necessary to have a perfect symmetric distribution of the z-planes and also to have a plane in the best image plane. But it is an advantage and should be approximated, if possible.

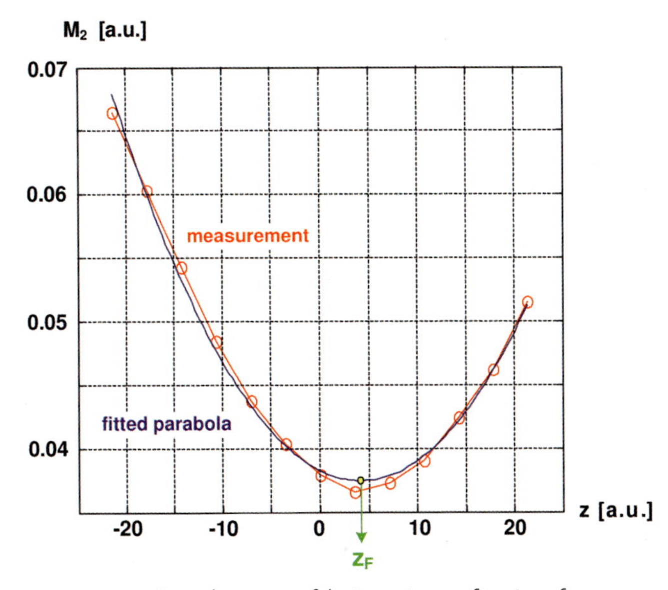

Figure 47-74: Second moment of the intensity as a function of the defocus z from the image plane. The red circles indicate real measurement data points, the blue line represents the parabola, which fits the data points.

Phase aberration causes a perturbation in the quadratic behavior of the second moment curve. In principle, it is also possible to fit a cubic polynomial, but this gives no relevant information about the appropriate setup of the measurement. If the numerical aperture is not known, an estimation or consistency check can be performed with the parameters of the parabola. This follows from the equation

$$NA = \frac{\sqrt{A_2}}{2} \tag{47-97}$$

47.7.5 Model Assumptions

One of the major aspects to consider in obtaining high accuracy for point spread function phase retrieval is the use of an appropriate physical diffraction model for the generation of the image stack. Usually, the measuring planes are quite near to the image plane and therefore the Fraunhofer approximation of far field diffraction is applicable. If the numerical aperture of the objective lens under test is not too high, the classical paraxial calculation of the diffraction point spread function, as described in chapter 20 and the appendix A1 of volume 2 of this book series, will be adequate. For the defocusing planes a quadratic phase term is assumed. This model delivers good results for Fresnel numbers above $N_F > 20$ and numerical apertures below $NA < 0.5$.

If the numerical aperture is high or quite low, there are ranges which need a more sophisticated model for the calculation of the point spread function. If the numerical aperture is in the range between the values $NA = 0.5$ and 0.8, a scalar extension of the Kirchhoff diffraction integral can be used, which can be written as [47-97]

$$E(x', y', z) = \int_{-\infty}^{\infty} \int_{-\infty}^{\infty} \frac{P(x_p, y_p)}{\sqrt[4]{1 - r_p^2 (NA/n)^2}} e^{-\frac{i\pi \cdot z \cdot \sin^2 u}{\lambda} \cdot \frac{1-\sqrt{1-r_p^2 NA^2}}{1-\sqrt{1-NA^2}}} e^{-\frac{ik}{z_{image}} \cdot \left[x' x_p + y' y_p\right]} dx_p \, dy_p \tag{47-98}$$

In this formulation, $P(x_p,y_p)$ is the pupil function, which incorporates apodization effects and special boundary shapes of the aperture. In the definition of the apodization function, a fulfilment of the Abbe sine condition of the optical system is assumed. z is the defocus relative to the image plane while z_{image} is the distance between the exit pupil and the image plane. In this formulation, the defocusing term is expressed as an exact spherical wave without the parabolic approximation. Figure 47-75 shows the effect of this correction as a function of the radial distance in the image plane for different values of the numerical aperture NA. From this representation the size of the deviations from the paraxial model can be roughly estimated.

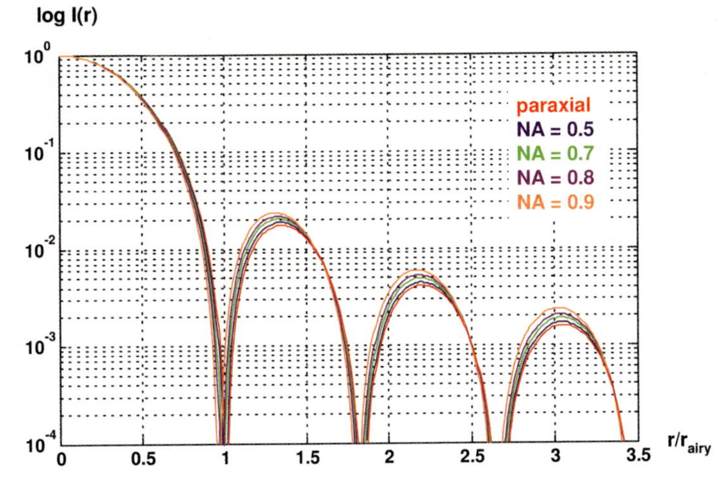

Figure 47-75: Cross-section through the radial intensity distribution of the point spread function in the extended scalar model for different values of the numerical aperture.

If the numerical aperture is further increased, the scalar consideration is no longer valid. In this case, a vector model must be taken and all the three complex field components must be considered in detail [47-84]–[47-87]. If the illumination field in the pupil is assumed to be linearly polarized

$$\vec{E}_p = \begin{pmatrix} E_{px} \\ E_{py} \\ 0 \end{pmatrix} \tag{47-99}$$

the field in the image plane can be written in components as

$$E'_x = \iint \frac{1}{\sqrt[4]{1-p_x^2-p_y^2}} \left[E_{px} \frac{p_x^2\sqrt{1-p_x^2-p_y^2}+p_y^2}{p_x^2+p_y^2} + E_{py} \frac{-p_x p_y\left(1-\sqrt{1-p_x^2-p_y^2}\right)}{p_x^2+p_y^2} \right] \times e^{2\pi i\cdot\left(xp_x+yp_y+z\sqrt{1-p_x^2-p_y^2}\right)} dp_x dp_y \tag{47-100}$$

$$E'_y = \iint \frac{1}{\sqrt[4]{1-p_x^2-p_y^2}} \left[E_{px} \frac{-p_x p_y\left(1-\sqrt{1-p_x^2-p_y^2}\right)}{p_x^2+p_y^2} + E_{py} \frac{p_y^2\sqrt{1-p_x^2-p_y^2}+p_x^2}{p_x^2+p_y^2} \right] \times e^{2\pi i\cdot\left(xp_x+yp_y+z\sqrt{1-p_x^2-p_y^2}\right)} dp_x dp_y \tag{47-101}$$

$$E'_z = \iint \frac{-1}{\sqrt[4]{1-p_x^2-p_y^2}} \left[E_{px}p_x + E_{py}p_y \right] e^{2\pi i\cdot\left(xp_x+yp_y+z\sqrt{1-p_x^2-p_y^2}\right)} dp_x dp_y \tag{47-102}$$

In these equations, the components p_x, p_y represent the direction cosines of the inclined incoming wave

$$\vec{k} = k_o \begin{pmatrix} p_x \\ p_y \\ p_z \end{pmatrix} \tag{47-103}$$

From the three field components, the intensity in the image is obtained by the equation

$$I'(x,y,z) = \left|E'_x(x,y,z)\right|^2 + \left|E'_y(x,y,z)\right|^2 + \left|E'_z(x,y,z)\right|^2 \tag{47-104}$$

Figure 47-76 shows in the image plane the relative deviation of the intensity values between the paraxial approximation and the full vector model as a function of the numerical aperture. In principle, there are changes in a new generated axial component E'_z and differences in the correct lateral components E'_x and E'_y. The axial and the transverse effects are separated in the figure. It can be seen that the axial error dominates the deviation from the scalar paraxial model. Depending on the desired accuracy of the phase retrieval, these vector aspects have to be incorporated into the propagation model for systems with higher aperture angles [47-102].

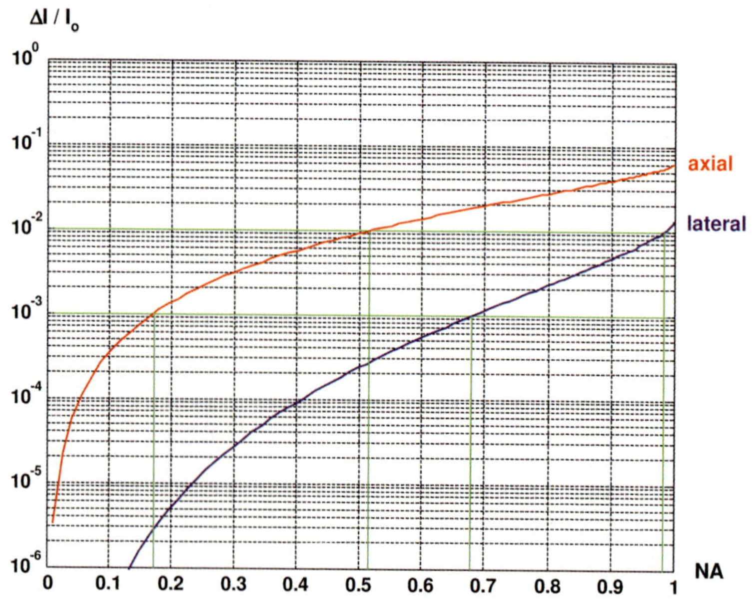

Figure 47-76: Relative error in the lateral and axial intensity distribution contributions of the point spread function in the paraxial scalar model for different values of the numerical aperture.

A third extension of the classical paraxial model is the case of low Fresnel numbers of the system, which occurs in systems with a large magnification and therefore a very small image-sided numerical aperture [47-103]. In this case the symmetry of the intensity profiles around the image plane vanishes. This is discussed in section 42.7.5 in detail and will not be repeated here.

Figure 47-77 shows a rough representation of the different models and their errors as a function of the numerical aperture, for comparison.

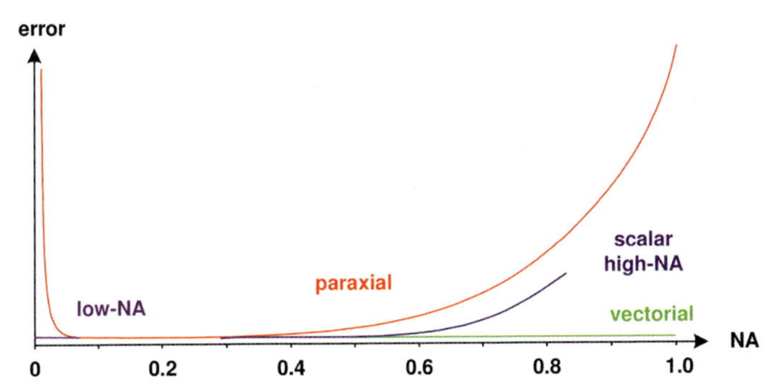

Figure 47-77: Validity ranges of the different diffraction calculation models as a function of the numerical aperture. The error axis is not to scale.

The errors in the approximated models change with defocusing. Therefore, an exact consideration must take the relevant defocus range into account. For illustration, figure 47-78 shows the point spread function for the numerical aperture $NA = 0.98$ in five focusing planes for the three models: paraxial, scalar extended and vector. It can be seen clearly that the differences are considerably larger in the defocus planes. But it should also be noticed that the intensity profiles are renormalized in every plane in this representation.

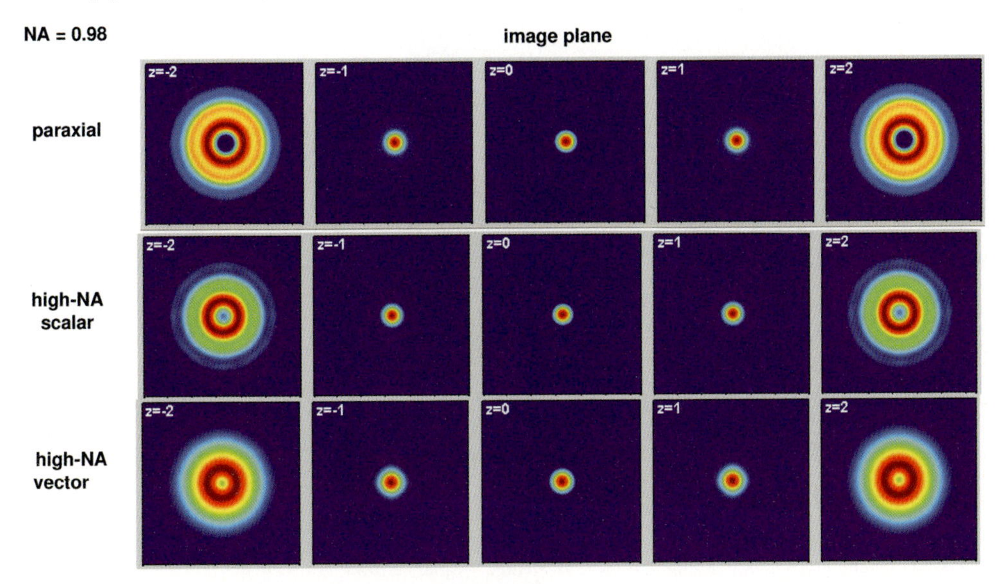

Figure 47-78: Point spread function for the five defocusing planes $z/R_u = -2, -1, 0, +1, +2$ for the three model calculations: paraxial, scalar extended and vector exact, for a focusing setup with numerical aperture $NA = 0.98$.

47.7.6
Image Processing

To obtain accurate results using the phase retrieval method, at first the signal intensity distribution has to be extracted from the measured image stack. Thus, before retrieval, an image processing must be carried out considering the following aspects.

1. Selection of a region of interest in the camera image, which contains the relevant signal data.
2. Centering of the images.
3. Removal of white noise.
4. Subtraction of the underground signal.
5. Normalization of the intensities over the whole stack.
6. Filtering of frequencies, which lie above the cut-off frequency of the optical system.
7. Accurate weighting of the measured data.

If the parameters are not chosen properly and the image processing is not performed well to deliver good data to the iterative algorithm, the result generally is not satisfactory. If a large range of the phase space rotation angle is chosen so as to get a high accuracy, the dynamic range of the beam diameters is also large. At the focal point, which usually lies near to the image plane, the spot has its smallest extension. In the images around this critical regime, the transversal discretization and the finite size of the camera pixel is most critical and must considered with care to fulfil the sampling theorem [47-104], [47-105].

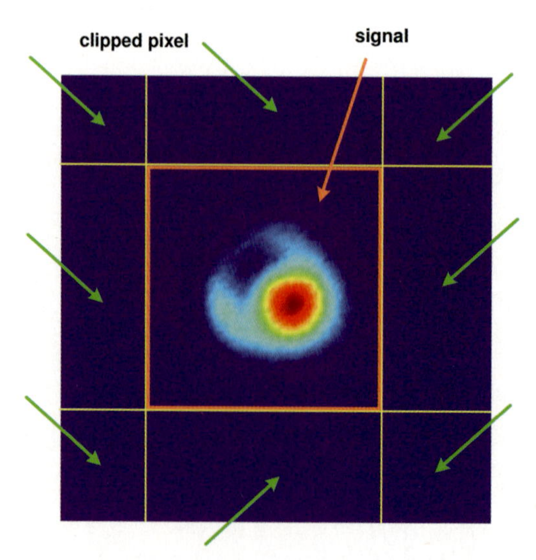

Figure 47-79: Definition of the region of interest in an arbitrary z-plane. All the pixels with only background and noisy content are removed by clipping. The regions marked by green arrows are removed.

The complete measurement of a z-stack contains two-dimensional point spread function images $I(x,y)$ for several z-planes. Therefore, a large data set is covered and evaluated by the algorithms. To reduce all the amount of data without significant signal values, an image processing is necessary and helps to speed up the calculation. Figure 47-79 illustrates this clipping process, which helps to select the relevant data. Since the size of the camera is constant, but the extent of the point spread function changes considerably with defocusing, this procedure must be done individually in every z-plane. In the region of the corners of the pictures the noisy background can be detected and analyzed. This intensity bias is often a critical parameter for the strongly focused measuring planes [47-106]

Usually the noise behavior is somewhat different in the image plane and in the strongly defocused planes. If a histogram is considered, which indicates the probability of the intensity values in every pixel, a picture like figure 47-80 is obtained. In the defocused planes, the intensity is renormalized to unity. A curve near the image plane shows a very small probability in the peak region, but a large probability of the low-level intensities in the noisy background. In the defocused plane, the signal strength is in the same range as the noise. Therefore the intensity scale is stretched out.

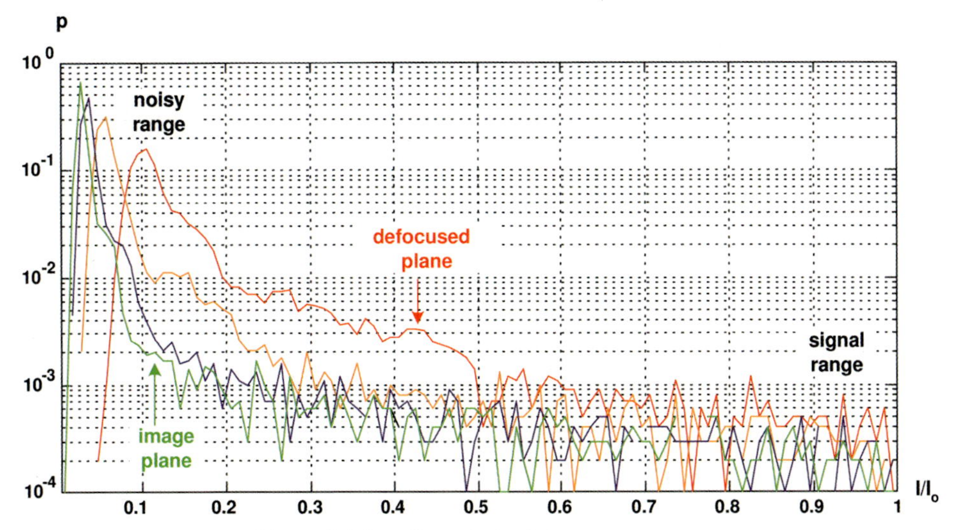

Figure 47-80: Histogram of the probability of intensity levels in every pixel of a complete z-stack with defocused planes.

In most practical applications, the determination of the Zernike coefficients is the major target required in order to qualify the performance of an objective lens. Therefore, a reconstruction of the pupil phase in every pixel as in the usual Gerchberg–Saxton algorithm is not required. If the Zernike polynomials are evaluated, they act as a low-pass filter. Therefore, experience shows that the white noise in the measured images of the z-stack is not a severe problem for the evaluation. The pixelation and quantization of camera gray levels is also uncritical and an unwrapping is not necessary. This modal approach is strongly recommended particularly for segmented pupils [47-107], [47-108].

However, in practice, beneath the white noise there is also Poisson noise, if the fluence of the photons is small. Figure 47-81 shows a perfect gaussian spot profile (a) together with 10% of noise of the Poisson type (b) or white behavior (c). The white noise is uniform for all pixels and therefore is low-pass filtered by a modal Zernike evaluation. The Poisson noise is largest in the peak region of a spot and therefore can influence the strongly weighted pixels significantly. This shows that the Poisson noise affects the accuracy much more than does the white noise.

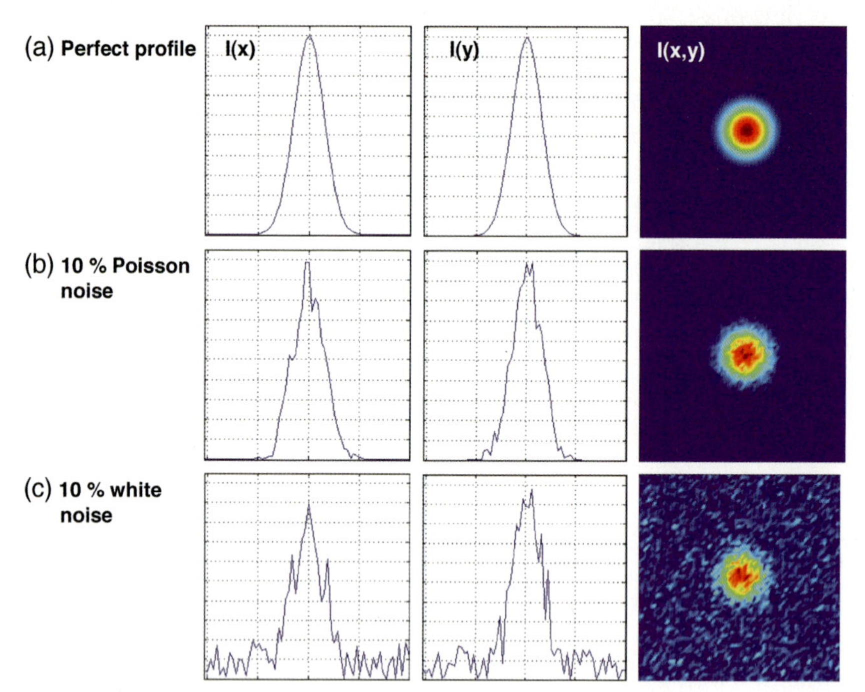

Figure 47-81: Illustration of the change in: (a) a perfect gaussian profile; (b) in the case of 10% Poisson noise and; (c) for 10% white noise, respectively.

The normalization of the image stack is also an important task in the evaluation of phase retrieval. In principle, there are two different ways of dealing with this problem. In the first case, the peak intensity is normalized to unity for every z-plane. This is shown in figure 47-82 in the upper row. The second possibility is a normalization for constant energy of the wave during propagation. The lower row of figure 47-76 shows the stack in this case.

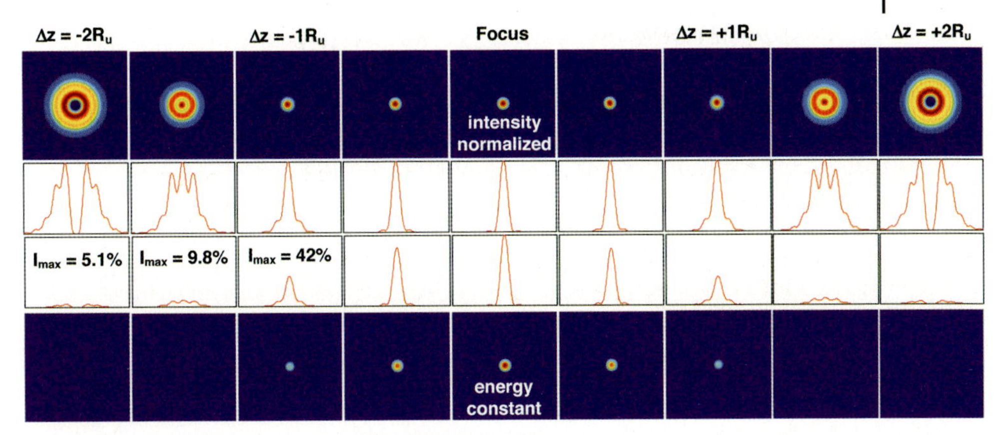

Figure 47-82: Comparison of ideal z-stack intensity profiles in the defocus range $-2\ R_u \ldots 2\ R_u$ for intensity and energy normalization, respectively.

The influence of this scaling onto the results of the retrieval has two aspects:

1. The information about the aberrations is coded more significantly in the defocused planes, but here the intensity level is low and the signal to noise ratio is usually worse.
2. If a non-least-squares fitting procedure is used to compute the parameters of the pupil function, it is possible to use weighted target values of the given pixel information. This weighting is therefore influenced by the scaling of the images.

Experience with experimental data shows that, in the combination of these two features, the best choice is to use an intensity normalization in every separate plane.

The centering of the images in the stack is one of the most important tasks in image processing. There are several reasons for a deviation in the lateral position of the intensity centroids from the optical axis of the system. Uneven coma-like phase aberrations or an asymmetrical apodization of the pupil intensity cause a characteristic transverse offset. Statistical errors in the measurement or an oblique movement of the camera have similar effects. For example, figure 47-83 shows the configuration of an asymmetrical intensity profile and its influence on the centroid.

For an accurate determination of the intensity centroid, it should be taken into account that the detector noise influences the centroid position. Therefore, careful noise removal is necessary and the signal should not be truncated. To illustrate the effect of a wrong choice of the region of interest, figure 47-84 shows the resulting error in the determination of the centroid for a z-position in the focal region and a large defocus. If truncation takes place, the centroid typically lies on a curve and causes residual errors as shown in images on the right.

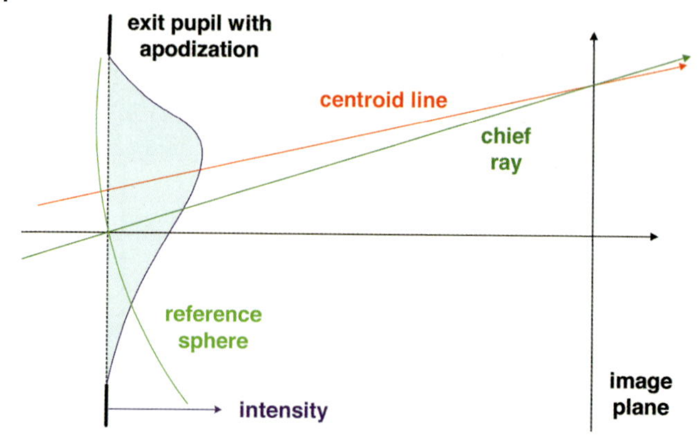

Figure 47-83: Lateral deviation of the intensity centroid of a light beam in the case of an asymmetrical intensity distribution of the pupil.

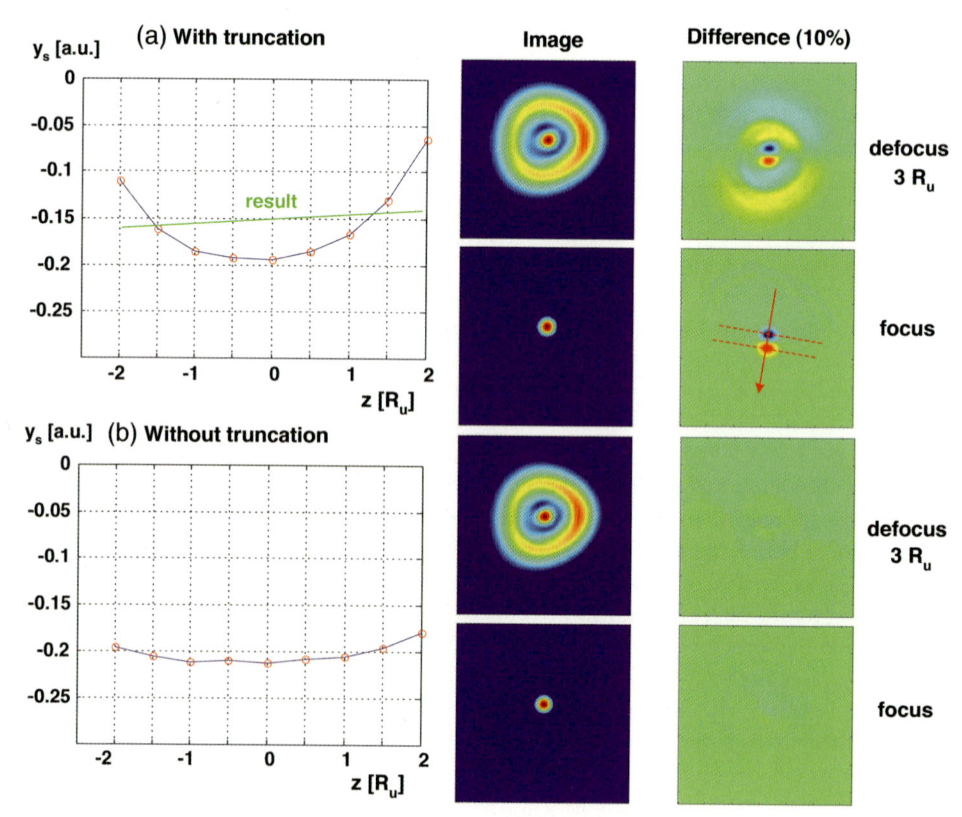

Figure 47-84: Effect of signal truncation on the determination of the centroid and the residual error differences in the retrieval results. In a) a case with truncation is shown, b) indicates an example without truncation and considerable smaller error.

Figure 47-85 shows the two major contributions from centering errors in a complete intensity z-stack. A linear drift of the blue centroid line can be obtained if apodization takes place and the wave suffers from coma-like aberrations also when the mechanical camera movement is not perfectly aligned to the optical axis. The second contribution is a statistical fluctuation of the exact centroid position due to noise effects or a large mechanical play of the movement.

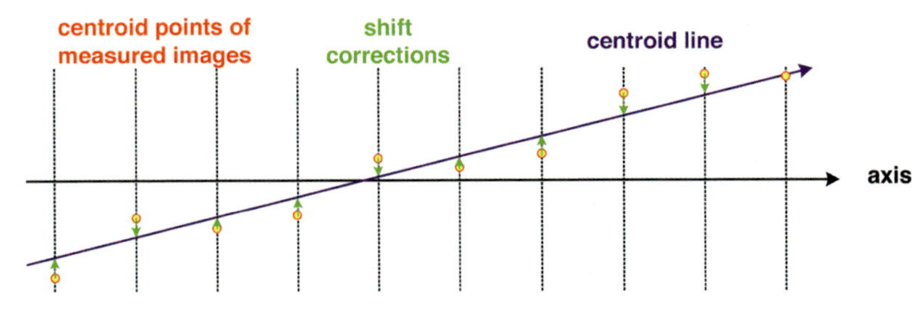

Figure 47-85: Schematic picture of the lateral drift of the centroid line in combination with statistical fluctuations.

47.7.7
Pinhole Deconvolution

If a physical pinhole is used as an artificial point source, the finite size of the circular opening must be taken into account for the evaluation of the image space data if it is larger than approximately 0.3 D_{airy}. If incoherent illumination is assumed, the image intensity distribution is given by the convolution of the pinhole object intensity and the point spread function of the system according to the well-known equation

$$I_{\text{image}}(x', y', z) = I_{\text{psf}}(x, y, z) * I_{\text{ph}}(x, y) \tag{47-105}$$

If the illumination uses a very small circular pinhole with a diameter much smaller than the Airy size, in practice, the signal to noise ratio is too low for a successful retrieval computation. Therefore, an enlarged pinhole is used to obtain more signal power at the detector. Thus, the observed images are the result of the point spread function convolved with a finite circular pinhole. As input for the retrieval algorithm the measured intensity distribution needs to be deconvolved to get the point spread function. If the PSF of the system is shift invariant, the deconvolution can be performed in the easiest way in the Fourier space according to Wiener with the help of a regularized inversion of the corresponding product of spectra [47-109]

$$I_{\text{psf}}(\nu) = \frac{I_{\text{ph}}^{*}(\nu) \cdot I_{\text{image}}(\nu)}{\left|I_{ph}(\nu)\right|^{2} + \mu} \tag{47-106}$$

where μ is the Tikhonov regularization parameter which helps to avoid singularities at the zeros of the pinhole spectrum [47-109]. Figure 47-86 shows image stacks for different pinhole sizes with an arbitrary constant set of Zernike coefficients for the wavefront. In the example a z-interval of four Rayleigh units is used for the stack. The diffraction fringes and the fine structure of the images vanish if the pinhole size is enlarged. The size of the pinhole diameter is scaled in Airy diameters.

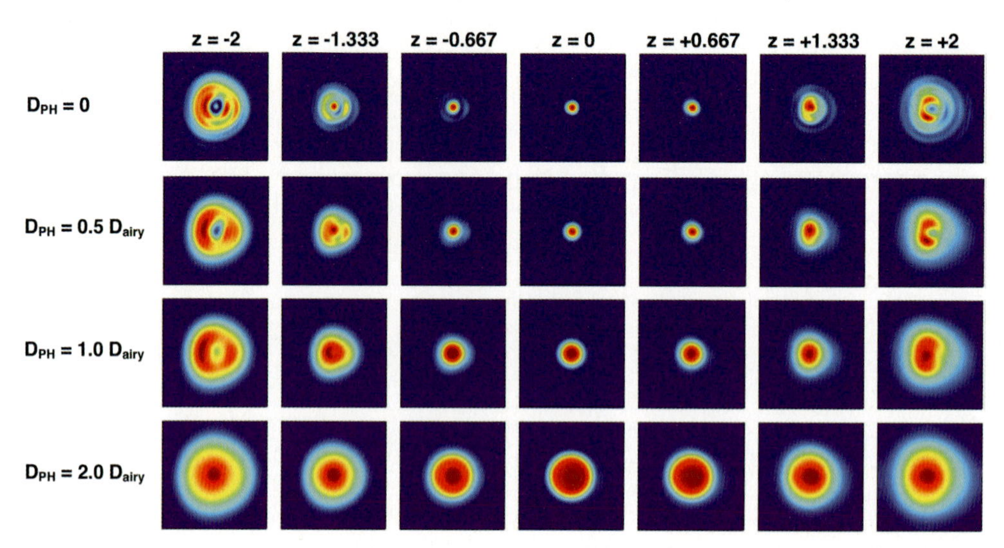

Figure 47-86: Variation in the pinhole image through focus as a function of the pinhole size. Defocus positions *z* are given in Rayleigh units.

Figure 47-87 shows the results of phase retrieval of these stacks [47-74]. The absolute mean error of the reconstructed Zernike coefficients in units of wavelength and the errors of spherical components are plotted as a function of pinhole size. With a well-conditioned deconvolution the accuracy of the Zernikes is nearly independent of the pinhole size in this interval. Without deconvolution, the uncertainty increases with pinhole size. As the spectrum of the circular pinhole shows a significant ring structure, the influence on pupil aberration is of rotational symmetry. Thus errors of the corresponding Zernikes C_{20}, C_{40} and C_{60} (defocus, primary and secondary spherical aberration) are especially large.

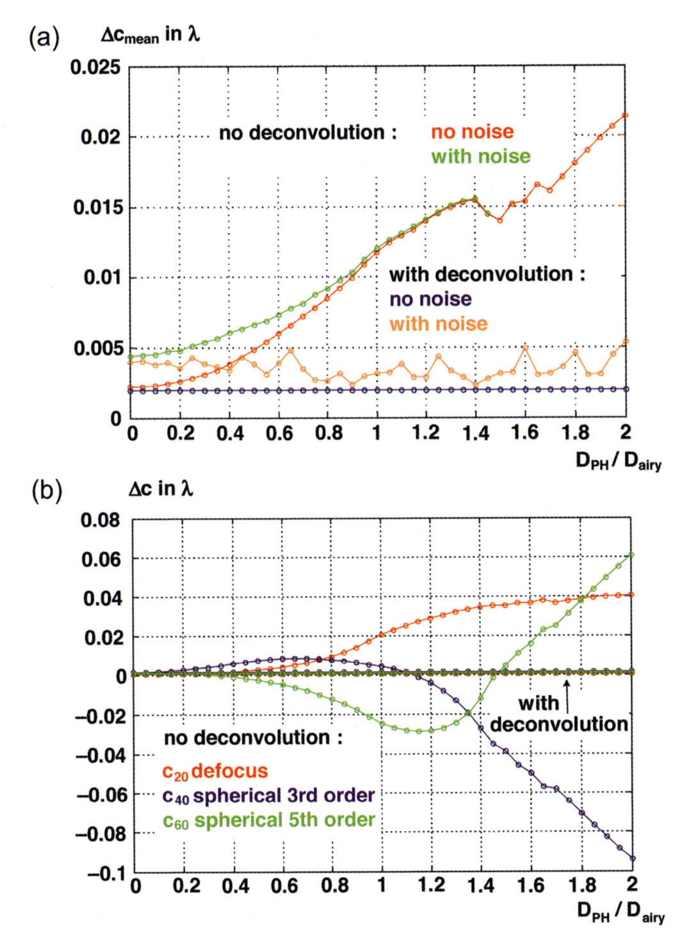

Figure 47-87: Error in the reconstructed Zernike coefficients, when the deconvolution of the finite pinhole size is either taken into account or neglected. Part a) shows the mean error of all coefficients and part b) the error in Zernike coefficients with rotational symmetry.

For pinholes size below 0.3 D_{Airy}, the approximation of a quasi point source seems to be acceptable.

A second effect arises due to the coherence of the pinhole illumination. Figure 47-88 shows the rms error of the image stack intensity for a pinhole of increasing normalized diameter $d = D_{ph} \, / \, D_{airy}$ for a coherent and an incoherent illumination, respectively. In the case of incoherence, the above estimated largest diameter of 0.3 D_{Airy} seems to be the limit to produce deconvolution with an approximately 1% residual error. If the illumination is coherent, the size of the pinhole influences the image stack much less and here a limit of 0.7 D_{Airy} is suggested.

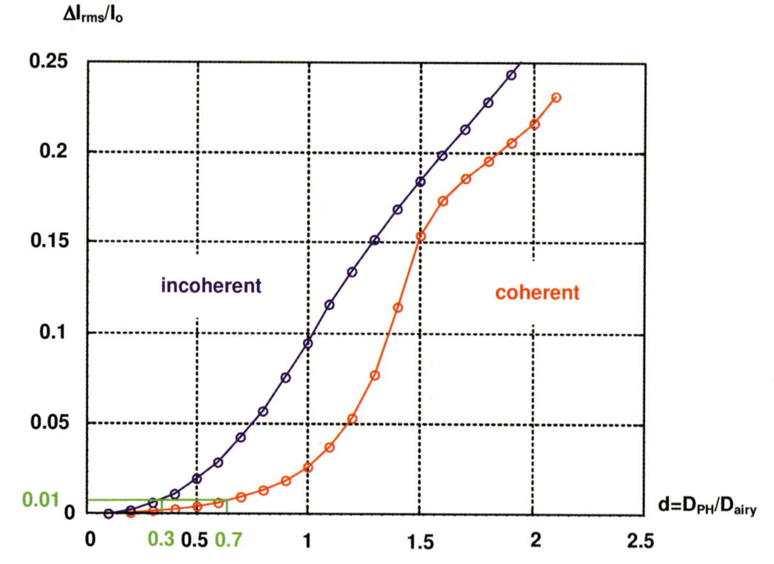

Figure 47-88: Comparison of the retrieval error for increasing size of the normalized pinhole diameter *d* for coherent and incoherent illumination.

47.7.8
Numerical Evaluation Algorithms

There are many algorithms which are used for iterative phase retrieval [47-71], [47-72], [47-110]–[47-112]. Most of them are based on the well-known IFTA or Gerchberg–Saxton algorithm [47-69], [47-70]. The most important proposed algorithms are the following

1. **Gerchberg–Saxton algorithm**
   The standard Gerchberg–Saxton or IFTA (iterative Fourier transform algorithm) is a brute force error reduction algorithm. It is described later in this section in more detail.
2. **Non-least-square algorithm**
   If the desired phase surface is described by a functional expansion like the Zernike representation, the retrieval calculation can be performed as a modal algorithm with the help of a non-least-square fit. The main advantage of this method is a low-pass filtering of the smooth modal functions, which significantly reduces the problems involved with measuring noise.
3. **Fourier algorithm**
   The transport of intensity equation can be used directly and can be solved very efficiently using Fourier methods.
4. **Yang–Gu algorithm**
   This is a special formulation of the IFTA algorithm, which suppose a non-

unitary transform between the pupil and the image plane. This allows to use the calculation scheme in cases of non-constant energy due to truncation or absorption effects.

5. **Fienups algorithms**
   There are several possible ways to refine the standard Gerchberg-Saxton algorithm, which were mostly developed by Fienup. The changes are found in the mathematical use of the boundary conditions. The importation of the known data into the image stack planes in the algorithms is properly weighted. This delivers a better convergence and control of the algorithm.

The standard version of the IFTA or Gerchberg–Saxton Algorithm is depicted in figure 47-89 and allows to calculate the field in the pupil from the intensity in one image plane only. This standard version only uses one image plane.

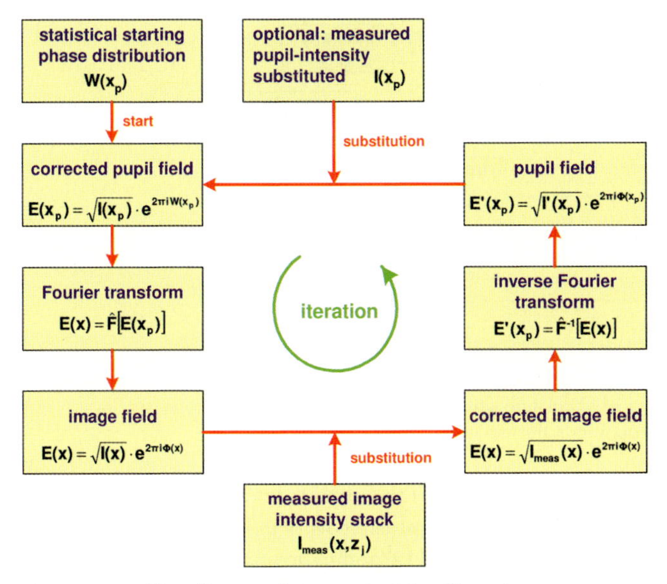

Figure 47-89: Flow diagram of the standard Gerchberg–Saxton algorithm. For simplicity, the formulas contain only one transverse coordinate.

In the basic algorithm the field is propagated back and forth between image and pupil, while in the image plane the calculated intensity is substituted by the measurement. The main disadvantage of this algorithm is the stagnation of the iteration [47-113]–[47-115]. This stagnation is avoidable by taking several z-planes into account. This is illustrated in figure 47-90.

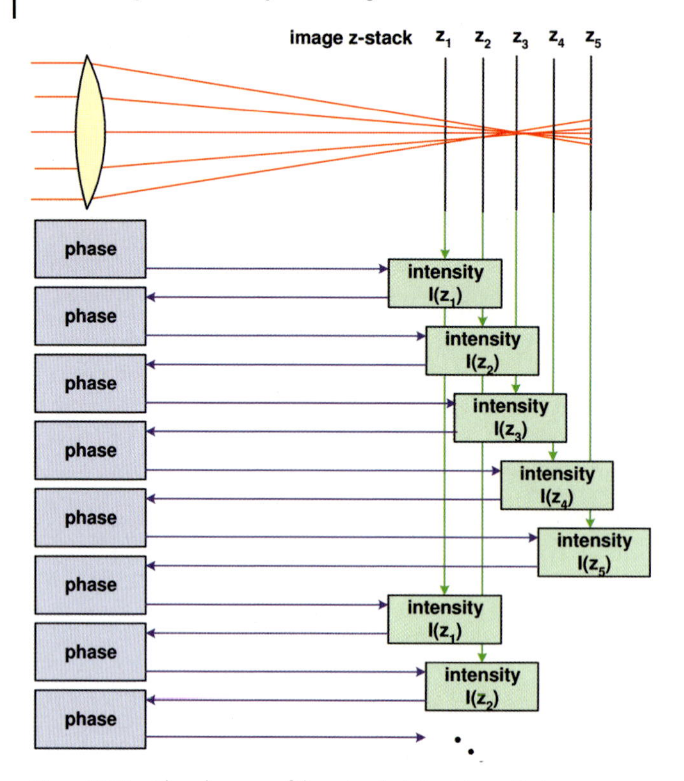

Figure 47-90: Flow diagram of the extended Gerchberg–Saxton algorithm for several planes in the image space. The blue arrows indicate one possible choice of calculation workflow. There are some other possibilities proposed in the literature.

If, in addition, the intensity distribution in the pupil is known, the algorithm can be accelerated significantly. There are many possible ways to improve the numerical performance of this algorithm [47-116]–[47-118] and to generalize the calculation. If the phase in the pupil is described by a finite number of Zernike polynomials, it is possible to calculate the gradient of the merit function for the coefficients directly. This modal-type algorithm shows good stability against disturbing noise. Other methods use a generalized non-least-square algorithm to fit intensity distributions to the measured stack. For example, a modified damped least-square method according to Marquardt performs well. In this case the model allows one to incorporate more details of the real system and therefore it is valid under more general conditions.

For example, the deconvolution operation described above is critical, since noise affects the result, and the exact value of the microscopic small pinhole is only known approximately. Therefore, an alternative computational scheme is used, which incorporates the convolution in the forward calculation inside the non-least-square algorithm. In this case, the merit function is defined at the image level and not as a deviation of the point spread functions. Figure 47-91 shows both alternatives for comparison. This kind of algorithm is much more stable and allows one to define the pinhole size and the intensity bias [47-106] as a variable of the fit, if its diameter is only roughly known.

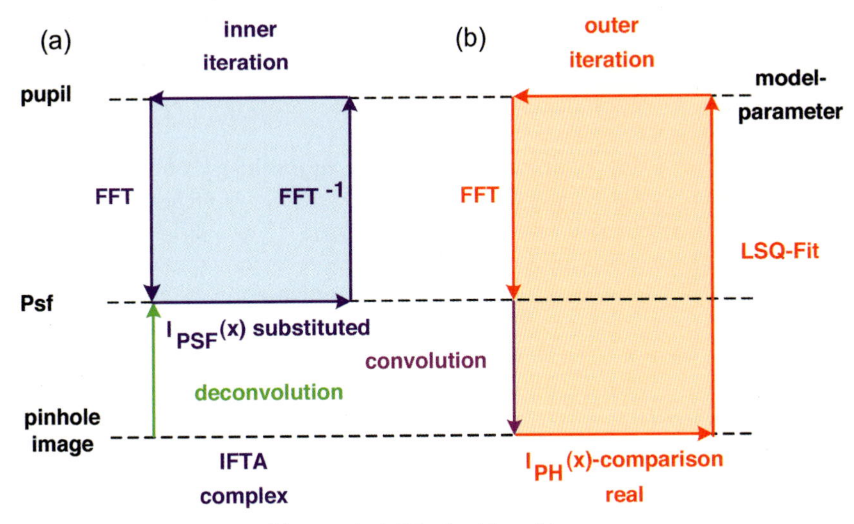

Figure 47-91: Comparison of the standard IFTA algorithm with pre-deconvolution (a) and the generalized non-least-square algorithm, which incorporates the finite-size pinhole convolution (b). Here only forward propagation steps are performed.

If the PSF phase retrieval is performed with the help of a non-least-square fit algorithm, a metric must be defined as a a suitable criterion. There are several possible ways to define the best adaptation of the fit onto the measuring data [47-78], [47-119]–[47-122]. One quite simple possibility is the standard deviation between the measured data and the model data with the help of

$$I_{\text{rms}} = \sqrt{\frac{1}{N} \sum_j \left[I_j^{(\text{model})} - I_j^{(\text{meas})} \right]^2} \qquad (47\text{-}107)$$

Here the sum goes over all pixel signal values that are used for the evaluation. The sum of all pixels is N. Another possibility is to use a normalized correlation between the data set and the model stacks

$$C = \frac{\iiint I_{\text{model}} \, I_{\text{meas}} \, dx dy dz}{\sqrt{\iiint I_{\text{model}}^2 dx dy dz} \sqrt{\iiint I_{\text{meas}}^2 dx dy dz}} \qquad (47\text{-}108)$$

where, in reality, the integrals are evaluated by finite sums over the pixels.

A third criterion is proposed in the literature [47-101], which uses the entropy of the intensity distribution to optimize the result of the fit

$$S = -\sum_j \left(I_j^{(model)} - I_j^{(meas)} \right) \ln \left(I_j^{(model)} - I_j^{(meas)} \right) \quad (47\text{-}109)$$

Practical experience shows that all three possible choices for the metric work well and can be used without problems.

47.7.9
Apodization

The intensity distribution in the image is more strongly influenced by phase errors in the pupil than by the amplitude. But if a high accuracy is required, the effect should also be taken into account, because phase errors show a different influence on the defocused intensity profiles. The phase produces asymmetrical effects in the z-direction around the image plane, while the apodization normally generates symmetrical distortions in z. In respect to the iteration there is no perfect decoupling, so one gets an incorrect phase result when the amplitude variation is neglected.

In the non-least-square algorithm, a more general model is also possible which allows one to describe apodized pupil functions. Figure 47-92 shows an example used to demonstrate the influence of apodization. The residual errors of the simple model, without taking an existing apodization into account, are shown in row c) for all z-positions. If the apodization is performed exactly, the residual errors are much better, as indicated in row e).

Figure 47-93 shows the correlation between the original data and the calculated model data for the two cases as a function of the z-position. It can be seen that row d) has a very good correlation in the range of $C = 0.99$. If the simpler model is considered, a typical asymmetrical decrease in the correlation with the z-position is shown.

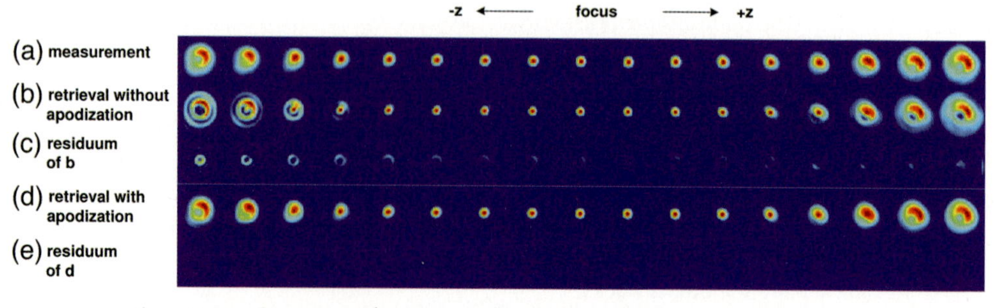

Figure 47-92: Reconstructed intensity profiles of an apodized pupil with and without taking apodization into account.

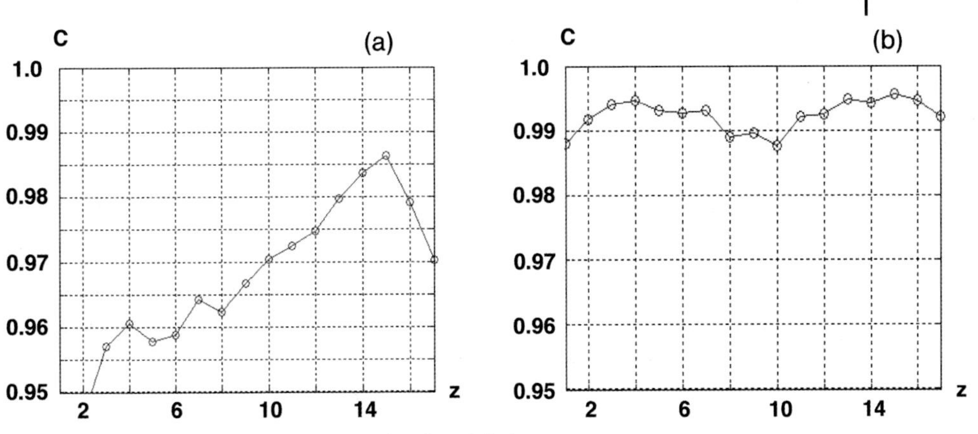

Figure 47-93: Correlation between measured and modelled image stacks in the two cases b) and d) of figure 47-92. The residual errors are large especially in the intrafocal defocussing part.

Segmented pupil shapes, as a special type of apodization, have already been mentioned. In practice, they occur in very large mirror telescopes, as a spider for the secondary telescope mirror or in a special case as phase contrast rings in microscopic lenses.

47.7.10 Object Space Defocusing

In order to measure through the focus a series of optical systems at high magnification and to cover a z-interval of four Rayleigh units in the image space, detector movement over large distances is required. On the other hand, defocusing of the object plane can be done very easily by the usual object stage mechanics in microscopy. Therefore, it is very appealing to perform the measurement with a fixed camera and a moving object instead of moving the image plane. One problem arises in this setup, and it is shown in figure 47-94. Since the ray paths through the lens under test are different for every z-plane in the object space, the wavefront in the exit pupil is no longer invariant. This fact must be taken into account and causes a more complicated retrieval calculation than in the case of an image-side defocusing.

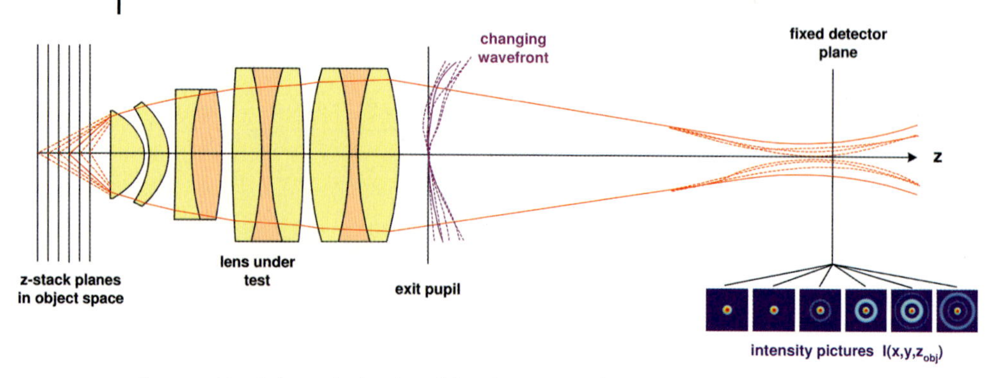

Figure 47-94: Schematic drawing of the phase retrieval setup in the case of object space defocusing. The wavefront in the exit pupil is not constant but changes its shape from z-plane to z-plane.

Thus the lens is not used at a constant working distance and the wave aberrations in the exit pupil depend on the modified z-values in the object plane. The conventional algorithm cannot be applied, but for small defocusing the Zernike coefficients are linearly dependent on defocus [47-74], [47-78]

$$c_j(z) = c_{jo}\left(1 + \Delta c_{\text{lin}} \frac{\Delta z}{R_u}\right) \tag{47-110}$$

The major part of this linear change described by the linear constants Δc_{lin} has its reason in the transformation of the Zernike coefficients from the high-NA side to the low-NA side of the system. A corrected system with a defocussed spherical wavefront corresponds to an expansion into Zernikes, because the quadratic term of c_{20} describes only the parabolic approximation. For higher numerical apertures, the expansion of the sphere needs higher order spherical aberration terms. The second minor contribution depend on the individual correction of the lens under test due to a retrace difference and have to be precalculated from the lens data or can alternatively be determinated by the algorithms. A third contribution to the change of the Zernike coefficients has its reason in a violation of the sine condition for defocussed object distances. Then the modal non-least-square algorithm must be used, which allows a more general physical model. Figure 47-95 shows the typical behavior of equation (47-110) for a real objective lens. As can be seen, linearity is quite a good assumption. Figure 47-96 shows an example for an original stack, simulated by defocusing the object plane, and reconstructed after retrieval with and without taking (47-110) into account. The improvement of the fit is clearly visible.

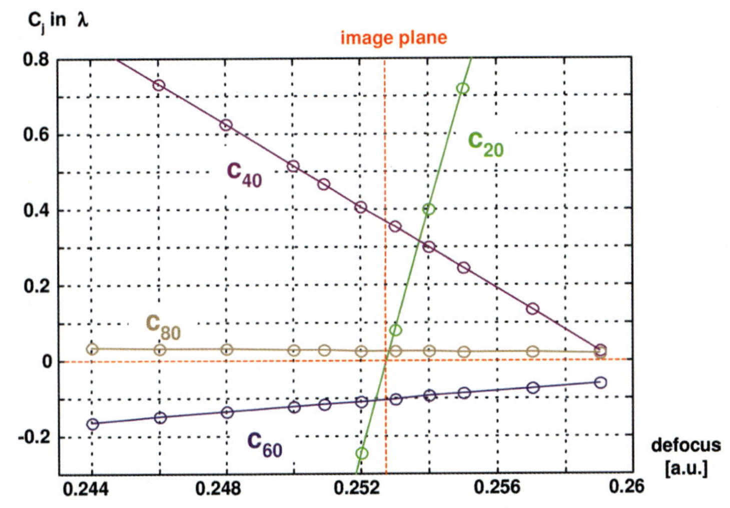

Figure 47-95: Change in the axisymmetrical Zernike coefficients when the system is defocused in the object space.

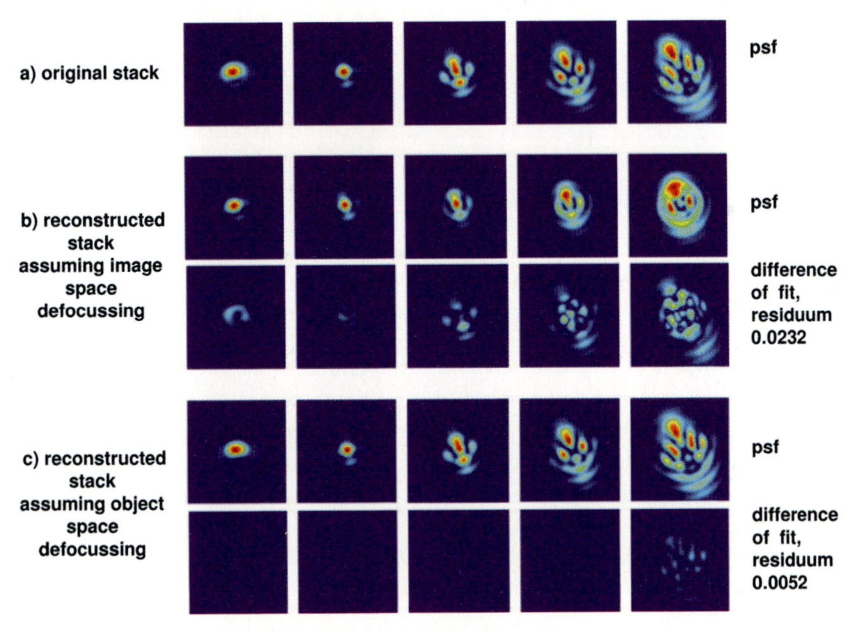

Figure 47-96: Comparison of the reconstructed intensity stack shown in a), obtained by defocusing the object plane, and either taking this into account (c) or not (b).

47.7.11
Accuracy of Phase Retrieval

Figure 47-97 shows a typical result for the phase surface in the pupil of a retrieval measurement for the optic under test. Figure 47-98 represents the expansion coefficients of the Zernike polynomials for the same result. The error bars indicate the relative uncertainty of the values. They are determined from the correlation matrix of the fitting procedure. If a parameter is independent and influences the result very strongly, the corresponding error bar is short. Large bars are obtained for those coefficients, which cannot be determined very accurately. This shows that the determination of the precise value for the accuracy of the phase retrieval is quite a complicated task. The precision depends on many parameters of the setup, the measuring process and the control of the algorithm.

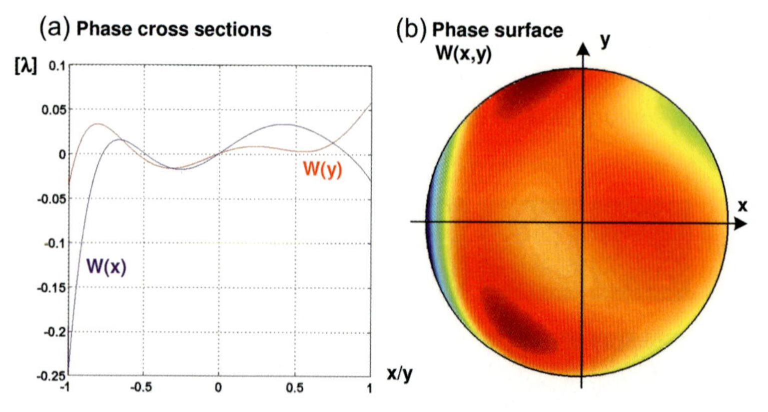

Figure 47-97: Typical result for the phase $W(x,y)$ of a retrieval measurement (b) together with the cross- sections through this surface (a).

The main advantage of phase retrieval for aberration measurements is the fact that the lens can be used under working conditions. According to the analysis of point spread images, the method is well suited for diffraction-limited systems with nearly ideal point spread images such as, for example, microscopic lenses or astronomical devices. For high magnifying systems in particular, the requirements on equipment for measurement must be balanced with the requirements on the accuracy of the retrieved wavefront. It should be noted that a finite image location is necessary in order to perform retrieval measurements.

In addition to the purely numerical task and the above-discussed difficulties, the following problems arise in practice.

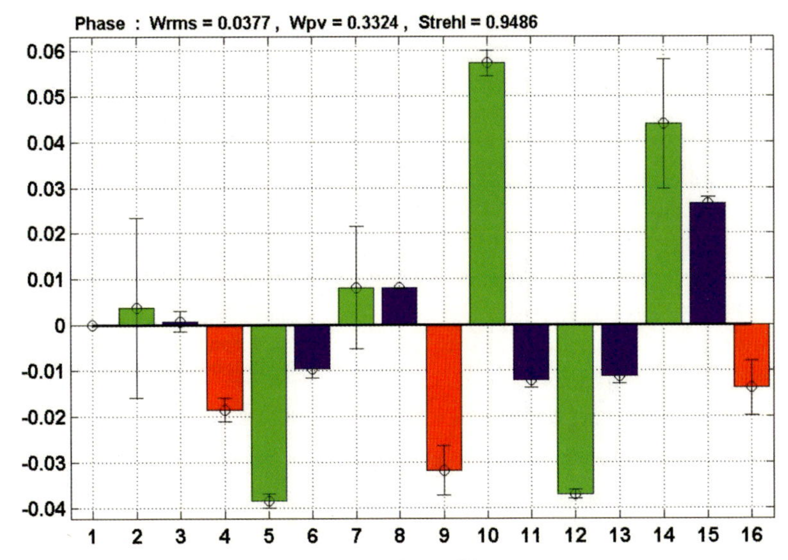

Figure 47-98: Zernike spectrum of the wavefront of figure 47-97. The colors indicate the symmetry of the corresponding Zernike polynomial. The axisymmetrical components are red, the cosine terms blue and the sine terms green. The error bars indicate the relative uncertainty of the retrieval, estimated by the condition number of the least square fit.

1. There is no ideal pinhole in an infinite thin layer, but remaining transmission and thickness of surrounding blocking layer.
2. In addition to monochromatic aberrations the point spread function is influenced by lateral and axial chromatic aberrations. For a state-of-the art lens design, all types of aberrations are balanced, e.g., there is no dominant error.
3. For microscopic lenses phase retrieval describes the imaging from the object to the image plane, which is performed by the microscopic lens as well as by the tube optics. The influence of tube optic and alignment cannot be neglected.
4. There is no simple way to check the results. Comparisons with other methods of wavefront measurement need further assumptions. These are, for instance, known numerical aperture, apodization, knowledge of additional components and their interfaces.

Finally, the result of phase retrieval will be satisfactory if the measured point spread image and the recalculated image match each other. For good results the normalized correlation between the measured and model data should be at least 95%, good results are in the range of 97–99%.

As described above, the accuracy of retrieval is mainly restricted by model assumptions and careful conditions of measurement. That is why the results of retrieval were checked by analysis of data with different retrieval algorithms for two repeated measurements. In figure 47-99 the Zernike coefficients and their mean errors are plotted. The optimization was done with and without apodization and with different settings of image processing, concerning the centering. The best correlation of 98% between measured and recalculated images was achieved with optimized apodization [47-98].

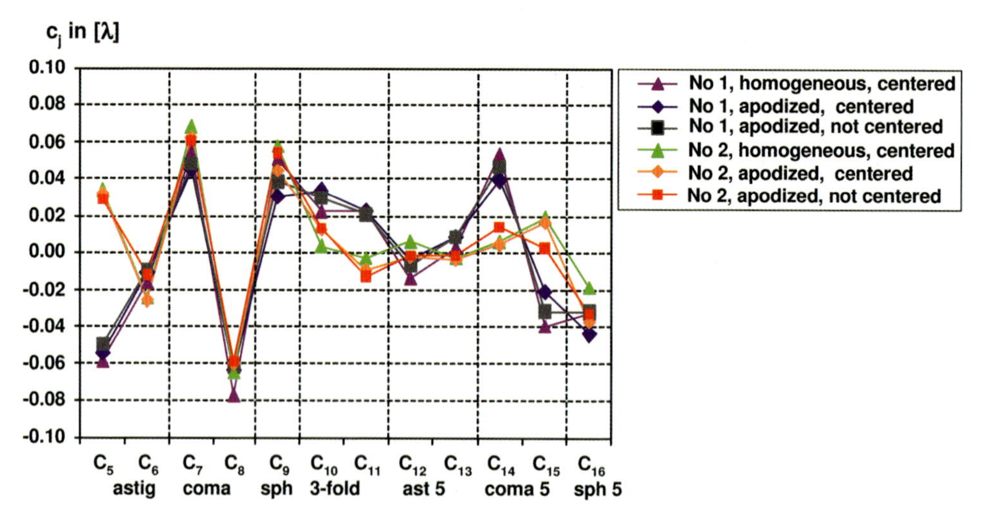

Figure 47-99: Repeatability of phase retrieval with different model assumptions and settings in image processing. No 1 / No 2 indicate two different measurements of the same lens. The c_j are the Zernike coefficients with index j in the usual Fringe definition.

For a more rigorous evaluation of retrieval a comparison between interferometry and retrieval at different setups was made. Two microscopic lenses were measured with interferometry and phase retrieval at two different setups.

The interferometer was of Fizeau type with a double pass of light, which was reflected at a reference mirror. The light directed onto the lens was collimated. The phase retrieval was done in ordinary microscopic setups with a tube lens, where the light leaving the lens is not collimated exactly. The different interfaces are taken into account by subtracting the corresponding Zernike coefficients from the interferometric data.

The repeatability of the interferometric data was higher than 0.005 λ per Zernike coefficient and the precision of phase retrieval with equal analysis parameters as well. The detailed comparison between interferometry and retrieval is shown in figure 47-100. As the angles of rotation were not recorded, only the numbers of Zernike coefficients are shown.

Except for spherical aberrations C_9 and C_{16} the results of retrieval and interferometry differ by about 0.02–0.04 λ. This is also valid for the two setups of the retrieval.

The difference in the spherical aberration between retrieval and interferometry and also between retrieval with samples 1 and 2 is supposedly caused by the tube lens or a weak difference in wavelength. This assumption is emphasized by the agreement of the differences in the spherical aberrations C_9 and C_{16} for both samples. Furthermore, the lenses need not exactly conform to the design, and non-ideal uniform illumination has to be taken into account.

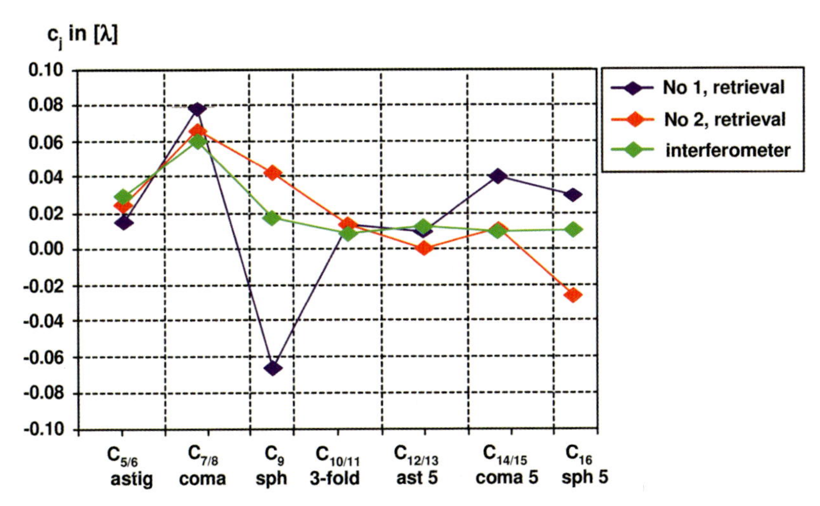

Figure 47-100: Comparison between the results of interferometry and retrieval with two lenses (samples 1 and 2). The largest deviations are observed for the axisymmetrical coefficients c_9 and c_{16} of the Zernike coefficients.

Figure 47-101 shows a second comparison of the retrieval results using another measurement method. The results of a Hartmann test and a retrieval for coma and astigmatism at various field positions of an objective lens are shown for comparison. A similar estimation for the accuracy in the region of 0.01 λ can be seen. The large difference in astigmatism at the positive field position results from a known measuring problem. The shaded coloured stripes indicate the estimated uncertainties of the various measurement setups.

Thus the uncertainty of the Zernike coefficients according to model assumptions is about $\lambda/100$. The repeatability by measurement only, is less. Considering the fact that the image is influenced by the wavefront as well as apodization, and apodization is not taken into account by interferometry, this result is pretty good.

The greatest problem in an accurate determination of the Zernike polynomials via phase retrieval is the problem of contributions with axisymmetric symmetry. There are three quite different reasons for an influence of the z-stack with rotational symmetry:

1. Spherical aberration.
2. Apodization with axisymmetrical distribution.
3. Effects of a finite-size circular pinhole, if the deconvolution is not performed well.

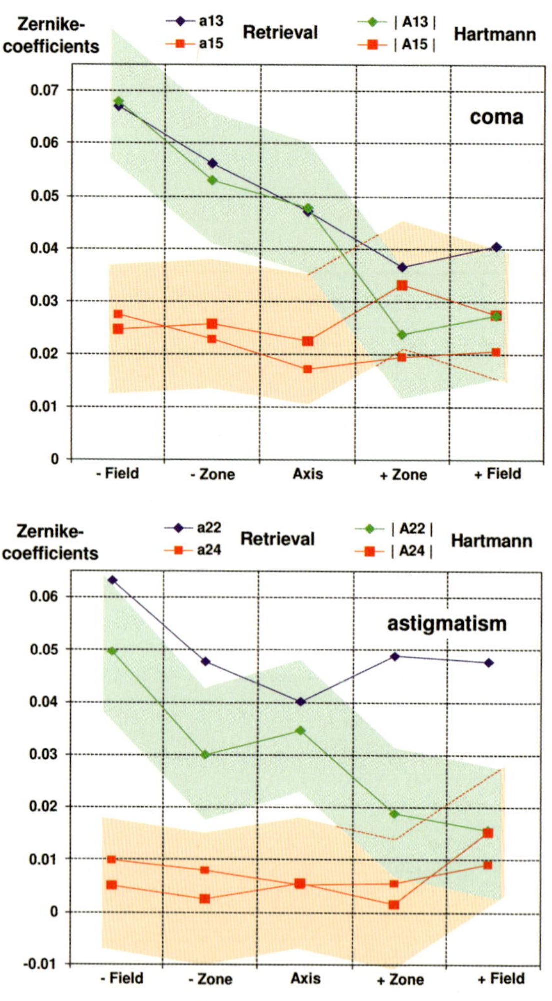

Figure 47-101: Comparison between the results of a Hartmann measurement and retrieval for coma and astigmatism at different field positions for a sample lens. The aberation coefficients are denoted by a and A in this figure.

As can be seen in figure 47-102, these three aspects all have a rather different influence on the intensity caustic in the image space. Spherical aberration causes asymmetrical behavior around the image plane. Apodization, with the usual decrease in intensity towards the rim of the pupil, causes smaller diameters in the defocus regime and an enlarged diameter in the focal region. The convolution of the ideal point spread function with the finite pinhole size, on the other hand, causes a quite uniform broadening of the caustic independently of the z-position. These, quite different, mechanisms of change allow the algorithm to separate all these effects with sufficient accuracy.

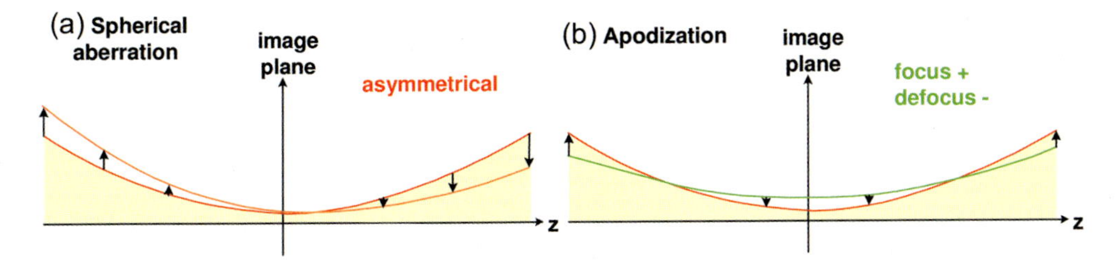

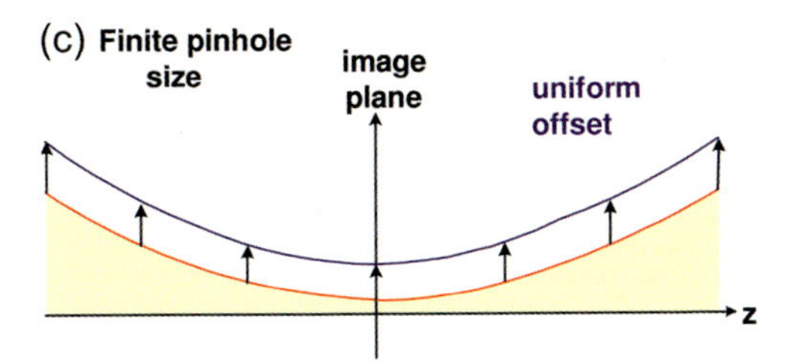

Figure 47-102: Different influences on the width of the caustic in the image space with axisymmetrical behavior.

47.8 Calculation of Wavefront and Zernike Coefficients

47.8.1 Introduction

In the different metrology methods used to measure the quality of an optical system by evaluating the wave aberration in the exit pupil of the system, the primary information is usually not the wavefront directly. In the Hartmann and Hartmann–Shack setups, the derivative components of the wavefront in the lateral directions x and y are measured in a scaled version. To obtain the wavefront itself, an integration of the gradient field components is necessary [47-124], [47-125].

A complete wavefront is a two-dimensional data set in the form of a surface on discrete grid points. This information is not very helpful for the certification of the system performance. A better way is to expand the wavefront into Zernike polynomials. This delivers a smaller set of coefficients, which are scaled onto the wavelength and can be compared and assessed more easily. This means that, if the wavefront is evaluated from a metrology tool, very often the calculation of the Zernike representation is a necessary second step.

This section deals with the methods used to reconstruct the wavefront from its gradient values and the determination of the Zernike coefficients. There are some metrology methods such as point spread function retrieval, which allow these steps to be combined into one.

47.8.2
Zonal Methods

If the measurement data are gathered onto a rectangular grid with elementary pixel sizes Δx and Δy respectively, we obtain the situation shown in figure 47-103.

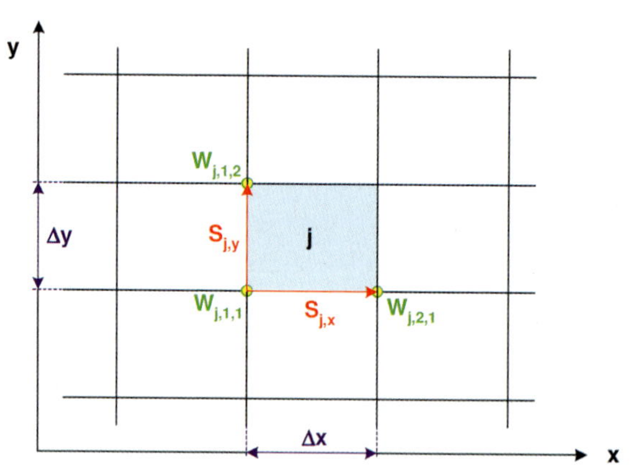

Figure 47-103: Calculation of the gradient components of the wavefront by finite differences in one cell number, *j*, of the grid.

If a cell with index j is considered, the relation between the wave aberration data at the corner points of the cell and the gradient components S can be approximated by finite differences using the equations

$$S_{j,x} = \frac{W_{j,2,1} - W_{j,1,1}}{\Delta x} \qquad S_{j,y} = \frac{W_{j,1,2} - W_{j,1,1}}{\Delta y} \tag{47-111}$$

This corresponds to a local integration of the gradient field. If the corresponding relations are now written for all mesh grid cells, a linear system of equations is obtained, which can be expressed in matrix form as

$$\vec{S} = \underline{B} \cdot \vec{W} \tag{47-112}$$

where the matrix $\underline{B}$ is a sparse band matrix, which combines the corresponding mesh points according to (47-111) for every cell and contains the coefficients of the finite difference formula used. The solution of this equation can be written in a formal way as

$$\vec{W} = \left(\underline{B}^T \cdot \underline{B}\right)^{-1} \cdot \underline{B}^T \cdot \vec{S} \tag{47-113}$$

and is performed in practice by an SVD algorithm. This is called the zonal method, in which the integration is calculated locally in every zone of the grid [47-126]–[47-131].

The computation and implementation of the zonal method is generally very easy to achieve, but is very sensitive to any noise influence on a single grid point. Furthermore, an integration constant must be added to the matrix in an additional row. There are several versions used when formulating the finite difference, because the exact spatial allocation of the wavefront and the gradient values inside the grid can be chosen in different ways.

47.8.3 Modal Methods

The sensitivity of the zonal method to any noise or perturbation leads to other calculation schemes, which have better performance. The so-called modal methods use an orthogonal function expansion for the wavefront and deal with the expansion coefficients [47-132], [47-133]. The advantage of these methods is a low-pass filtering influence of the modal functions, which decreases the sensitivity as desired. Depending on the geometrical shape of the aperture, there are several possible ways to take the choice of the modal function. The most important function systems are:

1. Elementary Polynomials
   This is, in principle, a Taylor expansion with the usual polynomials of increasing order in two variables x and y.
2. Zernike polynomials
   This is the most frequently used choice, it is well adapted to the case of a circular pupil shape.
3. Legendre polynomials
   These functions are used in the case of rectangular pupil shapes. They are described in appendix A.8 of volume 2 of this book series.
4. Exponential functions
   This special choice is discussed in section 47.8.4 in more detail. It is of high interest, because the fast Fourier transform and a low-pass filtering performance of the algorithm can both be used to produce a very stable method.

If the modal functions are denoted by $F_n(x,y)$, the expansion of the wavefront with coefficients a_n can be written in the form

$$W(x,y) = \sum_n a_n F_n(x,y) \tag{47-114}$$

If the derivatives of this equation in x and y are calculated by means of analytical formulas, the following two equations are obtained

$$\frac{\partial W}{\partial x} = \sum_n a_n \frac{\partial F_n(x,y)}{\partial x} \qquad \frac{\partial W}{\partial y} = \sum_n a_n \frac{\partial F_n(x,y)}{\partial y} \tag{47-115}$$

If a metrology tool is used, which primarily delivers the gradient of the wavefront, then these two equations can be used to compute a fit of the measured data points. This corresponds to a solution of the linear system of equations

$$\vec{S} = \underline{A} \cdot \vec{a} \tag{47-116}$$

with the vector of the modal coefficients $\vec{a}$ and the matrix of the derivatives

$$\underline{A} = \begin{pmatrix} \left.\frac{\partial F_1(x,y)}{\partial x}\right|_{j=1} & \left.\frac{\partial F_2(x,y)}{\partial x}\right|_{j=1} & \cdots \\ \left.\frac{\partial F_1(x,y)}{\partial x}\right|_{j=2} & \left.\frac{\partial F_2(x,y)}{\partial x}\right|_{j=2} & \cdots \\ \cdots & \cdots & \cdots \end{pmatrix} \quad (47\text{-}117)$$

In a similar way to the previous section, the solution can be found in the best way by using the SVD method according to the equation

$$\vec{a} = \left(\underline{A}^T \cdot \underline{A}\right)^{-1} \cdot \underline{A}^T \cdot \vec{S} \quad (47\text{-}118)$$

The condition matrix and the parameter of the SVD reduction appear very clearly, when the choice and number of modal functions are fitted well to the task.

47.8.4 Modal Fourier Reconstruction

One special choice of the modal functions is the use of plane waves or Fourier basis functions

$$Z_{pq}(m,n) = \frac{1}{N}\, e^{\frac{2\pi i}{N}(pm - qn)} \quad (47\text{-}119)$$

see [47-134], [47-135], [47-136]. The expansion of the wavefront can be written as

$$W(m,n) = \sum_{p,q} a_{pq}\, Z_{pq}(m,n) \quad (47\text{-}120)$$

The finite differences are determined, as explained above, by

$$\nabla_x W(m,n) = W(m,n+1) - W(m,n) \qquad n = 1, N-1;\, m = 1, N \quad (47\text{-}121)$$

$$\nabla_y W(m,n) = W(m+1,n) - W(m,n) \qquad m = 1, N-1;\, n = 1, N \quad (47\text{-}122)$$

Where N is the number of grid points in one direction. The boundary values at $n = N$ and $m = N$ are chosen with the help of the equations

$$\nabla_x W(m,N) = -\sum_{n=1}^{N-1} \nabla_x W(m,n) \quad (47\text{-}123)$$

$$\nabla_y W(N,n) = -\sum_{m=1}^{N-1} \nabla_y W(m,n) \quad (47\text{-}124)$$

This ensures that there is no global tilt of the wavefront. The expansion of (47-120) corresponds to a discrete Fourier transform, which can be computed using fast algorithms. The coefficients are then given by

$$a_{pq} = 0 \qquad if\ p = q = 0 \tag{47-125}$$

$$a_{pq} = \frac{\left[e^{-\frac{2\pi iq}{N}} - 1\right] \hat{F}[\nabla_x W(p,q)] + \left[e^{-\frac{2\pi ip}{N}} - 1\right] \hat{F}\left[\nabla_y W(p,q)\right]}{4\left[\sin^2\left(\frac{\pi p}{N}\right) + \sin^2\left(\frac{\pi q}{N}\right)\right]} \tag{47-126}$$

for all other values of the indices p, q. The synthesis of the wavefront can be written with the formula

$$W(m,n) = \sum_{p,q} \frac{a_{pq}}{N}\, e^{\frac{2\pi i}{N}(pm-qn)} \tag{47-127}$$

Since the Fourier transform has a filtering effect, the noise influence is suppressed and the algorithm is very stable. Some problems with this algorithm are as follows.

1. The finite differences are formulated asymmetrically. This can be solved by an iterative correction.
2. The margin of the pupil is mostly circular in shape. The corresponding steps must be handled with care.

47.8.5 Direct Determination of Zernike Coefficients from Slope Measurements

If the gradients of the wavefront are measured directly, the Zernike coefficients can be calculated from these primary data by a fitting procedure of the analytical functions of the derivatives of the Zernike polynomial functions. In this case, the explicit wavefront $W(x,y)$ is not obtained. The expansion of the wavefront into Zernike Polynomials can be written as [47-137]

$$W(x,y) = \sum_{n,m} c_{nm} Z_n^m(r,\varphi) = \sum_{n,m} c_{nm} R_n^m\left(\frac{r}{a}\right) e^{im\varphi} \tag{47-128}$$

where a is the normalization radius of the pupil. The derivatives of this function in cylindrical coordinates can be obtained by

$$\frac{\partial W}{\partial \varphi} = \sum_{n,m} i\, m\, c_{nm} R_n^m\left(\frac{r}{a}\right) e^{im\varphi} \tag{47-129}$$

and

$$\frac{\partial W}{\partial r} = \sum_{n,m} c_{nm} \frac{d\, R_n^m\left(\frac{r}{a}\right)}{dr} e^{im\varphi} \tag{47-130}$$

The following equations transform these derivatives into the cartesian representation

$$x' = -\frac{R}{n'}\frac{\partial W}{\partial x} = \frac{R}{n'}\left(\frac{\partial W}{\partial r}\cos\varphi - \frac{\partial W}{\partial \varphi}\frac{\sin\varphi}{r}\right) \tag{47-131}$$

$$y' = -\frac{R}{n'}\frac{\partial W}{\partial y} = \frac{R}{n'}\left(\frac{\partial W}{\partial r}\sin\varphi + \frac{\partial W}{\partial \varphi}\frac{\cos\varphi}{r}\right) \tag{47-132}$$

From these formulas the coefficients can be obtained by

$$c_{nm} = \frac{-i(n+1)}{\pi\, m a^2}\int\limits_0^{2\pi}\int\limits_0^{a}\left(\frac{\partial W}{\partial x}\sin\varphi - \frac{\partial W}{\partial y}\cos\varphi\right) R_n^m\left(\frac{r}{a}\right) e^{-im\varphi}\, r^2\, dr\, d\varphi \tag{47-133}$$

for m # 0 and, in particular, the axisymmetrical terms with $m = 0$

$$c_{n0} = \frac{n+1}{\pi\, n(n+2)}\int\limits_0^{2\pi}\int\limits_0^{a}\left(\frac{\partial W}{\partial x}\cos\varphi + \frac{\partial W}{\partial y}\sin\varphi\right)\frac{dR_n^0\left(\frac{r}{a}\right)}{d\left(\frac{r}{a}\right)}\left[1 - \left(\frac{r}{a}\right)^2\right]\left(\frac{r}{a}\right) dr\, d\varphi \tag{47-134}$$

The explicit terms of the derivatives occurring in this equation can be found in the literature [47-138].

47.8.6
Calculating the Zernike Coefficients of a Wavefront

The Zernike polynomials are orthogonal functions on a circular area with a constant weighting function of 1. Therefore, they are a good description for the wavefront aberrations in systems with circular pupil shape and uniform illumination. The orthogonality of the Zernike function is a great advantage for the calculation of the corresponding expansion coefficients [47-137].

If the above-mentioned assumptions are valid, from (47-128) it follows immediately that

$$c_j = \int\limits_0^{1}\int\limits_0^{2\pi} W(r,\varphi)\, Z_j^*(r,\varphi)\, d\varphi\, r\, dr \tag{47-135}$$

One possibility is the direct computation of the coefficients using this formula. The use of this integral representation delivers arbitrary coefficients and can be easily extended to higher indices j if necessary. But there are some disadvantages involved in using this equation.

1. In practice, the orthogonality is slightly disturbed. The reasons for this are the discrete computational grid and the zig-zag boundary line for a cartesian grid.
2. The computational effort is quite high as, for every coefficient of the expansion, it is necessary to calculate the double integral of (47-135).

The second way to calculate the coefficients is to formulate the fit as a non-least-square problem in the following equation [47-139], [47-140]

$$\sum_{i=1}\left[W_i - \sum_{j=1}^{N} c_j Z_j(r_i)\right]^2 = \min \qquad (47\text{-}136)$$

In this formula, the i-sum is over all grid points with the relevant data. The solution of this problem can be found using linear algebra as

$$\vec{c} = \left(\underline{Z}^T \underline{Z}\right)^{-1} \underline{Z}^T \vec{W} \qquad (47\text{-}137)$$

which can be solved with an SVD algorithm. The advantages of this method are as follows.

1. All coefficients are calculated in one step simultaneously.
2. The approximation is slightly better since, due to the perturbation of the orthogonality, some mixing effects generate a better description of the wavefront.

However, the lack of exact orthogonality also has the effect of correlating the coefficients to a certain extent. If the calculation is performed with different numbers of coefficients, the values of the lower coefficients are not perfectly constant.

Usually, the calculation of Zernike polynomials with high radial order is critical using the original formulation of the polynomials $R_n{}^m(r)$

$$R_n^m(r) = \sum_{q=0}^{\frac{n-m}{2}} (-1)^q \frac{(n-q)!}{q!\left(\frac{n+m}{2}-q\right)!\left(\frac{n-m}{2}-q\right)!} r^{n-2q} \qquad (47\text{-}138)$$

because the factorials cause an overflow in the computation for higher orders. Therefore it is strongly recommended to use the recursion formula

$$R_{n+2}^m = \frac{n+2}{(n+2)^2 - m^2}\left\{\left[4(n+1)r^2 - \frac{(n+m)^2}{n} - \frac{(n-m+2)^2}{n+2}\right] R_n^m - \frac{n^2-m^2}{n} R_{n-2}^m\right\} \qquad (47\text{-}139)$$

to calculate the radial polynomial for higher radial orders.

In reference [47-141] a method is proposed, which is much more stable and is recommended for high radial orders n. The performance near the rim of the pupil circle is particularly critical. With Chebyshev polynomials of the second kind

$$V_n(x) = \frac{\sin[(n+1)\ \arccos(x)]}{\sin[\arccos(x)]} = \frac{x\ T_n(x) - T_{n+1}(x)}{1-x^2},\ |x| \le 1 \tag{47-140}$$

the radial polynomials can be written using the equation

$$R_n^m(r) = \frac{1}{N} \sum_{j=0}^{N-1} V_n\left(r \cos \frac{2\pi j}{N}\right) \cos \frac{2\pi jm}{N} \tag{47-141}$$

which is a cosine transform of a series of Chebychev polynomials and can be calculated by fast methods. N is a number with the condition $N > n+m$.

47.8.7
Zernike Calculation via Fourier Transform

It is also possible to calculate the Zernike polynomial coefficients with the help of the Fourier transform [47-141]. For a circular pupil, the Fourier transform of the Zernike polynomials

$$U_{nm}(k,\theta) = \iint Z_{nm}(r,\varphi)\, e^{-2\pi i \cdot \vec{r}\vec{k}} d^2\vec{r} \tag{47-142}$$

can be written explicitly as

$$U_{nm}(k,\theta) = (-1)^{n/2+|m|} \sqrt{n+1} \frac{J_{n+1}(2\pi k)}{k} \psi_m(\theta) \tag{47-143}$$

with the auxiliary function

$$\psi_m(\theta) = \begin{cases} \sqrt{2}\cos(m\theta) & m > 0 \\ 1 & m = 0 \\ -\sqrt{2}\sin(m\theta) & m < 0 \end{cases} \tag{47-144}$$

If the Fourier transform of a wavefront is determined as

$$A(k,\theta) = \iint_{r \le 1} W(r,\varphi)\, e^{-2\pi i \cdot \vec{r}\vec{k}} d^2\vec{r} \tag{47-145}$$

the Zernike coefficients can be calculated using the equation

$$c_{nm} = \frac{1}{\pi} \sum_{l=0}^{N^2-1} A_l(k,\theta)\ U_{nm}^*(k,\theta) \tag{47-146}$$

from the discrete Fourier coefficients. The summation extends over all N^2 sampling points of the square grid of the fast Fourier transform.

47.8.8
Influence of Normalization Radius on Zernike Coefficients

The Zernike coefficients are defined on a unit circle with normalized radius. If, in the practical evaluation of the metrology measurement, the exact radius of the pupil is not known or is badly defined due to a smooth decrease of the illumination, the determination of the Zernike coefficients shows some errors [47-143]–[47-150]. There are some typical errors, which can be understood easily and which will be discussed in the following. In this consideration, the shrunk pupil is considered to be concentric with the original circle.

If the radius of the pupil is reduced from a value ρ to a smaller value r with a scaling factor ε according to

$$r = \varepsilon\rho \qquad \text{where} \qquad 0 \le \varepsilon \le 1 \tag{47-147}$$

we obtain the scenario shown in figure 47-104.

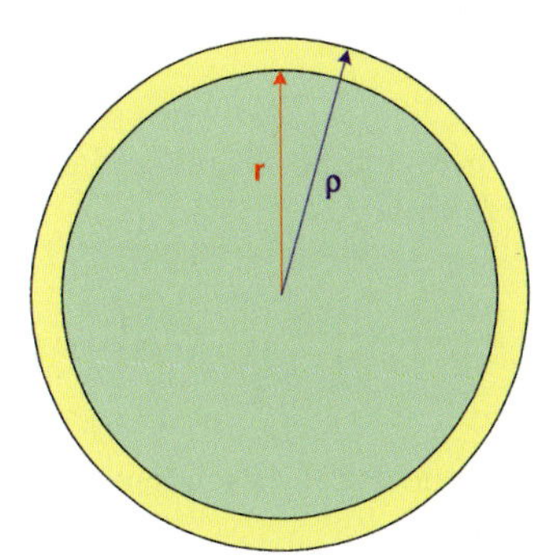

Figure 47-104: Reduced pupil radius.

A spherical aberration of for example third order with a Zernike polynomial Z_{40} changes in this case to

$$\begin{aligned} W(\rho) &= \bar{c}_{40}\, Z_{40}(\rho) + \bar{c}_{20}\, Z_{20}(\rho) + \bar{c}_{00}\, Z_{00}(\rho) \\ &= \frac{c_{40}}{\varepsilon^4}\,\overline{Z_{40}}(r) + c_{20}\,\frac{3(1-\varepsilon^2)}{\varepsilon^4}\,\overline{Z_{20}}(r) + c_{00}\,\frac{\varepsilon^4 - 3\varepsilon^2 + 2}{\varepsilon^4}\,\overline{Z_{00}} \end{aligned} \tag{47-148}$$

As can be seen, the coefficients change according to

$$\bar{c}_{40} = \frac{c_{40}}{\varepsilon^4},\; \bar{c}_{20} = \frac{3(1-\varepsilon^2)}{\varepsilon^4}\,c_{20},\; \bar{c}_{00} = \frac{\varepsilon^4 - 3\varepsilon^2 + 2}{\varepsilon^4}\,c_{00} \tag{47-149}$$

If the relative change of the coefficients is calculated as a function of the scaling parameter $\varepsilon < 1$, the curves shown in figure 47-105 are obtained. The piston term is not considered here. An error of 1% in the value of the pupil radius causes an error of 5% in the Zernike coefficient for spherical aberration.

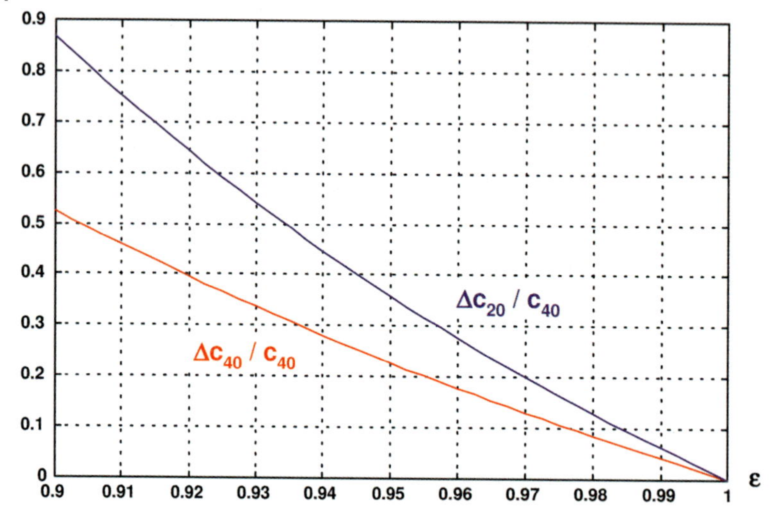

Figure 47-105: Relative changes in the Zernike coefficients for a decreased pupil radius.

In a more general formulation, with the equivalence between the shrunk and the new renormalized pupil radial coordinate, it is possible to write

$$W(\varepsilon\rho,\theta) = \sum_{n=0}^{N}\sum_{m=-n}^{n} a_{nm} Z_{nm}(\varepsilon\rho,\theta) = \sum_{n'=0}^{N}\sum_{m'=-n'}^{n'} c_{n'm'} Z_{n'm'}(\rho,\theta) \tag{47-150}$$

If the explicit formulas are inserted into this equation, one obtains a relationship, which is used to calculate the new coefficients [47-146]

$$c_{n'm'} = \varepsilon^{n'} \sum_{i=0}^{\frac{k-n'}{2}} a^{m}_{n'+2i} \sqrt{(n'+1)(n'+2i+1)} \sum_{j=0}^{i} (-1)^{j+i} \varepsilon^{2j} \frac{(n'+i+j)!}{(i-j)!(n'+j+1)!j!} \tag{47-151}$$

This can also be expressed by a single sum and the hypergeometric function.

47.8.9
Change in the Zernikes for Decentered, Rotated and Stretched Pupils

If an error in the centering of the pupil by a value a in the x-direction is assumed, a similar calculation leads to changes in several Zernike coefficients with different symmetry [47-143], [47-144]. The shifted coordinate is defined by

$$\bar{x} = x - a \tag{47-152}$$

and is illustrated in figure 47-106.

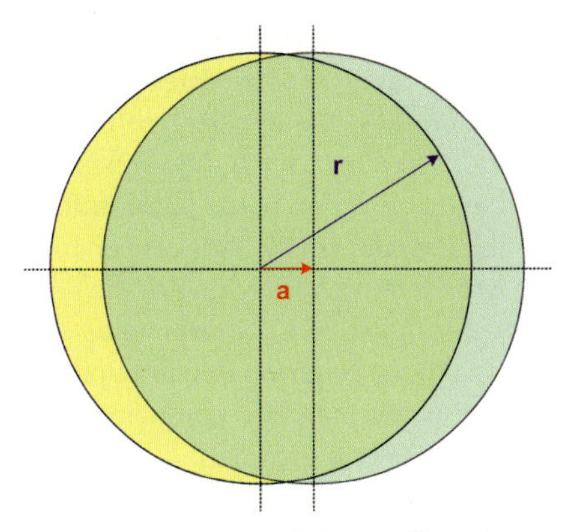

Figure 47-106: Pupil with lateral offset a.

If the spherical aberration is again considered in cartesian coordinates

$$Z_{40}(\bar{x}, y) = 6\,(\bar{x}^2 + y^2)^2 - 6\,(\bar{x}^2 + y^2) + 1 \tag{47-153}$$

the modified polynomial can be written as

$$\bar{Z}_{40} = Z_{40} + 8a\,Z_{31} + 12a^2\,Z_{20} - 12a^2\,Z_{22} + (24a^3 + 4a)\,Z_{11} + 12a^2\,Z_{00} \tag{47-154}$$

It is apparent that the coefficient of the spherical aberration is constant, and contributions of defocus, astigmatism, coma and tilt are generated. The corresponding behavior is shown in figure 47-107 as a function of the offset value a. The increase in the tilt and coma with a is linear, while the defocus and astigmatism increases quadratically.

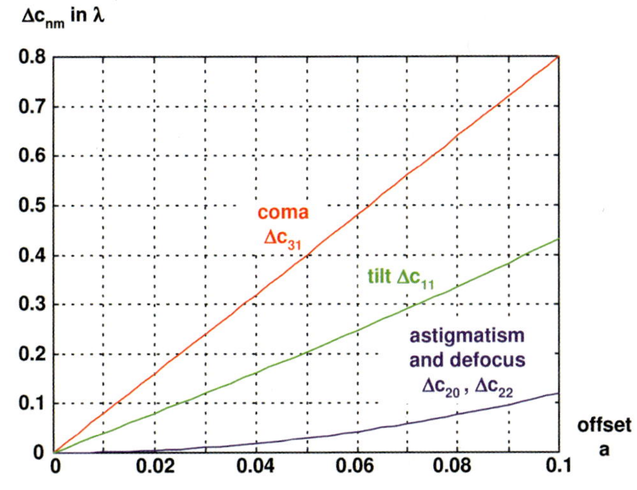

Figure 47-107: Relative changes in the Zernike coefficients due to a lateral offset a. The offset is scaled on the normalized radius of the pupil.

These simple examples show that an accurate knowledge of the centering position and the radius of the pupil circle is very important in order to obtain precise values of the Zernike coefficients. If there are slight differences, the Zernike coefficients change and contributions of lower order can be generated. Characteristic couplings and excited Zernike modes occur. Which correspond to the direction of the pupil shift.

If the pupil is rotated around the azimuth, the Zernike coefficients also change [47-144]. If the shape of the pupil changes due to an elliptical stretching or dilation, the change in the Zernike coefficients can be determined by a corresponding formula given in [47-144]. For these two modifications in the original circular pupil the explicit formulas for the coupling of the Zernike coefficients are complicated. It seems to be more practical to use numerical calculations in these cases.

Stretching the pupil circle into an elliptical shape has been used in the past to approximately describe the altered pupil geometry if an optical system shows a weak vignetting of the pupil. However, due to the mixing of defocus and astigmatic terms, an assessment of the system performance is critical for this approach.

47.8.10
Propagation Changes in the Zernike Coefficients

In practice, it is very often not possible to obtain access to the system pupil. One prominent example is the human eye, where the pupil plane lies inside the system. One possible way to derive the correct Zernike coefficients of the system is to transfer the pupil to an available location with a 4f-system as described in section 47.2.3 for the Hartmann–Shack sensor. Another possibility is to measure the Zernike polynomials as near as possible to the pupil location and to calculate the change in the coefficients during propagation over the difference distance z [47-151], [47-152]. The gradients in a non-spherical wavefront cause a change in the shape as a function of the propagation distance. This is shown in figure 47-108.

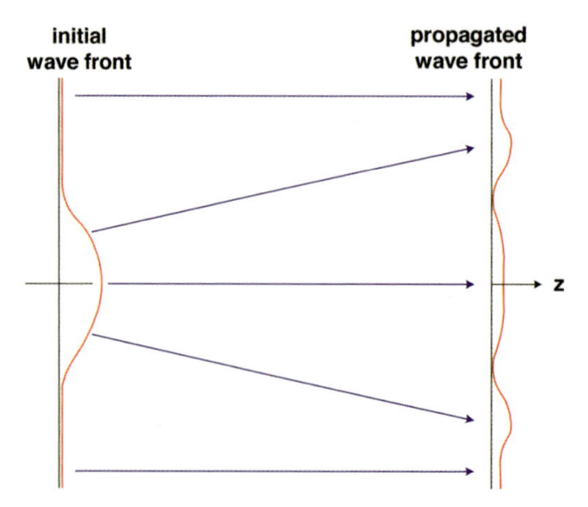

Figure 47-108: Relative change in the wavefront shape during propagation.

For the correct determination of the new wavefront, it needs to be taken into account, that the outer diameter of the beam usually changes, when a defocusing term exists in the Zernike expansion.

If a wavefront is described as a Taylor monomial expansion in the form

$$W(x,y) = \sum_{p=0}^{\infty}\sum_{q=0}^{p} a_{pq}\, P_p^q(x,y) = \sum_j a'_j P_j(x,y) = \sum_{p=0}^{\infty}\sum_{q=0}^{p} a_{pq}\, x^q y^{p-q} \tag{47-155}$$

and r is the radius of the pupil, a propagation of the wave changes the expansion coefficients. Here a single index j or a double index p,q can be used. With the direction factor

$$s(x,y) = \left[\frac{\partial W(x,y)}{\partial x}\right]^2 + \left[\frac{\partial W(x,y)}{\partial y}\right]^2 \tag{47-156}$$

then using the geometrical means of raytracing we get the transformation

$$\begin{aligned} s(x,y) &= \sum_{p,q}\sum_{p',q'} a_{pq}a_{p'q'}\, P_{p+p'-2}^{q+q'-2}(x,y) + \sum_{p,q}\sum_{p',q'} a_{pq}a_{p'q'}\,(p-q)(p'-q')\, P_{p+p'-2}^{q+q'}(x,y) \\ &= \sum_j b_j P_j(x,y) \end{aligned} \tag{47-157}$$

This simple approach is only valid when the aberrations are small and no ray crossover takes place. The change in the wavefront with radius of curvature R reads

$$\begin{aligned} W'(x',y') &= W(x,y) + \frac{|z|}{2R^2}\, s(x,y) \\ &= \sum_{j=1}\left(a_j + \frac{|z|}{2R^2} b_j\right) P_j(x,y) \end{aligned} \tag{47-158}$$

The new radial degree changes from p to $p' = 2p - 2$. Therefore, the propagation also induces higher radial orders. It should be noted that the normalization radius r also changes during the propagation. If only those contributions are considered, which influence the axisymmetric geometry, we will obtain those polynomials with quadratic terms. The radius changes according to

$$r' = r + \frac{z}{r}\sum_{j=1}^{N} 2c_{2j,0}\sqrt{2j+1)}\sum_{s=0}^{j-1}\frac{(-1)^s(j-s)(2j-s)!}{s![(j-s)!]^2} \tag{47-159}$$

If the well-known conversion between the Zernike coefficients and the monomial expansion coefficients is used, the change in the Zernike polynomial coefficients during the propagation can be calculated. The analytical formulas for these calculations are lengthy.

In figure 47-109 the result of a numerical calculation is shown. A collimated beam with wavelength $\lambda = 0.5$ μm and diameter 1 mm has an initial aberration of secondary spherical aberration $c_{60} = 1\ \lambda$. If the beam is propagated over a distance of 2 mm, the overall shape of the wavefront remains qualitatively equal, but the numbers change significantly, as can be seen in the figure.

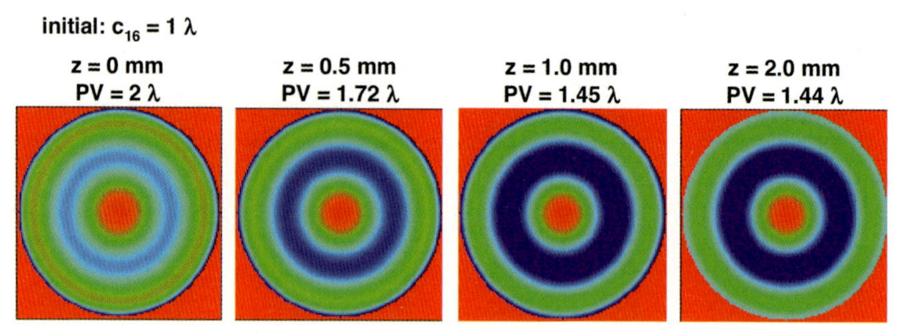

Figure 47-109: Change in the wavefront shape during propagation.

Figure 47-110 shows the changes in the first four axi-symmetrical Zernike coefficients of this example as a function of the propagation distance z. It can be seen that higher and lower Zernike modes are generated. The change in the normalization radius is negligible in this case.

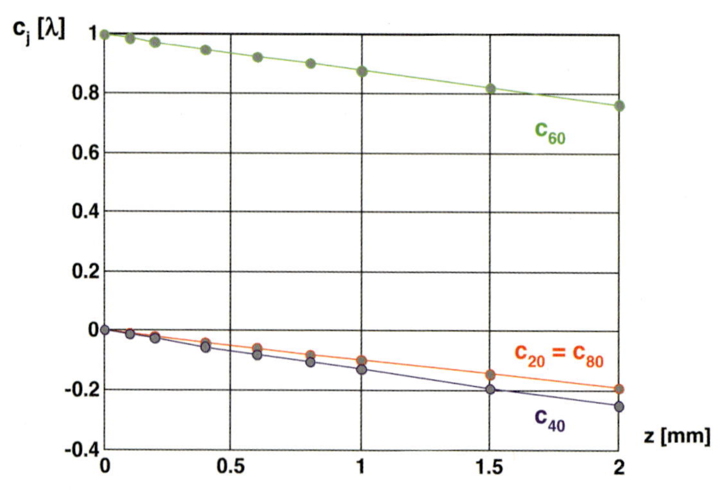

Figure 47-110: Change in the wavefront shape during propagation.

47.9 Literature

47-1 J. Novak, P. Novak and A. Miks, Application of Shack–Hartmann wavefront sensor for testing optical systems, Proc. SPIE **6609**, 660915 (2007).

47-2 B. Platt and R. Shack, Lenticular Hartmann Screen, Opt. Sci. Cent. Newsletter **5**, 15 (1971).

47-3 D. Malacara (Ed.), Optical Shop Testing (Wiley-Interscience, Hoboken, 2007).

47-4 J. M. Geary, Wavefront Sensors (SPIE Press, Bellingham, 1995).

47-5 C. Li, M. Xia, Z. Liu, D. Li and L. Xuan, Optimization for high precision Shack–Hartmann wavefront sensor, Opt. Comm. **282**, 4333 (2009).

47-6 J. Primot, Theoretical description of Shack–Hartmann wave-front sensor, Opt. Comm. **222**, 81 (2003).

47-7 Y. Dai, F. Li, X. Cheng, Z. Jiang and S. Gong, Analysis on Shack–Hartmann wave-front sensor with Fourier optics, Opt. and Laser Tech. **39**, 1374 (2007).

47-8 J. Novak, P. Novak and A. Miks, Application of Shack–Hartmann wavefront sensor for testing optical systems, Proc. SPIE **6609**, 15 (2007).

47-9 C. Canovas and E. N. Ribak, Comparison of Hartmann analysis methods, Appl. Opt. **46**, 1830 (2007).

47-10 L. Seifert, H. H. Tiziani and W. Osten, Wavefront reconstruction with the adaptive Shack–Hartmann sensor, Opt. Comm. **245**, 255 (2005).

47-11 R. Freimann and H. Gross, Propagation of the phase distribution through a double telecentric optical system, Optik **105**, 69 (1997).

47-12 P. Wu, E. DeHoog and J. Schwiegerling, Systematic error of a large dynamic range aberrometer, Appl. Opt. **48**, 6376 (2009).

47-13 C. Curatu, G. Curatu and J. Rolland, Fundamental and specific steps in Shack–Hartmann wavefront sensor design, Proc. SPIE **6288** (2006).

47-14 J. E. Greivenkamp, D. G. Smith and E. Goodwin, Calibration Issues with Shack–Hartmann sensors for metrology applications, Proc. SPIE **5252**, 372 (2004).

47-15 D. G. Smith, E. Goodwin and J. E. Greivenkamp, Important considerations when using the Shack–Hartmann method for testing highly aspheric optics, Proc. SPIE **5180**, 323 (2003).

47-16 J. Greivenkamp, D. Smith and E. Goodwin, Calibration issues with Shack–Hartmann sensors for metrology applications, Proc. SPIE **5252**, 372 (2004).

47-17 G. Cao and F. Yu, Accuracy analysis of a Hartmann–Shack wavefront sensor operated with a faint object, Opt. Eng. **33**, 2331 (1994).

47-18 D. M. Topa, Optimized methods for focal spot location using center of mass algorithms, Proc. SPIE **4769**, 116 (2002).

47-19 H. Li, H. Song, C. Rao and X. Rao, Accuracy analysis of centroid calculated by a modified center detection algorithm for Shack–Hartmann wavefront sensor, Opt. Comm. **281**, 750 (2008).

47-20 N. Zon, O. Srour and E. N. Ribak, Hartmann–Shack analysis errors, Opt. Express **14**, 635 (2006).

47-21 K. L. Baker and M. M. Moallem, Iteratively weighted centroiding for Shack–Hartmann wave-front sensors, Opt. Express **15**, 5147 (2007).

47-22 R. Rammage, D. Neal and R. Copland, Application of Shack–Hartmann wavefront sensing technology to transmissive optic metrology, Proc. SPIE **4779**, 27 (2002).

47-23 J. Pfund, N. Lindlein and J. Schwider, Misalignment effects of the Shack–Hartmann sensor, Appl. Opt. **37**, 22 (1998).

47-24 R. Irwan and R. G. Lane, Analysis of optimal centroid estimation applied to Shack–Hartmann sensing, Appl. Opt. **38**, 6737 (1999).

47-25 A. Zhang, C. Rao, Y. Zhang and W. Jiang, Sampling error analysis of Shack–Hartmann wavefront sensor with variable subaperture pixels, J.Mod. Opt. **51**, 2267 (2004).

47-26 J. Yang, L. Wei, H. Chen, X. Rao and C. Rao, Absolute calibration of Hartmann–Shack wavefront sensor by spherical wavefronts, Opt. Comm. **283**, 910 (2010).

47-27 A. Chernyshov, U. Sterr, F. Riehle, J. Helmcke and J. Pfund, Calibration of a Shack–Hartmann sensor for absolute measurement of wavefronts, Appl. Opt. **44**, 6419 (2005).

47-28 J. Pfund, N. Lindlein and J. Schwider, Dynamic range expansion of a Shack–Hartmann sensor by use of a modified unwrapping algorithm, Opt. Lett. **23**, 995 (1998).

47-29 N. Lindlein, J. Pfund and J. Schwider, Algorithm for expanding the dynamic range of a Shack–Hartmann sensor by using a spatial light modulator array, Opt. Eng. **40**, 837 (2001).

47-30 G. Yoon, S. Pantanelli and L. Nagy, Large-dynamic-range Shack–Hartmann wavefront sensor for highly aberrated eyes, J. Biomed. Opt. **11**, 030502 (2006).

47-31 N. Lindlein, J. Pfund and J. Schwider, Expansion of the dynamic range of a Shack–Hartmann sensor by using astigmatic microlenses, Opt. Eng. **39**, 2220 (2000).

47-32 N. Lindlein and J. Pfund, Experimental results for expanding the dynamic range of a Shack–Hartmann sensor using astigmatic microlenses, Opt. Eng. **41**, 529 (2002).

47-33 L. Zhao, W. Guo, X. Li and Z. Zhong, Sensing performance of a Shack–Hartmann wavefront sensor versus the properties of the light beam, Proc. SPIE **7390** (2009).

47-34 M. Ares, S. Royo and J. Caum, Shack–Hartmann sensor based on a cylindrical microlens array, Opt. Lett **32**, 769 (2007).

47-35 S. Rios and D. Lopez, Modified Shack–Hartmann wavefront sensor using an array of superresolution pupil filters, Opt. Express **17**, 9669 (2009).

47-36 X. Cui, J. Ren, G. Tearney and C. Yang, Wavefront image sensor chip, Opt. Express **18**, 16685 (2010).

47-37 J. Pfund, N. Lindlein, J. Schwider, R. Burrow, T. Blümel and K.-E. Elssner, Absolute sphericity measurement: a comparative study of the use of interferometry and a Shack–Hartmann sensor, Opt. Lett. **23**, 742 (1998).

47-38 J. Hartmann, Bemerkungen über den Bau und die Justierung von Spektrographen, Zeits. Instrumentenkunde **20**, 47 (1900).

47-39 W. Freitag, W. Grossmann and R. Wendler, Verbesserter Hartmanntest zur Wellenflächenanalyse von Hochleistungsoptik, Feingerätetechnik **33**, 554 (1984).

47-40 E. A. Vitrichenko, Methods of studying astronomical optics. Limitations of the Hartmann method, Sov. Astron. **20**, 373 (1976).

47-41 A. F. Brooks, T.-L. Kelly, P. J. Veitech and J. Munch, Ultra-sensitive wavefront measurement using a Hartmann sensor, Opt. Express **15**, 10370 (2007).

47-42 S. Rios, E. Acosta and S. Bara, Hartmann sensing with Albrecht grids, Opt. Comm. **133**, 443 (1997).

47-43 V. Voitsekhovich, L. Sanchez, V. Orlov and S. Cuevas, Efficiency of the Hartmann test with different subpupil forms for the measurement of turbulence-induced phase distortions, Appl. Opt. **40**, 1299 (2001).

47-44 B. Wells and J. Myrick, Revisiting the Hartmann test, Proc. SPIE **5180**, 340 (2003).

47-45 D. P. Salas-Peimbert, D. Malacara-Doblado, V. M. Duran-Ramirez, G. Trujillo-Schiaffino and D. Malacara-Hernandez, Wave-front retrieval from Hartmann test data, Appl. Opt. **44**, 4228 (2005).

47-46 J. Primot and N. Guerineau, Extended Hartmann test based on the pseudoguiding property of a Hartmann mask completed by a phase chessboard, Appl. Opt. **39**, 5715 (2000).

47-47 R. Diaz-Uribe, F. Granados-Agustin and A. Cornejo-Rodrigues, Classical Hartmann test with scanning, Opt. Express **17**, 13959 (2009).

47-48 N. Hodgson, T. Haase, R. Kosta and H. Weber, Determination of laser beam parameters with the phase space beam analyzer, Opt. and Quant. Electr. **24**, S926 (1992).

47-49 H. Weber, Wave optical analysis of the phase space analyzer, J. Mod. Opt. **39**, 543 (1992).

47-50 F. Zernike, Diffraction theory of the knife-edge test and its improved form, the phase contrast method, Physica I, 689 (1934).

47-51 G. Settles, Schlieren and Shadowgraph Techniques (Springer, Berlin, 2006).

47-52 D. Malacara, Analysis of the interferometric Ronchi test, Appl. Opt. **29**, 3633 (1990).

47-53 R. Ragazzoni, E. Diolaiti and E. Vernet, A pyramid wavefront sensor with no dynamic modulation, Opt. Comm. **208**, 51 (2002).

47-54 C. Verinaud, On the nature of the measurements provided by a pyramid wave-front sensor, Opt. Comm. **233**, 27 (2004).

47-55 X. Wei, T. van Heugten and L. Thibos, Validation of a Hartmann–Moire wavefront sensor with large dynamic range, Opt. Express **17**, 14180 (2009).

47-56 G. Andersen, L. Dussan, F. Ghebremichael and K. Chen, Holographic wavefront sensor, Opt. Eng. **48**, 085801 (2009).

47-57 M. Booth, M. Neil and T. Wilson, New modal wave-front sensor: application to adaptive confocal fluorescence microscopy and two-photon excitation fluorescence microscopy, JOSA A **19**, 2112 (2002).

47-58 M. Neil, M. Booth and T. Wilson, New model wave-front sensor: a theoretical analysis, JOSA A **17**, 1098 (2000).

47-59 C.Guo, K. Liang, X. Zhang and H. Wang, Real-time coherent diffractive imaging with convolution-solvable sampling array, Opt. Lett. **35**, 850 (2010).

47-60 O. Kafri and I. Glatt, The Physics of Moire Metrology (Wiley Interscience, New York, 1990), Ch. 6+7.

47-61 C. Hou and J. Bai, Wavefront measurement for long focal large aperture lens based on Talbot effect of Ronchi grating, J. of Phys. **48**, 1037 (2006).

47-62 L. Liu, Talbot and Lau effects on incident beams of arbitrary wavefront and their use, Appl. Opt. **28**, 4668 (1989).

47-63 S. Yokozeki and T. Sizuki, Shearing Interferometer using the grating as the beam splitter. Part 2, Appl. Opt. **10**, 1690 (1971).

47-64 S. Yokozeki and K. Ohnishi, Spherical aberration measurement with Shearing interferometer using Fourier imaging and Moire method, Appl. Opt. **14**, 623 (1975).

47-65 J. Quiroga, D. Crespo and E. Bernabeu, Fourier transform method for automatic processing of moire deflectograms, Opt. Eng. **38**, 974 (1999).

47-66 R. Sekine, T. Shibuya *et. al.*, Measurement of wavefront aberration of the human eye using Talbot Image of two-dimensional grating, Opt. Rev. **13**, 207 (2006).

47-67 Y. Nakano and K. Murata, Measurement of phase objects using the Talbot effect and Moire techniques, Appl. Opt. **23**, 2296 (1984).

47-68 D. Silva, A simple interferometric method of beam collimation, Appl. Opt. **10**, 1980 (1971).

47-69 R. W. Gerchberg and W. O. Saxton, Phase determination from image and diffraction plane pictures, Optik **34**, 275(1971).

47-70 R. W. Gerchberg and W. O. Saxton, A practical algorithm for the determination of phase from image and diffraction pictures, Optik **35**, 237 (1972).

47-71 J. R. Fienup, Phase retrieval algorithms: a comparison, Appl. Opt. **21**, 2758(1982).

47-72 J. R. Fienup, Phase-retrival algorithms for a complicated optical system, Appl. Opt. **32**, 1737 (1993).

47-73 B. M. Hanser, M. D. Gustafsson, D. A. Agard and J. W. Sedat, Phase retrieval for high-numerical aperture optical systems, Opt. Lett. **28**, 801 (2003).

47-74 B. Moeller and H. Gross, Characterization of complex optical systems based on wavefront retrieval from point spread function, Proc. SPIE **5965** (2005).

47-75 J. Wesner, J. Heil and T. Sure, Reconstructing the pupil function of microscope objectives from the intensity PSF, Proc. SPIE **4767** (2002).

47-76 R. M. von Buenau, H. Fukuda and T. Tersawa, Phase Retrieval from defocused Images and Its Applications in Lithography, Jpn. J. Appl. Phys. **36**, 7494 (1997).

47-77 J. L. Allen and M. P. Oxley, Phase retrieval from series of images obtained by defocus variation, Opt. Comm. **199**, 65 (2001).

47-78 H. Gross, A. Krause, L. Stoppe and B. Böhme, Characterization of optical systems based on wavefront retrieval from point spread function, Proc. EOS Topical Meeting, Lille, 2007.

47-79 R. Barakat and G. Newsam, Numerically stable iterative method for the inversion of wave-front aberrations from measured point-spread-function data, JOSA **70**, 1255 (1980).

47-80 A. J. Janssen, Extended Nijboer–Zernike approach for the computation of optical point-spread functions, JOSA **A 19**, 849 (2002).

47-81 P. Dirksen, J. J. Braat, A. J. Janssen, C. Juffermans and A. Leeuwestein, Experimental determination of lens aberrations from the intensity point-spread function in the focal region, Proc. SPIE **5040**, 1 (2003).

47-82 P. Dirksen , J. J. Braat , A. J. Janssen and C. Juffermans, Aberration retrieval using the extended Nijboer–Zernike approach, J. Microlith. **2**, 61 (2003).

47-83 J. J. Braat, P. Dirksen and A.J. Janssen, Assessment of an extended Nijboer–Zernike approach for the computation of optical point-spread functions, JOSA **A 19**, 858 (2002).

47-84 D. M. Paganin, Coherent X-Ray Optics (Oxford Science Publication, Oxford, 2006), Chapter 4.

47-85 V. V. Volkov, Y. Zhu and M. De Graef, A new symmetrized solution for phase retrieval using the transport of intensity equation, Micron **33**, 411 (2002).

47-86 T. E. Gureyev and K. A. Nugent, Phase retrieval with the transport-of-intensity equation. II. Orthogonal series solution for nonuniform illumination, JOSA **A 13**, 1670 (1996).
47-87 T. E. Gureyev, A. Roberts and K. A. Nugent, Phase retrieval with the transport-of-intensity equation. Matrix solution with the use of Zernike polynomials, JOSA **A 12**, 1932 (1995).
47-88 M. R. Teague, Image formation in terms of the transport equation, JOSA **A 2**, 2019 (1985).
47-89 A. Semichaevsky and M. Testorf, Phase space interpretation of deterministic phase retrieval, JOSA **A 21**, 2173 (2004).
47-90 C. Falldorf, M. Agour, C. Kopylow and R. Bergmann, Phase retrieval by means of a spatial light modulator in the Fourier domain of an imaging system, Appl. Opt. **49**, 1826 (2010).
47-91 Z. Li, Accurate optical wavefront reconstruction based on reciprocity of an optical path using low resolution spatial light modulators, Opt. Comm. **283**, 3646 (2010).
47-92 J. Rodrigo, H. Duadi, T. Alieva and Z. Zalevsky, Multi-stage phase retrieval algorithm based upon the gyrator transform, Opt. Express **18**, 1510 (2010).
47-93 B. H. Dean and C. W. Bowers, Diversity selection for phase-diverse phase retrieval, JOSA **A 20**, 1490 (2003).
47-94 Y. Han and C. Zhang, Reconstructing unknown wavefronts by use of diffraction-intensity pattern subtraction, Opt. Lett. **35**, 2115 (2010).
47-95 M. Li and X.-Y. Li, Linear phase retrieval with a single far-field image based on Zernike polynomials, Opt. Express **17**, 15257 (2009).
47-96 N. Hodgson and H. Weber, Optical Resonators (Springer, Berlin, 1997), Chapter 23.
47-97 E. Wolf (Ed.), Progress in Optics Vol. **51** (Elsevier, Amsterdam, 2008), p. 349, J. J. Braat, S. van Haver, A. J. Janssen and P. Dirksen, Assessment of optical systems by means of point-spread functions.
47-98 N. B. Voznesenskii and T. V. Ivanova, Mathematical modeling of the light-field distribution close to the focus of a high-aperture optical system, J. Opt. Techn. **65**, 787 (1998).
47-99 D. G. Flagello, T. Milster and A. E. Rosenbluth, Theory of high-NA imaging in homogeneous thin films, JOSA **A 13**, 53 (1996).
47-100 M. Mansuripur, Certain computational aspects of vector diffraction problems, JOSA **A 6**, 786 (1989).
47-101 M. Mansuripur, Distribution of light at and near the focus of high-numerical-aperture objectives, JOSA **A 3**, 2086 (1986).
47-102 N. Nakajima, Phase retrieval from a high-numerical-aperture intensity distribution by use of an aperture-array filter, JOSA A **26**, 2173 (2009).
47-103 Y. Li and E. Wolf, Three-dimensional intensity distribution near the focus in systems of different Fresnel numbers, JOSA A **1**, 801 (1984).
47-104 G. Brady and J. Fienup, Measurement range of phase retrieval in optical surface and wavefront metrology, Appl. Opt. **48**, 442 (2009).
47-105 S. Thurman and J. Fienup, Complex pupil retrieval with undersampled data, JOSA A **26**, 2640 (2009).
47-106 S. Thurman and J. Fienup, Phase retrieval with signal bias, JOSA A **26**, 1008 (2009).
47-107 H. Mao and D. Zhao, Alternative phase-diverse phase retrieval algorithm based on Levenberg–Marquardt nonlinear optimization, Opt. Express **17**, 4540 (2009).
47-108 H. Mao, X. Wang and D. Zhao, Application of phase-diverse phase retrieval to wavefront sensing in non-connected complicated pupil optics, Chin. Opt. Lett. **5**, 397 (2007).
47-109 M. Bertero and P. Boccacci, Introduction to Inverse Problems in Imaging (Institute of Physics Publishing, Bristol, 1998).
47-110 T. M. Jeong, D.-K. Ko and J. Lee, Method of reconstructing wavefront aberrations from the intensity measurement, Opt. Lett. **32**, 3507 (2007).
47-111 V. Y. Ivanov, V. P. Sivokon and M. A. Vorontsov, Phase retrieval from a set of intensity measurements: theory and experiment, JOSA **A 9**, 1515 (1992).
47-112 S. Marchesini, A unified evaluation of iterative projection algorithms for phase retrieval, Rev. Sci. Instr. **78**, 011301 (2007).
47-113 R. W. Gerchberg, The lock problem in the Gerchberg–Saxton algorithm for phase retrieval, Optik **74**, 91 (1986).
47-114 H. Kim and B. Lee, Optimal nonmonotonic convergence of the iterative Fourier-transform algorithm, Opt. Lett. **30**, 296 (2005).

47-115 J. R. Fienup and C. C. Wackerman, Phase-retrieval stagnation problems and solutions, JOSA A **3**, 1897 (1986).

47-116 H. Takajo, T. Takahashi, R. Ueda and M. Taninaka, Study on the convergence property of the hybrid input-output algorithm used for phase retrieval, JOSA A **15**, 2849 (1998).

47-117 R. G. Lane, Phase retrieval using conjugate gradient minimization, J. Mod. Opt. **38**, 1797 (1991).

47-118 H. H. Bauschke, P. L. Combettes and D. R. Luke, Phase retrieval, error reduction algorithm and Fienups variants, A view from convex optimization, JOSA A **19**, 1334 (2002).

47-119 D. S. Barwick, Efficient metric for pupil-phase engineering, Appl. Opt. **46**, 7258 (2007).

47-120 J. R. Fienup, Invariant error metrics for image reconstruction, Appl. Opt. **36**, 8352 (1997).

47-121 R. W. Deming, Phase retrieval from intensity-only data by relative entropy minimization, JOSA A **24**, 3666 (2007).

47-122 S. Thurman, R. DeRosa and J. Fienup, Amplitude metrics for field retrieval with hard-edged and uniformly illuminated apertures, JOSA A **26**, 700 (2009).

47-123 H. Shioya and K. Gohara, Generalized phase retrieval algorithm based on information measures, Opt. Comm. **266**, 88 (2006).

47-124 A. Talmi and E. N. Ribak, Wavefront reconstruction from its gradients, JOSA **A 23**, 288 (2006).

47-125 W. Zhou and Z. Zhang, Generalized wavefront reconstruction algorithm applied in a Hartmann–Shack test, Appl. Opt. **39**, 250 (2000).

47-126 W. Zou and J. P. Rolland, Iterative zonal wave-front estimation algorithm for optical testing with general-shaped pupils, JOSA A **22**, 938 (2005).

47-127 E. P. Wallner, Optimal wave-front correction using slope measurements, JOSA **73**, 1771 (1983).

47-128 J. Herrmann, Least-squares wavefront errors of minimum norm, JOSA **70**, 28 (1980).

47-129 W. H. Southwell, Wave-front estimation from wave-front slope measurements, JOSA **70**, 998 (1980).

47-130 B. R. Hunt, Matrix formulation of the reconstruction of phase values from phase differences, JOSA **69**, 393 (1979).

47-131 D. L. Fried, Least-square fitting a wave-front distortion estimate to an array of phase-difference measurements, JOSA **67**, 370 (1977).

47-132 S. Rios and E. Acosta, Orthogonal modal reconstruction of a wavefront from phase difference measurements, J. Mod. Opt. **46**, 931 (1999).

47-133 G. Dai, Modal wave-front reconstruction with Zernike polynomilas and Karhunen–Loeve functions, JOSA A **13**, 1218 (1996).

47-134 H. Guo and Z. Wang, Wavefront reconstruction using iterative discrete Fourier transforms, Optik, **117**, 77 (2006).

47-135 K. R. Freischlad and C. L. Koliopoulos, Modal estimation of a wavefront from difference measurements using the discrete Fourier transform, JOSA A **3**, 1852 (1986).

47-136 K. Freischlad and C. L. Koliopoulos, Wavefront reconstruction from noisy slope or difference data using the discrete Fourier transform, Proc. SPIE **551**, 74 (1985).

47-137 S. Bezdidko, Study of the properties of Zernikes's orthogonal polynomials, Proc. SPIE **5174**, 227 (2003).

47-138 C. Zhao and J. H. Burge, Orthonormal vector polynomials in a unit circle, Part I: basis set derived from gradients of Zernike polynomials, Opt. Express **15**, 18014 (2007).

47-139 D. Malacara, M. Carpio-Valadez and J. Sabchez-Mondragon, Wavefront fitting with discrete orthogonal polynomials in a unit radius circle, Opt. Eng. **29**, 671 (1990).

47-140 B. Qi, H. Chen, J. Ma and N. Dong, Regression analysis for wavefront fitting with Zernike polynomials, Proc. SPIE **5180**, 429 (2003).

47-141 A. Janssen and P. Dirksen, Computing Zernike polynomials of arbitrary degree using the discrete Fourier transform, JEOS **2**, 07012 (2007).

47-142 G. Dai, Zernike aberration coefficients transformed to and from Fourier series coefficients for wavefront representation, Opt. Lett. **31**, 501 (2006).

47-143 S. A. Comastri, L. Perez, G. D. Perez, G. Martin and K. Bastida, Zernike expansion coefficients: rescaling and decentering for different pupils and evaluation of corneal aberrations, J. Opt. A **9**, 209 (2007).

47-144 L. Lundström and P. Unsbo, Transformation of Zernike coefficients: scaled, translated and rotated wavefronts with circular and elliptical pupils, JOSA A **24**, 569 (2007).

47-145 J. Schwiegerling, Scaling Zernike expansion coefficients to different pupil sizes, JOSA A **19**, 1937 (2002).

47-146 J. Diaz, J. Fernandez-Dorado, C. Pizarro and J. Arasa, Zernike coefficients for concentric, circular scaled pupils: an equivalent expression, J. Mod Opt. **56**, 131 (2009).

47-147 H. Shu and L. Luo, General method to derive the relationship between two sets of Zernike coefficients corresponding to different aperture sizes, JOSA A **23**, 1960 (2006).

47-148 S. Comastri, K. Bastida, A. Bianchetti, L. Perez, G. Perez and G. Martin, Zernike aberrations when pupil varies: selction rules, missing modes and graphical method to identify modes, Pure Appl. Opt. **11**, 085302 (2009).

47-149 G. Dai, Scaling Zernike expansion coefficients to smaller pupil sizes: a simpler formula, JOSA A **23**, 539 (2006).

47-150 C. Campbell, Matrix method to find a new set of Zernike coefficients from an original set when the aperture radius is changed, JOSA A **20**, 209 (2003).

47-151 G. Dai, C. Campbell, L. Chen, H. Zhao and D. Chernyak, Wavefront propagation from one plane to another with the use of Zernike polynomials and Taylor monomials, Appl. Opt. **48**, 477 (2009).

47-152 G. Dai, Wavefront Optics for Vision Correction (SPIE Press, Bellingham, 2008).

48
Radiometry

Handbook of Optical Systems: Vol. 5. Metrology of Optical Components and Systems. First Edition.
Edited by Herbert Gross.

48.1 Introduction

The common task required of all optical systems is to transfer optical radiation energy. It is necessary to produce a desired amount of energy, with the correct spatial distribution within the spectral range needed and with a defined spectral distribution and this is dependent on the particular application. In the case of imaging systems, e.g., the brightness distribution of an illuminated object generally has to be transferred into the desired similar image distribution. The amount of energy must be sufficient to generate the signal needed for the sensor used. For example, the human eye as a natural sensor demands a minimum of about 1 lux in order to generate a signal by the eye receptors, making an object just visible. The total dynamic range covered by the human eye is about $1{:}10^5$. The technical sensors, which can measure the lowest radiation energy, are generally limited by sensor noise and sensitivity. The task of a modern photographic lens system, either for professional or leisure use, is to provide an image similar to the object and also to transfer an amount of energy per time unit, permitting illumination times which are short enough for convenient use. For conventional photography the amount of energy needed depends on the sensitivity of the film material and, for new digital photography, on the sensitivity of modern CCD imagers. Furthermore, enough grey levels must be resolved for the imaging purpose, and colours should be transferred sufficiently well. Microlithographic illumination and imaging lens systems provide an example for high significance of the system transmission: The higher the transmission the shorter the illumination times that can be chosen for evaluation of the resist structures and finally more chips can be produced in a shorter time. This small selection of examples shows the basic importance of radiometric quantities for optical systems.

In general, radiometry is the measurement of the energy or power (energy per time unit) content of electromagnetic radiation fields. This includes the transfer of energy from a source through a medium to a detector. All optical measurement techniques need at least a light source, a sensor and optical components which will bring the light to the sample under test and to the sensor. We can consider two categories of energy or power measurement:

1. Direct Measurement of power or energy as a primary property of an optical component and
2. Measurement of power or energy as a base quantity, from which other properties are derived.

Relative Measurements of the reflectivity of surfaces and coatings or transmission of optical components and systems belong within the first category. Other examples for category 1 are the measurement of absolute power and spatial distribution of a light source or the sensitivity of a detector. Often the motivation for this kind of measurement is to find the suitable source or detector for an optical measurement task within the second category.

The techniques included in category 2 do not measure the absolute or relative power as a property of the component under test but need the quantity of power or energy in order to evaluate other quantities. So interferometric methods evaluate the phase information of interferograms from the irradiance distribution in a plane (for example, a detector array plane) to get a relative surface form deviation or information about homogeneity. In a similar way, methods characterizing imaging quality use the primary irradiance of patterns transformed by an optical system in order to obtain phase and amplitude information.

This chapter introduces the fundamental requirements for measuring radiometric quantities with arbitrary sensors and gives an overview of the basic measuring principles. For the case involving the human eye so called photometric quantities are used instead of the radiometric ones.

48.2 Basic Principles of Radiometry

48.2.1 Energy Transport by Electromagnetic Fields

Fundamental to the understanding of energy or power measurement is energy transport as a principal property of electromagnetic waves.

In the case of light, the source of radiation is generally the rearrangement of charge distribution within the outer electron clouds of atoms. Accelerated charges produce electric fields and every electric field is accompanied by a magnetic field and vice versa. The electromagnetic field exists outside and independent of its primary sources and transports energy through space with the velocity of light. This process lasts eternally if there is no interaction with material – even if the source of this process no longer exists. By the interaction with any material the energy of light radiation can provide work at the electrons of the material. Electric charges are generated, which in turn generate signals by means of electric currents or voltages. This is the basic principle for any natural or technical radiometric sensor. Signals produced by the human eye in this way e.g. allow us to observe stars the light from which has been travelling for millions of years.

The description of light propagation as an electromagnetic wave is evaluated from Maxwell's equations. We will not derive the basic mathematic relations here, but present several results as far as they are important for the understanding of energy or power measurements (see for example [48-1] and [48-2], volume 1 of this handbook).

The energy of an electromagnetic field is described by the energy density u, which is the energy within an infinitesimal volume element. The energy density is distributed within the electric and magnetic fields of the travelling wave. In a homogenous, isotropic linear dielectric (non-conducting) medium the energy density of the electromagnetic field is

$$u = u_E + u_B \qquad (48\text{-}1)$$

where

$$u_E = \frac{1}{2}\varepsilon_r\varepsilon_0\vec{E}^2 = \frac{1}{2}\varepsilon\vec{E}^2 \qquad (48\text{-}2)$$

and

$$u_B = \frac{1}{2\mu_r\mu_0}\vec{B}^2 = \frac{1}{2}\mu_0\mu_r\vec{H}^2 = \frac{1}{2}\mu\vec{H}^2 \qquad (48\text{-}3)$$

and where
$\vec{E}$ is the electric field strength,
$\vec{B}$ is the magnetic induction,
ε_0 is the absolute constant of dielectricity, $\varepsilon_0 = 8.8542 \times 10^{-12}\,\frac{\text{As}}{\text{Vm}}$,
μ_0 is the absolute magnetic permeability, $\mu_0 = 4\pi \times 10^{-7}\,\frac{\text{Vs}}{\text{Am}}$.

The material constants ε_r and μ_r are the relative constant of dielectricity and the relative magnetic permeability, respectively, with the value 1 for vacuum.

The vectors $\vec{E}$ and $\vec{B}$ of the electric field strength and the magnetic induction are orthogonal to each other.

The energy carried by the electromagnetic field causes an energy flow perpendicular to the electric and magnetic field strength vectors. This is described by the Poynting vector

$$\vec{S} = \vec{E} \times \vec{H}. \qquad (48\text{-}4)$$

The direction of the Poynting vector can differ from the propagation direction of the electromagnetic wave, but in the case of electromagnetic plane waves in an isotropic medium, the propagation direction is also the direction of energy flow.

Figure 48-1 shows this schematically for the vector $\vec{E}$ parallel to the axis x. The Poynting vector can be understood as the energy of a volume element $v\Delta t\Delta A$ flowing through the area element ΔA within the time interval Δt:

$$\left|\vec{S}\right| = \frac{uv\Delta t\Delta A}{\Delta t\Delta A} = uv \qquad (48\text{-}5)$$

The time and space dependence of the electric field strength vector of an arbitrary harmonic monochromatic plane wave is described by

$$\vec{E}(\vec{r}, t) = \vec{E}_0 \cos(\vec{k} \cdot \vec{r} - \omega t). \qquad (48\text{-}6)$$

Where $\vec{k}$ is the wave vector, which determines the direction of wave propagation, $\vec{r}$ is the space vector, $\omega = 2\pi f$ isthe circle frequency of the frequency f for the variation of the vectors $\vec{E}$ and $\vec{B}$, and t is the time.

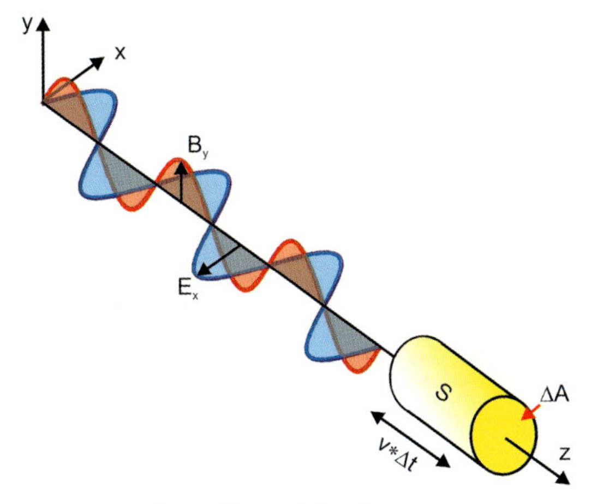

Figure 48-1: Energy Flow and Poynting vector.

Considering (48-1), (48-2), and (48-3) we obtain for the absolute value of the Poynting vector

$$\left|\vec{S}\right| = vu = \frac{c}{n}u = \sqrt{\frac{\varepsilon_0\varepsilon_r}{\mu_0\mu_r}}\left|\vec{E}\right|^2. \tag{48-7}$$

The phase velocity of the wave in the medium with the refraction index n is v. The refraction index is defined as

$$n \equiv \frac{c}{v} = \sqrt{\varepsilon_r\mu_r} \tag{48-8}$$

and depends on the light frequency f (for the phenomenon of dispersion see, for example, [48-1] and [48-2], and chapter 4 of volume 1 of this handbook).

In vacuum the phase velocity has the constant value

$$c = \frac{1}{\sqrt{\varepsilon_0\mu_0}}, \qquad c = \frac{1}{\sqrt{\varepsilon_0\mu_0}} = 2.99792458 \times 10^8\ \mathrm{ms}^{-1}. \tag{48-9}$$

Since both the electric and magnetic field strength vary with the same high frequency the Poynting vector and the energy flux density also vary with this frequency. According to the correlation

$$c = f\lambda = \frac{\omega}{2\pi}\lambda \tag{48-10}$$

between
the frequency f, respectively the circle frequency ω, the vacuum wavelength λ, and the vacuum velocity of light c, the frequencies are e.g. about 659–385 THz for the visible wavelength range from 455 nm (violet) to 780 nm (red).

The human eye and technical sensors used for measuring the energy characteristics of light are insensitive to the rapid variations of optical electromagnetic waves. Hence, one obtains a measurable quantity by using the irradiance I of the electromagnetic plane wave. I is the time averaged energy flux through an area unit perpendicular to the propagation direction. By using (48-6), (48-7) and (48-8) the relation (48-11) can be derived:

$$I = \varepsilon_0 \varepsilon_r v \left\langle \vec{E}^2 \right\rangle = \varepsilon_0 \varepsilon_r \frac{c}{n} \left\langle \vec{E}^2 \right\rangle \tag{48-11}$$

In the case of a harmonic plane wave (48-11) results in

$$I = \frac{1}{2} \varepsilon_0 \varepsilon_r v \left| \vec{E}_0 \right|^2 = \frac{1}{2} \varepsilon_0 \varepsilon_r \frac{c}{n} \left| \vec{E}_0 \right|^2 \tag{48-12}$$

where $|\vec{E}_0|$ is the absolute value of the electric field amplitude according (48-6).

The quantity I should not be mistaken for the radiant intensity I, the radiant flux per solid angle, which will be defined in the next paragraph. In (48-11) and (48-12) the symbol I is used instead of the usual radiometric symbol E for the quantity "irradiance" to avoid confusion, since the symbol E is generally used for the electric field strength. The standard radiometric symbol E for the irradiance will be introduced in section 48.2.2.

48.2.2
Radiometric and Photometric Quantities

In principle, every setup used for measuring the characteristics of optical components, a radiation source or a detector involves the radiation source, the detector and an optical system, which brings the radiation to the sensor. The optical system influences the radiation. This influence can itself be used to measure certain characteristics of the system. Alternatively, the characteristic of the source or the sensor can be measured and the optical system is used only to bring the radiation from the source to the sensor. In this case the influence of the system on the quantity to be measured must be eliminated. Figure 48-2 shows schematically the radiant flux Φ_{in} coming from one point P_S of the surface area element dA_S on the radiation source and the radiant flux Φ_{out} after passing through the lens and reaching one point P_D of the detector area element dA_D. For a given detector area size the optical system determines the amount of radiation reaching the detector as well as the source area size contributing to the radiation reaching the detector area (for more detail see section 48.2.4).

For radiometric and photometric measurements, the assumption is made that one can neglect polarization, diffractive or interference effects. A geometric treatment is appropriate when describing the correlations of radiant transfer.

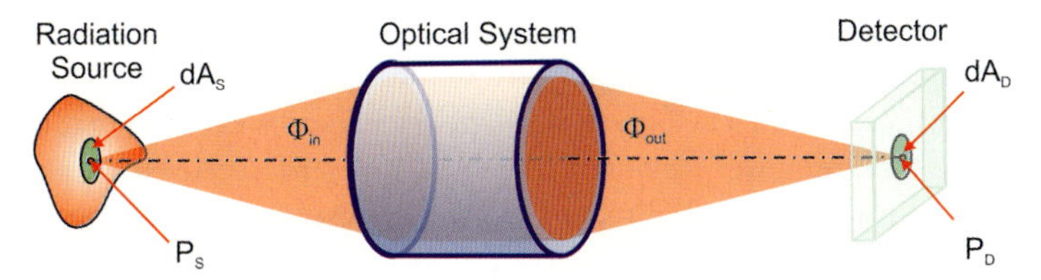

Figure 48-2: Basic components of an optical test setup.

Irradiance

The irradiance is defined as the radiant power (or radiant flux) $d\Phi$ of a source or a radiation beam incident on an infinitesimal surface element dA at a specified point P of the surface. The surface A can be a real detector element or the element of any virtual surface within a radiation field in space.

The formula for the irradiance is

$$E = \frac{d\Phi}{dA} \tag{48-13}$$

and the dimensions are $\frac{\mathrm{W}}{\mathrm{m}^2}$.

The quantity E is the area density of the radiant flux. It does not take into account the radiation direction, unlike the irradiance of an electromagnetic wave. The flux incident on the surface point considered can come from any direction of the hemispherical space above this surface, and may even come from all directions (see figure 48-3). The irradiance is the essential quantity used to determine the detector response of a detector array sensor. Those involved with optical measuring, traditionally use the term "intensity", which has another meaning in radiometric terminology (see below), rather than the term "irradiance" .

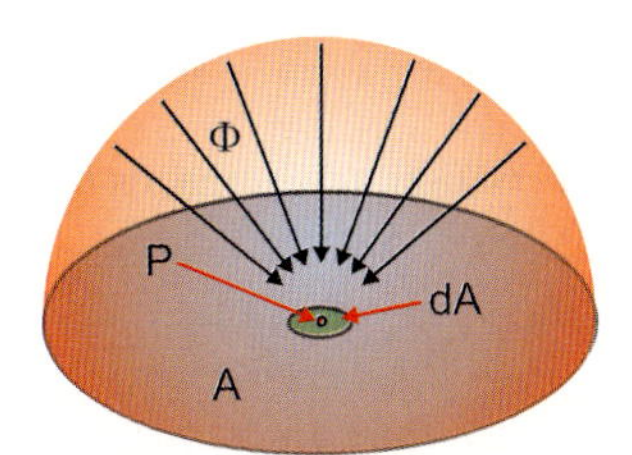

Figure 48-3: Irradiance of a surface element illuminated by radiation of the whole hemisphere.

Sometimes the irradiance is defined dependent on the direction of incident radiation as the power per projected area element, with the area element perpendicular to the direction of radiation [48.3]. In this book only the definition according (48-13) is used.

Radiant Exitance

The irradiance can also characterize the radiant flux area density leaving or passing through a specified surface. When leaving a surface the irradiance is called the radiant exitance. It is the radiant power leaving an infinitesimal surface element at a specified point in any direction or in all directions in the hemispherical solid angle. The surface of any radiation source can be characterized by this quantity.

Projection of Irradiance

For the special idealized case of a collimated beam falling on a surface A the irradiance is equal at every point of A and depends on the tilt angle θ between the surface normal and the direction of radiation. It is

$$E = E_0 \cos\theta \qquad (48\text{-}14)$$

where $E_0 = \dfrac{d\Phi}{dA_0} = \dfrac{\Phi}{A_0}$

is the irradiance for normal incidence ($\theta = 0°$) and $E = d\Phi/dA = \Phi/A$ is the irradiance on the area A with the normal deviating from the radiation direction by the angle θ according to definition (48-13). Figure 48-4 shows this situation schematically.

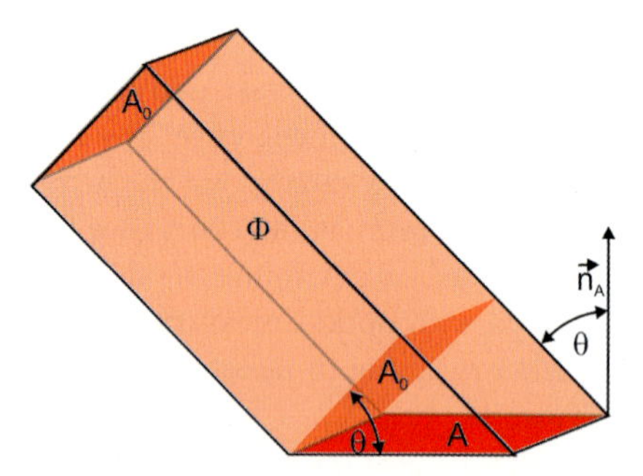

Figure 48-4: Irradiance of a collimated beam.

The relation (48-14) is derived from the law of energy conservation, which means for this case that the same amount of flux as passes through A_0 also flows through A.

Radiance

Usually radiant power is not homogeneously distributed over a solid angle, but varies with the direction. Furthermore, the radiant exitance of light sources varies over the surface. Hence, when characterizing radiation it is useful to define another more general radiometric quantity; the radiance (see figure 48-5).The radiance of a specified point P of area A in space is defined as the solid angle and projected area density of the radiant power Φ. This refers to the amount of radiant power per infinitesimal solid angle element $d\Omega$ and per infinitesimal surface element dA projected onto the plane perpendicular to the direction of radiation considered. The radiation direction is described by the angle θ to the surface normal of the surface element in point P. The radiance can be incident on, passing through or emerging from the point P considered.

It is

$$L = \frac{d^2\Phi}{d\Omega dA \cos\theta} \tag{48-15}$$

with dimensions $\frac{\text{W}}{\text{m}^2\ \text{srad}}$,

where "srad" is the measuring unit for the solid angle. An area of one square metre on the surface of a sphere with a radius of 1 metre forms a solid angle of 1 srad (see also below).

The quantity L is not only important for characterizing light sources but is also a determining quantity for another parameter in optical measuring tasks – the throughput of an optical system. Usually the throughput depends on the power per emitting surface element of the radiation source and the distribution over the solid angle, which is defined by the distance of the source and the aperture of the optical system used for conducting the radiation to a detector.

Irradiance and radiant power can be calculated from the radiance by integrating over a finite area A and a finite solid angle Ω.

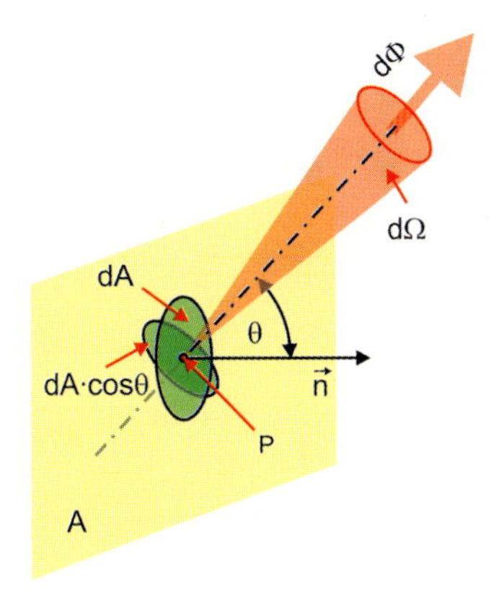

Figure 48-5: Definition of radiance.

Radiant Intensity

The radiant intensity I is the amount $d\Phi$ of radiant power Φ per infinitesimal solid angle element. Normally it varies with the direction and point of the light source area considered. It is

$$I = \frac{d\Phi}{d\Omega}. \tag{48-16}$$

For distances, which are large compared with the size of a radiation source, it makes sense to regard the source as a point source and characterize it by its radiant intensity (figure 48-6).

The dimensions of the radiant intensity are W/srad.

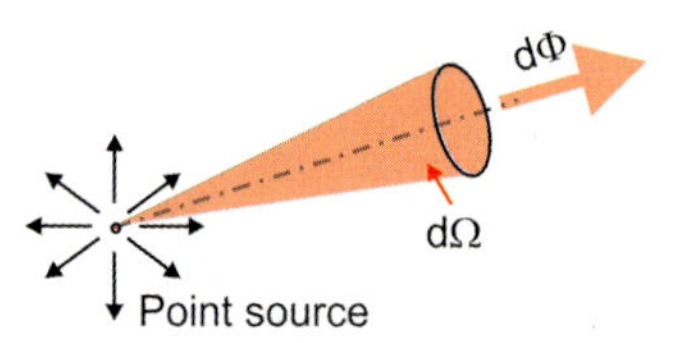

Figure 48-6: Definition of radiant intensity of a point source.

An example of those sources which can be described as point sources, would be a star observed from earth. For this case one can consider the star as a point source and the eye of the observer without the use of any optical instrument or telescope would define the solid angle for the observation.

Radiant Exposure

If the irradiance is integrated over time t an exposure takes place (e.g., exposure of a film, a detector array or of photoresist in microlithography procedures). During the integration time a specified amount of energy per area unit is generated by the incident radiant power flux. Hence this radiometric quantity is called the radiant exposure with

$$H = \int E dt \tag{48-17}$$

as the definition equation. The dimensions are $\mathrm{Ws/m^2}$.

Radiant Energy

The integration of the radiant flux of any source or beam over time t is the radiant energy

$$Q = \int \Phi dt. \tag{48-18}$$

This radiometric quantity is used, for example, for pulsed radiation sources, in which the energy of the pulses is important. As an amount of energy it is measured in the units Ws.

Solid Angle

The solid angle is important for the geometric definition of radiometric quantities. In analogy with the two-dimensional angle definition, the solid angle in three-dimensional space is defined as

$$d\Omega = \frac{dA \cos\theta}{r^2} \tag{48-19}$$

which is the infinitesimal surface element dA in space, the distance r of this element from a specified point (e.g., a source point or detector point) P and the angle θ between the surface normal $\vec{n}$ of dA and the vector $\vec{r}$, the connecting line of the surface element and point P. Figure 48-6 shows that, for the general case, the size of the area element can be defined by a closed curve C of arbitrary form in space. The solid angle is the projection of the curve onto a sphere with radius r_S and P as the center of curvature. From this an equivalent definition of the solid angle $d\Omega = dA_S/R^2$ can be found.

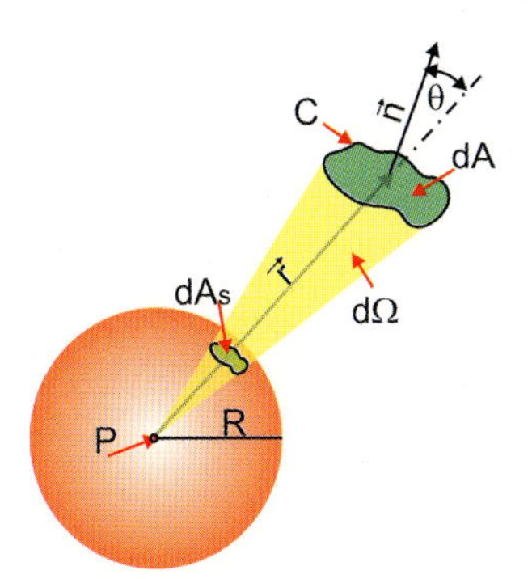

Figure 48-7: Definition of the solid angle for the general case and for spherical coordinates.

The solid angle of the complete space is 4π and for the hemisphere it is 2π.

In optical systems, often simple optical shapes of solid angles occur, which can be found analytically. For many optical applications the solid angle of a cone can be used. The solid angle of a cone with cone angle (half-aperture angle) σ is found by integrating over the solid angle element defined by (48-19) using spherical coordinates:

$$\Omega = \int_{\theta=0}^{\sigma} \int_{\varphi=0}^{2\pi} \sin\theta d\theta d\varphi = 2\pi(1 - \cos\sigma) \tag{48-20}$$

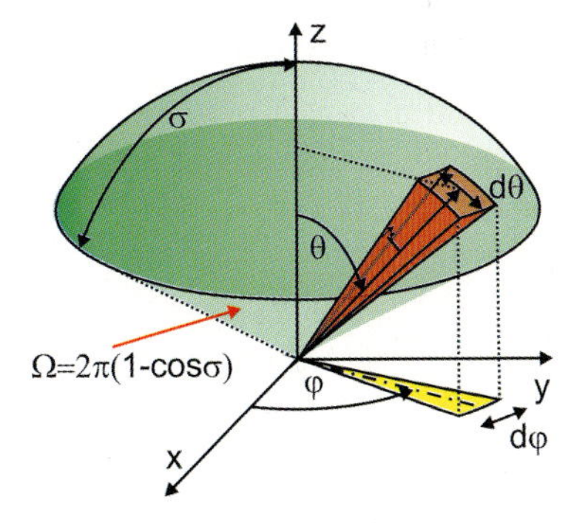

Figure 48-8: Solid angle of a cone.

Spectral Dependence of Radiometric Quantities

In general the radiometric quantities defined above depend on the wavelength of the radiation. Hence one has to define and measure the quantities for an infinitesimal interval of the wavelength range of interest, and all the quantities defined above have to be defined dependent on the wavelength considered.

Instead of the quantities defined above over the whole spectral range it is generally more useful to look at the amount per infinitesimal spectral unit. The *spectral radiant flux* is

$$Q_\lambda = \frac{dQ}{d\lambda}. \tag{48-21}$$

The total radiant flux over a finite spectral range of interest between λ_1 and λ_2 is calculated by integrating the spectral quantity over this range:

$$Q = \int_{\lambda_1}^{\lambda_2} Q_\lambda d\lambda. \tag{48-22}$$

The other quantities are generated in the same way:

$$L_\lambda = \frac{dL}{d\lambda} = \frac{d^3\Phi}{d\Omega dA \cos\theta d\lambda} \text{ and } L = \int_{\lambda_1}^{\lambda_2} L_\lambda d\lambda, \tag{48-23}$$

$$E_\lambda = \frac{dE}{d\lambda} = \frac{d^2\Phi}{dA d\lambda} \text{ and } E = \int_{\lambda_1}^{\lambda_2} E_\lambda d\lambda, \tag{48-24}$$

$$I_\lambda = \frac{dI}{d\lambda} = \frac{d^2\Phi}{d\Omega d\lambda} \text{ and } I = \int_{\lambda_1}^{\lambda_2} I_\lambda d\lambda. \tag{48-25}$$

Photometric Quantities

The radiometric quantities introduced above are general physical quantities. For a definition of similar quantities with respect to human vision the spectral response of the biological sensors has to be taken into account.

Light entering the eye, first passes the biological optical system of cornea, aqueous humour, the iris, the lens and the vitreous humour, before it is received by the receptors of the retina. There are two types of receptor – the cones and the rods. In their outer segments these receptors contain photo pigments, which absorb the light energy. Within the receptors the absorbed energy is converted into neural electrochemical signals, which are transmitted to the neurons and along the optic nerve to the brain.

The relative spectral sensitivity of this complex biological vision system is influenced by the spectral transmittance of the optical components in front of the retina and the spectral absorption sensitivity of the rods and cones. The cones are responsible for day vision and the observance of color, and the rods are for vision at low illumination levels. The spectral sensitivity of the rods is shifted to shorter wavelengths (into the blue region) compared with that of the cones. So the individual spectral response not only depends on the spectrum entering the eye but also on the light level. Vision at moderately high light levels, for which the cones dominate the sensitivity, is called *photopic vision*. The vision dominated by rods at low light levels is called *scotopic vision*. There is a wide range of vision with cones as well as rods involved, which is called *mesopic vision*.

In order to make the parameters involved with human vision comparable and measurable by technical sensors, standardized relative spectral eye sensitivity functions $V(\lambda)$ for photopic and $V'(\lambda)$ for scotopic vision had been introduced as international standards by CIE (*Commission Internationale de l'Eclairage*, the International Commission of Illumination) and CIPM (*Comité International des Poids et Mesures*, the International Committee on Weights and Measures) – see, for example, [48-4]. The $V(\lambda)$ function is mostly used for photopic vision, which is normalized to its maximum at 555 nm, the wavelength for maximum sensitivity of the human eye at moderately high light levels (see figure 48-9). It is defined for the *CIE standard photometric observer for photopic vision*. Additivity of sensation and a 2° field of view are assumed.

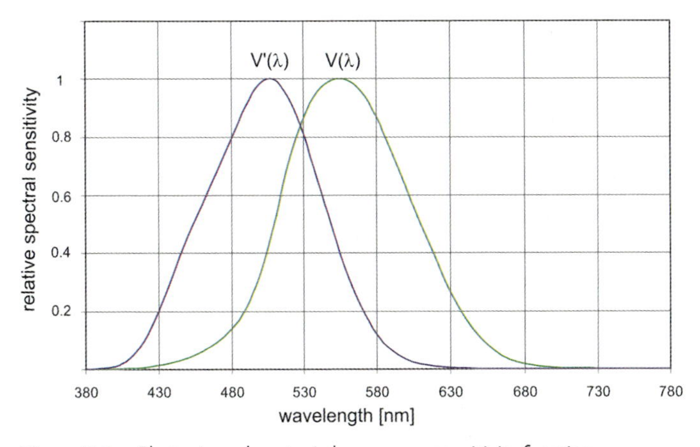

Figure 48-9: Photopic and scotopic human eye sensitivity function.

Between 360 and 830 nm the $V(\lambda)$ function is non-zero. This range of the electromagnetic spectrum is therefore called the visible spectrum range.

The sensation of a special color can be produced in different ways – by monochromatic radiation and by different spectral distributions over the visible range. Any color sensation can result from a variety of different distributions. In the case of monochromatic radiation the principal colors of the spectrum are defined for normal human eyes as follows:

Purple: 360 – 450 nm,
Blue: 450 – 500 nm,
Green: 500 – 570 nm,
Yellow: 570 – 591 nm,
Orange: 691 – 610 nm,
Red: 610 – 830 nm.

With the relative sensitivity function $V(\lambda)$, a photometric quantity X_V integrated over the visible spectrum range is defined in relation to its corresponding spectral radiometric quantity $X_{e,\lambda}$ by the equation

$$X_V = K_m \int_{\lambda=360\,\text{nm}}^{\lambda=830\,\text{nm}} X_{e,\lambda} V(\lambda) d\lambda. \tag{48-26}$$

The constant K_m is the *maximum spectral luminous efficacy of radiation for photopic vision*. Its value is derived from the definition of the photometric base unit candela (see below). K_m can be calculated as

$$K_m = 683 \frac{V(\lambda = 555\,\text{nm})}{V(\lambda = 555.016\,\text{nm})} \frac{\text{lm}}{\text{W}} = 683.002 \frac{\text{lm}}{\text{W}} \tag{48-27}$$

for photopic vision at luminance values of $> \approx 1\,\text{cd/m}^2$. $V(\lambda = 555\,\text{nm})$ is the maximum (value 1) of the relative spectral sensitivity function. According to definition (48-26) only a monochromatic source with $\lambda = 555$ nm can reach the value K_m. For other sources, which radiate in the spectrum range $\lambda_2 - \lambda_1$, the *spectral luminous efficacy of radiation* K_r is

$$K_r = \frac{\Phi_V}{\Phi_e} = \frac{K_m \int_{\lambda=360\,\text{nm}}^{\lambda=830\,\text{nm}} \Phi_{e,\lambda} V(\lambda) d\lambda}{\int_{\lambda=\lambda_1}^{\lambda=\lambda_2} \Phi_{e,\lambda} d\lambda} \tag{48-28}$$

with $K_r < K_m$ (see also table 48-2,which gives examples for different light sources).

For scotopic vision – valid for luminance values of $< \approx 10^{-3}\,\text{cd/m}^2$ – the relative sensitivity function $V'(\lambda)$ with its maximum at 507 nm must be used. In this case the maximum spectral luminous efficacy is calculated as

$$K'_m = 683 \frac{V'(\lambda = 507\text{ nm})}{V'(\lambda = 555,016\text{ nm})} \frac{\text{lm}}{\text{W}} \cong 1700 \frac{\text{lm}}{\text{W}}. \tag{48-29}$$

In this way all basic photometric quantities are generated. Table 48-1 gives a summarising overview of the radiometric and corresponding photometric quantities und measuring units.

Table 48-1: Radiometric and photometric quantities and units.

Quantity Description	Correlations	Radiant Term	Dimensions	Photometric Term	Dimensions
Energy	Q	Radiant Energy	Ws	Luminous energy	lms
Power	$\Phi = \frac{dQ}{dt}$	Radiant power, radiant flux	W	Luminous flux	lumen = lm
Incident power per area element	$E = \frac{d\Phi}{dA} = \int_\Omega L\cos\theta d\Omega$	Irradiance	$\frac{\mathrm{W}}{\mathrm{m}^2}$	Illuminance	$\mathrm{lux} = \frac{\mathrm{lm}}{\mathrm{m}^2}$
Emitted power per area element	$E = \frac{d\Phi}{dA} = \int_\Omega L\cos\theta d\Omega$	Exitance	$\frac{\mathrm{W}}{\mathrm{m}^2}$	Luminous Exitance	$\frac{\mathrm{lm}}{\mathrm{m}^2} = \mathrm{lux}$
Power per projected area element and solid angle element	$L = \frac{d^2\Phi}{\cos\theta dA d\Omega}$	Radiance	$\frac{\mathrm{W}}{\mathrm{m}^2\mathrm{sr}}$	Luminance	$\frac{\mathrm{lm}}{\mathrm{m}^2\mathrm{sr}} = \frac{\mathrm{cd}}{\mathrm{m}^2}$
Power per solid angle element	$I = \frac{d\Phi}{d\Omega} = \int_A L dA\cos\theta$	Radiant intensity	$\frac{\mathrm{W}}{\mathrm{sr}}$	Luminous Intensity	$\frac{\mathrm{lm}}{\mathrm{sr}} = \mathrm{cd}$
Time integral over power per area element	$H = \int E dt$	Radiant exposure	$\frac{\mathrm{Ws}}{\mathrm{m}^2}$	Luminous exposure	$\frac{\mathrm{lms}}{\mathrm{m}^2} = \mathrm{luxs}$

Photometric Base Unit Candela

The measuring unit candela cd is the unit for luminous intensity, the quantity corresponding to radiant intensity. For a given direction, a monochromatic light source has luminous intensity $I_V = 1cd = 1\,\mathrm{lm/srad}$ if it emits at 540×10^{12} Hz (555.016 nm in standard air) and has a radiant intensity of $I_e = \frac{1}{683}\frac{\mathrm{W}}{\mathrm{srad}}$.

This definition allows one to calculate the constants K_m and K'_m according (48-27) and (48-29) as follows:

$$I_V = K_m \int_{\lambda=360\,\mathrm{nm}}^{\lambda=830\,\mathrm{nm}} I_{e,\lambda=555.016\,\mathrm{nm}} \frac{V(\lambda = 555.016\,\mathrm{nm})}{V(\lambda = 555\,\mathrm{nm})} d\lambda. \tag{48-30}$$

This is

$$I_V = K_m I_e \frac{V(\lambda = 555.016\,\mathrm{nm})}{V(\lambda = 555\,\mathrm{nm})} \tag{48-31}$$

for the monochromatic source at 555.016 nm under standard photopic conditions and

$$I_V = K_m' I_e \frac{V(\lambda = 507\,\text{nm})}{V(\lambda = 555\,\text{nm})} \tag{48-32}$$

under standard scotopic conditions.

By the insertion of $I_V = 1\frac{\text{lm}}{\text{srad}}$ and $I_e = \frac{1}{683}\frac{\text{W}}{\text{srad}}$ according to the above definition of the unit candela, one gets (48–27) and (48–29).

In table 48-2 some representative examples of radiometric and photometric quantities are listed [48-5]).

Table 48-2: Examples of radiometric and photometric quantities.

Quantity	Value
Spectral luminous efficacy $K_r$of radiation sources	
Monochromatic radiation with $\lambda = 555$	$683\frac{\text{lm}}{\text{W}}$
White light, constant spectral radiant power over visible spectrum	$220\frac{\text{lm}}{\text{W}}$
Extraterrestrial direct sunlight beam	$99.3\frac{\text{lm}}{\text{W}}$
Direct sunlight beam, midday	$90 - 120\frac{\text{lm}}{\text{W}}$
Typical tungsten filament lamp	$15\frac{\text{lm}}{\text{W}}$
Phosphorescence of cool white fluorescence lamp	$348\frac{\text{lm}}{\text{W}}$
Radiometric quantities	
Extraterrestrial solar irradiance at mean earth orbit	$1367\frac{\text{W}}{\text{m}^2}$
Terrestrial direct normal solar irradiance, clear sky, winter, solar noon, south eastern U.S.	$852\frac{\text{W}}{\text{m}^2}$
Terrestrial global (hemispherical) solar irradiance on horizontal plane, clear sky, winter, solar noon, south eastern U.S.	$686\frac{\text{W}}{\text{m}^2}$
Radiance of the sun at its surface	$2.3 \times 10^7 \frac{\text{W}}{\text{m}^2\text{sr}}$
Apparent radiance of the sun from earth's surface	$1.4 \times 10^7 \frac{\text{W}}{\text{m}^2\text{sr}}$
Total radiant flux of a typical 100W tungsten filament lamp	$82\frac{\text{W}}{\text{m}^2}$

Quantity	Value
Photometric quantities	
Extraterrestrial solar luminance at mean earth orbit	$133,000\,\mathrm{lx}$
Terrestrial direct normal solar luminance, clear sky, winter, solar noon, south eastern U.S.	$94,600\,\mathrm{lx}$
Terrestrial global (hemispherical) solar luminance on horizontal plane	$78,900\,\mathrm{lx}$
Average solar luminance on its surface	$2.3\times10^{9}\,\frac{\mathrm{cd}}{\mathrm{m}^2}$
Blackbody luminance at 6,500 K	$3\times10^{9}\,\frac{\mathrm{cd}}{\mathrm{m}^2}$
Tungsten filament luminance inside 100 W light bulb	$1.2\times10^{7}\,\frac{\mathrm{cd}}{\mathrm{m}^2}$

Quantum Characteristics of Electromagnetic Radiation

Sometimes it is useful to consider the quantum characteristics of electromagnetic radiation.

The energy of one photon, the smallest quantum of radiation energy, depends on the frequency of light as follows:

$$Q_{\mathrm{Photon}} = hf \tag{48-33}$$

where
Planck's constant $h = 6.6260693\times10^{-34}$ Js.

Measuring the energy transported by radiation in units of photons is called *actinometry* [48-2], [48-6], [48-7].

Using (48-33), the relation (48-10) between the radiant frequency, the wavelength and velocity of light and the radiant flux – defined as energy flow per time unit – according to (48-18) the number of photons per time unit of a monochromatic radiation beam with wavelength λ and radiant power Φ_λ is

$$N_{\mathrm{Photon/t}} = \frac{d}{dt}\frac{Q_\lambda}{Q_{\mathrm{Photon}}} = \frac{\lambda\Phi_\lambda}{hc}. \tag{48-34a}$$

For Φ_λ measured in W and λ in nm the number of photons per second is

$$N_{\mathrm{Photon/s}} \approx 5.0341\cdot10^{15}\frac{\lambda\Phi_\lambda}{\mathrm{nmWs}} \tag{48-34b}$$

in vacuum.

In a medium with refraction index n the number of photons per time unit is

$$N_{\text{Photon/t}} = \frac{n\lambda\Phi_\lambda}{hc} \tag{48-35a}$$

or, for Φ_λ measured in W and λ in nm, the number of photons per second

$$N_{\text{Photon/s}} \approx 5.0341 \cdot 10^{15} \frac{n\lambda\Phi_\lambda}{\text{nmWs}}. \tag{48-35b}$$

If an *irradiance* E_λ generated by monochromatic radiation, with wavelength λ, is measured, the number of photons per time and surface unit is calculated by analogy with (48-34), respectively (48-35). It is

$$\frac{dN_{\text{Photon/s}}}{dA} = \frac{\lambda E_\lambda}{hc} \tag{48-36}$$

in vacuum.

For example, a He–Ne laser with a vacuum wavelength of about 633 nm illuminates a surface with about $3.2 \times 10^{18} E_\lambda(1/\text{Ws})$ photons per surface element per second, according to (48-36) for irradiance measured in W/m^2. If the laser beam has a total radiant power of 3 mW it carries about 10^{16} photons per second, calculated by using (48-34b). By using the energy unit *electron volt* , where $1\,\text{eV} = 1.60217653 \times 10^{-19}$ Ws the energy of 1 photon of the He–Ne laser beam is about 2 eV according to (48-33). A photon of monochromatic radiation with 248 nm wavelength, for example, has energy of about 5 eV, and a beam of the same power as the He–Ne laser beam with wavelength 248 nm only carries about 0.4×10^{16} photons per second.

The total number of photons impinging on a surface is measured by the quantity *dose* U. A monochromatic beam with constant radiant power Φ over a time interval Δt causes a dose of

$$U = N_{\text{Photon}} \Delta t, \tag{48-37}$$

where

$$U \approx 8.359 \times 10^{-9} \frac{\lambda\Phi_\lambda\Delta t}{\text{nmWs}} \text{ Einsteins}$$

for Φ_λ measured in W, Δt in seconds and λ in nanometers with 1 Einstein = the amount of energy in 1 mole of photons (1 mole = 6.0222×10^{23} particles = *Avogadro number*) on the irradiated surface.

48.2.3
Fundamentals of Radiation Transfer

Lambertian Radiator

Radiation sources can be characterized by their radiance or radiant intensity. An idealized model for a light source is the Lambertian radiator, a source with radiance which is independent of the direction considered.

From the correlations between (48-15) and (48-16) and using the condition of angular independent radiance of a Lambertian radiator, we find that

$$L = L(\theta) = \text{const} \tag{48-38}$$

and the radiant intensity is

$$I(\theta) = \int_A L dA \cos\theta = LA \cos\theta = I_0 \cos\theta \tag{48-39}$$

Real sources like the sun and a black body radiator often come very close to this model and so it is useful to look at this idealized case. However, illuminated surfaces generally show a dependence on the direction – usually they have an intensified radiance around the specular reflectance angle.

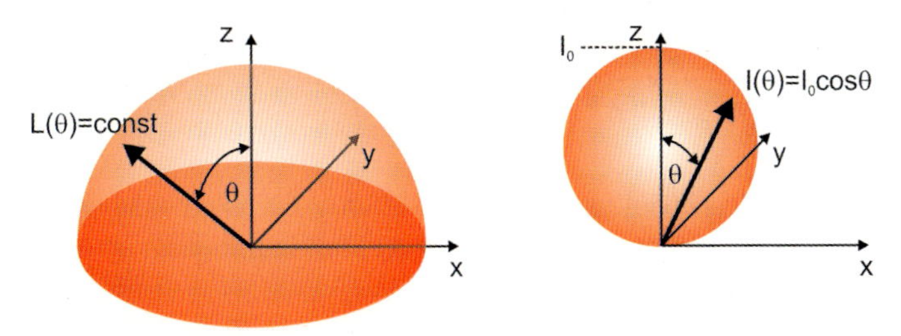

Figure 48-10: Radiance and radiant intensity of a Lambertian radiator.

Fundamental Law of Radiometry

In general, radiometric and photometric measurement tasks involve the exchange of radiation between source, sensor and optical components. In free space, the radiation transfer between an infinitesimal element dA_S at a point P_S of a source surface and the infinitesimal surface element dA_D of a detector surface at the point P_D at a distance r from point P_S can be calculated using (48-15). Neglecting all interactions except the radiation transfer between source and detector one obtains for the differential radiant power of the source

$$d^2\Phi = L d\Omega dA_S \cos\theta_S \tag{48-40}$$

where θ_S is the angle between the source surface normal at point P_S and the connecting line between the points P_S and P_D. According to (48-19) the solid angle for the radiation reaching the infinitesimal detector surface element dA_D at the point P_D is given by

$$d\Omega = \frac{dA_D \cos\theta_D}{r^2}. \tag{48-41}$$

where θ_D is the angle between the detector surface normal and the connecting line between the points P_S and P_D.

From this we get the *fundamental law of radiometry* by the insertion of (48-41) into (48-40):

$$d^2\Phi = L\frac{dA_S dA_D \cos\theta_S \cos\theta_D}{r^2}. \tag{48-42}$$

Figure 48-11 shows the geometric correlations.

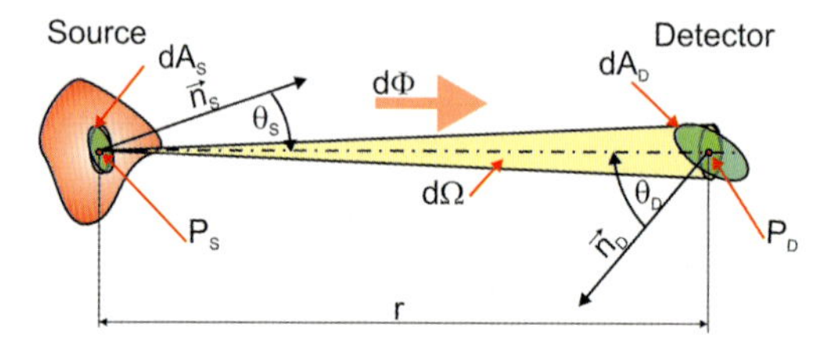

Figure 48-11: Fundamental law of radiometry.

By integration over the differential flux according to (48-42) one obtains the radiation power exchanged between source and detector for real configurations.

$$\Phi = \int_{A_S}\int_{A_D} L\frac{dA_S dA_D \cos\theta_S \cos\theta_D}{r^2}. \tag{48-43}$$

The correlation (48-43) can become very complex since, in general, L, r, θ_S and θ_D are functions of the position of the source and detector.

There are two different general cases:

1. The radiance depends neither on the direction considered nor on the coordinates of the emitting surface point. This means that the source is a Lambertian radiator with constant value L over the source surface. For this case only the geometric conditions have to be considered for integration.
2. The radiance is a function of both direction and position on the source surface. This dependence has to be taken into account for the integration (48-43).

Furthermore, for real configurations, the spectral dependence is also important because not only does the source have their specific spectral characteristics, but the transmission of an optical system is also dependent on the wavelength, and the sensor has a specific spectral sensitivity. The relevant wavelength ranges must therefore be correctly included in the integration.

Elementary Beams of Radiation and Radiance Invariance

According to (48-15) and figure 48-5, the radiance, being the most general quantity is related to a point $P(x, y, z)$ of an infinitesimal surface elementdA in space and the direction of flux flow determined by the angles θ and φ. Hence the constant radiance of an idealized ray of flux is defined as

$$L(x, y, z, \theta, \varphi) \equiv \frac{d^2\Phi(x, y, z, \theta, \varphi)}{d\Omega dA \cos\theta}. \tag{48-44}$$

Since this definition uses an infinitesimally small area element as well as an infinitesimally small solid angle element, no finite amount of flux can be carried. In real radiometry, however, the propagation of finite measurable amounts of flux is of interest. For this, the following concept of an elementary beam is useful [48-5].

In the direction of propagation of a ray, two infinitesimal areas with their normal vectors $\vec{n}_1$ and $\vec{n}_2$ are assumed. Coming from point P_1 in the center of dA_1 the idealized ray intersects the elemental area dA_2 at P_2 in its center. The elemental solid angles $d\Omega_1$ and $d\Omega_2$ at the points P_1 and P_2 are

$$d\Omega_1 = \frac{dA_2 \cos\theta_2}{r^2} \quad \text{and} \tag{48-45}$$

$$d\Omega_2 = \frac{dA_1 \cos\theta_1}{r^2}. \tag{48-46}$$

with a distance r between the area elements.

As shown schematically in figure 48-12, even very small finite surface elements would allow an infinite number of rays of radiation to propagate between them, if the area elements were finite.

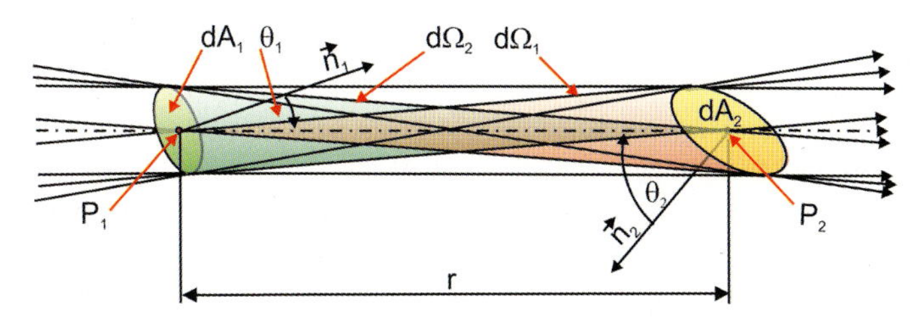

Figure 48-12: Concept of an elementary beam of radiation.

On this basis, the model of an elementary beam of radiation is considered as the collection of all these ray directions along which radiation flux propagates between the two area elements. The integration even over very small but finite surface elements delivers finite solid angles, through which finite amounts of flux can flow.

On the other hand, for sufficiently small surface elements ΔA_1 and ΔA_2, between which radiation flux flows, the difference in radiance for all points in ΔA_1 as well as for all points in ΔA_2 can be neglected and the differential equations can be used for a mathematical description, even though finite amounts of flux flow. All points in ΔA_1 (dA_1 in the infinitesimal limit) emit radiation of an equal amount L_1, and all points in ΔA_2 (dA_2 in the infinitesimal limit) emit radiation of an equal amount L_2.

Within a lossless isotropic medium, which is a good approximation for most optical materials, L_1 is equal to L_2. This can be proved by the application of the fundamental law of radiation transfer according to (48-42) and also (48-44), (48-45) and (48-46) and regarding figure 48-12 as follows.

The flux leaving surface element dA_1 from point P_1 is

$$d^2\Phi_1 = L_1 \frac{dA_1 \cos\theta_1 dA_2 \cos\theta_2}{r^2} \tag{48-47}$$

while the flux arriving at dA_2 is

$$d^2\Phi_2 = L_2 \frac{dA_2 \cos\theta_2 dA_1 \cos\theta_1}{r^2} \tag{48-48}$$

From the principal law of conservation of energy the relation $d^2\Phi_1 = d^2\Phi_2$ must be valid. From this one gets the radiance invariance (also valid for luminance)

$$L_1 = L_2 \tag{48-49}$$

within a lossless homogenous and isotropic medium.

Interface Processes

Processes at interfaces between optical media with different refractive indices are central to radiometric and photometric tasks for optical components. The incoming radiant flux can be reflected, scattered, transmitted and absorbed as a result of interaction with the electrons of the material. The model of the interface between two media assumes that there is no influence from other boundaries. Theoretically, the dimensions of the two media perpendicular to the interface are infinite. Although practical optical components have finite dimensions, the study of the radiation effects at idealized interfaces is useful for practical applications. Also, the influence of real boundaries can be considered on the basis of this model.

An idealized smooth interface without absorption and scattering is a useful model for a real smooth optical surface within the wavelength range of interest. For radiant flux flowing through such a lossless interface, the general principle of radiance invariance is valid in a modified form.

According to figure 48-13, an elementary beam of radiation with the solid angle $d\Omega_1$ transports the flux $d\Phi_1$ through a medium with refractive index n_1. At the interface the elementary beam is refracted as it enters the medium with refractive index n_2, and a part $Rd\Phi_1$ is reflected. According to (48-40) the incoming and the refracted flux are

$$d^2\Phi_1 = L_1 \cos\theta_1 dA d\Omega_1 \tag{48-50a}$$

and

$$d^2\Phi_2 = L_2 \cos\theta_2 dA d\Omega_2 \tag{48-50b}$$

where dA is an infinitesimal interface element with the input point centered in it.

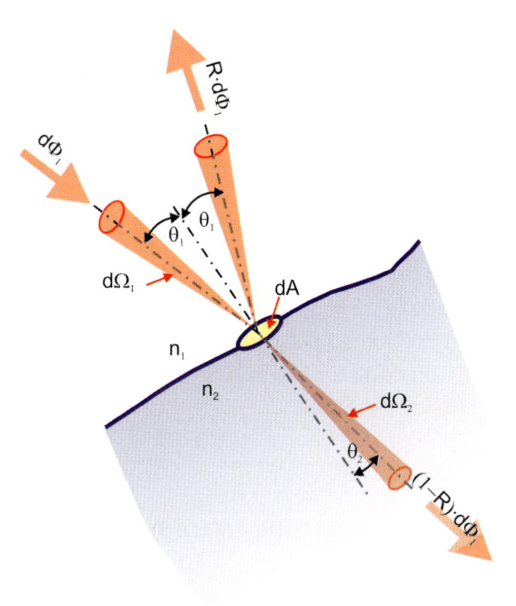

Figure 48-13: Propagation of radiation through an optical interface.

For an idealized non-absorbing, non-scattering material, the law of energy conservation is valid and one obtains

$$L_2 \cos\theta_2 dA d\Omega_2 = (1 - R)L_1 \cos\theta_1 dA d\Omega_1 \tag{48-51}$$

where the reflectivity R is the ratio of the reflected to incoming flux within the solid angle $d\Omega_1$, θ_1 is the angle of incidence, and θ_2 is the angle of the refracted beam in the medium with refractive index n_2.

The solid angle elements can be expressed by definition (48-19) with spherical coordinates as

$$d\Omega_1 = \sin\theta_1 d\theta_1 d\varphi \tag{48-52a}$$

and

$$d\Omega_2 = \sin\theta_2 d\theta_2 d\varphi. \tag{48-52b}$$

Using these expressions (48-51) is transformed to

$$\frac{(1-R)L_1 \cos\theta_1 d\theta_1 \sin\theta_1}{L_2 \cos\theta_2 d\theta_2 \sin\theta_2} = 1. \tag{48-53}$$

Using the law of refraction and differentiating, one finds

$$\frac{n_2}{n_1} = \frac{\sin\theta_1}{\sin\theta_2} = \frac{\cos\theta_1 d\theta_1}{\cos\theta_2 d\theta_2}. \tag{48-54}$$

Using (48-54) the equation (48-53) delivers the radiance transformation for transmission through a lossless interface between lossless media as

$$L_2 = \frac{n_2^2}{n_1^2}(1-R)L_1 \tag{48-55}$$

Equation (48-55) can also be applied to the flux transmission through a lens with two interfaces. The flux with radiance L_1 entering an idealized lens with refractive index n_2 embedded in a medium with refractive index n_1 is

$$L_2 = (1-R)^2 L_1 \tag{48-56}$$

and after passing the lens or – in a generalized form – with the transmittance T of the lens

$$L_2 = TL_1. \tag{48-57}$$

The radiance (and also the luminance) is an invariant for flux propagation through an optical system when there are no absorption or scattering losses. It is only the system transmission caused by reflection, which has to be considered. This also means that the radiance entering an optical system cannot be increased.

Influence of Angle of Incidence, Refraction Index and Polarization on Reflection and Transmission of Radiation Flux at Interfaces – Fresnel Equations

The fraction of flux, which is reflected at, and transmitted through, an ideal lossless optical interface depends on the polarization state of the electromagnetic waves entering and also on the angle of incidence and the refraction indices of the two media separated by the interface.

The relations are described by Fresnel equations for the amplitude and phase coefficients of the electric and magnetic field strength (see, for example, [48-1] and [48-2], and chapter 3 of this handbook volume 1). A plane wave of arbitrary polarization can be described by its electric and magnetic linearly polarized field components with polarization perpendicular to the plane of incidence E_S, B_S (TE-, σ- or s-polarization) and parallel to the plane of incidence E_P, B_P (TM-, π- or p-polarization).

The Fresnel equations describe the amplitude and phase relations between the incident and the reflected and transmitted components.

In the case of a monochromatic plane wave incident at the interface of two linear, homogenous, isotropic media with the refractive indices n_i and n_t the amplitude coefficients for the electric field strength, which is the determining quantity for most cases of optical radiation, are

$$r_S = \left(\frac{E_{0R}}{E_{0i}}\right)_S = \frac{\frac{n_i}{\mu_i}\cos\theta_i - \frac{n_t}{\mu_t}\cos\theta_t}{\frac{n_i}{\mu_i}\cos\theta_i + \frac{n_t}{\mu_t}\cos\theta_t} \tag{48-58a}$$

$$r_P = \left(\frac{E_{0R}}{E_{0i}}\right)_P = \frac{\frac{n_t}{\mu_t}\cos\theta_i - \frac{n_i}{\mu_i}\cos\theta_t}{\frac{n_t}{\mu_t}\cos\theta_i + \frac{n_i}{\mu_i}\cos\theta_t} \tag{48-58b}$$

for reflection of the s- and p-polarized components and

$$t_S = \left(\frac{E_{0T}}{E_{0i}}\right)_S = \frac{2\frac{n_i}{\mu_i}\cos\theta_i}{\frac{n_i}{\mu_i}\cos\theta_i + \frac{n_t}{\mu_t}\cos\theta_t} \tag{48-59a}$$

$$t_P = \left(\frac{E_{0T}}{E_{0i}}\right)_P = \frac{2\frac{n_i}{\mu_i}\cos\theta_i}{\frac{n_t}{\mu_t}\cos\theta_i + \frac{n_i}{\mu_i}\cos\theta_t} \tag{48-59b}$$

for transmission of the s- and p-polarized components.

In the four expressions, the material constants of the two media are the magnetic permeabilities $\mu_i = \mu_0\mu_{ri}$ and $\mu_t = \mu_0\mu_{rt}$ and the constants of dielectricity $\varepsilon_i = \varepsilon_0\varepsilon_{ri}$ and $\varepsilon_t = \varepsilon_0\varepsilon_{rt}$, which were introduced in section 48.2.1.

The irradiance I of a plane wave is the radiant flux through a surface element perpendicular to the propagation direction of the wave (see section 48.2.1). According to the law of energy in a lossless medium, the radiant flux $d\Phi_i = I_i dA_i$ of the incident plane wave with irradiance I_i remains constant until it reaches the interface surface element dA_B. From this surface element the part $d\Phi_R$ is reflected, and the part $d\Phi_t$ is transmitted through it (figure 48-14). The ratios of the reflected and transmitted perpendicular and parallel polarized radiant flux components to the conjugated incident components are called the reflectivities R_S and R_P and the transmissivities T_S and T_P respectively. According to figure 48-14 one finds

$$R_{S,P} = \left(\frac{d\Phi_R}{d\Phi_i}\right)_{S,P} = \left(\frac{I_R dA_B \cos\theta_r}{I_i dA_B \cos\theta_i}\right)_{S,P} = \left(\frac{I_R}{I_i}\right)_{S,P}, \tag{48-60}$$

and

$$T_{S,P}=\left(\frac{d\Phi_t}{d\Phi_i}\right)_{S,P}=\left(\frac{I_T dA_B\cos\theta_t}{I_i dA_B\cos\theta_i}\right)_{S,P}=\left(\frac{I_T\cos\theta_t}{I_i\cos\theta_i}\right)_{S,P}. \tag{48-61}$$

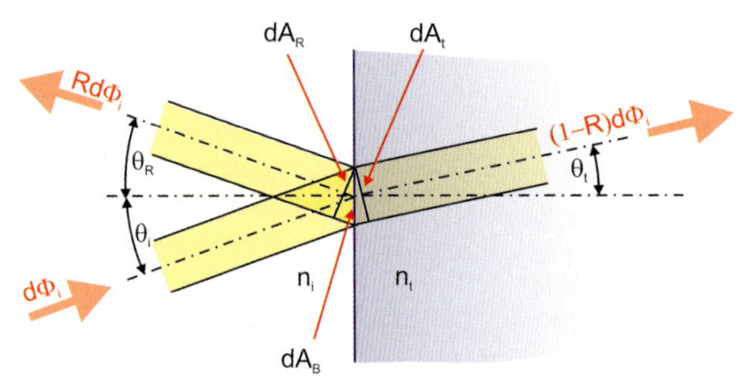

Figure 48-14: Flux and irradiance at an interface.

The area element dA_B in figure 48-14 is the boundary surface element, through which the flux within the "light pipe" characterizing the energy flow of the incoming wave is passing.

As was briefly discussed in section 48.2.1 the irradiance I of the electromagnetic wave is the time-averaged Poynting vector and fulfils the relation (48-12) in a linear, homogenous, isotropic medium for a monochromatic plane wave. Hence the irradiance components from (48-12) are found using the relation (48-7) for the refractive index as

$$(I_i)_S=\frac{1}{2}\sqrt{\frac{\varepsilon_0}{\mu_0}\frac{n_i}{\mu_{ri}}}(E_{0i})_S^2 \tag{48-62a}$$

$$(I_i)_P=\frac{1}{2}\sqrt{\frac{\varepsilon_0}{\mu_0}\frac{n_i}{\mu_{ri}}}(E_{0i})_P^2 \tag{48-62b}$$

$$(I_R)_S=\frac{1}{2}\sqrt{\frac{\varepsilon_0}{\mu_0}\frac{n_i}{\mu_{ri}}}(E_{0R})_S^2 \tag{48-63a}$$

$$(I_R)_P=\frac{1}{2}\sqrt{\frac{\varepsilon_0}{\mu_0}\frac{n_i}{\mu_{ri}}}(E_{0R})_P^2 \tag{48-63b}$$

for the incident and reflected components and as

$$(I_t)_S=\frac{1}{2}\sqrt{\frac{\varepsilon_0}{\mu_0}\frac{n_t}{\mu_{rt}}}(E_{0T})_S^2 \tag{48-64a}$$

$$(I_t)_P=\frac{1}{2}\sqrt{\frac{\varepsilon_0}{\mu_0}\frac{n_t}{\mu_{rt}}}(E_{0T})_P^2 \tag{48-64b}$$

for the transmitted components.

Using (48-62), (48-63) and (48-64) together with (48-60) and (48-61) the reflectivity and transmissivity of the p- and s-polarized components of the harmonic monochromatic plane wave are

$$R_S = r_S^2 \tag{48-65a}$$

$$R_P = r_P^2 \tag{48-65b}$$

and

$$T_S = \frac{n_t \cos\theta_t \mu_{ri}}{n_i \cos\theta_i \mu_{rt}} t_S^2 \tag{48-66a}$$

$$T_P = \frac{n_t \cos\theta_t \mu_{ri}}{n_i \cos\theta_i \mu_{rt}} t_P^2 \tag{48-66b}$$

where the amplitude coefficients r_S, r_P, t_S, t_P are from the Fresnel equations (48-58) and (48-59).

Taking into account the law of energy conservation for lossless media and a lossless interface furthermore, the relationships

$$R_S + T_S = 1 \tag{48-67a}$$

$$R_P + T_P = 1 \tag{48-67b}$$

are valid.

The reflectivity and transmissivity of a plane wave with an arbitrary polarization direction can be calculated from the components R_S, R_P, T_S, T_P using the azimuth angle τ_i between the polarization plane and the plane of incidence as follows:

$$R = R_P \cos^2\tau_i + R_S \sin^2\tau_i \tag{48-68}$$

and

$$T = T_P \cos^2\tau_i + T_S \sin^2\tau_i . \tag{48-69}$$

In the case of unpolarized or circularly polarized radiation, the total reflectivity and transmissivity are the averages

$$R = \frac{R_S + R_P}{2} \tag{48-70}$$

and

$$T = \frac{T_S + T_P}{2}. \tag{48-71}$$

If the interface separates two dielectric media (for example, air and glass) the magnetic permeabilities are $\mu_i \approx \mu_t \approx \mu_0$ and $\mu_{ri} \approx \mu_{rt} \approx 1$.

Under these conditions the Fresnel equations (48-58) and (48-59) and the relations for reflectivity and transmissivity are simplified. Using the law of refraction with the help of some trigonometric transformations we obtain the following relations for the reflectivity and transmissivity of the polarized components of the optical radiation perpendicular and parallel to the plane of incidence, dependent on the angle of incidence:

$$R_S = \left(\frac{d\Phi_R}{d\Phi_i}\right)_S = \left[\frac{n_i \cos\theta_i - \sqrt{n_t^2 - n_i^2 \sin^2\theta_i}}{n_i \cos\theta_i + \sqrt{n_t^2 - n_i^2 \sin^2\theta_i}}\right]^2 \tag{48-72a}$$

$$R_P = \left(\frac{d\Phi_R}{d\Phi_i}\right)_P = \left[\frac{n_t^2 \cos\theta_i - n_i\sqrt{n_t^2 - n_i^2 \sin^2\theta_i}}{n_t^2 \cos\theta_i + n_i\sqrt{n_t^2 - n_i^2 \sin^2\theta_i}}\right]^2 \tag{48-72b}$$

$$T_S = \left(\frac{d\Phi_t}{d\Phi_i}\right)_S = \frac{4n_i \cos\theta_i \sqrt{n_t^2 - n_i^2 \sin^2\theta_i}}{\left[n_i \cos\theta_i + \sqrt{n_t^2 - n_i^2 \sin^2\theta_i}\right]^2} \tag{48-73a}$$

$$T_P = \left(\frac{d\Phi_t}{d\Phi_i}\right)_P = \frac{4n_i n_t^2 \cos\theta_i \sqrt{n_t^2 - n_i^2 \sin^2\theta_i}}{\left[n_t^2 \cos\theta_i + n_i\sqrt{n_t^2 - n_i^2 \sin^2\theta_i}\right]^2} \tag{48-73b}$$

For perpendicular incidence ($\theta_i = 0$) there is no difference between the reflectivity and transmissivity for parallel and perpendicular polarized components:

$$R_S(\theta_i = 0) = R_P(\theta_i = 0) = \left(\frac{n_i - n_t}{n_i + n_t}\right)^2 \tag{48-74}$$

$$T_S(\theta_i = 0) = T_P(\theta_i = 0) = \frac{4n_i n_t}{(n_i + n_t)^2} \tag{48-75}$$

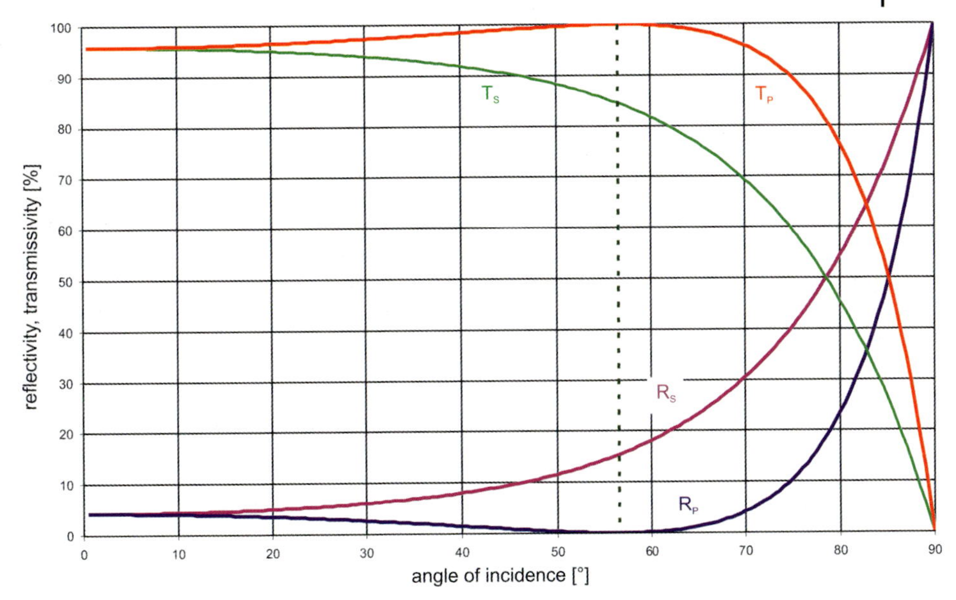

Figure 48-15: The reflectivity and transmissivity of the radiant flux for monochromatic light of wavelength 632.8 nm at an interface between air and the optical glass BK7 with refractive index 1.51509.

Knowing the refractive indices, then using (48-72) and (48-73), the reflectivity and transmissivity for monochromatic plane waves, dependent on the angle of incidence at the interfaces of lossless linear isotropic homogenous dielectric materials, can be calculated. Figure 48-15 shows an example of incident helium–neon laser light of wavelength 632.8 nm. The intersections of the dashed line with the transmission and reflection curves deliver the transmissivity and reflectivity values for the Brewster angle. For this angle of incidence the reflectivity of the component with parallel polarization is zero and the reflected light is completely polarized.

The reflectivities and transmissivities of ideal optical interfaces are the same for light coming from the other side of the interface at an angle of incidence θ_t due to the principle of optical reversibility (see, for example, [48-1] or [48-5]). Practically, measurements of the reflectivity and transmissivity usually involve measuring the incident, reflected and transmitted flux directly. The theoretically expected results can be found only approximately. Real absorption effects of the material and interface (e.g., contamination on a glass surface) do have an influence. The measuring conditions such as spectral bandwidth, accuracy of angle and wavelength setting must be carefully controlled.

Absorption Away from Interfaces

Optical materials are generally transparent over the wavelength range for which they are optimized, but a small part of the incident radiation is converted into other forms of energy, mostly heat. This can be caused by several processes, such as intrinsic (dissipated) absorption, the conversion of absorbed flux into vibration energy by the interaction of the glass atoms and molecules, selective absorption by impurities and scattering at bubbles, striae, and the inclusion of other material.

Even within the range of normal refractive indices, and neglecting reflection losses, at the interfaces of bulk material the radiance is attenuated by the absorption and scattering processes. The wavelength-dependent attenuation can be specified by a spectral volume absorption coefficient a, which is defined as

$$a(\lambda) \equiv -\frac{1}{L_\lambda}\frac{dL_\lambda}{dz} \tag{48-76}$$

Here L_λ is the spectral radiance as defined in (48-23). According to (48-76) the absorption coefficient is defined as the relative loss of radiance per unit length. An elementary beam of radiation in the propagation direction z, within a homogeneous medium away from interfaces, passes through the volume element $dV = dAdz$ of thickness dz and limited by two plane elements of area size dA perpendicular to the direction z. Since the input spectral radiance $L_\lambda(z)$ is attenuated by absorption loss along the path dz through the volume element, the output radiance $L_\lambda(z + dz)$ will be reduced. Figure 48-16 shows this schematically.

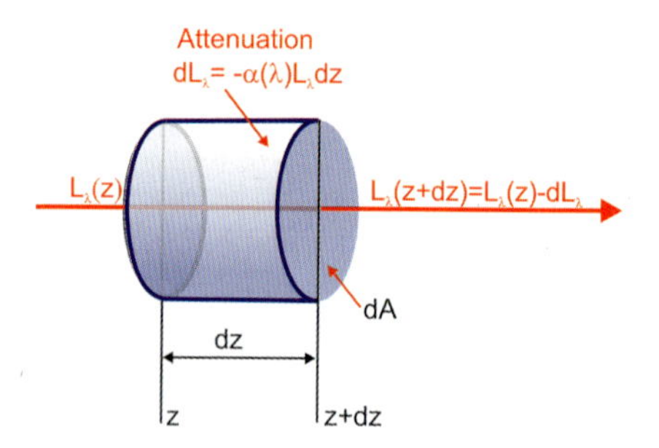

Figure 48-16: Radiance attenuation by absorption and single scattering along the path *dz* away from interfaces.

After integration of the relation (48-76) over the distance z, the *Lambert–Beer law* (48-77) is obtained for the radiance, dependent on the thickness of the material through which the radiation passes neglecting reflection losses at the interfaces:

$$L_\lambda(z) = L_\lambda(0)e^{-a\cdot z} \tag{48-77}$$

The Lambert–Beer law (48-77) in this form is only valid as long as the absorption coefficient does not depend on the position (homogenous material). Otherwise the more general form

$$L_\lambda(z) = L_\lambda(0)e^{-\int_z a(z)dz} \tag{48-78}$$

has to be used.

Other conditions for the validity of the Lambert–Beer law are:

- There are no interactions between the absorption centers.
- There are no non-linear effects.
- There is no saturation.

The Lambert–Beer law in the form of (48-77) can mostly be used for normal optical materials.

For a collimated beam of radiation it is valid in the same form for the spectral irradiance:

$$E_{\lambda}(z) = E_{\lambda}(0)e^{-\alpha \cdot z} \tag{48-79}$$

In simulations of optical components, the absorption coefficient defined in (48-76) can be considered using the complex refractive index $\tilde{n}$ of the optical material.

It is

$$\tilde{n} = n + i n_{\text{im}} = n + i \frac{\lambda_0}{4\pi} a = n(1 + i\kappa) \tag{48-80}$$

with the real part n as the normal refractive index according to (48-8), the attenuation coefficient $\kappa = (\lambda_0/4\pi n)a$ and the imaginary refractive index $n_{\text{im}} = (\lambda_0/4\pi)a$ describing the attenuation dependent on the wavelength of the radiation.

The attenuation of the spectral radiance can be considered as a combination loss of the absorption alone and the single scattering alone along the path of the ray. It is

$$dL_{\lambda} = dL_{a,\lambda} + dL_{b,\lambda} = L_{\lambda} dz[a(\lambda) + b(\lambda)] \tag{48-81}$$

the variation of the radiance along the infinitesimal path dz with the pure absorption loss $dL_{a\lambda}$ and the loss by single scattering $dL_{b,\lambda}$ alone. The sum of the coefficients, which are defined as

$$a(\lambda) \equiv -\frac{1}{L_{\lambda}} \cdot \frac{dL_{a,\lambda}}{dz} \tag{48-82a}$$

and

$$b(\lambda) \equiv -\frac{1}{L_{\lambda}} \frac{dL_{b,\lambda}}{dz} \text{ respectively,} \tag{48-82b}$$

is

$$\alpha(\lambda) = a(\lambda) + b(\lambda) \tag{48-83}$$

The spectral scattering coefficient $b(\lambda)$ describes the relative loss of radiance per length unit by single scattering. No part of the radiant flux is scattered back into the ray, which is a suitable model for optical materials. The amount of scattering loss according to (48-82b) depends on the scattering behavior of the material, which is described by the material-specific *volume scattering function VSF*. The VSF $\beta(\lambda)$ is defined as the ratio of the scattered amount of spectral flux $d\Phi_\lambda^2$ in the direction θ considered and the solid- angle element $d\Omega$ of the direction θ, the volume element dV and the irradiance of the beam propagating into dV. Figure 48-17 illustrates the definition of the VSF.

It is

$$\beta(\lambda,\theta) \equiv \frac{d^2\Phi_\lambda(\theta)}{E_\lambda d\Omega dV} \tag{48-84}$$

This definition presumes, as a suitable assumption for homogenous media, that the scattered radiation flux is rotationally symmetrically distributed around the propagation direction of a collimated beam. If there are other correlations they have to be considered by a modified definition.

For a thickness dz of the infinitesimal volume element dV an equivalent definition of the VSF is

$$\beta(\lambda,\theta) \equiv \frac{d^3\Phi_\lambda(\theta)}{E_\lambda dA d\Omega dz} \tag{48-85}$$

The VSF can also be expressed dependent on other radiometric quantities. Equivalent expressions are

$$\beta(\lambda,\theta) \equiv \frac{dI_\lambda(\theta)}{E_\lambda dV} \tag{48-86}$$

where the spectral intensity I_λ is defined by (48-25) and

$$\beta(\lambda,\theta) \equiv \frac{dL_\lambda(\theta)\cos\theta}{E_\lambda dz} \tag{48-87}$$

where the spectral radiance L_λ is defined by (48-23).

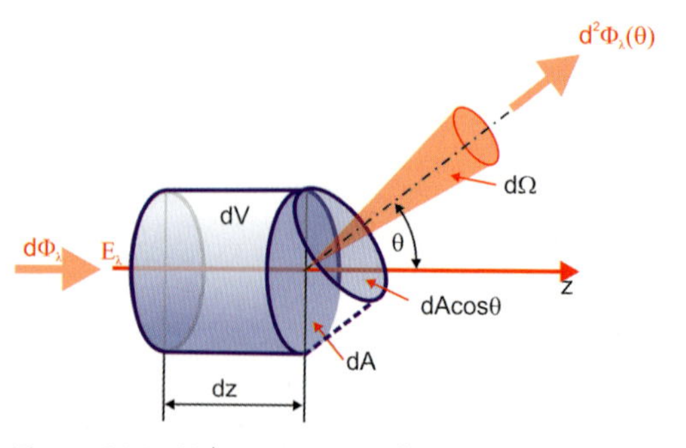

Figure 48-17: Volume scattering function.

The spectral scattering coefficient is obtained by integrating the VSF over the full space solid angle

$$b(\lambda) = \int_0^{4\pi} \beta(\lambda, \theta) d\Omega. \tag{48-88}$$

Since the VSF only depends on the angle θ the integral (48-88) can be transformed to (48-89):

$$b(\lambda) = 2\pi \int_0^{\pi} \beta(\lambda, \theta) \sin\theta d\theta. \tag{48-89}$$

Internal Transmittance

Suppliers of optical material usually specify the internal spectral transmittance, the transmittance excluding reflection losses, to characterize the internal losses by absorption and scattering. For a beam incident upon a sample with plane end faces the *internal transmittance* is obtained from the Lambert–Beer law (48-77) as (48-90):

$$\tau_i = \frac{\Phi_e}{\Phi_i} = e^{-a \cdot d} \tag{48-90}$$

for a sample of thickness d.
The internal spectral transmittance over relevant spectral ranges is specified for a reference sample thickness. In the catalogue of the company Schott, for example, the typical reference thicknesses for optical glass are 10 mm and 25 mm (see [48-8] and [48-9]) and for glass color filters 1 mm, 2 mm or 3 mm. In the glass data sheets the internal transmission data relate to the corresponding indicated reference thickness. If the internal transmission is known for thickness d_1 it can be calculated for another sample thickness d_2 by

$$\tau_i d_2 = (\tau_i d_1)^{\frac{d_2}{d_1}} \tag{48-91}$$

using the relation (48-90).
The transmission of optical material can also be specified by

$$D = \log\left(\frac{1}{\tau_i}\right), \tag{48-92}$$

which is referred to as the optical density.

Transmittance of Flux through a Plane Parallel Plate

Measurements of bulk material or coating properties are often carried out using a plane parallel plate. The flux transmitted through such a plate can be calculated as follows, if incoherent radiation is presumed. According to (48-60)–(48-75) the radiant flux of an incident beam is partly reflected at the interface, and the remaining part is transmitted depending on the angle of incidence, the refractive indices and the polarization state. If the absorption within the plate material is also included, then the flux or radiance of the transmitted beam is reduced by the multiplier τ_i, the internal transmittance according to (48-90). The reduced radiance is again partly reflected and transmitted at the second interface, and the reflected part is again reduced by absorption and reflected and transmitted at the entrance interface. There are multiple reflections between the two interfaces. The overall transmission can be calculated by summarizing all parts leaving the plate at the exit side. The overall reflectance is obtained as the sum of all parts reflected at the left entrance interface. Figure 48-18 shows this schematically for a plate with refractive index n_2 within another medium (for example, a glass plate in air). For this case the refractive index n_1 on the left side is the same as n_3 on the exit side, and the angle θ_e of the exit rays is the same as the angle θ_i of the input ray.

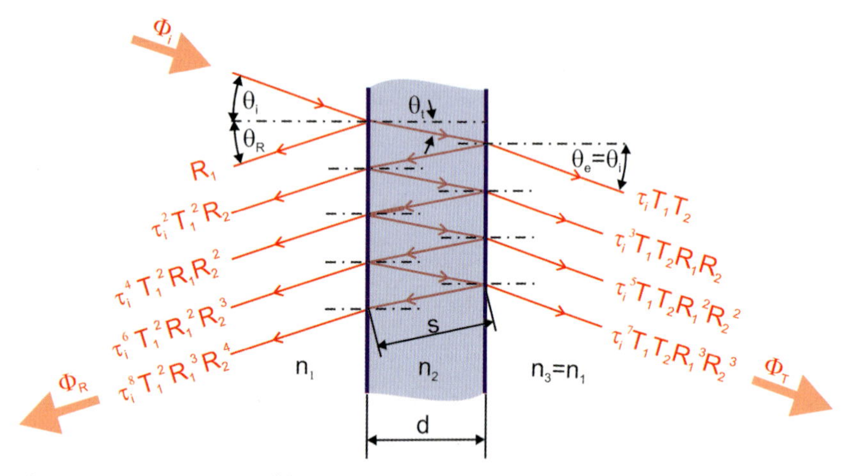

Figure 48-18: Transmission of flux through a plane parallel plate.

In this way the overall transmittance is found to be

$$\tau = T_1 T_2 \tau_i + T_1 T_2 \tau_i^3 R_1 R_2 + T_1 T_2 \tau_i^5 (R_1 R_2)^2 + ... \tag{48-93}$$

and the overall reflectance is

$$\rho = R_1 + \tau_i^2 T_1^2 R_2 + \tau_i^4 T_1^2 R_1 R_2^2 + ... \tag{48-94}$$

In (48-93) and (48-94) R_1 and R_2 are the reflectivity and T_1 and T_2 the transmissivity of the two interfaces according to (48-60)–(48-75). The absorption of the material along the single path s of the beam through the sample with thickness d is described by the absorption coefficient a. It is

$$\tau_i = \frac{L(s)}{L(0)} = e^{-as} \tag{48-95}$$

with the absorption coefficient α according to the Lambert–Beer law defined in (48-77) and the geometric path length

$$s = \frac{d}{\cos \theta_t} \tag{48-96}$$

through the plane parallel plate of thickness d.
The sums of (48-93) and (48-94) for the overall transmittance and reflectance can be transformed for infinite reflections to

$$\tau = \tau_i T_1 T_2 \sum_{j=0}^{\infty} \left(\tau_i \sqrt{R_1 R_2} \right)^{2j} \tag{48-97}$$

and

$$\rho = R_1 + \tau_i^2 T_1^2 R_2 \sum_{j=0}^{\infty} \left(\tau_i \sqrt{R_1 R_2} \right)^{2j} \tag{48-98}$$

For optical materials $\left(\tau_i \sqrt{R_1 R_2}\right)^2 < 1$ applies, and the sums of (48-97) and (48-98) can be calculated as the sums of an infinite geometric series.
Therefore the results are

$$\tau = \frac{\tau_i T_1 T_2}{1 - \tau_i^2 R_1 R_2} \tag{48-99}$$

and

$$\rho = R_1 + \frac{\tau_i^2 T_1^2 R_2}{1 - \tau_i^2 R_1 R_2} = R_1 + \tau \tau_i \frac{T_1}{T_2} R_2 \tag{48-100}$$

The overall absorptance is

$$A = 1 - \rho - \tau \tag{48-101}$$

due to the law of energy conservation.
If the plane parallel plate is in a homogeneous medium with $n_1 = n_3$, $n_2 = n$ then $R_1 = R_2 = R$, $T_1 = T_2 = T$ and $T_1 T_2 = T^2 = (1 - R)^2$.
For this case (48-99) and (48-100) can be simplified according to (48-102) and (48-103):

$$\tau = \frac{\tau_i T^2}{1 - \tau_i^2 R^2} = \frac{\tau_i (1 - R)^2}{1 - \tau_i^2 R^2} \tag{48-102}$$

$$\rho = R(1 + \tau_i \tau) \tag{48-103}$$

48.2.4
Fundamentals of Flux Detection

Radiometric measuring tasks usually require optical concentration facilities for flux detection. One reason is to obtain a sufficient signal-to-noise ratio, since the flux to be measured can be very small, when measured spectrally resolved and within small solid angles.

Irradiance on a Detector

The irradiance generated on a detector by radiation with a given radiance L_1 at the lens entrance pupil depends on the lens aperture at the detector side and the transmission of the lens. To obtain a general expression the following simplified model can be used [48-5]:

After passing through the lens with transmittance T the radiance of the flux is $L_2 = TL_1$. It is assumed to be coming from the exit pupil of the lens with area A_1. The radiance L_2 is assumed to be constant over A_1. The sensitive area of the detector is A_2, the distance between surfaces A_1 and A_2 is s. Surfaces A_1 and A_2 are perpendicular to the axis through their center points. The irradiation at the center point P of A_1 is considered. Figure 48-19 shows the principle correlations.

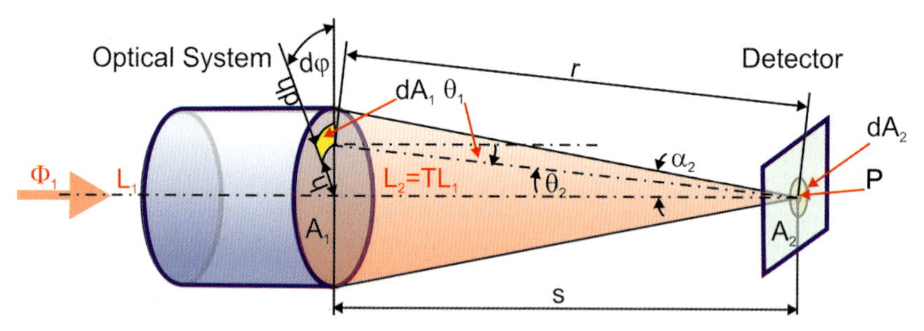

Figure 48-19: Calculation for the irradiance concentrated on a detector surface by a lens.

Using the *fundamental law of radiometry* (48-42) and the definition of irradiance (48-13) one gets for the differential of the irradiance at point P

$$dE_2 = \frac{d^2\Phi}{dA_2} = L_2 \frac{dA_1 \cos\theta_1 \cos\theta_2}{r^2} \tag{48-104}$$

This is the irradiance generated at point P in the infinitesimal detector element dA_2 by the flux emitted from the area element dA_1.

To obtain the total irradiance E_2 at point P the differential of (48-104) has to be integrated over A_1:

$$E_2 = L_2 \int_{A_1} \frac{\cos\theta_1 \cos\theta_2}{r^2} dA_1 . \tag{48-105}$$

Using the relations $\theta_1 = \theta_2 = \theta$, $dA_1 = hd\varphi dh$ and $r^2 = s^2 + h^2$ for this idealized setup, the integral (48-105) is transformed into (48-106):

$$E_2 = L_2 \int_0^{2\pi} d\varphi \int_0^{h_L} \frac{hdh \cos^2\theta}{s^2 + h^2} \tag{48-106}$$

where h_L is the radius of the exit pupil of the lens.

The quantity h can be expressed by θ and s, and using this, after integration φ over the range 0–2 π and θ over 0–a_2 the result is

$$E_2 = \pi L_2 \sin^2 a_2 = \pi T L_1 \sin^2 a_2. \tag{48-107}$$

The angle a_2 in (48-107) is the aperture angle of the lens on the detector side.

The fundamental relation (48-107) can be used to determine the radiance entering the lens from the flux $d\Phi_2$ measured by the detector.

Again using the definition (48-13) for the irradiance one gets

$$d\Phi_2 = E_2 dA_2 \tag{48-108}$$

for non-varying irradiance over the infinitesimal detector surface element dA_2.

Using (48-107) for the irradiance, the incoming radiance can be found from (48-108) as

$$L_1 = \frac{d\Phi_2}{\pi T dA_2 \sin^2 a_2}. \tag{48-109}$$

Following the general law of energy conservation the flux through a lossless optical system has to be conserved as well as the radiance. Only the reflection loss at the interfaces between the media with different refraction indexes has to be considered. This is

$$d\Phi_2 = T d\Phi_1. \tag{48-110}$$

For practical measurement tasks the actual dependence of the detector sensitivity on the input angle has to be considered. According to the general geometric relation (48-14) the irradiance should have a cosine dependence on the input angle. This can be obtained by limiting the flux incidence angles by the sensor optic.

Geometric Light Guide Value of an Optical System

Using the relation (48-107) for the irradiance on the detector surface as well as on the source surface, together with definition (48-13) for the irradiance and (48-110) for an optical illumination system with entrance aperture angle a_1 and surface element dA_1 of a light source, which is imaged onto the detector or into the pupil of a more complex optical imaging system, the general relation (48-111) is found:

$$dA_1 \sin^2 a_1 = T dA_2 \sin^2 a_2 \tag{48-111}$$

The surface element dA_2 is the image of dA_1, and α_2 the aperture angle on the image side (figure 48-20).

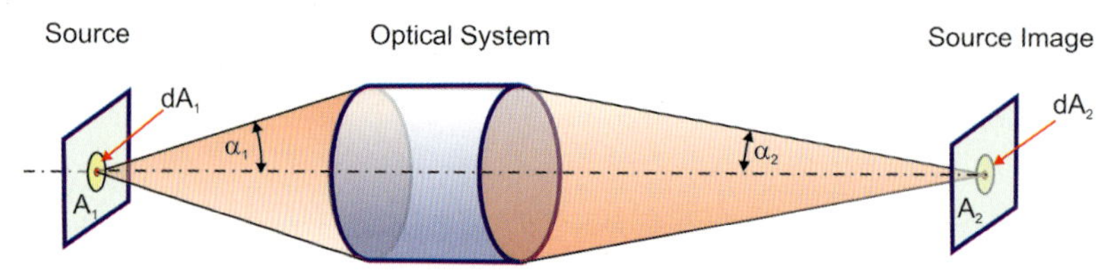

Figure 48-20: The light guide value of an optical system.

The value $dA_1 \sin^2\alpha_1$ and $dA_2 \sin^2\alpha_2$, respectively, is called the *geometric light guide value* of an optical system. According to (48-111) neither the light guide value nor the radiance can be increased by an optical system, but is constant for an idealized lossless system.

Increasing Irradiance for Radiometric Measurements

A typical measurement task is to measure the radiant flux within a defined small solid angle, for example, to measure the angular distribution of the radiance of a light source. A convenient radiometric setup is a detector in the focal plane of a concentration lens. Figure 48-21 shows the principal relations.

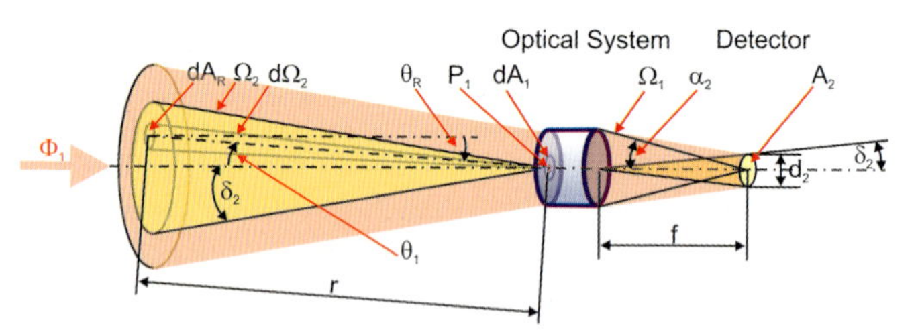

Figure 48-21: Calculating the irradiance on a detector in the focal plane of a lens.

The setup determines a solid angle Ω_2, a cone with half-angle δ_2 dependent on the sensitive detector surface dimensions and the focal length f of the lens as $\tan\delta_2 = d_2/2f$ where d_2 is the diameter of the sensitive detector surface assumed as a circular area.

Only the radiant flux Φ_1 within the solid angle Ω_2 can be focussed onto the sensor by the lens. The flux Φ_1 causes the irradiance E_1 at point P_1 of the entrance pupil. From the *fundamental law of radiometry* (48-42) the differential irradiance at point P_1 can be found according to (48-112):

$$dE_1 = \frac{d\Phi_1}{dA_1} = L_1 \cdot \frac{dA_R \cos\theta_R \cos\theta_1}{r^2}, \tag{48-112}$$

where dA_1 indicates the infinitesimal surface element in the lens entrance pupil on the optical axis centered around point P_1 and where dA_R is an infinitesimal radiating element (of a real source surface or free space) within the solid angle Ω_2 at a distance r from point P_1.

For the idealized configuration of figure 48-21, the angles θ_1 and θ_R are identical -$\theta_1 = \theta_R = \theta$.

With this assumption a solid angle element at point P_1 defined by dA_R is

$$\frac{dA_R \cos\theta_R}{r^2} = d\Omega_2 = \sin\theta d\theta d\varphi. \tag{48-113}$$

Substituting this expression into (48-112) the irradiance at point P_1 is calculated by integration over the solid angle Ω_2:

$$E_1 = \int_{\Omega_2} L_1 \cos\theta d\Omega_2 = \int_0^{2\pi}\int_0^{\delta} L_1 \cos\theta \sin\theta d\theta d\varphi. \tag{48-114}$$

The integration limit for θ is the acceptance angle δ of the detector.

Assuming the radiance L_1 to be constant over the solid angle Ω_2 one obtains the correlation

$$E_1 = \pi L_1 \sin^2\delta. \tag{48-115}$$

for the entrance irradiance.

After propagation through the optical system the radiation has the radiance L_2 dependent on the transmission of the optical system as described in (48-57). Assuming L_2 again to be constant over the lens exit pupil within solid angle Ω_1 dependent on the lens aperture, then (48-107) can be applied to obtain the irradiance E_2 caused by focussing on the detector. The increase of E_2 relative to the irradiance at the entrance pupil is

$$\frac{E_2}{E_1} = T \frac{\sin^2\alpha_2}{\sin^2\delta_2}. \tag{48-116}$$

Relation (48-116) shows that the irradiance on the detector can be increased by increasing the aperture angle α_2 for a given acceptance angle δ of the detector.

48.3 Monochromators

48.3.1 Introduction

The radiometric quantities of optical materials, components and systems generally have to be measured dependent on of the wavelength. Instruments which are used to measure in this way are called *spectrometers*.

The main component of all types of spectrometer is the *monochromator*, which generates radiation of small spectral bandwidth with different central wavelengths from radiation with a broader spectral range.

In principle, this can be done on the illumination side or on the sensor side of a spectrometric measurement system.

An *optical filter* can be considered as a simple monochromator. The use of a number of optical filters with different central wavelengths together with a light source of a suitable broader spectral bandwidth allows the sequential measurement of spectral radiant parameters.

A *Fabry–Pérot etalon* can be considered as a special form of monochromator allowing configurations with very high spectral resolution of < 0.1 pm over small spectral ranges. Typical applications of the etalon are wavelength stabilizator (one rigid component) and spectrum analyzer configurations.

The common, flexible solution for a spectrometer monochromator is a *dispersive component*, which delivers a spatial decomposition of light with broad spectral bandwidth into quasi-monochromatic fractions of small spectral bandwidths. Prism and grating monochromators are in common use. They deliver a wavelength-dependent angular distribution of the input radiation simultaneously. If this type of monochromator is arranged on the illumination side of a spectrometer, then quasi-monochromatic radiation can be generated. The central wavelength of the quasi-monochromatic radiation can be controlled by setting the angle of the exit radiation. The monochromator on the spectrometer's illumination side can be used to illuminate an optical component in order to measure its spectral reflection and transmission properties.

In the case of the dispersive component on the sensor side of the spectrometer the spectrum can be measured sequentially by using a single sensor, or simultaneously, if a multi-channel sensor is used. This is the usual configuration for spectral analysis of a light source, but can also be used for the spectrometric measurement of optical components, which have to be illuminated in this configuration with a light source of suitable spectral distribution.

Electronic tunable filters can, in principle, replace complete prism or grating monochromators. Their advantage is the ability to tune the wavelength without moving the filter. Common types are acousto-optic tunable filters and liquid crystal tunable filters. The latter allow large tuning ranges of some hundred nanometers within a few seconds, thus having the potential for real-time measurement applications.

In the next section the basic functions and inter-relationships of monochromators are described.

48.3.2 Optical Absorption Filters

One simple way to form a spectrometer is to use a broad-band source and a number of optical spectral filters transmitting radiance within a wavelength interval around the central wavelength for which they are designed. The light source should emit radiation within the spectral range for which the optical component has to be tested, and the filters must be chosen to allow spectral intervals within this range to pass through. Figure 48-22 shows this type of arrangement with a filter wheel allowing the analyzing wavelength to be changed sequentially by rotating the wheel manually – in a laboratory setup – or it could be motorized.

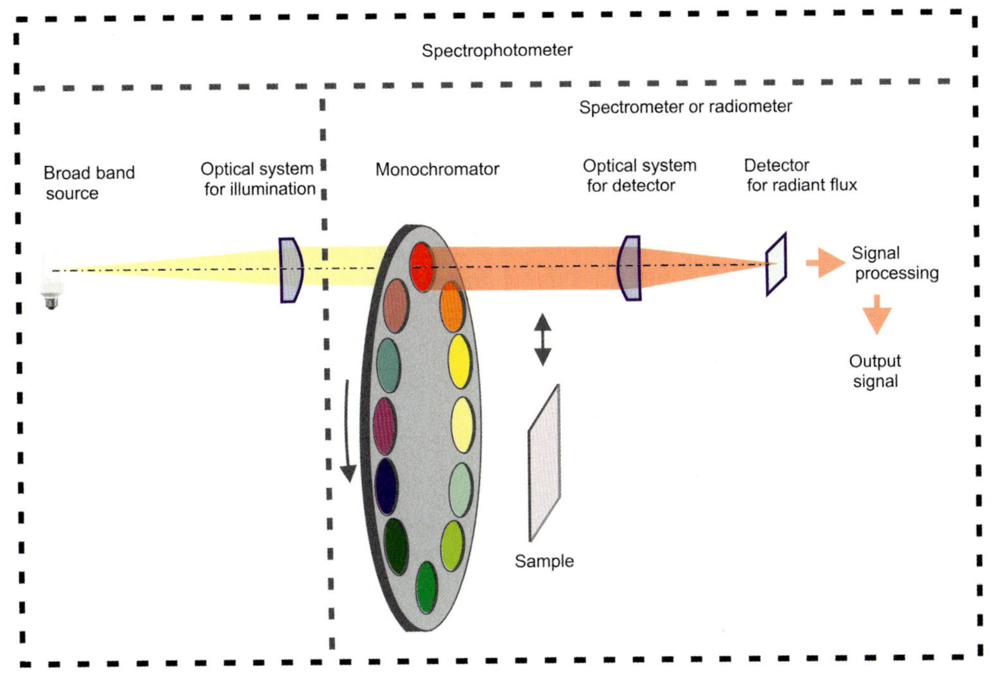

Figure 48-22: Simple spectrometer concept with spectral filters mounted on a filter wheel for use as as a monochromator.

As shown schematically in figure 48-22, the combination of light source, filter monochromator and detector can provide a simple spectrometer which can be used, for example, for measuring the transmission of optical samples by measuring the radiant power before and after placing the sample into the beam path between the monochromator (filter) and the detector.

Optical filters are designed using different physical-technical principles.

Classical absorption filters are based on the spectrally sensitive absorption of colored glass filters. Single filters can block defined short (long-pass filter) or long (short-pass filter) wavelengths. Band-pass filters of some hundred nanometers range are usual. By combination of long and short-pass filters, the band-pass configurations of defined spectral bands smaller than single filters, can be generated. The filter type defines the relative spectral transmission function, while the absolute value of the spectral transmission of colored glass filters depends on their thickness. A typical application for this filter type of spectral test equipment is the pre-filtering of a broad band light source in spectrometer test equipments.

Absorption filters (colored glass filters) are made of parallel plates of optical glass containing embedded molecules and/or ions that produce a strong wavelength-dependent behavior in the natural transmittance of the glass.

Ions of heavy metals or rare earths in solution act as colorants of the ionically colored glasses. The colorants of colloidally colored glasses are usually rendered effective by heat treatment of the, initially nearly colorless, glass [48-9]. Processes involved in the spectral dependent transmission are ionic absorption and colloidal scattering. The ionic material which is in solution in glass can be considered as being in a very viscous liquid. The attenuation effect on the transmitted radiant flux can be described by the Lambert–Beer law according to (48-77), if the reflection losses at the interfaces are neglected. The transmission can be adjusted by changing the thickness of the glass plate. In the catalogues of filter suppliers the spectral transmissions are listed for a defined reference thickness. For different filter thicknesses the internal transmission can be calculated by applying (48-91).

The overall transmittance, after considering the reflection losses at the two plate interfaces, can be calculated from (48-102) in the case of a plane parallel plate in air and normal incidence of incoherent radiation (see e.g. [48-9] and [48-10]). In this way, the overall transmittance of a plane parallel plate in air is calculated according to (48-117):

$$\tau = \frac{\tau_i (1 - R)^2}{1 - \tau_i^2 R^2} \tag{48-117}$$

where

τ_i is the internal transmission

and

$$R = \left(\frac{n - 1}{n + 1}\right)^2 \tag{48-118}$$

is the reflectivity of the plate interfaces in air according to (48-74) where the refractive index of air $n_i = 1$ and the refractive index of the glass $n_t = n$.

The values for n and τ_i can be found in filter supplier catalogues (see, for example, [48-9]). Usually these are reference values. The supplier values of τ_i are related to a reference thickness. For other thicknesses, the relative spectral dependence of τ_i remains the same, but the absolute transmission values vary with filter thickness. If the internal transmission τ_{id_1} is known for a reference thickness d_1 the required thickness d_2 to reach a desired internal transmission τ_{id_2} can usually be calculated from (48-91), since the Lambert–Beer law is valid for absorption glass filters. By generating the logarithm of (48-91) the thicknessd_2 can be found according to (48-119):

$$d_2 = d_1 \frac{\log \tau_{id_2}}{\log \tau_{id_1}} \tag{48-119}$$

If the absorption filters are combined, the total transmission characteristic can be obtained by multiplication of the single filter transmission values. Figure 48-23 shows an example of the combination of a short-pass and a long-pass filter from the Schott catalogue [48-9]. This combination is a band-pass filter with a maximum transmission of about 80%. At the short-wave limit and also at the long-wave limit it allows the passage of more than 1% of the incident radiant flux for wavelengths longer than about 480 nm and shorter than about 850 nm, respectively.

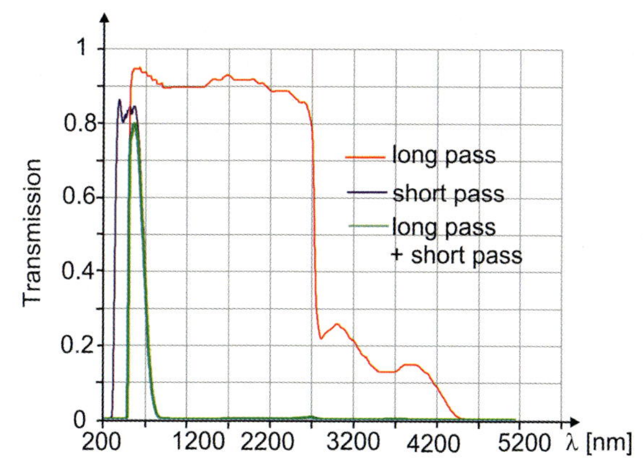

Figure 48-23: Combination of a long-pass filter GG 495 of reference thickness 3 mm and a short-pass filter KG5 of reference thickness 2 mm supplied by the Schott Company.

Specification of Absorption Filters

The filter characteristic, together with the spectral characteristic of the light source and the spectral sensitivity of the sensor, mainly determine the overall spectral responsivity of a spectrometer (see section 48.4).

The reference values for the spectral distribution of τ_i are not guaranteed absolute values, but only act as guidelines for the user. The spectral performance is specified by the suppliers by different parameters which are dependent on the filter type. In common use are the following spectral specifications for long-pass, short-pass and band-pass filters, which can be provided by absorption glass filters.

Short and Long-pass Filters

As shown schematically in figure 48-24 for the example of a long-pass filter, the transmission range of these filter types are specified by the three characteristic wavelength values: the edge wavelength λ_c; the limit wavelength of the blocking range λ_b; the limit of the pass range λ_p.

- Edge wavelength λ_c:
  At $\lambda = \lambda_c$ the spectral internal transmittance $\tau_i(\lambda)$ is one-half of the maximum value of the pass range.
- Limit of blocking range λ_b:
  For $\lambda < \lambda_b$ (long pass) or $\lambda > \lambda_b$ (short pass) respectively, the internal spectral transmission $\tau_i(\lambda)$ does not exceed a specified value τ_{ib} within a defined cer-

tain spectral range. The relation $\tau_i(\lambda) \leq \tau_{ib}$ is valid within this range.

– Limit of pass range λ_p:
  For $\lambda > \lambda_p$ (long pass) $\lambda < \lambda_p$ (short pass), respectively, the internal spectral transmission $\tau_i(\lambda)$ does not fall below a specified value τ_{ip} within a defined certain spectral range. The relation $\tau_i(\lambda) \geq \tau_p$ is valid within this range. The pass range can be divided into sub-ranges with different values τ_{ip}.

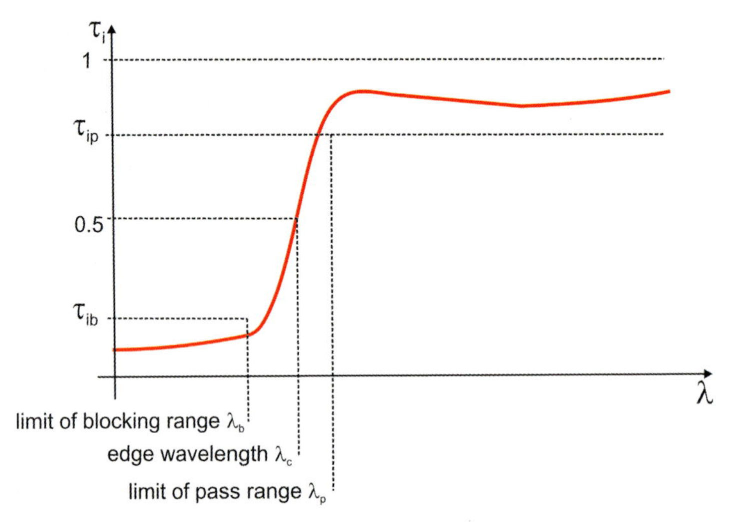

Figure 48-24: Parameters of a long-pass filter.

Band-pass Filters

Band-pass filters connect a spectral range of specified high-transmission (the so-called pass band) with adjacent spectral ranges of shorter and longer wavelengths and specified low transmission. According to figure 48-25, the main specification parameters are the maximum spectral transmission τ_{max} or the internal spectral transmission $\tau_{i\,max}$ in the pass range, the central wavelength λ_m of the pass range and the width of the spectral distribution curve. This width is also called the full width of half maximum FWHM.

The central wavelength is defined as the arithmetic mean of the two wavelengths, for which the transmission reaches one half of the maximum value τ_{max} according to (48-120):

$$\lambda_m = \frac{\lambda_{1\frac{\tau_{max}}{2}} + \lambda_{2\frac{\tau_{max}}{2}}}{2} \tag{48-120}$$

where

$\lambda_{1\frac{\tau_{max}}{2}}$, $\lambda_{2\frac{\tau_{max}}{2}}$ are the lower and upper wavelengths, respectively, at which the transmission falls to one-half of the maximum transmission τ_{max}.

With these two wavelengths the FWHM is defined according to (48-121):

$$FWHM = \lambda_{2\frac{\tau_{max}}{2}} - \lambda_{1\frac{\tau_{max}}{2}} \qquad (48\text{-}121).$$

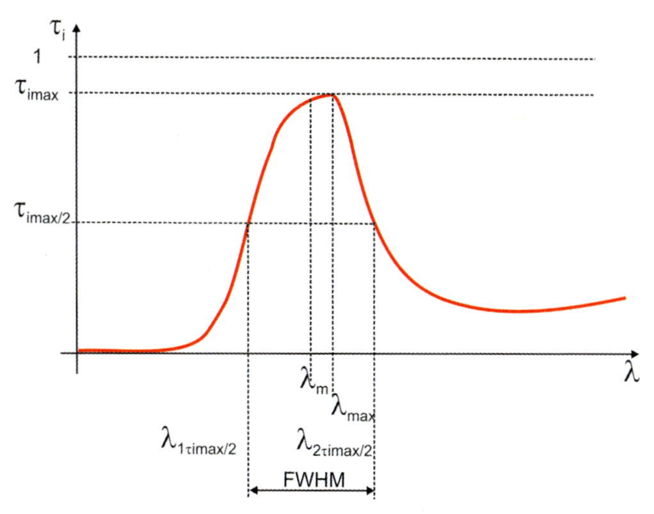

Figure 48-25: Parameters of a band-pass filter.

Suitable spectral filter characteristics as described above are the necessary conditions required for any measurement application. However, a variety of other material parameters have to be checked for their relevance to the special application considered.

These parameters can also be found in the catalogues. In this context, only a short overview of typical parameters is mentioned. For more details see, for example, [48-9].

Temperature Dependence of Transmission

Spectral characteristics can shift with changes in the temperature. Usually, for band-pass filters and filters with flat curve flanks, the changes in transmission as a function of temperature are relatively small. However, the edge of long-pass filters shifts toward higher wavelengths with increasing temperature. For this reason temperature coefficients are provided in catalogues produced by filter suppliers.

Transformation Temperature

If it is expected that the filters will be exposed to temperatures higher than 200°C, one consideration would be that the glass filters will be transformed through a brittle stage into a liquid stage with the increasing temperature. The transformation range between these two behaviors is described by the transformation temperature T_g measured in °C according to ISO 7884-8. The filter temperature should not exceed the temperature $T = T_g - 200°C$ to avoid any permanent change in the filter characteristics [48-9].

Thermal Expansion

The change in length of glass filters caused by a temperature change is characterized by the temperature dependent linear thermal expansion coefficient. Thermal, expansion will change both the filter thickness and outer dimensions of the filter faces. This has to be taken into consideration for filter mounting, but a change in the optical path length can also be important.

Solarization

Irreversible transmission changes can be caused by ultraviolet radiation. This solarization effect depends on the wavelength and irradiance. The shorter the wavelength the higher the influence. Typical effects are a reduction of the pass-band transmission and a shift in the short-wave edge to longer wavelengths. Depending on the duration, spectral distribution and irradiance, a saturation state will be reached. Filters can be artificially aged by pre-radiation. In this way, stable spectral transmission characteristics can be generated.

Luminescence

Filter glasses can generate luminescence radiation for wavelengths longer than the primary illumination wavelength. Usually this is not a problem for spectrometric applications because the luminescence radiation can be blocked.

Chemical Resistance

The surface quality of polished filter glasses can be changed by several chemical influences. In supplier catalogues the resistance classes are specified.

Acid and alkali resistance are usually specified according ISO documents. A stain resistance can also be specified [48-9].

Long-term Surface Changes

Filter surfaces are more or less sensitive dependent on the type of glass. Even without any direct strong chemical influence long-term changes in the polished filter surfaces can occur due to the normal influence of the surrounding air. Usually such processes are favored by humidity. Dependent on the application, clarification should be obtained from the supplier, whether a protective coating is necessary [48-9].

Internal Material Quality

Filter glass and optical glass are both characterized by internal properties such as bubbles, striae and the homogeneity of the refractive index (see also chapter 52).

The internal material quality of filter glass is not as good as that of high-quality optical glasses. Therefore optical filter glasses cannot generally be used in precise imaging systems.

48.3.3
Fabry–Pérot Etalons

In section 46.2.14 we have shown the basic laws for the transmission and reflection of light at two partially reflecting plane-parallel interfaces. An element of this kind is called a *Fabry–Pérot etalon* or *Fabry–Pérot interferometer.* In the following we describe the basic principles and characteristics of this type of monochromator.

The original Fabry–Pérot etalon consists of two parallel reflecting surfaces on glass carriers separated by a distance d, as shown in figure 48-26. The carrier plates are slightly wedged to prevent the outer surfaces from generating erroneous reflections. The medium with refractive index n' within the resonator can be air or any other gas, fluid or any solid transparent material.

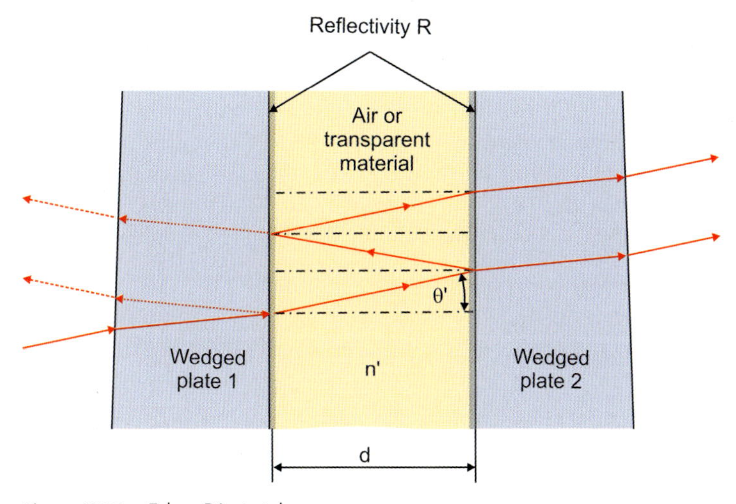

Figure 48-26: Fabry–Pérot etalon.

The transmissivity as the ratio of exit and incident irradiance can be calculated assuming that no absorption loss occurs. According to (46-100) the transmissivity T of a Fabry–Pérot etalon with resonator length d and reflectivity R of both reflecting surfaces is given by (48-122):

$$T_{FP} = \frac{(1-R)^2}{1+R^2-2R\cos\phi} = \frac{1}{1+F\sin^2\frac{\phi}{2}} \tag{48-122}$$

where

$$\phi = \frac{2\pi}{\lambda} 2n'd\cos\theta' \tag{48-123}$$

F is called the *coefficient of finesse* and is defined by

$$F = \frac{4R}{(1-R)^2} \tag{48-124}$$

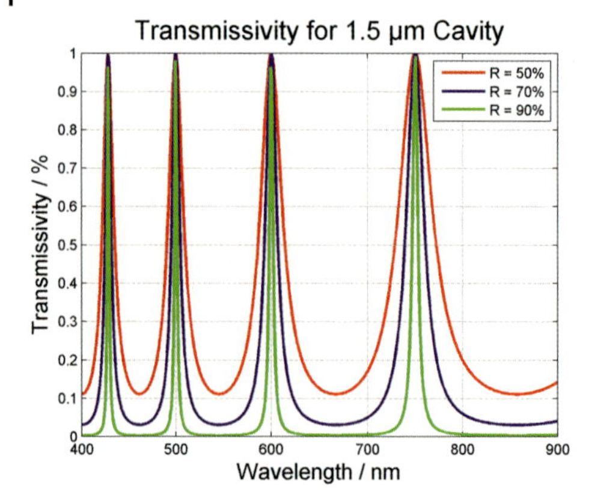

Figure 48-27: Transmissivity of a Fabry–Pérot etalon as a function of wavelength. Higher reflectivities of the resonator plates generate peaks with smaller FWHM.

Figure 48-27 shows the transmissivity of a Fabry–Pérot etalon as a function of wavelength λ. The full width of half maximum $\delta\lambda$ of the typical peaks in the spectrum depends on the reflectivity R of the resonator plates. The wavelength separation $\Delta\lambda$ between adjacent peaks in the spectrum is called the *free spectral range* (FSR) and is given by (48-125):

$$\Delta\lambda = \frac{\lambda_0^2}{2n'd\cos\theta' + \lambda_0} \approx \frac{\lambda_0^2}{2n'd\cos\theta'} \tag{48-125}$$

where λ_0 denotes the central wavelength of the nearest transmission peak in the spectrum.

The *finesse* $\hat{F}$ characterizing FSR/FWHM of the Fabry–Pérot etalon is defined by (48-126):

$$\hat{F} = \frac{\Delta\lambda}{\delta\lambda} = \frac{\pi}{2\arcsin\left(\frac{1}{\sqrt{F}}\right)} = \frac{\pi}{2\arcsin\left(\frac{1-R}{2\sqrt{R}}\right)} \tag{48-126}$$

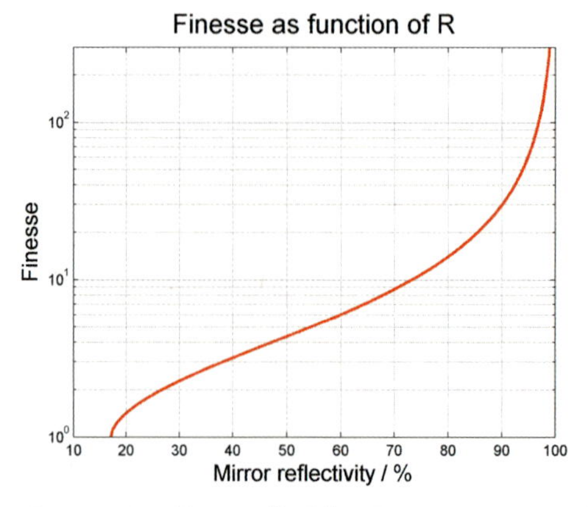

Figure 48-28: Finesse of a Fabry–Pérot resonator as a function of mirror reflectivity R.

For reflectivities > 0.5 (48-126) can be approximated by (48-127) as

$$\hat{F} \approx \frac{\pi\sqrt{F}}{2} = \frac{\pi\sqrt{R}}{1-R} \qquad (48\text{-}127)$$

Figure 48-28 shows $\hat{F}(R)$ according to (48-126) as a graphical presentation. For $R = 0.1716$ the finesse $\hat{F} = 1$. If R becomes smaller, the finesse will become complex. For $R = 0.7323$ a finesse $\hat{F} = 10$ is achieved, for $R = 0.9691$ a finesse of $\hat{F} = 100$ will result. A very high finesse with $\hat{F} > 10^6$ can be achieved for $R > 0.999997$ by using dielectric supermirrors.

A high finesse can be useful for optical spectrum analysis, because it allows the combination of a large free spectral range with a small resonator bandwidth. Therefore, a high spectral resolution in a wide spectral range is possible.

The FWHM $\delta\lambda$ of a peak in the spectrum will define the potential resolution of the Fabry–Pérot etalon. In figure 48-29 $\delta\lambda$ is shown as a function of the resonator length d for a finesse $\hat{F} = 100$ and wavelengths of 400 nm, 600 nm and 800 nm. For short resonator lengths in the μm range resolutions of nm will result, whereas for long resonators up to 1 m, resolutions of $\delta\lambda = 10^{-6}$ nm can be achieved for short visible wavelengths.

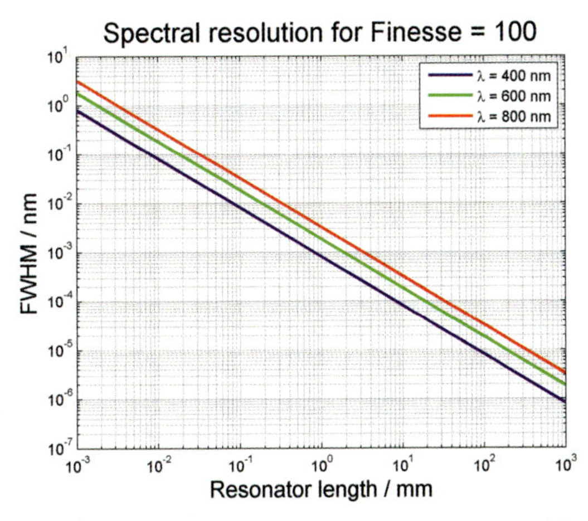

Figure 48-29: $\delta\lambda$ as a function of the resonator length d for a finesse of $\hat{F} = 100$ and wavelengths 400 nm, 600 nm and 800 nm.

Multiple Fabry–Pérot Etalons

It is possible to extend the FSR of a high-resolution Fabry–Pérot etalon by using multiple resonators in tandem, each with a different FSR [48-11]. The arrangement will exhibit high transmission only at those wavelengths which are simultaneously resonant in each of the resonators. For instance, if two resonators differing by 10% in thickness are combined, the resulting FSR will be ten times that of the longer resonator.

A combination of N resonators with individual cavities i specified by ϕ_i and individual resonator reflectivities R_i will have a final transmissivity of

$$T_{FP} = \prod_{i=1}^{N} \frac{(1-R_i)^2}{1+R_i^2+2R_i\cos\phi_i} \tag{48-128}$$

Tuning of Fabry–Pérot Etalons

Considering (48-122) and (48-123), the wavelength of the transmission peaks changes linearly with d and n'. For etalon tuning, both parameters can be used.

The most common way to change the spectral transmissivity of a Fabry–Pérot etalon is to change the spacing of the resonator plates. This can be accomplished by introducing piezo-electric elements which vary the spacing from the nanometer to the micrometer range. It is also possible to vary the pressure of the gas between the resonator plates in order to introduce changes in the refractive index. Solid-cavity etalons can be tuned by introducing specific temperature changes. When electro-optic material is used in the cavity, its refractive index can be changed by applying an electric field in the direction of the crystal's optical axis.

48.3.4
Interference Filters

Interference filters are used to reach small spectral pass bands and steep, short and long wavelength edges with high spectral resolution. The basic principle of an interference filter is identical to that of the Fabry–Pérot element described in the previous section. Interference filters consist of thin films deposited on a substrate of sufficient thickness. The thin film structure is composed of two partially transmitting layers separated by a dielectric layer of optical thickness $\lambda/2$ with λ as the specific design wavelength. The filter should ideally have maximum transmission at this wavelength. An incident beam is partially reflected and transmitted as shown in figure 48-30.

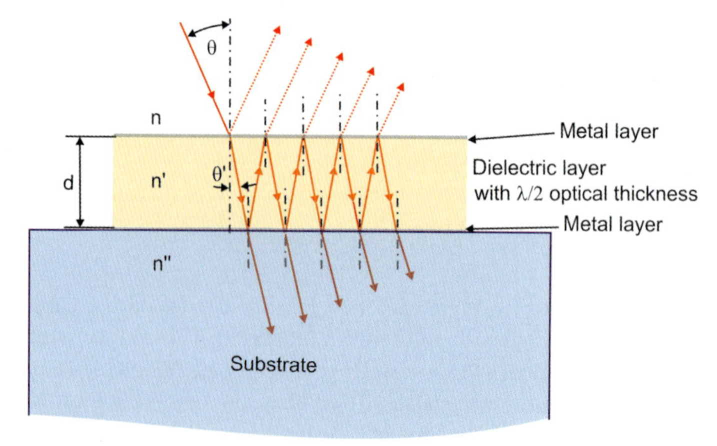

Figure 48-30: Basic configuration of an interference filter.

The transmitted and reflected parts from one direction are coherent and interfere constructively with each other if their optical path difference is a whole number multiple of the wavelength λ. In this case the phase difference of the interfering beams is a whole number multiple of 2π.

For the general case of an angle of incidence θ the transmitted beam components are in phase if equation (48-129) is fulfilled:

$$2\,n'\,d\,\cos\theta' = m\,\lambda \tag{48-129}$$

where

λ is the vacuum wavelength of the incident radiance,
n' is the refractive index of the dielectric layer,
d is the thickness of the dielectric layer,
m is the order.

The optical path difference for constructive interference according to (48-129) is found by a geometrical method as shown in figure 48-31 for two adjacent transmitted beams.

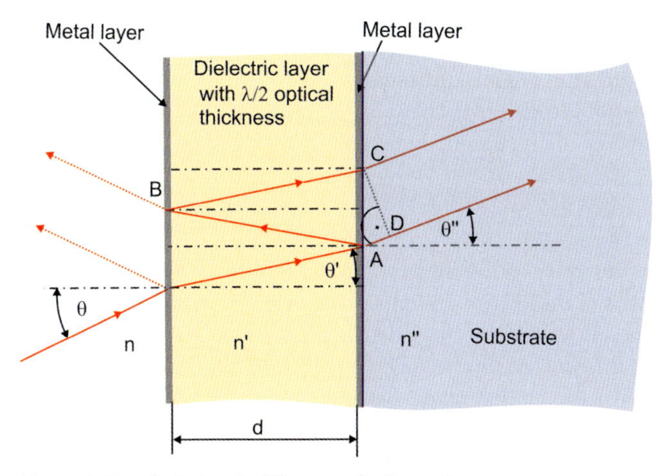

Figure 48-31: Optical path difference of adjacent beams transmitted through, and partly reflected within, the dielectric layer of an interference filter.

In figure 48-31 one transmitted beam is the directly transmitted component, while the adjacent beam is twice partly reflected before it is transmitted. The optical path difference is calculated from

$$\mathrm{OPD} = l_O = n'\left(\overline{AB} + \overline{BC}\right) - n''\overline{AD} \tag{48-130}.$$

The metal layers are considered as so thin that their influence on the optical path is negligible. After some trigonometric transformations and using the law of refraction $\frac{\sin\theta'}{\sin\theta''} = \frac{n''}{n'}$ where n'' is the refractive index of the substrate, one finds the optical path difference according to (48-131) as:

$$l_O = 2\,n'\,d\,\cos\theta' \qquad (48\text{-}131)$$

where
$\theta' = \arcsin\left(\frac{n}{n'}\sin\theta\right)$ is the inner angle of refraction. The general relation (48-129) is valid for arbitrary orders. According to (48-129) the longest wavelength transmitted is $\lambda = 2\,n'\,d\cos\theta'$, but transmission peaks also occur at the higher harmonics λ/m. For $\theta' = 0$ the wavelength of peak transmission is $\lambda = 2\,n'\,d$. The peak wavelength can be tuned by rotating the interference filter around an axis perpendicular to the plane of incidence. The peak transmission wavelength shifts to shorter wavelengths with increasing angle θ'. If the angle θ' is expressed by the angle of incidence θ, then using the law of refraction, relation (48-132) is valid for the peak transmission wavelength λ_θ at an incident angle θ as a function of the peak transmission wavelength λ_0 at normal incidence and the angle of incidence:

$$\lambda_\theta = \lambda_0\sqrt{1 - \left(\frac{n}{n'}\sin\theta\right)^2} \qquad (48\text{-}132)$$

where
λ_θ is the peak wavelength of zeroth order at an angle of incidence θ,
λ_0 is the peak wavelength at an angle of incidence $\theta = 0$,
n' is the refractive index of the dielectric layer,
n is the refractive index of the medium which the radiation passes through before entering the filter.

All-dielectric Interference Filters (ADI)
To obtain optimum performance for different applications there is a large variety of rather complex structures of multi-layer systems, which can by considered as extensions of the basic interference filter. For more details see [48-12].

Within the scope of this book only the following short overview is possible. In Table 48-3, the extended basic configuration types of interference filter with their advantages and targets are listed.

As an example, figure 48-32 shows the scheme of a simple one-cavity *all-dielectric interference filter* (ADI). The metal layers of the basic interference filter are exchanged for reflector layers composed of alternating high and low refractive index layers with optical thicknesses of $\lambda/4$. These layers have an anti-reflection function and therefore improve the transmission compared with the metal layers of the simple Fabry–Pérot system. Furthermore, the bandwidth is controlled by these layers, while the dielectric layer of optical thickness $\lambda/2$ still controls the peak wavelength as for the basic interference filter.

A better uniformity over the desired band pass and a sharper cutoff can be provided by an extension of the all-dielectric cavity configuration to more than one cavity. An *all-dielectric multicavity filter* can be configured by combining more than one cavity. Each cavity has the structure of a one-cavity all-dielectric filter. Between two cavities there is a coupling layer. In all types of ADI the primary spacer layers with their optical thickness of $\lambda/2$ control the peak wavelength, whereas the $\lambda/4$ reflector layers, with alternating high and low refractive index, control the pass band. With increasing number of cavities the pass band becomes more rectangular and the cut-off sharper.

The induced transmission filter, which is the third extended configuration type, uses metal layers for absorption at longer wavelengths additional to $\lambda/2$ spacer layers and $\lambda/4$ stacks. The dielectric layer combination has an anti-reflection function at the peak wavelength.

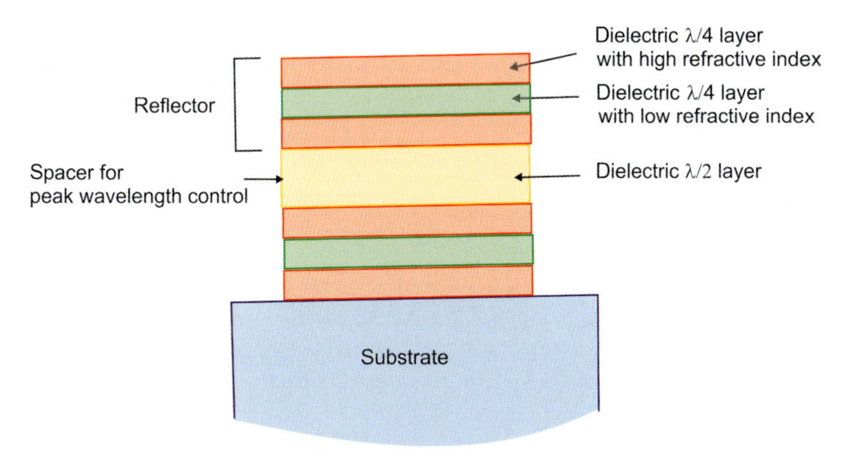

Figure 48-32: One-cavity-all-dielectric filter.

Table 48-3: Basic types of interference filter configurations.

Interference filter type	**Number of $\frac{\lambda}{2}$ dielectric layers**	**Reflector layers**	**Coupling layers**
Fabry–Pérot	1	2 metal layers	no
All-dielectric	1	Dielectric $\frac{\lambda}{4}$ layer stacks	no
All-dielectric multi-cavity	>1	Dielectric $\frac{\lambda}{4}$ layer stacks	1 between 2 cavities
Induced transmission	>1	Metal layers and dielectric $\frac{\lambda}{4}$ layer stacks	no

The influence of the number of cavities in the case of ADI filters is shown in figure 48-33 schematically for the filter transmission dependent on wavelength with the ordinate $\tau(\lambda)$ in diabatic scale $1 - \lg[\lg(1/\tau(\lambda))]$. As the number of cavities increases the pass band becomes more rectangular with a sharper cutoff.

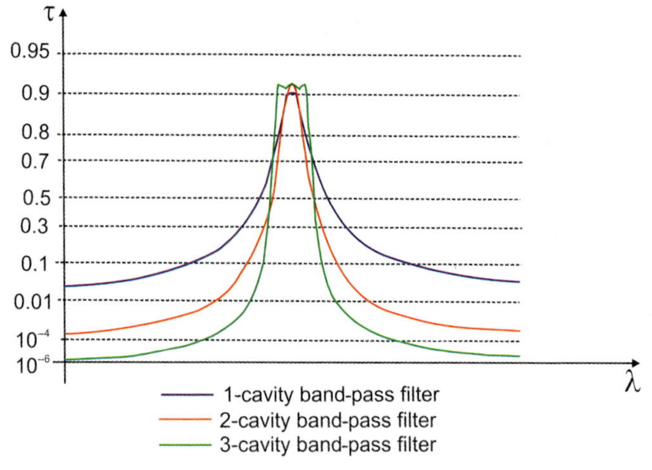

Figure 48-33: Multi-cavity filter of ADI type.

Also, in the case of more complex filter configurations, the transmission peak wavelength, dependent on the angle of incidence, can be described by analogy with (48-132) using an approximating formula with an effective index of refraction according to (48-133):

$$\lambda_\theta = \lambda_0 \sqrt{1 - \left(\frac{n}{n^*} \sin\theta\right)^2} \tag{48-133}$$

where
n^* is the effective refractive index of the total interference filter.

The effective refraction index is mainly a function of the refractive index of the $\lambda/2$ spacer layer. For spacer layers with index of refraction higher than the adjacent dielectric layers, the effective index of refraction is calculated according to (48-134):

$$n^* = \sqrt{\frac{n_H}{n_L}} \tag{48-134}$$

where
n_H is the index of refraction of the high refractive index layers,
n_L is the index of refraction of the low refractive index layers.

In the case of a spacer layer with lower refractive index the effective refractive index is approximately calculated according to (48-135):

$$n^* = \frac{n_L}{\sqrt{1 - \frac{n_L}{n_H} + \left(\frac{n_L}{n_H}\right)^2}} \tag{48-135}$$

For a convergent or divergent beam a weighted average of (48-132) or (48-133) over the range of angles of incidence must be used [48-12].

Often absorption filters are used as substrates for interference thin-film coatings. The combination of absorption filters and interference filters allows more design freedom and is used, for example, to reduce the transmission at harmonic wavelengths and flux outside the normal frequency band.

Linear and Circular Wedge Interference Filters

Circular and linear wedge filters have been developed, together with the interference filters (see e.g. [48-13]–[48-15]). For these filter types the thickness of all layers of a band pass varies in proportion across the surface of the substrate thus causing a variation in the central wavelength [48-16]. The layer wedge is generated by the coating process along a line or a circle. In particular, the circle wedge filter type can be used to generate a small lightweight scanning monochromator with moderate resolution and the capacity for fast central wavelength tuning by rotation of the filter disc. This can be used for example, for fluorescence imaging micro-spectrometers [48-17].

Specification of Interference Filters

In the same way as absorption glass filters, interference filters are characterized by their spectral transmission parameters. Representatively, a band-pass interference filter is considered here.

Band-pass Interference Filters

An interference band-pass filter is usually specified by the central avelength λ_m according to (48-120), the spectral width FWHM according to (48-121), the maximum transmission τ_{max} of the pass band and several wavelength values characterizing the transmission within blocking ranges. The latter are specified for different transmissions. As an example, figure 48-34 shows the specification of a Schott interference band-pass filter. Here the additional spectral widths for a transmission of $\tau_{max}/10$ and $\tau_{max}/1000$ are specified. Furthermore, the upper limits τ_{b1-2}, τ_{b3-4}, τ_{b4-5} etc., of spectral transmission within different blocking ranges from λ_{b1} to λ_{b2}, λ_{b3} to λ_{b4}, λ_{b4} to λ_{b5} etc., are specified.

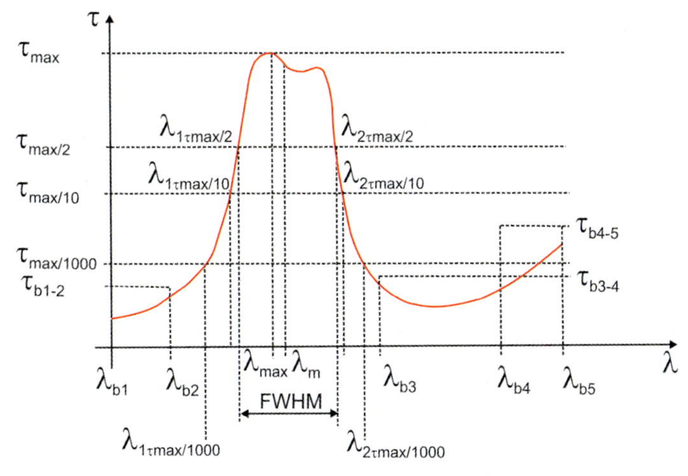

Figure 48-34: Parameters of a band-pass interference filter.

Influence of Temperature and Humidity

The spectral characteristic of interference filters is influenced by temperature changes. Usually with increasing temperature the central wavelength of band-pass filters is shifted to higher wavelengths. Depending on the type of coating the wavelength shift is in the range of 0.001–0.1 mm/K (see [48-18]). The temperature-shift coefficient depends on the type of coating.

Usually the so-called soft coatings, produced by classical physical vapor deposition, absorb moisture from the environment thus increasing the effective refractive index. Changes in the optical path are caused directly by thermal expansion and by changing the refractive index of the layers. Furthermore, the refractive index is influenced by temperature-induced variations of the water fraction within thin-film voids. So-called hard coatings are produced by ion-assisted evaporation or reactive ion plating processes. The layers of hard coatings have compact structures with higher density than is the case for soft coating layers. Therefore, they absorb almost no moisture from the environment. This results in a better lifetime and lower sensitivity against temperature changes. The temperature and moisture sensitivities are specified in filter suppliers' catalogues.

Maximum Temperatures and Resistance to Radiation

In a similar way to absorption color glass filters, there are maximum operating temperatures which have to be considered for the optical and mechanical configuration. For example, radiation of broad-band light sources can be partly reflected from mirror layers, back into the filter, and then absorbed. Therefore, filters with reflecting sides should be arranged with these sides to the light source to avoid heating. Generally, resistance to radiation should be considered in the case of high irradiance on the filter, such as intensive UV radiation or high-powered laser radiation of other wavelengths.

Thermo-optic Tuning of Multiple-cavity Filters

The dependence of the spectral characteristics of multiple-cavity interference filters on temperature is usually undesirable, but can, in fact, be utilized for wavelength tuning or switching. One example would be microscopic filters with thin-film combinations of amorphous silicon and silicon nitrides, which allow tuning ranges of up to about 60 nm to 1550 nm, caused by local microscopic temperature changes from room temperature to about 400°C. Heating is accomplished in packaged chips by special internal heating layers of electrically conductive polysilicon or by a doped conductive region on a crystalline silicon substrate. These filters can be useful for wavelength division multiplexing (WDM) network architectures in telecommunication applications [48-19].

Mechanical Stress

In contrast to colored glass filters, consisting of one glass plate, interference filters with many layers on a substrate or even more substrates cemented together, will be more sensitive against stress. This has to be considered by finding a suitable mounting. Stress can also be caused by temperature gradients generated by high local irradiance, such as from a laser source with an unexpanded beam of small diameter.

48.3.5
Electronically Tunable Filters

The most common electronically tunable filters are acousto-optical tunable filters, abbreviated as AOTF, and also liquid crystal tunable filters, abbreviated as LCTF. AOTFs allow random access and fast wavelength tuning with response times as short as several μs. Therefore, this filter type can be advantageous for spectrometric metrology systems, when random access and very fast measurement data generation are required. Real time *in situ* spectral measurement during the thin-film deposition process is one application – the rapid spectral measurement can be used to control the thickness of the single layers of complex thin-film multi-layers. Several types of AOTFs are commercially available.

LCTFs, which are a special type of birefringent filter, are of increasing interest especially for multi-spectral imaging applications such as fluorescense microscopy, because they can be used with high optical apertures and imaging angles. Compared to AOTFs they have lower transmission caused by the need for several polarizers and often longer response times of tens of milliseconds, dependent on the type of liquid crystal used. Nevertheless, LCTFs are of interest for use in other optical metrology tasks because of their low operation voltages, the lack of currents involved and ease of construction.

Whereas classical LCTFs are polarization dependent and therefore need polarizers which cause insertion losses dependent on the polarization state of the incident light, a newer type of LCTF development is based on a stack of switchable liquid crystal polarization gratings, abbreviated as LCPG. This filter type allows polarization-independent tuning with a high throughput of up to 100%. In the following we will give an overview of the main electronically tunable filter types.

Acousto-optic Filters

Principal Function of an AOTF

The basic physical mechanism exploited for acousto-optical filters is the elasto-optic effect. This effect occurs in many materials. As will be shown later in this chapter optical birefringence is an additional requirement for tunable filters. A common material used in AOTFs is Tellurium Dioxide TeO_2. A piezoelectric transducer generates an acoustic wave with a frequency in the radio frequency (RF) region. The acoustic wave travels through the crystal thus changing the refractive index periodically by generating periodic regions of compression and expansion. This acts as a traveling diffraction grating along the path of the acoustic wave. An incident beam of light can be diffracted in the interaction region with small spectral bandwidth, when certain conditions are fulfilled [48-20]. The peak wavelengths of the diffracted radiation depend on the frequency of the acoustic wave. Thus, by tuning the RF frequency, the filter band-pass can be tuned very quickly within microseconds (see, for example, [48-20] and [48-21]). By applying multiple-frequency RF signals even parallel multiplexed filtering is possible.

Anisotropic birefringent crystal materials are normally used for AOTF configurations. There are two basic types of birefringent crystal AOTF – the collinear beam AOTF with the incident optic beam collinear to the acoustic wave and the non- collinear AOTF type with the incident beam nearly perpendicular to the acoustic wave.

Figure 48-35 shows the function of a non-collinear AOTF schematically.

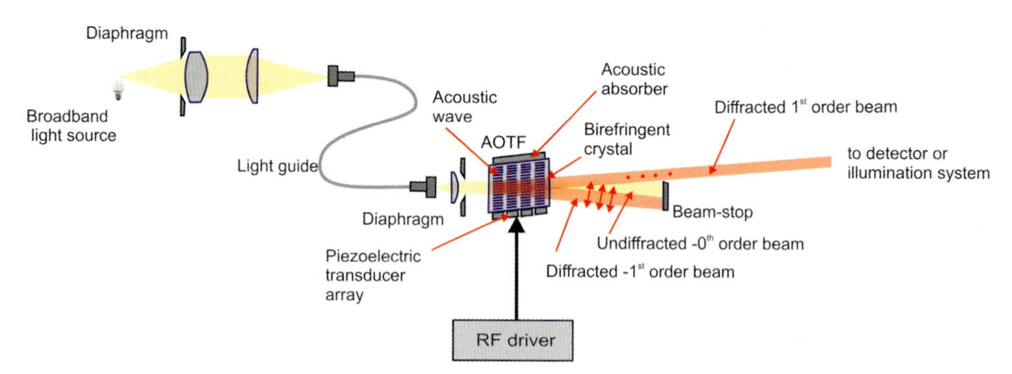

Figure 48-35: Configuration of a non-collinear AOTF monochromator.

After generation by an RF source the acoustic wave travels through the AOTF crystal. The RF energy is coupled into the crystal via a piezoelectric transducer bonded onto one crystal face. At the opposite side the acoustic energy is absorbed by an acoustic absorber. The incident optic beam from a light source of appropriate broad bandwidth, adapted to the spectrometric measuring task, travels through the crystal with a defined angle to the acoustic wave path. AOTFs work in the Bragg regime, in which the incident light is diffracted into the predominant first orders with high efficiency. Theoretically 100% efficiency can be achieved. For a fixed RF frequency, only narrow band light with one peak wavelength is diffracted. The acoustic frequency determines the peak wavelength of the diffracted light. Thus the wavelength of the diffracted light wave can rapidly be tuned by tuning the frequency of the RF source. The complete tunable filter is configured by blocking the undiffracted beam and the second beam of 1st or -1st order. The diffracted beams are linearly polarized with the polarization directions perpendicular to each other.

In case of the collinear AOTF the path of the optical beam is the same as the path of the acoustic beam along the principal crystal axis as shown schematically in figure 48-36.

At a fixed acoustic frequency, which is generated by a RF source, only a narrow spectral band with a peak wavelength determined by the RF frequency will be diffracted in the same direction but with orthogonal polarization. This is schematically shown in figure 48-36 as yellow dots for the linear polarization direction of the incident beam perpendicular to the drawing plane, which is the plane of incidence, and red double arrows for the polarization direction of the filtered diffracted beam in the drawing plane. The different polarization states enable the separation of the filtered

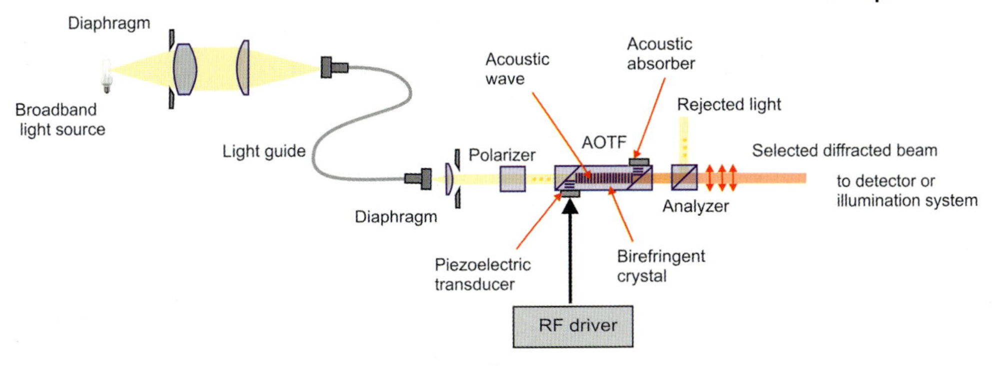

Figure 48-36: Configuration of a collinear AOTF monochromator.

beam from the primary beam by means of a second polarizer. The input beam has to be linearly polarized, and the required filtered monochromatic beam is selected by a second polarizer with crossed direction of transmission.

The main advantage of a collinear AOTF compared with the non-collinear is that a higher spectral resolution can be achieved due to longer collinear interaction ranges (see below). Despite this, collinear AOTFs are used commercially only for very special applications over small tuning ranges, whereas non-collinear AOTFs are already widely commercially available as filter components and recently also for spectrometers in the IR range (see [48-22]) due to the following main advantages as compared with the collinear type:

1. More highly efficient acousto-optic materials can be used (e.g. TeO_2).
2. No polarization is needed.
3. Fabrication is easier and more cost efficient, due to the separated paths of the acoustic and optic beam and the incident and diffracted output beam.

The typical RF frequencies needed to obtain optical wavelengths in the UV–VIS range from about 200 nm – 700 nm are about 200 – 50 MHz, whereas in the IR range from about 800 nm up to 4.5 μm RF frequencies of about 130 – 20 MHz are necessary. The RF power needed can reach several Watts dependent on the material and optical aperture [48-22] and [48-23].

Bragg Diffraction at an Acoustic Wave as a Basic Mechanism of an AOTF

The relations for the principle function of an AOTF are found from plane-wave analysis of the acoustic-optic interaction (see [48-24]).

Within the scope of this book, only important relations are presented, which are necessary for understanding the AOTF function.

Significant Bragg diffraction occurs only if the momentum matching between the acoustic and diffracted optic wave is achieved according to (48-136), the so-called *phase matching condition*:

$$\vec{k}_d = \vec{k}_i + \vec{k}_a \qquad (48\text{-}136)$$

where
$\vec{k}_d$ is the wave vector of the diffracted optic wave,
$\vec{k}_i$ is the wave vector of the incident optic wave,
$\vec{k}_a$ is the wave vector of the acoustic wave.

In the general case of an acousto-optical interaction in an anisotropic medium the magnitudes of the above wave vectors are given according to (48-137)–(48-139):

$$k_d \approx \frac{2\pi n_d}{\lambda} \tag{48-137}$$

$$k_i = \frac{2\pi n_i}{\lambda} \tag{48-138}$$

$$k_a = \frac{2\pi}{\Lambda_a} = \frac{2\pi f_a}{v_a} \tag{48-139}$$

where
λ is the vacuum wavelength of the incident optic wave,
n_i is the refractive index for the incident optic wave,
n_d is the refractive index for the diffracted optic wave,
Λ_a is the acoustic wavelength,
v_a is the acoustic velocity,
f_a is the frequency of the acoustic wave = RF frequency

In (48-137) the very small difference in wavelength due to acoustic frequency shift is neglected.

Figures 48-37 and 48-38 show the wave vector diagrams or *index ellipsoids* of a non- collinear and collinear AOTF configuration for a birefringent uniaxial crystal.

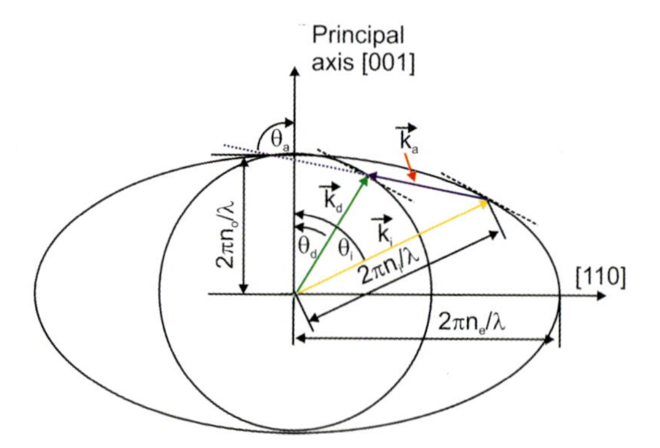

Figure 48-37: Wave vector diagram for a non-collinear AOTF.

Both configurations meet the phase-matching condition (48-136). In the non-collinear AOTF configuration of figure 48-37 the incident beam is extraordinarily polarized. The acousto-optic interaction in the uniaxial birefringent crystal generates an ordinarily polarized diffracted wave. The angle of incidence θ_i of the optic wave equals the polar angle θ_e of an extraordinarily wave with the principal axis, i.e. a wave polarized in the principal plane spanned by principal (optical) axis and wave vector $\vec{k}_i$.The angle θ_d of the diffracted beam equals the polar angle θ_o of the ordinary wave, and θ_a is the angle of the acoustic wave vector to the principal axis.

In the general case of an arbitrarily polarized incident light wave, two diffracted waves are generated [48-21], [48-24] and [48-25].

The non-collinear AOTF of figure 48-37 meets the phase-matching condition (48-136) and also the so-called *non-critical phase matching condition.* This means that the tangents to the loci of the incident light wave vector $\vec{k}_i$ and the diffracted light wave vector $\vec{k}_d$ are parallel. In this case a phase mismatch due to a change in the angle of incidence is compensated by the angular change of birefringence, within a certain angle range. In this angle range the non-collinear AOTF is insensitive to changes in the angle of incidence.

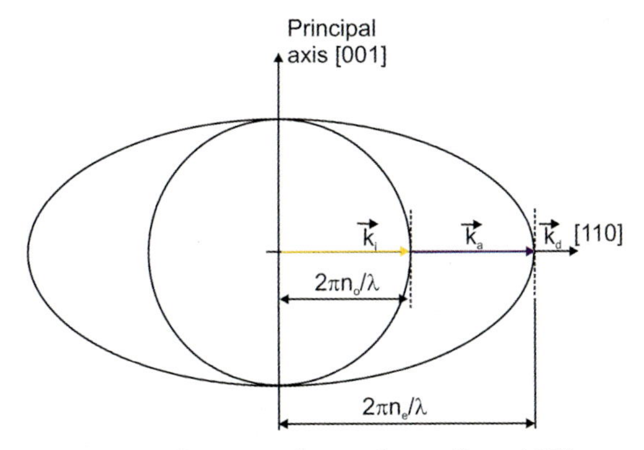

Figure 48-38: Wave vector diagram for a collinear AOTF.

Between the peak wavelength of the diffracted narrow band beam and the acoustic frequency, the general relation (48-140) is valid:

$$\lambda = \frac{v_a}{f_a} F(n_i, n_d) \tag{48-140}$$

$F(n_i, n_d)$ is a function of the refractive indices valid for incident and diffractive light. The function $F(n_i, n_d)$ is calculated from the phase-matching condition (48-136) using (48-137)–(48-139) for the vector magnitudes and considering the geometric relations of the index ellipsoid.

In the case of a collinear configuration, F is the crystal's birefringence, i.e., the difference between the extraordinary and the ordinary refractive index, according to (48-141). The incident light beam enters the crystal along the direction of a principal crystal axis. The refractive indices n_i and n_d of the incident beam and the beam diffracted in the orthogonal polarization, respectively, are the refractive indices n_o and n_e of the ordinary and the extraordinary beam, respectively. The ordinary and extraordinary beam are orthogonally polarized to each other and travel along the same direction.

$$\lambda = \frac{v_a}{f_a}(n_e - n_o) \tag{48-141}$$

For the non-collinear configuration the more complex relation (48-142) is valid. It is found from the geometric relations of the index ellipsoid of figure 48-37 by applying the cosine law:

$$\lambda = \frac{v_a}{f_a}\sqrt{n_o^2 + n_{ee}^2 - 2\,n_o\,n_{ee}\cos(\theta_o - \theta_e)} \tag{48-142}$$

In (48-142) θ_o and θ_e are the *polar angles* to the optical axis for the ordinary and extraordinary beam and equal θ_d and θ_i respectively. The effective refractive index n_{ee} ($= n_i$ for the example of figure 48-37) for the extraordinary beam with the polar angle θ_e to the optical axis, has a value lying between the crystal's ordinary and extraordinary refractive indices n_o and n_e dependent on the *polar angle* θ_e according to (48-143):

$$n_{ee} = \left(\frac{\cos^2\theta_e}{n_o^2} + \frac{\sin^2\theta_e}{n_e^2}\right)^{-\frac{1}{2}} \tag{48-143}$$

One obtains (48-143) by applying the geometric relations of an ellipse to the index ellipsoid of figure 48-37 with the lengths $2\pi n_e/\lambda$ and $2\pi n_o/\lambda$ for the large and small half-axes, respectively.

Furthermore, the *non-critical phase-matching condition* requires (48-144) for the polar angles of the ordinarily and extraordinarily polarized diffracted wave:

$$\frac{n_e}{n_o}\tan\theta_o = \frac{n_o}{n_e}\tan\theta_e \tag{48-144}$$

In figure 48-37, the tangents (black dashed lines) to the incident and diffracted light beam loci are parallel to each other, thus meeting this condition here as well as in the special case of the collinear AOTF according to figure 48-38.

The appropriate acoustic angle for an AOTF configuration is also found from the interaction geometry according to figure 48-37 by applying the sine law. Relation (48-145) for the acoustic angle θ_a dependent on the refractive indices and the polar angles θ_o and θ_e is valid:

$$\tan\theta_a = \frac{n_{ee}\sin\theta_e - n_o\sin\theta_o}{n_{ee}\cos\theta_e - n_o\cos\theta_o} \tag{48-145}$$

After substituting (48-143) and (48-144), relation (48-145) results in an exact expression for the acoustic angle, which fulfills the *non-critical phase matching condition*. Thus this exact value of the acoustic angle delivers the largest angular aperture for a given incident beam. Solving (48-142) for the exact acoustic angle will result in an exact tuning relationship for arbitrary acoustic angle and polarization [48-21] and [48-24].

In the case of small birefringence $(n_e - n_o)$ with the incident as the extraordinary ray an approximate tuning relation for the non-collinear AOTF configuration is given by (48-146):

$$\lambda = \frac{v_a}{f_a}(n_e - n_o)\sqrt{\sin^4\theta_e + \sin^2 2\theta_e} \tag{48-146}$$

For a *polar angle* $\theta_e = 90°$, i.e., the optic wave propagates along the extraordinary axis, this approximation delivers the collinear relation (48-141). For details see [48-24] and [48-26].

Spectral Resolution of an AOTF

The bandwidth of the diffracted radiant flux for one peak wavelength of the AOTF is determined by the transmission characteristic, which is a Sinc^2 function of the frequency according to (48-147):

$$\tau(\lambda) = \tau_0 \,\mathrm{Sinc}^2(\Delta k\, L) = \tau_0\left(\frac{\sin(\Delta k\, L)}{\Delta k\, L}\right)^2 \tag{48-147}$$

Here we have τ_0 as the peak transmission at exact phase matching,
L as the interaction length,
Δk as the momentum mismatch.

The spectral bandwidth is defined as the full width at half maximum (FWHM) of the central lobe of the transmission function. It can be approximated by relation (48-148) [48-24] and [48-25]:

$$\Delta\lambda = \frac{0.9\lambda^2}{b(\lambda)\, L \sin^2\theta_i} \tag{48-148}$$

where
λ is the vacuum wavelength,
$\Delta n = n_e - n_o$ is the birefringence,
θ_i is the angle of the incident beam with the principal crystal axis.

In relation (48-148) the dispersion of birefringence is described by the term $b(\lambda)$, which is defined as:

$$b(\lambda) = \Delta n - \lambda\frac{\partial \Delta n}{\partial \lambda} \tag{48-149}$$

The interaction length L is determined by the geometric AOTF configuration and can be calculated as follows:

$$L = \frac{d_a}{\cos\left(\theta_i + \theta_a - \frac{\pi}{2}\right)} \tag{48-150}$$

where θ_a is the angle of the acoustic beam to the principal axis according to figure 48-37 and d_a the piezoelectric transducer width, which determines the width of the acoustic beam. Figure 48-39 shows these geometric relations.

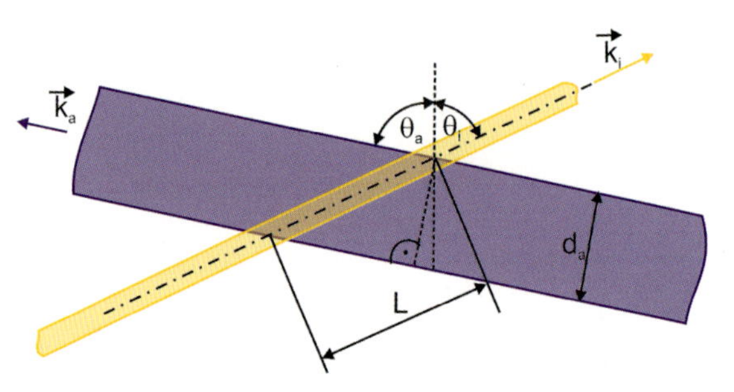

Figure 48-39: Interaction length of a non-collinear AOTF.

Relation (48-148) shows that the spectral bandwidth can be reduced by increasing the interaction length. As a rule of thumb, the spectral bandwidth increases as λ^2 and the resolving power $\lambda/\Delta\lambda$ decreases as $1/\lambda$, if the angle and interaction length remain constant. However, since the dispersion term $b(\lambda)$ decreases with increasing wavelength the bandwidth increases more rapidly than as λ^2, and the resolving power $\lambda/\Delta\lambda$ decreases more rapidly than as $1/\lambda$.

In the UV-VIS range from 250 nm to about 700 nm, commercially available AOTFs have spectral resolutions of about 0.05–7 nm dependent on the material, the tuning range for which they are configured and the acceptance incident angle [48-22] and [48-23].

As an example, the TeO_2 (Tellurium Dioxide) crystal as a typical material for non-collinear AOTFs, is considered. Using material specific dispersion formulas for room temperature, the ordinary and extraordinary refractive indices can be calculated [48-27]. Figure 48-40 shows the values of the refractive indices and the birefringence as well as the dispersion term $b(\lambda)$ of the birefringence at room temperature. The material coefficients are calculated over the optical transparency range from 0.34–4.5 μm. The transparency range is defined by its lower and upper limit each with an absorption coefficient of 1 cm^{-1} in this case. As shown in section 48.2.3 this means that, at the limits of the optical transparency range, the input irradiance is attenuated by the factor $\frac{1}{e}$ (about 0.37) over a length of $1cm$ according to (48-79).

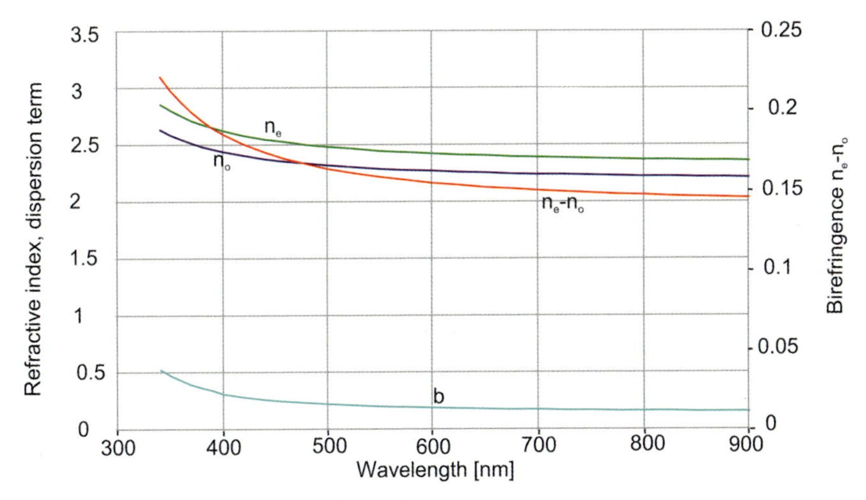

Figure 48-40: Ordinary and extraordinary refractive index, birefringence and dispersion term $b(\lambda)$of TeO_2 within the transparency range.

Figure 48-41 shows the spectral resolution for two different transducer widths and an optical wavelength of 500 nm. A typical angle $\theta_a = 80°$ is chosen for the acoustic wave vector. Note that the spectral resolution is increased by increasing the angle of incidence according to (48-148) and by increasing the transducer width, which increases the interaction length according to (48-150).

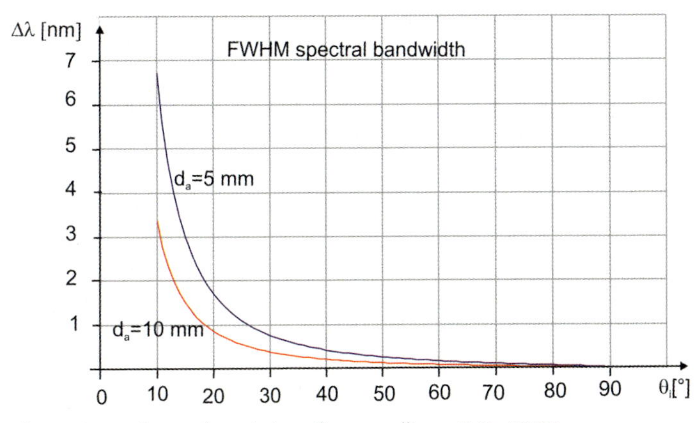

Figure 48-41: Spectral resolution of a non-collinear TeO_2 AOTF at an optical wavelength $\lambda_o = 500$ nm for an acoustic beam angle $\theta_a = 80°$ and two different transducer widths d_a dependent on the angle of incidence θ_i

Side Lobe Suppression and Out-of-band Rejection

In a similar way to absorption and interference filters, the ability of spectral discrimination by an AOTF is determined by the ratio of peak transmission and out-of-band rejection. Adjoining to the primary small band is of major importance for this filter parameter. In the case of uniform acoustic excitation the AOTF transmission has the $Sinc^2$ characteristic according to (48-145) with a ratio of only about 20 of the main lobe and nearest side lobe (13 dB suppression of the side lobe). Using amplitude apodization techniques of the acoustic beam, better suppression can be reached, e.g., with a non-collinear AOTF about 26 dB (ratio about 400). The apodization in the case of a non-collinear AOTF can be carried out by means of weighted acoustic excitation at a transducer array, as shown schematically in Figure 48-35 [48-24].

For collinear AOTFs apodization techniques with an acoustic pulse, which is apodized in time, have been developed.

Another factor which determines the out-of-band rejection is the contrast ratio; the fraction of undiffracted light. The contrast ratio is determined by the polarizer distinction in the case of a collinear AOTF, but can also be limited by residual strain in the filter material. Generally, a better contrast ratio is obtainable with a non collinear AOTF, because incident and diffracted light are spatially separated.

Transmission and Drive Power

The peak transmission in (48-147) is

$$\tau_0 = \sin^2\left(\frac{\pi^2}{2\lambda^2} M_2\, P_{ad}\, L^2\right)^{\frac{1}{2}} \tag{48-151}$$

where

M_2 is the material specific figure of merit

P_{ad} is the acoustic power density, which equals the RF power density,

L is the interaction length,

λ is the peak vacuum wavelength.

The figure of merit M_2 is defined according to(48-152):

$$M_2 = \frac{n^6\, p_{oe}^2}{\rho\, v_a^2} \tag{48-152}$$

where

n is therelevant refractive index

p_{oe} is the elasto-optic coefficient,

ρ is the material density,

v_a is the acoustic velocity.

According to (48-151) the maximal reachable peak transmission requires a certain value of the acoustic power density P_{ad}, which depends on the incident acoustic power P_a (= RF power) and the optical aperture A according to (48-153):

$$P_{ad} = \frac{P_a}{A} \tag{48-153}$$

The transmission function T_0 becomes maximal, if the relation $\left(\frac{\pi^2}{2\lambda^2} M_2\, P_{ad} L^2\right)^{\frac{1}{2}} = \frac{\pi}{2}$ is fulfilled. For given λ, M_2, L the necessary acoustic power density P_{ad} to reach the maximum peak transmission follows according to (48-154):

$$P_{ad} = \frac{\lambda^2}{2\, M_2\, L^2} \tag{48-154}$$

As an example, the crystal TeO_2 is considered again as appropriate material for a non-collinear AOTF. The typical design parameters $L = 2$ cm and $A = 5 \times 5$ mm^2 are assumed.

The figure of merit M_2 for TeO_2 is found in material tables to be about 10^{-12}m^2/W [48-27]. For a peak wavelength $\lambda = 500$ nm the necessary acoustic power density to reach the maximum transmission is calculated from (48-154) to be about 0.125 W/cm^2. The necessary RF power is about 0.03 W for the example's optical aperture of 25 mm^2.

Tuning Range

The tuning range of an AOTF depends on the pass band of the optical crystal, the material-dependent acoustic velocity and the acoustic drive power required to reach the wavelength-dependent maximum transmission as shown above. Commercial AOTFs are usually configured for specified spectral ranges, e.g., with crystal quartz for 250–650 nm or with TeO_2 for different ranges between 360 and 4500 nm [48-22].

Response Time

The time needed for switching from one wavelength to another, the response time, is limited by the acoustic transit times over the optical aperture. A few microseconds would be typical. These short response times allow much faster sequential spectrometer measurements than when using classical grating or prism spectrometers [48-25] and [48-26].

LCTF: Tunable Birefringent Filters with Liquid Crystal Layers as Tunable Components

Principal Function of an LCTF as a Special Type of Birefringent Filter

Liquid crystal tunable filters are birefringent filters. Primary birefringent filters had been developed as narrow band filters for single wavelengths for astronomical purposes (see for example [48-28]). A specific arrangement of birefringent plates and linear polarizers causes a transmission spectrum of a series of spaced spectral bands. Also, tunable birefringent filters on the basis of different phase-shifting technologies such as rotating half-wave plates and electro-optical components had been studied for many years [48-28].

A typical novel LCTF is based on a stack of fixed filters consisting of interwoven birefringent crystal and liquid-crystal combinations and linear polarizers. The function is explained as follows for the example of a Lyot–Ohman type of LCTF. This filter type consists of N birefringent crystal layers (usually uniaxial crystals, but configurations with biaxial crystals are also possible) with their optical axes in the layer plane perpendicular to the direction of incident light. Each birefringent element is arranged between two parallel or crossed polarizers. In figure 48-42 the principle is shown for the widespread configuration with parallel polarizers, but configurations with crossed polarizers are also possible (see [48-29]). The polarizer axes are oriented at 45° to one optic axis in the case of a uniaxial crystal and to two axes for a biaxial crystal. In this way the polarized light is split into two components, the extraordinary and the ordinary ray with perpendicular polarization planes traveling in parallel distribution directions but with their individual different velocities through the crystal layer. The optical path difference OPD of the two rays is given by the product of the layer thickness d and the difference of the refractive indices Δn of the two polarization directions. In the case of uniaxial crystals $\Delta n = n_e - n_o$ is the difference between the extraordinary and ordinary refractive index. After passing the second polarizer behind the birefringent layer the two components interfere with a phase difference (retardation) $\Delta\phi$ determined by the OPD and wavelength λ of the incident light according to (48-155):

$$\Delta\phi = \frac{2\pi}{\lambda} d\,\Delta n \tag{48-155}$$

If the phase difference is a multiple of 2π, light with the wavelength λ is again linearly polarized in the direction of the input and exit polarizers and in this way is transmitted with the lowest loss. The principle of the Lyot–Ohman filter is based on this relationship.

From (48-155), together with the full wave condition, the thickness d of the thinnest layer is found according to (48-156):

$$d = m\frac{\lambda}{\Delta n} = d_1 \tag{48-156}$$

where m is the order number and is a positive integer.

The thicknesses of the other layers of the arrangement increase in powers of 2 thus always delivering the full-wave condition. Equation (48-157) is the general relation for the thickness of an arbitrary layer:

$$d_j = 2^{j-1} d \tag{48-157}$$

with

$$j = 1, 2, 3 \ldots N.$$

The whole filter consists of N stages, every stage consisting of one birefringent layer, one input and one exit polarizer. The exit polarizer of one stage acts as the input polarizer of the following stage.

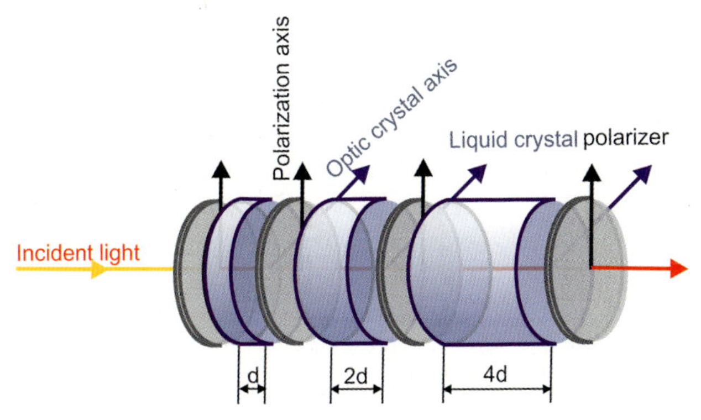

Figure 48-42: Configuration of a liquid crystal Lyot–Ohman filter of three stages.

The individual transmission function of the birefringent liquid crystal layer number j between two polarizers depends on the retardation $\Delta\phi$ according to (48-158):

$$\tau_j = \cos^2\left(2^{j-2}\Delta\phi\right). \tag{48-158}$$

In this way a cascaded relative transmission function for N stages is generated according to (48-159):

$$\tau = \prod_{j=1}^{N} \cos^2\left(2^{j-1}\Delta\phi\right) = \left(\frac{\sin\left(2^{N-1}\Delta\phi\right)}{2^N \sin\dfrac{\Delta\phi}{2}}\right) \tag{48-159}$$

where

N is the number of birefringent stages,
$\Delta\phi = \frac{2\pi}{\lambda} d\Delta n$ is the phase difference between the ordinary and extraordinary rays.

The transmission function does not include reflection and absorption losses due to interfaces and material of the birefringent elements and polarizers. The real world transmission is obtained by multiplication of (48-159) with all the individual transmission factors.

Due to this transmission characteristic, incident light can be filtered with a desired central wavelength by designing the suitable OPD of the birefringent layers. Using liquid crystal layers as birefringent layers allows one to tune the central wavelength continuously by electric voltage, because the difference $\Delta\phi$ changes when the birefringence is changed due the voltage tuning. Generally, the variation in the birefringence with the voltage is non-uniform over the layer thickness, and the overall retardation has to be calculated by integration along the light direction over the thickness (see [48-30]).

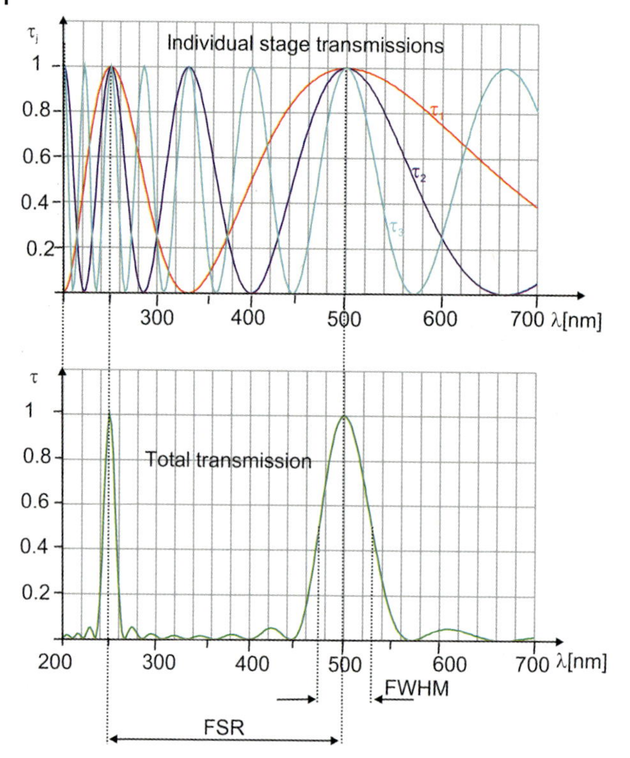

Figure 48-43: Theoretical transmission functions of a three-stage Lyot–Ohman filter for a peak wavelength of 500 nm.

As an example, the theoretical transmission functions of a three-stage filter for a central wavelength 500 nm is shown in figure 48-43. The individual transmission functions for the first, second and third stages between the polarizers are τ_1, τ_2, τ_3.

The transmittance characteristic according to (48-159) generates transmittance peaks at wavelengths according to (48-160) for incident white light, if the material is transparent for the spectral range.

$$\lambda_{\text{peak}} = \frac{d\Delta n}{m} \tag{48-160}$$

The *free spectral range* FSR between two adjacent peaks is determined by the thickness d of the thinnest birefringent layer according to (48-161):

$$\text{FSR} = \frac{\lambda_{\text{peak}}}{m+1} = \frac{d\Delta n}{m(m+1)}. \tag{48-161}$$

In figure 48-43 this parameter is shown as well as the *full width at half maximum* FWHM, which determines the *spectral resolution* according to (48-162) (see below).

Spectral Resolution

The spectral resolution of an LCTF is characterized by the FWHM depending on the peak wavelength, the order and the number of birefringent stages (see (48-162) and figure 48-43):

$$\mathrm{FWHM} = \frac{\lambda_{\mathrm{peak}}}{2^N m} = \frac{d\Delta n}{m^2 2^N}. \tag{48-162}$$

Improving the spectral resolution by enlarging the numbers of stages is limited because the transmission losses increase with the number of polarizers, and the tuning speed is reduced with the number of stages. Therefore, the possibility of increasing the spectral resolution by designing the filter for an order > 1 according to (48-162) can be advantageous, if the smaller FSR according to (48-161) does not matter for the application of interest. Commercially available filters reach FWHMs from 0.1 nm to some namometers dependent on spectral range [48-31].

Tuning Range

The tuning range of a LCTF is primarily limited by the free spectral range according to (48-161). By combination with additional blocking filters for unwanted orders the tuning range can be expanded to ranges larger than the FSR. For example, tuning ranges up to 800 nm with wavelength-dependent FWHMs of about 50–130 nm have been achieved [48-29].

Typical common commercially available configurations are configured for limited spectral ranges in the VIS, IR and NIR. The tuning ranges are about 250–300 nm with FWHMs down to 0.25 nm in the VIS range and can reach up to > 1200 nm in the NIR with FWHMs down to some nanometers (see [48-31] and [48-32]).

Response Time

The lower limit of the response time for switching from one wavelength to another is limited by the liquid crystal relaxation time depending on the liquid crystal type and ambient temperature. Commonly used nematic crystals have response times of the order of 100 ms, while ferroelectric liquid crystals can reach response times of < 20 µs [48-32]–[48-34].

Other LCTF Configurations

The retarders of the type shown schematically in figure 48-42 can be liquid crystal cells, e.g., of nematic type. In commercial instruments a combination of fixed retarders and liquid crystal cells has been successfully used ([48-34], [48-31]). For these configurations each stage of a Lyot filter consists of a fixed retarder of suitable thickness and an additional liquid crystal waveplate. The wavelength tuning is controlled by rotating the liquid crystal axis electronically, thus continuously changing the retardation.

Configurations with retarders between crossed polarizers work in a similar way to configurations of the above type, with retarders between parallel polarizers [48-29].

New developments concern larger tuning ranges by extension of the free spectral range [48-30].

These examples should be sufficient for the scope of this book.

LCPG: Liquid Crystal Polarization Gratings

The function of this quite new filter type is the same as described above for the usual LCTF: by using a stack of liquid crystal layers of suitable different thicknesses a desired peak wavelength according to the transmission function (48-159) and figure 48-43 is generated. By applying individually controlled voltages the birefringence of the layers is changed, thus changing the phase difference between ordinary and extraordinary rays and, in this way, tuning the transmission characteristics.

The difference between these and common LCTFs is the use of polarization gratings. Polarization gratings (abbreviation PG) are thin-film diffractive elements, which cause spatially varying periodic profiles of optical anisotropy. These profiles are generated by modulating the polarization state of the wavefront passing through the layers. Thus the incident radiation is diffracted, and the zero-order transmission can be controlled from about zero to about 100% independent of the input polarization state. Polarizers are not needed for a filter on this basis. The unwanted wavelengths are diffracted in an angle range around the axis and are blocked simply by an aperture stop, through which the zero order passes. This concept is shown schematically in figure 48-44 for a stack of three filters [48-35] and [48-36].

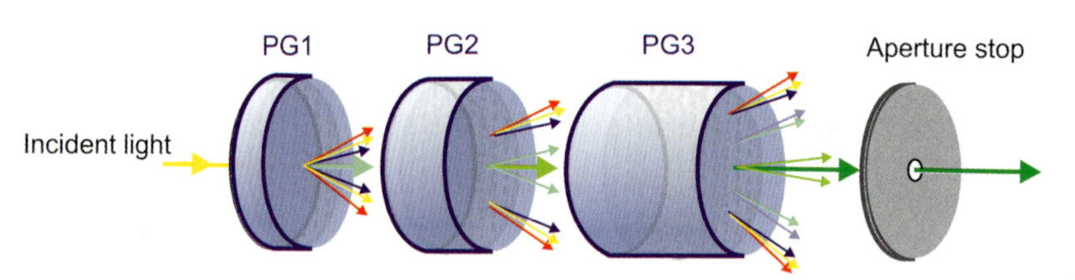

Figure 48-44: Schematic configuration of an LCPG with three PGs.

The transmission function is obtained by analogy with (48-159) using the individual diffraction efficiency $\eta_{0,j}$ instead of the individual transmission according to (48-158) of stage j. The diffraction efficiency and the complete relative transmission are described by relations (48-163) and (48-164) for the phase difference $\Delta\phi$ according to (48-155):

$$\eta_{0,j} = \cos^2(\Delta\phi/2) = \cos^2\left(\frac{\pi}{\lambda} d\,\Delta n\right) \tag{48-163}$$

$$\tau = \prod_{j=1}^{N} \eta_{0,j} = \prod_{j=1}^{N} \cos^2\left(\frac{\pi}{\lambda} d\,\Delta n\right) \tag{48-164}$$

Due to the use of diffraction and no need or polarizers between the retarder elements this filter type can achieve higher real transmission with a more compact configuration than traditional LCTF-types. Thus higher numbers of stages make sense, which makes better resolutions achievable. Figure 48-45 demonstrates this for three and five stages.

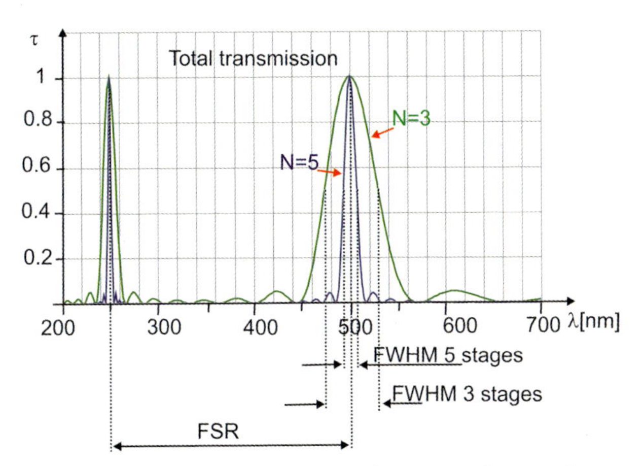

Figure 48-45: Theoretical transmission functions in the case of three and five stages for a 500 nm peak wavelength.

48.3.6
Prism Monochromators

Principle Function of a Prism Monochromator

A prism monochromator utilizes the wavelength dependent refraction of a prism. The principle configuration is shown in figure 48-46.

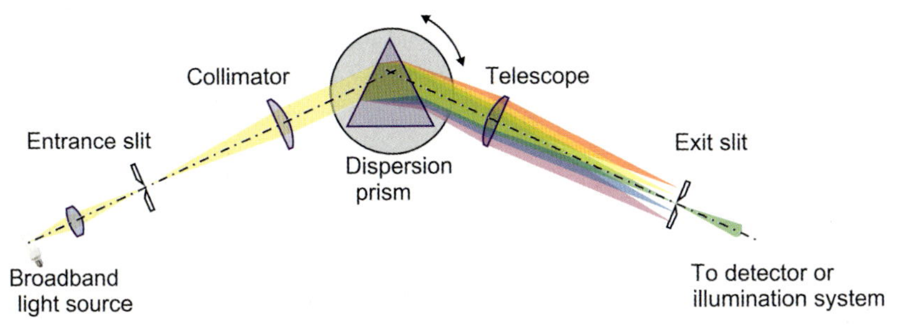

Figure 48-46: Principle of a prism monochromator.

The divergent radiance from the illuminated entrance slit is collimated by the collimator, which can be a lens system or mirror. The collimated beam enters the dispersion prism. At the prism's input face, and also at its output face, wavelength dependent refraction occurs and generates dispersing beams of different wavelengths which leave the prism with wavelength-dependent output angles. In the case of usual optical materials with normal dispersion the index of refraction increases with decreasing wavelength in the visual spectral range. Thus the deflection angle increases with decreasing wavelength. In the image plane of the following telescope system this wavelength-dependent angular distribution is transformed into a series of images of the entrance slit. The angular distribution of the beam fractions of different wavelengths is a nonlinear function of the wavelength. By rotating the prism around an axis perpendicular to the plane of incidence, flux quantities of different wavelengths sequentially pass the exit slit.

Deflection of a Monochromatic Beam by a Prism

The relations for the deflection of a single monochromatic beam by the prism refraction are shown in figure 48-47.

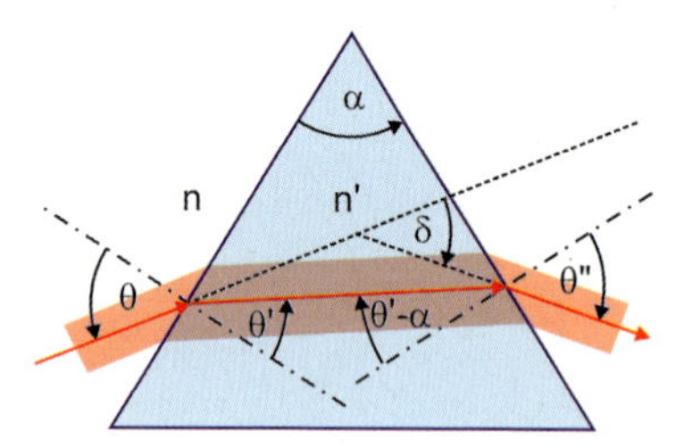

Figure 48-47: Deflection of a monochromatic beam deflected by a dispersion prism.

A monochromatic input beam impinges at an angle of incidence θ and is refracted at an angle θ' at the first interface of the dispersion prism with the prism angle a. The prism material has refractive index n' for the frequency of the input beam and is embedded in a medium with refractive index n. The output angle after refraction at the second prism interface is θ''. The direction of the output beam with regard to the input direction is described by the deflection angle δ according to (48-165). The sign of the angles is chosen according to figure 48-47 dependent on the direction of the axis of incidence, which has to be mentally rotated to obtain the same direction as the beam. In the example of figure 48-47, the sign of the deflection angle is negative.

$$\delta = \theta'' - \theta + a \tag{48-165}$$

When taking the absolute values of the angle, the relation $\delta = \theta'' + \theta - a$ will be valid.

By applying the law of refraction at both interfaces and using trigonometric transformation, the dependence of the deflection angle on the angle of incidence, the prism angle, the refractive index of the prism and that of the surrounding medium is found according to (48-166).

$$\delta = \arcsin\left(\sin a \sqrt{\left(\frac{n'}{n}\right)^2 - \sin^2\theta} - \sin\theta\cos a \right) - \theta + a \tag{48-166}$$

Due to the wavelength dependence of the prism's refractive index n' the deflection varies with the wavelength according to (48-164). This enables the monochromator function. The deflection of a monochromatic beam of light varies only with the angle of incidence. In a similar way, the relation for the refractive index of the prism can be described as a function of the angle of incidence and the deflection angle by (48-167) (see [48-1]).

$$n' = n\sqrt{\left(\frac{\sin\theta\cos\theta - \sin(\delta+\theta-a)}{\sin a}\right)^2 + \sin^2\theta} \tag{48-167}$$

Dispersion Prisms for Minimum Deflection

Usually the dispersion prisms of prism monochromators are used for *minimum deflection,* because this optical arrangement simplifies the technical configurations. The deflection angle according to (48-166) reaches a minimum absolute value for one specific angle of incidence. If the prism is used for this case of minimum deflection, (48-166) becomes the simplified equation (48-168).

$$\delta_{\min} = a - 2\theta \tag{48-168}$$

This can be found from (48-164) by considering the conditions for which the derivation $d\delta/d\theta = 0$ is valid: this is the case for a symmetric pass through the prism with $\theta'' = -\theta$ as shown in figure 48-48.

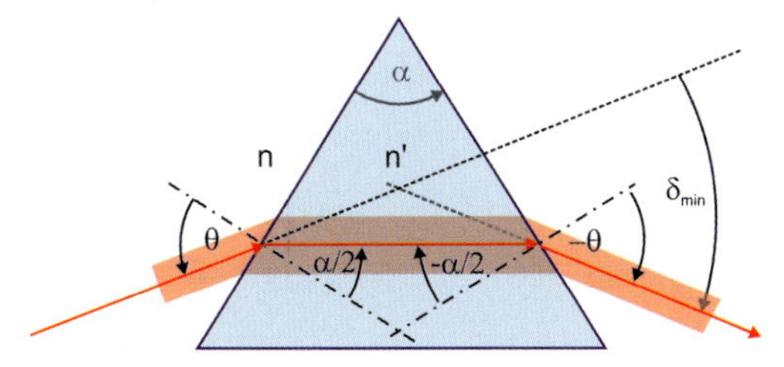

Figure 48-48: Dispersion prism used for minimum deflection.

For minimum deflection the refractive index n' of the prism can be calculated according to (48-169) from the prism angle a and the deflection angle $\delta_{\min}$.

$$n' = n\frac{\sin\dfrac{a-\delta_{\min}}{2}}{\sin\dfrac{a}{2}} \tag{48-169}$$

The angle of incidence for minimum deflection depends on the wavelength. By rotating the prism around an axis perpendicular to the plane of incidence the angle of incidence for minimum deflection is obtained sequentially for different wavelengths. The angle of deflection also varies with the wavelength. To obtain the sequential spectrum, the input collimator and slit and also the output collimator and slit would have to be rotated.

Relations (48-167) and (48-169), respectively, can be used to measure the refractive index of optical glass precisely: the prism angle, the angle of incidence and the deflection angle have to be measured using monochromatic light of known wavelength (see section 53.2).

Dispersion Prisms for Constant Deflection

Using a dispersion prism of constant deflection is a useful technical concept for prism monochromator setups. A prism of constant deflection delivers a constant angle of deflection independent of the wavelength. The prism only has to be rotated around an axis perpendicular to the plane of incidence to obtain the sequential spectrum. The entrance slit and the collimator lens and also the output slit and the output telescope lens are fixed.

To obtain a prism with constant deflection, the prism is used symmetrically for minimum deflection, and the input beam undergoing the refraction at the two-prism plane interfaces has also to be reflected once at a plane mirror perpendicular to the plane of incidence. Figure 48-49 shows this principle for a reflection plane outside the prism. The whole arrangement of prism and mirror has to be moved together around an axis perpendicular to the plane of incidence in order to obtain the spectral components for constant deflection.

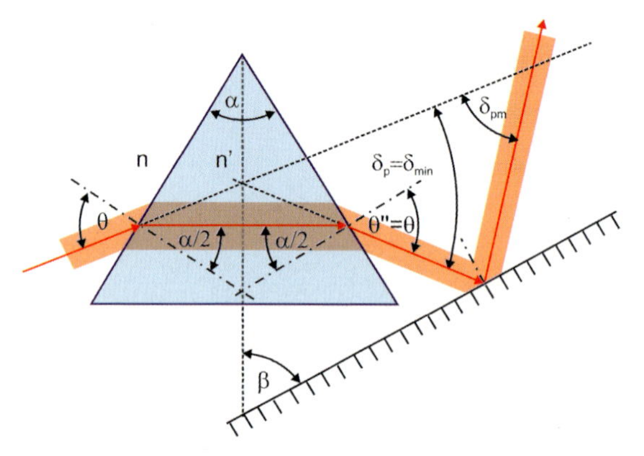

Figure 48-49: Constant angle of deflection for a prism with additional outside plane mirror.

The absolute (here unsigned) value of the deflection angle is

$$\delta_{pm} = 180^\circ - 2\beta \tag{48-170}$$

where β is the absolute value of the angle between the symmetry axis of the dispersion prism and the reflection plane. At an angle $\beta = 90^\circ$ one gets the special case for no deflection, which is known as the Wadsworth arrangement (see [48-37]).

The common version of a dispersion prism with constant deflection, as used in spectrometer instruments, is a prism with one internal reflection plane as part of the prism. This type of prism can be thought of as being composed of two separated prisms with the prism angle $\alpha/2$ (half of the prism angle of the comparable symmetric dispersion prism) used for minimum deflection. A quasi-lossless reflection can be achieved by choosing suitable material and prism angle in order to obtain total reflection. Ernst Abbe has already described this dispersion prism type for a spectral device as a microscope component [48-38], [48-1]. In figure 48-50 an Abbé prism for a deflection angle of 90° for $\beta = 45°$ is shown. This special prism type is also known as a *Pellin–Broca prism* [48-1].

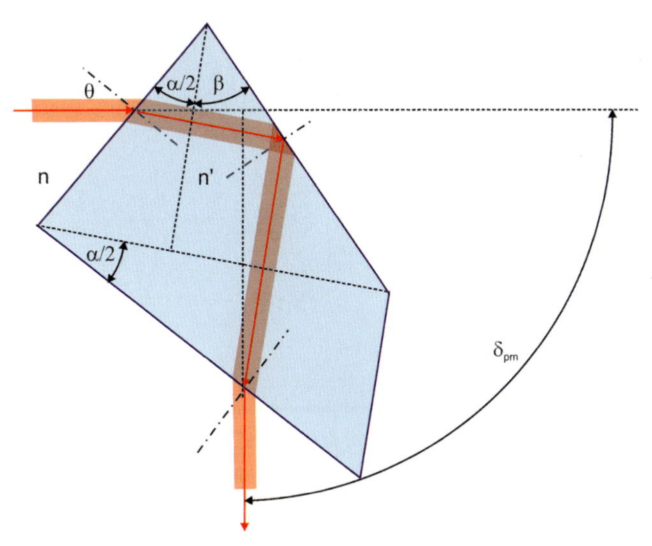

Figure 48-50: Abbe prism with a constant deflection angle of 90°.

For goniometric spectrometric measuring devices, e.g. for refractive index measurement, the *auto collimation prism* with a deflection angle of 180° is useful (see section 53.2).

A constant angle of deflection can also be achieved using two identical prisms without an additional reflection. The prisms have to be rotated in opposite directions around a common axis. For this arrangement the input collimator lens and the output (telescope) lens have each to be rotated with the dedicated prism. This principle is shown schematically in figure 48-51.

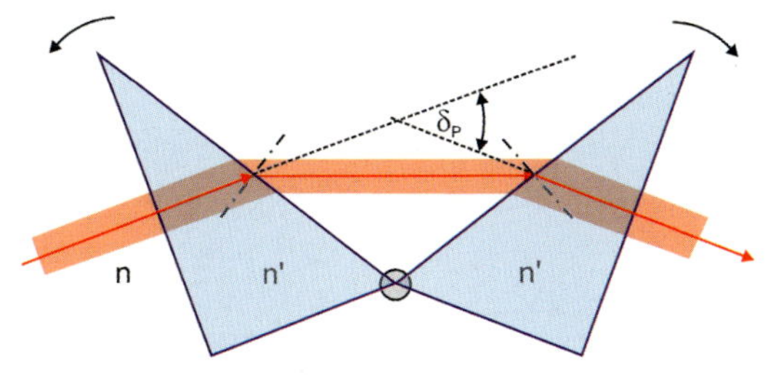

Figure 48-51: Constant deflection achieved using two prisms without reflection.

Figure 48-52 shows an example of the combination of an Abbé prism with a constant deflection angle of 60° with an autocollimation prism.

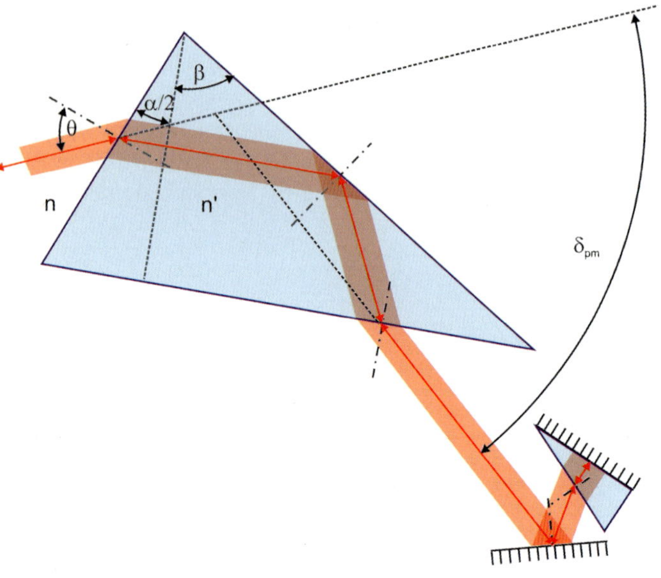

Figure 48-52: Combination of Abbe prism with 60° deflection angle and autocollimation prism.

To obtain the constant angle of deflection for the sequential spectrum, the prisms in this combination have to be rotated around a common axis perpendicular to the plane of incidence, while the mirror is fixed.

Angular Dispersion and Spectral Resolution of Prism Monochromators

The main functional parameter of a monochromator of arbitrary type is the spectral resolution $\Delta\lambda$; the smallest spectral distance for which two wavelengths can be distinguished [48-12]. The reciprocal of its normalized value is the usual parameter resolving power R according to (48-171) for determining the spectral resolution of a monochromator.

$$R = \frac{\lambda}{\Delta\lambda} \tag{48-171}$$

The resolvable wavelength difference $\Delta\lambda$ is determined by the angular dispersion of the prism. In the case of a prism used in minimum deflection, as described above, the angular dispersion is calculated as follows.

Considering a small wavelength interval, the rate of change of the refractive angle with the wavelength, the angular dispersion at one face $d\theta/d\lambda_{1face}$, is described according to (48-172):

$$\frac{d\theta}{d\lambda_{1face}} = \frac{d\theta}{dn'}\frac{dn'}{d\lambda} \qquad (48\text{-}172)$$

Since the refractive index is a function of wavelength, the angular dispersion $d\theta/d\lambda$ is separated into two contributions: the rate of change of the refractive angle with refractive index $d\theta/dn'$, which only depends on prism geometry; and the dispersion $dn'/d\lambda$, a characteristic of the prism material.

From (48-172) the *overall angular dispersion* at both prism faces is found according to (48-173):

$$\frac{d\theta}{d\lambda} = 2\frac{d\theta}{d\lambda_{1face}} = 2\frac{d\theta}{dn'}\frac{dn'}{d\lambda} \qquad (48\text{-}173)$$

Applying the law of refraction to one prism face one obtains relation (48-174) between the angle of incidence θ or the exit angle $\theta'' = -\theta$ respectively, the prism angle α, the refractive index n' of the prism material and the refractive index n of the medium in which the prism is embedded:

$$n' \sin\frac{\alpha}{2} = n \sin\theta \qquad (48\text{-}174)$$

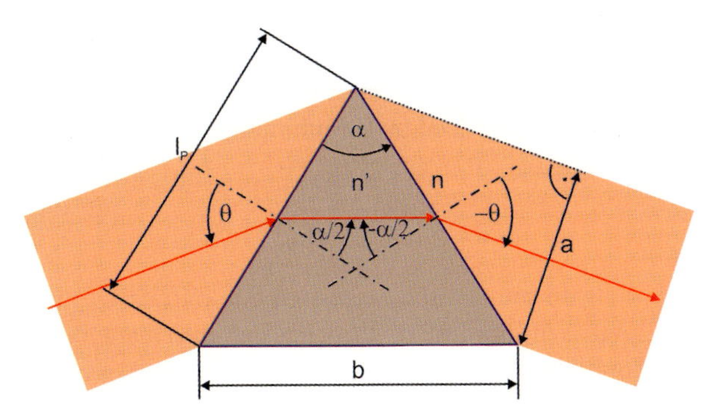

Figure 48-53: The prism dimensions determine the exit beam width.

From (48-174) the angle θ is obtained as a function of α, n', n, and one finds, for both prism faces, the relation (48-175):

$$\frac{d\theta}{dn'} = \frac{1}{n}\frac{\sin\frac{\alpha}{2}}{\cos\theta} \qquad (48\text{-}175)$$

Assuming the prism faces are the limiting apertures for the entrance and exit beam as shown in figure 48-53, relation (48-175) can be expressed in terms of the prism geometry according to (48-176):

$$\frac{d\theta}{dn'} = \frac{1}{2}\frac{b}{n}\frac{b}{a} \tag{48-176}$$

Since (48-176) and (48-173) give the angular dispersion at both prism faces, then the prism's *angular dispersion $d\theta/d\lambda$, the rate of change of the refractive angle with the wavelength, is found to fulfill relation (48-177):*

$$\frac{d\theta}{d\lambda} = 2\frac{d\theta}{dn'}\frac{dn'}{d\lambda} = \frac{1}{n}\frac{b}{a}\frac{dn'}{d\lambda} \tag{48-177}$$

where

n is the refraction index of the medium in which the prism is embedded,
b is the base length of the prism,
a is the beam width in the plane of incidence perpendicular to the exit beam,
$\frac{dn'}{d\lambda}$ is the material dispersion of the prism with refraction index n'.

Relation (48-177) is valid for a prism with refractive index n' embedded in a medium with refractive index n. According to (48-177) the angular resolution depends on the prism material dispersion $dn'/d\lambda$ and also the geometric prism parameters base length b and maximum beam width a in the plane of incidence, limited by the length l_P of the prism faces (figure 48-53).

Theoretical Maximum Resolving Power

The theoretically minimum resolvable angle $d\theta$ between two beams is limited by diffraction at the prism exit pupil. It can be described by Fraunhofer diffraction as

$$d\theta_0 = \frac{\lambda}{a} \tag{48-178}$$

where $d\theta_0$ is the infinitesimal angle distance in the case when one diffraction maximum falls into the first diffraction minimum of the adjacent diffracted beam [48-12]. With (48-178), definition (48-171) of the resolving power and relation (48-177) for the angular dispersion, the theoretical (diffraction-limited) minimum resolvable wavelength interval $\Delta\lambda_0$ and the maximum resolving power can be expressed according to (48-179) and (48-180) as:

$$\Delta\lambda_0 \approx d\lambda_0 = d\theta_0 \frac{1}{\frac{d\theta}{d\lambda}} = \frac{n}{b}\frac{\lambda}{\frac{dn'}{d\lambda}} = \frac{\lambda}{R_0} \tag{48-179}$$

$$R_0 = \frac{\lambda}{\Delta\lambda_0} = \frac{1}{n}b\frac{dn'}{d\lambda} \tag{48-180}$$

In the case of a prism in air $R_0 = \frac{\lambda}{\Delta\lambda_0} = b\frac{dn'}{d\lambda}$ is valid.

The resolving power R_0 is the theoretical maximum limit in the case of infinitesimal slit widths, whereas the actual resolving power can only yield values

$0 < R < R_0$. Relations (48-179) and (48-180) show that the resolution depends on the prism's material dispersion $dn'/d\lambda$ and the prism base length b.

Figure 48-54 shows the dispersion versus wavelength for some optical glass types. For heavy flint glass SF, e.g., the dispersion for 500 nm is found to be approximately 10^{-4}/nm. A prism in air made of this glass with a base length of 100 mm has a maximum diffraction limited resolving power $\lambda/d\lambda$ of approximately 10 000 at wavelength 500 nm according to the calculation from (48-180). The minimum resolvable wavelength difference $\Delta\lambda_0$ is about 0.05 nm.

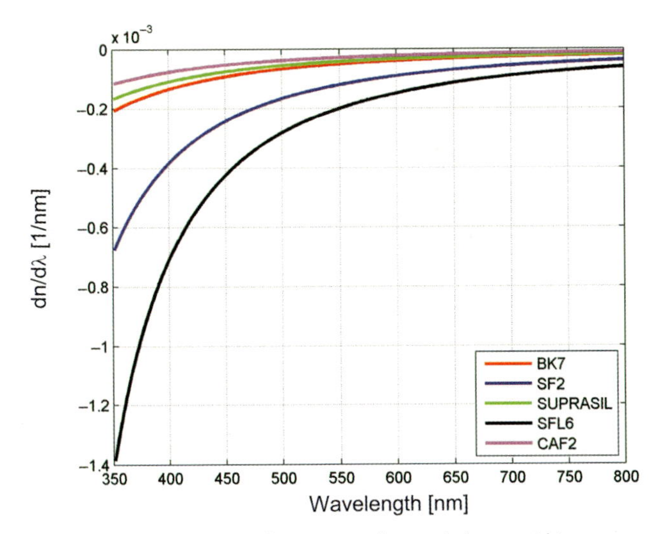

Figure 48-54: Dispersion of some typical optical glasses within the visual spectral range.

The real resolving power $R = \lambda/\Delta\lambda$ is usually much lower. It is characterized by the real bandwidth $\Delta\lambda$, which is mainly determined by the slit width but also by such influences as optical aberrations and stray light.

Linear Dispersion

The prism monochromator's dispersion can also be expressed as the linear dispersion $(d\lambda/dx)$ in the focal plane of the monochromator telescope in the direction x perpendicular to the slit length. It defines the linear extension of a spectral interval in the focal plane of the telescope as shown in figure 48-55. An angle interval of the exit beam can be expressed as $d\theta = dx/f_T$ where f_T is the focal length of the telescope. Using this and relation (48-177) for the angle dispersion the linear dispersion can be described according to (48-181):

$$\frac{d\lambda}{dx} = \frac{1}{\left(\frac{d\theta}{d\lambda}\right)} \frac{1}{f_T} = n \frac{a}{b} \frac{1}{f_T} \frac{1}{\frac{dn'}{d\lambda}} \qquad (48\text{-}181)$$

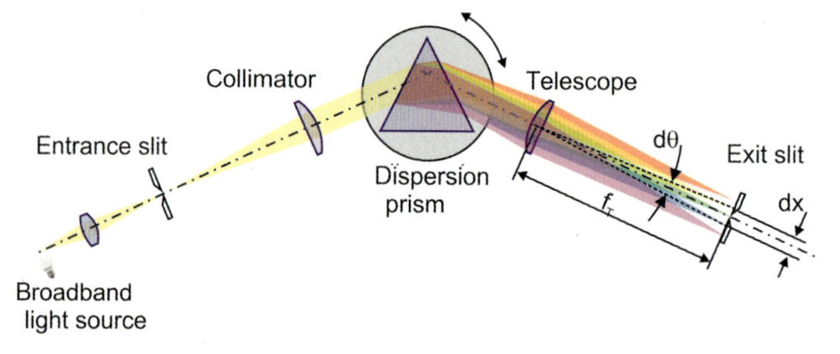

Figure 48-55: Linear dispersion in the focal plane of a prism monochromator.

Slit Width and Spectral Bandwidth

Finite widths of the entrance and exit slit are needed to get measurable quantities of radiant flux. The slit widths influence the transmission of the monochromator and also the effective bandwidth and the real resolution. As shown schematically in figure 48-55, the slit images for different wavelengths are distributed in the focal plane of the monochromator telescope. The distance between two slit images produced by the difference in the dispersion angle between the beams of two wavelengths separated by $d\lambda$ is approximately

$$\Delta x \approx dx = d\theta f_T \tag{48-182}$$

where f_T is the focal length of the monochromator telescope. The minimum useful exit slit width a_{S0} is the distance between two just-resolvable slit images in the focal plane of the telescope for $d\theta = d\theta_0$, the diffraction-limited angle interval according to (48-178). Thus from (48-182) the minimum useful slit width is found for $dx = a_{S0}$ and $d\theta = d\theta_0 = \frac{\lambda}{a}$ according (48-183):

$$a_{S0} \approx \lambda \frac{f_T}{a} = \lambda \ F/\# \tag{48-183}$$

where $F/\#$ is the telescope's f-number.

For a given slit width a_S and with the real angular dispersion $\frac{d\theta}{d\lambda}$ of the prism monochromator, the effective bandwidth $\Delta\lambda_S$ after the exit slit is

$$\Delta\lambda_S \approx d\lambda_S = \frac{a_S}{f_T} \frac{1}{\left(\frac{d\theta}{d\lambda}\right)} = \frac{a_S}{f_T}\left(\frac{d\lambda}{d\theta}\right) \tag{48-184}$$

In the case of a given linear dispersion $(d\lambda/dx)$ analogous to (48-182) the resulting bandwidth is calculated according to (48-185):

$$\Delta\lambda_S \approx d\lambda_S = a_S\left(\frac{d\lambda}{dx}\right) \tag{48-185}$$

Throughput of a Prism Monochromator

For practical applications involving a spectrometric measurement, one important parameter is the useable radiant flux at the exit slit for a given input radiance. According to figure 48-56, the spectral throughput is the spectral radiant flux $\Phi_{\lambda E}$ after the exit slit for a given radiance L_λ of the entrance slit. The spectral radiant flux at the output can be calculated approximately using the fundamental law of radiation (48-42). According to this general relation, the infinitesimal spectral radiant flux from a surface element dA_1 with spectral radiance L_λ at another surface element dA_2 located at a distance r from dA_1 is given by (48-186), if transmission losses are neglected.

$$d^2\Phi_\lambda = L_\lambda \frac{dA_1 dA_2 \cos\theta_1 \cos\theta_2}{r^2} \tag{48-186}$$

As shown in figure 48-11, θ_1 and θ_2 are the angles between the connecting line of the two surface elements and their respective surface normal. The total spectral radiant flux between the two surfaces A_1 and A_2 is calculated by integrating (48-186).

For the approximate case of constant and uniform spectral radiance L_λ and separable integrals over the surfaces A_1 and A_2 the spectral radiant flux emitted from surface A_1 with spectral radiance L_λ and received by surface A_2 is

$$\Phi_{\lambda_{12}} = L_\lambda \int_{A_1} dA_1 \int_{A_2} \frac{\cos\theta_1 \cos\theta_2}{r^2} dA_2 \tag{48-187}$$

By defining the projected solid angle Ω_P subtended by surface A_2 as

$$\Omega_P = \int_{A_2} \cos\theta_1 \frac{dA_2 \cos\theta_2}{r^2} = \int_{A_2} \cos\theta_1 \, d\Omega_2 \tag{48-188}$$

the approximation formula for the total spectral radiant flux according to (48-187) becomes

$$\Phi_{\lambda_{12}} = L_\lambda A \, \Omega_P \tag{48-189}$$

where

L_λ is the spectral radiance from a surface A,

Ω_P is the projected solid angle subtended by a second surface.

The approximation (48-189) can be applied in both directions.

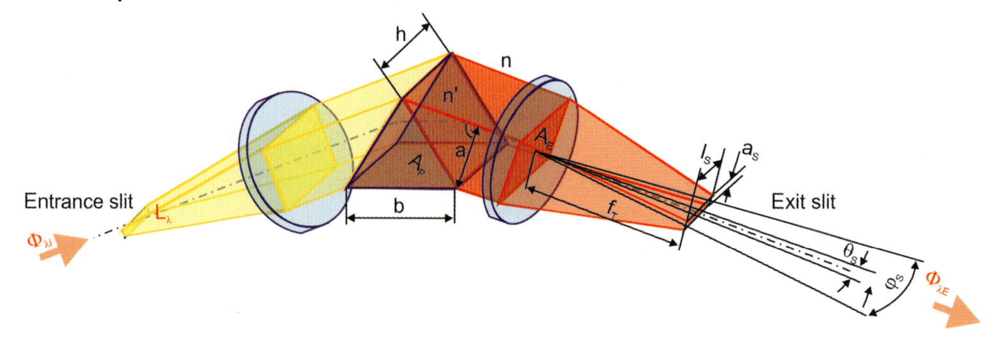

Figure 48-56: Radiant flux through a prism monochromator.

According to figure 48-56, the input and the output spectral radiant flux of a prism monochromator can be calculated approximately in this way when the diffraction at the exit slit is negligible. Applying relation (48-189) the spectral radiant flux Φ_{λ_E} through the exit slit with slit length l_S and the slit width a_S is

$$\Phi_{\lambda_E} = L_\lambda \, A \, \Omega_P = L_\lambda \, \tau(\lambda) \, A_E \frac{a_S \cdot l_S}{f_T^2} \tag{48-190}$$

where L_{λ_i} is the spectral radiance of the entrance slit, A_E the prism's exit aperture (when the telescope lens does not limit the exit beam then A_E would be the free collimator lens area), f_T is the focal length of the telescope. The term $\tau(\lambda)$ is the overall transmission from the entrance slit to the exit slit.

An approximated correlation between the monochromator's resolving power and the throughput is found from (48-190) by using the horizontal and vertical exit slit angles φ_S and θ_S seen from a point on the prism's exit surface A_E. The spectral radiant flux using these angles is

$$\Phi_{\lambda_E} = L_\lambda \, \tau(\lambda) \, A_E \, \varphi_S \, \theta_S \tag{48-191}$$

The small vertical angle $\theta_S = a_S/f_T$ and the angular dispersion $(d\theta/d\lambda)$ determine the resolvable spectral interval $\Delta\lambda_S$ according to (48-184). Thus the small angle θ_S can be expressed as $\theta_S = \Delta\lambda_S \, d\theta/d\lambda$. Applying this to (48-189) and the definition of resolving power R according to (48-171) the approximated calculation of the spectral radiant flux delivers (48-192):

$$\Phi_{\lambda_E} = L_\lambda \, \tau(\lambda) A_E \, \varphi_S \frac{\lambda}{R} \left(\frac{d\theta}{d\lambda} \right) \tag{48-192}$$

The radiant flux within a small spectral interval $\Delta\lambda_S$, for which Φ_λ can be considered as constant is found by multiplying (48-192) by $\Delta\lambda_S$ and using the resolving power definition (48-171) once more:

$$\Phi_E = \Phi_{\lambda_E} \, \Delta\lambda_S = L_\lambda \, \tau(\lambda) A_E \, \varphi_S \left(\frac{\lambda}{R} \right)^2 \left(\frac{d\theta}{d\lambda} \right) \tag{48-193}$$

The quantities in (48-192) and (48-193) are:

L_λ the spectral radiance of the entrance slit,
$\tau(\lambda)$ the transmission of the optical path from the entrance slit to the exit slit,
$\varphi_S = \frac{l_S}{f_T}$ the slit angle in the plane of the slit length direction and the optical axis
l_S the exit slit length,
a_S the width of the exit slit,
f_T the focal length of the telescope,
A_E the exit aperture area of the monochromator, perpendicular to the telescope axis,
λ the wavelength considered,
$R = \frac{\lambda}{\Delta\lambda}$ the resolving power of the monochromator,
$\left(\frac{d\theta}{d\lambda}\right)$ the angular dispersion of the monochromator.

The equations (48-192) and (48-193) for the throughput of the spectral radiant flux and the throughput of the radiant flux within a spectral interval $\Delta\lambda$, respectively, are based on the geometrical optical approximation, for which R is much smaller than R_0. Otherwise a diffraction correction factor must be applied (see [48-12]).

This approximated throughput calculation is valid for a prism monochromator and also for a grating monochromator (see also section 48.3.7). For a prism monochromator the throughput relations can be expressed by specific instrument parameters as follows.

Applying the general relation (48-177) for the prism's angular dispersion together with the expressions $A_E = a\,h$ and $A_P = b\,h$ for the exit aperture perpendicular to the telescope axis and the prism base area, respectively, (figure 48-56), the spectral radiant flux and the radiant flux within the spectral interval $\Delta\lambda_S$ through the exit slit are expressed by prism parameters according (48-194) and (48-195):

$$\Phi_{\lambda_E} = L_\lambda\,\tau(\lambda) A_P \frac{l_S}{f_T}\frac{\lambda}{R}\left(-\frac{1}{n}\frac{dn'}{d\lambda}\right) \tag{48-194}$$

$$\Phi_E = L_\lambda\,\tau(\lambda) A_P \frac{l_S}{f_T}\left(\frac{\lambda}{R}\right)^2\left(-\frac{1}{n}\frac{dn'}{d\lambda}\right) \tag{48-195}$$

where

n is the refractive index of the medium in which the prism is embedded ($n \approx 1$ for air)
n' is the refractive index of the prism material,
A_P is the prism base area.

48.3.7 Grating Monochromators

Principle Function of a Grating Monochromator

The dispersive element of a grating monochromator is a diffraction grating. Dispersion is caused by diffraction at the grating structure. In an analogous way to a prism monochromator, a collimated beam illuminates the grating, and the illuminated entrance slit in the focal plane of the collimator is imaged into the focal plane of the telescope. The grating, in principle, can be a transmission or reflection grating, but usual commercially available monochromators use reflection-grating configurations, which allow compact technical solutions. As shown in figure 48-57, typical configurations use mirrors as the collimator and telescope. To obtain appropriate technical configurations, the entrance and exit slits are fixed, which allows for the collimator and telescope also to be fixed. Only the grating rotates around an axis through the center of its face and perpendicular to the plane of incidence thereby generating different input and output angles. In the case of a light source with continuous spectrum every output angle represents another wavelength. The radiation leaving the output slit has a small spectral bandwidth depending on the monochromator's spectral resolution (see below).

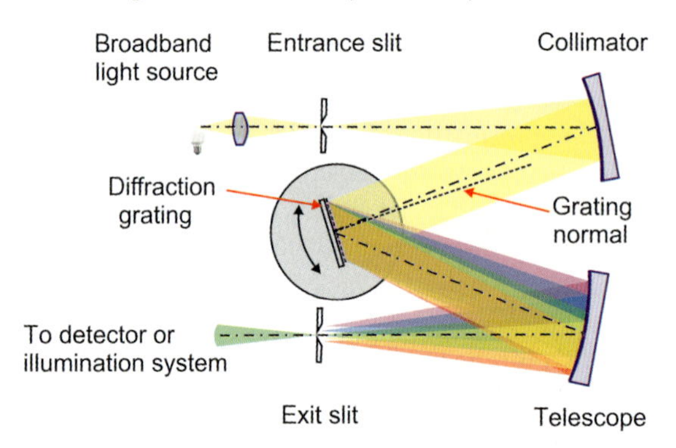

Figure 48-57: The principle of a grating monochromator using a plane diffraction grating.

Diffraction at Gratings

Modern gratings are manufactured holographically or by mechanical ruling. For the classical ruling method, equally spaced grooves are generated mechanically on mainly plane and concave surfaces. They are manufactured on ruling machines with a diamond stylus.

Modern holographic methods use equally spaced grooves as well as grooves of unequal distribution, which can be generated on many different surfaces such as plane, spherical or toroidal. Computer-generated holographic structures offer the possibility of optimizing the optical system performance by generating special wavefront forms, e.g., in order to correct aberrations of the optical images generated by the monochromator mirrors (see also below).

Independent of special groove forms some general relations are valid for wavelength dependent deflection caused by diffraction at a reflection diffraction grating. The fundamental one is the grating equation (48-196):

$$\sin\theta_m + \sin\theta_i = \frac{m\,\lambda}{g} \quad (48\text{-}196)$$

In (48-196) θ_i is the angle of incidence of an incident ray, θ_m the angle of the diffraction order m, λ the vacuum wavelength of the incident ray, g the grating constant, i.e., the distance between two adjacent grooves. The angles are measured in relation to the normal of the grating face (see figure 48-58). The grating constant g is often expressed by the groove density $1/g$ (see [48-39]). As shown schematically in figure 48-58, for a monochromatic ray, the angle θ_0 of the 0th order is the angle of reflection.

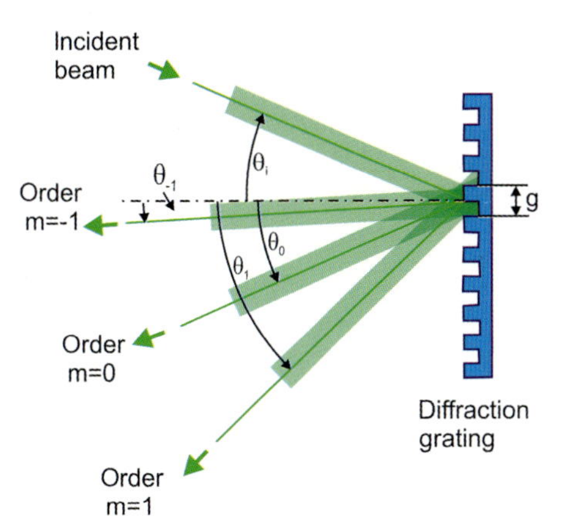

Figure 48-58: Grating dispersion caused by diffraction.

Equation (48-196) can be found by applying the diffraction theory at multiple slits (see [48-1] and [48-2]). A simple phenomenological approach is to consider plane waves which are diffracted at adjacent grating grooves as shown schematically in figure 48-59, the diffracted parts of the incident beam interfere with each other thus modulate the exit flux. If the optical path difference of the diffraction directions considered is an integer multiple of the wavelength, then maximum spectral flux is generated along this direction thus forming one diffraction order.

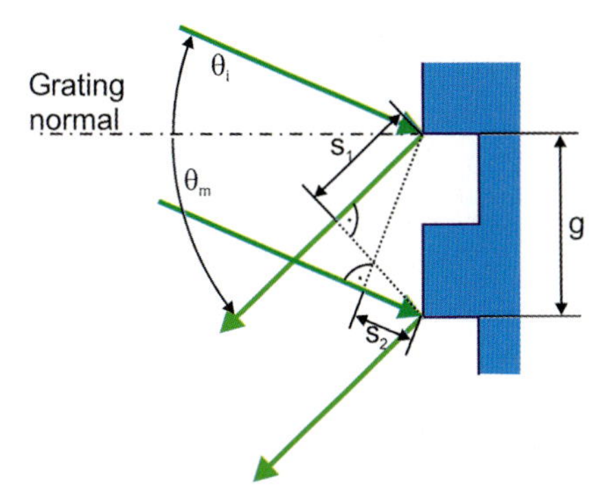

Figure 48-59: Optical path difference of elementary plane waves diffracted at adjacent grating grooves.

The optical path difference $s_1 - s_2$ between two waves, which are diffracted from adjacent grating grooves in the same direction, of diffraction order m, is $g(\sin\theta_m - \sin\theta_i)$ for the absolute values of the angle of incidence θ_i and a diffractive angle θ_m of diffraction order m. Thus one obtains the grating equation (48-196).

Common monochromator configurations with fixed entrance and output slits, according to figure 48-57, have a given fixed deflection angle $\delta = \theta_m - \theta_i$. The angle of incidence θ_i and the diffractive angle θ_m for the diffraction order m are determined by the vacuum wavelength λ of the incident beam and the grating constant g. The angle relations are shown in figure 48-60.

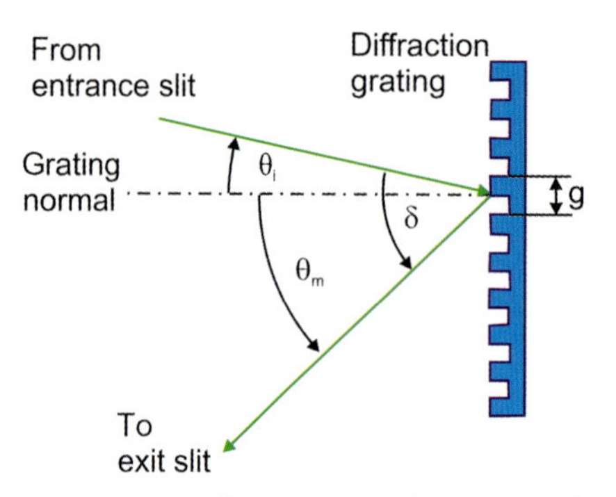

Figure 48-60: Schematic monochromator configuration with fixed entrance and exit slits and a given constant deflection angle δ.

Assuming that

$$\delta = \theta_m - \theta_i = \text{const.} \tag{48-197}$$

the angles θ_m and θ_i can be calculated for a particular diffraction order from the fundamental grating equation (48-196). By applying the trigonometric addition rule (48-196) is transformed into (48-198):

$$2\sin\frac{\theta_m + \theta_i}{2}\cos\frac{\theta_m - \theta_i}{2} = \frac{m\lambda}{g} \tag{48-198}$$

Applying (48-197) and (48-198) θ_i and θ_m are calculated as follows:

$$\theta_m = \arcsin\left(\frac{1}{2\cos\frac{\delta}{2}}\frac{m\lambda}{g}\right) + \frac{\delta}{2} \tag{48-199}$$

$$\theta_i = \arcsin\left(\frac{1}{2\cos\frac{\delta}{2}}\frac{m\lambda}{g}\right) - \frac{\delta}{2} \tag{48-200}$$

where

θ_m is the angle of diffraction with diffraction order m,
θ_i is the angle of incidence,
δ is the angle of deflection,
g is the grating constant.

As an example table 48-4 shows the variation in the angle of incidence and the associated angle of diffraction θ_1 with the deflection angle δ for a typical grating with 1200 grooves per mm set to diffract at a wavelength of 600 nm.

Table 48-4: The angle of incidence and angle of diffraction for first order, dependent on the angle of deflection for a wavelength of 600 nm and a grating with 1200 lines/mm.

Deflection angle δ [°]	Angle of diffraction θ_1 [°] 1st	angle of incidence θ_i [°]
0	21.100	21.100
5	23.621	18.621
10	26.185	16.185
15	28.791	13.791
20	31.442	11.442
25	34.138	9.138
30	36.882	6.882
35	39.677	4.677
40	42.526	2.526
45	45.433	0.433
50	48.404	–1.596

Free Spectral Range

In contrast to prisms, diffraction gratings generate several spectra according to the grating equation (48-196). Common diffraction angles for wavelengths of different diffraction orders can occur. Figure 48-61 demonstrates this for an example grating of table 48-4 with a fixed deflection angle δ of 30°:

At 600 nm the first diffraction order has a diffraction angle $\theta_1 = 36.882°$. Calculation using (48-197) shows that second and third orders with $\lambda = 300$ nm and $\lambda = 200$ nm respectively, are diffracted by the same angle. The diffraction angle $\theta_1 = 58.101°$ for the first diffraction order at 1100 nm occurs also for the second ($\lambda = 550$ nm), third ($\lambda = 366.667$ nm) and fourth ($\lambda = 275$ nm) orders, if the spectrum of the input radiation is broad enough. The overlapping of the orders causes an ambiguity in the correlation between the diffraction angle and wavelength.

An unambiguous correlation, as needed for spectral measurements, is given within the *dispersion range or free spectral range*. From (48-196) it is found that one wavelength λ_2 of order m has the same angle of diffraction as the wavelength λ_1 of order $m+1$ according to (48-201):

$$\frac{m\,\lambda_2}{g} = \frac{(m+1)\,\lambda_1}{g} \tag{48-201}$$

The free spectral range is given by (48-202):

$$\Delta\lambda_{\text{free}} = \lambda_2 - \lambda_1 = \frac{\lambda_1}{m} \tag{48-202}$$

where λ_2 and λ_1 are the upper and lower borders.

For the example, the free spectral range at $\lambda_2 = 600$ nm at the first diffraction order is 300 nm with the lower border $\lambda_1 = \frac{m}{m+1}\lambda_2 = 300\,\text{nm}$ according to (48-202). To measure the undisturbed spectrum within the free spectral range, the radiation for $\lambda \leq 300\,\text{nm}$ has to be blocked. For spectral measurements between 550 nm and 1100 nm the radiation for wavelengths $\lambda \leq 550\,\text{nm}$ has to be blocked.

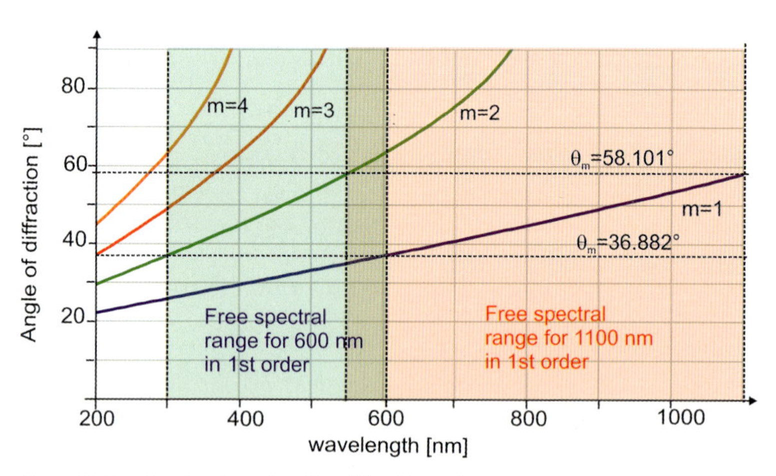

Figure 48-61: Overlapping of grating diffraction orders.

Blazed Diffraction Gratings

Blazed diffraction gratings have special saw tooth profiles enabling them to generate a concentration of nearly 100% of diffracted radiant flux into any one diffraction order other than the zeroth. This is a benefit for technical configurations, because in the zeroth diffraction order a spectral separation is not possible and therefore this part of the reflected flux is generally lost with the monochromator.

The blaze principle is shown in figure 48-62. At the face of a single saw tooth, the specular reflection generates radiance, which is reflected with angle θ_r determined by the blaze angle α. A coincidence of the specular reflection direction with a higher diffraction order direction is then generated. The preferred reflection direction can be controlled independently of the diffraction angle, which is determined by the grating constant. Theoretically, linearly blazed gratings can deflect 100% into the -first diffraction order for one wavelength and no flux in the zeroth diffraction order.

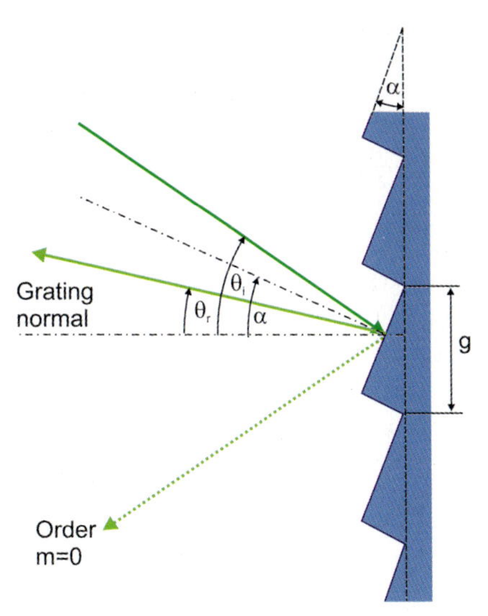

Figure 48-62: Preferred specular reflection at a single groove of a diffraction grating of the phase reflection type.

Normally, the triangular blaze profiles have an apex angle of 90°. However, higher angles of up to 110° are produced in order to avoid shadowing effects in the case of gratings used for higher orders, especially for holographic gratings [48-39].

For an incident beam parallel to the normal of the grating face, the angle of incidence is zero, and the angle of specular reflection is $\theta_r = -2\,a$. According to the grating equation (48-196) a coincidence with a higher diffraction order for a particular wavelength can be achieved when the condition (48-203) is fulfilled:

$$\sin(-2\,a) = \frac{m\,\lambda_b}{g} \tag{48-203}$$

where

a is the blaze angle
λ_b is the wavelength for which the grating is blazed with order m
m is the diffraction order
g is the grating constant.

This case is shown in schematically in figure 48-63.

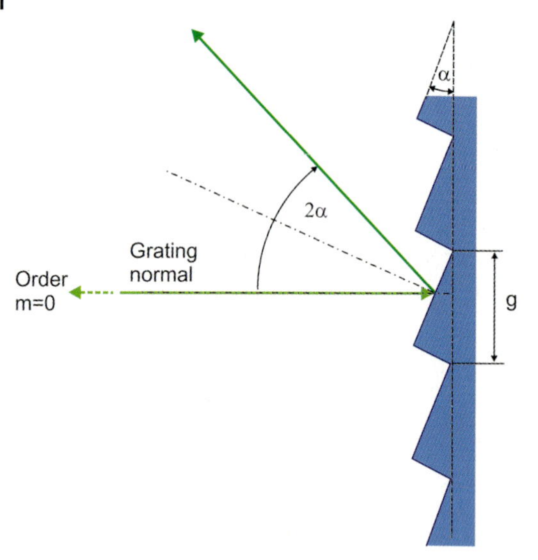

Figure 48-63: Preferred specular reflection at a single groove for incidence normal to the grating face.

Figure 48-64 shows the special case of a phase reflection grating of Littrow type. The incident beam is perpendicular to the face of the saw teeth of the blazed grating, and the reflected beam is in autocollimation with the incident beam. Following the grating equation (48-196) with the presumption $\alpha = \theta_i = \theta_m$ the relation (48-204) is valid for a Littrow grating:

$$2\sin\alpha = \frac{m\,\lambda_b}{g} \tag{48-204}$$

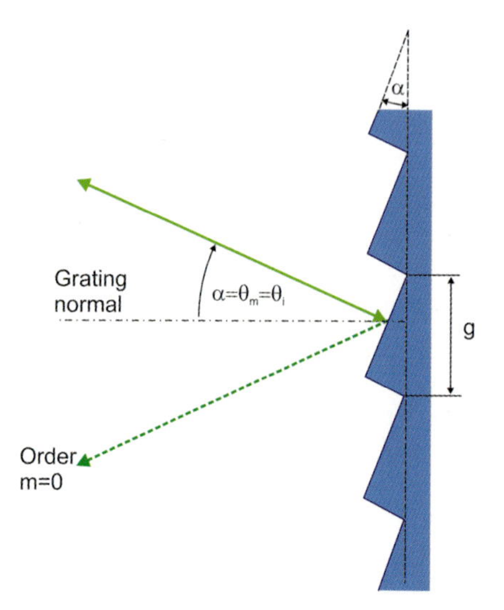

Figure 48-64: Littrow condition for a single groove of a blazed grating.

A grating blazed in first order is equally blazed in the higher orders. For example – a grating blazed in first order at 800 nm is also blazed at 400 nm in the second and at 266.667 nm in the third order.

Diffraction Efficiency and Order

The overall pattern of diffraction gratings mainly determines the shape of the wavefront, whereas the structure of the grooves, such as the groove depth and the duty cycle, determine the diffraction efficiency in the different orders.

Within the scope of this chapter the following general further remarks concerning the efficiency should suffice.

Blazed gratings are optimized to produce the maximum efficiency at desired wavelengths. Usually commercially available gratings are therefore described as blazed for this wavelength, but have a good efficiency over a specific, limited spectral range. The spectral efficiency follows a particular function over the spectral range for which the grating can be used. As a rule of thumb one can expect that the efficiency will be reduced to 50% that of the blaze wavelength for $\lambda \approx \frac{2}{3}\lambda_b$ as well as for $\lambda \approx 1.8\,\lambda_b$.

For higher orders the spectral efficiency variation with wavelength is usually comparable with that in first order. If the grating is blazed in first order the maximum efficiency decreases with increasing order number.

For further details see the overview in section 53.4.2 figure 53-46 and table 53-4 of this book or special publications, such as [48-39].

Angular Dispersion and Spectral Resolution of Grating Monochromators

The spectral resolution of a grating monochromator is characterized by its resolving power according to (48-171). In analogy with prism monochromators, the resolvable spectral interval $\Delta\lambda$ is determined by the angular dispersion $d\theta/d\lambda$. In the case of a grating monochromator, the angular dispersion is calculated according to (48-205) for the diffraction angle θ_m from the grating equation (48-196) by differentiating and considering the geometrical grating relations shown in figure 48-65;

$$\frac{d\theta}{d\lambda} = \frac{d\theta_m}{d\lambda} = \frac{m}{g}\frac{1}{\cos\theta_m} = \frac{m}{g}\frac{b}{a} = \frac{1}{\lambda}\frac{\sin\theta_m + \sin\theta_i}{\cos\theta_m} = \frac{1}{\lambda}\frac{b}{a}(\sin\theta_m + \sin\theta_i) \tag{48-205}$$

where

m is the diffraction order,
g is the grating constant,
b is the ruled grating width,
a is the beam width in the dispersion plane perpendicular to the diffracted beam,
θ_m is the angle of diffraction for order m,
θ_i is the angle of incidence,
λ is the wavelength considered.

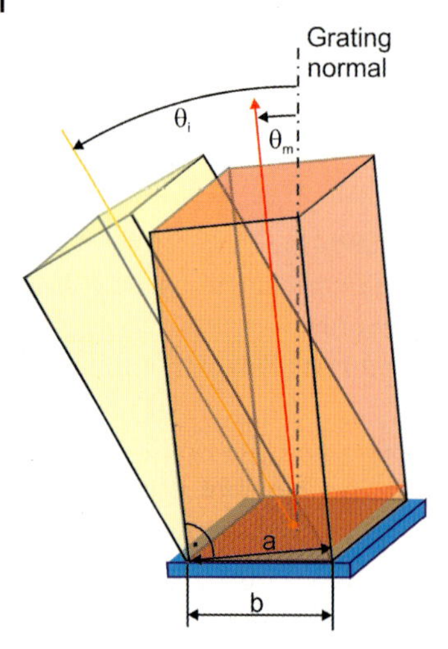

Figure 48-65: Grating dimensions and beam width.

For the Littrow configuration with $\theta_m = \theta_i$, relation (48-205) becomes (48-206):

$$\frac{d\theta}{d\lambda} = \frac{d\theta_m}{d\lambda} = \frac{2}{\lambda}\tan\theta_m = \frac{2}{\lambda}\frac{b}{a}\sin\theta_m \tag{48-206}$$

Equations (48-205) and (48-206) show that the angle dispersion of a grating monochromator, for a fixed grating length, depends only on the angle of incidence and diffraction.

Theoretical Maximum Resolving Power

In a similar way to prism monochromators, the theoretical maximum resolving power of a grating monochromator can be found from definition (48-171) for a resolving power of $R = \lambda/\Delta\lambda$ by considering the diffraction limited angle resolution $d\theta_0 = \lambda/a$ according to (48-178). In the case of the grating monochromator, a is the beam width in the dispersion plane perpendicular to the diffraction direction considered. The resolvable wavelength difference, dependent on the diffraction-limited minimum angle $d\theta_0$ and the angular dispersion (48-205), is calculated according to (48-207):

$$\Delta\lambda_0 \approx d\lambda_0 = d\theta_0 \frac{1}{\frac{d\theta}{d\lambda}} = \frac{\lambda}{a}\frac{g}{m}\cos\theta = \frac{\lambda}{b}\frac{g}{m} = \frac{\lambda^2}{b}\frac{1}{\sin\theta_m + \sin\theta_i} = \frac{\lambda}{R_0} \tag{48-207}$$

Using (48-207) and considering the grating equation (48-196) the theoretical maximum resolving power becomes

$$R_0 = \frac{\lambda}{\Delta\lambda_0} = \frac{b}{\lambda}(\sin\theta_m + \sin\theta_i) = b\frac{m}{g} = m\,N \tag{48-208}$$

The quantities in (48-207) and (48-208) are

g the grating constant
b the ruled width of the grating,
$N = \frac{b}{g}$ the total number of grooves,
a the beam width in the dispersion plane perpendicular to the diffracted beam direction,
m the diffraction order,
λ the wavelength considered,
θ_m the angle of diffraction for order m,
θ_i the angle of incidence.

As an example, a grating with a length b of 100 mm – the same as the prism base length in the example for the prism resolving power in section 48.3.6 – is considered. For 1200 grooves per mm the calculation according to (48-208) delivers a maximum resolving power of about 120 000 at a wavelength of 500 nm which is about 127 682 for $\theta_m = 34.297°$, $\theta_i = 4.297°$ according to table 48-4 for an angle of deflection of 30° in the first diffraction order. The minimum resolvable wavelength difference $\Delta\lambda_0$ at $\lambda = 500\,\text{nm}$ is about 4 nm.

For a Littrow configuration, the theoretical maximum resolving power (48-208) for $\theta_m = \theta_i$ is valid:

$$R_{0\text{Littrow}} = \frac{2\,b}{\lambda} \sin\theta_m \tag{48-209}$$

$(R_{0\text{Littrow}})_{\text{limit}} = \frac{2\,b}{\lambda}$ is the theoretical maximum $R_{0\text{Littrow}}$ can reach $(\sin\theta_m = 1)$, and the limit-resolvable wavelength difference $\Delta\lambda_0 = \lambda/R_0$ becomes $\Delta\lambda_{0\text{Littrow}} = \lambda^2/2\,b$. For a Littrow grating with 100 nm base length the theoretical minimum resolvable wavelength $\Delta\lambda_0$ at 500 nm therefore almost reaches 1.25 nm.

Linear Dispersion

The linear extension of a spectral interval in the focus plane of the monochromator telescope is defined as the linear dispersion $(d\lambda/dx)$ in an analogous way to the prism monochromator. The dependence on grating parameters and the focal length of the monochromator telescope is found by differentiating the wavelength, which depends on the angle of incidence, the angle of diffraction, the grating constant and the order of diffraction according to the grating equation (48-196):

$$\frac{d\lambda}{dx} = \frac{g}{m} \cos\theta_m \frac{d\theta_m}{dx} = \frac{g}{m} \cos\theta_m \frac{1}{f_T} \tag{48-210}$$

where

$$\frac{dx}{f_T} = d\theta_m.$$

Slit Width and Spectral Bandwidth

As for the prism monochromator, the minimal useful slit width $a_{S0} \approx \lambda\,(f_T/a) = \lambda\ F/\#$ is valid according to (48-183).

For a given linear dispersion $(d\lambda/dx)$ the resulting bandwidth is calculated by $\Delta\lambda_S \approx d\lambda_S = a_S(d\lambda/dx)$ according to (48-185).

Throughput of a Grating Monochromator

Equation (48-190) $\Phi_{\lambda_E} = L_\lambda\,\tau(\lambda) A_E\,\varphi_S\,\frac{\lambda}{R}(d\theta/d\lambda)$ for the approximated calculation of spectral radiant flux throughput is also valid for a grating monochromator.

With the beam area $A_E = A_g \cos\theta_m$ according to figure 48-66 and the angular dispersion (48-205) the throughput for the spectral radiant flux of a grating monochromator becomes (48-211):

$$\Phi_{\lambda_E} = L_\lambda\,\tau(\lambda) A_g\,\varphi_S\,\frac{1}{R}(\sin\theta_i + \sin\theta_m) \tag{48-211}$$

For a blazed grating with the blaze angle α according to figure 48-62 and $\theta_r = \theta_m$, equation (48-211) is transformed into (48-212):

$$\Phi_{\lambda_E} = L_\lambda\,\tau(\lambda) A_g\,\varphi_S\,\frac{1}{R} 2\sin\alpha \cos\frac{\theta_i - \theta_m}{2} \tag{48-212}$$

Expressing the horizontal angle φ_S by slit length l_S and the telescope focal length f_T, equation (48-212) becomes:

$$\Phi_{\lambda_E} = L_\lambda\,\tau(\lambda) A_g \frac{l_S}{f_T}\frac{1}{R} 2\sin\alpha \cos\frac{\theta_i - \theta_m}{2} \tag{48-213}$$

The total radiant flux is obtained by multiplying the expression (48-211) for the spectral radiant flux by the spectral interval $\Delta\lambda$ and applying definition (48-171), $R = \lambda/\Delta\lambda$ for the resolving power:

$$\Phi_E = \Phi_{\lambda_E}\Delta\lambda = L_\lambda\,\tau(\lambda) A_g \frac{l_S}{f_T}\frac{\lambda}{R^2} 2\sin\alpha \cos\frac{\theta_i - \theta_m}{2} \tag{48-214}$$

The quantities in relations (48-211)–(48-214) are:

L_λ	the spectral radiance of the entrance slit,
$\tau(\lambda)$	the transmission of the optical path from the entrance slit to the exit slit,
$\varphi_S = \frac{l_S}{f_T}$	the slit angle in the plane of the slit length direction and the optical axis,
l_S	the length of the exit slit,
a_S	the width of the exit slit
A_g	the used grating area
λ	the wavelength considered,
$R = \frac{\lambda}{\Delta\lambda}$	the resolving power of the monochromator with $\Delta\lambda$ as the resolvable spectral interval,
$\left(\frac{d\theta}{d\lambda}\right)$	the angular dispersion of the monochromator.

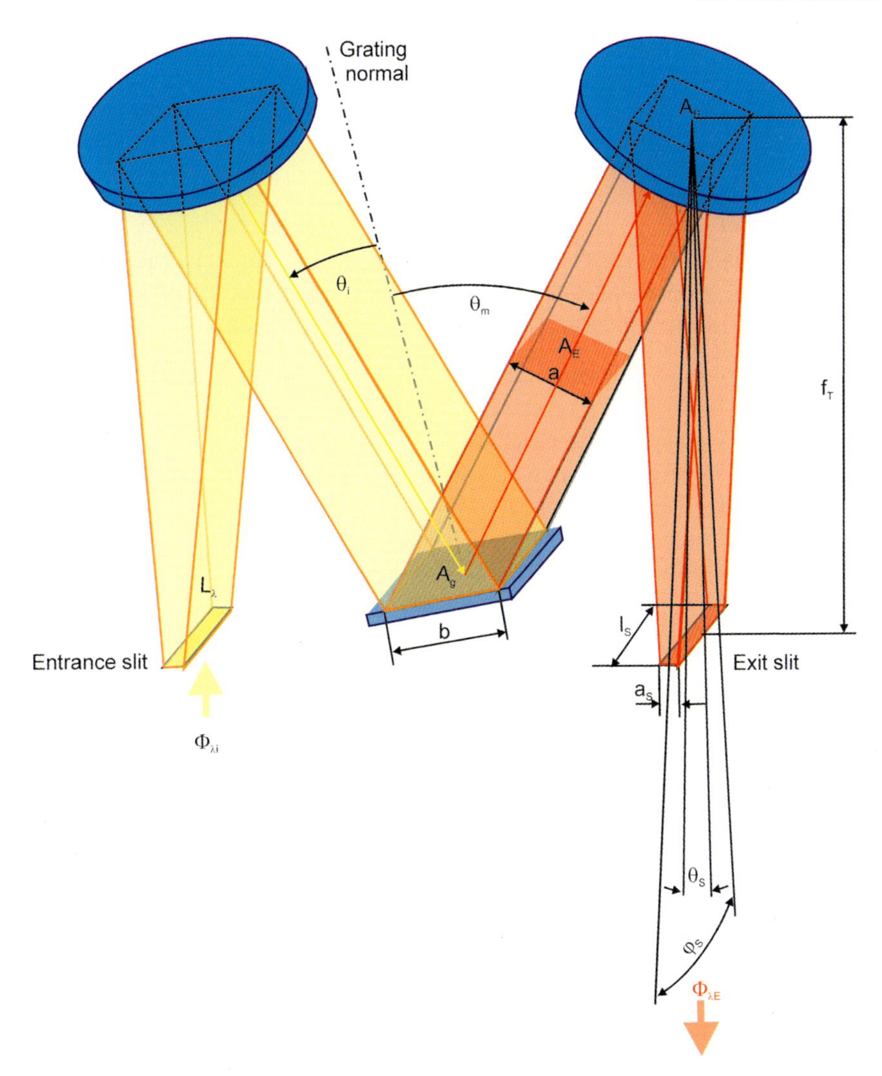

Figure 48-66: The radiant flux through a grating monochromator.

Equation (48-212) for the throughput of a grating monochromator can be compared with the corresponding equation (48-192) for a prism monochromator. Assuming a prism monochromator in air ($n \approx 1$) with equal resolving power R, and also equal parameters l_S, a_S, f_T, τ_λ and with a prism base area $A_P = A_g$ the ratio $\frac{\Phi_{\lambda_E \text{prism}}}{\Phi_{\lambda_E \text{grating}}}$ of the transmitted spectral radiance becomes (48-215):

$$\frac{\Phi_{\lambda_E \text{prism}}}{\Phi_{\lambda_E \text{grating}}} = \frac{\lambda\left(-\frac{dn'}{d\lambda}\right)}{2 \sin \alpha \cos \frac{\theta_i - \theta_m}{2}} \tag{48-215}$$

The ratio (48-215) becomes a maximum in the case of a Littrow grating with $\theta_i = \theta_m$:

$$\frac{\Phi_{\lambda_E \text{prism}}}{\Phi_{\lambda_E \text{grating}}} = \frac{\lambda\left(-\frac{dn'}{d\lambda}\right)}{2 \sin \alpha} \tag{48-216}$$

Equation (48-216) shows that a grating monochromator with typical blaze angles of about 30° is, in principle, able to reach higher throughputs than a comparable prism monochromator of equal resolving power in the case of normal optical glasses in the visual spectral range with values $\lambda\left(-\frac{dn'}{d\lambda}\right) < 0.5$ (see figure 48-54).

Basic Grating Monochromator Configurations

There are numerous grating monochromator configurations. Within the scope of this book only the most common basic types can be discussed. These can be classified by their grating types into plane grating systems and aberration-corrected holographic grating systems.

One of the most common basic plane grating monochromator configurations is the Czerny–Turner monochromator. Its principle is shown schematically in figure 48-67. This monochromator type uses two separate spherical mirrors as the collimator and telescope, respectively.

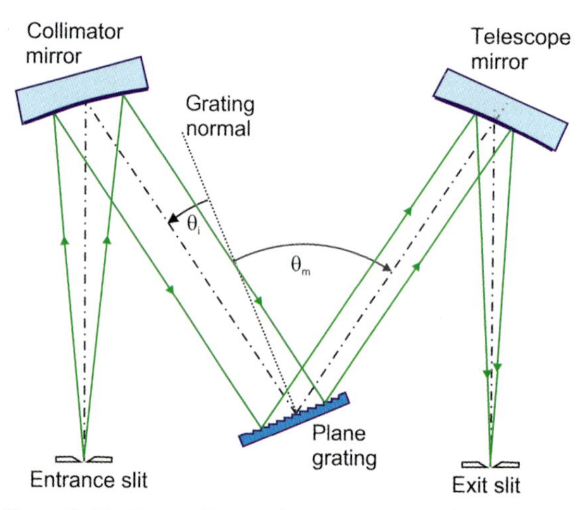

Figure 48-67: Czerny–Turner plane grating monochromator configuration.

The two separate mirrors enable a high flexibility for the geometrical correction of optical aberrations (see below).

The Fastie–Ebert plane grating monochromator, as the second basic type, uses one common spherical mirror for the collimator and telescope as shown in figure 48-68. This principle allows for simple, inexpensive designs but has the disadvantage of having limited possibilities for the correction of aberrations. Therefore, this very common concept makes sense for single detector spectrometers only, but not for multi-channel spectrometers, which need a flattened image plane (see section 48.4.4).

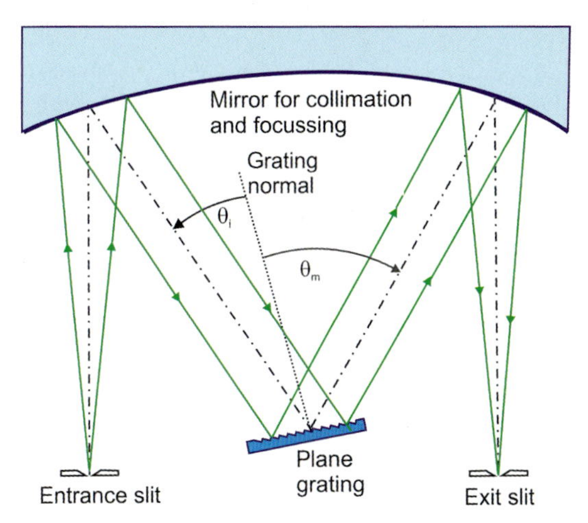

Figure 48-68: Schematic Fastie–Ebert plane grating monochromator configuration.

Plane grating monochromator configurations have several aberrations varying with the image plane, because the mirrors are used off-axis. All aberration types cause a degradation of the band width and the optical SNR. The size of the different aberration types depends on the numerical aperture (NA) at which the monochromator mirrors are used. For more details see [48-39].

Aberration-correcting Plane Gratings

With modern holographic grating technology all aberrations present in a spherical-mirror- based Czerny–Turner monochromator configurations can be corrected at one wavelength with very good reduction over a wide wavelength range (see [48-40]).

Concave Aberration-corrected Holographic Gratings

Monochromators of this type use a single holographic grating with no additional optics. The grating both focuses and diffracts the incident light as shown schematically in figure 48-69. In these configurations the direction of the grating normal and also the input angle and diffraction angles vary with the impact point. The principal grating relations discussed above are still valid, but have to be considered locally for the optical design.

Having only one optical component, this configuration type allows very compact spectrometer design and is therefore the preferred monochromator configuration used in multi-channel spectrometers (see section 48.4.4).

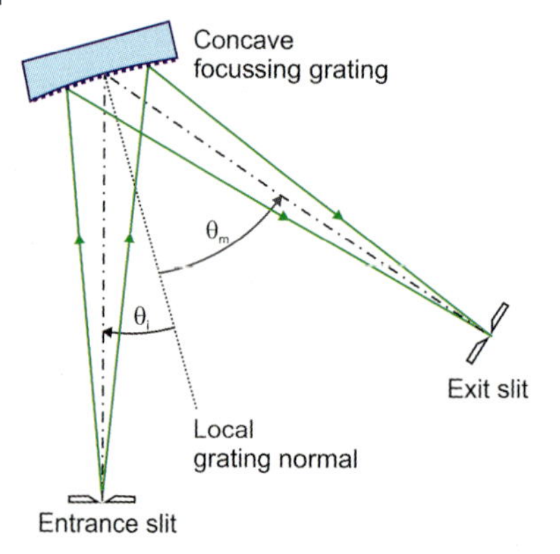

Figure 48-69: Schematic monochromator configuraton with one concave grating used for diffraction and focussing.

48.4
Spectrometers

48.4.1
Introduction

A *spectrometer* is an instrument which measures spectrally resolved radiometric quantities. Spectrometers can be used to measure radiometric properties of optical materials, components and systems. Monochromatic applications (for example, laser optics) can be considered as a special case.

Instruments which are designed to measure the spectrum in absolute units rather than relative units are called *spectroradiometers*. They are able to measure the absolute spectral characteristics of light sources and sensors and to calibrate them. Calibration is mostly carried out at national standard laboratories.

Spectrophotometers are instruments used to measure the ratio of two values of a radiometric quantity at the same wavelength in defined wavelength intervals. One value is taken from the impinging radiation, the other, after the radiation has been transmitted through or reflected from a material sample, an optical component or system in the test chamber within the instrument (the term *photometer* may be misleading, because the sensitivity function of the human eye is generally not involved). Spectrophotometers are equipped with artificial radiation sources and sensors as integrated components.

Spectrophotometers used to measure radiometric quantities in transmission are also called *transmittometers*, those measuring in reflection are called *reflectometers*.

In general, any spectrometer will operate over a small portion of the total electromagnetic spectrum due to the different techniques used to measure different portions of the spectrum. There is a large variety of radiometric or spectrometric arrangements which are adapted to special measuring tasks concerning optical components. Most of the commercially available complete spectrometers are *spectrum analyzers*, which can be used to measure the relative spectral distribution of the radiant flux of any radiation source.

Different types of spectrometer, which are able to measure spectrally resolved optical characteristics such as transmission and reflection of small test probes, are commercially available. This section gives a short overview of the basic configurations and concepts.

48.4.2
Basic Principles of Spectrometers

A spectrometer consists of:

1. a light source unit,
2. a monochromator unit proving at least one dispersive element,
3. a detector unit consisting of one or several single detectors or detector arrays,
4. some auxiliary optics.

A spectrometer can be arranged in two different basic ways:

- The monochromator is implemented in the illumination path (figure 48-70). The exit slit of the monochromator is used as the light source of the spectrophotometer arrangement in order to measure the spectral radiant quantity of interest, e.g., the transmission or reflection of an optical component or system. The test piece is illuminated sequentially by radiation of small bandwidth and continuously varying central wavelength, and the spectrally resolved radiant flux is measured by a single detector after having passed the test piece. Any fluctuation in the radiant flux of the light source can be measured and compensated for by using a reference sensor in the illumination path.
- The spectrometer is placed in the detector path of the setup. The test piece is illuminated by the radiant flux of the total spectrum of a suitable light source. Spectral analysis is carried out either sequentially or instantaneously depending on the type of spectrometer used. Figure 48-71 shows an example in which a multi-channel spectrometer provides a detector array. The array is illuminated by the spread spectrum and delivers the signals simultaneously over the whole spectral range (see below).

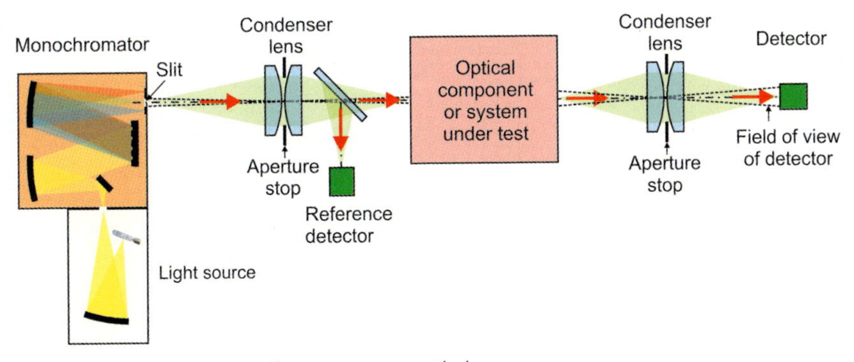

Figure 48-70: Arrangement of a spectrometer with the monochromator in the illumination path.

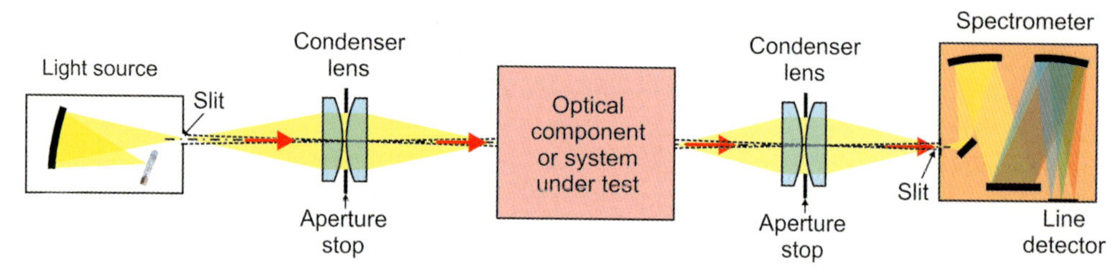

Figure 48-71: Arrangement of a test setup with the spectrometer in the detector path.

Optical components might be tested either:

- in reflection to measure the spectral reflectance of mirrors or the residual spectral reflectance of anti-reflection coatings (figure 48-72),
- in transmission to measure the spectral transmittance of optical materials, components or systems (figure 48-73).

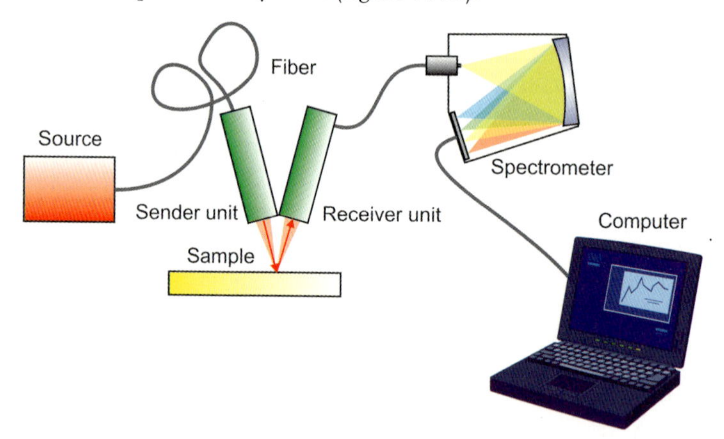

Figure 48-72: Measurement of the spectral reflectance of an optical component.

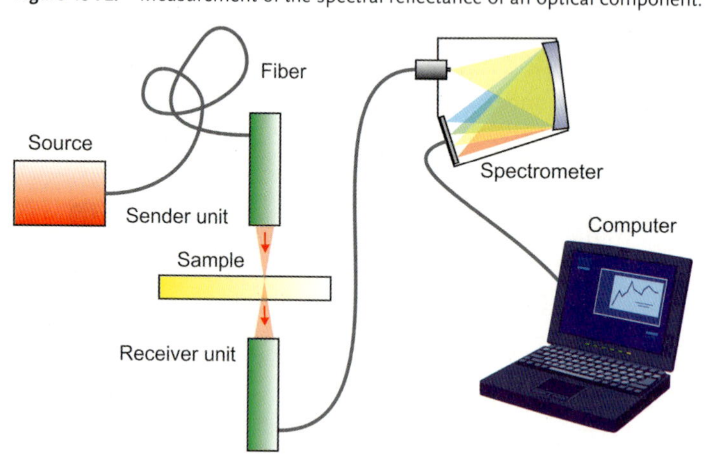

Figure 48-73: Measurement of the spectral transmittance of an optical component.

48.4.3
Single-Channel Spectrometers

Basic Principles

Spectrometers using single sensors (single-channel spectrometers) measure the spectrum sequentially. The monochromator being the basic element of the spectrometer must be tuned over the required spectral range to acquire the total spectrum from a light source or from light reflected by, or transmitted through, a test element.

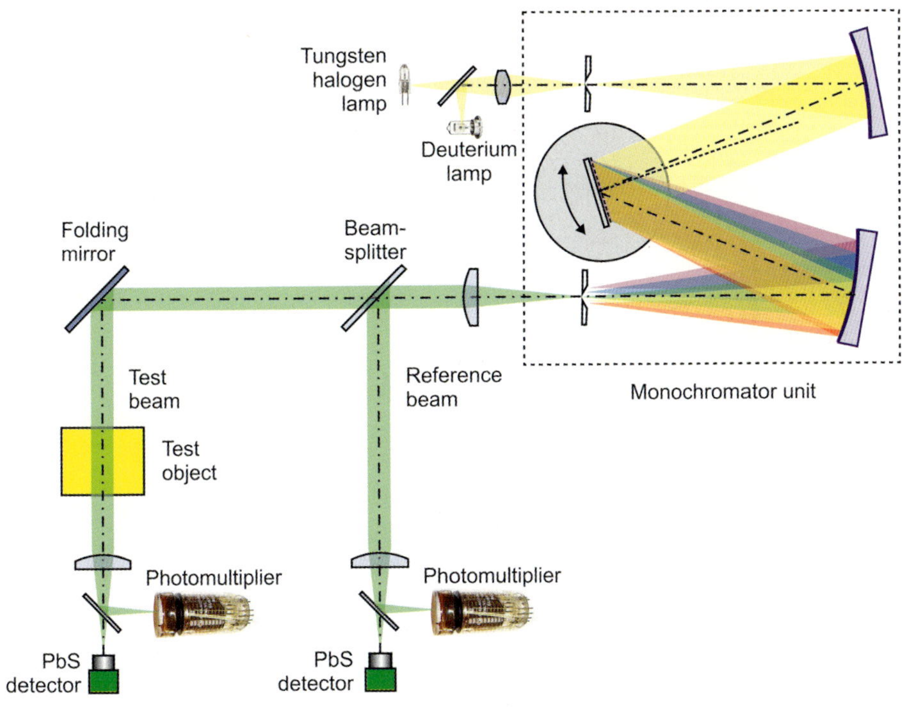

Figure 48-74: Schematic view of a double-beam spectrometer with separated test and reference beams and two detector sets.

Spectrometer configurations with single sensors allow the use of different sensors suitable for different spectral ranges. Furthermore, important sensor parameters such as the dynamic range, sensitivity and linearity can mainly be optimized and controlled better than in the case of sensor arrays. As examples, PbS detectors and photomultipliers which form a sensor set are shown in figures 48-74 and 48-75. A PbS detector (a lead sulfide integrating detector) is a photoconductive sensor with a resistance which decreases with increasing infrared light. It operates from 1.3–3.2 μm with a peak sensitivity around 2.2 μm. Photomultipliers are vacuum tubes which are extremely sensitive in the ultraviolet, visible, and near-infrared ranges, depending on the photocathode and window material. In the examples, the light sources consist of a combination of a tungsten halogen and a deuterium lamp. In this way a spectrum from 112 nm to > 3 μm can be covered.

A spectrometer can be either *single-beam* or *double-beam*. In a single-beam instrument, all of the light is reflected from or transmitted through the test piece. The flux calibration is achieved by removing the test piece or replacing it with a well-known test object.

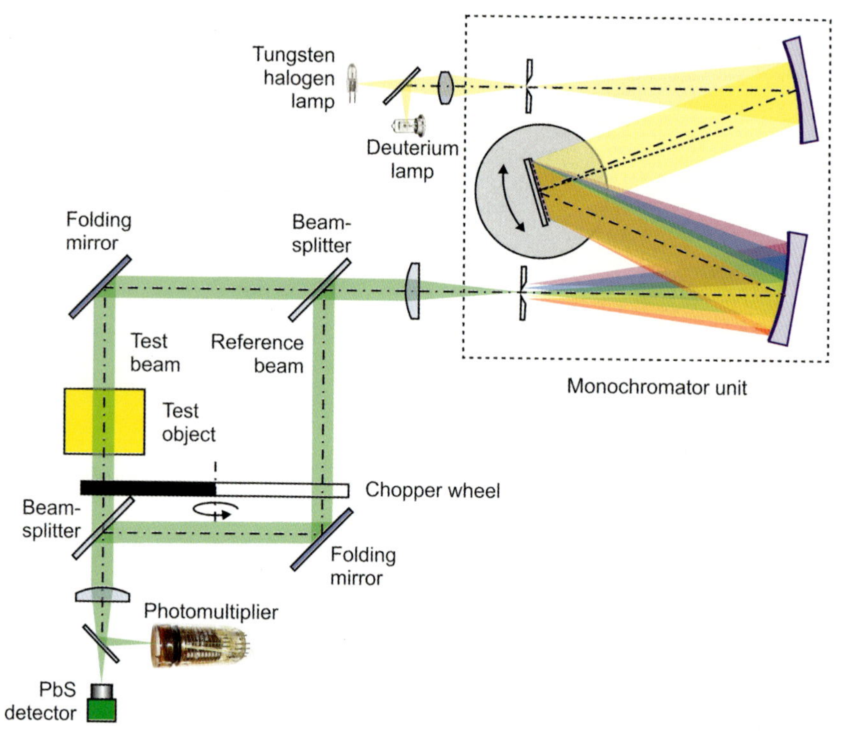

Figure 48-75: Schematic view of a double-beam spectrometer with a chopper and one detector set using sequential acquisition of the dark signal, test and reference signals.

In a double-beam instrument, the light is split into two beams before it reaches the test piece. One beam is used as the reference, the other beam passes the test piece. In a transmittance measurement the reference beam intensity is taken as 100% transmission, and the measurement displayed is the ratio of the two beam intensities. Some double-beam instruments have two detector sets measuring the test and reference beam at the same time (figure 48-74). In other instruments, the two beams pass through a beam-chopper, which blocks one beam at a time. The detector alternates between measuring the test beam and the reference beam which is synchronized with the chopper. The dark signal of the detector may be detected by introducing one or more dark intervals in the chopper cycle. In this way, drifts of the sensor and the light source are permanently being observed and compensated for by appropriate normalization (figure 48-75).

Devices

Most of the commercial spectrometers with single sensors are used in the UV and visible regions of the spectrum. Some of them also operate in the near-infrared region. Many UV/VIS instruments are used extensively in colorimetry, for example, by ink manufacturers, printing companies and textile vendors, who need to know the spectral reflectance curves of their products. Other users need to know the transmittance of fluid or solid materials. Fluid samples are usually prepared in cuvettes constructed of glass, plastic, or quartz, depending on the application and spectral region of interest.

In the following, we describe a commercial instrument as an example of a versatile spectrophotometer using a single sensor unit, a detector module combining a PbS detector with a photomultiplier. Figure 48-76 gives a schematic view of the arrangement.

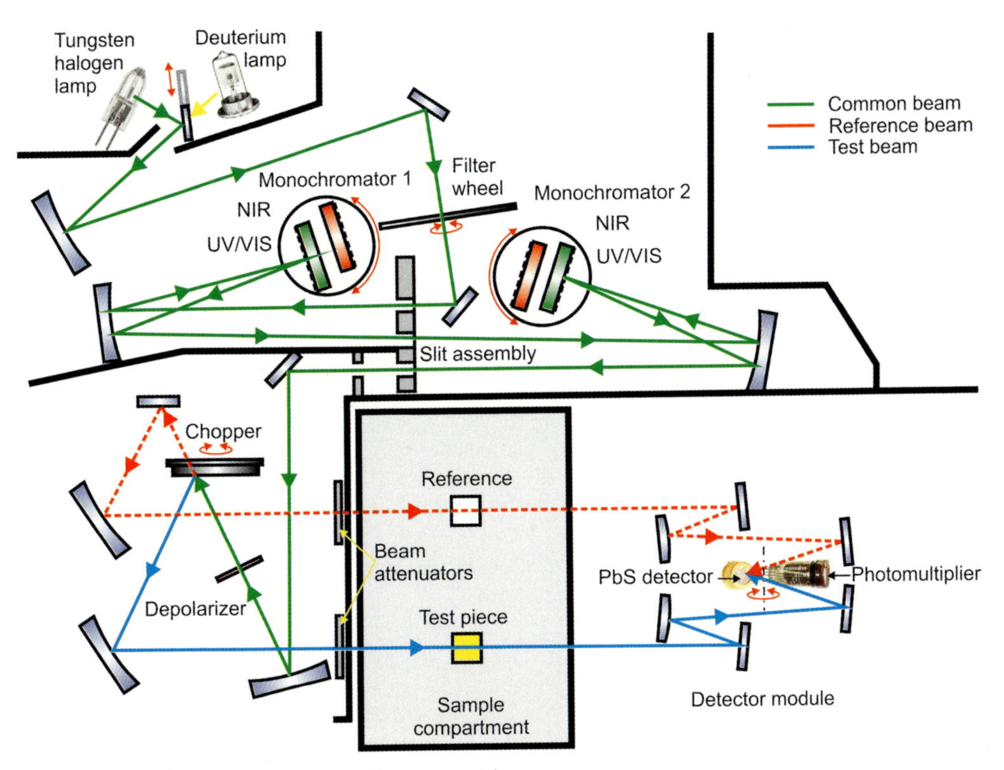

Figure 48-76: Schematic view of a versatile commercial spectrophotometer using a single sensor unit (detector module) [48-41].

The instrument is a double-beam, double-monochromator spectrophotometer covering the UV, VIS and NIR spectral region from 175 to 3300 nm.

The light source module provides a tungsten halogen lamp for the VIS/NIR region and a deuterium lamp for the UV/VIS region.

A reflecting chopper wheel switches between the reference beam (open), test beam (reflecting) and dark signal (closed) with a frequency of 46 Hz. The UV/VIS region is detected by a photomultiplier, while the NIR spectrum is detected by a Peltier cooled PbS detector.

Both monochromators are holographic grating monochromators, each holding two gratings, which can be selected depending on the spectral region. For the UV/VIS region, a grating of 1440 lines/mm blazed at 240 nm, and for the NIR region a grating of 360 lines/mm blazed at 1100 nm are required.

The total spectrum is covered by three instrument configurations which switch automatically during the wavelength scan. Table 48-5 gives an overview of the main specifications and the instrument's wavelength accuracy, reproducibility and resolution.

Table 48-5: Instrument configuration during a total wavelength scan [48-41].

Configuration	Scan 1	Scan 2	Scan 3
Wavelength	175 nm – 319 nm	319 nm – 860 nm	860 nm – 3300 nm
Grating	UV/VIS	UV/VIS	NIR
Lamp	deuterium	halogen	halogen
Detector	photomultiplier	photomultiplier	PbS detector
Wavelength accuracy / nm	± 0.08	± 0.08	± 0.30
Wavelength reproducibility / nm	≤ 0.01	≤ 0.01	≤ 0.04
Wavelength resolution / nm	≤ 0.05	≤ 0.05	≤0.20

In the deep UV region the instrument must be purged by nitrogen (N_2). The constituents of air start to absorb below 200 nm. The UV cutoff of ozone starts at 295 nm, while that of oxygen starts at 200 nm and that of nitrogen at 170 nm.

48.4.4 Multi-channel Spectrometers

Basic Principles

The advantage of multi-channel spectrometers is their ability to measure the total spectrum simultaneously. The lack of moving parts for spectrum scanning allows small and rigid setups which are ideal for in situ measurements and the control of optical coating processes.

In section 48.3 we discussed the basic characteristics of monochromators, i.e., the free spectral range (FSR), diffraction efficiency, angular and linear dispersion, maximum resolving power and throughput. These can be applied to any multi-channel spectrometer using prisms, diffraction gratings, Fabry–Pérot cavities, etc., as dispersive elements.

The sensor of a multi-channel system (for example, a diode array) will be characterized by

1. the number of pixels N,
2. the pixel width p,
3. the spectral responsivity $S(\lambda)$ of the detector.

The monochromator will be characterized by

1. the linear dispersion $d\lambda/dx$,
2. the maximum resolving power $R_0 = \lambda/\Delta\lambda_0$
3. the total spectral transmissivity $T_M(\lambda)$ of the monochromator,

The *spectral pixel pitch* is then given by

$$\Delta\lambda_{\text{pixel}} = \frac{\partial\lambda}{\partial x}p \tag{48-217}$$

As a rule of thumb $\Delta\lambda_{\text{pixel}}$ should be approximately 3x the spectral resolution of the monochromator:

$$\Delta\lambda_{\text{pixel}} \approx \frac{\Delta\lambda_0}{3} \tag{48-218}$$

From (48-217) and (48-218) it follows that, for a suitable pixel width:

$$p \approx \frac{\lambda}{3R_0\frac{\partial\lambda}{\partial x}} \tag{48-219}$$

Assuming a radiant exitance $E_0(\lambda)$of the light source and a transmittance (or reflectance) $T_T(\lambda)$ of the test element, the detected irradiance signal I_p at a single pixel is then given by

$$I_p(\lambda) \propto \int\limits_{\lambda}^{\lambda+\Delta\lambda_{\text{pixel}}} \frac{E_0(\lambda)T_T(\lambda)T_M(\lambda)S(\lambda)}{\frac{\partial\lambda}{\partial x}p} d\lambda \tag{48-220}$$

It is important to select the light source, monochromator and detector so that together they deliver a signal I_p which is sufficiently high above the noise level to achieve good quality results. In cases where this is not possible, filters are to be introduced in order to dampen spectral ranges with high signals. In this way the spectrum can be equalized to improve the S/N ratio of the multi-channel system.

Devices

In principle, multi-channel spectrometers can be arranged using any suitable type of monochromator. The great majority of commercially available multi-channel spectrometers, however, use diffractive monochromators providing diffraction gratings as dispersive elements. As an example, figure 48-77 shows the basic arrangement using a blazed holographic mirror as the dispersive element. Light from a suitable light source (halogen lamp, xenon lamp, deuterium lamp, etc.) is transmitted through or reflected from the test object and is then coupled into a fiber bundle. The fiber bundle may consist of around 30 fibers, which are arranged in a round circular area at one end. At the other end, the fibers are arranged along a line to form the entrance slit for the spectrometer device.

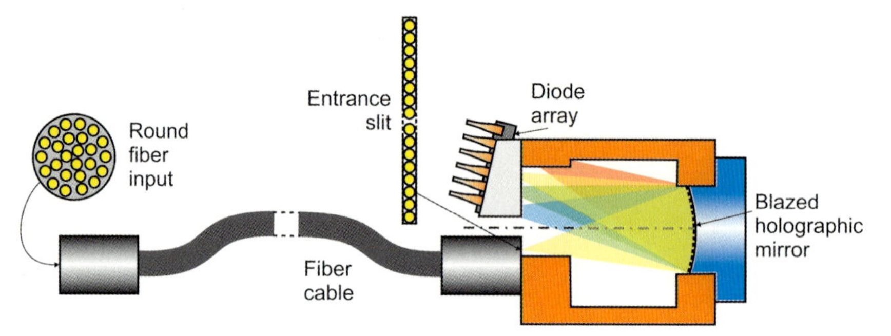

Figure 48-77: Basic arrangement of a multi-channel spectrometer using a blazed holographic mirror as the dispersive element.

The entrance slit is located at a distance r from a concave mirror where r is the radius of curvature of the mirror. The mirror surface carries a blazed diffraction grating etched into the substrate material. The blaze is optimized for a certain wavelength and the selected diffraction order projected onto the line detector (diode array). The line detector is located about a distance r from the mirror and tilted in a specific way to collect sharp images of the entrance slit for every color of the spectrum. In a first approximation, the grating provides a 1:1 ratio image. To obtain optimum sensitivity the entrance slit should be 2–3 pixels wide (see above). If more pixels are illuminated, the signal-to-noise ratio and the sensitivity will decrease. If fewer pixels are illuminated, the wavelength accuracy will decrease. For example, if a pixel width of 25 μm is selected, the effective slit width should be between 50 and 75 μm.

Table 48-6 shows typical specifications for three multi-channel spectrometers used for different spectral ranges.

Table 48-6: Typical specifications of multi-channel spectrometers for three different wavelength ranges [48-42].

Spectrometer type	UV – VIS – NIR	VIS	NIR
Spectral range / nm	190 – 1015	360 – 780	910 – 2200
Light source	deuterium / halogen	halogen / pulsed xenon	halogen
Pixel number of diode	1024	512	256
Mean spectral pixel pitch / nm	0.8	0.8	6
Spectral resolution (half width at 1/10 maximum) / nm	2.4	2.4	18
Wavelength accuracy / nm at 500 nm	1.0	1.0	–
Wavelength reproducibility / nm	0.05	0.05	0.10
Minimum integration time / ms	12	6	0.1

48.4.5
Fourier Spectrometers

A *Fourier spectrometer* or *Fourier transform spectrometer (FTS)* (sometimes referred to as a *multiplex spectrometer*) is an interferometer which can be used to measure the temporal autocorrelation of radiation passing through the interferometer. The autocorrelation function is represented by the visibility of the interference fringes. The Fourier transform of the fringe visibility leads to the spectrum of the radiation, which is of interest to the user.

In many cases two-beam interferometers of the Twyman–Green, Michelson or Mach–Zehnder type (see section 46.3) are used. In a *temporally modulated FTS* a single detector at the exit of the interferometer measures the time-averaged irradiance while the optical path difference between both interferometer arms is monotonously changed. This spectrometer has the advantage that the total spectrum is projected onto a single detector, making it suitable for very weak light sources.

In a *spatially modulated FTS* a line detector or detector array measures the irradiance of the interference fringes generated by tilting one of the interferometer mirrors. The advantage is a very rapid collection of the total autocorrelation signal at the cost of spreading the light over many detector pixels and thus losing the ability to analyze weak signals.

In the following we will describe the basic principles of an FTS, various different configurations and their limitations.

Basic Principles

The *temporally modulated FTS* of the Michelson interferometer type consists of:

1. a light source providing the unknown radiation,
2. a beam-splitter,
3. a fixed mirror,
4. a mirror movable in the beam direction,
5. a linear encoder used to determine the relative position of the movable mirror,
6. a photodetector which detects the interference signal at the exit of the interferometer.

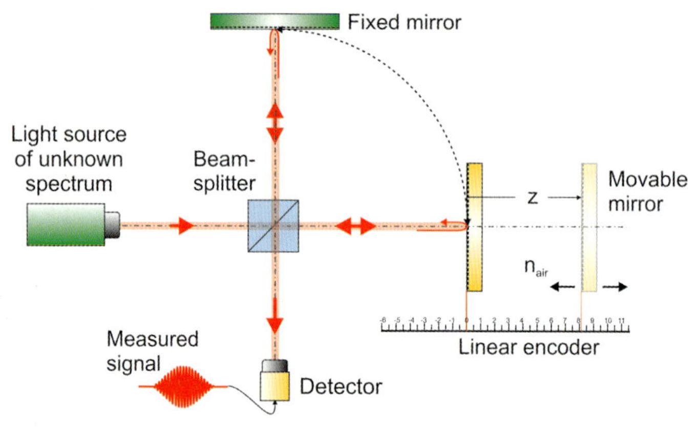

Figure 48-78: Scheme of a temporally modulated FTS of Michelson interferometer type used to detect the autocorrelation function of an unknown radiation.

Figure 48-78 shows the scheme of a temporally modulated FTS of Michelson interferometer type. The radiation of unknown spectral distribution is incident on a beam-splitter, where it is divided into a transmitted and a reflected beam. The beams are reflected from plane mirrors, one of which is in a fixed position, while the other can be moved in the beam direction. The position z of the movable mirror can be measured by a position-sensing device (a linear encoder or displacement-measuring interferometer). The zero position of the movable mirror coincides with the position at which both beams have zero optical path difference. The two beams interfere at a detector at the exit of the interferometer [48-43].

A temporally modulated FTS can collect a series of irradiance measurements, while the movable mirror is moved over a defined travel range and its actual position is recorded.

We assume a spectral density $B'(\sigma)$ at the entrance of the FTS as a function of the wavenumber $\sigma = \lambda^{-1}$. Effects of polarization and absorption are neglected.

The measured irradiance $I(z)$ at the detector is expressed as

$$I(z) = I_0 + \int_0^{\infty} B(\sigma)\cos(2\pi\sigma z)d\sigma \tag{48-221}$$

where

$$I_0 = \int_0^{\infty} A(\sigma)d\sigma \tag{48-222}$$

and

$$A(\sigma) = B'(\sigma)S(\sigma)(T_1(\sigma) + T_2(\sigma)) \tag{48-223}$$

$$B(\sigma) = 2B'(\sigma)S(\sigma)\sqrt{T_1(\sigma)T_2(\sigma)} \tag{48-224}$$

and with the following denotations:

$S(\sigma)$ is the spectral responsivity of the detector,

$T_1(\sigma)$ is the total spectral transmissivity of the optical system for the beam reflected from the fixed mirror,

$T_2(\sigma)$ is the total spectral transmissivity of the optical system for the beam reflected from the movable mirror.

Since the interferometer is not perfect we have to introduce phase errors due to unequal path length, for instance, within the beam-splitter. We therefore define a complex spectrum

$$\tilde{B}(\sigma) = 2B(\sigma)e^{i\phi(\sigma)} \tag{48-225}$$

which satisfies the reality condition

$$\tilde{B}(-\sigma) = \tilde{B}^*(\sigma) \tag{48-226}$$

(48-221) can then be rewritten as

$$I(z) - I_0 = \int_{-\infty}^{+\infty} \tilde{B}(\sigma)e^{i2\pi\sigma z}d\sigma \tag{48-227}$$

The complex spectrum $\tilde{B}(\sigma)$ can therefore be calculated by the inverse Fourier transform of the measured signal:

$$\tilde{B}(\sigma) = \int_{-\infty}^{+\infty} (I(z) - I_0)e^{-i2\pi\sigma z}dz \tag{48-228}$$

The magnitude of (48-228) leads to the unknown spectrum

$$B(\sigma) = \left|\tilde{B}(\sigma)\right| = \sqrt{\tilde{B}_{\mathrm{Re}}^2 + \tilde{B}_{\mathrm{Im}}^2} \tag{48-229}$$

whereas the unknown phase is calculated from

$$\phi(\sigma) = \arctan \frac{\tilde{B}_{\mathrm{Im}}}{\tilde{B}_{\mathrm{Re}}} \tag{48-230}$$

From (48-225) – (48-230) we can finally determine the unknown spectral density at the entrance of the FTS [48-44]:

$$B'(\sigma) = \frac{1}{2S(\sigma)\sqrt{T_1(\sigma)T_2(\sigma)}} \left| \int_{-\infty}^{+\infty} (I(z) - I_0) e^{-i2\pi\sigma z} dz \right| \tag{48-231}$$

For the evaluation of (48-231) the spectral sensor responsivity and the spectral transmissivities of the device must be known a priori. $S(\sigma)$ must be calibrated in advance by an apropriate method, either by comparing with a calibrated sensor or by using the radiation of a well-known source.

$T_1(\sigma)$ and $T_2(\sigma)$ can be determined by measuring the spectral transmissivities of the individual beam paths separately with a suitable spectrometer.

A *spatially modulated FTS* works very similar to the principle described above, except that both mirrors are fixed and one mirror is tilted to introduce a defined number of fringes at the interferometer's exit. A line or array sensor detects the interference fringes by a single shot. The interference fringes represent the spatial autocorrelation function of the unknown radiation (figure 48-79).

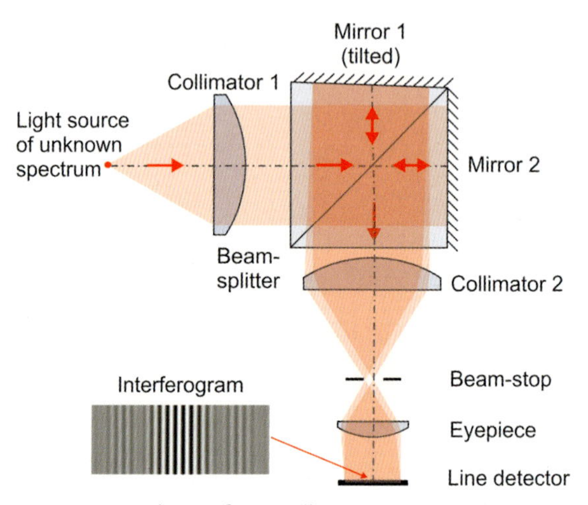

Figure 48-79: Scheme of a spatially modulated FTS of Michelson interferometer type used to detect the autocorrelation function of unknown radiation.

Figure 48-80 shows different examples, in which a measured interference signal was simulated and the corresponding spectrum was calculated according to (48-231). A displacement of the movable mirror over 20 μm in steps of 10 nm has been introduced. The spectrum is a function of the wavenumber σ expressed in μm^{-1}.

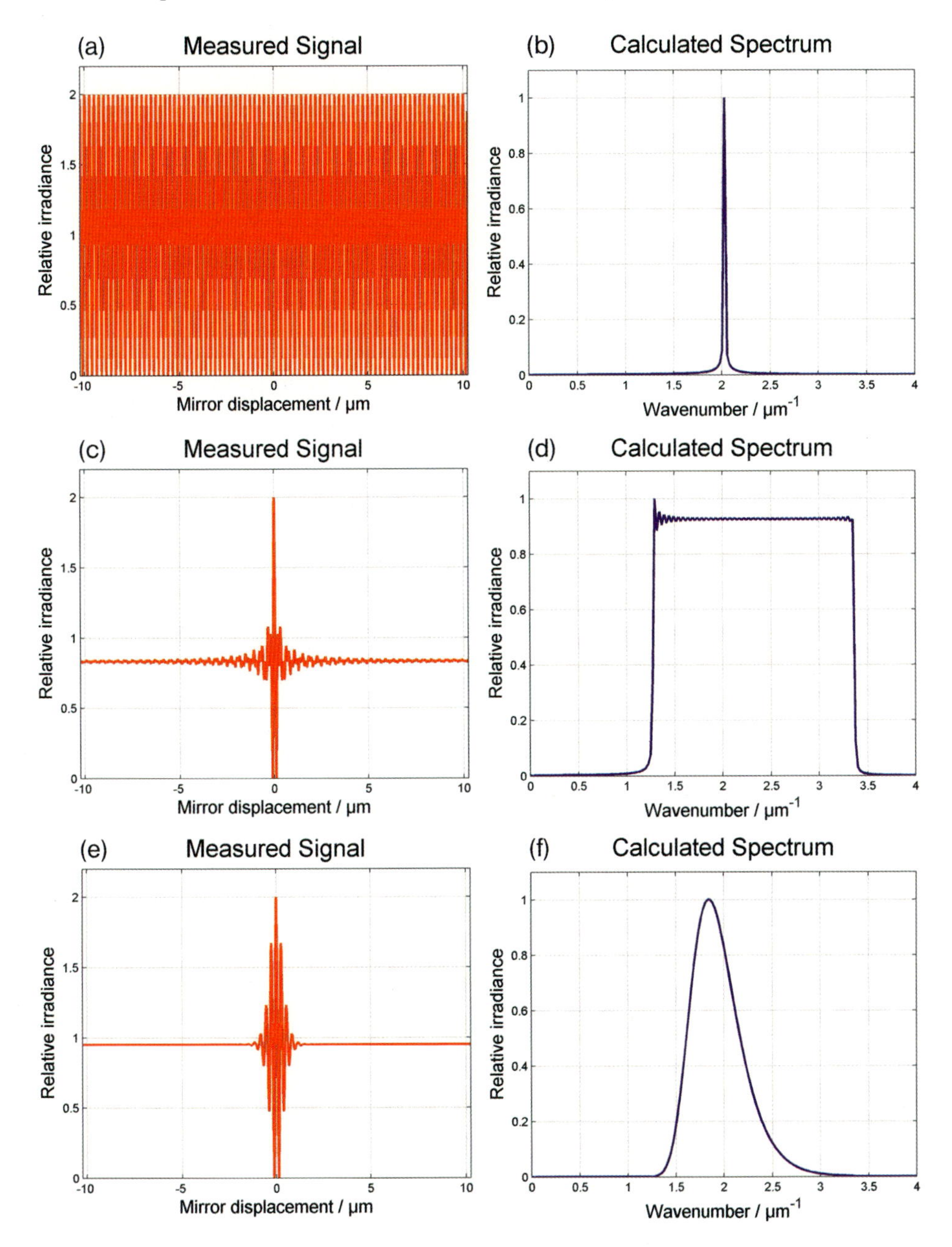

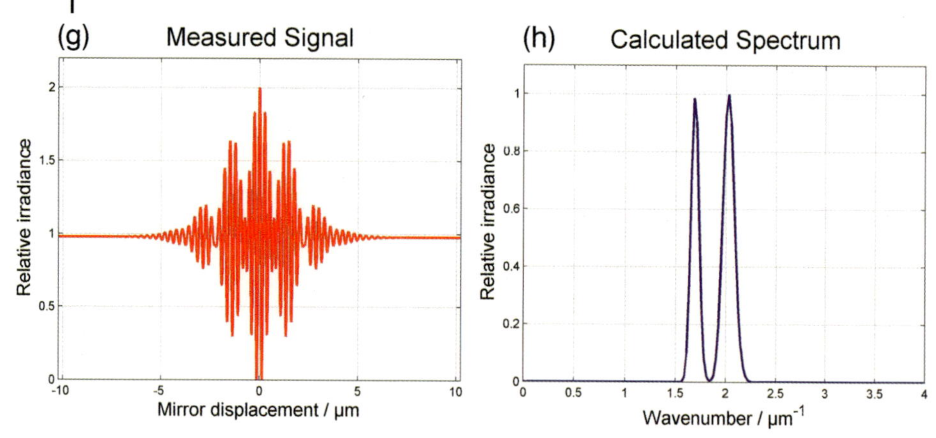

Figure 48-80: Measured interference signal of an FTS and the calculated spectra: (a)+(b) narrow line spectrum; (c)+(d) rectangular spectrum from 300–800 nm wavelength; (e)+(f) Gaussian spectrum with FWHM = 166 nm, maximum at 550 nm wavelength; (g)+(h) two Gaussian spectra with FWHM = 33 nm separated by 100 nm.

The resolution of an FTS depends on the maximum optical path difference OPD achievable by the scanning mechanism of the interferometer. In practice, the integral over z in (48-228) can then only be carried out by replacing infinity with OPD [48-45] and [48-46]. The Fourier transform of a sinusoidal signal then leads to a sinc function, whose central maximum shows a width of 1/OPD (figure 48-81).

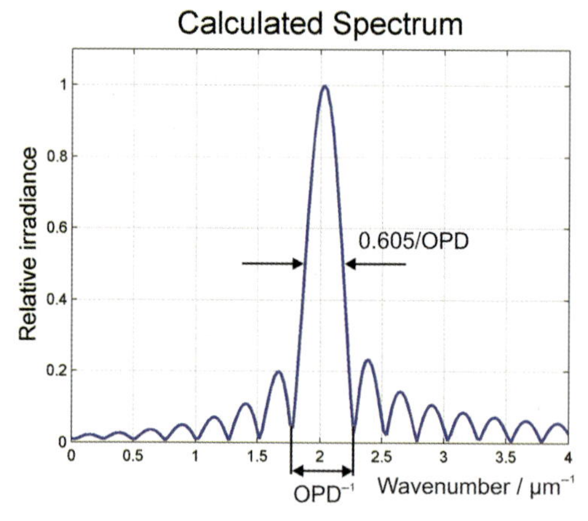

Figure 48-81: Spectrum of monochromatic radiation sampled over an optical path difference of OPD.

The theoretical full-width-half-maximum (FWHM) resolution of an FTS in wave-numbers σ is then given by

$$\Delta\sigma \approx \frac{0.605}{\text{OPD}} \tag{48-229}$$

or, expressed as wavelength λ

$$\Delta\lambda \approx \frac{0.605\lambda^2}{\text{OPD}} \tag{48-230}$$

The resolution of an FTS is therefore wavelength-dependent. Shorter wavelengths are resolved much more distinctly than longer wavelengths.

For a mirror scan range of 0.5 mm in a Michelson interferometer with resulting OPD of 1 mm, the resolutions as shown in table 48-7can be achieved.

Table 48-7: Wavelength resolutions for OPD = 1 mm, equivalent to a wave number resolution of $\Delta\sigma = 6.05\ \text{cm}^{-1}$.

λ/ nm	$\Delta\lambda$ / nm
500	0.15
1000	0.61
1500	1.36
2000	2.42
2500	3.78
3000	5.45
3500	7.41
4000	9.68
4500	12.25

For scanning systems of the temporally modulated FTS type, larger OPDs can be achieved, leading to very high resolutions. Mirror displacements of 150 mm are reported, leading to a resolution of $\Delta\sigma = 0.02\ \text{cm}^{-1}$ (0.5 pm at 500 nm wavelength) [48-55].

In a spatially modulated FTS type, the maximum OPD achievable is restricted by the number of pixels N of the line sensor. Considering the Nyquist limit, the maximum number of fringes that can be resolved is given by $N/2$. Using (48-230) the resolution of a spatially modulated FTS is then approximated by

$$\Delta\lambda \approx \frac{1.21\lambda}{N} \tag{48-231}$$

Since the total spectrum is simultaneously projected onto the detector, the signal-to-noise ratio (SNR) of the detector will limit those spectral regions that are weakly represented in the original light-source spectrum. For instance, when using a halogen lamp, the shorter visible spectral regions are much lower than the infrared regions. The detector might be saturated by the IR, whereas the VIS portion itself would be close to the noise level. In this case it is appropriate to introduce well-defined low-pass or band-pass filters to allow the FTS to measure the specified spectra very precisely.

Devices

There are many different FTS-methods which are used in a variety of fields. The types of FTS are manifold. In addition to the Michelson and Mach–Zehnder interferometer arrangements, there are also Fabry–Perot arrangements [48-48], lamellar grating interferometers [48-49], Sagnac interferometers [48-50] and more.

One main issue of contemporary development is the miniaturization of FTS techniques which makes instruments and sensors easier to handle, faster and cheaper. Compact spectrometers are used in various applications, including color measurement, quality and process control, gas detection and chemical analysis. Their fields of application cover environmental monitoring, the food and beverage industry, imagery, telecommunication, life science and medical diagnostics. Microelectromechanical system technology (MEMS) [48-51]–[48-53] and silica-based planar waveguide technology [48-50], [48-54] and [48-55] provide suitable elements for miniaturization.

A further issue is the development of *spectral imaging*, providing techniques that are capable of measuring individual spectra for each pixel of an array sensor [48-48], [48-49], [48-56], enabling the spatially resolved color measurements of objects.

Contemporary commercial FTS use fiber arrangements for measuring the transmitted or reflected spectra of samples in a convenient way as shown in figures 48-82 and 83.

Wavelength ranges cover the visible range (VIS) and the near-infrared range (NIR) from 400–2600 nm. Other instruments also cover the mid-infrared range up to a wavelength of 4500 nm. Fibered systems use NIR multi-mode fibers for illumination and FTS connection. Spectral resolutions depend on the wavelength range and can be expected to be between < 1 nm for 400 nm and <10 nm at a wavelength of 2600 nm. Scanning of the total spectrum takes several seconds [48-45].

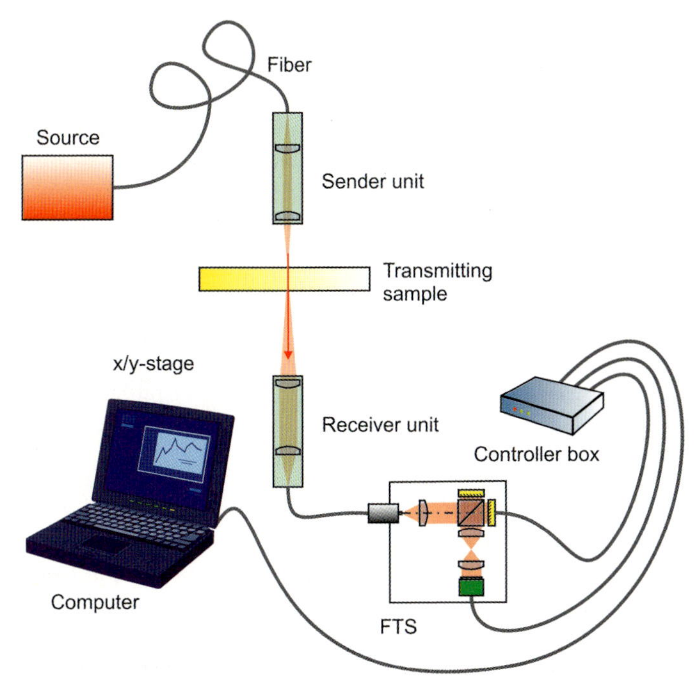

Figure 48-82: Schematic fiber arrangement used to measure spectra of transmitting samples with an FTS.

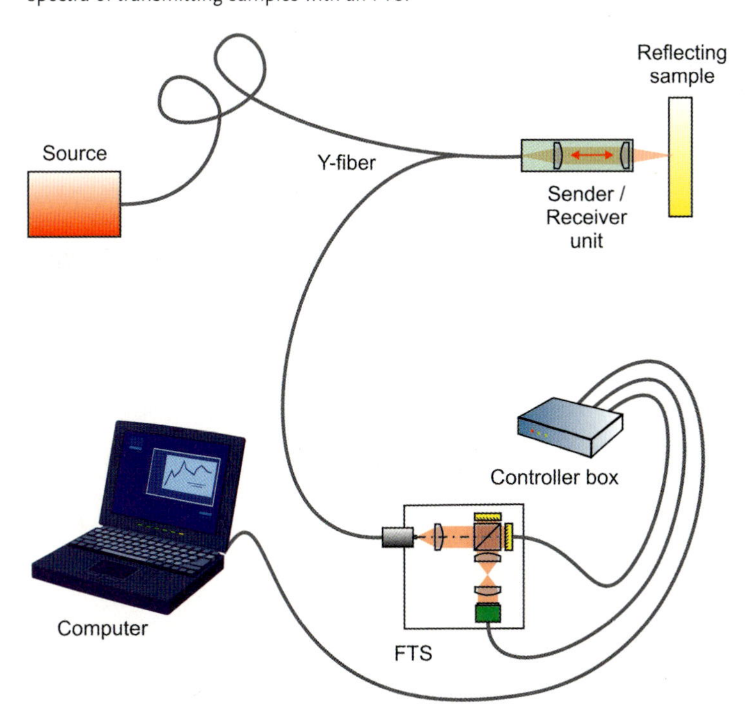

Figure 48-83: Schematic fiber arrangement used to measure spectra of reflecting samples with an FTS.

48.4.6
Accuracy and Error Sources

The following main parameters determine the accuracy of measurement for spectral transmission and reflection measurements of optical components and systems with spectrometers and spectrophotometers:

- the spectral resolution and band pass (dependent on wavelength range),
- the wavelength accuracy and reproducibility,
- the photometric accuracy and reproducibility.

Spectral Resolution and Band Pass

The spectral resolution is determined by the monochromator resolution as discussed in section 48.3. Usually, commercially available, single-channel spectrophotometers allow the adjustment of the slit width to meet the special measurement requirements. The smallest band pass specified for commercial spectrometers is their real spectral resolution.

The spectral resolution of a multi-channel spectrometer is limited by the monochromator and the spectral pixel pitch as defined in (48-217).

The spectral resolution of a Fourier spectrometer depends on the maximum optical path difference OPD achievable by the scanning mechanism of the interferometer.

Wavelength Accuracy and Reproducibility

The wavelength accuracy of a spectrometer is one of the most important parameters. Calibration of the wavelength scale is achieved by reference measurements with well-known spectral lines as will be described in section 48.4.7. After calibration, the spectral reproducibility will mainly determine the wavelength accuracy. The spectral reproducibility of a system depends on the stability of the monochromator under changing environmental conditions (temperature, air pressure, humidity, mechanical vibrations, air turbulences, etc.).

In a Fourier spectrometer, systematic wavelength errors can be introduced by the following.

- Errors in the displacement measurement: the absolute error in length metrology directly effects the determination of the wavelength scale. If the relative displacement error is, for example, 10^{-5}, then the wavelength is determined with an accuracy of $\Delta\lambda > \lambda \times 10^{-5}$.
- Nonlinearities in the total movable-mirror system, including the displacement metrology will broaden the spectrum. A sinusoidal signal will be recorded as a “chirped” signal, thus the line spectrum will be affected and broadened.

Photometric Accuracy and Reproducibility

The absolute value of the radiant flux or derived quantities is usually not necessary when testing optical components and systems. It is only the relative radiance parameters which have to be considered when measuring the transmittance and reflectance of components or coatings. The absolute radiant flux or derived quantities are of interest for characterizing sensors and light sources. In this case additional absolute calibration techniques must be considered [48-57].

To characterize the accuracy of the spectrometric transmission and reflection measurements, traditionally the photometric accuracy is specified. This is often specified in *spectral absorbance units*. The spectral absorbance $A(\lambda)$ is the logarithm to the base ten of the reciprocal of the spectral transmittance $T(\lambda)$:

$$A(\lambda) = \log\frac{1}{T(\lambda)} \qquad (48\text{-}232)$$

The reading displayed by commercially available photometers is usually the absorbance, because it is proportional to the concentration according to the *Lambert–Beer law.* Since the resolution is generally good in the high-transmittance range, in practice, a range of 0–0.6 A (corresponding to T from 100% to 25 %) is preferred.

Photometric accuracy is critical for many applications of UV-VIS spectrophotometers. For example, when comparing measurements taken on different instruments, it is necessary to know the accuracy of each instrument before a true comparison can be made [48-60].

The main systematic error source degrading the photometric accuracy is light, other than the wavelength of interest, reaching the detector. Usually this disturbance light is caused by stray light. Stray light can have several origins:

- scattered light due to any optical surfaces within the spectrometer,
- back reflections in the case of transparent samples,
- ghost images in the dispersion plane caused by non-periodic errors in the ruling of the grating grooves.

Scattered light can occur in any spectrometer type. For grating spectrophotometers the grating can add a dominant part to the stray light. The type of grating determines the level of grating stray light in that stray light from a holographic grating is normally less than that of a mechanically ruled grating (up to a factor of 10). Moreover, holographic gratings do not have periodic ruling errors and therefore no ghosts in the dispersion plane. Typically, the stray light from holographic gratings is not focussed.

The photometric accuracy is further limited by the linearity of the detector. To linearize the detection system, calibration methods can be applied as will be described in section 48.4.7. One very accurate method is the *double aperture method.*

The photometric reproducibility is mainly limited by detector noise and fluctuations in the light source. The latter is mainly compensated for by implementing reference detectors, which normalize the measured signal.

In a Fourier spectrometer, statistical errors in the interferogram detection will result in higher orders affecting the spectrum by erroneous spectra. For instance, a line spectrum at λ will contain an erroneous spectrum at $\lambda/2$.

Determining the photometric accuracy of a spectrophotometer has been traditionally performed by using solutions of high-purity compounds prepared by the operator, or by measuring the absorbance of calibrated neutral density filters, issued by one of the national standards organizations. The former method is limited by the volumetric accuracy of the solution preparation and the measurement technique. The latter method relies upon the filters being kept clean, free from scratches and at a constant temperature. The transmittance of the filters can also change by as much as ± 1% of the value over the first year [48-58].

Examples of wavelength accuracy and reproducibility, photometric reproducibility, linearity deviation and photometric accuracy, for commercial spectrophotometers are shown in table 48-8.

Table 48-8: Wavelength accuracy and reproducibility, photometric reproducibility, linearity deviation and accuracy of commercial spectrophotometers (from [48-59]).

Wavelength accuracy and reproducibility	Wavelength error	
Accuracy UV/VIS	± 0.08 nm	
Accuracy NIR	± 0.3 nm	
Reproducibility UV/VIS	± 0.01 nm	
Reproducibility NIR	± 0.04 nm	
Absorbance and transmittance range	**Absorbance reproducibility**	**Transmittance reproducibility**
1.0 A ⇔ T = 10.0% at 546.1 nm	± 0.00016 A	± 0.0037%
0.5 A ⇔ T = 31.6% at 546.1 nm	± 0.00008 A	± 0.0058%
0.3 A ⇔ T = 50.1% at 546.1 nm	± 0.00008 A	± 0.0092%
Absorbance and transmittance range	**Absorbance linearity deviation**	**Transmittance linearity deviation**
1.0 A ⇔ T = 10.0% at 546.1 nm	± 0.006 A	± 0.139%
2.0 A ⇔ T = 1.0% at 546.1 nm	± 0.017 A	± 0.040%
3.0 A ⇔ T = 0.1% at 546.1 nm	± 0.020 A	± 0.0047%
1.0 A ⇔ T = 10.0% at 1200 nm	± 0.0005 A	± 0.012%
2.0 A ⇔ T = 1.0% at 1200 nm	± 0.001 A	± 0.0023%

Wavelength accuracy and reproducibility	Wavelength error	
Calibration method and absorbance and transmittance range	**Absorbance accuracy**	**Transmittance accuracy**
Double Aperture Method 1 A ⇔ T = 10%	± 0.0003 A	± 0.0069%
Double Aperture Method 0.5 A ⇔ T = 31.6%	± 0.0003 A	± 0.022%
NIST 1930D Filters 2 A ⇔ T = 1%	± 0.003 A	± 0.0069%
NIST 930D Filters 1 A ⇔ T = 10%	± 0.003 A	± 0.069%
NIST 930D Filters 0.5 A ⇔ T = 31.6%	± 0.002 A	± 0.15%

48.4.7 Calibration Techniques

For the calibration of light sources and detectors so-called *electrical substitution radiometers* (ESR), also referred to as absolute radiometers, are used. They are able to scale the measured radiometric quantities such as radiant flux, radiant energy, irradiance, radiance, etc., to absolute units within the *International System of Units* (SI) [48-57].

For the measurement of the spectral reflectance and transmittance of optical components and systems, an absolute calibration of the light source and detector is generally not necessary and is therefore beyond the scope of this book. Spectrometers and spectrophotometers are calibrated to relative units, setting the response with no light through the sample beam to be zero and the response with no sample in the beam to be 100%. The calibration process must then measure and compensate for the deviations from linearity.

The (relative) calibration process for spectrometers and spectrophotometers includes:

a) the measured radiant flux and
b) the associated wavelength scale.

In the following we will describe both calibration principles.

Calibration of Radiant Flux

All instruments are subject to non-linearity errors that result in a systematic error in the measured parameter. For example, a measured transmittance value T_m must be corrected by subtracting $\Delta T(T)$ to obtain the true transmittance T [48-60]:

$$T_m(T) = T + \Delta T(T) \tag{48-233}$$

The standard procedure used to determine non-linearities in spectrometers and spectrophotometers is to use a set of filters of known transmittance. The national standards organizations, such as NIST in the USA, supply customers with standards of spectral transmittance for checking the photometric scale of spectrophotometers. The standards are polished glass disks of defined size, thickness and spectral transmittance over a defined spectral range at a specified temperature. The uncertainty ranges from approximately 0.2 % to 0.3 % of the specified value. The uncertainty includes the effects of random and systematic errors of the calibration procedure, as well as estimated systematic errors associated with alignment of the filters and material properties [48-61].

In the following, we describe a suitable technique based on the *method of light addition* to calibrate spectrophotometers without the need to introduce calibrated filter sets. The method assumes the behavior of linear systems, in which readings $I(A)$ and $I(B)$ for radiant fluxes A and B will give a reading $I(A+B) = I(A) + I(B)$, when the two fluxes are added incoherently. The method described in the following is called the *double aperture method* and utilizes pairs of apertures that are opened and closed separately or in combination.

A photometric transmittance scale is arbitrarily defined by setting the response with no light through the sample beam to be zero and the response with no sample in the beam to be 100%. If the response is linear, then the scale is absolutely accurate. The double-aperture method can be used to measure deviations from the linearity independent of wavelength, spectral bandwidth or temperature.

The equipment for the double-aperture method consists of a plate with two separated holes that can be individually covered or uncovered to produce two independent light fluxes onto the detector. The two aperture areas are adjusted, so that the fluxes passing both apertures are equal. The beam passing the double aperture can also be attenuated by a wedged neutral filter or by decreasing the current supplying the light source.

Following (48-233) we define a function $E(T_1, T_2)$:

$$\begin{aligned} E(T_1, T_2) &= {}^1\!/\!_2(T_m(T_1) + T_m(T_1) - T_m(T_1 + T_2)) \\ &= {}^1\!/\!_2(\Delta T(T_1) + \Delta T(T_1) - \Delta T(T_1 + T_2)) \end{aligned} \tag{48-234}$$

If $T_1 = T_2 = T$ (48-234) can be simplified to (48-235):

$$\begin{aligned} E(T) &= {}^1\!/\!_2(2T_\mathrm{m}(T) - T_m(2T)) \\ &= {}^1\!/\!_2(2\Delta T(T) - \Delta T(2T)) \end{aligned} \tag{48-235}$$

The calibration process now includes the following steps as illustrated in figure 48-84.

1. Let the unattenuated beam pass both apertures and set $T_\mathrm{m}(100\%) = 100\%$ with $\Delta T(100\%) = 0$.
2. Let the unattenuated beam path a single aperture and measure $T_\mathrm{m}(50\%)$, then calculate from (48-233): $\Delta T(50\%) = T_\mathrm{m}(50\%) - 50\%$.

3. Attenuate the beam to 50% and repeat the measurements with both apertures open and a single aperture open to obtain $T_m(50\%)$ and $T_m(25\%)$. Use the result from step 2 and calculate from (48-235):

$$\Delta T(25\%) = \tfrac{1}{2}(2T_m(25\%) - T_m(50\%) + \Delta T(50\%))$$

4. Attenuate the beam to 25% and repeat the measurements and calculations according to steps 1–3. Then attenuate to 12.5%, 6.25% etc. and repeat the process, respectively.

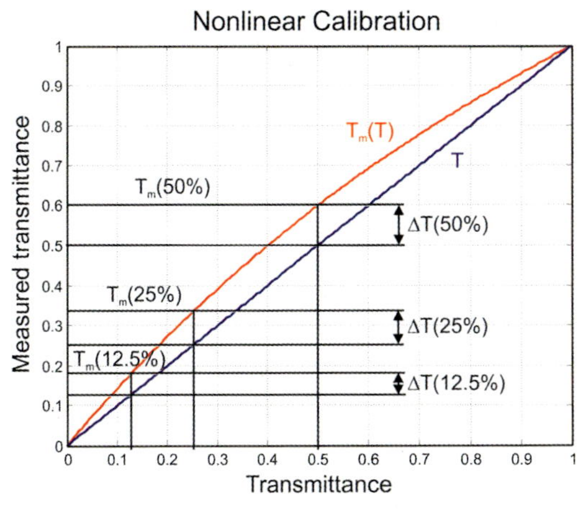

Figure 48-84: Applying the double-aperture method to measure a spectrophotometer's transmittance deviation from linearity.

The described 50% method has some disadvantages, namely:

1. the measured corrections are unevenly distributed along the T-axis, with wide gaps between them in the regions of most interest,
2. most measurements are made at low light levels, where the correction is small and the experimental error is large,
3. each measured point depends on the previous ones, so that the errors accumulate as T is decreased [48-60].

A method of extending the calibration curve is to use the method of *addition of filters*. It is capable of measuring relative linearity errors. When coupled with the *double aperture method*, absolute errors can be measured [48-58] and [48-62].

Calibration of Wavelength Scale

Spectrometers are commonly calibrated by using well-known wavelengths. Usually two or more wavelengths are necessary. In a Fourier spectrometer only one well-known wavelength is necessary, if the displacement of the movable mirror is controlled by a precise displacement-measuring instrument. However, if the stage adjustment or the travel direction are not quite accurate, it is wise to use a further or several well-known wavelengths that are spread over the total spectral range.

An intracavity, iodine-stabilized helium–neon laser operating at 633 nm is an excellent standard of wavelength [48-63]. Its laser line accuracy $\Delta\lambda/\lambda$, when locked on hyperfine transitions of iodine, is $< 10^{-9}$. The absolute wavelength of this line is 632 991.398 ± 0.003 pm. Other reference lines to be used for calibration in the visible spectrum are shown in table 48-9 [48-64].

Table 48-9: Reference lines used for calibration (from [48-64]).

Reference line	Wavelength [nm]
Fourth anti-Stokes of YAG	410.8300 ± 0.0004
Third anti-Stokes of YAG	435.6778 ± 0.0003
Second anti-Stokes of YAG	463.7247 ± 0.0003
Ar^+ 476	476.622 ± 0.006
Ar^+ 488	488.122 ± 0.006
First anti-Stokes of YAG	495.6311 ± 0.0003
Ar^+ 497	496.645 ± 0.004
Ar^+ 514	514.676 ± 0.004
He–Ne green	543.5159 ± 0.0005
Ne	572.094 ± 0.004
First Stokes of YAG	574.7177 ± 0.0002
U	576.20331 ± 0.00005
Second Stokes of YAG	624.5463 ± 0.0002
He–Ne red	632.991398 ± 0.000003
Rb D_2 d/f	780.2462916 ± 0.0000008
Rb D_2 b/f	780.246450 ± 0.000002
Rb D_1 c′	794.981364 ± 0.000002
Rb D_1 d′	794.974964 ± 0.000003
Cs	852.33512 ± 0.00004

For very high-precision purposes, the national standards laboratories such as PTB, NPL, and NIST provide very accurately derived radio frequencies from atomic standards, such as the cesium atomic clock, by coupling them in a phase-locked manner to optical frequencies, preserving the accuracy from the atomic standard for optical measurements. Using frequency multiplication and frequency division in so-called frequency chains, very precise measurements of the order of $\Delta\lambda/\lambda=10^{-14}$ for optical emission wavelengths or optical transitions within atoms can be achieved.

An improvement with a potential of up to $\Delta\lambda/\lambda=10^{-17}$ is possible, when frequency combs from a femtosecond laser are used. In frequency space, their emission spectrum consists of an equidistant, phase-rigid multitude of spectral lines, with a frequency spacing equal to the repetition frequency. The frequency and phase of this laser light can also be stabilized to radio frequency sources using atomic standards [48-65].

48.5 Literature

48-1 E. Hecht and A. Zajac, Optics (Pie) (Addison-Wesley Longman, Amsterdam, 4th edn International, 2003).

48-2 M. Born and E. Wolf, Principles of Optics, (Cambridge University Press, Cambridge, 1999).

48-3 E. F. Zalewski, Radiometry and Photometry, Handbook of Optics, Volume II, Devices, Measurements and Properties, edited by M. Bass, editor in chief, E. W. van Stryland, D. R. Williams, W. L. Wolfe, associate editors, chap. 24, p. 24.3–24.51 (McGraw-Hill, Inc., New York, San Francisco, Washington D.C., Auckland, Bogotá, Caracas, Lisbon, London, Madrid, Mexico City, Milan, Montreal, New Delhi, San Juan, Singapore, Sydney, Tokyo, Toronto, 1995).

48-4 Y. Ohno, Photometry, in Optical Radiometry, ed. by A. C. Parr, R. U. Datla, J. l. Gardner, Volume 41 of "Experimental Physical Sciences", treatise editors R. Celotta, T. Lucatorto (Elsevier Academic Press, Amsterdam, Boston, Heidelberg, London, New York, Oxford, Paris, San Diego, San Francisco, Singapore, Sydney, Tokyo, 2005), chap. 7, p.327–31.

48-5 R. McCluney, Introduction to Radiometry and Photometry (Artech House, Inc., Norwood, MA, 1994).

48-6 P. J. Mohr and B. N. Taylor, CODATA recommended values of the fundamental physical constants, Rev. Mod. Phys. 77, 1 (2005), http://physics.nist.gov/cuu/Constants (2010).

48-7 Maßstäbe, H.01 (2001), http://www.ptb.de/cms/publikationen (2010).

48-8 http://www.schott.com/advanced_optics/english/tools_downloads/download/ Datasheets (2010).

48-9 Optical Glass Filters, catalogue, data sheets, calculation program, http://www.schott.com/advanced_optics/english/our_products/filters/index.html (2010).

48-10 Transmittance of Optical Glass, Technical Information TIE 35, http://www.schott.com/advanced_optics/english/tools_downloads/download/ (October 2005).

48-11 W. Gunning, Double-cavity electrooptic Fabry–Perot tunable filter, Applied Optics, Vol. 21, No. 17, p. 3129–31 (1982).

48-12 F. Grum and R. J. Becherer, Optical Radiation Measurements, vol. 1 Radiometry (Academic Press, New York, San Francisco, London, 1979).

48-13 V. L. Yen, Circular variable filters, Opt. Spectra, p.78–83 (1969).

48-14 A. Thelen, Circularly wedged optical coatings. I. Theory, Applied Optics, Vol. 4, No. 8, p. 977–81 (1965).

48-15 A. Thelen, Circularly wedged optical coatings. II. Experimental, Applied Optics, Vol. 4, No. 8, p. 983–85 (1965).

48-16 S. Yang, Circular, variable, broad-bandpass filters with induced transmission at 200–1100 nm, Applied Optics, Vol. 32, No. 25, p. 4836–42 (1993).

48-17 D. C. Youvan, William J. Coleman, Chris M. Silva, Julien Petersen, Edward J. Bylina and Mary M. Yang, Fluorescence Imaging Micro-Spectrophotometer (FIMS), Biotechnology et alia, p. 1–16, www.et-al.com (1997).

48-18 Interference Filters and Special Filters Catalogue, http://www.schott.com/advanced_optics/english/our_products/filters/interference_datasheets.html (2010).

48-19 L. Domash, M. Wu, N. Nemchuk and E. Ma, Tunable and switchable multiple-cavity thin film filters, Journal of Lightwave Technology, vol. 22, no. 1, p. 126–35 (2004).

48-20 N. Gupta and V. B. Voloshinov, Development and characterization of two-transducer imaging acousto-optic tunable filters with extended tuning range, Applied Optics, Vol. 46, No. 7, p. 1081–88 (2007).

48-21 P. A. Gass and J.R. Sambles, Accurate design of a noncollinear acousto-optic tunable filter, Optics Letters, Vol. 16, No.6, pp. 429–31 (1991).

48-22 Brimrose, AOTF Model Number Guide, www.brimrose.com (2010).

48-23 Crystal Technology, Inc., Acousto-Optic Tunable Filters, www.crystaltechnology.com (2010).

48-24 I. C. Chang, Acousto-Optic Devices and Applications, in Handbook of Optics, Volume II, Devices, Measurements and Properties, edited by M. Bass, editor in chief, E. W. van Stryland, D. R. Williams, W. L. Wolfe, associate editors, chap. 12, p. 12.1–12.54 (McGraw-Hill, Inc., New York, San Francisco, Washington D.C., Auckland, Bogotá, Caracas, Lisbon, London, Madrid, Mexico City, Milan, Montreal, New Delhi, San Juan, Singapore, Sydney, Tokyo, Toronto, 1995).

48-25 G. Georgiev, D. A. Glenar and J. J. Hillman, Spectral characterization of acousto-optic filters used in imaging spectroscopy, Applied Optics, Vol. 41, No. 1, p. 209–17 (2002).

48-26 J. Zhu and A. Y. S. Cheng, A high-speed automatic spectrometer based on a solid state non-collinear acousto-optic tunable filter, Chinese Optics Letters, Vol.1, No. 2, p. 85–87 (2003).

48-27 W. J. Tropf, M. E. Thomas and T. J. Harris, Properties of Crystals and Glasses, in: Handbook of Optics, Volume II, Devices, Measurements and Properties, edited by M. Bass, editor in chief, E. W. van Stryland, D. R. Williams, W. L. Wolfe, associate editors, chap. 33 (McGraw-Hill, Inc., New York, San Francisco, Washington D.C., Auckland, Bogotá, Caracas, Lisbon, London, Madrid, Mexico City, Milan, Montreal, New Delhi, San Juan, Singapore, Sydney, Tokyo, Toronto, 1995).

48-28 J. W. Evans, The Birefringent Filter, JOSA, Vol. 39, No. 3, p. 229–42 (1949).

48-29 O. Aharon and I. Abdulhalim, Tunable optical filter having a large dynamic range, Optics Letters, Vol. 34, No. 14, p. 2114–16 (2009).

48-30 O. Aharon and I Abdulhalim, Liquid crystal tunable Lyot filter with extended free spectral range, Optics Express, Vol. 17, No. 14, p. 11426–33 (2009).

48-31 VariSpec Liquid Crystal Tunable Filters, Brochure Channel Systems and CRI, www.spectralcameras.com (2010).

48-32 Meadolawk Optics Selectable Bandwidth Tunable Optical Filters. Brochure, www.meadolark.com (2010).

48-33 H. J. Masterson, G. D. Sharp and K. M. Johnson: Ferroelectric liquid-crystal tunable filter, Optics Letters, November 15, Vol. 14, No. 22, p. 1249 (1989).

48-34 G.D. Sharp, K. M. Johnson and D. Doroski, Continuously tunable smectic A* liquid-crystal color filter, Optics Letters,Vol. 15, No. 10, p. 523–25 (1990).

48-35 Elena Nicolescu and Michael J. Escuti, Polarization-independent tunable optical filters using bilayer polarization gratings, Applied Optics, Vol. 49, No. 20, p. 3900–04 (2010).

48-36 Elena Nicolescu and Michael J. Escuti, Compact Spectrophotometer using Polarization-independent Liquid Crystal Tunable Optical Filters, Proceedings of SPIE, vol. 6661, no. 666105 (2007).

48-37 H. Haferkorn, Optik (Wiley-VCH, 4. Auflage, 2003).

48-38 E. Abbe, Gesammelte Abhandlungen, Band I, p. 3–6 (Georg Olms Verlag Hildesheim, Zürich, New York 1989).

48-39 Diffraction Gratings Ruled and Holographic Handbook, published by HORIBA Jobin Yvon Inc, Edison, New Jersey 08820 USA and HORIBA Jobin Yvon Div. d'Instruments SA, 1618 Rue du Canal 91160, Longjumeau, France (1988).

48-40 A. Thevenon, J. Flamand, J.P. Laude, B. Touzet and J.M. Lerner, Aberration Corrected Plane Gratings, SPIE Proc 815, 136145 (1987).

48-41 PerkinElmer, Inc., Technical information on LAMBDA series instruments, Waltham, MA, USA (2004).

48-42 Carl Zeiss MicroImaging GmbH, MCS 600 – Technical Specifications, Doc. Nr. 72-1-0001/e, Jena, Germany (2008).

48-43 J. E. Chamberlain, The Principles of Interferometric Spectroscopy, Wiley, New York (1979).

48-44 W. Otto, Ein Fourierspektrometer für das Sichtbare und Ultraviolette, Dissertation at the Universität-Gesamthochschule-Siegen, Germany (1997).

48-45 Arcoptix S. A, Product Brochure on Ultra-High-Throughput Fourier Transform Spectrometer, Neuchâtel, Switzerland (2010).

48-46 V. Saptari, Fourier Transform Spectroscopy Instrumentation Engineering (SPIE Press, Bellingham, Washington, 2004).

48-47 V. V. Arkhipov, Scanning systems of rapid-scanning Fourier spectrometers, J. Opt. Technol. 77, p. 435–41 (2010).

48-48 M. Pisani and M. Zucco, Compact imaging spectrometer combining Fourier transform spectroscopy with a Fabry-Perot interferometer, Optics Express 17, p. 8319–31 (2009).

48-49 Y. Garini, I. T. Young and G. McNamara, Spectral imaging: Principles and applications, Cytometry Part A, 69A(8), p. 735–47 (2006).

48-50 E. Le Coarer, S. Blaize, P. Benech, I. Stefanon, A. Morand, G. Lérondel, G. Leblond, P. Kern, J. M. Fedeli and P. Royer, Stationary waves integrated Fourier transform spectrometry (SWIFTS): towards an ultimate wavelength scalespectrometer, Nature Photonics 8, 473–78 (2007).

48-51 L. Wu, A. Pais, S. R. Samuelson, S. Guo and H. Xie, A miniature Fourier transform spectrometer by a large-vertical-displacement microelectromechanical mirror, in Fourier Transform Spectroscopy, OSA Technical Digest (CD) (Optical Society of America, 2009), paper FWD4.

48-52 O. Manzardo, R. Michaely, F. Schädelin, W. Noell, T. Overstolz, N. De Rooij and H. P. Herzig, Miniature lamellar grating interferometer based on silicon technology, Optics Letters 29, p. 1437–39 (2004).

48-53 O. Manzardo, H. P. Herzig, C. R. Marxer and N. F. de Rooij, Miniaturized time-scanning Fourier transform spectrometer based on silicon technology, Optics Letters 24, p. 1705–07 (1999).

48-54 M. Froggat and T. Erdogan, All-fiber wavemeter and Fourier-transform spectrometer, Optics Letters 24, p. 942–44 (1999).

48-55 K. Okamoto, H. Aoyagi and K. Takada, Fabrication of Fourier-transform, integrated-optic spatial heterodyne spectrometer on silica-based planar waveguide, Optics Letters, Vol. 35, Issue 12, pp. 2103–05 (2010).

48-56 J. Li and R. K. Y. Chan, Imaging Fourier transform spectrometer based on a beam-folding position-tracking technique, in Fourier Transform Spectroscopy, OSA Technical Digest (CD) (Optical Society of America, 2009), paper FWB4.

48-57 A. C. Parr, A National Measurement System for Radiometry, Photometry, and Pyrometry Based upon Absolute Detectors, NIST Technical Note, 1421 (2009).

48-58 R. Francis, Measuring photometric accuracy using the double aperture method, Technical information from Varian Australia Pty Ltd, Mulgrave, Victoria, Australia (1993).

48-59 Perkin Elmer, Inc., Technical Specifications for the LAMBDA 1050 UV/Vis/NIR and LAMBDA 950 UV/Vis/NIR, Waltham, MA, USA (2007).

48-60 K. D. Mielenz and K. L. Eckerle, Spectrophotometer linearity testing using the double-aperture method, Appl. Opt. 11, p. 2294–303 (1972).

48-61 NIST Special Publication SP 260-116 – Glass Filters as a Standard Reference Material for Spectrophotometry: Selection, Preparation, Certification and Use of SRM 930 and SRM 1930 (March 1994).

48-62 R. C. Hawes, Technique for measuring photometric accuracy, Appl. Opt. 10, p. 1246–53 (1971).

48-63 H. P. Layer, A portable iodine stabilized helium-neon laser, IEEE Trans. Instrum. Meas. IM-29, p. 358–61 (1980).

48-64 P. S. Bhatia, C. W. McCluskey and J. W. Keto, Calibration of a computer-controlled precision wavemeter for use with pulsed lasers, Appl. Opt. 38, p. 2486–98 (1999).

48-65 T. Fischer and W. Kaenders, Wavelength meters: Solid-state etalons improve wavelength measurement, Laser Focus World, June 2004.

49
Image Analysis

Handbook of Optical Systems: Vol. 5. Metrology of Optical Components and Systems. First Edition.
Edited by Herbert Gross.

49.1
Introduction

The traditional way to inspect optical systems is the use of a specially designed and well known test object for image analysis of the system under test. Depending on the test object all important aberrations can be detected and – by means of a suitable analysis method – can be qualified and quantified. Aberrations include those affecting the geometry of an image such as magnification, distortion and field curvature as well as those affecting the image resolution and sharpness.

Image analysis methods have also been used to inspect optical surfaces. The basic principle is to let the surface under test reflect a specially designed object and to measure the geometry of the distorted image in order to determine the radius of curvature or the general shape of the test surface. The technique is generally known as *deflectometry* and can be applied only to specular reflecting surfaces.

In an early fabrication state, after grinding or lapping, the surface might not have sufficient specular reflectivity. Deflectometry can then be replaced by methods where specially designed patterns are projected onto the matt, opaque surfaces under test. The surface is observed under an angle which is different from the projection angle. The lateral displacement and distortion of the pattern is then analyzed and used to determine the shape of the matt test surface. The technique is generally known as *triangulation* or *pattern projection* and can be applied only to surfaces which scatter enough light in the direction of the observer.

49.2
Basic Principles of Image Analysis

49.2.1
System Setup

Figure 49-1 shows the arrangement to test the performance of an optical system. A mask carrying a special pattern is illuminated by a light source and imaged onto a suitable camera by the test system. The camera is connected to a computer via a frame-grabbing system. Image analysis is carried out by a specially designed software program.

When deflectometry or pattern projection are applied, the optical test system can be thought of as an optical arrangement containing the mask-projecting system, the test surface and the imaging system. In each case, the principle is the same as that illustrated in Figure 49-1.

In image analysis techniques the pattern of the mask plays a major role. The pattern is selected according to the system characteristic to be investigated. Table 49-1 shows a variety of patterns in use for different test methods. Graytone patterns are shown as well as their binary relatives and their mathematical description. Depending on the application T_b either describes the transmissivity or the reflectivity of a binary test pattern, whereas T_g describes that of a graytone test pattern.

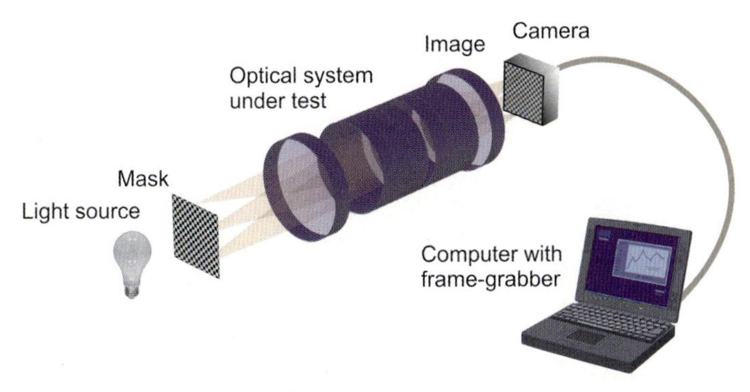

Figure 49-1: Testing the performance of an optical system by image analysis.

Table 49-1: Test patterns used in image analysis for various qualification applications.

	Binary	Graytone	
Single circular spot a half width $T_g = e^{-\left(\frac{r}{a}\right)^2}$ $T_b = \begin{cases} 0 \to r > a \\ 1 \to r \le a \end{cases}$			Used as "star test" with a spot as small as possible. The "Point Spread Function" (PSF) results from the measured irradiance distribution in the image plane.
Single slit a half width $T_g = e^{-\left(\frac{x}{a}\right)^2}$ $T_b = \begin{cases} 0 \to \lvert x\rvert > a \\ 1 \to \lvert x\rvert \le a \end{cases}$			In many situations it is easier to use a very thin slit instead of a star to derive the PSF. The slit then has to be measured in at least two orthogonal orientations.
Uniform equidistant grid a cycle period, b threshold $T_g = \frac{1}{2}\left(1 + \cos\left(2\pi \frac{x}{a}\right)\right)$ $T_b = \begin{cases} 0 \to T_g > b \\ 1 \to T_g \le b \end{cases}$			Used to measure • modulation transfer value for one spatial frequency, • distortion of optical systems. Needs to be measured in at least two orthogonal directions.
Checkerboard a cycle period $T_g = \frac{1}{4} \times$ $\left(2 + \cos\left(2\pi \frac{x+y}{a}\right) + \cos\left(2\pi \frac{x-y}{a}\right)\right)$ $T_b = \begin{cases} 0 \to T_g > \frac{1}{2} \\ 1 \to T_g \le \frac{1}{2} \end{cases}$			Used to measure • modulation transfer value for one spatial frequency, • distortion of optical systems. Measurement contains information in all lateral directions.

	Binary	Graytone	
Circular spot grid a cycle period, b width, c threshold $T_g = e^{-\left(\frac{\cos\left(2\pi\frac{x}{a}\right)\cos\left(2\pi\frac{y}{a}\right)}{b}\right)^2}$ $T_b = \begin{cases} 0 \to T_g > c \\ 1 \to T_g \le c \end{cases}$	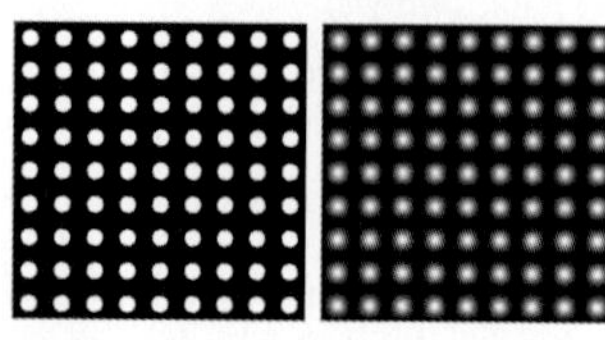		Used to measure • Point Spread Functions (PSF) in the field, • distortion of optical systems. Measurement contains information in all lateral directions.
Concentric equidistant rings a radial cycle period $T_g = \frac{1}{2}\left(1 + \cos\left(2\pi\frac{r}{a}\right)\right)$ $T_b = \begin{cases} 0 \to T_g > \frac{1}{2} \\ 1 \to T_g \le \frac{1}{2} \end{cases}$	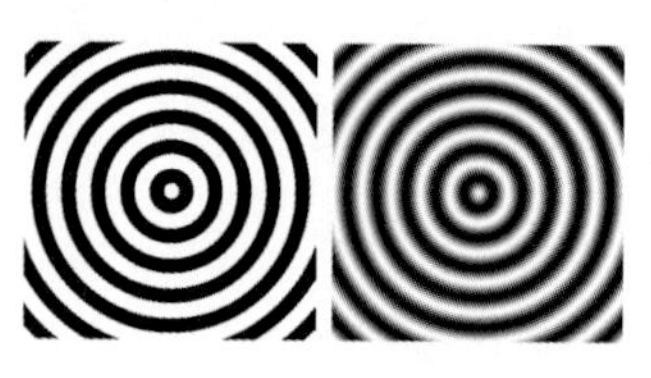		Used to measure • modulation transfer value for one spatial frequency in azimuthal direction, • distortion in radial direction.
Siemens star a azimuthal cycle period $T_g = \frac{1}{2}\left(1 + \cos\left(2\pi\frac{\varphi}{a}\right)\right)$ $T_b = \begin{cases} 0 \to T_g > \frac{1}{2} \\ 1 \to T_g \le \frac{1}{2} \end{cases}$			Used to measure • Modulation Transfer Function (MTF) in radial direction, • distortion in azimuthal direction.
Chirped circular grid a cycle period, r_0 coordinate start $T_g = \frac{1}{2}\left(1 + \cos\left(2\pi\left(\frac{r - r_0}{a}\right)^2\right)\right)$ $T_b = \begin{cases} 0 \to T_g > \frac{1}{2} \\ 1 \to T_g \le \frac{1}{2} \end{cases}$	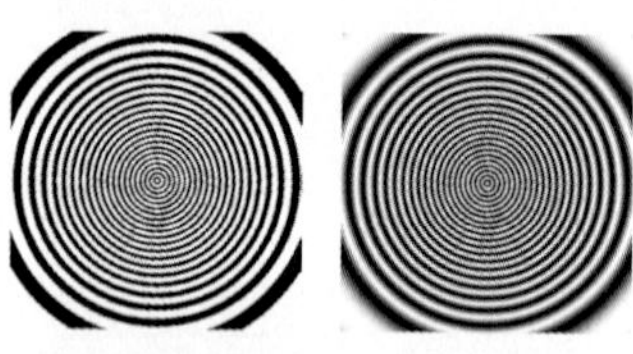		Used to measure Modulation Transfer Function (MTF) in azimuthal direction, distortion in radial direction.
Chirped linear grid a cycle period, x_0 coordinate start $T_g = \frac{1}{2}\left(1 + \cos\left(2\pi\left(\frac{x - x_0}{a}\right)^2\right)\right)$ $T_b = \begin{cases} 0 \to T_g > \frac{1}{2} \\ 1 \to T_g \le \frac{1}{2} \end{cases}$	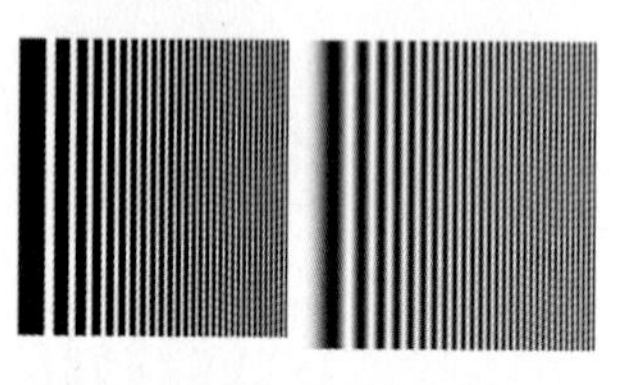		Used to measure Modulation Transfer Function (MTF) of optical systems in one direction. Needs to be measured in at least two orthogonal directions.
Random pattern Arbitrary distribution $T_g = random(0...1)$ $T_b = \begin{cases} 0 \to T_g > \frac{1}{2} \\ 1 \to T_g \le \frac{1}{2} \end{cases}$	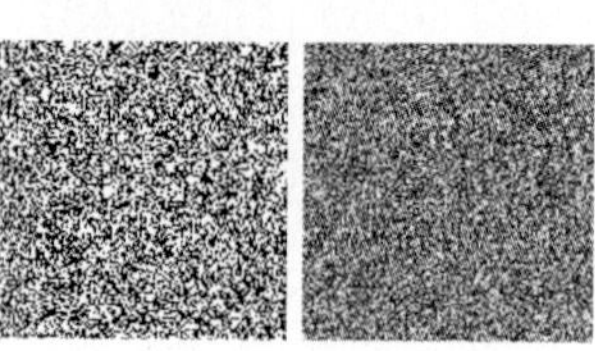		Used to measure distortion of optical systems. Measurement contains information in all lateral directions.

Depending on the application either the distortion or the "sharpness" (PSF, MTF, etc.) of the imaged pattern is analyzed.

49.2.2
Calibration Principles

In image analysis the measured irradiance distribution at the sensor or camera for at least one optical configuration is used to determine the optical performance of an optical system or a component under test.

Calibrations can be carried out in three different ways.

1) Determination of optical properties and behaviour of all critical parts in the setup as basis for a mathematical model.
2) Measurements of well-known objects (objectives, aspheres, spheres, plano surfaces) within the range of the test system as basis for a linear or nonlinear approximating model.
3) A mixture of 1 and 2, applied to subsystems of the test arrangement.

The critical parts in an image analysis test system are:

- sensor or camera,
- test target,
- light source,
- additional optical subsystems and elements.

To achieve a perfect mathematical model the following properties have to be determined.

1) The sensor or camera properties such as fixed pattern noise, bias, amplification, linearity of each pixel. Also helpful is information on dark noise, shot noise, saturation, blind pixels, etc.
2) The test target geometry, local reflectivities or transmissivities.
3) The light source spatial and angular distribution, spectral distribution and stability in time.
4) Additional optical element geometries, their relative positions, spatial and spectral reflectivities or transmissivities.

Once the optical properties and behaviour of all critical setup parts are determined, a perfect mathematic simulation can be established to determine the performance of the system under test from the measured irradiance distribution.

This is the most universal calibration method, however it is not always possible to determine properties and positions of all individual components within the test setup. A mixture, as indicated in 3, could therefore be a compromise in the case when substantial subsystems of the total test setup are well known to make a partial mathematical model, whereas the unknown parts are calibrated by means of well-known calibration elements.

The method denoted as 2 assumes that the setup is basically known and results can be produced, but with unknown accuracy. Measuring one or several well-known calibration elements distributed over the range of the measurement system then gives an opportunity to correct individual results by providing interpolated correction terms.

The following chapters will describe:

- basis techniques and setups,
- analysis methods and algorithms,
- calibration techniques and
- accuracy and error sources

which are related to the following image analysis techniques:

- star test, slit test,
- distortion metrology,
- deflectometers,
- pattern and fringe projectors.

49.3 Star Test, Slit Test

49.3.1 Basic Setups

A conceptually simple method used to test the image quality of optical systems is the so-called star test, in which the image of a point source is analyzed visually or by means of an electronic detector and a computer [49-1]–[49-10]. Figure 49-2 schematically shows a setup to test photographic lenses with a point source at infinity.

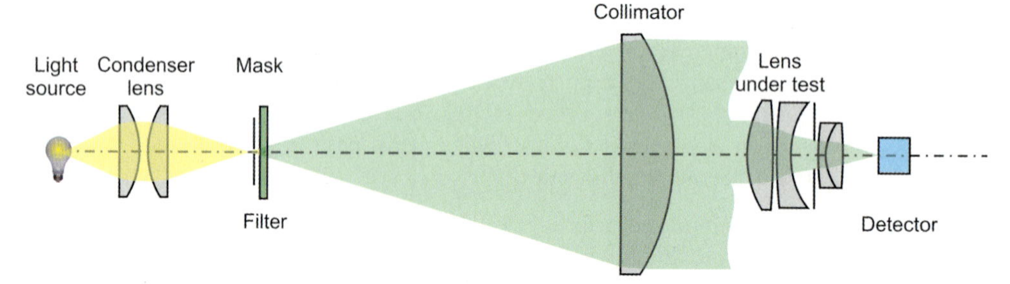

Figure 49-2: Star test setup with electronic detection to test photographic lenses.

The light from a suitable light source is imaged onto a mask, in this case an opaque screen carrying a round hole of very small size. The transmitted light then passes a filter, which transmits a spectrum adapted to the spectrum for which the test lens is designed. The filter also compensates for spectral sensitivities of the detector system. A collimator images the mask to infinity. The collimator must be corrected for all relevant aberrations including color aberrations. Parallel light then enters the lens under test. In its focal plane the image of the mask can be observed and analyzed.

For visual inspection the star image is observed using an eyepiece of suitable magnification. Due to the pupil size of the human eye, a rule of thumb can be found, in which the eyepiece's focal length in mm should be equal to or smaller than the f-number of the system under test. A test lens with $F/\# = 10$ would therefore need an eyepiece with $f' \leq 10$ mm.

For photoelectronic detection we need the following considerations:

Dimensions of Star Mask

Very well corrected lenses perform near the diffraction limit. The star image can be described by the irradiance distribution $I(x)$ of an Airy disc pattern, which can be calculated for a circular aperture from the Fraunhofer diffraction pattern given by

$$I(x) = I(0)\left(\frac{2J_1(x)}{x}\right)^2 \tag{49-1}$$

with $J_1(x)$ as the Bessel function of the first kind of order one, and x given by

$$x = 2\pi\frac{a\,q'}{2\,\lambda\,r} \tag{49-2}$$

with the following denotations:

- a is the diameter of the exit pupil diameter of the imaging optics,
- q' is the radial distance from optical axis in the image plane,
- r is the observation distance from the center of the exit pupil,
- λ is the wavelength.

In practical situations $q' << r$ and (49-2) then transforms to

$$x = 2\pi\frac{q'}{2\,\lambda\,F/\#} = 2\pi\frac{NA}{\lambda}q' \tag{49-3}$$

where F/# denotes the F-number or NA denotes the numerical aperture of the optical system.

For an annular aperture the Airy pattern can be calculated from

$$I_{\text{annular}}(x) = I(0)\left(\frac{2(J_1(x) - \varepsilon J_1(\varepsilon x))}{(1-\varepsilon^2)x}\right)^2 \tag{49-4}$$

where ε is the obscuration ratio with $0 \leq \varepsilon \leq 1$, and where $\varepsilon = 0$ means an unobscured aperture. Figure 49-3 explains the obscuration ratio.

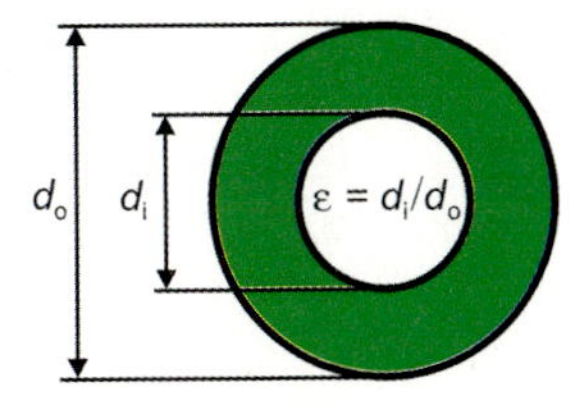

Figure 49-3: Definition of obscuration ratio.

Figure 49-4 shows the Airy patterns for obscurations $\varepsilon = 0$ and $\varepsilon = 0.5$. Figure 49-5 shows a central cross-section through both patterns in a logarithmic scale.

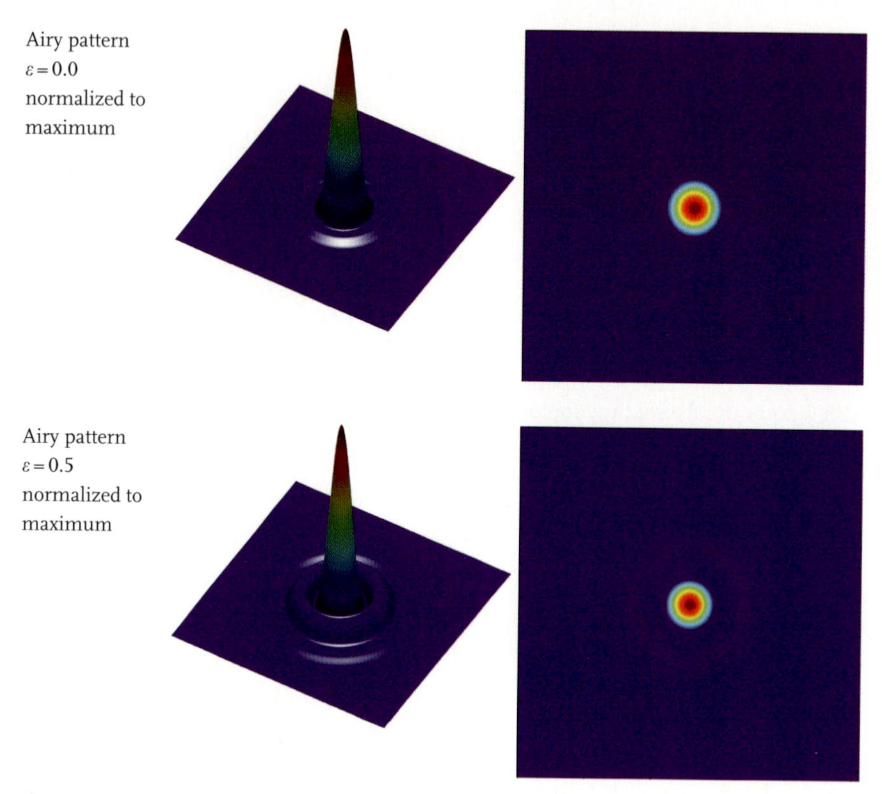

Figure 49-4: Airy pattern normalized to maximum for (a) an unobscured circular aperture, (b) a circular aperture with a central obscuration of $\varepsilon = 0.5$.

The locations of the maxima and minima in the Airy pattern up to the fourth ring and their relative irradiances are shown in table 49-2. The diameter of the first ring is approximately given by $2.44 \cdot \lambda \cdot$ F/#. For green light we therefore can assume the diameter of the Airy disc as F/# $\cdot$ 1.2 µm.

For the projection of the star mask, a collimator with a large F/# must be chosen to enable the fabrication of a suitable hole size. Since the smallest necessary diameter of the collimator is determined by the largest entrance pupil diameter to be tested, the focal length must be as large as possible. For a collimator of $\varnothing = 100$ mm and $f' = 1000$ mm, the hole size must be < 10 µm for the visible spectrum.

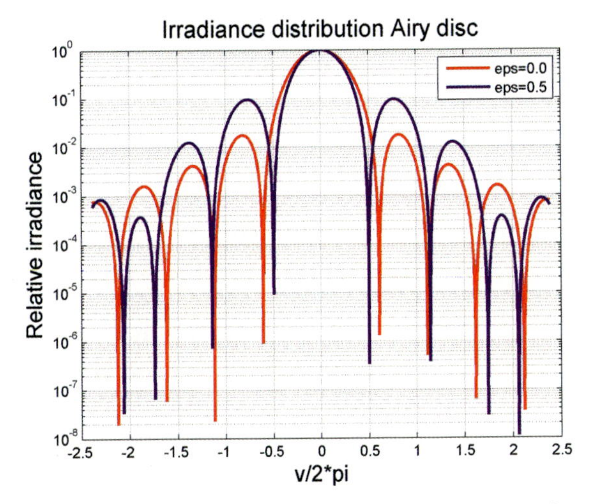

Figure 49-5: Cross-sections through Airy pattern for obscured and unobscured aperture in logarithmic scale.

Table 49-2: Normalized half diameter and relative irradiation of diffraction rings in an Airy disc.

	$\frac{NA}{\lambda} \cdot q'$	$I\left(\frac{NA}{\lambda} \cdot q'\right) / I(0)$
Central maximum	0	1.00000
1st minimum	0.6098	0.00000
1st maximum	0.817	0.01749
2nd minimum	1.117	0.00000
2nd maximum	1.340	0.00416
3rd minimum	1.619	0.00000
3rd maximum	1.849	0.00160
4th minimum	2.121	0.00000
4th maximum	2.355	0.00078

Detecting the Star Image Irradiance Distribution

There is a variety of techniques which are known to detect star images for different optical systems under test. Some are describe in the following.

A detection unit must be able to resolve much less than 1 μm if high-aperture lenses need to be tested. If a scanning single-channel unit is used to scan the *x*/*y*-image plane directly as shown in figure 49-6, a very small aperture stop, preferably < 1 μm, must be used. The scanning field must typically cover a range of

250 μm × 250 μm. If the depth of focus needs to be investigated, scanning in z-direction should be provided additionally. Figure 49-7 shows examples for perfect star images and those shaped by spherical aberration and coma wavefronts of 0.5 λ (Zernike).

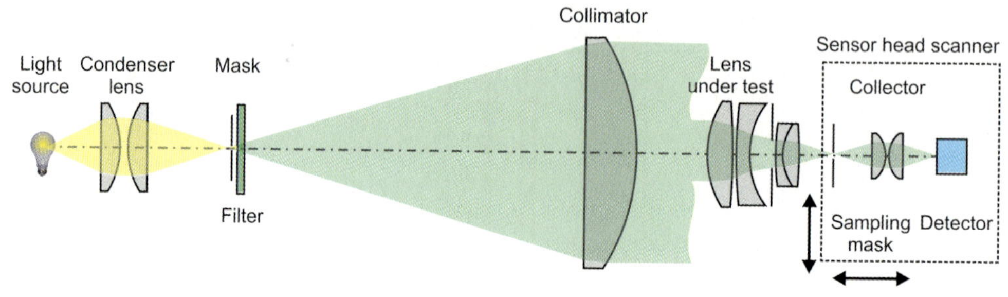

Figure 49-6: Star test setup with scanning sensor head to test photographic lenses.

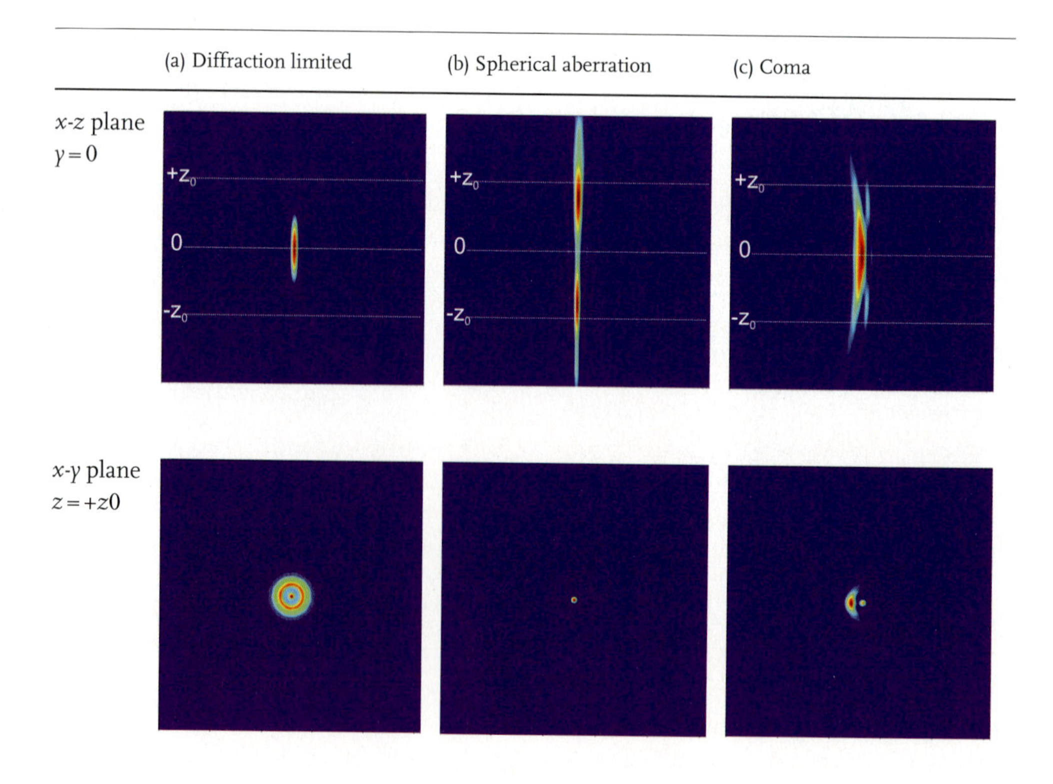

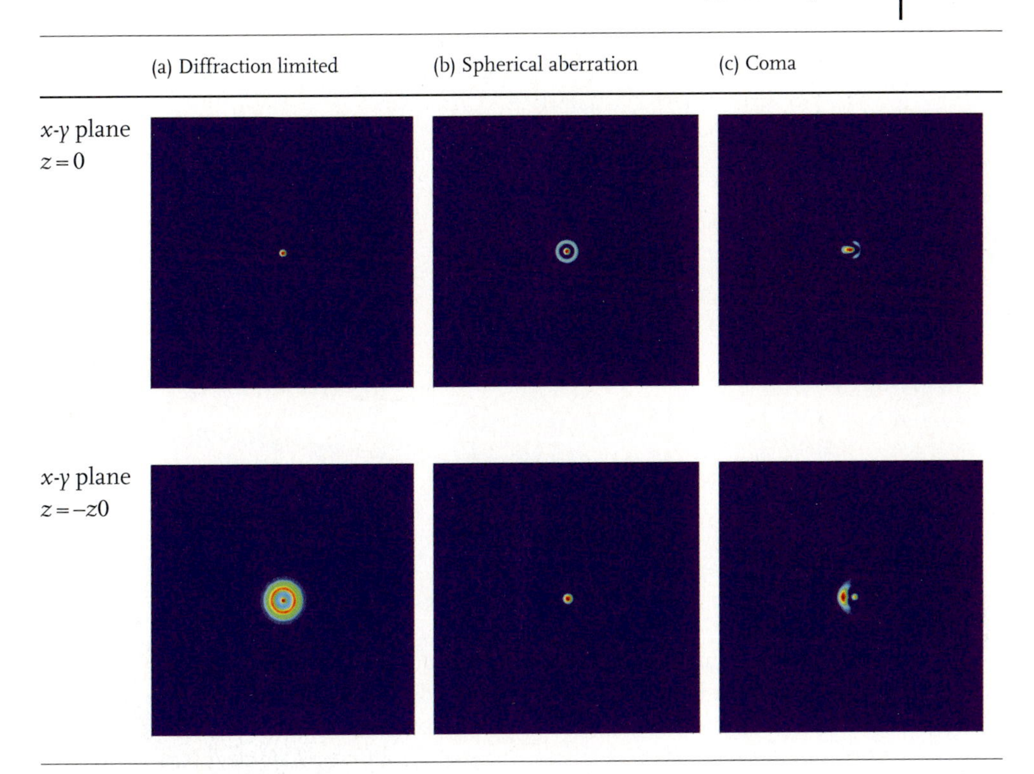

Figure 49-7: Star images along the *x-z* plane at $y = 0$ and along the *x-y* plane at $z = 0, \pm z_0$ for (a) perfect imaging, (b) spherical aberration and (c) coma (0.5 λ Zernike, respectively).

Direct imaging onto a CCD camera is not possible, because usually the pixel size is too large. The image has to be enlarged by a microsope objective of NA higher than that of the lens under test and well corrected for the wavelength spectrum.

One problem for a CCD camera is the necessary dynamic range when the 2D point spread function is to be measured. If the fourth ring has to be resolved a 12 bit dynamic range is mandatory, which usually cannot be provided by a standard CCD camera. By combining images taken with different exposure times the range could be extended.

When testing afocal systems such as telescopes or binoculars, a second collimator has to be introduced behind the eyepiece of the test system [49-13], [49-14]. The collimator can be selected with high F/# to project the star image directly onto the CCD sensor (figure 49-8).

To investigate the object point within the total field of view, a mechanism has to be found where the system under test can be tilted relative to the projecting collimator. The observing collimator then also has to be tilted to catch the star image (figure 49-9).

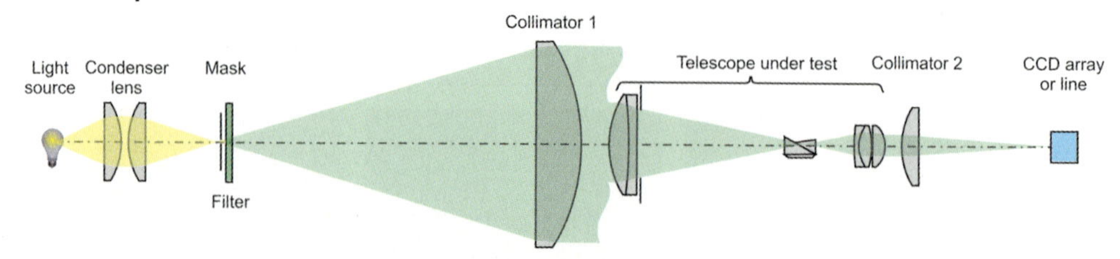

Figure 49-8: Star test setup to test afocal systems using two collimators, with the object point on the optical axis.

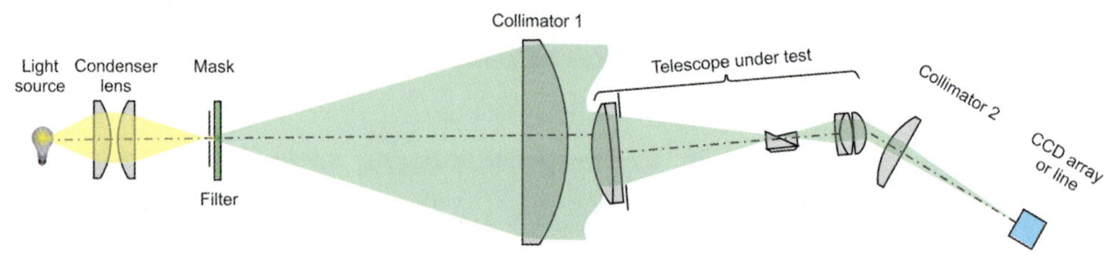

Figure 49-9: Star test setup to test afocal systems using two collimators, with the object point at the limit of the field of view.

The star test can be applied in many modifications. Instead of a single star, an array of stars can be imaged simultaneously onto a sensor system to investigate the total field or parts of it. The star array can be generated by a set of optical fibers attached to a common carrier plate. The use of a diffractive element that generates a set of spherical waves of appropriate NA from an incident plane wave (figure 49-10) is very convenient.

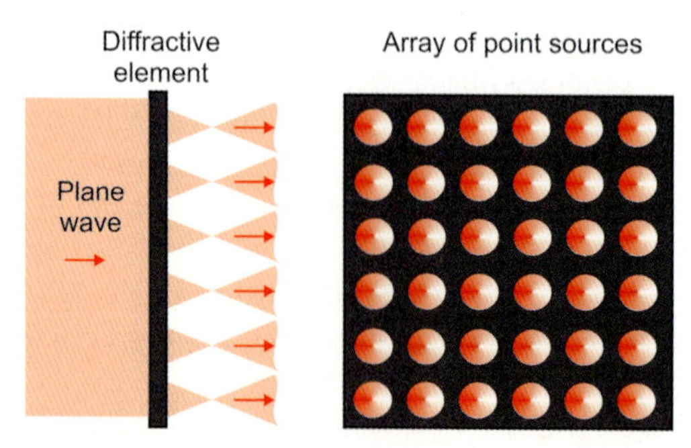

Figure 49-10: Array of point sources generated by means of a specially designed diffractive element.

In many applications the star can be replaced by a slit source of adequate width [49-8]. In this case the detector collects more light, which makes the measurement more stable and efficient. A CCD line perpendicular to the slit can then be used to detect the image. To obtain full information on the system aberrations the slit must be rotated by 90° for a second measurement. The descent of the slit signal is weaker than a cross-section through the point spread function (figure 49-11).

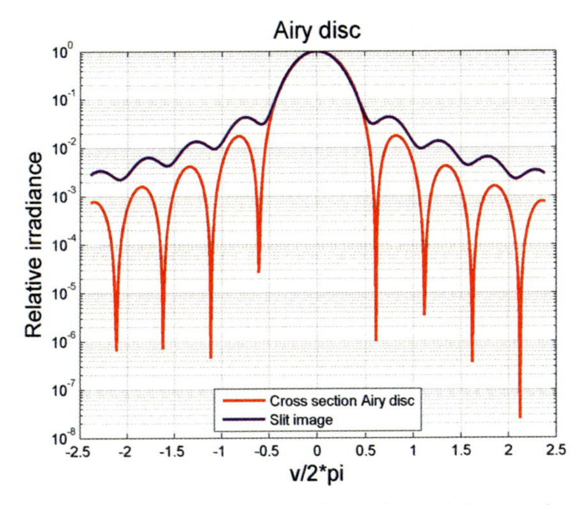

Figure 49-11: Cross-section through Airy disc and slit image for sampled Airy disc on a logarithmic scale.

49.3.2
Image Deconvolution

Qualifying the "sharpness" of an image usually leads to the determination of the point spread function (PSF) of the system under test. The PSF of an optical device is the image of a single point object normalized so that its integral over the total space is equal to one. From the PSF we can derive other system characteristics such as the optical transfer function (OTF) or the encircled energy (EE) [49-15]–[49-17].

In the following we consider incoherent imaging systems where the image formation process is linear. We refer to linear systems, when the images of several objects do not interfere with each other when simultaneously imaged, their sum being equal to the sum of the independently imaged objects. This is different from coherent imaging where the different images interfere while simultaneously imaged.

The OTF of an optical system can be derived from the complex pupil function P of an optical system by autocorrelation. The Fourier transform of the pupil function is called the coherent PSF, its squared magnitude is called the incoherent PSF [49-15]. Figure 49-12 gives an overview of the relationships between P, OTF, coherent and incoherent PSF.

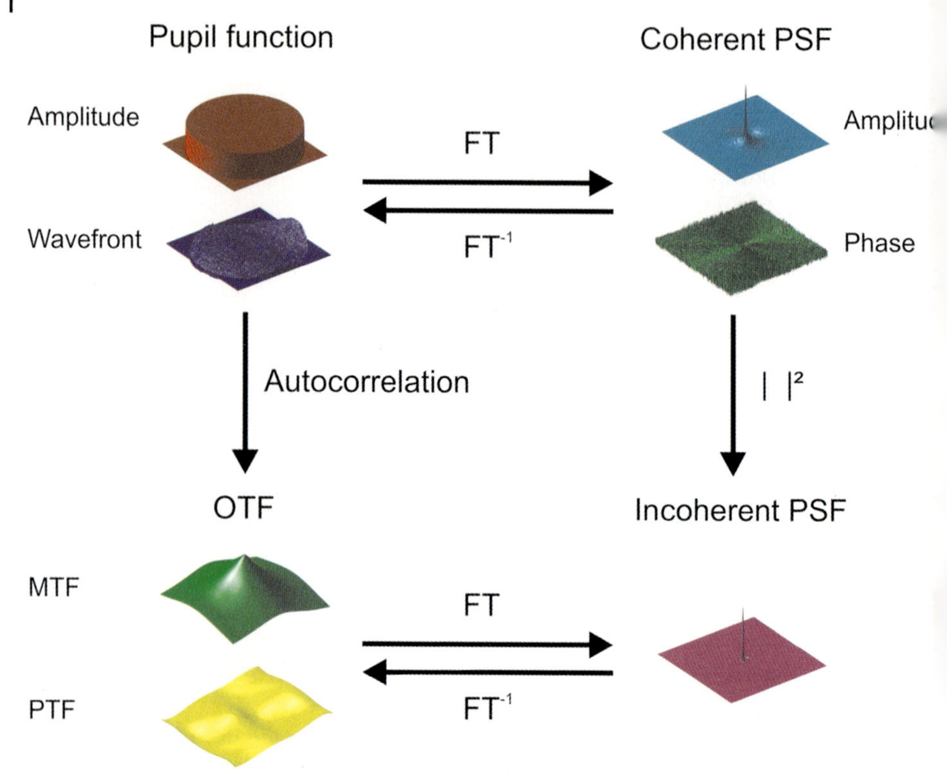

Figure 49-12: Relationship between pupil function P, coherent PSF, OTF and incoherent PSF.

In incoherent imaging, the final image $O'(x,y)$ can be thought of composed from individual PSFs, each shifted to the location and scaled according to the irradiance $O(x,y)$ of the corresponding object point. This is mathematically represented by a convolution equation

$$O'(x,y) = O(x,y) * PSF(x,y) \tag{49-5}$$

When the image is sampled by an image detecting device, its sampling function $S(x,y)$ needs to be considered. The final signal $I(x,y)$ as an image of the system under test and the result of the detection process is therefore given by

$$I(x,y) = (O(x,y) * PSF(x,y)) * S(x,y) \tag{49-6}$$

Assuming that $O(x,y)$ and $S(x,y)$ are well known, we can determine the PSF by transforming (49-6) into the Fourier domain und dividing by the spatial spectra of $O(x,y)$ and $S(x,y)$.

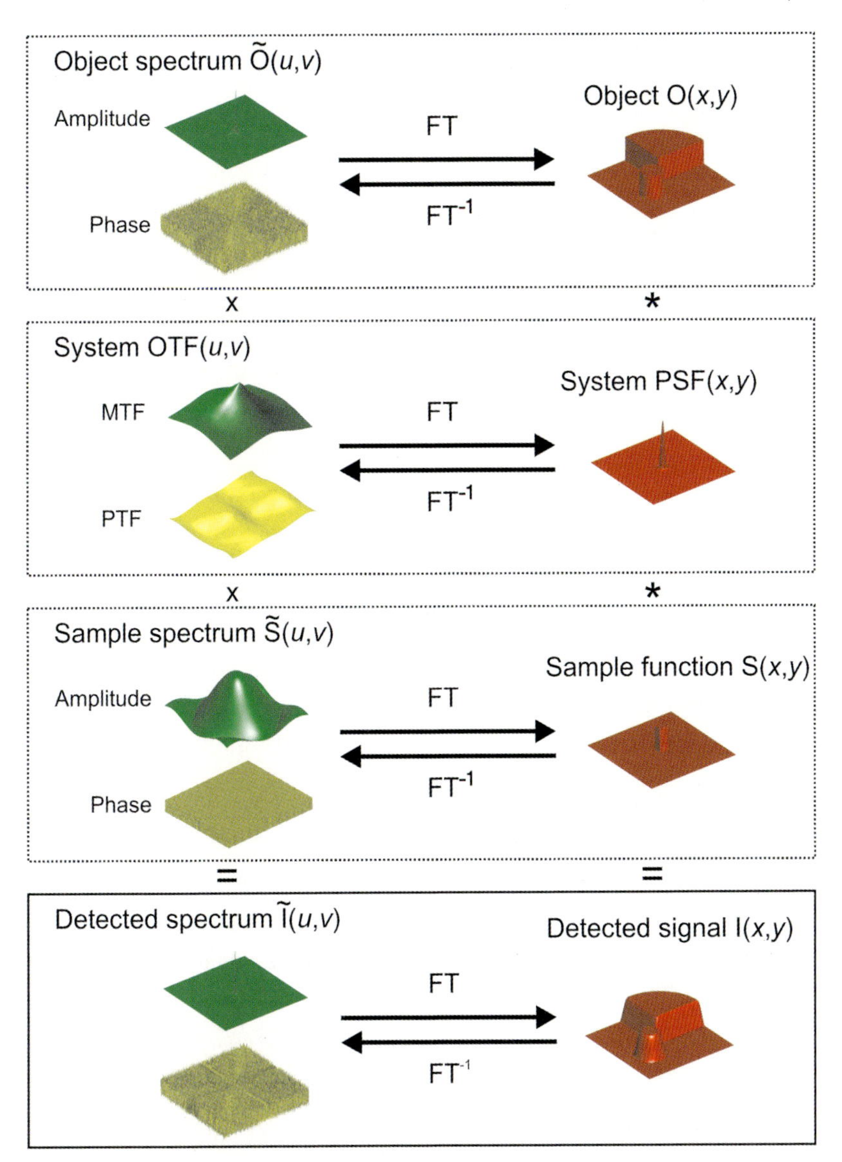

Figure 49-13: The detected signal $I(x,y)$ as result of convolutions with system PSF(x,y) and sample function S(x,y) as well as the corresponding spectra combined by complex multiplication.

$$\mathrm{OTF}(u,v) = \mathrm{FT}\{\mathrm{PSF}(x,y)\} = \frac{\tilde{I}(u,v)}{\tilde{\mathrm{O}}(u,v)\cdot\tilde{\mathrm{S}}(u,v)} \tag{49-7}$$

where u and v denote the spectrum coordinates and "~" denotes the corresponding Fourier transformed field of the object and the sample function. The Fourier transform of the PSF is called the optical transfer function OTF.

The PSF is therefore determined by a reverse Fourier transform of (49-7):

$$\mathrm{PSF}(x,y) = \mathrm{FT}^{-1}\left\{\frac{\tilde{I}(u,v)}{\tilde{O}(u,v)\cdot\tilde{S}(u,v)}\right\} \tag{49-8}$$

Figure 49-13 shows the corresponding relationships. We refer to image deconvolution if the object $O(x,y)$ is reconstructed by convolving the measured image $I(x,y)$ with the inverted spectra of PSF and sample function S as shown in (49-9).

$$O(x,y) = I(x,y) * \mathrm{FT}^{-1}\left\{\frac{1}{\mathrm{OTF}(u,v)}\right\} * \mathrm{FT}^{-1}\left\{\frac{1}{\tilde{S}(u,v)}\right\} \tag{49-9}$$

(49-8) can then also be written as a convolution

$$\mathrm{PSF}(x,y) = I(x,y) * \mathrm{FT}^{-1}\left\{\frac{1}{\tilde{O}(u,v)}\right\} * \mathrm{FT}^{-1}\left\{\frac{1}{\tilde{S}(u,v)}\right\} \tag{49-10}$$

In order to solve (49-7)–(49-10) properly, the spectra of the object and sample function should not show any zero crossings, otherwise the solutions become indeterminate.

49.3.3
Calibration

In order to determine the OTF or PSF of an optical system under test, it is necessary to solve (49-7) or (49-8) respectively. Three data fields are combined to give the following results.

- The measured irradiance distribution $I(x,y)$.
- The irradiance sensitivity over the sample window, the sample function $S(x,y)$.
- The relative irradiance distribution over the test target $O(x,y)$.

In order to measure $I(x,y)$ properly, the sensor signal has to be compensated for nonlinearities (single-channel detector). If a multi-channel sensor such as a CCD camera is used, the spatial variations in bias (fixed pattern noise) and amplification have to be corrected [49-19], [49-20]. We assume that variations which occur over the incident angle are negligible and behaviour is stable over time.

When color sensitive measurements are being made, the sensor signal must be compensated for varying spectral sensitivity.

The active sensor area generally has a spatially varying sensitivity defined by $S(x,y)$. When using a CCD camera this variation is usually not known exactly. The so-called fill factor is specified, which is a measure of how much of the light being directed at the sensor actually strikes the photosensitive diodes in each pixel. Frame-transfer CCDs have architectures where nearly all of the pixel area is photosensitive, leading to a 100% fill factor. Interline transfer CCDs, which are found in most industrial cameras, must divide the pixel area between photodiodes, transfer gates and shift register circuitry, reducing the fill factor to less than 30%. To compensate for this, interline transfer CCDs typically have small lenslet arrays positioned above the pixel to gather and focus as much light as possible on the photodiode portion of the pixel, thus reaching fill factors of 60–70%.

$S(x,y)$ must also include the sample width, which is automatically given by the pixel width and height, when using a CCD camera. If a scanning single-channel detector is employed, the scanning step must be used to define the grid for the necessary Fourier transformation.

Finally, the original relative irradiance distribution over the test target $O(x,y)$ must be determined. Depending on the type of test target this can be quite tedious and time consuming. For extended test targets an illumination system must be found that homogeneously illuminates the target area either in transmission or in (diffuse) reflection. If illumination is critical then reference measurements have to be made with blank targets of uniform transmission or reflection. The setup can then be compensated for non-uniform illumination.

The geometry of the test target has to be checked when specified to give quantitative results. Circular spot masks and arrays, slits and uniform grids, checkerboards, circular equidistant rings, etc., can be measured by means of comparators, microscopes and coordinate measurement machines, depending on their overall size. For ultra-high precision lithographic mask metrology there are machines that measure registration (overlay on reticles) as well as critical dimensions (CD) in transmitted and reflected light [49-21]. Structures of 45 nm and below can be resolved on 6″ or 9″ masks.

49.3.4
Accuracy and Error Sources

The accuracy denotes the variation of a measurement process around the true value. It includes the repeatability of the system and also the reproducibility of the measurement process as well as the systematic offset between the measured results and the true value. The latter is very dependent on the calibration process being applied.

Like all measurement processes, star or slit tests are limited by changes in environmental conditions that either influence the test sample itself or the measuring instrument.

Environmental issues have been discussed in more detail in section 46.8.1 covering mechanical and acoustic vibrations, radiation, convection, air turbulence, changes in temperature, humidity, pressure, etc.

With relation to a star test we will discuss limitations caused by sensor noise or unstable light sources as well as deviations from the exact sensor positions. Using

(49-1) and (46-210) for the origin of shot noise we can define irradiance measurement errors ΔI:

$$\Delta I(x, \Delta x, t) = d(t) + \frac{2J_1(x + \Delta x(t))}{x + \Delta x(t)} \sqrt{\left(\frac{I(0)\, h\nu}{t A \eta}\right)} \qquad (49\text{-}11)$$

where

$I(0)$ is the irradiance at the center of the star image in Wm^{-2}
h is the Planck constant $= 6.62606896 \times 10^{-34}$ J s,
ν is the radiation frequency in Hz,
A is the sensor area in m^2,
t is the exposure time in s,
η is the quantum efficiency,
$d(t)$ is the dark current noise added to the measured signal,
$\Delta x(t)$ is the placement error (jitter) due to deviations from the ideal sensor position while scanning,
$J_1(x)$ is the Bessel function of the first kind as described in (49-1).

In most cases the modulation transfer function MTF is calculated from the acquired system point spread function PSF in order to describe the system performance. In the following we will discuss impacts from measurement errors on the MTF. Figure 49-14 shows simulated results for (a) 10% shot noise relative to the signal maximum, (b) 2% additive noise, like dark current noise, relative to the signal maximum, (c) x/y jitter of 1/500 of the scanned field.

Results:

a) Shot noise has a negligibly small influence on the MTF. This is due to the fact that only the top regions of the detected peak are disturbed, whereas the areas of low irradiance are undisturbed. The MTF describing the resolution of the system remains unchanged at lower spatial frequencies, whereas slight changes show up at high spatial frequencies.
b) Additive noise considerably influences the MTF in all spatial frequencies. Dark current noise causes the main limitations to the star test device.
c) The x/y jitter has the greatest impact in regions with strong signal gradient. It therefore behaves similar to shot noise ending up with a negligible change in MTF over the total spectrum.

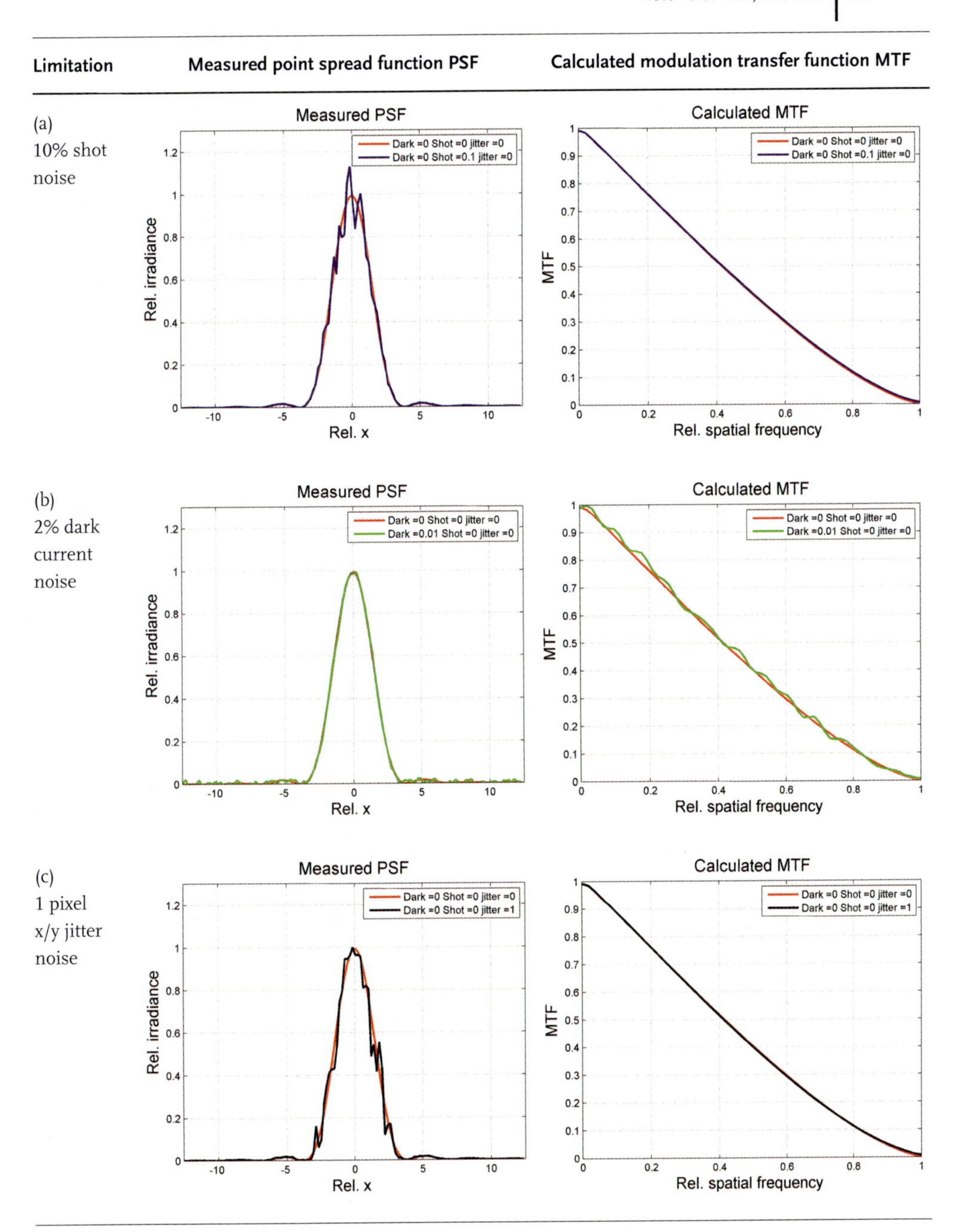

Figure 49-14: Measured PSFs and the corresponding calculated MTFs for a diffraction limited system. The red curves show the perfect results for comparison. The following errors have been introduced: (a) 10% multiplicative noise (shot noise), (b) 2% additive noise (dark current noise), (c) *x/y* jitter 1/500 of the scanned field.

49.4
Test Targets, Visual Inspection

A variety of test targets are available which are used to check the resolution and contrast transfer of optical systems [49-21]–[49-23]. In most cases the targets are used for visual inspections. However, they can also be used to qualify systems in a quantitative way by letting a computer evaluate and quantify the captured test images.

Figure 49-15 shows the USAF target as a resolution test example with varying frequencies. The target is divided into four groups, each group consisting of six elements. The group number is indicated by a red number, the element number by a green number. Each element is composed of three horizontal and three vertical equally spaced bars. Each element within a group corresponds to an associated spatial frequency R in linepairs/mm, which is defined by (49-12).

$$R[\mathrm{lp/mm}] = 2^{\text{Group No.}+\frac{\text{Element No.}}{6}} \tag{49-12}$$

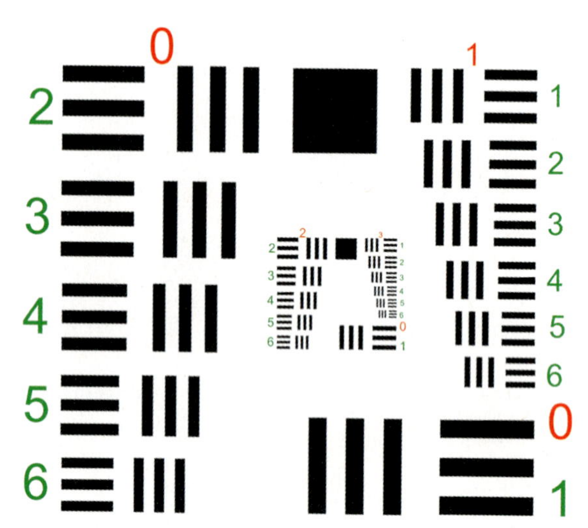

Figure 49-15: USAF target to test the resolution of optical systems.

Test targets can be presented to the total field of the test system or sequentially to different small ranges within the field of view. The latter is necessary if the image quality is expected to change substantially within the field. Figure 49-16 shows three different test targets, rings and spokes, a Siemens star target, a USAF target, when imaged by a perfect system, a defocussed system and an astigmatic system.

In the defocussed image all structures become equally blurred. The rings and spokes are equally broadened, the Siemens star shows a uniform radial drop in image contrast, the USAF target is equally resolved in the horizontal and vertical directions.

In the astigmatic image all structures in orthogonal directions become differently blurred. In the example horizontal lines are sharply imaged while vertical lines are broadened. Thus, the Siemens star shows drops in image contrast depending on the radial direction. The USAF target shows different resolutions in the horizontal and vertical directions.

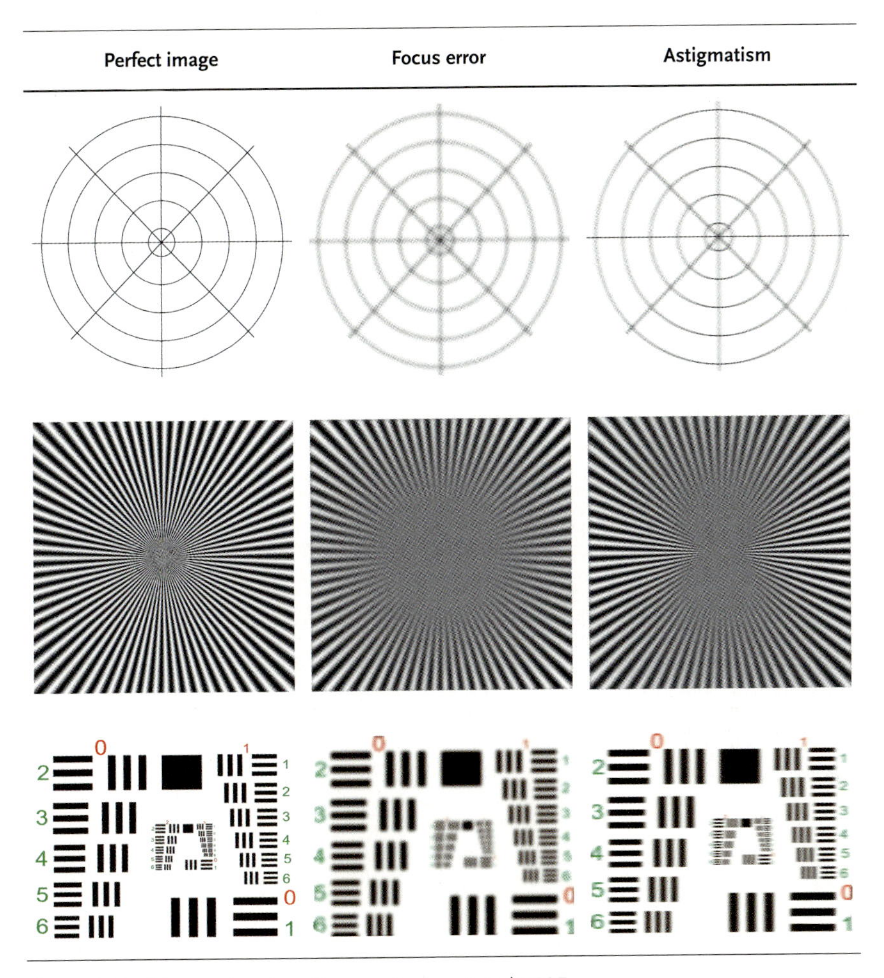

Figure 49-16: Rings and spokes, Siemens star target, and USAF target in perfect, defocussed and astigmatic imaging.

49.5
Distortion Metrology

49.5.1
Basic Setups

When an object plane is imaged by an optical system the image differs from the object in different ways: the image

- is partially or totally blurred (defocus, spherical aberration, astigmatism, coma, ...),
- lies on a curved surface (field curvature),
- is distorted (distortion).

The previous chapter covered the investigation on "blurred" images. We now consider the case where the image lies in an image plane but is distorted. Distortion means that the coordinates x' and y' in the image plane are received by nonlinearly transformation of the object coordinates x and y. In general the transformation can be described by using x-y polynomials as given in (49-13) and (49-14).

$$x' = \sum_{i=0}^{M} \sum_{j=0}^{N} a_{ij} x^i y^j \tag{49-13}$$

$$y' = \sum_{i=0}^{M} \sum_{j=0}^{N} b_{ij} x^i y^j \tag{49-14}$$

Most optical systems are rotationally symmetric, which simplifies the transformation to (49-15) and (49-16).

$$x' = x \sum_{i=1}^{M} c_{2i-1} (x^2 + y^2)^{2(i-1)} \tag{49-15}$$

$$y' = y \sum_{i=1}^{M} c_{2i-1} (x^2 + y^2)^{2(i-1)} \tag{49-16}$$

The coefficient c_1 defines the magnification of the system. If all other coefficients are zero the system will be undistorted. If the coefficient c_3 is not equal to zero and has the same sign as c_1 we refer to cushion shaped distortion. If the coefficient c_3 is not equal to zero and has a different sign from c_1 we refer to barrel shaped distortion. Figure 49-17 shows barrel and cushion distorted images of an equidistant dot grid, a checkerboard grid and a linear grid.

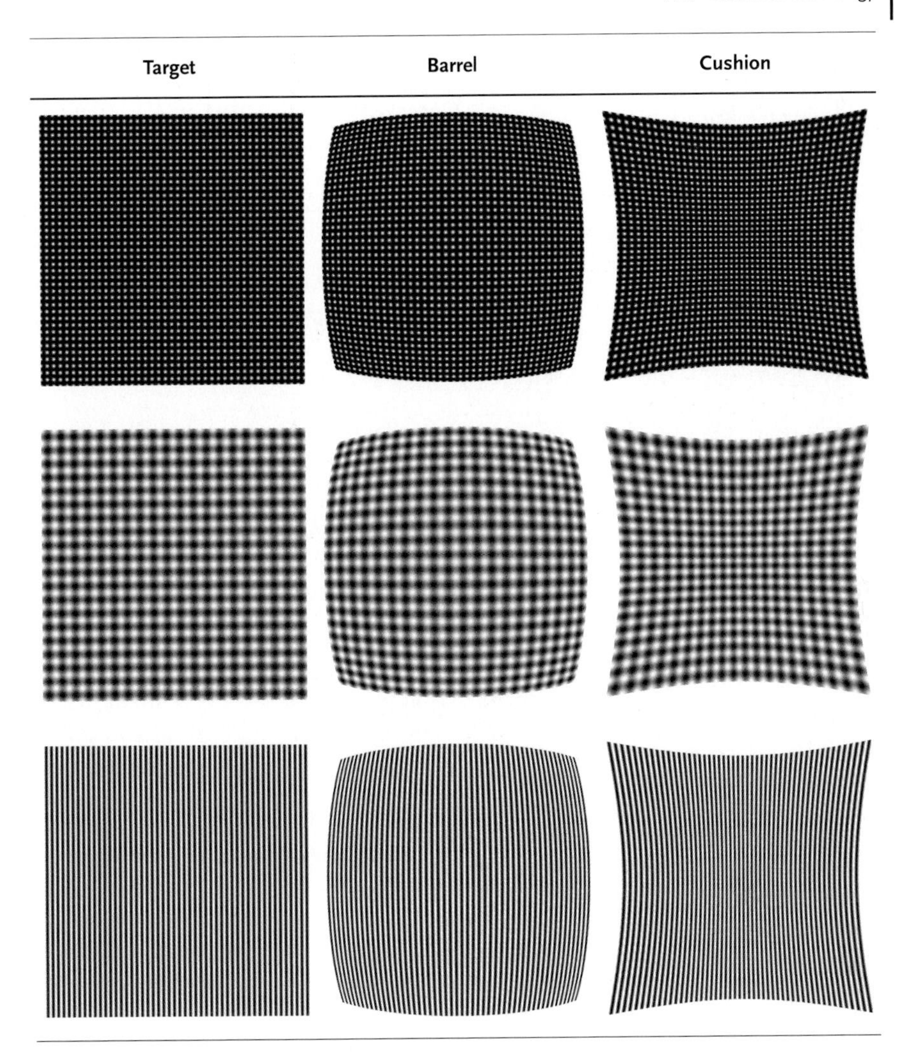

Figure 49-17: Equidistant dot grid and linear grids imaged by barrel and cusion shaped distortion.

In a test setup a detector, i.e., a CCD camera, captures the image directly at the image plane. If this is not possible an additional converting system must be added to fit the size of the sensor area. The converting system must be carefully calibrated to compensate for its own residual distortion errors [49-25].

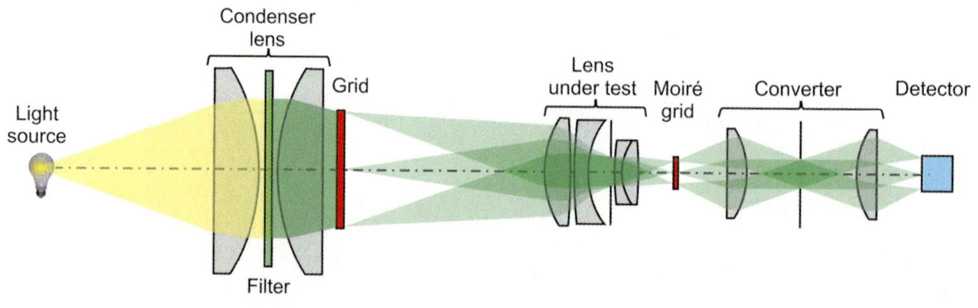

Figure 49-18: Test setup to measure distortion using a grid target and a Moiré grid target in an intermediate image plane.

Note that a linear grid in one position can only be used to determine rotational symmetric distortions. For verification the grid should be rotated by 90° for an additional measurement. For unsymmetrical distortions either a linear grid in orthogonal positions or an equally spaced two-dimensional pattern as the dot grid, must be used as a target.

The direct detection of the distorted image by means of a CCD camera can typically resolve 0.1% distortion. When subpixel resolving algorithms are used 0.01% – 0.0005% can be detected.

Moiré Technique

If higher resolution is necessary a Moiré technique can be applied, where an extremely fine structured target is used, which the test system still can resolve, but not necessarily the camera [49-26–49-28]. A second target is then placed in the intermediate image or directly in front of the detector, matching the image structure (Figure 49-18). The detector then mainly sees the Moiré pattern of the image and the second target. As an example we consider a sinusoidal grid target whose image is superimposed with a second sinusoidal Moiré grid. The detected signal is given by the product of the irradiance distribution in the image and the transmissivity of the Moiré grid as shown in (49-17)

$$\begin{aligned} I(x,y) &= T_T(1+\cos(2\pi x'))T_M(1+\cos(2\pi c_M x)) \\ &= T_T T_M(1+\cos(2\pi x')+\cos(2\pi c_M x)+\cos(2\pi x')\cos(2\pi c_M x)) \\ &= T_T T_M\left(1+\cos(2\pi x')+\cos(2\pi c_M x)+\frac{1}{2}\cos(2\pi(x'+c_M x))+\frac{1}{2}\cos(2\pi(x'-c_M x))\right) \end{aligned} \qquad (49\text{-}17)$$

where T_T denotes the maximum irradiance of the intermediate target image and T_M denotes the maximum transmissivity of the Moiré target. c_M is the spatial frequeny of the Moiré grid.

Since the CCD camera cannot resolve high spatial frequencies, the detected camera signal I_{CCD} will be given by (49-18).

$$I_{\mathrm{CCD}}(x, y) = T_{\mathrm{T}} T_{\mathrm{M}} \left(1 + \frac{1}{2} \cos\left(2\pi (x' - c_{\mathrm{M}} x) \right) \right) \tag{49-18}$$

Replacing x' by the expression in (49-15) we get (49-19).

$$I_{\mathrm{CCD}}(x, y) = T_{\mathrm{T}} T_{\mathrm{M}} \left(1 + \frac{1}{2} \cos\left(2\pi x \left(\sum_{i=1}^{M} c_{2i-1} (x^2 + y^2)^{2(i-1)} - c_{\mathrm{M}} \right) \right) \right) \tag{49-19}$$

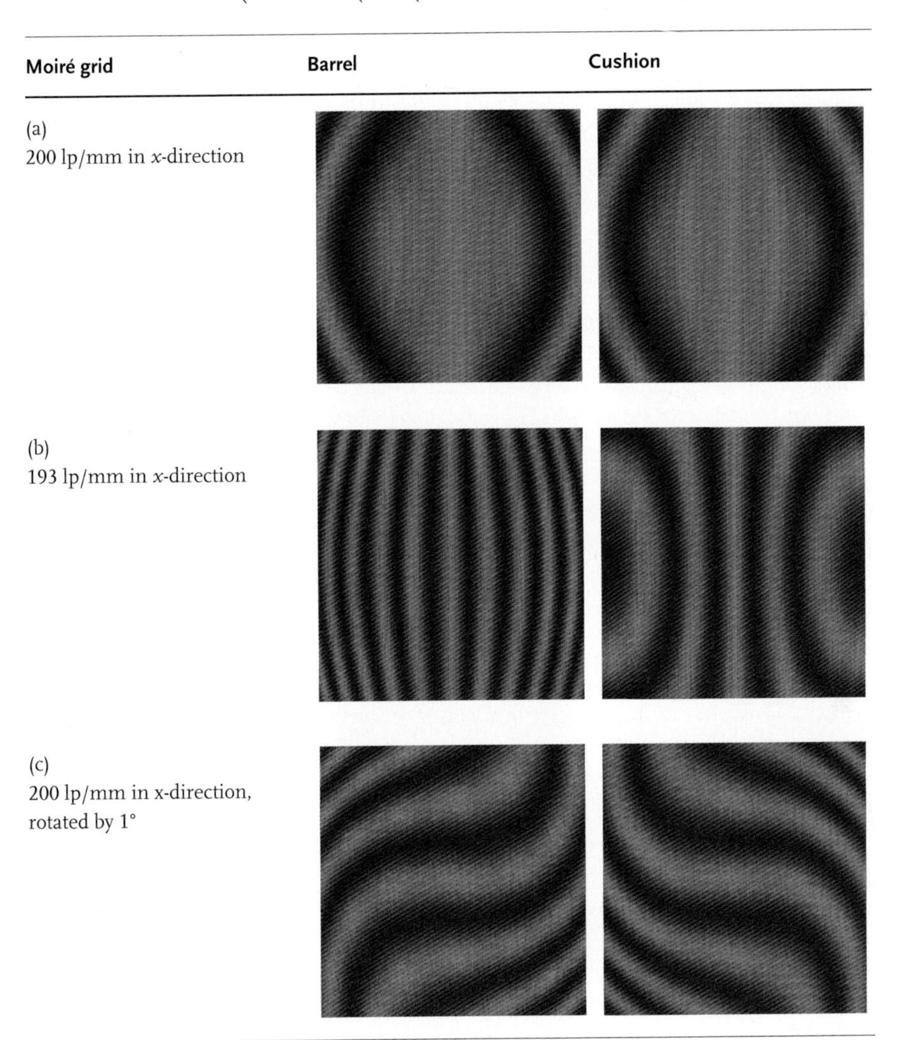

Figure 49-19: Distortion detection by applying a Moiré grid in an intermediate image plane.

The magnification factor c_1 of the test system can be sensitively resolved, because the camera now only sees a spatial frequency of $\mathbf{c}_1$–c_M or any slight deviations from the linearity of the image.

Figure 49-19 shows examples in which a linear grid image was superimposed by a linear Moiré grid. If $c_1 = c_M$ the images for $+c_3$ and $-c_3$ distortion are identical, barrel and cushion distortion cannot be distinguished. If the Moiré grid is replaced by a grid with $c_M \neq c_1$, distortions show different Moiré patterns. If the Moiré grid is rotated, different distortions can also be distinguished.

If a computer is used for image analysis the Moiré grid can be displaced perpendicular to the grid lines by steps of $2\pi/4$ to introduce phase shifts between the image and the Moiré grid. Standard phase-shift algorithms can then be applied to recover the cosine argument in (49-18) which carries all the information on the sign and magnitude of the distortion.

49.5.2 Correlation Method

The general approach to identifying distortion in an image is to apply cross-correlation methods using suitable masks [49-29]–[49-31]. If the mask and the pattern being sought are similar the cross-correlation will be high. The mask is usually a synthetic model of an image which needs to have the same functional appearance as the pattern to be found.

Cross-correlation and General Masks

Figure 49-20 shows a pixel grid on which a rectangular mask with M pixels in i and N pixels in j is positioned with its center at pixel i,j. The unnormalized correlation coefficient $c_{i,j}$ is then calculated by (49-20)

$$c_{i,j} = \sum_{j'=1}^{N} \sum_{i'=1}^{M} \left(Q_{i',j'} - \overline{Q}\right) \cdot \left(I_{i+i'-M/2,j+j'-N/2} - \overline{I_{i,j}}\right) \tag{49-20}$$

where $\overline{Q}$ denotes the mean of the masks pixels, $\overline{I}_{i,j}$ denoting the mean of the image pixels covered by the mask. The mathematical process is also called convolution, with the mask often denoted by the convolution kernel.

The best matches of mask and image are shown by peaks in the resulting 2D cross-correlation function $c_{i,j}$, which needs to be calculated for every pixel in the image (Figure 49-21). The mask size should be chosen to be as small as practicable, otherwise the process might be extremely time consuming. To find the position of the peaks in the cross-correlation function the definition of a threshold might be necessary.

Digital image correlation (DIC) is a general method used to identify displacements of object textures [49-32]–[49-34]. A widely used application is non-contact strain measurement in which the image of an object is captured before and after stress application. The resulting local displacements in the texture are detected by DIC.

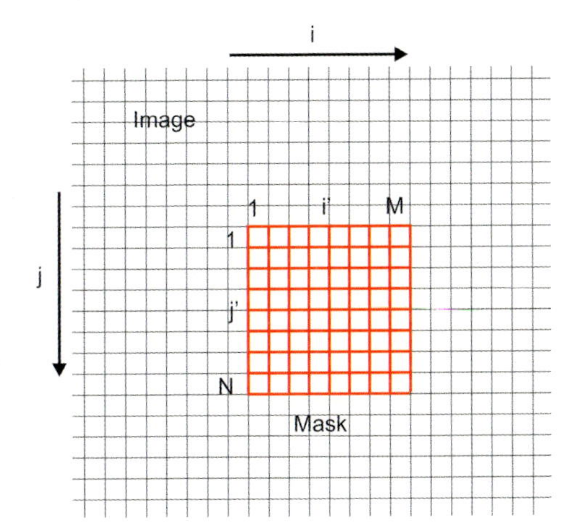

Figure 49-20: Pixel grid of image and mask.

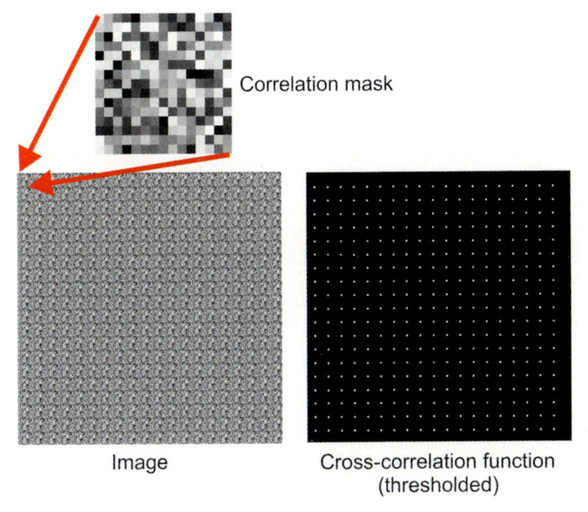

Figure 49-21: 2D cross-correlation of an image using a 2D correlation mask (convolution kernel) to calculate the thresholded cross-correlation function.

In DIC a set of neighboring points in an undeformed state are assumed to remain as neighboring points after deformation. Figure 49-22 illustrates schematically the deformation process of a planar object. Quadrangle S represents a sub-image of the undistorted image, quadrangle S′ the corresponding distorted area in the distorted image. In order to obtain the displacement vectors u_O and v_O of point O, sub-image S is matched with sub-image S′ using a correlation operation. If subset S is sufficiently small, the coordinates of points in S′ can be approximated by first-order Taylor expansion as shown in (49-21) and (49-22).

$$x_{P'} = x_O + u_O + \left(1 + \left.\frac{\partial u}{\partial x}\right|_O\right)\Delta x + \left.\frac{\partial u}{\partial y}\right|_O \Delta y \tag{49-21}$$

$$y_{P'} = y_O + v_O + \left.\frac{\partial v}{\partial x}\right|_O \Delta x + \left(1 + \left.\frac{\partial v}{\partial y}\right|_O\right)\Delta y \tag{49-22}$$

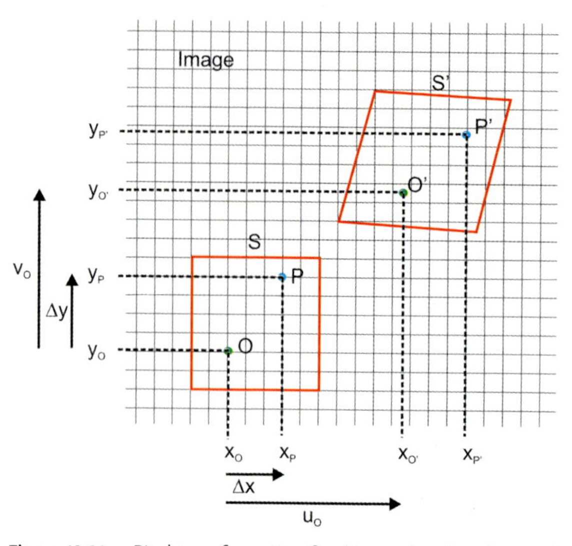

Figure 49-22: Pixel transformation for distorted and undistorted image.

Let $I(x,y)$ and $I'(x,y)$ be the gray value distributions of the undistorted and distorted image, respectively. For a subset S, a correlation coefficient c is defined as shown in (49-23).

$$c = \frac{\sum_{i \in S} \left(I(x_i, y_i) - I'(x_{i'}, y_{i'})\right)^2}{\sum_{i \in S} \left(I(x_i, y_i)\right)^2} \tag{49-23}$$

where (x_i, y_i) is a point in subset S in the undistorted image, whereas $(x_{i'}, y_{i'})$ is a corresponding point in subset S' in the deformed image as defined by (49-21) and (49-22). u_O and v_O are the displacements, $\left.\frac{\partial u}{\partial x}\right|_O, \left.\frac{\partial u}{\partial y}\right|_O, \left.\frac{\partial v}{\partial x}\right|_O, \left.\frac{\partial v}{\partial y}\right|_O$ are the displacement derivatives of point O. The correlation coefficient c becomes zero if the approximation is equal to the real displacement. Hence, minimization of the coefficient c would provide the best estimates of the parameters.

Minimization of the correlation coefficient c is a nonlinear optimization process, and Newton–Raphson or Levenburg–Marquardt iteration methods [49-33], [49-34] are usually used in the implementation of the process.

Subpixel accuracy can be achieved if interpolation schemes are implemented to reconstruct a continuous gray value distribution in the deformed images. Higher order interpolation would provide more accurate results at the cost of more computation time. Bi-cubic and bi-quintic spline interpolation schemes are widely used.

Cross-correlation and Sinusoidal Masks

Sinusoidal mask structures play a particular role in cross-correlation techniques, because it is easily possible to determine the relative shift of the image portion for each pixel within the image, whereas in the general case only the locations of correlation maxima are determined.

When periodic patterns such as grids or parallel fringes are projected, evaluation techniques like sequential phase shifting or spatial carrier frequency analysis can be applied (figure 49-23). The latter is a special form of spatial cross-correlation, whereas phase shifting is a special form of cross- correlation in time.

We consider two periodic patterns $I_x(x,y)$ and $I_y(x,y)$ to be projected or imaged by an optical system under test:

$$I_x(x,y) = a + b\cos(2\pi c_x x) \tag{49-24}$$

$$I_y(x,y) = a + b\cos(2\pi c_y y) \tag{49-25}$$

Considering linear behavior in the imaging process the corresponding images $I_x'(x,y)$ and $I_y'(x,y)$ can then be described by

$$I'_x(x,y) = a' + b'\cos\left(2\pi\left(c'_{x0} + c'_{x1}x + c'_{x2}x^2 + c'_{x3}x^3 + ...\right)\right) \tag{49-26}$$

$$I'_y(x,y) = a' + b'\cos\left(2\pi\left(c'_{y0} + c'_{y1}x + c'_{y2}x^2 + c'_{y3}x^3 + ...\right)\right) \tag{49-27}$$

c_{x0}' and c_{y0}' denote the lateral displacement of the fringe patterns, whereas c_{x1}' and c_{y1}' denote the magnification factors in x and y direction, which in some applications might be of no interest. The other coefficients denote the different distortion factors in the x and y directions, respectively.

To determine the individual coefficients, the fringe patterns detected by an image analysis system can be evaluated by standard phase-shifting techniques or spatial carrier frequency analysis as known from interferometry and as described in section 46.5.

After sequential projection and analysis of the individual patterns in the x and y directions the results of the phase functions are available, from which the unknown coefficients can be determined by standard least-squares minimization.

Note that, generally, the offsets c_{x0}' and c_{y0}' are not known from fringe detection techniques, because the fringes itself can not be distinguished from each other. In practice they are set to zero when possible. In cases they are important for the results, at least one fringe must be identified by additionally projecting a marker pointing onto the zeroth fringe.

Figure 49-23: Cross-correlation techniques in space (spatial carrier frequency analysis) or time (phase shifting): (a) image of fringe modulation in x, (b) image of fringe modulation in y, (c) wrapped phases in x after correlation in x, (d) wrapped phases in y after correlation in y, (e) unwrapped phases in x, (f) unwrapped phases in y, (g) calculated distortion including linear terms from unwrapped phases in x and y. The arrows describe the pixel displacement without linear terms.

If the sequential projection of grids in x and y is not possible, a sinusoidal pattern periodic in both directions is an alternative (figure 49-24) as described by (49-28).

$$I(x, y) = a + b\cos(2\pi cx) + b\cos(2\pi cx) \tag{49-28}$$

After projection a distorted irradiance distribution $I'(x,y)$ can be detected with distortions in the x and y directions as defined by (49-26) and (49-27).

For the 2D evaluations four correlation masks have to be defined as described by (49-29)–(49-32).

$$Q_{x0}(x, y) = I(x, 0) \tag{49-29}$$

$$Q_{x1}(x, y) = I\left(x + \frac{1}{4c}, 0\right) \tag{49-30}$$

$$Q_{y0}(x, y) = I(0, y) \tag{49-31}$$

$$Q_{y1}(x, y) = I\left(0, y + \frac{1}{4c}\right) \tag{49-32}$$

where $0 \leq x \leq \frac{1}{c}$ and $0 \leq y \leq \frac{1}{c}$.

The lateral displacements in x and y can then be calculated by three correlation procedures using well-known phase evaluation formulas as shown in (49-33) and (49-34).

$$x'(x, y) = \frac{c}{2\pi}\arctan\frac{Q_{x0}(x, y) * I'(x, y)}{Q_{x1}(x, y) * I'(x, y)} \tag{49-33}$$

$$y'(x, y) = \frac{c}{2\pi}\arctan\frac{Q_{y0}(x, y) * I'(x, y)}{Q_{y1}(x, y) * I'(x, y)} \tag{49-34}$$

where "*" denotes the correlation procedure as defined in (49-20).

Figure 49-24 shows an example using a periodic pattern modulated in two dimensions (sinusoidal checkerboard structure).

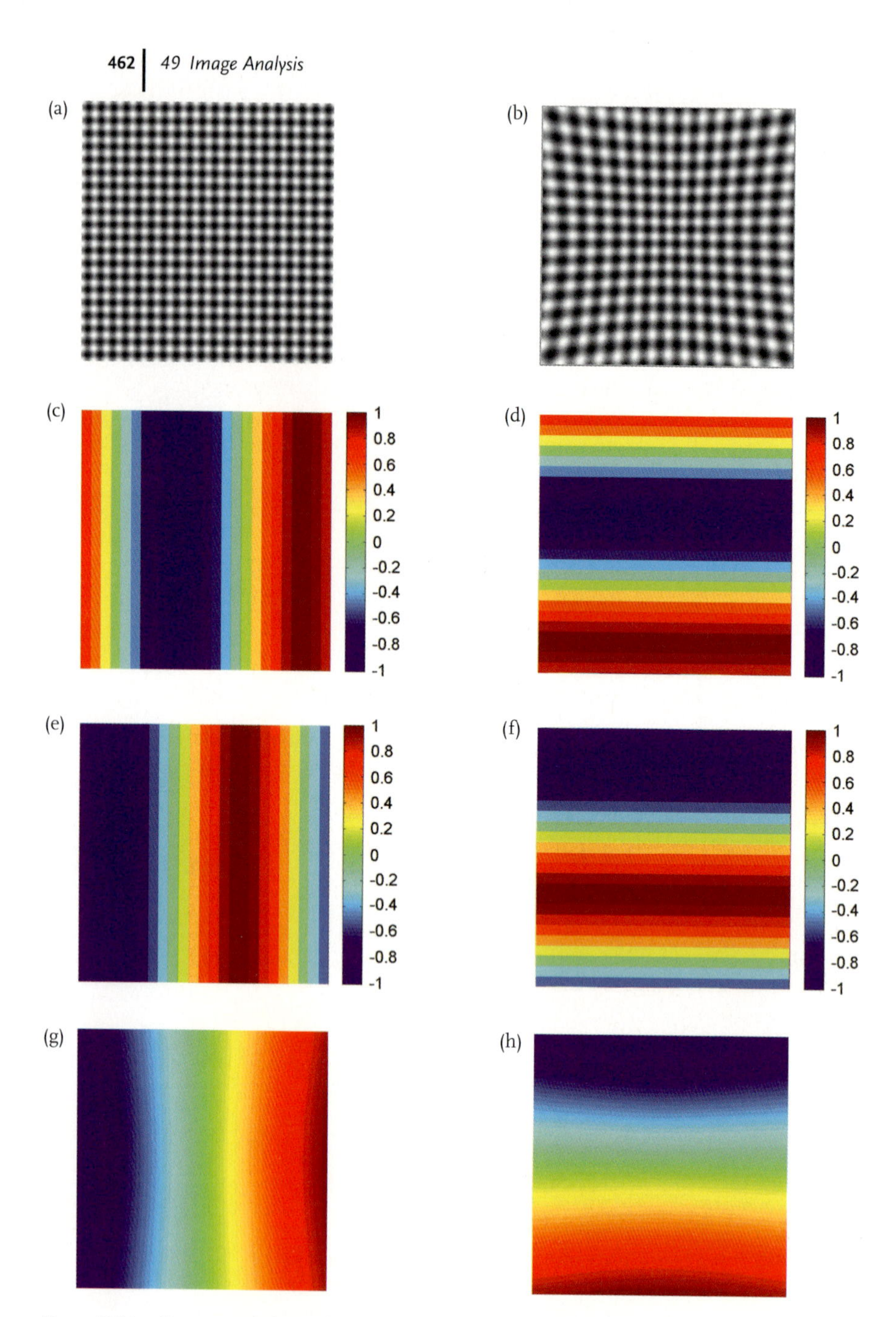

Figure 49-24: Cross-correlation techniques in space using a 2D periodic pattern: (a) projected pattern, (b) image of pattern, (c) correlation mask Q_{x0}, (d) correlation mask Q_{y0}, (e) correlation mask Q_{x1}, (f) correlation mask Q_{y1}, (g) calculated magnification and distortion in x, (h) calculated magnification and distortion in y.

49.5.3
Calibration

A distortion-measuring instrument can be calibrated by using either:

a) a well-known, high-precision reference object or objective, or by
b) a calibration method compensating for systematic errors of the mask.

If a suitable reference objective is not available, a regular, periodic mask pattern can be at least partially calibrated when measured in different positions and orientations. Rotating by 90°, 180° and 270° is possible for 2D patterns with 90° symmetry. Additionally lateral shifts can be made, making use of the periodicity of the patterns.

49.5.4
Accuracy and Error Sources

Besides the influence of environmental conditions which were discussed in section 46.8.1 the accuracy of distortion metrology is mainly limited by the accuracy of the mask and the capability of the detector.

The most accurate masks are produced by lithographic mask writers. These machines are known as electron-beam writers (E-beam writers) or laser writers. They sequentially expose photoresist on a glass plate which later turns into a mask with a pattern. Typically mask substrates have sizes up to 300 × 300 mm^2. The lateral machine resolution in x/y is in the order of 20 nm with pixel sizes of the same magnitude. The placement errors of mask writers can be detected by special inspection systems capable of detecting misplacements of a few nanometers.

When using the mask in high-precision applications, the following issues play a major role:

- the support and bending of the mask,
- the environmental conditions like temperature, air pressure, humidity,
- the wavelength,
- the position and adjustment of the mask within the test setup.

Taking high-precision production and calibration techniques into account, a typical mask has a dynamic range DR_{mask} of approximately

$$DR_{mask} = \frac{\text{max dimension}}{\text{smallest Feature}} \leq 10^7 = 140\,\text{dB} \approx 23\,\text{bit}$$

Simpler masks that have been produced by mechanical tools have DR_{mask} of 10^5–10^6.

The second limiting device is the detector system. Using a CCD camera and a pattern that the camera still can resolve by making the pattern period larger than the Nyquist limit, a camera can have an approximate dynamic range comparable to that of a mask. Selecting a CCD camera with 1000 x 1000 pixels that capture 500 fringes with a maximum resolution of 10^{-4} fringe will lead to a dynamic range DR_{camera} of

$$DR_{camera} = \frac{\text{max Fringes}}{\text{smallest Fringe Fraction}} \leq 0.5 \times 10^7 = 134\,\text{dB} \approx 22\,\text{bit}$$

To receive a fringe resolution of 10^{-4} fringes, many frames have to be captured and the mask has to be moved into several positions to average out local mask defects. In a simple setup a DR_{camera} of 10^3–10^4 is more likely to be achieved.

Scanning single-detector systems have comparable dynamic ranges if their positioning systems are based on accurate coordinate metrology. Laser displacement metrology can theoretically have huge dynamic ranges of > 10^9 with resolutions of a fraction of a nanometer. However, due to environmental influences and the deformation of the target mirror, real resolutions are in the order of 0.3 μm. Depending on the field to be scanned, a CCD camera with pixel sizes of a few μm might be more adequate.

When Moiré techniques are necessary, the errors of the additional Moiré mask and the additional imaging system will add to the errors of the primary mask and the detector. The total error budget must be calculated by raytracing from the system parameters and the individual geometry aberrations of the mask patterns.

In general, the distortion of optical systems can be detected with accuracies ranging from 0.1% to 0.0005% depending on the accuracies of the mask and the detection system, the compensation for environmental influences and the compensation for setup design deviations by raytracing.

49.6 Deflectometers

49.6.1 Basic Setups

The aim of deflectometry is to determine the shape of a specular reflecting test surface. A general setup consists of an appropriate mask of known geometry, the test surface and a well-known imaging optics to image the mask onto a CCD target after reflection at or transmission through the test object [49-35]–[49-41]. Figure 49-25 shows the arrangement for a reflective test surface.

The imaging optics can be composed of a single lens unit or a collimator – eyepiece arrangement as shown. The latter has the advantage of introducing a stop at the intermediate focus to establish a telecentric setup making the numerical evaluation easier.

The illumination optics in deflectometry is always spatially incoherent. Monochromatic sources are often useful in order to keep the optical design simple. Figure 49-26 shows a test arrangement for a transmissive object including the extended light source and a condenser lens at a distance close to the mask. The aperture angle θ of the extended source must be larger than the largest expected deflection angle ε of the test object.

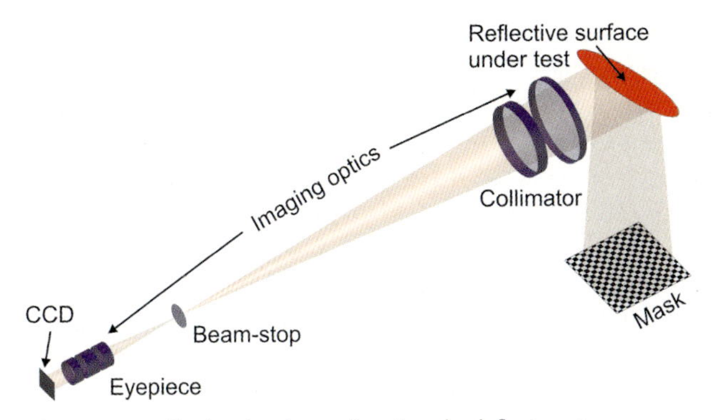

Figure 49-25: Testing the shape of a mirror by deflectometry.

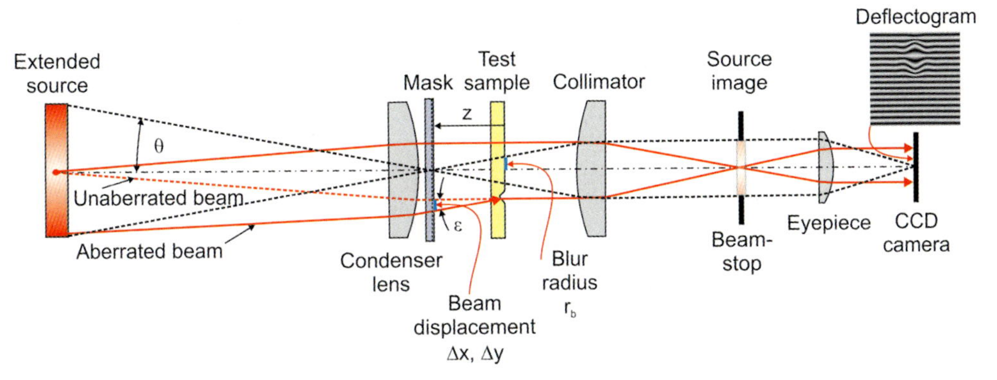

Figure 49-26: Deflectometer arrangement to test transmissive test samples.

The mask is imaged onto the CCD camera by means of a collimator and an eyepiece. A beam-stop is placed, where the back focal point of the collimator coincides with the front focal point of the eyepiece. This location also coincides with the image of the extended light source. Since the camera cannot focus onto the mask and the test sample simultaneously, either the mask image or the test sample image will be blurred, the radius r_b of a blurred object point depending on the distance z between mask and sample and θ. By decreasing the beam-stop size the blur radius is also decreased, at the same time restricting the maximum detectable deflection angle.

To study the sensitivity of deflectometry we will look at the simple arrangement shown in figure 49-27. The mask is located at a distance z from the vertex of the test surface which is at distance b from the principal plane H′ of the imaging optics. The CCD target is located at distance a in front of the principal point H. The imaging optics has a focal length f', the CCD target has a diameter d. We assume that the mask is sharply imaged onto the CCD target.

When a small surface element of the test surface alters its slope by an angle ε the corresponding pixel of the CCD target will look at a different portion of the mask.

We suppose that the CCD target has N pixels over its diameter d and that the lateral resolution of the image analysis algorithm is capable of resolving a fraction δ of a pixel. The smallest slope angle deviation ε_{res} that the system can resolve is then given by

$$\varepsilon_{res} = \frac{1}{2}\arctan\frac{d\delta}{N z\,|m|} \quad (49\text{-}35)$$

with m denoting the lateral magnification for the mask image at the CCD sensor. If $z \rightarrow \infty$ and $m \rightarrow 0$ the expression simplifies to

$$\varepsilon_{res} \approx \frac{1}{2}\arctan\frac{d\delta}{N f'} \quad (49\text{-}36)$$

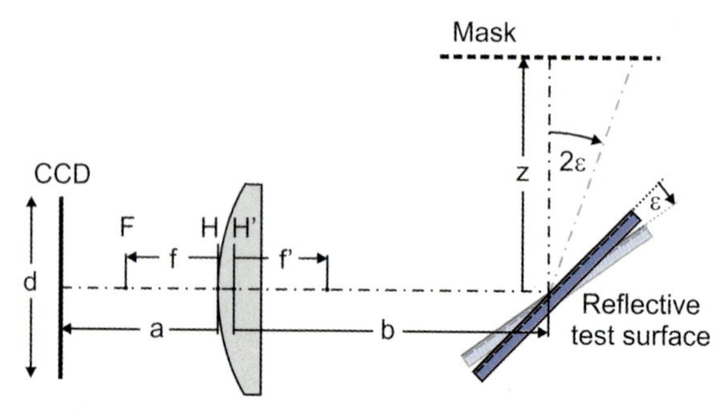

Figure 49-27: Geometric parameters for a deflectometry arrangement to test the shape of a mirror.

When z becomes large, deflectometry becomes extremely sensitive and even comparable to interferometry. If z is in the range of several meters, nanometer deviations of a test surface become visible for high spatial frequency deviations. Note, however, that the method is sensitive to the gradient of the surface rather than to the sagittal height.

Figure 49-28, 49-29 and 49-30 show a telescopic arrangement using a checkerboard mask and the related CCD images when convex, plane and concave mirrors are tested. Since the setup is made telecentric geometric relations are mainly independent of the distance of the test mirror from the telescope. Note that the angle-resolving sensitivity is different for each point on the test mirror, because the distance from the mask is different for each point.

The mask must carry a structure enabling the algorithms to derive surface gradients in the x and y directions. Checkerboard, spot grid or cross-hair grid patterns are suitable. If a mask with parallel fringes is used, two consecutive measurements have to be carried out with the mask rotated by 90°.

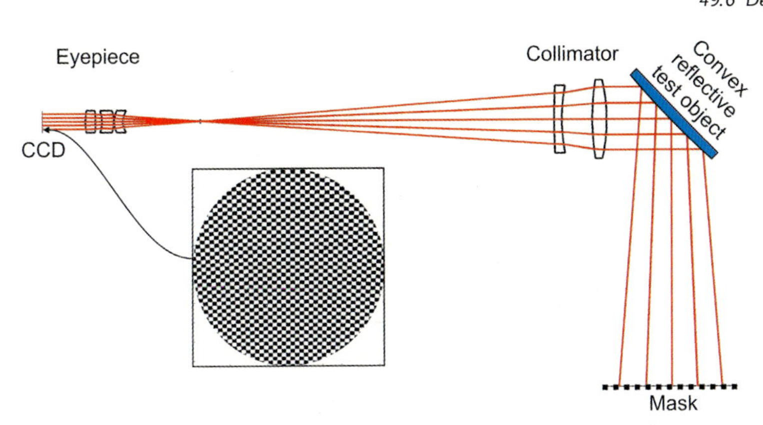

Figure 49-28: Deflectometric arrangement using a telescope and a checkerboard mask to test a convex mirror.

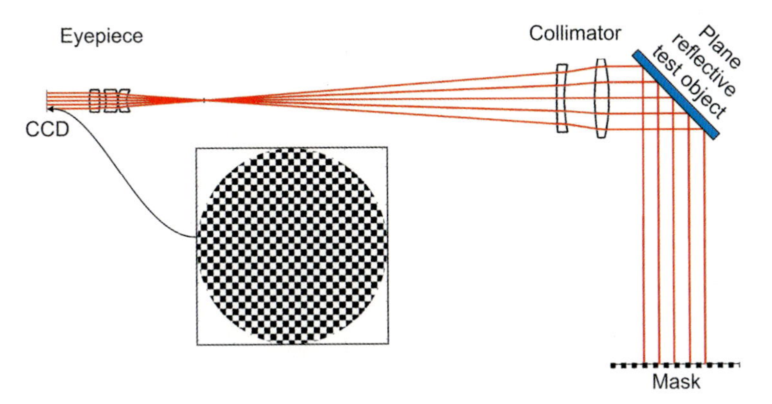

Figure 49-29: Deflectometric arrangement using a telescope and a checker board mask to test a plane mirror.

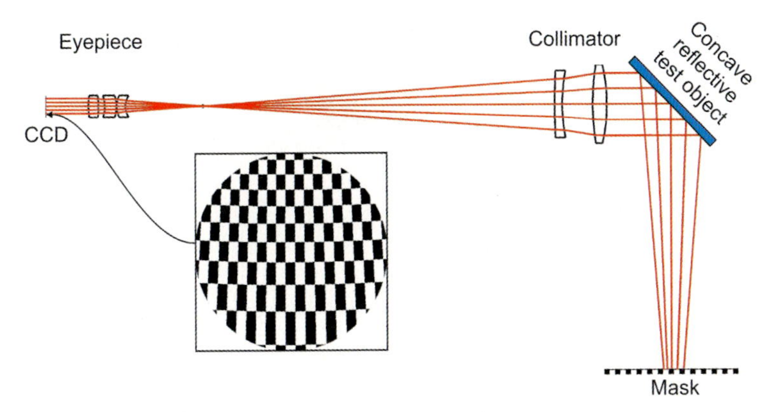

Figure 49-30: Deflectometric arrangement using a telescope and a checkerboard mask to test a concave mirror.

One general problem is the testing of strongly curved surfaces, because the angle range of the reflected beams can cover up to 360° if, for instance, a semi-sphere is observed in parallel light (Figure 49-31). In this case a plane mask is not sufficient.

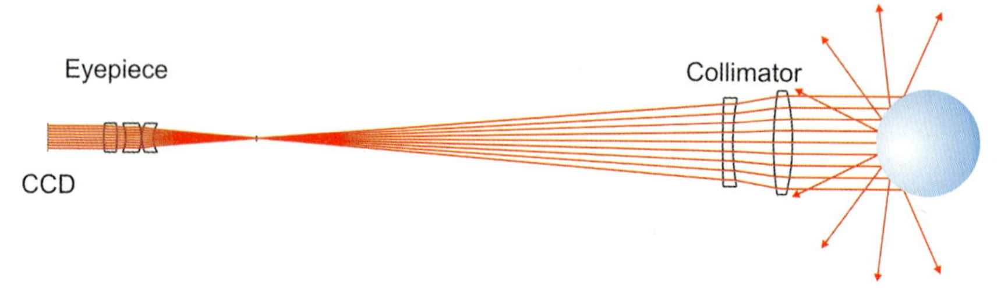

Figure 49-31: Deflectometric arrangements at strongly curved surfaces need masks that surround the reflecting test object.

For strongly curved surfaces the mask must be placed on a curved surface partially surrounding the test surface. The surrounding surface must carry a bore to allow a complete view of the test surface. The shape of the surrounding surface can be, for instance, a cone (Figure 49-32), a sphere, a cylinder or any other shape suitable for carrying the mask pattern.

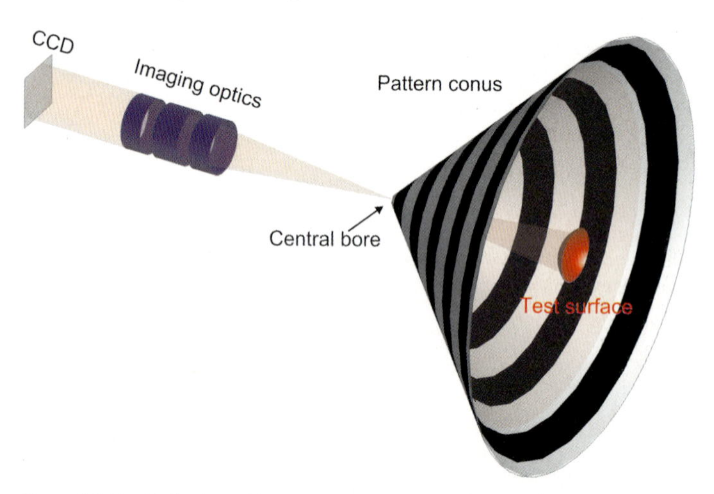

Figure 49-32: Deflectometric arrangement at strongly curved surfaces using a conical pattern carrier with a central bore for observation.

The illumination of a hollow pattern carrier is non-trivial. Light must be applied from the outside by using a transparent pattern carrier or from the inside without being shaded by the test object.

49.6.2 Algorithms

Deflectometric patterns can be analyzed by the same cross-correlation methods as described for distortion measurement systems in section 49.5.2.

If grids with sinusoidal patterns are used, evaluation techniques such as sequential phase shifting or spatial carrier frequency analysis can be applied, as described above. Since deflectometry mainly delivers gradients of surfaces, measurements must be made with grids oriented in at least two different directions to generate full surface deviation.

49.6.3 Calibration

In image geometry measuring devices one of the main issues is the quality and position of the mask. Main quality errors are deviations from:

- the plane nature of the mask, or – when masks are spherical or bent – the exact shape of the substrate,
- the exact placement of the structure.

Deflectometers generally use camera lenses and sometimes additional folding mirrors or collimation optics that need to be included in the calibration. The easiest method is calibration with known calibration elements like plano or spherical surfaces. The advantage is that absolute testing procedures such as shift-rotation techniques, three-flat testing or the rotating sphere test (see section 46.6.2) can be applied. Note, however, that the shape and position of the test objects need to be very close to those of the calibration elements, because:

- the beams returning from the test object might follow very different paths from the calibration process leading to so-called retrace errors,
- the camera pixels will look at a very different portion of the mask, so that the quality of the mask will limit the accuracy.

In cases where mask quality errors are sufficiently small and the camera lens optics is well-known, calibration by a mathematical model is preferred over calibration using perfect spheres or flats in different positions. In this case the total range of the measuring device can be used without recalibration. The evaluation software of the deflectometer should preferably be combined with raytracing software to approximate the simulated and measured irradiance distributions by iteration. For the initial setup process unknown parameters like the position of the mask and camera can also be found by iteration in an optimization run.

49.6.4
Accuracy and Error Sources

Since deflectometry mainly detects the slope of a surface under test, the resolving ability of a surface deviation $z(x)$ is dependent on the spatial frequency f of the deviating structure. Surface structures of the type (Figure 49-33)

$$z(x) \propto \frac{\sin(2\pi f x)}{f} \tag{49-37}$$

have the same maximum slope and can equally be detected by deflectometers.

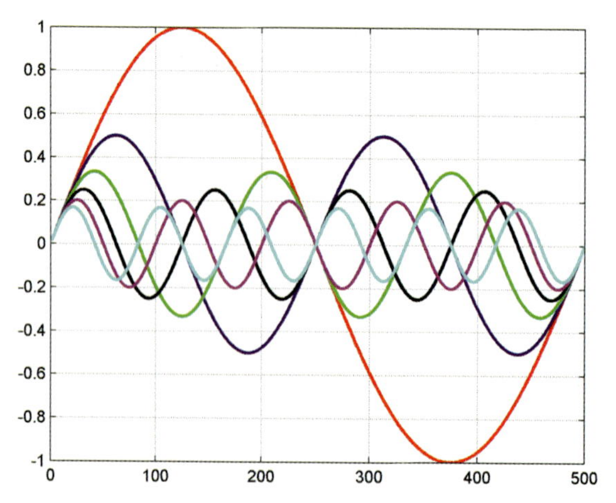

Figure 49-33: Sine functions of equal maximum slope.

Therefore a deflectometer resolves high spatial frequency structures much better than low spatial frequency structures, in contrast to an interferometer which mainly makes no difference between different spatial frequencies. Figure 49-34 shows simulated examples of a camera with 1000 × 1000 pixels of 14 μm size, a lens with $f' = 100$ mm and a plane surface under test at a distance of $b = 500$ mm. It is assumed that the convolution algorithm is able to resolve 1/30 of a pixel. Differently scaled masks of the same pattern were placed at different distances z from the test surface such that their image fills the imager of the camera. The results show that sub-nanometer deviations are resolved for structures finer than 1 mm with mask distances $z > 1$ m.

The accuracy of deflectometers is mainly limited by the following factors.

- The accuracy of the mask including its plane nature and the placement of the structure.
- The position of the mask relative to the test surface, camera and optics.
- The magnification of the imaging optics.
- The calibration process for the setup.

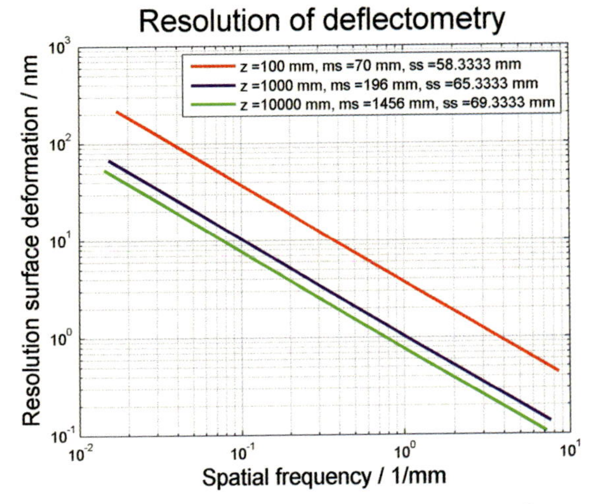

Figure 49-34: Resolution of a deflectometer for plane surfaces with the following parameters: d=14 mm, f'=100 mm, b=200 mm, N=1000 pixels, δ=1/30 pixel. Distance of mask z=100 mm / 1000 mm / 10000 mm, mask size=70 mm / 196 mm, 1456 mm, test surface size=58 mm / 65 mm / 69 mm.

For high-precision applications the following factors play an additional role.

- The support and bending of the mask.
- The environmental conditions like temperature, air pressure, humidity.
- The mechanical stability of all the components.

49.7 Pattern and Fringe Projectors

49.7.1 Basic Setups

The process of finding the coordinates of a point C in space relative to two known observation points A and B is called *triangulation*. If the distance c between the observation points and the viewing angles α and β from each observation point are known, the coordinate of the unknown point C can be determined (figure 49-35).

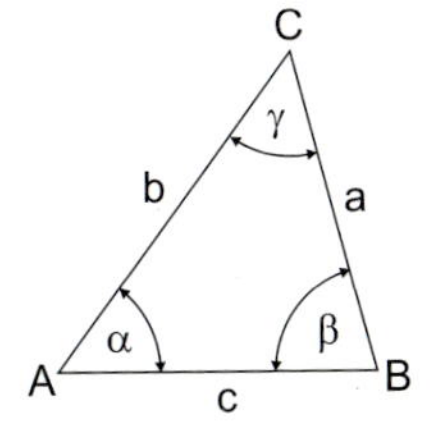

Figure 49-35: Parameters of a triangle.

Triangulation can be performed either by

- combining a projection system in A with an observation system in B (pattern projection, fringe projection), or
- placing two observation systems in A and B, respectively (photogrammetry, videogrammetry).

Pattern Projection

Pattern projection makes use of the triangulation principle and can be applied to rough objects, i.e., objects that generate enough scattered light in the direction of observation, preferably those objects which have a bright basic color. Dark or black colors will generally absorb too much of the illuminating light [49-42]–[49-55].

We assume a light beam projected onto a plane test object. We observe the light spot on the object surface from a great distance. The line of projection and the line of observation include an angle ϑ_P and coincide at the light spot at x. The direction of observation is identical with the z-axis. When the object moves in the z direction by Δz the light spot moves in the lateral direction by Δx according to (49-38) (figure 49-36).

$$\Delta x = \Delta z \tan \vartheta_P \tag{49-38}$$

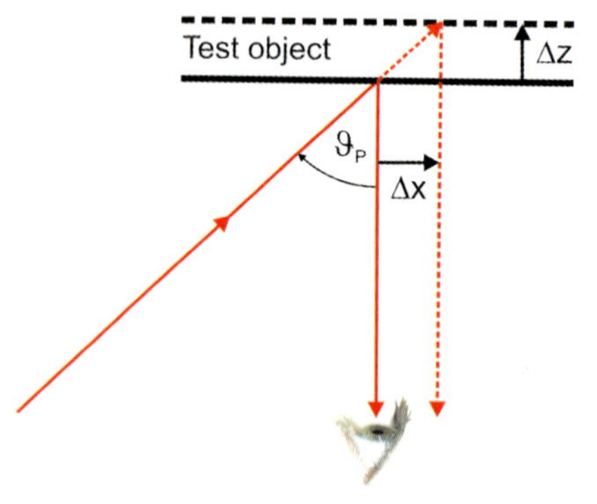

Figure 49-36: Triangulation by beam projection onto a plane test surface.

The sensitivity of the test increases with the angle between projection and observation.

From the projection of a single light spot, only a single coordinate point of the object can be determined. To obtain more coordinate points either the projection beam has to be scanned over the test surface or the test object must be moved in a suitable way.

When a light line is projected onto a test surface, coordinate points along the line can be detected (figure 49-37). Thus, the type and magnitude of deformations along the line can be determined by the single shot of a camera. The line must be oriented perpendicular to the plane defined by the principal line of projection and the principal line of observation.

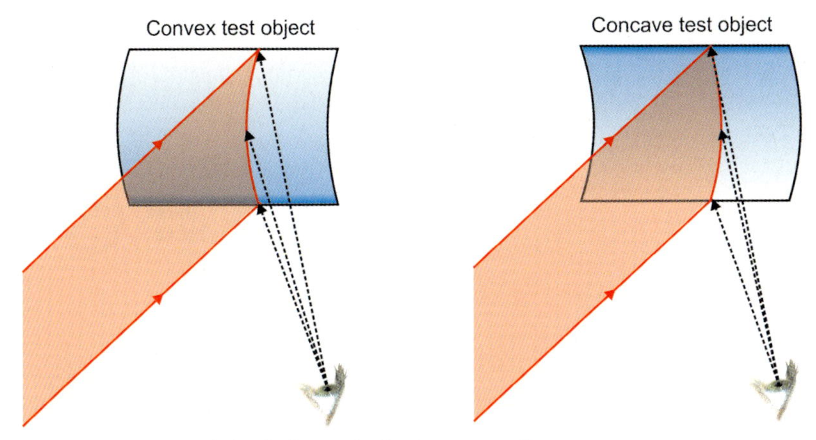

Figure 49-37: Triangulation by line projection to measure the deformations of objects along a line.

When many light lines (fringes) are projected onto a test surface, coordinate points across the projection surface can be detected in shape and magnitude by the single shot of a camera (figure 49-38). The fringes must be oriented perpendicular to the plane defined by the principal line of projection and the principal line of observation.

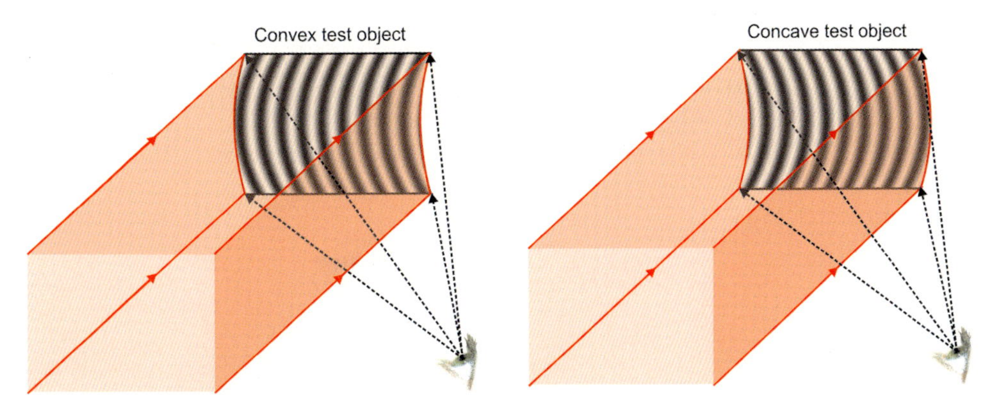

Figure 49-38: Triangulation by fringe projection to measure the deformation of objects across their surface.

The general setup for a pattern projection system is shown in figure 49-39. The mask carrying the projection pattern is imaged onto the surface under test via a suitable projection system. The test object must be a rough surface, which generates enough scattered light in the direction of observation. In the case of high residual specular reflectivity directly reflected light should be prevented from entering the observing aperture.

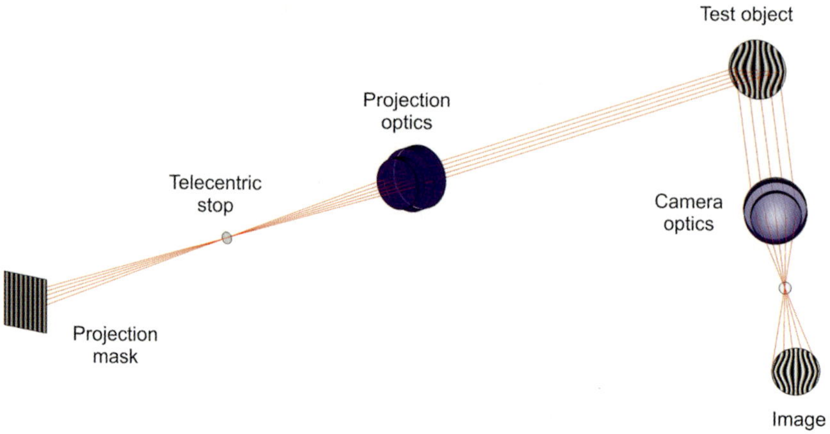

Figure 49-39: General optical setup of a pattern projection (fringe projection) system.

The projection system and the camera system can be designed either as a telecentric system, in which parallel projection or observation is possible (figure 49-40), or as a central projection or observation system (figure 49-41). For parallel projection or observation a telecentric stop must be placed in the focal plane of the projector or camera optics, respectively. For central projection or observation the aperture stop is near the principal planes of the optics.

Telecentric projection and observation has the following advantages.

- The measurement is independent of the distance from the test object.
- The resolution is constant over the field of view.
- The calibration and evaluation of the measurement signal is easy compared to central projection and observation.

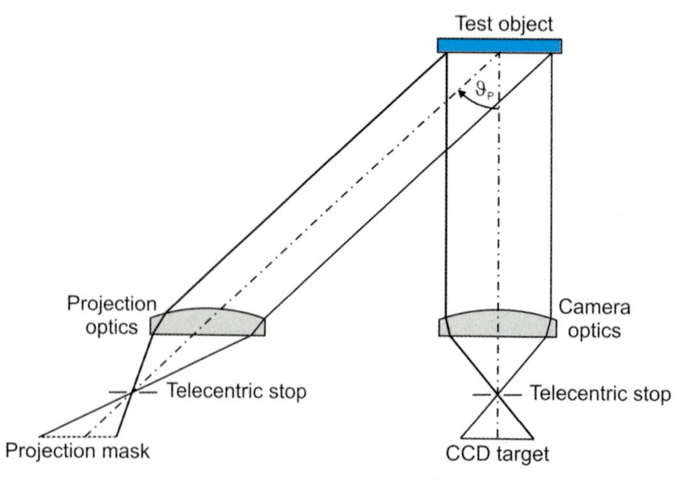

Figure 49-40: Pattern projection arrangement with telecentric projection and camera optics enabling parallel projection and observation.

Telecentric projection and observation is, however, limited to smaller testpieces, because the optics must be larger than the test object itself. The diameter of the telecentric lens affects the depth of focus, which has to be optimized to give enough lateral resolution for projection and observation.

The resolution of a telecentric arrangement, as shown in figure 49-40, is given by

$$\Delta z_{\text{res}} = \frac{\Delta s \delta}{\sin \vartheta_{\text{P}}} \tag{49-39}$$

where Δs denotes the smallest lateral pattern dimension being projected. In the case of fringe projection, Δs denotes the lateral fringe spacing perpendicular to the projection direction. ϑ_{P} is the angle between projection and observation. We assume that the camera optics resolves the projected pattern and that the image analysis algorithm is capable of resolving a fraction δ of the smallest pattern dimension Δs.

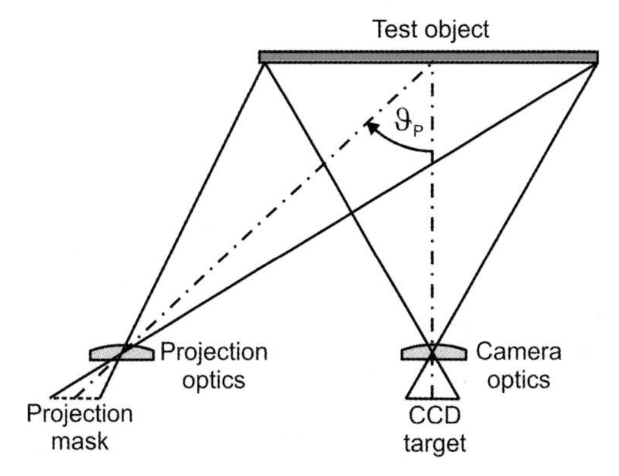

Figure 49-41: Pattern projection arrangement with aperture stops near the principal planes of projection and camera optics (central projection and observation).

Central projection and observation as shown in figure 49-41 has the advantage that there is theoretically no limitation on the size of the object. However, in practice, the size is limited because the irradiation of the projected pattern decreases by the illuminated area of the test surface. The problem is worse when scattering is low and the surface has a dark color.

Central projection and observation has the following complications.

- The measurement is dependent on the distance of the test object, therefore, an additional absolute distance-sensing system or feature has to be combined with the projection and camera system.
- The resolution changes over the field of view.

For large projection angles ϑ_{P} sharp imaging of the pattern across the total test object might become difficult, because there is no projection optics with a sufficiently large field. It will not be possible to align the mask, principal plane of projection optics and mean surface plane so that they are all parallel, in order to establish a sharp pattern image onto the test surface. In those cases all three elements can be arranged according to the *Scheimpflug condition*, in which the mask plane, the principal plane of projection optics and the mean test piece plane cross at a common straight line (figure 49-42). The mask will then be sharply imaged onto the test piece. Note, however, that the mask pattern will be distorted when projected in the Scheimpflug condition.

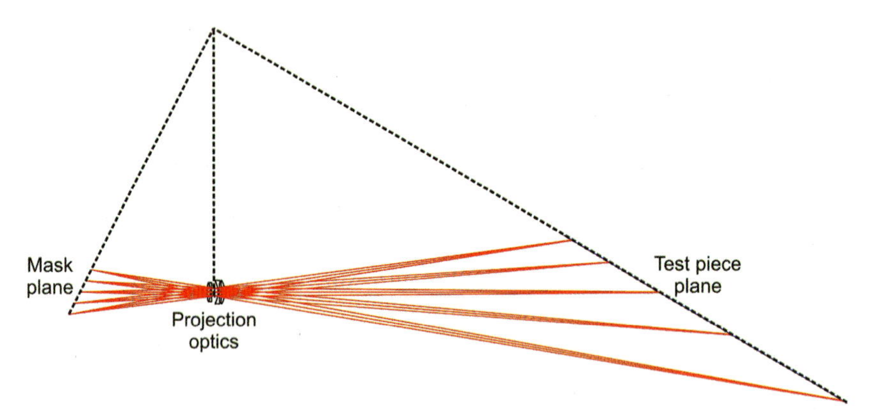

Figure 49-42: Alignment of mask, projection optics and mean test piece plane according to the *Scheimpflug condition*.

When strongly curved surfaces are tested by means of central projection, the testable area might be limited by the perspective of the camera and projector with respect to the maximum tilt of a surface element. Figure 49-43 shows an example in which a spherical surface of radius r is viewed by a camera at distance d and illuminated by a projector translated by t with respect to the camera. The limiting angle α is due to the perspective view of the camera given by (49-40). The angle β is limited by the perspective view of the projector given by (49-41).

$$\alpha = \arccos \frac{r}{r+d} \tag{49-40}$$

$$\beta = \arccos \frac{r}{\sqrt{t^2 + (r+d)^2}} - \arctan \frac{t}{r+d} \tag{49-41}$$

In the following we will derive some equations for central pattern projection from simple vector relations. Figure 49-44 shows an object with surface point P, two center points C_1 and C_2 denoting the principal points of the camera lens and the projection lens, respectively, and two planes perpendicular to the vectors $\vec{c}_1$ and $\vec{c}_2$ pointing to C_1 and C_2, respectively. The vectors $\vec{q}_1$ and $\vec{q}_2$ define points within the two planes. We assume that $\vec{q}_1$ describes the camera plane, whereas $\vec{q}_2$ describes the mask plane.

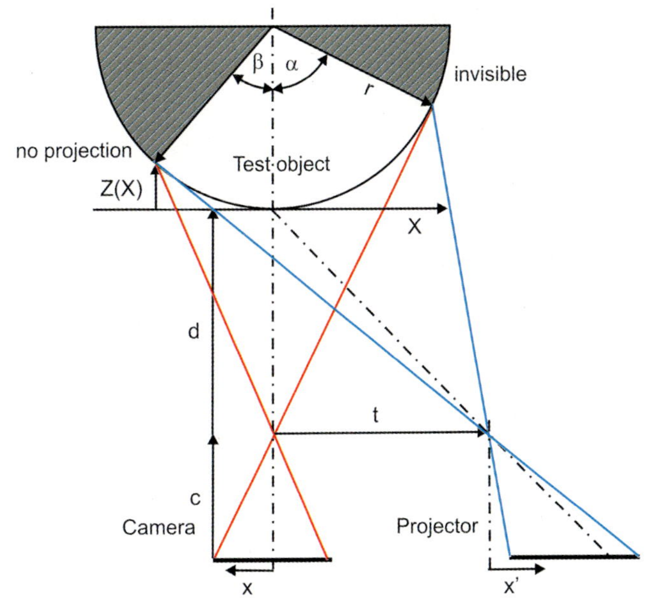

Figure 49-43: Visible and invisible regions of a spherical test surface in a pattern projection setup.

We assume that the coordinate transformation from the camera plane to the mask plane $x \leftrightarrow x'$, $y \leftrightarrow y'$, $z \leftrightarrow z'$ is known including the separation of the local origins $\vec{t}$. We also assume that the coordinates of the center points C_1 and C_2 are known, described by the vectors $\vec{c}_1$ and $\vec{c}_2$.

A point P(x,y,z) on the unknown object will then be displayed on the camera plane in relative coordinates at $\vec{q}_1$ and will be illuminated by a mask point from $\vec{q}_2'$, the latter becoming $\vec{q}_2$ after transformation into the coordinate system of the camera.

The unit vectors pointing in the direction of P(x,y,z) can then be calculated from

$$\vec{a}_1 = \frac{\vec{c}_1 - \vec{q}_1}{|\vec{c}_1 - \vec{q}_1|} \tag{49-42}$$

$$\vec{a}_2 = \frac{\vec{c}_2 - \vec{q}_2}{|\vec{c}_2 - \vec{q}_2|} \tag{49-43}$$

The coordinate of point P(x,y,z) is then calculated from

$$P(x, y, z) = p_1 \vec{a}_1 + \vec{q}_1 \tag{49-44}$$

with p_1 given by

$$p_1 = \frac{(\vec{q}_1 - \vec{q}_2 - \vec{t})(\vec{a}_2(\vec{a}_1 \cdot \vec{a}_2) - \vec{a}_1)}{1 - (\vec{a}_1 \cdot \vec{a}_2)^2} \tag{49-45}$$

The tasks for the pattern projection algorithms to solve are then:

1. to measure the irradiance distribution $I(\vec{q}_1) = I(x, y)$ within the image of the test object,
2. to identify the related pattern point $\vec{q}_2'$ from $I(x,y)$,
3. to transform $\vec{q}_2'$ to $\vec{q}_2$,
4. to calculate the coordinates of P(x,y,z) from (49-42), (49-43) and (49-44).

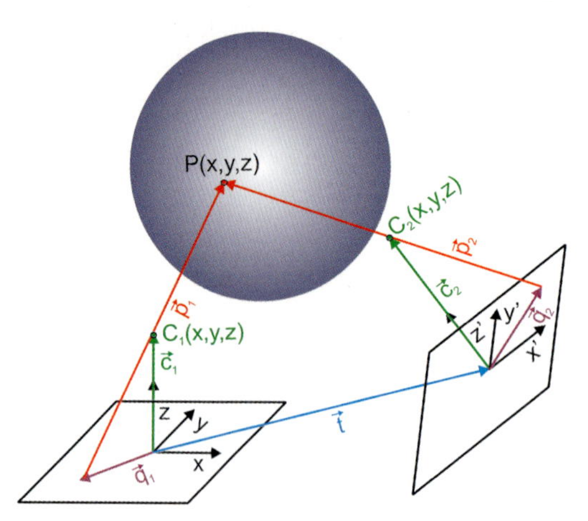

Figure 49-44: Vector relations describing central pattern projection (see text).

The main problem is to identify $\vec{q}_2'$ from $I(x,y)$. When a periodic pattern is projected, as in the case of fringe projection, the 0th fringe must be identified and the assumption is made that the object does not show any discontinuity that would produce ambiguities in the pattern interpretation.

Thus, for smooth, continuous surfaces $\vec{q}_2'$ is calculated from $I(x,y)$ by correlation or phase-deriving algorithms that are able to calculate the associated x'/y' coordinates of the mask.

For discontinuous surfaces there are techniques where patterns of different periodicities are projected and detected sequentially to solve the ambiguity problem in interrupted periodic patterns. Alternatively, periodic patterns can be projected or detected sequentially from different directions relative to the test object. For the measurement of optical components, however, we assume that the objects have smooth and continuous surfaces such that a single projection and observation direction is sufficient.

Photogrammetry

Photogrammetry is mainly based on the same triangulation algorithms as pattern projection [49-56]–[49-60]. The main differences are that:

1. the projector is replaced by a second camera,
2. the object must carry a texture.

The texture is needed to identify corresponding points in the two images. For all identified correlated surface points their coordinates can be calculated from (49-42), (49-43) and (49-44).

Testing ground or lapped optical surfaces with photogrammetry would not be convenient, because the surface texture is generally not suitable for image correlation. However, a suitable texture can be projected onto the surface under test acting as the structure needed for the image correlation process. The method would also work if fiducials were painted onto the surface. However, this is not very practical in an optical shop.

49.7.2 Algorithms

The patterns from pattern or fringe projectors can be analyzed by the same cross-correlation methods as described for distortion measurement systems in section 49.5.2. As in deflectometry, sinusoidal patterns preferably use well-known fringe evaluation techniques like sequential phase shifting or spatial carrier frequency analysis. Unlike deflectometry, pattern projection mainly measures surface shapes rather than their gradients. Therefore a single measurement with a grid in one orientation is sufficient when contiguous surfaces without discontinuations are being tested.

49.7.3 Calibration

Pattern projection and photgrammetry depend on calibration by mathematical model. The main issue is the camera calibration which results in determining the relation between the three-dimensional object coordinates X,Y,Z and the two-dimensional pixel coordinates i,j. The general relation neglecting any distortion effects from the camera lens is given by the matrix equation (49-46) where s denotes an arbitrary scaling factor.

$$\begin{pmatrix} s \cdot i \\ s \cdot j \\ s \end{pmatrix} = \begin{pmatrix} q_{11} & q_{12} & q_{13} & q_{14} \\ q_{21} & q_{22} & q_{23} & q_{24} \\ q_{31} & q_{32} & q_{33} & 1 \end{pmatrix} \begin{pmatrix} X \\ Y \\ Z \\ 1 \end{pmatrix} \tag{49-46}$$

The 11 unknown parameters q_{11} ... q_{33} contain five camera intrinsic parameters

1. the projection point distance c from the CCD array
2. the pixel width p_i and height p_j,
3. the pixel number i_c, j_c for a pixel on the optical axis,

as well as six extrinsic parameters defining the position of the camera relative to the object coordinate system (figure 49-45)

4. translations in X, Y, Z,
5. rotations relative to the X, Y, Z axes.

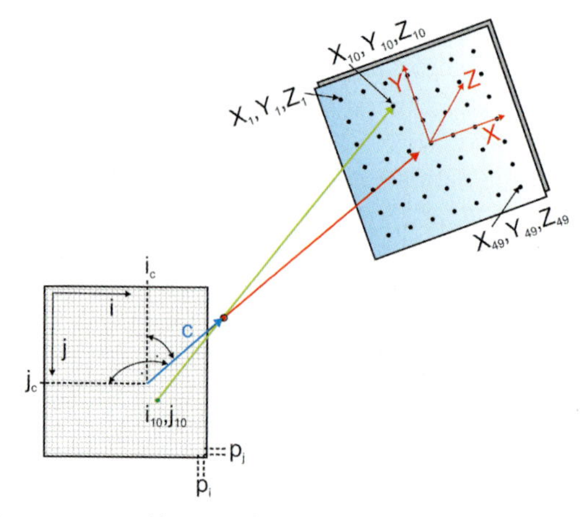

Figure 49-45: Calibration of intrinsic and extrinsic parameters of a camera – object arrangement.

For N measurement points a set of equations can be derived from (49-46) leading to (49-47).

$$\begin{pmatrix} X_1 & Y_1 & Z_1 & 1 & 0 & 0 & 0 & 0 & -i_1X_1 & -i_1Y_1 & -i_1Z_1 \\ 0 & 0 & 0 & 0 & X_1 & Y_1 & Z_1 & 1 & -j_1X_1 & -j_1Y_1 & -j_1Z_1 \\ \dots & \dots & \dots & \dots & \dots & \dots & \dots & \dots & \dots & \dots & \dots \\ X_N & Y_N & Z_N & 1 & 0 & 0 & 0 & 0 & -i_NX_N & -i_NY_N & -i_NZ_N \\ 0 & 0 & 0 & 0 & X_N & Y_N & Z_N & 1 & -j_NX_N & -j_NY_N & -j_NZ_N \end{pmatrix} \begin{pmatrix} q_{11} \\ q_{12} \\ q_{13} \\ q_{14} \\ q_{21} \\ q_{22} \\ q_{23} \\ q_{24} \\ q_{31} \\ q_{32} \\ q_{33} \end{pmatrix} = \begin{pmatrix} i_1 \\ j_1 \\ \dots \\ i_N \\ j_N \end{pmatrix} \tag{49-47}$$

In a short form we can write (49-47) as

$$\overline{\overline{M}} \cdot \vec{Q} = \vec{N} \tag{49-48}$$

The solution for $\vec{Q}$ is then given by

$$\vec{Q} = \left(\overline{\overline{M^T}}\,\overline{\overline{M}}\right)^{-1} \overline{\overline{M^T}}\,\vec{N} \tag{49-49}$$

To determine the unknown parameters in $\vec{Q}$ a least-squares estimate is achieved when N known object points X_k, Y_k, Z_k with $k = 1 \dots N$ and $N \geq 6$ with their associated pixels i_k, j_k are used as input to (49-49).

The N object points have to be chosen carefully to make the matrix inversion well-conditioned.

The calibration technique can be applied to pattern projection or photogrammetry. A calibration object like a plane plate or spherical surface with well-defined marks and fiducials must be observed by the camera under calibration (figure 49-45).

In photogrammetry, calibration must be provided for two or more cameras depending on the technique. In the case of static objects, photogrammetry can be applied using a single camera sequentially in different well-defined positions. Calibration must then also be carried out for the six extrinsic parameter sets in the further camera positions.

In pattern projection the projected mask is well known. It is projected onto the calibration object and then observed by the calibrated camera. The projected pattern is then observed along with the well-defined fiducials of the calibration object (figure 49-46). Equations (49-47) and (49-49) can then be applied to determine the unknown parameters in $\vec{Q}$ for the projector.

Once calibration is done, the parameters in $\vec{Q}$ for different cameras or the projector are defined and unknown objects can be measured. The unknown object coordinates X,Y,Z are then determined from a combination of (49-46) applied to the individual cameras as shown in (49-50) as an example for two cameras. The parameters of the second camera are marked by a single prime.

$$\begin{pmatrix} s\cdot i \\ s\cdot j \\ s \\ s'\cdot i' \\ s'\cdot j' \\ s' \end{pmatrix} = \begin{pmatrix} q_{11} & q_{11} & q_{11} & q_{11} \\ q_{21} & q_{11} & q_{11} & q_{11} \\ q_{31} & q_{11} & q_{11} & 1 \\ q'_{11} & q'_{11} & q'_{11} & q'_{11} \\ q'_{21} & q'_{11} & q'_{11} & q'_{11} \\ q'_{31} & q'_{11} & q'_{11} & 1 \end{pmatrix} \begin{pmatrix} X \\ Y \\ Z \\ 1 \end{pmatrix} \tag{49-50}$$

Scaling factors s and s′ must be determined by measuring the absolute distance of at least one object point from the camera. (49-50) can then be inverted using least-squares methods to give the unknown object coordinates from the measured pixels i, j and i', j' of both cameras.

In a short form we can write (49-50) as

$$\vec{I'} = \overline{\overline{Q'}} \cdot \vec{X} \tag{49-51}$$

The solution for $\vec{X}$ denoting the unknown object coordinates is then given by

$$\vec{X} = \left(\overline{\overline{Q'^T Q'}}\right)^{-1} \overline{\overline{Q'^T}}\,\vec{I'} \tag{49-52}$$

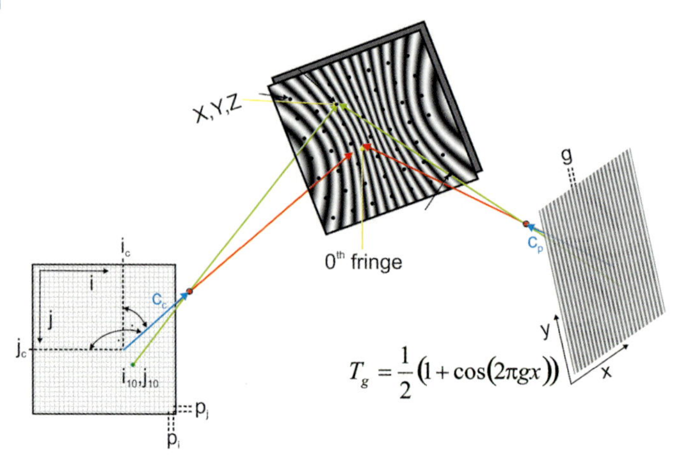

Figure 49-46: Calibration of a fringe projection arrangement by means of a calibration plate with well-known fiducials.

49.7.4 Accuracy and Error Sources

The accuracy of pattern projection devices is mainly limited by the following factors.

- The accuracy of the mask.
- The pattern recognition and evaluation algorithm.
- The stability of the position of the mask, the projecting optics and the camera including the optics to each other and relative to the test surface,
- The calibration process for the setup.

As stated in section 49.5.4 the precision of masks lies between 10^{-5} and 10^{-7} depending on the manufacturing process.

Pattern recognition and evaluation processes can be estimated to sub-pixel resolutions. They resolve between 1/10 and 1/1000 of a pixel depending on the amount of data collected and averaged.

The influence of the stability of the individual components within the setup can be estimated from the general geometric configuration. We consider a fringe projection setup as shown in Figure 49-47. The grid and camera target share the same plane, and the camera and projection optics have the same focal length. The measurable volume within the overlapping views of observation and projection is indicated in yellow, the camera and projector are focussed to a distance d. The back focal length is denoted by c. The entrance pupils are separated by t. Coordinates of the camera target are denoted by x, those of the projection grid by x'. The coordinates X, Z of a point P in the measurable volume is then identified by its camera and grid coordinate x, x' as indicated in (49-53) and (49-54).

$$Z = \frac{ct - d(x + x')}{x + x'} \tag{49-53}$$

$$X = \frac{x}{c}\left(d + \frac{ct - d(x + x')}{x + x'}\right) \tag{49-54}$$

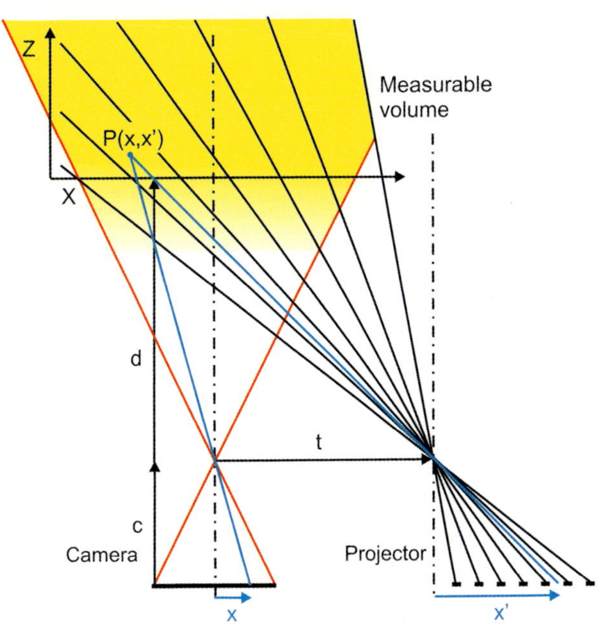

Figure 49-47: Fringe projection setup with camera and grid in the same plane.

The resolution of the system in Z is dependent on the location of the object within the measurable volume. It is proportional to t and c as shown in (49-55).

$$\frac{\partial Z}{\partial x'} = \frac{-ct}{(x + x')^2} \tag{49-55}$$

As an example, Figure 49-48 shows the projection range in Z along the optical axis of the observing camera ($x = 0$) as a function of the mask coordinate x'. The resolution gets worse with increasing object distance from the camera and with decreasing projection angles. The example shows results for a configuration, where a CCD camera with chip size 10 mm observes an object of 100 mm size. With 1000 pixels along x the chip can resolve a projection grid with 350 fringes to be projected. If the fringe detection algorithm can resolve 1/100 of a fringe spacing it is possible to resolve 1–2 μm in Z with projection angles > 60°.

To estimate the requirements for stability (49-52) must be differentiated by the system parameters c, t and d to give the translatoric sensitivities. Additionally, the tilts of the camera, grid and optics have to be investigated.

The calibration process influences the systematic part within the accuracy error budget. Since it generally consists of a combination of different measurements carried out with a calibration object (gauge), the reproducibility of the setup and the measurement process will influence the quality of the calibration. Of course, the quality of the gauge itself will also limit the accuracy.

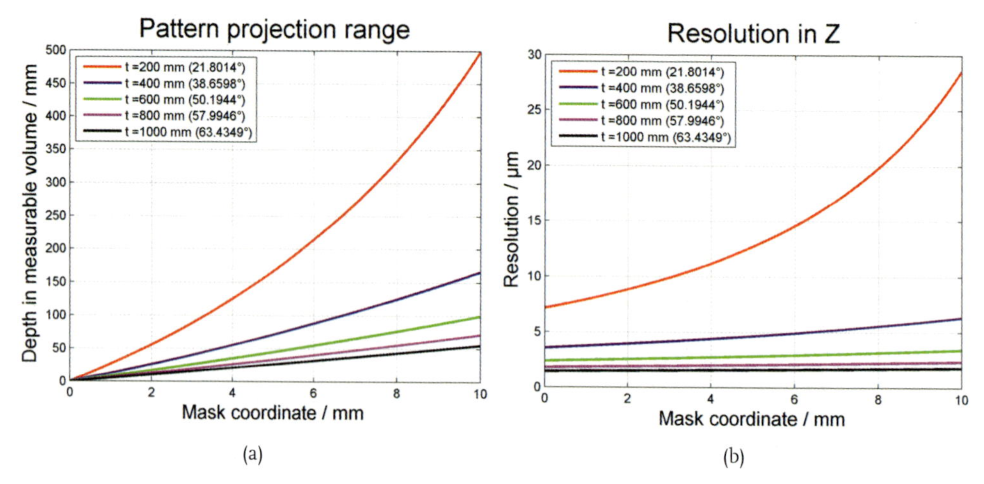

Figure 49-48: (a) Z as a function of x' for $x = 0$ according to (49-53) for $d = 500$ mm, $c = 50$ mm and $t = 200$ ... 1000 mm corresponding to projection angles between 21.8° and 63.4°, (b) resolution in Z according to (49-55) for 350 projected fringes and an assumed fringe resolution of 1/100 fringe.

Changes in environmental conditions are usually not primarily critical to changes in the optical path, but they might disturb the relative positions of the camera, grid and optics to make recalibration necessary.

49.8 Literature

49-1 D. Malacara (Editor), Optical Shop Testing, 3rd Edition (John Wiley & Sons, Inc., Hoboken, New Jersey, 2007).

49-2 C. D. Hause, J. G. Woodward and C. E. Clellan, Direct measurement of the intensity distribution of Fresnel diffraction patterns, J. Opt. Soc. Am. **29**, 147–51 (1939).

49-3 L. A. Jones and R. N. Wolfe, A method for the measurement of the energy distribution in optical images, J. Opt. Soc. Am. **35**, 559–69 (1945).

49-4 W. Herriot, A photoelectric lens bench, J. Opt. Soc. Am. **37**, 472–74 (1947).

49-5 J. M. Naish and P. G. Jones, Direct measurement of intensity distributions in star images, Nature 173, pp. 1241–242 (1954).

49-6 R. E. Hopkins, H. Kerr, T. Lauroesch and V. Carpenter, Measurement of Energy Distribution in Optical Images, NBG Circular 526, pp. 183–204 (1954).

49-7 R. E. Hopkins, S. Oxley and J. Eyer, The problem of evaluating a white light image, J. Opt. Soc. Am. **44**, 692–98 (1954).

49-8 V. Carpenter and R. E. Hopkins, Comparison of three photoelectric methods of image evaluation, J. Opt. Soc. Am. **46**, 764–67 (1956).

49-9 W. T. Welford, On the limiting sensitivity of the star test for optical instruments, J. Opt. Soc. Am., 50, p. 21 (1960).

49-10 V. A. Zverev, N. I. Boldyrev and M. N. Sokol'skii, Photoelectric photometer for measuring the energy concentration in the point spred function, Sov. J. Opt. Technol. **44**, 278–79 (1977).

49-11 J. M. Beckers, Interpretation of out-of-focus star images in terms of wave-front curvature, J. Opt. Soc. Am. A., 11, p. 425–27 (1994).

49-12 C. Loebich, D. Wueller, B. Klingen and A. Jaeger, Digital Camera Resolution Measurement Using Sinusoidal Siemens Stars IS+T, SPIE Electronic Imaging Conference 2007.

49-13 H. R. Suiter, Star Testing Astronomical Telescopes: A Manual for Optical Evaluation and Adjustment, Willmann-Bell; 1st English Ed edition (1994).

49-14 S. Monteil, M. Romero and D. Gale, Evaluation of the image quality of telescopes using the star test, Proceedings of the SPIE – The International Society for Optical Engineering (2004).

49-15 H. Gross, Handbook of Optical Systems, Vol. 1: Fundamentals of Technical Optics, Wiley-VCH, Weinheim (2005).

49-16 C. Williams and O. Becklund, Introduction to the Optical Transfer Function, Wiley-Interscience, New York (1989).

49-17 R. R. Shannon and A. H. Newman, An instrument for measurement of the optical transfer function, Appl. Opt. **2**, 365–69 (1963).

49-18 K. Murata, Instruments for the measuring of optical transfer functions, Progress in Optics **5**, 201–44 (1966).

49-19 G. Healey and R. Kondepudy, CCD camera calibration and noise estimation, Proceedings of IEEE Computer Society Conference on Computer Vision and Pattern Recognition (1992).

49-20 G. Healey and R. Kondepudy, Radiometric CCD camera calibration and noise estimation, IEEE Transactions on Pattern Analysis and Machine Intelligence, **16**, no.3, 267–76 (1994).

49-21 C. Enkrich, G. Antesberger, O. Loeffler, K.-D. Roeth, F. Laske and D. Adam, Registration Measurement Capability of VISTEC LMS IPRO4 with Focus on Small Features, Photomask Japan (2008).

49-22 G. Smith and D. A. Atchison, The Eye and Visual Optical Instrument, Cambridge University Press, New York (1997).

49-23 AIR FORCE MIL-STD-150A CANC NOTICE 5: Photographic Lenses (2006).

49-24 International Council of Ophthalmology, Visual Functions Committee, Visual Acuity Measurement Standard, Italian Journal of Ophthalmology II/I 1988, 1–15 (1988).

49-25 W. Grossmann, H. Nehler and J. Müller, Distortion measurement on high-performance objectives, Feingeraetetechnik **30** (10), 454–56 (1981).

49-26 O. Kafri and I. Glatt, The Physics of Moiré Metrology, John Wiley & Sons, New York, NY (1990).

49-27 L. Zimmermann, A Method for measuring the distortion of photographic objectives, Appl. Opt. **2**, 759–60 (1963).

49-28 G. Oster et al., Theoretical interpretation of Moiré patterns, JOSA **54**, 169-75 (1964).

49-29 M. A. Sutton, M. Cheng, W. H. Peters, Y. J. Chao and S. R. McNeill, Application of an optimized digital correlation method to planar deformation analysis, Image and Vision Computing **4** (3), 143–51 (1986).

49-30 H. W. Schreier, J. R. Braasch and M. A. Sutton, Systematic errors in digital image correlation caused by gray-value interpolation, Opt. Eng. **39** (11), 2915–21 (2000).

49-31 J. D. Helm, S. R. McNeill and M. A. Sutton, Improved 3-D image correlation for surface displacement measurement, Opt. Eng. **35** (7), 1911–20 (1996).

49-32 C. J. Tay, C. Quan, Y. H. Huang and Y. Fu, Digital image correlation for whole field out-of-plane displacement measurement using a single camera, Optics Communications **251, 23–36 (2005).**

49-33 H. A. Bruck, S. R. McNeill, M. A. Sutton and W. H. Peters, Digital image correlation using Newton–Raphson method of partial differential correction, Experimental Mechanics **29**, 261 (1989).

49-34 G. Vendroux and W. G. Knauss, Submicron deformation field measurements: Part 2. Improved digital image correlation, Exp. Mech. **38**, 86 (1998).

49-35 Y. Surrel, N. Fournier, M. Grediac and P. Paris, Phase-stepped deflectometry applied to shape measurement of bent plates. Experimental Mechanics, **39** (1), 66–70 (1999).

49-36 R. D. Geckeler and I. Weingärtner, Sub-nm topographie measurement by deflectometry: Flatness standard and wafer nanotopographie, Proc. SPIE **4779**, 1–12 (2002).

49-37 T. Bothe, W. Li, C. von Kopylow and W. Jüptner, High-resolution 3D shape measurement on specular surfaces by fringe reflection, Proceedings of the SPIE – The International Society for Optical Engineering 5457 (1), 411–22 (2004).

49-38 M. C. Knauer, J. Kaminski and G. Häusler, Phase measuring deflectometry: a new approach to measure specular free-form surfaces, Proceedings of the SPIE – The International Society for Optical Engineering **5457** (1), 366–76 (2004).

49-39 I. Scheele and S. Krey, Measurement of Aspheric Surfaces with 3D-Deflectometry, DGaO Proceedings 2005 – http://www.dgao-proceedings.de – ISSN: 1614–8436.

49-40 R. D. Geckeler, Optimal use of pentaprisms in highly accurate deflectometric scanning, Meas. Sci. Technol. 18, 115–25 (2007).

49-41 R. D. Geckeler, Shearing deflectometry as a flatness standard: Comparison with a mercury mirror and absolute interferometry, Proc. 6th EUSPEN Internat. Conf., ed. H. Zervos, Vol. 1, 390–93 (2006).

49-42 M. Takeda and K. Mutoh, Fourier transform profilometry for the automatic measurement of 3D object shapes, Appl. Opt. **22**, 3977–982 (1983).

49-43 M. Küchel and A. Hof, Device and Procedure to Measure without Contact the Surface-Contour of an Object, EP 445618 (1992).

49-44 E. Klaas and B. Breuckmann, Dynamisches Verfahren zur Erfassung von Form und Formaenderungen, DE 4023368 (1992).

49-45 G. Häusler, D. Ritter, Parallel three-dimensional sensing by color-coded triangulation, Appl. Opt. **32**, 7162–69 (1993).

49-46 W. Schreiber, G. Notni, P. Kuhmstedt, J. Gerber and R. Kowarschik, Optical 3D-coordinate measuring system using structured light, Proceedings of the SPIE – The International Society for Optical Engineering **2782**, 620–27 (1996).

49-47 V. Kirschner, W. Schreiber, R. Kowarschik and G. Notni, Self-calibrating shape-measuring system based on fringe projection, Proceedings of the SPIE - The International Society for Optical Engineering **3102**, 5–13 (1997).

49-48 H. Steinbichler, Verfahren und Vorrichtung zur Bestimmung der Absolut-Koordinaten eines Objektes, EP 534284 (1997).

49-49 P. S. Huang, Q. Y. Hu, F. Jin and F. P. Chiang, Color-encoded digital fringe projection technique for high-speed three-dimensional surface contouring, Opt. Eng. 38, 1065–71 (1999).

49-50 J. Gerber, R. M. Kowarschik, G. Notni and W. Schreiber, Adaptive optical 3D measurement with structured light, Proceedings of the SPIE – The International Society for Optical Engineering **3749**, 222–23 (1999).

49-51 R. Kowarschik, P. Kuhmstedt, J. Gerber, W. Schreiber and G. Notni, Adaptive optical three-dimensional measurement with structured light, Optical Engineering **39** (1),150–58 (2000).

49-52 F. Chen, G. M. Brown and M. Song, Overview of three-dimensional shape measurement using optical methods, Opt. Eng. **39**, 10–22 (2000).

49-53 P. Kuhmstedt, J. Gerber, M. Heinze and G. Notni, Analysis of the measurement capability of optical fringe projection systems, Proceedings of the SPIE – The International Society for Optical Engineering **5144**, 728–36 (2003).

49-54 P. S. Huang, C. P. Zhang and F. P. Chiang, High-speed 3-D shape measurement based on digital fringe projection, Opt. Eng. **42**, 163–68 (2003).

49-55 M. Steinbichler, A. Maidhof, M. Prams and M. Leitner Markus, Method for determining the 3D coordinates of the surface of an object, US 20060265177 (2006).

49-56 S. F. El-Hakim, Real-time image metrology with CCD cameras, Photogrammetric Engineering and Remote Sensing, 52 (11), 1757–66 (1986).

49-57 C. S. Fraser, State of the art in industrial photogrammetry, International Archives of Photogrammetry and Remote Sensing. XXVII(V), 166–81 (1988).

49-58 K. Wong, Machine vision, robot vision, computer vision, and close-range photogrammetry, Photogrammetric Engineering & Remote Sensing, 58(8): 1197–98 (1992).

49-59 M. R. Shortis, T. A. Clarke and T. Short, A comparison of some techniques for the subpixel location of discrete target images, Videometrics III. Boston. SPIE Vol. 2350. pp. 239–50 (1994).

49-60 T. A. Clarke, M. A. R. Cooper, J. Chen and S. Robson, Automated 3-D measurement using multiple CCD camera views. Photogrammetric Record. Vol. XV. No 86. pp. 315–22 (1994).

50
Distance and Angle Metrology

Handbook of Optical Systems: Vol. 5. Metrology of Optical Components and Systems. First Edition.
Edited by Herbert Gross.

50.1 Introduction

One of the main subjects in optical metrology is the measurement of the basic geometrical parameters *distance* and *angle*. When an optical instrument or testing device is set up, the relative distance and angular position of optical and mechanical components must be measured and adjusted. Often element positions are referenced against specific datum points and must be specifically displaced during the alignment process. We then refer to *displacement* metrology, because it is the distance between two specific element positions which is being measured. We refer to *thickness* metrology when determining the optical path length between optical surfaces. We refer to *profile* metrology, when a pointwise measuring sensor is scanned over the surface under test in order to determine its surface profile.

In the following, we will distinguish between long-range and short-range displacement metrology. Long-range techniques cover distances of up to several tens of meters, while short-range techniques may have ranges up to tens of millimeters.

We make a similar distinction when measuring rotations. Angle metrology denotes measurement techniques including ranges with full 360° rotations, while tilt metrology describes techniques covering ranges which are usually smaller than 1°.

In the following, we have concentrated on those optical techniques which deliver the resolutions necessary for optical setup purposes, i.e., < 1 μm resolution for displacement metrology and < 1 arcsec resolution for angle and tilt metrology, and have omitted those of coarser resolution.

The following sections will describe the basic principles of the techniques, different implementations of the devices and the accuracy which is achievable. Calibration techniques and sources of error will also be discussed.

At the end of this chapter there is a short section on optical profile metrology including a local displacement and a local angle technique used to reconstruct the profile of an optical surface.

50.2 Long-range Displacement Metrology

50.2.1 Displacement-measuring Interferometer

A displacement-measuring interferometer is a two-beam interferometer of the Twyman–Green or Michelson type (see section 46.3.3) used to measure displacement. It is used for high- precision displacement measurements ranging from nanometers to almost 100 m, but can also be applied to measure angles, flatness, straightness, velocity and acceleration (vibrations). As a non-contact high-precision metrology its main use is for:

- precision stage metrology, for instance, in semiconductor metrology tools and lithography steppers,

- the measurement of length standards and the calibration of measurement tools in standard laboratories,
- the calibration of machine tools to check whether they can produce parts within the required specifications (microns),
- the calibration of piezoelectric transducers, encoders, grid plates, etc.

In the following we will describe the basic principles, the different configurations and their limitations.

Basic Principles

The displacement laser interferometer consists of:

1. a laser,
2. a beam-splitter,
3. a fixed reflector component,
4. a reflector component attached to the moving object, whose displacement is to be measured,
5. a photodetector placed in the path of the interfering beams at the output of the interferometer.

The beam reflected from the fixed reflector is called the *reference beam*, while the beam reflected from the moving reflector is called the *object beam*. For high accuracy, it is essential to know the precise wavelength of the laser light [50-2] and [50-6].

There are two major classifications of displacement laser interferometers: *homodyne* and *heterodyne*.

A *homodyne* interferometer utilizes a single-frequency laser source. The reference beam and the object beam use identical wavelengths. A change in the optical-path difference (OPD) between the reference and object beams results in an increase or decrease of the signal detected at the photodetector.

Heterodyne interferometry utlizes a stabilized dual-frequency laser with a fixed beat frequency $\Delta\omega$ which is defined by the frequency split between the two polarization states [50-3]. The measurement signal is generated from a frequency-based measurement which uses phase-detection techniques. Displacement in a heterodyne system is derived from the change in phase which occurs with a change in the optical path of the object beam with respect to the reference beam. This Doppler-shifted signal is compared to the reference signal and converted into precise time-based velocity and position-displacement information.

Let $I(t)$ be the irradiance of the interference signal detected at the photodetector:

$$I(t) = A_R + A_O + 2A_R A_O \cos \Delta\phi(t) \tag{50-1}$$

where

A_R is the amplitude of the reference wave,
A_O is the amplitude of the object wave,
$\Delta\phi(t)$ is the phase difference between the object and the reference wave given by:

$$\Delta\phi(t) = k_O d_O - k_R d_R - \Delta\omega t \tag{50-2}$$

with

$$\Delta\omega = \omega_O - \omega_R \tag{50-3}$$

and

$$k_O = \frac{2\pi n_O}{\lambda_O} \text{ and } k_R = \frac{2\pi n_R}{\lambda_R} \tag{50-4}$$

and with

$$\omega_O = 2\pi \frac{c}{\lambda_O} \text{ and } \omega_R = 2\pi \frac{c}{\lambda_R} \tag{50-5}$$

where

λ_O, λ_R are the vacuum wavelengths of the object and the reference wave,
d_O, d_R are the geometrical paths of the object and the reference wave,
n_O, n_R are the refractive indices of the media transmitted by the object and the reference wave, and
c is the speed of light in vacuum.

Assuming a fixed optical reference path with $n_R d_R = const$, the time dependency of the signal is then given by the beat frequency $\Delta\omega$ and the change of the optical path $n_O(t)d_O(t)$ of the object beam. A homodyne signal is achieved when $\Delta\omega = 0$. A heterodyne signal results when $\Delta\omega \neq 0$.

When the object reflector is moved by a velocity v in the direction of the beam propagation, the beat frequency $\Delta\omega$ is altered by a Doppler shift $\delta\omega$:

$$\Delta\omega' = \Delta\omega + \delta\omega = \omega_O - \omega_R - 2\frac{v}{c}\omega_O = 2\pi\left(\frac{c - 2v}{\lambda_O} - \frac{c}{\lambda_R}\right) \tag{50-6}$$

The number of fringes detected in an interval of time is therefore dependent on the wavelength difference and the velocity of the moving object.

To detect the signal, the photodetector will have to integrate over a certain integration time T. After normalization we obtain for the detected signal

$$I(t, T) = A_R + A_O + 2A_R A_O \frac{2\sin\left(\frac{\Delta\omega' T}{2}\right)}{\Delta\omega' T} \cos\left(k_O d_O - k_R d_R - \Delta\omega'\left(t + \frac{T}{2}\right)\right) \tag{50-7}$$

The integration over a time interval T leads to a decrease in the contrast of the fringes as well as in a constant phaseshift of $\Delta\omega' T/2$.

For the determination of displacement in the system electronics

1. count the number of beats (waves),
2. derive the phase from the signal given by (50.7)

while the object is moved.

Neither the object nor the reference beam should be interrupted during the measurement process, otherwise the beat count will contain uncertain information. Initializing the system and repeating the measurement process is then unavoidable.

The accuracy of the measurement is dependent on the stability of the laser frequencies. The two frequencies are usually generated by a two-mode stabilized laser, a Zeeman laser, an acousto-optic modulator (AOM) or a pair of AOMs [50-1], [50-3], [50-4] and [50-6]. Typical beat frequencies and the resulting maximum object velocities v are:

600 MHz	± 12.5 m/s	for a stabilized Helium–Neon two-mode laser,
20 MHz	± 2.1 m/s	for an AOM arrangement.
3–4 MHz	± 0.5 m/s	for a Zeeman laser.

The faster the beat frequency the faster will be the maximum allowed velocity of the object displacement.

Heterodyne Interferometer Using Corner Cubes

The principle of a heterodyne displacement-measuring interferometer using corner cubes as reflectors is shown in figure 50.1. Corner cubes are useful because they reflect the beams back in the same direction with a defined lateral displacement, thus avoiding feedback to the laser. The corner cubes must carry a metallic coating on the reflecting surfaces, otherwise the total internal reflection of the glass material will change the polarization state of the beams considerably.

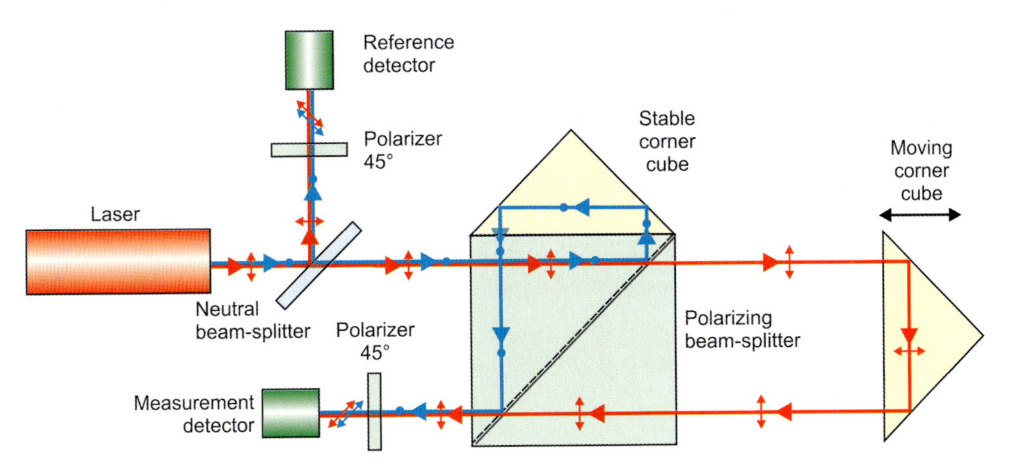

Figure 50-1: Principle of a heterodyne displacement-measuring interferometer using corner cubes as reflectors.

A laser generates two frequencies, which are linear polarized and orthogonal to each other. To generate a reference beat frequency and to compensate for any flux fluctuations, a reference signal is reflected from a neutral beam-splitter and detected by a reference detector after passing a polarizer positioned at an angle of 45°.

The light transmitting the beam-splitter is incident on a polarizing beam-splitter reflecting the s-polarization (reference beam) and transmitting the p-polarization (object beam). A stable corner cube reflects the s-polarized beam to return it to the beam-splitter surface, where it is reflected towards a 45° rotated polarizer and then on to a measurement signal detector. The second corner cube is attached to the object, which is moved by an appropriate amount in the beam direction. Its displacement is then measured by the device. The object beam is returned by the corner cube to transmit the polarizing beam-splitter and then interfere with the reference beam after passing the 45° polarizer. The phase difference between the object beat signal and the reference beat signal indicates the displacement of the moving object corner cube.

Heterodyne Interferometer Using Plane Mirrors

There are many applications requiring plane mirrors as reflectors. For example, x-y stages often carry plane, reflecting reference surfaces, from which the measuring beam must be reflected for position control.

The scheme of a heterodyne displacement-measuring interferometer using a plane object mirror as a reflector is shown in figure 50.2.

The fundamental difference from figure 50.1 is the use of a quarter-waveplate which forces the object beam to traverse the measurement path twice, leading to a doubled resolution.

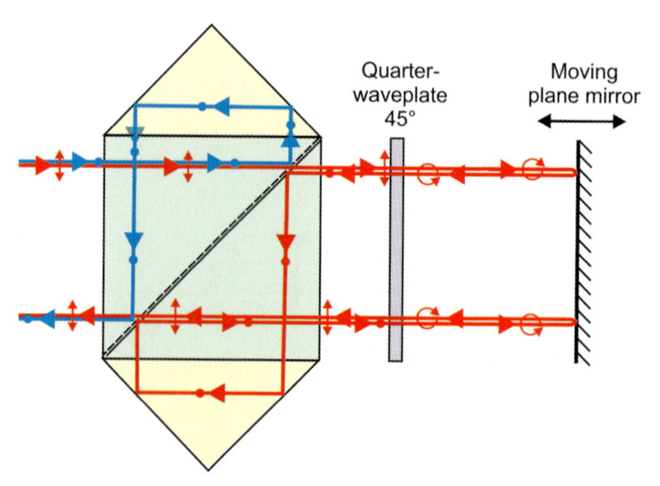

Figure 50-2: Principle of a heterodyne displacement-measuring interferometer using a plane mirror as the object reflector (laser and detectors are omitted).

The linear and orthogonal polarized beams with different frequencies enter the polarized beam-splitter. The p-polarized component acts as the object beam and transmits the beam-splitter and a following quarter-waveplate with axis at 45°. After reflection from the movable object mirror and transmission through the quarter-waveplate, the beam is converted to s-polarization and is therefore reflected from the beam-splitter surface. A corner cube at the lower end of the beam-splitter cube reflects the object beam back to the beam-splitter, where it is again reflected towards the object mirror, passing the quarter-waveplate. After the second reflection from the object mirror and transmission through the quarter-waveplate, the beam is reconverted to p-polarization and transmits the beam-splitter surface to enter the detector channel.

The s-polarized beam entering the beam-splitter acts as the reference beam and is reflected towards an upper corner cube. After reflection the reference beam returns to the beam-splitter surface, where it is reflected to enter the detector channel.

The reference and object beams interfere with each other after passing a 45° polarizer, forming the object beat signal, which is detected by the measurement detector. The phase difference between the object beat signal and the reference beat signal indicates the displacement of the moving object mirror.

Heterodyne Interferometer Using Balanced Optical Paths

Optical media generally change their refractive index as the temperature changes. The refractive index temperature dependence of glass material is much larger than that of air. For high-accuracy measurements it is therefore mandatory to find arrangements in which the optical paths of the reference and object beams through the glass material are balanced. Figure 50-3 shows an example in which the upper corner cube from figure 50-2 has been exchanged by a second quarter-waveplate and

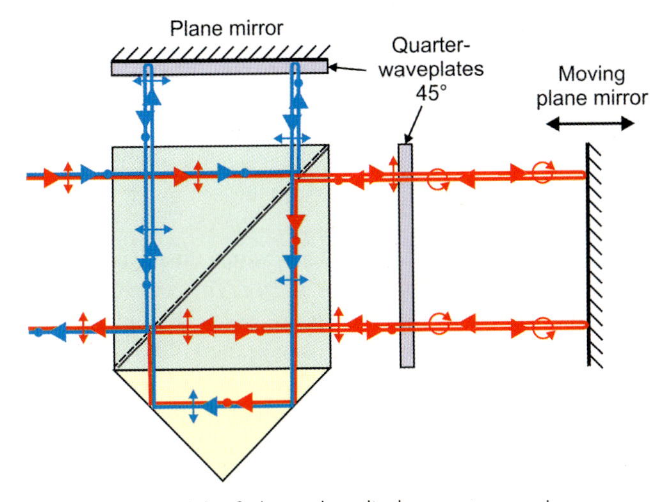

Figure 50-3: Principle of a heterodyne displacement-measuring interferometer with balanced glass paths using plane mirrors as reflectors (laser and detectors are omitted).

a second plane mirror. The s-polarized reference beam now performs the analog double passage through the reference arm as does the object beam through the object arm. Reference and object beam interfere with each other after passing a 45° polarizer, forming the object beat signal, which is detected by the measurement detector. The phase difference between the object beat signal and the reference beat signal indicates the displacement of the moving object mirror.

Accuracy and Error Sources

Error sources in displacement interferometry can be classified as being caused by:

1. the environment,
2. imperfect components,
3. misalignment of the components,
4. instrument errors.

In most cases environmental influences and object mirror surface deviations are the greatest contributors to measurement errors.

A typical accuracy achieved in air at 20°± 0.5°C is ± 500 nm per meter displacement. A typical accuracy achieved in vacuum at 20°± 0.5°C is ± 20 nm per meter displacement. The position resolution of displacement interferometers is typically between 0.5 and 2.5 nm.

In the following we discuss the error sources in more detail.

Environmental Errors

Environmental errors are usually the largest contributor to the error budget of displacement-measuring interferometers. The index of refraction for air changes with deviations in temperature, pressure and humidity, thus altering the optical path difference between the object and reference beams. It is essential to control or monitor the environment in order to maintain control of the measurement accuracy.

An optical wavelength compensator is often used to measure the change in the refractive index of the air. Since the effective wavelength changes with the refrative index, it is important to measure the temperature, pressure and humidity to calculate the index of refraction of the air from Edlén's formula (46-205) (see chapter 46). Environmental compensators or wavelength trackers can be set up in different arrangements [50-1] and [50-8].

Other compensator setups are used as refractometers, in which the measurement and reference beams travel across the same nominal distance. The reference beam travels through a vacuum sealed tube while the object beam travels through air. The optical path difference between the two can be used to indicate the change in the index of refraction of the air during the time of the measurement [50-8].

Air turbulence is caused by a local movement of the thermal gradients in the air. When located along the beam path the magnitude of the air turbulence can considerably affect the measurement. Precautions can be taken to avoid this by placing tubes around the actual dead path outside the range of the object reflector motion. The dead path is the difference in distance in air between the reference path R and measurement path M of an interferometer configuration (figure 50-4). To minimize the deadpath distance M–R the reflectors should be located as close to the beam-splitter as possible. The dead path error is effectively reduced by minimizing environmental changes during the time of the measurement. If necessary, all optical paths must operate within a helium atmosphere or in a vacuum.

When operating in air the environmental errors lead to the greatest significant limitations in accuracy. Depending on the dead path and the environmental conditions of the object beam path a displacement error of several hundreds of nanometers can result.

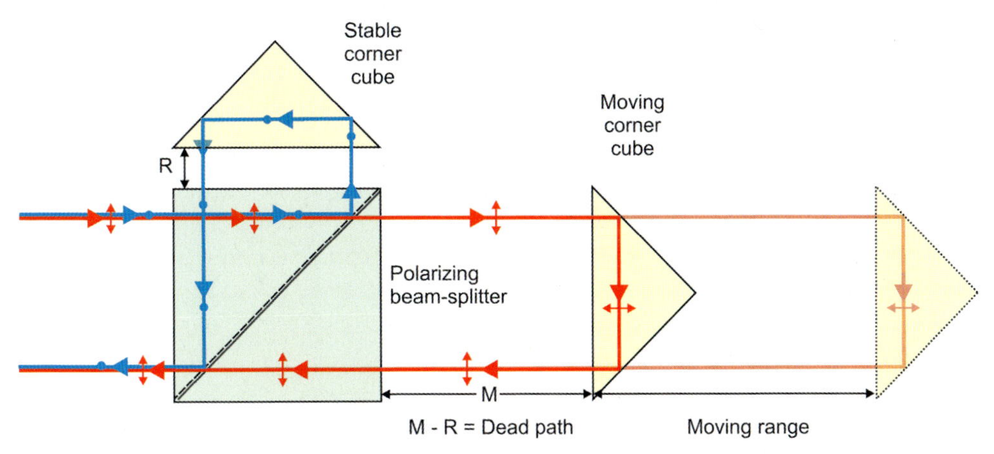

Figure 50-4: Dead path M–R in a displacement-measuring interferometer.

Imperfect Components

In a measurement configuration where the object beam always reflects from the same location on the object reflector, the geometrical error in the components will play a minor role. If, however, the object reflector consists of a laterally moving mirror surface, the *figure error of the mirror surface* will strongly influence the displacement result (figure 50-5). As in the case of a travelling stage, with reflecting sides which act as the target mirror, a surface deviation $\Delta S(x_1,x_2)$ between the two reflecting points x_1 and x_2 will result in an optical path difference of $2\Delta S$. If the surface deviations are known by calibration they can be compensated for by means of a software look-up table.

Imperfections in AR-coatings on the optical components may introduce erroneous reflections which disturb the interference signal from the object and reference beams and lead to significant errors. Usually a small tilt of the beam-splitter is introduced to direct erroneous reflections out of the field of view of the detector [50-3].

The *polarization leakage error* is caused by an unwanted polarization mixing of the laser's frequency components [50-5] and [50-7]. It causes a periodic, nonlinear relationship between the measured displacement and the actual displacement.

Polarization mixing is caused by misaligned or imperfect optical polarization components such as the polarizing beam-splitter or the quarter-waveplates. Corner cubes must have a proper metallic coating on the outer surfaces to prevent the polarization-sensitive inner total reflection. The rotational misalignment of the components about the optical axis should be limited to less than 1° to minimize polarization errors.

For high-quality systems the residual imperfection of components, including figure errors, imperfect coatings and polarization leakage, can lead to displacement errors of several tens of nanometers.

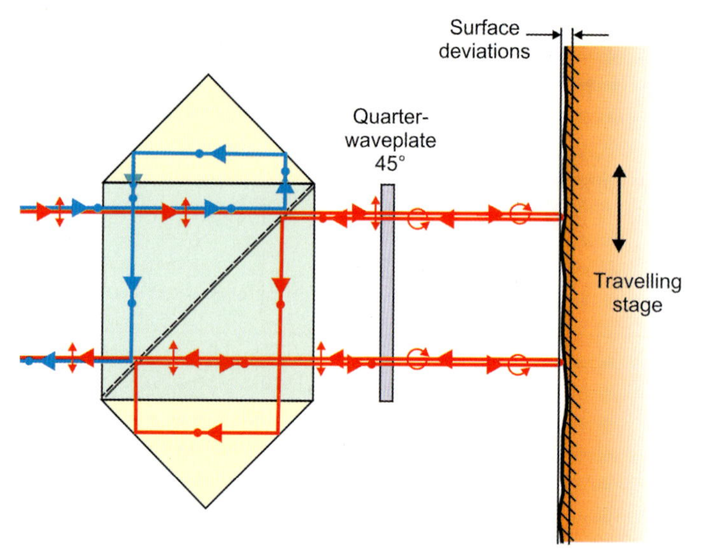

Figure 50-5: Heterodyne displacement-measuring interferometer using the reflecting side of a traveling stage as object reflector.

Misalignment of Components

Misalignments of the optical components could lead to measurement errors. Some typical misalignments carry specific names such as *cosine error* and *Abbe error.*

A *cosine error* results from an angular misalignment between the object beam and the axis of motion. The cosine error is generally negligible until the angle becomes quite large. It will then cause the interferometer to measure a displacement shorter than the actual distance traveled. As a result of a larger tilt between the optical and mechanical axis, the object and reference beams will shear and the signal efficiency will decrease. We then refer to an *overlap error.* In a single-axis system, a minimum beam overlap of 50% is sufficient. However, with an increasing number of axes using the same laser, the efficiency of the interferometer channels decreases and the overlap will be nearly 100% [50-8].

The cosine error is minimized by proper adjustment. The object corner cube is moved over the total range, and optical and mechanical axes are carefully aligned, until the overlap error is no longer visible.

An *Abbe error* results when the axis of measurement is offset from the axis of interest. If the object beam and the moving axis are separated by a distance d and the translation stage carrying the corner cube is tilted by θ during the measurement, the displacement result will carry an error of $d \sin \theta$. To avoid the Abbe error, the object beam and the moving axis must coincide. If this is impossible, a separate tilt control must be provided.

Mechanical stability of the interferometer arrangement is mandatory. If the physical attachment between the object reflector and the point of interest changes during the measurement time, this is indistinguishable from actual motion. To minimize vibration effects, several measurements can be taken at one position and then averaged.

After careful adjustment and using the best mechanical design, residual misalignment errors may be only of the order of a few nanometers.

Instrument Errors

The wavelength of the laser source is the metrology basis for displacement-measuring interferometry. Opto-electronic circuitry within the laser head must ensure that the output frequency of the laser tube is kept at a fixed value. The electronic resolution must be well-balanced with the optical resolution of the interferometer.

To control the motion of moving stages precisely, it is not sufficient to provide position and time data alone. The delay between the time of measurement and the time when the control system obtains the position data is also relevant, especially in high-velocity applications. We refer to *data age uncertainty*, which is defined as the maximum variation in the data age in a multi-axis system, due to process variations in the electronics.

Instrument errors, including wavelength variations, electronic and optical resolution as well as data age uncertainty can be in the order of a few nanometers.

50.2.2 Low-coherence Interferometers

A *low-coherence interferometer* (sometimes referred to as a *partial coherence interferometer*) is a two-beam interferometer of the Twyman–Green or Michelson type (see section 46.3.3) which can be used to measure relative distances between reflecting surfaces. In this context, the term "coherence" refers to the temporal coherence of the light.

Usually, the reflecting surfaces belong to transparent objects such as lenses or plane plates. The low-coherence interferometer is used for high-precision measurements from nanometers to several tens of centimeters, depending on the length of the delay line (see below). Its main application is the measurement of the thickness of transparent optical elements (see section 53.3) and the magnitude of air gaps between optical elements within an optical system [50-9]–[50-11].

In the following we will describe the basic principles, configurations and their limitations.

Basic Principles

The low-coherence interferometer consists of:

1. a low-coherent light source,
2. a beam-splitter,
3. collimators to generate collimated reference and object beams,
4. a reference mirror which can be moved in the beam direction,
5. a linear encoder, used to determine the relative position of the reference mirror,
6. the test element or system carrying the optical surfaces, whose relative distance is to be determined,
7. a photodetector detecting the interference signal of reference and the object beam at the exit of the interferometer.

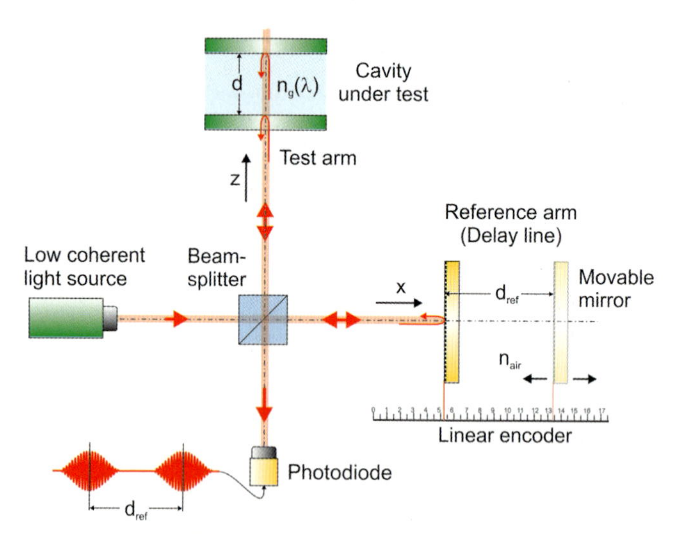

Figure 50-6: Principle of a low-coherence interferometer used to detect the relative distance between optical surfaces (cavities) within an optical system.

Figure 50-6 shows the principle of a low-coherence interferometer. It works as a comparator of cavities. The cavity length in the test arm is given by the distance d between the reflecting surfaces multiplied by the refractive group index $n_g(\lambda)$ of the medium between the surfaces. The group index is given by

$$n_g(\lambda) = n_p(\lambda) - \frac{\partial n_p}{\partial \lambda}\lambda \tag{50-8}$$

with $n_p(\lambda)$ as the refractive phase index of the medium at the wavelength λ. For air the dispersion is very small, in which case we can set $n_g(\lambda)= n_p(\lambda)= n_{air}$.

The reference arm consists of a mirror which can be linearly displaced on a translation stage. The position of the stage is measured by a linear encoder. During the displacement of the mirror the interference signal is detected by the photodiode. There are two distinct positions of the reference mirror in which the contrast of the detected interference signal is maximum. They are separated by a distance

$$d_{ref} = \frac{n_g(\lambda)}{n_{air}} d \tag{50-9}$$

as will be shown below.

Following the laws of three-beam interference, the normalized irradiance detected by the photodiode over a spectrum, assuming an equally dividing beam-splitter, is given by

$$\begin{aligned} I(x) &= R_{O_1} + R_{O_2} + R_R + 2\gamma(d_{ref})\sqrt{R_{O_1} R_{O_2}} \cos\phi(d_{ref}) \\ &+ 2\gamma(z-x)\sqrt{R_{O_1} R_R} \cos\phi(z-x) \\ &+ 2\gamma(z + d_{ref} - x)\sqrt{R_{O_2} R_R} \cos\phi(z + d_{ref} - x) \end{aligned} \tag{50-10}$$

with the reference mirror positioned at x and the test cavity positioned at z and with R_{O_1}, R_{O_2} denoting the reflectivities of the two surfaces in the test cavity and R_R denoting the reflectivity of the movable mirror.

Furthermore, with

$$\phi(x) = \frac{2\pi x}{\lambda} \tag{50-11}$$

and also with the envelopes $\gamma(z)$ desribing the modulation variation in the movable mirror's axial position z.

For a rectangular spectrum between λ_0 and λ_1 the envelope is given by:

$$\gamma(z) = \sin c\left\{ 2\pi z \left(\frac{n(\lambda_1)}{\lambda_1} - \frac{n(\lambda_0)}{\lambda_0} \right) \right\} \tag{50-12}$$

The corresponding coherence length l_c is determined by

$$l_c \approx \frac{\lambda_1 \lambda_0}{|\lambda_1 - \lambda_0|} \tag{50-13}$$

and must be considerably smaller than d_{ref}. In this case $\gamma(d_{ref})$ will be close to zero. During the displacement of the movable mirror there will then be modulation maxima of the signal $I(x)$ at $x = z$ and at $x = z + d_{ref}$.

d_{ref} can be measured by finding the exact positions of the modulation maxima, which will be described in the following section.

Envelope Detection

The envelope detection is performed by phase shifting techniques. The movable mirror in the delay line travels at a constant velocity v_M while the interferometric signal is sampled at a given acquisition frequency f_S to obtain N irradiance values

$$I_i = I\left(i\frac{v_M}{f_S}\right) \text{ with } i = 1...N \tag{50-14}$$

The phase step Δ between adjacent samples is then given by

$$\Delta = \frac{4\pi v_M}{\lambda_c f_S} \tag{50-15}$$

and should be unequal to an integer number of π.

λ_c denotes the center wavelength of the instrument.

The envelope of a fringe packet can then be computed using Kieran Larkin's Five-Sample-Adaptive (FSA) nonlinear algorithm [50-12]. The envelope value γ_i for the sample irradiance I_i is calculated from the five adjacent samples (figure 50-7)

$$\gamma_i = \sqrt{(I_{i-1} - I_{i+1})^2 - (I_{i-2} - I_i)(I_i - I_{i+2})} \tag{50-16}$$

The two maxima within the N γ_i-values are determined to find the related distance d_{ref}.

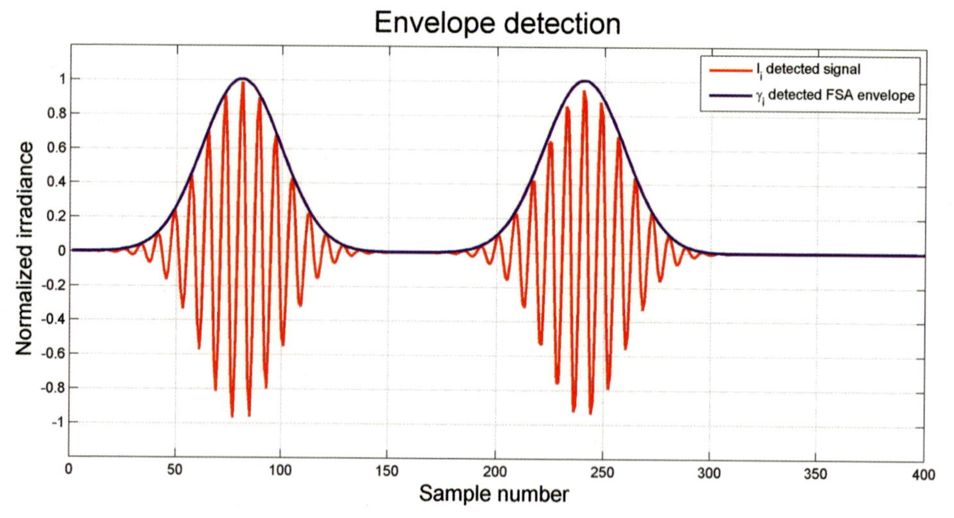

Figure 50-7: Envelope detection using the Kieran Larkin's Five-Sample-Adaptive (FSA) algorithm.

Devices

In many applications, the beam-splitter, light source, photodetector and collimators are connected by optical fibers, forming a very stable, compact unit [50-9]–[50-13]. The principle of operation is presented in figure 50-8. The light source might be a super luminescent diode (SLD) with a (typical) center wavelength $\lambda_c = 1310$ nm, a Gaussian spectral distribution and a spectral bandwidth ranging from 30 nm to 60 nm FWHM (full width at half maximum) leading to coherence lengths of 25.2–12.6 μm. SLDs are similar to laser diodes, in that they contain an electrically driven p-n junction and an optical waveguide, but lack optical feedback, so that only weak coherent amplification can occur. Optical feedback is suppressed by tilting the output facet relative to the waveguide, and can be suppressed further with anti-reflection coatings.

In figure 50-8 the light is split by a coupler into the reference and test arm. Adjustable collimators are mounted at the exits of each fibre to collimate the beams or to introduce a well-defined focus. When testing optical systems a focus is often necessary, in order to maximize the signal reentering the fiber. The reference arm contains the delay line where the light is reflected from a moving mirror and coupled back into the fiber. The mirror is mounted on a motorized translation stage. Its position is measured by an internal linear encoder.

The working distance (the distance between the collimator of the test arm and the reflecting test surface) is typically in the cm range up to 1 m, depending on the flux coupled back into the collimator.

The time for a measurement run depends on the necessary scanning range of the translation stage. For a 600 mm range a measurement run takes approximately 30 seconds.

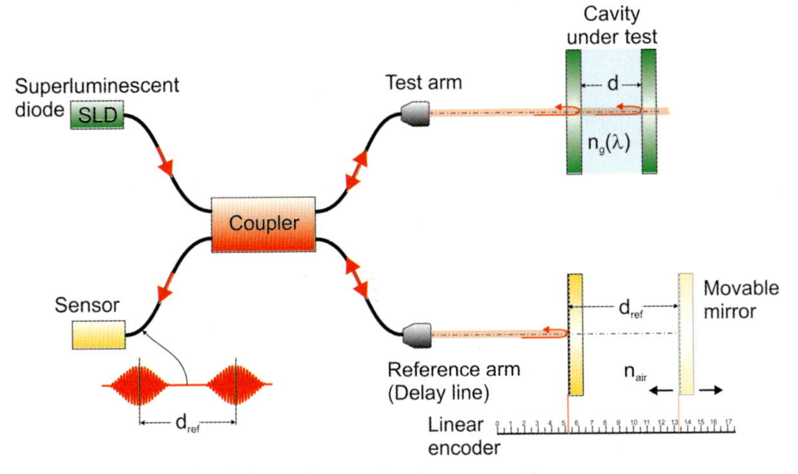

Figure 50-8: Principle of a low-coherence interferometer with optical fibers, coupler, SLD, sensor and collimators forming a compact unit.

The absolute accuracy after calibration and with stabilization of environmental parameters is ± 0.1 μm for d_{ref} in air up to 100 mm with a repeatability of 50 nm RMS.

The absolute accuracy for glass thicknesses depends on the precise knowledge of the dispersion curve in the spectral band around the center wavelength λ_{SLD}.

Calibration Techniques

A correct distance measurement relies on the accuracy of the linear encoder and its appropriate alignment. For calibration purposes a well-known airgap in a spacer block made from a material with ultral-low thermal expansion coefficient is measured and compared to a nominal, precisely known distance. Such a calibration piece can be made from a block of Zerodur® with a central bore drilled through and with two Zerodur® plates attached to both sides of it. The two plates are fixed to the central piece by molecular adhesion. The length of the airgap can be determined to an accuracy of about 50 nm by other methods (for instance by high-precision tactile coordinate metrology).

The environmental conditions (temperature, air pressure and relative humidity) must be known in order to compensate for the refractive index of the air. Assuming linear behavior, a linear correction can be introduced to calibrate the system.

In a stable environment with temperature fluctuations of less than 1°C, a monthly or longer calibration time period can be chosen.

50.2.3 Femtosecond Frequency Combs

A low-coherence interferometer using *femtosecond frequency combs* – provided by a femtosecond pulse laser – is a two-beam interferometer of the Twyman–Green or Michelson type (see section 46.3.3) which can be used to measure relative distances between reflecting surfaces over ranges of tens of meters with the resolution of a displacement-measuring interferometer. In the following, we will describe the basic principles, an appropriate setup and its limitations.

A displacement-measuring interferometer, as covered in section 50.2.1, is an *incrementally* measuring instrument based on counting wave increments. If the reference or object beams are interrupted, the system must be initialized and the measurement must be repeated.

A low-coherence interferometer, as discussed in the previous section, is an instrument for measuring absolute displacements without the need for counting increments. Its range, however, is limited by the length of the delay line.

The use of a frequency-stabilized femtosecond pulse laser enables a length metrology which measures absolute displacements comparable to the low-coherence interferometer. The advantage is that a continuous train of coherent pulses is provided, such that low coherence interferometry can be carried out by comparing different pulses. In this way, the measurement range can be extended to very large displacements without the need for a long delay line.

Absolute measuring systems are comparable to displacement-measuring interferometers in that they can be used for high-precision displacement measurements from nanometers to extremely large distances, but can also be applied to measure angles, flatness, and straightness.

Several methods for the measurement of length, using a frequency comb, have been investigated [50-14], [50-26], for instance:

a) using a perfect Fabry–Perot interferometer [50-15];
b) using a combination of time-of-flight measurements and fringe-resolved interferometry [50-16];
c) using a combination of spectrally resolved interferometry, synthetic wavelength interferometry, and time-of-flight measurements [50-17];
d) using multiple wavelength interferometry with several cw lasers referenced to a frequency comb [50-18];
e) detecting the high harmonics of the frequency comb's repetition rate and deriving the distance by measurement of the phase shift of the returning radiation [50-74].

Basic Principles

A mode-locked femtosecond laser is able to generate a phase-stabilized frequency comb [50-21]–[50-24]. As an example, the Ti:sapphire (titanium-sapphire) laser is mentioned, emitting a spectrum in the range 650 – 1100 nanometers. The laser emits a periodic train of pulses equivalent to a comb of equidistant modes in the frequency domain with a mutual separation equal to the repetition frequency f_r (figure 50-9). The difference between the group velocity and phase velocity gives rise to a pulse-to-pulse phase shift $\Delta\phi$ between the carrier wave and the envelope (figure 50-10). This phase shift results in an offset frequency f_0, also called the *carrier-envelope offset frequency (CEO)*, which is related to the repetition frequency by

$$f_0 = \frac{\Delta\phi f_r}{2\pi} \tag{50-17}$$

In a phase-locked femtosecond laser both f_r and f_0 are stabilized to a reference value provided, for instance, by a cesium atomic clock [50-25].

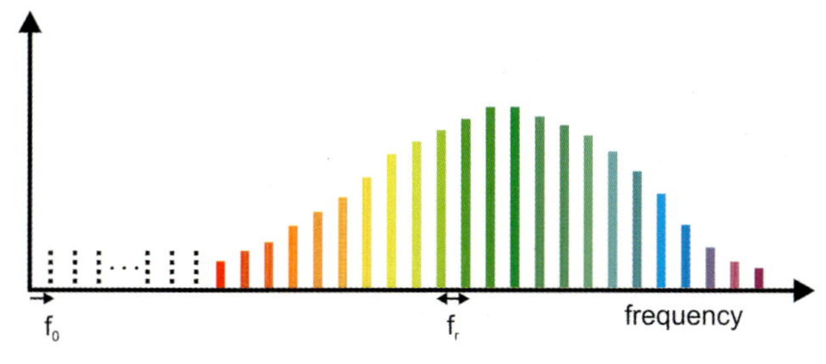

Figure 50-9: Frequency spectrum of a femtosecond laser.

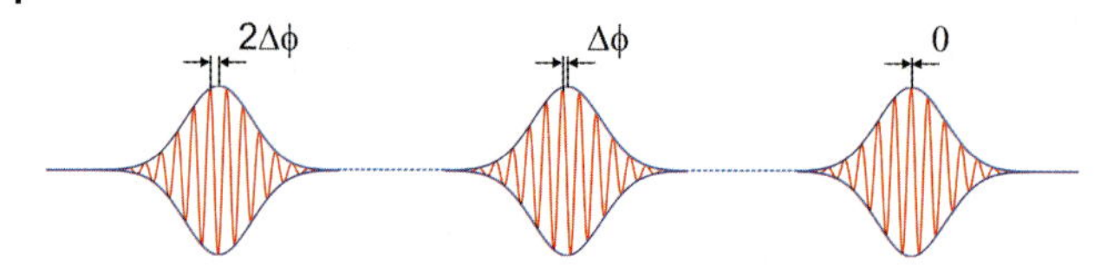

Figure 50-10: Pulse-to-pulse phase shift $\Delta\phi$ due to the difference between phase velocity and group velocity.

As a result of this, the phase relation between subsequent pulses is conserved. Interferometry between different pulses is then possible. The pulse train may be sent into an interferometer, as shown in figure 50-11 [50-14]. When the path-length difference in the interferometer arms is a multiple of the interpulse distance l_{pp}, an interferogram can be measured. l_{pp} is given by the repetition frequency f_r and the effective group refractive index of air n_g through the relation

$$l_{pp} = \frac{c}{n_g f_r} \qquad (50\text{-}18)$$

where c denotes the speed of light in vacuum.

Devices

A frequency comb usually has a repetition rate f_r of 50–1000 MHz, which leads to an interpulse distance of 30–600 cm. f_0 is fixed at a value in the range of 100–200 MHz. Pulse durations are in the order of 10–200 femtoseconds, leading to pulse width of 3–60 μm. A Ti:sapphire laser operates most efficiently at a center wavelength near 800 nm with FWHM (full width at half maximum) of about 20 nm.

Two different approaches using a frequency comb as a displacement measuring instrument will be explained in the following: 1) frequency comb as an incoherent series of pulses (phase-measuring time-of-flight metrology), and 2) frequency comb as a coherent signal (absolute displacement-measuring interferometer).

1) Phase-measuring Time-of-flight Method

In this method a femtosecond laser is used as an amplitude modulated source. As an example, figure 50-11 shows a specific arrangement [50-75]. The modulated beam is split into two by means of a polarizing beam-splitter. The reference beam is directed onto the high-speed reference detector near the beam-splitter. The test beam is transmitted, passes a quarter-waveplate and is directed onto the target reflector, usually a corner cube attached to the moving object, the displacement of which is to be measured. The reflected light is reflected from the polarizing beam-splitter, transmits through a half-waveplate and a second polarizing beam-splitter and impinges on the test detector.

As the reference and test beams travel different paths a phase difference $\Delta\phi$ is introduced between them. Since the detected signals contain many phase-locked modulation frequencies ranging from ~100–1000 MHz to tens of THz, a discrete beat frequency f_b can be selected and filtered. The phase difference is determined by mixing down the detected signals with fixed frequencies by means of a lock-in amplifier. The non-ambiguity range Λ is given by half the corresponding wavelength of the modulation wave:

$$\Lambda = \frac{c}{2n_g f_b} \tag{50-19}$$

If the repetition rate f_r of a 100 MHz frequency comb is selected as the beat frequency f_b, the corresponding non-ambiguity range is 1.5 m.

To extend the ambiguity range other different beat frequencies can be chosen and combined.

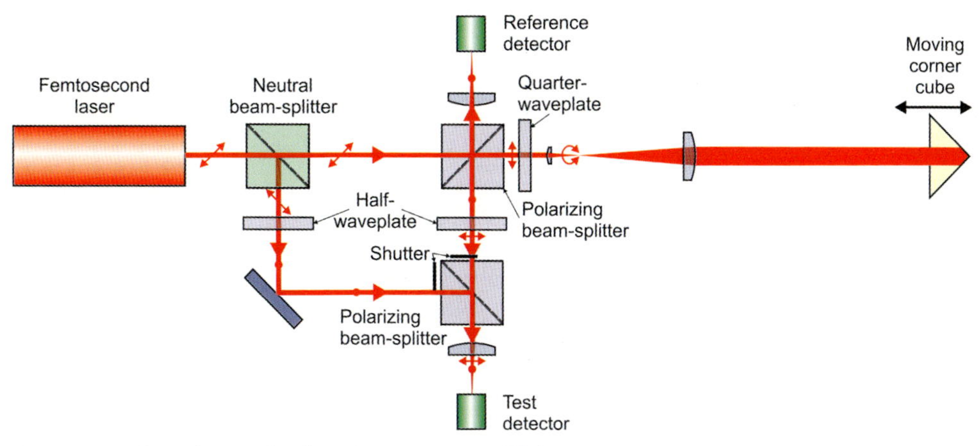

Figure 50-11: Optical setup of a phase-measuring time-of-flight arrangement using a femtosecond laser as a modulated source.

In figure 50-11 a neutral beam-splitter next to the femtosecond laser reflects a portion of the emitted light, via the polarizing beam-splitter, directly onto the test detector. This signal serves as a further reference signal to compensate for any drifts in the HF-electronics that could severely disturb the phase measurement. The signal is sequentially separated from the test signal by means of a shutter. Arrangements similar to the one shown can provide accuracies in the order of 10^{-6}–10^{-7}.

2) Absolute Displacement-measuring Interferometer

In a displacement-measuring interferometer, the coherent reference and object beams must be directed onto a detector to interfere with each other. In order to accomplish spatial overlap between the different pulses of a frequency comb in the reference and object arms, the following condition has to be fulfilled:

$$m \equiv n_g (d - d_{ref}) \frac{f_r}{c} \tag{50-20}$$

where

m denotes an integer number,
d denotes the geometrical traveling distance of the measuring beam,
d_{ref} denotes the geometrical traveling distance of the reference beam.

In order to satisfy (50-20) two alternative methods can be applied:

a) altering d_{ref} by use of a mechanical delay line [50-19],
b) altering f_r by use of a femtosecond laser with tunable repetition rate [50-20].

Method a) is shown in figure 50-12. The reference arm can be adjusted within the range $l_{pp}/2$ by means of a translation stage. An additional piezoelectric transducer (PZT) is used to obtain the first-order cross-correlation functions by modulating the PZT at around 50 Hz with a sine modulation within a range of 50–100 μm.

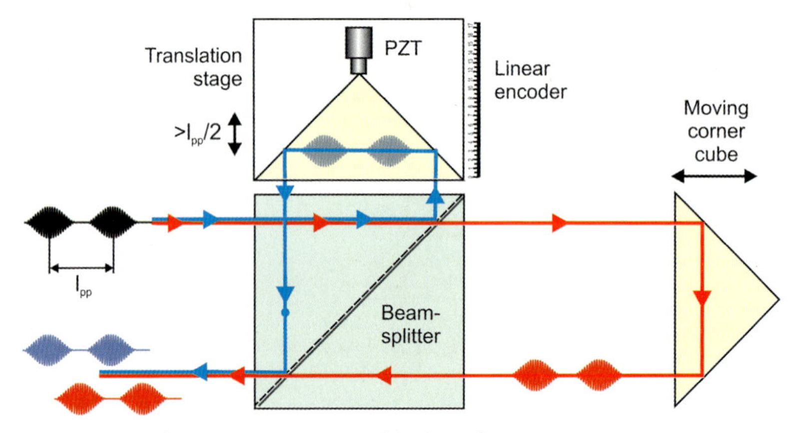

Figure 50-12: Schematic representation of the setup for interferometric absolute distance measurement with a frequency comb using an adjustable delay line.

l_{pp} defines the range of non-ambiguity of the distance measurement. To extend the measurement range over l_{pp} an initial measurement by other methods, which have a better accuracy than l_{pp}, can be carried out.

For an absolute distance measurement, the object reflector is placed in the zero position and the translation stage with the reference reflector is adjusted until the coherence maximum is achieved. In this position the cross-correlation pattern is recorded by scanning the PZT.

The object reflector is then moved to the target position and the reference reflector is again adjusted to coherence maximum and the cross-correlation pattern is recorded by scanning the PZT. The displacement of the translation stage is measured by an auxiliary displacement measuring device such as a counting interferometer or a calibrated linear encoder. While the object reflector is moved, the measurement beam may be interrupted. There is no need for continuous measurement while moving the object reflector.

Method b) is shown in figure 50-13. The reference reflector is now held in a steady position. The repetition rate can be altered until the pulses of the reference and object beams overlap. In order to fulfill (50-20) the repetition rate must have a range of

$$\Delta f_r = f_r' - f_r = \frac{c}{n_g(d - d_{ref})} \tag{50-21}$$

For a cavity length $d - d_{ref}$ of 1 m the tuning range must be 300 MHz.

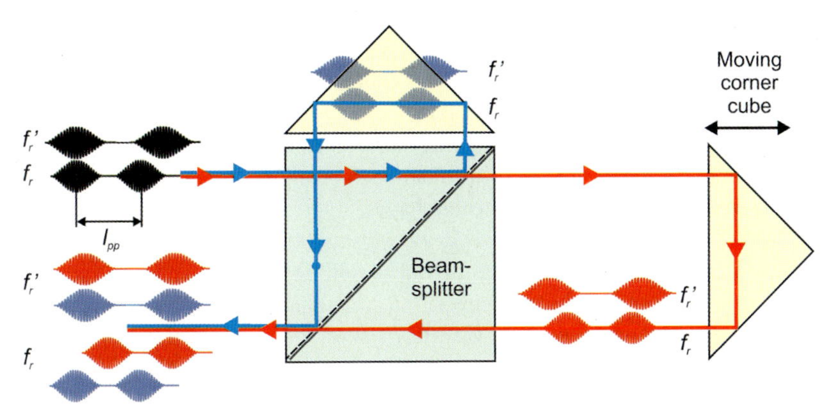

Figure 50-13: Principle of the setup for interferometric absolute distance measurement with a frequency comb using a tunable repetition rate f_r.

The accuracy of the displacement results obtained with a mode-locked femtosecond laser is comparable to that of a counting displacement-measuring interferometer. Comparisons are reported [50-14], [50-19], in which an agreement over 50 m within 2 μm has been achieved (4×10^{-8}).

As for any counting instrument, the error sources can be classified as being caused by:

1. the environment,
2. imperfect components,
3. misalignment of components,
4. instrument errors.

For error groups 2–4. see section 50.2.1.

As in the case of any instrument which measures optical path differences, the uncertainty is mainly dominated by variations in the air's temperature, pressure and humidity. For example, a typical uncertainty of 0.2°C in the temperature or of 0.5 hPa in the pressure already leads to an uncertainty of the refractive index of air of about 5×10^{-7}, which corresponds to 25 μm at 50 m.

50.2.4
Linear Encoders

Linear encoders are displacement-measuring sensors comprising:

1. a scale that encodes the position (main scale);
2. a reading head attached to the moving object reading the scale;
3. a digital readout or motion controller.

The field of application for linear encoders is immense and includes servo-controlled motion systems, metrology instruments, high-precision machining tools, such as digital calipers or coordinate measuring machines, or lithographic wafer steppers or scanners and inspection tools for lithographic masks or wafers, to name but a few.

Linear encoders can be either incremental or absolute, as will be explained below.

The linear motion is determined by the change in position over time.

There is a variety of linear encoder technologies, such as optical, magnetic, inductive, or capacitive [50-1]. In the following, we will concentrate on optical technologies. Optical linear encoders dominate the high-resolution market and may employ Moiré, diffraction or holographic principles.

Typical incremental scale periods vary from several hundreds down to a few micrometers and – depending on the detection technique – can provide resolutions down to a nanometer. Commonly used light sources include infrared LEDs, visible LEDs, miniature light bulbs and laser diodes [50-29].

Basic Principles

The main scale of a linear encoder consists of a glass or steel substrate onto which a periodic structure, known as graduation, is applied. Glass scales are used in encoders up to 4 m length, whereas steel scale tapes are used on encoders for lengths up to 30 m. Optical precision graduations may consist of chromium lines manufactured by various photolithographic processes [50-30].

The basic type of linear encoder is the image transmission (shutter) type shown in figure 50-14. Light from a light source (LED, light bulb or laser diode) is collimated and transmits both the index scale and the main scale. The transmitted irradiance is detected by a set of detectors, depending on the detection method. The light source, collimator, index scale and photodetectors are contained in the reading head and are moved along with the displaced object, whereas the main scale is fixed in a steady position.

Devices

There are two types of linear encoder:

a) the incremental type and
b) the absolute type.

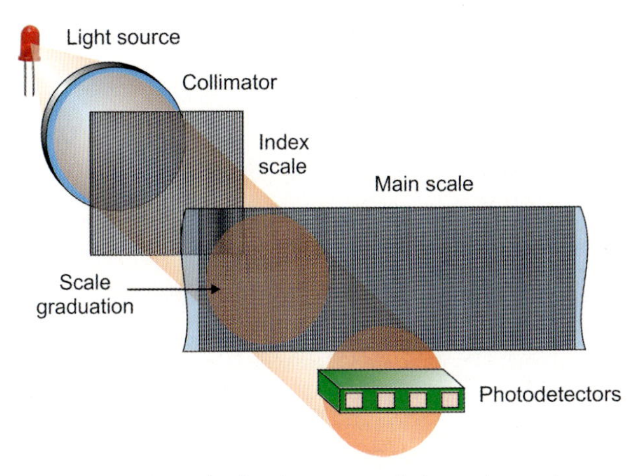

Figure 50-14: Principle of an image transmission scale encoder.

Incremental Encoders

The *incremental* type of encoder carries a single periodic graduation on the main scale and a second periodic graduation on the index scale. The detector in the reading head counts the number of periods (marks) through which the reading head passes. The width of a period is well known, so that the displacement is calculated from the number of periods passed by the reading head. In order to interpolate between adjacent marks and to detect the direction of travel, a sinusoidal signal is generated by the light passing through the main and index scale and detected by four photodiodes, which simultaneously detect four signals phase-shifted by 0°, 90°, 180° and 270°. The four photodiodes could just as easily be replaced by a line sensor covering a signal period. The sensor signals may be approximated by:

$$I_{0^\circ} = a + b\cos\left(2\pi\frac{z}{p}\right) \tag{50-22}$$

$$I_{90^\circ} = a + b\cos\left(2\pi\frac{z}{p}+\frac{\pi}{2}\right) = a - b\sin\left(2\pi\frac{z}{p}\right) \tag{50-23}$$

$$I_{180^\circ} = a + b\cos\left(2\pi\frac{z}{p}+\pi\right) = a - b\cos\left(2\pi\frac{z}{p}\right) \tag{50-24}$$

$$I_{270^\circ} = a + b\cos\left(2\pi\frac{z}{p}+\frac{3\pi}{2}\right) = a + b\sin\left(2\pi\frac{z}{p}\right) \tag{50-25}$$

where z denotes the position of the reading head, p the width of the scale period,

a is the irradiance bias and b is the modulation depth of the signal.
The interpolated position within a period is then calculated from

$$z = \frac{p}{2\pi} \arctan \left(\frac{I_{270^\circ} - I_{90^\circ}}{I_{0^\circ} - I_{180^\circ}} \right) \tag{50-26}$$

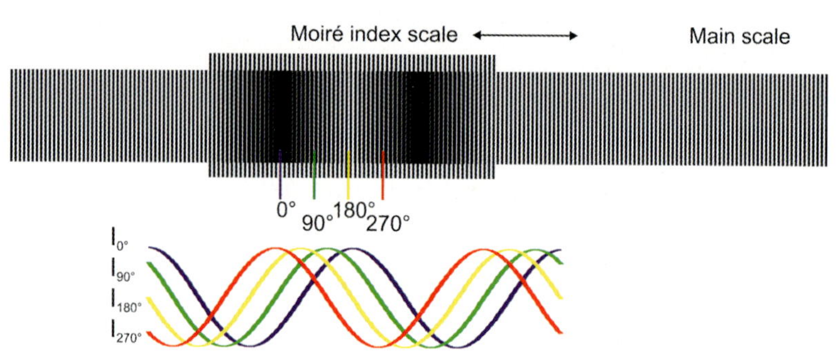

Figure 50-15: The Moiré method used to generate a sinusoidal signal from slightly different graduation periods.

There are different methods used to generate a four-field signal from the main scale of a linear encoder. For the *Moiré method* an index scale with a graduation period slightly different from that of the main scale is introduced into the reading head. Moiré fringes are generated along a cross-section in the direction of travel (figure 50-15). Four photodiodes (alternatively a line sensor) are placed equally spaced at the 0°, 90°, 180° and 270° positions, each one detecting a signal phase-shifted accordingly.

It is also possible to use four index scales with graduation periods identical to that of the main scale but with its relative position shifted such that shifts of 0, $p/4$, $p/2$ and $3p/4$ occur (figure 50-16).

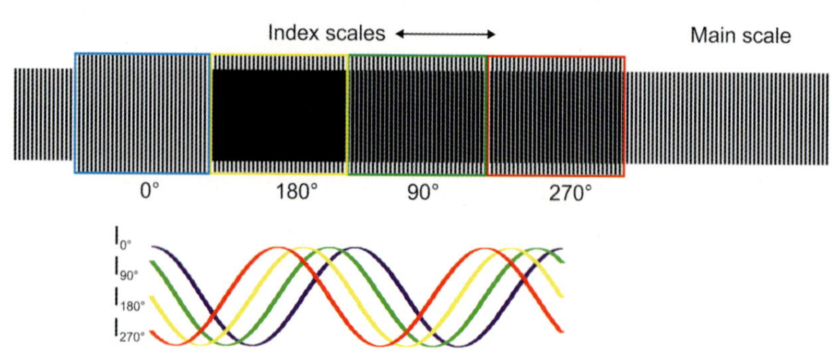

Figure 50-16: The Moiré method used to generate a four-field signal from four index scales with relative displacements of 0, $p/4$, $p/2$ and $3p/4$.

Most incremental encoder scales carry an index or *reference mark* used to find the zero position of the scale at power-up or following a loss of power. The reference mark may comprise a single feature on the scale, an autocorrelator pattern or a chirp pattern.

Incremental encoders may also have limit switches, such that on power-up the controller can determine if the encoder is at an end-of-travel position.

Diffraction scale encoders use a diffraction grating on the main scale or index scale or both scales. Their basic function is identical to that of Moiré-based methods, except that the scale pitch is smaller (5–8 μm instead of ≥ 10 μm for Moiré encoders).

Hologram scale encoders use a hologram diffraction grating on the main scale. The index scale is avoided and replaced by a polarization sensitive projection and detection systems as shown in figure 50-17 [50-31].

The scale is made from a low thermal expansion ceramic. The grating is a volumetric phase hologram providing a very high diffraction efficiency. In this arrangement the primary signal period is 1/4 of the scale period.

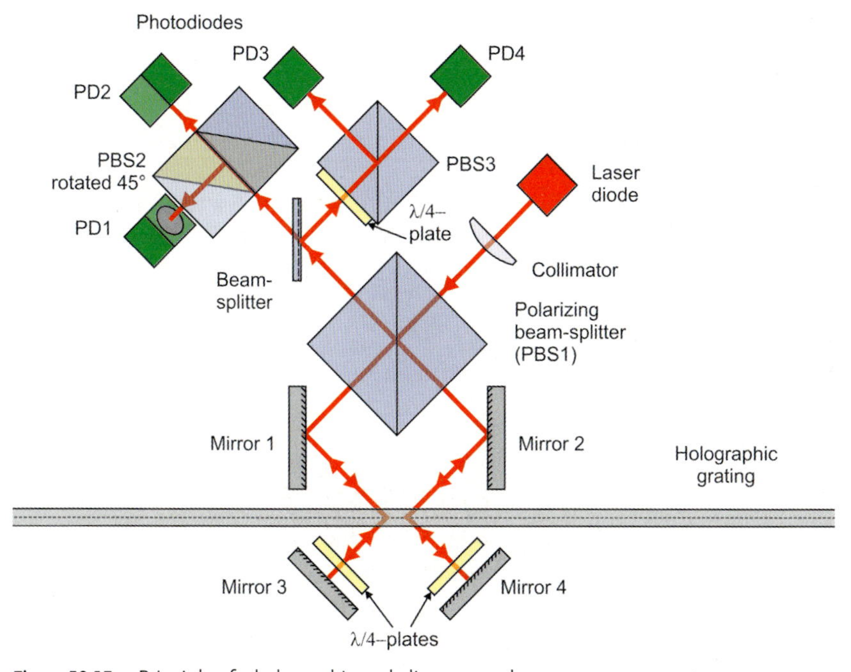

Figure 50-17: Principle of a holographic scale linear encoder.

The measuring beam of 790 nm wavelength is provided by a laser diode and passes through a polarized beam-splitter (PBS1), which divides it into two beams of different linear polarization (p- and s-polarization). Both beams are deflected by folding mirrors 1 and 2 to be directed onto the hologram lattice of period 550 nm. Both beams are diffracted and then impinge on a mirror with a quarter wave-plate in front. After reflection the s-polarized beam is converted to a p-polarized beam and the p-polarized beam is converted to an s-polarized beam. The beams returning to the grating are diffracted a second time and proceed to the detector systems transmitting PBS1. A beam-splitter divides both beams, so that one portion enters detector system 1 consisting of a polarized beam-splitter (PBS2) and two photodiodes PD1 and PD2, all rotated around the axis of beam propagation by 45°. The second portion transmits a quarter wave-plate before entering detector system 2. It then enters a polarized beam-splitter PBS3, to be divided and deflected towards the photodiodes PD3 and PD4. In this arrangement, the photodiodes detect signals shifted by 0°, 90°, 180° and 270°.

Absolute Encoders

The *absolute-type* of encoder carries several graduations or code tracks forming a serial absolute code structure. With an absolute measuring method, the position value is available at any time without the need to find a reference position upon switch-on. For the highest resolution a separate incremental track or the track with the finest graduation structure is interpolated to generate an incremental signal [50-34].

An example of a binary multi-track code is shown in figure 50-18.

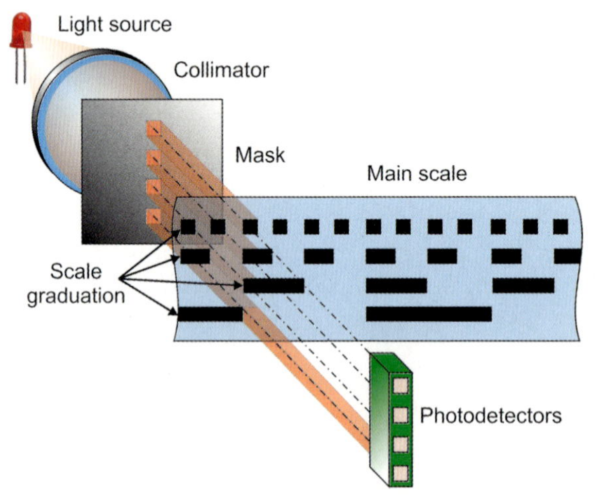

Figure 50-18: Principle of an absolute-type encoder using a binary multi-track code.

Accuracy and Error Sources

The overall *accuracy* of a linear encoder is a combination of the scale accuracy and errors introduced by the reading head. The scale contributes errors in linearity and slope (scaling factor error) to the error budget. Errors of the reading head are usually described as cyclic errors or sub-divisional errors, since they repeat every scale period [50-31].

In comparison to displacement-measuring interferometers linear encoders are difficult to use in line. Also their linearity is inferior to that of interferometers due to bending of the scale and due to grating errors occurring during production. However, solid scales of low-expansion material are resistant to environment influences and provide a supreme *stability* (*precision*) appropriate to their resolution. To minimize drifts, the detector head, which is a source of heat, is designed for minimal power consumption. If the main components of the system are mounted with adhesives, humidity might become a problem in the nanometer range.

Stabilities within ± 1 nm for measurements taken over a 40-day period have been achieved. Over an eight-hour period, stabilities within ± 0.1 nm have been reported [50-31].

Due to their supreme stability linear encoders are preferred over interferometers in systems demanding extremely high resolutions, such as semiconductor manufacturing equipment.

Calibration Techniques

Linear encoders are generally calibrated by a comparative calibration method, in which the reading head of the encoder to be calibrated is attached to a movable linear stage, whose position is measured by a second encoder or displacement-measuring interferometer. In this configuration Abbe's principle cannot be satisfied. The resulting Abbe error must then be compensated by considering the distance between the encoder axis and the axis of the linear stage and the straightness of the movable stage.

The use of a displacement-measuring interferometer, which measures the position of a reflector attached to the reading head, is preferable. The Abbe error is then avoided. If not operated in vacuum, air temperature, air pressure, humidity and the density of carbon oxide must be measured for the compensation of the refractive index of air. In air, accuracies of 50 nm per 350 mm scale have been achieved. In a vacuum accuracies of 10 nm for a 600 mm scale have been reported [50-1].

50.3 Short-range Displacement and Thickness Metrology

50.3.1 Triangulators

A triangulator is a sensor measuring the coordinate of a single surface point of an object by using triangulation (see section 49-7). Figure 50-19 shows the basic principle. A light beam from a laser or laser diode is projected onto an object. To keep the light spot on the target small a collimator can be used. A projection lens images the spot and its neighborhood onto a line sensor. The laser beam and the axis of the projection lens cross each other at an angle of ϑ. For a distortion-free projection lens and a mean magnification of β the z-coordinate of the object point is obtained from

$$z = \frac{z'}{\beta \sin \vartheta} \tag{50-27}$$

where z' is the coordinate along the line sensor with $z' = 0$ and $z = 0$ at the points where the optical axis crosses the line sensor or the laser beam, respectively.

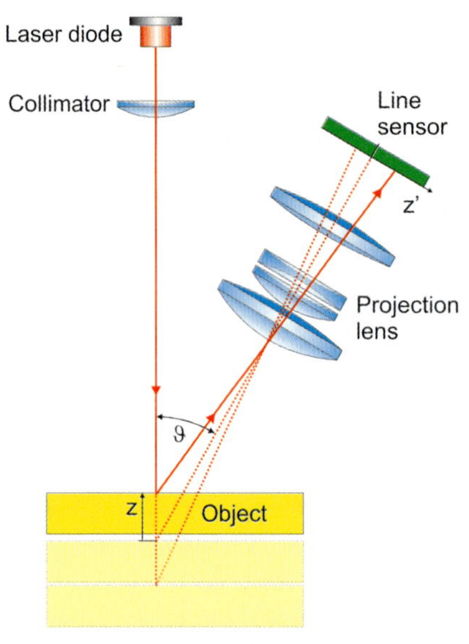

Figure 50-19: Principle of an optical triangulator.

Range and resolution of a triangulator are determined by:

- β and ϑ,
- the length and number of pixels of the line sensor,
- the size of the projected spot,
- the correlation algorithm of the signal detection (usually a hundredth to a thousandth of a pixel can be resolved).

Triangulators are commercially available for short-range displacement or thickness measurements. They use visible light with powers < 1 mW. Sampling rates range between 1 kHz and 50 kHz. Some typical measurement ranges and repeatabilities in z are shown in table 50-1.

Table 50-1: Some typical specifications of optical triangulators [50-35].

Mean free working distance [mm]	Spot size [μm]	Measurement range [mm]	Repeatability [μm]
10	20	2	0.01
30	30	10	0.05
80	70	30	0.2
150	120	80	0.5
400	290	200	2

50.3.2 Confocal Sensors

A confocal sensor is a displacement-measuring sensor working according to the confocal principle shown in figure 50-20. The confocal principle is based on light focussed through a lens and transmitted through the aperture of a pinhole or stop. A photodetector measures the flux transmitted through the pinhole, which is related to the distance z between the pinhole and the focal point F' of the focussing lens.

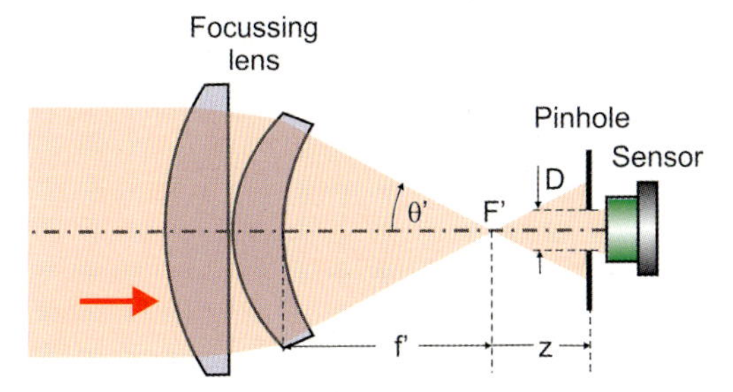

Figure 50-20: Principle of a confocal sensor.

Using simple geometrical considerations and assuming a focussing lens of numerical aperture $NA = \sin\theta'$ and homogeneous illumination, the measured flux I is given by

$$I = I_0 \frac{D^2(1 - NA^2)}{4NA^2 z^2} \tag{50-28}$$

where D denotes the diameter of the pinhole and I_0 denotes the total flux.

(50-28) is valid only for ranges

$$|z| > \frac{D\sqrt{1 - NA^2}}{2NA} \tag{50-29}$$

There are several short-range displacement sensor types which use the confocal principle combined with additional optical principles. These will be explained in the following sections.

Monochromatic Confocal Sensors

A simple monochromatic confocal sensor uses light from a laser diode focussed onto an object, whose displacement is to be measured (figure 50-21 (a) and (b)). After reflection from the object, the light re-enters the focussing lens and is directed towards a beam-splitter to be deflected onto a pinhole. The transmitted light is detected by a sensor [50-37] and [50-38].

When the focus of the focussing lens coincides with the object surface, the flux transmitting the pinhole is maximal (figure 50-21 (a)). When the object surface is moved out of focus, the transmitted flux decreases (figure 50-21 (b)).

According to (50-28) it is not possible to detect the sign of z in a steady position. The measured flux is also dependent on the object color, surface roughness and incident angle of the measurement beam. Therefore, monochromatic confocal sensors use vibrating focussing lenses, the vibration provided by tuning fork elements vibrating in the kHz range. The oscillating signal is then analyzed with respect to bias and amplitude of first and second harmonic contributions.

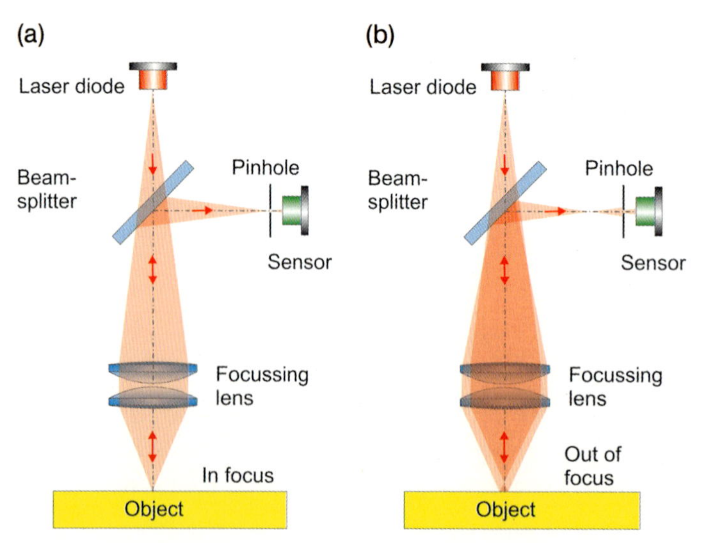

Figure 50-21: Principle of a monchromatic confocal displacement sensor, (a) object in focus, (b) object out of focus.

Monochromatic confocal sensors are commercially available for short-range displacement or thickness measurements. They use visible light with powers < 1 mW. Sampling rates range between 3 Hz and 1.5 kHz when tuning fork arrangements are used. Some typical measurement ranges and repeatabilities in z are shown in table 50-2.

Table 50-2: Two typical specifications of monochromatic confocal sensors [50-36].

Mean free working distance [mm]	Focus spot size [μm]	Measurement range [mm]	Repeatability [μm]
6	2	0.6	0.01
30	7	2	0.1

Chromatic Confocal Sensors

A chromatic confocal sensor is based on a combination of chromatic coding and the confocal principle. It uses the wavelength dependence of longitudinal chromatic aberration of the focus lens and obtains measurement data corresponding either to the distance of the measuring sensor from the object surface or the thickness of optically transparent layers [50-39]–[50-42].

The physical principle of a confocal chromatic sensor system is shown in figure 50-22. Polychromatic light from a broad-band light source (for instance a halogen lamp) is coupled into an optical fiber which is connected to a chromatic focussing lens. A chromatic focussing lens is a lens with large longitudinal chromatic aberration (hyperchromatic lens). Alternatively, a diffractive focussing element can be used, introducing specially designed longitudinal chromatic aberration by diffraction [50-41].

The opposite end of the fiber is imaged onto the sample surface by means of the focussing lens. Because of its chromatism only one color is in focus, while all other colors are out of focus. The reflected light is coupled back to the fiber, the end of the fiber acting as a pinhole to introduce the confocal effect differently with respect to different colors. By using a fiber optic splitter the returning light is sent to a spectrometer to determine its spectral components.

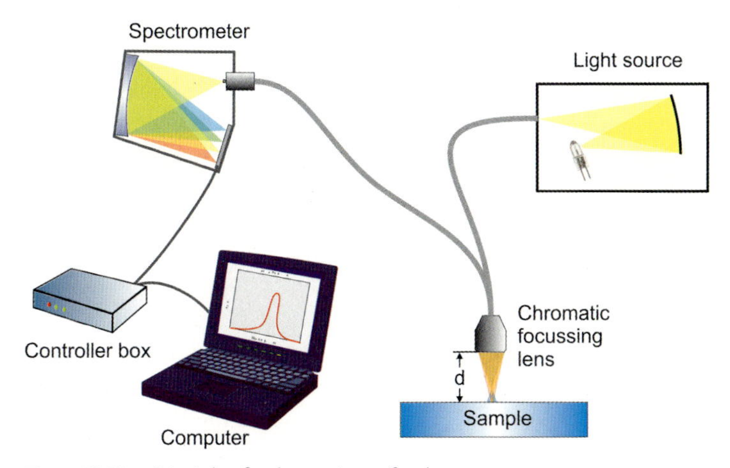

Figure 50-22: Principle of a chromatic confocal sensor system.

As the distance of the sample surface from the focussing lens is altered, the spectral distribution of the light entering the spectrometer changes. A calibration of the spectral maximum wavelength as a function of the sample's displacement is necessary to establish a proper displacement-measuring system.

Chromatic confocal sensors are commercially available for short-range displacement or thickness measurements. They use LEDs, SLDs, halogen or xenon lamps with powers from 10 to 150 W. Sampling rates range between 2 kHz and 15 kHz. Some typical measurement ranges and repeatabilities in z are shown in table 50-3.

Table 50-3: Some typical specifications of chromatic confocal sensors [50-44].

Mean free working distance [mm]	Spot size [μm]	Measurement range [mm]	Repeatability [μm]
1.85	3.5	0.1	0.003
15.3	4	0.4	0.014
20.8	3.5	1	0.035
70	24	10	0.3
76.5	25	25	0.8

50.3.3
Coaxial Interferometric Sensors

A coaxial interferometric sensor is used to measure the optical path difference (OPD) of small cavities by means of spectral analysis of the reflected or transmitted polychromatic light. Its main application is the thickness measurement of thin layers or the displacement measurement of reflecting samples in ranges from several μm up to 2 mm with nanometer resolutions [50-43].

The OPD of the unknown cavity is measured by analysis of the interferometric signal in the polychromatic spectrum, described in the following. The term "coaxial" is used to describe the fact that the measurement beam is applied normal to the cavity, leading to reflected beams which are coaxial to the illuminating beam.

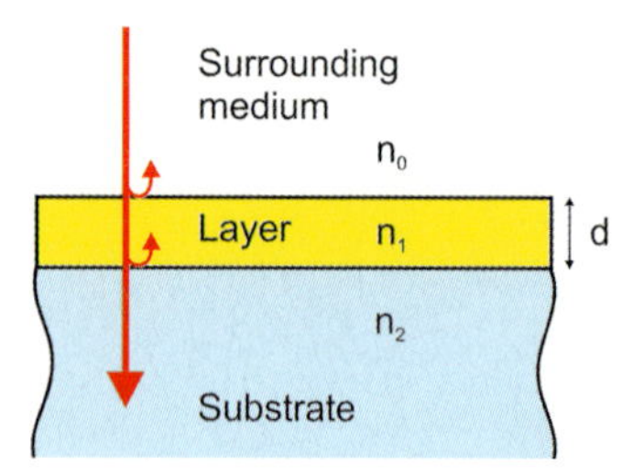

Figure 50-23: Transparent layer on a sample illuminated by a light beam in a direction normal to the layer surfaces.

We assume a transparent layer of thickness d and refractive index n_1 on a substrate material of index n_2, the surrounding medium having the index n_0, as shown in figure 50-23. All refractive indices are assumed to be functions of the wavelength λ.

Applying Fresnel's formulas for normal incidence and neglecting higher reflection orders we can describe the reflectivity R of the layer by

$$R = r_1^2 + r_2^2 + 2r_1 r_2 \cos\phi \tag{50-30}$$

with

$$\phi = \frac{4\pi n_1 d}{\lambda} = \frac{4\pi}{c} n_1 d\nu \tag{50-31}$$

and

$$r_1 = \frac{n_1 - n_0}{n_1 + n_0} \tag{50-32}$$

$$r_2 = \frac{n_2 - n_1}{n_2 + n_1} \tag{50-33}$$

where c denotes the speed of light in vacuum and ν denotes the frequency of the light. Equation (50-31) shows a linear dependence of ϕ from the wavenumber λ^{-1} or from ν with $n_1 d$ defining the oscillation frequency of the fringes in the spectrum. Figure 50-24 shows examples for two thin plates of fused silica in air with thicknesses 5 μm and 10 μm. Figure 50-24 (a) shows the spectral reflectivities as functions of the wavelength in the visible range. The fringe density changes with the wavelength. Figure 50-24 (b) shows the corresponding spectrum with a wavenumber abscissa. If dispersion effects are neglected, the fringes will show a constant density over the spectrum. Figure 50-24 (c) shows the magnitude of the Fourier spectrum of (b). Assuming that the refractive index $n_1(\lambda)$ is known, the abscissa of (c) can be normalized to thickness values. The position of the first-order peak then gives the thickness of the corresponding layer.

It is possible to use the method to analyze the individual thicknesses of multilayer systems of N layers. In this case a peak for every individual cavity within the system will occur in the Fourier spectrum, leading to $N(N+1)/2$ peaks. The magnitude of the individual peaks is limited by the reflectivities at the corresponding interfaces given by the corresponding refraction index differences.

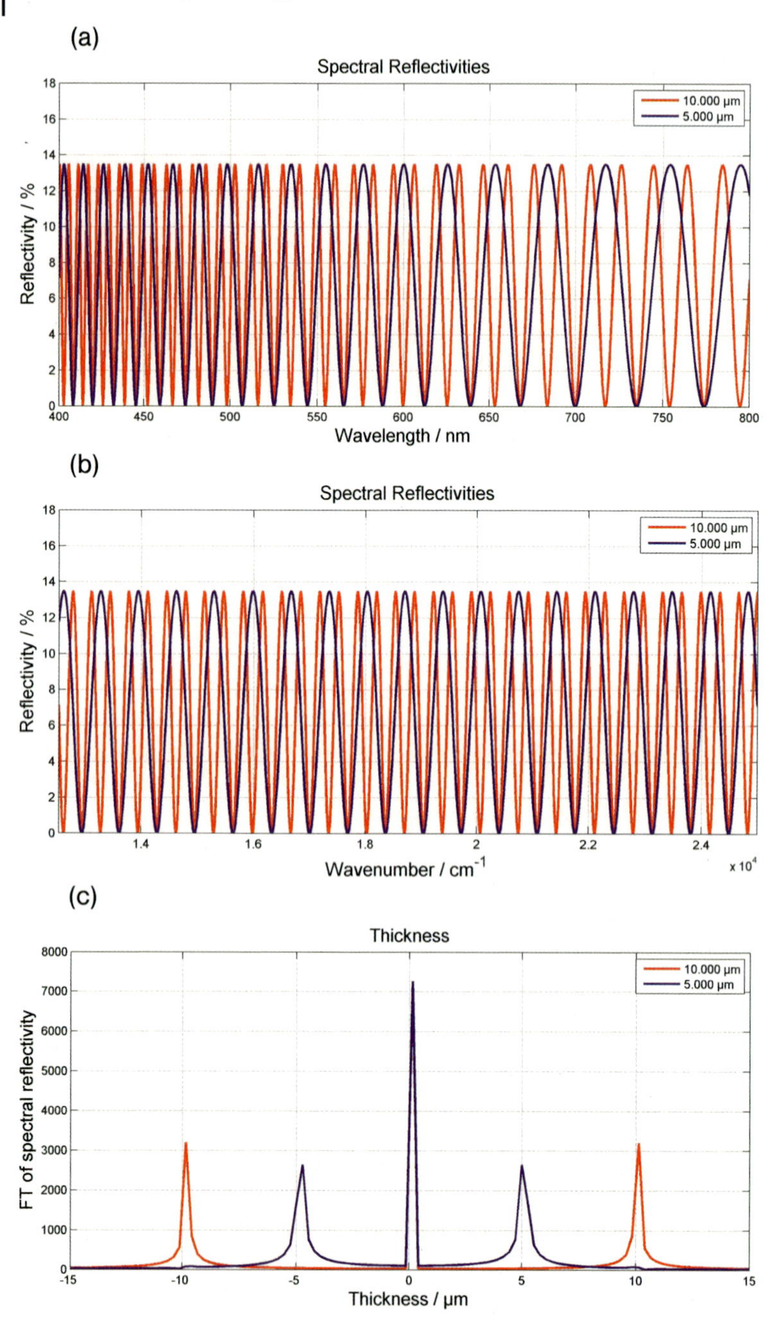

Figure 50-24: Spectral reflectivities of thin plates made from fused silica of thickness 5 µm and 10 µm: (a) reflectivity as a function of wavelength; (b) reflectivity as a function of wavenumber; (c) the Fourier spectrum of (b).

The physical principle of a coaxial interferometric sensor system is shown in figure 50-25. Polychromatic light from a broad-band light source (for instance a halogen lamp) is coupled into an optical fiber which is connected to a low-aperture focussing lens. The focussing lens generates an image of the fiber end near the layer surfaces. The reflected light is coupled back to the fiber. A fiber optic splitter sends the returning light to a spectrometer to determine its spectral components. The ν-abscissa of the spectrum must be calibrated and is then normalized to $n_1(\nu)$. The normalized spectrum is Fourier transformed. The position of the first-order peak gives the geometrical thickness d of the layer under test.

The measurement range is given by the spectrometer range. A broad spectrum is used for small ranges with high resolutions, a narrow spectrum is used for long ranges with low resolutions. Resolutions are limited mainly by the spectrometer configuration as well as spot size and aperture of the focussing lens.

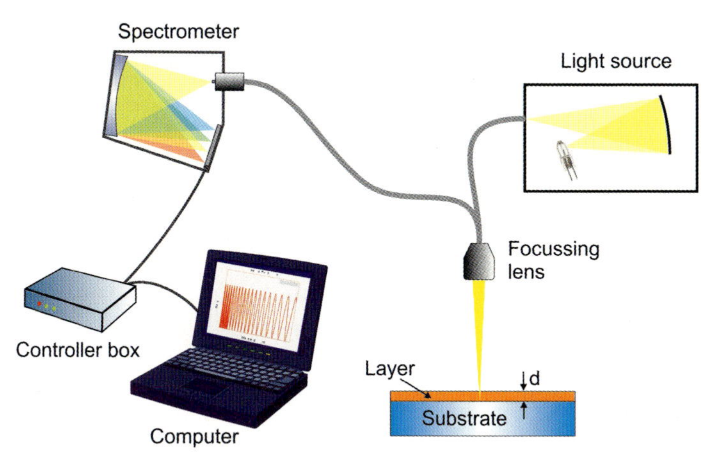

Figure 50-25: Principle of a coaxial interferometric sensor system used to measure the thickness of thin layers.

A coaxial interferometric sensor system can also be used to measure small displacements of reflecting surfaces as shown in figure 50-26. In this case a reference surface must be provided and attached to the focus lens along with a beam-splitter to form a rigid reference unit. The position of the reference surface determines the free working distance of the displacement sensor, which can freely be selected. The displacement measurement range is given by the spectrometer range as in the case of layer thickness measurements.

Coaxial interferometric sensors are commercially available for short-range displacement or thickness measurements. They use LEDs, SLDs, halogen or xenon lamps with powers from 10 to 150 W. Sampling rates range up to 4 kHz. Some typical measurement ranges and repeatabilities in z are shown in table 50-4.

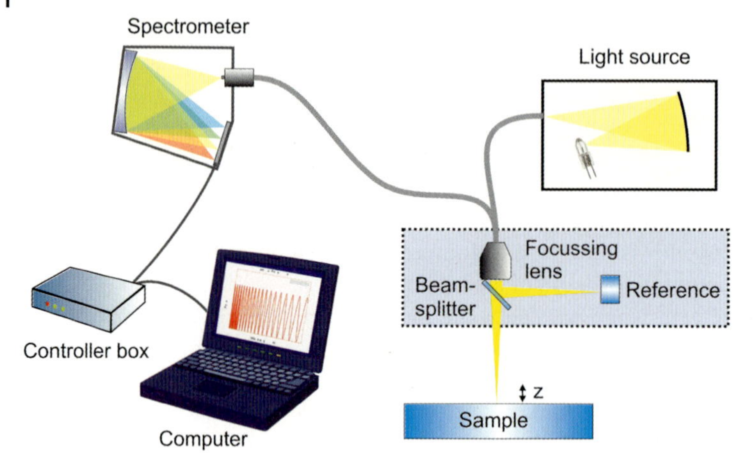

Figure 50-26: Principle of a coaxial interferometric sensor system to measure small displacements of reflecting surfaces.

Table 50-4: Some typical specifications of chromatic confocal sensors [50-44].

Mean free working distance [mm]	Spot size [μm]	Measurement range [mm]	Repeatability [μm]
10 – 100	10 – 50	0.003 – 0.18	0.01
10 – 100	10 – 50	0.002 – 0.25	0.01
19 – 23	13	0.028 – 1.1	0.075
19 – 23	13	0.034 – 1.9	0.1
19 – 23	13	0.060 – 3.5	0.2

50.4 Angle and Tilt Metrology

50.4.1 Angle Encoders

The instrument most frequently used to measure rotational positions and angles in optics is the *goniometer*. The instrument comprises an object table that carries an optical component which can be rotated or positioned in a specific rotational position. The rotational position is generally measured by means of an *angle* or *rotary encoder*.

The term angle encoder is used to describe encoders that have an accuracy of typically better than ± 5″, which is the usual application in precision optics. Rotary encoders typically have an accuracy of more than ± 10″.

Angle encoders are angle-measuring sensors comprising:

1. a scale that encodes angular position (main scale),
2. a reading head attached to the rotating object table and indicating the scale,
3. a digital readout or rotation controller.

Angle encoders can be either incremental or absolute, as will be explained in the following. Although a variety of different techniques exist, we will concentrate on optical technologies, because optical angle encoders dominate the high-resolution market.

Typical angle encoders provide system accuracies from ± 1″ to ± 5″ while permitting a maximum angular speed of up to 20 000 min^{-1}, depending on the specific opto-mechanical layout [50-33].

Basic Principles

The main scale of an angle encoder consists of a glass or steel disk onto which a periodic structure, known as graduation, is applied (figure 50-27). Glass scales are used primarily in encoders for speeds of up to 10 000 min^{-1}. For higher speeds of up to 20 000 min^{-1}, steel drums are used. For drums with large diameters, steel tapes carrying the graduation are applied (figure 50-28).

Precision graduations are manufactured in various photolithographic processes. Graduations are fabricated from:

- extremely hard chromium lines on glass or gold-plated steel drums,
- matte-etched lines on gold-plated steel tape,
- three-dimensional structures etched into quartz glass.

The basic type of angle encoder is the image transmission (shutter) type shown in figure 50-27. Light from a light source (LED, light bulb or laser diode) is collimated and transmits the index scale and the main scale. The transmitted irradiance is detected by a set of detectors, depending on the detection method. The light source, collimator, index scale and photodetectors are contained in the reading head and are attached to the stator of the rotary table, whereas the main scale rotates along with the rotator of the rotary table.

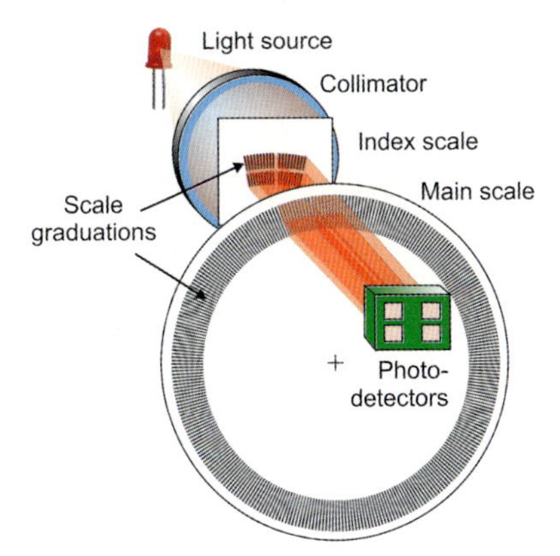

Figure 50-27: Principle of an image transmission scale angle encoder.

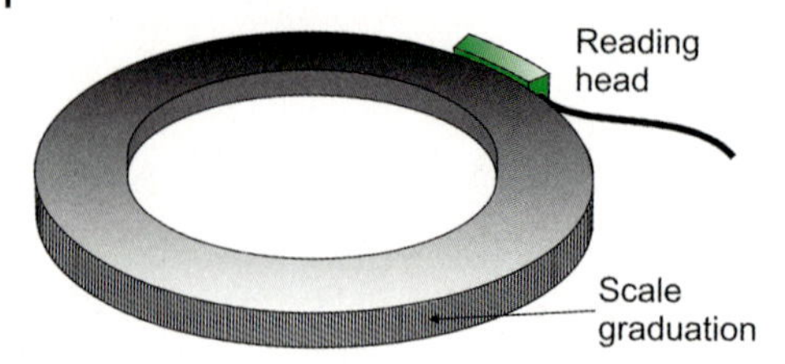

Figure 50-28: Principle of a reflective scale angle encoder using a steel tape as graduation.

There are two types of angle encoder:

a) the incremental type and
b) the absolute type.

Incremental Angle Encoders

The *incremental type* of angle encoder carries a single periodic graduation on the main scale and one or several periodic graduations on the index scale. The detectors in the reading head count the number of periods (marks) through which the graduation on the main scale passes. The width of a period is already known, so that the relative angle of rotation can be calculated from the number of periods passed by the reading head. In order to interpolate between adjacent marks and to detect the direction of travel, a sinusoidal signal is generated by the light passing through the main and index scales and detected by four photodiodes, which simultaneously detect four signals, which are phase-shifted by 0°, 90°, 180° and 270° as described in section 50.2.4 by (50-22)–(50-25).

The interpolated rotational position ϑ within an angular period ρ is then calculated from

$$\vartheta = \frac{\rho}{2\pi} \arctan\left(\frac{I_{270^\circ} - I_{90^\circ}}{I_{0^\circ} - I_{180^\circ}}\right) \tag{50-34}$$

Analogue to linear encoders. There are different methods used to generate a four-field signal from the main scale of a linear encoder. For the Moiré method an index scale with a graduation period slightly different from that of the main scale is introduced into the reading head. Alternatively, the use of an index scale with four graduations identical to that of the main scale but with its relative position shifted such that shifts of 0, $\rho/4$, $\rho/2$ and $3\rho/4$ occur.

Most incremental angle encoder scales carry an index or *reference mark,* which is used to find the zero position of the scale at power-up or following a loss of power. The reference mark may comprise a single feature on the scale, an autocorrelator pattern or a chirp pattern.

Absolute Encoders

The *absolutetype* of angle encoder carries several graduations or code tracks forming a serial absolute code structure [50-32]. With an absolute measuring method, the position value is available at any time without the need to find a reference position upon switch-on. For the highest resolution a separate incremental track or the track with the finest graduation structure is interpolated to generate an incremental signal.

An example of a binary multi-track code is shown in figure 50-29.

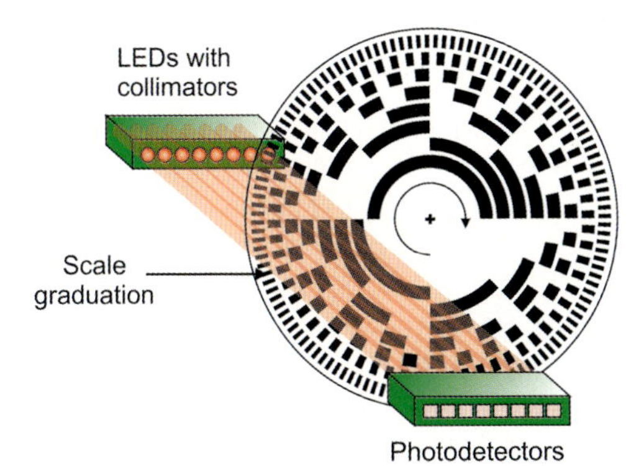

Figure 50-29: Principle of an absolute type of angle encoder using a binary multi-track code.

Accuracy and Error Sources

The overall *accuracy* of an angle encoder is a combination of its resolution, repeatability and systematic deviation (absolute error).

The resolution of an angle encoder is a measure of how many counts per angle the encoder generates. It can be expressed in counts per revolution, often specified as a binary number. For example, a 16-bit encoder generates $2^{16} = 65536$ counts per revolution. By interpolation a resolution of 1% of a count can be achieved leading to resolutions of better than 0.2 arsec for a 16-bit encoder.

The repeatability is given by the sum over all statistical errors including the following.

- The *quantization error,* which is the error given by the resolution of the scale, when no interpolation is applied. It then corresponds to ± 1/2 count cycle of the main scale.
- The *quadrature error* denoting the combined effect of phasing and duty-cycle tolerances and other variables in the basic analog signals. This error applies to data taken at all four transitions within an optical cycle,
- The *interpolation error, which is* the error arising when the resolution has been electronically increased to more than four measuring steps per optical cycle. It is the sum of all the tolerances in the electronic interpolation circuitry.
- *The environmental error* which consists of errors caused by changes in temperature, pressure and humidity as well as shaft loading.

The systematic error is mainly given by:

- the directional deviation of the radial grating;
- the eccentricity of the graduated disk to the bearing. Disk eccentricity can also lead to vibrations during high rotational speeds, resulting in early encoder failure;
- the radial deviation of the bearing;
- the error resulting from the connection with a shaft coupling.

For incremental angle encoders and absolute angle encoders with complementary incremental signals having line counts up to 5000, a typical accuracy at 20 °C ambient temperature and scanning frequency of up to 1 kHz is approximately ± 1/20 of the grating period. A 5000-count encoder therefore has an estimated accuracy of ± 13 arcsec.

The encoder accuracy is measured by rotating the encoder at a very precisely controlled speed, and then measuring the time interval between successive transitions of the encoder's output. A very accurate method, which can be used to calibrate the total goniometer system, is covered in the following section.

Calibration Techniques

Angle encoders can be calibrated by mounting a multi-faceted mirror of known geometry and the best quality, on the shaft of the encoder under test. An autocollimator (see next section) is used to align each facet and compare the known rotation angle with the encoders read-out [50-45]. In this way a calibration for as many angles as there are facets can be made. By applying appropriate interpolations, the calibration is extended to any angle of rotation. The method is, however, very slow and labor-intensive and only allows for a limited number of angle positions.

An improved calibration method can be set up by using a high-precision master encoder equipped with multiple reading heads reading the graduation of a large diameter disk mounted onto an air-bearing spindle. The encoder under test is coupled to the master encoder by precision fixturing. Both the master and test encoder are rotated together, and the relative deviation of both read-outs is stored for calibration purposes. In this way, fast, automated calibrations can be carried out [50-46].

50.4.2 Autocollimators

An autocollimator is an optical instrument used for non-contact measurement of small tilts in the reflecting plane elements. They are typically used to align components, especially plane mirrors, or to check the parallelism of optical windows and wedges. They are also useful for measuring deflections in optical systems. A typical application is also to measure the straightness of a surface plate onto which a plane mirror is mounted and moved.

An autocollimator works by projecting an image onto a target mirror, and measuring the deflection of the returned image against a scale. The measurement can be carried out either visually or by means of an electronic detector. A visual autocollimator can measure angles as small as 0.5 arcsec, while an electronic autocollimator can be up to 100 times more accurate. Electronic and digital autocollimators are often used for monitoring angular movement over long periods of time and for checking angular position repeatability in mechanical systems [50-47] and [50-48].

Basic Principles

An autocollimator comprises a projection beam and an observation beam, which use a common collimator. Both beam paths are separated by means of a beam-splitter. Figure 50-30 is a diagram showing the principle of a visual autocollimator. An object reticle is projected to infinity by means of the collimator. The reticle is illuminated by an illumination system consisting of an appropriate light source, a condenser, a filter and a diffusor. The object reticle is placed at the back focus of the collimator in the reflected arm of the beam-splitter. The collimated beam is reflected from the plane target mirror to re-enter the collimator. After transmission through the beam-splitter, the image of the object reticle is projected onto an eyepiece reticle. An observer looking through the eyepiece sees both the object and eyepiece reticle simultaneously, their separation being a measure of the tilt of the target mirror.

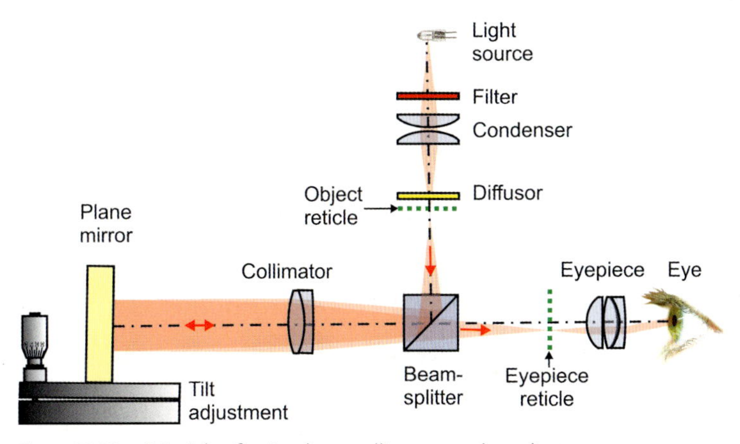

Figure 50-30: Principle of a visual autocollimator used to adjust a plane mirror.

The tilt a of an object mirror leads to a lateral displacement d of the object reticle image:

$$d = f'_{\text{coll}} \tan 2a \tag{50-35}$$

where f'_{coll} denotes the back focal length of the collimator.

Assuming small tilt angles, the object mirror angle is directly proportional to the measured shift in the image plane. The resolution of an autocollimator increases proportionally and the angular field of view reciprocally with the focal length of the collimator lens.

The tilt resolution limit α_{res} of a visual autocollimator can be estimated considering the angle resolution of diffraction-limited optical systems:

$$\sin 2\alpha_{res} = 1.22 \frac{\lambda}{D_{\text{coll}}} \tag{50-36}$$

where D_{coll} denotes the diameter of the collimator and λ denotes the wavelength of the illumination. For green light with $\lambda = 500$ nm and a 50 mm collimator aperture, an object mirror tilt of 1.26 arcsec can be resolved. Note that the resolution only limits the separation of the object and eyepiece reticle images during observation. When using electronic collimators the relative displacement of the object reticle images is recorded and this is not limited by diffraction.

Devices

While a visual inspection is sufficient for most system alignment purposes, a higher resolution is necessary for high-precision inspection and alignment. In many cases drifts in optical setups must be observed and measured to define improvement procedures. There are electronic autocollimators commercially available which can be connected to a computer for high-precision analysis and automatic sampling.

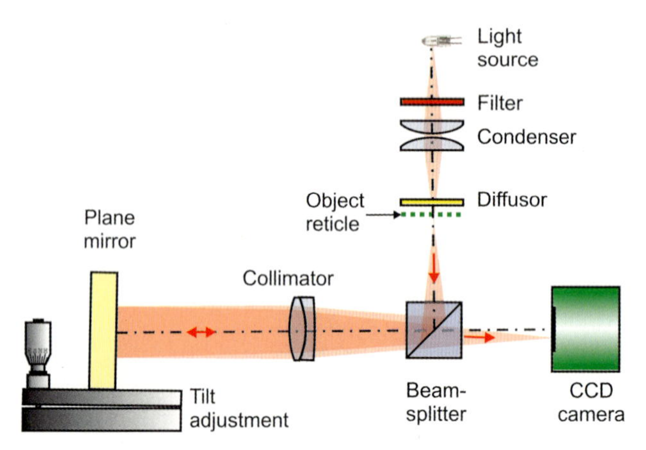

Figure 50-31: Principle of an electronic autocollimator using a CCD camera as 2D array detector to measure the tilt of a plane mirror.

In an electronic autocollimator, the eyepiece is replaced by an electronic digital camera with discrete sensor pixels, for example, a CCD or CMOS sensor type (figure 50-31). In the general case a 2D array type, allowing angular measurements in two directions, is preferred over a 1D line-scan sensor for single-axis measurements. The digital camera is usually connected to a computer which calculates the measured angle from the image by using image analysis software. The computer-assisted measurement guarantees much higher resolution, accuracy and repeatability of the results compared to the visual inspection, since it does not depend on the operator's experience or attention.

For high-speed applications, the CCD or CMOS sensor is replaced by a high-bandwidth position-sensitive photo detector (PSD) for rates up to 10 kHz. The analog signal output of the angular data enables a closed-loop control operation for fast adjustments or angular control operations [50-47].

Accuracy and Error Sources

The resolution of an autocollimator using digital image processing is limited by the sub-pixel resolution of the analysis algorithm and the S/N ratio of the sensor. Generally 1/100 of a pixel can be resolved but, in cases where many images are collected and averaged and environmental conditions are carefully controlled, 1/1000 of a pixel might be resolved.

Assuming a quadratic sensor array with diameter D_{CCD} and $N \times N$ pixels, the angular resolution can be calculated from

$$\alpha_{\mathrm{res}} = \frac{1}{2} \arctan \frac{D_{\mathrm{CCD}} \Delta p}{N f'_{\mathrm{coll}}} \tag{50-37}$$

where Δp denotes the pixel fraction of the resolution capability, usually between 10^{-2} and 10^{-3}.

For a collimator with 1000 mm focal length and a sensor with 2k × 2k pixels and a diameter of 25 mm a resolution of between 0.026 arcsec and 0.0026 arcsec can be achieved.

The autocollimator's field of view is limited by the maximum tilt angle $\pm \alpha_{\max}$ given by

$$\alpha_{\max} = \frac{1}{2} \arctan \frac{D_{\mathrm{CCD}}}{2 f'_{\mathrm{coll}}} \tag{50-38}$$

For the example above the field of view would be $\pm 0.36°$.

The repeatability of an electronic autocollimator is limited by the stability of the total setup. The main reasons for unstable results are as follows.

1. Thermal drifts in the optical setup, which can either influence the beam paths in air or the position and optical function of the individual components.
2. Mechanical drifts causing small misalignments of the optical elements, light source and sensor.
3. Drifts in the electronic sensor system.

Note that, to achieve a repeatability of 1/1000 of a pixel, the alignment of the reticle, the collimator and the sensor must fulfill a stability of better than 10 nm–15 nm. This is an extremely challenging task and demands a granite table setup, a vibration-isolation system, a thermally stabilized environment, to better than 50 mK, and acoustic shielding.

In order to estimate the overall accuracy of the measurement, the thermal and mechanical drifts of the plane target mirror and its mechanical mount system must also be considered and must be added to the system's repeatability.

Note that, in an electronic autocollimator, the quality of the optical components does not play as important a roll as it does in other optical instruments. Aberrations in the system can be neglected to a certain extent, since it is the change in position of the reticle image, which is measured, rather than the image quality.

Calibration Techniques

To calibrate and verify the relation between the pixel coordinate and tilt angle of an electronic autocollimator, a plane mirror must be provided, placed in a series of known angular positions. A high-precision goniometer providing an object table onto which the plane mirror is mounted, will be particularly suitable. While the goniometer is rotated to a set of discrete positions, the reticle image position is analyzed and the relation is recorded for calibration purposes. If possible, the autocollimator should be rotated by 90° around its optical axis to repeat the calibration procedure. Thus, asymmetric aberrations or misalignments of the collimator can be detected and compensated by software.

If only small-angle deviations near the optical axis are to be detected, a full calibration is not necessary. It is then sufficient to use the relation (50-35) together with the collimator's focal length and the sensor's pixel size to calculate the appropriate angle deviations.

50.4.3
Surface-measuring Interferometers

As an alternative to an autocollimator, a surface-measuring interferometer can be used to measure small tilts of plane surfaces. Since the tilt of a surface under test leads to a certain number of fringes which will then be detected by the interferometer camera, the analysis of the measured wavefront only requires determination of the linear terms in the x- and y-direction, which are proportional to the tilt of the test mirror.

The resolution and the field of view of interferometers are different from those of autocollimators, but are not necessarily more accurate. While the autocollimator's resolution is increased by the collimator having a long focal length, the interferometer's resolution is increased by the collimator having a large diameter.

In the following we briefly describe the principle of a Fizeau arrangement to detect small tilts.

Basic Principles

Similar to an autocollimator, the interferometer comprises a projection beam and an observation beam, which use a common collimator. Both beam paths are separated by means of a beam-splitter. Figure 50-32 shows the arrangement for a Fizeau interferometer (see section 46.3.2), in which the transmission flat holding the reference surface is placed in front of the collimator. A pinhole is projected to infinity by means of the collimator. The pinhole is illuminated by a laser and an appropriate objective. The collimated beam is partially reflected from the reference surface and re-enters the collimator. The transmitted portion is reflected from the plane target mirror and also re-enters the collimator. After transmission through the beam-splitter, both beams transmit an eyepiece before they overlay and interfere at the CCD sensor.

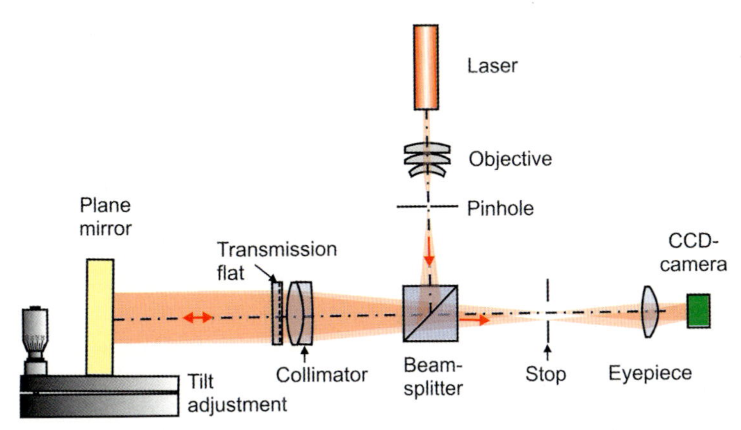

Figure 50-32: Principle of a Fizeau interferometer used to measure the tilt of a plane mirror.

Devices

In principle, any type of interferometer capable of measuring plane mirrors is appropriate for use as a tilt-measuring instrument. Besides the Fizeau type the Twyman–Green type of interferometer is also in widespread use. The Fizeau interferometer has the advantage that the reference flat can be freely positioned in front of the collimator. In this way the cavity can be shortened, leading to a very robust setup system. The Twyman–Green interferometer (see section 46.3.3) has a separate reference arm, therefore the effective cavity is much longer than in the Fizeau arrangement. The setup might be more affected by vibrations and air turbulences.

Accuracy and Error Sources

The tilt resolution of an interferometer can be expressed as

$$a_{res} = \frac{1}{2} \arctan \frac{\lambda \Delta p}{D_{coll}} \tag{50-39}$$

where Δp denotes the phase resolution of linear wavefront terms in fringes.

λ is the wavelength of the light source. D_{coll} denotes the diameter of the collimator.

Precision interferometers have linear phase resolutions of between 10^{-3} and 10^{-4}.

For a collimator with 100 mm diameter and $\lambda = 633$ nm a tilt resolution of between 6.5×10^{-4} arcsec and 6.5×10^{-5} arcsec can be achieved.

By applying the Nyquist limit the interferometer's field of view is then limited by the maximum tilt angle $\pm \alpha_{max}$ given by

$$\alpha_{max} = \frac{1}{2} \arctan \frac{N\lambda}{2D_{coll}} \tag{50-40}$$

where N denotes the number of sensor pixels across one dimension.

For the example of $D_{coll} = 100$ mm and red laser light the field of view for a $2k \times 2k$ camera would be $\pm 0.18°$.

In a similar way to electronic autocollimators, the repeatability of an interferometer is limited by the stability of the total setup, which is mainly dominated by thermal, mechanical and electronic drifts. Note, however, that the main major influence in a Fizeau interferometer comes from the relative instability of the reference surface and target mirror. If it is possible to make the cavity length short and provide a common stable mechanical basis (i.e., a granite table), a vibration-isolation system, a thermal stabilized environment to better than 50 mK and some acoustic shielding, then the precision will be extremely high.

Calibration Techniques

To calibrate and verify the relation between the measured wavefront tilt and the tilt of the test mirror, a plane mirror, placed in a series of known angular positions is required, as described in the previous section (50.4.2). However, in most cases, it is sufficient to measure the effective diameter of the collimator D_{coll} and calculate the related tilt angle according to

$$\alpha = \frac{1}{2} \arctan \frac{\lambda W_{tilt}}{D_{coll}} \tag{50-41}$$

where W_{tilt} denotes the measured linear wavefront term over the collimator aperture measured in fringes.

50.4.4
Differential Heterodyne Laser Interferometer

Heterodyne displacement-measuring interferometers, as described in section 50.2.1, can be used to measure tilts very accurately. Since they are two-beam interferometers of the Twyman–Green or Michelson type (see section 46.3.3) they can be easily modified to form a tilt-measuring instrument, when their reference beam is directed towards the test object being laterally displaced with respect to the measurement beam. The interference signal then indicates the displacement difference between the measurement and reference beam, which contains the information concerning the tilt in one direction [50-6] and [50-49].

In the following, the principle and some technical configurations are shown.

Basic Principles

In oder to determine the tilts α_x, α_y, α_z of a surface with respect to the x, y and z-axis of a coordinate system, the coordinate measurement of at least three surface points $P_1(x_1, y_1, z_1)$, $P_2(x_2, y_2, z_2)$, $P_3(x_3, y_3, z_3)$ is necessary (see figure 50-33).

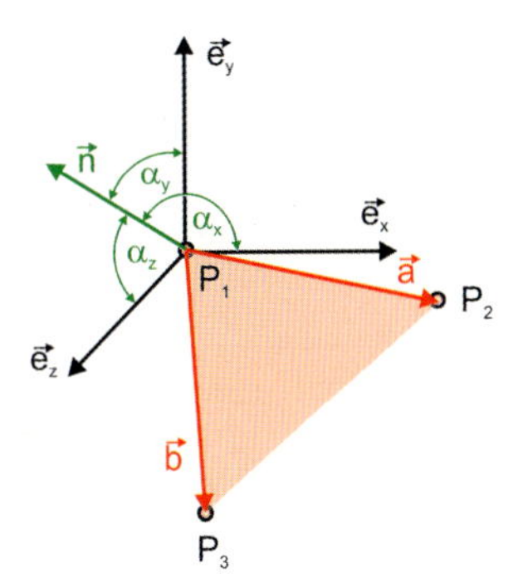

Figure 50-33: A surface determined by three points in a coordinate system.

The tilts can then be calculated from

$$\alpha_x = \arccos\left(\vec{n} \cdot \vec{e_x}\right) = \arccos n_x \tag{50-42}$$

$$\alpha_y = \arccos\left(\vec{n} \cdot \vec{e_y}\right) = \arccos n_y \tag{50-43}$$

$$\alpha_z = \arccos\left(\vec{n} \cdot \vec{e_z}\right) = \arccos n_z \tag{50-44}$$

where

$$\vec{n} = \begin{pmatrix} n_x \\ n_y \\ n_z \end{pmatrix} = \frac{\vec{a} \times \vec{b}}{\left|\vec{a}\right|\left|\vec{b}\right|} \tag{50-45}$$

denotes the unit vector normal to the surface and
$\vec{e_x}$, $\vec{e_y}$, $\vec{e_z}$ denotes the unit vectors in the x, y and z direction.

Inserting the point coordinates into (50-42)–(50-45) the tilt angles can then be calculated from

$$\alpha_x = \arccos\frac{(y_2 - y_1)(z_3 - z_1) - (y_3 - y_1)(z_2 - z_1)}{A} \tag{50-46}$$

$$\alpha_y = \arccos\frac{(x_3 - x_1)(z_2 - z_1) - (x_2 - x_1)(z_3 - z_1)}{A} \tag{50-47}$$

$$\alpha_z = \arccos\frac{(x_2 - x_1)(y_3 - y_1) - (x_3 - x_1)(y_2 - y_1)}{A} \tag{50-48}$$

with

$$A = \sqrt{(x_2 - x_1)^2 + (y_2 - y_1)^2 + (z_2 - z_1)^2}\sqrt{(x_3 - x_1)^2 + (y_3 - y_1)^2 + (z_3 - z_1)^2} \quad (50\text{-}49)$$

The usual way to measure tilts is to define the coordinate system with $\vec{e_z}$ in the initial normal direction of the surface under test and to select three points on the surface forming a right-angled triangle, its legs representing the x and y-axis. To achieve a high resolution the points should be widely separated. The separations $x_2 - x_1$, $x_3 - x_1$, $y_2 - y_1$ and $y_3 - y_1$ must be known.

When small tilts are introduced, a_z will remain practically unchanged, whereas a_x and a_y will change according to (50-46) and (50-47) with the actually measured values for z_1, z_2 and z_3.

Devices

A very simple device, used to measure small tilts, consists of three short-range displacement sensors, for instance, confocal sensors, arranged in a triangle, each sensor pointing to the surface under test. The displacements in z are recorded and the relative tilt of the surface is determined using the equations from the previous section.

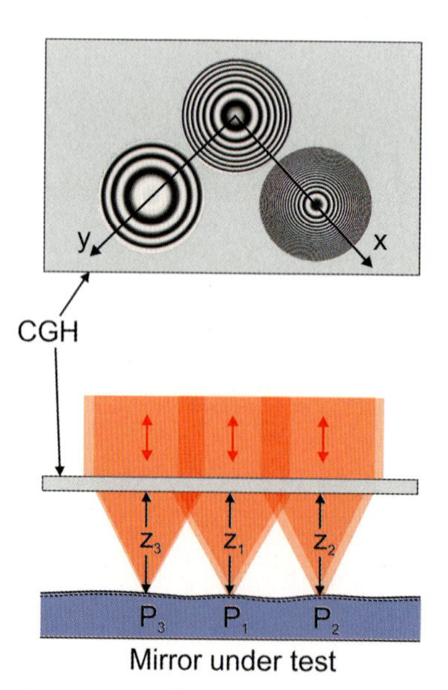

Figure 50-34: Tilt-measuring device using a computer-generated hologram (CGH) forming three spherical wavefronts focussed onto the surface under test.

An interferometer, when used to measure plane surfaces, can be combined with a special computer-generated hologram (CGH) to form three spherical wavefronts focussed onto the surface under test (figure 50-34). From the interferograms in the three apertures, the distance of the focal points from the test surface, and therefore z_1, z_2 and z_3 can be determined. The arrangement is useful for test surfaces which are not plane enough to be tested in a surface-measuring interferometer, for instance, free-form surfaces.

The most common method used to measure small tilts very precisely is by means of the heterodyne displacement-measuring interferometer as described in section 50.2.1.

The laser generates two frequencies, which are linearly polarized and orthogonal to each other. One frequency is directed to the first location, the other to the second location of the object under test. After recombination of both beams their optical path difference is detected by the system. Tilts in one direction can be detected very accurately. In order to measure the tilt in the orthogonal direction, an appropriate second system or channel must be set up.

In figure 50-35 an arrangement is shown in which two corner cubes are attached to an object to measure its tilt in one direction [50-50] and [50-51].

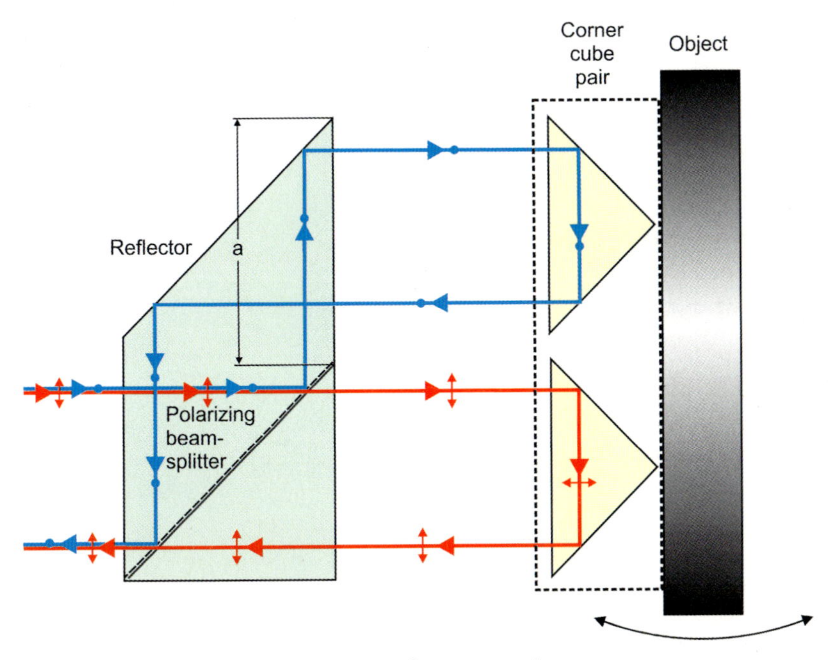

Figure 50-35: Displacement-measuring interferometer used to measure the tilt of an object in one direction by means of two corner cubes attached to the object.

The optical arrangement comprises a polarizing beam-splitter and a reflector, both integrated in a single cemented module. The p-polarized beam transmitting the beam-splitter is incident on the first corner cube reflector attached to the object. After reflection the beam transmits again through the beam-splitter and exits the module heading for the detector. The s-polarized beam is reflected by the beam-splitter and again by total reflection at the surface parallel to the beam-splitting surface of the rhomboidal component. The length a of this component provides the separation of both beams necessary for tilt detection. After reflection from the second corner cube, the s-polarized beam re-enters the rhomboid and leaves the module after two reflections to overlay with the p-polarized beam.

The optical path difference OPD measured by the system is given by

$$\mathrm{OPD} = a \sin 2\alpha \approx 2a\alpha \tag{50-50}$$

where a denotes the length of the rhomboid defining the separation of the beams and α denotes the tilt angle of the object.

In figure 50-36 an arrangement is shown in which the tilt of a mirror is directly measured, avoiding the use of retroreflectors.

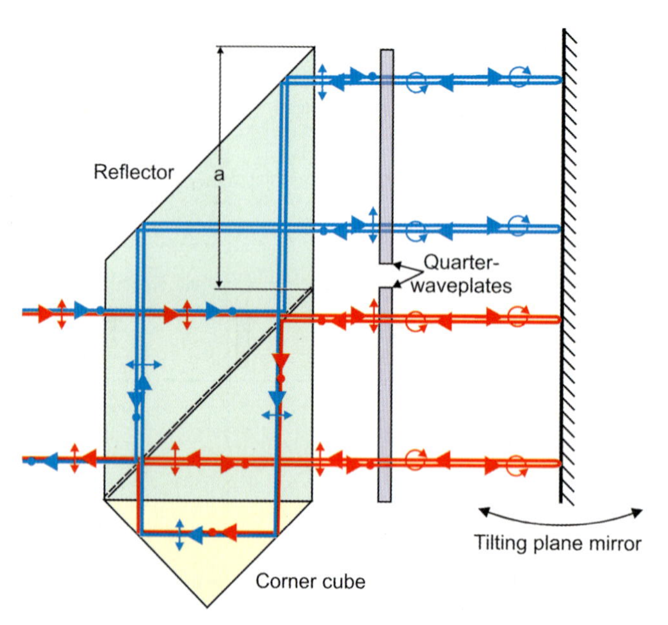

Figure 50-36: Displacement-measuring interferometer used to measure the tilt of a plane mirror in one direction.

The optical arrangement comprises a polarizing beam-splitter, a reflector and a corner cube retroreflector, all integrated into a single cemented module. Additionally two quarter-waveplates are required.

The p-polarized beam transmitting the beam-splitter passes a quarter-waveplate before it is reflected from the test mirror. On its return the p-polarization is converted to s-polarization, such that the beam is now reflected from the beam-splitter, then reflected by the corner cube and again reflected by the beam-splitter to be directed towards the test mirror a second time. On its return, its polarization changes again from s to p-polarization, and the beam transmits the beam-splitter and exits the module heading for the detector.

The s-polarized beam is reflected by the beam-splitter and again by total reflection at the surface parallel to the beam-splitting surface of the rhomboidal component. The length a of this component provides the separation of both beams necessary for tilt detection. The s-polarized beam then follows a path parallel to the p-polarized beam and is reflected from the test mirror twice in the same way, but laterally separated by the length of the rhomboid. At the exit of the module the s and p-polarized beams overlay to form the detection signal.

The optical path difference OPD measured by the system is given by

$$\mathrm{OPD} = 2a\sin 2\alpha \approx 4a\alpha \tag{50-51}$$

where a denotes the length of the rhomboid defining the separation of the beams and α denotes the tilt angle of the mirror.

In figure 50-37 an alternative arrangement is shown, in which the tilt of a mirror is directly measured [50-6].

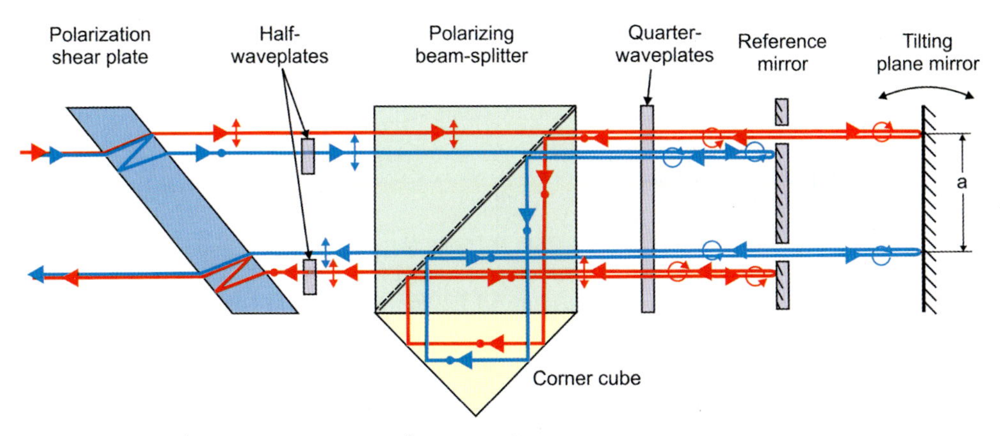

Figure 50-37: Displacement-measuring interferometer using a shear plate and reference mirror to measure the tilt of a plane mirror in one direction.

The optical arrangement comprises a polarizing shear plate, a polarizing beam-splitter, a corner cube retroreflector, a quarter-waveplate, two half-waveplates and a reference mirror carrying two holes.

The polarization shear plate divides the beam into an s- and p-polarized part, both laterally separated. The s-polarized beam is converted to a p-polarized beam by means of a half-waveplate. Both beams transmit the polarizing beam-splitter and transmit a quarter-waveplate. The first beam passes the reference mirror through a bore and is reflected from the test mirror, the second is reflected from the reference mirror. On their return both beams are converted to s-polarization and are reflected from the beam-splitter. After their return from a corner cube retroreflector, they are again reflected from the beam-splitter to pass the quarter-waveplate. This time the first beam is reflected from the reference mirror, while the second beam is reflected from the test mirror. On their return both beams transmit the beam-splitter. The first beam is converted to s-polarization by means of a half-waveplate, before both beams re-enter the shear plate to be recombined to exit the module.

The separation a of the beams is limited by the size of the beam-splitter and the corner cube. The optical path difference OPD measured by the system is given by (50-50).

Accuracy and Error Sources

The angle resolution of a differential heterodyne laser interferometer can be calculated from (50-50). Since a displacement interferometer has a typical resolution of 1 nm, the angle resolution for a basis of $a = 10$ mm would be 0.01 arcsec.

The repeatability of a measurement with a differential heterodyne laser interferometer is limited by the stability of the total setup. The main reasons for unstable results are described in detail in section 50.2.1. An optimization problem is caused by the separation a of the beams. If a is too large, air turbulences and thermal and mechanical drifts will influence both beam paths differently and the repeatability will worsen. If a becomes too small, the resolution will drop accordingly.

Assuming a reasonable accuracy of 10 nm for the differential measurement, an accuracy of 0.1 arcsec for a basis of $a = 10$ mm will result. Of course, for measurements in vacuum with mechanically stable setups and larger separations, accuracies can be below 0.01 arcsec.

Calibration Techniques

To calibrate and verify the relation between the tilt of the test mirror and the OPD measured by the system, a plane mirror in a series of known angular positions must be provided, as described in section 50.4.2. However, in most cases it is sufficient to measure the effective separation a of the beams and calculate the related tilt angle according to (50-50) or (50-51), depending on the related type of setup.

50.5
Combined Distance and Angle Metrology

The alignment of opto-mechanical systems is often carried out with the help of instruments which measure the distances and angles of components with reference to each other. The alignment of complex multi-axis systems needs instruments which can provide 3D data of component positions. When multi-axis stages are involved, their relative orientations with reference to given datum axes are required. In these cases combined distance and angle metrology is necessary.

A *theodolite* is an optical instrument used to measure horizontal and vertical axis orientations. A *total station* is a theodolite with additional distance metrology.

A *laser tracker* or *laser tracer* is a total station with the capability of following a moving target by providing a beam-steering mechanism and a position control feedback system.

In the following we will introduce the theodolite, the total station and the laser tracker.

50.5.1 Theodolites and Total Stations

Basic Principles

A theodolite is a precision instrument for measuring angles in the horizontal and vertical planes. Theodolites are mainly used for surveying and geodetic applications, but have been adapted for alignment purposes in complex optical setups. A theodolite consists of a movable telescope mounted within two perpendicular axes – the horizontal axis and the vertical axis (figure 50-38). The telescope provides a reticle, which is needed to point it to special datum points within the setup to be aligned. The angle of both axes for each datum point is measured with a typical precision of an arcsecond or better. In principle, the telescope can be used to measure distances, if the reticle has a known scale. When objects of known height are observed, the object distance from the objective's principle point can be determined.

Instead of using known targets to measure distance, modern instruments are equipped with absolute distance meters (ADM), which measure the optical path length between the instrument and the object. The preferred techniques are based on the *time-of-flight* or *amplitude modulation* method [50-26]. Instruments combining two-axis-angle metrology of a theodolite with absolute distance metrology are called *total stations*.

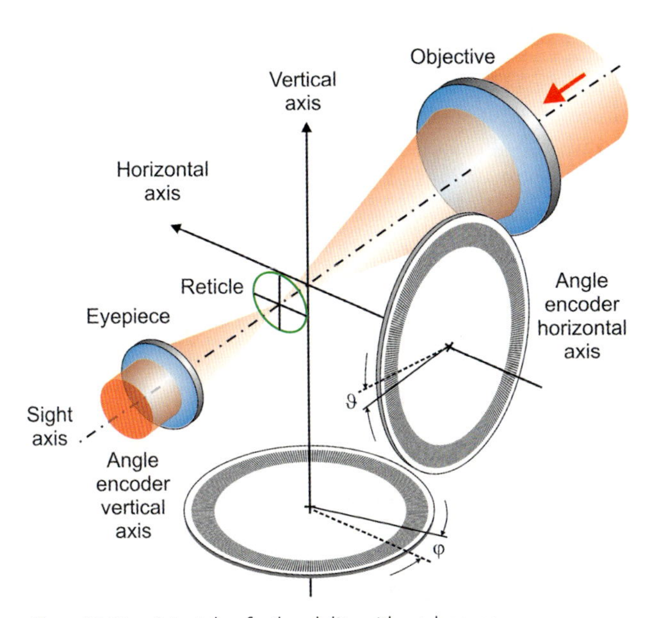

Figure 50-38: Principle of a theodolite with a telescope mounted within a horizontal and a vertical axis.

Devices

Modern theodolites use digital angle encoders to determine the rotational position around the horizontal and vertical axes. For convenience, absolute encoders are preferred. Some instruments have a CCD sensor positioned at the focal plane of the telescope objective allowing both auto-targeting and the automated measurement of residual target offset. All this is implemented by embedded software [50-52].

Total stations are equipped with integrated electro-optical absolute distance-measuring devices (ADM), generally infrared-based, allowing the measurement of the object distance. A total station can therefore collect 3D coordinates simultaneously.

Among the various ways used to measure an absolute distance, the time-of-flight method is conceptually the easiest. A pulse is sent out to a reflective object and the time it takes for the pulse to travel back to the source is recorded. The object distance d from the instrument is then obtained by

$$d = \frac{c\tau}{2n} \tag{50-52}$$

where
τ denotes the measured round-trip time τ,
c denoting the speed of light in vacuum,
n denotes the refractive index of the medium between the object and the instrument, usually air.

The accuracy at large lengths is limited by the relative stability of the clock. When an atomic clock is available, the results are extremely stable and are mainly limited by the refractive index of the air. The resolution is usually limited to around a millimeter [50-26].

An improvement over the time-of-flight method as far as resolution is concerned is the *amplitude modulation* method. Instead of sending out individual pulses, the measurement beam intensity is modulated with a sinusoidal pattern as a function of time. The analysis of the signal then changes from a simple time measurement to a phase measurement. The irradiance measured at the detector is given by

$$I(t) = I_0(1 + a\sin(2\pi ft - \phi)) \tag{50-53}$$

where
f denotes the modulation frequency,
a denotes the modulation amplitude,
I_0 denotes the mean signal at the detector.

The phase shift ϕ contains the information about the unknown distance d:

$$\phi = 2\pi f\tau = 2\pi f\frac{2nd}{c} = 2\pi\frac{2nd}{\Lambda} \tag{50-54}$$

where Λ denotes the so-called synthetic wavelength describing the range of unambiguity given by

$$\Lambda = \frac{c}{f} \tag{50-55}$$

The unambiguity range can be increased by lowering the modulation frequency. However, the resolution will drop if a constant error in the phase measurement is assumed. By using a pseudo-random bit pattern to modulate the amplitude, the unambiguity range can essentially be extended [50-55].

An alternative to amplitude modulation is the use of *polarization modulation*. The principle is analogue, but any change in the polarization status of the laser beam that is un-accounted for can create an error source. By selecting a superluminescent light-emitting diode (SLD) with a broadband spectrum rather than a laser diode, the sensitivities to polarization changes can be reduced significantly [50-53].

The highest resolution and fastest acquisition time is achieved when an ADM and a displacement-measuring interferometer (section 50.2.1) are combined [50-52]. The latter only measures displacements rather than absolute distances, but has a rapid acquisition time (in the MHz regime), while an ADM measures absolute distances but needs a longer integration time for high resolutions (typically 10 kHz). By combining both and making use of their benefits, a rapid absolute distance metrology can be achieved [50-53].

Accuracy and Error Sources

The overall *accuracy* of a theodolite is a combination of the resolution, repeatability and systematic deviation of its angle encoders and their relative adjustment to each other.

In section 50.4.1 we have discussed the accuracy of incremental and absolute angle encoders. We stated that for 20 °C ambient temperature and a scanning frequency of up to 1 kHz, the error is approximately ± 1/20 of the grating period. A 5000-count encoder has an estimated accuracy of ± 13 arcsec. High-precision angle encoders provide system accuracies from ± 1 to ± 5 arcsec. 1 arcsec is equivalent to a lateral displacement of 4.8 μm at a distance of 1 m.

For the accuracy of total stations, the accuracy of the ADM has also to be considered.

The resolution of a time-of-flight instrument is limited to approximately 1 mm, which is usually unsuitable for the adjustment of optical setups. For amplitude or polarization- modulated sytems a resolution of several μm can be achieved over long distances, for example, 240 m [50-55]. The accuracy, however, is very limited by the environmental conditions.

The highest resolution is achieved with a heterodyne displacement-measuring interferometer, typically specified between 0.5 and 2.5 nm.

Typical accuracies achieved with displacement interferometers in air at 20°C are ± 0.5 μm per meter displacement.

Systematic errors occur when the horizontal and vertical axes are not perpendicular to each other or do not cross each other or the line of sight of the telescope. A calibration process can determine the relative position of the axes. The results can be used to compensate for systematic errors by means of an appropriate analysis software.

Calibration Techniques

Theodolites can be calibrated using targets with known geometry in different positions and orientations. In most cases, however, it is useful to calibrate the rotation axes in a similar way to that in which angle encoders are calibrated. For the calibration of one rotation axis the theodolite is mounted on top of a multi-faceted mirror of known geometry and of the best quality. Both objects are carried by a high-precision rotary table. An autocollimator is used to align each facet. For the calibration process the rotary table is rotated until a facet is oriented perpendicular to the axis of the autocollimator. The theodolite is then aligned to observe a fixed target several meters away. For each facet the known rotation angle is compared with the read-out of the theodolite. In this way a calibration for as many angles as there are facets can be made [50-45], [50-54], [50-56], [50-63]. By applying appropriate interpolations, the calibration can be extended to any angle of rotation.

The multi-faceted mirror can also be replaced by a high-precision master encoder equipped with multiple reading heads reading the graduation of a large diameter disk mounted on the rotary table. The theodolite is coupled to the master encoder with precision fixturing. While the rotary table is moved to any arbitrary position, the theodolite is aligned to observe a fixed target. The read-outs of both the theodolite and the master encoder are then stored for calibration purposes [50-46].

For total stations, it is necessary also to calibrate their ADM. This is done on long calibration benches, of about 50 m length [50-54]. A displacement-measuring interferometer is installed on a concrete pillar at one end while the total station is installed on a concrete pillar at the other end. Retro-reflectors for the interferometer and for the total station are installed on the motorized carriage which is moved along the bench. During the carriage movement both interferometer and total station read-outs are recorded for calibration purposes. The calibration can be repeated for different angle adjustments of the total station.

50.5.2 Laser Trackers

A laser tracker is a three-dimensional measurement system which uses two angle encoders and a laser interferometric length measurement to determine the target point coordinates. Basically, it is identical to a total station providing an additional tracking function, which enables the system to follow a moving retro-reflector.

A laser tracker points the laser beam emitted by a displacement-measuring interferometer or an absolute distance meter (ADM) with the help of a beam-steering mechanism to a retro-reflector attached to the moving object. Together with the two angle encoders for the horizontal and vertical angular rotations of the beam-steering mechanism, 3D coordinates of the target can be determined [50-57].

Laser trackers with displacement-measuring interferometers must track the path of the retro-reflector continuously without interruption of the beam. Once the tracker station loses the target, the interferometer will lose its reference distance and will not be able to determine the current distance to the target.

Basic Principles

A typical commercial laser tracker consist of the following main components.

1. A laser-based distance-measuring device.
2. A beam-steering mechanism with two angle encoders.
3. An optical position-sensitive diode (PSD) sensor.
4. A beam-splitter.
5. A retro-reflector attached to the moving object.
6. A control unit.
7. Software for system control and analysis.

These components work together and are used both to track the target and to measure the coordinates of the target (see figure 50.39).

Devices

Figure 50-39 shows a laser tracker system, also called a laser tracer [50-58] and [50-59]. A particular feature is a reference sphere with its center at the crossing point of both rotation axes. The whole interferometer block is rotatable about this sphere.

The interferometer beam consists of two orthogonal polarizations, which are separated by a polarizing beam-splitter. The p-polarization passes straight through the beam-splitter and acts as the reference beam, while the s-polarized beam is reflected in a direction towards the reference sphere and acts as the measurement beam.

Before being reflected by the sphere's surface the measurement beam transmits a quarter-waveplate, which changes its polarization to circular. A collimator then focusses the beam onto the surface of the reference sphere, from which it is reflected, transmitting through the collimator and quarter-waveplate to be changed to p-polarization. The beam passes the beam-splitter and transmits a second quarter-waveplate changing the polarization to circular. A portion of the beam then transmits a neutral beam-splitter and aims at the retro-reflector attached to the object. In figure 50-39 a corner cube is shown as the retro-reflector, but it is also quite common to use a cat's eye type of retro-reflector, which is usually made from two glass hemispheres with the same refractive index, but different radii, with their flat sides joined together to form the target [50-60].

If the measurement beam hits the retro-reflector not exactly in the center, a laterally displaced beam will return to the beam-splitter. A portion of it will be directed towards a position-sensitive diode (PSD), which generates two electrical signals proportional to the relative x and y position of the beam on the PSD detector surface. The other portion transmits the beam-splitter and is converted to s-polarization after passing the quarter-waveplate a second time. The measurement beam is finally reflected from the polarizing beam-splitter to enter the receiver unit of the displacement interferometer, together with the reference beam.

The two PSD signals are used by the motion-controller system to rotate the interferometer block around the reference sphere and thereby track the target.

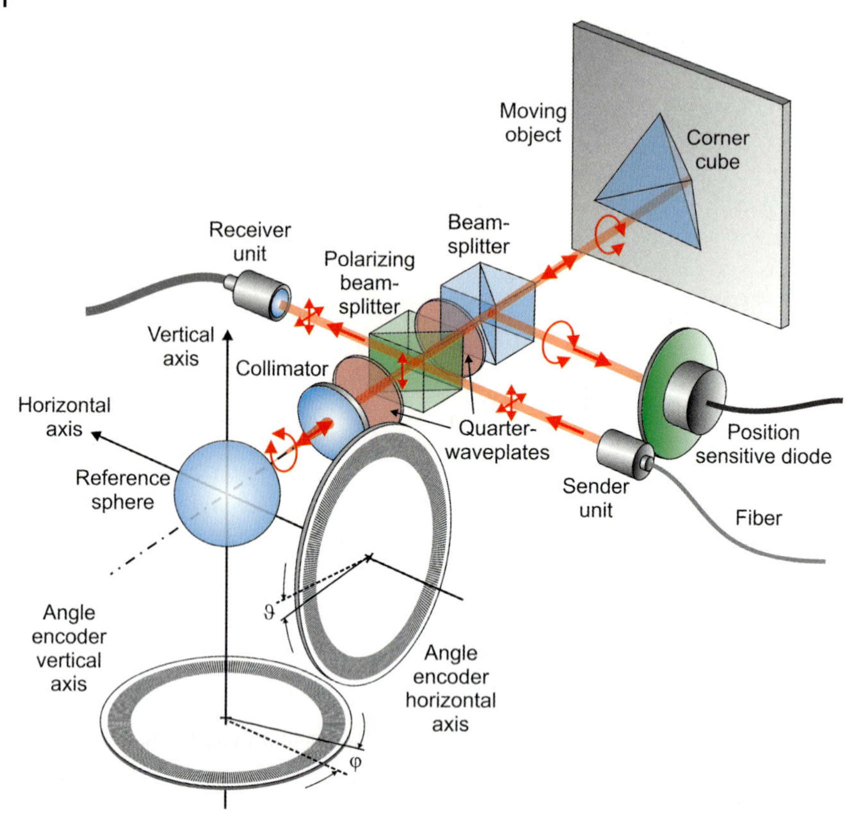

Figure 50-39: Laser tracker using a displacement-measuring interferometer to measure the optical path between a stationary reference sphere and a corner cube attached to the moving object.

The tracking is achieved by minimizing the measured displacement between the outgoing and returning measurement beams detected by the PSD after calibration.

The controller system uses the detected beam-offsets to minimize the tracking error by rotating the interferometer block to point the measurement beam towards the center of the target. When this is done continuously, the target will be tracked, while the relative distance of the retro-reflector is simultaneously measured by the displacement-measuring interferometer. The coordinates of the retro-reflector are finally completed by means of the two encoder readings.

Accuracy and Error Sources

The overall *accuracy* of a laser tracker is a combination of the resolution, repeatability and systematic deviation of its angle encoders, the displacement-measuring interferometer and their relative adjustment to each other [50-61], [50-62].

In section 50.4.1 we discussed the accuracy of incremental and absolute angle encoders. We stated that, for 20°C ambient temperature and a scanning frequency of up to 1 kHz, the error is approximately ± 1/20 of the grating period. A 5000-count encoder has an estimated accuracy of ± 13 arcsec. High-precision angle encoders provide system accuracies from ± 1 to ± 5 arcsec. 1 arcsec is equivalent to a lateral displacement of 4.8 μm at a distance of 1 m.

In section 50.2.1 we discussed the accuracy of displacement-measuring interferometers. Their accuracy is extremely limited by the environmental conditions. The resolution is usually specified between 0.5 and 2.5 nm. Typical accuracies achieved with displacement interferometers in air at 20°C are ± 0.5 μm per meter displacement.

Systematic errors in laser trackers can be caused by:

- the angle encoders,
- the displacement-measuring interferometer,
- the tracking system,
- beam misalignment,
- misalignment of the axes.

Systematic errors from the displacement-measuring interferometer are caused by imperfect components or an insufficient installation process leading to misaligned components. Typical errors are the cosine error and the Abbe error as discussed in section 50.2.1.

Misalignment of axes occurs when:

- the reference sphere is decentered,
- the horizontal and vertical axes are not perpendicular to each other,
- the horizontal and vertical axes and the measurement axis of the interferometer do not cross each other.

An extended calibration process can determine all relative misalignments, so that this can be compensated for by means of the appropriate analysis software.

Calibration Techniques

Laser trackers are calibrated in a similar way to total stations. A retro-reflector can be moved to known positions within a measurement volume and the positions then compared with the results from the laser tracker.

To calibrate the individual rotation axes the laser tracker can be mounted on top of a multi-faceted mirror of known geometry, which is mounted on top of a rotary table (see previous section on total stations). The multi-faceted mirror can be avoided if a rotary table with a high-precision master encoder is used. The laser tracker observes a retro-reflector fixed at a certain distance. For the calibration process the rotary table is rotated to previously known rotary positions, while the tracker compensates for the rotation of the rotary table. For each facet the known rotation angle is compared with the laser tracker's read-out. In this way a calibration over many angles can be made.

50.6 Optical Profile Metrology

An instrument used to measure the topography or profile of a surface is called a profilometer. A profilometer can measure the form, also called the figure, of the surface. If designed for small test areas, it can also quantify the roughness of the surface. In section 54.2 we have described the different instruments used to measure surface texture and roughness, including the tactile instruments.

In section 53.4 we have covered the different interferometer configurations used to measure surface form and figure. However, we have excluded the tactile coordinate measuring machines (CMM), since they are unsuitable for the measurement of surface figure with nanometer resolution.

There are applications in which the use of a CMM to measure optical figure is preferable; for example, in an early state of fabrication or in the case of a single piece production, where the use of dedicated optical test equipment would be too expensive.

The disadvantage of the surface-covering interferometers is their lack of universality. In order to measure a variety of radii and apertures of spherical surfaces, a set of expensive transmission spheres is necessary. When measuring aspheres, a computer-generated hologram is required, which can be used only for a single type of asphere.

More universality is given by subaperture-measuring interferometeric devices. However, their asphericity range is limited and the stitching technique is susceptible to stitching errors.

Optical profile metrology is used to measure optical surfaces pointwise with nanometer resolutions without contacting the test surface. In the following we describe:

- a CMM with an optical sensor (a short-range displacement measuring sensor) used to measure a variety of spherical and aspherical surfaces and
- profilers using slope (tilt) measuring sensors to determine the surface profiles of large flat optical surfaces with nanometer accuracy.

50.6.1 CMMs with Optical Sensors

Basic Principles

Conventional CMMs move a probe along three linear and orthogonal axes to touch a surface under test and take its 3D coordinates for an arbitrary number of points. The probe usually consists of a ball-shaped stylus, which contacts the surface. For optical surfaces in a final state of production, contacting methods should be avoided. The CMM must then carry a contactless or optical probe. For surfaces with large surface slopes, the probe should be oriented perpendicular to the surface. Therefore, CMMs must be equipped with at least two additional rotational axes.

To position an optical probe (small-distance sensor) perpendicular to a freeform surface, at least four axes have to be moved. The fifth axis is provided by the range of the optical probe, while the sixth axis can be avoided, if surfaces close to rotational-symmetry are to be tested. For the general machine setup a Cartesian (orthogonal), cylindrical or polar setup can be chosen. Since optical freeform surfaces are often more or less rotationally symmetric within several millimeters PV, a cylindrical setup is preferable.

In the following we describe a cylindrical CMM carrying an optical probe to measure the figure of spherical, aspherical and freeform surfaces.

Devices

In a cylindrical machine setup the surface under test is mounted on a rotating spindle providing the θ-axis, while the optical probe is translated in the radial (r) and vertical (z) direction and rotated around the ψ-axis (figure 50-40) [50-64].

The optical probe measures the distance between the ψ-axis rotor and the surface under test. A special dual stage design is implemented in the example, offering a 5 mm range, nanometer resolution and 5° unidirectional acceptance angle. This is necessary to enable the measurement of freeform surfaces, where exact normal adjustment is not possible [50-64].

The result for a single sample point is provided by the optical probe, consisting of a value for the normal surface height along the optical axis of the probe, and the position of the four axes of the system. Sampling of the total surface is done on consecutive circular tracks. During the sampling of a circular track only the test surface is rotated, while all other stages are locked. To obtain redundant data for averaging, the track can be measured several times. The stages then move the probe to the next circular track, and the process is repeated. This sampling strategy leads to measurement times in the order of 15 minutes for large surfaces [50-64]–[50-66].

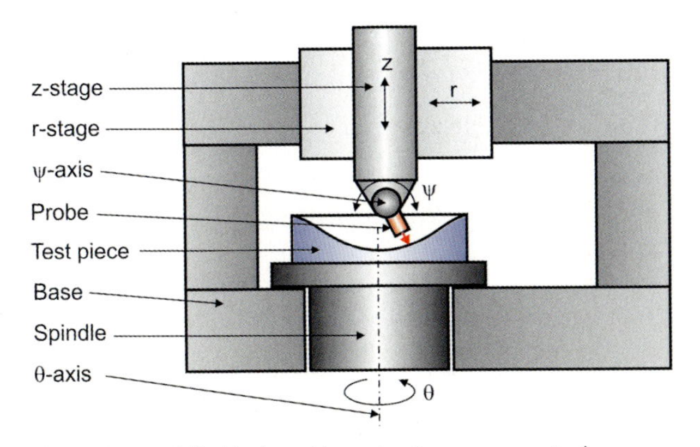

Figure 50-40: Cylindrical machine setup to measure optical freeform surfaces [50-64].

Accuracy and Error Sources

The factors limiting the system precision are given by:

- the *motion system* consisting of all four axis and their controller systems,
- the *metrology system* including the optical probe and all related auxiliary systems which measure the distance between the surface under test and the ψ-axis,
- the *environmental conditions* including
  - mechanical and acoustic vibrations,
  - radiation from light or heat sources,
  - changes in the temperature, humidity, or pressure of ambient air, or
  - heat conduction.

The measurement uncertainty is furthermore dependent on:

- the probe position,
- the position of the test surface,
- the form of the test surface,
- the form stability of the test surface,
- the data-acquisition and the data-processing software.

The CMM is generally able to measure freeform optics with diameters of up to 500 mm with an uncertainty of 30 nm (2σ).

Repeatabilities of < 10 nm RMS along circular tracks with locked probe-moving axes can be achieved. Measurements performed on a high-quality optical flat with a diameter of 100 mm lead to repeatabilities of 2 nm RMS.

The machine's accuracy depends on the quality of the calibration and the reproducibility of the measurement process, including the machine's repeatability and the handling and thermal adaptation of the test piece.

Calibration Techniques

To calibrate the CMM, a set of known spherical surfaces can be used, which cover the total measuring volume of the machine. Spherical surfaces can be measured by high-precision interferometers with uncertainties better than a nanometer, except for the quadratic term. The quadratic term related to the error in the radius of curvature can be measured to an accuracy of between 50 and 100 nm PV. This limits the total system accuracy after calibration.

Calibration can also be provided by using a laser tracker in different rigid positions following the path of a retro-reflector attached to the probe-holder. In this case, all axes of the system can be calibrated individually with μm accuracy.

50.6.2
Devices Using Angle Sensors

Basic Principles

An alternative method for collecting the coordinates of a test surface is to collect its derivatives. The technique is called *deflectometry*. For flat surfaces an autocollimator can be used to measure the inclination of surface subapertures by collecting the beam deflected by the test surface. To establish a scanning process, a pentaprism can be used, which is moved parallel to the surface under test at a certain distance [50-67].

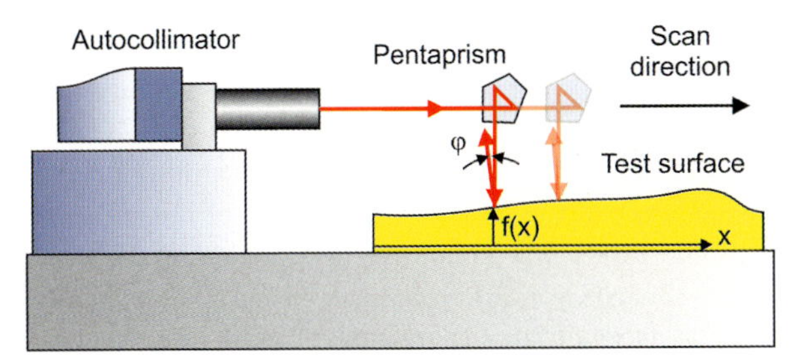

Figure 50-41: Principle of deflectometry using an autocollimator and a scanning pentaprism [50-71].

The advantage in using a pentaprism for scanning is that the deflection angle of 90° remains unaltered even though the prism might tilt slightly during the scanning process.

The principle of deflectometry by means of a pentaprism is shown in figure 50-41. The straightness reference is represented by the autocollimator, which is rigidly mounted on a granite table. The light from the autocollimator is deflected at a right angle by a pentaprism and propagates towards the surface under test. After reflection from the test surface, the returning beam is inclined by an angle φ, which is detected by the autocollimator. The pentaprism is scanned along the x-direction to collect a series of sample points. From the local tilt values, the topography $f(x)$ of the test surface is calculated by integration with appropriate algorithms [50-67] and [50-68].

Devices

Extended Shear Angle Difference Deflectometer (ESAD)

For high-precision measurements of flat surfaces with nanometer and subnanometer accuracy, the *Extended Shear Angle Difference method* (ESAD) can be used. This method utilizes angle difference information as a primary measurement property. The angle difference for two points on the surface – separated by the shear distance – is not influenced by small misalignments of the test surface during the scanning process. Large test surfaces, requiring long measurement times, can thus be tested very accurately.

The surface profile is reconstructed from two sets of angles measured by the autocollimator. A sophisticated integration process is necessary for the mathematical solution of the shearing problem. Multiple shears have to be introduced to avoid the blindness of the process for the periodicities corresponding to a single shear [50-67].

Two-dimensional topographies can be combined from profiles in different directions. The lateral resolution is limited by the beam diameter used for scanning the surface. It can be as low as 1 mm [50-69] independent of the size of the test surface. The cost for measuring systems needed for large test surfaces is dependent only on the size of the scanning unit.

Traceable Multi-sensor Technique (TMS)

The ESAD principle can be extended by a number of consecutive sensors arranged in a multiple-distance sensor module (MDS) and an additional measurement of the module's tilt. Figure 50-42 shows the arrangement of the measuring system. The tilt of the MDS is measured by a rigidly mounted autocollimator. The scanning process is selected such that the distance between the sensors is a multiple of the scanning step [50-70]. As a sensor array, the pixels of a small interferometer may be used, either directly or in combinations. By using an interferometer with a small aperture, typically in the range of 1 mm, surfaces of larger curvature can also be measured. The technique is called the *Traceable Multi-Sensor Technique* (TMS).

The surface profile under the MDS module can be modeled according to

$$f(p_i + s(j)) = -d_{ij} + \varepsilon_j + a_i + b_i s(j) \qquad (50\text{-}56)$$

where

- d_{ij} denotes the distance between the j-th sensor at the i-th scanning position and the surface under test,
- p_i is the position of the first sensor of the multiple-distance sensor array at scanning step i,
- $s(j)$ is the relative position of the j-th sensor,
- ε_j is the sensor offset error, and
- a_i, b_i are the positioning offset and tilt errors of the scanning stage at scanning step i.

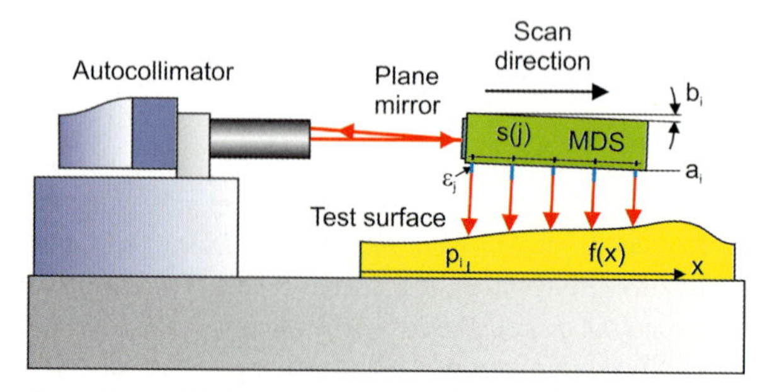

Figure 50-42: TMS measuring system. An autocollimator measures the tilt angle of a small mirror attached to the multiple-distance sensor (MDS) scanned over the test surface [50-71].

From (50-56) topography profiles can be reconstructed along an arbitrary straight line.

During the scanning process small tilts b_i of the MDS module are detected by the autocollimator, while the MDS module delivers results for the sensor distances d_{ij}.

The sensor offset errors ε_j, which represent the systematic errors of the MDS module, must be determined by a calibration process.

Although, for nearly flat surfaces, uncertainties in the nanometer range can be achieved, for topographies with larger peak-to-valley heights, the uncertainty will increase, especially if the lateral position of the MDS module is not exactly determined. For accuracy improvement a displacement interferometer can be introduced, measuring the lateral position p_i of the MDS module very precisely [50-72].

Long Trace Profiler (LTP)

The *long trace profiler* (LTP) is another instrument to measure the local slope of a surface under test. Its principle is shown in figure 50-43. A collimator generates a narrow parallel beam from linear polarized laser light sent through a polarization-preserving optical fiber. A linear polarizer is used to adjust the total intensity. A neutral beam-splitter splits the beam and sends both parts to right-angle prisms (roof prisms). When decentered, a roof prism returns a beam laterally displaced by twice the decentration. In this way a pair of collinear beams with zero optical path difference is generated. The beam pair is further split into two beam pairs, a test beam pair and a reference beam pair, by means of a polarized beam-splitter. A half-waveplate between the neutral and the polarized beam-splitter is used to balance the intensity of the reference and the test pair. Both beam pairs transmit a quarter-waveplate when leaving the polarized beam-splitter to change the polarization from linear to circular. The reference pair is reflected from a plane reference mirror, while the test pair is deflected by a pentaprism towards the surface under test. After reflection both beam pairs return to the quarter-waveplate and the polarized beam-splitter, where they are recombined and then deflected towards a Fourier transform lens. Since the reference and test beams are perpendicularly polarized, they do not interfere with each other. Interference occurs only between the components of the beam pairs. A line sensor is positioned at the focus of the Fourier transform lens, where the spot image interference pattern is detected.

As the test surface is scanned by the pentaprism, the angle of the reflected test beam pair depends on the local slope of the test surface. The collected slope profile can be integrated to obtain the height profile of the surface under test along the scanning line.

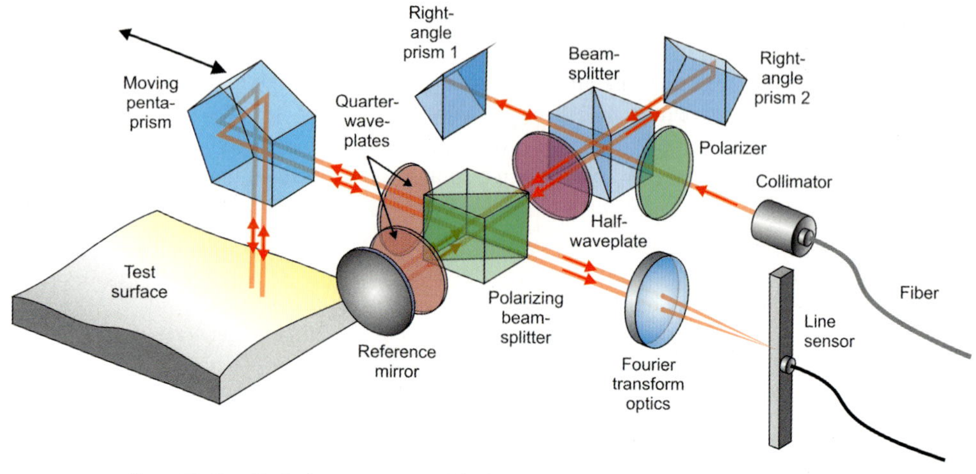

Figure 50-43: Optical arrangement of the long trace profiler (LTP).

Accuracy and Error Sources

Unlike interferometers, ESAD, TMS and LTP systems do not depend on the stability of a reference surface. Their precision limiting factors are given by:

- the precision and resolution of the angle sensor,
- the spatial resolution of the angle sensor,
- the quality of the pentaprism or MDS unit,
- the motion system for the pentaprism or MDS unit,
- the metrology system for the lateral position of the pentaprism or MDS unit,
- the integration algorithm used to calculate the topography from the local slopes,
- the environmental conditions including:
  - mechanical and acoustic vibrations,
  - radiation from light or heat sources,
  - changes in temperature, humidity, or pressure of ambient air, or
  - heat conduction.

Although the repeatability of the angle-sensor-based systems can be in the sub-nanometer region, their accuracy is limited by the knowledge of the exact shear distance. For instance, if the size of the pixel array used in the TMS system is known to 10^{-3}, the reconstructed topography is known with similar precision. Therefore, topographies with millimeter sagittal heights can only be measured with μm accuracy. High-precision measurements in the nanometer region are restricted to surfaces with deviations smaller than 100 μm from a perfect plane [50-73].

The accuracy can be improved by applying a calibration process from which a more reliable knowledge of the shear distance can be achieved.

Calibration Techniques

To calibrate ESAD, TMS or LTP, surfaces with well-known deformations can be used, which cover the total measuring volume needed for a required test surface spectrum. Deformations in the opticals surfaces can be qualified by high-precision interferometers with uncertainties better than a nanometer. Excluded are quadratic terms covering the total surface related to the error in the radius of curvature, which can be measured to an accuracy of between 50 and 100 nm PV. Local deformations of small size but steep slope, which the interferometer can still resolve, are more suitable for calibration. The interferometer results can be used to analyze and calibrate systematic shear errors and other systematic deviations.

50.7 Literature

50-1 T. Yoshizawa (Ed.), Handbook of Optical Metrology, CRC Press, Taylor & Francis Group LLC, Boca Raton (2009).

50-2 F. C. Demarest, High-resolution, high-speed, low data age uncertainty, heterodyne displacement measuring interferometer electronics, Meas. Sci. Technol. **9**, p. 1024–30 (1998).

50-3 C. M. Wu, J. Lawall and R. D. Deslattes, Heterodyne interferometer with subatomic periodic nonlinearity, Appl. Opt. **38**, p. 4089 (1999).

50-4 J. Lawall and E. Kessler, Michelson interferometry with 10 pm accuracy, Rev. Sci. Instrum. **71**, p. 2669 (2000).

50-5 P. de Groot, Jones matrix analysis of high-precision displacement measuring interferometers, Proc. 2nd Topical Meeting on Optoelectronic Distance Measurement and Applications (ODIMAP II), p. 9–14 (1999).

50-6 G. E. Sommargren, A new laser measurement system for precision metrology, Precis. Eng. **9** (4), p. 179–84 (1987).

50-7 J. Hu, H. Hu and Y. Ji, Detection method of nonlinearity errors by statistical signal analysis in heterodyne Michelson interferometer, Optics Express **18**, No. 6, p. 5831–39 (2010).

50-8 Zygo Corp., Brochure on error sources, Middlefield, Ct., USA (2008).

50-9 R. Wilhelm, A. Courteville, F. Garcia and F. de Vecchi, On-axis, non-contact measurement of glass thicknesses and airgaps in optical systems with submicron accuracy, Proceedings of the SPIE – The International Society for Optical Engineering, vol. 6616, 66163P (12 pp.) (2007).

50-10 R. Wilhelm and A. Courteville, Dimensional metrology for the fabrication of imaging optics using a high accuracy low coherence interferometer, Optical Measurement Systems for Industrial Inspection, Proceedings of SPIE 5856 (2005).

50-11 A. Courteville and R. Wilhelm, Contact-free on-axis metrology for the fabrication and testing of complex optical systems, Optical Fabrication, Testing, and Metrology II, Proceedings of SPIE 5965 (2005).

50-12 K. G. Larkin, Efficient nonlinear algorithm for envelope detection in white light interferometry, J. Opt. Soc. Am. A **13**, pp. 832–43 (1996).

50-13 P. Pavlíèek and G. Häusler, White-light interferometer with dispersion: an accurate fiber-optic sensor for the measurement of distance, Appl. Opt. **44**, p. 2978–83 (2005).

50-14 P. Balling, P. Køen, P. Mašika and S. A. van den Berg, Femtosecond frequency comb based distance measurement in air, Optics Express Vol. **17**, No. 11, pp. 9300–13 (2009).

50-15 E. V. Baklanov and A. K. Dmitriev, Absolute length measurements with a femtosecond laser, Quantum Electron. **32**, p. 925–28 (2002).

50-16 J. Ye, Absolute measurement of a long, arbitrary distance to less than an optical fringe, Opt. Lett. **29**, p. 1153–55 (2004).

50-17 K.-N. Joo, Y. Kim and S.-W. Kim, Distance measurements by combined method based on a femtosecond pulse laser, Opt. Express **16**, 19799–806 (2008).

50-18 Y. Salvadé, N. Schuhler, S. Lévêque and S. Le Floch, High-accuracy absolute distance measurement using frequency comb referenced multiwavelength source, Appl. Opt. **47**, 2715–20 (2008).

50-19 M. Cui, M. G. Zeitouny, N. Bhattacharya, S. A. van den Berg, H. P. Urbach and J. J. M. Braat, High-accuracy long-distance measurements in air with a frequency comb laser, Optics Letters **34**, pp. 1982–84 (2009).

50-20 T. Hochrein, R. Wilk, M. Mei, R. Holzwarth, N. Krumbholz and M. Koch, Optical sampling by laser cavity tuning, Optics Express **18**, pp. 1613–17 (2010).

50-21 M. Cui, R. N. Schouten, N. Bhattacharya and S. A. van den Berg, Experimental demonstration of distance measurement with a femtosecond frequency comb laser, J. Eur. Opt. Soc. Rapid Publ. **3**, 08003 (2008).

50-22 D. Jones, S. A. Diddams, J. K. Ranka, A. Stentz, R. S. Windeler, J. L. Hall and S. T. Cundiff, Carrier-envelope phase control of femtosecond mode-locked lasers and direct optical frequency synthesis, Science **288**, 635 (2000).

50-23 R. Holzwarth, T. Udem, T. W. Hänsch, J. C. Knight, W. J. Wadsworth and P. S. J. Russell, Optical frequency synthesizer for precision spectroscopy, Phys. Rev. Lett. **85**, 2264 (2000).

50-24 T.Yasui, K. Minoshima and H. Matsumoto, Stabilization of femtosecond mode-locked Ti:sapphire laser for high-accuracy pulse interferometry, IEEE J. Quantum Elect. **37**, No.1 (2001).

50-25 Sangwon Hyun, Young-Jin Kim, Yunseok Kim, Jonghan Jin and Seung-Woo Kim, Absolute length measurement with the frequency comb of a femtosecond laser, Meas. Sci. Technol. **20** 095302 (6pp) (2009).

50-26 B. L. Swinkels, High-accuracy absolute distance metrology, Ph.D. Thesis, Technische Universiteit Delft, The Netherlands (2006).

50-27 H. Walcher, Position Sensing: Angle and Distance Measurement for Engineers, Butterworth Heinemann (1994).

50-28 D. S. Nyce, Linear Position Sensors: Theory and Application, John Wiley & Sons Inc., New Jersey (2003).

50-29 Dr. Johannes Heidenhain GmbH, General Catalog, 350 457-2D · 30 · 10/2009, Traunreut, Germany (2009).

50-30 Dr. Johannes Heidenhain GmbH, Technical information: Linear Encoders with Single-Field Scanning, 510 431–21 · 20 · 11/2006, Traunreut, Germany (2006).

50-31 Sony Manufacturing Systems Corporation, Measuring Systems Division, Product Information: Achieving 17-Picometer Resolution –Sony Laserscale Products for the Next Generation of Innovative Manufacturing (2008).

50-32 Dr. Johannes Heidenhain GmbH, Technical information: Optimized Scanning in Absolute Rotary Encoders, 606 132–21 · 10 · 10/2008, Traunreut, Germany (2008).

50-33 Dr. Johannes Heidenhain GmbH, Technical information: Rotary Encoders, 349 529–2A · 40 · 8/2010, Traunreut, Germany (2010).

50-34 K. Engelhardt and P. Seitz, Absolute, high-resolution optical position encoder, Appl. Opt. **35**, p. 201–08 (1996).

50-35 Keyence Corp., Laser Measurement Product Brochure, LKLJ-General-KA-C-E 0129-2 611115, Osaka, Japan (2008).

50-36 Keyence Corp., High-Accuracy Surface Scanning Method, LT9-KA-C-E 0019-3 611001, Osaka, Japan (2009).

50-37 Joon Ho Bae, Ki Hyun Kim, Mun Heon Hong, Chang Hun Gim and Wonho Jhe, High-resolution confocal detection of nanometric displacement by use of a 2 × 1 optical fiber coupler, Optics Letters **25**, p. 1696–98 (2000).

50-38 E. Shafir and G. Berkovic, Expanding the realm of fiber optic confocal sensing for probing position, displacement, and velocity, Appl. Opt. **45**, pp. 7772–77 (2006).

50-39 A. Miks, J. Novak and P. Novak, Analysis of method for measuring thickness of plane-parallel plates and lenses using chromatic confocal sensor, Appl. Opt. **49**, pp. 3259–64 (2010).

50-40 A. K. Ruprecht, K. Körner, T. F. Wiesendanger, H. J. Tiziani and W. Osten, Chromatic confocal detection for high speed micro-topography, Proc. SPIE 5302-6, p. 53–60 (2004).

50-41 C. Pruss, A. Ruprecht, K. Körner, W. Osten and P. Lücke, Diffractive Elements for Chromatic Confocal Sensors, DGaO Proc. ISSN 1614–8436 (2005).

50-42 B. Michelt and J. Schulze, The spectral colours of nanometers, Mikroproduktion 3 (2005).

50-43 E. Papastathopoulos, K. Körner and W. Osten, Chromatic confocal spectral interferometry, Appl. Opt. **45**, p. 8244–52 (2006).

50-44 Precitec Optronik GmbH, Technical information: Optical Probes and Sensors, Rodgau, Germany (2010).

50-45 J. Jeko, Calibration and Verification of Horizontal Circles of Electronic Theodolites, Slovak Journal of Civil Engineering, p. 32–38 (2007).

50-46 G. S. Gordon, Let's Talk Accuracy, Technical information from Gurley Precision Instruments, Troy, N.Y., USA (2002).

50-47 Trioptics GmbH, Technical information: Electronic Autocollimators for Precise Angle Measurement, Wedel, Germany (2010).

50-48 Trioptics GmbH, Technical information: TriAngle® UltraSpec - Precision Electronic Autocollimator, Wedel, Germany (2010).

50-49 C.-M. Wu, S.-T. Lin and J. Fu, Heterodyne interferometer with two spatial-separated polarization beams for nanometrology, Optical and Quantum Electronics **34**, p. 1267–76 (2002).

50-50 JENAer Meßtechnik GmbH, Technical information: ZLM 700 / 800 / 900 Dual – Frequency Laser Interferometer, Jena, Germany (2010).

50-51 Zygo Corp., Technical information: ZMI Differential Plane Mirror Interferometer (DPMI), Middlefield, Ct., USA (2008).

50-52 Leica Geosystems, Product Brochure: Leica TDRA6000, Unterentfelden, Switzerland (2009).

50-53 Leica Geosystems, Product Brochure: Leica Absolute Tracker AT401 – white paper, Unterentfelden, Switzerland (2010).

50-54 D. Martin and G. Gatta, Calibration of Total Stations Instruments at the ESRF, XXIII FIG Congress, Munich, Germany (2006).

50-55 R. D. Peterson and K. L. Schepler, Timing modulation of a 40-MHz laser-pulse train for target ranging and identification, Appl. Opt. **42**, pp. 7191–96 (2003).

50-56 D. Martin, The Analysis of Parasitic Movements on a High Precision Rotation Table, MEDSI 2006, Himeji, Japan (2006).

50-57 G. P. Greeff, A Study for the Development of a Laser Tracking System Utilizing Multilateration for High Accuracy Dimensional Metrology, MScEng (Mechatronic), Department of Mechanical and Mechatronics Engineering, University of Stellenbosch, South Africa (2010).

50-58 E. Hughes, A. Wilson and G. Peggs, Design of a high-accuracy CMM based on multilateration techniques. CIRP Annals – Manufacturing Technology **49**, pp. 391–94 (2000).

50-59 C. Schneider, Lasertracer – a new type of self tracking laser interferometer. Technical report, IWAA, Geneva, Switzerland (2004).

50-60 L. Yongbing, Z. Guoxiong and L. Zhen, An improved cat's-eye retro-reflector used in a laser tracking interferometer system. Measurement Science and Technology **14**, pp. 36–40 (2003).

50-61 P. Teoh, B. Shirinzadeh, C. Foong and G. Alici, The measurement uncertainties in the laser interferometry-based sensing and tracking technique. Measurement **32**, pp. 135–50 (2002).

50-62 B. B. Gallagher, Optical shop applications for laser tracking metrology systems. Master's thesis, Department of Optical Sciences, University of Arizona (2003).

50-63 B. H. Walser, 'Development and calibration of an image assisted total station', dissertation submitted to the Swiss Federal Institute of Technology Zurich (2004).

50-64 R. Henselmans, Non-contact Measurement Machine for Freeform Optics, Ph. D. Thesis, Eindhoven University of Technology, The Netherlands (2009), ISBN 978-90-386-1607-0.

50-65 R. Henselmans, G. Gubbels and C. van Drunen, Application of the NANOMEFOS Non-contact Measurement Machine in Asphere and Freeform Optics Production, OSA/IODC/OF&T (2010).

50-66 R. Henselmans, L. A. Cacace, G. F. Y. Kramer, P. C. J. N. Rosielle and M. Steinbuch, Nanometer level freeform surface measurements with the NANOMEFOS non-contact measurement machine, Proc. of SPIE, Optical Manufacturing and Testing VIII, Vol. 7426 (2009).

50-67 C. Elster, I. Weingärtner, High-accuracy reconstruction of a function f(x) when only $df(x)/dx$ or $d^2f(x)/dx^2$ is known at discrete measurement points, Proc. SPIE. 4782, 12–20 (2002).

50-68 R. D. Geckeler and I. Weingärtner, Sub-nm topography measurement by deflectometry: Flatness standard and wafer nanotopography, Proc. SPIE 4779, 1–12 (2002).

50-69 F. Siewert, H. Lammert and T. Zeschke, The Nanometer Optical component measuring Machine – NOM, Modern Developments in X-Ray and Neutron Optics, Springer Series in Optical Sciences **137**, Springer-Verlag, Berlin/Heidelberg (2008).

50-70 M. Schulz and C.Elster, Traceable multiple sensor system for measuring curved surface profiles with high accuracy and high lateral resolution, Opt. Eng. **45** 060503 (2006).

50-71 M. Schulz, A. Wiegmann, A. Márquez and C. Elster, Optical flatness metrology: 40 years of progress, Opt. Pura Apl. **41** (4) 325–31 (2008).

50-72 A. Wiegmann, M. Schulz and C. Elster, Improving the lateral resolution of a multi-sensor profile measurement method by nonequidistant sensor spacing, Optics Express Vol. **18**, p. 15807–819 (2010).

50-73 A. Wiegmann, Multiple Sensorsysteme zur Topographiebestimmung optischer Oberflächen, Ph. D. Thesis, Technische Universität Berlin, Germany (2009).

50-74 K. Minoshima and H. Matsumoto, High-accuracy measurement of 240 m distance in an optical tunnel by using of a compact femtosecond laser, Appl. Opt. **39**, p. 5512–17 (2000).

50-75 N. R. Doloca, M. Wedde, K. Meiners-Hagen and A. Abou-Zeid, Femtosekundenlaserbasierendes Messsystem für geodätische Längen, PTB-Mitteilungen 120, p. 120–23 (2010).

51
Polarimetry

Handbook of Optical Systems: Vol. 5. Metrology of Optical Components and Systems. First Edition.
Edited by Herbert Gross.

51.1 Introduction

Polarimetry is the measurement of the polarization state of light. The instrument built to measure the specific characteristics of polarization is called a polarimeter. Polarimetry of thin films and surfaces is commonly known as ellipsometry.

Polarimetry can be used to measure the polarizing properties of materials, optical components and systems such as:

- **Retardance**, which represents the phase shift between two eigenpolarizations. Eigenpolarizations are those polarization states which propagate unchanged through optically anisotropic materials. In homogeneous polarization elements eigenpolarizations are orthogonal.
- **Diattenuation**, which represents the property of materials and elements with transmittances depending on the incident polarization state. Diattanuation occurs in materials with absorption coefficients depending on the polarization (dichroism) or at interfaces. The eigenpolarizations have principal transmittances T_{min} and T_{max}.
- **Depolarization**, which represents the property of optical components and devices to convert incident polarized light into a state of less polarized light up to completely unpolarized light. Examples are rotating ground glass plates or integrating spheres.

To measure these properties, there have been many different polarimeter designs. Most conventional polarimeters are based on arrangements of polarizing filters and wave plates, while others use electro-optic modulators and other devices.

To describe polarization states and polarizing properties of optical materials, components and systems mathematically two methods are mainly used [51-1].

1. The two-element **Jones vector** which describes the two-element electric field vector combined with the **Jones matrix** describing the polarization properties of a material, component or system. Jones calculus, however, is not able to describe depolarization effects, but it retains phase information so that coherent beams can be combined properly.
2. The four-element **Stokes vector** which describes the polarization state of light by four measurable intensities combined with the **Mueller matrix** describing the polarization properties of a material, component or system.

In the following we will describe the Jones as well as the Stokes/Mueller calculus as a basis for the description of the functionality of polarimetric devices.

51.2
Basic principles of Polarimetry

51.2.1
Jones Calculus

Jones Vector

Let us assume a monochromatic plane wave as described by (51-1) [Handbook of Optical Systems **2**, 26.2.1].

$$\vec{E} = \begin{pmatrix} E_x \\ E_y \end{pmatrix} = \begin{pmatrix} A_x e^{i\phi_x} \\ A_y e^{i\phi_y} \end{pmatrix} \tag{51-1}$$

$\vec{E}$ denotes the polarization state of a completely polarized field in isotropic media and is called the Jones vector. It also contains the amplitude and phase of each component.

We introduce the following replacements:

$$\delta = \phi_y - \phi_x \tag{51-2}$$

$$A_x = \sqrt{A_x^2 + A_y^2} \cos \psi \tag{51-3}$$

$$A_y = \sqrt{A_x^2 + A_y^2} \sin \psi \tag{51-4}$$

and obtain (51-5).

$$\vec{E} = e^{i\frac{\phi_x + \phi_y}{2}} \sqrt{A_x^2 + A_y^2} \begin{pmatrix} \cos \psi \cdot e^{-i\frac{\delta}{2}} \\ \sin \psi \cdot e^{i\frac{\delta}{2}} \end{pmatrix} \tag{51-5}$$

If we omit the complex factor in front of the vector we obtain the so-called phase-reduced Jones vector $\vec{E}_{red}$, which will be the basis for the description of the retardance and the diattanuation states of light (51-6). In the following we will consider the Jones vector, referring to the phase-reduced form. Figure 51-2 shows an overview of Jones vectors with different ψ- and δ-values.

$$\vec{E}_{red} = \begin{pmatrix} A_x e^{-i\frac{\delta}{2}} \\ A_y e^{i\frac{\delta}{2}} \end{pmatrix} = \sqrt{A_x^2 + A_y^2} \begin{pmatrix} \cos \psi \cdot e^{-i\frac{\delta}{2}} \\ \sin \psi \cdot e^{i\frac{\delta}{2}} \end{pmatrix} \tag{51-6}$$

The ratio ρ will be defined as:

$$\rho = \frac{E_{red}y}{E_{red}x} = \tan \psi \cdot e^{i\delta} \tag{51-7}$$

The irradiance corresponding to a Jones vector is

$$I = n|E_{\text{red}}x|^2 + n|E_{\text{red}}y|^2 = n\left(A_x^2 + A_y^2\right) \qquad (51\text{-}8)$$

where n is the refractive index of the medium in which the wave is propagating.

Unpolarized or partially polarized light cannot be represented by a single Jones vector. However, it can be represented in the Jones vector calculus by the incoherent superposition of two orthogonal polarization states:

$$I = \frac{1+g}{2}\left|\vec{E}\right|^2 + \frac{1-g}{2}\left|\vec{E}_{\text{orth}}\right|^2 \qquad (51\text{-}9)$$

$$\text{with } g = \frac{I_{\text{pol}}}{I_{\text{pol}} + I_{\text{unpol}}} \qquad (51\text{-}10)$$

as the degree of polarization given by the ratio of the irradiance of the polarized light and the total irradiance.

Ellipsometry generally concentrates on measuring ψ and δ. It is usual to take the so-called *p*- and *s*-directions as the two orthogonal basis vectors, where the *p*-direction takes the part of the *x*-coordinate and is defined as the plane that contains the incident ray, the reflected ray and the vector normal to the reflecting surface. The *s*-direction lies perpendicular to the *p*-direction and takes the part of the *y*-coordinate (Figure 51-1).

Figure 51-2 shows examples of different polarization states and their associated δ- and ψ-values, Jones and Stokes vectors.

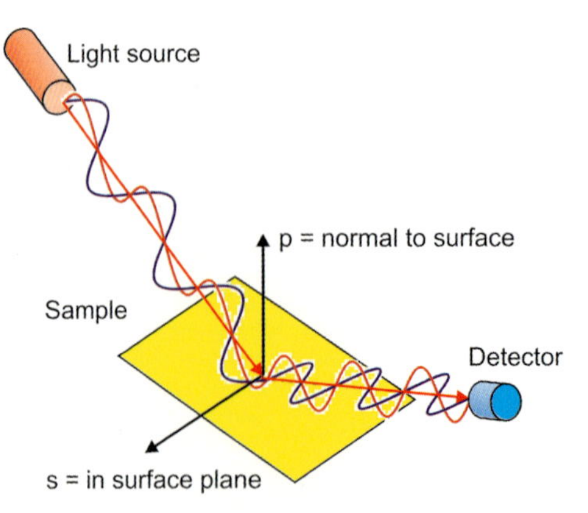

Figure 51-1: The coordinate system in ellipsometry: the *p*-direction coincides with the plane which contains the incident and reflected ray and the sample normal (plane of incidence), the *s*-direction is perpendicular to the plane of incidence.

Jones Matrix

The polarizing properties of optical components and systems can be represented as complex 2 × 2 Jones matrices **J**. The matrix describes actions of non-depolarizing elements on polarized fields. The output polarization state $\vec{E'}$ is then given by

$$\vec{E'} = \mathbf{J} \cdot \vec{E} \tag{51-11}$$

with $\mathbf{J} = \begin{pmatrix} J_{xx} & J_{xy} \\ J_{yx} & J_{yy} \end{pmatrix}$ denoting the Jones matrix of the element.

Polarimetry on surfaces is most conveniently described in the sp-coordinate system. In the following we therefore give some general rules for the transmission of a polarized beam through a system of polarizing elements.

If a beam travels through a series of m optical components with individual Jones matrices J_i, a single common matrix **J** can be constructed according to

$$\mathbf{J} = \prod_{i=1}^{m} \mathbf{R}(-a_i) \cdot \mathbf{J}_i \cdot \mathbf{R}(a_i) \tag{51-12}$$

where

$$\mathbf{R}(a) = \begin{pmatrix} \cos a & \sin a \\ -\sin a & \cos a \end{pmatrix} \tag{51-13}$$

denotes the rotation matrix of the element i rotated by a_i around the axis of propagation.

Since the matrices are not commutative they have to be arranged in a sequential order with the rightmost matrix representing the first element upon which the light is incident. Each element can be rotated by $\mathbf{R}(a_i)$such that its eigenpolarization coincides with the p- or s-axis and back rotated by $\mathbf{R}(-a_i)$. We can write explicitly

$$\mathbf{J}' = \mathbf{R}(-a) \cdot \mathbf{J} \cdot \mathbf{R}(a) = \begin{pmatrix} J'_{xx} & J'_{xy} \\ J'_{yx} & J'_{yy} \end{pmatrix} \tag{51-14}$$

with

$$J'_{xx} = J_{xx} \cos^2 a - (J_{xy} + J_{yx}) \sin a \cos a + J_{yy} \sin^2 a \tag{51-15}$$

$$J'_{xy} = J_{xy} \cos^2 a - (J_{yy} - J_{xx}) \sin a \cos a - J_{yx} \sin^2 a \tag{51-16}$$

$$J'_{yx} = J_{yx} \cos^2 a - (J_{yy} - J_{xx}) \sin a \cos a - J_{xy} \sin^2 a \tag{51-17}$$

$$J'_{yy} = J_{yy} \cos^2 a + (J_{xy} + J_{yx}) \sin a \cos a + J_{xx} \sin^2 a \tag{51-18}$$

The general form of the Jones matrix of a polarizing component with its eigenpolarization axes at 0° and 90° is shown in (51-19) [51-2].

$$\mathbf{J}_0 = \frac{1}{1+\tan^2\varepsilon}\begin{pmatrix} e^{\gamma}+\tan^2\varepsilon & -i(e^{\gamma}-1)\tan\varepsilon \\ i(e^{\gamma}-1)\tan\varepsilon & e^{\gamma}\tan^2\varepsilon+1 \end{pmatrix} = \begin{pmatrix} e^{\gamma}\cos^2\varepsilon+\sin^2\varepsilon & -i(e^{\gamma}-1)\sin\varepsilon\cos\varepsilon \\ i(e^{\gamma}-1)\sin\varepsilon\cos\varepsilon & e^{\gamma}\sin^2\varepsilon+\cos^2\varepsilon \end{pmatrix} \tag{51-19}$$

with

$$\gamma = -i\cdot\frac{2\pi d}{\lambda}(n_1-n_2-i(k_1-k_2)) \tag{51-20}$$

and the following notation:
d is the thickness of the material,
n_1, n_2 are the refractive indices of the material in both eigenpolarizations,
k_1, k_2 are the extinction coefficients of the material in both eigenpolarizations,
$\tan\varepsilon$ denotes the eigenpolarization of the element,
$\varepsilon = 0$ describes elements with orthogonal linear eigenpolarizations which cover most polarization elements such as linear polarizers, diattenuators and retarders.
$\varepsilon = \pi/4$ describes orthogonal circular eigenpolarizations. In this case, the left- and right-circular polarizations represent the eigenpolarizations of the element or system. Circular eigenpolarization covers elements showing optical activity leading to optical rotation and circular dichroism. In natural media the effect is caused by their helical molecule structure. The Faraday effect induces circular birefringence when optically isotropic media are placed in strong magnetic fields.

We introduce the abbreviations
$\Delta\phi$ for relative retardation (retardance) or birefringence,
κ for dichroism
with

$$\Delta\phi = \frac{2\pi d}{\lambda}(n_1-n_2) \tag{51-21}$$

$$\kappa = \frac{2\pi d}{\lambda}(k_1-k_2) \tag{51-22}$$

We can now write

$$e^{\gamma} = e^{-i\Delta\phi-\kappa} = \frac{e^{-\frac{2\pi d}{\lambda}k_1}}{e^{-\frac{2\pi d}{\lambda}k_2}}e^{-i\Delta\phi} = \frac{t_1}{t_2}e^{-i\Delta\phi} \tag{51-23}$$

where t_1/t_2 denotes the ratio of the amplitude transmittances. We can then rewrite (51-19) neglecting a normalizing factor:

$$\mathbf{J}_0 = \begin{pmatrix} t_1 e^{-i\Delta\phi}\cos^2\varepsilon + t_2\sin^2\varepsilon & -i(t_1 e^{-i\Delta\phi}-t_2)\sin\varepsilon\cos\varepsilon \\ i(t_1 e^{-i\Delta\phi}-t_2)\sin\varepsilon\cos\varepsilon & t_1 e^{-i\Delta\phi}\sin^2\varepsilon + t_2\cos^2\varepsilon \end{pmatrix} \tag{51-24}$$

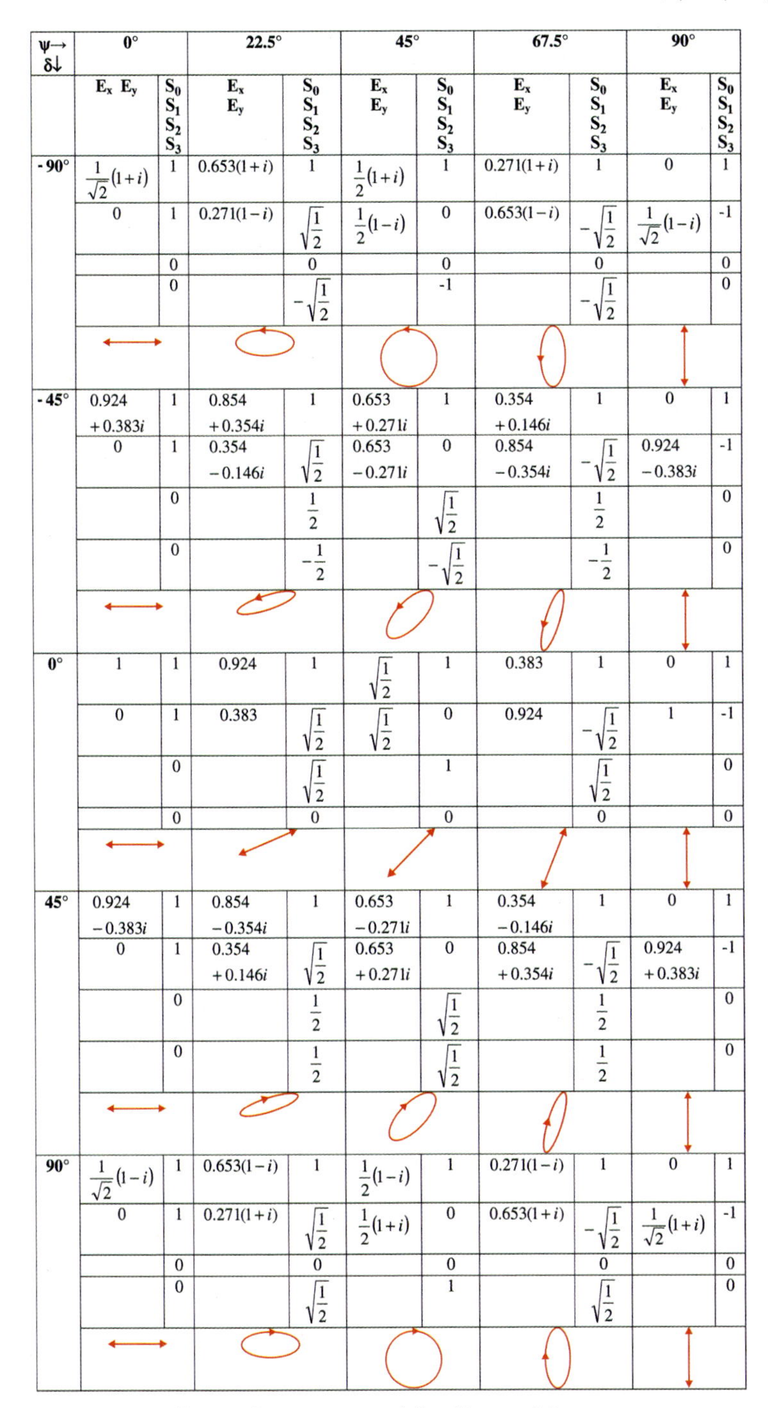

ψ→ δ↓	0°		22.5°		45°		67.5°		90°	
	E_x E_y	S_0 S_1 S_2 S_3	E_x E_y	S_0 S_1 S_2 S_3	E_x E_y	S_0 S_1 S_2 S_3	E_x E_y	S_0 S_1 S_2 S_3	E_x E_y	S_0 S_1 S_2 S_3
-90°	$\frac{1}{\sqrt{2}}(1+i)$	1	$0.653(1+i)$	1	$\frac{1}{2}(1+i)$	1	$0.271(1+i)$	1	0	1
	0	1	$0.271(1-i)$	$\sqrt{\frac{1}{2}}$	$\frac{1}{2}(1-i)$	0	$0.653(1-i)$	$-\sqrt{\frac{1}{2}}$	$\frac{1}{\sqrt{2}}(1-i)$	-1
		0		0		0		0		0
		0		$-\sqrt{\frac{1}{2}}$		-1		$-\sqrt{\frac{1}{2}}$		0
-45°	$0.924+0.383i$	1	$0.854+0.354i$	1	$0.653+0.271i$	1	$0.354+0.146i$	1	0	1
	0	1	$0.354-0.146i$	$\sqrt{\frac{1}{2}}$	$0.653-0.271i$	0	$0.854-0.354i$	$-\sqrt{\frac{1}{2}}$	$0.924-0.383i$	-1
		0		$\frac{1}{2}$		$\sqrt{\frac{1}{2}}$		$\frac{1}{2}$		0
		0		$-\frac{1}{2}$		$-\sqrt{\frac{1}{2}}$		$-\frac{1}{2}$		0
0°	1	1	0.924	1	$\sqrt{\frac{1}{2}}$	1	0.383	1	0	1
	0	1	0.383	$\sqrt{\frac{1}{2}}$	$\sqrt{\frac{1}{2}}$	0	0.924	$-\sqrt{\frac{1}{2}}$	1	-1
		0		$\sqrt{\frac{1}{2}}$		1		$\sqrt{\frac{1}{2}}$		0
		0		0		0		0		0
45°	$0.924-0.383i$	1	$0.854-0.354i$	1	$0.653-0.271i$	1	$0.354-0.146i$	1	0	1
	0	1	$0.354+0.146i$	$\sqrt{\frac{1}{2}}$	$0.653+0.271i$	0	$0.854+0.354i$	$-\sqrt{\frac{1}{2}}$	$0.924+0.383i$	-1
		0		$\frac{1}{2}$		$\sqrt{\frac{1}{2}}$		$\frac{1}{2}$		0
		0		$\frac{1}{2}$		$\sqrt{\frac{1}{2}}$		$\frac{1}{2}$		0
90°	$\frac{1}{\sqrt{2}}(1-i)$	1	$0.653(1-i)$	1	$\frac{1}{2}(1-i)$	1	$0.271(1-i)$	1	0	1
	0	1	$0.271(1+i)$	$\sqrt{\frac{1}{2}}$	$\frac{1}{2}(1+i)$	0	$0.653(1+i)$	$-\sqrt{\frac{1}{2}}$	$\frac{1}{\sqrt{2}}(1+i)$	-1
		0		0		0		0		0
		0		$\sqrt{\frac{1}{2}}$		1		$\sqrt{\frac{1}{2}}$		0

Figure 51-2: Different polarization states defined by ψ and δ and their associated Jones and Stokes vectors.

Combining (51-14) with (51-24) we arrive at (51-25–51-28) for the general Jones matrix elements with arbitrary eigenpolarization axes.

$$J'_{xx} = \frac{1}{2}\left(t_1 e^{-i\Delta\phi} + t_2 + \left(t_1 e^{-i\Delta\phi} - t_2\right)\cos 2\varepsilon \cos 2a\right) \quad (51\text{-}25)$$

$$J'_{xy} = \frac{1}{2}\left(t_1 e^{-i\Delta\phi} - t_2\right)\left(-i\sin 2\varepsilon + \cos 2\varepsilon \sin 2a\right) \quad (51\text{-}26)$$

$$J'_{yx} = \frac{1}{2}\left(t_1 e^{-i\Delta\phi} - t_2\right)\left(i\sin 2\varepsilon + \cos 2\varepsilon \sin 2a\right) \quad (51\text{-}27)$$

$$J'_{yy} = \frac{1}{2}\left(t_1 e^{-i\Delta\phi} + t_2 - \left(t_1 e^{-i\Delta\phi} - t_2\right)\cos 2\varepsilon \cos 2a\right) \quad (51\text{-}28)$$

Table 51-1 shows a variety of elements and their associated a, ε, t_1, t_2 and $\Delta\phi$.

Table 51-1: Parameters of Jones matrices for different polarizing elements.

Component	a	ε	t_1	t_2	$\Delta\phi$
Nonpolarizing element	0	0	1	1	0
Absorber with amplitude transmittance $0<t<1$	0	0	t	t	0
Linear polarizer at 0°	0	0	1	0	0
Linear polarizer at 90°	$\pi/2$	0	0	1	0
Linear diattenuator at 0°	0	0	$0<t_1<1$	$0<t_2<1$	0
General linear retardance by $\Delta\phi$, fast axis at 0°	0	0	1	1	$\Delta\phi$
Quarter-wave ($\Delta\phi=\pi/2$) linear retarder, fast axis at 0°	0	0	1	1	$\pi/2$
Half-wave ($\Delta\phi=\pi$) linear retarder, fast axis at 0°	0	0	1	1	π
Mirror	0	0	1	–1	0
Right circular polarizer	0	$\pi/4$	1	0	0
Left circular polarizer	0	$-\pi/4$	1	0	0
Right circular diattenuator when $t_1>t_2$ Left circular diattenuator when $t_2>t_1$	0	$\pi/4$	$0<t_1<1$	$0<t_2<1$	0
General circular retardance by $\Delta\phi$	0	$\pi/4$	1	1	$\Delta\phi$
Faraday mirror	0	$\pi/4$	1	–1	0

Determination of Jones Matrix from Measured Jones Vectors

It is the aim of polarimetry to determine the polarization properties of an optical component or system, when the elements of the Jones matrix are determined. If the elements of at least two different input Jones vectors $\vec{E}_1$, $\vec{E}_2$ and the corresponding output Jones vectors $\vec{E}'_1$, $\vec{E}'_2$ are known, the elements of J can be determined according to (51-29)–(51-32).

$$J_{xx} = \frac{E_{2y}E'_{1x} - E_{1y}E'_{2y}}{E_{1x}E_{2y} - E_{2x}E_{1y}} \tag{51-29}$$

$$J_{xy} = \frac{E_{1x}E'_{2x} - E_{2x}E'_{1x}}{E_{1x}E_{2y} - E_{2x}E_{1y}} \tag{51-30}$$

$$J_{yx} = \frac{E_{2y}E'_{1y} - E_{1y}E'_{2y}}{E_{1x}E_{2y} - E_{2x}E_{1y}} \tag{51-31}$$

$$J_{yy} = \frac{E_{1x}E'_{2y} - E_{2x}E'_{1y}}{E_{1x}E_{2y} - E_{2x}E_{1y}} \tag{51-32}$$

If a number $m > 2$ of different input Jones vectors $\vec{E}_i$ and their associated output vectors $\vec{E}'_i$ are known, the elements of the Jones matrix can be determined by least-square fitting according to (51-33)–(51-36).

$$J_{xx} = \frac{\left(\sum_{i=1}^{m} E'_{ix}E_{ix}\right)\left(\sum_{i=1}^{m} E^2_{iy}\right) - \left(\sum_{i=1}^{m} E'_{ix}J_{iy}\right)\left(\sum_{i=1}^{m} E_{ix}E_{iy}\right)}{\left(\sum_{i=1}^{m} E^2_{ix}\right)\left(\sum_{i=1}^{m} E^2_{iy}\right) - \left(\sum_{i=1}^{m} E_{ix}E_{iy}\right)^2} \tag{51-33}$$

$$J_{xy} = \frac{\left(\sum_{i=1}^{m} E'_{ix}J_{iy}\right)\left(\sum_{i=1}^{m} E^2_{ix}\right) - \left(\sum_{i=1}^{m} E'_{ix}J_{ix}\right)\left(\sum_{i=1}^{m} E_{ix}E_{iy}\right)}{\left(\sum_{i=1}^{m} E^2_{ix}\right)\left(\sum_{i=1}^{m} E^2_{iy}\right) - \left(\sum_{i=1}^{m} E_{ix}E_{iy}\right)^2} \tag{51-34}$$

$$J_{yx} = \frac{\left(\sum_{i=1}^{m} E'_{iy}E_{ix}\right)\left(\sum_{i=1}^{m} E^2_{iy}\right) - \left(\sum_{i=1}^{m} E'_{ix}E_{iy}\right)\left(\sum_{i=1}^{m} E_{ix}E_{iy}\right)}{\left(\sum_{i=1}^{m} E^2_{ix}\right)\left(\sum_{i=1}^{m} E^2_{iy}\right) - \left(\sum_{i=1}^{m} E_{ix}E_{iy}\right)^2} \tag{51-35}$$

$$J_{yy} = \frac{\left(\sum_{i=1}^{m} E'_{iy}E_{iy}\right)\left(\sum_{i=1}^{m} E^2_{ix}\right) - \left(\sum_{i=1}^{m} E'_{iy}E_{ix}\right)\left(\sum_{i=1}^{m} E_{ix}E_{iy}\right)}{\left(\sum_{i=1}^{m} E^2_{ix}\right)\left(\sum_{i=1}^{m} E^2_{iy}\right) - \left(\sum_{i=1}^{m} E_{ix}E_{iy}\right)^2} \tag{51-36}$$

Once the elements of the Jones matrix are known, the unknown angle of rotation a, the eigenpolarization ε, the birefringence $\Delta\phi$ and the amplitude transmittances t_1, t_2 can be calculated from the Jones matrix elements using (51-25)–(51-28):

$$\tan 2a = \frac{J_{xy} + J_{yx}}{J_{xx} - J_{yy}} \tag{51-37}$$

$$\tan 2\varepsilon = i\frac{J_{xy} - J_{yx}}{J_{xy} + J_{yx}} \sin 2a \tag{51-38}$$

$$t_1 = \sqrt{\left(J_{xx} + J_{yy} + \frac{J_{xy} + J_{yx}}{\cos 2\varepsilon \sin 2a}\right)\left(J^*_{xx} + J^*_{yy} + \frac{J^*_{xy} + J^*_{yx}}{\cos 2\varepsilon \sin 2a}\right)} \tag{51-39}$$

$$t_2 = \frac{1}{2}\left(J_{xx} + J_{yy} - \frac{J_{xy} + J_{yx}}{\cos 2\varepsilon \sin 2a}\right) \tag{51-40}$$

$$\Delta\phi = i \cdot \ln\left(\frac{1}{t_1}\left(J_{xx} + J_{yy} + \frac{J_{xy} + J_{yx}}{\cos 2\varepsilon \sin 2a}\right)\right) \tag{51-41}$$

where * denotes the conjugate complex element.

51.2.2
Stokes/Mueller Calculus

The Jones calculus formalism is not sufficient to describe the depolarizing effects of optical elements and systems. We will use the so-called Mueller calculus to cover polarizing and depolarizing effects [51-1], [51-5] and [51-8].

The basis of Mueller calculus is the four-element Stokes vector $\vec{S}$ representing the polarization state of light. The elements S_0, S_1, S_2, S_3 of the Stokes vector are related to the quasi-monochromatic field representation by (51-42)–(51-45), where E_x and E_y refer to (51-1).

$$S_0 = \langle E_x E^*_x \rangle + \langle E_y E^*_y \rangle = A^2_x + A^2_y \tag{51-42}$$

$$S_1 = \langle E_x E^*_x \rangle - \langle E_y E^*_y \rangle = A^2_x - A^2_y \tag{51-43}$$

$$S_2 = \langle E_x E^*_y \rangle + \langle E_y E^*_x \rangle = 2A_x A_y \cos\left(\phi_y - \phi_x\right) \tag{51-44}$$

$$S_3 = i\left(\langle E_y E^*_x \rangle - \langle E_x E^*_y \rangle\right) = 2A_x A_y \sin\left(\phi_y - \phi_x\right) \tag{51-45}$$

The angle brackets denote time averaging, the asterisk denotes the conjugate complex value.

Each Stokes parameter can be determined experimentally by introducing ideal polarizers and measuring the irradiance in the following way:

$$\vec{S} = \begin{pmatrix} S_0 \\ S_1 \\ S_2 \\ S_3 \end{pmatrix} = \begin{pmatrix} I_{0^\circ} + I_{90^\circ} \\ I_{0^\circ} - I_{90^\circ} \\ I_{+45^\circ} - I_{-45^\circ} \\ I_{rcp} - I_{lcp} \end{pmatrix} \tag{51-46}$$

with the following notation:

I_{0° is the irradiance measured through an ideal polarizer with transmittance in 0°,

I_{90° is the irradiance measured through an ideal polarizer with transmittance in 90°,

I_{+45} is the irradiance measured through an ideal polarizer with transmittance in +45°,

I_{-45° is the irradiance measured through an ideal polarizer with transmittance in –45°,

I_{rcp} irradiance measured through a circular ideal polarizer with transmittance for right circular polarization,

I_{lcp} irradiance measured through a circular ideal polarizer with transmittance for left circular polarization.

Note that $I_0 = I_{0^\circ} + I_{90^\circ} = I_{+45^\circ} + I_{-45^\circ} = I_{rcp} + I_{lcp}$. Because of these identities the Stokes vector is actually determined by four irradiance measurements, for instance I_{0°, I_{90°, I_{-45° and I_{lcp} as shown in (51-47).

$$\vec{S} = \begin{pmatrix} I_{0^\circ} + I_{90^\circ} \\ I_{0^\circ} - I_{90^\circ} \\ I_{0^\circ} + I_{90^\circ} - 2I_{-45^\circ} \\ I_{0^\circ} + I_{90^\circ} - 2I_{lcp} \end{pmatrix} \tag{51-47}$$

The degree of polarization g is expressed by

$$g = \frac{\sqrt{S_1^2 + S_2^2 + S_3^2}}{S_0^2} \tag{51-48}$$

For unpolarized light, all three differences in (51-47) vanish and only S_0 is other than zero.

Inserting (51-2)–(51-4) into (51-42)–(51-45) gives the Stokes vector of a totally elliptical polarized beam defined by its ellipsometric angles δ and ψ

$$\vec{S} = S_0 \begin{pmatrix} 1 \\ \cos 2\psi \\ \sin 2\psi \cos\delta \\ \sin 2\psi \sin\delta \end{pmatrix} \tag{51-49}$$

The Stokes vector for partially polarized light ($g<1$) can be considered as the superposition of a totally polarized Stokes vector $\overrightarrow{S_P}$ and an unpolarized Stokes vector $\overrightarrow{S_U}$. Their relationship is expressed as:

$$\vec{S} = \overrightarrow{S_P} + \overrightarrow{S_U} = \begin{pmatrix} S_0 \cdot g \\ S_1 \\ S_2 \\ S_3 \end{pmatrix} + \begin{pmatrix} S_0 \cdot (1-g) \\ 0 \\ 0 \\ 0 \end{pmatrix} \tag{51-50}$$

An optical element changing the polarization state of light represented by a Stokes vector can be expressed by its Mueller matrix $\mathbf{M}$ carrying 4×4 real elements as shown in (51-51) and (51-52).

$$S' = \mathbf{M} \cdot S \tag{51-51}$$

where

$$\mathbf{M} = \begin{pmatrix} M_{00} & M_{01} & M_{02} & M_{03} \\ M_{10} & M_{11} & M_{12} & M_{13} \\ M_{20} & M_{21} & M_{22} & M_{23} \\ M_{30} & M_{31} & M_{32} & M_{33} \end{pmatrix} \tag{51-52}$$

A Stokes vector can be transferred to a coordinate system which is rotated around its direction of propagation by the rotation matrix $\mathbf{R}$:

$$\mathbf{R}(a) = \begin{pmatrix} 1 & 0 & 0 & 0 \\ 0 & \cos 2a & \sin 2a & 0 \\ 0 & -\sin 2a & \cos 2a & 0 \\ 0 & 0 & 0 & 1 \end{pmatrix} \tag{51-53}$$

If a beam travels through a series of m optical components with individual Mueller matrices $\mathbf{M}_i$, a single common matrix $\mathbf{M}$ can be constructed in the same way as described for Jones calculus according (51-12):

$$\mathbf{M} = \prod_{i=1}^{m} \mathbf{R}(-a_i) \cdot \mathbf{M}_i \cdot \mathbf{R}(a_i) \tag{51-54}$$

In the case of a non-depolarizing optical element or system, the Mueller matrix can be calculated from the elements of the Jones matrix. From the 16 elements then only 7 are independent.

The conversion is carried out by

$$\mathbf{J} = \mathbf{A} \cdot (\mathbf{J} \otimes \mathbf{J}^*) \cdot \mathbf{A}^{-1} \tag{51-55}$$

with

$$\mathbf{A} = \begin{pmatrix} 1 & 0 & 0 & 1 \\ 1 & 0 & 0 & -1 \\ 0 & 1 & 1 & 0 \\ 0 & i & -i & 0 \end{pmatrix}.$$

The symbol $\otimes$ indicates the Kronecker product of the Jones matrix **J** with its complex conjugate **J***.

Using the elements J_{ij} of the Jones matrix we then can write explicitly [51-2], [51-10]:

$$M_{00} = \frac{1}{2}\left(|J_{xx}|^2 + |J_{yy}|^2 + |J_{xy}|^2 + |J_{yx}|^2\right) \tag{51-56}$$

$$M_{01} = \frac{1}{2}\left(|J_{xx}|^2 - |J_{yy}|^2 - |J_{xy}|^2 + |J_{yx}|^2\right) \tag{51-57}$$

$$M_{02} = \mathrm{Re}\left(J_{xx}J_{xy}^*\right) + \mathrm{Re}\left(J_{yx}J_{yy}^*\right) \tag{51-58}$$

$$M_{03} = -\mathrm{Im}\left(J_{xx}^*J_{xy}\right) - \mathrm{Im}\left(J_{yx}^*J_{yy}\right) \tag{51-59}$$

$$M_{10} = \frac{1}{2}\left(|J_{xx}|^2 - |J_{yy}|^2 + |J_{xy}|^2 - |J_{yx}|^2\right) \tag{51-60}$$

$$M_{11} = \frac{1}{2}\left(|J_{xx}|^2 + |J_{yy}|^2 - |J_{xy}|^2 - |J_{yx}|^2\right) \tag{51-61}$$

$$M_{12} = \mathrm{Re}\left(J_{xx}J_{xy}^*\right) - \mathrm{Re}\left(J_{yx}J_{yy}^*\right) \tag{51-62}$$

$$M_{13} = -\mathrm{Im}\left(J_{xx}^*J_{xy}\right) + \mathrm{Im}\left(J_{yx}^*J_{yy}\right) \tag{51-63}$$

$$M_{20} = \mathrm{Re}\left(J_{xx}J_{yx}^*\right) + \mathrm{Re}\left(J_{xy}J_{yy}^*\right) \tag{51-64}$$

$$M_{21} = \mathrm{Re}\left(J_{xx}J_{yx}^*\right) - \mathrm{Re}\left(J_{xy}J_{yy}^*\right) \tag{51-65}$$

$$M_{22} = \mathrm{Re}\left(J_{xx}J_{yy}^*\right) + \mathrm{Re}\left(J_{xy}J_{yx}^*\right) \tag{51-66}$$

$$M_{23} = -\mathrm{Im}\left(J_{xx}^*J_{yy}\right) + \mathrm{Im}\left(J_{xy}^*J_{yx}\right) \tag{51-67}$$

$$M_{30} = \mathrm{Im}\left(J_{xx}^*J_{yx}\right) + \mathrm{Im}\left(J_{xy}^*J_{yy}\right) \tag{51-68}$$

$$M_{31} = \mathrm{Im}\left(J_{xx}^*J_{yx}\right) - \mathrm{Im}\left(J_{xy}^*J_{yy}\right) \tag{51-69}$$

$$M_{32} = \mathrm{Im}\left(J_{xx}^*J_{yy}\right) + \mathrm{Im}\left(J_{xy}^*J_{yx}\right) \tag{51-70}$$

$$M_{33} = \mathrm{Re}\left(J_{xx}J_{yy}^*\right) - \mathrm{Re}\left(J_{xy}J_{yx}^*\right) \tag{51-71}$$

When optical elements or systems exhibit depolarization, then (51-56)–(51-71) do not apply. The Jones formalism does not possess a method for depolarization. In the case of depolarization, more than from 7 up to 16 real elements of the Mueller matrix can be independent of each other.

51.3 Polarizing Elements

Polarimetry relies on the use of polarizing elements which change the state of polarization of a source or of light returning from a sample in a specific way. We can distinguish between polarizers and retarders which, in general, damp or retard right-elliptically polarized light in a different way from left-elliptically polarized light. There are special cases of linear polarizers and retarders which damp and retard orthogonal linearly polarized light differently, or circular polarizers and retarders, when referring to right- and left-circular polarized light [51-3] and [51-8]. Since linear polarizers and retarders are the most common the word "linear" is usually omitted.

51.3.1 Polarizers

An ideal polarizer transmits one unique state of polarization totally, while all other states are blocked. In practice, polarizers are not ideal, as other states will be transmitted. The transmittivity T for the radiant power of an imperfect linear polarizer is given by

$$T = (T_{\max} - T_{\min}) \cos^2\beta + T_{\min} \tag{51-72}$$

with $T_{\min}$ and $T_{\max}$ as the principal transmittances of the polarizer and β as the angle between the polarization azimuth of the incident beam and the polarizer's transmission axis. Referring to the Jones formalism in (51-39) and (51-40) $T_{\min}$ and $T_{\max}$ are identical to t_1^2 and t_2^2. We can define the diattenuation ΔT_D by

$$\Delta T_D = \frac{T_{\max} - T_{\min}}{T_{\max} + T_{\min}} \tag{51-73}$$

Ideal polarizers have $\Delta T_D = 1$, and the diattenuations of real polarizers should approach 1. Most optical interfaces exhibit linear diattenuation, because their Fresnel reflection and transmission coefficients for non-normal incidence are different for s- and p-polarization. High-quality polarizers make use of these effects and achieve high diattenuations. Most commercial polarizers, however, exploit the dichroism of optical materials as a result of polarization-dependent reflection or refraction in birefringent crystals or dielectric thin-film structures.

Polarization by Reflection

According to Fresnel's formulas, light is polarized when reflected or transmitted at an optical surface. Full polarization is achieved when light is reflected at the Brewster angle for which the refracted and reflected beams are perpendicular to each other. In this case no p-polarization is reflected.

A simple polarizer can be made by tilting a stack of glass plates [51-11]. Depending on the incidence angle, transmitted and reflected light are more or less polarized. Figure 51-3 shows the arrangement.

When tilted near the Brewster angle each reflection depletes the incident beam of s-polarized light, leaving a greater fraction of p-polarized light in the transmitted beam. To achieve a high degree of polarization for the transmitted light it takes at least ten plates. The reflected beam will be fully polarized, but it will be spread out and may not be very useful.

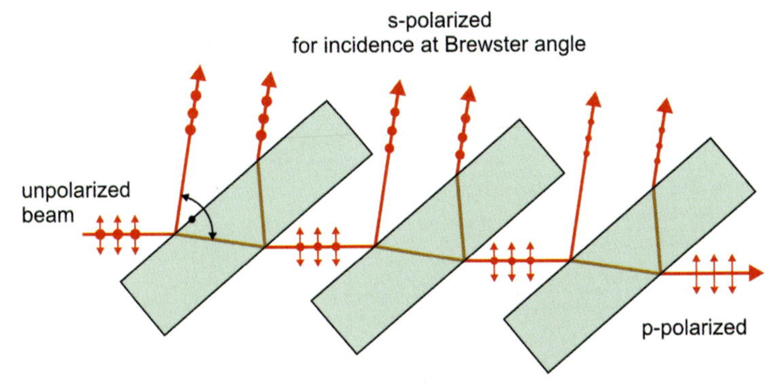

Figure 51-3: A stack of tilted glass plates used to generate light of high diattenuation (degree of polarization) in transmission and reflection.

A better diattenuation for transmitted light is obtained by tilting the stack to steeper angles at the expense of losing overall transmission (figure 51-4 (a), (b)). When piling up 20 plates, 50% of the incident light is transmitted thereby achieving nearly perfect polarization at the Brewster angle.

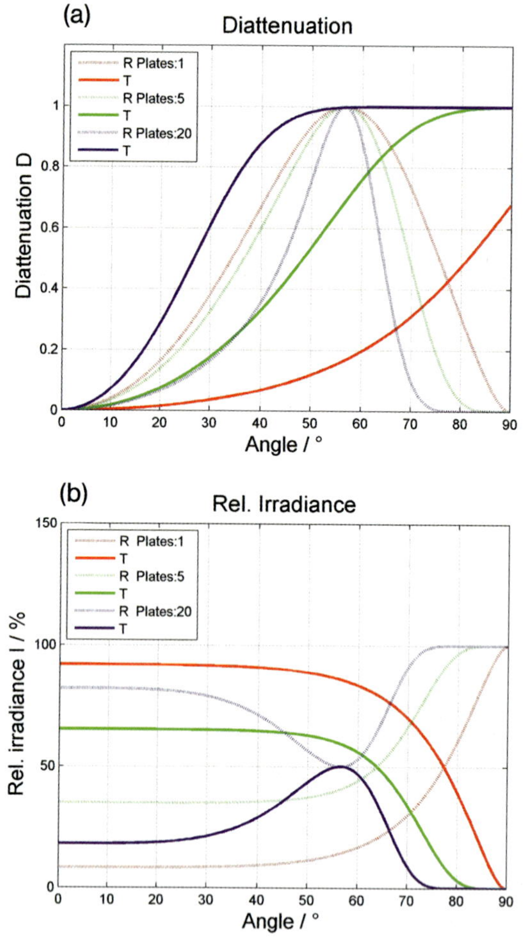

Figure 51-4: (a) diattenuation and (b) relative irradiance for reflected (dotted curve) and transmitted (solid curve) light at a stack of plates consisting of 1 (red), 5 (green) and 20 (blue) individual plates. The tilted glass plates are used to generate light of high diattenuation (degree of polarization) during transmission and reflection.

Polarizing Beam-Splitters

Beam-splitting polarizers split the incident unpolarized beam into two beams of different polarization. In the ideal case a polarizing beam-splitter would polarize totally, generating beams of orthogonal polarizations. Depending on the polarization principle and the optical arrangement sometimes only one of the two output beams is fully polarized, whereas the other contains a mixture of polarization states [51-4] and, [51-12].

Beam-splitting polarizers theoretically do not absorb light in the same way as dichroitic polarizers. They are therefore more suitable for use with high-intensity beams. Polarizing beam-splitters are also useful in applications, where the two orthogonal polarization states are needed for simultaneous analysis. The large group of birefringent beam-splitting polarizers is discussed in the next section.

Birefringent Polarizers

A linearly birefringent crystal, such as calcite, will divide an entering beam of monochromatic light into two beams with orthogonal polarizations. The beams will usually propagate in different directions and have different propagation speeds. Depending on whether the crystal is uniaxial or biaxial, there will be one or two directions along which the beams will travel at the same speed. These directions define the optical axes of the crystal. If a beam enters a birefringent plane-parallel plate in a direction other than the optical axis, the beam will split and emerge as two separate, orthogonally polarized beams. The two beams are referred to as the ordinary and extraordinary beam, respectively. The refractive index experienced by the ordinary beam is denoted by n_o, the most extreme extraordinary beam index by n_e. The direction of the lower index is called the fast axis, that of the higher index is called the slow axis. The polarization of the extraordinary beam lies in the plane containing the direction of propagation and the optical axis, and the polarization of the ordinary beam is perpendicular to this plane [51-3] and [51-4].

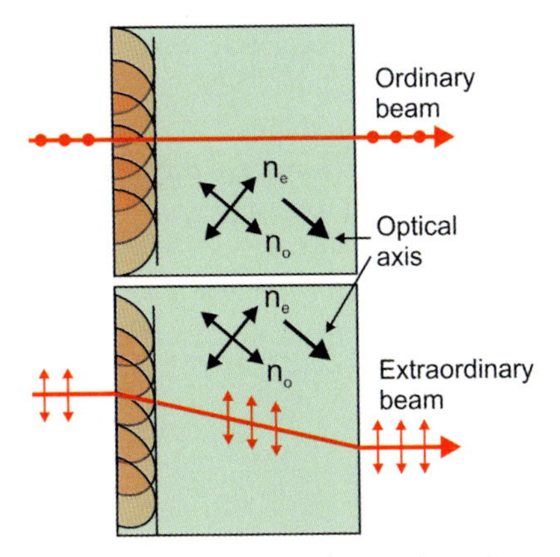

Figure 51-5: Ordinary and extraordinary beam generated at a birefringent crystal with oblique optical axis.

There are a variety of polarizers made of birefringent crystals which are used for various applications. Some of them are mentioned in the following section. A particular form of birefringent polarizer is made using birefringent prisms. Many polarizing prisms are made from calcite, a rhombohedral crystalline form of calcium carbonate. Since calcite is a naturally occurring material, imperfections are quite usual. The highest quality materials are difficult to find and are more expensive than those with some defects. Calcite transmits from 250 nm to 2 μm wavelength. Magnesium fluoride can be used for wavelengths from 140 nm to 7 μm.

Among the different prism polarizers the most useful are those where the entrance and exit faces are perpendicular to the direction of use.

Nicol Prism

A Nicol prism (figure 51-6 (a)) consists of a pair of identical prisms, which have been cut from calcite and rejoined with Canada balsam. The crystal is cut such that the o- and e-beams are in orthogonal linear polarization states. The o-beam is totally reflected at the balsam interface, since it experiences a larger refractive index in calcite than in the balsam. It is then deflected to the side of the crystal. The e-beam "sees" a smaller refractive index in the calcite and is transmitted through the interface without deflection. Nicol prisms produce polarized light of very high purity [51-4].

In modern times these have been replaced by Glan-type prisms, which have entrance and exit faces normal to the transmitted beam, although they are not true polarizing beam-splitters because only the transmitted beam is fully polarized.

Glan–Taylor Prism

A Glan–Taylor prism (figure 51-6 (b)) transmits the extraordinary beam, while the ordinary beam is deflected. It consists of two birefringent prisms with parallel optical axes. The axes are parallel to the entrance and exit faces but oblique to the inner hypotenuse surface. The prism elements are separated by an air space, which makes them useful for higher power applications but limits the angular field to less than 8°. The angular field of view is asymmetrical about the mechanical axis and is a function of wavelength. The usual mechanical mounts and housings are designed to absorb the deflected ordinary beam. Uncoated Glan–Taylor prisms have a useful wavelength range of 250–2300 nm. Most prisms are supplied with anti-reflection coatings on the entrance and exit faces. They generate a high degree of polarization purity and high total transmission. Extinction ratios of 10^{-5}–10^{-7} are typical, extensions below 10^{-7} are possible [51-4] and[51-11].

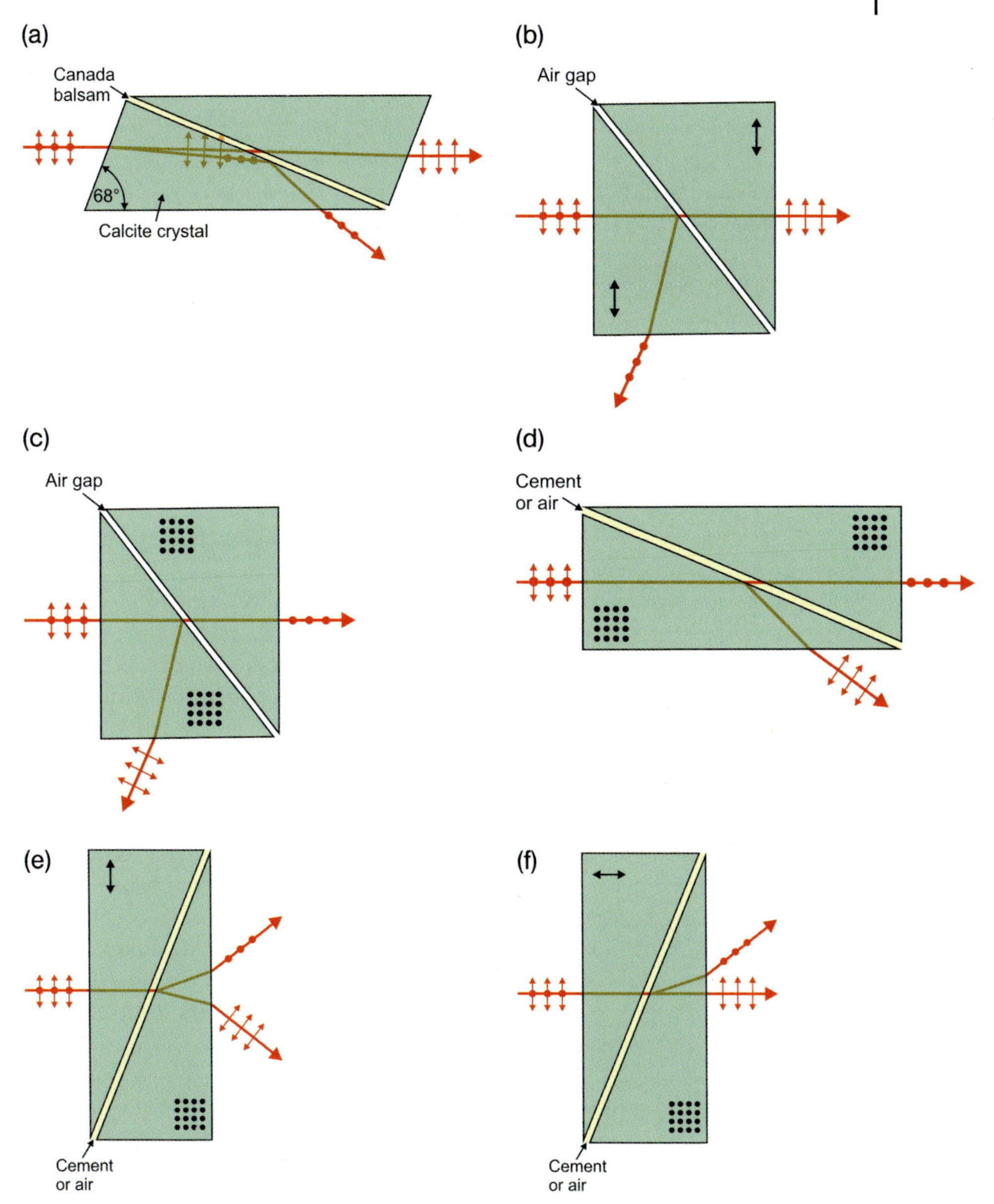

Figure 51-6: Various prism polarizers: (a) Nicol, (b) Glan–Taylor, (c) Glan–Foucault, (d) Glan–Thompson, (e) Wollaston, (f) Rochon.

Glan–Foucault Prism

A Glan–Foucault prism (figure 51-6 (c)) transmits the ordinary beam, while the extraordinary beam is deflected. It consists of two birefringent prisms with parallel optical axes. Contrary to the Glan–Taylor prism, the axes are parallel to the entrance and exit faces and also parallel to the inner hypotenuse surface. Prism elements are also separated by an air space. The characteristics are similar to those of the Glan–Taylor prism [51-4].

Glan–Thompson Prism

A Glan-Thompson prism (fig. 51-6 (d)) transmits the ordinary beam, while the extraordinary beam is deflected. It consists of two birefringent prisms with parallel optical axes. Contrary to the Glan–Foucault prism, the prism elements are cemented together at the hypotenuse surfaces. The optical axes are parallel to the entrance and exit faces and also parallel to the inner hypotenuse surface.

Glan–Thompson prisms are recommended for low-to-medium power application requiring a large field of view and a high degree of polarization purity. The asymmetrical field of view is enlarged to approximately 30° depending on the length/aperture ratio and the wavelength.

The usual mechanical mounts and housings are designed either to absorb the deflected extraordinary beam or to let it escape through an appropriate aperture. Uncoated Glan–Thompson prisms have an effective wavelength range of 250 nm–2300 nm. Most prisms are supplied with anti-reflection coatings on the entrance and exit faces. They generate a high degree of polarization purity and high total transmission. Extinction ratios of 10^{-5}–10^{-6} are typical [51-3] and [51-4].

Wollaston Prism

A Wollaston prism (figure 51-6 (e)) transmits ordinary and extraordinary beams and deflects the ordinary beam in a direction different from the incident beam. It consists of two triangular calcite prisms with orthogonal crystal axes that are cemented together. The axis of the first prism is perpendicular to the direction of the incident beam. At the internal interface, an unpolarized beam splits into two linearly polarized beams, which leave the prism at a divergence angle of 15°–45° depending on the prism angle and wavelength. Both beams deviate from the axis of the incoming beam.

Wollaston prisms provide a simple way of splitting a beam of light into two mutually orthogonal, linearly polarized beams that are separated by an angle. Uncoated Wollaston prisms have an effective wavelength range of from 250 nm–2300 nm. Most prisms are supplied with anti-reflection coatings on the entrance and exit face.

The Wollaston prism is used to compare the intensities of both polarization states by measuring irradiances in both the exiting beams. The irradiance of the two output beams will be equal if the input beam is unpolarized, but they will differ if the input beam is polarized. Extinction ratios of 10^{-5} are typical [51-3] and [51-4].

Rochon Prism

A Rochon prism (figure 51-6 (f)) transmits ordinary and extraordinary beam and deflects them in different directions. It consists of two triangular calcite prisms with orthogonal crystal axes that are cemented together. The axis of the first prism shows into the direction of the incident beam. In the first Rochon prism element, both the ordinary and the extraordinary beams travel with the same velocity. In the prism element the ordinary beam continues at the same speed, whereas the extraordinary beam travels more rapidly and therefore is deviated by an amount that depends on the angle of the interface. Characteristics are similar to those of the Wollaston prism [51-4].

Thin-Film or Interference Polarizer

Dielectric stacks of thin-film coatings made of alternating high and low refractive index layers with quarter-wave optical thickness can be designed to provide reflectances and transmittances with large diattenuations. When positioned on suitable plane plates or wedged glass substrates, which are inclined with respect to the incident beam, they are known as thin-film polarizers or interference polarizers (figure 51-7) [51-12].

Sometimes the dielectric stack is deposited onto a wedge of glass that is cemented to a second wedge to form a cube with the stack cutting diagonally across the center (polarizing beam-splitter cube, see next section).

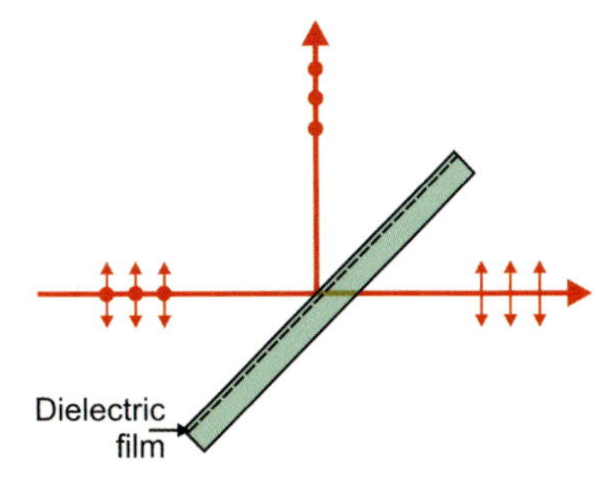

Figure 51-7: Polarizing thin-film beam-splitter plate.

Thin-film polarizers generally do not perform as well as Glan-type polarizers, but they are inexpensive and provide two beams that are fairly equally well-polarized. The polarizing beam-splitter cubes generally perform better than the plate polarizers. The extinction of thin-film polarizers is limited by the defects in the coating layers and the quality of the substrate material. Typical extinction ratios for reflectance or transmittance are 10^{-2}.

Polarizing Beam-Splitter Cubes

A polarizing beam-splitter cube is a special application of a thin-film polarizer. It consists of a pair of right-angled prisms made from optical glass or quartz cemented together at their hypotenuse surface with a special multi-layer dielectric film inbetween. They separate monochromatic unpolarized light into two highly polarized output beams separated by an almost 90° angle. The extinction ratios for reflectance and transmittance are 10^{-2}–10^{-3}. Figure 51-8 shows the principle.

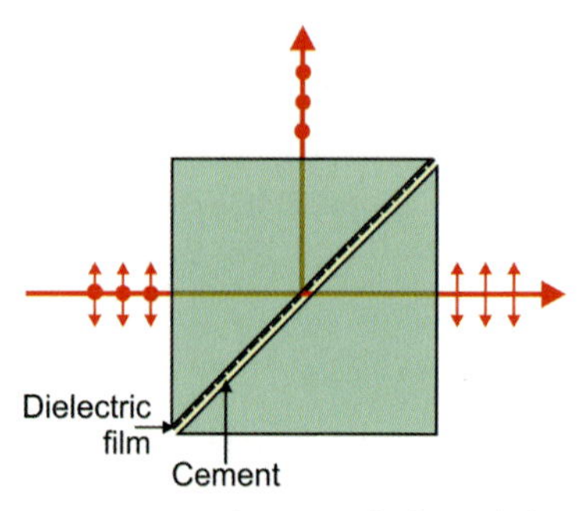

Figure 51-8: Polarizing cube beam-splitter.

Since the thickness of the multi-layer beam division film is only a few wavelengths, there are no ghost images disturbing the application. Attention is necessary, however, especially in interferometric applications, concerning the cement, which might have a slight different index of refraction sometimes causing unwanted interference effects. In these cases the user has to properly decide, which side the dielectric film must face.

Polarizing beam-splitters usually are supplied with antireflection coatings at their entrance and exit faces [51-11].

Wire-Grid Polarizer

A wire-grid polarizer (figure 51-9) consists of a series of fine parallel metallic lines which are coated onto a transparent material, usually glass. The wire-grid polarizer acts like a linear polarizer, when the wire spacings are smaller than half the wavelength and when the duty cycle of the wire grid is near 0.5. Light polarized parallel to the wires (s-polarization) is reflected from the surface, whereas light polarized perpendicular to the wires (p-polarization) is mostly transmitted [51-16].

If the wire grid period is longer than twice the wavelength, the polarizing effect disappears and the wire grid acts as a regular diffraction grating.

The parameters determining the quality of the wire-grid polarizer are:

- the wire-grid period (which should be $< \lambda/2$),
- n and k for the wire metal,
- the wavelength of the light,
- n for the substrate and the material between the wires (usually air).

Usually aluminum or gold are used as metals for the wires. When used as a polarizing beam-splitter cube, the polarization directions selected are set by the orientation of the wires and not by the orientation of the plane of incidence. Consequently, there is no rotation of the plane of polarization for skew rays when the polarizer is used in an imaging system. The polarizer can thus split along any orthogonal direction designated by the wire-grid direction.

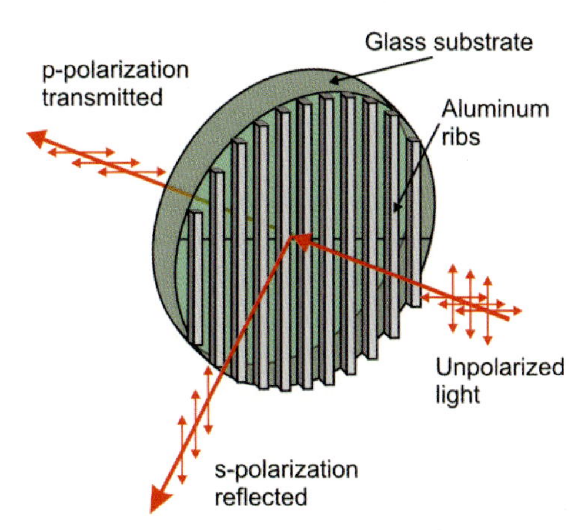

Figure 51-9: Wire-grid polarizer.

Wire-grid polarizers achieve purities similar to those of thin-film polarizers for wavelengths longer than about 700 nm up to at least 1550 nm. The transmitted beam can be used over a large range of angles, i.e., ± 30° in air. Absorption of the incident light is approximately 10%–20% which makes the elements critical for high-power applications or those depending on high efficiencies.

Dichroic Polarizers

When a light beam passes through a dichroic polarizer, one of the orthogonal polarization components of the light is strongly absorbed while the other passes across with only weak absorption (figure 51-10).

Certain crystals show dichroism and can therefore be used as polarizers. The best- known crystal of this type is tourmaline, which is, however, seldom used as a polarizer, since the dichroic effect is strongly wavelength-dependent and the crystal appears colored. Herapathite is also dichroic, and is not strongly colored, but is difficult to grow in large crystals.

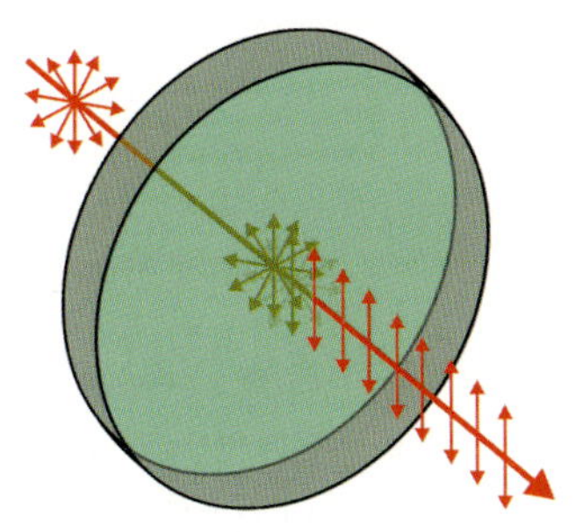

Figure 51-10: Dichroic polarizer converting random polarized light to linear polarized light.

Dichroic polarizers are made of materials, which consist of optically anisotropic molecules or crystals that are oriented in a defined direction within a plastic or glass matrix. The most famous dichroic polarization filter is the Polaroid filter invented by Edwin Land, which consists of many microscopic crystals of iodoquinine sulfate (herapathite) embedded in a transparent nitrocellulose polymer film. The needle-like crystals are aligned during the manufacture of the film by stretching or by applying electric or magnetic fields, making the filter dichroic.

An improved type of absorptive polarizer is made of elongated silver nanoparticles embedded in glass. These polarizers are more durable and can polarize light to a higher degree with low absorption of correctly-polarized light. Such glass polarizers are widely used in optical fiber communications.

Dichroic sheet polarizers are used to convert random polarization light to linear polarized light. Transmission for unpolarized light is typically 48%. Since dichroic polarizers operate by selective absorption they cannot be used in high-power applications. Compared with polarization prisms, dichroic sheet polarizers offer a much larger diameter and field angles of up to ± 20°. Extinction ratios between 10^{-2} and 10^{-5} are possible [51-4].

51.3.2 Retarders

Retarders are optical components which induce an optical path difference between the orthogonally polarized components of a wave. The path difference is called retardation. If the path difference occurs between linearly polarized components we refer to linear retardation (figure 51-11 (b)), if it occurs between right and left-circular polarized components we refer to circular retardation (figure 51-11 (a)). The general case is elliptical retardation with a path difference between the right and left-elliptically polarized components.

Since retarders can be used to convert an arbitrary polarization state to any other state, they are used for altering and controlling the polarization of a wave [51-4].

Linear Retarders

Waveplates

Waveplates are linear retarders made from birefringent materials, such as crystals or plastic sheets. The latter are stretched to produce an anisotropy that gives rise to birefringence. There are a variety of birefringent (uniaxial) crystals within which crystalline quartz and mica are the materials commonly used for waveplates. Crystalline quartz has a difference in refractive index of $\Delta n = n_e - n_o \approx 0.009$ for red light, whereas that for mica is $\Delta n \approx 0.033$.

Mica is a mineral that can be split into very thin sheets. Clear mica is called Muscovite because it is found near Moscow in Russia. It is used for broadband waveplates, because its principal indices of refraction vary slowly across the visible spectrum. A retarder made for 550 nm and normal incidence will produce approximately the same retardation in the spectrum from 400 nm–700 nm. Quartz elements are recommended for higher-power applications.

Usually a thin mica or quartz disk is sandwiched between optical glass disks for protection on both sides and is oriented so that the optical axes are orthogonal.

Note that Muscovite is, in fact, biaxial or tri-refringent, whereas crystalline quartz is optically active, which means that the plane of linear polarization rotates with the distance through the material. However, neither of these effects will usually disturb their function as proper thin waveplates.

From (51-21) we can determine the thickness of a birefringent plate to introduce a certain retardation. If the retardation is π the plate is called a zero-order half-waveplate (sometimes called a first-order half-waveplate). If the retardation is $\pi/2$ the plate is called a zero-order quarter-waveplate. If the phase difference at emergence is some multiple of π or $\pi/2$ the plate is called a multi-order or higher-order waveplate. The thickness d of a plate of mth-order is determined from

$$d = \left(\frac{\Delta\phi}{2\pi} + m\right)\frac{\lambda}{(n_1 - n_2)} \qquad (51\text{-}74)$$

A zero-order quarter-waveplate for red light made from quartz would be only 18 μm thick, which is quite difficult to fabricate and handle. Common practice is to fabricate a "pseudo" or compound zero-order waveplate from two thicker quartz plates with their optical axis rotated by 90° around the axis of transmission such that the second plate compensates for the phase shift introduced by the first plate. In this case, only the thickness difference of the two plates must be equal to 18 μm to give a quarter-waveplate, whereas the overall thickness of the element is several mm.

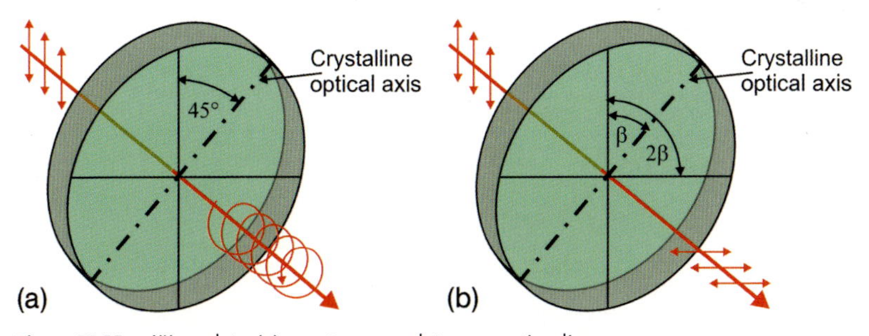

Figure 51-11: Waveplate (a) quarter-waveplate converting linear polarized light to circular polarized light; (b) half-waveplate rotating linear polarized incident light by 2β.

True zero-order retarders are generally required for applications with high retardance stability since they are generally the least sensitive to variations in wavelength, angle of incidence and temperature. Compound zero-order elements are more sensitive to variations in wavelength and angle of incidence. They show the same sensitivity to angle of incidence as multiple-order retarders of the same thickness. Multi-order elements have a poor wavelength and thermal stability and variations in the angle of incidence are critical.

The retardation of waveplates will change with temperature caused by a change in thickness of the retarder and a change in the birefringence of the material. For quartz the change in retardation is approximately 0.0001 δ per K. Therefore, a compound zero-order retarder, made of two plates of the same material, will show the same thermal shift as a true zero-order retarder of the same material [51-4].

Rhombs

Retarders can also be made without using birefringent materials. Taking Fresnel's equations into account, a phase shift between the s- and p-polarized components of a wave's total internal reflection can be used to create optical elements with well-defined retardation. When light is incident at an angle θ_i larger than the critical angle, the retardance $\Delta\phi$ at the reflection is calculated from

$$\Delta\phi = 2\arctan\left(\frac{\cos\theta_i\sqrt{\sin^2\theta_i - \left(\frac{n_t}{n_i}\right)^2}}{\sin^2\theta_i}\right) \tag{51-75}$$

where n_i denotes the refractive index of the inner material and n_t denotes that of the outer medium and where $n_i > n_t$. Figure 51-12 shows $\Delta\phi$ as a function of the incident angle for 632.8 nm wavelength and three different glass materials.

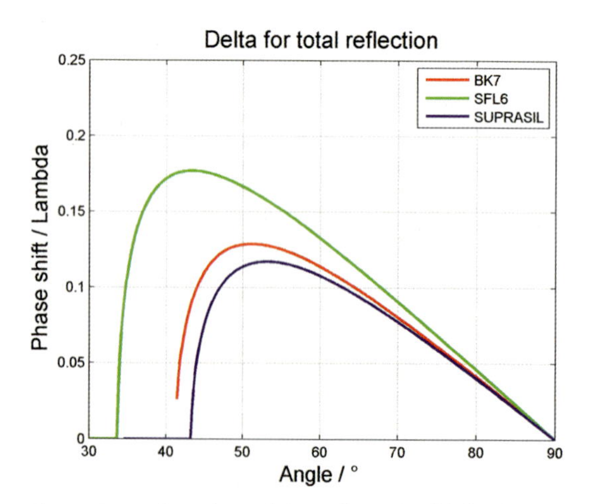

Figure 51-12: Retardance $\Delta\phi$ as a function of θ_i for glasses BK7, SFL6 and Suprasil at $\lambda = 632.8$ nm.

A Fresnel rhomb (figure 51-13) is a solid parallelogram with entrance and exit surfaces normal to the incident and emerging beam and two surfaces at which total internal reflections occur. If the rhomb angle is chosen such that the phase shift between s- and p-polarization after one reflection is equal to $\pi/2$, the Fresnel rhomb has a total retardation of π and thus can be used as a quarter-wave retarder. However, the element is rather sensitive to the angle of incidence and laterally displaces the beam, which makes it unsuitable for applications, where the retarder must be rotated.

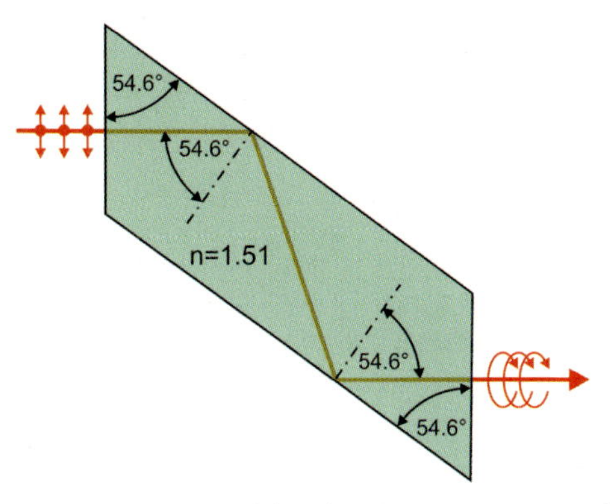

Figure 51-13: Fresnel rhomb as a quarter-wave retarder for red light, made from BK7.

The retardance sensitivity to variations in the incident angle can be greatly reduced when two identical Fresnel rhombs are concatenated to provide a collinear output beam. Figure 51-14 shows a Fresnel double rhomb used to produce a quarter-wave retarder, while figure 51-15 shows an example for a half-wave retarder.

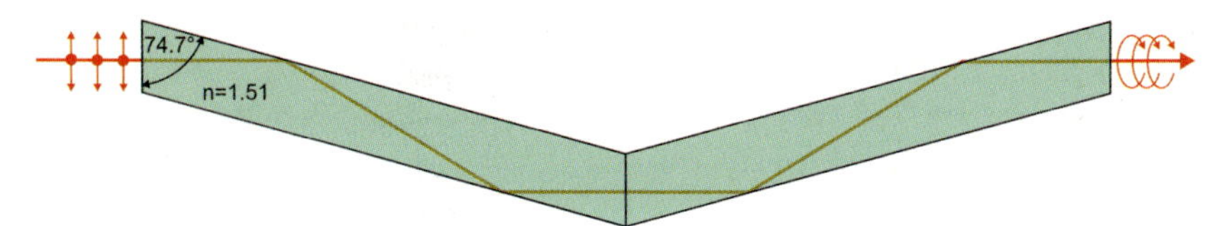

Figure 51-14: Fresnel double rhomb designed as a quarter-wave retarder for red light, made from BK7.

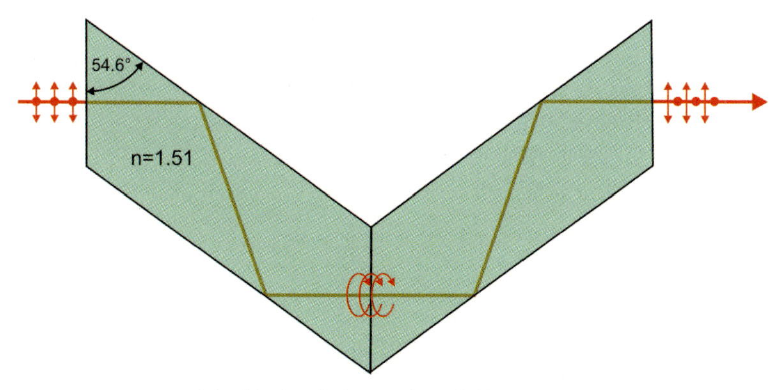

Figure 51-15: Fresnel double rhomb designed as a half-wave retarder for red light, made from BK7.

Since the retardance of total-internal-reflection retarders is independent of the optical path length within the element, they are less sensitive to wavelength variations than waveplates. Their dimensions scale proportionally with the required cross-sections of the entrance and exit surfaces in all three dimensions, which makes them unsuitable when small, compact dimensions are required. When the dimensions are large, birefringence in the bulk glass or that introduced by fabrication or mounting can disturb the retardance and limit the stability of the element for its final function [51-3].

Circular Retarders

Some materials, such as crystalline quartz, exhibit circular birefringence in which the eigenpolarizations are left- and right-circular and the retardance is the phase shift between the eigenpolarizations. The phenomenon is also called *optical activity* and the elements are called optical rotators, because the incident linear polarization will be rotated during passage through the optical element (figure 51-16). Equation (51-76) defines the angle of rotation β for an optical element with thickness d and the refractive indices n_l and n_r for left-circular and right-circular polarizations, respectively [51-9].

$$\beta = \frac{\pi\, d\,(n_l - n_r)}{\lambda} \tag{51-76}$$

A material with $n_l > n_r$ rotates linearly polarized light in a clockwise direction when viewed facing the light source. It is called right-handed or dextro-rotary, whereas a material with $n_l < n_r$ causes a counterclockwise rotation and is called left-handed or levo-rotary. The circular retardance per unit length is called the specific rotary power SRP and is defined as

$$\mathrm{SRP}(\lambda) = \frac{\beta}{d} = \frac{\pi(n_l - n_r)}{\lambda} \tag{51-77}$$

The rotary power and orientation of a material is independent of the propagation direction. A beam reflected back after passage through an optically active material will be restored to the initial azimuth.

Optical activity occurs in materials with asymmetrical molecules in which the electrons are more easily accelerated in one orientation than another. There are many optically active materials in nature due to the fact that almost all amino acids (the building blocks of life) are left-handed.

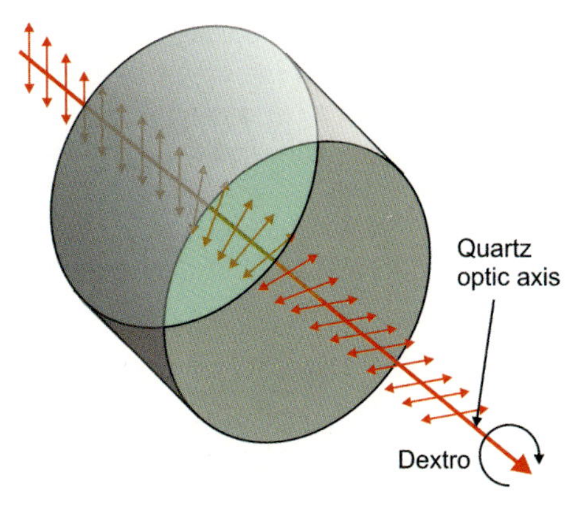

Figure 51-16: Optically active component rotating the polarization clockwise (= dextro-rotary or right-handed).

Electro-optic and Magneto-optic Effects

There are materials in which retardance can be induced by an electric or magnetic field. These elements are exploited to be used as electrically controllable retardance elements.

In non-centrosymmetric crystals such as lithium niobate or gallium arsenide the *Pockels effect* is observed, where retardance can be induced by a constant or varying electrical field proportional to the exciting electrical field. Field application causes isotropic crystals to become uniaxial, and uniaxial crystals to become biaxial crystals. Crystal symmetry determines the direction of the optical axes. The induced retardation depends on the material, the polarization of the incident beam, and the magnitude and direction of the applied electrical field [51-9].

Pockels cells are devices consisting of an electro-optic crystal with some electrodes attached to it, through which a light beam can propagate. The birefringence can be modulated by applying a variable electric voltage to make it a voltage-controlled waveplate. Pockels cells are the basic components of electro-optic modulators used, e.g,. for Q switching lasers.

Common nonlinear crystal materials for Pockels cells are:

- potassium di-deuterium phosphate (KD*P = DKDP),
- potassium titanyl phosphate (KTP),
- β-barium borate (BBO),
- lithium niobate ($LiNbO_3$),
- lithium tantalate ($LiTaO_3$),
- ammonium dihydrogen phosphate ($NH_4H_2PO_4$, ADP).

Pockels cells are available as *longitudinal devices* in which the electric field is applied in the direction of the light beam, or as *transversal devices* in which the electrical field is applied perpendicular to the light beam. They are characterized by their half-wave voltage V_π which must be applied to induce a birefringence of 180°.

The **Kerr effect** is an electro-optic effect occurring in solids, liquids and gases, in which birefringence is excited proportional to the square of the applied electrical field. It needs no requirements concerning the crystal structure of the material, and the induced optical axis is parallel to the field direction. The change in refractive index is given by

$$\Delta n = K \lambda E^2 \tag{51-78}$$

where

K is the Kerr constant.

λ is the wavelength,

E is the amplitude of the applied electrical field,

causing the material to act as a waveplate when the field is applied in a direction perpendicular to the incident light.The effect is obeserved in all materials but is typically smaller than the Pockels effect and is often negligible in Pockels materials [51-9].

Very large Kerr constants can be found in some polar liquids, such as nitrotoluene ($C_7H_7NO_2$) and nitrobenzene ($C_6H_5NO_2$). A Kerr cell is obtained when a glass cell is filled with one of these liquids and a high-voltage field up to 30 kV is applied. Kerr cells respond very quickly to changes in the electric field and can be driven up to 10 GHz.

The *Faraday effect* is an induced circular birefringence proportional to an applied magnetic field B. It can occur in all materials. The sensitivity of a material to the induction of circular birefringence is described by the *Verdet constant V* (in units of radians per tesla per meter). It varies with wavelength and temperature and is tabulated for various materials [51-9].

The induced angle of rotation β of an element of thickness d to which a magnetic flux density B_z in the direction of the beam propagation is applied, is defined by

$$\beta = V\, d\, B_z \tag{51-79}$$

Figure 51-17 shows the arrangement.

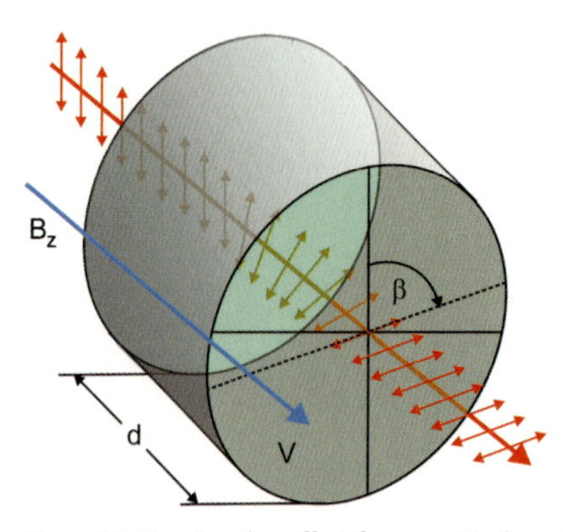

Figure 51-17: Faraday effect for an optical material of thickness d and Verdet constant V induced by a magnetic flux density B_z in the direction of beam propagation.

The sense of Faraday rotation is determined by the direction of the magnetic field. A positive Verdet constant corresponds to left-handed rotation, when the direction of propagation is parallel to the magnetic field and to right-handed rotation when the direction of propagation is anti-parallel. Thus, if a light beam passes through a material and is reflected back through it, the rotation doubles. Note that this is different from circular birefringent materials, where the rotary power and orientation of a material is independent of the propagation direction.

This property is exploited in optical isolators and components that transmit light in only one direction, for instance the Faraday isolator.

Faraday mirrors can be made by combining a 45° Faraday rotator and a plane mirror. Polarized light passing through an arbitrary retarder, then being reflected by a Faraday mirror and repassing the arbitrary retarder will always exit with a fixed polarization. For instance, if the entering light is linearly polarized, the emerging light will be orthogonally linearly polarized, independent of the arbitrary retardation during its passage. Faraday mirrors are implanted in fiber optic systems to control bend-induced retardances in ordinary optical fibers [51-16].

51.3.3
Compensators

Compensators are variable linear retarders, which can be adjusted over a continuous range of retardations. The Babinet compensator (figure 51-18 (a)) is a device containing two opposed quartz wedges of equal angle with an air gap in between. One wedge can be moved along its length by a micrometer screw or a motoric drive. The wedges are cut so that the axis of the first wedge is oriented along, the axis of the second wedge is perpendicular to the incident beam. The individual wedges introduce opposite signs of retardance such that the resulting retardance is equal to the difference between the individual retardances. The magnitudes depend on the thickness of each wedge being traversed by the light beam. The moving wedge provides a variable thickness and therefore a variable retardance. The Babinet compensator has the advantage of leaving the input and output beam collinear; however, the retardance varies across the optical beam, making it suitable only for beams with small diameters.

The Babinet–Soleil compensator (figure 51-18 (b)) contains two opposed quartz wedges of equal angle with the optical axes oriented parallel to each other and perpendicular to the incident beam. The moving wedge changes the total thickness and retardance of the combined retarder. The total thickness of the two wedges is constant over the full aperture. To improve performance, a compound zero-order retarder is placed in front of the wedges with its optical axis perpendicular to those of the wedges [51-3].

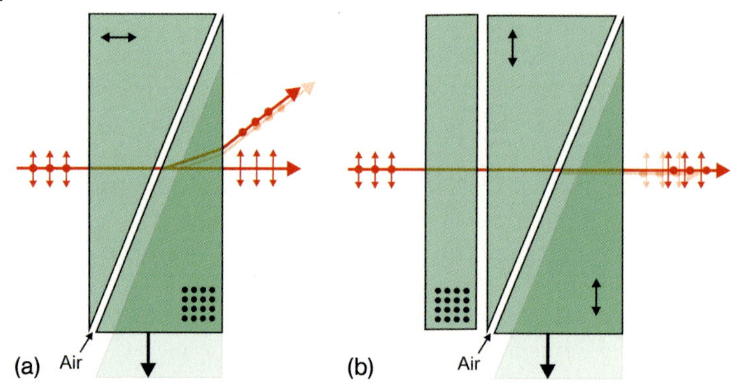

Figure 51-18: Compensators for variable adjustment of retardance: (a) Babinet compensator, (b) Babinet–Soleil compensator.

Photoelastic Modulators

A photoelastic modulator (PEM) is a compensator with variable retardance caused by mechanical stress applied to a normally isotropic material. An isotropic material, such as fused quartz, will become anisotropic when stressed und thus will induce birefringence as an anisotropic crystal such as calcite. Figure 51-19 shows the principle of a PEM consisting of a piezoelectric transducer and a block of fused quartz cemented to each other. The piezoelectric transducer is a block of crystalline quartz cut in a specific orientation with metal electrodes deposited on each of the two sides. It is cut in such a way that it resonates when a 50 kHz electric field is applied. The resonance in the transducer causes a periodic strain in the longitudinal direction of the fused quartz block which is cemented to it. A light beam entering from the side experiences a change in index of refraction in the horizontal (longitudinal) direction, whereas the vertical direction remains unchanged. The perpendicular polarization components traversing the fused quartz block are therefore periodically phase shifted. The phase shift is positive when the fused quartz block is compressed, and negative when it is in tension. The frequency of oscillation is determined by the size and shape of the crystals and the amplitude by the voltage applied at the resonant frequency [51-3], [51-17] and [51-18].

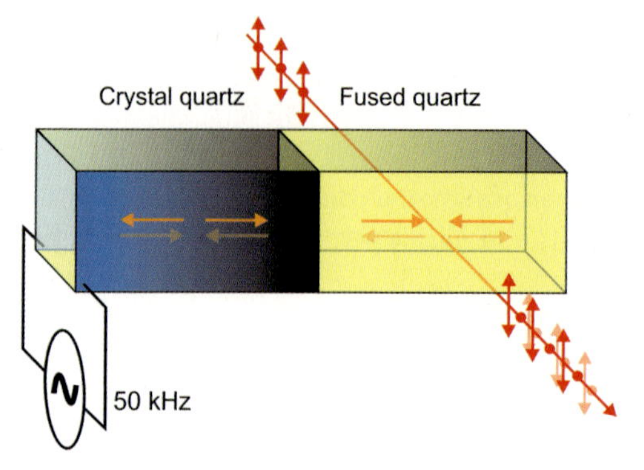

Figure 51-19: Photoelastic modulator (PEM) used to provide periodically varying retardance.

51.3.4 Depolarizers

True depolarization of a polarized light beam is, in practice, difficult to achieve. Common methods such as using an integrating sphere often result in an unacceptable loss in irradiance, especially when collimated light is necessary.

There are a variety of pseudo-depolarizers that vary polarization states over time, wavelength or the beam cross-section. For example, the retardance across a wedged waveplate varies with the local thickness. An incident beam will exit with spatially varying polarization. The method is quite simple and often satisfies the requirement for unpolarized light for an application.

In some applications it is possible to vary the polarization rapidly so that a detector integrates over many different states of polarization. For instance, a rotating half-waveplate produces polarization, which is periodic in time to, effectively, scrambled polarization states for sufficiently slow detectors.

Polarization-maintaining fibers have about one wavelength of retardation per several millimeters length. A fiber bundle of statistically varying fibers can be used as a depolarizer with laterally varying polarization.

In the following, the main pseudo-depolarizer elements are briefly explained.

Cornu Depolarizer

The Cornu depolarizer consists of a pair of 45° crystal quartz prisms optically in contact at their hypothenuse surface to form a cuboid. The fast axes are perpendicular to each other and inclined at 45° relative to the sides of the depolarizer (figure 51-20). Each prism acts as a waveplate with retardance variation in the y-direction across the beam aperture [51-19]. The resulting retardance variation $\Delta\phi$ is expressed by

$$\Delta\phi(y) = 2y \tan a \frac{2\pi}{\lambda}(n_1 - n_2) \tag{51-80}$$

where n_1, n_2 denote the birefringence of the crystal quartz and a denotes the wedge angle. For the Cornu depolarizer, $a = 45°$.

An input beam of uniform polarization will be periodically modulated in the y-direction. Due to dispersion the phase shift is also dependent on the wavelength. The use of two prisms has the advantage that input and output beams are mainly collinear; however, a separation takes place at the 45°-tilted interface because the refractive indices for the ordinary and extraordinary beam are exchanged.

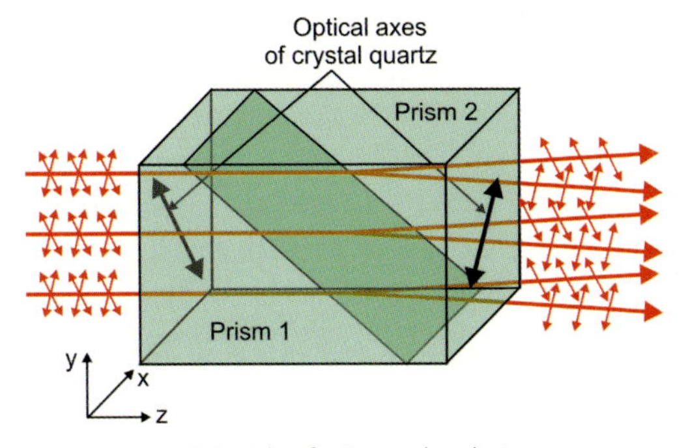

Figure 51-20: Principle of a Cornu depolarizer.

Alternatively, a Cornu depolarizer can comprise two prisms of optically active material having polarization with opposite directions of rotation. For example, one prism might consist of left-handed quartz, and the other of right-handed quartz. In both prisms the main axis extends parallel to the propagation direction of the incident light. Each prism acts as an optical rotator with a rotation variation in the y-direction across the beam aperture. The resulting rotation variation $\beta(y)$ is expressed by

$$\beta(y) = 2y \tan\alpha \frac{\pi}{\lambda}(n_l - n_r) \tag{51-81}$$

with refractive indices n_l and n_r for left-circular and right-circular polarizations, respectively, and where α denotes the wedge angle.

An input beam of uniform polarization will be differently rotated in the y-direction. Due to dispersion the rotation is also dependent on the wavelength.

Lyot Depolarizer

The Lyot depolarizer consists of two waveplates with their fast axes crossing at 45°. The second plate has twice the thickness of the first (figure 51-21). The polarization of the emerging beam is periodic in wavelength, which is why the depolarizer is only suitable for use in white light where it performs a temporal decorrelation of any input polarization state [51-20].

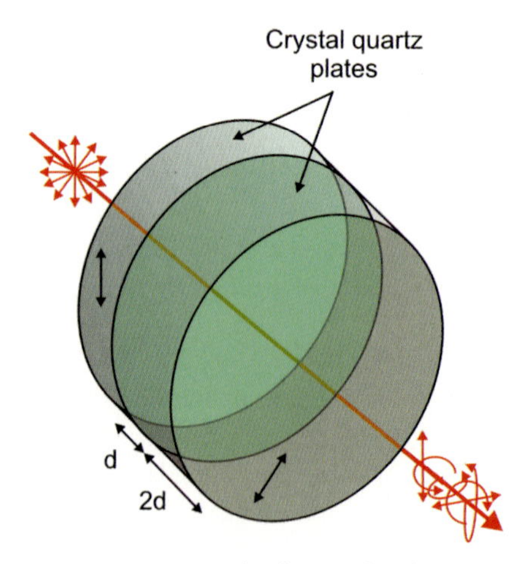

Figure 51-21: Principle of a Lyot depolarizer.

Quartz-Silica Wedge Depolarizer

The quartz-silica wedge depolarizer is based on concatenated wedges as in the case of the Cornu depolarizer. The angle of the two wedges is much smaller, typically around 2°, which makes it very compact (figure 51-22). Only the first component is birefringent. The second component is made of fused silica having a similar refractive index, which, however, does not introduce additional retardation. Generally the fast axis of the quartz element is rotated by 45° with respect to the wedge.

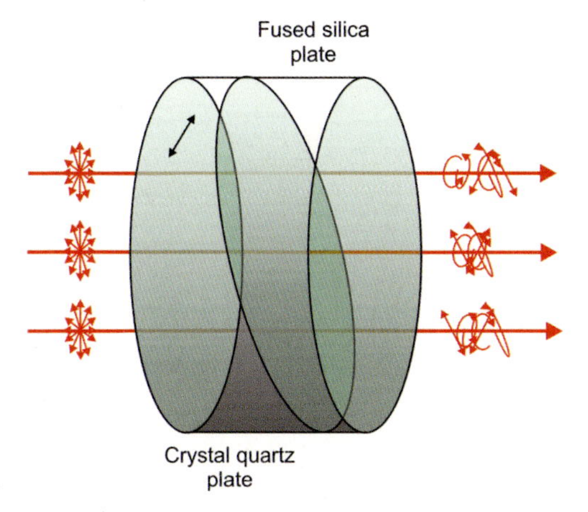

Figure 51-22: Principle of a Quartz–Silica wedge depolarizer.

As with the Cornu depolarizer, there is some separation of the emerging beams depending on the entering polarization. There is also some beam deviation in the order of 10 arcmin due to the imperfect match in refractive index of the cement between the quartz and silica. The cement will usually also limit the broad-band transmission of the depolarizer.

As with the Cornu depolarizer the input beam of uniform polarization will be periodically modulated in the y-direction. But because the wedge angle is so much smaller than in a Cornu depolarizer, the period is larger, usually around several mm. The depolarizer has a preferred orientation because of its single defined fast axis, which is usually marked [51-21].

51.3.5 Jones and Mueller Matrix Representations of Selected Optical Components

The following table 51-2 gives an overview of selected polarizing components and their Jones and Mueller matrices.

Table 51-2: Selected polarizing components and their Jones and Mueller matrices.

Component	Jones Matrix	Mueller Matrix
Nonpolarizing element	$\begin{pmatrix} 1 & 0 \\ 0 & 1 \end{pmatrix}$	$\begin{pmatrix} 1 & 0 & 0 & 0 \\ 0 & 1 & 0 & 0 \\ 0 & 0 & 1 & 0 \\ 0 & 0 & 0 & 1 \end{pmatrix}$
Absorber with amplitude transmittance $0<t<1$	$t\begin{pmatrix} 1 & 0 \\ 0 & 1 \end{pmatrix}$	$t^2\begin{pmatrix} 1 & 0 & 0 & 0 \\ 0 & 1 & 0 & 0 \\ 0 & 0 & 1 & 0 \\ 0 & 0 & 0 & 1 \end{pmatrix}$
Linear polarizer at 0°	$\begin{pmatrix} 1 & 0 \\ 0 & 0 \end{pmatrix}$	$\frac{1}{2}\begin{pmatrix} 1 & 1 & 0 & 0 \\ 1 & 1 & 0 & 0 \\ 0 & 0 & 0 & 0 \\ 0 & 0 & 0 & 0 \end{pmatrix}$
Linear polarizer at angle a	$\begin{pmatrix} \cos^2 a & \sin a \cos a \\ \sin a \cos a & \sin^2 a \end{pmatrix}$	$\frac{1}{2}\begin{pmatrix} 1 & \cos 2a & \sin 2a & 0 \\ \cos 2a & \cos^2 2a & \sin 2a \cos 2a & 0 \\ \sin 2a & \sin 2a \cos 2a & \sin^2 2a & 0 \\ 0 & 0 & 0 & 0 \end{pmatrix}$
Linear diattenuator at 0°	$\begin{pmatrix} t_1 & 0 \\ 0 & t_2 \end{pmatrix}$	$\frac{1}{2}\begin{pmatrix} t_1^2 + t_2^2 & t_1^2 - t_2^2 & 0 & 0 \\ t_1^2 - t_2^2 & t_1^2 + t_2^2 & 0 & 0 \\ 0 & 0 & 2t_1t_2 & 0 \\ 0 & 0 & 0 & 2t_1t_2 \end{pmatrix}$
Quarter-wave $\left(\Delta\phi = \frac{\pi}{2}\right)$ linear retarder, fast axis at 0°	$\frac{1}{\sqrt{2}}\begin{pmatrix} 1+i & 0 \\ 0 & 1-i \end{pmatrix}$	$\begin{pmatrix} 1 & 0 & 0 & 0 \\ 0 & 1 & 0 & 0 \\ 0 & 0 & 0 & 1 \\ 0 & 0 & -1 & 0 \end{pmatrix}$

Component	Jones Matrix	Mueller Matrix
Half-wave ($\Delta\phi = \pi$) linear retarder, fast axis at 0°	$\begin{pmatrix} i & 0 \\ 0 & -i \end{pmatrix}$	$\begin{pmatrix} 1 & 0 & 0 & 0 \\ 0 & 1 & 0 & 0 \\ 0 & 0 & -1 & 0 \\ 0 & 0 & 0 & -1 \end{pmatrix}$
General linear retardance by $\Delta\phi$, fast axis at 0°	$\begin{pmatrix} e^{i\frac{\Delta\phi}{2}} & 0 \\ 0 & e^{-i\frac{\Delta\phi}{2}} \end{pmatrix}$	$\begin{pmatrix} 1 & 0 & 0 & 0 \\ 0 & 1 & 0 & 0 \\ 0 & 0 & \cos\Delta\phi & \sin\Delta\phi \\ 0 & 0 & -\sin\Delta\phi & \cos\Delta\phi \end{pmatrix}$
General linear retardance by $\Delta\phi$, fast axis at a	$\begin{pmatrix} e^{i\frac{\Delta\phi}{2}}\cos^2 a + e^{-i\frac{\Delta\phi}{2}}\sin^2 a & i\sin\frac{\Delta\phi}{2}\sin 2a \\ i\sin\frac{\Delta\phi}{2}\sin 2a & e^{i\frac{\Delta\phi}{2}}\sin^2 a + e^{-i\frac{\Delta\phi}{2}}\cos^2 a \end{pmatrix}$	$\begin{pmatrix} 1 & 0 & 0 & 0 \\ 0 & \cos 4a\sin^2\frac{\Delta\phi}{2} + \cos^2\frac{\Delta\phi}{2} & \sin 4a\sin^2\frac{\Delta\phi}{2} & -\sin 2a\sin\Delta\phi \\ 0 & \sin 4a\sin^2\frac{\Delta\phi}{2} & -\cos 4a\sin^2\frac{\Delta\phi}{2} + \cos^2\frac{\Delta\phi}{2} & \cos 2a\sin\Delta\phi \\ 0 & \sin 2a\sin\Delta\phi & -\cos 2a\sin\Delta\phi & \cos\Delta\phi \end{pmatrix}$
Mirror	$\begin{pmatrix} 1 & 0 \\ 0 & -1 \end{pmatrix}$	$\begin{pmatrix} 1 & 0 & 0 & 0 \\ 0 & 1 & 0 & 0 \\ 0 & 0 & -1 & 0 \\ 0 & 0 & 0 & -1 \end{pmatrix}$
Right circular polarizer	$\frac{1}{2}\begin{pmatrix} 1 & -i \\ i & 1 \end{pmatrix}$	$\frac{1}{2}\begin{pmatrix} 1 & 0 & 0 & -1 \\ 0 & 0 & 0 & 0 \\ 0 & 0 & 0 & 0 \\ -1 & 0 & 0 & 1 \end{pmatrix}$
Left circular polarizer	$\frac{1}{2}\begin{pmatrix} 1 & i \\ -i & 1 \end{pmatrix}$	$\frac{1}{2}\begin{pmatrix} 1 & 0 & 0 & 1 \\ 0 & 0 & 0 & 0 \\ 0 & 0 & 0 & 0 \\ 1 & 0 & 0 & 1 \end{pmatrix}$

Component	Jones Matrix	Mueller Matrix
Right circular diattenuator when $t_1>t_2$	$\frac{1}{2}\begin{pmatrix} t_1+t_2 & (t_2-t_1)\cdot i \\ (t_1-t_2)\cdot i & t_1+t_2 \end{pmatrix}$	$\frac{1}{2}\begin{pmatrix} t_1^2+t_2^2 & 0 & 0 & t_1^2-t_2^2 \\ 0 & 2t_1t_2 & 0 & 0 \\ 0 & 0 & 2t_1t_2 & 0 \\ t_1^2-t_2^2 & 0 & 0 & t_1^2+t_2^2 \end{pmatrix}$
Left circular diattenuator when $t_2>t_1$		
General circular retardance by $\Delta\phi$	$\begin{pmatrix} \cos\frac{\Delta\phi}{2} & \sin\frac{\Delta\phi}{2} \\ -\sin\frac{\Delta\phi}{2} & \cos\frac{\Delta\phi}{2} \end{pmatrix}$	$\begin{pmatrix} 1 & 0 & 0 & 0 \\ 0 & \cos\frac{\Delta\phi}{2} & \sin\frac{\Delta\phi}{2} & 0 \\ 0 & -\sin\frac{\Delta\phi}{2} & \cos\frac{\Delta\phi}{2} & 0 \\ 0 & 0 & 0 & 1 \end{pmatrix}$
Faraday mirror	$\begin{pmatrix} 0 & 1 \\ -1 & 0 \end{pmatrix}$	$\begin{pmatrix} 1 & 0 & 0 & 0 \\ 0 & -1 & 0 & 0 \\ 0 & 0 & -1 & 0 \\ 0 & 0 & 0 & 1 \end{pmatrix}$
Ideal depolarizer	None	$\begin{pmatrix} 1 & 0 & 0 & 0 \\ 0 & 0 & 0 & 0 \\ 0 & 0 & 0 & 0 \\ 0 & 0 & 0 & 0 \end{pmatrix}$
Partial depolarizer	None	$\begin{pmatrix} 1 & 0 & 0 & 0 \\ 0 & d & 0 & 0 \\ 0 & 0 & d & 0 \\ 0 & 0 & 0 & d \end{pmatrix}$

51.4 Polarimeters

51.4.1 Measurement of the Jones vector

Since only irradiances can directly be measured instead of phases and amplitudes, an unknown Jones vector can be determined by letting the beam pass a rotating perfect polarizer and then measuring the irradiance on a polarization neutral detector. Figure 51-23 shows the main setup [51-3]

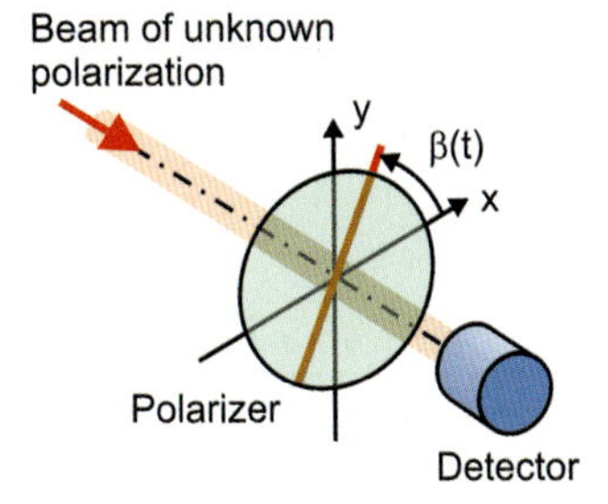

Figure 51-23: Detecting an unknown Jones vector by means of a rotating ideal linear polarizer.

The Jones vector will be described by (51-6), the rotating polarizer by its Jones matrix **M(t)** as indicated in table 51-2. Using addition theorems leads to

$$\mathbf{M}(t) = \frac{1}{2}\begin{pmatrix} 1 + \cos 2\beta(t) & \sin 2\beta(t) \\ \sin 2\beta(t) & 1 - \cos 2\beta(t) \end{pmatrix} \tag{51-82}$$

For a rotational position $\beta(t) = \omega_P t$ of the linear polarizer the time-dependent measured irradiance $I(t)$ is given by

$$I(t) = \frac{I_0}{2}\left(1 + a\cos\left(2\beta(t) + \varphi\right)\right) \tag{51-83}$$

where I_0 denotes the irradiance of the incident beam and

$$a = \sqrt{1 - \sin^2 2\psi \sin^2\delta} \tag{51-84}$$

$$\varphi = -\arctan\left(\tan 2\psi \cos\delta\right) \tag{51-85}$$

I_0, a and φ can be determined by measuring at least three irradiances at different rotation angles, for instance $\beta(t_1) = 0$, $\beta(t_2) = \frac{\pi}{4}$ and $\beta(t_3) = \frac{\pi}{2}$. Better accuracy is achieved by using sophisticated phase-shift algorithms in combination with irradiances acquired at many different rotation angles.

Once I_0, a and φ are determined, ψ and δ can be calculated from (51-86) and (51-87).

$$\psi = \frac{1}{2} \arcsin \sqrt{1 - a^2 \cos^2 \varphi} \tag{51-86}$$

$$\delta = \arccos \left(\frac{-a \cdot \sin \varphi}{\sqrt{1 - a^2 \cos^2 \varphi}} \right) \tag{51-87}$$

Figure 51-24 shows irradiances at the detector for various Jones vectors while the ideal polarizer is moving through a full rotation. For linear polarized light (figure 51-24 (a)) the sinusoidal function shows a periodicity of $2\beta(t)$, unity amplitude and a phase varying with ψ. If δ is varied while ψ is set to $\pi/4$, the amplitude varies with δ while the phase stays constant (figure 51-24 (b)). Figures 51-24 (c) and (d) show the limitation of the rotating polarizer method: the curves for $+\delta$ and $-\delta$ cannot be distinguished due to the restricted range of the arccos-function between 0 and π.

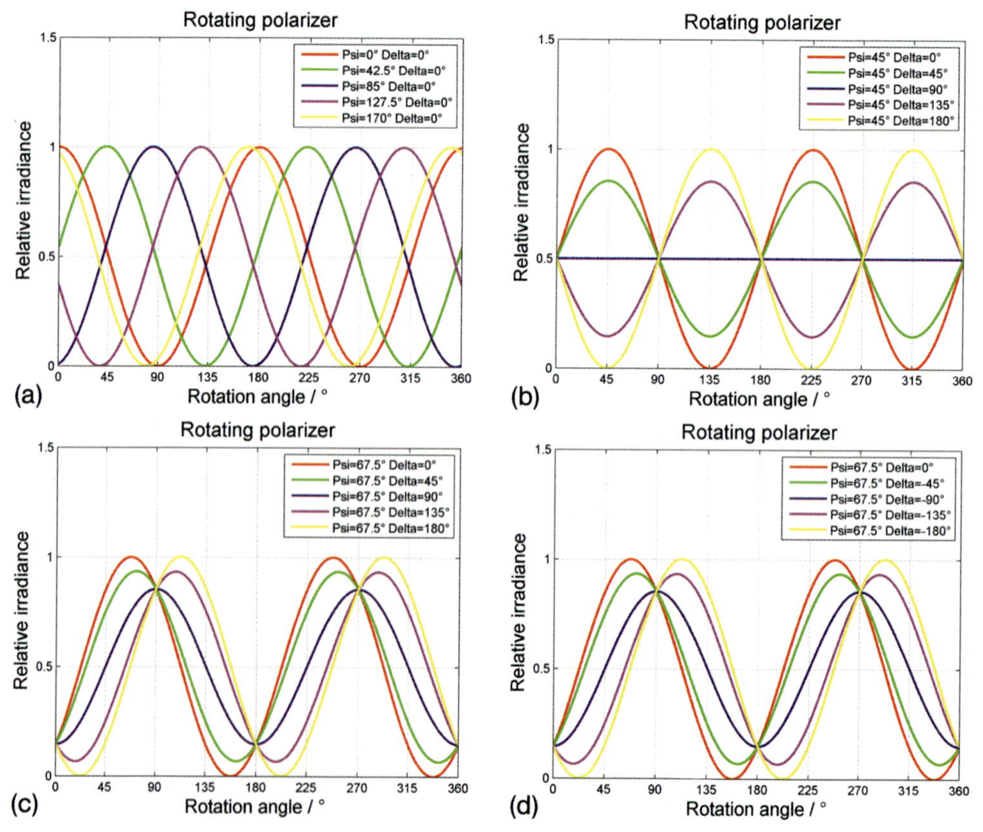

Figure 51-24: Detector irradiances for various Jones vectors measured with rotating ideal polarizer: (a) variations in ψ when δ=0, (b) variations in δ, when $\psi=\pi/4$, (c) variations in δ from 0 to π, when $\psi=3\pi/8$, (d) variations in δ from 0 to $-\pi$, when $\psi=3\pi/8$.

To extend the measuring range for δ to $\pm\,\pi$ an additional fixed compensator is necessary [51-3]. It can be introduced in front of the rotating polarizer as shown in figure 51-25. The compensator is a retarder with its fast axis at 45° and a retardation of $\pi/4$ (quarter-waveplate). By introducing a retarder with retardance $\Delta\phi$ the irradiance is calculated from (51-88)–(51-90).

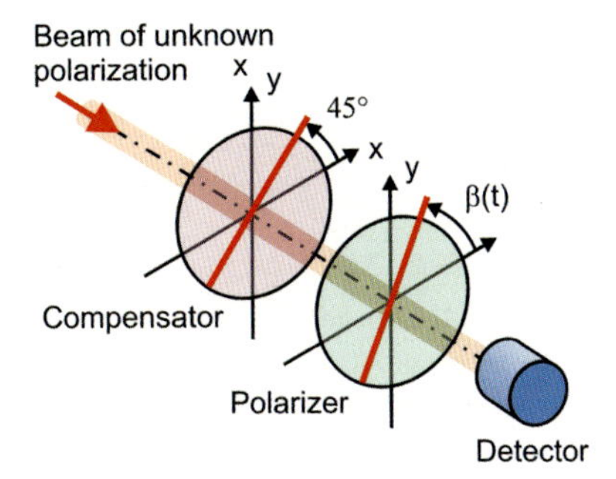

Figure 51-25: Detecting an unknown Jones vector by means of a compensator (quarter-waveplate with fast axis at 45°) and rotating ideal linear polarizer.

$$I(t) = \frac{I_0}{2}\left(1 + a\sin\left(2\beta(t) + \varphi\right)\right) \tag{51-88}$$

with I_0 denoting the irradiance of the incident beam and

$$a = \sqrt{(\cos 2\psi \cos\Delta\phi - \sin 2\psi \sin\Delta\phi \sin\delta)^2 + \sin^2 2\psi \cos^2\delta} \tag{51-89}$$

$$\varphi = \arctan\left(\frac{\cos\Delta\phi}{\tan 2\psi \cos\delta} - \sin\Delta\phi \tan\delta\right) \tag{51-90}$$

For $\Delta\phi = \pi/2$ the unknown δ and ψ can be calculated according to

$$\psi = \frac{1}{2}\arcsin(a) \tag{51-91}$$

$$\delta = -\varphi \tag{51-92}$$

The measuring range for δ is now extended to $\pm\,\pi$

Figure 51-26 shows irradiances at the detector for various Jones vectors with the introduced quarter-waveplate while the ideal polarizer is moving through a full rotation. For linear polarized light (figure 51-26 (a)) the sinusoidal function shows a periodicity of $2\beta(t)$, a zero phase and an amplitude varying with ψ. If δ is varied while ψ is set to $\pi/4$, the phase varies with $-\delta$ while the amplitude stays constant (figure 51-26 (b)). Figures 51-26 (c) and (d) show that the sign of δ can now be clearly identified.

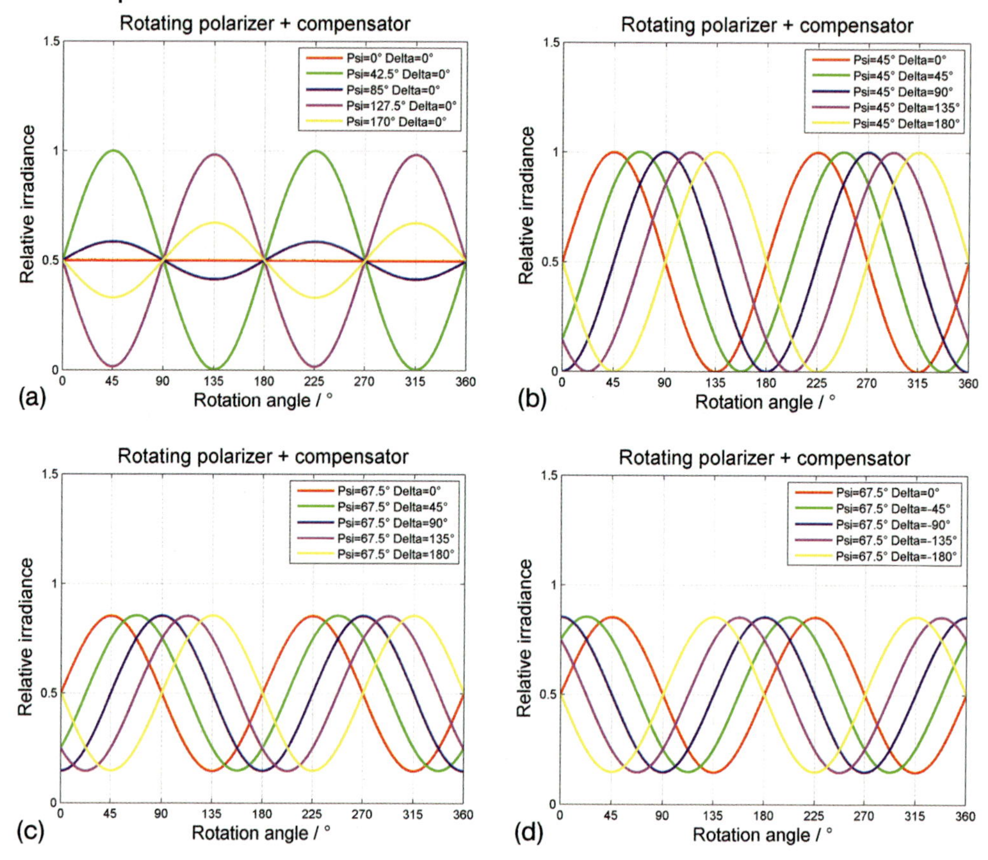

Figure 51-26: Detector irradiances for various Jones vectors measured with quarter-waveplate and rotating ideal polarizer: (a) variations in ψ when δ=0, (b) variations in δ, when $\psi=\pi/4$, (c) variations in δ from 0 to π, when $\psi=3\pi/8$, (d) variations in δ from 0 to $-\pi$, when $\psi=3\pi/8$.

51.4.2
Measurement of the Stokes Vector

An unknown Stokes vector can be determined by measuring the flux transmitted through a set of N polarizing analyzers with Mueller matrices $\mathbf{M}^{(i)}$, $i=1, 2, \ldots N$. An analyzer either consists of a linear polarizer in rotary position β or a combination of a linear $\lambda/4$-retarder in rotary position a and a following linear polarizer in rotary position β. Each analyzer determines the flux of one polarization component in the incident beam [51-5] and [51-25]. The analyzer transforms the unknown Stokes vector $\vec{S}$ to a new Stokes vector $\vec{S}_i'$ according to

$$\vec{S}_i' = \mathbf{M}^{(i)} \cdot \vec{S} \tag{51-93}$$

A detector measures the overall irradiance I_i transmitted through the analyzer, which is equal to the element $S'_{i,0}$ of the transmitted Stokes vector

$$I_i = S'_{i,0} = M_{00}^{(i)} S_0 + M_{01}^{(i)} S_1 + M_{02}^{(i)} S_2 + M_{03}^{(i)} S_3 \tag{51-94}$$

If the analyzer consists of a linear polarizer only, then the transmitted irradiance is given by

$$I_{\text{Pol}}(\beta) = \frac{1}{2}(S_0 + S_1 \cos 2\beta + S_2 \sin 2\beta) \tag{51-95}$$

Since S_3 has no influence, a polarizer alone is not able to discriminate between all four components of a Stokes vector.

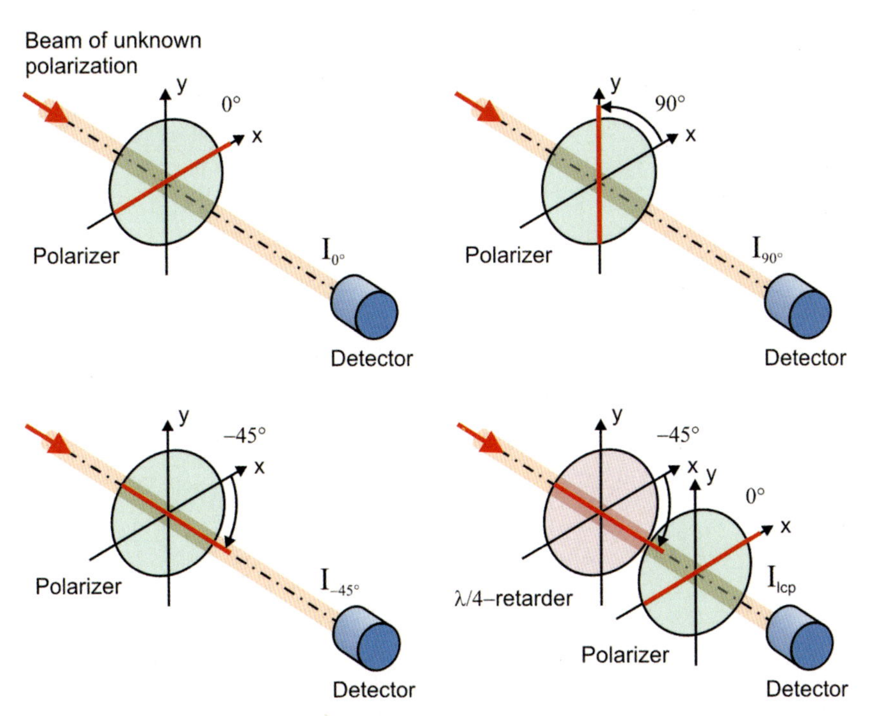

Figure 51-27: Measurement of an unknown Stokes vector by four analyzer states : detection of $I_{0°}$, $I_{90°}$, and $I_{-45°}$ by means of an ideal polarizer, detection of I_{lcp} by means of an additional quarter-waveplate in the –45° rotational position.

If the analyzer consists of a linear retarder of retardance $\Delta\phi$ and a linear polarizer, then the transmitted irradiance is given by

$$I_{\text{Ret+Pol}}(\alpha, \beta, \Delta\phi) = \frac{1}{2}\left\{\begin{array}{l} S_0 \\ +S_1\left[\cos 2\beta(1 - \sin^2 2\alpha(1 - \cos\Delta\phi)) + \sin 2\beta \sin 2\alpha \cos 2\alpha(1 - \cos\Delta\phi)\right] \\ +S_2\left[\cos 2\beta \sin 2\alpha \cos 2\alpha(1 - \cos\Delta\phi) + \sin 2\beta(1 - \cos^2 2\alpha(1 - \cos\Delta\phi))\right] \\ +S_3 \sin 2(\beta - \alpha) \sin\Delta\phi \end{array}\right. \tag{51-96}$$

In this combination, all four Stokes vector components can be discriminated.

Figure 51-27 shows an arrangement used to measure an unknown Stokes vector by four analyzer states : detection of $I_{0°}$, $I_{90°}$, and $I_{-45°}$ by means of an ideal polarizer in positions 0°, 90° and –45°; detection of I_{lcp} by means of an additional quarter-waveplate in the –45° rotational position.

Using (51-95) and (51-96) then leads to the solutions:

$$S_0 = I_{0°} + I_{90°} \tag{51-97}$$

$$S_1 = I_{0°} - I_{90°} \tag{51-98}$$

$$S_2 = I_{0°} + I_{90°} - 2I_{-45°} \tag{51-99}$$

$$S_3 = 2I_{lcp} - I_{0°} - I_{90°} \tag{51-100}$$

If a series of N measurements, with altered analyzers but constant Stokes vector are taken, a measurement matrix using (51-94) can be created

$$\vec{F} = \begin{pmatrix} I_1 \\ I_2 \\ \ldots \\ I_N \end{pmatrix} = \mathbf{W} \cdot \vec{S} = \begin{pmatrix} M_{00}^{(1)} & M_{01}^{(1)} & M_{02}^{(1)} & M_{03}^{(1)} \\ M_{00}^{(2)} & M_{01}^{(2)} & M_{02}^{(2)} & M_{03}^{(2)} \\ \ldots & \ldots & \ldots & \ldots \\ M_{00}^{(N)} & M_{01}^{(N)} & M_{02}^{(N)} & M_{03}^{(N)} \end{pmatrix} \cdot \begin{pmatrix} S_0 \\ S_1 \\ S_2 \\ S_3 \end{pmatrix} \tag{51-101}$$

The Stokes vector is then determined by least-squares fitting according to

$$\vec{S} = \left(\mathbf{W}^T\mathbf{W}\right)^{-1} \cdot \mathbf{W}^T \cdot \vec{F} \tag{51-102}$$

with

$$\mathbf{W}^T\mathbf{W} = \begin{pmatrix} \sum_{i=1}^{N} M_{00}^{(i)2} & \sum_{i=1}^{N} M_{00}^{(i)} M_{01}^{(i)} & \sum_{i=1}^{N} M_{00}^{(i)} M_{02}^{(i)} & \sum_{i=1}^{N} M_{00}^{(i)} M_{03}^{(i)} \\ \sum_{i=1}^{N} M_{01}^{(i)} M_{00}^{(i)} & \sum_{i=1}^{N} M_{01}^{(i)2} & \sum_{i=1}^{N} M_{01}^{(i)} M_{02}^{(i)} & \sum_{i=1}^{N} M_{01}^{(i)} M_{03}^{(i)} \\ \sum_{i=1}^{N} M_{02}^{(i)} M_{00}^{(i)} & \sum_{i=1}^{N} M_{02}^{(i)} M_{01}^{(i)} & \sum_{i=1}^{N} M_{02}^{(i)2} & \sum_{i=1}^{N} M_{02}^{(i)} M_{03}^{(i)} \\ \sum_{i=1}^{N} M_{03}^{(i)} M_{00}^{(i)} & \sum_{i=1}^{N} M_{03}^{(i)} M_{01}^{(i)} & \sum_{i=1}^{N} M_{03}^{(i)} M_{02}^{(i)} & \sum_{i=1}^{N} M_{03}^{(i)2} \end{pmatrix} \tag{51-103}$$

and

$$\mathbf{W}^T \cdot \vec{F} = \begin{pmatrix} \sum_{i=1}^{N} M_{00}^{(i)} I_i \\ \sum_{i=1}^{N} M_{01}^{(i)} I_i \\ \sum_{i=1}^{N} M_{02}^{(i)} I_i \\ \sum_{i=1}^{N} M_{03}^{(i)} I_i \end{pmatrix} \tag{51-104}$$

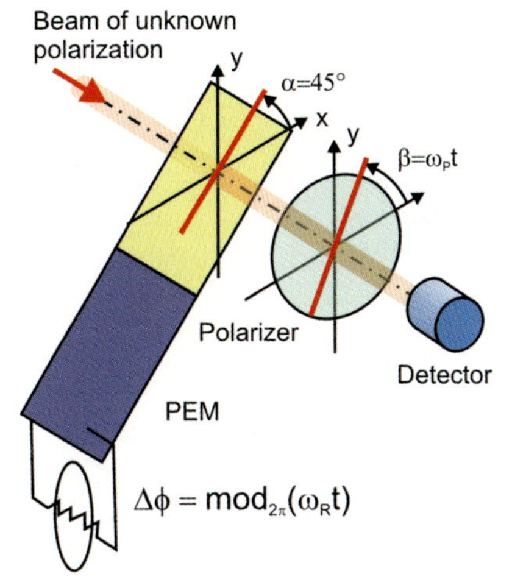

Figure 51-28: Measurement of an unknown Stokes vector by means of a PEM (photoelastic modulator) and a rotating polarizer. The PEM is modulated by a ramp sequence, covering a retardance of $-\pi \le \delta \le +\pi$.

In an alternative setup a rotating polarizer and a modulated PEM can be used (figure 51-28) to determine an unknown Stokes vector [51-22] and [51-23]. We let the polarizer rotate in time by $\beta(t) = \omega_P t$ and set the orientation a of the PEM to 45°. The phase modulation of the PEM is varied in time by $\Delta\phi(t) = \mathrm{mod}_{2\pi}(\omega_R t)$. Using (51-96) the transmitted irradiance is then given by

$$I(t) = \frac{1}{2}\{S_0 + S_1 \cos 2\omega_P t \cos \omega_R t + S_2 \sin 2\omega_P t - S_3 \cos 2\omega_P t \sin \omega_R t\} \tag{51-105}$$

Using addition theorems leads to

$$I(t) = \frac{1}{2}\left\{\begin{array}{l} S_0 \\ +S_1 \frac{1}{2}[\cos(2\omega_P - \omega_R)t + \cos(2\omega_P + \omega_R)t] \\ +S_2 \sin 2\omega_P t \\ +S_3 \frac{1}{2}[\sin(2\omega_P - \omega_R)t - \sin(2\omega_P + \omega_R)t] \end{array}\right\} \tag{51-106}$$

The signal contains the DC term and the three frequencies $2\omega_P$, $2\omega_P - \omega_R$ and $2\omega_P + \omega_R$. From their amplitudes the unknown Stokes vector components S_0 S_1, S_2 and S_3 can be determined.

The modulation frequency ω_R of the retardation must be selected to be different from the rotational frequency ω_P of the polarizer, for instance $\omega = \omega_P = m \cdot \omega_R$.

Figure 51-29 shows examples for $\Delta\phi(t) = \omega t$ and $\beta(t) = 3\omega t$ for different sets of S_1-, S_2- and S_3-variations. Shown are the measured signals $I(t)$ over a full $\Delta\phi$-cycle and the amplitude and phase of the signal spectrum. It can be seen that S_2 can easily be determined from the amplitude spectrum at 6ω. S_1 and S_3 are detected at 5ω and 7ω; however, their phase has to be analyzed to separate the terms clearly.

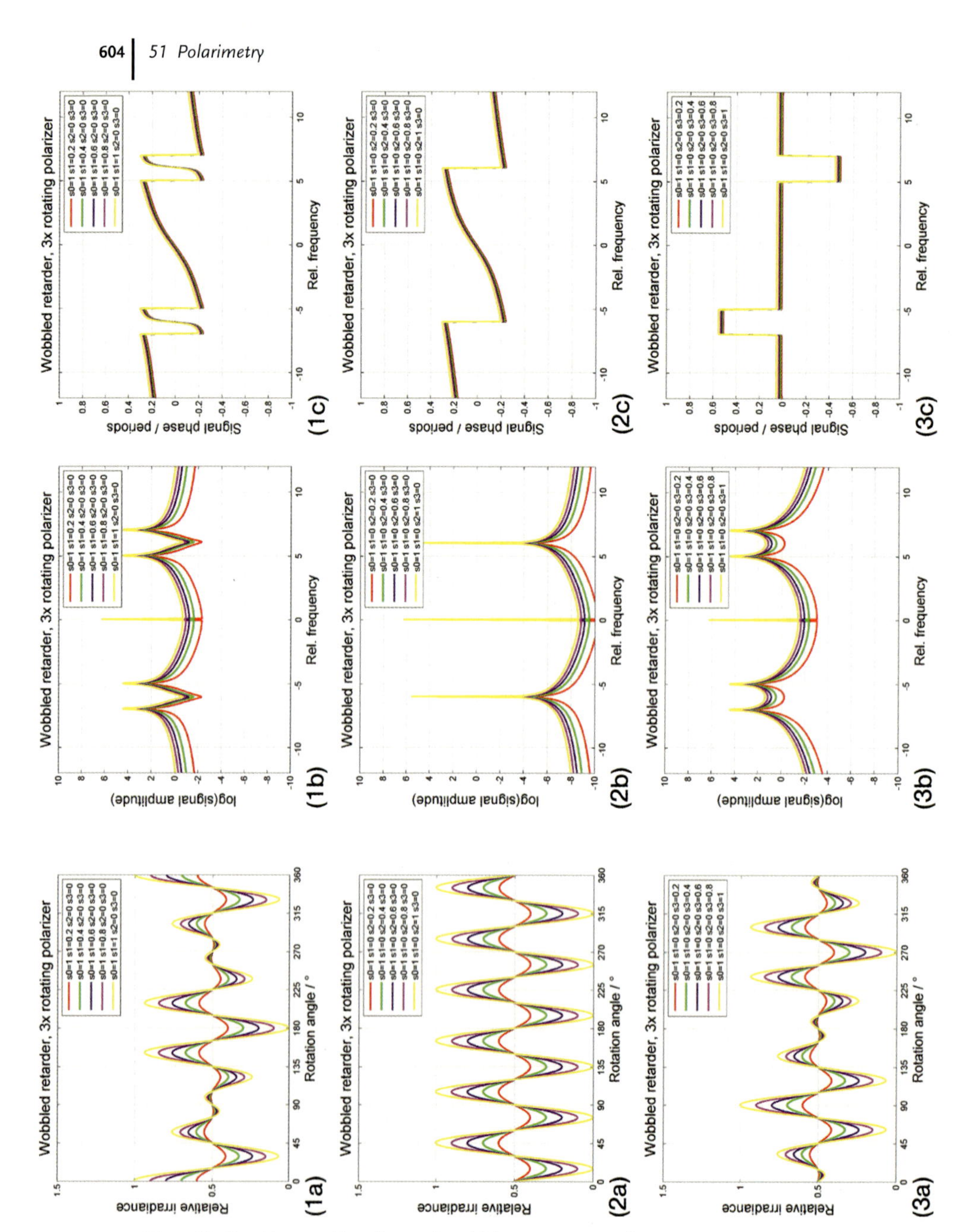

Figure 51-29: Measured irradiance signal (a), spectrum amplitude (b) and spectrum phase (c) for an arrangement in figure 51-28 with varying S_1 (1), S_2 (2) and S_3 (3). The variations are from 0.2 to 1.0 with $S_0 = 1$ and the other $S_i = 0$. The retardance modulation $\Delta\phi(t) = \omega t$ and the angle of the polarizer $\beta(t) = 3\omega t$.

51.4.3
Measurement of the Jones Matrix

The principle of polarimetry is based on the measurement of irradiances achieved by transmitting or reflecting light of a specific polarization through, or from, sample elements and observing through well-defined polarizing elements.

The optical configuration of a polarimeter is characterized by a sequence of letters denoting the individual polarizing components from source to detector. In this notation P, C, S and A designate the polarizer, compensator, sample and analyzer, respectively [51-3]. A rotating element is indicated by the subscript 'r' followed by the rotational frequency in brackets. A modulated element like a PEM is indicated by the subscript 'm' For instance, a $\text{PSC}_\text{r}(\omega)\text{A}$ configuration consists of a fixed polarizer, a fixed sample followed by a rotating compensator rotating with ω, and a fixed analyzer. A configuration $\text{PC}_\text{m}(\omega)\text{SC}_\text{m}(5\omega)\text{A}$ consists of a fixed polarizer, a PEM modulated by ω, a fixed sample, a PEM modulated by 5ω, and a fixed analyzer.

We will denote the Jones vector of the light, when it has passed the polarizer, by $\vec{P}$, the Jones matrix of the unknown sample by **J** and the Jones matrix of the analyzing polarization element (analyzer) in front of the detector, by **A**. If a compensator element is used, we will denote it by **C**.

For a general PSA-arrangement the Jones vector $\vec{E}$ arriving at the detector is defined by

$$\vec{E} = \mathbf{A}\cdot\mathbf{J}\cdot\vec{P} = \begin{pmatrix} A_{xx}\left(J_{xx}P_x + J_{xy}P_y\right) + A_{xy}\left(J_{yx}P_x + J_{yy}P_y\right) \\ A_{yx}\left(J_{xx}P_x + J_{xy}P_y\right) + A_{yy}\left(J_{yx}P_x + J_{yy}P_y\right) \end{pmatrix} \tag{51-107}$$

The related irradiance I measured by the detector is then defined by

$$I = \vec{E}\cdot\vec{E}^{*} = \left(\mathbf{A}\cdot\mathbf{J}\cdot\vec{P}\right)\cdot\left(\mathbf{A}\cdot\mathbf{J}\cdot\vec{P}\right)^{*} \tag{51-108}$$

For a general PCSA-arrangement the Jones vector $\vec{E}$ arriving at the detector is defined by

$$\vec{E} = \mathbf{A}\cdot\mathbf{J}\cdot\mathbf{C}\cdot\vec{P} = \begin{pmatrix} A_{xx}\left(J_{xx}P'_x + J_{xy}P'_y\right) + A_{xy}\left(J_{yx}P'_x + J_{yy}P'_y\right) \\ A_{yx}\left(J_{xx}P'_x + J_{xy}P'_y\right) + A_{yy}\left(J_{yx}P'_x + J_{yy}P'_y\right) \end{pmatrix} \tag{51-109}$$

where

$$\vec{P'} = \begin{pmatrix} C_{xx}P_x + C_{xy}P_y \\ C_{yx}P_x + C_{yy}P_y \end{pmatrix} \tag{51-110}$$

The related irradiance I measured by the detector is then defined by

$$I = \vec{E}\cdot\vec{E}^{*} = \left(\mathbf{A}\cdot\mathbf{J}\cdot\mathbf{C}\cdot\vec{P}\right)\cdot\left(\mathbf{A}\cdot\mathbf{J}\cdot\mathbf{C}\cdot\vec{P}\right)^{*} \tag{51-111}$$

For a general PSCA-arrangement the Jones vector $\vec{E}$ arriving at the detector is defined by

$$\vec{E} = \mathbf{A} \cdot \mathbf{C} \cdot \mathbf{J} \cdot \vec{P} = \begin{pmatrix} A'_{xx}\left(J_{xx}P_x + J_{xy}P_y\right) + A'_{xy}\left(J_{yx}P_x + J_{yy}P_y\right) \\ A'_{yx}\left(J_{xx}P_x + J_{xy}P_y\right) + A'_{yy}\left(J_{yx}P_x + J_{yy}P_y\right) \end{pmatrix} \tag{51-112}$$

where

$$\mathbf{A}' = \mathbf{A} \cdot \mathbf{C} = \begin{pmatrix} A_{xx}C_{xx} + A_{xy}C_{yx} & A_{xx}C_{xy} + A_{xy}C_{yy} \\ A_{yx}C_{xx} + A_{yy}C_{yx} & A_{yx}C_{xy} + A_{yy}C_{yy} \end{pmatrix} \tag{51-113}$$

The related irradiance I measured by the detector is then defined by

$$I = \vec{E} \cdot \vec{E}^{*} = \left(\mathbf{A} \cdot \mathbf{C} \cdot \mathbf{J} \cdot \vec{P}\right) \cdot \left(\mathbf{A} \cdot \mathbf{C} \cdot \mathbf{J} \cdot \vec{P}\right)^{*} \tag{51-114}$$

In the following we will calculate the irradiances for the following specific arrangement examples:

1. P_rSA
2. P_rCSA
3. PS_rA
4. P_rSA_r

Example 1: Rotating Linear Polarizer, Fixed Sample and Linear Analyzer (P_rSA)

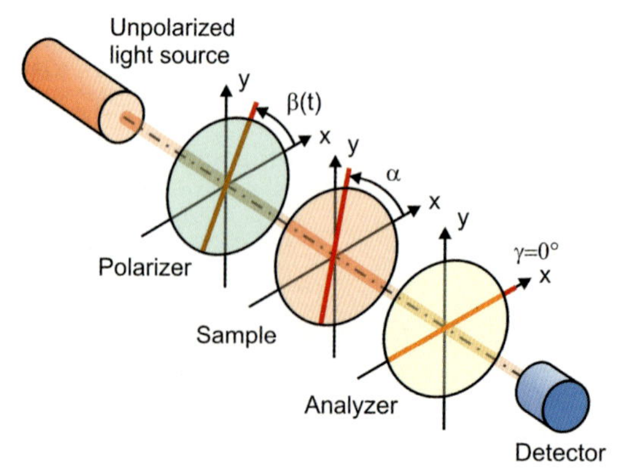

Figure 51-30: Polarimeter arrangement (P_rSA) with rotating polarizer, fixed sample and analyzer.

$$\mathbf{J} = \begin{pmatrix} J_{xx} & J_{xy} \\ J_{yx} & J_{yy} \end{pmatrix} \qquad \vec{P} = \begin{pmatrix} \cos\beta(t) \\ \sin\beta(t) \end{pmatrix} \qquad \mathbf{A} = \begin{pmatrix} 1 & 0 \\ 0 & 0 \end{pmatrix}$$

For a rotational position $\beta(t) = \omega_p t$ of the incident linear polarization the time-dependent measured irradiance $I(t)$ is given by

$$
\begin{aligned}
I(t) &= \left(J_{xx}\cos\beta(t) + J_{xy}\sin\beta(t)\right)\left(J_{xx}^{*}\cos\beta(t) + J_{xy}^{*}\sin\beta(t)\right) \\
&= \frac{1}{2}\left(J_{xx}J_{xx}^{*} + J_{xy}J_{xy}^{*} + \left(J_{xx}J_{xx}^{*} - J_{xy}J_{xy}^{*}\right)\cos 2\beta(t) + 2\mathrm{Re}\left(J_{xx}J_{xy}^{*}\right)\sin 2\beta(t)\right) \\
&= a + b\sin\left(2\omega_p t + \varphi\right)
\end{aligned}
\tag{51-115}
$$

with the following notation

$$
a = \frac{J_{xx}J_{xx}^{*} + J_{xy}J_{xy}^{*}}{2} \tag{51-116}
$$

$$
b = \sqrt{\left(\frac{J_{xx}J_{xx}^{*} - J_{xy}J_{xy}^{*}}{2}\right)^2 + \left(\mathrm{Re}\left(J_{xx}J_{xy}^{*}\right)\right)^2} \tag{51-117}
$$

$$
\varphi = \arctan\frac{J_{xx}J_{xx}^{*} - J_{xy}J_{xy}^{*}}{2\mathrm{Re}\left(J_{xx}J_{xy}^{*}\right)} \tag{51-118}
$$

Introducing the elements of (51-25) and (51-26) leads to

$$
a = \frac{1}{4}\left(t_1^2 + t_2^2 + (t_1^2 - t_2^2)\cos 2\varepsilon\cos 2\alpha\right) \tag{51-119}
$$

$$
b = \sqrt{b_1^2 + b_2^2} \tag{51-120}
$$

$$
\varphi = \arctan\frac{b_1}{b_2} \tag{51-121}
$$

with

$$
b_1 = \frac{1}{2}\left(2t_1t_2\cos\Delta\phi + (t_1^2 - t_2^2)\cos 2\varepsilon\cos 2\alpha + (t_1^2 - 2t_1t_2\cos\Delta\phi + t_2^2)\cos^2 2\varepsilon\cos^2 2\alpha\right) \tag{51-122}
$$

$$
b_2 = \frac{1}{4}\left((t_1^2 - t_2^2)\cos 2\varepsilon\sin 2\alpha + 2t_1t_2\sin\Delta\phi\sin 2\varepsilon + (t_1^2 - 2t_1t_2\cos\Delta\phi + t_2^2)\cos^2 2\varepsilon\cos 2\alpha\sin 2\alpha\right) \tag{51-123}
$$

The irradiance is therefore a sinusoidal function with frequency $2\omega_P$ from which the offset a, the amplitude b and the phase φ can be derived, to determine the unknown parameters of the Jones matrix. However, since the general Jones matrix carries the five unknowns t_1, t_2, $\Delta\phi$, ε and α, it will not be possible to obtain a complete solution from a, b and φ. Therefore, the method of polarizer rotation is only suitable for special sample elements, for instance, if the diattenuation and eigenpolarization are known.

In the case of plain linear birefringent samples ($t_1 = t_2 = 1$, $\varepsilon = 0$) b_1 and b_2 become

$$b_1 = \frac{1}{2}\left(\cos^2\frac{\Delta\phi}{2} + \sin^2\frac{\Delta\phi}{2}\cos 4\alpha\right) \tag{51-124}$$

$$b_2 = \frac{1}{2}\sin^2\frac{\Delta\phi}{2}\sin 4\alpha \tag{51-125}$$

Solutions for $\Delta\phi$ and α are then calculated from

$$\Delta\phi = 2\arcsin\sqrt{\frac{1-4b_1+4b_1^2+4b_2^2}{1-2b_1}} = 2\arcsin\sqrt{\frac{1-4b\sin\varphi+4b^2}{1-2b\sin\varphi}} \tag{51-126}$$

$$\alpha = \frac{1}{2}\arctan\frac{1-2b_1}{2b_2} = \frac{1}{2}\arctan\frac{1-2b\sin\varphi}{2b\cos\varphi} \tag{51-127}$$

For very small $\Delta\phi$-values (51-126) becomes undetermined $\left(\frac{0}{0}\right)$. In this case, a quarter-waveplate (a compensator with its fast axis at 45°) has to be introduced (see next section).

Figure 51-31 shows examples of different samples in a polarimeter arrangement with rotating polarizer, fixed sample and fixed analyzer. When ideal polarizing elements are used the signal is plainly sinusoidal with a frequency of $2\omega_P$.

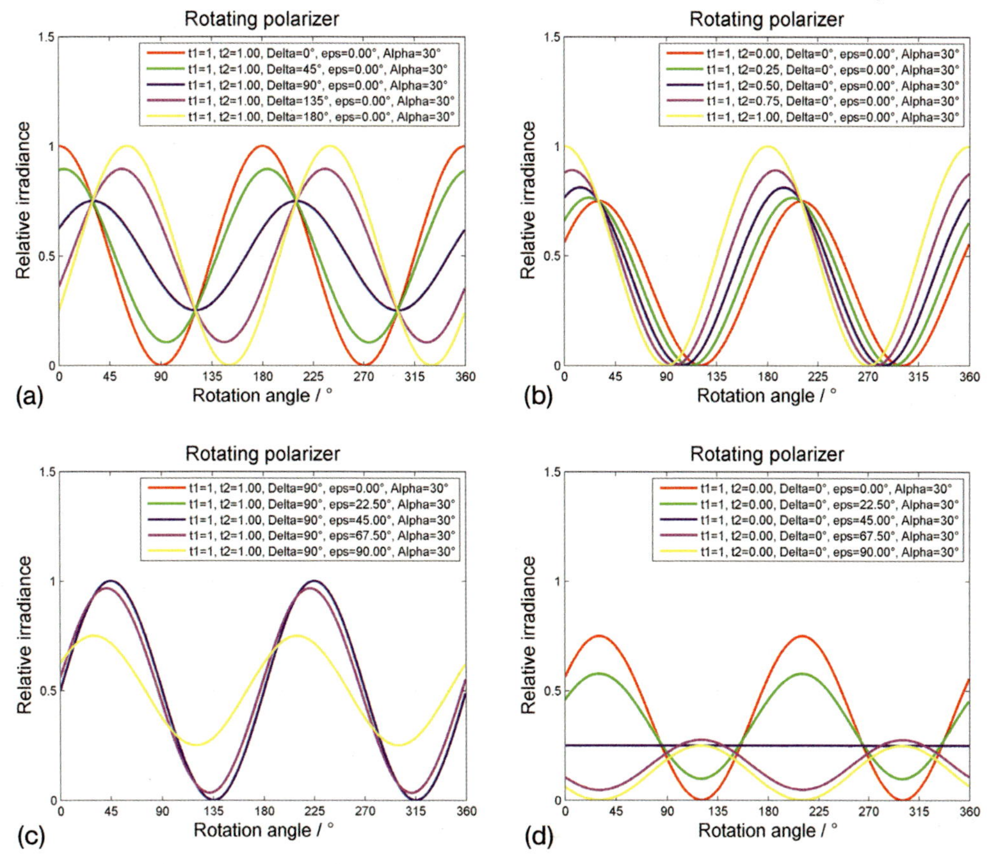

Figure 51-31: Relative irradiance at the detector for a polarimeter arrangement with rotating polarizer, fixed analyzer and sample. Samples: (a) linear birefringent with t_1=1, t_2=0, $\Delta\phi$=0°...180°, ε = 0°, α =30°; (b) linear diattenuating with t_1=1, t_2=0 ... 1, $\Delta\phi$=0°, ε = 0°, α=30°; (c) ellipsoidal birefringent with t_1=1, t_2=0, $\Delta\phi$=90°, ε = 0°...90°, α = 30°; (d) ellipsoidal diattenuating with t_1=1, t_2=0, $\Delta\phi$=0°, ε = 0°...90°, α =30°.

Example 2: Rotating Linear Polarizer, Fixed Compensator, Sample and Linear Analyzer (P_rCSA)

$$\mathbf{J} = \begin{pmatrix} J_{xx} & J_{xy} \\ J_{yx} & J_{yy} \end{pmatrix} \qquad \vec{P} = \begin{pmatrix} \cos\beta(t) \\ \sin\beta(t) \end{pmatrix} \qquad \mathbf{C} = \frac{1}{\sqrt{2}}\begin{pmatrix} 1 & i \\ i & 1 \end{pmatrix} \qquad \mathbf{A} = \begin{pmatrix} 1 & 0 \\ 0 & 0 \end{pmatrix}$$

For a rotational position $\beta(t) = \omega_P t$ of the incident linear polarization the time-dependent measured irradiance $I(t)$ is given as:

$$\begin{aligned} I(t) &= \frac{1}{2}\big(J_{xx}(\cos\beta(t) + i\sin\beta(t)) + J_{xy}(\sin\beta(t) + i\cos\beta(t))\big)\cdot \\ &\quad \big(J^*_{xx}(\cos\beta(t) - i\sin\beta(t)) + J^*_{xy}(\sin\beta(t) - i\cos\beta(t))\big) \\ &= \frac{J_{xx}J^*_{xx} + J_{xy}J^*_{xy}}{2} + \mathrm{Im}\left(J_{xx}J^*_{xy}\right)\cos 2\beta(t) + \mathrm{Re}\left(J_{xx}J^*_{xy}\right)\sin 2\beta(t) \\ &= a + b\sin(2\omega_P t + \varphi) \end{aligned} \tag{51-128}$$

with the following notation

$$a = \frac{J_{xx}J^*_{xx} + J_{xy}J^*_{xy}}{2} \tag{51-129}$$

$$b = \sqrt{\left(\mathrm{Im}\left(J_{xx}J^*_{xy}\right)\right)^2 + \left(\mathrm{Re}\left(J_{xx}J^*_{xy}\right)\right)^2} = \sqrt{b_1^2 + b_2^2} \tag{51-130}$$

$$\varphi = \arctan \frac{\mathrm{Im}\left(J_{xx}J^*_{xy}\right)}{\mathrm{Re}\left(J_{xx}J^*_{xy}\right)} = \arctan \frac{b_1}{b_2}$$

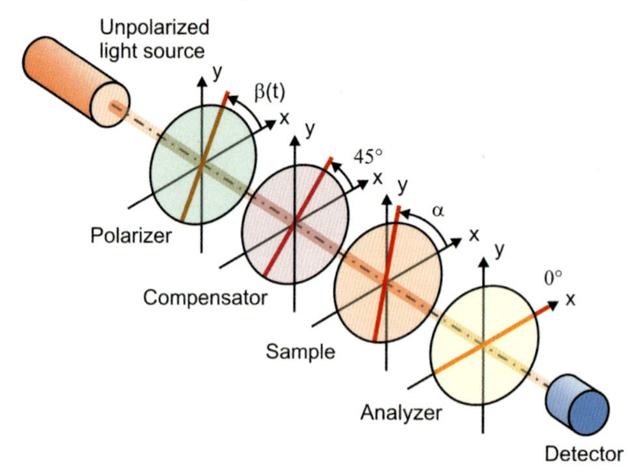

Figure 51-32: Polarimeter arrangement (P_rCSA) with rotating polarizer, fixed compensator, sample and analyzer.

In the case of plain linear birefringent samples ($t_1 = t_2 = 1$, $\varepsilon = 0$) b_1 and b_2 become

$$b_1 = \sin \frac{\Delta\phi}{2} \cos \frac{\Delta\phi}{2} \sin 2\alpha \tag{51-131}$$

$$b_2 = \sin^2 \frac{\Delta\phi}{2} \sin 2\alpha \cos 2\alpha \tag{51-132}$$

Solutions for $\Delta\phi$ and α are then calculated from

$$\Delta\phi = 2 \arctan \frac{b_1}{1 - b_2} = 2 \arctan \frac{b \sin \varphi}{1 - b \cos \varphi} \tag{51-133}$$

$$\alpha = \frac{1}{2} \arccos \frac{b_2(1 - b_2)}{b_1^2} = \frac{1}{2} \arccos \frac{\cos \varphi (1 - b \cos \varphi)}{b \sin^2 \varphi} \tag{51-134}$$

Figure 51-33 shows examples of different samples in a polarimeter arrangement with rotating polarizer, fixed sample and fixed analyzer. When ideal polarizing elements are used the signal is plainly sinusoidal with a frequency of $2\omega_P$.

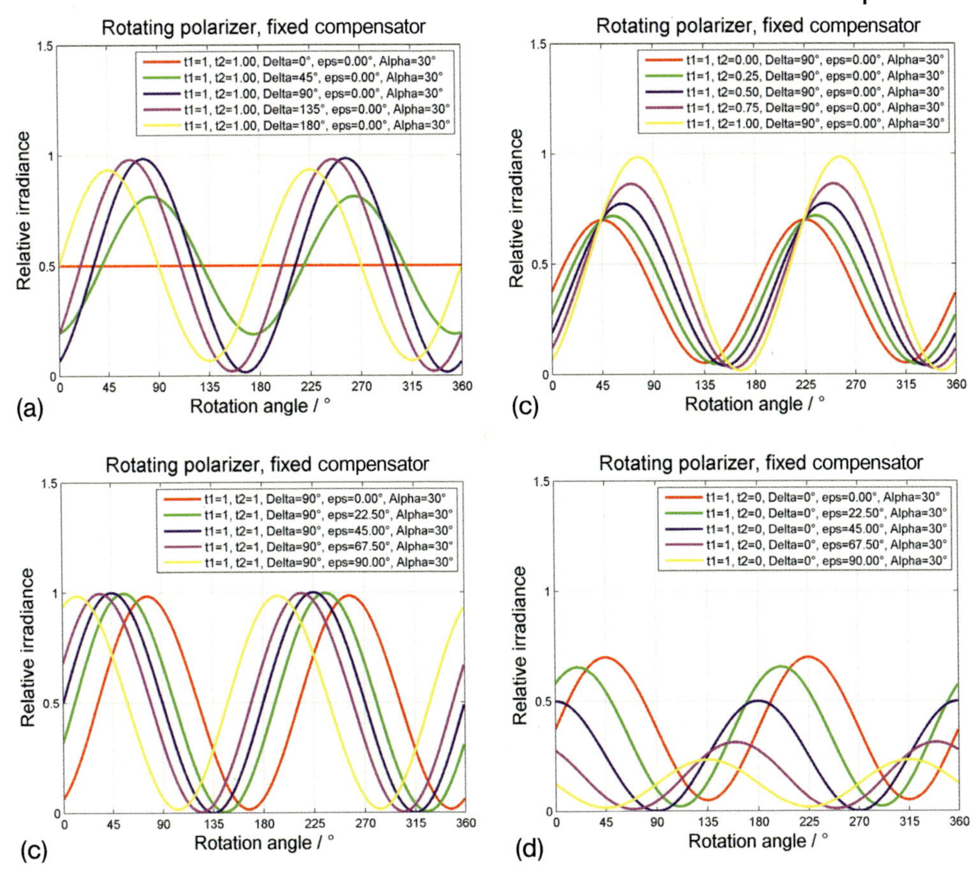

Figure 51-33: Relative irradiance at detector for a polarimeter arrangement with rotating polarizer, fixed compensator, analyzer and sample. Samples: (a) linear birefringent with t_1=1, t_2=0, $\delta = 0°...180°$, $\varepsilon = 0°$, $\alpha = 30°$; (b) linear diattenuating with t_1=1, t_2=0 ... 1, $\Delta\phi$=0°, $\varepsilon = 0°$, $\alpha = 30°$; (c) ellipsoidal birefringent with t_1=1, t_2=0, $\Delta\phi$=90°, $\varepsilon = 0°...90°$, $\alpha = 30°$; (d) ellipsoidal diattenuating with t_1=1, t_2=0, $\Delta\phi$=0°, $\varepsilon = 0°...90°$, $\alpha = 30°$.

Example 3: Rotating Sample, Fixed Linear Parallel Polarizer and Analyzer (PS_rA)

$$\mathbf{M}(t) = \begin{pmatrix} J'_{xx} & J'_{xy} \\ J'_{yx} & J'_{yy} \end{pmatrix} \qquad \vec{P} = \begin{pmatrix} 1 \\ 0 \end{pmatrix} \qquad \mathbf{A} = \begin{pmatrix} 1 & 0 \\ 0 & 0 \end{pmatrix}$$

M(t) refers to (51-14).

For a rotational position $\theta(t) = \omega_S t$ of the sample, the time-dependent measured irradiance $I(t)$ is given by

$$I(t) = a + b\cos(2\theta(t) + \alpha) + c\cos(4\theta(t) + \alpha) \tag{51-135}$$

with

$$a = \frac{3}{8}\left(J_{xx}J_{xx}^* + J_{yy}J_{yy}^*\right) + \frac{1}{8}\left(J_{xx}J_{yy}^* + J_{xx}^*J_{yy}\right) \tag{51-136}$$

$$b = \frac{1}{2}\left(J_{xx}J_{xx}^* - J_{yy}J_{yy}^*\right) \tag{51-137}$$

$$c = \frac{1}{8}\left(J_{xx} - J_{yy}\right)\left(J_{xx}^* - J_{yy}^*\right) \tag{51-138}$$

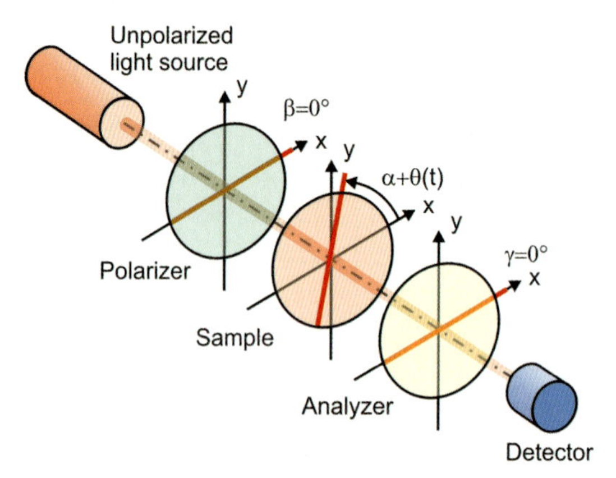

Figure 51-34: Polarimeter arrangement PS_rA with rotating sample, fixed polarizer and analyzer.

Introducing the elements of (51-25) and (51-28) leads to

$$a = \frac{1}{8}\left(\left(t_1^2 + t_2^2\right)\left(2 + \cos^2 2\varepsilon\right) + 2t_1t_2\left(2 - \cos^2 2\varepsilon\right)\cos\Delta\phi\right) \tag{51-139}$$

$$b = \frac{1}{2}\left(t_1^2 - t_2^2\right)\cos 2\varepsilon \tag{51-140}$$

$$c = \frac{1}{8}\left(\left(t_1^2 + t_2^2\right)\cos^2 2\varepsilon - 2t_1t_2\cos 2\varepsilon\cos\Delta\phi\right) \tag{51-141}$$

We notice that a periodicity of 2θ only ocurs in the case of linear or elliptical dichroism when $t_1 \neq t_2$; however, it is not due to a birefringence $\Delta\phi$. The 2θ periodicity also vanishes for left- or right-circular diattenuation with $\varepsilon = \pm\pi/4$.

The rotational offset α of the sample's eigenpolarization axis is easily detected from the phase shifts of the cosine functions in (51-135).

We are able to calculate three unknowns from (51-139)–(51-141). If we consider the sample as being linear in diattenuation and birefringence ($\varepsilon = 0$), we can determine t_1, t_2 and $\Delta\phi$ according to

$$t_1 = \sqrt{2(a+b+c)} \tag{51-142}$$

$$t_2 = \sqrt{2(a-b+c)} \tag{51-143}$$

$$\Delta\phi = \arccos\frac{3c-a}{\sqrt{a^2-2ac-b^2+c^2}} \tag{51-144}$$

Due to the arccos function $\Delta\phi$ being only defined in the interval 0 to π.

Figure 51-35 shows examples of different samples in a polarimeter arrangement with rotating sample, fixed polarizer and analyzer. When ideal polarizing elements are used, the signal is a mixture of sinusoidal functions with frequencies at $2\omega_P$ and $4\omega_P$. Note that all signals carry the same phase defining the sample's eigenpolarization axis.

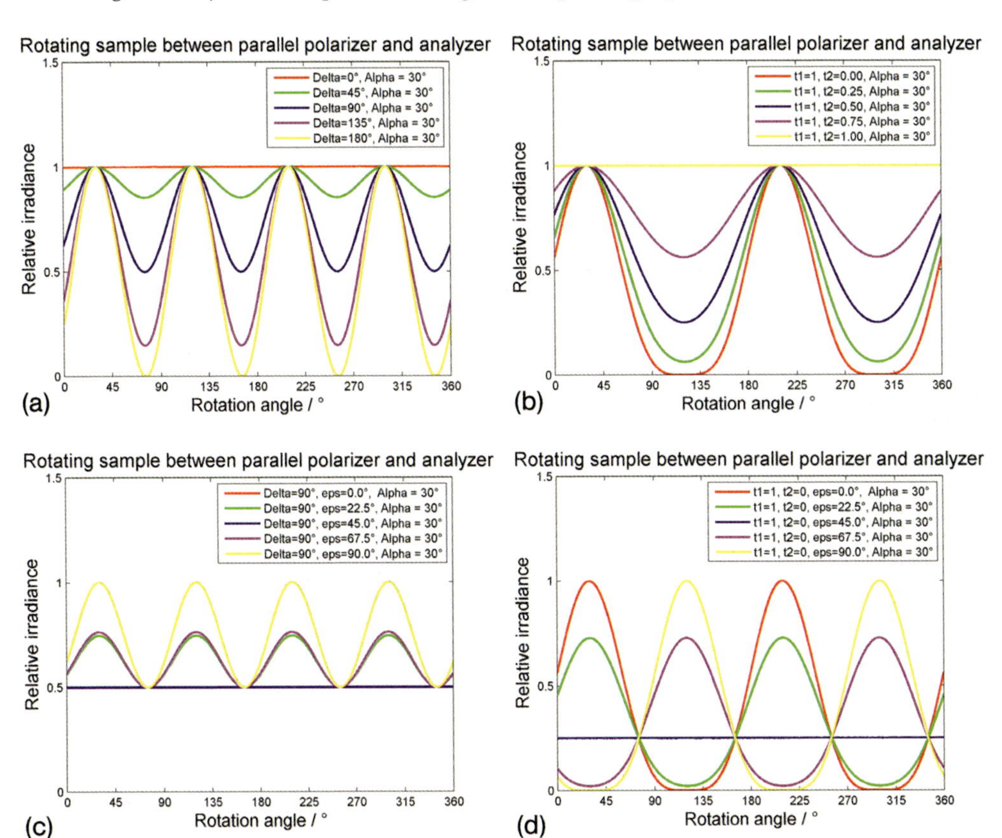

Figure 51-35: Relative irradiance at detector for a polarimeter arrangement with rotating sample, fixed polarizer and analyzer. Samples: (a) linear birefringent with t_1=1, t_2=0, $\Delta\phi$=0°...180°, ε = 0°, α =30°; (b) linear diattenuating with t_1=1, t_2=0 ... 1, $\Delta\phi$=0°, ε = 0°, α =30°; (c) ellipsoidal birefringent with t_1=1, t_2=0, $\Delta\phi$=90°, ε = 0°...90°, α =30°; (d) ellipsoidal diattenuating with t_1=1, t_2=0, $\Delta\phi$=0°, ε = 0°...90°, α =30°.

Example 4: Fixed Sample, Rotating Linear Polarizer, Rotating Linear Analyzer (P_rSA_r)

$$\mathbf{J} = \begin{pmatrix} J_{xx} & J_{xy} \\ J_{yx} & J_{yy} \end{pmatrix} \quad \vec{P} = \begin{pmatrix} \cos\beta(t) \\ \sin\beta(t) \end{pmatrix} \quad \mathbf{A} = \begin{pmatrix} \cos^2\gamma(t) & \sin\gamma(t)\cos\gamma(t) \\ \sin\gamma(t)\cos\gamma(t) & \sin^2\gamma(t) \end{pmatrix}$$

For a rotational position $\beta(t) = \omega_P t$ of the incident linear polarization and a rotational position $\gamma(t) = \omega_A t$ of the analyzer, the measured irradiance $I(t)$ is given by

$$\begin{aligned} I(t) &= \frac{1}{2}(AA^* + BB^* + (AA^* - BB^*)\cos 2\gamma(t) + (AB^* + A^*B)\sin 2\gamma(t)) \\ &= \frac{1}{2}\left(AA^* + BB^* + \sqrt{(AA^* - BB^*)^2 + (AB^* + A^*B)^2}\sin\left(2\gamma(t) - \arctan\frac{AB^* + A^*B}{AA^* - BB^*}\right)\right) \end{aligned} \tag{51-145}$$

where

$$A = (J_{xx}\cos\beta(t) + J_{xy}\sin\beta(t)) = \sqrt{J_{xx}^2 + J_{xy}^2}\sin\left(\beta(t) - \arctan\frac{J_{xy}}{J_{xx}}\right)$$

$$B = (J_{yx}\cos\beta(t) + J_{yy}\sin\beta(t)) = \sqrt{J_{yx}^2 + J_{yy}^2}\sin\left(\beta(t) - \arctan\frac{J_{yy}}{J_{yx}}\right)$$

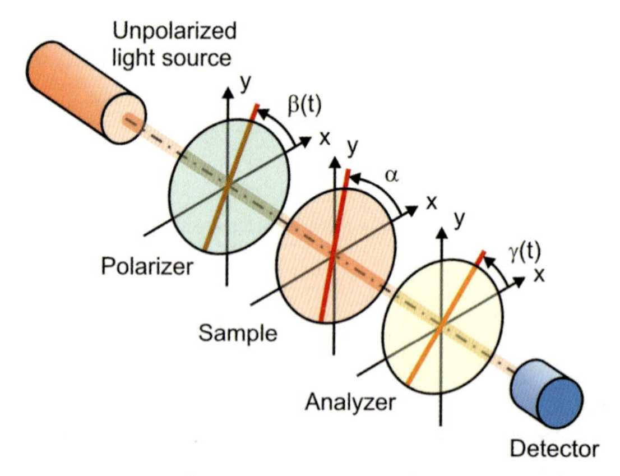

Figure 51-36: Polarimeter arrangement $P_r(\omega)SA_r(5\omega)$ with rotating polarizer, differently rotating analyzer and sample with fixed axis at a.

If we assume periodical rotations of the polarizer and analyzer with frequencies ω_P and ω_A, we will expect signals in the frequency domain at $\omega = 0, \omega_P, \omega, \omega_p + \omega_A, \omega_P - \omega_A$. Figure 51-37 shows time-dependent irradiances at the detector in an arrangement with $\omega_A = 5\omega_P$ for different samples. Figure 51-38 shows frequency spectra for a variety of ellipsoidal diattenuations. Significant signals are found at $\omega = 0, \omega_P, 4\omega_P, 5\omega_P, 6\omega_P$. From the phase and amplitude at the five frequencies, all the unknowns t_1, t_2, $\Delta\phi$, ε and a can be determined. The solution is conveniently achieved by numerical iteration rather than analytical expression, which is generally difficult to derive.

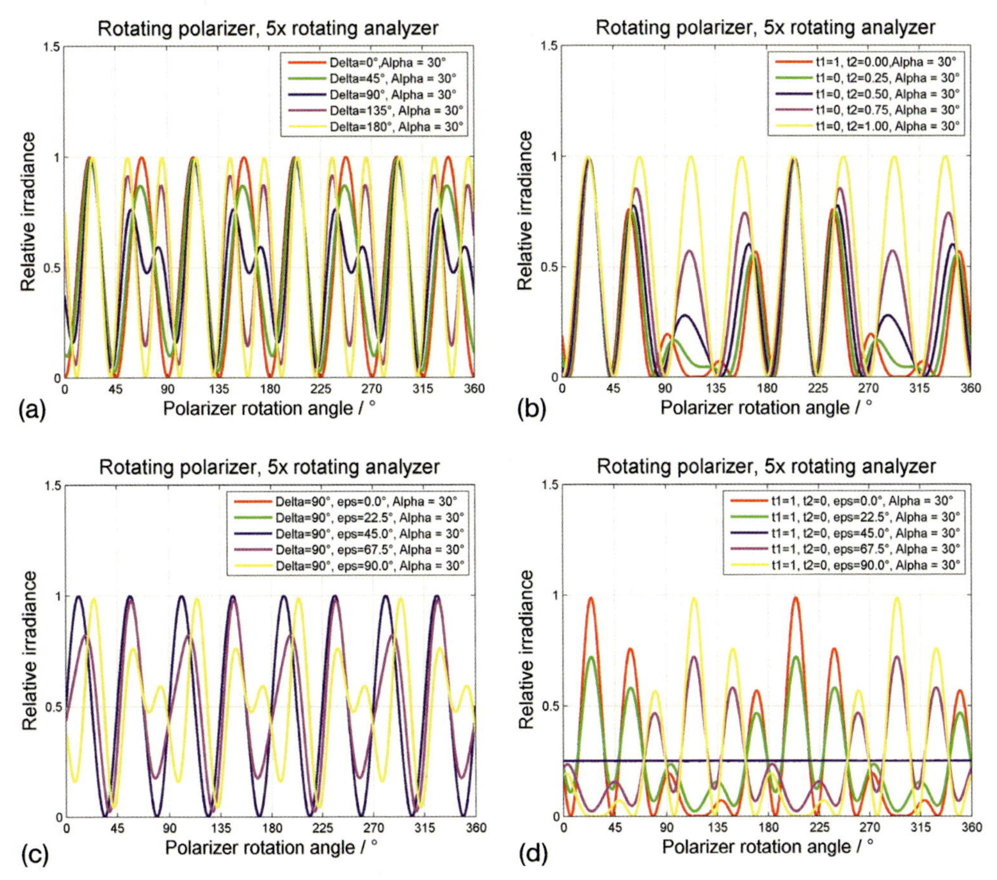

Figure 51-37: Relative irradiance at the detector for a polarimeter arrangement with rotating polarizer, a 5x faster rotating analyzer and a fixed sample. Samples: (a) linear birefringent with $t_1=1$, $t_2=0$, $\Delta\phi=0°...180°$, $\varepsilon=0°$, $\alpha=30°$; (b) linear diattenuating with $t_1=1$, $t_2=0 ... 1$, $\Delta\phi=0°$, $\varepsilon=0°$, $\alpha=30°$; (c) ellipsoidal birefringent with $t_1=1$, $t_2=0$, $\Delta\phi=90°$, $\varepsilon=0°...90°$, $\alpha=30°$; (d) ellipsoidal diattenuating with $t_1=1$, $t_2=0$, $\Delta\phi=0°$, $\varepsilon=0°...90°$, $\alpha=30°$.

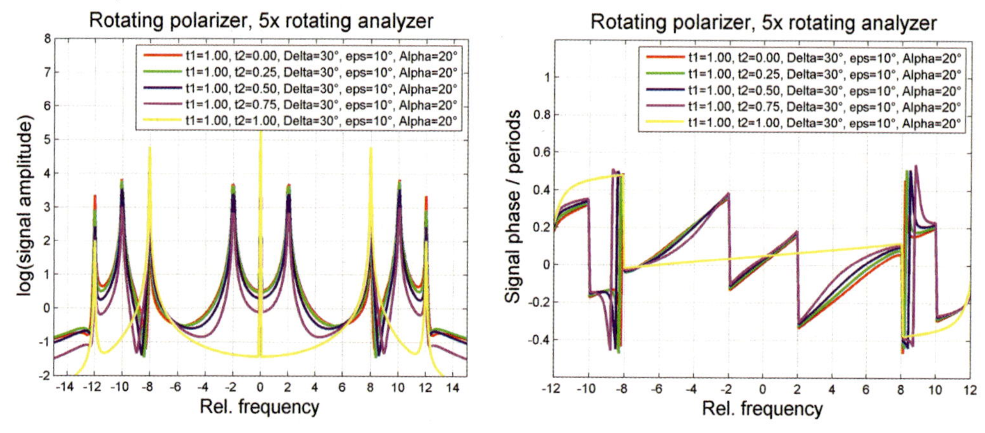

Figure 51-38: Fourier spectrum of the detector irradiance signal in a polarimeter arrangement with rotating polarizer, fixed analyzer and sample. Sample: ellipsoidal diattenuating with t_1=1, t_2=0...1, $\Delta\phi$=30°, $\varepsilon = 10°$, α =20°. (a) Log of frequency signal amplitude; (b) frequency signal phase.

51.4.4
Measurement of the Mueller Matrix

As in the case of the Jones matrix determination, the Mueller matrix elements are derived from the measurement of irradiances achieved by transmitting or reflecting light of specific polarization through or from sample elements while observing through well-defined polarizing elements.

We will denote the Stokes vector of the incident unpolarized light by $\vec{S}$, the Mueller matrix of the polarizer by **P**, the Mueller matrix of the unknown sample by **M** and the Mueller matrix of the analyzing polarization element (analyzer) in front of the detector by **A**. Compensator elements will be denoted by **C**.

Since the irradiance at the detector is described by the first element S_0' of the arriving Stokes vector $\vec{S}'$, the following considerations will concentrate on S_0'.

For a general PSA-arrangement, the Stokes vector $\vec{S}'$ arriving at the detector is defined by

$$\vec{S'} = \mathbf{A} \cdot \mathbf{M} \cdot \mathbf{P} \cdot \vec{S} \tag{51-146}$$

The related irradiance I measured by the detector is then defined by

$$I = S_0' = \sum_{i=0}^{3} \sum_{j=0}^{3} A_{0i} M_{ij} P_{j0} \tag{51-147}$$

For a general PCSA-arrangement the Stokes vector $\vec{S}$ arriving at the detector is defined by

$$\vec{S'} = \mathbf{A} \cdot \mathbf{M} \cdot \mathbf{C} \cdot \mathbf{P} \cdot \vec{S} = \mathbf{A} \cdot \mathbf{M} \cdot \vec{P'} \tag{51-148}$$

where

$$\vec{P'} = \begin{pmatrix} \sum_{i=0}^{3} C_{0i} P_{i0} \\ \sum_{i=0}^{3} C_{1i} P_{i0} \\ \sum_{i=0}^{3} C_{2i} P_{i0} \\ \sum_{i=0}^{3} C_{3i} P_{i0} \end{pmatrix} \tag{51-149}$$

The related irradiance I measured by the detector is then defined by

$$I = S'_0 = \sum_{i=0}^{3} \sum_{j=0}^{3} A_{0i} M_{ij} P'_j = \sum_{i=0}^{3} \sum_{j=0}^{3} \sum_{k=0}^{3} A_{0i} M_{ij} C_{jk} P_{k0} \tag{51-150}$$

For a general PSCA-arrangement, the Stokes vector $\vec{S}$ arriving at the detector is defined by

$$\vec{S'} = \mathbf{A} \cdot \mathbf{C} \cdot \mathbf{M} \cdot \mathbf{P} \cdot \vec{S} = \mathbf{A}' \cdot \mathbf{M} \cdot \mathbf{P} \cdot \vec{S} \tag{51-151}$$

where

$$\mathbf{A}' = \mathbf{A} \cdot \mathbf{C} \tag{51-152}$$

The related irradiance I measured by the detector is then defined by

$$I = S'_0 = \sum_{i=0}^{3} \sum_{j=0}^{3} \sum_{k=0}^{3} A_{0k} C_{ki} M_{ij} P_{j0} \tag{51-153}$$

The arrangements shown above measure only a subset of the 16 Mueller matrix elements. In the following we will give an example of a PCSCA-arrangement, which is able to measure all 16 elements.

An unknown complete Mueller matrix **M** can be determined by measuring the flux transmitted through a set of N ($N \geq 16$) polarizer units P_i and N analyzers units A_i, $i = 1, 2, \ldots N$. In each configuration i the flux transmitted through the total setup is measured. A setup suitable to measure all 16 Mueller matrix elements is shown in Figure 51-39.

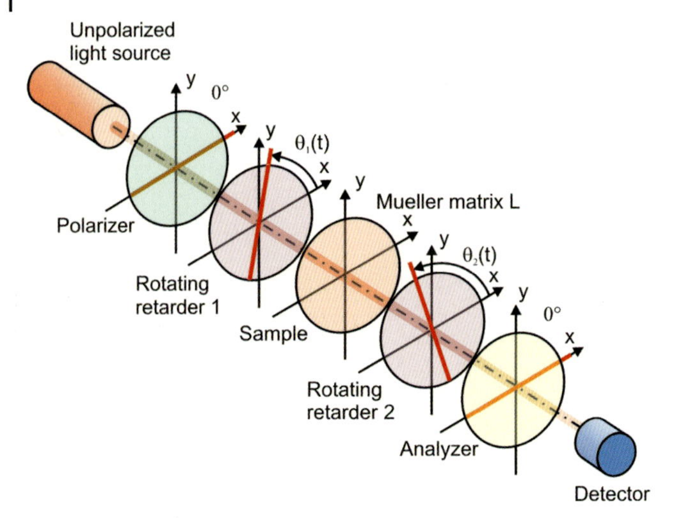

Figure 51-39: Polarimeter arrangement $PC_r(\omega_1)SC_r(\omega_2)A$ suitable to measure the total Mueller matrix of an unknown optical sample in transmission.

Light from an unpolarized source passes the polarizing unit which consists of a fixed linear polarizer at 0° and a rotating retarder of retardation $\Delta\phi_1$ and angular position $\theta_1(t)$. The light then transmits through the sample and enters the following analyzer unit consisting of a rotating retarder of retardation $\Delta\phi_2$ in angular position $\theta_2(t)$ and a fixed linear analyzer at 0° [51-26]–[51-30].

The Stokes vector reaching the detector is expressed by

$$\vec{S'} = \mathbf{A} \cdot \mathbf{C}^{(2)} \cdot \mathbf{M} \cdot \mathbf{C}^{(1)} \cdot \mathbf{P} \cdot \vec{S} \tag{51-154}$$

where $\vec{S}$ denotes the Stokes vector of the unpolarized incident light,

P denotes the Mueller matrix of the polarizer,

$\mathbf{C}^{(1)}$ and $\mathbf{C}^{(2)}$ are the Mueller matrices of the first and second retarder,

M is the Mueller matrix of the unknown sample and

A is the Mueller matrix of the analyzer.

The irradiance I is expressed by the first element S_0 of the Stokes vector $\vec{S}$. After solving (51-154) the irradiance can be expressed as

$$I = \frac{1}{2}\sum_{i=0}^{3}\sum_{j=0}^{3} M_{ij}\left(C_{j0}^{(1)} + C_{j1}^{(1)}\right)\left(C_{0i}^{(2)} + C_{1i}^{(2)}\right) \tag{51-155}$$

Rearranging (51-155) to form a 16-element-vector equation leads to

$$I=\frac{1}{2}\overrightarrow{W}\cdot\overrightarrow{M}=\frac{1}{2}\begin{pmatrix}\left(C^{(1)}_{00}+C^{(1)}_{01}\right)\left(C^{(2)}_{00}+C^{(2)}_{10}\right)\\ \left(C^{(1)}_{10}+C^{(1)}_{11}\right)\left(C^{(2)}_{00}+C^{(2)}_{10}\right)\\ \left(C^{(1)}_{20}+C^{(1)}_{21}\right)\left(C^{(2)}_{00}+C^{(2)}_{10}\right)\\ \left(C^{(1)}_{30}+C^{(1)}_{31}\right)\left(C^{(2)}_{00}+C^{(2)}_{10}\right)\\ \left(C^{(1)}_{00}+C^{(1)}_{01}\right)\left(C^{(2)}_{01}+C^{(2)}_{11}\right)\\ \left(C^{(1)}_{10}+C^{(1)}_{11}\right)\left(C^{(2)}_{01}+C^{(2)}_{11}\right)\\ \cdots\\ \left(C^{(1)}_{30}+C^{(1)}_{31}\right)\left(C^{(2)}_{03}+C^{(2)}_{13}\right)\end{pmatrix}\cdot\begin{pmatrix}M_{00}\\ M_{01}\\ M_{02}\\ M_{03}\\ M_{10}\\ M_{11}\\ \cdots\\ M_{33}\end{pmatrix}=\frac{1}{2}\begin{pmatrix}W_0\\ W_1\\ W_2\\ W_3\\ W_4\\ W_5\\ \cdots\\ W_{15}\end{pmatrix}\cdot\begin{pmatrix}M_{00}\\ M_{01}\\ M_{02}\\ M_{03}\\ M_{10}\\ M_{11}\\ \cdots\\ M_{33}\end{pmatrix} \tag{51-156}$$

Table 51-3 shows the individual elements for W_i for a first retarder at a_1 with retardation $\Delta\phi_1$ and second retarder at a_2 with retardation $\Delta\phi_2$.

When a set of N measurements I_i, $i=1, 2, \ldots N$, is taken, the following matrix equation can be generated

$$\overrightarrow{I}=\begin{pmatrix}I_1\\ I_2\\ \cdots\\ I_N\end{pmatrix}=\frac{1}{2}\mathbf{W}\cdot\overrightarrow{M}=\begin{pmatrix}W_0^{(1)} & W_1^{(1)} & \ldots & W_{15}^{(1)}\\ W_0^{(2)} & W_1^{(2)} & \ldots & W_{15}^{(2)}\\ \cdots & \cdots & \cdots & \cdots\\ W_0^{(N)} & W_1^{(N)} & \ldots & W_{15}^{(N)}\end{pmatrix}\cdot\begin{pmatrix}M_{00}\\ M_{01}\\ \cdots\\ M_{33}\end{pmatrix} \tag{51-157}$$

If $N>16$ is selected, (51-157) can be solved for $\overrightarrow{M}$ by applying least-squares fitting and matrix inversion:

$$\overrightarrow{M}=\left(\mathbf{W}^{\mathrm{T}}\mathbf{W}\right)^{-1}\cdot 2\overrightarrow{I} \tag{51-158}$$

M can therefore be detected by measuring the irradiance for a variety of discrete rotations of both retarders. **M** can also be detected, when the retarders are uniformly rotated with, for instance, $a_1(t)=\omega t$ and $a_2(t)=5\omega t$ [51-5]. Significant information can then be found in the Fourier spectrum of the time signal $I(t)$ at the following frequencies: DC-term and 2, 4, 6, 8, 10, 12, 14, 16, 18, 20, 22, 24 times the base frequency ω (table 51-4). To determine the unknown Mueller matrix elements the amplitudes and also the phases of the frequency orders have to be investigated.

Table 51-3: 16-element vector components for complete Mueller polarimeter arrangement with first retarder at a_1 with retardation $\Delta\phi_1$ and second retarder at a_2 with retardation $\Delta\phi_2$.

Element from $\overrightarrow{W}$	Expression
W_0	1
W_1	$\cos 4a_1 \sin^2\frac{\Delta\phi_1}{2} + \cos^2\frac{\Delta\phi_1}{2}$
W_2	$\sin 4a_1 \sin^2\frac{\Delta\phi_1}{2}$
W_3	$-\sin 2a_1 \sin\Delta\phi_1$
W_4	$\cos 4a_2 \sin^2\frac{\Delta\phi_2}{2} + \cos^2\frac{\Delta\phi_2}{2}$
W_5	$\left(\cos 4a_1 \sin^2\frac{\Delta\phi_1}{2} + \cos^2\frac{\Delta\phi_1}{2}\right)\left(\cos 4a_2 \sin^2\frac{\Delta\phi_2}{2} + \cos^2\frac{\Delta\phi_2}{2}\right)$
W_6	$\sin 4a_1 \sin^2\frac{\Delta\phi_1}{2}\left(\cos 4a_2 \sin^2\frac{\Delta\phi_2}{2} + \cos^2\frac{\Delta\phi_2}{2}\right)$
W_7	$-\sin 2a_1 \sin\Delta\phi_1\left(\cos 4a_2 \sin^2\frac{\Delta\phi_2}{2} + \cos^2\frac{\Delta\phi_2}{2}\right)$
W_8	$\sin 4a_2 \sin^2\frac{\Delta\phi_2}{2}$
W_9	$\left(\cos 4a_1 \sin^2\frac{\Delta\phi_1}{2} + \cos^2\frac{\Delta\phi_1}{2}\right)\sin 4a_2 \sin^2\frac{\Delta\phi_2}{2}$
W_{10}	$\sin 4a_1 \sin^2\frac{\Delta\phi_1}{2}\sin 4a_2 \sin^2\frac{\Delta\phi_2}{2}$
W_{11}	$-\sin 2a_1 \sin\Delta\phi_1 \sin 4a_2 \sin^2\frac{\Delta\phi_2}{2}$
W_{12}	$-\sin 2a_2 \sin\Delta\phi_2$
W_{13}	$-\left(\cos 4a_1 \sin^2\frac{\Delta\phi_1}{2} + \cos^2\frac{\Delta\phi_1}{2}\right)\sin 2a_2 \sin\Delta\phi_2$
W_{14}	$-\sin 4a_1 \sin^2\frac{\Delta\phi_1}{2}\sin 2a_2 \sin\Delta\phi_2$
W_{15}	$\sin 2a_1 \sin\Delta\phi_1 \sin 2a_2 \sin\Delta\phi_2$

Table 51-4: 16-element vector comonents for complete Mueller polarimeter arrangement with first $\pi/2$-retarder rotating at ωt and second $\pi/2$-retarder rotating at $5\omega t$.

Element from $\overrightarrow{W}$	Expression
W_0	1
W_1	$\frac{1}{2}(\cos 4\omega t + 1)$
W_2	$\frac{1}{2}\sin 4\omega t$
W_3	$-\frac{1}{\sqrt{2}}\sin 2\omega t$
W_4	$\frac{1}{2}(\cos 20\omega t + 1)$
W_5	$\frac{1}{8}(2 + 2\cos 4\omega t + \cos 16\omega t + 2\cos 20\omega t + \cos 24\omega t)$
W_6	$\frac{1}{8}(2\sin 4\omega t - \sin 16\omega t + \sin 24\omega t)$
W_7	$-\frac{1}{4}(2\sin 2\omega t - \sin 18\omega t + \sin 22\omega t)$
W_8	$\frac{1}{2}\sin 20\omega t$
W_9	$\frac{1}{8}(2\sin 20\omega t + \sin 16\omega t + \sin 24\omega t)$
W_{10}	$\frac{1}{8}(\cos 16\omega t - \cos 24\omega t)$
W_{11}	$-\frac{1}{4}(\cos 18\omega t - \cos 22\omega t)$
W_{12}	$-\sin 10\omega t$
W_{13}	$-\frac{1}{4}(2\sin 10\omega t + \sin 6\omega t + \sin 14\omega t)$
W_{14}	$-\frac{1}{4}(\cos 6\omega t - \cos 14\omega t)$
W_{15}	$\frac{1}{2}(\cos 8\omega t - \cos 12\omega t)$

51.4.5 Polarimeters and Ellipsometers

Polarimeters are generally used to measure and interpret the polarization state of electromagnetic waves. Typically, they inspect waves that have traveled through or have been reflected, refracted, or diffracted from an object in order to characterize that object. In optics, polarimetry is used to measure properties such as linear and circular birefringence, linear and circular dichroism and scattering of materials and components. Since the polarizing characteristics of samples are represented by a 4 × 4 Mueller matrix, a polarimeter is called a *general* or *complete Mueller polarimeter,* when all 16 elements of the Mueller matrix can be determined. Otherwise, it is called an *incomplete Mueller polarimeter.*

Ellipsometers are special polarimeters that are dedicated to the measurement and characterization of thin films and optical surfaces. The goal is to determine the refractive indices n and the extinction coefficients k of individual layers or substrate materials. To achieve this goal sophisticated iterative algorithms have to be applied to the polarization- dependent measured reflective irradiances. For reliable and accurate n- and k-results, measurements with multiple wavelengths and multiple incident angles must be carried out. We then refer to spectroscopic ellipsometers.

A general polarimeter consists of

1. a light source,
2. a polarization generator,
3. a sample,
4. a polarization analyzer,
5. a detector.

Polarization generators and analyzers consist of optical components, which manipulate the polarization: polarizers, compensators (retarders), and phase modulators (modulated retarders). Polarimeter and ellipsometer configurations include the rotating analyzer (RAE), the rotating polarizer (RPE), the rotating compensator (RCE), and phase modulation (PME) [51-36]–[51-38].

In the following we will explain and give examples of polarimeter and ellipsometer arrangements.

Polarimeter used to Measure Birefringence

The traditional way to determine birefringence of a material is the PS_rA-configuration, i.e., by passing a light beam through a sample which is placed between crossed polarizers. The sample is rotated through 360° and the transmitted irradiance is recorded. The magnitude of birefringence is related to the difference between the maximum and minimum signals. In section 52.5.2 we have shown a setup comprising a rotating polarizer used to measure the magnitude and local orientation of the birefringence of a sample. The setup is only suitable for measurement of plane optical elements. Since a CCD camera observes the entire sample, local scanning is not necessary. A general drawback is that retardance can only be detected within 0 and π.

Senarmont Method

The classical method of stress mapping is the Senarmont method, in which the test sample is placed between a 90°-polarizer and a 0° oriented quarter-waveplate, followed by a rotatable analyzer as shown in figure 51-40 [51-47]. The configuration is P(90°)SC(0°)A_r. When the retardance axis of the sample is oriented at ± 45°, a simple quantitative visual inspection can be made as shown below.

Following (51-153) the irradiance $I(x,y)$ at the CCD camera for a rotation γ of the analyzer can be calculated as

$$\begin{aligned} I(x,y) &= \frac{1}{4}\{M_{00} - M_{01} + (M_{10} - M_{11})\cos 2\gamma + (M_{30} - M_{31})\sin 2\gamma\} \\ &= A - B \cdot \cos(\varphi - 2\gamma) \end{aligned} \tag{51-159}$$

with M_{ij} as the appropriate Mueller matrix elements of the sample and

$$A = \frac{M_{00} - M_{01}}{4} \tag{51-160}$$

$$B = \frac{1}{4}\sqrt{(M_{10} - M_{11})^2 + (M_{30} - M_{31})^2} \tag{51-161}$$

$$\varphi = \arctan\frac{M_{30} - M_{31}}{M_{10} - M_{11}} \tag{51-162}$$

The signal is therefore sinusoidal in 2γ and its amplitude B and its phase contain information about the unknown Mueller matrix elements.

For a birefringent sample with retardance $\Delta\phi(x,y)$ and local orientation, $a(x,y)$ the irradiance becomes

$$I(x,y) = \frac{1}{4}\left(1 - \left(\cos^2 2a + \sin^2 2a\cos\Delta\phi\right)\cos 2\gamma - \sin\Delta\phi\sin 2\gamma\right) \tag{51-163}$$

and therefore

$$A = \frac{1}{4} \tag{51-164}$$

$$B = \frac{1}{4}\sqrt{\left(\left(\cos^2 2a + \sin^2 2a\cos\Delta\phi\right)^2 + \sin^2\Delta\phi\right)} \tag{51-165}$$

$$\varphi = \arctan\left(\frac{\sin\Delta\phi}{\cos^2 2a + \sin^2 2a\cos\Delta\phi}\right) \tag{51-166}$$

B and φ can be determined by regular phase-shifting techniques, in which the rotational position γ of the analyzer is altered, for instance, in 45° steps introducing $\pi/2$-phase steps in the signal. The results are then given by

$$\Delta\phi = \arcsin 4B \cdot \tan\varphi \tag{51-167}$$

$$a = \frac{1}{2} \arcsin \sqrt{\frac{1 - 4B}{1 - \sqrt{1 - 16B^2 \tan^2 \varphi}}} \tag{51-168}$$

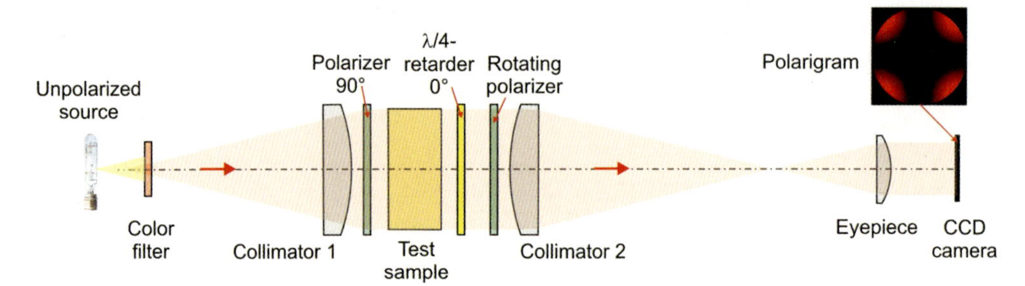

Figure 51-40: Senarmont arrangement used for stress mapping of optical elements.

For the special cases $a = \pm 45°$ the irradiance is

$$I_{\pm 45°}(x, y) = \frac{1}{4}(1 - \cos(\Delta\phi(x, y) - 2\gamma)) \tag{51-169}$$

While the analyzer rotates with $\gamma = \omega t$, the phase shift of the signal at the location x,y is equal to the retardance $\Delta\phi(x, y)$ of the sample. A dark fringe occurs, when $\Delta\phi = 2\gamma$. In this case, the rotational angle of the analyzer is identical to half the retardance. The range for $\Delta\phi$ is 0–2π. In this sense the method is still used for simple quantitative inspection of stressed optical components.

Polarimeters Using PEMs

There are many situations, in which local scanning is necessary, for instance if the sample is not plane or is too large to get observed completely. In the following, we describe PC_mSA and PC_mSC_mA configurations comprising photoelastic modulators (PEM) as they are used in industrial applications.

PEM technology offers an appropriate alternative to the crossed polarizer technique [51-3], [51-31] and [51-32]. Figure 51-41 shows a PC_mSA-arrangement to measure birefringence of a transmitting sample in which the light of a Helium-Neon laser is polarized in +45° direction before being modulated by a PEM oriented in 0°. The PEM modulates the light polarization at a frequency around 50 kHz, which then passes through the birefringent sample. Before the light enters the detector it passes another polarizer oriented at –45°.

Following the general expression (51-150) for the measured irradiance in a *PCSA*-arrangement we have

$$\begin{aligned} I(t) &= \frac{1}{4}\{M_{00} - M_{20} + (M_{02} - M_{22})\cos(\Delta\phi_{PEM}(t)) + (M_{23} - M_{03})\sin(\Delta\phi_{PEM}(t))\} \\ &= A - B \cdot \cos(\varphi - \Delta\phi_{PEM}(t)) \end{aligned} \tag{51-170}$$

with M_{ij} as the appropriate Mueller matrix elements of the sample and

$$A = \frac{M_{00} - M_{20}}{4} \tag{51-171}$$

$$B = \frac{1}{4}\sqrt{(M_{02} - M_{22})^2 + (M_{23} - M_{03})^2} \tag{51-172}$$

$$\varphi = \arctan\frac{M_{23} - M_{03}}{M_{02} - M_{22}} \tag{51-173}$$

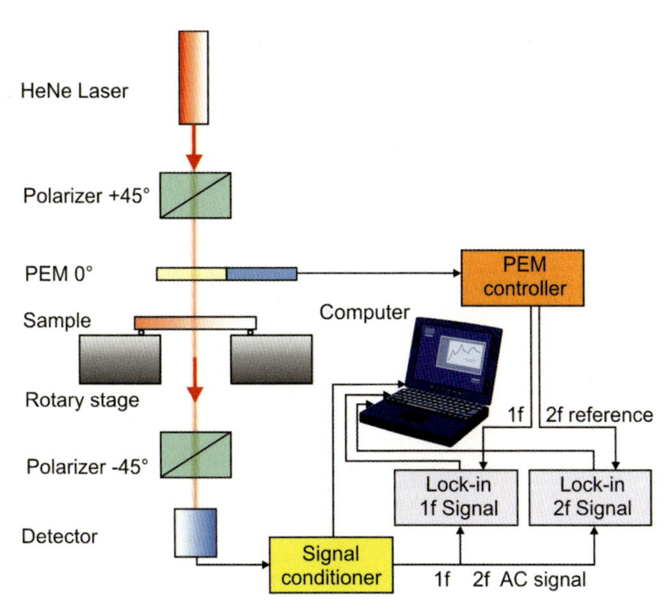

Figure 51-41: Polarimeter setup PC_mSA using a PEM (photoelastic modulator) to measure the birefringence magnitude of optical elements in transmission.

For a birefringent sample with retardance $\Delta\phi$ and the local orientation α, the irradiance becomes

$$I(t) = \frac{1}{4}\left\{1 - \left(\sin^2 2\alpha + \cos^2 2\alpha\cos\Delta\phi\right)\cos\left(\Delta\phi_{\mathrm{PEM}}(t)\right) + \cos 2\alpha\sin\left(\Delta\phi_{PEM}(t)\right)\right\} \tag{51-174}$$

and therefore

$$A = \frac{1}{4} \tag{51-175}$$

$$B = \frac{1}{4}\sqrt{\left(\left(\sin^2 2\alpha + \cos^2 2\alpha\cos\Delta\phi\right)^2 + \cos^2 2\alpha\right)} \tag{51-176}$$

$$\varphi = \arctan\left(\frac{\cos 2a}{\sin^2 2a + \cos^2 2a \cos\Delta\phi}\right) \tag{51-177}$$

Usually the PEM is modulated by a sinusoidal retardance modulation $\Delta\phi_{\mathrm{PEM}} = \Delta\phi_{\mathrm{PEM},0} \sin 2\pi ft$. The orientation of the retardance axis of the sample should either be known or should be measured by rotating the sample until a maximum signal is observed. The magnitude of the birefringence can then be determined from the lock-in outputs (1f and 2f) and the average signals. The arrangement can be used for measuring small residual birefringence of optical materials and for measuring the retardance of waveplates.

Since the PEM solution measures the modulated signals very quickly it can average over light intensity fluctuation which otherwise would disturb the birefringence measurement.

Typical birefringence repeatabilities for visible and ultraviolet wavelengths are ± 0.1 nm with a retardation range of 0.5 nm–80 nm. Spot size is typically 3 mm. The typical measurement time is less than 3 seconds per data point.

To measure the birefringence magnitude and orientation of an arbitrarily oriented sample the setup in Figure 51-41 must be modified by allowing the second polarizer to be oriented in at least two different positions, for instance, –45° and 0°. Sequential measurements with both orientations must then be carried out and analyzed to give the birefringence magnitude and angular position.

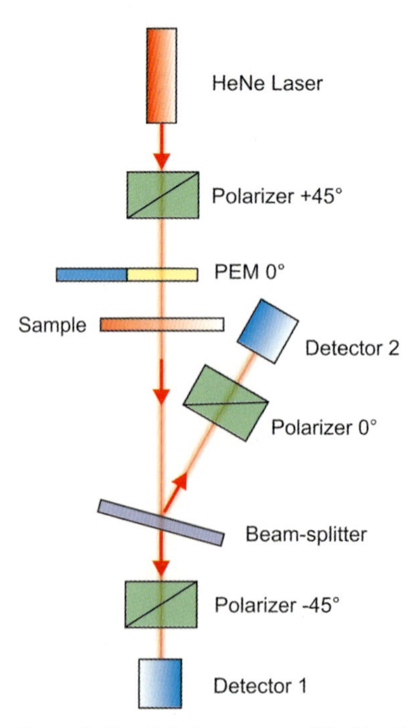

Figure 51-42: Polarimeter setup PC_mSA using a PEM to measure the birefringence magnitude and the angular position of arbitrarily oriented optical elements in transmission.

To enable rapid acquisition without moving parts, the arrangement in Figure 51-42 can be used in which a beam-splitter is used to supply two detector channels with different polarizer orientations [51-33] and [51-34]. After passing the sample, the modulated beam is divided to pass a 45°- and a 0°-oriented polarizer. The electronic signals are processed through a lock-in amplifier and some analysis software to determine the retardation magnitude and axis angle.

When the PEM is modulated by a sinusoidal retardance modulation the retardance magnitude $\Delta\phi$ and the angle α of the fast axis of a linear birefringent sample are calculated from

$$\Delta\phi = \sqrt{R_1^2 + R_2^2} \quad (51\text{-}178)$$

$$\alpha = \frac{1}{2} \arctan \frac{R_2}{R_1} \quad (51\text{-}179)$$

with

$$R_1 = \frac{V_1^{(1f)}\left(1 - J_0(\Delta\phi_{\mathrm{PEM},0})\right)}{\sqrt{2} D_1 J_1(\Delta\phi_{\mathrm{PEM},0})} \quad (51\text{-}180)$$

$$R_2 = \frac{V_2^{(1f)}}{\sqrt{2} D_2 J_1(\Delta\phi_{\mathrm{PEM},0})} \quad (51\text{-}181)$$

and D_1, D_2 are the average signals from the detectors 1 and 2, respectively, $V_1^{(1f)}$, $V_2^{(1f)}$ as the corresponding modulated 'AC' signals at the modulation frequency f, J_0 and J_1 as the 0th and 1st orders of the Bessel functions, respectively. The factor $\sqrt{2}$ is due to the fact that the output of a lock-in amplifier measures the root-mean-square instead of the signal amplitude.

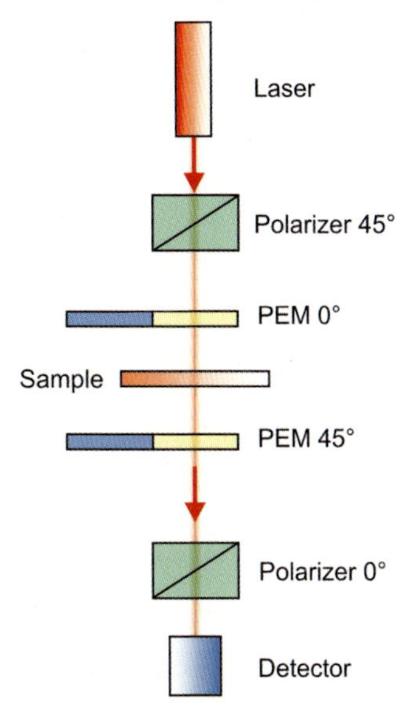

Figure 51-43: Dual-PEM setup PC_mSC_mA to measure the birefringence magnitude and angular position of arbitrarily oriented optical elements in transmission.

A single-channel polarimeter can be achieved by introducing two PEMs oriented at 0° and 45° as indicated in Figure 51-43. Both PEMs are operated at different frequencies f_1 and f_2 by sinusoidal retardance modulations $\Delta\phi_{\mathrm{PEM1}} = \Delta\phi_{\mathrm{PEM1,0}} \sin 2\pi f_1 t$ and $\Delta\phi_{\mathrm{PEM2}} = \Delta\phi_{\mathrm{PEM2,0}} \sin 2\pi f_2 t$. A polarizer at 45° is followed by the first PEM oriented with its fast axis at 0°, which is followed by the optical sample. A second PEM behind the sample oriented at 45° is followed by an analyzer at 0° and the detector [51-35].

This type of instrument will measure 9 of the 16 Mueller matrix elements of the sample. When the orientation of the polarizers and PEMs are varied, all 16 Mueller matrix elements of the sample can be measured to make it a complete Mueller polarimeter.

R_1 and R_2 in (51-180) and (51-181) are then calculated from

$$R_1 = \frac{V_1^{(1f)}}{D J_0(\Delta\phi_{\mathrm{PEM1,0}}) J_1(\Delta\phi_{\mathrm{PEM2,0}})} \tag{51-182}$$

$$R_2 = \frac{V_2^{(1f)}}{D J_0(\Delta\phi_{\mathrm{PEM2,0}}) J_1(\Delta\phi_{\mathrm{PEM1,0}})} \tag{51-183}$$

where D is the average signal measured by the detector, $V_1^{(1f)}$, $V_2^{(1f)}$ are the corresponding modulated "AC" signals at the modulation frequency f_1 and f_2, J_0 and J_1 are the 0th and 1st orders of the Bessel functions, respectively.

Measuring systems of this kind can give accuracies of less than 0.005nm. They are used to measure linear birefringence of fused silica or CaF_2 elements for microlithography like photomask blanks or optical lens elements or plane plates of the projection optics as well as low retardation wave-plates.

Ellipsometer to Measure Complex Reflectivities

The Jones matrix $\mathbf{J_R}$ of a reflective sample in an aligned ellipsometer is defined by

$$\mathbf{J_R} = \begin{pmatrix} \cos\psi & 0 \\ 0 & \sin\psi e^{i\Delta\phi} \end{pmatrix} = \begin{pmatrix} |r_s| & 0 \\ 0 & |r_p| e^{i\Delta\phi} \end{pmatrix} \tag{51-184}$$

with the complex reflectance ratio ρ as introduced in (51-7)

$$\rho = \frac{r_p}{r_s} = \tan\psi \, e^{i\Delta\phi} \tag{51-185}$$

where the sample's complex reflectances are r_p in the plane of incidence and r_s perpendicular to the plane of incidence. Here $\tan\psi$ denotes the diattenuation ratio, $\Delta\phi$ the retardance of the sample for a given incidence angle. Usual ellipsometers measure the full Jones matrix of a reflective sample. The sample is characterized by the parameters ψ and $\Delta\phi$ as a function of the incident angle and wavelength.

In the following, different types of ellipsometers are explained.

Null Ellipsometer

Early ellipsometric techniques were based on so-called nulling techniques, in which the azimuthal angles of the polarizer, retarder and analyzer were altered until the irradiance at the detector was minimal. The position of the polarizing elements determined the result of the sample's polarizing characteristics. Figure 51-44 shows an example in which the polarizer and the analyzer are rotated while the quarter-wave retarder behind the polarizer is held fixed at 45°.

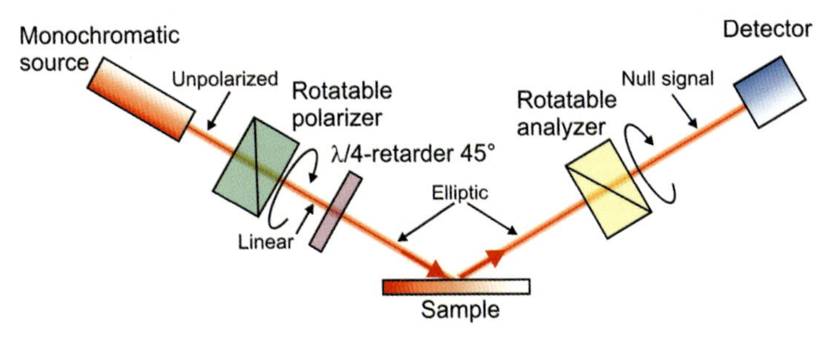

Figure 51-44: Configuration P_r CSA_r of a nulling single-wavelength ellipsometer.

Following (51-111) the irradiance at the detector for a P_r CSA_r configuration, as shown in figure 51-44, is given by

$$I = \frac{I_0}{4}\{2 + \cos 2(\psi - \gamma)[1 + \sin(2\beta - \Delta\phi)] + \cos 2(\psi + \gamma)[1 - \sin(2\beta - \Delta\phi)]\} \tag{51-186}$$

where β denotes the rotational position of the polarizer,
γ is the rotational position of the analyzer,
ψ and $\Delta\phi$ are the polarizing characteristics of the sample
and I_0 is the irradiance of the incident light.
The irradiance becomes zero for the conditions:

$$\beta = \frac{\Delta\phi}{2} \pm 45^\circ \text{ and } \gamma = \pm\psi \mp 90^\circ,$$

which have to be fulfilled simultaneously.

When connected to appropriate driving units and to a computer, some software finds the angles of minimum irradiance. The process, however, is rather slow [51-3].

Rotating Analyzer Ellipsometer (RAE)

An improvement with respect to speed is obtained by use of the rotating analyzer ellipsometer in which the polarizer is fixed at an angle of 45°. A quarter-waveplate with its fast axis at 0° can be inserted into or removed from the beam between the-sample and polarizer (Figure 51-45). The analyzer rotates continuously so that the detector receives a sinusoidally varying irradiance [51-3].

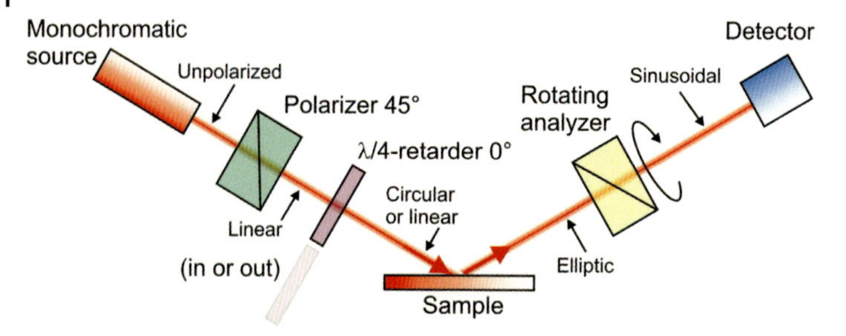

Figure 51-45: Configuration $PC_x SA_r$ (where C_x denotes an exchangable compensator) of a single-wavelength rotating analyzer ellipsometer.

Following (51-145) the irradiance $I(t)$ at the detector is given by

$$\begin{aligned} I(t) &= \frac{I_0}{4}\{1 + \cos 2\psi \cos 2\gamma(t) + \sin 2\psi \sin 2\gamma(t) \cos \Delta\phi\} \\ &= \frac{I_0}{4}\{1 + A \cos 2\gamma(t) + B \sin 2\gamma(t)\} \end{aligned} \tag{51-187}$$

where $\gamma(t)$ denotes the rotational position of the analyzer at time t and I_0 is the irradiance of the incident light.

When the rotation of the analyzer is periodic with $\gamma(t) = \omega t$ the signal can be Fourier analyzed at the second harmonic 2ω to determine the ellipsometric parameters ψ and $\Delta\phi$ according to

$$\tan \psi = \sqrt{\frac{1+A}{1-A}} \tag{51-188}$$

$$\cos \Delta\phi = \frac{B}{\sqrt{1-A^2}} \tag{51-189}$$

When the emerging light is linearly polarized ($\Delta\phi$ of the sample is near 0° or 180°) the uncertainty in the determination of $\Delta\phi$ is greatest, because the signal's phase modulation by $\cos \Delta\phi$ is very insensitive. In this case the quarter-wave plate is inserted to shift the phase by 90°. $\cos \Delta\phi$ in (51-187) is then replaced by $\sin \Delta\phi$.

Spectroscopic Ellipsometer

Since the main purpose of ellipsometry is to characterize thin films, ellipsometers have to respect the dispersion characteristics of the optical constants n and k of the material. Therefore measurements at multiple wavelengths are unavoidable. Lasers are not usually capable of performing spectroscopic ellipsometry measurements, unless they are tuneable over a considerable wavelength range [51-36]–[51-40].

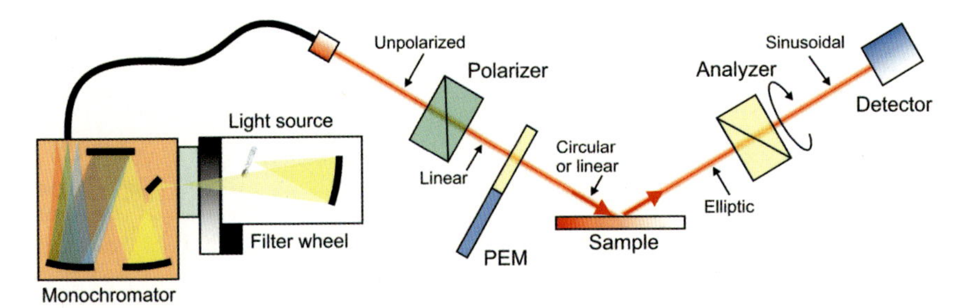

Figure 51-46: Configuration $PC_m SA_r$ of a spectroscopic rotating analyzer ellipsometer using a monochromator as the light source.

In Figure 51-46 the laser source is replaced by a wavelength-tunable source using a white light source with a monochromator to select narrow-band spectra over a wide spectral range. Light from the monochromator is transferred to the ellipsometer by means of a fiber. The basic setup is the same as in Figure 51-45. Note, however, that a quarter-waveplate would have a retardation of 90° only for a specific wavelength. When a wider spectrum is investigated, the quarter-waveplate is usually no longer needed, since a lot of data is acquired from which the cases with $\Delta\phi$ near 0° or 180° are removed. When necessary, its spectral retardance must be determined and can then be considered in cases where it enhances the accuracy. Alternatively, a PEM can be inserted into selected individual retardances for each wavelength and sample as shown in Figure 51-46.

When the light level becomes very low the white light source can be chopped to remove influences of the room light and also any stray light from other sources. In this case, an additional sinusoidal signal is superimposed onto the modulation by the rotating analyzer. The ratio of the modulation frequencies should be selected appropriately so that their separation is easily achieved.

In a spectroscopic ellipsometer it is convenient to keep the fixed polarizer adaptable to other azimuthal angles. Since the uncertainty in the data is significantly reduced when the axis of the polarizer has a value near to the ψ of the sample, the polarizer can be adjusted prior to the measurement, which is referred to as "polarizer tracking" [51-3].

Ellipsometers with a monochromator in the sending channel are slow, because the monochromator has to be tuned to the individual wavelengths. A much more efficient arrangement is to use a spectrometer in the receiving channel and replacing the single channel detector by a CCD line or matrix which detects the total spectrum simultaneously. Figure 51-47 shows a rotating-polarizer multichannel ellipsometer in the $P_r(\omega)SA$ optical configuration used for spectroscopic ellipsometry, where three parameters are determined in real time [51-36] and [51-37].

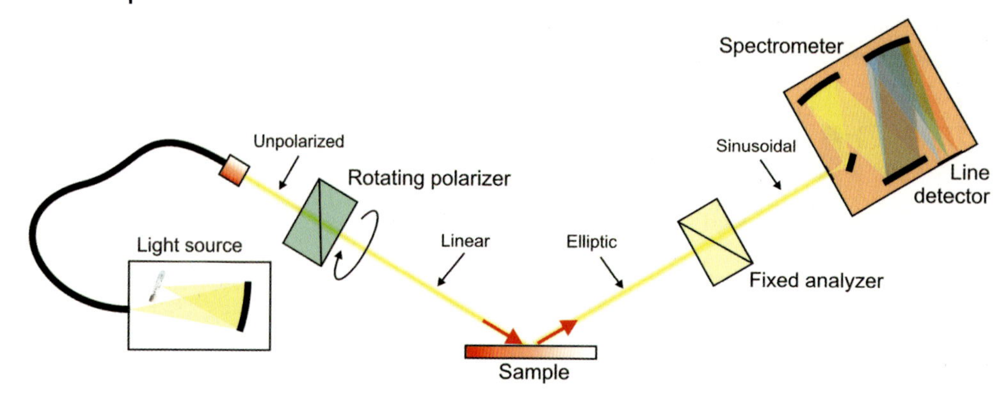

Figure 51-47: Configuration P_r SA of a spectroscopic rotating-polarizer multichannel ellipsometer using a line detector spectrometer in the receiving channel.

The rotating-polarizer multichannel ellipsometer is best suited for measuring strongly-absorbing material structures, i.e., semiconductor or metal thin films on semiconductor substrates. The surfaces should be isotropic, homogeneous and non-depolarizing over the incident beam. The benefits of the arrangement are:

- a potentially wide spectral range,
- ease of calibration,
- high speed,
- good precision and accuracy.

Since only three parameters can be detected simultaneously, the instrument is an incomplete Mueller ellipsometer and only capable of partially determining the Stokes vector components. When the reflected light is nearly linearly polarized, the accuracy is not sufficient.

Figure 51-48 shows a single rotating-compensator multichannel ellipsometer in the $PSC_r(\omega)A$ optical configuration used for spectroscopic ellipsometry, where four unnormalized parameters of the Stokes vector are determined in real time.

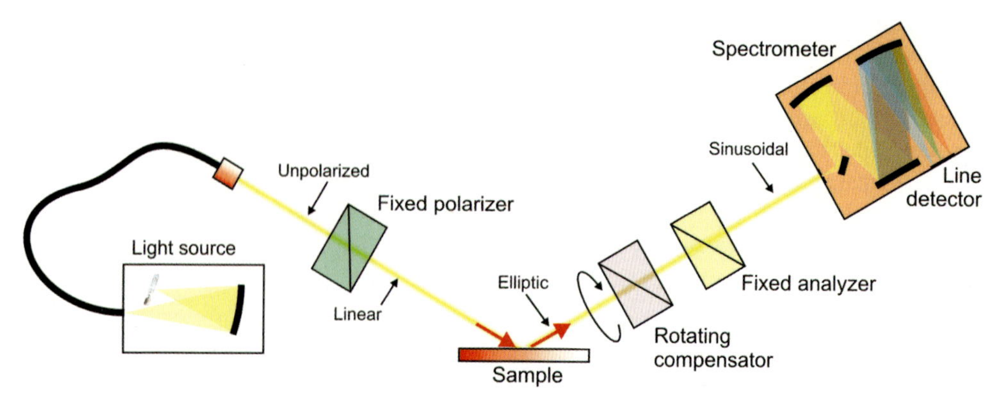

Figure 51-48: Configuration PSC_rA of a single rotating-compensator multichannel ellipsometer using a line detector spectrometer in the receiving channel.

The single rotating-compensator multichannel ellipsometer is able to detect the full Stokes vector of a reflected beam from a polarizing sample surface. It is best suited for weakly absorbing layers or transparent substrates. In addition, it can handle simple inhomogeneities of a sample. It should be noted that the rotating compensator must keep a retardance near the quarterwave value over the entire spectrum, otherwise the precision and accuracy will decrease substantially.

Figure 51-49 shows a dual rotating-compensator multichannel ellipsometer in the $PC_r(5\omega)SC_r(3\omega)A$ optical configuration used for spectroscopic ellipsometry to determine all 16 elements of the Mueller matrix in real time [51-41].

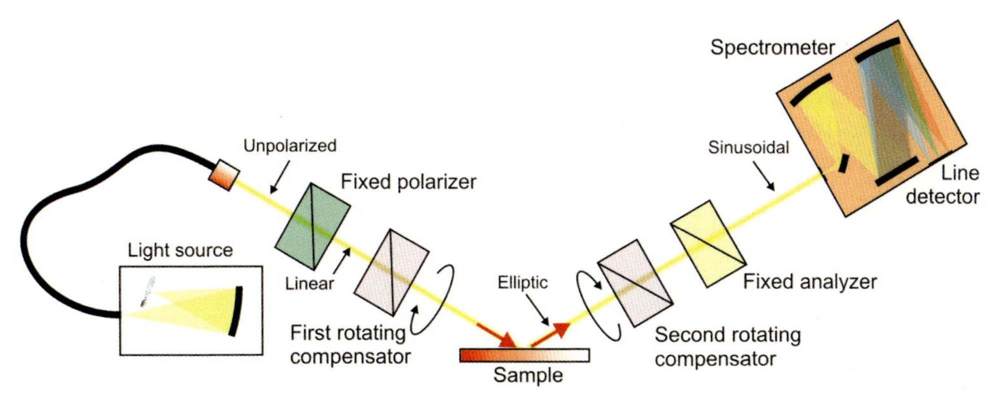

Figure 51-49: Configuration $PC_rSC_r\,A$ of a dual rotating-compensator multichannel ellipsometer using a line detector spectrometer in the receiving channel.

The instrument is suitable for the analysis of all kinds of strongly or weakly absorbing samples, and also for the analysis of samples exhibiting both heterogeneity and anisotropy. Potential applications are in the study of anisotropic thin film growth.

51.5 Calibration Techniques

In the previous chapters we have assumed that the angular positions of the polarizing elements are well-defined and known. During the setup of real systems these positions are usually only known approximately. The orientation is indicated by scales and marks on the mounts of the components. Rotating elements are driven, for example, by stepping motors with integrated angle encoders. Their exact physical zero position relative to the nominal values acquired by the control computer might be arbitrary. For precise measurements this is, of course, insufficient.

Before the setup of a new system, the calibration method should be defined. The angular adjustment of the polarizing elements is best achieved in the straight-through configuration without a sample. However, it should also be possible to calibrate the system with the sample in place, using exactly the same configuration and alignment as used in the measurement. The calibration sample should be selected appropriately to the main polarimeter task.

In the calibration process the angle offsets of the fixed elements and the phase offsets of the rotating elements must be adjusted and determined, to be considered in the final evaluation of the data. The general approach is to alter the angular position of the fixed elements by appropriate mechanical (motor driven) rotators, while the rotating element moves and the modulated signal is acquired. The Fourier coefficients at various relevant frequencies are traced and compared with theoretical values. The correct positions of the fixed elements as well as the phase offset of the rotating element can then be determined by interpolation from the acquired curves [51-3].

In phase modulated ellipsometers (PME) the angle and also the dynamic and static retardation of the individual PEMs need to be calibrated, which is similar in complexity to the calibration of RPEs and RAEs. However, with the help of computer control both calibration techniques are comparable and involve only little user action [51-42]. Each PEM needs four individual calibrations:

1) the azimuthal angle of the PEM with respect to the polarizer,
2) the static strain of the modulator,
3) the modulation amplitude relative to the driving voltage,
4) the azimuthal angle of the PEM with respect to the plane of incidence at the sample.

Calibrations 2) and 3) are wavelength dependent, while the azimuthal angles are not. If the instrument is equipped with mechanics for sample adjustment, the PEM needs to be adjusted only once.

Calibration 1) is best done in the straight-through configuration after polarizer and analyzer have been crossed perfectly by nulling the transmitted signal. The PEM is then set to the 0° or 90° position with respect to the polarizer, in which no modulation is detected, and afterwards moved in the ± 45° position by means of a precision rotator.

Calibration 2) is determined in the same position. With the modulator turned off, the wavelength dependent strain induced retardation is determined and approximated by polynomials of low order.

Calibration 3) is the most crucial and difficult process. The modulator is turned off and the polarizer is rotated in the 0° position with respect to the analyzer, where the transmitted intensity is $2I_0$, whereas it is 0, when the polarizer and analyzer are crossed. By adjusting the modulator amplitude, a point can be found where the DC component of the transmitted intensity is I_0. At this specific point the modulation amplitude is 2.4048, because $J_0(2.4048)=0$.

Calibration 4) is carried out with a sample in place, since it defines the plane of incidence. Depending on whether the PEM is used in the PSC_mA or the PC_mSA configuration, it is rotated together with the analyzer or polarizer, respectively, until nulling is achieved. The components are then rotated back to their specified positions by means of accurate motoric rotators.

For dual PEM systems, the calibration procedure is essentially the same as for single PEM systems, except that both PEM pairs need to be calibrated. The four calibration steps for a single PEM system must be performed for each PEM.

Once the instruments are adjusted and calibrated, their ψ- and $\Delta\phi$-results can be checked with the help of commercially available standards for ellipsometers. For the visible wavelength they usually consist of silicon wafers with a thermally grown silicon dioxide film. ψ and $\Delta\phi$ are specified as well as the derived thickness and refractive index of the silicon dioxide layer on the silicon wafer.

51.6 Accuracy and Error Sources

As in any other measuring device, errors can be classified as random or systematic. Random errors can be reduced by collecting and averaging more data. The standard deviation of the resulting random error is then reduced by a factor of $N^{-\frac{1}{2}}$ where N is the number of collected samples.

Random error sources can be any of the following.

- Fluctuation from the light source.
- Irregularities in the rotation of the polarizing components.
- Detector noise.
- Shot noise.

To minimize random errors, various preventative measures can be taken as follows.

- Collecting a reference signal from the light source to normalize the signal to the actual dc level.
- Using precision rotary stages.
- Using low noise detector systems.
- Generating and collecting as much light as necessary for a sufficiently low shot noise level.

Since the individual random error contributions are independent, their standard deviations can be added quadratically to calculate the total random error.

Changes in the environmental conditions such as temperature, pressure and humidity lead to drift errors, which are slowly varying random errors. They have an impact on the retardance and diattenuation of the components depending on their type. For instance, zero order or compound zero order retarders have much higher thermal stabilities than multi-order retarders.

Wavelength changes and mechanical drifts within the optical setup might also lead to slowly varying effects which depend on the optical configuration and the selected components. For instance, if zero-order or compound zero-order retarders are preferred over multi-order retarders, any mechanical drifts causing altered angles of incidence can be neglected.

Systematic errors are the dominating errors in polarimetry because they are difficult to identify. Taking more data generally does not improve the quality of the results. The strategy is to identify and minimize systematic errors and then calculate a budget for the residuals.

Systematic errors are caused by the following.

- Misalignments of the polarizing components (including the sample) relative to each other.
- Erroneous calibration of the zero position of the rotating polarizing elements.
- Erroneous calibration of the PEM's phase modulation amplitude.
- Erroneous calibration of the PEM's static phase bias.
- Diattenuation error of the polarizer or analyzer.
- Unwanted retardance of the polarizer or analyzer.
- Retardance error of the compensator.
- Unwanted diattenuation of the compensator.
- Wavelength calibration error of the spectrometer or monochromator.
- Wavelength spread error of the spectrometer or monochromator.
- Sample depolarization effects.
- Angle of incidence error at the sample.

In order to estimate systematic errors, simulations are necessary including probable misalignments, erroneous calibrations and the estimated retardance and diattenuation deviations of all setup components [51-43]–[51-46].

In the following we give examples for P_rSA and P_rCSA arrangements measuring the birefringence $\Delta\phi$ and azimuth angle α of a given retardation element. The simulations refer to figures 51-30 and 51-32, where the fixed analyzer is nominally set to 0°, while the polarizer is rotating with ω_P. In the P_rCSA arrangement of figure 51-32 the compensator is set to 45°.

Following (51-115) and (51-128), the irradiance at the detector is given for both setups by

$$I(t) = a + b\sin\left(2\omega_P t + \varphi\right) \tag{51-190}$$

The parameters a, b and φ of the signal are detected by the instrument and the results for $\Delta\phi$ and α are calculated for the P_rSA arrangement from

$$\Delta\phi = 2\arcsin\sqrt{\frac{1-4b\sin\varphi+4b^2}{1-2b\sin\varphi}} \tag{51-191}$$

$$\alpha = \frac{1}{2}\arctan\frac{1-2b\sin\varphi}{2b\cos\varphi} \tag{51-192}$$

and for the P_rCSA arrangement from

$$\Delta\phi = 2\arctan\frac{b\sin\varphi}{1 - b\cos\varphi} \tag{51-193}$$

$$a = \frac{1}{2}\arccos\frac{\cos\varphi(1 - b\cos\varphi)}{b\sin^2\varphi} \tag{51-194}$$

Estimations of the error behavior can be made by calculating the total derivatives

$$\delta\Delta\phi = \left|\frac{\partial\Delta\phi}{\partial b}\right|\delta b + \left|\frac{\partial\Delta\phi}{\partial\varphi}\right|\delta\varphi \tag{51-195}$$

$$\delta a = \left|\frac{\partial a}{\partial b}\right|\delta b + \left|\frac{\partial a}{\partial\varphi}\right|\delta\varphi \tag{51-196}$$

by using the expressions from (51-191) to (51-194) to calculate the partial derivatives. Then estimates for δb and $\delta\varphi$ must be found assuming different parameters that lead to an erroneous signal $I(t)$.

Figure 51-50 compares the error behavior of a P_rSA and a P_rCSA arrangement for a misalignment of the fixed polarizer relative to the perfectly aligned rest of the system. A birefringent sample with $\Delta\phi = 5°$ has been introduced. It is obvious that the P_rSA arrangement is extremely sensitive to small polarizer misalignments when $\Delta\phi$ and a must be determined. A P_rCSA arrangement measures a small $\Delta\phi$ precisely; however, a becomes rather uncertain when a polarizer misalignment is introduced.

Figure 51-51 makes the same comparison with $\Delta\phi = 85°$. The determination of $\Delta\phi$ and a is now much less sensitive to polarizer misalignment errors also for the P_rSA arrangement.

In the following example, samples of $\Delta\phi = 5°$ and 85° have been simulated in a P_rCSA arrangement. Figure 51-52 shows the deviations of birefringence $\Delta\phi$ and azimuth angle a as a function of a compensator retardance error. The deviations for $\Delta\phi$ are comparable for small and large birefringence; however, the azimuth angle becomes quite uncertain when small birefringence is to be analyzed. Due to the fact that, for small birefringence, the signal $I(t)$ is very weakly modulated (figure 51-52 (a)), the results become very sensitive to noise. The S/N of 4096 in the example already leads to significant noise in the results for a. When the birefringence approaches 90°, the sinusoidal signal is large and is therefore quite insensitive to noise.

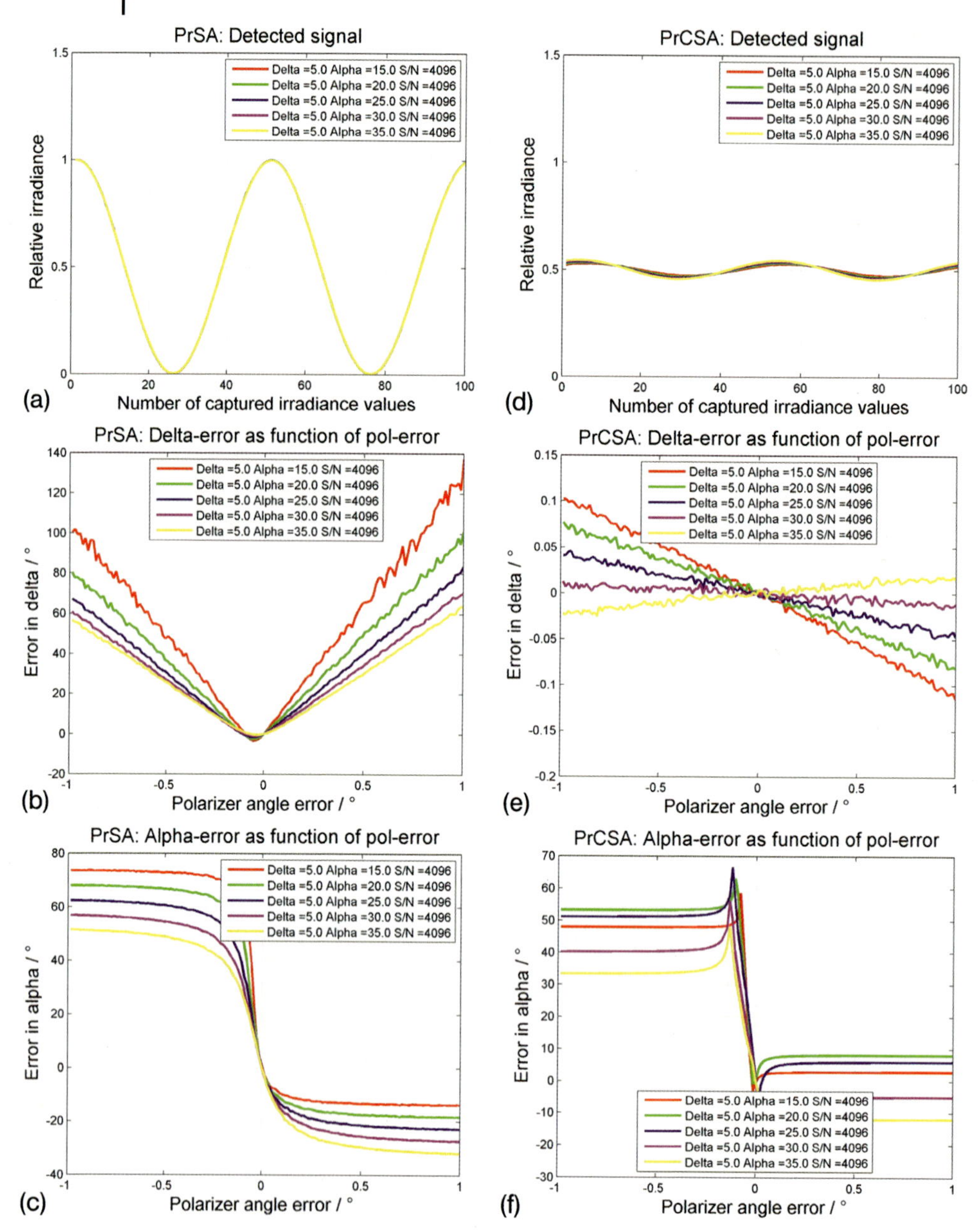

Figure 51-50: The influence of a polarizer angle error on the result for birefringence $\Delta\phi$ and azimuth angle α in a P_rSA and P_rCSA arrangement. An S/N ratio of 4096 caused by shot noise has been introduced. Simulated are $\Delta\phi = 5^\circ$ and $\alpha = 15^\circ - 35^\circ$: (a) + (d): detected signal for P_rSA and P_rCSA; (b) + (e) $\Delta\phi$-error as function of polarizer angle error for P_rSA and P_rCSA; (c) + (f) α-error as function of polarizer angle error for P_rSA and P_rCSA.

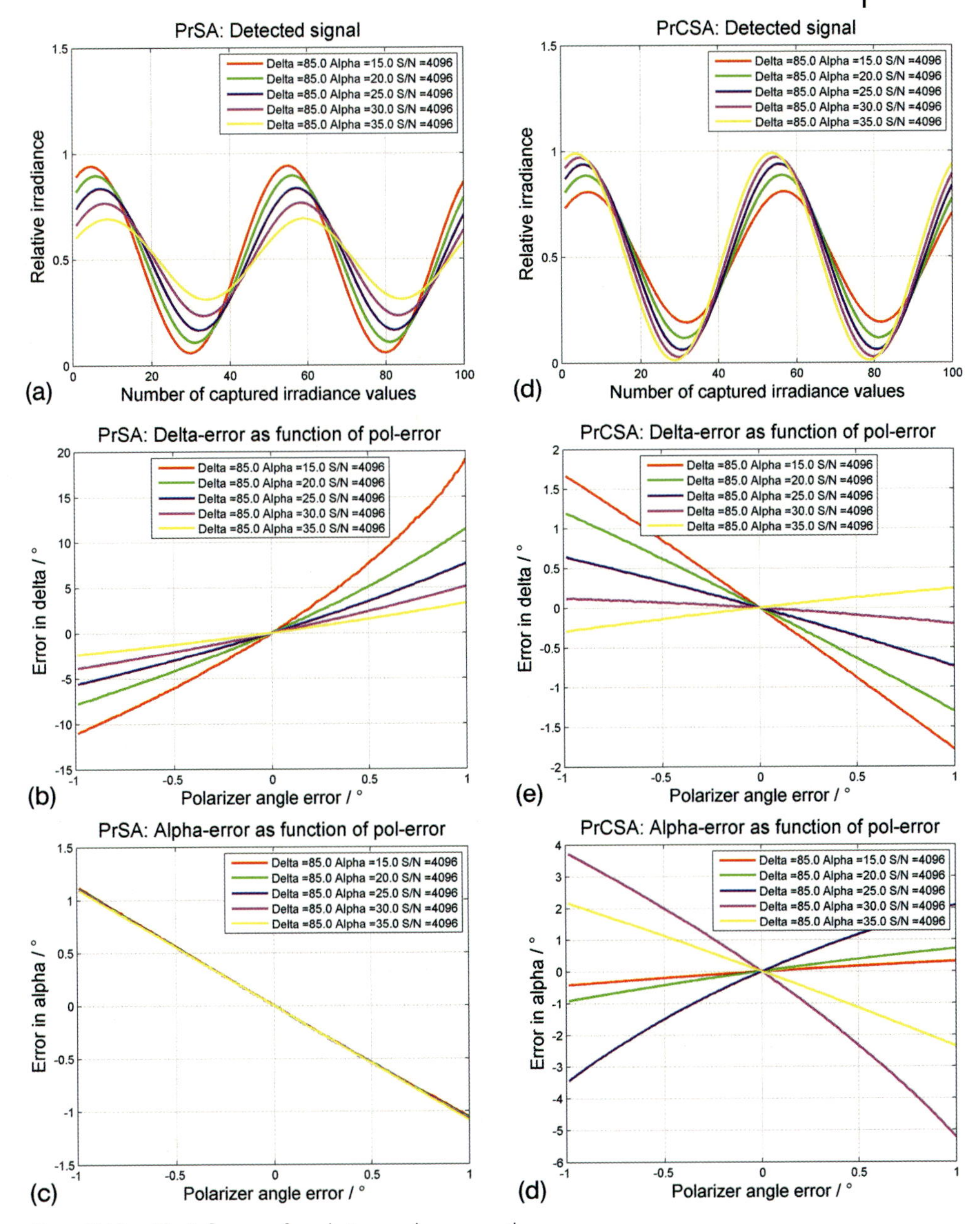

Figure 51-51: The influence of a polarizer angle error on the result for birefringence $\Delta\phi$ and azimuth angle α in a P_rSA and P_rCSA arrangement. An S/N ratio of 4096 caused by shot noise has been introduced. Simulated are $\Delta\phi = 85^\circ$ and $\alpha = 15^\circ - 35^\circ$: (a) + (d) detected signal for P_rSA and P_rCSA; (b) + (e) $\Delta\phi$-error as a function of polarizer angle error for P_rSA and P_rCSA; (c) + (f) α-error as a function of polarizer angle error for P_rSA and P_rCSA.

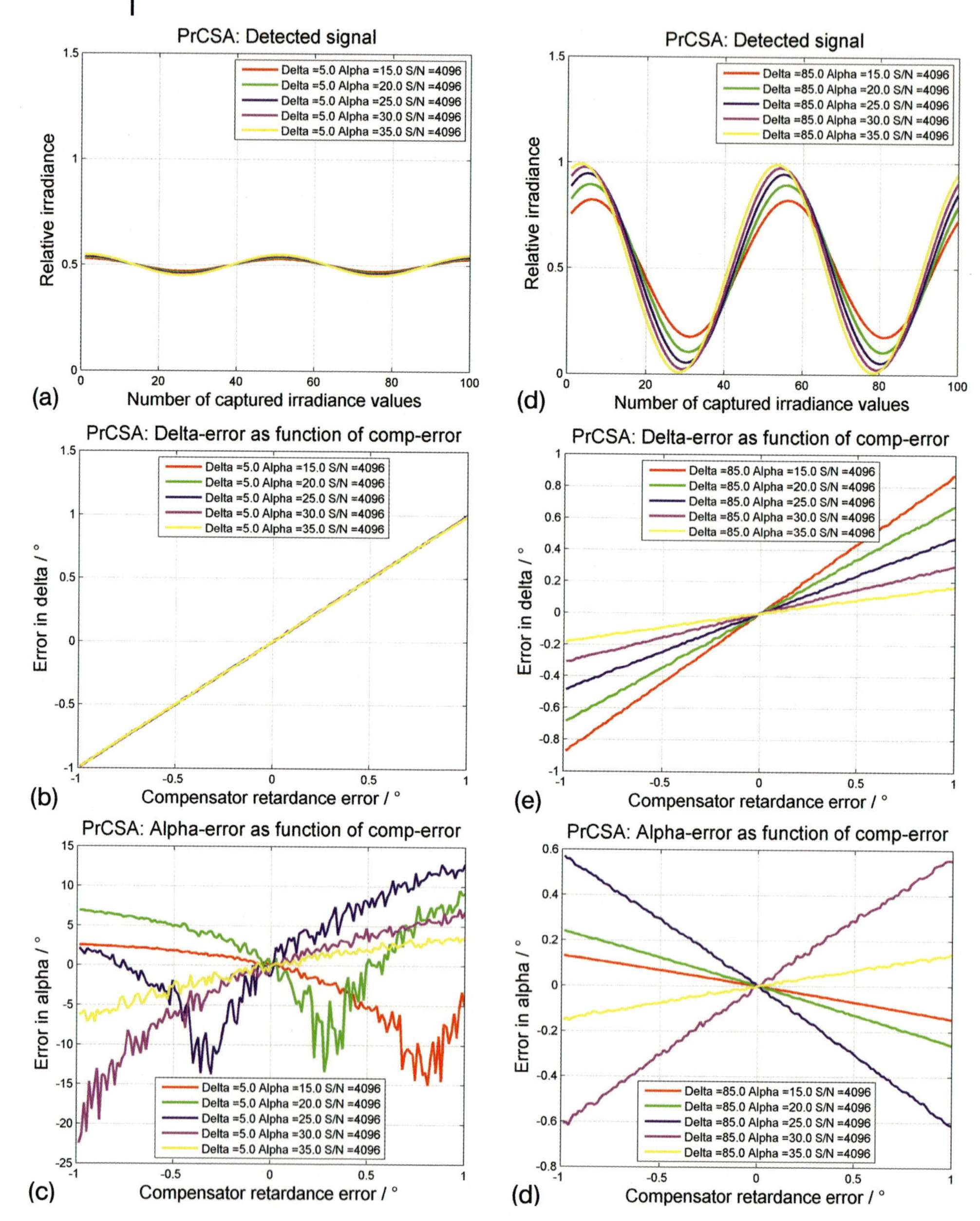

Figure 51-52: The influence of a compensator retardance error on the result for birefringence $\Delta\phi$ and azimuth angle α in a P_rCSA arrangement. An S/N ratio of 4096 caused by shot noise has been introduced. Simulated are $\Delta\phi = 5°$ and 85° and $\alpha = 15° - 35°$: (a) + (d) detected signal for $\Delta\phi = 5°$ and 85°; (b) + (e) $\Delta\phi$-error as a function of compensator retardance error for $\delta = 5°$ and 85°; (c) + (f) α-error as a function of compensator retardance error for $\Delta\phi = 5°$ and 85°.

51.7 Literature

51-1 K. Rochford, Polarization and Polarimetry, Encyclopedia of Physical Science and Technology, 3rd Edition, Vol. 12, Academic Press, San Diego (2001).

51-2 R. M. A. Azzam and N. M. Bashara, Ellipsometry and Polarized Light, Elsevier Science B.V., Amsterdam (1987).

51-3 H. G. Tompkins and E. A. Irene, Handbook of Ellipsometry, William Andrews Publishing, Norwich, NY (2005).

51-4 M. Bass (Ed.), Handbook of Optics, Vol. II, Chapter 3: Polarizers, McGraw-Hill, New York (1995).

51-5 M. Bass (Ed.), Handbook of Optics, Vol. II, Chapter 22: Polarimetry, McGraw-Hill, New York (1995).

51-6 D. S. Kliger, J. W. Lewis and C. E. Randall, Polarized Light in Optics and Spectroscopy, Academic Press (1990).

51-7 C. Brosseau, Fundamentals of Polarized Light, John Wiley & Sons, New York (1998).

51-8 D. Clarke and J. F. Grainger, Polarized Light and Optical Measurement, Pergamon Press, Oxford (1971).

51-9 E. Hecht, Optics, 3rd Ed., Addison-Wesley (1998).

51-10 D. G. M. Anderson and R. Barakat, Necessary and sufficient conditions for a Mueller matrix to be derivable from a Jones matrix, J. Opt. Soc. Am. 11, 2305–19 (1994).

51-11 W. A. Shurcliff, Polarized Light: Production and Use, Harvard University Press, Cambridge, Mass. (1966).

51-12 J. L. Pezzaniti and R. A. Chipman, Angular dependence of polarizing beam-splitter cubes, Appl. Opt. 33, 1916–29 (1994).

51-13 L. Li and J. A. Dobrowolski, High-performance thin-film polarizing beam splitter operating at angles greater than the critical angle, Appl. Opt. 39, 2754–71 (2000).

51-14 M. G. Robinson, J. Chen and G. D. Sharp, Wide field of view compensation scheme for cube polarizing beam splitters, SID03 Digest (2003).

51-15 Thomas Baur, A new type of beam-splitting polarizer cube, Proc. SPIE, Vol. 5158, 135 (2003).

51-16 P. Drexler and P. Fiala, Utilization of Faraday mirror in fiber optic current sensors, Radioengineering, Vol. 17, No. 4 (2008)

51-17 J. C. Kemp, Piezo-optical birefringence modulators: new use for a long-known effect, J. Opt. Soc. Am. 59, 950–54 (1969).

51-18 J. C. Kemp, Basic laboratory set-up for various measurements possible with the photoelastic modulator, Application note, Hinds Instruments, Inc. (1975).

51-19 N. Hodgson and H. Weber, Laser Resonators and Beam Propagation: Fundamentals, Advanced Concepts and Applications (2nd ed.) Springer, Berlin (2005).

51-20 K. Mochizuki, Degree of polarization in jointed fibers: The Lyot depolarizer, Appl. Opt. 23, 3284 (1984).

51-21 W. Hanle, Messung des Polarisationsgrades von Spektrallinien, Zeitschrift für Instrumentenkunde 51, p. 488 (1931).

51-22 T. C. Oakberg, Measurement of wave-plate retardation using a photoelastic modulator, SPIE, 3121, p. 19–22 (1997).

51-23 J. P. Badoz, M. P. Silverman and J. C. Canit, Wave propagation through a medium with static and dynamic birefringence: theory of the photoelastic modulator, JOSA A, Vol. 7, 672–82 (1990).

51-24 R. Anderson, Measurement of Mueller matrices, Appl. Opt. 31, 11 (1992).

51-25 R. M. A. Azzam, Division-of-amplitude photopolarimeter DOAP for the simultaneous measurement of all four Stokes parameters of light, Opt. Acta 29, 767–77 (1985).

51-26 G. E. Jellison, Jr and F. A. Modine, Two-modulator generalized ellipsometry: experiment and calibration, Appl. Opt. 36, 8184 (1997).

51-27 G. E. Jellison, Jr. and F. A. Modine, Two-modulator generalized ellipsometry: theory, Appl. Opt. 36, 8190 (1997).

51-28 G. E. Jellison, Jr and F. A. Modine, Two Modulator Generalized Ellipsometer for Complete Mueller Matrix Measurement, US patent No. 5,956,147 (1999).

51-29 D. H. Goldstein, Mueller matrix dual-rotating retarder polarimeter, Appl. Opt. 31, 6676–83 (1992).

51-30 R. C. Thompson, J. R. Bottinger and E. S. Fry, Measurement of polarized light interactions via the Mueller matrix, Appl. Opt. 19, 1323–32 (1978).

51-31 J. C. Kemp, Detecting polarized light at levels below 1 ppm, Proc. SPIE Int. Soc. Opt. Eng., Vol. 891, 79 – 83 (1988).

51-32 B. L. Wang, T. C. Oakberg and P. Kadlec, Industrial applications of a high-sensitivity linear birefringence measurement system, polarization: measurement, analysis and remote sensing, SPIE, Vol. 3754, p.197 (1999).

51-33 T. Oakberg, Stokes Polarimetry, PEMlabs Application Note, Hinds Instruments Inc. (2007).

51-34 T. Oakberg, Birefringence Measurement, PEMlabs Application Note, Hinds Instruments Inc. (2007).

51-35 B. Wang and T. Oakberg, Dual PEM Systems: Polarimentry Applications, PEMlabs Application Note, Hinds Instruments Inc. (2005).

51-36 J. A. Woollam, B. Johs, C. Herzinger, J. Hilfiker, R. Synowicki and C. Bungay, Overview of Variable Angle Spectroscopic Ellipsometry (VASE), Part I: Basic Theory and Typical Applications, Optical Metrology, Vol. CR72, 3–28. SPIE, Bellingham, Washington (1999).

51-37 J. A. Woollam, B. Johs, C. Herzinger, J. Hilfiker, R. Synowicki and C. Bungay, Overview of Variable Angle Spectroscopic Ellipsometry (VASE), Part II: Advanced Applications, Vol. CR72, p. 29–58. SPIE, Bellingham, Washington (1999).

51-38 J. A. Woollam, Ellipsometry, Variable Angle Spectroscopic, in: J.G. Webster (ed.) Wiley Encyclopedia of Electrical and Electronics Engineering, John Wiley & Sons, New York, p. 109–16 (2000).

51-39 W. A. McGahan, B. Johs and J. A. Woollam, Techniques for ellipsometric measurement of the thickness and optical constants of thin absorbing films, Thin Solid Films, 234, 443–46 (1993).

51-40 G. E. Jellison, Jr., The calculation of thin film parameters from spectroscopic ellipsometry data, Thin Solid Films 290–91, p. 40–45 (1996).

51-41 G. E. Jellison, Jr and F. A. Modine, Two-modulator generalized ellipsometry: experiment and calibration, Appl. Opt. 8184–89 (1997).

51-42 G. E. Jellison, Jr and F. A. Modine, Accurate calibration of a photoelastic modulator in a polarization modulation ellipsometry experiment, in Polarization Considerations for Optical Systems II, R. A. Chipman, ed., Proc. Soc. Photo-Opt. Instrum. Eng. 1166, 231–41 (1989).

51-43 P. S. Hauge, Mueller matrix ellipsometry with imperfect compensators, J. Opt. Soc. Am. 68, 1519–28 (1978).

51-44 D. H. Goldstein and R. A. Chipman, Error analysis of a Mueller matrix polarimeter, J. Opt. Soc. Am. 7, 693–700 (1990).

51-45 F. A. Modine, G. E. Jellison, Jr and G. R. Gruzalski, Errors in ellipsometry measurements made with a photoelastic modulator, J. Opt. Soc. Am. 73, 892–900 (1983).

51-46 F. A. Modine and G. E. Jellison, Jr, Errors in polarization measurements due to static retardation in photoelastic modulators, Appl. Phys. Commun. 12, 121–39 (1993).

51-47 T. Yoshizawa (Ed.), Handbook of Optical Metrology, CRC Press, Taylor & Francis Group, LLC, Boca Raton (2009).

52
Testing the Quality of Optical Materials

Handbook of Optical Systems: Vol. 5. Metrology of Optical Components and Systems. First Edition.
Edited by Herbert Gross.

52.1 Specifications

The specification of optical materials covers a number of optical, mechanical, thermal and chemical properties which an engineer has to consider when selecting a material for a special application. The optical properties include [52-1]:

- Refractive index
- Transmittance
- Homogeneity
- Striae
- Birefringence
- Bubbles and Inclusions

The mechanical properties include:

- Density
- Knoop hardness
- Grindability [52-2]
- Viscosity

The thermal properties include:

- Coefficient of linear thermal expansion
- Thermal conductivity
- Heat capacity

The chemical properties include:

- Climatic resistance [52-3]
- Stain resistance
- Acid resistance [52-4]
- Alkali resistance [52-5] and phosphate resistance [52-6]
- Environmental aspects
- Hazardous substances

In the following we will discuss how optical properties can be measured and quantified.

52.2 Refractive Index

52.2.1 Basics

The most important optical parameters characterizing an optical glass are the refractive index n_d in the middle range of the visible spectrum and the Abbe number

$$\nu_d = \frac{n_d - 1}{n_F - n_C} \tag{52-1a}$$

as a measure for dispersion. The difference $n_C - n_F$ is called the principal dispersion.

If not specified as a number in nanometers the letter subscripts as d, F, C, etc., refer to the wavelengths shown in table 52-1 [52-1], [52-7].

Table 52-1: Subscripts for refractive index and their associated wavelength

Subscript	Wavelength [nm]
t	1013.98
s	852.11
r	706.5188
C	656.2725
C'	643.8469
D	589.2938
d	587.5618
e	546.0740
F	486.1327
F'	479.9914
g	435.8343
h	404.6561
i	365.0146

It is also very common to specify an Abbe number based on the e-line as shown in (52-1b).

$$\nu_e = \frac{n_e - 1}{n_{F'} - n_{C'}} \tag{52-1b}$$

The suppliers of optical materials offer different optical qualities of refractive index and Abbe numbers specified in steps. As an example, Schott offers four steps of tolerances shown in table 52-2.

Table 52-2: Tolerances for refractive index and Abbe number.

	n_d	ν_d
Step 4	–	± 0.8 %
Step 3	± 0.0005	± 0.5 %
Step 2	± 0.0003	± 0.3 %
Step 1	± 0.0002	± 0.2 %

For a given wavelength spectrum the relation between refractive index and wavelength is approximated by the Sellmeier equation shown in (52-2).

$$n^2(\lambda) = 1 + \sum_{i=1}^{M} \frac{B_i \lambda^2}{\lambda^2 - C_i} \quad (52\text{-}2)$$

B_i and C_i are the experimetally determined Sellmeier coefficients usually quoted for λ in micrometers.

λ is the vacuum wavelength. In the usual form of (52-2) M is set to 3, therefore 6 Sellmeier coefficients are supplied to describe the refracting properties of the material. For common optical glasses the refractive index deviates from the actual refractive index by less than 5×10^{-6} over the wavelengths range of 365 nm to 2.3 μm, which is of the order of the homogeneity of a glass sample.

The achieved refractive index is usually specified by the supplier with an accuracy up to $\pm 3 \times 10^{-5}$ for refractive index and up to $\pm 2 \times 10^{-5}$ for dispersion. The numerical data usually are listed to five decimal places.

52.2.2
Metrology

The refractive index of the material is usually measured before the component is manufactured. A dispersion prism has to be made from a sample of the material to be tested by suitable test equipment. A refractive index measurement from a complete lens component is difficult to achieve and can only be carried out indirectly by determining, for instance, the focal length, radius of curvature and thickness of the component, in order to calculate the actual refractive index.

It is also possible to place the component in a glass tank filled with immersion liquid to neutralize the refractive effect of the lens. The refractive index of the liquid is changed by adding portions of different ingredients until a test beam remains undeflected by the component (figure 52-1). The process is awkward and complicated and should be used only exceptionally.

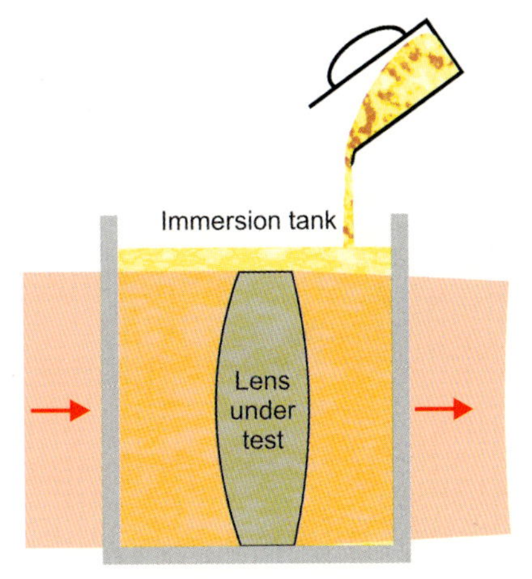

Figure 52-1: Determining the refractive index of a component by immersion compensation.

A precise measurement of the refractive index $n(\lambda)$ of a transparent material is carried out by means of a goniometer spectrometer setup shown in figure 52-2 [52-8]–[52-10], [52-34]. The goniometer spectrometer usually consists of:

1. a spectral lamp generating a multiline spectrum,
2. a small slit in the back focus of a collimator,
3. a precision rotary stage carrying the sample prism,
4. a precision rotary stage carrying a telescope,
5. a telescope comprising a crosshair reticle in the intermediate focus.

The angle between the axis of the collimator and that of the telescope can be precisely adjusted by means of a rotary stage to which the telescope is attached. The sample prism can be rotated individually and relatively to the telescope axis and that of the collimator. By adjusting the telescope rotary angle the slit images of different spectral lines can be brought into coincidence with the reticle crosshair. Rotating the prism to different measurement positions and adjusting the telescope to the appropriate observation positions for each spectral line leads to the results for refractive indices of the prism under test as described in the following.

Figure 52-3 shows a beam being refracted by a prism of prism angle a. The incidence angle is denoted by θ, the refraction angle at the second surface by θ''. The total angle of deflection δ is given by (52-3).

$$\delta = a - \theta + \theta \tag{52-3}$$

θ'' is calculated from a and θ according to (52-4) and (52-5).

$$\theta' = \arcsin\left(\frac{n}{n'}\sin\theta\right) \tag{52-4}$$

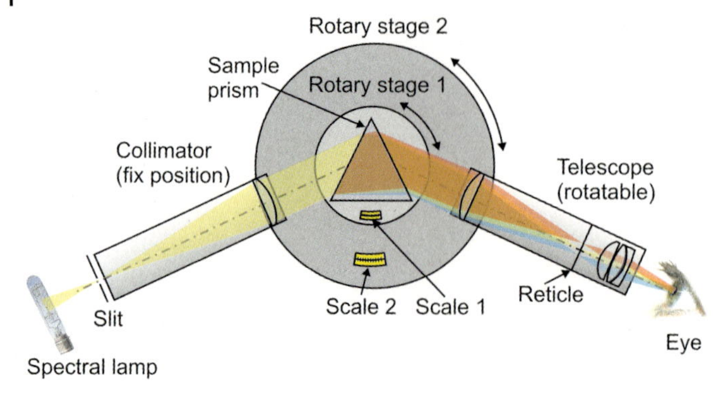

Figure 52-2: Goniometer spectrometer setup to measure the index of refraction of a sample prism.

$$\theta'' = \arcsin\left(\frac{n'}{n}\sin(\theta' - a)\right) \tag{52-5}$$

The refractive index of the prism is then calculated from (52-6).

$$n' = n\sqrt{\left(\frac{\sin\theta\cos a - \sin\theta'}{\sin a}\right)^2 + \sin^2\theta} \tag{52-6}$$

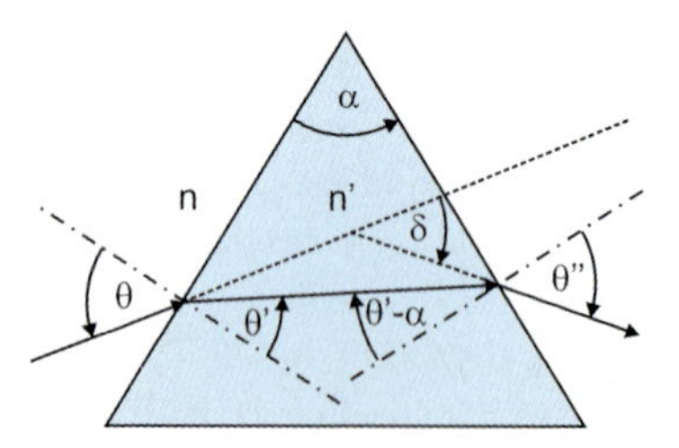

Figure 52-3: Beam being refracted by a prism.

The deflection angle δ can be measured directly by the goniometer spectrometer. The relative position of the prism can also be determined. Figure 52-4 shows the deflection angles as a function of the incidence angles for 45° prisms made of BK7 and SFL6 for three different wavelengths.

The curves generally have two characteristic points:

a) the minimum deflection angle at $\theta = -\theta''$ leading to $\delta_{\min} = a - 2\theta_{\min}$,
b) the maximum deflection angle at $\theta'' = -90°$ leading to $\delta_{\max} = a - \theta_{\max} - 90°$.

If the refractive index of the surrounding medium (air) is known, $n(\lambda)$ and a can be determined from the measured deflection angles and the relative incidence angles. For best results measurements are made for many rotating positions of the prism and the best fitting results are determined by least-square routines.

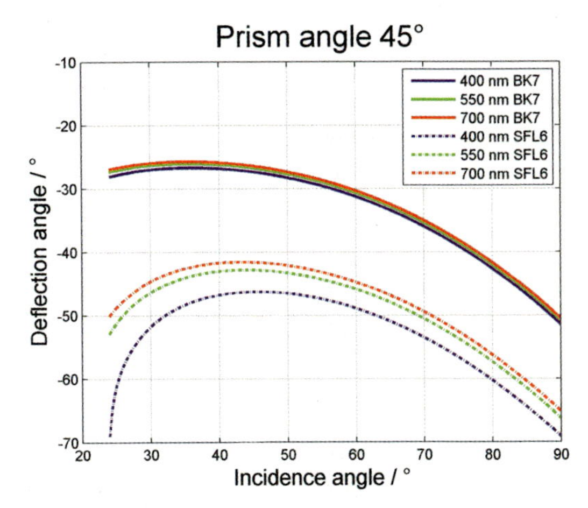

Figure 52-4: Deflection angles as functions of incidence angles for wavelengths 400 nm, 550 nm and 700 nm for 45° prisms made of BK7 and SFL6.

Errors in the refractive index determination are mainly caused by errors from the angle measurements. Common precision rotary stages have resolutions of less than 1 arcsec. The results are affected by the measurement errors of the prism angle and the deflection angle. Figure 52-5 shows the resulting measurement errors in the refractive index when an angle error of 0.1 arcsec is assumed for the corresponding examples in figure 52-4 and prism angles of 30°, 45° and 60°. The calculated deviations from the true refractive index are less than 2×10^{-6}. It is obvious that larger prism angles lead to smaller refractive index errors.

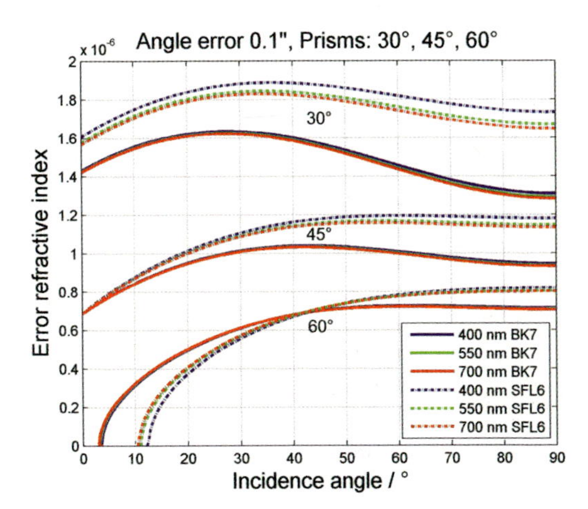

Figure 52-5: Refractive index errors caused by 0.1 arcsec angle errors as a function of incidence angles for wavelengths 400 nm, 550 nm and 700 nm for 30°, 45° and 60° prisms made of BK7 and SFL6.

There is an alternative spectrometer arrangement according to Abbe where an autocollimator telescope is used to observe the slit image after reflection at the back surface of a sample prism of angle α. The prism can be rotated by a rotary stage, and an autocollimator provides the slit-image projection and the observation of the reflected image at the reticle plane via its telescope arrangement (figure 52-6).

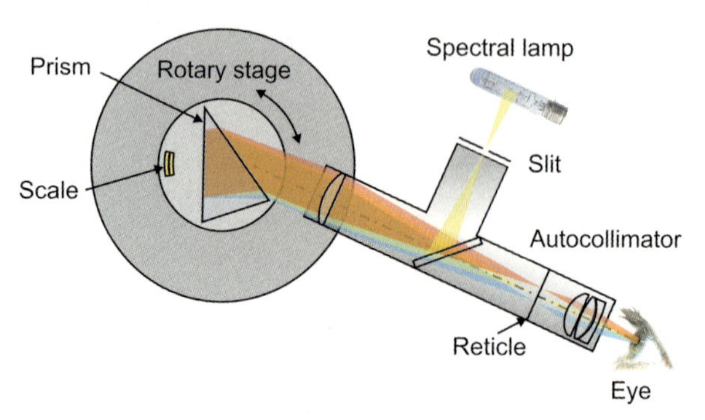

Figure 52-6: Goniometer spectrometer setup (Abbe arrangement) to measure the index of refraction of a sample prism.

Prism angles must be smaller than $\arcsin\left(\frac{n}{n'}\right)$, otherwise no reflection at the back surface can be observed. From figure 52-6 we derive the relation of (52-7),

$$\theta' = \alpha = \arcsin\left(\frac{n}{n'}\sin\theta\right) \tag{52-7}$$

from which the unknown refractive index can be calculated (52-8).

$$n' = n\frac{\sin\theta}{\sin\alpha} \tag{52-8}$$

While the prism is rotated, reflections at the prism surfaces directly and after refraction at the first surface, can be observed and the associated rotary angles are recorded. The four reflections can be used to calculate $\theta(\lambda)$ and α which lead to the unknown $n'(\lambda)$ as long as the n of the surrounding medium is known.

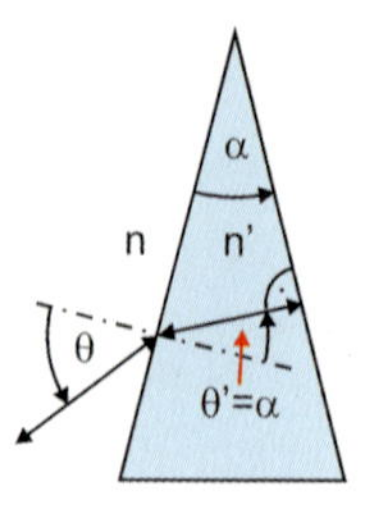

Figure 52-7: Beam being refracted and reflected at the back surface of a prism.

Angle errors lead to refractive index errors that can be estimated by (52-9).

$$\Delta n'(\lambda) = \frac{n(\lambda)}{\sin^2 a}\sqrt{\sin^2\theta(\lambda)\cdot\cos^2 a + \cos^2\theta(\lambda)\cdot\sin^2 a\cdot\Delta\varphi} \qquad (52\text{-}9)$$

$\Delta\varphi$ denotes the accuracy of the angle measurement in radians. Note that a usually must be smaller than 40°. The results from (52-9) are comparable to those achieved with the arrangement in figure 52-2.

52.3 Transmittance

52.3.1 Basics

The transmittance T of an optical material is the fraction of incident light at a specified wavelength that passes through a sample, usually a plane parallel plate,

$$T = \frac{I}{I_0} = P\cdot T_i = P\frac{I_i}{I_{i0}} \qquad (52\text{-}10)$$

where I_0 is the irradiance of the incident light and I is the irradiance of the light coming out of the sample. Due to the reflections at the surfaces of the plane parallel sample the transmitted light is reduced by a factor P, the so called "reflection factor", derived from Fresnel's formula for normal incidence given by (52-11).

$$P = \frac{2nn'}{n'^2 + n^2} \qquad (52\text{-}11)$$

n is the refractive index of the surrounding medium, n' the refractive index of the sample material.

T_i is the internal transmittance of the material. I_i is called the internally transmitted irradiance, whereas I_{i0} is the irradiance in the sample after the light has entered the first surface. While traveling through the material a distance d the irradiance is attenuated according to the exponential formula given in (52-12),

$$T_i = \frac{I_i}{I_{i0}} = e^{-k\cdot d} \qquad (52\text{-}12)$$

where k is the absorption coefficient of the material. Once the internal transmittance T_{i1} of a material is specified for a thickness d_1, it can easily be calculated for a thickness d_2 according to (52-13).

$$T_{i2} = T_{i1}^{\frac{d_2}{d_1}} \qquad (52\text{-}13)$$

Figure 52-8 shows examples of internal transmittances between 250 nm and 2500 nm wavelength for a 25 mm sample thickness of some optical glasses [52-1].

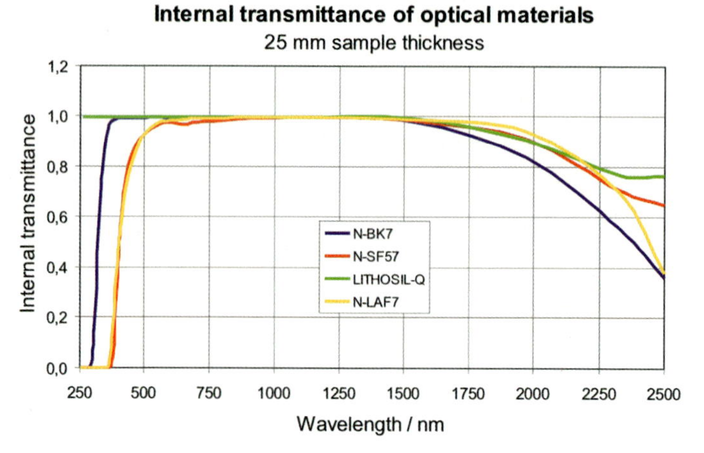

Figure 52-8: Internal transmittances between 250 nm and 2500 nm of some optical materials for a 25 mm sample thickness.

52.3.2
Metrology

The measurement of internal transmittances is usually carried out by spectral photometer setups that comprise the following subsystems (figure 52-9) [52-12], [52-13]:

1. a broadband light source,
2. a filter wheel,
3. a monochromator,
4. a reference signal generator,
5. a sample chamber,
6. a detection system.

The suitable light source has to be selected to cover the necessary wavelength range. Examples are shown in table 52-3.

Table 52-3: Some lamp types suitable for transmittance photometers.

Lamp type	Spectrum	Specification
Xenon (Xe)	250 nm – 800 nm	continuous spectrum
Mercury (Hg):	240 nm – 600 nm	strong lines with continuum
Low pressure Mercury (Hg)	240 nm – 600 nm	sharp discrete peaks for calibration
Halogen	250 nm – 2700 nm	continuous spectrum
Quartz Tungsten Halogen (QTH):	300 nm – 2700 nm	very smooth continuum
Deuterium	160 nm – 400 nm	smooth continuum

The selection of the detector follows the selection of the light source. The following considerations have to be made:

- the total wavelength region required for the sample measurements,
- the expected level of sample absorption,
- the light source intensity over the entire wavelength region for the sample measurements.

Photomultiplier tubes (PMT) are normally required for UV measurements, while silicon detectors typically work well over the range of 400 nm to 1100 nm. InGaAs and other appropriate detectors are recommended for measurements above 800 nm.

The light source module comprises the lamp as well as a focusing mirror to illuminate a slit at the entrance of the monochromator module. A filter wheel contains filters to block umwanted partial spectra. The monochromator projects the slit image of a preselected wavelength onto the exit slit of the monochromator. A beam divider projects a portion of the monochromatic light onto a reference diode in order to compensate for any irradiance variations due to lamp instabilities. A telescope optic projects the slit image onto a suitable detector to measure the amount of transmitted light. The sample plate is placed in between the telescopic elements where it is traversed by parallel light of a certain wavelength. A computer collects the measured irradiance values of the sample detector and the reference diode. It also selects the wavelength of the monchromator to acquire the total spectrum of the sample transmittance.

A computer program corrects the measured irradiances concerning light source variations and normalizes the irradiances to the lamp spectrum measured in the absence of a sample to obtain the transmittance values $T(\lambda)$. Considering the reflection losses the internal transmittance $T_i(\lambda)$ is then calculated from (52-14).

$$T_i(\lambda) = \frac{T(\lambda)}{P(\lambda)} = e^{-k(\lambda)\cdot d} \tag{52-14}$$

The absorption coefficient of the sample material is obtained by (52-15).

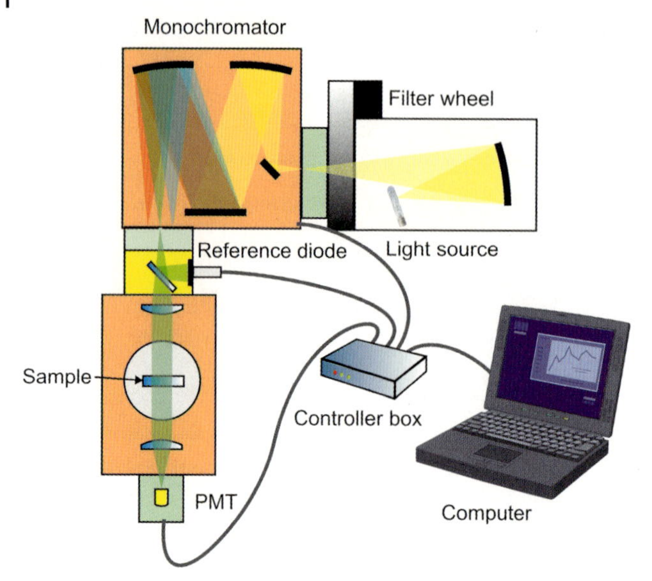

Figure 52-9: Principle of spectral photometer setup to measure the transmittance of optical materials.

$$k(\lambda) = -\frac{1}{d} \ln\left(\frac{T(\lambda)}{P(\lambda)}\right) \tag{52-15}$$

Typical measurement accuracies over a wavelength range from 250 nm to 2500 nm are about ± 0.5%, for the visible spectrum the accuracy reaches ± 0.3%. High-precision setups can obtain accuracies of ± 0.08% in the visible and UV range (200 nm–850 nm). Typical wavelength accuracies are ± 0.2 nm to ± 0.8 nm. The usual standard measurement sample thickness is 25 mm.

Note that, when measuring in air, there are several absorption bands of oxygene especially in the UV range, as well as of water vapor in the infrared range, which might prevent exact measurements or any measurements at all. In these cases the samples have to be placed into a chamber that can be evacuated or filled with nitrogen. The optical path from the light source to the detector then also has to be evacuated or purged with nitrogen.

In cases where rapid transmittance changes are to be recorded, a scanning monochromator is not suitable for sequential spectrometric measurements. A diode array spectrometer with a diode line or array used in place of the exit slit simultaneously records a complete spectrum within a fraction of a second and makes moving components superfluous.

An application example is shown in figure 52-10, where a photochromic sample is excited by means of a UV light source. Photochromic materials reversibly change their transmittance spectrum with changes in the sourrounding light irradiance.

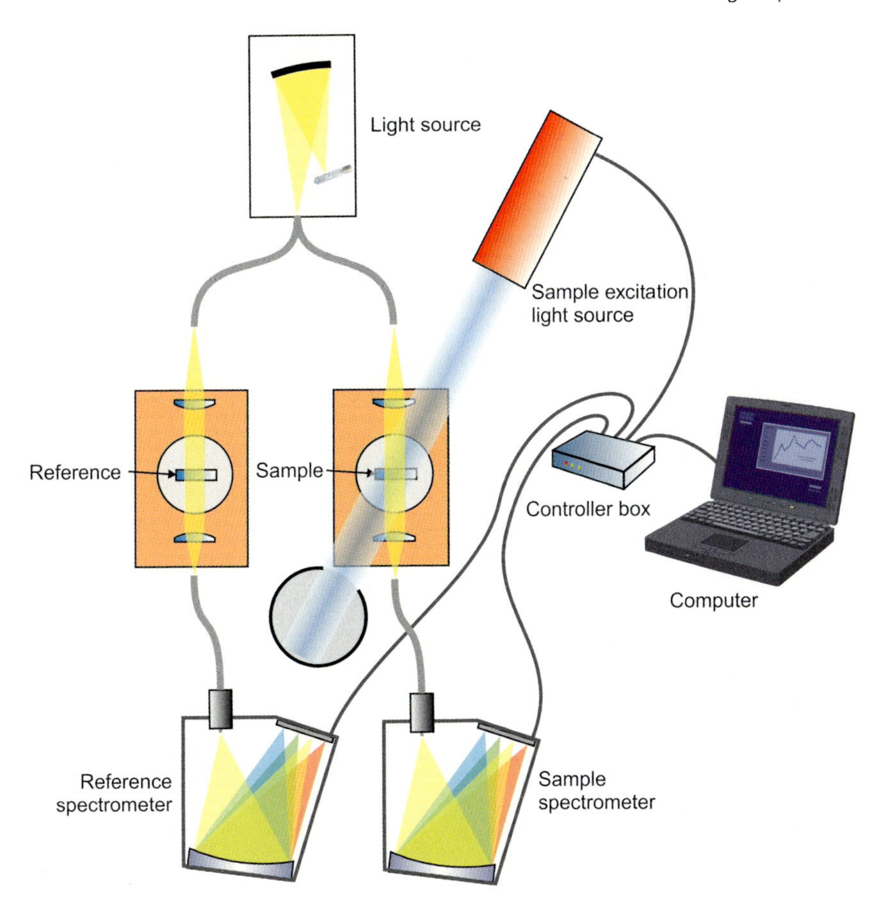

Figure 52-10: A spectrometric setup to measure the change in transmittance of a photochromic material during UV excitation.

For calibration and normalization purposes a second spectrometer can be used to simultaneously measure a reference sample which remains unexcited. In this case it is possible, for example, to heat up or cool down the sample with reference to the same levels without the need for recalibration of the total system.

52.4
Inhomogeneity and Striae

52.4.1
Basics

The refractive index inhomogeneity is a measure used to designate deviations of the refractive index within the material of an optical element. The cause of inhomogeneity is a variation in the chemical composition and other effects within the bulk material.

The glass manufacturer can obtain pieces of glass with refractive index of high homogeneity by melting and fine annealing. The achievable refractive index inhomogeneity depends on the glass type, the volume and the shape of the individual glass pieces.

Inhomogeneity is defined as the difference between the maximum and minimum values of the refractive index within the element. In many cases it is acceptable to subtract certain aberration terms with negligible impact on the application. For example, focal aberrations (expressed by the rotational symmetric quadratic term in the transmitted wavefront of a plane parallel sample) can often be corrected by adapting the geometry of the final part.

According to the standard ISO 10110 Part 4 [52-14] inhomogeneity classes are defined to characterize the different quality ranges. They correspond to the allowable variation of the refractive index within the optical element.

Table 52-4: Inhomogeneity of optical elements as defined by ISO 10110 Part 4.

Inhomogeneity class	Maximum variation of refractive index within a part 10^{-6}
0	± 50
1	± 20
2	± 5
3	± 2
4	± 1
5	± 0.5

Striae are inhomogeneities of refractice index with small spatial extent. They resemble bands in which the refractive index deviates with a typical period of 0.1 mm to several millimeters. It can appear in the form of sharply defined cordlike regions, especially when the glass was made by the clay-pot melting process. The tank melting process, which can cause band-like striae structures, is more common today in the production of optical glass.

The standard ISO 10110 Part 4 contains a classification with reference to striae. It refers to finished optical components and is therefore only conditionally applicable to optical glass in its original form of supply, because the tested bulk material is usually much larger than the finished optical component.

ISO 10110 Part 4 divides striae quality into five classes according to their area based on the optically effective total surface of the component. For classes 1 to 4, striae are considered only if they cause an optical path difference Δs of at least 30 nm. Under this provision striae can be tested and classified by their projected area perpendicular to the optical path through the element. Class 5 also allows the specification of tolerances for striae causing an optical path difference of less than

30 nm. Classes 1 to 4 are related to the density of striae, which is defined as the ratio of the effective projected area of striae to the area of the test region. The values are shown in table 52-5. Class 5 applies to optical elements with the highest quality requirements. The restriction to striae exceeding a 30 nm optical path difference does not apply to this class.

Table 52-5: Classes of striae as defined by ISO 10110 Part 4.

Striae class	Density of striae causing an optical path difference of at least 30 nm in%
1	≤ 10
2	≤ 5
3	≤ 2
4	≤ 1
5	Extremely free of striae. Restriction to striae exceeding 30 nm does not apply. Further information to be supplied in a note to the drawing.

According to ISO 10110 Part 4, inhomogeneity and striae are indicated in drawings by the code number 2, followed by a slash, and the class numbers for inhomogeneity and striae.

The indication has the following form:

2/A;B

where A is the class number for inhomogeneity according to table 52-4 and B is the class number for striae according to table 52-5. If no specifications are needed, A and B are replaced by dashes, respectively. Figure 52-11 shows an example of an appropriate indication with inhomogeneity class 3 and striae class 2.

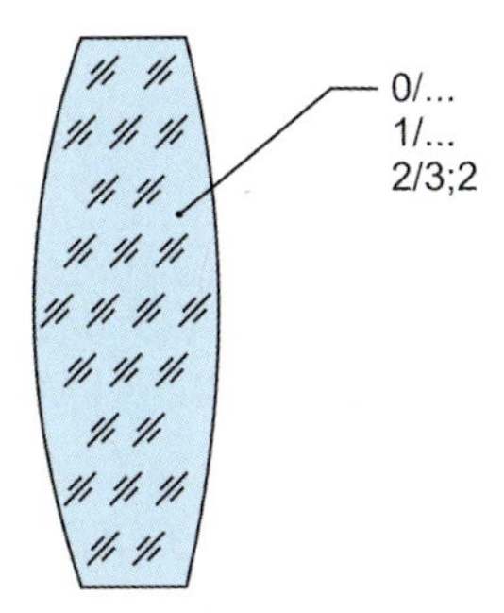

Figure 52-11: Indication of drawing for inhomogeneity and striae.

52.4.2
Metrology

Inhomogeneity

When light travels through a plane plate in air the optical path length OPD varies over the aperture of the beam due to:

1. the surface deviations $\Delta z_1(x,y)$, $\Delta z_2(x,y)$ from a perfect plane,
2. inhomogeneities of refractive index $\Delta n(x,y,z)$ of the optical material,
3. temperature inhomogeneities $\Delta T(x,y,z)$ within the sample.

If the influence of the surrounding medium is neglected, the OPD variation $\Delta\mathrm{OPD}(x,y)$ is then expressed by (52-16) for a single pass through the sample.

$$\Delta\mathrm{OPD}(x,y) = (d + \Delta z_1(x,) + \Delta z_2(x,y)) \cdot \left(\Delta n(x,y) + \left(a(n + \Delta n(x,y) - 1) + \frac{\mathrm{d}n}{\mathrm{d}T}\right)\Delta T(x,y)\right) \tag{52-16}$$

where:
d is the mean thickness of the sample,
n is the mean refractive index of the sample,
a is the coefficient of linear thermal expansion,
$\mathrm{d}n/\mathrm{d}T$ is the relative temperature coefficient of the refractive index.

Since we cannot detect local variations $\Delta n(x,y,z)$ *and* $\Delta T(x,y,z)$ in the volume by a single pass arrangement, we will assume that the average variations $\Delta n(x,y)$ and $\Delta T(x,y)$ represent the mean variations along a ray at x,y.

In order to measure the sample's inhomogeneities $\Delta n(x,y)$, we must be sure that:

1. the surface deviations are clearly compensated,
2. any effects from temperature variations $\Delta T(x,y)$ can be neglected.

Effects from temperature variations in a sample are in the order of 10^{-5}/K, which means that for higher inhomogeneity classes samples must have homogeneous temperatures of better than 0.01 K in order to be neglected during homogeneity measurements. This means that samples need to be adapted to a constant temperature and should not be touched several hours before the measurement. They must usually be mounted and adjusted by remotely controlled devices in a climatically stable environment of $\Delta T < 0.05$ K.

The general test setup for homogeneity measurements is a Twyman–Green interferometer [52-16] or a Fizeau interferometer [52-17]. The latter is shown in figure 52-12. A test sample with plane polished surfaces is placed between a flat mirror and the transmission flat, carrying the reference surface. The test sample is slightly wedged in order to avoid disturbing interferences between the surfaces of the test sample. The sample is then adjusted so that reflections from the test sample are blocked by the beam-stop in the interferometer.

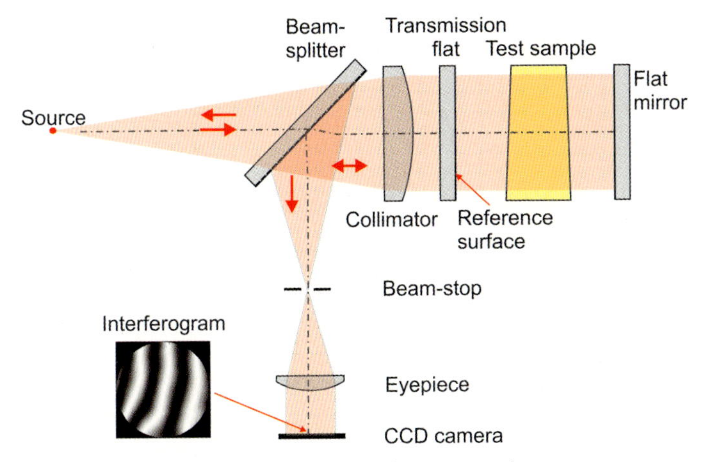

Figure 52-12: A Fizeau interferometer used to measure the inhomogeneities of a glass sample.

The homogeneity measurement consists of a sequence of four individual measurements as shown in figure 52-13:

W_1 first surface of the sample compared to reference surface,

W_2 sample in transmission with reflection at second surface compared to reference surface,

W_3 sample in transmission with reflection at flat mirror compared to reference surface,

W_4 empty cavity with reflection at flat mirror compared to reference surface.

Assuming that the interferometer is calibrated before the measurements are taken (the deviations of the reference surface are known and subtracted from the measurements), the inhomogeneity is then calculated from (52-17).

$$\Delta n(x,y) = \frac{1}{2d}[(1-n)(W_1(x,y) - W_2(x,y)) - n(W_4(x,y) - W_3(x,y))] \tag{52-17}$$

The reproducibility of the measurement process can be used to estimate the accuracy of the inhomogeneity measurement. For stable setups and a carefully designed masurement process the reproducibility of a wavefront $\Delta W_{\mathrm{repro}}$ lies in the range of a few nanometers peak to valley. The mesurement error $\delta\Delta\mathrm{n}$ can be estimated from (52-18).

$$\delta\Delta n \approx \frac{\Delta W_{\mathrm{repro}}[\mathrm{nm}]}{d[\mathrm{mm}]} 10^{-6} \tag{52-18}$$

The sensitivity of the measurement increases with increasing sample thickness d. For class 5 inhomogeneity the sample needs to be at least 10–20 mm thick to guarantee the accuracy.

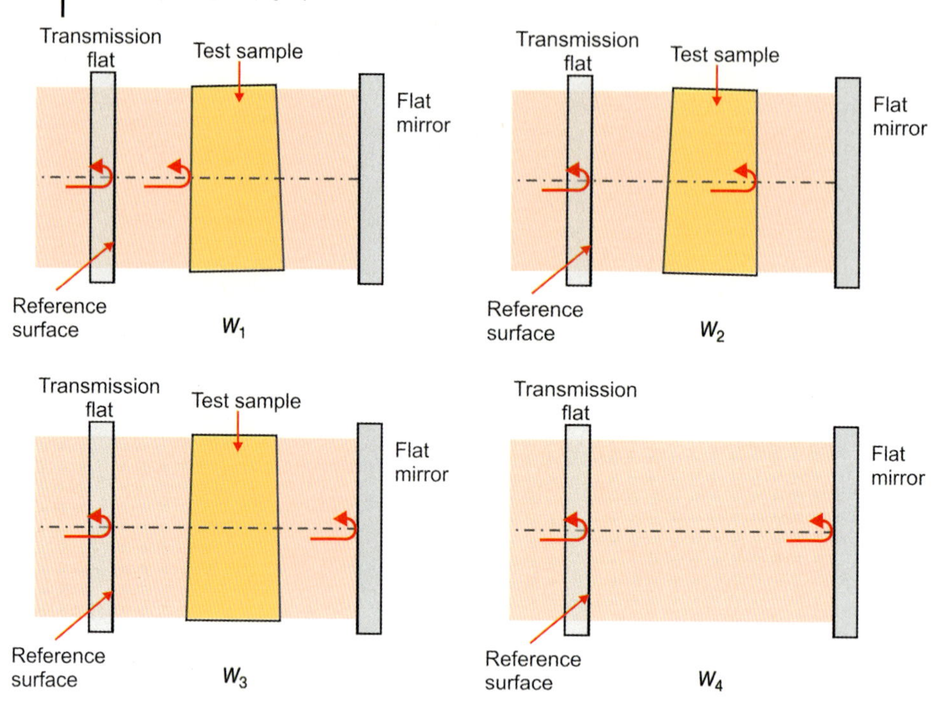

Figure 52-13: Measurements needed to determine the inhomogeneity of a polished sample.

A drawback of the method described above is that each sample has to be polished to a sufficient quality, i.e., a few fringes. This takes quite a considerable effort which makes the sample expensive.

An alternative method, where the samples only have to be lapped to a roughness of approximately 3 μm RMS, is the so-called "oil-on-plates sandwich" method (figure 52-14) [52-18]. For this method the rough sample is placed between two glass plates, which exhibit accurate polished surfaces. The glass plates are connected with the samples using an immersion oil liquid that has the same refractive index as the sample.

Two measurements have to be carried out:

1. a first measurement by measuring the oil-on-plates sandwich alone without the sample,
2. a second measurement with the sample in between the oil-on-plates.

The result for the inhomogeneity is obtained by subtracting the first from the second mesurement.

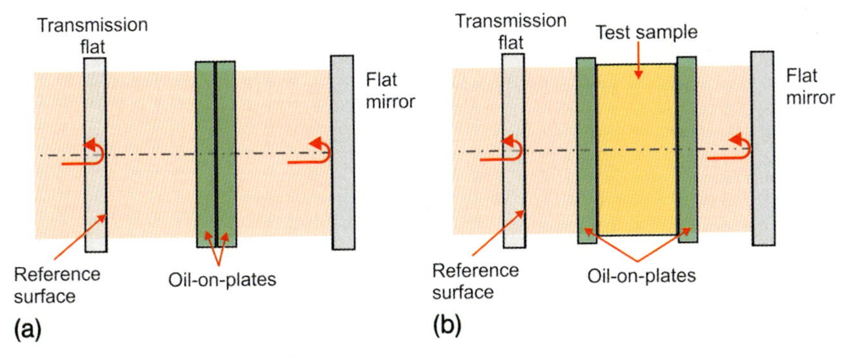

Figure 52-14: Oil-on-plates sandwich method to measure the inhomogeneity of a glass sample: a) measurement of oil-on-plates alone; b) measurement of oil-on-plates with sample.

The accuracy of the method is limited by the accuracy of the refractive index of the immersion oil. Deviations from the refractive index of the sample should be less than 10^{-4}. Mixtures of two immersion oils are used to cover a range of indices between 1.473 and 1.651 [52-17].

To measure the inhomogeneity of optical components the immersion tank method can be applied (figure 52-15). The tank is filled with immersion oil that matches that of the component under test. The resulting accuracy is, however, neither comparable to that of the oil-on-plate sandwich method or to the polished sample method. The limitation is caused by the imhomogeneity of the liquid which tends to form layers of different density and temperature.

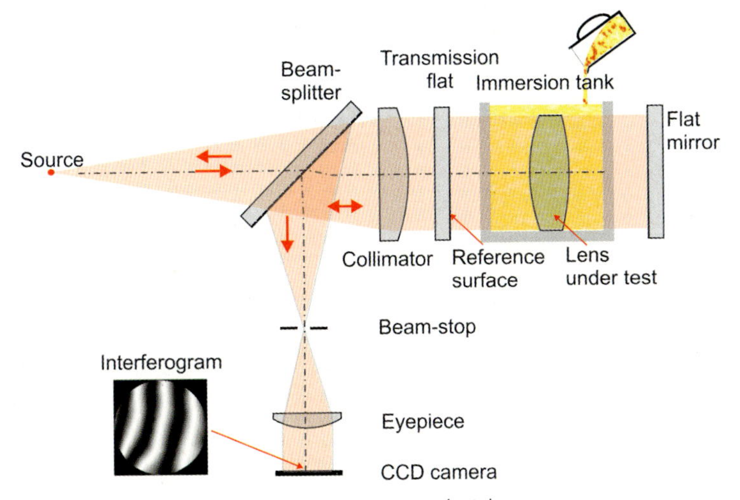

Figure 52-15: Fizeau interferometer to measure the inhomogeneity of an optical component in an immersion tank.

Striae

The usual metrology to detect and qualify striae is the shadowgraph method as shown in figures 52-16 and 52-17 [52-19]–[52-21]. It consists of the following elements:

1. a high-pressure short arc lamp (for example 100 W Mercury) with a pinhole,
2. a sample holder on a turn table,
3. a white non-transparent projection screen.

There is no further imaging optics in the setup. The light emitted from the lamp is divergent and partly coherent. Without the sample the projection screen will show a constant bright area. A sample with striae placed between the light source and the screen will locally deflected the beam, so that the striae will become visible on the screen as gray or dark areas. If the sample was produced by the clay-pot melting process, striae will appear in the form of sharp, cord-like regions. When produced by the tank melting process striae structures usually look more band-like.

The glass sample must have plane, almost parallel, polished surfaces. Since the method is sensitive to large inhomogeneity gradients, it will not show any large-scale homogeneity changes.

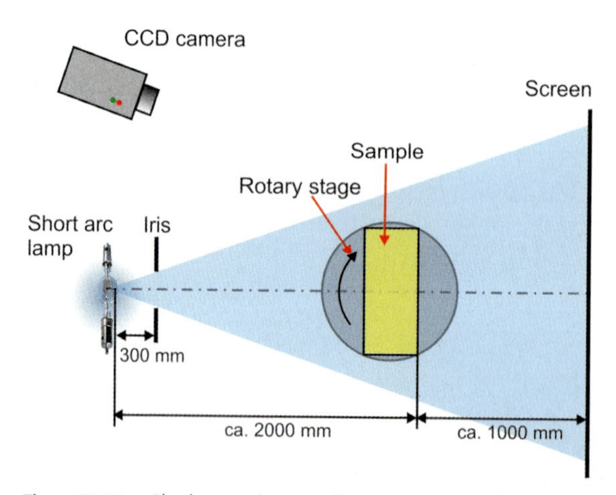

Figure 52-16: Shadowgraph setup for striae testing of glass samples.

The sample must be rotated by ± 45° into the position where the striae are most prominent. The rotation also allows discrimination between surface flaws and internal striae, because the amplitude of their lateral movement on the screen is different.

The observed striae structures are documented by an operator either by drawing or by additional CCD camera documentation.

The sensitivity of the shadowgraph method depends merely on the distances of the sample from the screen and from the light source. In order to detect striae of 10 nm optical path differences, the sample should be placed approximately in the middle between light source and screen which should be at a distance of at least 3 m from each other.

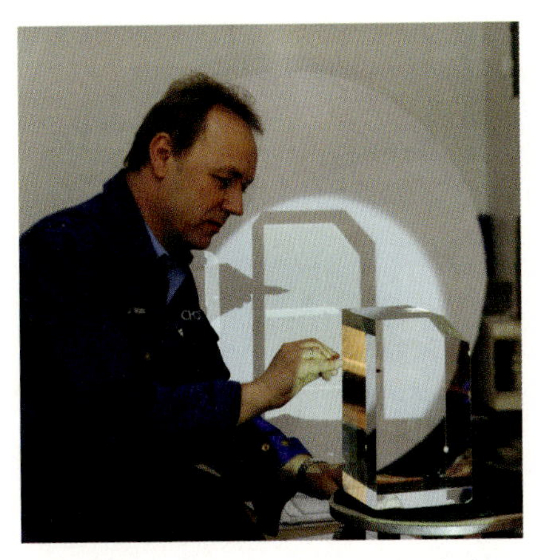

Figure 52-17: Striae testing using the shadowgraph method (Schott).

A simple striae model might explain the general phanomenon as shown in figure 52-18. A thin layer of refractive index n_2 extends in a glass block of refractive index n_1. The layer's thickness is denoted *a*, its size $b \times c$.

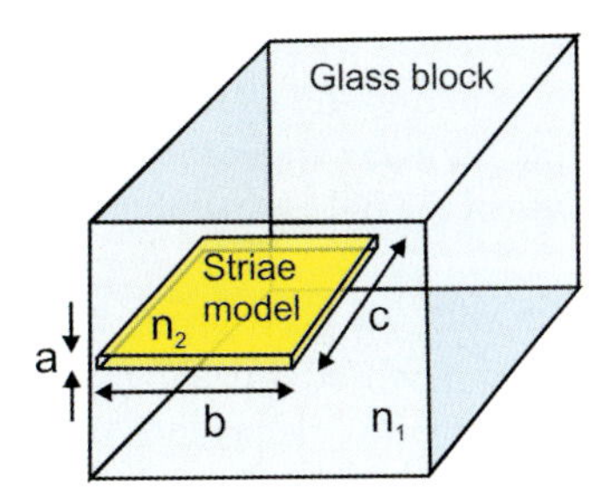

Figure 52-18: A simple striae model.

The optical path differences in different observation directions are

$$\mathrm{OPD}_a = (n_2 - n_1) \cdot a \tag{52-19a}$$

$$\mathrm{OPD}_b = (n_2 - n_1) \cdot b \tag{52-19b}$$

$$\mathrm{OPD}_c = (n_2 - n_1) \cdot c \tag{52-19c}$$

As an example we assume $n_2 - n_1 = 3 \times 10^{-7}$ and a = 2 mm, b = 100 mm and c = 200 mm. The resulting OPDs are then

$$\mathrm{OPD}_a = 6\,\mathrm{nm},$$
$$\mathrm{OPD}_b = 30\,\mathrm{nm},$$
$$\mathrm{OPD}_c = 60\,\mathrm{nm}.$$

Shadowgraph images can be simulated by Fourier transform methods using a defocused projection technique as shown in figure 52-19. The resulting images for thin striae lines of OPDs from 5 nm to 160 nm show that striae below 10 nm can hardly be detected. Striae should always be observed along its b and c dimensions. To find these directions the glass sample must be rotated in the shadowgraph setup.

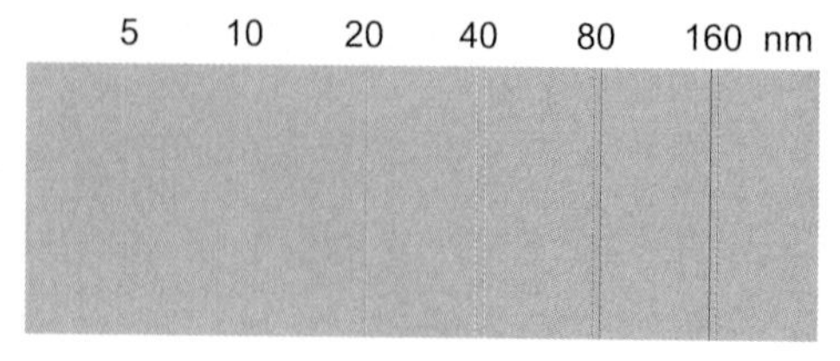

Figure 52-19: Simulated shadowgraph striae lines of different OPD.

52.5 Birefringence

52.5.1 Basics

Birefringence produces a difference in the index of refraction in the glass for light polarized parallel or perpendicular to the residual stress. In many applications this will affect the wavefront quality or optical path difference of the light transmitted by the optical element.

Stress birefringence is the result of residual stresses within a glass blank. The main causes are the differential cooling during the forming and/or annealing process, or certain fabrication processes carried out on the optical element [52-23], [52-26]. ISO 10110 Part 2 [52-22] describes the permissable stress birefringence in the following way:

A measure of birefringence is the optical path difference (OPD) Δs between orthogonal polarizations of transmitted light over the thickness of the sample. It is given in nanometres by (52-20).

$$\Delta s = a \cdot \sigma \cdot K \tag{52-20}$$

where:

a is the sample path length in centimetres,

σ is the residual stress in newtons per square millimeter,

K is the difference between the photoelastic constants in units of 10^{-7} square millimeters per newton ($10^{-7}\text{mm}^2 \cdot \text{N}^{-1}$).

The residual stress-induced birefringence is specified in terms of OPD per unit path length in nanometers per centimeter. A retardation of more than 20 nm cm^{-1} sample thickness generally corresponds to “coarse” annealed glass while a retardation of less than 10 nm cm^{-1} sample thickness refers to a "fine" anneal and is usually specified for precision optical elements.

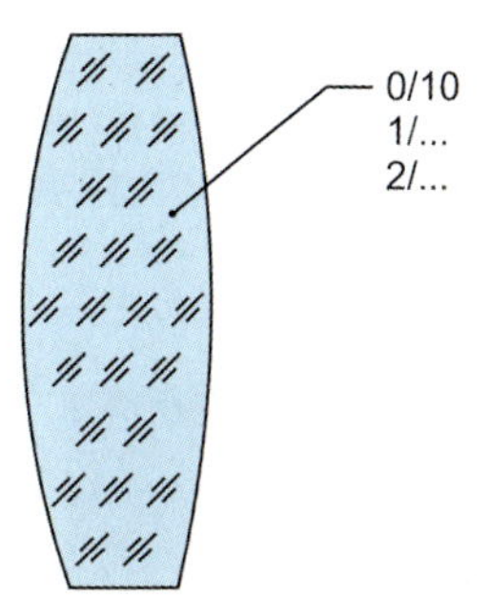

Figure 52-20: Indication of drawing for stress birefringence.

According to ISO 10110 Part 2, stress birefringence is indicated in drawings by the code number 0, followed by a slash, and a value for the maximum permissable OPD per unit path length. The indication has the following form:

$0/A$

where A is the maximum permissible stress birefringence in nanometres per centimeter of the optical path length. Figure 52-20 shows an example.

ISO 10110 Part 2 gives some typical examples of birefringence tolerances and their corresponding typical applications as shown in table 52-6.

Table 52-6: Examples of birefringence tolerances and typical applications as shown in ISO 10110 Part 2.

Permissible optical path difference (OPD) per cm glass path	Typical applications
< 2 nm cm^{-1}	Polarization instruments Interference instruments
5 nm cm^{-1}	Precision optics Astronomical optics
10 nm cm^{-1}	Photographic optics Microscope optics
20 nm cm^{-1}	Magnifying glasses Viewfinder optics

52.5.2 Metrology

A very simple test to make birefringence visible is to place a sample between two polarizers that are either crossed or parallel. Figure 52-21 shows an example where a plastic ruler has been used as a sample with the sun as the light source. To make the test quanitative we will use the Jones formalism which will be described in the following.

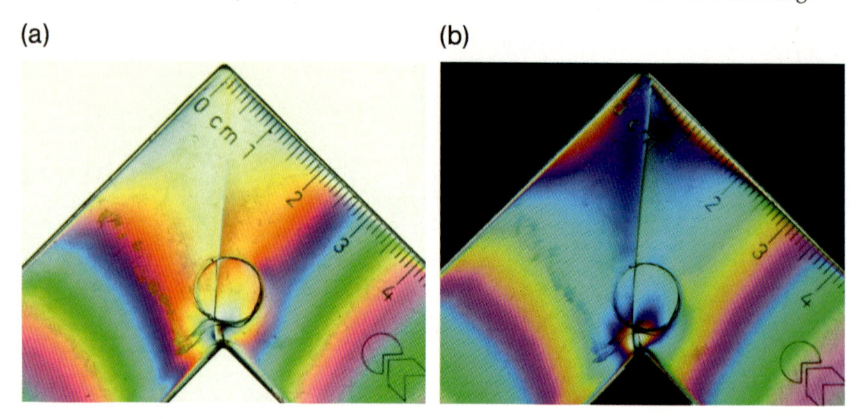

Figure 52-21: Plastic ruler observed in white light in between two polarizers in: a) parallel; b) crossed orientation.

We assume a setup as shown in figure 52-22, where a sample is placed between two polarizers. The light of an unpolarized light source passes a suitable color filter and a collimator. The parallel light then passes both polarizers and the sample. An observation telescope consisting of a collimator and an eyepiece projects an image of the sample onto a CCD camera. When the polarizers are crossed no light will arrive at the camera as long as the sample is free of birefringence. Birefringence of the sample will change the state of the incoming light from linear to elliptical, circular or rotated linear polarized light, after passage, such that a portion of the light will be transmitted by the second polarizer depending on the birefringence magnitude and orientation [52-24], [52-25], [52-27].

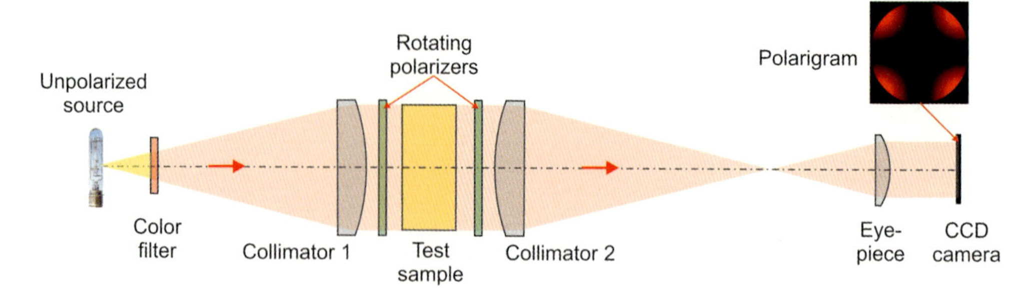

Figure 52-22: Setup used to measure birefringence of a test sample by rotating polarizers.

The polarizing characteristic of a transmitting sample will be described by a matrix $S(x,y)$ as shown in (52-21a–e)

$$S(x,y) = \begin{pmatrix} S_{11}(x,y) & S_{12}(x,y) \\ S_{21}(x,y) & S_{22}(x,y) \end{pmatrix} \tag{52-21a}$$

with

$$S_{11}(x,y) = t_p(x,y)e^{i\delta_p(x,y)}\cos^2\alpha(x,y) + t_s(x,y)e^{i\delta_s(x,y)}\sin^2\alpha(x,y) \tag{52-21b}$$

$$S_{12}(x,y) = \left(-t_p(x,y)e^{i\delta_p(x,y)} + t_s(x,y)e^{i\delta_s(x,y)}\right)\sin\alpha(x,y)\cos\alpha(x,y) \tag{52-21c}$$

$$S_{21}(x,y) = S_{12}(x,y) \tag{52-21d}$$

$$S_{22}(x,y) = t_p(x,y)e^{i\delta_p(x,y)}\sin^2\alpha(x,y) + t_s(x,y)e^{i\delta_s(x,y)}\cos^2\alpha(x,y) \tag{52-21e}$$

where:

$t_p(x,y)$, $t_s(x,y)$	local parallel (p) and perpendicular (s) transmittence,
$\delta_p(x,y)$, $\delta_s(x,y)$	local parallel (p) and perpendicular(s) phase shift,
$\alpha(x,y)$	local orientation of birefringence.

Following the Jones formalism we will describe a polarizer by its Jones matrix P (52-22) and the rotation matrix by R (52-23).

$$P = \begin{pmatrix} 1 & 0 \\ 0 & 0 \end{pmatrix} \tag{52-22}$$

$$R(\beta) = \begin{pmatrix} \cos\beta & \sin\beta \\ -\sin\beta & \cos\beta \end{pmatrix} \tag{52-23}$$

If an unpolarized light source is assumed, the electromagnetic field passing through the sample and the two rotating polarizers is described by (52-24).

$$\vec{E}_{out} = R(-\beta_2)\cdot P\cdot R(\beta_2)\cdot S(x,y)\cdot \vec{E}_{in}(\beta_1) \tag{52-24}$$

with $\vec{E}_{in}(\beta_1) = \begin{pmatrix} \cos\beta_1 \\ \sin\beta_1 \end{pmatrix}$

and β_1 and β_2 denoting the rotational positions of both polarizers. For parallel polarizers we set $\beta_1 = \beta_2$, for crossed polarizers we set $\beta_1 = \beta_2 + \pi/2$.

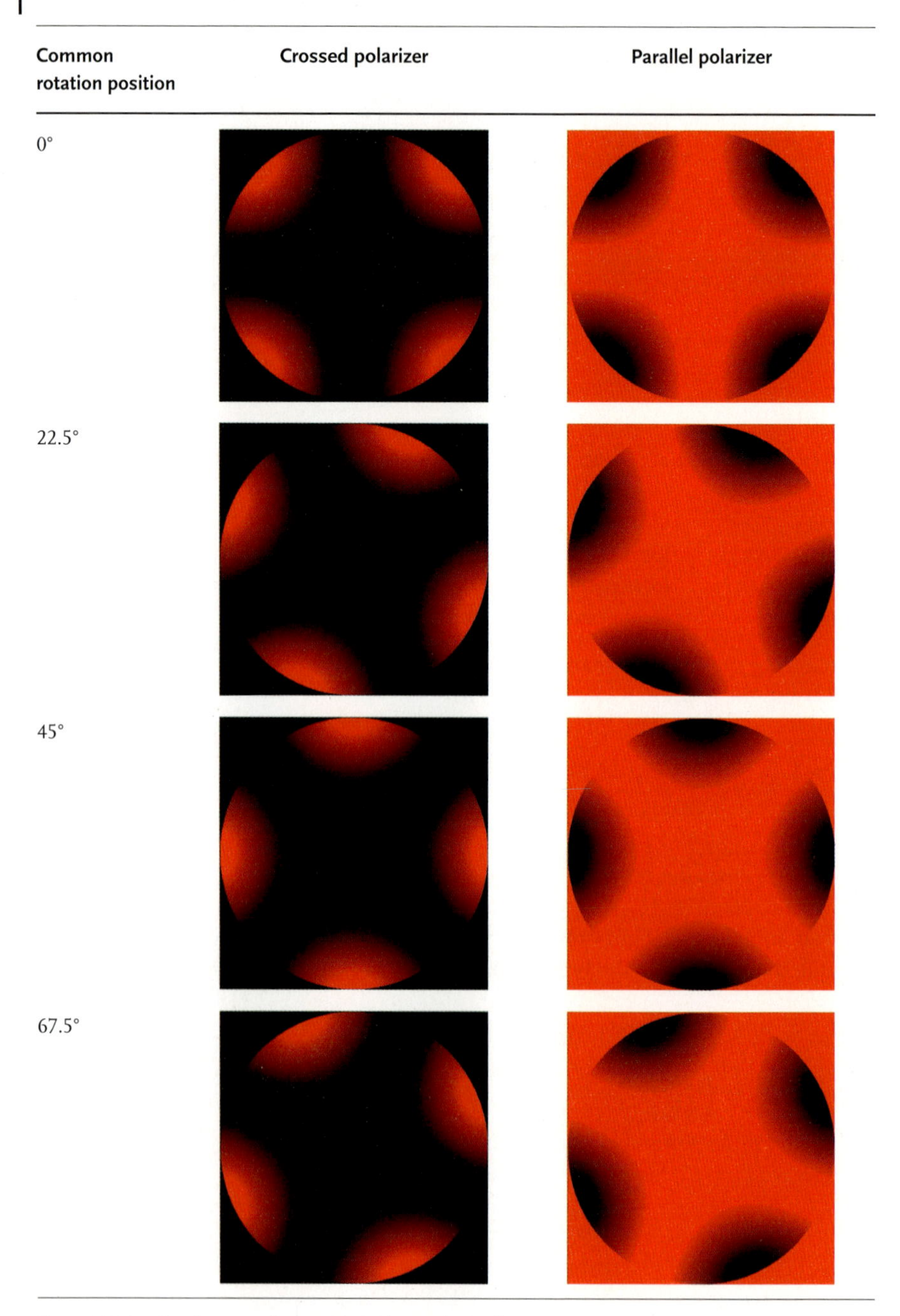

Figure 52-23: Irradiances of a sample with quadratically rising radial brefringence and radial orientation of local axis between crossed and parallel polarizers.

Figure 52-23 shows irradiances in red light of a sample with quadratically rising radial birefringence and a radial orientation of the local axis:, $\delta_s - \delta_p = \pi \cdot \frac{2\sqrt{x^2+y^2}}{d^2}$, $a = \arctan\left(\frac{y}{x}\right)$ with x, y denoting the local coordinates within the sample and d denoting the sample diameter. The polarizers were rotated together by steps of $\pi/8$ (22.5°) for the cases "crossed" and "parallel".

While the irradiance modulation with respect to the rotation angle is a measure of the birefringence magnitude, the phase is a measure of its orientation. Equations (52-25a,b) show the irradiance as a function of birefringence for crossed and parallel polarizers.

$$I_{\text{crossed}} = \frac{1}{4} I_0 \cdot \left[1 - \cos\left(\delta_s - \delta_p\right)\right] \cdot \left[1 - \cos 4(a - \beta)\right] \quad (52\text{-}25a)$$

$$I_{\text{parallel}} = I_0 \cdot \left\{1 - \frac{1}{4}\left[1 - \cos\left(\delta_s - \delta_p\right)\right] \cdot \left[1 - \cos 4(a - \beta)\right]\right\} \quad (52\text{-}25b)$$

with I_0 denoting the irradiance incident on the first polarizer and β denoting the rotation angle of the polarizers. Figure 52-24a,b shows relative irradiances for three birefringence magnitudes $\delta_s - \delta_p = 60°$, 120° and 180° and for two orientations $a = 30°$ and 60° as a function of the polarizer rotation angle.

A measurement of the birefringence magnitude and orientation is carried out by rotating the polarizers simultaneously to at least three different rotational positions β and applying regular phase-shift algorithms to determine $\delta_s - \delta_p$ and a. For instance, if β is rotated to 0°, 22.5° and 45.0°, the results are determined by (52-26a,b).

$$a = \frac{1}{4} \arctan \frac{I_{0°} - 2I_{22.5°} + I_{45°}}{I_{0°} - I_{45°}} \quad (52\text{-}26a)$$

$$\delta_s - \delta_p = \arccos\left(1 - \frac{2}{I_0}\sqrt{\left(I_{0°} - 2I_{22.5°} + I_{45°}\right)^2 + \left(I_{0°} - I_{45°}\right)^2}\right) \quad (52\text{-}26b)$$

Birefringence can also be measured in an interferometric arrangement, where the test sample is transmitted by linear polarized light whose orientation can be manipulated [52-28]. Figure 52-25 shows an example where a plane plate is tested in transmission in a Fizeau arrangement. A rotatable polarizer is placed before the transmission flat such that linear polarized light enters the cavity. When the light returns its polarization state is altered by the birefringence of the test sample. After passing the polarizer, on its return the unmodulated reference wave interferes with the modulated test wave having transmitted the test sample twice. The arrangement behaves in the same way as two parallel polarizers and a test sample of double thickness. The irradiance of the test wave is modulated over the size of the sample according to (52-25b), whereas the wavefronts ϕ measured from the interference patterns are altered while the polarizer is rotated by $\beta = 0°$, 22.5° and 45°. The birefringence is determined in an analogous way to (52-26a,b) as shown in (52-27a,b).

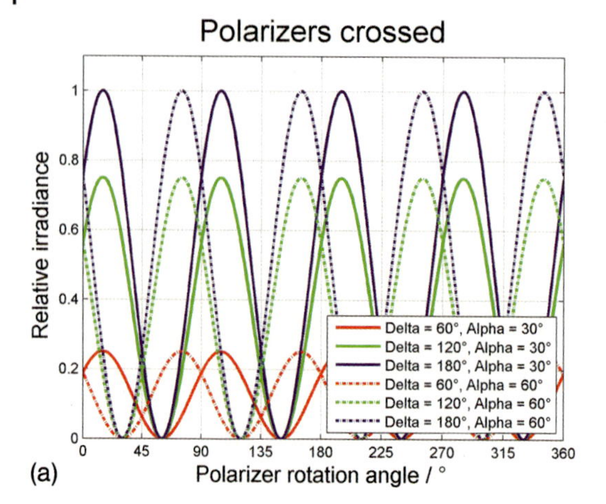

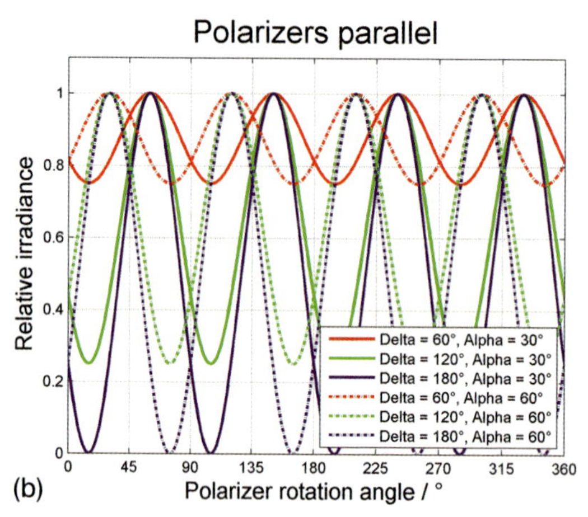

Figure 52-24: Irradiances at detector for three birefringence magnitudes $\delta_s - \delta_p = 60°$, 120° and 180° and for two orientations $\alpha = 30°$ and 60° as a function of the polarizer rotation angle: a) crossed polarizers; b) parallel polarizers.

$$\alpha = \frac{1}{4} \arctan \frac{\phi_{0^\circ} - 2\phi_{22.5^\circ} + \phi_{45^\circ}}{\phi_{0^\circ} - \phi_{45^\circ}} \tag{52-27a}$$

$$\delta_s - \delta_p = \arccos\left(1 - \frac{1}{\pi}\sqrt{(\phi_{0^\circ} - 2\phi_{22.5^\circ} + \phi_{45^\circ})^2 + (\phi_{0^\circ} - \phi_{45^\circ})^2}\right) \tag{52-27b}$$

The setups shown above can measure birefringence magnitude $\delta_s - \delta_p$ with a reproducibility of approximately 0.1 nm RMS, if the environmental conditions are sufficient for interferometric high precision measurements. The reproducibility of the orientation α depends strongly on the current birefringence magnitude. Low $\delta_s - \delta_p$ will lead to unstable results for α.

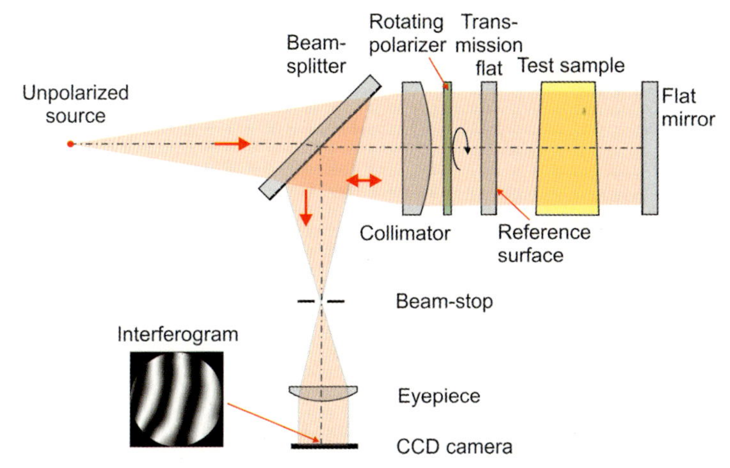

Figure 52-25: Fizeau interferometer to measure transmitting wavefronts and birefringence of a test sample by introducing a rotating polarizer in front of the transmission flat.

For single-channel birefringence measurements, polarimeter units consisting of a sender and a receiver unit can be applied. They measure birefringence magnitude and orientation integrated along an optical path through the optical sample. Units with rotating polarizing elements are available as well as those avoiding moving parts [52-29], [52-30]. The latter use photoelastic modulators (PEM) whose birefringence can be modulated rapidly to change the polarization state of an incident linear polarized beam. After passing through the sample the modulated beam enters the receiver unit, where it transmits further polarizing elements or a PEM. The electronic signals might be processed through a lock-in amplifier to determine the phase and amplitude of the modulated signal until a computer program converts the signal levels into birefringence parameters.

Figure 52-26 shows an arrangement used to measure the birefringence of plane samples. A sample is placed on an x/y-translation stage to be measured in arbitrary positions, while polarimeter sender and receiver units are fixed to the setup. A monochromator in front of a broadband light source selects the appropriate mean wavelength. The computer controls the x/y-stage and the polarimeter units and, if possible, the wavelength selection as well. Fully automatic birefringence measurements of plane samples for different wavelengths are possible.

Typical specifications for single-channel scanning devices are:

- Spot size: 1–3 mm,
- Time per single measurement: 1–3 s,
- Retardation range: 0–100 nm, depending on wavelength,
- Repeatability: 0.005–0.01 nm or 1%.

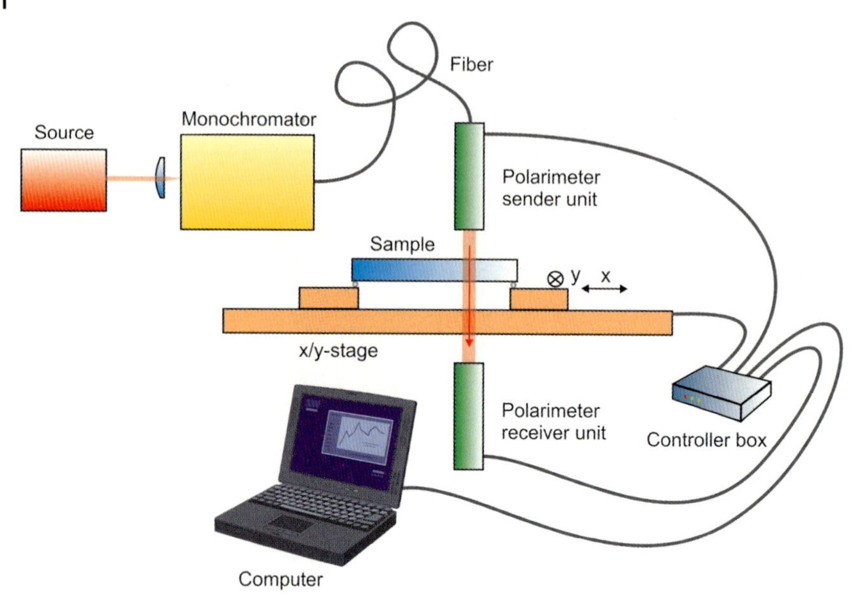

Figure 52-26: Setup to measure the birefringence of plane samples at different wavelengths by means of an electronic polarimeter. The sample is laterally translated by an *x*/*y*-stage to measure arbitrary points.

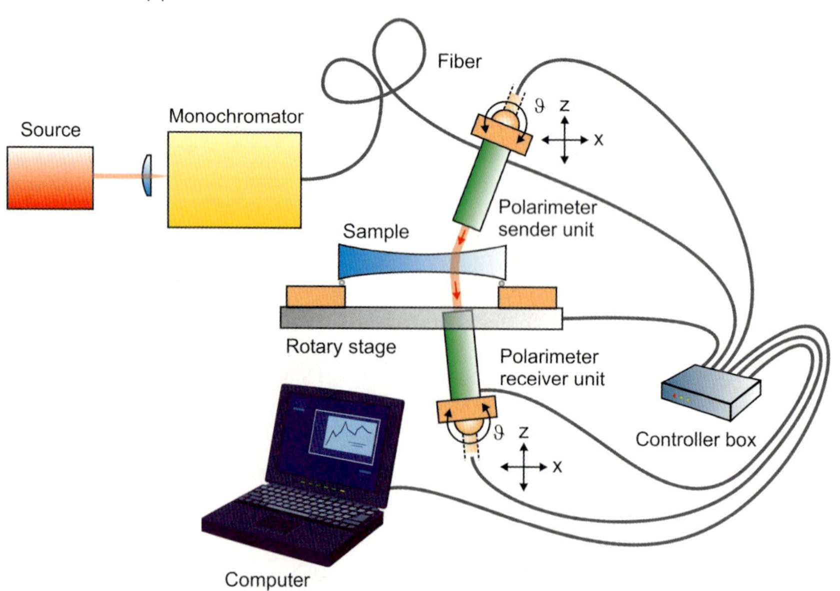

Figure 52-27: Setup to measure the birefringence of lens elements at different wavelengths by means of an electronic polarimeter. The polarimeter units are positioned by means of individual 4-axis positioners (robots), while the lens under test

is rotated on a rotary stage in order to measure arbitrary points.

For applications where it is necessary to measure birefringence of optical lens components, a setup as shown in figure 52-27 is appropriate. The polarimeter units are positioned by means of individual robot arms, which can move up to 6 axes. In the example, only 3- or 4-axis positioners are necessary, because the lens under test is rotated on a rotary stage. In this way arbitrary points of the sample can be accessed.

52.6 Bubbles and Inclusions

52.6.1 Basics

Bubbles are gaseous voids in the bulk material, generally of circular cross-section, which sometimes appear in glass as the result of the manufacturing process. Melting of some raw materials like carbonates or hydrogen-carbonates produces reaction gases forming bubbles in the melt. These bubbles will generally be removed by the refining process. However, some residual bubbles may remain.

Inclusions are solid particles within the glass which can be generated due to (Schott):

- remains from the batch that have not been melted completely,
- wall material with low solubility,
- particles from the outer surroundings,
- Platinum particles from the tank tools,
- devitrification processes, crystallization.

We will mention inclusions which refer to all localized bulk material defects of essentially circular cross-section, including striae knots, small stones, sand and crystals.

Bubbles and other inclusions usually appear in roughly constant numbers per unit volume of glass. Their number depends on the glass type and the manufacturing process. Optical elements in an optical system carrying bubbles and inclusions affect the optical performance in the following ways:

- the unwanted scattering of light,
- the visibility in the final image when located near an image plane.

When high-power light sources are used, for instance pulsed lasers, very high power densities can be achieved within an optical system. Under these conditions, damage to the optical elements is likely due to the presence of bubbles and other inclusions. Platinum particles in the glass are of special importance. The damage threshold for platinum particles extends from energy densities of 2.5 to 8 J cm^{-2}.

The resulting damage site can grow by continued radiation to sizes larger than 1 mm [52-31].

Since the harmfull effects of bubbles and inclusions are proportional to the area they fill, ISO 10110 Part 3 [52-28] defines gradings for individual optical elements in terms of apparent cross-sectional areas of bubbles and inclusions per unit volume.

According to ISO 10110 Part 3, bubbles and inclusions are indicated in drawings by the code number 1, followed by a slash, and a specification for the permissible bubbles in the element given in the form (figure 52-28):

$$1/N \times A$$

N is the number of bubbles and inclusions of the maximum permitted size. A is called the grade number, which is the measure of the size of the bubbles and inclusions. It is equal to the square-root of the projected area of the largest permissible bubble and/or inclusion, expressed in mm.

Preferred values for A are given in the first column of table 52-7.

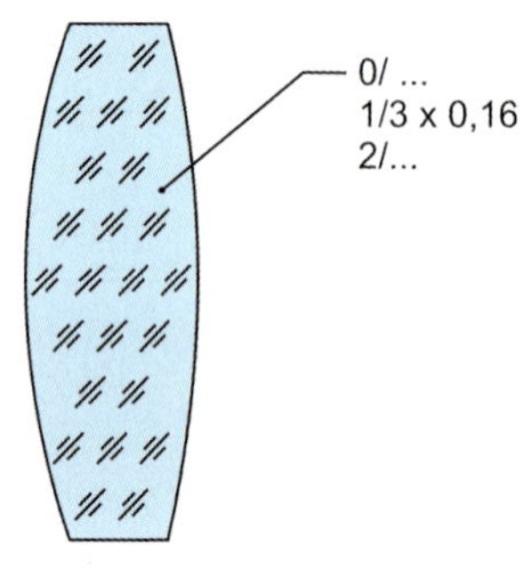

Figure 52-28: Example of tolerance indication for bubbles and other inclusions: a maximum of three bubbles or other inclusions of grade number 0,16 is permitted.

A larger number of bubbles and other inclusions with a smaller grade number is permitted if the sum of the projected areas of all bubbles and inclusions does not exceed

$$N \times A^2 \text{ (= maximum total area)}$$

Grade numbers are given in the columns of table 52-7, and the corresponding multiplication factors appear in the first line. The example at the bottom explains how to use the multiplication factors.

According to ISO 10110 Part 3 [52-33] concentrations of bubbles and inclusions are not allowed. A concentration occurs when more than 20 % of the number of allowed bubbles and other inclusions is found in any test region. If the total number of bubbles and other inclusions is less than 10, then 2 or more bubbles or other inclusions falling within a 5 % sub-area constitute a concentration.

The presence of very small inclusions with sizes < 0.01 mm and much smaller leads to an effect referred to as haze. The associated inclusions originate from phase separation or devitrification, i.e., small droplets or crystals with different refractive index from the surrounding glass matrix.

Table 52-7: Preferred size designation and factors for sub-division for bubbles and other inclusions (from ISO 10110 Part 3)

	Multiplication factors			
	1 (preferred values)	**2,5**	**6,3**	**16**
Grade numbers A mm	0,006			
	0,010	0,006		
	0,016	0,010	0,006	
	0,025	0,016	0,010	0,006
	0,040	0,025	0,016	0,010
	0,063	0,040	0,025	0,016
	0,10	0,063	0,040	0,025
	0,16	0,10	0,063	0,040
	0,25	0,16	0,10	0,063
	0,40	0,25	0,16	0,10
	0,63	0,40	0,25	0,16
	1,0	0,63	0,40	0,25
	1,6	1,0	0,63	0,40
	2,5	1,6	1,0	0,63
	4,0	2,5	1,6	1,0

EXAMPLE
If the indication is $1/2 \times 0{,}25$ (i.e. 2 bubbles of grade number 0,25), then $2 \times 2{,}5 \approx 5$ bubbles and/or other inclusions of grade number 0,16, or $2 \times 6{,}3 \approx 12$ bubbles of grade number 0,1 or $2 \times 16 \approx 32$ bubbles of grade number 0,063 are permissible. Alternatively, any corresponding combination of the above is permissible, provided that the total projected area of all bubbles and/or other inclusions with a grade number greater than $0{,}16 \times 0{,}25 = 0{,}04$ does not exceed $2 \times 0{,}25^2\ \text{mm}^2 = 0{,}125\ \text{mm}^2$.

When high-power light sources, such as pulsed lasers are used, very small amounts of microscopic platinum particles can lead to local damage in the glass. To specify glass with extra low platinum content Schott adds the expression “extra low platinum particle content” (EP) to the desired bubble class. In this case platinum particles ≥ 0.003 mm are examined independently from the usual bubble classification.

The quality grade "extra low platinum particle content" (EP) is characterized by table 52-8.

Table 52-8: Specification of platinum particles

	Max. amount of platinum particles in 100 cm³
≥0.05 mm	0
≥0.03 mm up to < 0.05 mm	5
≥0.003 mm up to < 0.03 mm	12

52.6.2
Metrology

Bubbles and other inclusions are mostly visually inspected. Figure 52-29 shows a suitable setup for inspection of bubbles and inclusions of larger grade numbers, usually > 0.03 mm. The glass is placed on a black screen and is illuminated from the side by means of a halogen lamp whose beam has been narrowed by a suitable stop and/or optics. The glass is viewed from above by looking through it toward the black background. Bubbles and inclusions become visible as bright spots. The classification of sizes is carried out either by comparison to standards or with the help of microscopes [52-32].

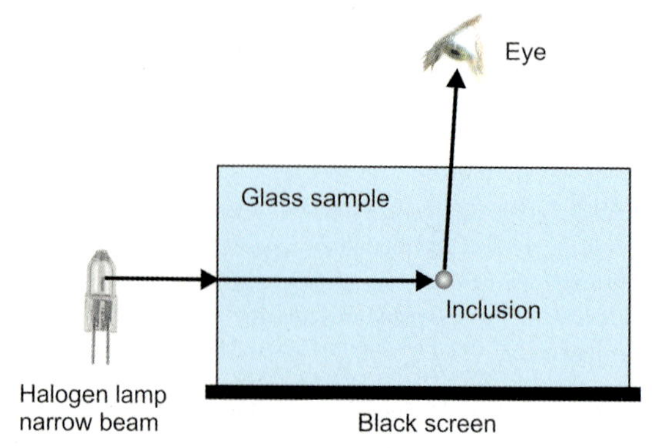

Figure 52-29: Setup to inspect bubbles and inclusions with diameter > 0.03 mm [Schott].

To inspect very small inclusions, referred to as haze, a similar setup is used (figure 52-30). However, the direction of the illumination is different. The glass sample under test will be illuminated just slightly differently from the direction of vision in front of a black screen. The irradiance of the scattered light is then a measure of the haze level. The arrangement could be made quantitative by either using standards for comparision or by measuring the haze level by means of suitable electronic detectors.

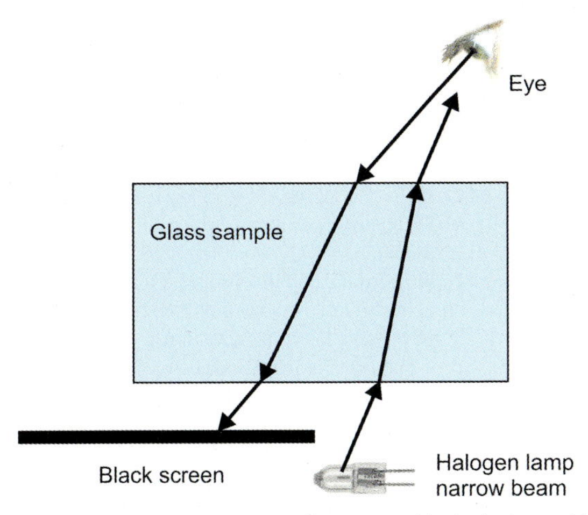

Figure 52-30: Setup to inspect haze caused by inclusions with diameters < 0.01 mm [Schott].

The inspection of platinum particles is carried out as shown in figure 52-29. The light of a strong halogen lamp must be collimated to a small parallel beam which must enter the sample from the side. The sample needs to be polished on four surfaces. The light is scattered by the platinum particles which appear as small bright spots. For particles much smaller than 0.03 mm high-resolution microscopes must be used for detection.

52.7 Literature

52-1 Schott catalogue, Optical Glass: Description of Properties (2007).

52-2 ISO 12844, Raw optical glass – Grindability with diamond pellets – Test method and classification (1999).

52-3 ISO/CD 13384 (proposed standard), Climatic Resistance.

52-4 ISO 8424: Raw optical glass – Resistance to attack by aqueous acidic solutions at 25 degrees C – Test method and classification (1996).

52-5 ISO 10629, Raw optical glass – Resistance to attack by aqueous alkaline solutions at 50 degrees C – Test method and classification (1996).

52-6 ISO 9689, Raw optical glass – Resistance to attack by aqueous alkaline phosphate-containing detergent solutions at 50 degrees C – Testing and classification (1990).

52-7 F. Wooten, Optical properties of solids, Academic Press, New York (1972).

52-8 Börsch, Spectral-Apparat und Reflexions-Goniometer, Annalen der Physik und Chemie, vol. 205, No. 11, p. 384–93 (1866).

52-9 E. Abbe, Gesammelte Abhandlungen, Zweiter Band: Wisenschaftliche Abhandlungen aus verschiedenen Gebieten, Patentschriften, Gedächtnisreden, Verlag Gustav Fischer, Jena (1906).

52-10 G.-J. Ulbrich, J. Trede, Goniometer-spectrometer for index of refraction measurements from the near UV through the near IR, Proc. SPIE vol. 1327, p. 32–39, Properties and Characteristics of Optical Glass II (1990).

52-11 M. V. R. K. Murty, R. P. Shukla, Simple method for measuring the refractive index of a liquid or glass wedge, Optical Engineering, vol.22, no.2, pp. 227-30 (1983).

52-12 W. R. McCluney, Introduction to Radiometry and Photometry, Artech House Inc., Boston, London (1994).

52-13 M. Bass (Editor), Handbook of Optics, vol. II, Devices, Measurements, & Properties, 2nd Edition, McGraw-Hill Inc., New York (1995).

52-14 ISO 10110-4, Optics and optical instruments – Preparation of drawings for optical elements and systems – Part 4: Material imperfections – Inhomogeneity and striae (1997)

52-15 D. W. Harper, G. B. Boulton, Measurement of the refractive index of optical glass, Optical instruments and techniques, pp. 177–88 (1970).

52-16 J. Schwider, R. Burow, K.-E. Elssner, R. Spolaczyk, J. Grzanna, Homogeneity testing by phase sampling interferometry, Appl. Opt. vol. 24, No. 18 p. 3059–61 (1985).

[52-17 Schoenfeld, Doerte, B. Kühn, W. Englisch, R. Takke, Interferometry at the physical limit: How to measure sub-parts-per-million optical homogeneity in fused silica, Appl. Opt. OT, vol. 42, Issue 10, pp.1814–19 (2003).

52-18 Schott Technical Information, TIE-26: Homogeneity of optical glass (2004).

52-19 Schott Technical Information, TIE-25: Striae in optical glass (2006).

52-20 H. Jebsen-Marwedel, R. Brückner (Editors), Optische Verfahren zur Erfassung von Schlieren, Glastechnische Fabrikationsfehler, 3rd Edition, p. 89 (1980).

52-21 H. Schardin, Glastechnische Interferenz- und Schlierenaufnahmen, Glastechnische Berichte 27, p. 1 (1954).

52-22 ISO 10110-2, Optics and optical instruments – Preparation of drawings for optical elements and systems –- Part 2: Material imperfections – Stress birefringence (1996).

52-23 Schott Technical Information, TIE-27: Stress in optical glass (2004).

52-24 ISO 11455, Raw optical glass – Determination of birefringence (1955).

52-25 H. W. Mc Kenzie, R. J. Hand, Basic Optical Stress Measurement in Glass, Society of Glass Technology (1999).

52-26 H. Bach, N. Neuroth (Editors), The Properties of Optical Glass, Springer Verlag (1998).

52-27 H. G. Tompkins, E. A. Irene, Handbook of Ellipsometry, William Andrew Inc., Norwich (2005).

52-28 M. Noguchi, T. Ishikawa, M. Ohno, S. Tachihara, Measurement of 2D birefringence distribution, Proceedings of the SPIE – The International Society for Optical Engineering, vol.1720, pp. 367–78 (1992).

52-29 B. Wang, T. C. Oakberg, A new instrument for measuring both the magnitude and angle of low level linear birefringence, Review of Scientific Instruments, vol.70, no.10, pp. 3847–54 (1999).

52-30 B. Wang, T. C. Oakberg, P. Kadlec, Industrial applications of a high-sensitivity linear birefringence measurement system, Proceedings of the SPIE – The International Society for Optical Engineering, vol.3754, pp. 197–203 (1999).

52-31 Schott Technical Information Nr. 21, 4 (1988).

52-32 Schott Technical Information, TIE-28: Bubbles and inclusions in optical glass (2004).

52-33 ISO 10110-3, Optics and optical instruments– Preparation of drawings for optical elements and systems – Part 3: Material imperfections – Bubbles and inclusions (1996)

52-34 V. Y. Demchuka: Measuring the refractive indices of optical glass by means of automatic goniometerspectrometers based on a ring laser, Opticheski Zhurnal 73, 51–56 (2006).

53
Testing the Geometry of Optical Components

Handbook of Optical Systems: Vol. 5. Metrology of Optical Components and Systems. First Edition.
Edited by Herbert Gross.

53.1 Specifications

The specification of the geometrical parameters of an optical component and their tolerances includes the following features:

1. Radius of curvature (spherical surfaces).
2. Central thickness.
3. Surface form (figure) irrigularities (ISO 10110-5).
4. Centering (ISO 10110-6).
5. Diameter and chamfer.

The International Organization of Standardization (ISO) provides the international standard ISO 10110 "Optics and optical instruments – preparation of drawings for optical elements and systems", which adresses some of the features mentioned above [53-1, 53-2]. As an example figure 53-1 shows an optical lens element carrying a spherical and an aspherical optical surface as well as a spherical chamfer which acts as a contact surface for the lens mount. The geometrical parameters to be toleranced and measured are indicated.

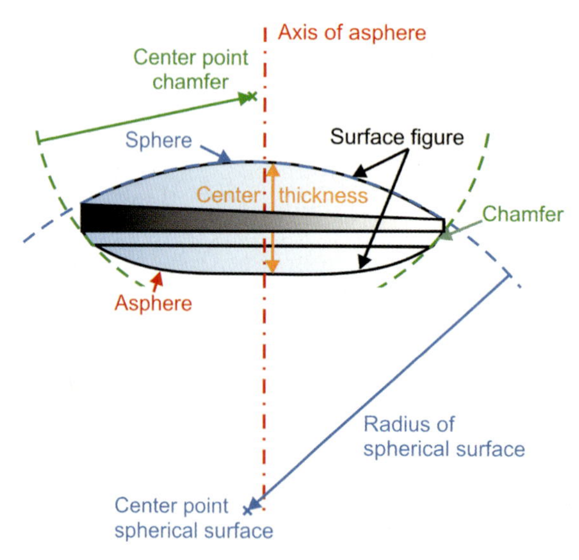

Figure 53-1: Lens element with a spherical and an aspherical optical surface as well as a spherical chamfer.

The following chapters cover basics and metrology techniques related to the lens features mentioned above.

53.2 Radius of Curvature

53.2.1 Basics

The radius of curvature of a fabricated spherical optical surface generally deviates from the ideal radius required by the optical design of the element. According to ISO 10110 Part 5 [53-1] the *total surface deviation function* denotes the difference between an optical surface and the nominal theoretical surface measured perpendicular to the theoretical surface.

The *spherical surface deviation* denotes the difference between the approximating sphere and the nominal theoretical surface (figure 53-2). The approximating sphere is the spherical surface for which the root-mean-square (rms) difference from the total surface deviation function is a minimum,.

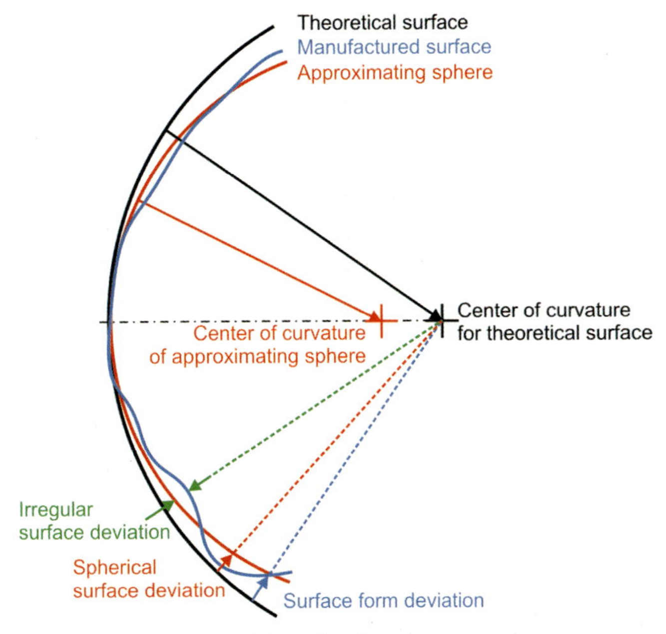

Figure 53-2: Definitions of the surface form deviation and related terms.

The *sagitta error* is the peak-to-valley difference between the spherical surface deviation and a plane – in other words: the approximating sphere's maximum deviation from the theoretical surface.

In practice, the theoretical surface is represented by a test glass or an interferometric reference surface or any other device of sufficient accuracy as descibed below.

Unless otherwise specified, the surface form deviation is measured in half the wavelength of 546.07 nm, which corresponds to the green spectral line of mercury (the e-line).

53.2.2
Metrology

The curvature of a spherical surface may be measured by mechanical or optical methods as described in this chapter. The methods can also be used to measure local curvatures of aspherical surfaces, either at their vertex or in any off-axis location.

Curvature measuring methods can be either *absolute* or *relative*. Absolute methods produce a direct result for the measured radius of curvature, whereas relative methods produce a curvature difference from a given reference.

We will concentrate on those methods that are able to test concave and convex surfaces of all radius ranges. There are several methods which are only suitable to test concave surfaces or surfaces with very long radii. We will mention them only briefly at the end of this chapter.

Test-glass Method

The test-glass (also called the test-plate) method is a relative method, where the surface under test is brought into contact with a corresponding test glass of opposite curvature, as described in section 46.3.1 (figure 53-3). In the case of a curvature deviation, Newton rings can be observed and their number m is, to a good approximation, related to the radius deviation ΔR as

$$\Delta R = R_S - R_C \approx 4m\lambda \frac{D^2}{R_C R_S} \approx 4m\lambda \left(\frac{D}{R}\right)^2 \qquad (53\text{-}1)$$

where λ is the wavelength, D is the diameter of the observed surface under test, R is the nominal radius of curvature, R_S is the radius of curvature of the surface under test and R_C is that of the test glass [53-3, 53-4, 53-8].

In the case when the nominal theoretical radius is infinite (plane surface), the measured radius of curvature R_S can be calculated from

$$R_S \approx \frac{D^2}{4m\lambda} \qquad (53\text{-}2)$$

The sign of m can be determined by applying slight pressure to the components in contact and observing the direction in which the fringes move.

A drawback of the test-glass method is the danger of surface damage as it is brought into contact. It is also necessary to have an inividual test glass for each radius to be fabricated. Using a Fizeau interferometer along with a suitable transmission sphere as described in section 46.3.2 gives much more flexibility and avoids the danger of damage. A lens mount will be necessary which can be adjusted to a fixed position relative to the reference surface of the transmission sphere. The interferometer arrangement can then be calibrated by means of a well-known test glass, which will be exchanged by the surface under test without altering the position of the lens mount (figure 53-4). Equations (53-1) and (53-2) apply in the same way as in the case for contacting test glasses.

Figure 53-3: Test-glass method showing colored Newton rings in white light.

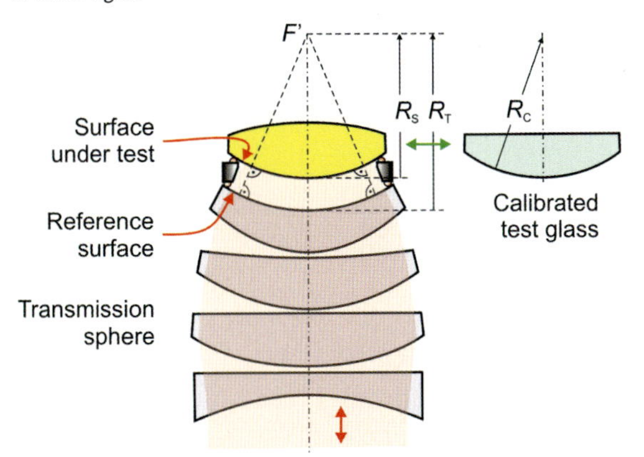

Figure 53-4: Test-glass method to measure radii of curvature in a Fizeau interferometer using a transmission sphere and a fixed lens mount, where a calibrated test glass and the surface under test are mounted for measurement.

Spherometers

The most popular mechanical device in use is the spherometer [53-3]–[53-5]. When calibrated, it measures the absolute radius of curvature of a spherical surface. It consists of three equally spaced feet, which form the vertices of an equilateral triangle, and a central moving plunger or a fine screw moving within a nut. The legs terminate in spheres or hemispheres, so that each rests on a point on the surface under test. The spherometer measures the sagitta, which is related to the position of the plunger, when placed onto the surface under test (figure 53-5). For calibration the spherometer is first placed on top of a known flat surface, then the scale connected to the plunger is nulled. When placed on top of an unknown spherical surface, the radius of curvature R is determined from

$$R = \frac{z}{2} + \frac{g^2}{2z} - r \tag{53-3}$$

where z is the plunger movement relative to the position for $R = \infty$, g is the spherometer leg length and r is the radius of the (hemi)spheres (see figure 53-5).

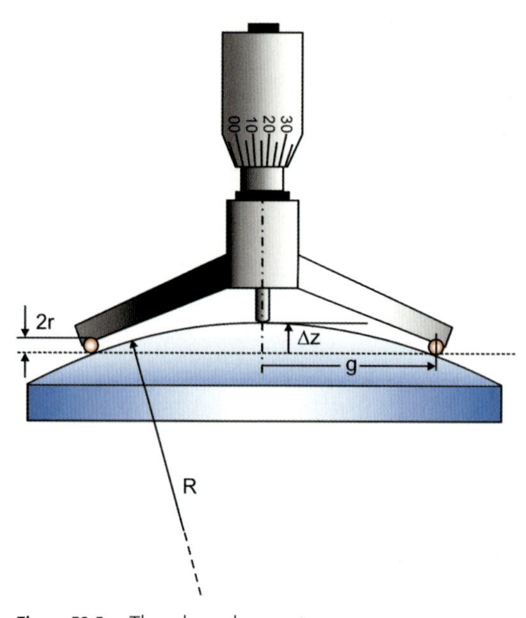

Figure 53-5: Three-leg spherometer.

The accuracy of the instrument is given by the total differential of (53-3) leading to

$$\Delta R = \left|\frac{g}{z}\right| \Delta g + \frac{1}{2}\left|1 - \left(\frac{g}{z}\right)^2\right| \Delta z \tag{53-4}$$

with

$$z = (R + r) \cdot \left(1 - \sqrt{1 - \left(\frac{g}{R + r}\right)^2}\right) \tag{53-5}$$

where Δg denotes the error in leg length and Δz denotes the error in the plunger position. Δg is a systematic error and limits accuracy, whereas Δz is mainly a statistical error limiting the precision of the measurement process.

As an example, we assume a spherometer leg length of $g = 50$ mm, a precision in z of 1 μm, 5 μm and 10 μm and a systematic error in g of the same magnitude; the resulting relative radius errors $\Delta R/R$ can be calculated from (53-4). The results for radii between 50 mm and 10000 mm are shown in figure 53-6.

For plane or very flat surfaces it may be useful to specify the curvature $\rho = \frac{1}{R}$ and evaluate the errors in the curvature rather than those for the radius of curvature.

$$\rho = \frac{1}{R} = \frac{2z}{z^2 - 2z\,r + g^2} \tag{53-6}$$

$$\Delta\rho = \left|\frac{4g\,z}{\left(z^2 - 2z\,r + g^2\right)^2}\right|\Delta g + \left|\frac{2(g^2 - z^2)}{\left(z^2 - 2z\,r + g^2\right)^2}\right|\Delta z \tag{53-7}$$

with

$$z = \frac{\rho\,g^2}{(1+\rho\,r)\left(1+\sqrt{1-\left(\frac{\rho\,g}{1+\rho\,r}\right)^2}\right)} \tag{53-8}$$

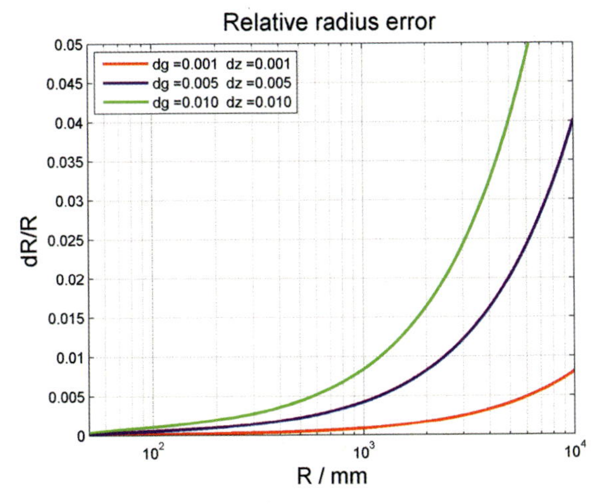

Figure 53-6: Relative radius error $\Delta R/R$ for $g = 50$ mm and $\Delta g = \Delta z = 1$ µm (red), $\Delta g = \Delta z = 5$ µm (blue), $\Delta g = \Delta z = 10$ µm (green).

There are different types of spherometers available; for instance, the so-called ring spherometer which has a cup instead of three legs. The cup is cylindrical with a flat surface at the end. Concave surfaces touch the cylinder at the outer edge, convex surfaces at the inner edge. Usually the cups are interchangeable so that cups with different diameters can be used to match different radii. Other than with the three legs the contact with the surface under test is undefined, because the rim of the cup strikes only the high spots of the surfaces, but an astigmatic surface is therefore easily detected [53-4].

The so-called *bar spherometer* has only two fixed contact points with a moveable plunger as the third contact point in the middle. It can be used to measure astigmatic surfaces, but its general accuracy is not as high as with a three-leg or ring spherometer due to the risk of tilting the instrument unintentionally.

Two-position Method

The two-position method is an absolute technique used to measure the radius of curvature of spherical surfaces with the highest accuracy [53-3]–[53-7]. It makes use of the fact that a convergent light beam with focus point F′ can be reflected by a spherical surface, when

a) F′ coincides with the center point C,
b) F′ coincides with the vertex point V

of the sphere. In both positions the light returns the same way as it arrived from the light source. The distance between V and C is equal to the surface's radius of curvature.

In the measurement process the surface under test is adjusted consecutively in both positions, until the perfect position is reached. In figure 53-7 a Fizeau interferometer is used as a measuring system equipped with a transmission sphere of back focal length R_T. In both perfect positions the Fizeau interferometer detects a wavefront free of quadratic terms (focus term). Any focus residuals can be taken into account for numerical position correction, when the F/# of the transmission sphere is considered (see section 53.4.2). The back focal length of the transmission sphere limits the test range for convex radii to $R_T > R_{cx}$. There is no such restriction for concave surfaces except for the maximum length of the test bench. The travel distance between both positions must be carefully measured by an adequate measuring system. It is the most critical accuracy-limiting issue in the two-position technique.

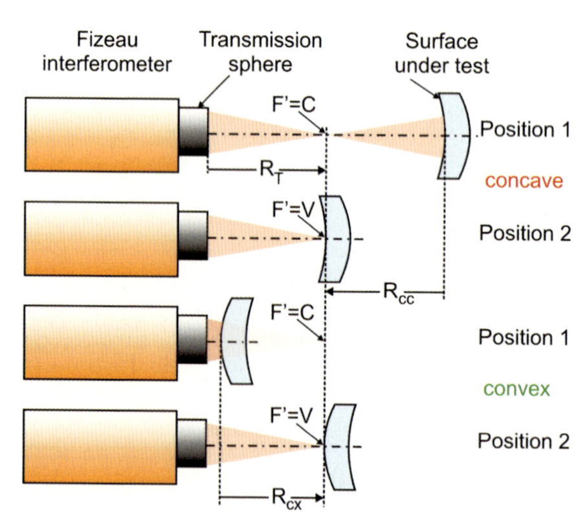

Figure 53-7: Two-position test using a laser interferometer and a transmission sphere to determine the radii of curvature R_{cc} and R_{cx} of concave and convex surfaces, respectively.

Figure 53-8 shows a horizontal setup used to measure radii of curvature for convex and concave surfaces with the two-position test by means of a Fizeau interferometer, a transmission sphere and laser distance metrology. The lens is kept in an adjustable lens mount, which is placed on a horizontally traveling sledge. Its position is detected by a heterodyne displacement interferometer, which is mounted on a second horizontally traveling sledge. To measure the radius of a surface, this second sledge is kept in a fixed position, minimizing the length of the measurement beam, and only the lens sledge is moved to the two specific positions.

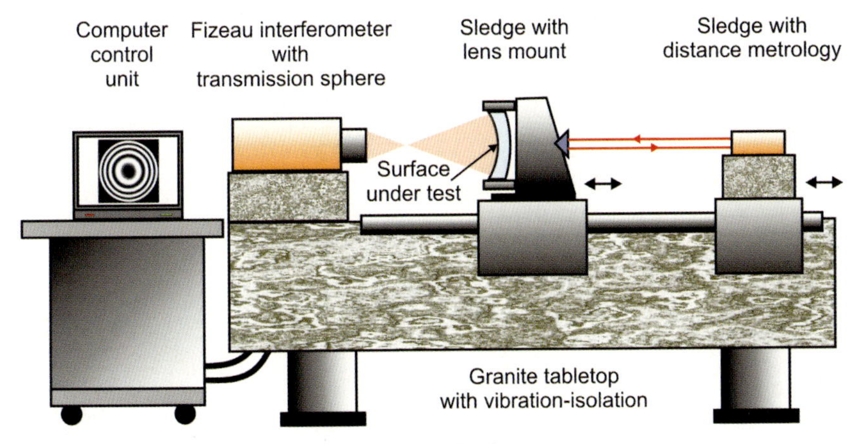

Figure 53-8: Horizontal setup used to measure radii of curvature for convex and concave surfaces using the two-position method.

A computer controls the Fizeau interferometer as well as the displacement interferometer and the sledge movements.

The following parameters have a major impact on the accuracy of the curvature results:

1. Lens-bending effects caused by gravity or stress from the lens mount.
2. Rotational aspherical deformation of the surface under test combined with partial coverage of the surface by the transmission sphere.
3. Environmental changes in temperature which deform the lens under test.
4. Environmental changes in temperature, air pressure and humidity which influence the displacement interferometer.
5. Abbe error (parallel misplacement of the Fizeau interferometer axis and the displacement interferometer axis).

Since the radius of curvature changes with temperature, depending on the material, careful temperature control is necessary for high-precision measurements. The measured radius of curvature is usually corrected to a nominal temperature of 20°C. Figure 53-9 shows the radius change with temperature for different materials.

Precision setups are able to determine radii of curvature with accuracies between 0.5 μm and 2 μm.

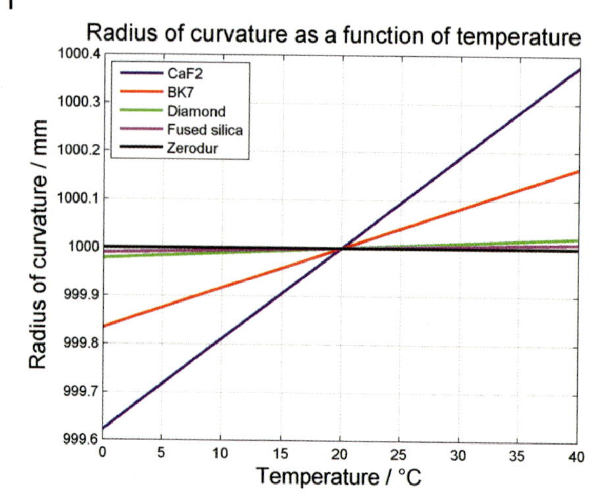

Figure 53-9: Radius of curvature as a function of temperature for different optical materials.

Other Methods

There are some curvature test methods, which are only suitable to test concave surfaces [53-10]–[53-13]. Among them are the following.

- The Foucault Test (knife-edge test), in which a point source and a knife-edge are brought to the center point of a spherical or nearly spherical concave surface under test. The distance from the knife-edge to the vertex of the surface under test is then measured. In special cases it is possible to measure the radius of curvature of the second lens surface indirectly through the material, when the radius of the first surface, the thickness and the refractive index of the lens are known.
- The Traveling Microscope Method, which is identical to the Two-Position Method, except that the interferometer is replaced by a microscope. A point source is produced at the front focus of the microscope objective, which is then brought to the vertex and the center of curvature of the concave surface under test while observing the image of the light source through the microscope. The travel distance of the microscope is measured by a suitable displacement metrology.

There are also some curvature test methods suitable for testing very large radii. These are as follows.

- Non-contacting interferometry (Fizeau or Twyman–Green interferometer) comparing the surface under test with a plane reference surface. The reference flat has to be calibrated carefully. The gradient of the resulting test wavefront must not exceed the Nyquist limit (> 2 pixels per fringe). For a camera with 1000 × 1000 pixels, the limit is 500 fringes per field of view diameter.
- Deflectometry using a reference pattern (checkerboard, grid, etc.) as a test mask. The surface under test reflects the mask image which is observed by a camera. The distortion of the image contains information about the radius of curvature, which is analyzed by a computer (see section 49.6).

53.3 Central Thickness

53.3.1 Basics

The central thickness *d* of a lens element is the geometrical distance between its vertex points (figure 53-10). Vertex points are surface points that coincide with the optical axis of the lens element. In the case of a spherical element the optical axis crosses the centers of curvature. In the case of a rotationally symmetric aspherical element the optical axis usually coincides with the axes of the aspheres.

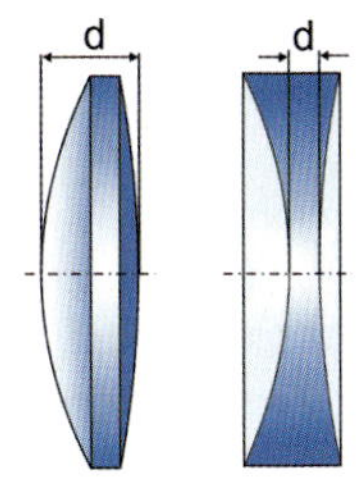

Figure 53-10: Central thickness *d* of convex and concave lens elements.

53.3.2 Metrology

There are contacting and non-contacting methods in use to measure the central thickness of optical elements. Contacting methods mainly refer to CMM-based techniques [53-15], whereas non-contacting methods include capacitive or optical sensors that either act as proximity sensors or displacement measurement systems. Optical methods include those that measure the thickness from one side of the element by using the partial reflections returning from both surfaces, as well as those that measure the position of the surface vertices from each side individually. In the following we give an overview of measurement methods for geometrical and optical central thickness, the latter defined by the product of refractive index and geometrical thickness.

Measurement of Geometrical Central Thickness

One type of earliest high-precision metrology for central thickness is the contacting method proposed by Abbe [53-14] as shown in figure 53-11. The lens element under test rests with its convex back side (figure 53-11 (a)) on a polished plane steel plate. Its optical axis is adjusted so that it coincides with the mechanical axis of a plunger movable in a direction perpendicular to the steel plate. The plunger is connected to a precision scale that – in earlier times – could be read by a microscope. Nowadays opto-electronic devices will detect the z-position of the plunger precisely.

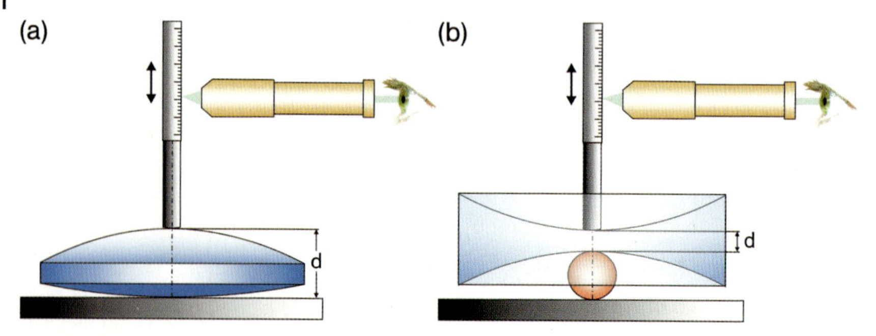

Figure 53-11: Central thickness measurement of : (a) a convex lens element by means of a contacting movable plunger; (b) a concave lens element by means of an additional steel ball and a contacting movable plunger.

The zero position is calibrated without the lens by putting the plunger in contact with the steel plate. When a lens element with two concave surfaces is being tested, an additional steel ball is placed between the steel plate and the lower surface (figure 53-11 (b)). The zero position is then found by letting the plunger contact the upper vertex of the steel ball.

Calibration is carried out by introducing at least one calibrated element of known thickness and correcting the scale accordingly. To compensate for non-linear deviations, several calibration elements of known thickness covering the total thickness range have to be measured.

Care has to be taken to correctly find the vertices of the lenses. Lateral displacement by Δx and tilting of the elements by α (in radians) will lead to a thickness measurement error Δd as shown in (53-9).

$$\Delta d = (r_1 + r_2 - d)(1 - \cos\alpha) + r_1\left(\sqrt{1 - \left(\frac{\Delta x + (r_1 + r_2 - d)\sin\alpha - r_2 \cdot \alpha}{r_1}\right)^2} - 1\right) \quad (53\text{-}9)$$

Here r_1 denotes the radius of the upper surface, r_2 that of the lower surface. Convex surfaces have $r_1, r_2 > 0$, concave surfaces have $r_1, r_2 < 0$.

An alternative to this method is to place the lens element onto a regular lens mount and let two individual plungers contact the lens surfaces as shown in figure 53-12 (a). The thickness is measured by two scales associated with the plungers. The zero position is found by letting the plungers contact each other. Calibration is carried out by introducing at least one calibrated element of known thickness and correcting the scale accordingly.

Placing the optical surfaces in contact with mechanical plungers always carries the risk of damaging the surfaces. It is therefore advisable to use a non-contacting proximity sensor, either an optical or a capacitve sensor which either shows the relative distance of the surface's vertex from the sensor, or which gives a signal when the vertex is positioned in its correct zero position (figure 53-12 (b)). The thickness range can be extended by attaching the sensors to motion systems traveling in the z-direction, which are connected to readable scales.

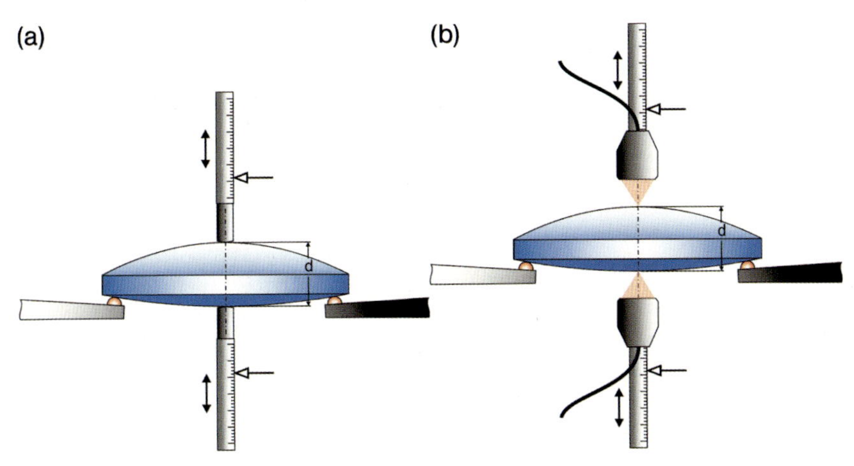

Figure 53-12: Central thickness measurement using: (a) two mechanical plungers; (b) two optical proximity sensors which are in contact with both surfaces of the lens element under test.

If the thickness ranges to be measured are fairly short, two optical displacement sensors kept in fixed positions relative to the lens mount are a suitable solution. Displacement systems with different principles are also in use. Figure 53-13 shows a thickness measurement arrangement using two chromatically coded confocal sensors [53-20]. Each color of a broadband light source is focussed at a different distance from the sensor, so that light returning from the lens vertices shows an individual color related to the distance from the sensor. The signal is analyzed by a spectrometer and the displacement is indicated precisely.

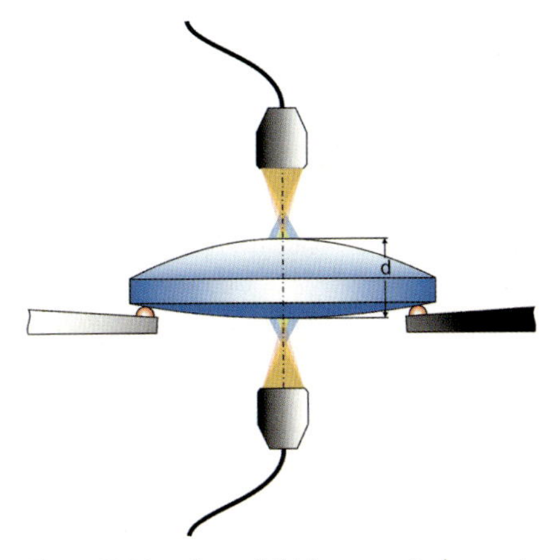

Figure 53-13: Central thickness metrology using two chromatically coded confocal sensors.

Measurement of Optical Central Thickness

For many applications it is useful to measure the central thickness of the optical elements from one side. The optical solution is to use the reflections from both surfaces and analyze their properties related to the applied method. In analogy to figure 53-12 (b) an optical proximity sensor can be placed in two positions, where its focus is placed first on the upper surface and then on the lower surface (figure 53-14 (a)). The traveling distance between these two positions is a, which is given by:

$$a = \frac{d\,r_1}{n\,r_1 - d(n-1)} \tag{53-10}$$

where r_1 denotes the radius of curvature of the upper surface, n denotes the refractive index of the element for the wavelength of the proximity sensor and d is the central thickness of the element. Note that in general there is spherical aberration, when the sensor spot is focussed onto the lower surface. Therefore, only low beam apertures may be used for this method. Note also that the refractive index of the element and the radius of curvature of the upper surface have to be known to determine the geometrical central thickness d of the element.

If the refractive index n is not known and if $r_1 \neq r_2$, a second measurement can be carried out with the lens element turned upside down, so that the measurement is repeated through the second surface of known radius r_2. The central thickness is then determined from

$$d = \frac{a_1\,r_1(r_2 - a_2) - a_2\,r_2(r_1 - a_1)}{a_1\,r_2 - a_2\,r_1} \tag{53-11}$$

where a_1 and a_2 denote the measured sensor displacement in both measurements.

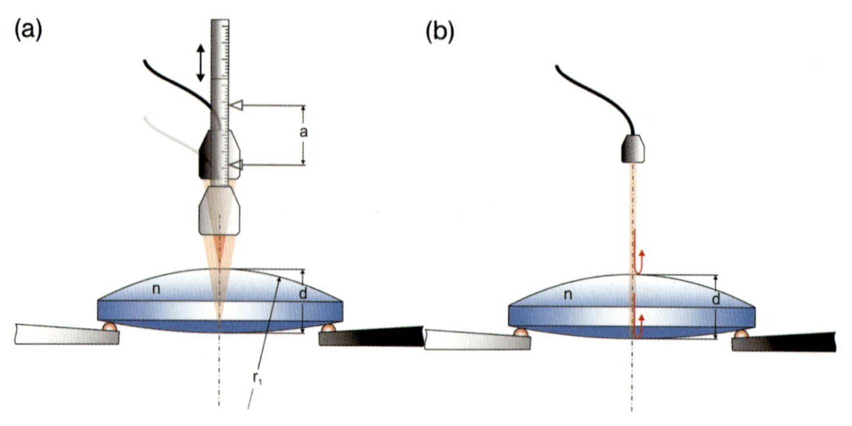

Figure 53-14: Central thickness measurement from one side: (a) detecting the proximity sensor displacement a when focussed on the upper and lower surfaces consecutively: (b) detecting the optical path difference of the two simultaneous reflections from the upper and lower surface vertices.

Another method uses short coherent light supplied through a fiber to a collimator, which generates a thin, almost parallel, beam directed along the optical axis of the lens element under test (figure 53-14 (b)) [53-16]–[53-19]. The reflections from the upper and lower vertex re-enter the collimator and return through the fiber for analysis. Both reflections have a relative optical path difference of $2nd$. The analysis of the optical path difference (OPD) can be carried out, for instance, by a Michelson-type interferometer, in which the interferometer arms are provided by fibers as shown in figure 53-15. Light from a short coherent superluminescent diode is fed into a fiber and split into two parts by a beam-dividing coupler element. The first part reaches the lens under test and returns to the coupler after reflection. The second part reaches a delay line, which consists of a movable plane mirror. The mirror can be moved continuously along the optical axis while its position is registered. The reflected light returns to the collimator and the coupler, where it is directed towards a sensor along with the returning light from the lens under test. The sensor detects the light level while the reference mirror travels along the optical axis. Interference modulation is detected when the OPD between the test arm and the reference arm is within the coherence length of the light. The distance of travel between the two modulation maxima being detected is equal to nd. When the refractive index of the lens material is known for the measurement light, the central thickness can be determined.

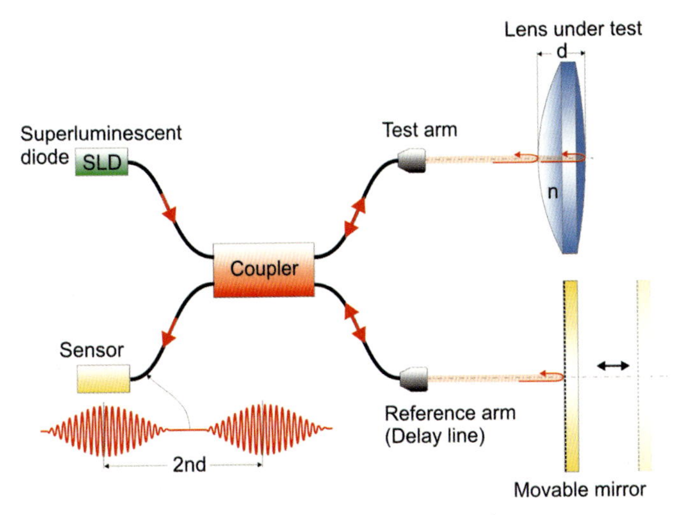

Figure 53-15: Michelson-type fiber interferometer used to measure the central thickness of an optical lens element.

If the refractive index n is unknown, which is quite likely, since most commercial fiber interferometers work at 1300 nm wavelength, then an air cavity can be placed around the lens under test, as shown in figure 53-16. The cavity consists of two partially reflecting plane mirrors. While the reference mirror travels, coherence bursts are detected whenever the OPD between the reference arm and the individual reflections from the cavity or the lens element surfaces is small enough. It is therefore possible to measure the air gaps d_1 and d_2 between the cavity elements and the lens element. A second measurement with the lens element removed measures the total length D of the empty cavity. The central thickness of the lens element is then calculated from $d = D - d_1 - d_2$, without the knowledge of the refractive index of the lens element.

The accuracy of central thickness measuring interferometers is limited to 0.5 μm – 1 μm when the instrument is kept within a mechanically and thermally stable environment.

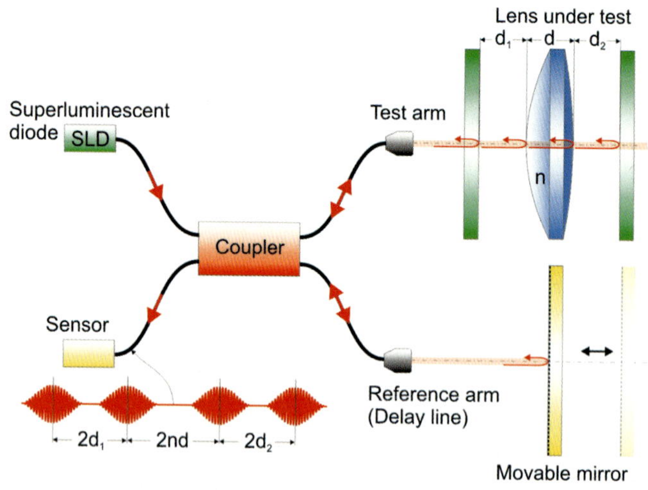

Figure 53-16: Michelson-type fiber interferometer used to measure the central thickness of an optical lens element by means of a surrounding air cavity.

53.4
Surface Form and Figure Irregularities

53.4.1
Basics

Definitions

The deviation of a fabricated optical surface from an ideal surface as specified by the optical design can be described by a number of terms, which are defined in ISO 10110 Part 5 [53-1]. In the follwing we will summarize the fundamental definitions. An example is given in figure 53-17.

1. Surface form deviation
   The distance between the optical surface under test and the nominal theoretical surface, measured perpendicular to the theoretical surface, which is nominally parallel to the surface under test.

2. Peak-to-valley (PV) difference
The maximum distance minus the minimum distance between two surfaces. If one of the surfaces is a theoretical surface, it is possible that the surfaces cross, in which case the minimum distance between the surfaces is a negative number; the sign must be taken into account in computing the PV difference.

3. Total surface deviation function
The theoretical surface defined by the difference between the actual surface and the desired theoretical surface.

4. Approximating spherical surface
The spherical surface for which the root-mean-square (rms) difference from the total surface deviation function is a minimum.

5. Sagitta error
The peak-to-valley difference between the approximating spherical surface and a plane. A sagitta error results from the test surface having a radius of curvature different from the specified radius.

6. Irregularity function
The theoretical surface defined by the difference between the total surface deviation function and the approximating spherical surface.

7. Irregularity
The peak-to-valley difference between the irregularity function and the plane which best approximates it. For nominally spherical surfaces, the irregularity represents the departure of the surface from sphericity. For aspherical surfaces, the irregularity represents the aspherical part of the total surface deviation function.

8. Approximating aspherical surface
The rotationally symmetric surface for which the rms difference from the irregularity function is a minimum.

9. Rotationally symmetric irregularity
The peak-to-valley difference between the approximating aspherical surface and the plane which best approximates it. The rotationally symmetric irregularity is the rotationally symmetric part of the *irregularity* defined above. Its value cannot exceed that of the irregularity.

10. Total rms deviation, RMSt
The root-mean-square difference (RMS) between the optical surface under test and the desired theoretical surface, without subtraction of any surface-form deviation types. The RMS value of a function f of two variables x and y over a given area A is given by

$$\mathrm{RMS} = \sqrt{\frac{\int_A (f(x,y))^2 dA}{\int_A dA}} \tag{53-12}$$

In practice, the integrals are replaced by a corresponding summation over the measured data points, provided that a sufficient number of data is used.

11. The rms irregularity, RMSi
    Root-mean-square value of the *irregularity function* defined above.
12. The rms asymmetry, RMSa
    Root-mean-square value of the difference between the irregularity function and the approximating aspherical surface.

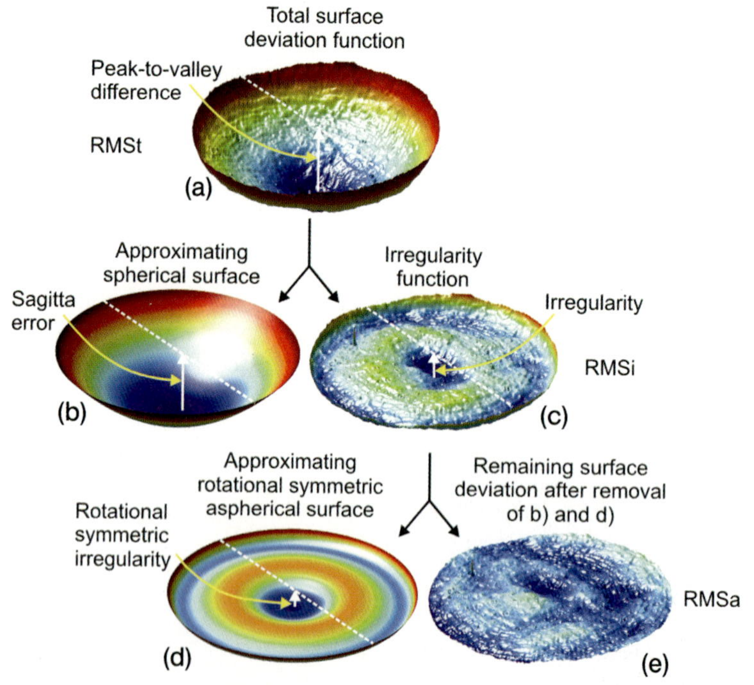

Figure 53-17: Example of a measured surface and its decomposition into surface error types.

All terms defined above are expressed in units of fringe spacing. When a surface is tested interferometrically, a surface-form deviation of one-half the wavelength of light causes an interference pattern in which the intensity varies from one bright fringe to the next, or from one dark fringe to the next – that is, one "fringe spacing" is visible. The expression "fringe spacing" does not refer to the transverse distance between fringes, but to the fact that the number of fringe spacings visible in the interference pattern corresponds to the number of half-wavelengths of surface-form deviation. Unless otherwise specified, the wavelength is 546.07 nm, which corresponds to the green spectral line of mercury (e-line).

Indication

On the drawing of an optical element, the surface-form tolerance is indicated by a code number followed by the tolerances for sagitta error, irregularity, rotationally symmetric irregularity and rms deviation types, as appropriate (figure 53-18) [53-1].

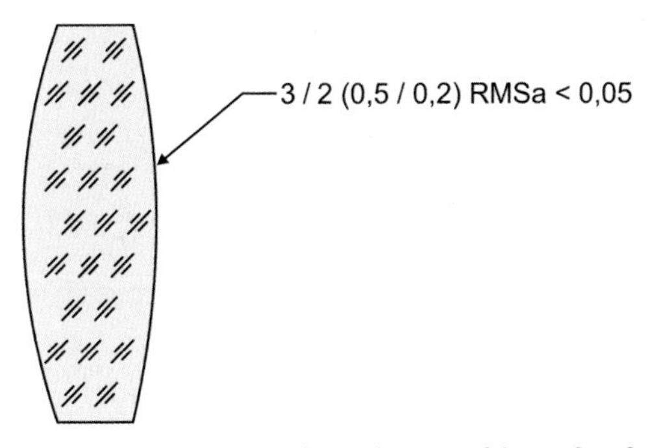

Figure 53-18: Example of an indication of the surface-form tolerances.

The code number for the surface-form tolerance is 3. The indication will therefore have one of three forms:

$3/A(B/C)$

or

$3/A(B/C)$ RMSx $< D$ (where x is one of the letters t, i or a)

or

3/–RMSx $<D$ (where x is one of the letters t, i, or a).

The quantity A is either:

1) the maximum permissible sagitta error, expressed in fringe spacings, or
2) a dash (–) indicating that the total radius of curvature tolerance is given in the radius of curvature dimension (not applicable for planar surfaces).

The quantity B is either:

1) the maximum permissible value of the irregularity expressed in fringe spacings, or
2) a dash (–) indicating that no explicit irregularity tolerance is given.

The quantity C is the permissible rotationally symmetric irregularity expressed in fringe spacings. If no tolerance is given, the slash (/) is replaced by the final parenthesis, i.e. 3/A(B).

If no tolerance is given for all three deviation types, then A, B, C, the slash (/) and the parentheses are replaced by a single dash (–), i.e. 3/–.

The quantity D is the maximum permissible value of the rms quantity of the type specified by x where x is one of the letters t, i or a. The specification of more than one type of rms deviation is allowed. These specifications are separated by a semicolon.

The surface-form tolerance indicated applies to the optically effective area, except when the indication is to apply to a smaller test field for all possible positions.

Decomposition

When testing curved surfaces interferometrically, the surface under test is compared with a reference surface. The resulting fringe pattern represents the difference between the surface under test and the projection of the reference surface into the location of the surface under test. This projected surface is called the *effective reference surface*. The measured wavefront $W(x,y)$, called the *wavefront error function*, therefore contains figure deviations as well as any deviations caused by misalignment of the surface under test relative to the reference surface.

For the decomposition of $W(x,y)$ into the different surface-error types, an approximation by polynomials is useful. The polynomials should be adequate to easily represent the piston, tilts and focus terms, because they are needed to express and subtract misalignment effects of the plane and low- and medium-aperture spherical surfaces. However, higher terms are also of great interest as they describe spherical aberration, coma, astigmatism or trefoil. Using x-y-monomials is easy, but they do have two main drawbacks:

a) since monomials do not form an orthogonal function system, the calculated monomial coefficients depend on the number of terms which were used for the polynomial fit;
b) the monomial coefficients have different units, which makes any reasonable comparison impossible.

It is therefore useful to select sets of orthonormal polynomials for wavefront decomposition, which avoid both drawbacks. As examples Zernike polynomials for round-shaped apertures and Chebyshev polynomials for rectangular-shaped are introduced later in this chapter.

The decomposition consists of the following steps:

1. Determine the *wavefront error function* $W(x,y)$.
2. Remove any terms introduced by misalignment of the surface under test to obtain the *total surface deviation function*.
3. Calculate the PV and RMSt.
4. Determine the *approximating spherical surface* with its PV value (= *sagitta error*) by fitting the rotationally symmetric quadratic polynomial term to the surface deviation function.
5. Subtract the rotationally symmetric quadratic term to obtain the *irregularity function* with its PV-value called the *irregularity* and its RMS-value called the RMSi.

6. Determine the *approximating rotationally symmetric aspherical surface* with its PV-value (= *rotationally symmetric irregularity*) by fitting the mean radial profile (MRP, see below) or a set of rotationally symmetric polynomial terms to the irregularity function.
7. Subtract the approximating rotationally symmetric aspherical surface to obtain the remaining surface deviation with its RMS called the RMSa.

Polynomial fitting is achieved by applying a standard least-squares fitting algorithm to the wavefront, in which the sum of the deviation-squares is minimized [53-4].

The removal of the misalignment terms is, in many cases, achieved by subtraction of the piston, tilt x and tilt y terms for plane surfaces, and additionally of the focus term (rotationally symmetric quadratic term) for low- and medium-aperture spherical surfaces. For high-aperture spherical surfaces an appropriate combination of rotationally symmetric polynomial terms must be found, which can then be subtracted. The appropriate combination can be found theoretically by raytracing or experimentally by determining the polynomial terms for a set of different axial positions of the surface under test.

If aspherical surfaces are tested, a compensating system is introduced to the interferometer to provide the necessary aspherical wavefront. When a compensating system is introduced, the interferometer camera in most cases sees a distorted image of the surface under test. In this case tilts and axial misalignments of the surface under test lead to higher polynomial terms which need to be removed. An experimental investigation would be the correct way to determine the necessary polynomial term combinations (see the section on *rotationally symmetric aspherical surfaces* later in this chapter).

Zernike Polynomials

Since most optical elements have a circular shape, Zernike polynomials are preferred [53-4]. They represent a complete set of characteristic wavefront deformations making it possible to classify, quantify and separate individual wavefront terms.

Zernike polynomials represent an orthonormal basis for a circular area at a constant weight function. They are basically defined in unified polar coordinates r $(0 \le r \le 1)$ and φ and are generated by the product of a radial term $R(r)$ and a term which is dependent on the azimuthal angle φ.

$$Z_n^m(r,\varphi) = R_n^m(r)\begin{cases} \sin(m\varphi) \text{ for } m > 0 \\ \cos(m\varphi) \text{ for } m \le 0 \end{cases} \tag{53-13}$$

where n denotes an integer number representing the radial order, $n \ge 0$, and m denotes an integer number representing the azimuthal order with $m = -n, -n+2, \cdots, n-2, n$.

The radial term can be generated by

$$R_n^m(r) = \sum_{k=0}^{\frac{n-m}{2}} \frac{(-1)^k (n-k)!}{k!\left(\frac{n+m}{2}-k\right)!\left(\frac{n-m}{2}-k\right)!} r^{n-2k} \tag{53-14}$$

It is very common to combine n and m to a single number j which indicates the individual terms. A widespread method for numbering is

$$j = \left(\frac{n+2+|m|}{2}\right)^2 - 2|m| + \frac{1}{2}(\text{sign}|m| + \text{sign}(m)) \tag{53-15}$$

with sign$(m) = 1$ for $m > 0$ and sign$(m) = -1$ for $m < 0$ and sign$(m) = 0$ for $m = 0$.

Rotationally symmetric terms with $m = 0$ are therefore numbered as 1, 4, 9, 16, 25, ...

Table 53-1 shows the first 16 terms.

Table 53-1: The first 16 Zernike polynomials in normalized r-φ-coordinates.

j	n	m	$Z_n^m(r,\varphi) = R_n^m(r)\begin{cases}\sin(m\varphi)\text{for} m > 0\\ \cos(m\varphi)\text{for} m \le 0\end{cases}$	Name	3D surface
1	0	0	1	Piston	
2	1	–1	$r\cos\varphi$	Tilt x	
3	1	1	$r\sin\varphi$	Tilt y	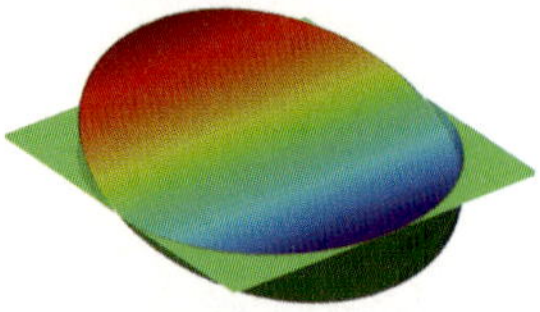
4	2	0	$2r^2 - 1$	Focus	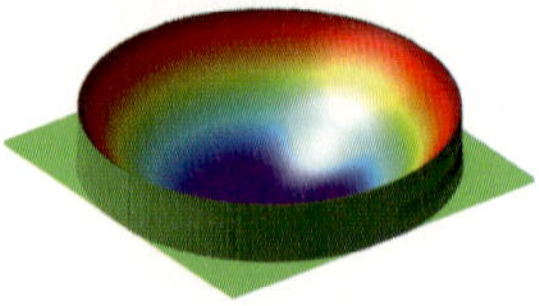

j	n	m	$Z_n^m(r,\varphi) = R_n^m(r)\begin{cases} \sin(m\varphi)\text{form} > 0 \\ \cos(m\varphi)\text{form} \le 0 \end{cases}$	Name	3D surface
5	2	−2	$r^2 \cos 2\varphi$	Astigmatism 0°/90° 2nd order	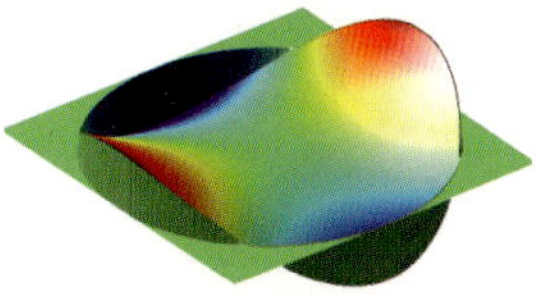
6	2	2	$r^2 \sin 2\varphi$	Astigmatism ± 45° 2nd order	
7	3	−1	$(3r^3 - 2r)\cos\varphi$	Coma x 3rd order	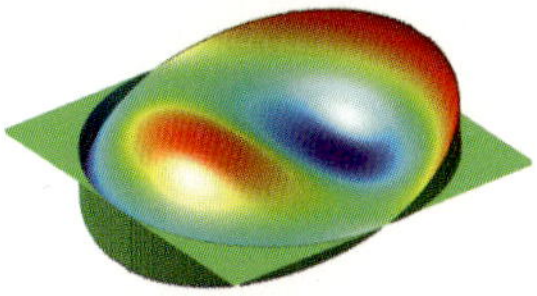
8	3	1	$(3r^3 - 2r)\sin\varphi$	Coma y 3rd order	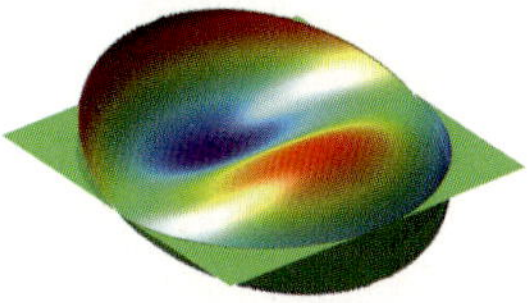
9	4	0	$6r^4 - 6r^2 + 1$	Spherical aberration 4th order	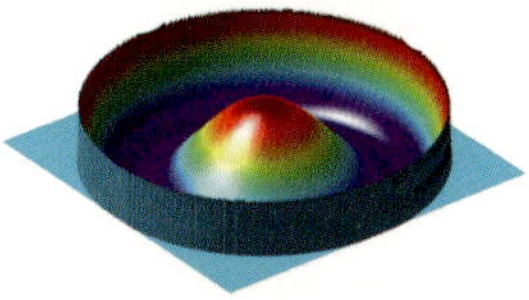
10	3	−3	$r^3 \cos 3\varphi$	Trefoil 0°/ 120°/240° 3rd order	
11	3	3	$r^3 \sin 3\varphi$	Trefoil 30°/ 150°/270° 3rd order	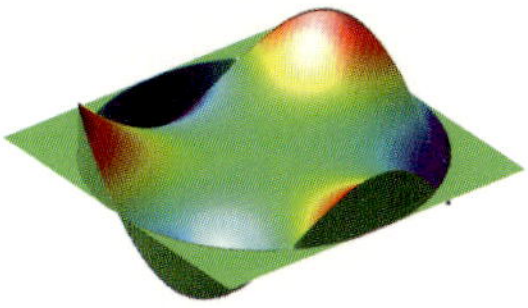

j	n	m	$Z_n^m(r,\varphi) = R_n^m(r)\begin{cases} \sin(m\varphi)\text{form} > 0 \\ \cos(m\varphi)\text{form} \le 0 \end{cases}$	Name	3D surface
12	4	−2	$(4r^4 - 3r^2)\cos 2\varphi$	Astigmatism 0°/90° 4th order	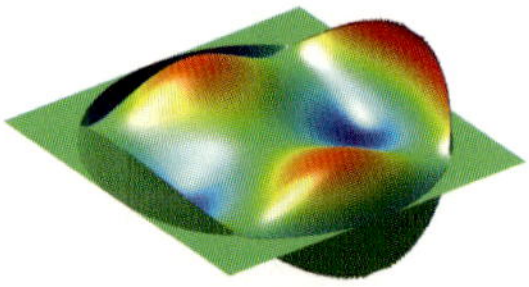
13	4	2	$(4r^4 - 3r^2)\sin 2\varphi$	Astigmatism ± 45° 4th order	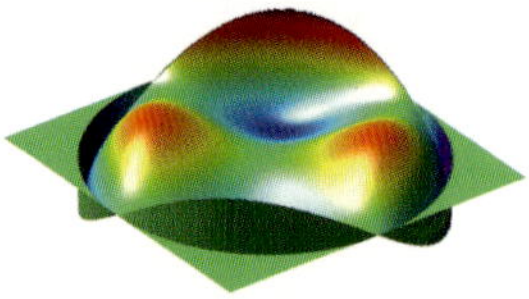
14	5	−1	$(10r^5 - 12r^3 + 3r)\cos\varphi$	Coma x 5th order	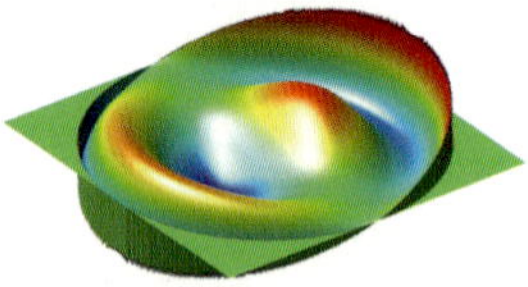
15	5	1	$(10r^5 - 12r^3 + 3r)\sin\varphi$	Coma y 5th order	
16	6	0	$20r^6 - 30r^4 + 12r^2 - 1$	Spherical aberration 6th order	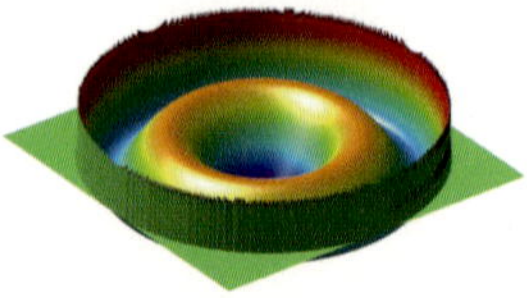

Chebyshev Polynomials

Measurements of optical elements having a rectangular shape cannot be treated by means of Zernike polynomials. Zernike polynomials are orthonormal only for circular-shaped apertures. A possible solution for rectangular-shaped surfaces are Chebyshev polynomials [53-21]. As an example we will select Chebyshev polynomials of the first kind applied independently to x- and y-coordinates. The resultant two-dimensional polynomials then represent an orthonormal basis for a rectangular area at a constant weight function. They are basically defined in unified cartesian coordinates x and y ($-1 \le x \le 1$ and $-1 \le y \le 1$) and are generated by the product of an x-term $T(x)$ and a y-term $T(y)$.

$$U_{n,m}(x,y) = T_n(x)\,T_m(y) \tag{53-16}$$

with the integer numbers $n \geq 0$ and $m \geq 0$.

A one-dimensional Chebyshev polynomial $T_n(x)$ is generated by

$$T_n(x) = \cos\left(n \arccos(x)\right) \tag{53-17}$$

It is also possible to generate individual terms by a recursion formula (53-17). Considering $T_0(x) = 1$ and $T_1(x) = x$ any higher term can be generated by

$$T_{n+1}(x) = 2x\,T_n(x) - T_{n-1}(x) \tag{53-18}$$

It is useful to combine n and m into a single number to indicate the individual terms. A possible numbering is given by

$$j = \frac{1}{2}\left((n+m)^2 + n + 3m + 2\right) \tag{53-19}$$

The first 15 Chebyshev terms are shown in table 53-2.

Table 53-2: The first 15 Chebyshev polynomials in normalized x-y-coordinates.

j	n	m	$U_{n,m}(x,y) = T_n(x)\,T_m(y)$	Name	3D surface
1	0	0	1	Piston	
2	1	0	x	Tilt x	
3	0	1	y	Tilt y	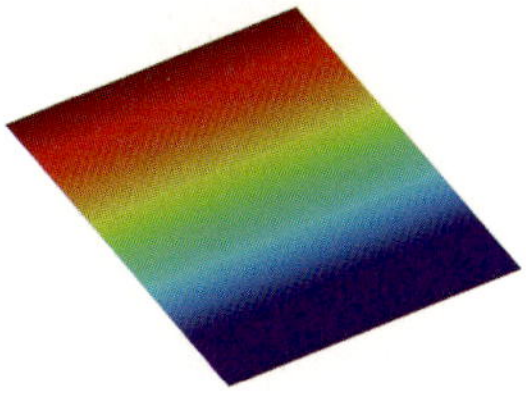

j	n	m	$U_{n,m}(x,y) = T_n(x)\,T_m(y)$	Name	3D surface
4	2	0	$2x^2 - 1$	Cylinder x	
5	1	1	xy	Saddle ± 45°	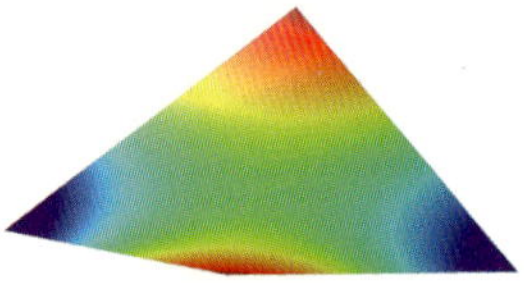
6	0	2	$2y^2 - 1$	Cylinder y	
7	3	0	$4x^3 - 3x$	–	
8	2	1	$(2x^2 - 1)y$	–	
9	1	2	$(2y^2 - 1)x$	–	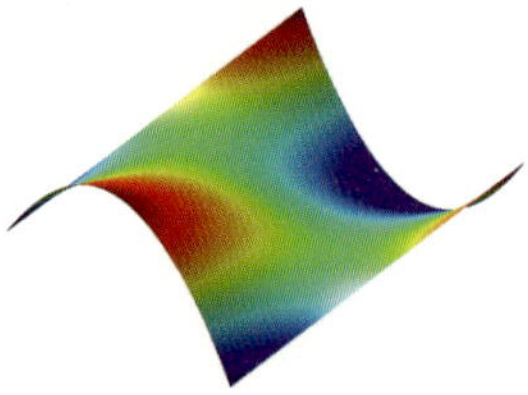

j	n	m	$U_{n,m}(x,y) = T_n(x)\,T_m(y)$	Name	3D surface
10	0	3	$4y^3 - 3y$	–	
11	4	0	$8x^4 - 8x^2 + 1$	–	
12	3	1	$(4x^3 - 3x)y$	–	
13	2	2	$(2x^2 - 1)(2y^2 - 1)$	–	
14	1	3	$(4y^3 - 3y)x$	–	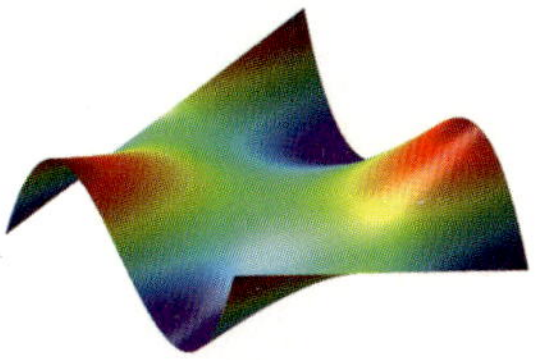
15	0	4	$8y^4 - 8y^2 + 1$	–	

Mean Radial Profile

To determine the *rotationally symmetric irregularity* from the *irregularity function* it is necessary to separate the *approximating rotationally symmetric aspherical surface* from the irregularity function. ISO 10110-5 suggests using rotationally symmetric Zernike terms for the approximation. For smooth irregularities containing only low-order rotationally symmetric terms this is the most appropriate way. However, in many applications, where aspherical surfaces are manufactured by means of computer-controlled polishing machines, rotationally symmetric deviations caused by machine errors are of higher order. These deviations would need very high polynomial terms for a good approximation. In these cases it is useful to calculate the so-called *Mean Radial Profile* (MRP), which will be explained in the following.

The MRP is defined by the mean integral over the circle of radius r around the center of the circular aperture with radius $r_{\max}$ (see figure 53-19).

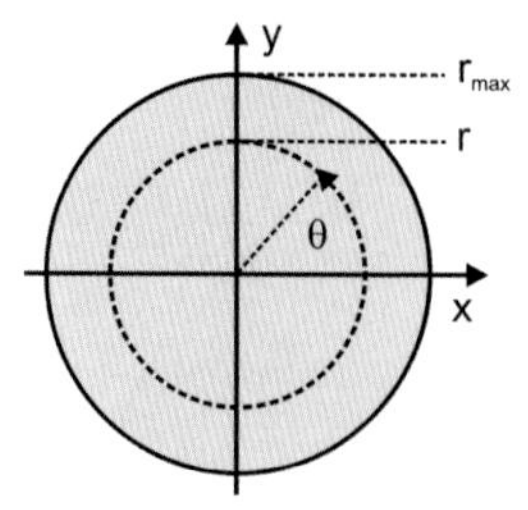

Figure 53-19: Coordinates for the calculation of the Mean Radial Profile.

$$\mathrm{MRP}(r) = \frac{1}{2\pi} \int_{0}^{2\pi} f(r,\theta) d\theta \tag{53-20}$$

where θ denotes the azimuthal angle, $f(r,\theta)$ denotes the irregularity function, of which the *approximating rotationally symmetric aspherical surface* – which is identical with the MRP(r) – is to be determined. Figure 53-20 compares the MRP with a Zernike approximation up to twelfth order. We can see that the Zernike approximation is not able to express the fine ring-shaped deviations of the original wavefront.

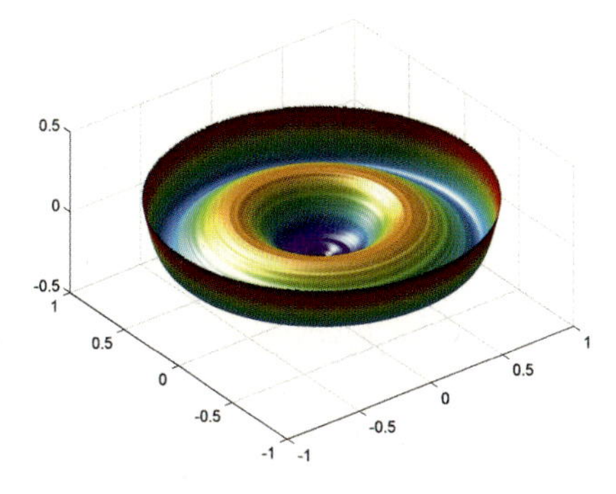

(a) Original wavefront

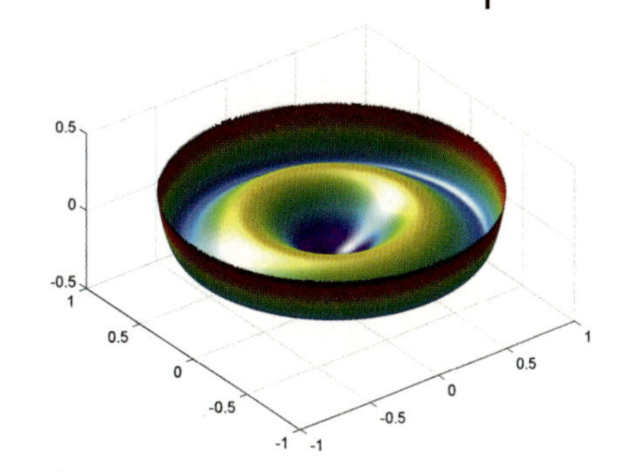

(b) Zernike approximation

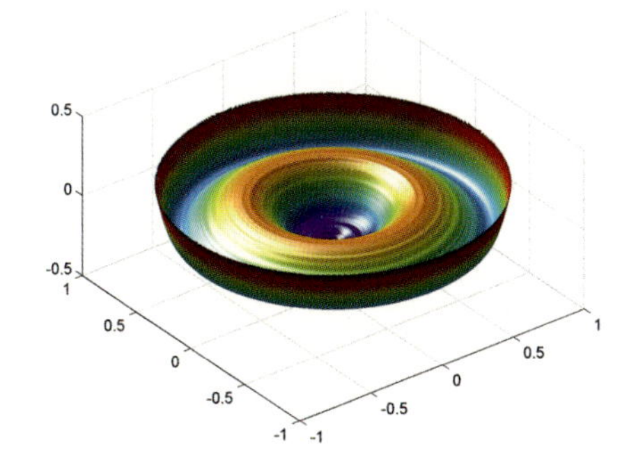

(c) Mean Radial Profile

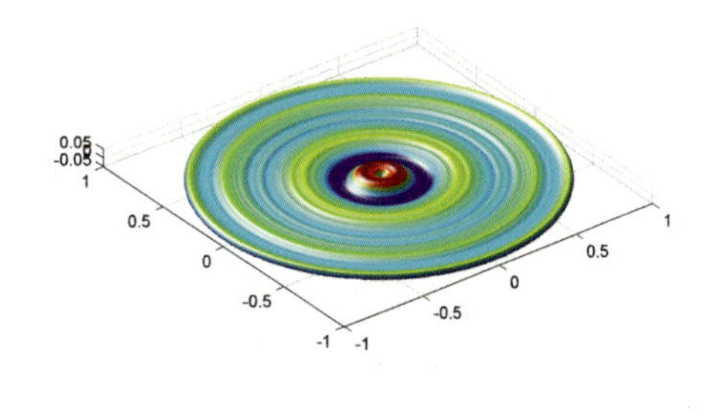

(d) Difference between Zernike approximation and MRP

Figure 53-20: Comparison of Zernike approximation and Mean Radial Profile for a measured wavefront: (a) original wavefront; (b) Zernike approximation with highest power = 12; (c) Mean Radial Profile; (d) difference between (b) and (c).

53.4.2 Metrology

Optical systems can be regarded as well corrected, if they fulfill the Maréchal criterion (see section 30.5.2 in Vol. 3). The Maréchal criterion demands that the system's rms wavefront error is smaller than $\lambda/14$ [53-22]. This is one of the reasons why the surface forms of optical surfaces are preferably measured by means of interferometers. Interferometers provide a surface-covering deviation map, generally of sufficient resolution.

Neglecting any wavefront error residuals due to the optical design, the surface-form deviations of individual surfaces i add up to a total RMS wavefront deviation error RMS_W for a system with N optical surfaces:

$$\mathrm{RMS_W} = \sqrt{\sum_{i=1}^{N} \left((n_i - n_{i-1})\, \mathrm{RMS_{t,i}} \right)^2} < \frac{\lambda}{14} \tag{53-21}$$

where n_i denotes the refractive indices between surface n_i and n_{i-1}. For example, a perfect three-element optical system with normal glasses suitable for green light must have RMS surface-form deviations smaller than 30 nm. To qualify the individual components, full surface-covering interferometers are necessary. The deviation maps they provide serve as qualification data as well as intermediate results for further polishing processes. Regular tactile coordinate measuring machines are able to qualify optical components in a ground state, when the surface form is still far from its final shape, but not in a polished state.

Interferometers measure surface forms relative to a reference surface or system. This makes it necessary to calibrate the reference system to an accuracy which is sufficient to produce the end quality required.

In the following, we will describe the general interferometric test setups used to measure plane, spherical, aspherical, cylindrical and free-form surfaces. We will mention the various calibration techniques and the limitations in the accuracy and repeatability.

Surface–Form Measuring Interferometers

Several main aspects must be considered for the concept of an interferometric setup:

1. The requirement specifications for the surfaces to be tested.
2. The location of test equipment and necessary environmental conditions (vibration, temperature).
3. The mounts of the test pieces during test and in its final lens mount.
4. The direction of gravitation during test and in the final application.
5. The environmental conditions during test and in the final application.
6. The handling of test pieces, tools and equipment.
7. The security aspects concerning laser safety and general safety aspects during usage and maintenance.
8. The capacity of test equipment considering the standard measurement process, the calibration process and down-time for maintenance.
9. The skill level of users.
10. The total costs of ownership (TCO).

In our further discussions we will concentrate on the technical aspects of the necessary test equipment rather than on the economic aspects. In most cases Fizeau-type interferometers are proposed. They are by far the most widely used interferometers because of their stability, flexibility and simplicity in setup and usage.

Gravity and Orientation

One of the basic questions in surface testing is how to treat the problem af gravity. The deformation of the optical element depends on its shape and material as well as on the position and number of contact points supporting the element. If the element can be tested in the final mount in the same orientation relative to gravity as in the final application, then no further consideration is necessary.

In the usual case, however, the optical element is tested in a test mount which is different from the later application. In this case the bending of the element due to gravitation must be estimated and subtracted from the measured surface form. The estimation is carried out by the *finite element method* (FEM), which is a numerical technique for finding approximate solutions of partial differential equations (PDE) as well as of integral equations. The solution approach is based either on eliminating the differential equation completely (steady state problems), or rendering the PDE into an approximating system of ordinary differential equations, which are then numerically integrated using standard techniques, such as Euler's method, Runge-Kutta, etc. [53-23]–[53-24]. Figure 53-21 shows an example of a gravitational surface-form deviation map for an optical lens element placed on a three-ball-mount as simulated by the FEM.

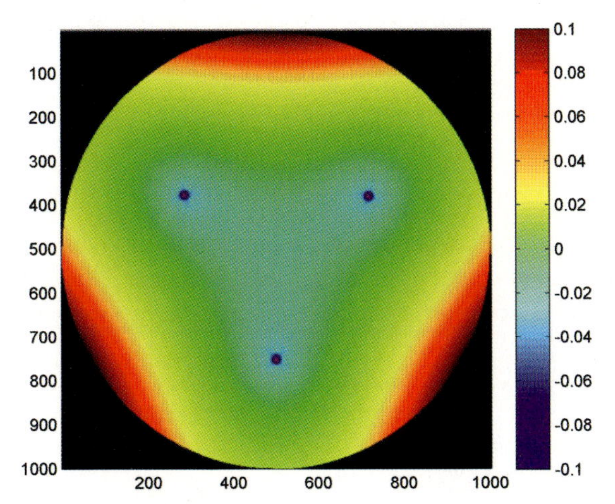

Figure 53-21: FEM simulation of a lens element's surface form deformation in µm when placed on a three-ball mount.

The interferometric system must be set up either with the optical axis of the element perpendicular (horizontal setup, figure 53-23) or parallel (vertical setup, figure 53-24) to the gravity vector.

A horizontal setup is easy to build up on a tabletop placed on a vibration isolation system to damp any mechanical vibrations from the ground. The element under test can be mounted on a V-block or two steel cylinders contacting the element at its outer rim. A simple method is to fix the lens in between three screws. For calibration purposes, the element should be rotatable around its optical axis, which is achieved, for instance, by making the mount cylinders rotatable and driving the rotation by means of a friction wheel. Heavy elements should be placed in a strap that supports and rotates the element when driven by the friction wheel (figure 53-22(a)–(d)) [53-25].

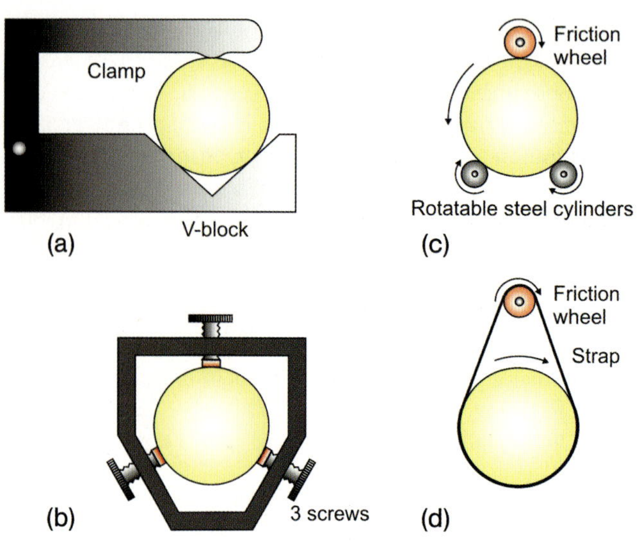

Figure 53-22: Different mount techniques for optical elements in a horizontal setup: (a) V-block with clamp; (b) three screws in a frame; (c) two rotatable cylinders with friction wheel; (d) strap support with driving friction wheel.

Horizontal setups are useful for laboratory purposes and in a production process with a small series of elements. In a production process with larger numbers of elements a vertical setup has more advantages in handling and adjustment, although the setup itself is more complicated and expensive.

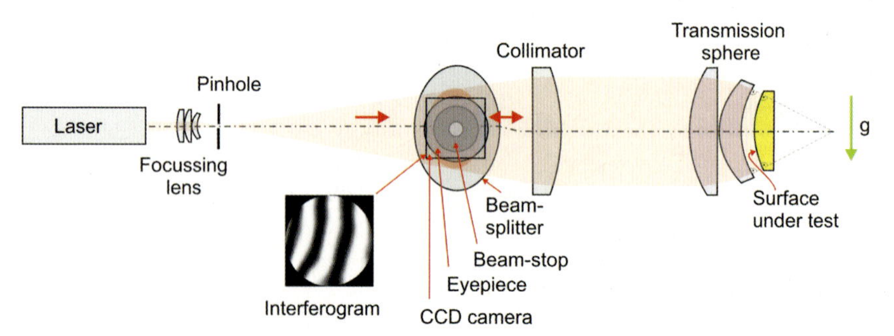

Figure 53-23: Horizontal interferometric setup to test spherical surfaces by means of a transmission sphere (the observation axis including eyepiece, beam-stop and camera is perpendicular to the plane of the drawing).

Figure 53-24 shows an example of a vertical setup in which the interferometer, including the collimator, is horizontally orientated and in which the parallel test beam is directed upward by a plane folding mirror. When testing spherical surfaces, a transmission sphere is introduced with its optical axis in the direction of gravitation as is the element under test. The element mount can be a simple three-ball mount (figure 53-25) on which the surface under test is placed and adjusted. A three-ball mount provides a well-defined support geometry which can easily be simulated by the FEM to subtract any gravitational bending effects. Once the three-ball mount is adjusted relative to the optical axis of the interferometer, any spherical element under test only needs to be adjusted in two degrees of freedom to bring it on axis. For calibration purposes and for high-precision measurements the lens mount should be rotatable around the optical axis of the transmission sphere.

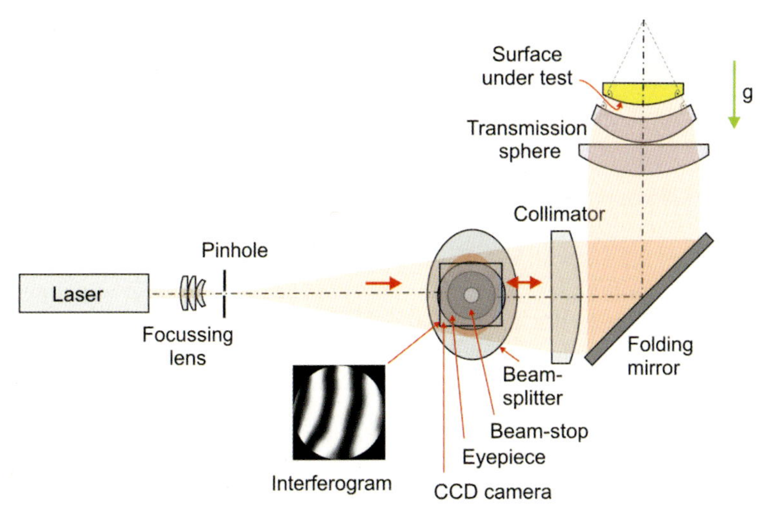

Figure 53-24: Interferometric setup with a horizontal interferometer and a vertical test beam to test spherical surfaces by means of a transmission sphere (the observation axis including eyepiece, beam-stop and camera is perpendicular to the plane of the drawing).

In a vertical setup, looking down, the supporting arms of the element mount might cover parts of the test area. This is a big disadvantage, especially when the surface-form deviation map is used to correct the surface using computer-controlled polishing machines. One solution is to support the surface in the outer freeboard area, which is not used for optical imaging. However, the lens element will then bend much more under gravity. For large lens elements this might be a problem. Another solution is to make a second measurement with the element lifted and rotated to make the hidden areas visible. Both measurements must then be combined by a "stitching" process.

A vertical setup can also be arranged with the surface under test looking upward. In this case the element is supported at its back side. Bending effects will be smaller and smooth and are less critical when the FEM simulation is subtracted. The disadvantage is that adjustment of the element is more complicated, because five degrees of freedom must be readjusted every time the element is exchanged.

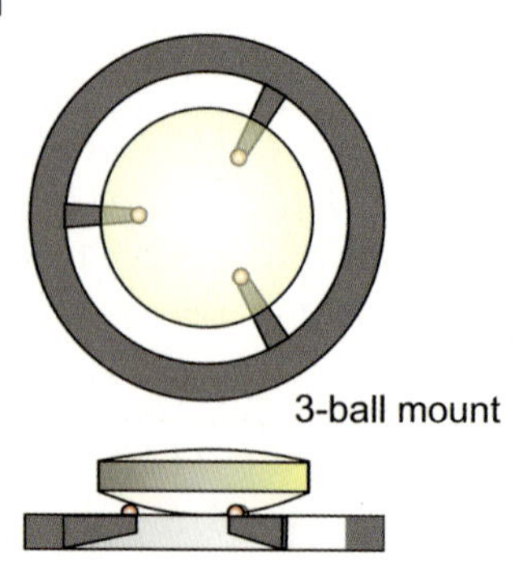

Figure 53-25: Three-ball mount to support lens elements in a vertical interferometric test setup.

Environmental Conditions

Interferometric setups in general need an environment with a low mechanical vibration level. Using red light a peak-to-peak element vibration of 300 nm would cause the interferogram to shift by one fringe. High frequent vibrations of this amplitude would diminish the interference contrast severely and would make measurements impossible. In practice, vibration amplitudes should be less than 30 nm. Interferometric setups should therefore preferrably be installed in building basements, near posts and on optical tables used for vibration–isolation systems [Section 46.8.1].

Acoustic excitation vibrations might also disturb phase-shift-based fringe analysis techniques. The disturbance is not caused by the compressed and rarefied air within the cavity, but by the eigenfrequency of the mechanical setup being excited by the acoustical source. Damping might be necessary if the source cannot be removed.

The requirement specifications for the elements to be tested determine how accurate the measurement has to be and therefore strongly determine the environmental requirements. For a $\lambda/30$ quality a temperature stability of better than 1 K will be sufficient for the interferometric inspection. For qualities better than $\lambda/100$ as needed, for example, in lithography projection optics, a flow box arrangement with temperature stabilities around 50 mK is mandatory [53-26]–[53-27].

Productivity

When mounted and adjusted in the final test position, lens elements must rest for a while to adapt to the test environment temperature, which might be around 20°C. The element has been exposed to the heat radiation of the human body and must now cool down. Also, mechanical stress introduced by friction of the three-ball lens mount at the contact points must be released. For high-precision measurements the element therefore needs to rest for more than one hour before the measurement begins. During this time the test equipment is idle if only one channel is provided. To keep productivity up, two or more channel arrangements (cascades) may be in use, where the channels can be connected to a single interferometer by movable folding mirrors. One channel can be used for the actual measurement, while a second element in the other channel adapts to temperature. Figure 53-26 shows a two-channel cascade, in which a movable folding mirror connects the left channel when in the 45° position and the right channel when in the 0° position. Figure 53-27 shows a two-channel cascade for high-precision measurements in a special housing to keep the measurement environment at a constant temperature level.

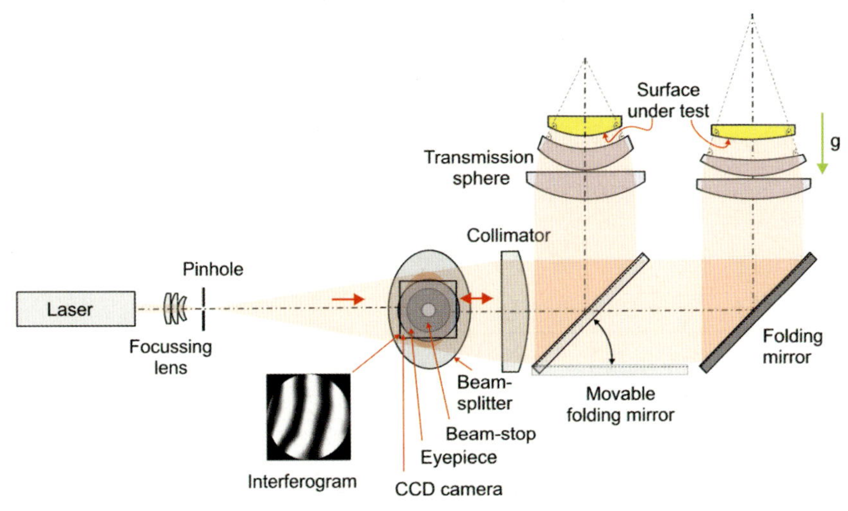

Figure 53-26: Interferometric setup with a horizontal interferometer and two vertical test beams. A movable folding mirror can be set to a 45° position to connect the left channel to the interferometer, or to a 0° position to use the right channel (the observation axis including eyepiece, beam-stop and camera is perpendicular to the plane of the drawing).

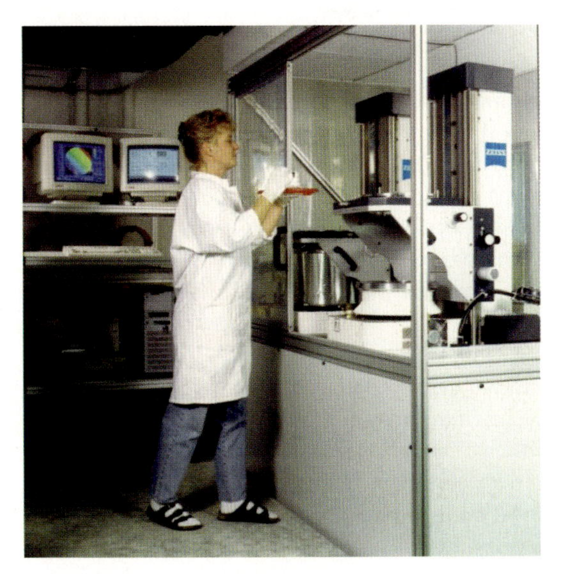

Figure 53-27: Interferometric setup with two vertical test towers (cascade) to test spherical surfaces by means of a transmission sphere in each tower (Courtesy of Carl Zeiss SMT GmbH).

Testing Plane Surfaces

To measure plane surfaces, the Fizeau interferometer must be equipped with a transmission flat carrying the reference surface. The transmission flat is a slightly wedged plane plate which is placed in front of the collimator of the instrument (figure 53-28). The transmission flat is made of fused silica or a low-expansion glass material like Zerodur® (glass ceramic) or ULE® (Ultra Low Expansion titanium silicate glass). The uncoated reference surface faces the surface under test, while the anti-reflective coated opposite surface is slightly tilted, so that the remaining reflection is blocked by the beam-stop.

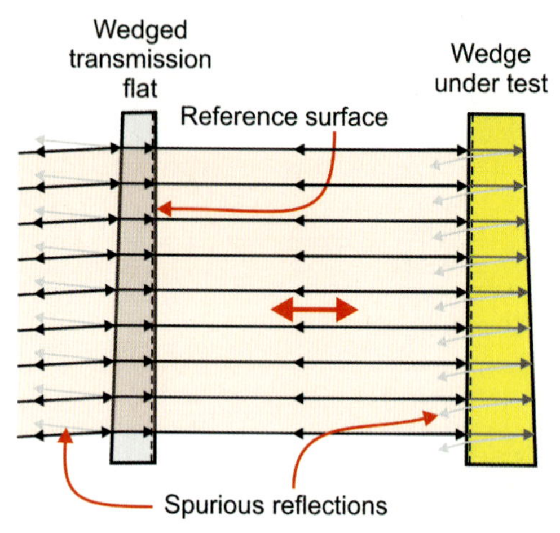

Figure 53-28: Reference flat and wedged plane plate under test in an interferometric test arrangement.

Optical elements with only one plane surface or with a second plane surface forming a wedge can be tested in a regular high-coherent interferometric arrangement. In this case the reflection from the back side is blocked by the interferometer's beam-stop.

Tests can also be performed when the back surface is rough (figure 53-29). In this case the diffusely scattered light partially reaches the interferometer's camera so that additional speckle effects will disturb the measurement. In general, however, these effects are small and can be minimized by using an interferometer with an extended light source (see section 46.4.2).

In the case of a plane plate with parallel front and back, surface measurements are not possible in a high-coherent interferometer. The reflection from the back surface must be suppressed considerably. The usual way is to cover the back surface with a black lacquer of refractive index very close to that of the test element. Transmitted light is then absorbed almost entirely with only a very small portion being reflected at the element's back surface (figure 53-30).

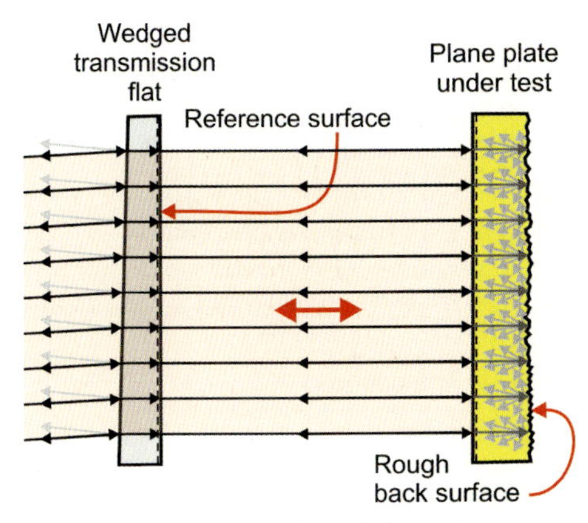

Figure 53-29: Reference flat and plane plate under test with a rough back surface in an interferometric test arrangement.

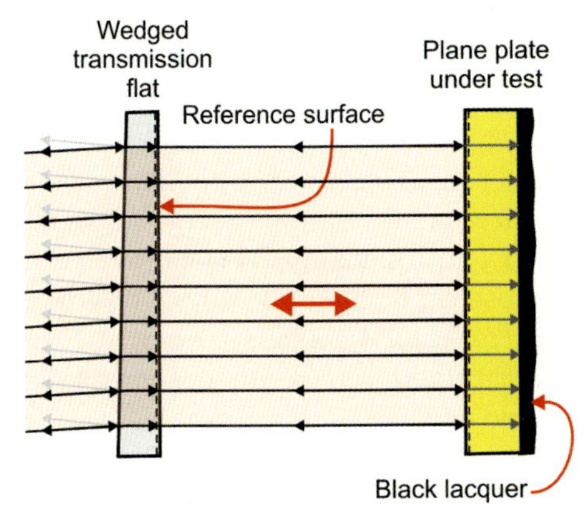

Figure 53-30: Reference flat and plane parallel plate under test with black lacquer on the back surface to suppress spurious reflections in an interferometric arrangement.

If coverage is not possible, the plane parallel element must be tested in a low-coherent interferometer, equipped, for instance, with a superluminescent diode. Superluminescent diodes utilize stimulated emission as their primary radiative mechanism, but do not exceed the threshold for oscillation. They provide a high power output of broad-band low-coherent radiation with coherence lengths of less than about 200 micrometers and typically about 50 micrometers. The plane plate under test is then placed in a Twyman–Green interferometer (section 46.3.3) and the reference arm is adjusted to match the optical path length of the test arm.

For high-precision measurements of large elements the use of a Twyman–Green interferometer might be cumbersome and unsufficient. In this case a Fizeau interferometer can be used in combination with a delay-producing Twyman–Green interferometer positioned between the light source and the focussing lens of the Fizeau interferometer (figure 53-31) [53-28].

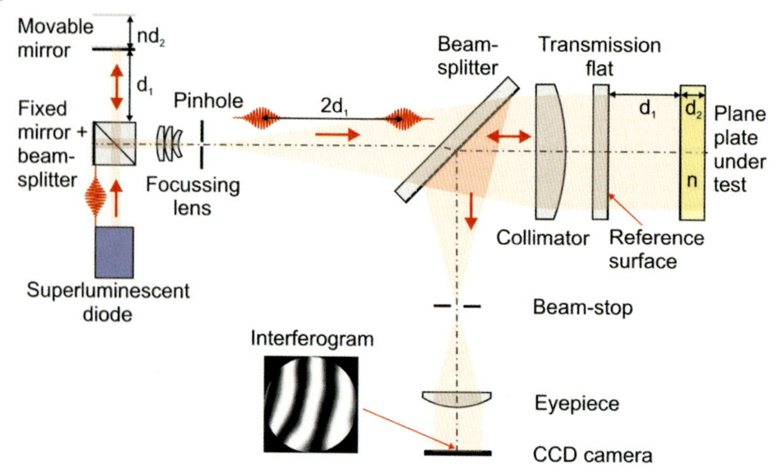

Figure 53-31: Interferometeric arrangement using a short coherent light source, a Twyman–Green interferometer as delay line and a Fizeau interferometer to test plane parallel elements.

If the cavity length between the transmission flat and the surface under test is d_1, then the path difference in the Twyman–Green is also adjusted to d_1. The result is that two waves delayed by $2d_1$ will enter the Fizeau interferometer. Each wave is split into a reference and a test portion. From the four resulting portions only two can interfere, because their optical paths are almost identical: the test surface reflection from the first wave with the reference surface reflection from the second wave. Note that the visibility of the fringes is 0.5 as a maximum, because the non-interfering portions of the waves reach the camera and produce an irradiance bias which diminishes the contrast of the interference fringes.

It is also possible to test the back surface through the plane parallel plate without changing its position. When the delay is set to $2(d_1 + n \cdot d_2)$, where n is the refractive index of the test element, the reflected portion from the back side of the plane plate element can interfere with the delayed reflected portion from the transmission flat. In this case surface deviations and also the inhomogeneity of the material will influence the result.

When a plane surface is measured, its tilt and distance from the transmission flat are generally arbitrarily adjusted and of no interest. Therefore, piston and linear tilt in x and y are removed from the measured wavefront by means of Zernike or Chebyshev polynomials or by monomials (see above).

In the case of very large plane plates it is possible to set up a stitching arrangement, in which the plane plate is moved laterally to capture subapertures in discrete, overlapping regions. The subapertures are then stitched together by means of a stitching software capable of removing the individual tilts and offsets in the subapertures [53-29]. Alternatively, large plane surfaces can be tested by means of the Hindle test or the Ritchey–Common test as described in section 46.6.2.

In the case of a plane mirror which is large in only one dimension, the mirror can be tested in a non-normal (oblique incidence) arrangement with the help of an additional plane mirror as shown in figure 53-32.

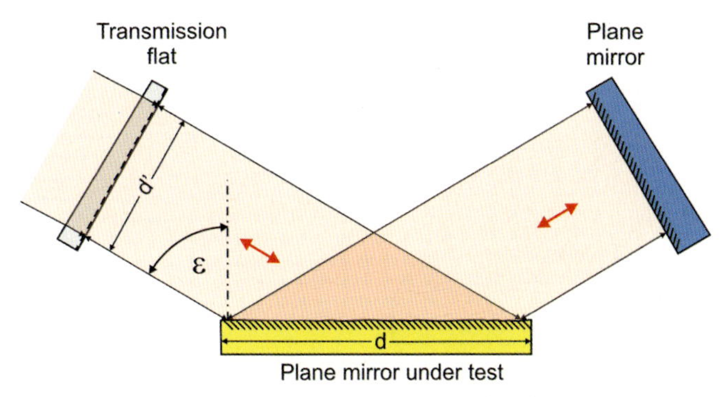

Figure 53-32: Testing a plane mirror at an angle of incidence ε against an additional plane mirror.

In this case the diameter d decreases to the observed diameter d' by a factor of $\cos\varepsilon$. The range and resolution of the interferometric detection normal to the test surface change by a factor of $1/2\cos\varepsilon$ in comparison to the regular normal incidence measurement of the surface. The factor 1/2 is due to the fact that the wavefront passes the surface under test twice. For $\varepsilon = 60^\circ$ the range and resolution are identical to the normal incidence arrangement. For angles larger than 60° the method can be used to measure aberrations that, at normal incidence, would exceed the detectable range. Even rough surfaces with a roughness of up to 1 μm can be measured with visible light, when the incidence angle approaches 90°.

A very common arrangement is to put a prism in contact with a semi-rough surface under test (figure 53-33). The surface can be lapped or fine-ground with hardly any visible specular reflection [53-30]–[53-31]. The contacting prism surface acts as the reference surface as in the case of a test glass (Newton interferometer, see section 46.3.1). The light from a collimated laser beam enters the prism from one side, but not normal to the prism surface to avoid retroreflection. The beam hits the reference surface at an angle ε and a portion of it is reflected to exit through the opposite prism surface. The portion transmitting through the reference surface exits the reference surface at an angle $\varepsilon' = \arcsin(n \sin\varepsilon)$, where n denotes the refractive index of the prism material. After reflection from the surface under test, the beam enters the prism again and mainly interferes with the reflected beam from the reference surface. The sensitivity is reduced by a factor of $1/\cos\varepsilon'$. To reduce the sensitivity by 0.1 an angle of $\varepsilon' = 84^\circ$ is necessary.

Calibration of the interferometric setup can be achieved by any of the following methods:

1. Using a liquid surface as calibration surface (section 46.6.1).
2. The shift-rotation technique (section 46.6.2).

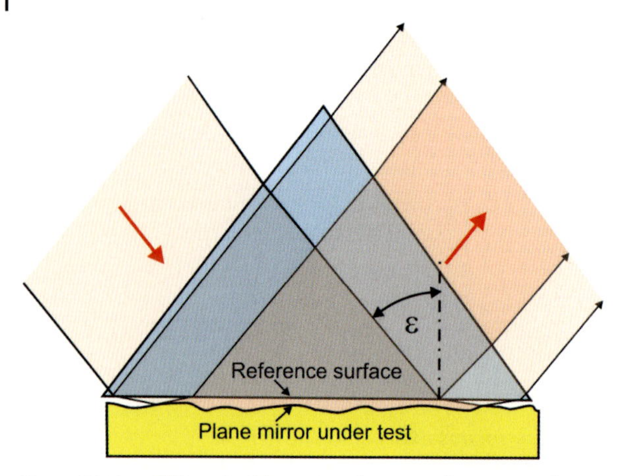

Figure 53-33: Oblique incidence interferometer by means of a prism brought into close contact with a semi-rough flat surface under test.

3. The three-flat test (section 46.6.2),
4. The Ritchey–Common test (section 46.6.2),
5. The Hindle test (section 46.6.2).

For most applications the shift-rotation technique is the most versatile and flexible method.

The *repeatability* of a measurement depends very much on the quality of the interferometric setup, the environmental conditions, the mechanical stability of the test tower and the lens mount. High-quality setups in a stable environment show repeatabilities of 0.1 nm RMS surface deviation.

The *reproducibility* of the measuring process, including the remounting and readjusting of the lens under test, depends on

- the repeatability of the interferometric setup,
- the geometry of the lens under test,
- the handling process,
- the time spent allowing the lens to adapt to the test environment temperature and to release any mechanical tension.

In a high-quality process reproducibilities of 0.5 nm RMS surface deviation are typical.

The *accuracy* of the measurement depends on

- the reproducibility of the measurement process and
- the quality of the calibration process.

The quality of the calibration defines the systematic error by which all further measurements will deviate from the ideal plane surface in addition to the repeatability of the measurement. When absolute calibration methods are carefully used, systematic deviations of a few nanometers RMS are possible, depending strongly on the stability of the calibration mirrors as well as the stability of the equipment and the environment.

Transmission Spheres

To measure spherical surfaces an interferometer must be equipped with a so-called *transmission sphere.*

Transmission spheres are special test objectives usually designed with an infinite object distance and a well-defined back focal length and numerical aperture for a well-defined wavelength. When used in a Fizeau-type interferometer, the transmission sphere carries a final uncoated reference surface called the *Fizeau surface,* which partially reflects the concentric exiting wavefront to produce the reference wavefront (figure 53-34). The transmitted portion converges to the focal point of the objective coinciding with the center of curvature of the Fizeau surface. To test large-radius concave surfaces there are also divergent transmission spheres in use which have a convex Fizeau surface. However, in this case the focal point is not accessible, which often makes calibration more difficult.

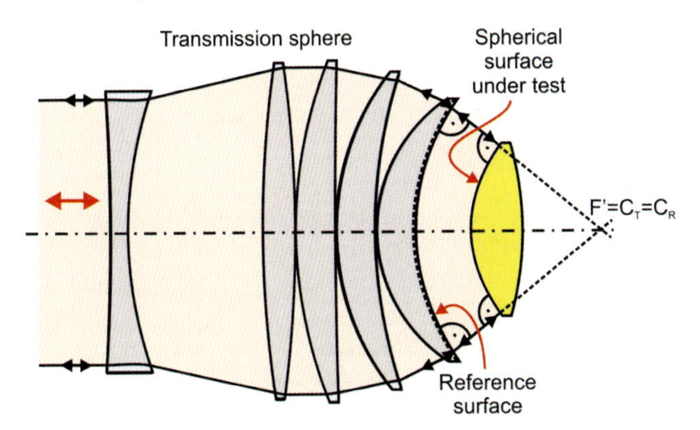

Figure 53-34: Transmission sphere used to test spherical surfaces in a Fizeau interferometer.

When used in a Twyman–Green type interferometer, transmission spheres do not carry a Fizeau surface, since the reference wavefront is supplied from the reference arm of the interferometer.

Transmission spheres are characterized by the radius of curvature of the reference surface R_T and the F-number F/# where

$$F/\# = \frac{R_T}{D_T} \qquad (53\text{-}22)$$

and D_T is the diameter of the converging test beam at the reference surface.

The F/# is related to the numerical aperture NA of the exiting beam by

$$F/\# = \frac{1}{2\,NA} \qquad (53\text{-}23)$$

where NA=$\sin\theta'$ when used in air, θ' being the half-cone angle of the exiting beam.

A spherical surface is tested with its center of curvature coinciding with the back focal point F′ of the transmission sphere. However, in order to cover the surface completely with the test wave, the F/# of the surface has to match that of the transmission sphere. Optimum coverage is obtained when the F/# of the transmission sphere is equal to or only slightly smaller than that of the surface under test. If it is larger, only partial coverage can be obtained. If it is smaller, the image of the test surface only covers a small portion of the camera sensor.

An additional limitation exists for convex surfaces. If the radius of curvature of the test surface is larger than or equal to that of the reference surface, then testing is not possible due to simple geometric limitations. Table 53-3 shows the limitations for convex and concave test surfaces when R_T and D_T for a transmission sphere are given. An optical shop would therefore need to have a set of transmission spheres to cover the total spectrum of lenses to be fabricated. A suitable set of transmission spheres can be selected for instance by starting with a high aperture (low F/#) like 0.67 and multiplying continuously by a factor of 1.5 until the lowest aperture is reached, i.e. F/# = 0.67, 1.0, 1.5, 2.3, 3.4, 5.1, 7.6, 11.4.

Figure 53-35 shows the arrangements for a variety of convex and concave surfaces which are to be tested with a transmission sphere.

Table 53-3: Limitations for convex and concave surfaces to be tested with a transmission sphere having a reference surface with radius of curvature R_T and a diameter of the exiting test wave of D_T.

	Full coverage of surface	**Partial coverage of surface**	**No testing possible**
Convex surfaces	$\frac{R_T}{D_T} \leq \frac{R_S}{D_S}$ and $R_S < R_T$	$\frac{R_T}{D_T} > \frac{R_S}{D_S}$ and $R_S < R_T$	$R_S \geq R_T$
Concave surfaces	$\frac{\lvert R_T \rvert}{D_T} \leq \frac{\lvert R_S \rvert}{D_S}$	$\frac{\lvert R_T \rvert}{D_T} > \frac{\lvert R_S \rvert}{D_S}$	

For practical use the D/R-diagram is used to show the ranges of radii and diameters that can be tested with existing transmission spheres (figure 53-36). In the D/R-diagram transmission spheres of a certain F-number are displayed as a triangle, the extension in R being defined by the radius of the reference surface R_T. Since, for practical reasons, a minimum airspace of several millimeters must be considered, the maximum radius actually used for convex surfaces is R_S. Spherical surfaces with R/D values within at least one of the triangles can be tested. Those belonging to the higher F/#, are preferable because they will appear larger on the interferometer sensor.

Transmission spheres are usually designed following the sine condition. In this case they are called *aplanatic*. The crossing points of the incoming and outgoing beams lie on a sphere concentric to the exiting spherical wavefront. The sum of all crossing points forms a surface called the *principal surface* of the objective (figure 53-37).

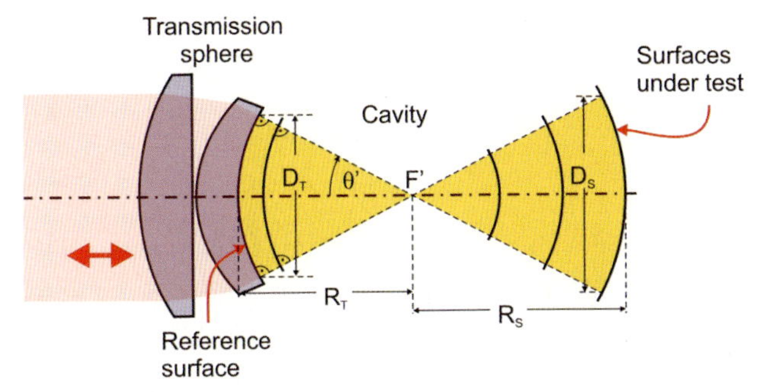

Figure 53-35: Arrangements for testing convex or concave spherical surfaces using a transmission sphere. The centers of curvature have to coincide with the back focal point F′ of the transmission sphere which is identical to the center of curvature of the reference surface. The yellow marked space between the reference and the test surface is called the *cavity*.

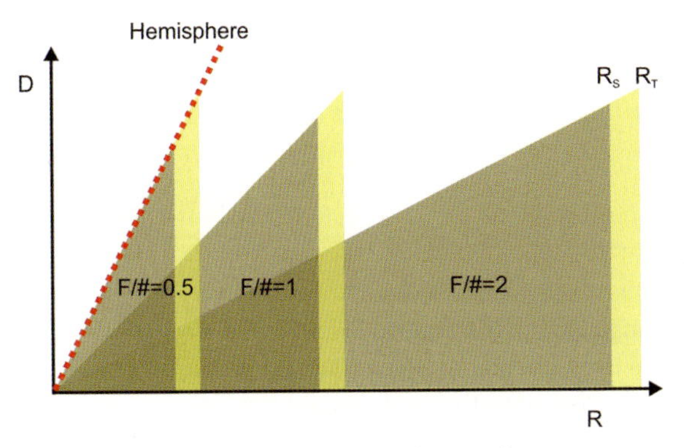

Figure 53-36: A *D/R*-diagram display of ranges for transmission spheres. Yellow triangles show the theoretical ranges, grey triangles the practical (normally used) ranges. F/# = 0.5 corresponds to the hemisphere.

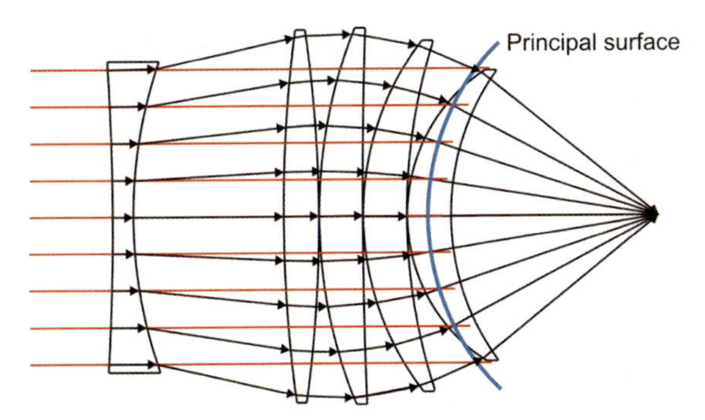

Figure 53-37: Principal surface (blue) of an aplanatic transmission sphere.

A beam of height h, when entering the transmission sphere, is deflected by an angle θ'. If the focal length of the transmission sphere is f', the relation for the sine condition is given by

$$h_S = f' \sin\theta' \tag{53-24}$$

Transmission spheres following the sine condition enable a camera perspective onto the surface under test appearing from an infinite distance. The consequence is that the arc on the surface under test covered by a pixel is proportional to $1/\cos\theta'$. Therefore pixels further away from the optical axis become stretched.

A transmission sphere can also be realized by a computer-generated hologram (CGH). In this case the principal surface is a plane and h and θ' are related by the tangent according to

$$h_T = f' \tan\theta' \tag{53-25}$$

Following the tangent condition the arc on the surface under test covered by a pixel is then proportional to $\cos^2\theta'$. Thus pixels further away from the optical axis become squeezed.

A transmission sphere where h is proportional to θ' is called an f-theta objective, and is described by the relation

$$h = f'\theta' \tag{53-26}$$

In this case the principal surface is an asphere which lies between the sphere of the exiting wavefront and its tangential plane (figure 53-38). With an f-theta transmission sphere each pixel covers the same arc on the surface under test. For computer-controlled polishing processes this might have a decisive advantage.

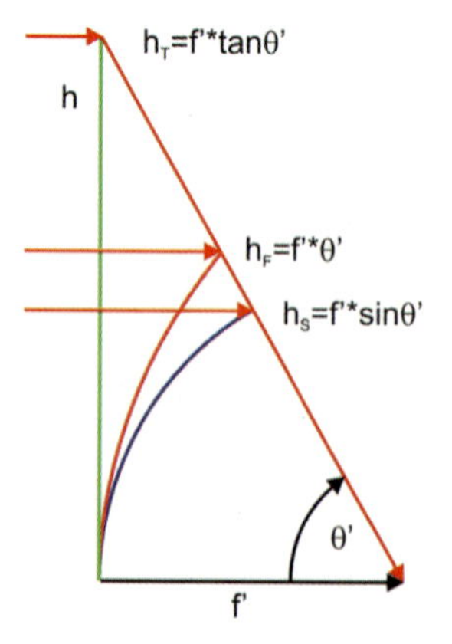

Figure 53-38: Principal surfaces of transmission spheres with a focal length f' and an aperture angle θ': CGH (green), f-theta (red), aplanatic (blue).

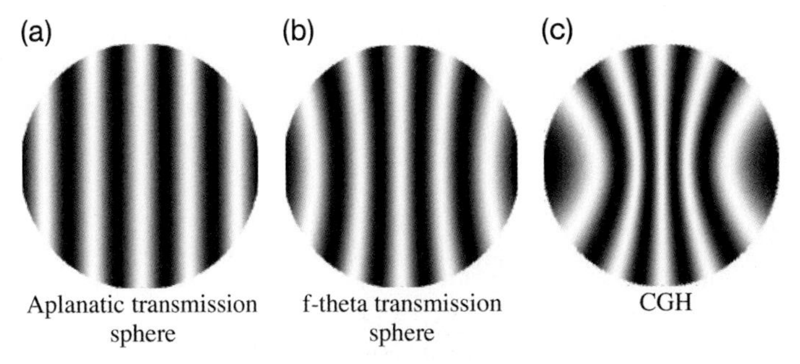

Figure 53-39: Interferograms for x decentered test surfaces using a transmission sphere of F/#=0.6 with different distortions: a) aplanatic; b) f-theta; c) CGH type.

The shape of the principal plane defines the distortion of the test surface's image. For an aplanatic transmission sphere only, will the interferometric fringes be straight, parallel and equidistant when the surface is tilted or decentered by a small amount. In any other case fringes will show coma due to a distorted test surface image (figure 53-39). If the image distortion is corrected by software procedure the fringes will again be straight, parallel and equidistant as in the aplanatic case.

Testing Spherical Surfaces

When testing convex surfaces, the reference surface is always larger in diameter than the surface under test. For large-lens testing the necessary transmission spheres are extremely expensive. Their production is challenging and time-consuming with long lead times for suitable glass material. Figure 53-40 shows a test tower arrangement equipped with a 12″-transmission sphere.

The surface under test must be adjusted so that its center of curvature C_T coincides with the back focal point F' of the transmission sphere. The center of curvature of the reference surface C_R also coincides with F′.

The detected wavefront error function $W(x,y)$ is composed of the total surface deviation function $W_S(x,y)$ and the misalignment wavefront $W_M(x,y)$ of the surface. Let the transmission sphere be aplanatic such that the sine condition is fulfilled. The interferometer camera then sees the surface under test as undistorted.

The misalignment wavefront $W_M(x,y)$ can be calculated from:

$$W_M(x,y) = 2\left(\sqrt{2\left(x_0 x + y_0 y + z_0\left(z_0 + \sqrt{r^2 - (x-x_0)^2 - (y-y_0)^2}\right) + \frac{1}{2}(r^2 - x_0^2 - y_0^2 - z_0^2)\right)} - r\right) \quad (53\text{-}27)$$

where x_0, y_0, z_0 denote the misalignments of the surface relative to the centered position and

r is the radius of curvature of the surface under test.

For large radii with $r >> x, y, z$ the following approximation is valid:

$$W_M(x,y) \approx 2\left(z_0 + x_0\frac{x}{r} + y_0\frac{y}{r} - \frac{z_0}{2r^2}(x^2+y^2)\right) \quad (53\text{-}28)$$

Figure 53-40: Interferometric test tower arrangement with a 12″-transmission sphere used to test large convex spherical surfaces (Courtesy of Carl Zeiss SMT GmbH).

Misalignments perpendicular to the optical axis z lead to linear terms in x and y proportional to x_0 and y_0, respectively. Misalignments in z lead to piston and quadratic terms proportional to z_0.

For short radii the exact expression (53-27) must be considered containing additionally all necessary higher than quadratic orders. This has to be considered when misalignment terms are to be removed from the wavefront error function $W(x,y)$ for further decomposition.

Figure 53-41 shows the remaining deviations occurring when only quadratic terms are removed to eliminate defocus effects. Figure 53-41 (a) shows an interferogram of 15 rings at $\lambda = 632.8$ nm with F/# = 0.5 for a defocus of 5 µm. The deviation from the quadratic approximation is then up to six rings as shown in figure 53-41 (b). For transmission spheres with F/# < 1.0 higher order rotationally symmetric terms might therefore become relevant for high-precision measurements. Deviations from the linearity in the case of decenters in x and y are extremely small and can be neglected. Figure 53-41 (c) shows a 30- fringe interferogram leading to deviations from linearity of less than 1/1000 fringe.

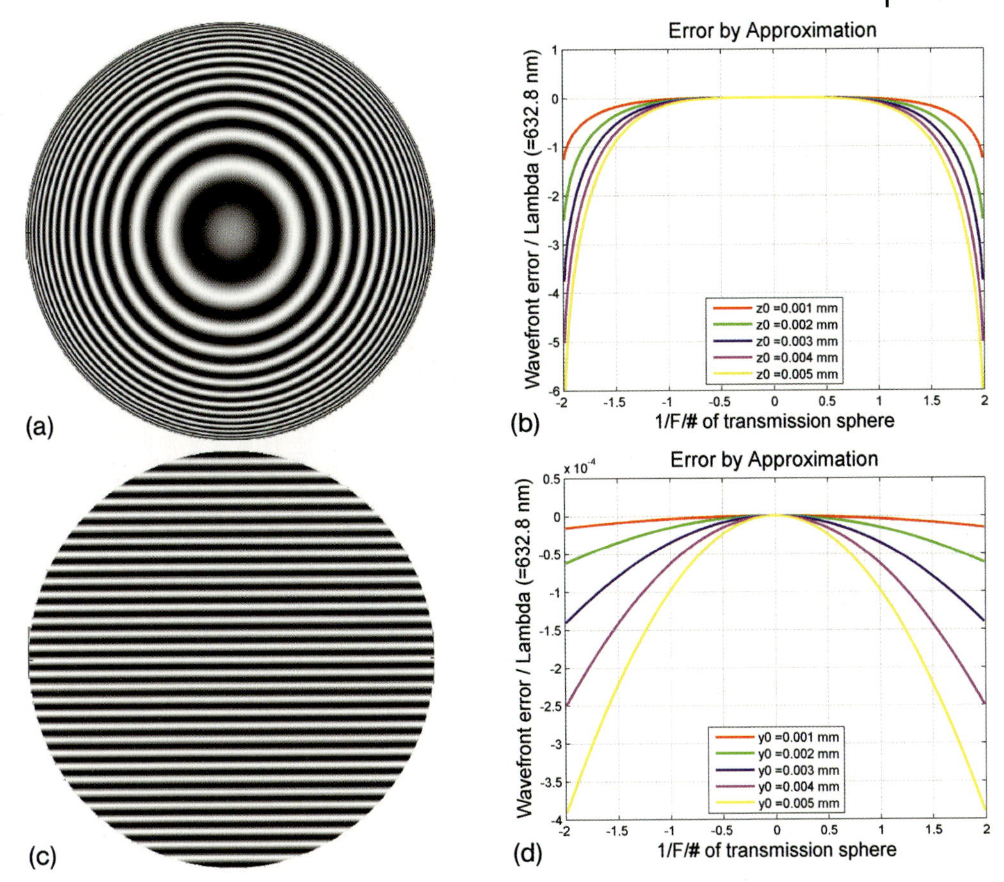

Figure 53-41: Deviations from linear and quadratic misalignment terms for test spherical surfaces with a transmission sphere F/# = 0.5 at $\lambda = 632.8$ nm: (a) interferogram for 5 μm defocus; (b) wavefront deviations from quadratic approximation for 1 ... 5 μm defocus; (c) interferogram for 5 μm decenter in y; (d) wavefront deviations in y from linear approximation for 1 ... 5 μm decenter.

Testing spherical lenses is mostly not affected by spurious reflections. However, there are some lens element configurations which could become problematic:

1. Concentric lens elements, in which the front and back surface have a common center point (figure 53-42 (a)),
2. Lens elements, in which a reflection from the back surface hits the front or back surface vertex (figure 53-42 (b), (c)).

The concentric lens element behaves in principle like the plane parallel plate from the previous chapter. Interferometric measurement is only possible with short coherent light or when the reflection from the back surface is diminished by black laquer. High coherent light will lead to three-beam interferograms, which cannot be analyzed (figure 53-42 (d)).

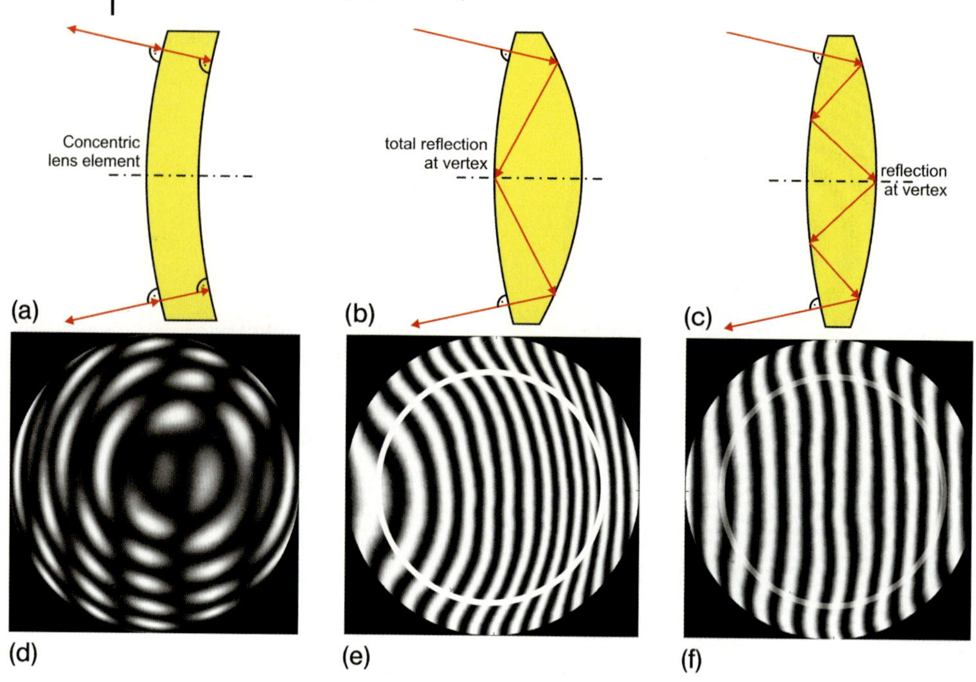

Figure 53-42: Spherical lens elements with problematic geometry and their corresponding interferograms: (a)+(d) concentric, (b)+(e) biconvex with total internal reflection hitting the front vertex for a certain incident height, (c)+(f) biconvex with internal reflections hitting the vertex for a certain incident height.

Elements in which beams at a certain incident height are reflected from the back surface to hit the front vertex (figure 53-42 (b)), will produce a bright ring within the interferogram (figure 53-42 (e)). For some geometries all reflection angles are above the angle of total internal reflection, resulting in more than 90% of the light returning to the interferometer. In this case, the ring area cannot be detected. Geometries in which only the vertex reflection is a total internal reflection lead to three-beam interference in a ring-shaped area, which is another problem to be analyzed.

Elements in which beams at a certain incident height are reflected in between the front and back surfaces to hit the back vertex (figure 53-42 (c)), will produce a weak ring within the interferogram (figure 53-42 (f)). In this case the wavefront can be measured, but the ring-shaped area suffers from three-beam interference artefacts, which have to be corrected for further usage.

In the case of very large spherical surfaces it is possible to set up a stitching arrangement, in which the surface is rotated and tilted in front of a transmission sphere to capture subapertures in discrete, overlapping regions (see section 46.7.4). The subapertures are then stitched together by means of a stitching software capable of removing the individual tilts and focus terms in the subapertures [53-32].

Calibration of the interferometric setup can be carried out by the following methods (see chapter 46.6.2):

1. The three-position technique.
2. The shift-rotation technique.
3. The calibration of cross-sections.
4. The rotating sphere.

For most applications the three-position technique is the most versatile and flexible method.

Considering *repeatability, reproducibility* and *accuracy* for interferometric measurements of spherical surfaces the discussion in the previous chapter on plane surfaces is valid. High-quality setups in a stable environment show

- *repeatabilities* of 0.1 nm RMS surface deviation,
- *reproducibilities* of 0.5 nm RMS surface deviation,
- *accuracies* of the measurement dependent on the reproducibility of the measurement process and the quality of the calibration process. When absolute calibration methods are used carefully, systematic deviations of a few nanometers RMS are possible, depending strongly on the stability of the calibration mirrors, the stability of the equipment and the environment.

Null Correctors

Null correctors or compensating systems are test objectives designed to test aspherical surfaces in an interferometer. Reference to aspherical surfaces in most cases means rotationally symmetric aspherical surfaces that have a specific deviation from a best-fitting spherical surface. Free-form aspheres describe a class of aspheres that do not have an axis of symmetry (see sections below).

Rotationally symmetric aspheres can be tested similar to spheres. The difference is that a null corrector is used instead of a transmission sphere, which is designed specifically for a certain asphere. In general a null corrector cannot be used for a different asphere, except for "similar" aspheres that correspond to the shape of the aspherical wavefront as it propagates after passing through the null corrector.

Null correctors are usually designed with an infinite object distance and without a reference surface. The beams transmitting through the null corrector are deflected so that they hit the surface under test in a normal direction when it is located exactly in the design position. When used in a Fizeau- type interferometer the null corrector is preceded by a transmission flat providing the reference surface (figure 53-43).

Since the null corrector is located within the cavity it therefore defines the quality of the test wavefront. Any deviation from the design parameter such as radii, surface shapes, thicknesses, air spaces and refractive indices, will directly influence the shape of the test wavefront and will lead to measurement errors. Unfortunately the errors can barely be detected by other methods on a nanometer scale, because no other metrology exists with a resolution comparable with interferometry.

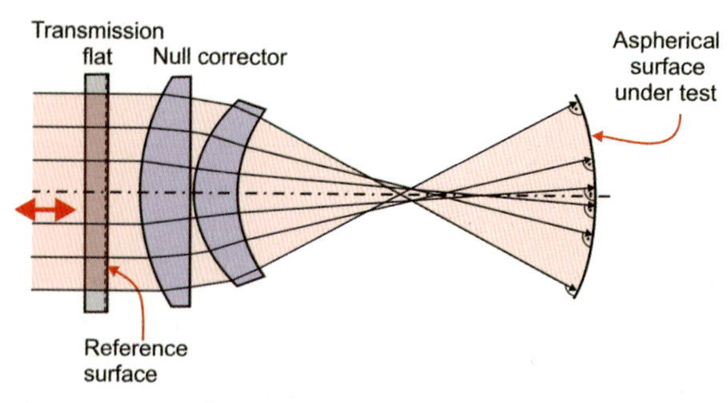

Figure 53-43: Null corrector in a Fizeau interferometer to test reflection in a rotationally symmetric asphere.

The position of the aspherical surface under test is critical, since the aspherical wavefront constantly changes its shape as it propagates. The test surface therefore has to be placed carefully into its design position, otherwise spherical aberration will be detected and falsely interpreted as aspherical shape deviation.

A null corrector can easily be used in a Twyman-Green type interferometer, where the reference wavefront is supplied from the reference arm of the interferometer.

Occasionally null correctors are designed to carry an aspherical last reference surface similar to the reference surface in a regular transmission sphere. In this case the null corrector is called a *matrix system* (figure 53-44). Deviations in the radii, thicknesses, distances, and refractive indices from their target values mostly do not influence the result because most of the matrix system is located outside the cavity. The critical element is the aspherical reference surface which has to be manufactured very carefully. For its production a further special null corrector is necessary. The benefit is, however, that null correctors for large concave surfaces are usually much easier to produce than for large convex surfaces. Once the matrix system is set up correctly, it is very stable when used in any production environment.

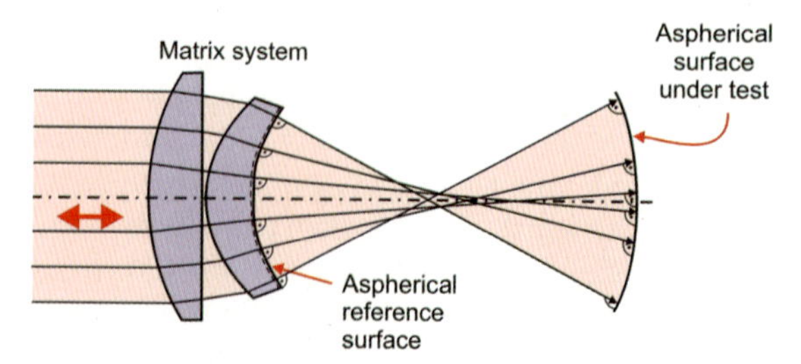

Figure 53-44: Matrix system in a Fizeau interferometer to test reflection in a rotationally symmetric asphere.

Null correctors are designed using the same criteria as those for transmission spheres. First, the designer has to find a refractive system consisting of spherical lens elements which will provide an aspherical wavefront matching the surface under test. In many cases this can be achieved only approximately. The designer can add lens elements to achieve a better approximation, but the error budget will increase when more lens parameters are involved if they have been manufactured with non-zero tolerances.

Secondly, the designer has to consider lateral resolution which is mainly influenced by field curvature and field astigmatism when the test surface is imaged onto the sensor. By checking spot diagrams at different heights from the optical axis during the design process, solutions for sufficient lateral resolution can be found.

Thirdly, distortion has to be considered in order to relate pixel positions to locations on the test surface in order to correct surface deviations by computer-controlled polishing methods. In the case of severe distortion the lateral resolution might be critical on those parts of the surface with large stretching factors.

For some aspheres it can be very difficult to find a refractive null corrector. Deviation from the best-fitting sphere, its derivative or its change in curvature mainly determine the feasibility of a null corrector. For those severe cases, the use of *computer-generated holograms* may be the only suitable solution.

Computer-generated Holograms

A computer-generated hologram (CGH) is a diffractive optical element designed by a computer and produced by a computer-controlled writing machine. There are different techniques for writing diffractive structures onto a substrate. Amongst these are laser pattern generation or e-beam writing of sub-μm structures into photoresist layers, followed by an etching process to copy the phase structures into the substrate material. Most common is the binary phase CGH. However, binary amplitude CGH can be manufactured also using laser beam writing onto chromium coated substrates.

The current pattern generators differ in their coordinate adressing methods which may be either *x-y- or r-φ*-adressing. Most high-precision *x-y*-adressing pattern generators have been developed to produce masks for optical lithography purposes. They adress *x-y*-positions with resolutions down to some tens of nanometers.

A binary CGH consists of many curved lines drawn onto or etched into glass substrates. The curved lines follow the shape of bright and dark interference fringes of two interfering wavefronts. When testing aspheres the two interfering wavefronts are formed by the incident wavefront from the interferometer, in most cases a spherical or plane wave, and the aspherical wavefront which is identical to the perfect asphere under test in the correct design position (figure 53-45).

The structure of the CGH is thus calculated from the optical paths of the spherical wavefront and the aspherical wavefront. The geometrical paths are expressed by

$$\vec{g}_1 = \vec{h} - \vec{z}_1 - \vec{l} \tag{53-29}$$

$$\vec{g}_2 = \vec{h} - \vec{z}_2 - \vec{s} \tag{53-30}$$

The transmittance $T(x,y)$ to be coded as diffracted structure is calculated for each pixel $h(x,y)$ of the CGH addressed by the pattern generator according to the interference of the two waves in the plane of the CGH (see equation (53-31)).

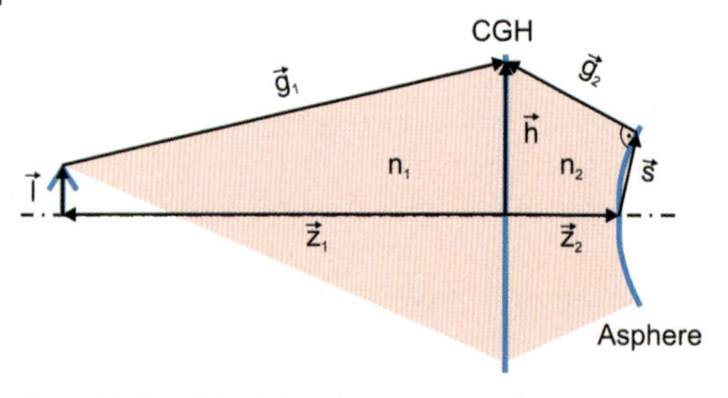

Figure 53-45: Calculating the structures of a computer-generated hologram: the OPD in the plane of the CGH is calculated from the optical path of beams normal to the aspherical surface and of beams from the incident wavefront illuminating the CGH.

$$I(x,y) = A_1^2 + A_2^2 + 2A_1A_2\cos(\phi_2(x,y) - \phi_1(x,y)) \tag{53-31}$$

with

$$\phi_1(x,y) = \frac{n_1|\vec{g}_1(x,y)|}{\lambda} \tag{53-32}$$

$$\phi_2(x,y) = \frac{n_2|\vec{g}_2(x,y)|}{\lambda} \tag{53-33}$$

Assuming that A_1 and A_2 are equal and homogeneous over the CGH, the transmittance $T(x,y)$ of the CGH is calculated according to

$$T(x,y) = 1 + \cos(\phi_1(x,y) - \phi_2(x,y)) \tag{53-34}$$

In practice, however, most pattern generators are unable to produce CGH with a continously varying transmittance or phase structure. Machines for mask coding produce binary transmissive or phase structures. Continuously varying structures can be approximated by multi-step processes, which require a much more complicated manufacturing process.Usually a choice must be made between an amplitude or a phase CGH. Figure 53-46 shows some examples.

For a binary amplitude CGH (53-34) transforms to

$$T(x,y) = \mathrm{rect}\left(\frac{1}{2}\{1 + \cos(\phi_1(x,y) - \phi_2(x,y))\} + \sin\frac{\pi}{2}(1-2a)\right) \tag{53-35}$$

with $0 < a < 1$ defining the duty cycle of the binary structure.

For a binary phase, CGH (53-35) can be modified to a phase profile distribution $W(x,y)$ where p denotes the phase step or peak-to-valley (PV) of the phase distribution:

$$W(x,y) = p\,\mathrm{rect}\left(\frac{1}{2}\{1 + \cos(\phi_1(x,y) - \phi_2(x,y))\} + \sin\frac{\pi}{2}(1-2a)\right) \tag{53-36}$$

The groove depth h to be etched into the substrate of refractive index n is calculated from

$$h = \frac{p}{n - n'} \tag{53-37}$$

where n' denotes the refractive index of the surrounding medium.

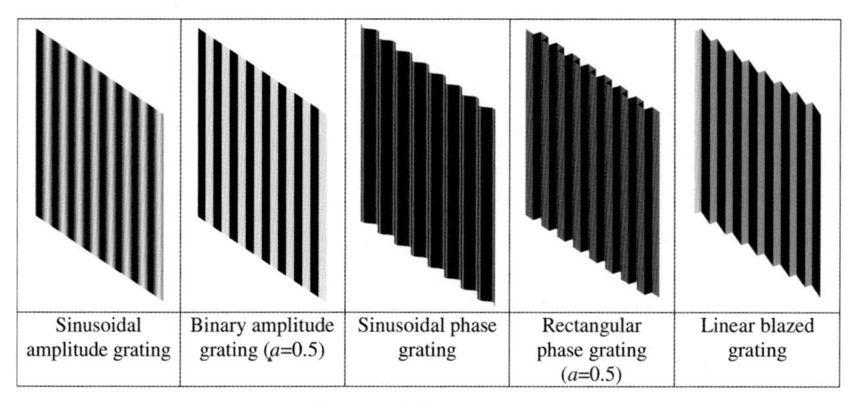

Figure 53-46: Various types of amplitude and phase gratings.

Amplitude and phase CGHs differ considerably in their diffraction efficiency, which is defined as the ratio of the intensity of the diffracted wavefront to that of the incident wavefront. While the overall pattern of lines mainly determines the shape of the wavefront, the structure of the lines, such as the groove depth h, or duty cycle a, determines the diffraction efficiency in the 0th, 1st and higher diffraction orders. In most applications the first diffraction order of the diffracting element is used.

Table 53-4 shows theoretical efficiencies for 0th and $\pm\, 1^{st}$ diffraction orders for various amplitude and phase gratings.

Note that the 0th order can be supressed totally with rectangular shaped phase gratings of 0.5λ phase step and duty cycle of 0.5, reaching a diffraction efficiency of 40,5% in the 1st diffraction orders. A linearly blazed phase grating theoretically deflects all illuminating light in direction of the 1st diffraction order.

In practice, a variety of effects diminishes the theoretical diffraction efficiencies, such as reflection at the optical interfaces, absorption within the substrates or coatings, scatter, and "rigorous" effects. The latter are due to the fact that the simple scalar diffraction theory is only valid for grating periods considerably larger than the wavelength used in the application. If the grating period approaches the wavelength a rigorous solution of Maxwell's equations is necessary to compute more accurate diffraction efficiencies and phase effects for differently polarized incident light. A good knowledge of the real phase profiles including the depth and shape of the grooves as well as their density and duty cycle is necessary as input for a rigorous computation.

Table 53-4: Various amplitude and phase gratings and their theoretical diffraction efficiencies in 0th and 1st order.

Amplitude profile	Phase profile	Duty cycle *a*	PV of phase profile *p*	0th diffraction order	$\pm 1^{st}$ diffraction order
Sinusoidal	Constant	–	–	25,0%	6,2%
Rectangular	Constant	0.3	–	8,9%	6,6%
		0.5	–	25,0%	10,1%
		0.7	–	49,2%	6,6%
Constant	Sinusoidal	–	0.5λ	22,4%	32,0%
		–	0.59λ	9,6%	33,8%
		–	0.75λ	0,0%	27,9%
Constant	Rectangular	0.3	0.5λ	16,2%	26,4%
		0.5	0.5λ	0,0%	40,5%
		0.7	0.5λ	16,2%	26,4%
Constant	Linear blaze	–	1.0λ	0,0%	100% in $+1^{st}$, 0% in -1st

The local spatial frequency of a CGH at a point *x,y* defines the angle of deflection for an incident beam. Since the CGH may in general be viewed as a collection of linear gratings with variable groove spacings, we can apply the grating formula to calculate the angle of deflection. The local grating period *g(x,y)* is defined as

$$g(x,y) = \frac{m\lambda}{\sin\theta(x,y) + \sin\theta'(x,y)} \tag{53-38}$$

where $\theta(x,y)$ is the local incidence angle, $\theta'(x,y)$ is the local diffraction angle, and *m* is the diffraction order (figure 53-47).

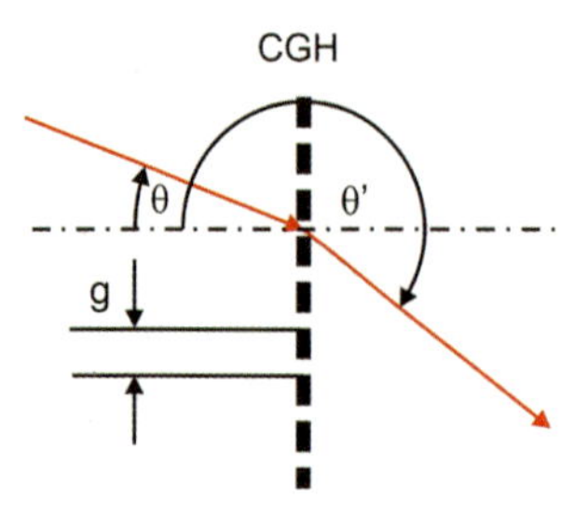

Figure 53-47: Deflection of an incident beam by a diffraction grating of groove period g.

Figure 53-48 shows examples of deflection angles θ' for the first three diffraction orders and normal incidence ($\theta = 0$) and $\lambda = 632.8$ nm as a function of the grating density.

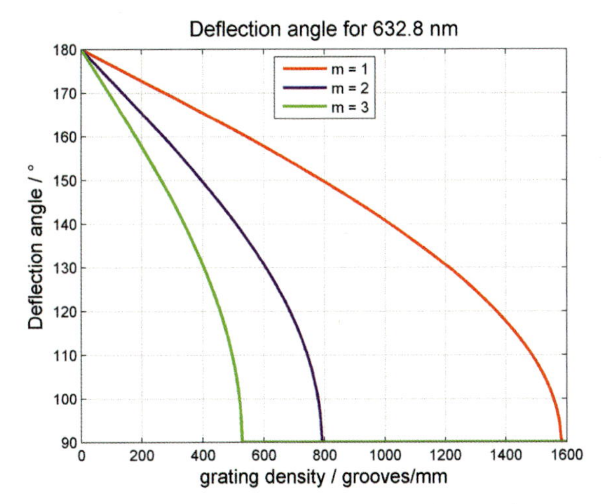

Figure 53-48: Deflection angle as a function of groove density for normal incidence and $\lambda = 632.8$ nm.

Sometimes CGHs are used in reflection. In this case $|\theta'| < 90°$. A special situation is the so-called *Littrow* arrangement, where the beam is redirected into the same direction from which it arrived ($\theta = \theta'$). (53-38) then transforms to

$$g(x, y) = \frac{m\lambda}{2 \sin \theta_{\text{Littrow}}} \tag{53-39}$$

Figure 53-49 shows examples of Littrow angles θ_{Littrow} for the first three diffraction orders and $\lambda = 632.8$ nm as a function of the grating density.

CGHs are often used as null correctors with an infinite object distance. The beams transmitting the CGH are deflected so that they hit the surface under test in a normal direction when it is located exactly in the design position. When used in a Fizeau-type interferometer the CGH is preceded by a transmission flat providing the reference surface.

Often the incident beam is inclined, making it necessary to use an off-axis CGH (figure 53-50). This shows a carrier frequency which moves the "pole" from the optical axis, possibly even outside the CGH region. The pole is the location where the grating density is zero. It lets all diffraction orders pass into the same direction and is therefore not suitable for metrological evaluation. Figure 53-51 shows a binary amplitude CGH with a pole in the upper center.

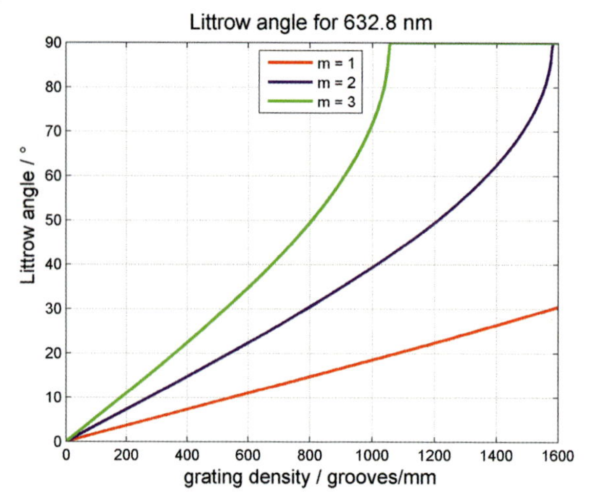

Figure 53-49: Littrow angle as a function of groove density for normal incidence and $\lambda = 632.8$ nm.

Since the CGH is located within the cavity it therefore defines the quality of the test wavefront. In addition to deviations of the groove location, any deviation of the CGH substrate from its perfect configuration will directly influence the shape of the test wavefront and will lead to measurement errors. As in the case of the refractive null corrector the errors of the CGH can barely be detected by other methods on a nanometer scale, because no other metrology exists with a resolution comparable to that of interferometry.

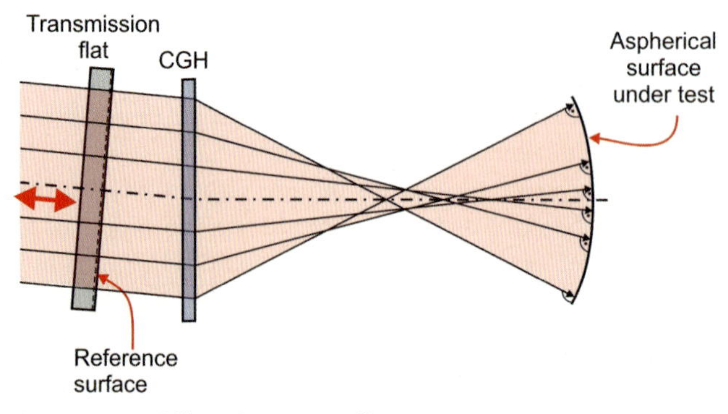

Figure 53-50: Off-axis CGH as a null corrector in a Fizeau arrangement to test reflection from an aspherical surface.

The aspherical surface under test has to be carefully aligned to avoid misinterpretation of focus, spherical aberration, and coma due to misalignment rather than to shape errors.

A CGH can of course also be used in a Twyman–Green type interferometer, where the reference wavefront is supplied from the reference arm of the interferometer.

Figure 53-51: Binary off-axis CGH with a pole in the upper center (fringes are enlarged 25 times).

Testing Rotationally Symmetric Aspherical Surfaces

In this chapter we discuss the metrology needed to measure rotationally symmetric aspherical surfaces, called aspherical surfaces for brevity. According to ISO 10110 Part 12 [53-33] we will describe the sagittal height $z(h)$ of an aspherical surface by:

$$z(h) = \frac{h^2}{R\left(1 + \sqrt{1 - (1+\kappa)\frac{h^2}{R^2}}\right)} + \sum_{i=2}^{M} A_{2i} h^{2i} \tag{53-40}$$

where

- h is the lateral coordinate perpendicular to the axis of symmetry (z-axis),
- R is the vertex radius of the aspherical surface,
- κ is the conic constant denoting the type of conic surface (second-order surface):
  - $\kappa > 0$ for an oblate ellipsoid,
  - $\kappa = 0$ for a sphere,
  - $-1 < \kappa < 0$ for a prolate ellipsoid,
  - $\kappa = -1$ for a paraboloid,
  - $\kappa < -1$ for a hyperboloid,
- A_{2i} are the coefficients of polynomial expansion (power series) with $i = 2, 3, ..., M$.

There are generally three different methods used to measure aspheres by means of interferometry: measurement with

a) full compensation,
b) partial compensation,
c) no compensation

of the asphericity of the surface under test [53-34].

In order to fullfill the requirements of a *fully compensated* setup, an aspherical wavefront has to be generated showing the shape of the ideal surface under test at its ideal location within the setup.

In a *partially compensated* setup an aspherical wavefront is generated, which deviates in a well-determined way from the ideal asphere. The deviation is small enough to be detected by the interferometrical setup and the applied fringe analysis system.

In an *uncompensated* setup the asphere is tested by means of a plane or spherical wavefront, which after reflection is aspherically deformed. The deformation must be small enough to be detected by the interferometrical setup and the applied fringe analysis system. Therefore, only very weak aspheres can be tested when uncompensated. However, there are various methods in which the asphere is tested sequentially by collecting subapertures, in which the aspherical deformation of the wavefront can be detected and evaluated. The total surface deformation is then determined by assembling and evaluating all subapertures.

In the following chapters we explain all three methods in more detail.

When testing aspheres we have to take account of the following facts.

- **Propagation.** An aspherical wavefront changes its shape while propagating. This is different from the propagation of a spherical wavefront. It is therefore mandatory to control the distance between the compensating optics and the surface under test. The accuracy of this distance is – among others – a main parameter to define the accuracy of the measurement.
- **Reference coordinate system.** The deviation of an aspherical surface from its ideal shape is generally measured with reference to a well-defined coordinate system. The coordinate system can be provided by reference marks (fiducials) or reference surfaces at the test piece, by the edges of the test piece, by the back surface of a test lens element or any other characteristics referenced during manufacture of the test piece. When an interferometric test setup is designed it is therefore also necessary to set up metrology enabling the detection of the reference system of the test piece. The coordinate system of the interferometer and that of the referencing metrology will then have to be calibrated against each other so that the measured deviations from the perfect asphere can be expressed in terms of the reference coordinate system.

a) Testing with Full Compensation

Fully compensated testing needs a compensating system, often called null system, to provide the perfect aspherical wavefront at the location of the surface under test. The null system can be refractive [53-35], deflective [53-38] or diffractive [53-39]. Figure 53-52 gives an overview.

Null systems for rotationally symmetric aspheres cannot be calibrated except for their non-rotationally symmetric deviations. By rotating the surface under test around the optical axis of the null system, non-rotational symmetries can be detected very accurately (see section 46.6.2). However, rotationally symmetric deviations remain uncalibrated. The shift-rotation technique is not applicable, because very small lateral shifts lead to many undetectable fringes. The only way to estimate rotationally symmetric deviations is to calculate an error budget by considering all possible parameter deviations influencing rotational symmetries. The error budget has to account for systematic deviations of radii of curvature, refractive indices of glass materials, central thicknesses and distances between components; but also for statistical variations caused by changes in temperature, air pressure and humidity.

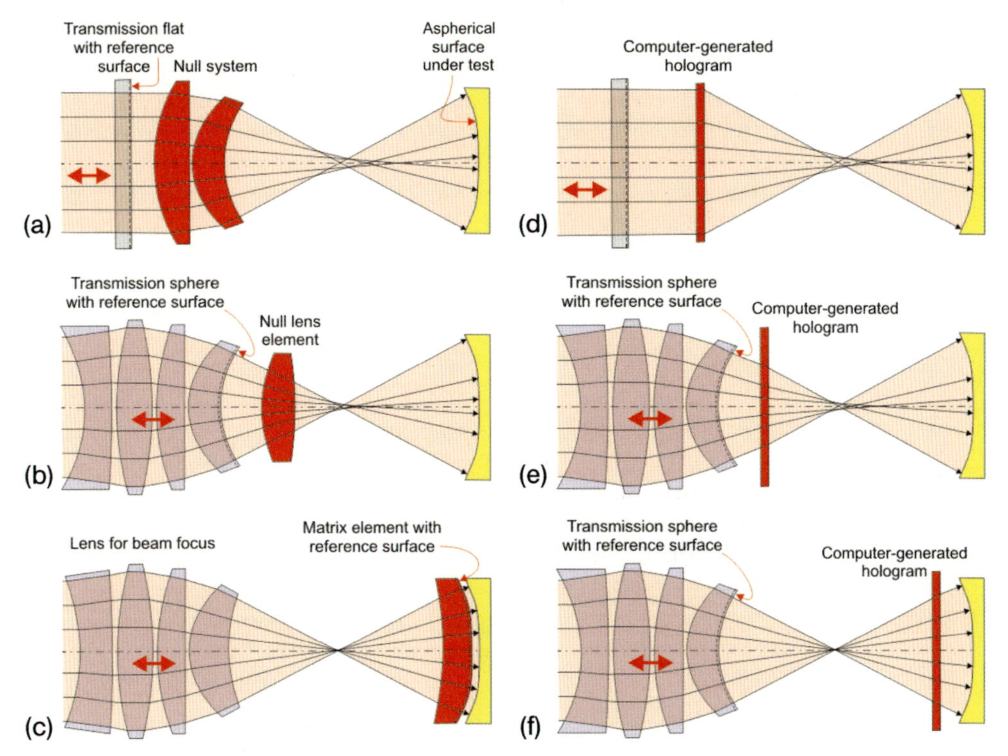

Figure 53-52: Various compensating or null systems used to measure aspherical surfaces in a Fizeau interferometer: (a) transmission flat with following refractive null system; (b) transmission sphere with following null lens element; (c) refractive system with following matrix element carrying the aspherical reference surface; (d) transmission flat with following computer-generated hologram; (e) transmission sphere with computer-generated hologram in intrafocal position; (f) transmission sphere with computer-generated hologram in extrafocal position.

Testing with Refractive Null Systems

Refractive null systems differ in the way in which the reference surface is chosen. If the reference surface is provided by a transmission flat (figure 53-52 (a)), then the null system has to provide the total necessary refraction including the basic spherical convergence or divergence and the asphericity. For high-aperture aspheres and strong asphericities a null system can have many lens elements [53-35]–[53-36]. Such complex null systems are not only expensive and cumbersome to manufacture, but they also have considerable uncertainty in their rotationally symmetric deviations. Additionally, they might be sensitive to changes in temperature and air pressure. Their design has to be checked not only for the provision of the exact aspherical wavefront, but also for the distortion and the lateral resolution when imaged onto the interferometer's camera.

Less critical is the combination of a transmission sphere and a single spherical lens element as the null lens element (figure 53-52 (b)). In this case the transmission sphere provides the basic spherical convergence or divergence and carries the spherical reference surface, which can be fully calibrated by a suitable method (see section 46.6). The null lens element mainly provides the asphericity. It can be placed in the intrafocal or extrafocal position relative to the transmission sphere. Its axial position as well as the position of the surface under test can easily be referenced against the focal point of the transmission sphere. The error budget can be kept small when both radii of curvature, the central thickness and the refractive index of the lens element are well known.

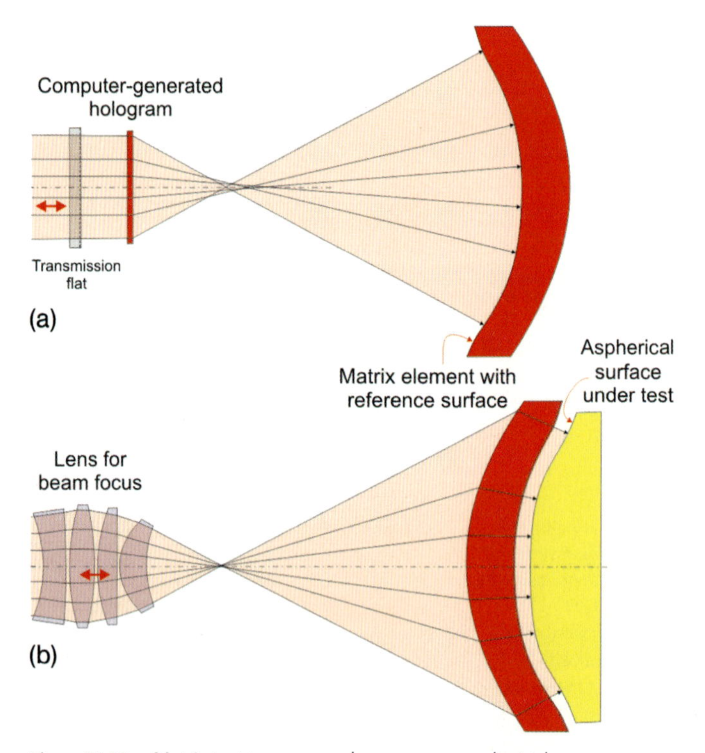

Figure 53-53: Matrix test to measure large convex aspherical surfaces: (a) testing a concave reference surface at the matrix element by means of a small null system (computer-generated hologram); (b) testing a convex surface by means of the final matrix element.

To keep the cavity length (the distance between the reference and test surfaces) small, a matrix lens element can be provided just in front of the surface under test. The matrix lens element carries an aspherical surface having the inverse shape of the asphere under test. Using exact ray-tracing the small change in shape due to propagation from the aspherical reference surface to the asphere under test may be investigated. One drawback is that the aspherical reference surface first has to be manufactured to a high accuracy. In practice, matrix testing is therefore only applied for very large convex aspherical surfaces, where the provision of a null system is impossible (figure 53-53 (b)). In this case the matrix element carries a concave aspherical reference surface which generally can be measured by means of a small null system (figure 53-53 (a)).

Testing with Diffractive Null Systems (CGHs)

Diffractive null systems are those in which at least one computer-generated hologram (CGH) shapes the test wave to a specific asphericity [53-39]–[53-54]. In most cases the reference surface is provided by a transmission flat (figure 53-52 (d)). The CGH is located within the cavity and provides the total wavefront shaping including the spherical convergence or divergence and the asphericity.

When using CGH the following characteristics have to be considered:

- **Efficiency**, in most cases, binary phase CGHs, generated by lithographic mask processes, are used in transmission in the first diffraction order, delivering a maximum efficiency of 40%. In double passage as in figures 53-52 (d),(e),(f), a maximum efficiency of 16% is achieved. To improve the fringe visibility the reference surface can be coated to balance the amplitudes of reference and test wave.
- **Spurious reflections, poles.** Due to light from other diffraction orders each configuration has to be checked for spurious reflections reaching the camera, which would severly disturb the measurement [53-51]. When the CGH is designed so-called poles, in which the wavefront has a zero slope, should be avoided. Light passing the pole is not diffracted, such that all the light reaches the camera causing a bright ring or spot, which cannot be evaluated. Poles can be avoided by adding a carrier frequency to the CGH structure shifting the pole outside the CGH plate.
- **Line density.** The higher the aperture of the surface under test (the smaller its F/#) the higher will be the grating density (1/groove spacing) on the CGH. Figure 53-54 shows the maximum grating density required to provide a transmission sphere with a certain F/# based on a CGH for wavelengths 400 nm, 600 nm and 800 nm. Modern e-beam writers are able to produce grating densities of 2000 grooves per mm and more. However, it must explicitly be noted that when groove spacing reaches the dimension of the wavelength, geometric calculations are unsufficient to describe deflection characteristics. This is the case, when groove spacings $< 2\ \mu m$ are used to deflect visible light. In this case *rigorous calculations* based on the explicit solution of Maxwell's equations must be carried out. A model of the 3D geometry of the grooves must be developed and the incidence direction as well as the polarization state of the incident light must be considered. Since for CGH the diffractive structure varies locally in density, duty cycle, depth and geometrical shape, rigorous calculations become very time-consuming and cumbersome even with powerful computers.

In order to reduce grating densities on CGHs in aspherical testing, a combination of CGH with refractive transmission spheres is quite common (figures 53-52(e) and (f)). In this case the CGH mainly carries the asphericity, while the transmission sphere provides the spherical part in the test wave. The CGH can be located at an intrafocal (figure 53-52 (e)) or an extrafocal (figure 53-52 (f)) position with respect to the transmission sphere.

One advantage of CGHs is their flexibility in implementing additional functional areas into their structure. It is therefore possible, for example, to add a ring-shaped area to the original test area, which provides a cat-eye wavefront with its focus point at the vertex of the surface under test (figure 53-55). This area surrounds the actual surface test area and can be controlled separately by the fringe analysis system to keep the surface at its correct distance from the CGH [53-55]–[53-56].

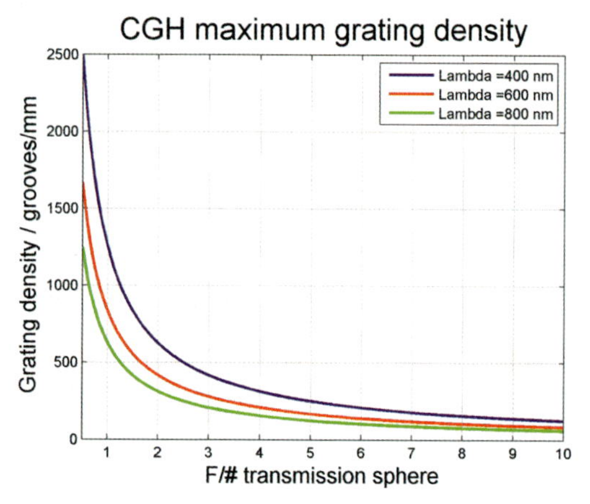

Figure 53-54: Maximum grating density in a CGH which acts as a transmission sphere with aperture F/# for wavelengths 400 nm, 600 nm and 800 nm.

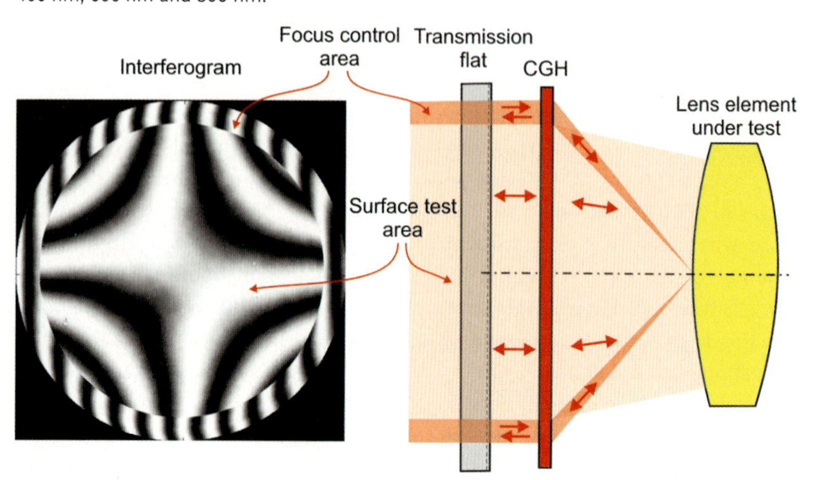

Figure 53-55: CGH with ring-shaped focus control zone, which can be analyzed separately by fringe analysis to keep the surface under test in the correct test position.

b) Testing with Partial Compensation

If a variety of rotationally symmetric aspheres which differ only slightly in shape need to be tested, it might be possible to use a single null system for all aspheres. The residual aspherical deviation must then be detected by the interferometrical test setup and the fringe analysis system [53-57]–[53-62].

One example is the testing of aspherical contact lens back surfaces, which differ in their curvature vertex radii and asphericity, as shown in the following (see also [53-63]–[53-64]). In the example, a movable lens element compensates for a substantial part of the asphericity, while the interferometer detects the aspherical residuals as a non-null signal.

The arrangement shown in figure 53-56 can be used to adapt to the necessary total range of radii and asphericities.

A spherical aplanatic–concentric lens element is placed between a transmissison sphere of suitable F/# and the contact lens element under test. When used in its "zero" position, the aplanatic–concentric element does not introduce asphericity to the transmitted wavefront. The aplanatic distance s_{apl} for a refracting surface of curvature of radius r_1 is given by

$$s_{apl} = r_1 \frac{n_1 + n_2}{n_1} \tag{53-41}$$

where n_1 and n_2 denote the refractive index of the surrounding medium and the lens material, respectively. An object point at a distance s_{apl} from a refracting spherical surface is imaged without introducing spherical aberrations to an image point at distance s'_{apl}

$$s'_{apl} = r_1 \frac{n_1 + n_2}{n_2} \tag{53-42}$$

The second radius r_2 of an aplanatic–concentric lens element is then calculated from (53-43) where d denotes the selected thickness of the element.

$$r_2 = r_1 \frac{n_1 + n_2}{n_2} - d \tag{53-43}$$

When the element is moved toward the surface under test, the transmitted wavefront changes from a spherical to an increasingly elliptical shape until a parabolical shape is reached. This range is usually sufficient to test the total variety of contact lens shapes. Compared to the required ideal surface shapes there remains an uncompensated asphericity, which can be detected by the interferometer as a non-null wavefront.

Partial compensating setups suffer from the so-called retrace error of the interferometer [53-65]. The retrace error is a usually unkown measurement error, which occurs, when a considerable non-null wavefront is detected. In this case the beams returning from the surface under test travel through different sections of the interferometer elements as during calibration. An estimation of the expected retrace error can be made when the quality of the interferometer elements is known. The calibration methods for the retrace error will be explained in the next section.

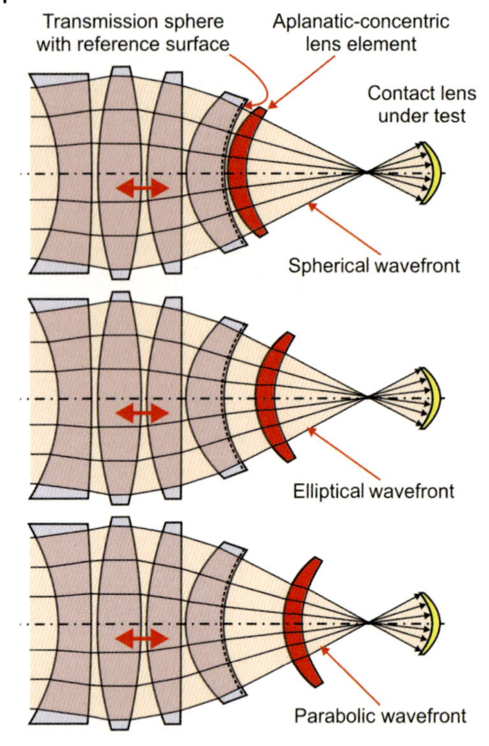

Figure 53-56: Interferometric setup to test a variety of aspherical contact lens back surfaces using a movable aplanatic–concentric spherical lens element.

c) Uncompensated Testing

In an *uncompensated* setup the asphere is tested by means of a plane or spherical wavefront, which after reflection is aspherically deformed. The deformation must be small enough to be detected by the interferometrical setup and the applied fringe analysis system. Therefore, only very weak aspheres can be tested uncompensated. However, there are various methods in which the asphere is tested sequentially by collecting subapertures, in which the aspherical deformation of the wavefront can be detected and evaluated. The total surface deformation is then determined by assembling and evaluating all subapertures.

Subaperture Testing

Subaperture testing of aspherical surfaces by means of a spherical or plane wavefront can be performed either by

a) moving the interferometer relative to the steady test piece;
b) moving the test piece relative to the steady interferometer; or
c) moving the interferometer as well as the test piece,

until the total surface is covered by overlapping apertures, which will be combined by a stitching algorithm to a total surface deviation map [53-66]–[53-67].

Figure 53-57 shows an example of a setup according to c) in which the interferometer is moved along a cross-section through the asphere's axis of symmetry while the lens element is rotated around the same axis. In this way subapertures are collected in overlapping ring zones sequentially as shown in figure 53-58.

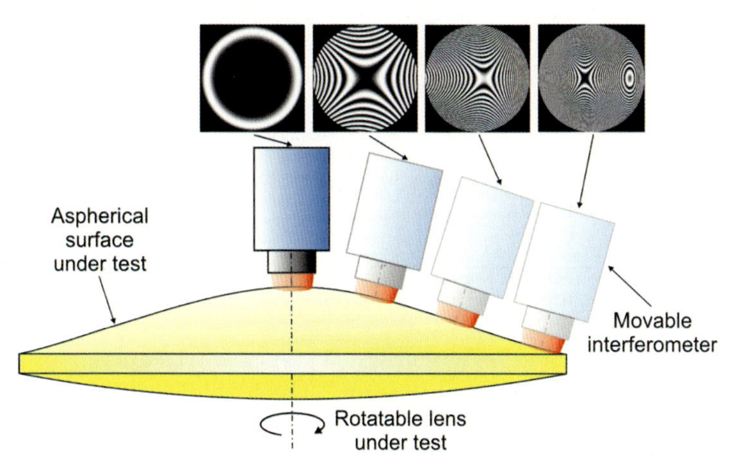

Figure 53-57: Interferometric subaperture testing of a rotationally symmetric asphere by moving the interferometer along a cross-section and rotating the lens element under test. Corresponding interferograms are shown for several interferometer positions.

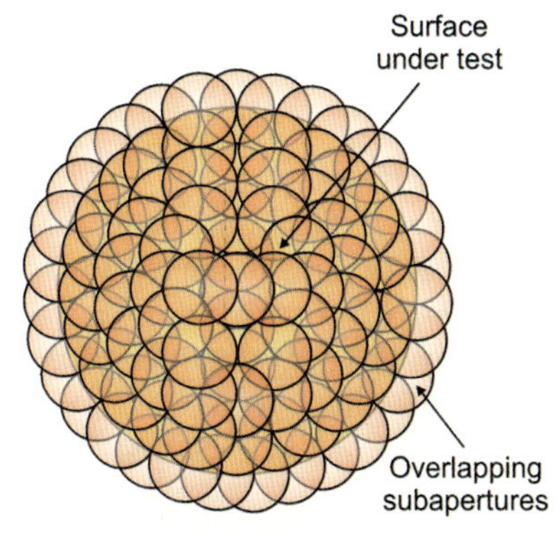

Figure 53-58: Overlapping subapertures to cover the total aspherical surface under test.

The width of a subaperture is selected to be as large as possible in order to minimize the number of subapertures and to stabilize the stitching process as much as possible. The maximum width is limited by the maximum fringe density within the subaperture, i.e., the fringe density must be smaller than 0.5 fringes per camera pixel. The largest possible subaperture width is estimated according to the procedure described below for which figure 53-59 shows the corresponding graphs.

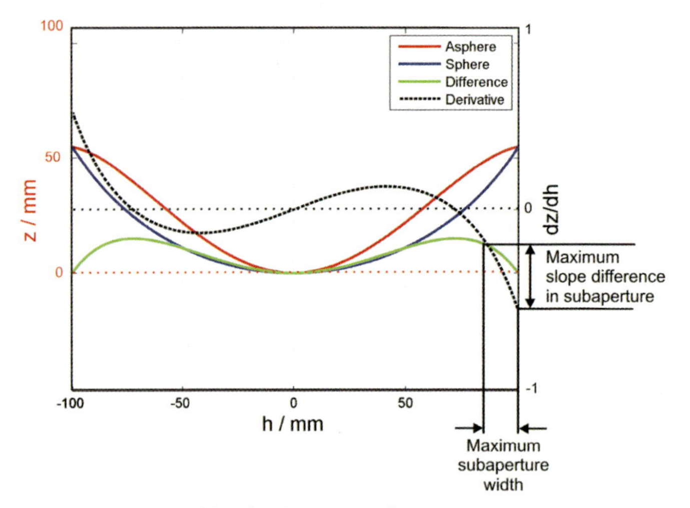

Figure 53-59: Sagittal height of a rotationally symmetric asphere (red), approximating sphere (blue), sagittal height difference (green) and slope of sagittal height difference (black). By varying the approximating sphere, different tilts of the black slope curve can be achieved in order to find the minimum fringe density.

1. Determine a sphere that contacts the asphere at the vertex and at the rim (also possible: calculate the best fitting sphere).
2. Subtract the sagittal heights of the sphere from those of the asphere: $\Delta z(h) = z_{\text{asph}}(h) - z_{\text{sph}}(h)$.
3. Calculate and plot the derivative function $\dfrac{d\Delta z(h)}{dh}$.
4. Determine the maximum acceptable derivative range which the interferometer can detect within a single aperture. The range is equal to the maximum slope difference = maximum slope – minimum slope.
5. Mark the maximum slope difference in the derivative function plot at the location with the steepest ascent or descent as shown in figure 53-59.
6. Estimate the corresponding maximum subaperture width at the *h*-axis.

The transmission sphere of the interferometer must carry a reference surface matching the curvature of the asphere's approximating sphere. For each subaperture the optical axis of the interferometer must be oriented normal to the surface under test. For a stable stitching process the adjacent subapertures should have enough overlap, for example around 50%.

The stitching algorithm must consider possible inaccuracies in positioning the interferometer relative to the test surface in all six degrees of freedom. The magnitude of the inaccuracies is defined by the precision of the mechanical positioning system and can be in the order of a few μm to 0.1 mm. To arrive at a nanometer scale for the surface deviations, the stitching algorithm therefore has to compensate for any inaccurate subaperture position.

Two critical points limit the accuracy of subaperture testing:

a) The absolute lateral scale.
b) The retrace error of the interferometer.

The absolute lateral scale is lost when the stitching algorithm is allowed to adjust all degrees of freedom. Therefore, the outer subapertures must be fixed to keep the scale at a well-defined magnitude. Generally, the local position of the subapertures is known to the degree of accuracy of the positioning system, which is a few μm in the best case. If improvement is necessary, a high resolution length metrology must be installed to measure the position of the outer subapertures as accurately as possible.

The retrace error in an interferometer occurs when rays returning from the test surface follow a path which is considerably different from the path in the calibration situation. A calibration is usually carried out by means of a good quality spherical surface which is well adjusted, so that only very few fringes are seen by the camera. Rays from reference and test surfaces then basically follow identical paths. In the case of subaperture testing of aspherical surfaces, each subaperture sees a different number and orientation of fringes, so the retrace error varies accordingly. In the following, a general method for compensation of the retrace error is described.

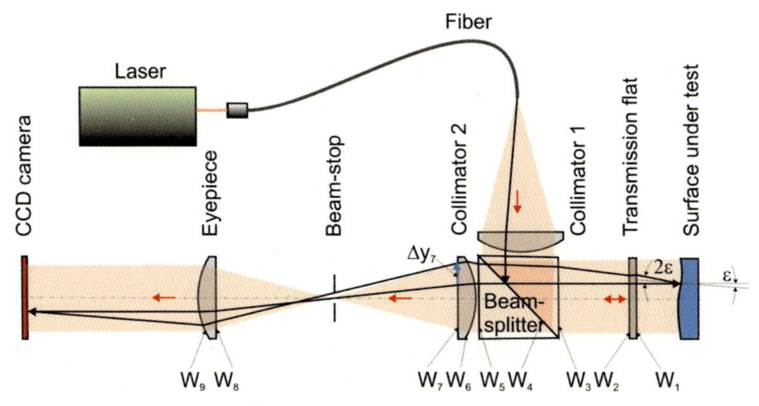

Figure 53-60: A Fizeau interferometer used to test an asphere without null compensation. The reflected ray returning from the surface under test follows a path which is different from the incident ray.

We consider a Fizeau interferometer as shown in figure 53-60. Light from a laser is brought to the back focal point of collimator 1 by means of a fiber. A beam-splitter directs the collimated light to a transmission flat (sphere) and the subsequent strong asphere under test. A ray hits the asphere at an incident angle ε other than zero. After reflection the ray follows a considerably different path as it returns to the beam-splitter than the corresponding ray reflected from the reference surface. On its way to the camera it passes collimator 2 and the eyepiece and crosses the individual surfaces with lateral displacement Δx_i, Δy_i ($i = 1 \ldots N$ = number of crossed surfaces) relative to a ray with incident angle $\varepsilon = 0$. Since calibration is usually carried out with $\varepsilon = 0$ for all incident rays, its validity is limited. The resulting systematic deviations mainly depend on the following.

1) The magnitude of Δx_i, Δy_i.
2) The optical design of the interferometer.
3) The surface deviations and inhomogeneities of the individual components.
4) The adjustment errors of the individual components.

If distortion of the optical design is small the resulting retrace error can be approximated by

$$\Delta W_R(x,y) = \sum_{i=1}^{N} \left(W_i(x + k_i\varepsilon_x, y + k_i\varepsilon_y) - W_i(x,y) \right) \tag{53-44}$$

where

ε_x, ε_y are the tilt of the surface under test in x and y at pixel x,y,
k_i is the shearing factor (lateral displacement per incident angle ε) for a displacement at surface number i,
$W_i(x,y)$ is the wavefront contribution at surface i when $\varepsilon = 0$.

For the compensation procedure of long spatial retrace errors, measurements must be carried out using either the aspherical test surface or a spherical calibration surface in at least three different alignments:

a) centered,
b) tilted in x,
c) tilted in y.

From the three measurements retrace differential quotients $G_x(x,y)$ and $G_y(x,y)$ are calculated using the first term of a Taylor series as an approximation:

$$\text{from a) and b)}: \qquad G_x(x,y) = \frac{\Delta W_R(x,y)}{\varepsilon_x} \tag{53-45}$$

$$\text{from a) und c)}: \qquad G_y(x,y) = \frac{\Delta W_R(x,y)}{\varepsilon_y} \tag{53-46}$$

When the aspherical surface is tested, the local incident angles $\varepsilon_x(x,y)$, $\varepsilon_y(x,y)$ are determined from the slopes of the measured wavefront $W(x,y)$:

$$\varepsilon_x(x,y) = \arctan\left(m \frac{\partial W(x,y)}{\partial x} \right) \tag{53-47}$$

$$\varepsilon_y(x,y) = \arctan\left(m \frac{\partial W(x,y)}{\partial y} \right) \tag{53-48}$$

where m is the appropriate scaling factor.

The local retrace error is then calculated from

$$\Delta W_R(x,y) = G_x(x,y)\varepsilon_x(x,y) + G_y(x,y)\varepsilon_y(x,y) \tag{53-49}$$

and can be calculated and subtracted from each individual measurement (see overview in figure 53-61).

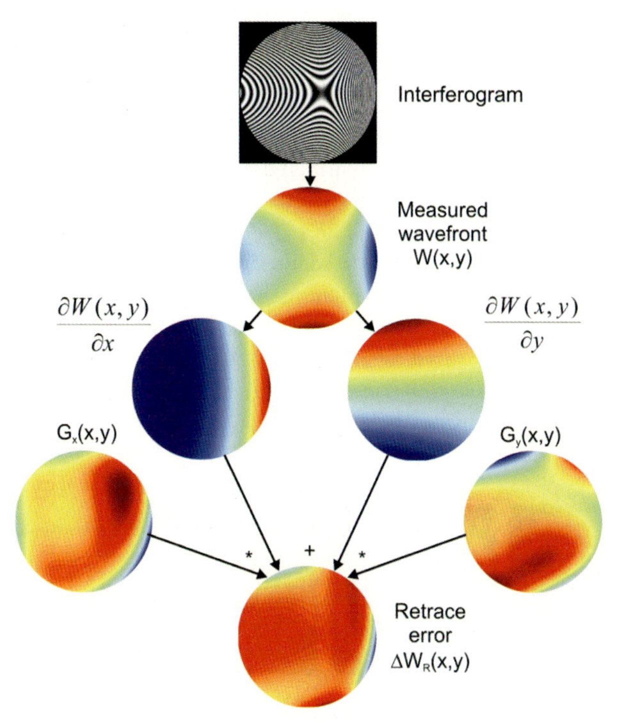

Figure 53-61: Retrace error compensation, schematic overview.

The calibration of mid-spatial and high-spatial retrace errors as well as non-linearly behaving retrace errors needs special iterative treatment which is beyond the scope of this chapter.

To decrease or avoid retrace errors, sometimes variable or exchangeable null optics are used in subaperture stitching interferometers. In this case fringe densities are kept to a minimum by selecting an appropriate null system for extensive compensation of the local asphericity [53-68]–[53-70].

The achievable results with subaperture stitching can be extremely accurate for mid-spatial and high-spatial frequencies; however, long-spatial frequencies which extend over many subapertures have the tendency to deviate increasingly from the true surface shape. When collecting many subapertures, the lateral resolution is very high. The high-frequent information could be used to qualify surface deviations contributing to flare of the optical system to which the optical element belongs.

Ring-Zone Testing

One of the earliest methods to test rotationally symmetric aspheres interferometrically was invented by Abbe [53-71]. He proposed to use a set of spherical test glasses that would sequentially produce ring zones in the vicinity of the contact zones when placed onto the asphere. The diameter of the contact zone would then be measured and the interference pattern analyzed and compared with the required specifications.

Using modern digital interferometers, sequential ring-zone testing can be implemented by moving the surface under test in the z-direction in front of a transmission sphere and recording the relative position z along with the interference pattern for each axial position [53-72]–[53-78]. Resolvable interference fringes are detected in the vicinity of those points, where the aspherical and spherical reference surfaces have common tangents, i.e., where the rays are normal to both surfaces. For aspherical surfaces this is a ring zone with a diameter of $2h$ and a zone around the vertex of the surface. At these points no retrace errors are induced, because the returning rays follow the same path as in the calibration situation. Thus, by scanning along the symmetry axis and collecting all necessary ringzones and the zone around the vertex, the complete surface is measured with interferometric resolution (figure 53-62). Overlapping of the individual ring zones is not necessary, since the shape of each ring zone is determined individually. In the following the algorithm is decribed in more detail [53-76].

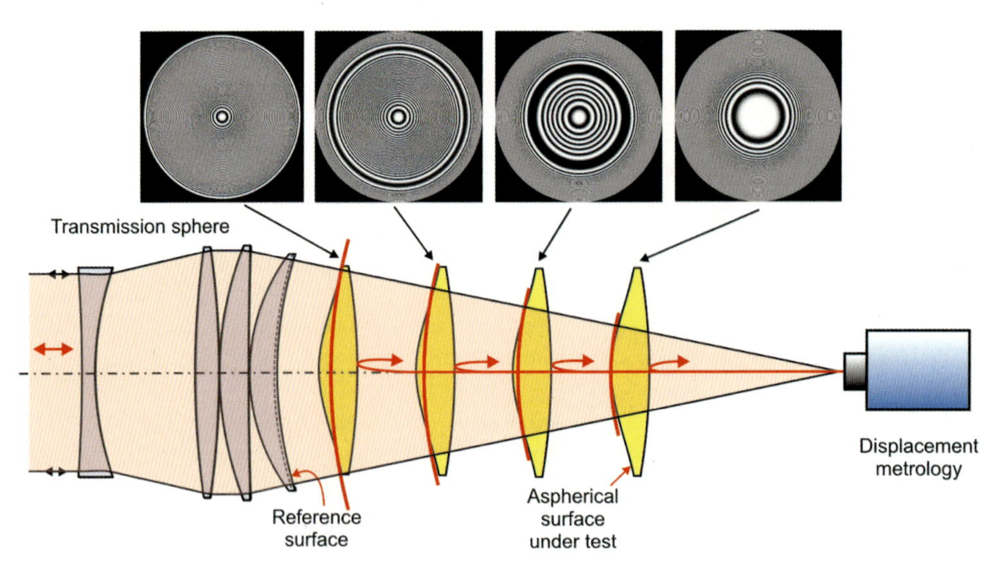

Figure 53-62: Ring-zone testing to sequentially measure the total shape of a rotationally symmetric surface by means of a transmission sphere, a test surface stage movable in the axial direction, and a displacement-measuring system.

Let d_A be the distance from the vertex of the reference surface to the vertex of the aspherical surface in an arbitrary scan position. Let d_Z be the distance between the reference surface and the aspherical surface normal to both surfaces at a zone height h (figure 53-63). Also let d_0 be the difference between the radius of curvature of the reference surface R_T and the vertex radius of the asphere R_0. Then a pair of coordinates v and p can be defined, which can be measured for any ringzone selected by moving the asphere along the axis of symmetry:

$$v = d_0 - d_A \tag{53-50}$$

$$p = d_Z - d_A \tag{53-51}$$

v and p can be measured directly interferometrically with phase resolution in the ring-zone setup. It can be shown ([53-75]–[53-76]) that v and p can be converted into h and z, which uniquely describe the surface under test. The necessary formulas are given in the following.

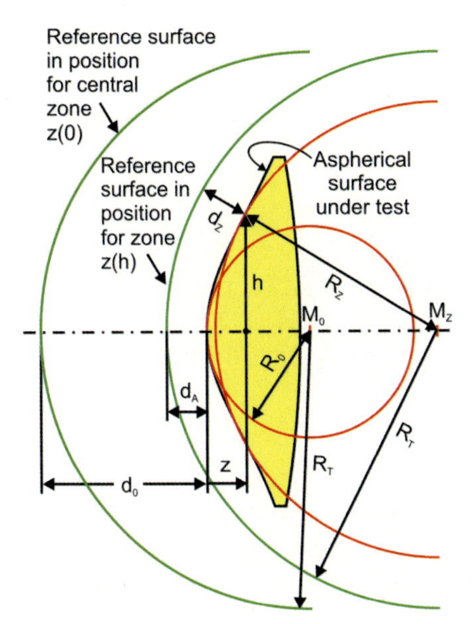

Figure 53-63: Geometrical parameters for ring-zone testing of aspherical surfaces.

From figure 53-63 we derive the relations:

$$v = z - R_0 + \frac{h}{z'} \tag{53-52}$$

$$p = z + \frac{1 - \sqrt{1 + z'^2}}{z'} h \tag{53-53}$$

where z' denotes the first derivative of the sagittal height z of the aspherical surface with respect to h as described by (53-40):

$$z'(h) = \frac{h}{R\sqrt{1-(1+\kappa)\frac{h^2}{R^2}}} + \sum_{i=2}^{M}\left(2i \cdot A_{2i}h^{2i-1}\right) \tag{53-54}$$

z' can be determined from the measured quantities v and p according to

$$z' = \frac{dz}{dh} = \frac{\sqrt{p'(2-p')}}{1-p'} \tag{53-55}$$

$p' \approx \Delta p/\Delta v$ is the differential change in the value for p by a differential change in the value for v.

$$p' = \frac{dp}{dv} = 1 - \frac{1}{\sqrt{1+z'^2}} \tag{53-56}$$

The quantity can either be deduced by differentiating the function $p(v)$ or can be directly measured. Then h and z can be determined according to

$$h = (R_0 + v - p)\sqrt{p'(2-p')} \tag{53-57}$$

$$z = p + (R_0 + v - p) \cdot p' \tag{53-58}$$

Before the scanning starts, the asphere is brought to "cat's eye position", where its vertex coincides with the center point of the reference surface. Then its position is axially changed by the amount R_0 to reach the starting position in which the central zone covers a circular area around the vertex. In this way R_0 is known absolutely, i.e., the surface is measured including the deviation of the vertex radius from the design value.

It should be noted that the lateral coordinate h of the location of a shape deviation is achieved from interferometric measurements rather than from pixel coordinates. The interferometer's actual magnification or its distortion are not needed.

Critical points in ring-zone testing are as follows.

1. The smallest possible zone width is limited by the lateral resolution of the interferometer, i.e., by the number of pixels of the implemented camera. The zone width is related to the derivative of the local tangential curvature of the asphere under test. This limits the dynamic of the testable aspheres.
2. The method described above depends essentially on the interferometric detection in the individual zones and the central vertex zone. The latter, however, is often disturbed by erroneous reflections from the surfaces in the transmission sphere and the collimator. Special care has to be taken regarding this point.
3. The transmission sphere must be perfectly calibrated. Its accuracy is directly

related to the accuracy of the ring-zone measurement. Rotating the surface under test while capturing individual zones in order to compensate for asymmetrical transmission sphere deviations is very complicated. Thus, the interferometer must be calibrated perfectly in advance and must be stable over the time of ring-zone capture.

Testing Generalized Second-Order and Toric Surfaces

In this chapter we discuss the metrology needed to measure generalized second-order surfaces and toric surfaces. Both types of surface may have different aspherical cross-sections in their perpendicular axes (principal sections) leading to a generalized astigmatic asphericity.

According to ISO 10110 Part 12 [53-33]we will describe the sagittal height $z(x,y)$ of a *generalized second-order surface with polynomial expansions* by:

$$z(x,y) = \frac{\frac{x^2}{R_x} + \frac{y^2}{R_y}}{1 + \sqrt{1 - (1+\kappa_x)\left(\frac{x}{R_x}\right)^2 - (1+\kappa_y)\left(\frac{y}{R_y}\right)^2}} + \sum_{i=2}^{M} \left(A_{2i}x^{2i} + B_{2i}y^{2i}\right) \tag{53-59}$$

where

x, y	are the lateral coordinates perpendicular to the axis of symmetry (z-axis),
R_x, R_y	are the vertex radii in the x- and y-direction,
κ_x, κ_y	are the conic constants in the x- and y-direction denoting the type of conic surface (second- order surface):
$\kappa_{x,y} > 0$	for an oblate ellipsoid,
$\kappa_{x,y} = 0$	for a sphere,
$-1 < \kappa_{x,y} < 0$	for a prolate ellipsoid,
$\kappa_{x,y} = -1$	for a paraboloid,
$\kappa_{x,y} < -1$	for a hyperboloid,
A_{2i}	are coefficients of polynomial expansion in the x-direction with $i = 2$, 3, ..., M,
B_{2i}	are coefficients of polynomial expansion in the y-direction with $i = 2$, 3, ..., M.

Some special cases follow:

a) $R_u \to \infty$ *Planotoric* or *cylindrical surface* with $u =$ either x or y

$$z(x,y) = \frac{\frac{u^2}{R_u}}{1 + \sqrt{1 - (1+\kappa_u)\left(\frac{u}{R_u}\right)^2}} + \sum_{i=2}^{M} A_{2i}u^{2i} \tag{53-60}$$

b) Cone with elliptical principal sections (a ≠ b)

$$z(x,y) = c\sqrt{\left(\frac{x}{a}\right)^2 + \left(\frac{y}{b}\right)^2} \tag{53-61}$$

c) V-edge with $u =$ either x or y

$$z(x,y) = c \cdot |u| \tag{53-62}$$

A *toric surface* is generated by the rotation of a defining curve, contained in a plane, about an axis which lies in the same plane.

If we define a curve $z = g(x)$ in the xz-plane and rotate around an axis parallel to the x-axis which intersects the z-axis at R_y, we can describe the sagittal height $z(x,y)$ according to ISO 10110 Part 12:

$$z(x,y) = R_y \mp \sqrt{(R_y - g(x))^2 - y^2} \tag{53-63}$$

$g(x)$ can generally be derived from (53-60) by setting $u = x$.

When $g(x)$ is selected as a circular shape with radius R_x, we have the special case of a toric surface:

$$z(x,y) = R_y - \sqrt{\left(R_y - R_x + \sqrt{R_x^2 - x^2}\right)^2 - y^2} \tag{53-64}$$

Figure 53-64 shows some examples for second-order surfaces, cone, V-edge and toric surfaces.

As in the case of rotationally symmetric aspheres, there are generally three different methods used to measure general second-order or toric surfaces by means of interferometry. These are measurement by:

a) full compensation,
b) partial compensation and
c) no compensation

of the astigmatic asphericity or toricity of the surface under test.

When measuring second-order or toric surfaces we have to note the following facts.

- A second-order or toric wavefront changes its shape while propagating, therefore the distance between the compensating optics and the surface under test must be controlled carefully.
- It is often useful and necessary to set up a detection system for the position of the test surface within the interferometric setup. The lack of rotational symmetry causes a strong sensitivity with respect to the rotational position. Therefore, at least an azimuthal mark (fiducial) must be provided with the element under test. If the cross-sections of the surface are aspherical, additional fiducials might be necessary for the measurement relative to a correct center position.

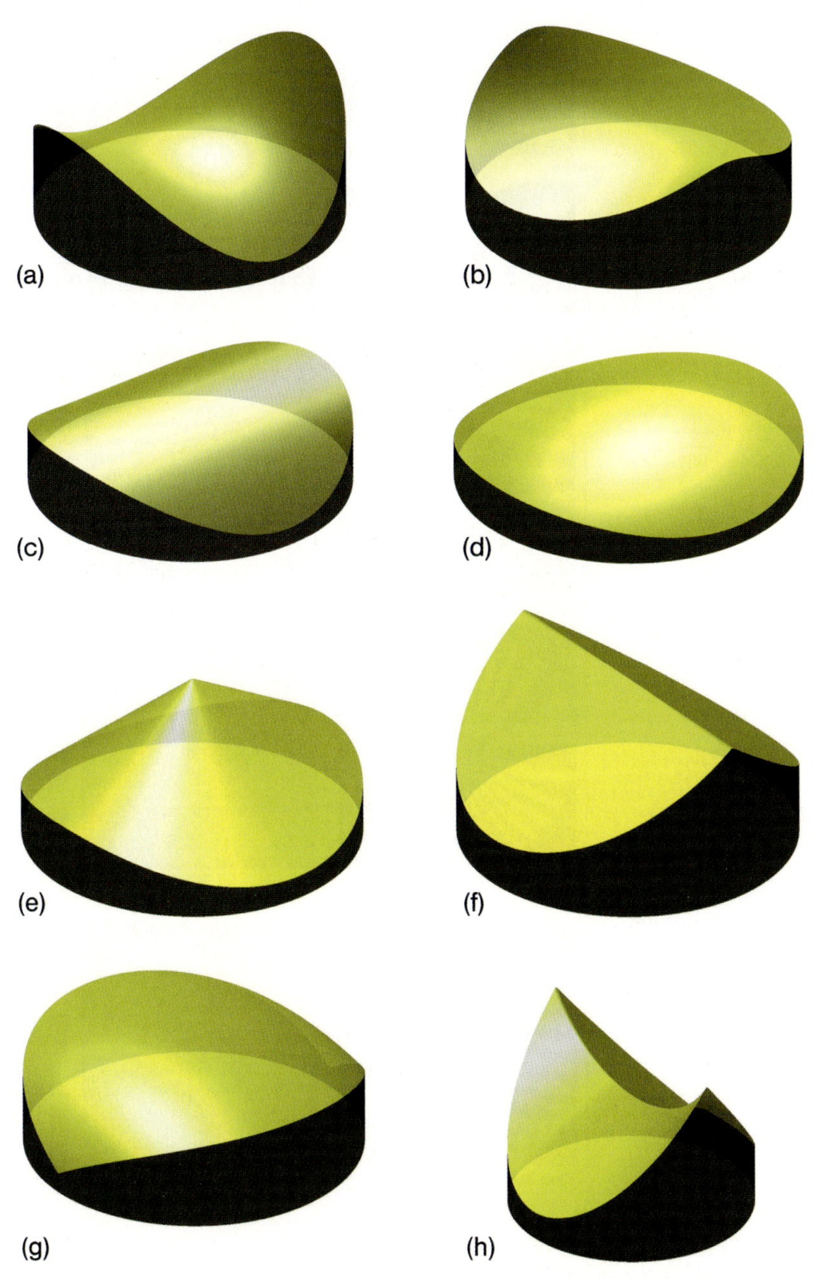

Figure 53-64: Examples of different second-order and toric surfaces: a) saddle-shaped second-order surface with circular principal sections of opposite sign; b) saddle-shaped second-order surface with aspherical principal sections of opposite sign; c) cylindrical second-order surface with circular principal sections; d) second-order surface with convex circular principal sections; e) cone with elliptical principal section; f) V-edge; g) toric surface with elliptical $g(x)$, h) toric surface with $g(x)$ = V-edge.

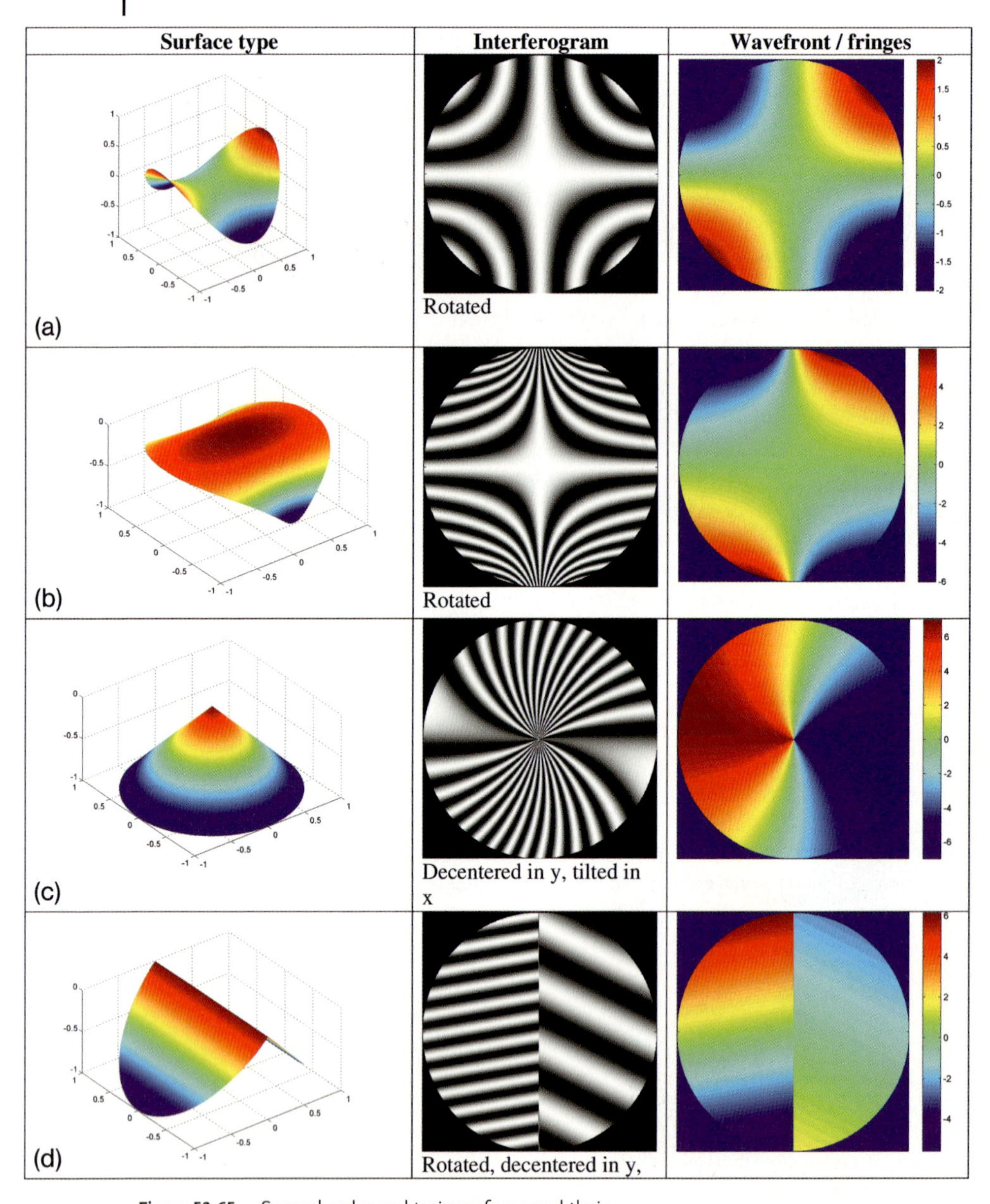

Figure 53-65: Second-order and toric surfaces and their misalignment effects in a fully compensated test setup: (a) saddle-shaped surface rotated around the optical axis; (b) toric surface with elliptical and hyperbolic principal sections rotated around the optical axis; (c) cone with circular principal section decentered in *y* and tilted in *x*; (d) V-edge rotated, decentered in *y* and tilted in *x* and *y*.

In figure 53-65 some misalignment effects for different second-order and toric surfaces are shown. When astigmatic or toric surfaces are misrotated in a fully compensated setup, the measured wavefront is also astigmatic (figure 53-65(a)). The astigmatism contains higher orders when the cross-sections of the test surface are non-circular (figure 53-65(b)). When testing cones, the decenters lead to wavefront discontinuities at the location of the tip (figure 53-65(c)), which must be covered by a mask, when conventional fringe analysis software is applied. When testing V-edge types of surfaces, there will be wavefront discontinuities along the edge, which – depending of the type of misalignment – become inconsistent for a conventional fringe analysis software, unless a mask divides both halves of the surface (figure 53-65(d)).

a) Testing with Full Compensation

As in the case of rotationally symmetric aspherical surfaces null systems are used for full compensation of a toric surface under test. Toric null systems can be refractive or diffractive [53-79]–[53-81]. In contrast to spherical or rotationally symmetric aspherical null systems toric null systems can not be calibrated. For error estimation an error budget must be calculated by considering all possible parameter deviations influencing the generated wavefront. The error budget has to account for systematic deviations of radii of curvature, rotational position of individual components, refractive indices of glass materials, central thicknesses, distances between components, but also for statistical variations caused by changes in temperature, air pressure and humidity.

Testing with Refractive Null Systems

In analogy to refractive null systems used in testing rotationally symmetric aspheres, it is possible to design and fabricate null systems for toric surfaces. An extreme example in shown in figure 53-66, in which the null system comprises lens elements with cylindrically shaped surfaces used to test cylindrical surfaces.

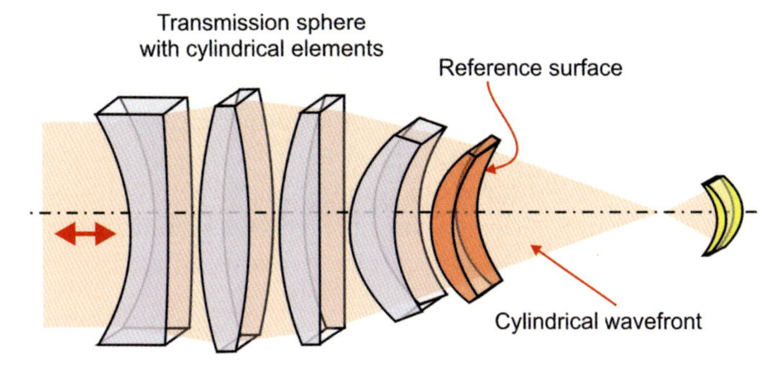

Figure 53-66: Example of a refractive null system to test cylindrical surfaces comprising cylindrical lens elements.

To test toric surfaces containing a mean spherical shape plus an astigmatic deviation, a combination of a transmission sphere and at least one toric element will be suitable. Preferably, the last refractive surface should act as the reference surface.

A general problem, depending on the toricity of the test surface, can be the fact that there will be no stigmatic imaging of the surface under test onto the CCD camera. The lateral resolution of the test surface will vary in azimuthal direction.

The fabrication of cylindrical or toric refractive null systems is cumbersome and time consuming. Unlike transmission spheres measuring a large variety of spherical surfaces, they cover only a small spectrum of astigmatic surfaces with equal toricity (the astigmatic focus lines must coincide).

An improvement in variability is introduced by a so-called Alvarez lens [53-82], [53-85]. It contains two transmissive refractive plates, each having a plane surface and a surface shaped in a two-dimensional cubic profile (figure 53-67 (a)). The two cubic surfaces are designes to be the inverse of each other, so that when both plates are placed with their vertices on the optical axis, the induced phase variations cancel out (figure 53-67 (d)).

If the two plates are laterally translated relative to each other, a phase variation is induced proportional to the derivative of the cubic surface profiles, resulting in a quadratic phase profile. The profile is selected such that a relative movement in the y direction induces a rotationally symmetric focus term, whereas a movement in the x direction induces a pure astigmatic term proportional to the magnitude of the movement.

The wavefront transmitted through a single Alvarez lens element is given by

$$W_{1/2}(x,y) = \pm a\left(x^2 y + \frac{1}{3}y^3\right) \tag{53-65}$$

where a denotes an aribtrary constant. The wavefront ΔW transmitted through the lens pair is then approximately given by

$$\Delta W(x,y) = a((2xy)\,dx + (x^2 + y^2)\,dy) \tag{53-66}$$

where dx and dy denote the relative displacement of the lens elements. Combined movements can therefore produce any quadratic wavefront shapes including focus, astigmatic and cylindrical phase profiles.

The use of refractive null systems to test toric or cylindrical surfaces will generally be the exception because of the tremendeous effort required in design, fabrication and qualification. It is much easier to use diffractive null systems as discussed in the next section.

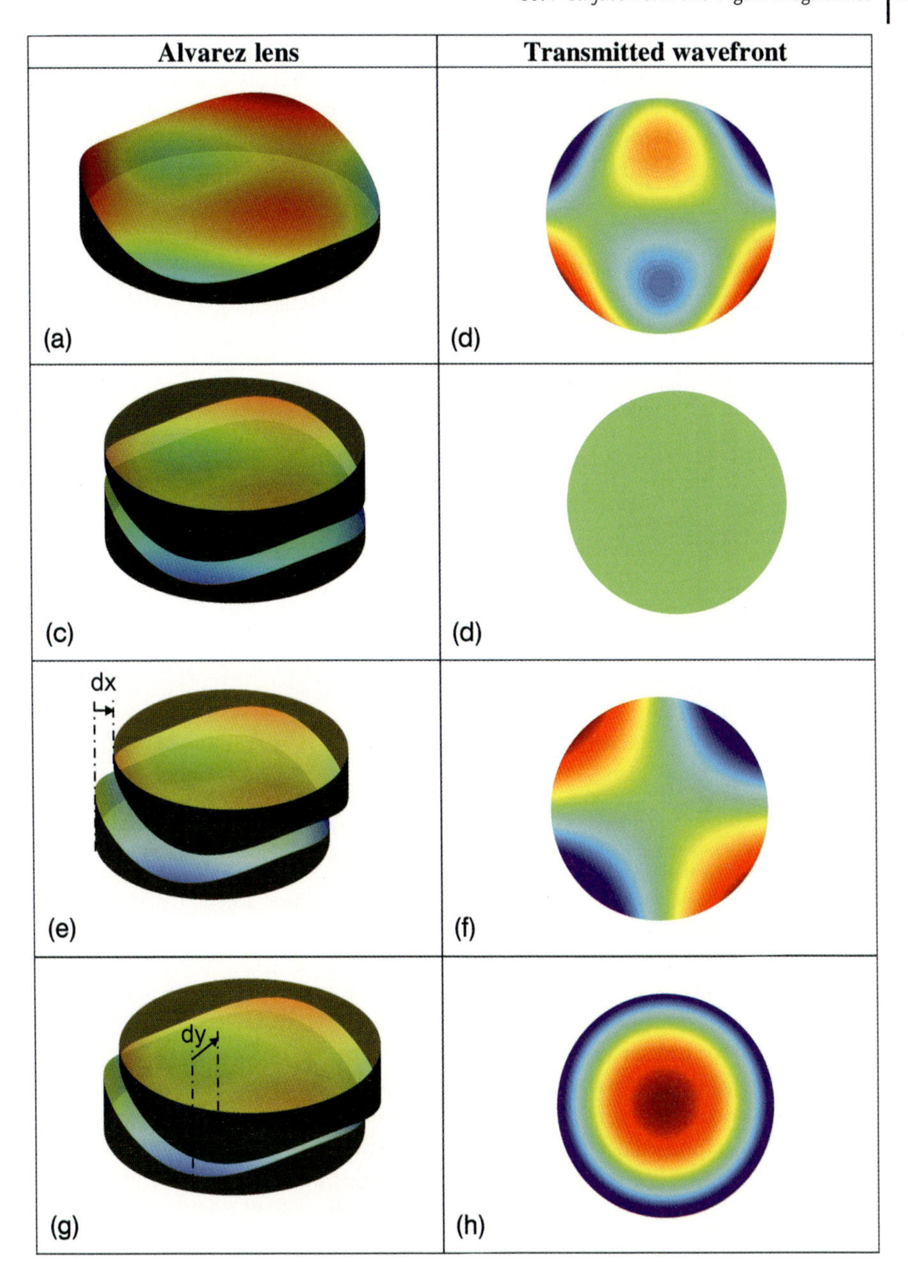

Figure 53-67: Alvarez lens and the transmitted wavefront: (a) single lens element; (b) transmitted wavefront for a single lens element; (c) lens pair in zero position; (d) transmitted wavefront for (c); (e) lens pair shifted dx; (f) transmitted wavefront for (e); (g) lens pair shifted by dy; (h) transmitted wavefront for (g).

Testing with Diffractive Null Systems

In analogy with the test of aspherical surfaces, diffractive null systems can preferably be used to test toric or cylindrical surfaces [53-81]–[53-82]. In most cases the reference surface is provided by a transmission flat (figure 53-68). The CGH is located within the cavity and provides the total wavefront shaping including the spherical convergence or divergence and the toricity.

When using CGH the characteristics concerning efficiency, spurious reflections and maximum line density must be considered as discussed above.

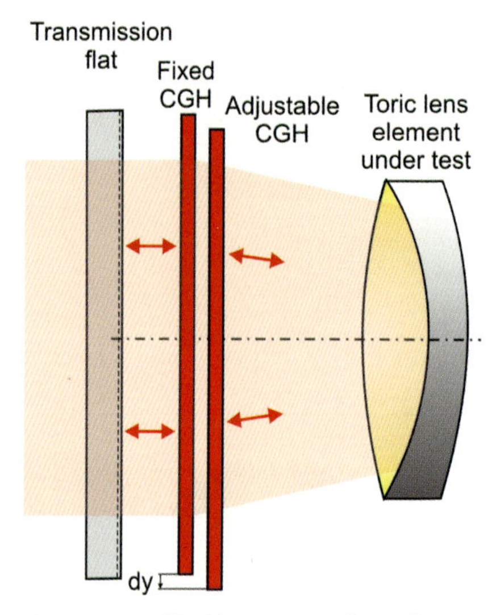

Figure 53-68: The Alvarez principle used to test toric surfaces provided by two CGHs of opposite cubic wavefront terms. The second CGH can be translated in the x- and y-direction to generate spherical and astigmatic wavefronts.

In oder to reduce grating densities on CGHs in toric testing, a combination of CGH with refractive spherical transmission spheres might be useful. In this case the CGH mainly carries the toricity, while the transmission sphere provides the spherical part in the test wave. The CGH can be located intrafocally or extrafocally with respect to the transmission sphere. It can be configured to carry additional functional areas for distance and adjustment control as shown in figure 53-55.

In order to improve the versatility of toric testing, the Alvarez principle can be applied as described in the previous section [53-83]–[53-84]. Two CGHs producing opposite cubic wavefronts are brought closely together. The first one is fixed while the second can be translated in the x- and y-direction to generate either focus or astigmatism (figure 53-68). The first CGH can additionally be equipped with a spherical or aspherical wavefront term to deliver a rotationally symmetric surface which best fits the surface under test.

b) Testing with Partial Compensation

Partial compensation makes sense if a variety of toric surfaces need to be tested covering a toricity range which is detectable by the interferometrical test setup and the applied fringe analysis system. By introducing, for example, a pair of Alvarez CGHs for compensation of the mean astigmatism of a surface under test, a variation in asphericity along the axes of symmetry might be covered by the range of the detection system.

It might be useful to provide partial compensation by means of a transmission sphere with a movable lens element as shown in figure 53-56 in combination with a fixed toric or cylindrical element or an adjustable element, such as in the Alvaraz arrangement.

As explained above, partial compensating setups suffer from the so-called retrace error of the interferometer. The beams returning from the surface under test travel through different sections of the interferometer elements in the same way as during calibration. Retrace errors can be calibrated by applying the procedure as explained above in the section on subaperture testing.

c) Uncompensated Testing

In an *uncompensated* setup the toric surface is tested by means of a plane or spherical wavefront, which after reflection is astigmatically deformed. The deformation must be small enough to be detected by the interferometrical setup and the applied fringe analysis system. Therefore, only weak toricities can be tested uncompensated in a single measurement. However, when the surface is tested sequentially by collecting subapertures with sufficient fringe densities, strong overall astigmatic deformations can be detected by assembling all subapertures to a total surface map.

The subaperture testing process has been described above in detail and can be fully applied to toric surfaces.

Testing Freeform surfaces

In the previous chapter on generalized second-order and toric surfaces we were already very close to what are known as freeform surfaces. While toric surfaces still show two axes of symmetry, a freeform surface does not generally show any. However, in many applications optical setups are designed in 2D rather than 3D space, leaving a remaining axis of symmetry for the freeform element comprising a quasi-freeform surface.

Applications of optical freeform surfaces are, for instance:

- progressive-addition ophthalmic lenses with freeform front surfaces,
- headup display systems in cars,
- headmounted devices,
- optical elements in automotive forward lighting applications,
- rear view mirrors,
- Alvarez lens element (see above),
- Telescopes in space optics.

The preferred testing methods for higher accuracies are the use of:

- CGH null systems in an interferometric setup,
- Uncompensated subaperture measuring devices.

All methods are described in detail in the section above and can be applied fully to freeform surfaces.

For systems and elements with lower accuracy specifications, uncompensated measurement might be possible over the entire element diameter. For instance, rear view mirrors with freeform shapes or optical elements in car lighting systems can be easily tested by deflectometers as described in section 49-6 [53-86].

Opthalmic lenses with ranges of ± 30 dpt should be qualified to a resolution of 0.01 dpt. The necessary dynamic is achieved by a deflectometric arrangement in transmission, using a CCD camera observing a suitable geometrical mask (checkerboard, circular spot grid) through the lens under test (see figure 49-26).

53.5 Centering

53.5.1 Basics

The centering state of an optical element and its related tolerances, as specified by the optical design, can be described by a number of terms, which are defined in ISO 10110 Part 6 [53-2]. In the following we will summarize the fundamental definitions.

1. *Optical axis*

The theoretical axis, about which the optical element or system is nominally rotationally symmetric. Exceptions are deflecting elements, such as plane mirrors, prisms, etc.

2. *Datum axis*

An axis selected after consideration of specific features of an optical system. It serves as a reference for the location of surfaces, elements and assemblies.

3. *Datum point*

A specified point on the datum axis. It serves as an additional reference for the location of an optical surface, element or optical system. For individual optical surfaces the datum point is identical with the vertex point. For optical elements the datum point is identical with the vertex point of the first surface.

4. *Tilt angle of a spherical surface*

The angle between the datum axis and the normal to the surface at its intersection point with the datum axis.

5. *Tilt angle of an aspherical surface*

The angle between the rotation axis of the aspherical surface and the datum axis of the part, subsystem, or system to which the aspherical surface belongs.

6. *Lateral displacement of an aspherical surface*

The distance from the point of rotational symmetry of the aspherical surface (vertex point) to the datum axis.

7. *Tilt angle of an optical element*

The angle between the datum axis of the element and the datum axis of the system of which the element is a part.

8. *Lateral displacement of an optical element or subsystem*

The distance between the datum axis of the element and the datum axis of the system of which the element is a part, measured at the datum point of the element or subsystem.

Centering Errors

- The centering error of an individual spherical surface is specified only by a surface tilt angle α (figure 53-69).
- The centering error of an aspherical surface is specified by the tilt angle of its rotation axis α and the lateral displacement d of its vertex point from the datum axis (figure 53-70).
- The centering error of a lens element or a subsystem is specified by the tilt angle α of its datum axis (in general the optical axis of the element or subsystem) and the lateral displacement d of its datum point from the datum axis of the total system (figure 53-71).

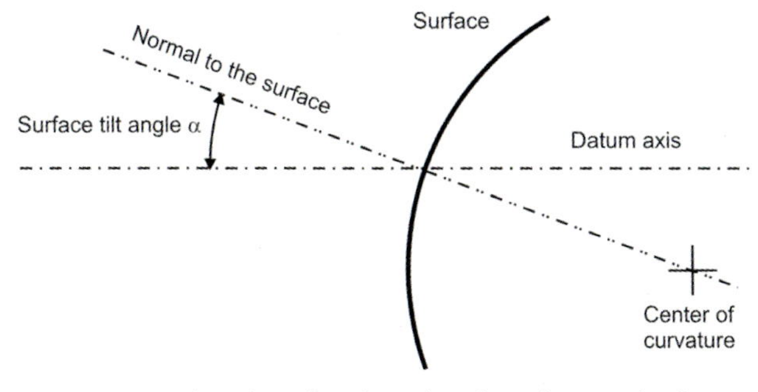

Figure 53-69: Tilt angle α of a spherical surface relative to the datum axis.

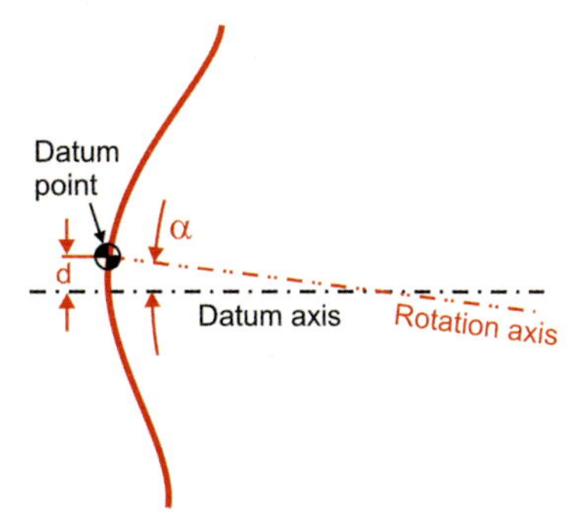

Figure 53-70: Tilt angle α and lateral displacement d of an aspherical surface relative to the datum axis.

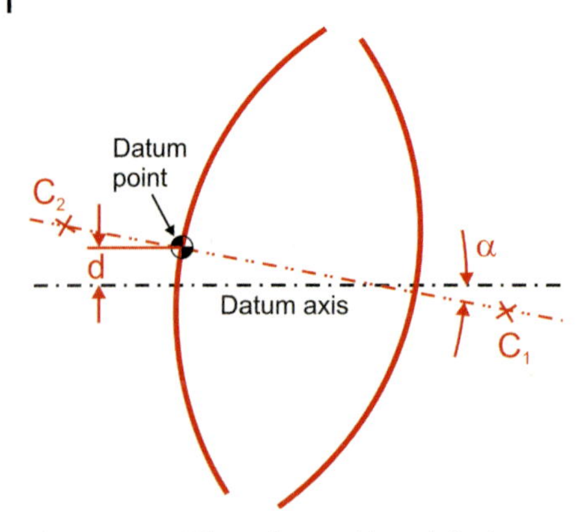

Figure 53-71: Tilt angle α and lateral displacement d of a lens element relative to the datum axis.

Inner and Outer Centering

It must be pointed out that there is a significat difference between the centering nature of a spherical lens element and an aspherical lens element.

The optical axis of a spherical lens element is always well defined by the line crossing the two centers of curvature, its centering property therefore depends on how well its axis coincides with the datum axis. The lens itself is perfect, it is only its position which must be corrected for perfect performance. The "inner" centering of a spherical lens is always perfect, while its "outer" centering can be achieved by adjustment.

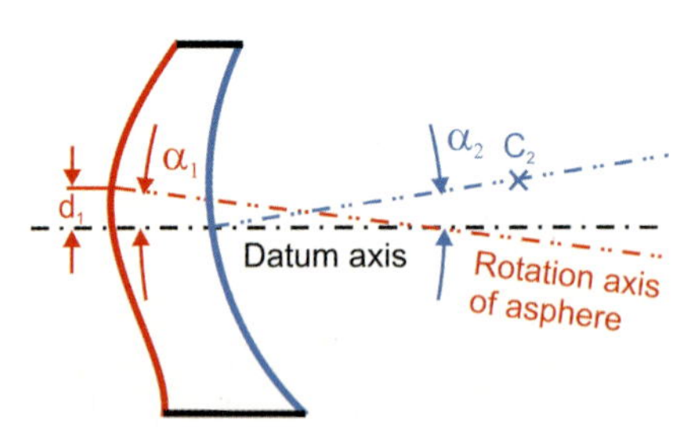

Figure 53-72: Lens element with aspherical first surface and spherical second surface showing inner decenter.

For a perfect aspherical lens element the rotation axis of the aspherical surface must coincide with the center of curvature of the opposite surface, if the latter is spherical. If the opposite surface is also aspherical, both rotation axes must coincide. If the misalignment of the opposite surfaces exeeds the specified tolerances, then the lens is unusable. It cannot be corrected just by mere processes. Figure 53-72 shows an example. The "inner" centering of an aspherical lens therefore depends on the relative centering of the optical surfaces to each other, while its "outer" centering is related to the misalignments referencing outer datum axes.

Indication of Drawings

The indication of centering tolerances on an optical drawing consists of the code number 4, one or two tolerance values and, if necessary, the appropriate reference to the elements of the datum axis.

For the indication of cement wedge angle tolerances, the triangular delta symbol (Δ) precedes the tolerance value (figure 53-76).

The structure of the indication has one of the following three forms:

$4/a$

or

$4/a(l)$

or

$4/\Delta\beta$

where a denotes the maximum permissible tilt angle,

l denotes the maximum permissible lateral displacement, and

β following the triangular symbol Δ denotes the maximum permissible cement wedge angle.

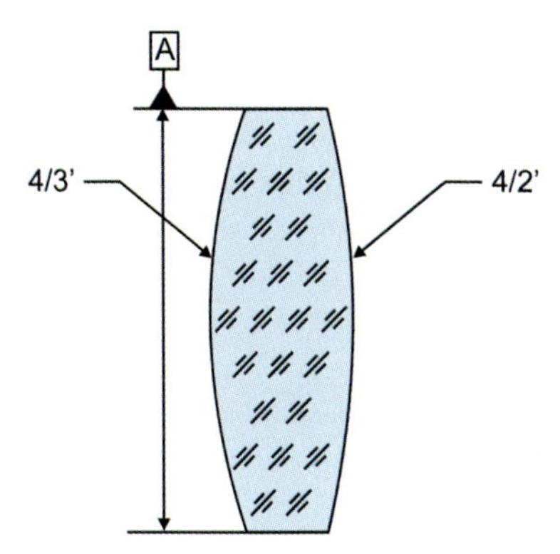

Figure 53-73: Indication of centering tolerances for a spherical lens element, datum axis referencing the outer cylinder.

The indicated centering tolerances refer to the datum axis of the optical element or subsystem. If more than one datum axis is indicated in the drawing, the reference letters of the appropriate datum axis are appended to the tolerance values.

The values for the tolerances are specified in minutes ['] or seconds ["] of arc for angular dimensions and in millimetres [mm] for linear dimensions.

The indication is shown in figures 53-73 to 53-75 with a line pointing to the surface or optical system to which it refers.

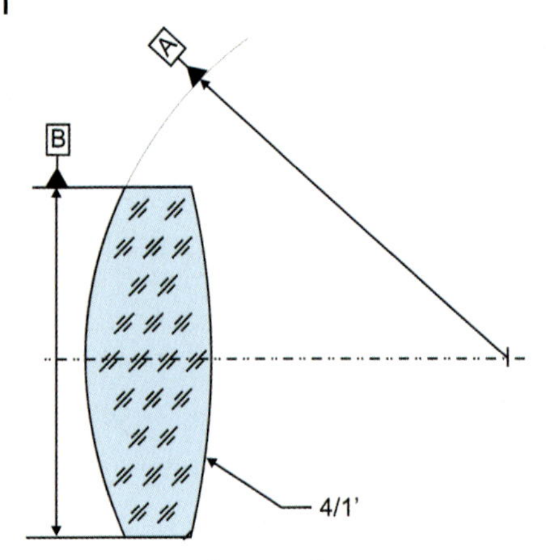

Figure 53-74: Indication of centering tolerances for a spherical lens element, datum axis referencing the center of curvature and the central point (vertex) of the first surface.

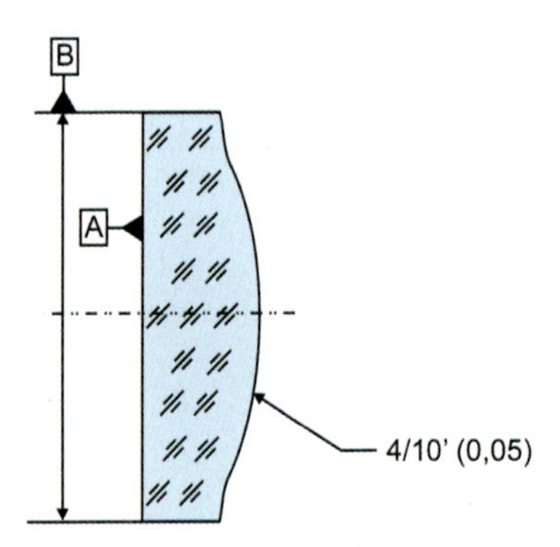

Figure 53-75: Indication of centering tolerances for an aspherical lens element, datum axis referencing the outer cylinder and the first (plane) surface.

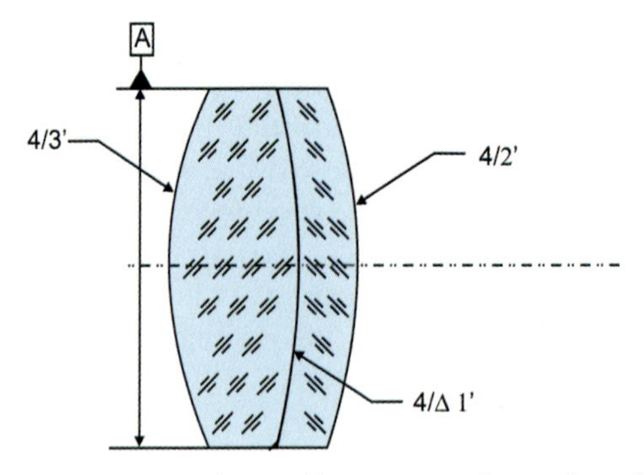

Figure 53-76: Subassembly consisting of two spherical lens elements including indication of cement wedge tolerance with datum axis referencing the outer cylinder.

53.5.2 Metrology

In this section we describe the metrology needed to measure the outer centering of spherical lens elements or subsystems and the inner centering of aspherical lens elements.

The general procedure used to measure centering errors is to rotate the lens under test in transmitted or reflected light [53-87]–[53-93]. A typical mechanical setup therefore consists of:

1. a precision rotary table carrying an adjustable lens mount;
2. a length gauge measuring the lateral displacement of the sample during rotation;
3. a length gauge measuring the precession movement of the sample during rotation.

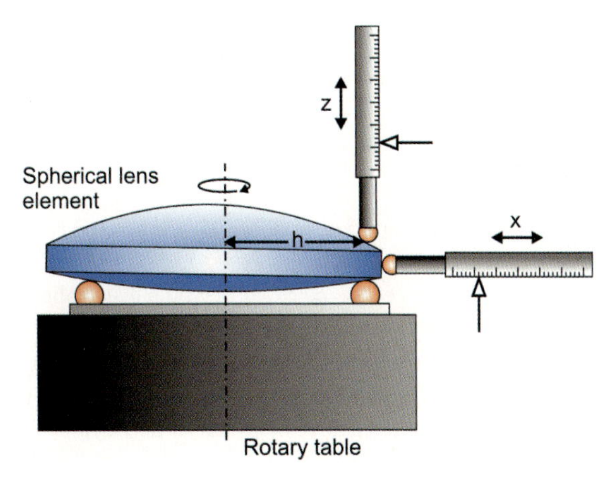

Figure 53-77: Mechanical testing of centering properties of a spherical lens.

Figure 53-77 shows a typical arrangement of a mechanical setup. If the lens support, represented for instance by a three-ball mount, is adjusted so that its axis coincides with the rotation axis of the rotary table, the relative movements of the length gauges in x and z are given by

$$\Delta x = l \cdot \cos(\omega t + \varphi_l) + \sqrt{\frac{D^2}{4} - l^2 \sin^2(\omega t + \varphi_l)} - \frac{D}{2} \tag{53-67}$$

$$\Delta z = \sqrt{r_2^2 - h^2 + 2h \sin(\omega t + \varphi_a) \sin a - (d - r_2)^2 \sin^2 a} - \sqrt{r_2^2 - h^2} + (d - r_2)(1 - \cos a) \tag{53-68}$$

where

D	is the diameter of the lens element,
l	is the lateral displacement of the lens edge relative to the optical axis,
φ_l	is the azimuthal direction of lateral displacement,
a	is the tilt angle of the optical axis,
φ_a	is the azimuthal direction of tilt,
r_2	is the radius of curvature of upper surface,
d	is the thickness of the lens element,
h	is the contact point of the z gauge from the rotation axis,
ω	is the rotation frequency,
t	is the time.

The decenter l of a spherical lens and its azimuthal direction is calculated from (53-67). The maximum of $\Delta x = l$ is found at $\omega t = -\varphi_l$.

The tilt angle a of the lens on the lens mount is calculated from (53-68). Usually the lens is iteratively centered while rotating until Δz is zero.

Adjustment of Lens Support

For a correct centering metrology it is mandatory that the lens support is centered relative to the rotation axis of the rotary table. To achieve the correct centering, two adjustments have to be carried out:

1. the lens support is tilted, until a contacting plane is perpendicular to the rotation axis of the rotary stage,
2. the lens support is centered, until the center point of a contacting sphere coincides with the rotation axis of the rotary stage.

While a plane plate is carried by the rotating lens support, the precession of its contacting surface is detected either by a mechanical length gauge or by observing reflected light by means of an autocollimator (figure 53-78). The lens support is tilted until the signal from the lower surface remains unchanged during rotation.

Once the support is perpendicular to the rotation axis, its centering is achieved by means of a spherical surface. The remaining rotation excentricity is detected either by a mechanical length gauge contacting the lower surface or by an autocollimator as shown in figure 53-79. The collimator has to be combined with a compensating lens, which focusses the parallel beams from the autocollimator to the center point of the spherical surface. The lens support is then centered in the lateral direction, until the reflection remains unchanged during rotation.

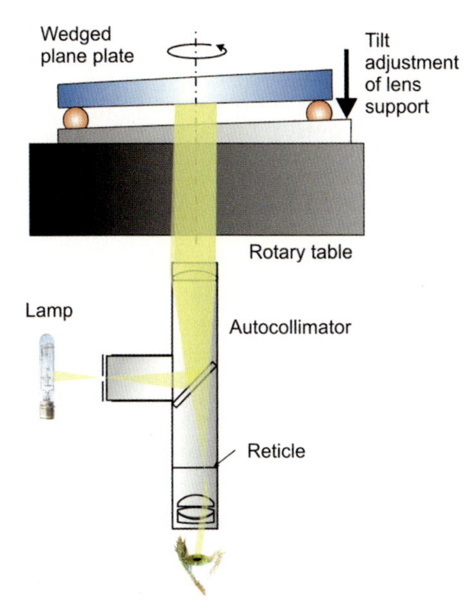

Figure 53-78: Tilt adjustment of lens support by means of an autocollimator and a plane plate.

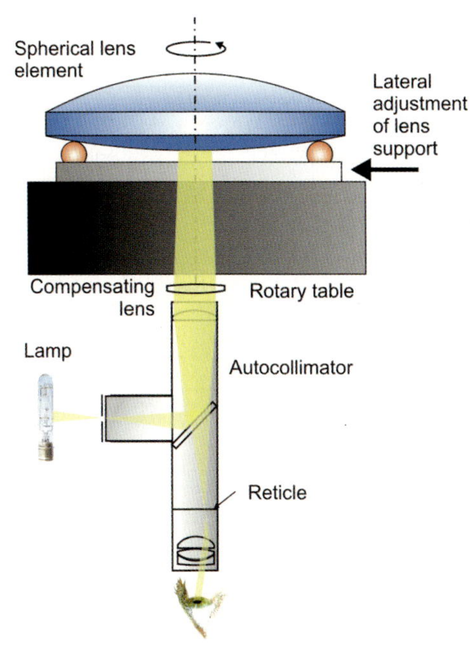

Figure 53-79: Lateral center adjustment of lens support by means of a collimator and a spherical surface.

Centering Metrology with Adjusted Lens Support

Once the lens support is properly adjusted, measurements of different lens elements can be made without additional adjustment of the support. In many cases a contactless optical detection method is preferred over a contacting mechanical gauge.

Optical centering methods are based on either transmitted or reflected light. A simple method of transmission is shown in figure 53-80. A laser beam is directed through the sample approximately coinciding with the rotation axis. It is detected by a position-sensitive detector (PSD). If the optical axis of the rotating lens coincides with the rotation axis of the rotary table, the detected spot remains in a steady position. If the lens is tilted, the light spot on the PSD rotates on a circle while the lens is rotating. When the lens has been correctly positioned ($a = 0$), the remaining decenter l of the lens is detected by a length gauge which detecs the edge of the lens. For small lenses a mechanical gauge should be replaced by a non-contacting length gauge to prevent unwanted displacements by the applied forces.

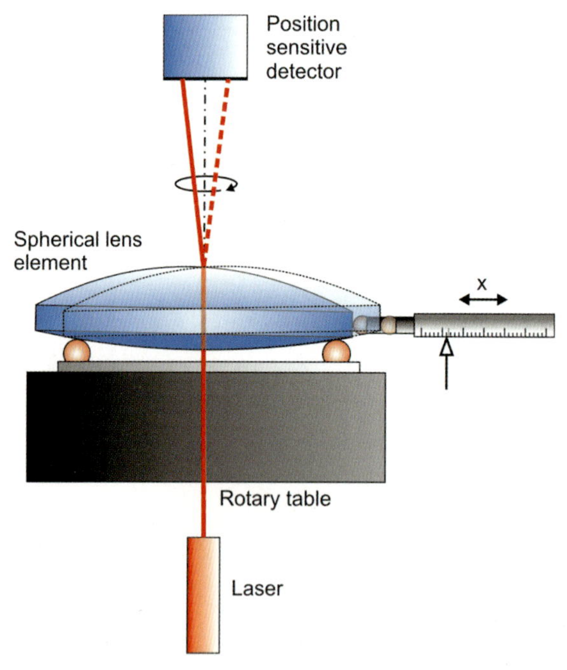

Figure 53-80: Method used to measure centering errors of a lens component or subsystem during transmission using a laser, a position-sensitive detector and a length gauge.

For high-precision decenter inspection during transmission a collimator which projects a cross-reticle through the lens under test can be used as shown in figure 53-81. A second collimator, in combination with a compensating lens, is used to project the cross-reticle image onto a CCD camera. The focus of the compensating lens should coincide with the focus of the lens under test. When the lens is centered correctly on the lens support ($a = 0$) the cross- reticle image remains unchanged while the lens is rotating. The decenter l of the lens is then detected by a length gauge which detects the edge of the lens.

For some requirements the transmission method might not have sufficient resolution. In this case it is useful to apply the method after reflection, which is approximately $2/(n\text{-}1)$ times more sensitive, where n denotes the refractive index of the element under test. For the reflection method an autocollimator projects a reticle to infinity (figure 53-82). A compensating lens projects the target image onto the center of curvature of the upper surface of the test lens. After reflection the target image is projected onto a CCD camera, its lateral position being proportional to the tilt of the upper test surface.

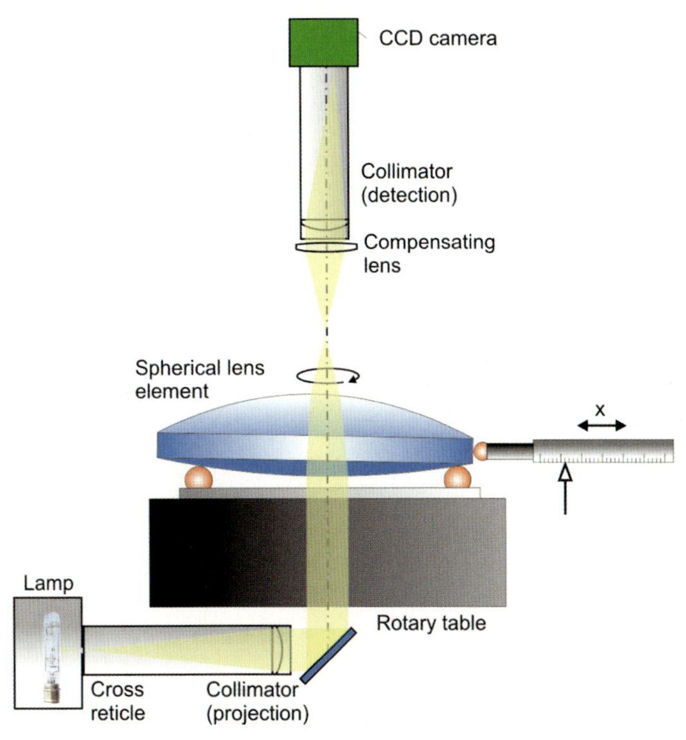

Figure 53-81: Method used to measure centering errors of a lens component or subsystem during transmission using a projection collimator and a detection collimator, a cross-reticle to be projected and a CCD camera for cross-reticle image detection and a length gauge.

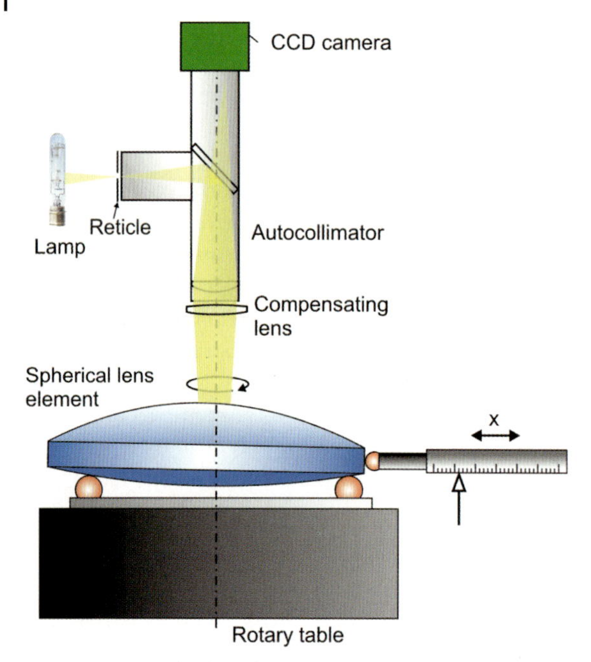

Figure 53-82: Method used to measure centering errors of a lens component or subsystem during reflection using an electronic autocollimator and a compensating lens to detect lens tilt and a length gauge to detect the decenter of the lens edge.

Centering Metrology with Unadjusted Lens Support

In situations where the lens support must frequently be exchanged to adapt to different test lens diameters, it will be necessary to readjust the lens support after each exchange. In this case it will be helpful to use a setup detecting the upper and lower surfaces of the test lens simultaneously, while the lens is rotating. Figure 53-83 shows a setup in which two electronic autocollimators with appropriate compensating lenses observe the reflections from the upper and lower surfaces during rotation.

Both autocollimators should be aligned, which can easily be checked by observing the images projected from the opposite autocollimator. As long as the optical axis of the test lens does not coincide with the rotation axis, the reflections C_1 and C_2 on the CCD cameras will move in circles, the diameter of the circles being proportional to the magnitude of the surface tilts α_1 and α_2 (figure 53-84). While the test lens rotates with ωt, the lateral positions x_1, y_1, x_2, y_2 of C_1 and C_2 on the CCD cameras will be recorded sequentially, given by

$$x_1 = 2f'_{\text{coll1}} \tan \alpha_1 \cos(\omega t + \varphi_1) \tag{53-69}$$

$$y_1 = 2f'_{\text{coll1}} \tan a_1 \sin(\omega t + \varphi_1) \tag{53-70}$$

$$x_2 = 2f'_{\text{coll2}} \tan a_2 \cos(\omega t + \varphi_2) \tag{53-71}$$

$$y_2 = 2f'_{\text{coll2}} \tan a_2 \sin(\omega t + \varphi_2) \tag{53-72}$$

where φ_1 and φ_2 denote the azimuthal surface tilt directions and f'_{coll1} and f'_{coll2} denote the back focal lengths of the autocollimators 1 and 2, respectively. From the data set (53-69)–(53-72) the unknown tilts and azimuths can be determined and the lens element as well as the lens support can be adjusted until the circular movements of the reflections C_1 and C_2 have shrunk to steady spots. C_1 can be used to align the lens support, the correct alignment being indicated by the minimized circle. C_2 is used to tilt the lens on the support to its centered position. In this position, the remaining decenter l of the lens edge and its azimuthal direction φ_l can be measured by the length gauge according to (53-67).

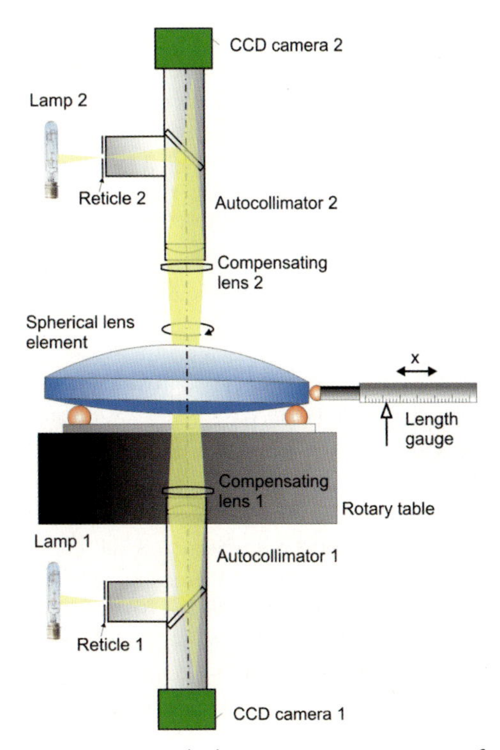

Figure 53-83: Method to measure centering errors of a lens component or subsystem in reflection using two electronic autocollimators and two appropriate compensating lenses to detect lens decenter and tilt and a length gauge to detect decenter of the lens edge.

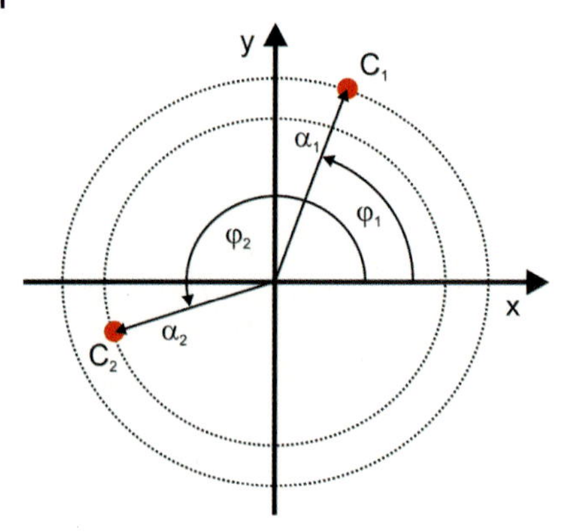

Figure 53-84: Position of the autocollimator reflections C_1 and C_2 during rotation of a tilted lens element on a decentered lens support (see text).

Centering Metrology for Aspherical Lens Elements

The sections above are related to the outer centering of lens elements or subsystems, in which the direction of their axis of symmetry and the centering of their edge cylinders are determined. In this section we describe the technique needed to measure the inner and outer centering of aspherical lens elements [53-94]–[53-97].

To determine the inner centering of an aspherical lens element, it is necessary to measure the distance of the spherical surface's center point from the symmetry axis of the asphere. If the element has two aspherical surfaces, then the relative inclination and separation of their axes must be measured.

To determine the rotation axis of an asphere it is not sufficient to measure the sagittal height of the surface or its inclination in a small surface region while the lens is rotating. Since the tilt and decenter of the surface must be clearly distinguished, it is necessary at least to measure the asphere at two different points, where the difference between the related tangential radii is large. Figure 53-85 shows a configuration in which two autocollimators are pointing at different regions of the surface, which represent areas, where the separation of the surface normals' cross-points with the symmetry axis is a maximum. Compensating lenses are added to adapt to the local curvature of the surface. Autocollimator 2 points towards a region where the local curvature is astigmatic. For strong aspheres an astigmatic compensating lens might be necessary to record a clear spot of the reflected signal.

An alternative method is shown in figure 53-86, in which two chromatically coded confocal sensors are used to detect the relative surface distance along the surface normal at two different points. While the autocollimator measures variations in the local surface angle, the confocal sensor measures variations in the local surface topometry.

During the rotation of the aspherical lens the centering of two spots on the asphere's symmetry axis can be detected by both methods, giving information about the inclination of the symmetry axis and its position relative to the rotary table axis.

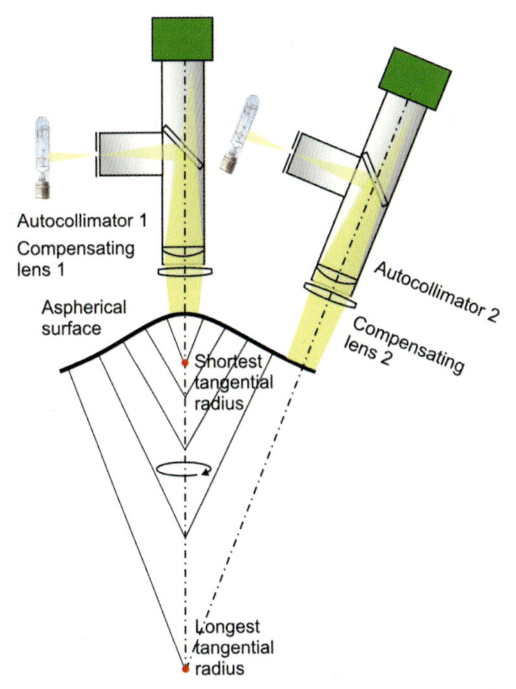

Figure 53-85: Method used to measure centering errors of an aspherical surface during reflection, using two electronic autocollimators with appropriate compensating lenses to detect the inclination and decenter of the symmetry axis.

For further improvement a surface-covering method can be used to replace the two autocollimators. Figure 53-87 shows a setup in which the aspherical surface is observed by a surface-covering laser interferometer. The interferometer is equipped with a diverging transmission sphere and a computer-generated hologram (CGH) as a compensating element. The opposite spherical surface is observed by an electronic autocollimator. As the lens element is rotated in discrete, well-defined positions, wavefronts from the aspherical surface and the relative tilt of the spherical surface are recorded. The individual wavefronts carry information about the axis position relative to the optical axis of the interferometer. Their evaluation in relation to the tilt of the sperical surface leads to the inner centering of the lens element.

Once the inner centering is determined and found to be in tolerance, the outer centering can be measured by means of a length gauge as described in the sections above.

To measure elements where both surfaces are aspherical, two by two autocollimators or two surface-covering instruments are needed to detect the relative inclination and separation of both axes. In most cases the use of two interferometers is beyond the user's technical and financial potential.

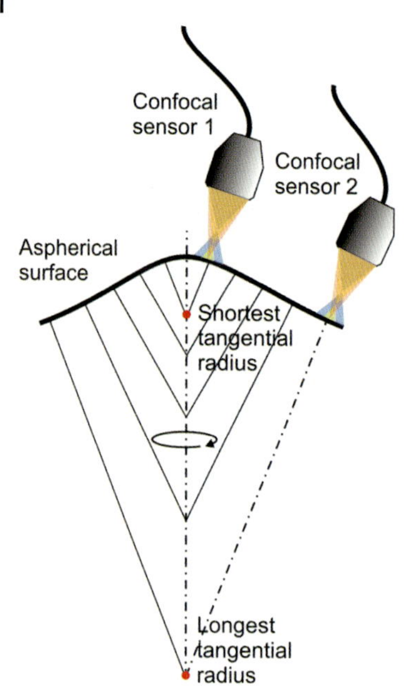

Figure 53-86: Method used to measure centering errors of an aspherical surface during reflection using two chromatically coded confocal sensors to detect the inclination and decenter of the symmetry axis.

Critical points in centering metrology are as follows.

1. **Rotary table**
   The quality of the measurement strongly depends on the quality of the rotary table. If the table's axis decenters or tilts in a statistical way during a measurement, effects from the table cannot be distinguished from those of the sample surfaces. It is therefore mandatory to select a table with tolerances for tilt and decenter less than 1/10 of the narrowest sample tolerances to be detected.
2. **Autocollimators and sensors**
   All collimators and sensors must have sensitivities of better than 1/10 of the narrowest sample tolerances to be detected. Their alignment is not critical when only the relative movement of the surface part is observed and the sample is aligned until the detected signals are constant. Length gauges have sensitivities from a few nanometers up to 1 μm, depending on the selected sensor type. Autocollimators have resolutions of below 1 arcsec with fields of view of up to 2°.
3. **Contacting mechanical sensors**
   For small lens elements or elements where scratches and dents are critical, contacting sensors must be avoided to prevent unwanted misalignment or damage of the optical surfaces. Using optical or capacitive position sensors avoids direct contact.

4. **Azimuth detection**
   For some applications the knowledge of the sample's decenterings relative to a given lens coordinate is important. In this case the centering has to be measured relative to a fiducial at the edge of the element, marking its azimuthal zero position. The fiducial then has to be detected by the length gauge, making it necessary to provide an appropriate notch at the edge of the element.

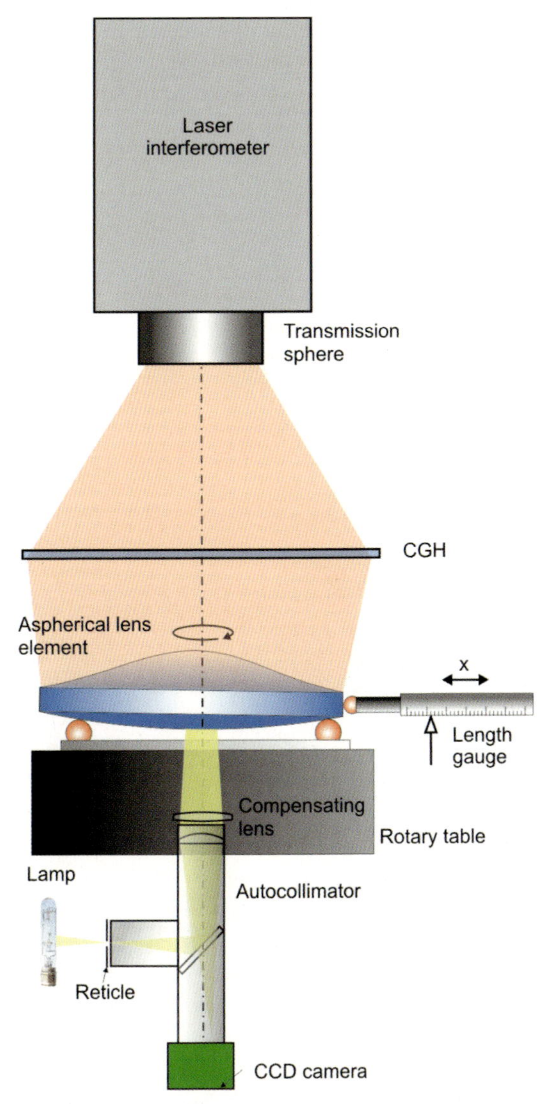

Figure 53-87: Method used to measure centering errors of an aspherical lens component during reflection using an interferometer with transmission sphere and CGH and an electronic autocollimator with an appropriate compensating lens to detect the inner centering of the element. The length gauge is used to detect the decenter of the lens edge.

53.6
Diameter and Chamfer

53.6.1
Basics

The diameter of a round lens element is the length of any straight line segment that passes through the center of the best-fitting circle through the outer edge of the element. Since the element's edge deviates from a perfect circle, there is a largest and a shortest diameter to be tolerated when the element is manufactured (figure 53-88).

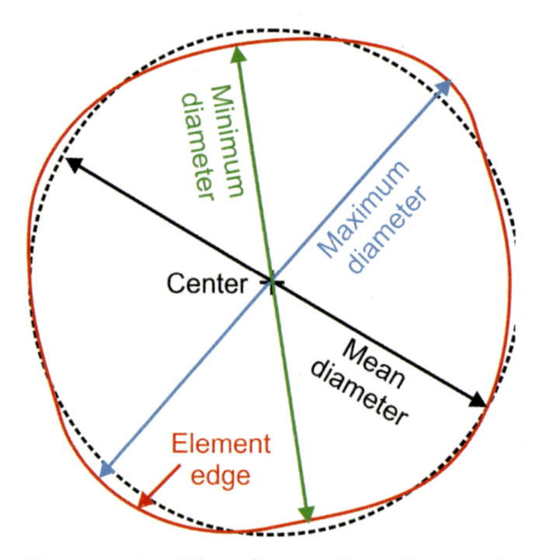

Figure 53-88: Edge of a round lens element showing its best-fitting circle and its mean, minimum and maximum diameter.

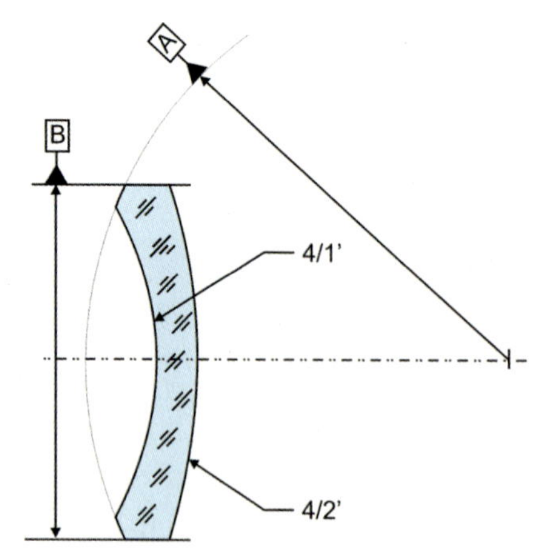

Figure 53-89: Radius chamfer as a third functional surface at the edge of a lens element.

A chamfer is a beveled edge connecting two surfaces. Usually lens elements are chamfered at their edges to prevent chipping. In some cases chamfers are also needed to provide a third functional surface used for mounting and centering; this is called a radius chamfer. In most cases the chamfer surface is then spherical and must be specified on a technical drawing including radius and centering tolerances (figure 53-89).

53.6.2
Metrology

The standard method of measuring the diameter of a round optical element is by use of a mechanical caliper (figure 53-90). The element is rotated to measure a variety of diameters so that the maximum and minimum diameters can be found. The usual accuracy of 0.05 mm is sufficient for most cases.

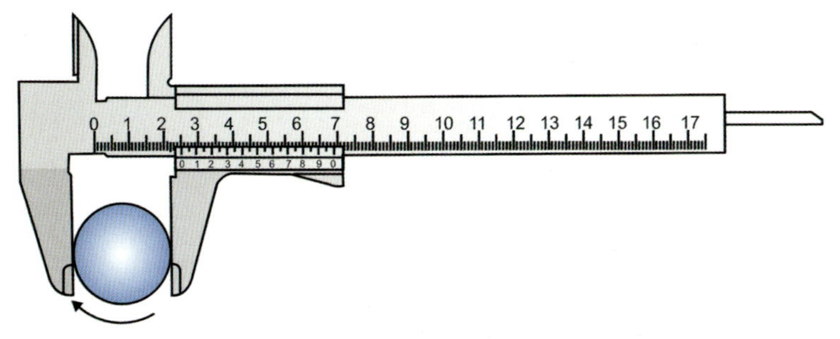

Figure 53-90: Use of a caliper to measure the diameter of a round optical component in different orientations.

In cases where higher precision is necessary, the use of a rotary table is adequate (figure 53-91). The component under test is mounted and centered on top of the table. While the component rotates, its edge is measured at two opposite sides simultaneously and the diameter variation as a function of azimuthal angle is found. Neither a residual mount decenter nor an inaccurate rotary table will influence the measurement. For small lenses the mechanical length gauges should be exchanged by non-contacting optical gauges. The gauge position must be calibrated by means of a well-known reference component to give the absolute diameter; otherwise only the diameter variations are detected. The method is capable of measuring diameters with accuracies to 1 μm.

The metrology for radius chamfers is equivalent to the centering measurement procedure. The lens element under test is mounted on a rotary stage, where the centering of its optical surfaces is inspected by autocollimators, interferometers or contactless or tactile length gauges. An additional length gauge is necessary to control the centering of the chamfer with respect to the optical axis of the lens element under test. Figure 53-92 shows an arrangement in which two autocollimators measure the centering of the optical surfaces while the lens is rotated on the rotary table.

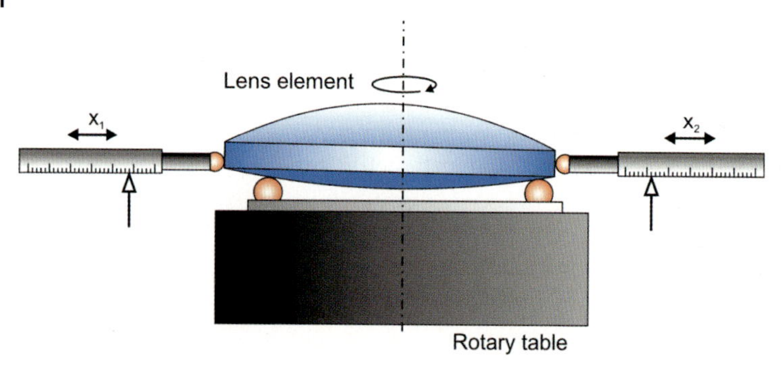

Figure 53-91: Measurement of the diameter of a round optical component in different azimuthal positions be means of a rotary table and two length gauges in opposing positions.

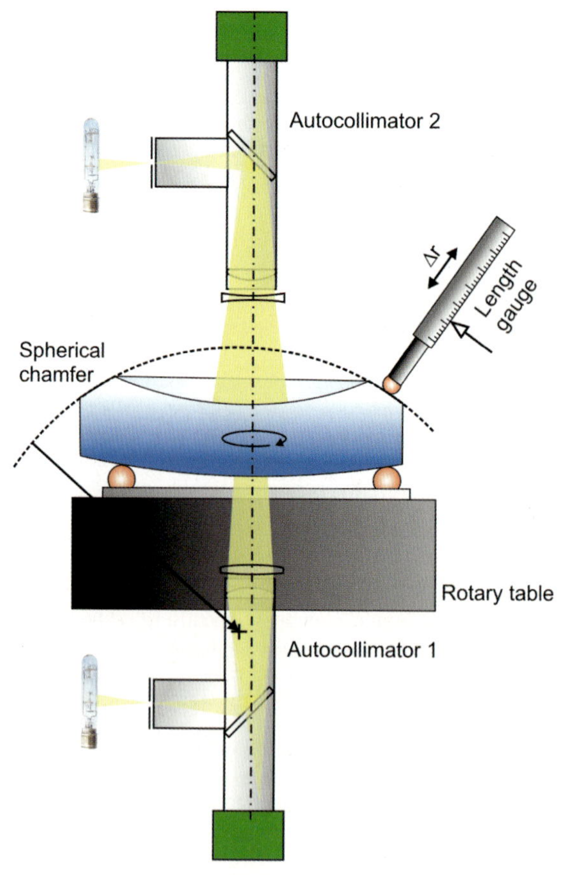

Figure 53-92: Method for measuring the centering error of a spherical chamfer using two autocollimators to control the centering of the optical surfaces and a length gauge to measure the centering of the chamfer simultaneously.

An additional tactile length gauge simultaneously measures the distance of a chamfer surface point in the normal direction as a function of the rotational position. The center point of the radius chamfer can then be calculated with respect to the rotation axis. Depending on the accuracy of the rotary table as well as the resolution of the autocollimators and the length gauge centering, accuracies to 1 μm can be achieved.

53.7 Literature

53-1 ISO 10110-5, Optics and optical instruments– Preparation of drawings for optical elements and systems – Part 5: Surface form tolerances (1996).

53-2 ISO 10110-6, Optics and optical instruments– Preparation of drawings for optical elements and systems – Part 6: Centering tolerances (1996).

53-3 M. Bass (Editor), Handbook of Optics, Vol. II, Devices, Measurements, & Properties, 2nd Edition (McGraw-Hill Inc., New York,1995).

53-4 D. Malacara (Editor), Optical Shop Testing, 3rd Edition (John Wiley & Sons, Inc., Hoboken, New Jersey, 2007).

53-5 G. Molesini, R. Regini and A. Capecchi, Measurement of radius of curvature with contact spherometers, Optik, vol. 82, no. 2, pp. 73–78 (1989).

53-6 L. A. Selberg, Radius measurement by interferometry, Optical Engineering, vol. 31, no. 9, pp. 1961– 6 (1992).

53-7 S. Y. El-Zaiat, L. Missotten and J. Engelen, Radius of curvature measurement by Twyman–Green's interferometer, Optics and Lasers in Engineering, vol. 15, no. 3, pp. 203–8 (1991).

53-8 S. Y. El-Zaiat, J. Engelen, L. Missotten, N. Barakat and M. Emarah, Comparative radius of curvature measurement by non-contact Newton's rings, Measurement Science & Technology, vol. 2, no. 8, pp. 780–4 (1991).

53-9 Y. Yang, Y. Zhou and H. Wu, Digitalization measurement of large radius of curvature by non-contact Newton's rings, Proceedings of the SPIE, vol. 4231, pp. 346–51 (2000).

53-10 J. C. Bhattacharya and A. K. Aggarwal, Finite fringe moire deflectometry for the measurement of radius of curvature: an alternative approach, Optics and Laser Technology, vol. 25, no. 3, pp. 167–9 (1993).

53-11 Hua Shiqun and Luo Ying, Measuring the radius of curvature of a spherical surface with diffraction method, Proceedings of the SPIE vol. 7375, 73754T (2008).

53-12 M. Spiridonov and D. Toebaert, Simple yet accurate noncontact device for measuring the radius of curvature of a spherical mirror, Applied Optics, vol. 45, no. 26, pp. 6805–11 (2006).

53-13 Z. Chunyu, R. Zehnder and J. H. Burge, Measuring the radius of curvature of a spherical mirror with an interferometer and a laser tracker, Optical Engineering, vol. 44, no. 9, pp. 90506-1–3 (2005).

53-14 E. Abbe, Gesammelte Abhandlungen, Zweiter Band, Wissenschaftliche Abhandlungen aus verschiedenen Gebieten. Patentschriften. Gedächtnisreden, Georg Olms Verlag, Hildesheim, Zürich, New York (1989).

53-15 G. Peggs, The dimensional metrology of large refractive optical components using non-optical techniques, Proceedings of the SPIE - The International Society for Optical Engineering, vol. 4411, pp. 171–6 (2002).

53-16 R. Wilhelm, A. Courteville, F. Garcia and F. de Vecchi, On-axis, non-contact measurement of glass thicknesses and airgaps in optical systems with submicron accuracy, Proceedings of the SPIE - The International Society for Optical Engineering, vol. 6616, 66163P (12 pp.) (2007).

53-17 R. Wilhelm and A. Courteville, Dimensional metrology for the fabrication of imaging optics using a high accuracy low coherence interferometer, Optical Measurement Systems for Industrial Inspection, Proceedings of SPIE 5856 (2005).

53-18 A. Courteville and R. Wilhelm, Contact-free on-axis metrology for the fabrication and testing of complex optical systems, Optical Fabrication, Testing, and Metrology II, Proceedings of SPIE 5965 (2005).
53-19 K. G. Larkin, Efficient nonlinear algorithm for envelope detection in white light interferometry, J. Opt. Soc. Am. A 13, pp. 832–43 (1996).
53-20 M. Kunkel and J. Schulze, Noncontact measurement of central lens thickness, Glass Science and Technology, vol. 78, no. 5, pp. 245–7 (2005).
53-21 G. Freud, Orthogonale Polynome, Birkhauser Verlag, Basel (1969).
53-22 A. Maréchal, Influence de faibles aberrations geometriques sur le maximum central de la tache de diffraction regles de correction, Revue d'Optique 26, 257 (1947).
53-23 P. Solin, K. Segeth and I. Dolezel, Higher-Order Finite Element Methods, (Chapman & Hall/CRC Press, 2003).
53-24 P. E. Lewis and J. P. Ward, The Finite Element Method: Principles and Applications (Modern Applications of Mathematics), (Addison Wesley Publishing Company, 1991).
53-25 P. R. Yoder, Mounting Optics in Optical Instruments, SPIE Press, Bellingham (2004).
53-26 B. Dörband, High-precision testing of optical components, OSA Technical Digest, International Optical Design Conference, Hawaii, p. 225–28 (1998).
53-27 B. Dörband, H. Müller and G. Seitz, High Precision Interferometric Measurements of Lens Elements, Proceedings FRINGE 97, (Akademie Verlag, Berlin, 1997).
53-28 M. Küchel, Interferometer for measuring optical phase differences, US Patent No. US4872755A (1988).
53-29 M. Bray, Stitching interferometer for large plano optics using a standard interferometer, Optical Manufacturing and Testing II, SPIE, vol. 3134, San Diego (1997).
53-30 T. Yoshizawa, Handbook of Optical Metrology, (CRC Press, Boca Raton, London, New York, 2009).
53-31 D. Boebel, B. Packross and H. J. Tiziani, Phase shifting in an oblique incidence interferometer, Optical Engineering, vol. 30, no. 12, pp. 1910–14 (1991).
53-32 P. Murphy, G. Forbes, J. Fleig, P. Dumas and M. Tricard, Stitching interferometry: A flexible solution for surface metrology, Optics & Photonics News 14, pp. 38–43 (2003).
53-33 ISO 10110-12, Optics and optical instruments– Preparation of drawings for optical elements and systems – Part 12: Aspheric surfaces (1997).
53-34 J. C. Wyant, Interferometric testing of aspheric surfaces, Proceedings of the SPIE, vol. 816, pp. 19–39 (1987).
53-35 P. L. Ruben, Refractive null correctors for aspheric surfaces, Appl. Opt., vol. 15, no. 12, pp. 3080–83 (1976).
53-36 P. Nageswara Rao, K. K. Banerjee and H. S. Singh, Null lenses for testing surface figure of aspheric mirrors, Journal of Optics, vol. 16, no. 3, pp. 58–66 (1987).
53-37 Sandri, P. and Pecchioli, E., The optical design of a null compensator for testing an aspheric lens, Atti della Fondazione Giorgio Ronchi, vol. 61, no. 2, pp. 267–71 (2006).
53-38 M. V. Mantravadi, V. Kumar and R. J. von Handorf, Aspheric testing using null mirrors, Proceedings of the SPIE, vol. 1332, pt. 1, pp. 107–14 (1990)
53-39 J. C. Wyant and P. K. O'Neill, Computer generated hologram: Null lens test of aspheric wavefronts, Appl. Opt., vol. 13, no. 12, pp. 2762–5 (1974)
53-40 J. Schwider, J. Grzanna, R. Spolaczyk and R. Burow, Testing aspherics in reflected light using blazed synthetic holograms, Optica Acta, vol. 27, no. 5, pp. 683–98 (1980).
53-41 Yu Tsu-liang, Wu Min-shen and Chin Kuo-fan, Two methods for optical testing aspherical surface by using a computer-generated hologram, Chinese Journal of Scientific Instruments, vol. 2, no. 4, pp. 64–70 (1981).
53-42 D. C. Smith, Testing diamond turned aspheric optics using computer-generated holographic (CGH) interferometry, Proceedings of the SPIE, vol. 306, pp. 112–21 (1981).
53-43 B. Dörband and H. J. Tiziani, Testing aspheric surfaces with computer-generated holograms: analysis of adjustment and shape errors, Appl. Opt., vol. 24, no. 16, pp. 2604–11 (1985).
53-44 S. M. Arnold, A. K. Jain, An interferometer for testing of general aspherics using com-

puter generated holograms, Proceedings of the SPIE, vol. 1396, pp. 473–80 (1991)

53-45 J. H. Burge and D. S. Anderson, Full-aperture interferometric test of convex secondary mirrors using holographic test plates, Proc. SPIE 2199, pp. 181–92 (1994).

53-46 S. M. Arnold and R. Kestner, Verification and certification of CGH aspheric nulls, Proceedings of the SPIE, vol. 2536, pp. 117–26 (1995).

53-47 J. H. Burge, Applications of computer-generated holograms for interferometric measurement of large aspheric optics, Proceedings of the SPIE, vol. 2576, pp. 258–69 (1995).

53-48 S. M. Arnold, A. P. Stuart and L. Koudelka, L, CGH-LUPI interferometer for aspheric figure metrology, Proceedings of the SPIE, vol. 3134, pp. 390–7 (1997).

53-49 E. O. Curatu, Min Wang, Tolerancing and testing of CGH aspheric nulls, Proceedings of the SPIE, vol. 3782, pp. 591–600 (1999).

53-50 H. J. Tiziani, S. Reichelt, C. Pruss, M. Rocktaschel and U. Hofbauer, Testing of aspheric surfaces, Proceedings of the SPIE, vol. 4440, pp. 109–19 (2001).

53-51 N. Lindlein, Analysis of the disturbing diffraction orders of computer-generated holograms used for testing optical aspherics, Appl. Opt. vol. 40, no. 16, pp. 2698–708 (2001).

53-52 Tae-hee Kim and Soon Cheol Choi, Measurement of highly aspherical surface using computer generated holograms, Journal of the Optical Society of Korea, vol. 6, no. 2, pp. 21–6 (2002).

53-53 Wang Chunxia, Wu Fan, Wu Shibin, Du Chunlei and Hou Desheng, Research on testing the null corrector using the computer-generated hologram, Acta Photonica Sinica, vol. 32, no. 5, pp. 592–4 (2003).

53-54 C. Pruss, S. Reichelt, H. J. Tiziani and W. Osten, Computer-generated holograms in interferometric testing, Optical Engineering, vol. 43, no. 11, pp. 2534–40 (2004).

53-55 R. Zehnder, J. H. Burge and Chunyu Zhao, Use of computer generated holograms for alignment of complex null correctors, Proceedings of the SPIE, vol. 6273, pp. 62732S-1–8 (2006).

53-56 A. G. Poleshchuk, V. P. Korolkov, R. K. Nasyrov and J.-M. Asfour, Computer generated holograms: fabrication and application for precision optical testing, Proceedings of the SPIE, vol.7102, pp. 710206–15 (2008).

53-57 A. E. Lowman and J. E. Greivenkamp, Modeling an interferometer for non-null testing of aspheres, Proceedings of the SPIE, vol. 2536, pp. 139–47 (1995).

53-58 R. O. Gappinger, J. E. Greivenkamp, Non-null interferometer for measurement of aspheric transmitted wavefronts, Proceedings of the SPIE, vol. 5180, no.1, pp. 307-18 (2004).

53-59 J. J. Sullivan and J. E. Greivenkamp, Design of partial nulls for testing of fast aspheric surfaces, Proceedings of the SPIE, vol. 6671, no.1, pp. 1–8 (2007).

53-60 Q. Hao and Q. D. Zhu, Aspheric surface testing using a partial compensation lens, Key Engineering Materials, vol. 381–82, pp. 263–6 (2008).

53-61 Dong Liu, Yongying Yang, Yongjie Luo, Chao Tian, Yibing Shen and Yongmo Zhuo, Non-null interferometric aspheric testing with partial null lens and reverse optimization, Proceedings of the SPIE, vol. 7426, 74260M (7 pp.) (2009).

53-62 Dong Liu, Yongying Yang, Chao Tian, Lin Wang and Yongmo Zhuo, Non-null interferometric system for general aspheric test, Proceedings of the SPIE, vol. 7283, 728305 (6 pp.) (2009).

53-63 B. Dörband and T. Pesler, Prüfung asphärischer Kontaktlinsenrückflächen mit dem Interferometer DIRECT 100 (Teil I), DOZ 8, p. 108–13 (1994).

53-64 B. Dörband and T. Pesler, Prüfung asphärischer Kontaktlinsenrückflächen mit dem Interferometer DIRECT 100 (Teil II), DOZ 9), p. 124 – 28 (1994).

53-65 N. Gardner and A. Davies, Retrace error evaluation on a figure-measuring interferometer, Proceedings of the SPIE, vol. 5869, 58690V (8 pp.) (2005).

53-66 P. Murphy, J. Fleig, G. Forbes, D. Miladinovic, G. DeVries and S. O'Donohue, Subaperture stitching interferometry for testing mild aspheres, Proceedings of the SPIE, vol. 6293, pp. 62930J-1–10 (2006).

53-67 Wang Xiao-kun, Zheng Li-gong, Zhang Bin-zhi, Li Rui-Gang, Zhang Zhong-yu, Zhang Feng and Zhang Xue-jun,Testing of large aspheric surfaces by subaperture stitching interferometry, Appl. Opt., vol. 30, no. 2, pp. 273–8 (2009).

53-68 E. Garbusi, C. Pruss and W. Osten, Interferometer for precise and flexible asphere testing, Optics Letters, vol. 33, no. 24, pp. 2973–5 (2008).

53-69 Chunyu Zhao and J.H. Burge, Stitching of off-axis sub-aperture null measurements of an aspheric surface, Proceedings of the SPIE, vol.7063, 706316 (7 pp.) (2008).

53-70 P. Murphy, G. DeVries, J. Fleig, G. Forbes, A. Kulawiec and D. Miladinovic, Measurement of high-departure aspheric surfaces using subaperture stitching with variable null optics, Proceedings of the SPIE, vol.7426, 74260P (9 pp.) (2009).

53-71 E. Abbe, Verfahren, sphäroide Flächen zu prüfen und Abweichungen von der vorgeschriebenen Gestalt nach Lage und Größe zu bestimmen, German Patent No. 131536, Klasse 42h (1899).

53-72 Ying-Moh Liu, G. N. Lawrence and C. L. Koliopoulos, Subaperture testing of aspheres with annular zones, Appl. Opt., vol. 27, no. 21, pp. 4504–13 (1988).

53-73 M. Melozzi, L. Pezzati and A. Mazzoni, Testing aspheric surfaces using multiple annular interferograms, Optical Engineering, vol. 32, no. 5, pp. 1073–9 (1993).

53-74 M. J. Tronolone, J. F. Fleig, C. H. Huang and J. H. Bruning, Method of Testing Aspherical Optical Surfaces with an Interferometer, U.S. Patent No. US5416586A (1993).

53-75 M. Küchel, Scanning interferometer for aspheric surfaces and wavefronts, US Patent Nos. US2003002049A, US2003043385A (2003).

53-76 M. Küchel, Absolute Measurement of Rotationally Symmetrical Aspheric Surfaces, OSA – Optical Fabrication and Testing (2006).

53-77 Xiao-kun Wang, Li-hui Wang, Li-gong Zheng, Wei-jie Deng and Xue-jun Zhang, Annular sub-aperture stitching interferometry for testing of large aspherical surfaces, Proceedings of the SPIE, vol.6624, pp. 66240A-1–8 (2007).

53-78 M. Küchel, Scanning interferometric methods and apparatus for measuring aspheric surfaces and wavefronts, US Patent No. US2008068613A (2008).

53-79 Der-Shen Wan and Ding-Tin Lin, Profile measurements of cylindrical surfaces, Appl. Opt., vol.32, no.7, pp. 1060–4 (1993).

53-80 S. Brinkmann, T. Dresel, R. Schreiner and J. Schwider, Axicon-type test interferometer for cylindrical surfaces, Optik, vol. 102, no. 3, pp. 106–10 (1996).

53-81 S. Brinkmann, R. Schreiner, T. Dresel and J. Schwider, Interferometric testing of plane and cylindrical workpieces with computer-generated holograms, Optical Engineering, vol. 37, no. 9, pp. 2506–11 (1998).

53-82 L. W. Alvarez, Two-element variable-power spherical lens, US Patent 3,305,294 (1967).

53-83 I. M. Barton, S. N. Dixit, L. J. Summers, K. Avicola and J. Wilhelmsen, Diffractive Alvarez lens, Opt. Lett. 25, 1–3 (2000).

53-84 J. Wallace, Alvarez lens enters the real world, Laser Focus World, Vol. 36 Issue 3, p15–16 (2000).

53-85 Y. Ohsaki, K. Saitoh and A. Suzuki, Interferometer and interference measurement method, US Patent No. US2002176090A (2002).

53-86 M. C. Knauer, J. Kaminski and G. Häusler, Phase measuring deflectometry: a new approach to measure specular free-form surfaces, Proceedings of the SPIE, vol.5457, no.1, pp. 366–76 (2004).

53-87 D. R. Herriott, J. H. Bruning and A. D. White, A zoom autocollimator for centering optical surfaces, Optical Soc. America, pp. 14, 88 pp. (1972).

53-88 V. Guyenot, Opto-electrical evaluation of reflex images in lens centering, Feingeraetetechnik, vol. 30, no. 3, pp. 121–3 (1981).

53-89 A. Zaltz and D. Christo, Methods for the control of centering error in the fabrication and assembly of optical elements, Proceedings of the SPIE, vol. 330, pp. 39–48 (1 982).

53-90 B. A. Chunin, F. I. Kalugin, M. I. Bakaev and Yu. B. Popov'Dyumin, Centering of small optical elements having an aspherical surface, Soviet Journal of Optical Technology, vol. 50, no. 11, pp. 730 (1983).

53-91 V. N. Senatorov, S. G. Korolev, A. A. Kuri'ko, V. S. Cherednik and N. S. Seleznez, Apparatus for centering lenses to be cemented, Soviet Journal of Optical Technology, vol. 57, no. 10, pp. 637–8 (1990).

53-92 J. Heinisch, E. Dumitrescu and S. Krey, Novel technique for measurement of centration errors of complex completely mounted multi-element objective lenses, Proceedings of the SPIE, vol. 6288, pp. 628810-1–7 (2006).

53-93 Xiang Li, Zhao Liping and Fang Zhong Ping, Inspection of misalignment factors in lens assembly, Proceedings of the SPIE, vol. 7390, 739006 (12 pp.) (2009).

53-94 Der-Shen Wan, Decenter and defocus for testing aspheric surfaces, Proceedings of the SPIE, vol. 1776, pp. 140–50 (1992).

53-95 R. H. Wilson, R. C. Brost, D. R. Strip, R. J. Sudol, R. N. Youngworth, P. O. McLaughlin, Considerations for tolerancing aspheric optical components, Appl. Opt., vol. 43, no. 1, pp. 57–66 (2004).

53-96 U. Birnbaum, H. Bernitzki, O. Falkenstorfer, H. Lauth, R. Schreiner and T. Waak, Manufacturing of high-precision aspheres, Proceedings of the SPIE, vol. 6149, pp. 61490H-1–-7 (2006).

53-97 Ma Zhen, Li Ying-cai, Fan Xue-wu, Chen Rong-li and Duan Xue-ting, Study on optical centering of aspheric mirror by interferometry, Acta Photonica Sinica, vol. 37, no. 7, pp. 1455–8 (2008).

54
Testing Texture and Imperfections of Optical Surfaces

Handbook of Optical Systems: Vol. 5. Metrology of Optical Components and Systems. First Edition.
Edited by Herbert Gross.

54.1
Specifications

The total topography of an optical surface can be classified by *form* and *finish*. In the previous chapters we have described in detail how to measure the geometry of an optical component including the form deviations of the optical surfaces. In this chapter we will cover test methods and devices for the measurement of finish characteristics. The finish of a surface can be further classified by *texture* and *imperfections*, as shown in figure 54-1.

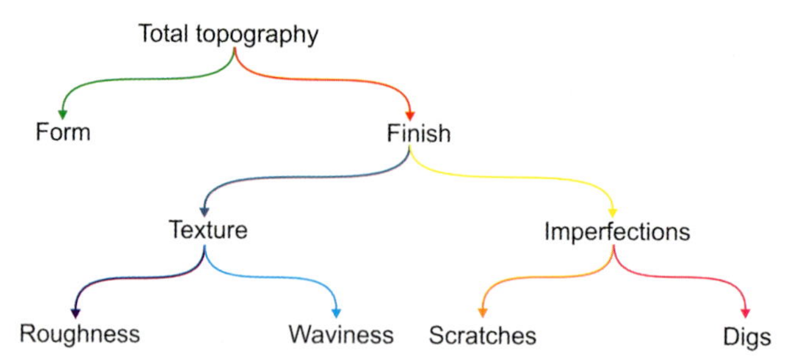

Figure 54-1: Classification of total topography of an optical surface.

The specification of texture and imperfections of an optical surface and their tolerances are part of a worldwide standard. The International Organization of Standardization (ISO) provides the international standard ISO 10110 "Optics and optical instruments – preparation of drawings for optical elements and systems" with the following parts:

- ISO 10110 – Part 7 Surface imperfection tolerances [54-1].
- ISO 10110 – Part 8 Surface texture [54-2].

This chapter makes use of the definitions given by ISO 10110 to define the basics and the metrology related to texture and imperfections.

54.2
Surface Texture

54.2.1
Basics

We define *surface texture* as the global statistical property relating to the profile of an optical surface. Localized defects, known as surface imperfections, are excluded as these are covered in the following chapter.

In most cases it is assumed that the character and magnitude of the texture in any one area of the surface is similar to that in all other areas of the same surface. In this case a measurement made in one part of the surface may be considered representative of the entire surface. Since the measured surface part is small, long spatial periods associated with form error are not included in the designation of texture.

A *matt surface* is an optical surface for which the height variation of the surface texture is not considerably smaller than the wavelength of the visible light. Matt surfaces are usually produced by brittle grinding of glass or other dielectric material, or by etching.

A *specular surface* is an optical surface for which the height variation of the surface texture is considerably smaller than the wavelength of the visible light. A specular surface is usually produced by polishing or moulding and is called optically smooth.

Microdefects are small irregularities (generally less than 1 μm in size) in a specular surface. They are pits remaining after an incomplete polish or may be caused by mishandling and contamination during polishing. Microdefects are not considered to be surface imperfections as described in the next chapter, because they are reasonably uniformly distributed over the surface and thus have a global characteristic associated with texture.

Surface texture specifications are applicable to matt surfaces, specular surfaces and microdefects. Depending on the application of a surface and the magnitude of the surface height variation, one or more methods outlined below may be appropriate for the numerical description of surface texture.

Specification of Optical Surface Texture

There are three main quantities specifying the textural quality of an optical surface:

1. Root-mean-square (r.m.s.) surface roughness R_q, depending on the sampling length.
2. Density of microdefects.
3. Power spectral density function (PSD).

We will define these in the following. For matt surfaces the specification of microdefects and PSD is not usually very useful.

Surface Roughness R_q

Texture quantities are derived from the amplitude parameters, which are based on the vertical deviations z_i of a roughness profile from the mean line. It is assumed that the data set is extracted from the global topography, i.e., form deviations are removed. The z_i values are the vertical distances from the mean line $\bar{z}$ to the ith data point. Height is assumed to be positive in the upward direction, away from the sample. Figure 54-2 shows an example.

If we assume a sample of N profile values z_i distributed at equal spaces along a line at the x-coordinates x_i, R_q is calculated from

$$R_q = \sqrt{\frac{1}{N}\sum_{i=1}^{N}(z_i - \bar{z})^2} \tag{54-1}$$

R_q is specified together with the indication of the lower and upper limits of the sampling length. $\Delta x_{\text{low}} = x_i - x_{i-1}$ and $\Delta x_{\text{up}} = x_N - x_1$.

In some cases, functional requirements may dictate a roughness criterion other than R_q. Examples are the arithmetic average R_a calculated from

$$R_a = \frac{1}{N}\sum_{i=1}^{N} |z_i - \bar{z}| \qquad (54\text{-}2)$$

or the maximum height or peak-to-valley (PV) value R_t calculated from

$$R_t = \max(z_i) - \min(z_i) \qquad (54\text{-}3)$$

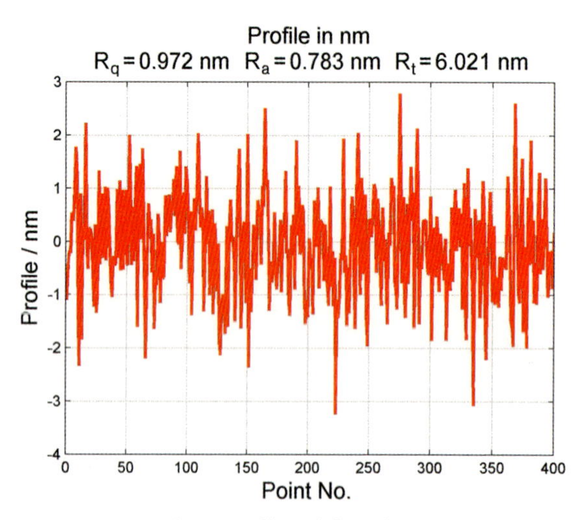

Figure 54-2: Roughness profile with form deviations and piston removed. Calculated roughness parameters are: $R_q = 0.972$ nm, $R_a = 0.783$ nm, $R_t = 6.021$ nm.

By convention every one-dimensional (1D) roughness parameter is a capital *R* followed by an additional character specifying the formula applied to a 1D profile. Different capital letters imply that the formula was applied to a different profile. For example, P specifies an unfiltered 1D raw profile set of data, S specifies a 2D profile data set. If the surface height variations obey certain statistical distribution properties, R_q can be related to the magnitude of the optical scattering (see below).

Microdefects

Microdefects are localized pits in a smooth surface, resulting from incomplete polishing. They are quantified by measuring the profile along a line of 10 mm length and counting the number of pits (figure 54-3). The metrology can be contacting by means of the sharp stylus of a mechanical profilometer, or by using a low-power microscope or optical profilometer. It is assumed that a better polish leads to fewer microdefects.

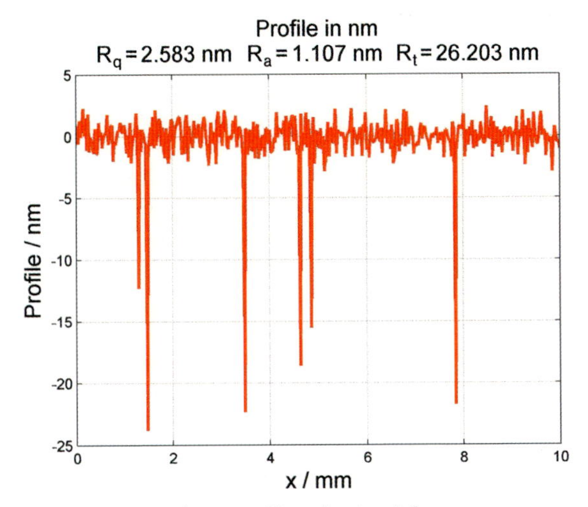

Figure 54-3: Roughness profile with microdefects over a sampling length of 10 mm.

Power Spectral Density (PSD)

The PSD function is the frequency spectrum of the surface roughness measured in inverse length units, for example, μm^{-1}. It is especially useful for specifying super-smooth surfaces used in high-precision optics, allowing a complete description of the surface texture characteristics.

The PSD can be applied to one- or two-dimensional profile data sets. The 2D *power spectral density function* $PSD_2(f_x,f_y)$ is defined as the squared modulus of the Fourier transform of the surface topography $z(x,y)$:

$$\mathrm{PSD}_2(f_x,f_y) = \lim_{L_x,L_y \to \infty} \frac{1}{L_x L_y} \left| \int_{-\frac{L_x}{2}}^{\frac{L_x}{2}} \int_{-\frac{L_y}{2}}^{\frac{L_y}{2}} z(x,y) e^{-i2\pi(f_x x + f_y y)} \, dx dy \right|^2 \tag{54-4}$$

where L_x and L_y are the tangential and sagittal dimensions of the measured surface region and f_x and f_y are the spatial frequency variables corresponding to the x and y coordinates.

The PSD expresses the power of different roughness components in terms of the spatial frequencies f_x and f_y of the surface topography. The units are μm^4 or any appropriate variation such as $\mu m^2 mm^2$ or $nm^2 mm^2$.

In many cases, optical surfaces have structures with polar symmetry due to the type of fabrication process such as grinding, polishing, etching or thin-film deposition. In this case the use of the one-dimensional *2D-isotropical* PSD is useful, which is calculated by averaging the 2D PSD over all azimuthal directions:

$$\mathrm{PSD}_{\mathrm{iso}}(f) = \frac{1}{2\pi} \int_0^{2\pi} \mathrm{PSD}_2(f, \varphi) d\varphi \tag{54-5}$$

with the transformations

$$f = \sqrt{f_x^2 + f_y^2} \quad \text{and} \quad \varphi = \arctan\left(\frac{f_y}{f_x}\right) \tag{54-6}$$

Figures 54-4 (a)–(d) show an example of a measured surface topography (a); and the related 2D power spectral density function (b); as well as the related one-dimensional 2D-isotropical PSD (c). It is often convenient to use the spatial wavelength – the inverse of the spatial frequency – as the abscissa in units of µm or mm. Figure 54-4 (d) shows the 2D-isotropical PSD equivalent to (c) but as a function of the spatial wavelength in mm.

To characterize the roughness of optical surfaces it is often useful to calculate a band limited roughness $R_q^{(f_{min}, f_{max})}$ expressed by

$$R_q^{(f_{min} f_{max})} = \sqrt{2\pi \int_{f_{min}}^{f_{max}} PSD_{iso}(f) f \, df} \tag{54-7}$$

with the spatial frequency band limits f_{min} and f_{max} depending on the spatial frequency range of the instrument and the related application.

The PSD is regarded as a very convenient description of the roughness properties of an optical surface. In practice, surface topography is sampled using a finite number of points. The integral expressions are then replaced by discrete sums, sometimes combined with apodization functions to suppress numerical artifacts [54-3], [54-4] and [54-17].

When the PSD has to be applied to a one-dimensional profile data set $z(x)$, the calculation will be made according to

$$\mathrm{PSD}_1(f_x) = \lim_{L_x \to \infty} \frac{1}{L_x} \left| \int_{-\frac{L_x}{2}}^{\frac{L_x}{2}} z(x) e^{-i2\pi f_x x} dx \right|^2 \tag{54-8}$$

using a one-dimensional Fourier transform. Units of the PSD_1 are μm^3 or compatible length unit^3 combinations.

For an easy analytical description and approximation the PSD_1 can be modeled by

$$\mathrm{PSD}_1(f) = \frac{A}{f^B} \quad \text{for} \quad \frac{1}{1000D} < f < \frac{1}{1000C} \tag{54-9}$$

where

f is the spatial frequency of the roughness, (in μm^{-1}),
B is the power to which the spatial frequency is raised,
C and D are the minimum and maximum spatial periods (sampling lengths) of the measurement, (in mm),
A is a constant, expressed in μm^{3-B}.

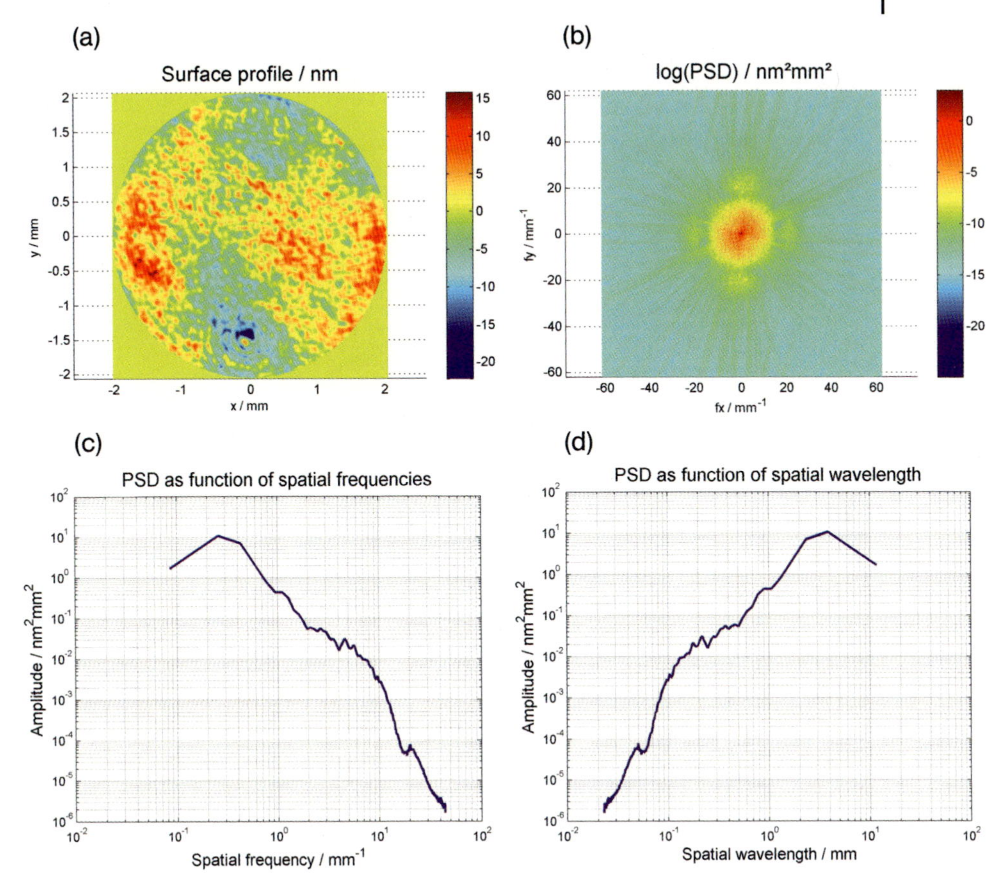

Figure 54-4: (a) The 2D surface profile of a polished surface showing roughness; (b) the related 2D power spectral density function $PSD_2(f_x,f_y)$ (logarithmic scale); (c) the related 2D-isotropical PSD as function of spatial frequency f; (d) the related 2D-isotropical PSD as a function of spatial wavelength f^{-1}.

The value of B should be greater than zero. It has been shown experimentally that most polished surfaces scatter light according to a power law [54-5]–[54-8]. For many real surfaces the value of B is between 1 and 3.

In this way, the surface texture requirement specification may be given by specifying the four values A, B, C and D of (54-9). Figure 54-5 shows some examples.

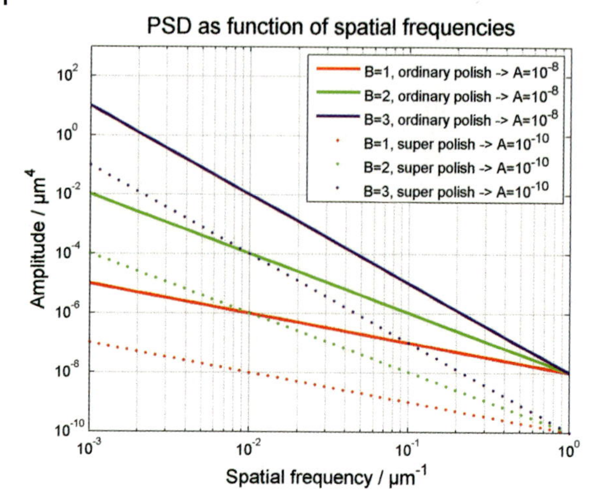

Figure 54-5: Examples of $PSD_1(f)$ with $C = 0.001$ mm and $D = 1$ mm and $B = 1$, 2 and 3 for ordinary polished and super-polished surfaces.

Indication in Drawings

The required quality of polished or matt surfaces is indicated on optical drawings by a triangle, a leader line and an alphanumeric description of the quality [54-9]. Depending on whether the surface is matt (ground) or polished, the letter “G” or “P” is placed above the horizontal line. Figure 54-6 shows an example for a ground surface.

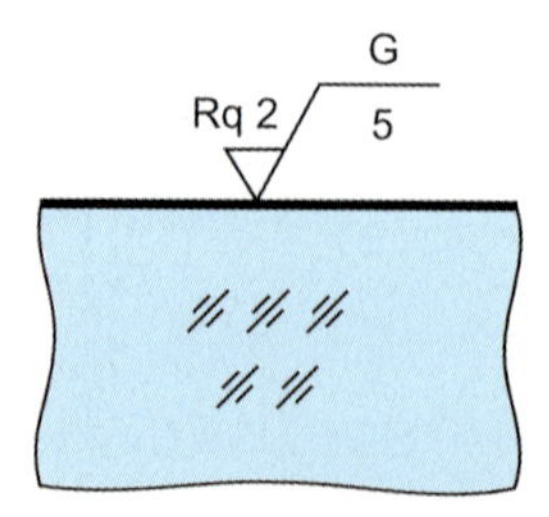

Figure 54-6: Indication for ground surface with $R_q = 2$ µm and minimum sampling length of 5 mm.

The maximum permissible surface roughness R_q max, in micrometers, is indicated above the triangle. When a single value of R_q is given, it represents the upper limit of the surface roughness parameter. If, in addition, the roughness is not permitted to lie below a certain value, a minimum surface roughness value R_q min, is indicated below the maximum value.

If desired, a lower limit of the sampling length in mm may be indicated under the horizontal line. If an upper limit is necessary, it is separated from the lower limit by a slash.

For polished surfaces, the number of allowed microdefects may be indicated by placing a grade number between 1 and 4 to the right of the letter "P". The range of the corresponding permissible number of microdefects is given in table 54-1. If the number is omitted, no specific microdefect specification is required.

Table 54-1: Grade numbers to describe the permissible number of microdefects on a polished surface.

Polishing grade designation	Number N of microdefects per 10 mm of sampling length
P1	$80 \leq N < 400$
P2	$16 \leq N < 80$
P3	$3 \leq N < 16$
P4	$N < 3$

Instead of R_q it is possible to indicate the maximum permissible value of the PSD function by placing the letters PSD and the values for A and B, as defined in (54-9) and separated by a slash, above the triangle in the texture symbol as shown in figure 54-7. The minimum and maximum spatial periods (sampling lengths), C and D, expressed in mm, are placed under the horizontal line and separated by a slash.

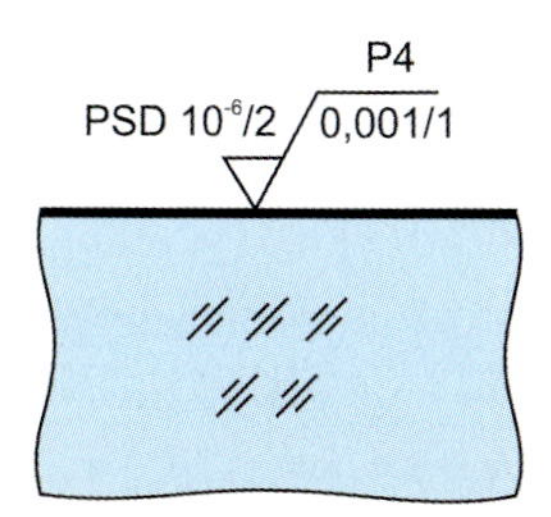

Figure 54-7: Indication for polished surface with < 3 microdefects per 10 mm scan and a PSD $< 10^{-6}/f^2$ (μm^3) between sampling lengths of 0.001 mm and 1 mm.

Specification of Optical Surface Waviness

Surface waviness is the periodic component of the surface texture. It arises most frequently from vibrations of a single-point surface generator such as a diamond turning lathe [54-10]. When Fourier analyzed, the surface profile is dominated by a spatial frequency band between the roughness and form. Such a surface causes unwanted diffraction when used in reflection or transmission, giving rise to multiple images which disturb the signal.

The preferred metric for measuring waviness is, the PSD. Significant deviations from the power law defined in (54-9) at specific frequency bands can be interpreted as waviness. Figure 54-8 gives an example of a smooth surface showing waviness at a specific spatial frequency.

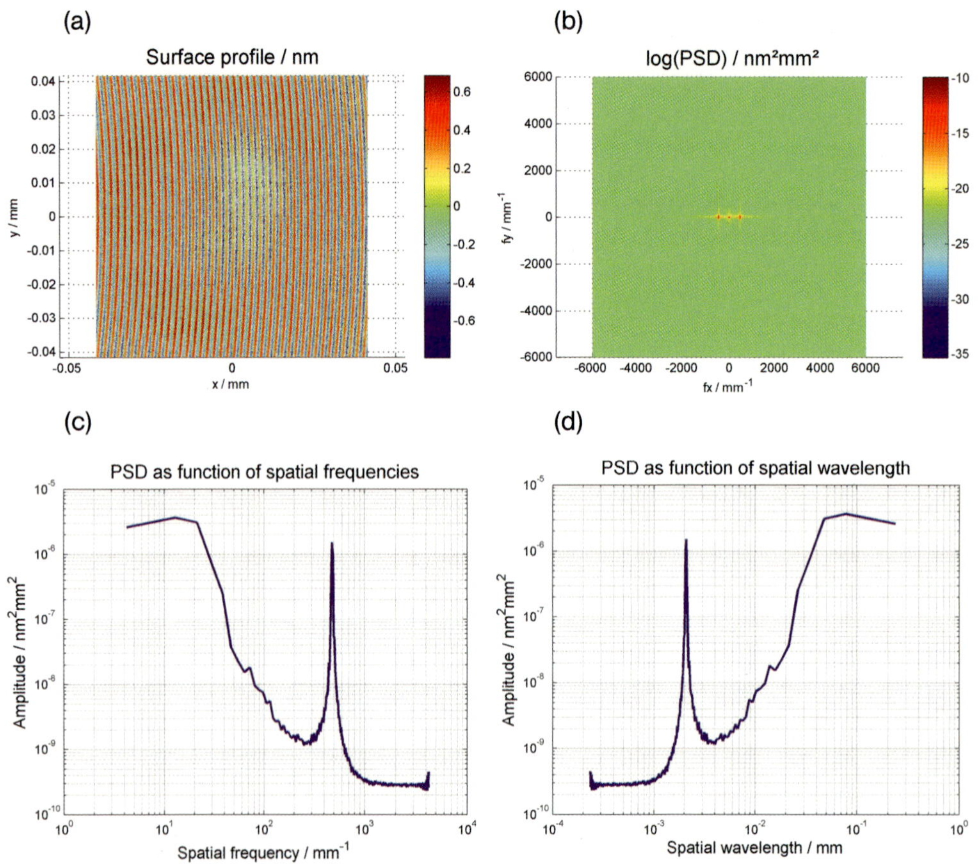

Figure 54-8: (a) A 2D surface profile of a polished surface showing waviness; (b) related 2D power spectral density function $PSD_2(f_x, f_y)$ (logarithmic scale); (c) related 2D-isotropical PSD as a function of spatial frequency f; (d) related 2D-isotropical PSD as a function of the spatial wavelength f^{1}.

Light Scattering by Optical Surfaces

The surface texture and the scattering behavior of an optical surface are strongly related. The texture profile can be described as a Fourier series of sinusoidal waves, as represented by the PSD. In the same way as diffraction at a grating, a certain waviness with spatial frequency f causes scattering at the wavelength λ into the angle θ_s:

$$\sin\theta_s = m\lambda f - \sin\theta_i \tag{54-10}$$

where θ_i denotes the incident angle and m denotes the diffraction order.

Since a surface with statistical roughness contains a large diversity of spatial frequencies, a correlation between PSD and the scattered light can be expected.

The scattering of an optical element can be described by the *bidirectional scatter distribution function* (BSDF), defined as the scattered flux Φ_s per solid angle Ω and per incident flux Φ_i weighted by an "obliquity" factor cos θ_s

$$\mathrm{BSDF} = \lim_{\Omega \to 0} \frac{\Phi_s}{\Phi_i\, \Omega \cos\theta_s} = \frac{1}{\Phi_i \cos\theta_s} \frac{\partial \Phi_s}{\partial \Omega} \tag{54-11}$$

where

θ_s is the scattering polar angle,
θ_i is the incident polar angle,
φ_s is the scattering azimuthal angle,
φ_i is the incident azimuthal angle,

as illustrated in figure 54-9.

An infinitesimal polar angle dΩ can be expressed as

$$d\Omega = \sin\theta_s\, d\theta_s\, d\varphi_s \tag{54-12}$$

The *Angular Resolved Scattering* (ARS) is defined as

$$\mathrm{ARS} = \mathrm{BSDF} \cos\theta_s = \lim_{\Omega \to 0} \frac{\Phi_s}{\Phi_i\, \Omega} = \frac{1}{\Phi_i} \frac{\partial \Phi_s}{\partial \Omega} \tag{54-13}$$

Ω is determined by the size of the detector aperture and its distance from the sample.

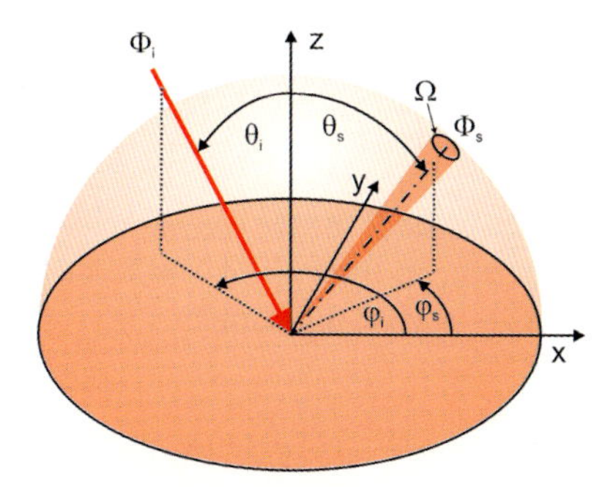

Figure 54-9: Geometrical parameters of an incident and scattered beam at an optical surface.

According to [54-11] the *Total Scatter* (TS) is defined as the backscattered or forward-scattered flux Φ_s divided by the incident flux Φ_i. TS is the scattering loss of the optical component and applies to transparent and opaque elements.

The *Total Integrated Scatter* (TIS) is defined as the backscattered or forward-scattered flux Φ_s divided by the flux of the total transmitted or reflected light including the specular beam. TS and TIS can be converted from one to the other provided that the reflectance or transmittance of the component is known.

TS is equal to the integral of the BSDF function over the hemisphere multiplied by a cosine obliquity factor:

$$\mathrm{TS} = \int_0^{2\pi}\int_0^{\frac{\pi}{2}} \mathrm{BSDF}\cos\theta_s \sin\theta_s d\theta_s d\varphi_s = \int_0^{2\pi}\int_0^{\frac{\pi}{2}} \mathrm{ARS}\sin\theta_s d\theta_s d\varphi_s \tag{54-14}$$

Scattering from rough and slightly rough surfaces can be described by *Vector Perturbation Theories* (VPT). The basic procedure of VPT is to solve Maxwell's equations for the ideally smooth surface and to replace the interface roughness, which induces the perturbation of the specular field, by a plane carrying surface currents, which act as sources of scattered plane waves [54-12].

The VPT result for the ARS of a slightly rough single surface is

$$\mathrm{ARS} = \frac{16\pi^2}{\lambda^4} Q\,\mathrm{PSD}(f_x, f_y)\cos\theta_i \cos^2\theta_s \tag{54-15}$$

where Q is the optical factor containing all information on the corresponding perfectly smooth surface such as the dielectric constant and the conditions of illumination and observation (angles of incidence and scattering, polarization states). Q can be interpreted as a generalized Fresnel reflectance of the scattering surface. The reflectance is a function of wavelength, incidence angle, and polarization. Explicit formulas are given in [54-13].

For polished surfaces with a surface roughnesses $R_q << \lambda$ the relationship between TIS, TS and R_q is given by

$$\mathrm{TIS} = \frac{\mathrm{TS}}{P} = \left(\frac{2\pi}{\lambda}(n_1 - n_2)\cos\theta_i\, R_q\right)^2 \tag{54-16}$$

where n_1 and n_2 denote the refractive indices on both sides of the surface and P denotes the reflectance or transmittance of the surface. In the case of a reflecting surface in air $n_1 = -n_2 = 1$ and P is the specular reflectance of the sample.

54.2.2
Metrology

The surface texture can be measured by a variety of microtopometric instruments. The way the instruments sample the microtopography of a surface can be classified as:

- tactile or non-contacting;
- pointwise, line- or surface-covering.

The instruments differ considerably in their range and resolution when collecting lateral and normal coordinates of a surface point. Figure 54-10 gives examples for the spatial bandwidth of different methods and instruments used to measure surface texture.

To generate a PSD of a sample with a sufficiently large spatial frequency range, it is often necessary to combine topometric measurements of different instruments to a unique PSD. An Atomic Force Microscope (AFM) [54-19], [54-20], for instance, might have a lateral range of 10 µm and a lateral resolution of 10 nm, whereas an interference microscope might use an objective providing a lateral range of 1 mm and a lateral resolution of 1 µm. Results from both instruments can be combined with the result from a form-measuring interferometer to deliver the total PSD, for instance, for a spatial frequency range of $10^{-5} - 50\ \mu m^{-1}$.

In the following, we discuss the important types of microtopometric instruments as well as different scatterometers to detect TIS and ARS distributions of optical surfaces [54-14].

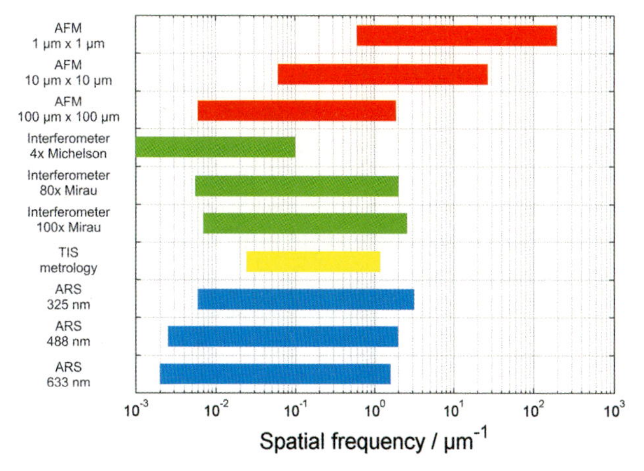

Figure 54-10: Bandwidths of some texture measuring instruments.

Tactile Profilometer

Tactile profilometers are pointwise measuring instruments used to sequentially acquire a 2D or 3D topography of a small surface. During operation a diamond stylus is brought vertically into contact with a sample using a specified contact force. It is then moved laterally across the sample for a specified distance. Small surface variations are detected as vertical stylus displacements during the scanning process. The coordinates of the sample points are referenced against an intrinsic coordinate system, which is supplied by the machine after a calibration process.

Tactile profilometry can be applied universally to optical surfaces. The aim is to measure the shape (figure) as well as the roughness of optical surfaces during different states of the production process. Both, polished and ground or lapped optical elements of different materials can be measured. No a priori information about the shape is necessary. Unknown elements can be measured as long as they fit into the measuring range of the instrument.

A typical profilometer can measure small vertical features ranging in height from 1 nm to several mm. The radius of the stylus tip ranges from 20 nm to 25 μm. The stylus tracking force can range from 10 μN to several tens of mN.

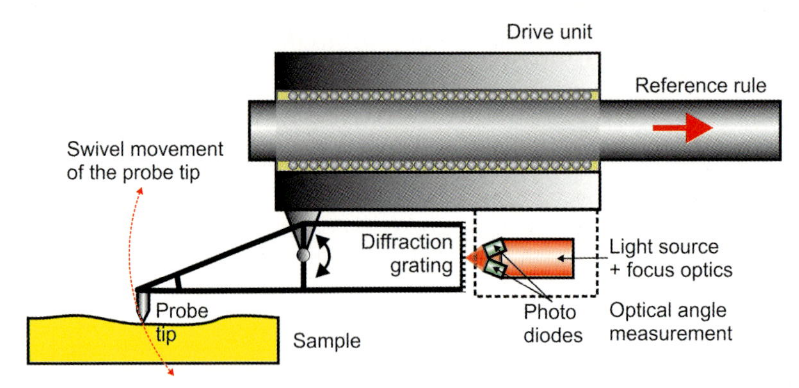

Figure 54-11: Principle of a tactile profilometer [54-15].

The height position of the diamond stylus is mainly detected by a rocker system with the stylus tip on one side and a detection system on the other side of the rotary axis. There are different detection systems in use. Figure 54-11 shows one system, which uses a diode combined with a lens to focus the light onto a diffraction grating. The grating is attached to the rocker system at the opposite end of the stylus tip. The diffraction grating divides the light into three partial components, thereby forming a light patch having three overlapping maxima: the 0th, $+1^{st}$ and -1st diffraction orders. In the overlap regions an interference pattern is formed. Photo diodes in the overlap region of the 0th and $+1^{st}$ diffraction order and of the 0th and -1st diffraction order are positioned on either side to register the irradiance of a fringe. When the grating moves, the interference fringes also move in the same direction. The signal detected by the diodes is evaluated in an evaluation circuit in order to provide precise detection of each movement of the diffraction grating.

For different measuring tasks, various tracing arm and stylus tip geometries are available. According to the measuring task, diamond tips with 2 μm radius for roughness measurements are in use. For contour measurements carbide tips with, for example, 25 μm radius as well as ruby balls are available. An adequate selection of the stylus tip enables the detection and evaluation of short-wave and long-wave profile components.

The physical geometry of the tactile profilometer may have a large effect on the data. When measuring smooth optical surfaces the most obvious problem is that the stylus may scratch the surface. In cases of higher roughness the stylus may be too blunt to reach the bottom of deep valleys or it may round the tips of sharp peaks. In this case the probe is a physical filter that limits the accuracy of the instrument (figure 54-12).

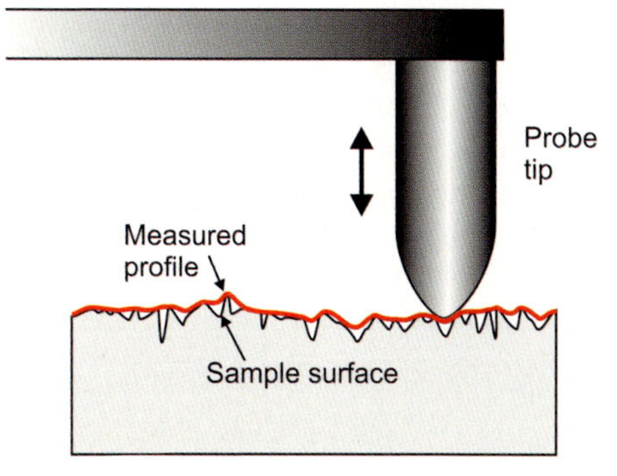

Figure 54-12: Low-pass filtering of a sample surface by the geometry of the probe tip: red curve represents the profile detected by the tip.

The advantages of a tactile profilometer are as follows.

- The profilometer is insensitive to surface reflectance or color, coating or contamination, because the stylus is in direct contact with the surface.
- With a stylus tip of 20 nm, the lateral resolution can be significantly better than optical profiling.
- Tactile profiling is a direct measuring technique, no additional null-systems or compensators are necessary to adapt to curved or aspherical surfaces.

Table 54-2 gives an overview of some typical parameters characterizing commercially available tactile profilometers.

Table 54-2: Some typical parameters of tactile profilometers.

Parameter	Minimum	Maximum
Sample diameter	1 mm	200 mm
Sag in z	0 mm	38 mm
Slope	0°	60°
Duration of measurement for ∅ = 100 mm	50 s	17 min
Measurement rate	0.1 mm/s	20 mm/s
Resolution in z	0.8 nm	20 nm
Repeatability in z	25 nm	50 nm
Range in z	6 mm	12.5 mm
Resolution in x	0.05 μm	1 μm
Measuring force	10 μN	30 mN

Scanning Probe Microscope

Scanning Probe Microscopy (SPM) is a relatively young branch of microscopy in which images of surfaces are formed by using a scanning physical probe. The image is obtained by mechanically moving the probe in a raster scan line by line, and recording the probe–surface interaction as a function of position. A large variety of SPM techniques are available, all differing in their physical type of probe–surface interaction. Many SPMs can image several interactions simultaneously [54-16]–[54-18].

The most commonly used techniques for the measurement of surface roughness are:

- STM – Scanning Tunneling Microscopy (which was the first SPM technique invented in 1981),
- AFM Atomic Force Microscopy

The SPM consists of a cantilever with an extremely sharp tip (probe) at its end which is used to scan the specimen surface. The cantilever is typically silicon or silicon nitride with a tip radius of curvature on the order of nanometers. As it is brought into the proximity of the sample surface, the forces between the tip and the sample cause a deflection of the cantilever. In an AFM these will be mechanical contact forces. In an STM a voltage difference is applied between the tip and the sample surface. When the tip is very close to the surface, electrons are allowed to tunnel through the vacuum between them. The resulting tunneling current is a function of tip position, which is monitored as the tip scans across the surface.

The deflection of the cantilever is typically measured using a laser beam reflected from the top surface of the cantilever into a position sensitive detector (PSD). In a simple case the array may consist of a differential or quadrant diode (figure 54-13). Other methods in use include optical interferometry and capacitive sensing. The system detects cantilever deflections smaller than 1 Å, limited mainly by thermal noise. A long beam path (several cm) amplifies changes in the laser beam angle.

The scanning of the sample is carried out by a piezoceramic scanner device, which holds the sample and positions it in the x-, y- and z-direction with very high precision (figure 54-13). Most SPMs use tube-shaped piezoceramics because they combine a simple one-piece construction with high stability and a large scanning range. Four electrodes cover the outer surface of the tube, while a single electrode covers the inner surface. Application of voltages to one or more of the electrodes causes the tube to bend or stretch, and hence the sample can be moved in all three dimensions (figure 54-14).

The SPM is different from a tactile scanner in that it uses a feedback loop between the positioning system and the sensor system, regulating the force on the sample during acquisition. As the cantilever deflection is measured, the feedback system attempts to keep it constant by varying the voltage applied to the scanner and hence adjusts the height of the sample. In this way, an SPM can measure sample topography either by recording the feedback output or the cantilever deflection.

Since the sample has to be moved in z-direction, the control parameters of the feedback system need to be adjusted every time the sample mass changes. If the sample is large and heavy, the tube scanner is not able to position the sample. For this case there are systems with the tube scanner driving the total probing unit with cantilever, laser and PSD, as shown in figure 54-15.

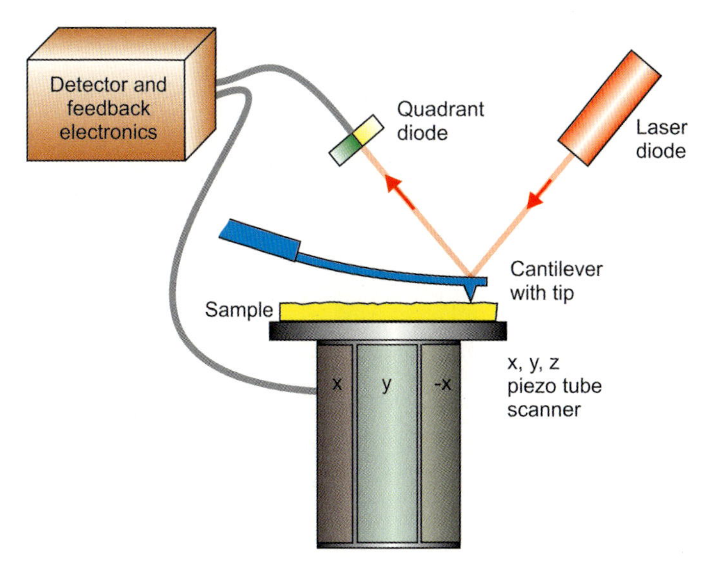

Figure 54-13: Principle of the technique used in an atomic force microscope (AFM). For small samples the tube scanner carries the sample.

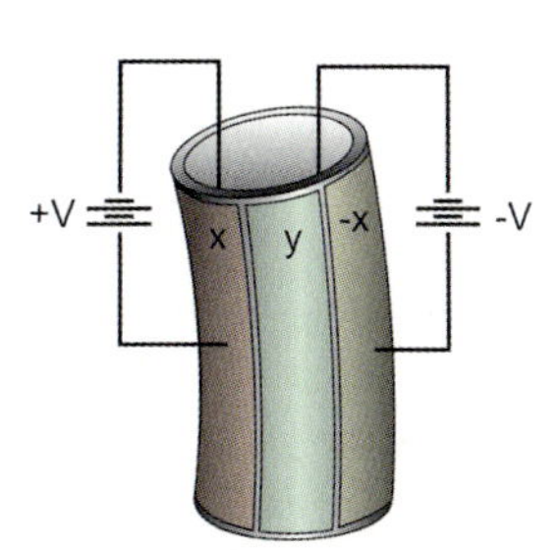

Figure 54-14: Piezoelectric tube scanner used in conventional AFM. When a mirror symmetric voltage is applied to the opposite electrodes, the tube bends sideways.

The resolution normal and parallel (lateral) to the sample surface depend on the applied forces technique, some of which reach atomic resolution. This is largely due to the resolution of piezoelectric actuators, which can execute motions with an accuracy in the region of the atomic level. The resolution is limited only by the size of the probe–sample interaction volume, which can be as small as a few picometers. Laterally the probe–sample interaction extends only across the tip atom or atoms involved in the interaction.

Unlike electron microscope methods, specimens in this case do not require a partial vacuum but can be observed in air at standard temperature and pressure or in a liquid environment.

A typical standard AFM head scans up to 100 μm in the lateral x-y direction and up to 6 μm in the normal direction z. The control system provides a linear scan motion in x and y by means of a closed loop system. The noise level is kept low so as to ensure a resolution in z, which is necessary for sub-Angstrom surface roughness on ultrasmooth surfaces.

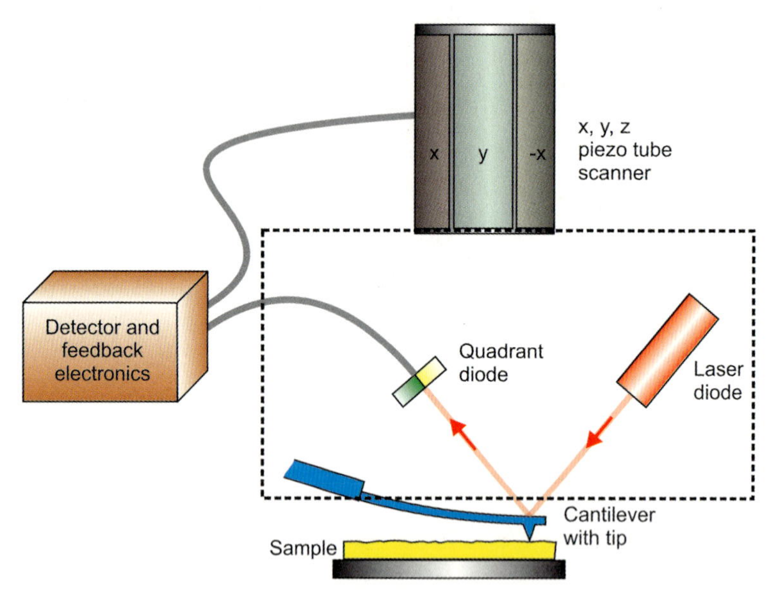

Figure 54-15: SPM for large samples: the tube scanner must carry the entire miniaturized probing unit.

Probe tips are normally made of platinum/iridium or gold. There are two main methods for obtaining a sharp probe tip:

a) *Acid etching* involves dipping a wire end first into an acid bath and waiting until it has etched through the wire and the lower part drops away. The remainder is then removed and the resulting tip is often one atom in diameter.
b) *Cutting* a thin wire with a pair of scissors or a scalpel. Fifty percent of the cuts lead to a sufficiently sharp probe, which is tested by means of a sample with a known profile.

Imaging Modes

The primary modes of operation of an AFM are:

a) *Contact mode* (static mode), in which the static tip deflection is used as the feedback signal. In the contact mode, the force between the tip and the surface is kept constant during scanning by maintaining a constant deflection. Low-stiffness cantilevers are used to receive a large deflection signal. The attractive forces can be quite strong close to the sample surface, causing the tip to "snap-in" to the surface, making scanning difficult. Therefore the contact mode is used in cases where the overall force is repulsive.

b) *Tapping mode* (dynamic mode), in which the "cantilever" is made to vibrate at close to its natural frequency by a small piezoelectric element mounted in the AFM tip holder. Typical vibration frequencies are between 50 kHz and 500 kHz. The proximity of the surface is determined by the damping of this oscillation, the amplitude of which is greater than 10 nm, typically 100–200 nm. It eliminates lateral shear forces on the tip and reduces the force normal to the tip and the surface, which can damage soft samples. Schemes for dynamic-mode operation include frequency and amplitude modulation. In frequency modulation, changes in the oscillation frequency provide information about tip–sample interactions. The use of very stiff cantilevers is possible. Stiff cantilevers provide stability very close to the surface and enable atomic resolution in ultra-high vacuum conditions [54-19].

c) *Non-contact mode,* in which the AFM derives topographic images from measurements of attractive forces. The tip does not touch the sample. The cantilever is instead oscillated at a frequency slightly above its resonant frequency. The amplitude of oscillation is typically below 10 nm. The van der Waals forces, which are strongest from 1 nm to 10 nm above the surface, or any other long-range force, which extends above the surface, all act to decrease the resonance frequency of the cantilever. This decrease is used as the input signal of a closed-loop system, which keeps the tip-to-sample distance constant. Measuring the feedback response at each data point produces the topographic image of the sample surface. An AFM in non-contact mode does not suffer from tip or sample degradation effects. This makes the non-contact mode preferable to the contact mode, when soft samples are measured. In the case of rigid samples, contact and non-contact images may look the same. However, if a few monolayers of adsorbed fluid are lying on the surface of a rigid sample, the images may look quite different. An AFM operating in contact mode will penetrate the liquid layer to image the underlying surface, whereas in non-contact mode an AFM will oscillate above the adsorbed fluid layer to image both the liquid and the surface.

Characteristics of SPMs

When using an SPM, some typical characteristics have to be considered, for example:

- The scanning technique is sequential and therefore slow, requiring several minutes for a typical scan.
- A sequential scanning process is sensitive to thermal and mechanical drifts of the sample, mechanical vibrations or feedback loop oscillations.
- SPM images can be affected by hysteresis of the piezo material and cross-talk between the x-, y- and z-axes which may require software filtering ("flattening"). Hysteresis causes the forward and reverse scans to behave differently, making it necessary to apply a non-linear voltage to the piezo electrodes and to calibrate the scanner accordingly.

- The sensitivity of piezoelectric material decreases exponentially with time. After a continuous 48 hour run, however, the piezos are relatively stable, seldom requiring recalibration.
- Due to the nature of SPM probes, they cannot normally measure steep walls or overhangs. Specially-made cantilevers can be used to modulate the probe sideways as well as up and to measure sidewalls. These will be more expensive and have lower lateral resolution and additional artifacts.

The test area selection is made by means of an x-y stage carrying the sample and a CCD camera built into the SPM unit for visual control of the test area. The video field can be adjusted by zoom optics and covers the SPM scan field as well as the cantilever and tip. Table 54-3 shows some typical data of a commercial AFM.

Table 54-3: Some typical data for range and resolution of an AFM.

Resolution in z	< 50 pm RMS with acoustic hood
Range in z	6 μm
Resolution in x and y	< 2 nm RMS
Range in x and y	90 μm x 90 μm
Accuracy in x and y	< 2%
Field size of CCD camera for tip and sample viewing	150 – 700 μm (zoom)
Resolution of CCD camera viewing	1.5 – 7 μm

Mirau and Michelson Interference Microscope

For the rapid acquisition of a 3D texture topometry, the interference microscope (or optical microscopic profiler) is the most popular instrument. It is basically a standard microscope equipped with a CCD camera and a piezo element to provide phase-shifting [54-22]–[54-24]. The interference microscope needs a special objective, which provides the reference surface to produce interference. Figure 54-16 shows the principle by which it works. A low- or high-coherent light source is imaged via a condenser system, a tube lens and a beam-splitter, onto the pupil of the microscope objective. An adjustable aperture-stop at the pupil location can be used to select different aperture settings. These have an impact on the lateral resolution, the depth of focus and the irradiance of the camera image. After reflection from the sample, the light transmits through the beam-splitter and is then imaged onto the camera by a second tube lens. A field-stop between the condenser unit and the first tube lens encircles the conjugate image of the sample.

To provide interference, special microscope objectives are available. For higher magnifications from 10x up to 100x (numerical apertures from 0.3 to 0.7) so called *Mirau objectives* are in use. In these objectives, a light beam passes through a beam-splitter, which directs the light to both the surface of the sample and a built-in reference mirror (figure 54-17). The light reflected from these surfaces recombines and a fringe interference pattern is formed.

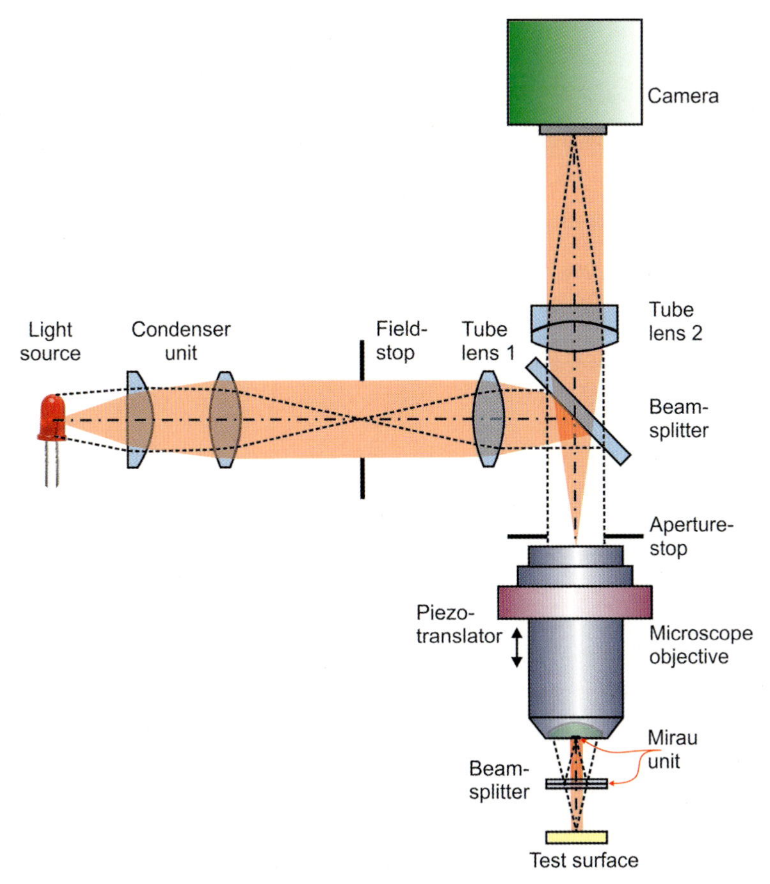

Figure 54-16: Principle of an interference microscope using a Mirau interference objective.

For lower magnification from 2.5 to 5 (numerical apertures from 0.07 to 0.13) so called *Michelson objectives* provide comparatively longer working distances, wider fields of view and a larger depth of focus (figure 54-18) [54-25]–[54-27]. Because of their low numerical aperture their lateral resolution is lower than those from Mirau objectives. Table 54-4 gives some examples of Mirau and Michelson objectives available for standard microscopes.

In a Michelson objective a beam-splitter is placed within the working distance of the objective, separating the light into a reference and a test beam. The test beam transmits the beam-splitter, whereas the reference beam is reflected towards a rigidly mounted reference surface. After reflection from the sample and the reference surface, the beams combine to produce the interferogram. The reference surface should have a reflectance similar to that of the typical samples. In cases where the sample surfaces are considerably curved, the reference surface should also carry a similar curvature to reduce the number of interference rings in the interferogram, so that the camera can easily resolve the fringes.

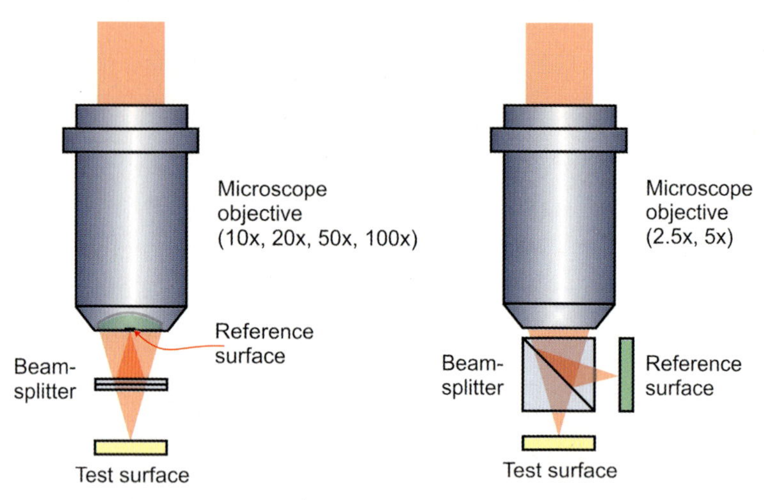

Figure 54-17: Mirau microscope objective.

Figure 54-18: Michelson microscope objective.

Table 54-4: Examples of interference microscope objectives.

Magnification	**2.5x**	**5x**	**10x**	**20x**	**50x**	**100x**
Interference type	Michelson	Michelson	Mirau	Mirau	Mirau	Mirau
Numerical aperture	0.075	0.13	0.30	0.40	0.55	0.7
Lateral resolution at $\lambda = 500$ nm [μm]	6.7	3.8	1.7	1.3	0.9	0.7
Working distance [mm]	10.3	9.3	7.4	4.7	3.4	2
Focal length [mm]	80.0	40.0	20.0	10.0	4.0	2
Depth of focus [μm]	48.6	416.2	3.04	1.71	0.90	0.56
Maximum field of view [mm]	10.0	5.0	2.5	1.25	0.50	0.25

To provide the phase-shifting necessary for the acquisition of several interferograms with well-defined relative fringe positions, a piezo translator can move the total Mirau or Michelson unit relative to the sample (figure 54-16). It is also possible to move the sample relative to the objective if the sample table can be lifted by a piezo transducer.

Linnik Interference Microscope

Mirau and Michelson objectives both need a sufficiently large working distance to enable the insertion of a beam-splitter and a reference surface. Naturally, objectives with high numerical apertures provide very small working distances, which are insufficient for use as interference objectives.

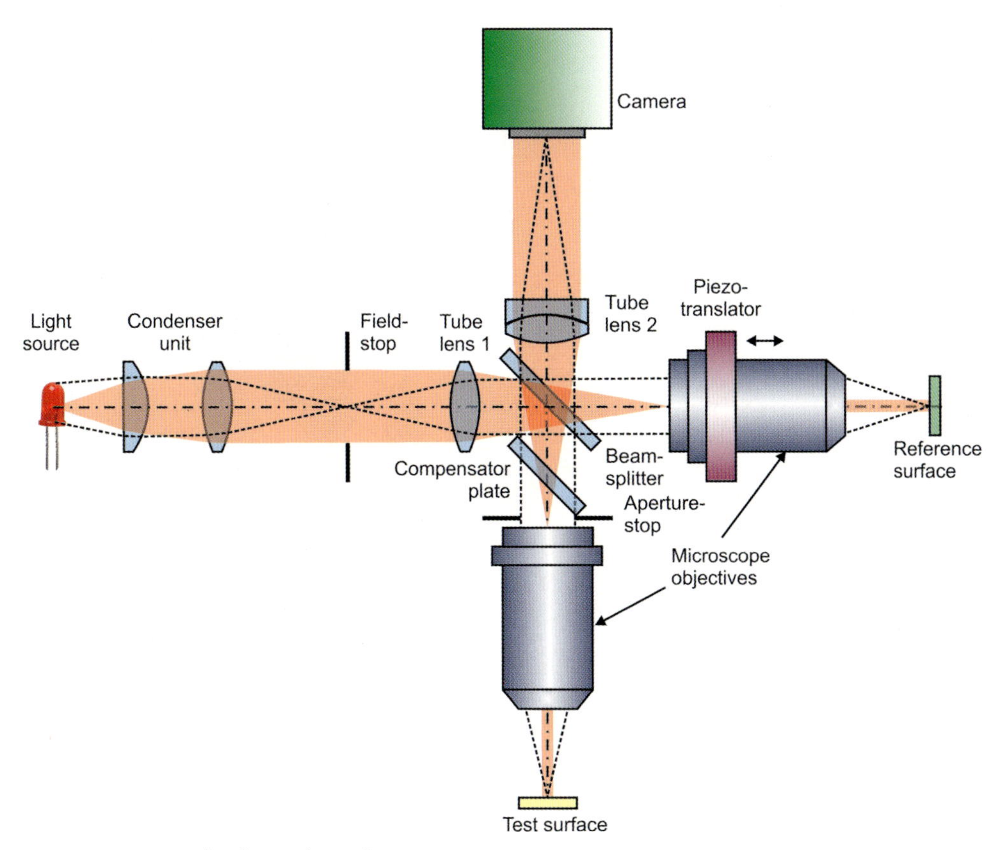

Figure 54-19: Principle of a Linnik interference microscope.

For high lateral resolution obtained with numerical apertures > 0.9 the so-called *Linnik interference microscope* is one solution (figure 54-19) [54-25] and [54-28]. A Linnik interferometer is basically the same as a Michelson interferometer. The difference is in the use of the optics in the reference arm, which essentially duplicates the microscope objective in the test arm. Besides the use of high-aperture objectives, the advantage is its ability to compensate for chromatic dispersion and monochromatic optical aberrations. The Linnik configuration therefore can be used with low-coherent light sources. In this case the use of an appropriate compensator plate is necessary along with the beam-splitter plate, unless a beam-splitter cube is used. In order to achieve phase-shifting, the objective in the reference arm is moved by a piezo-translator along with the rigidly connected reference surface.

Reference-free Interference Microscope

The drawback of a Linnik interference microscope is the requirement to use a second microscope objective in the reference arm, which is generally not compatible with commercially available microscopes. When testing high-quality polished samples, it is important that the reference surface is of the best quality otherwise small misalignments of the reference surface, after calibration, will limit the accuracy of the measurement.

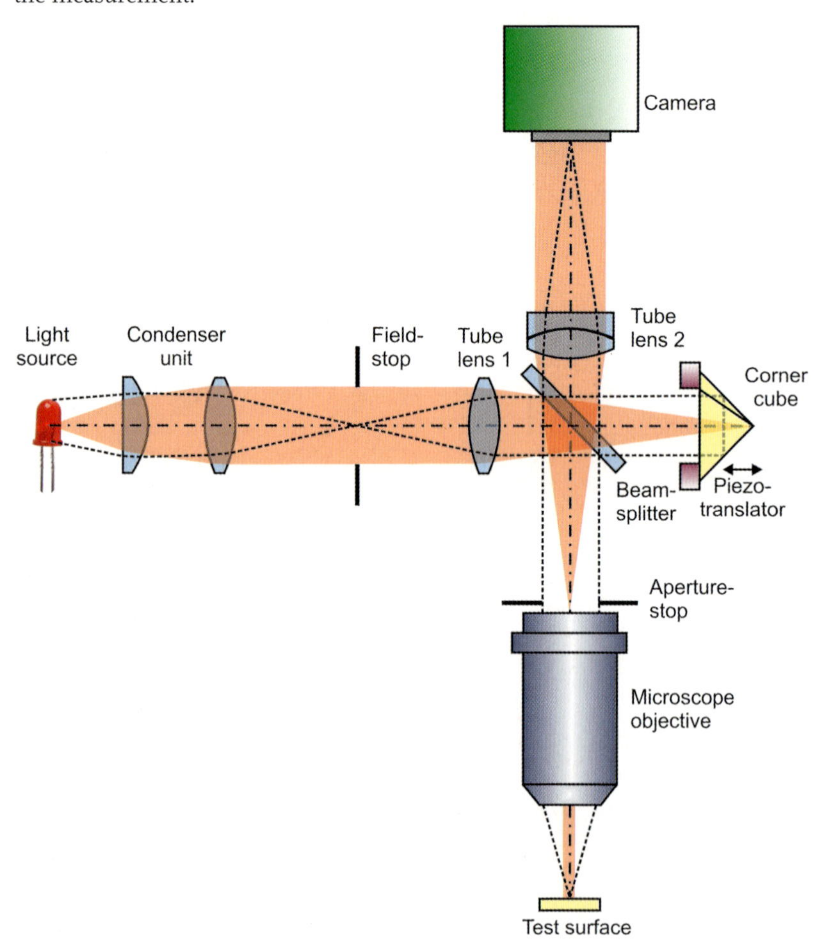

Figure 54-20: Principle of a reference-free interference microscope.

An improvement on the Linnik concept is the reference-free interference microscope [54-29], in which the reference objective and the reference surface are replaced by a corner cube (figure 54-20). The light source is imaged onto the tip of the corner cube, while its image is inverted. In this position it replaces the function of an objective and a sample as well as acting as a retroreflector.

For general-purpose applications the use of a laser light source is unavoidable, because optical paths and dispersions in the test and reference arms are quite different. Using a low-coherence light source would demand well-balanced optical paths in both arms.

Phase-shifting can be provided by a piezo-driven support ring for the corner cube as shown in figure 54-20.

Differential Interference Contrast Microscope (DIC)

A *Differential Interference Contrast microscope* (DIC), also known as a *Nomarski microscope*, uses a polarizing illumination technique to enhance the contrast in microscopy [54-30]–[54-32]. It works by separating a polarized light source into two laterally displaced beams, which take slightly different paths through the sample. After recombination, the two beams interfere showing a modulation in irradiance related to their optical path difference. The irradiance distribution carries the information of the OPD gradient of the sample.

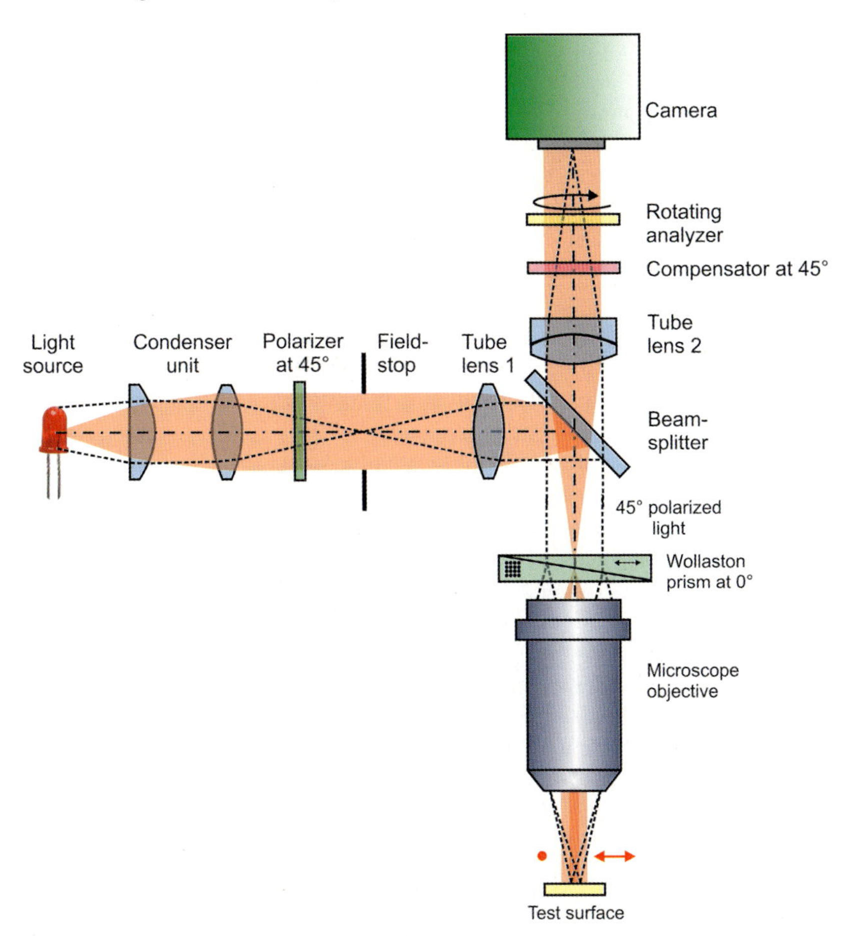

Figure 54-21: Principle of a differential interference contrast microscope (DIC), also known as a Nomarski microscope.

The Nomarski arrangement can also be used to measure the microtopography of a surface. Figure 54-21 shows an example setup, in which phase shifts between the two laterally displaced beams can be introduced to acquire a series of interferograms for quantitative analysis.

In figure 54-21 a linear polarizer is placed in the illumination arm at 45°. After reflection at the beam-splitter, the light transmits through a Wollaston prism, which splits the light into two angularly separated beams of orthogonal linear polarization. After passing the objective, two parallel beams, laterally displaced by Δs – usually a fraction of a μm – are incident onto the sample. Δs is called the lateral shear. After reflection and transmission through the Wollaston prism, both beams are recombined. However, the orthogonal polarization now carries a relative phase shift which is proportional to the surface gradient of the sample, resulting in elliptical polarization. A quarterwave-plate (compensator) at 45° in the camera arm adds an additional phase shift of $\lambda/4$ to the orthogonal components at 45° and 135°. A rotating linear polarizer (analyzer) in front of the camera introduces a sinusoidal irradiance modulation during rotation at each pixel. The phase of the sinusoidal modulation is proportional to the optical path difference of the laterally displaced measuring beams. The camera acquires a series of sample images synchronized with the analyzer rotation. The analysis of the phase at each pixel gives the gradient map of the sample surface in the 0° direction [54-33], [54-34].

The integration of the data set in 0° does not give the full topography of the sample. However, to analyze a single cross-section, the one-dimensional integration is sufficient.

To gain complete topometric information, a second measurement must be carried out with the Wollaston prism rotated through 90°. The resulting gradient map in 90° can then be used together with the 0° map to reconstruct the total topography of the sample.

Figure 54-22 gives an example of a surface profile measure with an 80x objective. The gradient in y – as measured with the Nomarski arrangement – shows different statistical behavior, because the lower frequency components in the signal are damped. Figures 54-22 (c) and (d) show the related 2D-PSD of the surface profile (a) and the gradient in y (b). In the latter the frequency bands in the x- and y-direction now differ considerably.

Note that shearing signals are “blind” for all spatial frequencies corresponding to the $N\,\Delta s^{-1}$, where $N = 1, 2, 3, \ldots$ and Δs denotes the lateral shear displacement. If a strong signal is required, the lateral shear must be larger. The shearing signal then loses information on spatial frequencies which might be the main interest. This is one of the principal reasons why a DIC microscope is rarely used for PSD measurement.

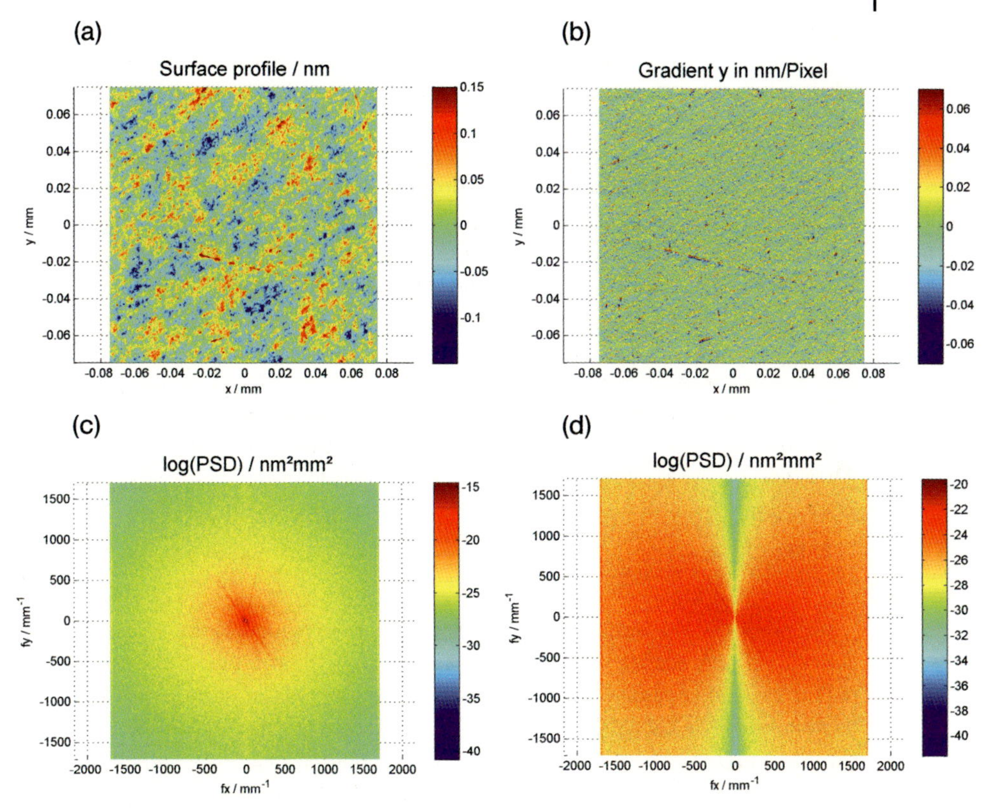

Figure 54-22: Microtopography of an optical surface measured using an 80x microscope objective: (a) regular topographic map; (b) gradient map in y as measured with the Nomarski arrangement; (c) PSD of (a) in a logarithmic scale; (d) PSD of (b) in a logarithmic scale.

Interferometer Microscopes for very Rough Surfaces

For very rough surfaces in which the roughness exceeds the wavelength, a high-coherent interferometer is not able to measure the topography. Using an extended low-coherence light source enables the instrument to measure deviations over several hundred microns of surface heights, if a vertical scanning technique is used.

The extended broadband light source is imaged at the pupil of a microscope objective equipped with a Mirau or Michelson objective. When the sample is in best focus, interference fringes are observed at the camera. Owing to the wide spectral range and spatial extent of the source the contrast of these fringes falls off rapidly for object points that are located above or below the best focus plane. A measurement is performed by moving the objective vertically while recording the intensity pattern at each pixel of the camera. A special algorithm analyzes the images taken sequentially to reconstruct the rough topography of the sample [54-35]–[54-38].

Scatterometers

As described in the sections above, the scattering behavior of an optical surface is strongly related to the surface texture. Instruments measuring scattered light are called *scatterometers.*

In the following we will describe scatterometers measuring *Angular Resolved Scattering* (ARS) or the *Bidirectional Scatter Distribution Function* (BSDF) as well as scatterometers measuring *Total Integrated Scatter* (TIS). BSDF contains the reflected portion, called the *Bidirectional Reflectance Distribution Function* (BRDF), as well as the transmitted portion of the scattered light, called the *Bidirectional Transmission Distribution Function* (BTDF).

BSDF and ARS Scatterometers

BSDF and ARS can be used to calculate the PSD of a sample and therefore characterize the roughness R_q for individual spatial frequency bands. Using (54-15) the expression for the 2D PSD can be found [54-39], [39-40]:

$$\mathrm{PSD}(f_x, f_y) = \frac{\lambda^4}{16\pi^2 Q \cos\theta_i \cos^2\theta_s} \mathrm{ARS} = \frac{\lambda^4}{16\pi^2 Q \cos\theta_i \cos\theta_s} \mathrm{BSDF} \tag{54-17}$$

Although Q – the optical factor containing all information on the corresponding perfectly smooth surface – might not be known explicitly, samples of the same material can be compared in their PSD, as long as the same angles of incidence and polarization states are used. When the 2D PSD is integrated over a spatial frequency region ($f_{x\min}$ to $f_{x\max}$ and $f_{y\min}$ to $f_{y\max}$), the corresponding roughness R_q can be calculated:

$$R_q = \int_{f_{y\min}}^{f_{y\max}} \int_{f_{x\min}}^{f_{x\max}} \mathrm{PSD}(f_x, f_y) df_x df_y \tag{54-18}$$

Note that all roughness metrology including scatterometers are limited in their spatial bandwidth.

A typical setup for a BSDF or ARS scatterometer is shown in figure 54-23 [54-41]–[54-44]. The light from a laser is focused onto beam dump 1 by a beam expander telescope. A receiver unit with a solid angle aperture Ω can be moved in a circle around the sample surface. Each receiver position is characterized by the scattering polar angle θ_s. The sample can be rotated to an incident polar angle θ_i . Since all axes are in the plane of incidence, all azimuthal angles are set to zero.

The light from the source is chopped to reduce optical and electronic noise, which is accomplished by means of lock-in detection in the electronics of the reference detector and the receiver unit, suppressing all signals except those at the chopping frequency. The reference detection is used to normalize the receiver signal Φ_s, thus making it insensitive to fluctuations in the incident flux Φ_i.

The spatial filter within the beam expander removes coherent noise and scatter from the source. It presents a point source, which is imaged at the aperture of beam dump 1 or at the entrance pupil of the receiver unit in the zero position. Beam dump 2 captures the images reflected from the sample surfaces. The source region is separated by a shield, which isolates stray light from the detector.

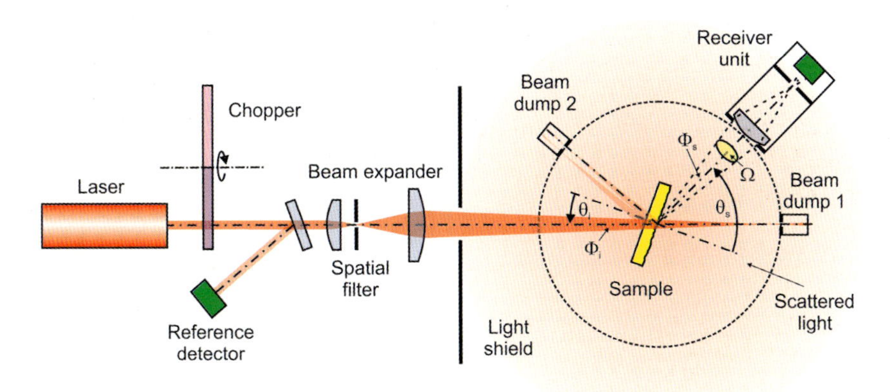

Figure 54-23: Scatterometer arrangement used to measure BSDF.

Laser light sources are convenient, but not necessary. For particular applications broadband sources are required. In order to select arbitrary wavelengths, monochromators and filters can be added to the light source. A tunable laser, for example a TiSa (Titanium–Sapphire) laser providing 650–1100 nm wavelength, is very convenient.

A professional scatterometer system uses motorized rotation stages controlled by a computer for the following axes:

- sample rotation,
- receiver unit rotation,
- beam dump 2 rotation (optional).

For special applications the receiver unit can often be equipped with changeable elements such as different aperture stops, bandpass filters, polarizers, lenses and field stops. The field stop in the receiver unit determines the field of view of the system, whereas the entrance pupil stop determines the solid angle Ω over which scatter is gathered. Any light entering the receiver unit from the field of view and within the solid angle aperture Ω will reach the detector and become part of the signal.

Figure 54-24 shows a receiver unit operating with a light source imaged at the sample (converging source), whereas figure 54-25 shows a receiver unit operating with a light source imaged to infinity (collimated source).

The collimated source arrangement is in better accordance with the strict definition of BSDF. The field stop is located in the back focal plane of the receiver lens. In this way, bundles of nearly parallel rays scattered from the sample are collected by the detector. The solid angle aperture Ω is now determined by the field stop, whereas the field of view is limited by the size of the aperture stop.

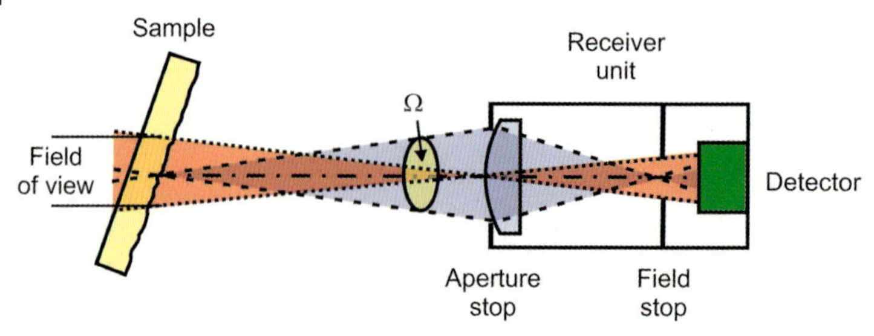

Figure 54-24: Receiver unit configuration of a scatterometer with converging source.

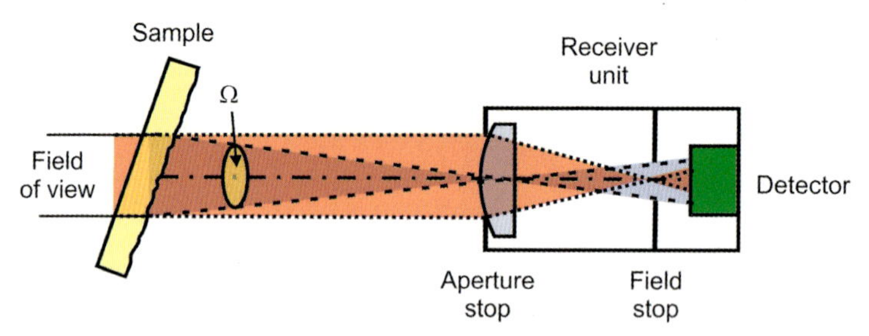

Figure 54-25: Receiver unit configuration of a scatterometer with collimated source.

A BSDF or ARS measurement near specular transmission or reflection becomes very difficult. The critical regions may be divided into two zones around the specular beam. The start angle of the outer critical zone is denoted by θ_N. Inside θ_N, scattered light from the focusing element may enter the FOV of the receiver unit (figure 54-26). The start angle of the inner critical zone is denoted by θ_{spec}. Inside θ_{spec}, portions of the specular beam from the focusing element enter the FOV of the receiver unit. The measurement is then dominated by the convolution of the receiver FOV and the illuminating aperture. Scatter measurement is not possible in this region. According to the geometry in figure 54-26 θ_N can be calculated using small-angle approximation:

$$\theta_N = \frac{2\alpha_R f_F' + D_R + D_F}{2(f_F' - R)} \tag{54-19}$$

In regular BSDF or ARS scatterometers, θ_N values of 1° – 10° are achieved. The specular zone angle θ_{spec} is calculated from

$$\theta_{spec} = \frac{D_R + D_{diff}}{2R} \approx \frac{1}{2R}(D_R + 4\lambda\, F/\#) \tag{54-20}$$

where D_{diff} denotes the spot diameter of the specular beam, $F/\#$ denotes the F-number of the focusing optics and λ denotes the wavelength.

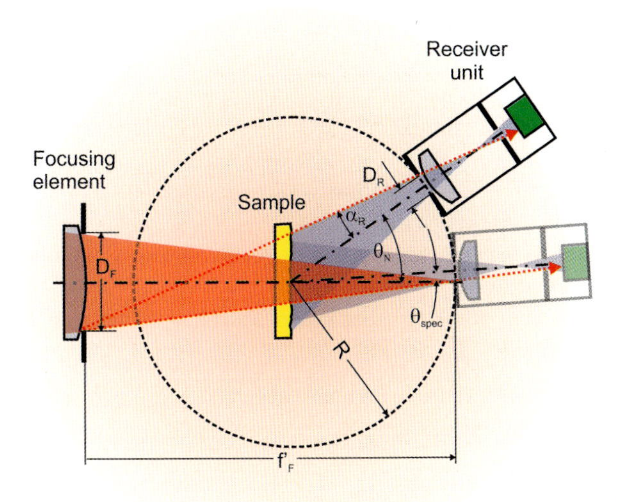

Figure 54-26: Near specular geometry of a BSDF scatterometer.

To make the near-specular regions small, a high *F/#* and a small D_R should be used.

θ_N, θ_{spec} and the so-called *Noise Equivalent BSDF* (NEBSDF) form the *instrument signature,* which characterizes the signal of the scatterometer with no sample (transmission) or a perfect sample (reflection). A schematic instrument signature as a function of the angle difference θ_s- θ_i is shown in figure 54-27.

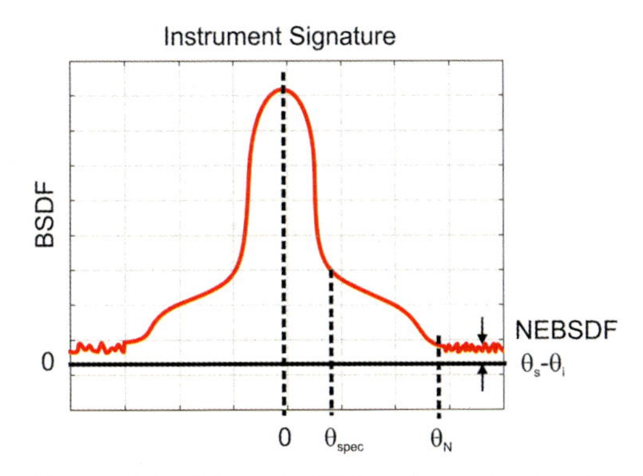

Figure 54-27: Schematic BSDF scatterometer signature.

The major requirements when characterizing low-scatter samples are a low NEBSDF and also small angles for θ_N and θ_{spec}. Taking (54-10) into account, spatial frequencies $f > \frac{0.0175}{\lambda}$ can be detected with a BSDF scatterometer that provides a signature with $\theta_N > 1°$. When visible light is used, spatial wavelengths smaller than approximately 40 μm can be detected.

NEBSDF is dominated by optical and electronic noise. Optical noise can be caused by the following factors.

- Scatter from optical components in the illumination path, especially from the focusing elements. Therefore they should be super-polished and extremely clean.
- Erroneous reflections from the illumination path. Therefore all mounts, stops or pinholes should be blackened.
- Erroneous reflections from the sample area. So all specular beams should be dumped by effective beam dumps.
- Dust in the air along the path of the specular beams leads to a scatter signal, which can be reduced by a filtered air supply.

The electronic contributions to NEBSDF can be detected by covering the receiver unit aperture during measurement. Photomultiplier tubes (PMT) are preferred detectors for low-level light detection.

BSDF and ARS can be measured as functions of:

- lateral position on a sample,
- incident angle,
- wavelength,
- polarization.

To characterize sample's uniformity, raster scans are taken by shifting the sample in the x- and y-directions.

Polarization can be fully detected using ellipsometric techniques. To determine the total Mueller matrix, a polarimeter arrangement, for example a $PC_r(\omega_1)SC_r(\omega_2)$ arrangement as described in section 51.4.4, must be installed.

When a laser source is used, speckle effects will generate optical noise, thus limiting the precision of the instrument. In this case, measurements can be averaged while the sample is rotated around its normal through the illuminating spot.

Calibration of a BSDF and ARS scatterometer includes the following steps.

- Determination and correction of the receiver linearity, which can be achieved, for example, by using a set of well-known neutral density filters.
- Determination of the incident flux Φ_i with the sample removed or with the help of a reference sample.
- Determination of the receiver unit's solid angle Ω by measuring the geometry of the arrangement.
- Calibration of the goniometers rotating the receiver unit and the sample, determination of the 0°-positions for both axes.

The stability of a scatterometer is checked by measuring the instrument signature or by measuring a reference sample. Recalibration is necessary if the signature or the result for the reference sample has changed significantly. The usual precision is between ± 1% and ± 10% in regions away from the near-specular and grazing directions.

TIS Scatterometers

Total Integrated Scatter (TIS) can be used to characterize the roughness R_q of polished surfaces. Using (54-16) the expression for normal incidence and surfaces in reflection leads to

$$R_q = \frac{\lambda}{4\pi}\sqrt{\mathrm{TIS}} \tag{54-21}$$

There are two main instruments used to measure TIS : the *Coblentz sphere* and the *Ulbricht sphere* (or *integrating sphere*) [54-44]–[54-49]. Their setups and characteristics are described in the following sections.

Coblentz Sphere

A typical setup for a TIS scatterometer is shown in figures 54-28 (backward scatter) and 54-29 (forward scatter). A hemispherical mirror, a so-called *Coblentz sphere*, gathers a large fraction of the light scattered from a sample and focuses it onto a detector [54-50]. By separately measuring both the diffuse and the specular part of the reflected light, the total integrated scatter TIS can be determined. It is the ratio of the diffuse reflectance to the total reflectance.

To measure backward scatter (figure 54-28) the sample is illuminated by light from a laser, which – after reflection from a beam-splitter – enters the Coblentz sphere through a small hole (port). The specular reflection leaves the sphere through the same hole. The diameter of that hole defines the near-specular limit of the instrument. The reflected beam – not the incident beam – should be centered in the hole, because the BSDF will be symmetrical about it. The transmitted beam leaves the Coblentz sphere and can be measured by another detector or can be absorbed by a beam dump.

To determine the specular reflectivity of the sample the specular reflection can be detected by a further detector.

To measure forward scatter (figure 54-29) the arrangement of the Coblentz sphere and the sample, including the mount, is rotated by 180°, so that the sample is illuminated from behind.

A reference signal is generated by the light transmitted through the beam-splitter which illuminates the reference detector. It is used to normalize the receiver signal Φ_s, to make it insensitive to fluctuations in the incident flux Φ_i.

TS and TIS are calculated from

$$\mathrm{TS} = \frac{\Phi_s}{\Phi_i} = \frac{\Phi_s}{k\,\Phi_0} \tag{54-22}$$

$$\mathrm{TIS} = \frac{\Phi_s}{\Phi_r} = \frac{\Phi_s}{R\,\Phi_i} = \frac{\Phi_s}{R\,k\,\Phi_0} \tag{54-23}$$

where R denotes the reflectance of the sample and k denotes a normalizing factor related to the reflectivity of the folding beam-splitter in figures 54-28 and 54-29.

As in other scatterometers, the light from the laser is chopped and filtered through the use of lock-in detection in the electronics of the reference detector and the receiver unit, suppressing all signals except those at the chopping frequency.

The Coblentz sphere must be adjusted so that most of the light scattered from the sample is collected by the inner reflecting surface and reflected onto the scatter detector, which is placed near the sample.

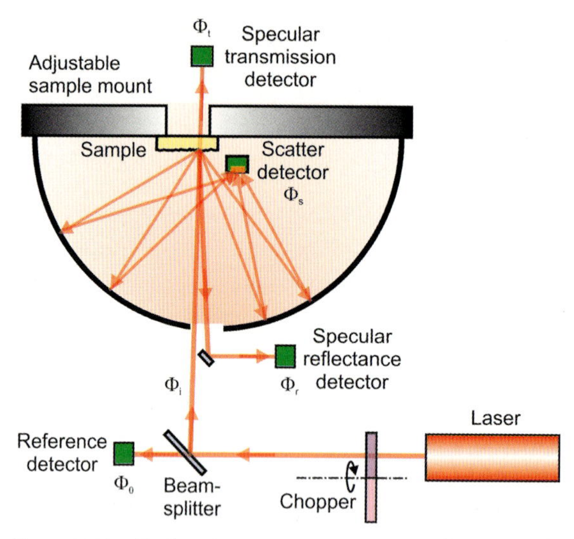

Figure 54-28: Backward scatter TIS measurement by means of a Coblentz sphere.

Making use of (54-16) the measured TIS result can be expressed as the R_q value. However, $R_q$values must be expressed with an associated spatial bandwidth to compare them with results measured by other instruments.

An *American Society for Testing and Materials* (ASTM) standard for the total integrated scatter measurement has been established [54-46]. The method described determines the integrated scattering for scattering angles between ~2.5° and ~70° from the surface normal. The alignment of the position of the sample, laser beam, and the scattered-light detector must be done carefully in order to meet the large angle collection efficiency limit of ~70°. Relatively small misalignments of the above parts of a TIS instrument can lead to an overfilling of the detector, especially if the detector is not of the appropriate size, and also to a loss of light scattered to large angles. The alignment of the TIS instrument can be facilitated, for example, by choosing an oversized detector (e.g., 100 mm^2) or by trying several detectors and choosing the optimal one. The largest scattered angles collected by TIS instruments are reported to be 85° [54-51].

In the design of a Coblentz sphere arrangement, all geometrical dimensions must be considered and optimized in order to determine the effective spatial bandwidth of the TIS measurement. The following dimensions should be considered.

- The radius of the Coblentz sphere.
- The port location and size (position and geometry of the entrance hole).
- The geometry and position (distance, tilt, rotation) of the sample.

- The position, size and angular (FOV), spectral and polarizing sensitivity of the detector unit.
- The characteristics of the illuminating beam (polarization, spectral bandwidth, direction, divergence, diameter, etc.),
- The angular and spectral reflectivity and the roughness of the Coblentz sphere mirror surface.

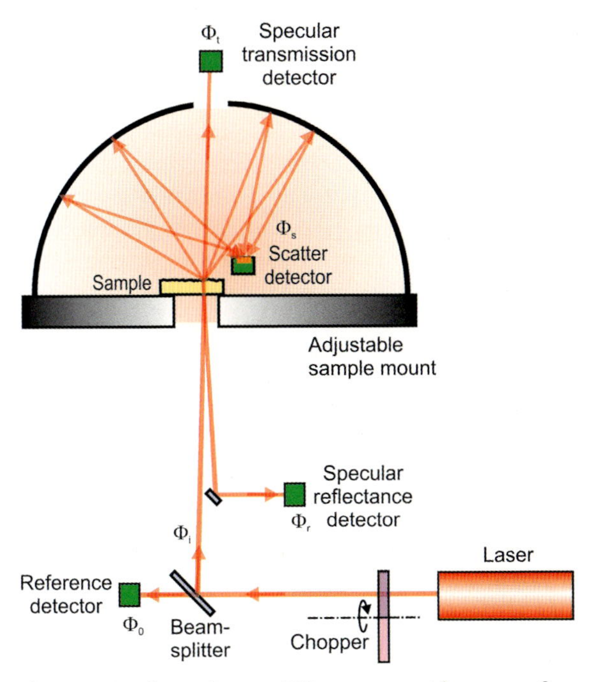

Figure 54-29: Forward-scatter TIS measurement by means of a Coblentz sphere.

With the help of ray-tracing calculations the system can be simulated and optimized. It is then possible to determine a spatial bandwidth transfer function characterizing the total system as a function of wavelength and polarization.

Modern Coblentz sphere scatterometers provide detection levels in the order of 1–10 ppm for forward and backward scatter, respectively. Calibration procedures also include the determination of the detector linearities and the determination of the system signature when the sample is removed. The use of a reference sample with a well-known TIS can be helpful.

Ulbricht Sphere (Integrating Sphere)

An alternative method used to measure TS or TIS is by means of an *integrating sphere*, also known as an *Ulbricht sphere* [54-44]. It is an optical component consisting of a hollow sphere with its interior coated for high diffuse reflectivity. Ideally, the coating on the inner side of the integrating sphere should have a very high reflectivity over the required wavelength range. Since its relevant property is uniform scattering, its entrance and exit holes (ports) are assumed to be small in order to keep a high degree of radiation uniformity within the cavity. Due to multiple-scattering reflections, light incident on any point on the inner surface is distributed equally to all other points. An integrating sphere may be thought of as a diffuser, which preserves power but destroys spatial information. If the optical losses in the sphere and through the small ports are low, the multiple reflections can lead to a fairly high optical intensity inside the sphere, even if the sphere is much larger than the light source and the detector.

Figure 54-30 shows a typical Ulbricht sphere setup used to measure TS and TIS of a sample. The field of view (FOV) of the recessed detector is limited to a section of the sphere that is not directly illuminated by the scatter from the sample. This is accomplished by blocking the direct light by means of appropriate baffles.

The specular reflection from the sample should be centered on the exit hole, because the BSDF will be symmetrical about it. A further detector measures the specular reflection to determine the specular reflectivity of the sample.

TS and TIS are calculated from

$$\mathrm{TS} = k_1 \frac{\Phi_\mathrm{d}}{\Phi_\mathrm{i}} = k_1 \frac{\Phi_\mathrm{d}}{k_2 \Phi_0} \tag{54-24}$$

$$\mathrm{TIS} = k_1 \frac{\Phi_\mathrm{d}}{\Phi_\mathrm{r}} = k_1 \frac{\Phi_\mathrm{d}}{R\,\Phi_\mathrm{i}} = k_1 \frac{\Phi_\mathrm{d}}{R\,k_2\,\Phi_0} \tag{54-25}$$

where R is the reflectance of the sample, k_1 is a normalizing factor related to the integrating sphere (see below) and k_2 is a normalization factor related to the reflectivity of the folding beam-splitter in figure 54-30.

As with a Coblentz sphere, a reference signal is generated by the light transmitted through the beam-splitter, illuminating the reference detector. The light from the source is also chopped and filtered by means of lock-in detection to suppress erroneous reflections and signals.

In the following we will discuss the general radiance equation for an integrating sphere. It is derived from [54-52] and [54-53].

The main contributions are derived from:

- the geometrical considerations of a sphere with given entrance and exit ports and
- multiple reflection theory within the sphere for a given Lambertian reflectance of the inner wall.

For an incident flux Φ_i the flux Φ_d detected at the scatter detector is then given by

$$\Phi_\mathrm{d} = \frac{R_\mathrm{w} a}{A(1 - R_\mathrm{w} a) - A_\mathrm{s} R_\mathrm{ss}} \frac{\Omega_\mathrm{d} A_\mathrm{d}}{4\pi} R_\mathrm{s} \Phi_\mathrm{i} = M \frac{\Omega_\mathrm{d} A_\mathrm{d}}{4\pi} R_\mathrm{s} \Phi_\mathrm{i} \tag{54-26}$$

where

$a = \dfrac{A - A_\mathrm{s} - A_\mathrm{d} - A_\mathrm{o}}{A}$ is the effective surface factor of the sphere

and

R_s is the diffuse reflectance of the sample for collimated irradiance.
R_{ss} is the diffuse reflectance of the sample for diffuse irradiance.
R_w is the diffuse reflectance factor of the sphere wall.
A_d is the area of the detector.
A_s is the area of the sample.
A_o is the area of any other aperture within the inner sphere wall.
A is the area of the inner surface of the total sphere.
Ω_d is the solid angle corresponding to the FOV of the detector.

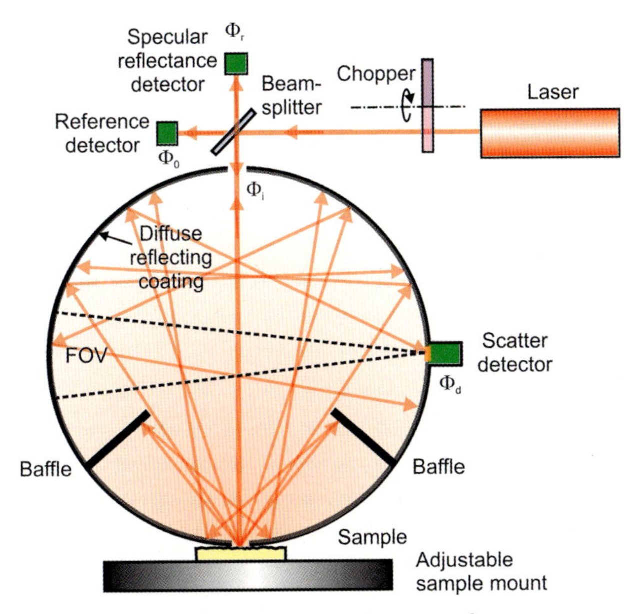

Figure 54-30: TS and TIS measurement by means of an integrating sphere (Ulbricht sphere).

M is a factor mainly driven by the areas of the sphere and its ports and by the reflectance of the inner wall. When $R_w a \rightarrow 1$ then Φ_d will approach infinity. In order to achieve a good S/N ratio the integrating sphere must have:

- a high reflectance R_w of the inner wall,
- small entrance and exit ports relative to the total sphere area to achieve $a \rightarrow 1$.

Commercially available integrating spheres are coated with a special paint based on Barium Sulphate ($BaSO_4$). The reflectance covering the UV-VIS-NIR wavelength region yields reflectances of 95% – 98% over the wavelength region 300–1200 nm (figure 54-31) [54-52]. Other materials based on thermoplastic resin (for example Spectralon®) achieve reflectances > 99% over a range of 400–1500 nm and > 95% for 250–2500 nm. All materials show highly Lambertian behavior.

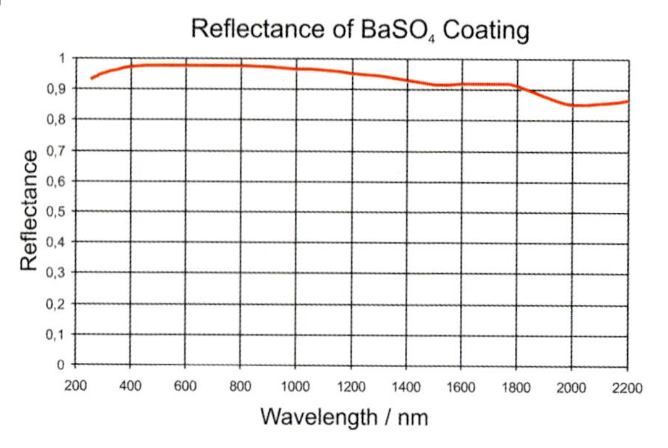

Figure 54-31: Spectral reflectance for a special $BaSO_4$ material used to coat the inner wall of an integrating sphere.

The effective surface factor a of commercially available integrating spheres range between 0.95 and 0.98. A rule of thumb for integrating spheres is that no more than 5% of the sphere surface area should be consumed by port openings. Port diameter are driven by both the size of devices, as well as the geometrical constraints required by a sphere system.

Diameters of commercially available integrating spheres range between 75 mm and 2000 mm. The detected flux of the scattered light is inversely proportional to the sphere diameter. Thus, the smallest sphere generally produces the highest radiance. However, since the integrating sphere is usually employed for its ability to spatially integrate input flux, a larger sphere diameter and smaller port fraction will improve the spatial performance. With a small sphere diameter the effective surface factor a becomes smaller, because the ports occupy a larger relative portion of the sphere surface. Figure 54-32 compares three different sphere diameters and their M factors as a function of the $R_w a$ factor. The areas of the sample have been neglected.

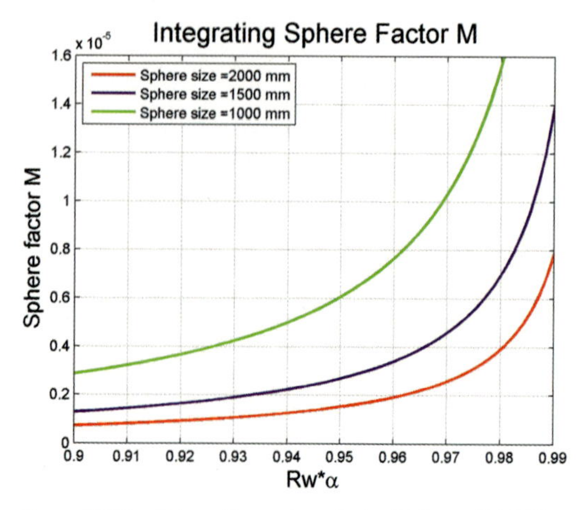

Figure 54-32: Integrating sphere factor M (equ. (54-26)) as a function of the factor $R_w a$ for sphere diameters 1000 mm, 1500 mm and 2000 mm.

54.3 Surface Imperfections

54.3.1 Basics

Surface imperfections are addressed in [54-1] and may comprise:

1. *Surface imperfections* described as "localized defects within the effective aperture of an optical surface produced by improper treatment during or after the fabrication process." Examples of such surface imperfections are scratches, pits, broken bubbles, sleeks, scuffs and fixture marks. Imperfections also include localized coating blemishes such as grey spots and color sites that absorb or reflect light differently from the bulk of the coating.
2. *Long scratches* described as thin surface imperfections longer than 2 mm.
3. *Edge chips* described as localized defects around the periphery of an element. Although they are usually located outside the optically used area, they may be a source of scattered light or may be the origin of crack propagation.

Since no surface can be fabricated perfectly, there is a need to define imperfection tolerances related to the application, which can be measured by an adequate metrology. The acceptance level for surface imperfections can take into account:

- *functional effects,* which affect the image formation or durability of the optical element,
- *cosmetic or aesthetic effects,* which do not affect the image formation but dissatisfy the user, leading to products which are hard to sell.

In the following we will apply imperfection tolerances to both transmissive and reflecting surfaces and to finished optical elements (including coating), but not to assemblies. Generally there are two different methods used to measure surface imperfections:

1. *Method I* also called the *obscured or affected area method,* involving a metrology in which individual surfaces can be measured.
2. *Method II* also known as the *visibility method,* involving a metrology in which an entire optical element is tested. The results contain the defects of all optically effective surfaces, as well as material defects such as bubbles and inclusions.

Indication in Drawings

On optical drawings, the required quality related to surface imperfections is indicated by a leader line and an alphanumeric description of the quality [54-9]. The code number for surface imperfections is 5. The indications take into account either test Method I or II which will be used for the inspection process.

According to Method I, the drawing indication for the number and size of general surface imperfections, which are permissible within the effective aperture of a surface, is

$5/\,N \, x \, A$

N x A specifies the number N and the grade number A of allowed maximal surface imperfections. A is equal to the square-root of the surface area of the maximum allowed defect, expressed in millimeters.

It is possible to specify the tolerances for coating blemishes (C), long scratches (L) and edge chips (E) separately:

$$5/\, N \times A;\, CN' \times A',\, LN'' \times A'';\, E\, A'''$$

Since A specifies the bare surface imperfections, then A' specified for an anti-reflection coating will be larger than the grade A, because of the difficulty in distinguishing between small coating blemishes and surface imperfections. For edge chips, any number of edge chips is permissible as long as their extent does not exceed A'''.

A larger number of general surface imperfections with a smaller grade number is permitted, if the sum of their areas does not exceed the maximum total area

$N\,A^2$ for general surface imperfections or

$N'\,A'^2$ for coating blemishes.

Table 54-5 shows the size designation for A according to [54-1] and the corresponding value for A^2.

Table 54-5: Preferred size designation A and the corresponding area A^2 for surface imperfections according to Method I.

Grade numbers A in mm	A^2 in mm^2
0,006	0,000036
0,01	0,0001
0,016	0,000256
0,025	0,000625
0,04	0,0016
0,063	0,003969
0,1	0,01
0,16	0,0256
0,25	0,0625
0,4	0,16
0,63	0,3969
1	1
1,6	2,56
2,5	6,25
4	16

Figure 54-33 shows an example of an indication specifying 10 surface imperfections of grade ≤ 0.1 mm and maximum edge chip grade of 0.4 mm. Imperfections of grade ≤ 0.1 mm are allowed, as long as the sum of their areas does not exceed 0.1 mm^2. Each occurring edge chip must cover a surface smaller than 0.16 mm^2.

Concentrations of surface imperfections are not allowed. A concentration occurs when more than 20 % of the number of allowed defects is found in any 5 % of the test region. If $N < 10$ we refer to a concentration, when two or more surface defects fall within a 5 % sub-area.

Method II expresses the visibility of surface imperfections. Although an entire optical element is tested, the imperfection tolerances are specified individually for each surface, because they provide guidelines for the surface fabrication.

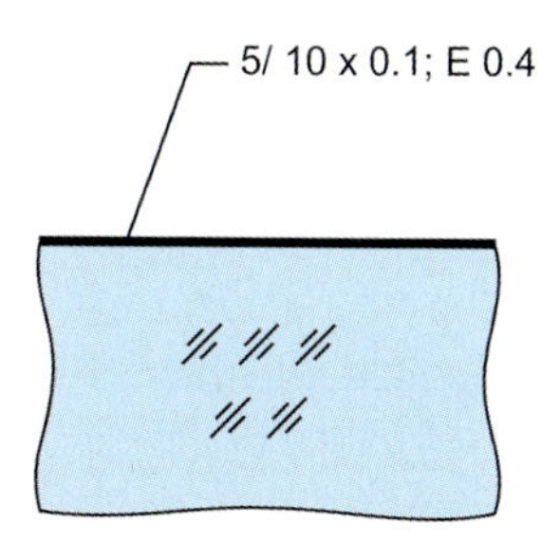

Figure 54-33: Indication for surface imperfections (maximum of 10 imperfections of maximum grade 0.1 mm) and edge chip specification (no maximum number specified, maximum edge chip size of grade 0.4 mm).

The drawing indication for the visibility of surface imperfections of an optical element, as determined by Method II, is expressed as

5/ TV

or as

5/ RV

The T or R in the indication specifies a transmissive or reflective test and the V is a visibility class number. It is expressed as an integer number from 1 to 5, where 1 indicates the most stringent and 5 the weakest requirements for surface imperfections (table 54-6). The illumination for different classes is defined within ± 5%. However, the illumination ratio between the classes should be within ± 2%.

In Method II no special distinction is made for long scratches and coating blemishes. Edge chips are specified in the same way as in Method I.

Table 54-6: Visibility class designations for transmitted and reflected light inspection.

Visibility class	Illumination of sample under test in lx (± 5% tolerance)	Standard background
T5 / R5	310	Adjusted
T4 / R4	625	Adjusted
T3 / R3	1250	Adjusted
T2 / R2	2500	Adjusted
T1 / R1	2500	Black

Method II is based on an instrument, in which light scattered from the sample's surface imperfections is compared with a reference background illumination. The level of sample illumination at which surface defects become visible determines the visibility class.

The following section gives an overview of the metrology used to detect and classify surface imperfections, in particular the instrument used in Method II.

54.3.2
Metrology

Imperfections on optical surfaces have for centuries been inspected by using the eye in association with a controlled test-piece movement by hand. The human eye has an unsurpassed light flux dynamic range of 10^{14}, very well suited for the detection of weak imperfections in bright-field or dark-field illumination [54-10]. The eye is also able to classify different imperfections, once they are detected on the entire surface under test.

Human hands can quickly change the position of small and medium-sized elements no matter what curvature or thickness they might have.

However, there are some disadvantages to an inspection process, which is based on the skill of the eye:

1. Suitable individuals must be selected and specially trained.
2. The ability to detect imperfections might change during the day due to fatigue.
3. Detection levels will only gradually improve over the long term.
4. Contact with the samples might occasionally give rise to further damage.
5. The inspection process is basically subjective, making it difficult to monitor quality changes in the production process objectively.

To change the inspection process from a subjective to a more objective and reproducible one, ISO 10110-7 [54-1] proposes two different methods of measurement.

Method I: Measurement of the imperfection's area of obscuration.
Method II: Measurement of the imperfection's visibility.

In the following we will give examples of a suitable metrology for both methods.

Method I

As discussed in the previous section, the tolerances of Method I are specified by the maximum number *N* of allowed tolerances and a grade number *A* representing the square-root of the surface area of the maximum allowed imperfection, expressed in millimeters.

A suitable metrology must be able to count the number of detected imperfections and evaluate their grade. In the following, a fully automatic metrology used to measure imperfections according to Method I is described for both a *bright-field* and a *dark-field* version.

A *bright-field* detector is shown in figure 54-34. Light from a small light source is collimated and reflected by a beam-splitter to pass a microscope objective of low numerical aperture (NA). The image of the light source coincides with the surface under test. The specular reflected light then re-enters the microscope objective, while a considerable portion of the scattered light misses the NA of the objective. The specular light is focused onto a detector by a collimator after transmitting through the beam-splitter. The irradiation at the detector therefore is related to the level of scattering at the inspected surface point. A smooth surface at the inspected area leads to a bright signal, an imperfection leads to a dark signal.

The principle can be inverted by using a *dark-field detector* as shown in figure 54-35. An aperture stop reduces the diameter of the parallel beam entering the microscope objective, which might have a large NA. The specular light re-entering the microscope objective is blocked by a dark-field stop behind the beam-splitter. Light scattered from an imperfection and re-entering the microscope objective passes the dark-field stop and illuminates the detector after being focused by the collimator. Again, the irradiation at the detector is related to the level of scattering at the inspected surface point. A smooth surface at the inspected area leads to a dark signal, an imperfection leads to a bright signal.

A metrology covering the total surface should provide solutions for the following difficulties:

1. A vast amount of data per inspected surface has to be collected and evaluated. Figure 54-36 gives examples for different pixel sizes needed to detect low-grade imperfections. For large optical elements the processing might be extremely slow.
2. A stray-light detector has to be precisely positioned in a direction normal to the surface under test. The positioning must be precise on a micrometer level. If only plane surfaces are to be inspected, a rotary table for the sample rotation and a motorized linear detector positioner are necessary. For spherically and rotationally symmetric aspherical surfaces, a four-axis system including sample rotation and tilt and x- and z-translation of the detector are necessary (figure 54-37). To measure freeform surfaces a further tilt axis will be necessary.

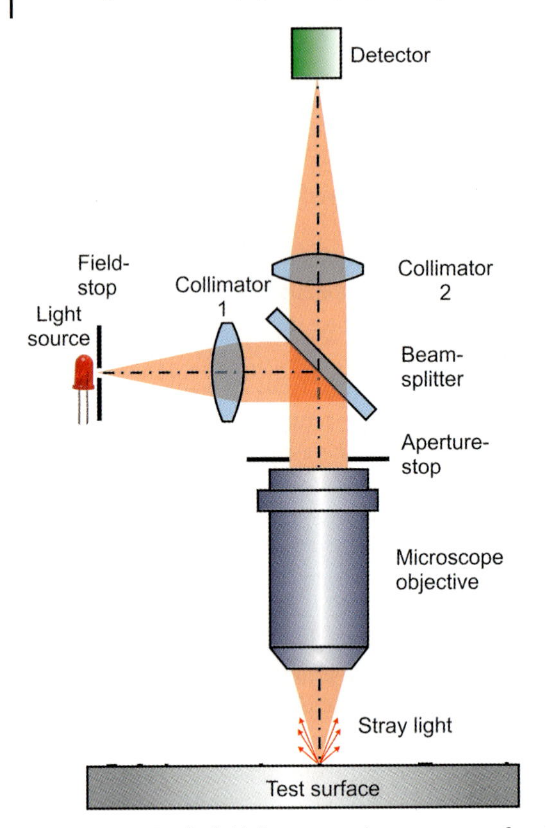

Figure 54-34: Bright-field detector used to measure surface imperfections.

Once a system is set up, calibration against a standard is necessary. Since the detector irradiation is a function of several system parameters such as spot size and NA of the microscope objective, calibration is made against a set of reference imperfections. There are reference graticules on plane substrates defined by national standard organizations and by ISO 10110-7 [54-10] suitable for use in transmission as well as in reflection. They consist of non-reflecting lines and dots, the latter having diameters between 4.5 μm and 450 μm. The lines representing standard scratches have sizes from $1 \times 16\ \mu m^2$ up to $100 \times 1600\ \mu m^2$.

Figure 54-38 gives an example of an inspected surface having a surface topography as shown in (a) measured by an interferometer. Figure 54-38 (b) shows the result of a bright-field inspection, and (c) shows that of a dark-field inspection. Using a calibrated system, the results can be classified according to the applied standards by an appropriate software analysis.

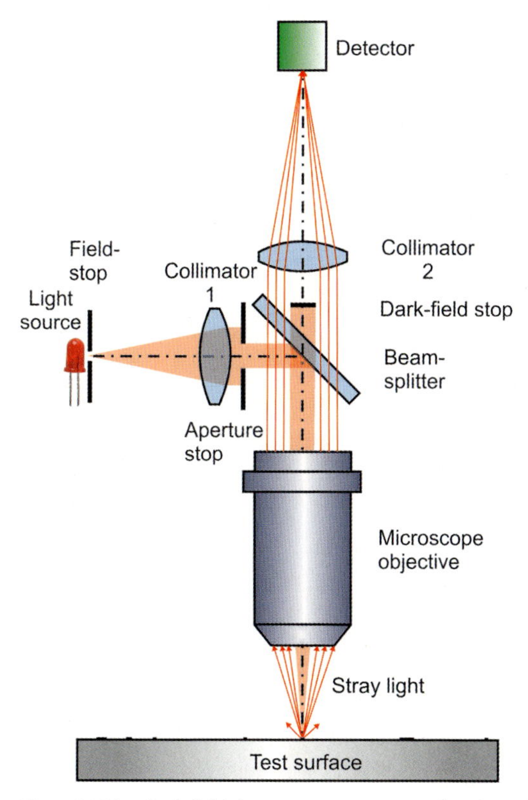

Figure 54-35: Dark-field detector to measure surface imperfections.

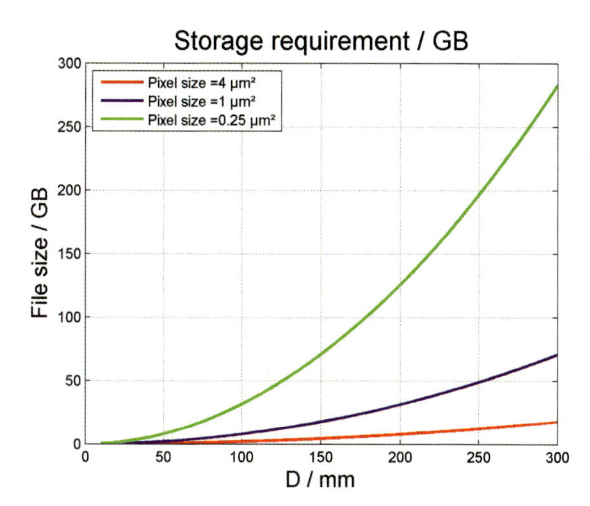

Figure 54-36: Required amount of data storage in Gigabytes for pixel sizes 0.5×0.5 mm^2, 1×1 μm^2 and 2×2 μm^2 as a function of surface diameter in mm.

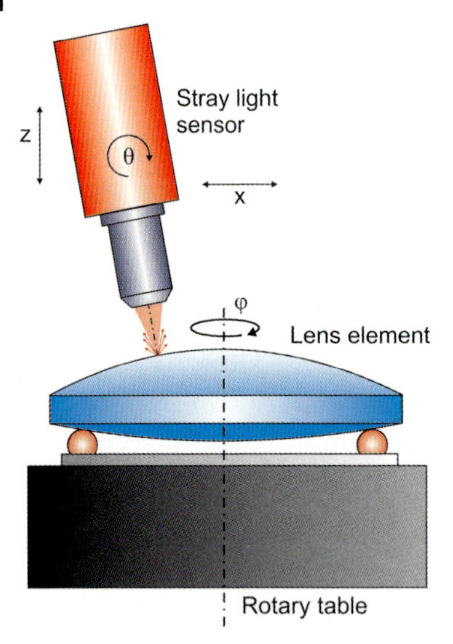

Figure 54-37: Metrology for surface imperfection inspection using a stray-light sensor and a rotary table.

The costs of development and fabrication of a surface covering imperfection metrology as shown in figure 54-37 are very high. That is why a much simpler technique is used for quality control in the many optical shops [54-54]–[54-56]. The techniques involve a person sequentially comparing sections of the surface under test with a reference graticule. For the comparison a so-called *analogue Microscope Image Comparator (MIC)* as shown in figure 54-39 can be used [54-10]. The example shown is a technique based on *Zernike phase contrast microscopy* in order to make pure phase structures visible.

Light from a tungsten lamp is focused through a polarizer (axis at 45°) onto an annular aperture by means of a condenser system. A further lens (2) images the annular aperture to infinity. The polarizing beam-splitter transmits the parallel and reflects the perpendicular polarized part of the parallel beam. The reflected beam enters the test arm, and the transmitted beam the reference arm, of the instrument. Each arm carries a quarter-wave plate at 45° to convert the light from linear to circular polarization. After reflection from the test and reference surfaces and transmission through the quarter-waveplates, the light is reconverted to linear polarized light, however parallel polarization is now changed to perpendicular polarization and vice versa. An objective images the surface under test onto a camera. A phase-retarding ring is located in its back focal plane followed by a rotatable analyzer. The phase-retarding ring coincides with the image of the annular stop. Thus, if the test surface shows only specular reflection, a homogeneous phase retardation is introduced, generating an unmodulated image at the camera.

The reference arm carries the graticule in a plane conjugate to the test surface. A retroreflective screen follows, producing scattering of light over a few degrees as it is hardly affected by the phase-retarding ring. The camera adds the incoherent images from the test and reference arms.

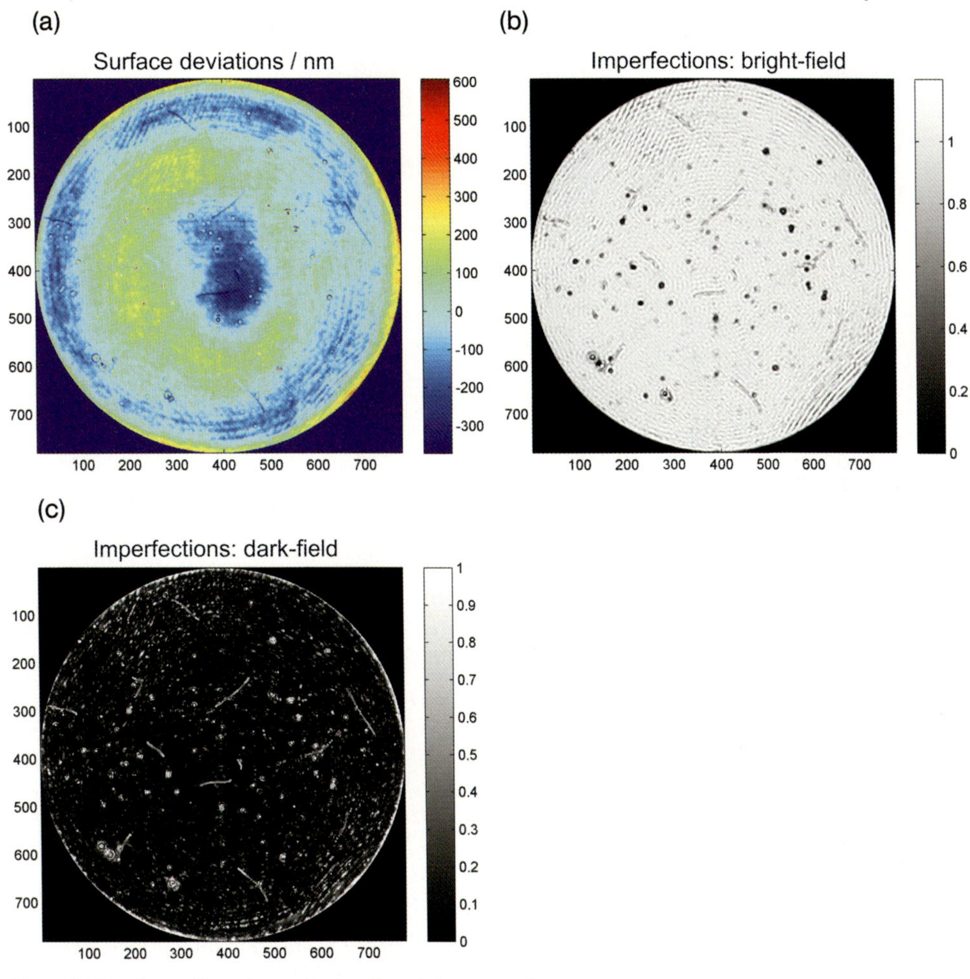

Figure 54-38: Inspection of an optical surface: (a) measured surface deviations in nm; (b) detected imperfections using a bright-field detector; (c) detected imperfections using a dark-field detector.

Let θ be the angular position of the rotatable analyzer in front of the camera. The detected irradiance $I(x,y)$ at the camera is given by

$$I(x,y,\theta) = I_{\mathrm{T}} R_{\mathrm{T}}(x,y)\cos^2\theta + I_{\mathrm{R}} R_{\mathrm{R}}(x,y)\sin^2\theta \tag{54-27}$$

where $R_{\mathrm{T}}(x,y)$ and $R_{\mathrm{R}}(x,y)$ denote the local reflectivities of the test surface and graticule, respectively. I_T and I_R are the irradiances generated by 100% reflecting reference and test surfaces in the test and reference arm, respectively. When the irradiance contributions from the reference and test arm are equal, the following condition is valid:

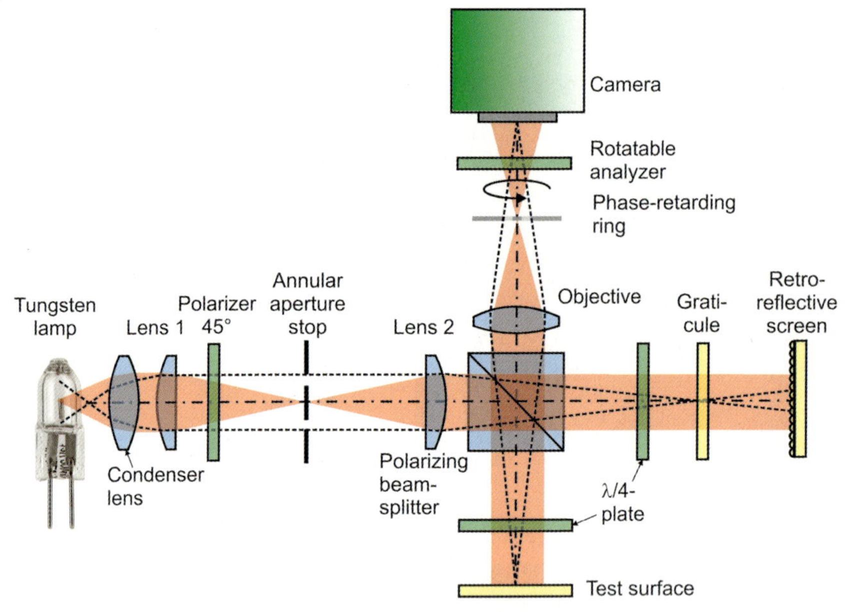

Figure 54-39: *Analogue microscope image comparator (MIC)* to compare surface imperfections with a reference graticule.

$$\frac{R_T(x,y)}{R_R(x,y)} = \frac{I_R}{I_T} \tan^2\theta \qquad (54\text{-}28)$$

Once I_T and I_R have been calibrated, the ratio of the reflectivities from a sample imperfection and the reference scratches from the graticule is expressed by $\tan^2\theta$.

The method is therefore suitable for use by hand adjustment of θ as well as by an instrument that continuously rotates the analyzer while recording images for different rotational positions.

Method II

This method requires an inspection station, in which illumination of the sample and its observation by the inspector is well defined and carefully controlled. Before a measurement the visual sensitivity of the inspector must be calibrated. This is done by viewing a calibrated reference defect and adjusting the illumination system accordingly, as described below.

Elements tested in transmission or reflection need separate stations. Their principle of operation will be described in the following.

Both versions need an *illumination module* consisting of:

1. a light source, preferably a Xenon lamp,
2. a beam-splitter separating the light coming from the source into a *background illumination arm* and a *sample illumination arm*,

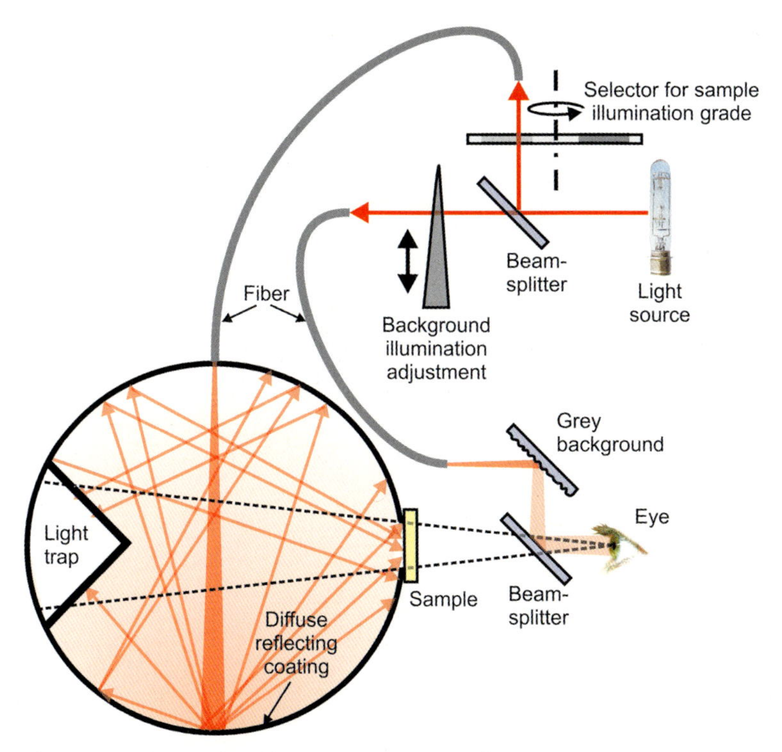

Figure 54-40: Integrating sphere setup to inspect surface imperfections in transmission according to Method II.

3. a background illumination arm providing an optical element to continuously adjust the background illumination (a linear-wedge neutral-density filter would be a suitable choice),
4. a sample illumination arm providing a filter device to select the sample illumination grade as specified in table 54-6,
5. two light-guidance devices, preferably provided by appropriate fibers,
6. an integrating sphere with
   a. an entrance port, into which light from the sample illumination arm is guided,
   b. an observation port, through which the sample can be observed.

The *observation module* consists of

7. a beam-splitter through which the observer can view the sample,
8. a ground glass diffusely reflecting light from the fiber connected to the background illumination arm.

For the transmission test samples are placed near the observation port, so that light exiting from the integrating sphere enters the observer's eye after transmission through the sample (figure 54-40). A light trap (beam dump) is necessary inside the integrating sphere to prevent direct light from entering the observer's eye. With the sample removed, the observer should see a perfect black observation port-hole.

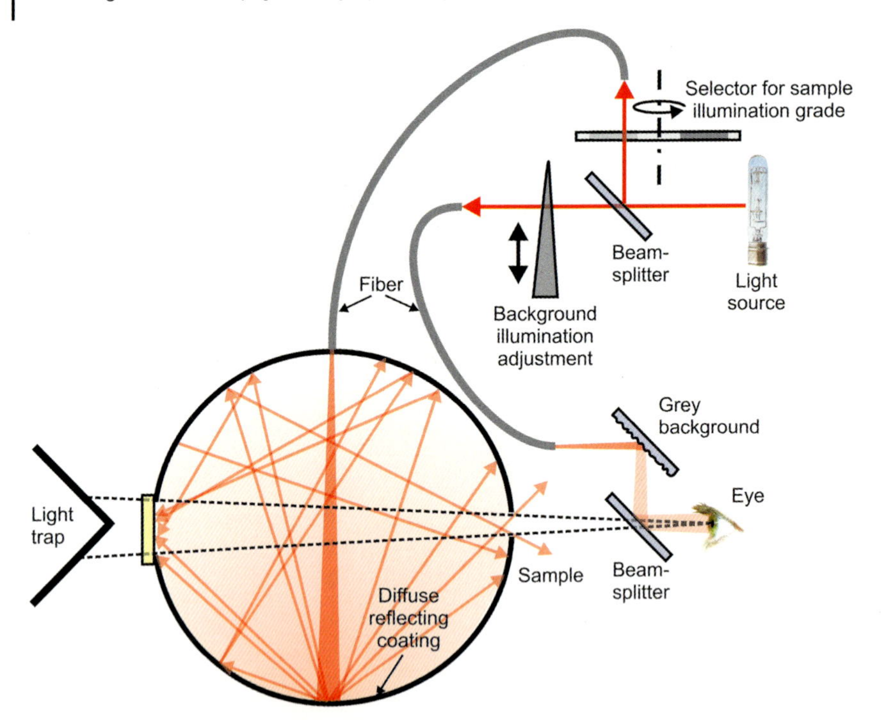

Figure 54-41: Integrating sphere setup to inspect surface imperfections in reflection according to Method II.

For the reflection test, samples are placed opposite to the observation port in front of an additional exit port. Light from the integrating sphere then illuminates the sample from all directions and enters the observer's eye (figure 54-41). A light trap (beam dump) is necessary outside the integrating sphere behind the sample to prevent direct light from the environment from entering the observer's eye. With the sample removed, the observer should see a perfect black exit port-hole.

The principle of operation is as follows:

The observer simultaneously observes light from the sample and the background illumination arm. Light scattered from the surface imperfections of the sample is compared with a reference background illumination. The visibility class is determined by the level of sample illumination at which surface defects become visible. Each visibility class corresponds to a particular sample illumination (table 54-6).

In order to guarantee a uniform level of sensitivity independent of the observer, the element in the background illumination arm is adjusted accordingly. For the adjustment process a standard sample carrying a well-defined standard defect is used [54-1]. The sample consists of a chrome reticle cross of certain size evaporated onto a polished plate. The luminance is adjusted until the standard defect is just visible to the observer when illuminated at a level of 2500 lx.

Comparison of Methods I and II

Both methods are more or less subjective since a human operator must determine the area of individual imperfections or must find a threshold of visibility. The results are dependent upon the degree of concentration and fatigue. Table 54-7 gives an overview of the main characteristics of both methods.

Table 54-7: Comparison of Method I and II.

Method I (Area method)	Method II (Visibility method)
Individual imperfections are observed and measured	Whole surface is observed
No optical significance measured	Optical significance is measured
Limited by the resolution of the microscope	Component size limited to geometry of the observation port
Time consuming	Quick test
Areas are measured precisely within the microscope resolution	Insensitive to small imperfections
Objective measure of imperfection area, however, depending on operator's fatigue	Subjective comparison by an operator, depending on operator's fatigue

Both methods have the potential to be operated by an automatic inspection system, replacing the operator by a camera system and replacing the human decision by an indefatigable image analysis system. In this case Method II could be established at much lower costs, because no complicated motorized positioning system would be necessary. Note, however, that the system would be insensitive to small imperfections.

54.4 Literature

54-1 ISO 10110-7, Optics and photonics – Preparation of drawings for optical elements and systems – Part 7: Surface Imperfection Tolerances (2008).

54-2 ISO 10110-8, Optics and photonics – Preparation of drawings for optical elements and systems – Part 8: Surface Texture (1997).

54-3 S. Schröder, S. Gliech and A. Duparré, Sensitive and flexible light scatter techniques from the VUV to IR regions, Proc. SPIE 5965, 424–32 (2005).

54-4 S. Schröder, S. Gliech and A. Duparré, Measurement system to determine the total and angle resolved light scattering of optical components in the deep-ultraviolet and vacuum-ultraviolet spectral regions, Appl. Opt. 44, 6093–107 (2005).

54-5 P. Croce and L. Prod'homme, Analyse des surfaces polies par la distribution spatiale de la lumiere diffusee, Opt. Commun. **35**, pp. 20–24 (1980).

54-6 P. Croce and L. Prod'homme, Sur les conditions d'application de la diffusion optique a la caracterisation des surfaces rugueuses, J. Opt. **15**, pp. 95–104 (1984).

54-7 E. L. Church, Fractal surface finish, Appl. Opt. **27**, pp. 1518–26 (1988).

54-8 D. J. Janeczko, Power spectrum standard for surface roughness: Part I, Proc. SPIE Vol. 1165, p. 175–83 (1990).

54-9 ISO 1302, Geometrical Product Specifications (GPS) – Indication of surface texture in technical product documentation (2002).

54-10 L. R. Baker, Metrics for High-Quality Specular Surfaces, SPIE Press, Bellingham (2004).

54-11 ISO 13696, Optics and optical instruments – Test methods for radiation scattered by optical components (2002).

54-12 P. Bousquet, F. Flory and P. Roche, Scattering from multilayer thin films: theory and experiment, J. Opt. Soc. Am. **71**, 1115–23 (1981).

54-13 J. C. Stover, Optical scattering: measurement and analysis (2nd ed.), Optical and Electro-Optical Engineering Series (McGraw-Hill, Inc. 1990).

54-14 B. W. Scheer, J. C. Stover, Development of a smooth-surface microroughness standard, SPIE Vol. 3141, 78-87 (1997).

54-15 US-Patent 2003117633A: Probe with a diffraction grating for +1,0 and –1 orders (2003).

54-16 E. Meyer, H. J. Hug and R. Bennewitz, Scanning Probe Microscopy : The Lab on a Tip, (Springer-Verlag, Berlin, Heidelberg, New York, 2004).

54-17 R. Wiesendanger, Scanning Probe Microscopy: Analytical Methods (Springer-Verlag, Berlin, Heidelberg, New York,1998).

54-18 B. Bhushan and H. Fuchs, Applied Scanning Probe Methods II: Scanning Probe Microscopy Techniques (Springer-Verlag, Berlin, Heidelberg, New York, 2006).

54-19 P. J. de Pablo, J. Colchero, M. Luna, J. Gómez-Herrero and A. M. Baró, Tip-sample interaction in tapping-mode scanning force microscopy, Phys. Rev. B 61, 14179–83 (2000).

54-20 J. Jahanmir and J. C. Wyant, Comparison of surface roughness measured with an optical profiler and a scanning probe microscope, SPIE Vol. 1720, 111–18 (1992).

54-21 Y. Namba, J. Yu, J. M. Bennett and K. Yamashita, Modeling and measurements of atomic surface roughness, Appl. Opt. Vol. **39** Issue 16, 2705–18 (2000).

54-22 B. Bhushan, J. C. Wyant and C. Koliopoulos, Measurement of surface topography of magnetic tapes by Mirau interferometry, Appl. Opt. 24, 1489–97 (1985).

54-23 J. C. Wyant, C. L. Koliopoulos, B. Bhushan and D. Basila, Development of a three-dimensional noncontact digital optical profiler, Journal of Tribology **108**, 1–8 (1986).

54-24 D. S. Mehta, S. Saito, H. Hinosugi, M. Takeda and T. Kurokawa, Spectral interference Mirau microscope with an acousto-optic tunable filter for three-dimensional surface profilometry, Appl. Opt. **42**, 1296–1305 (2003).

54-25 K. Creath, Calibration of numerical aperture effects in interferometric microscope objectives, Appl. Opt. 28, 3333–8 (1989).

54-26 E. Beaurepaire and A. C. Boccara, Full-field optical coherence microscopy, Optics Letters 23, 244–6 (1998).

54-27 P. de Groot, X. Colonna de Lega, J. Kramer and M. Turzhitsky, Determination of fringe

order in white-light interference microscopy, Appl. Opt. 41, 4571–8 (2002).
54-28 P. C. Montgomery, A. Benatmane, E. Fogarassy and J. P. Ponpon, Large area, high resolution analysis of surface roughness of semiconductors using interference microscopy, Materials Science and Engineering B **91-92**, 79–82 (2002).
54-29 H. Fritz, Über ein Interferenzmikroskop ohne körperliche Referenzfläche unter besonderer Beachtung der Genauigkeit der automatischen Interferenzbildauswertung, Dissertation, Friedrich-Schiller-Universität, Jena (1991).
54-30 S. Ruzin, Differential Interference Contrast, Plant Microtechnique and Microscopy, (Oxford University Press, New York, 20–23,1999).
54-31 F. Rost and R. Oldfield, Differential Interference Contrast (DIC) Microscopy, Photography with a Microscope (Cambridge University Press, Cambridge, United Kingdom, 136–40, 2000).
54-32 D. Murphy, Differential Interference Contrast (DIC) Microscopy and Modulation Contrast Microscopy, Fundamentals of Light Microscopy and Digital Imaging, (Wiley-Liss, New York, 153–68, 2001).
54-33 C. J. Cogswell, N. I. Smith, K. G. Larkin and P. Hariharan, Quantitative DIC microscopy using a geometric phase shifter, Proceedings of the SPIE, The International Society for Optical Engineering **2984**, 72–81 (1997).
54-34 M. Davidson and M. Abramowitz, Optical Microscopy, Encyclopedia of Imaging Science and Technology (Wiley-Interscience, New York, 1106–41, 2002).
54-35 T. Dresel, G. Häusler and H. Venzke, Three-dimensional sensing of rough surfaces by coherence radar, Appl. Opt. **31**, 919–25 (1992).
54-36 T. R. Thomas, Rough Surfaces (Imperial College Press, London, 1999).
54-37 P. J. Caber, Interferometric profiler for rough surfaces, Appl. Opt. **32**, 3438–41 (1993).
54-38 L. Deck and P. de Groot, High-speed non-contact profiler based on scanning white-light interferometry, Appl. Opt. **33**, 7334–38 (1994).
54-39 A. Duparré, J. Ferré-Borrull, S. Gliech, G. Notni, J. Steinert and J. M. Bennett, Surface characterization techniques for determining rms roughness and power spectral densities of optical components, Appl. Opt. **41**, 154–71 (2002).
54-40 J. M. Bennett and L. Mattsson, Introduction to Surface Roughness and Scattering, 2nd ed. (Optical Society of America, Washington, D.C., 1999).
54-41 A. Duparré, Light Scattering Techniques for the Inspection of Microcomponents and Microstructures in Optical Inspection of Microsystems, edited by Wolfgang Osten (CRC Press, New York,2007).
54-42 J. Neubert, T. Seifert, N. Czarnetzki and T. Weigel, Fully automated angle resolved scatterometer, Space Optics 1994: Space Instrumentation and Spacecraft Optics, T. M. Dewandre, J.J. Schulte in den Bäumen, E. Sein, (eds), Proc. SPIE 2210, 543–52 (1994).
54-43 S. Schröder, S. Gliech and A. Duparré, Measurement system to determine the total and angle resolved light scattering of optical components in the deep-ultraviolet and vacuum-ultraviolet spectral regions, Appl. Opt. **44**, 6093–107 (2005).
54-44 M. Bass (ed.), Handbook of Optics, 2nd ed. (McGraw-Hill, Inc., New York, 1995).
54-45 J. A. Detrio and S. M. Miner, Standardized total integrated scatter measurements of optical surfaces, Opt. Eng. **24**, 419–22 (1985).
54-46 Standard Test Method for Measuring the Effective Surface Roughness of Optical Components by Total Integrated Scattering, Vol. 10.05 of ASTM Standards, ASTM Doc. No. F1048-87, American Society for Testing Materials, Philadelphia, Pa., 418–21 (1996).
54-47 T. A. Leonard, Standardization of optical scatter measurements, Stray Radiation in Optical Systems, R.P. Breault, ed., Proc. SPIE **1331**, 188–94 (1990).
54-48 A. Duparré and S. Gliech, Quality assessment from supersmooth to rough surfaces by multiple wavelength light scattering measurement, Scattering and Surface Roughness, Z. Gu and A.A. Maradudin, (eds) Proc. SPIE **3141**, 57–64 (1997).
54-49 O. Kienzle, V. Scheuer, J. Staub and T. Tschudi, Design of an integrated scatter instrument for measuring scatter losses of superpolished optical surfaces, application to surface characterization of transparent fused quartz substrates, Optical Interference Coat-

ings, F. Abèles, (ed.) Proc. SPIE **2253**, 1131–42 (1994).

54-50 H. E. Bennett and J. O. Porteus, Relation Between Surface Roughness and Specular Reflectance at Normal Incidence, J. Opt. Soc. Am. **51**, 123–29 (1961).

54-51 J. Lorincik and J. Fine, Focusing properties of hemispherical mirrors for total integrating scattering instruments, Appl. Opt. **36**, 8270–74 (1997).

54-52 Labsphere, Inc., A guide to integrating sphere radiometry and photometry, North Sutton (2003).

54-53 M. W. Finkel, Integrating sphere theory, Opt. Comm. **2**, 25–8 (1970).

54-54 A. Huard, Visibility method for classifying microscopic surface defects for both reflection and transmission systems, Proc. SPIE **525**, 37–42 (1985).

54-55 J. P. Marioge, A. Huard, M. Munier and J. L. Hautcolas, Validation of a local defect classification procedure, Proc. SPIE **1009**, 218 (1988).

54-56 L. R. Baker, Inspection of surface flaws by comparator microscopy, Appl. Opt. **27**, 4620–5 (1988).

55
Testing the Quality of Coatings

Handbook of Optical Systems: Vol. 5. Metrology of Optical Components and Systems. First Edition.
Edited by Herbert Gross.

55.1 Introduction

In this section we give a brief introduction on the optical characteristics of thin-film coatings.

Optical elements are usually coated with one or more thin layers in order to achieve specific characteristics of spectral transmittance or reflectance. Optical coatings, which reduce unwanted reflections from optical surfaces, are called AR (Anti-Reflection) coatings. Optical coatings, which increase the reflectivity up to a required level, are called HR (High- Reflection) coatings. More complex optical coatings may exhibit high reflection over some range of wavelengths, and anti-reflection over another range. These elements are used as special dichroic filters.

ISO 9211-1 [55-1] defines coatings according to their function as shown in table 55-1.

Table 55-1: Definition of coatings by function.

Function	Definition	Example of application
Reflecting	Coating increases the reflectance of an optical surface over a specified wavelength and angle of incidence range	HR coatings, mirrors
Anti-reflecting	Coating reduces the reflectance of an optical surface over a specified wavelength range (usually related to an increase in transmittance)	AR coating limits parasitic reflections
Beam-splitting	Coating separates the incident flux into two beams of specified flux: one transmitted, the other reflected. The separation is made in a non-selective manner over a specified wavelength range	Neutral beam-splitters, partial reflectors
Attenuating	Coating reduces the transmittance of an optical surface in a non-selective manner over a specified wavelength range	Neutral density filters
Filtering	Coating modifies the spectral transmittance in a selective manner	Filters
Selecting or combining	Coating separates the incident flux into two or more beams of specified spectral regions, the beams being propagated either by reflection or transmission. The reverse path combines beams of different spectral regions	Dichroic mirrors, beam-combiners
Polarizing	Coating controls the state of polarization of the emergent electromagnetic radiation over a specified wavelength range	Polarizers, non-polarizers, polarizing beam-splitters

Function	Definition	Example of application
Phase-changing	Coating controls the phase change of the emergent electromagnetic radiation relative to the incident radiation in the s- and p-direction over a specified wavelength range	Phase-matching, phase-retarding coatings
Absorbing	Coating absorbs a specified value of the incident flux over a specified wavelength range	Sunglasses, light traps
Supplementary function	Coating can provide non-optical properties	Chemical or mechanical protection, electrical conduction

To provide a glass substrate with an HR coating a choice can be made between:

- a single metallic layer, such as aluminum, silver or gold layer, and
- a dielectric coating made of multiple layers of different transparent dielectric materials such as *Magnesium Fluoride* MgF_2 and *Tantalum Pentoxide* Ta_2O_5.

Metallic layers always absorb a significant portion of the incident light, while dielectric layers can be made to show very high reflectance (> 99.99%) with extremely low loss by absorption.

The general concept of the so-called *quarter-wave system,* based on the periodic layer system composed from two materials, one with a high index, such as Tantalum Pentoxide with $n_{550\,\mathrm{nm}} = 2.07$, and a low-index material, such as Magnesium Fluoride with $n_{550\,\mathrm{nm}} = 1.39$. This periodic system significantly enhances the reflectance of the surface in a certain wavelength range called the *band-stop,* the width of which is determined by the ratio of only the two indices used. The thicknesses d of the layers are generally

$$d = \frac{\lambda}{4n} \tag{55-1}$$

to maximize reflection and minimize transmission.

Figure 55-1 shows the normal incidence reflectances of gold, silver and aluminum coatings in comparison to an HR coating made of 10 MgF_2 layers of thickness 66.3 nm between 11 layers of Ta_2O_5 layers of thickness 98.9 nm.

Figure 55-2 shows the reflectances for s- and p-polarization of the same coatings as a function of the incidence angle θ_i at a wavelength of 550 nm. The reflectance of the HR coating is very high up to an angle of 45°, then the p-polarization drops rapidly to zero, while all metal coatings keep their reflectances at high levels up to 90°.

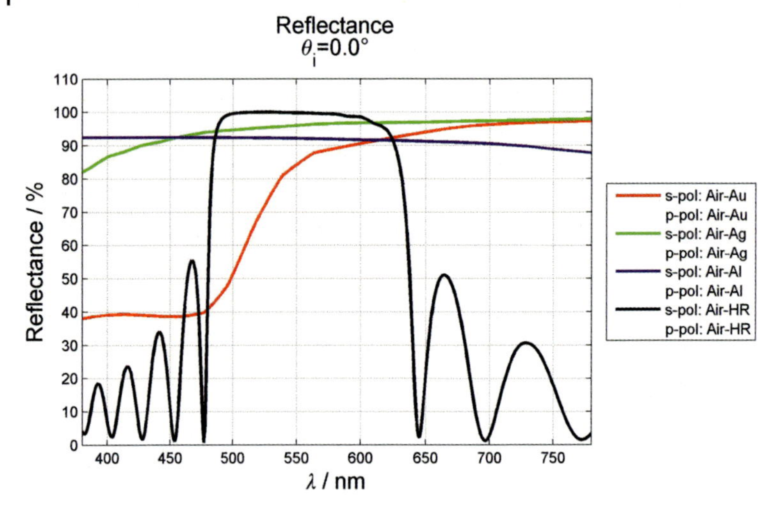

Figure 55-1: Spectral reflectances (VIS) in normal incidence of gold (red), silver (green), aluminum (blue) and a 21-layer MgF_2/Ta_2O_5 coating (black) on BK7 (s- and p-polarizations coincide for normal incidence).

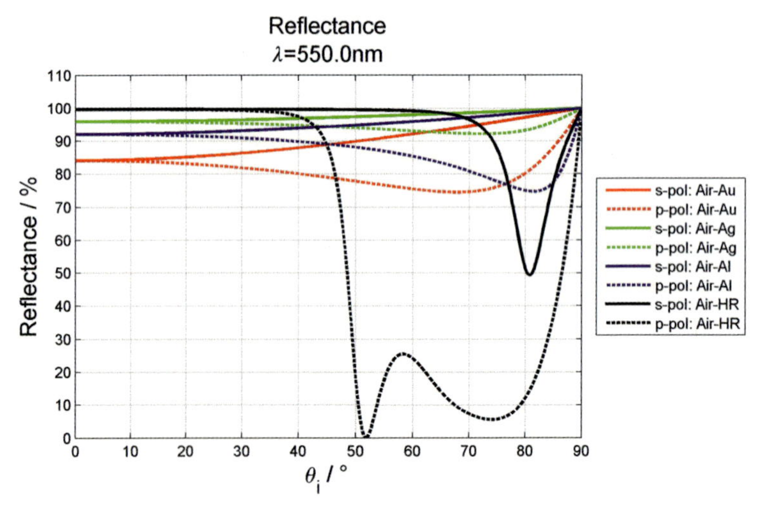

Figure 55-2: Reflectances (p- and s-polarization) as a function of incidence angle for $\lambda = 550$ nm for gold (red), silver (green), aluminum (blue) and a 21-layer MgF_2/Ta_2O_5 coating (black) on BK7.

To provide a glass substrate with an AR coating, a dielectric coating made of one or multiple layers of different transparent dielectric materials such as *Magnesium Fluoride* MgF_2 and *Tantalum Pentoxide* Ta_2O_5 may be used. A single layer should have a refractive index n_{AR} of

$$n_{AR} = \sqrt{n_0 n_{sub}} \tag{55-2}$$

where n_0 denotes the refractive index of the surrounding medium, usually air, and n_{sub} denotes the index of the glass substrate. The thickness d of the layer is calculated from (55-1) using n_{AR} as the refractive index of the layer.

Since there is no material matching the requirement of (55-2) exactly for all necessary wavelengths, a significant residual reflection will remain. Figure 55-3 shows normal incidence reflectances of a BK7 substrate compared with a single layer AR coating made of MgF_2 of thickness 98.9 nm on top of a BK7 substrate. The remaining reflectance is 1.5% at 550 nm wavelength; a third that of an uncoated substrate.

A significant improvement is achieved by a dielectric AR coating made of multiple layers of different transparent dielectric materials such as *Magnesium Fluoride* MgF_2 and *Tantalum Pentoxide* Ta_2O_5. Figure 55-3 shows an example of an AR coating made of 3 MgF_2 and 3 Ta_2O_5 layers in alternating sequence of optimized thickness. The remaining normal incidence reflectance is below 0.5% in the entire visible spectrum.

Alternatively to the alternating two-material coatings, a series of layers with small differences in refractive index can be used to create a broadband anti-reflective coating by means of a refractive index gradient.

Figure 55-4 shows the reflectances for s- and p-polarization of the uncoated BK7 substrate and the single and multi-layer AR coatings as a function of the incidence angle θ_i at a wavelength of 550 nm. The s-polarized parts increase rapidly, while the p-polarized parts drop to zero near the *Brewster angle*:

$$\theta_B = \arctan \frac{n_{sub}}{n_0} \tag{55-3}$$

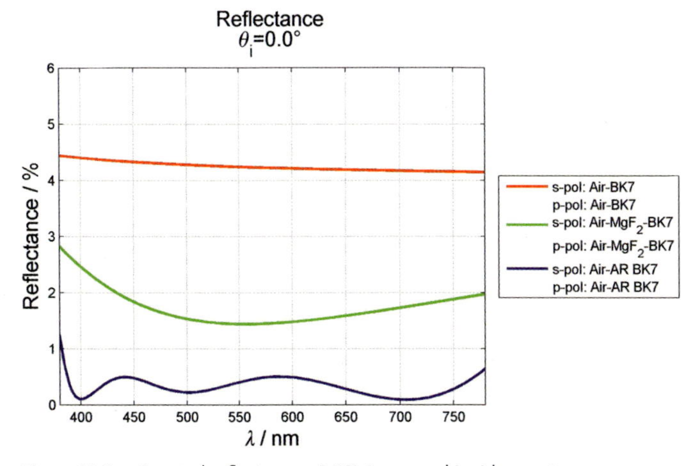

Figure 55-3: Spectral reflectances (VIS) in normal incidence at an uncoated BK7 substrate (red), a single-layer coating of MgF_2 on top of BK7 (green) and a 6-layer coating of alternating MgF_2 and Ta_2O_5-layers on BK7 (black) (s- and p-polarizations coincide for normal incidence).

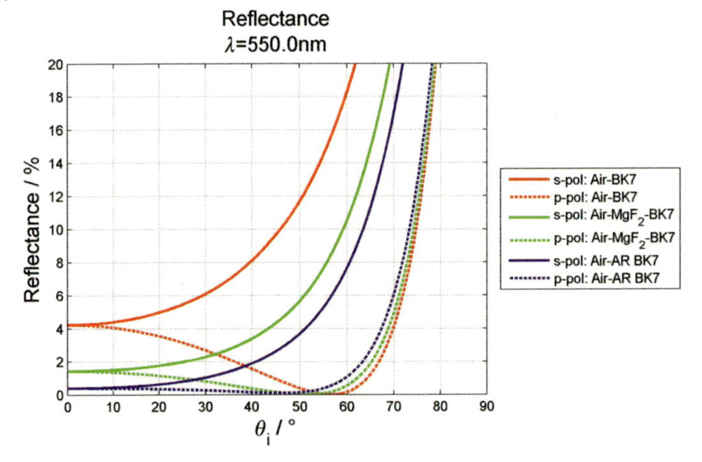

Figure 55-4: Reflectances (p- and s-polarization) as a function of incidence angle for $\lambda = 550$ nm at an uncoated BK7 substrate (red), a single-layer coating of MgF_2 on top of BK7 (green) and a six-layer coating of alternating MgF_2 and Ta_2O_5 layers on BK7 (blue).

In the EUV portion of the spectrum below a wavelength of 30 nm, as used in modern lithography, nearly all materials absorb strongly. The mirrors used for normal incidence reflections utilize multi-layer mirrors constructed of hundreds of alternating layers of a high-mass metal and a low-mass spacer. A suitable metal is molybdenum with a refractive index $n = 0.921$ and an extinction coefficient $k = 0.006$ at 13 nm wavelength. A suitable spacer material is silicon with $n = 0.999$ and $k = 0.002$. Each layer pair is designed to have a thickness equal to half the wavelength of the light to be reflected. Multi-layer mirrors in the EUV spectrum reflect up to 70% of the incident light at the particular wavelength for which they were optimized.

55.2 Specifications

An optical thin-film coating is described by a set of N layers, each layer i defined by a material of defined refractive index n_i, extinction coefficient k_i and a geometrical thickness d_i.

The optical constants of the surrounding medium are described by n_0 and k_0 and those of the substrate by n_{N+1} and k_{N+1}, accordingly (see figure 55-5).

The optical properties of a coated surface can be characterized by spectrophotometric values. They are related to the flux reflected from or transmitted through the coated surface, and also to the flux scattered from or absorbed by the coated surface.

The *spectral transmittance* $T(\lambda)$ is the ratio of the spectral concentration of the radiant or luminous flux transmitted to that of the incident radiation [55-5] and [55-6].

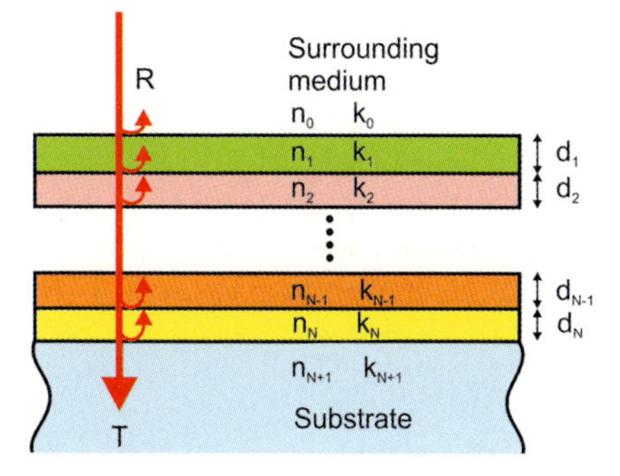

Figure 55-5: Thin-film coating on a substrate consisting of *N* individual layers.

The *spectral reflectance* $R(\lambda)$ is the ratio of the spectral concentration of the radiant or luminous flux reflected to that of the incident propagation.

The *spectral absorptance* $A(\lambda)$ is the ratio of the spectral concentration of radiant or luminous flux absorbed to that of the incident radiation. $A(\lambda)$ depends on the direction of the radiation.

The *spectral scatter* is the change in the spatial distribution of a radiation when it is deviated or scattered in many directions by a surface or by a medium without change of frequency of the monochromatic components from which the radiation is composed.

The metrology for transmittance, reflectance, absorptance and scatter are addressed in this chapter in more detail.

The following relations are valid:

$$1 = R + T + A \tag{55-4}$$

$$R = R_r + R_d \tag{55-5}$$

$$T = T_r + T_d \tag{55-6}$$

where:

R_r is the regular (specular) spectral reflectance,
R_d is the diffuse spectral reflectance,
T_r is the regular (specular) spectral transmittance,
T_d is the diffuse spectral transmittance.

The optical properties $R(\lambda)$, $T(\lambda)$ and $A(\lambda)$ generally vary with:

1) the wavelength λ,
2) the angle of incidence θ_i,
3) the state of polarization (s- or p-polarization),
4) the state of spatial distribution (direction)

of the incident light. 4) is related to the scatter behavior of the coating.

The optical properties may also vary with environmental parameters such as temperature, pressure and relative humidity.

The coating of an optical component is usually specified and tolerated by setting limits for the spectral reflectance $R(\lambda)$ and transmittance $T(\lambda)$, specifying the spectral range and the range of incident angles θ_i at which the component is to be used. Figure 55-6 shows an example. If polarization matters, the specification must express the range of polarization states at which the component is to be used.

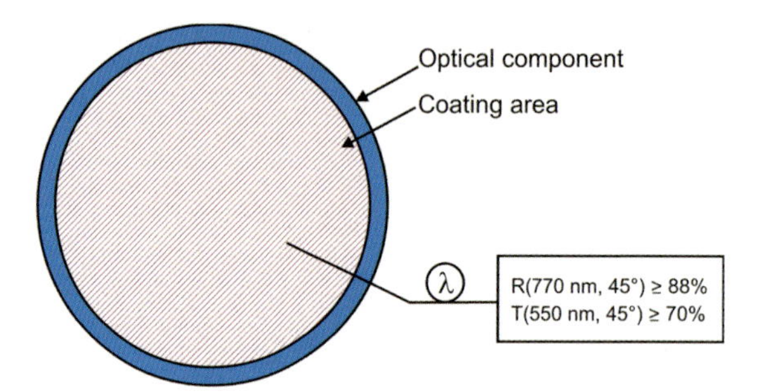

Figure 55-6: Tolerance for a coating on an optical surface specified by setting a lower limit for the reflectance and transmittance at $\lambda = 770$ nm and an angular inclination of 45°.

Specifications of optical coatings also include mechanical and chemical properties according to ISO 9211-4 [55-4]:

1) Abrasion resistance
   The specification limits the extent to which the optical and mechanical properties of optical coatings on components and substrates are affected when subjected to specific abrading conditions at ambient atmospheric conditions.
2) Adhesion
   The specification limits the extent to which the mechanical properties of optical coatings on components and substrates are affected when subjected to specific tensile or shear stress conditions at ambient atmospheric conditions.
3) Cross-hatch resistance
   The specification limits the extent to which the adhesion properties of optical coatings on components and substrates are affected after cutting the coating (distorting the stress and influencing the adhesion).
4) Solubility
   An appropriate specification limits the extent to which the optical and mechanical performance characteristics of optical coatings on components and substrates are affected after immersion in distilled or de-ionized water or a salt-water solution.

Tests related to the mechanical and chemical properties of coatings are beyond the scope of this book.

ISO 9211-4 also specifies point-like, line-like, area and volume imperfections on optical coatings such as those listed below.

1) Point-like imperfections
   a. Pinhole
   b. Spatter
   c. Particle
   d. Fine dust
   e. Nodule.
2) Line-like imperfections
   a. Scratches
   b. Hairline
   c. Crack
   d. Crazing
   e. Sleek.
3) Area imperfections
   a. Stain
   b. Abrasion
   c. Lint mark
   d. Coating void.
4) Volume imperfections
   a. Peeling
   b. Flaking
   c. Large spatter
   d. Large particle
   e. Blister and bubble.

The metrology for surface imperfections is covered in chapter 54.

55.3 Model Simulation

55.3.1 Transfer-matrix Method

The transfer-matrix method [55-7]–[55-10] is a mathematical procedure used to calculate the spectral reflectance and transmittance of a given thin-film coating defined by the individual layer thicknesses, refractive indices and extinction coefficients. The method is extremely helpful in cases when actual layer parameters must be determined from measured reflectance or transmittance results. The unknown parameters are then obtained by mathematical inversion.

In this section we describe the mathematical basics for the transfer-matrix forward calculation.

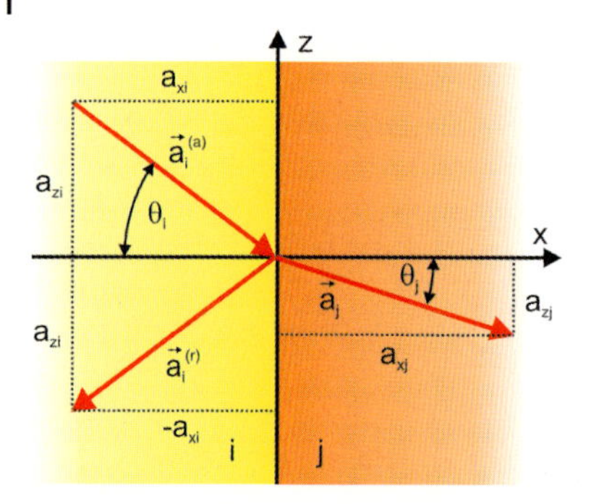

Figure 55-7: Transition of a plane wave from medium *i* to medium *j*.

We start with the transition of a plane wave from a medium i to a medium j as shown in figure 55-7. The transition is described by the *complex Fresnel equations* [46-2], which combine the transmitted and reflected complex field amplitudes with the amplitude of the incident wave.

A plane wave crossing the interface at an angle θ_i is partially reflected at an angle $-\theta_i$. The transmitted portion leaves the interface at an angle θ_j. $\vec{a}_i^{(a)}$ denotes the propagation vector of the incident wave, $\vec{a}_i^{(r)}$ denotes that of the reflected wave, $\vec{a}_j$ that of the transmitted wave.

The normal component of the incident propagation vector is described by

$$a_{xi} = k_0\sqrt{N_i^2 - N_{\text{eff}}^2} \tag{55-7}$$

where

$k_0 = \frac{2\pi}{\lambda}$ and

$N_{\text{eff}} = n_i \sin\theta_i = n_j \sin\theta_j$ describes *Snell's law of refraction*,

$N_i = n_i - ik_i$ denotes the complex refractive index of the medium i, (note: in physics the relation $N_i = n_i + ik_i$ is used)

n_i denotes the refractive index of medium i,

k_i denotes the extinction coefficient of medium i.

Since N_{eff} is equal for all interfaces within the thin-film stack, the normal component a_{xi} can easily be calculated for all interfaces by applying the incident angle θ_0 and the refractive index n_0 of the surrounding medium.

The tangential component of the incident propagation vector is equal for all interfaces and is described by

$$a_{zi} = k_0 N_{\text{eff}} \tag{55-8}$$

The Fresnel equations for the reflected field components are expressed as

$$r_{s,ij} = \frac{a_{xi} - a_{xj}}{a_{xi} + a_{xj}} \tag{55-9}$$

for s-polarization and

$$r_{p,ij} = \frac{N_j a_{xi} - N_i a_{xj}}{N_j a_{xi} + N_i a_{xj}} \tag{55-10}$$

for p-polarization.

The Fresnel equations for the transmitted field components are expressed as

$$t_{s,ij} = \frac{2a_{xi}}{a_{xi} + a_{xj}} \tag{55-11}$$

for s-polarization and

$$t_{p,ij} = \frac{2N_i N_j a_{xi}}{N_j a_{xi} + N_i a_{xj}} \tag{55-12}$$

for p-polarization.

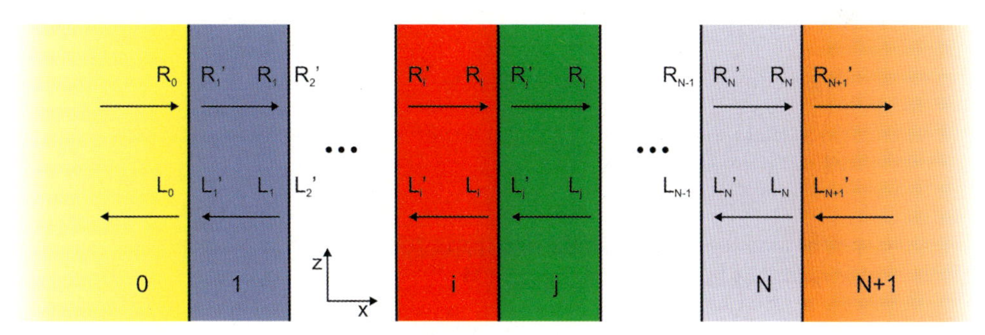

Figure 55-8: Scheme of a layer system consisting of *N* layers and the half-spaces 0 and *N*+1.

To fulfill the *stationary Helmholtz equation* [55-7], the solution for the electrical field in each layer is selected as a linear combination of a wave traveling in the positive and negative x-direction:

$$E_i(x) = R_i e^{ik_{xi}x} + L_i e^{-ik_{xi}x} \tag{55-13}$$

where R_i denotes the amplitudes of the waves travelling in the positive (to the right) and L_i denotes those travelling in the negative (to the left) x-direction. The s-polarized component of the electrical field vector is perpendicular to the drawing plane and is therefore only dependent on y, whereas the p-polarized component is dependent on x and z.

Figure 55-8 shows the scheme of a layer system consisting of N layers characterized by their complex refractive index N_i and their thickness d_i. The half-space to the left is characterized by the complex refractive index N_0, the one to the right by N_{N+1}. Partial waves to the right of an interface are primed, those to the left of an interface are unprimed. The relationship between the partial waves to the left and right of the interface ij is given by

$$\begin{pmatrix} R_i \\ L_i \end{pmatrix} = \frac{1}{t_{ij}} \begin{bmatrix} 1 & r_{ij} \\ r_{ij} & 1 \end{bmatrix} \begin{pmatrix} R_j' \\ L_j' \end{pmatrix} = \mathbf{D}_{ij} \begin{pmatrix} R_j' \\ L_j' \end{pmatrix} \tag{55-14}$$

where $\mathbf{D}_{ij}$ denotes the transition matrix for the interface ij, and r_{ij} and t_{ij} denote the reflected and transmitted field components – either for s- or p-polarization – as given in (55-9)–(55-12).

The primed and unprimed components L_i and R_i within a layer i differ by a phase ϕ_i calculated by

$$\phi_i = a_{xi} d_i = k_0 \sqrt{N_i^2 - N_{\text{eff}}^2}\, d_i \tag{55-15}$$

and implemented in a phase matrix $\mathbf{P}_i$

$$\begin{pmatrix} R_j' \\ L_i' \end{pmatrix} = \begin{bmatrix} e^{-i\phi_i} & 0 \\ 0 & e^{i\phi_i} \end{bmatrix} \begin{pmatrix} R_i \\ L_i \end{pmatrix} = \mathbf{P}_i \begin{pmatrix} R_i \\ L_i \end{pmatrix} \tag{55-16}$$

Combining (55-14) and (55-16) leads to

$$\begin{pmatrix} R_j' \\ L_i' \end{pmatrix} = \mathbf{P}_i \mathbf{D}_{ij} \begin{pmatrix} R_j' \\ L_j' \end{pmatrix} \tag{55-17}$$

The amplitude vectors in the half-spaces 0 and N+1 of an N-layer system can then be related by the total transfer matrix $\mathbf{M}$

$$\begin{pmatrix} R_0 \\ L_0 \end{pmatrix} = \mathbf{M} \begin{pmatrix} R_{N+1}' \\ L_{N+1}' \end{pmatrix} = \begin{bmatrix} M_{11} & M_{12} \\ M_{21} & M_{22} \end{bmatrix} \begin{pmatrix} R_{N+1}' \\ L_{N+1}' \end{pmatrix} \tag{55-18}$$

with $\mathbf{M}$ given by

$$\mathbf{M} = \begin{bmatrix} M_{11} & M_{12} \\ M_{21} & M_{22} \end{bmatrix} = \mathbf{D}_{01} \prod_{i=1}^{N} \mathbf{P}_i \mathbf{D}_{i,i+1} \tag{55-19}$$

(55-19) can be used to calculate a 2 × 2 matrix for any thin-film system consisting of N homogeneous layers of known thickness and complex refractive indices deposited on and surrounded by media of known complex refractive index.

Note that s- and p-polarizations need different transfer matrices $\mathbf{M}_s$ and $\mathbf{M}_p$, respectively, due to the fact that $\mathbf{D}_{ij}$ is formed by r_{ij} and t_{ij} expressing either s- or p-polarized components, as given by (55-9)–(55-12).

The resulting amplitudes of the reflected field components are then calculated from

$$r = \frac{L_0}{R_0} = \frac{M_{21}}{M_{11}} \tag{55-20}$$

whereas the amplitudes of the transmitted field components are calculated from

$$t = \frac{R'_{N+1}}{R_0} = \frac{1}{M_{11}} \tag{55-21}$$

r_s, r_p, t_s, and t_p are then calculated from (55-20) and (55-21) using either $\mathbf{M}_s$ or $\mathbf{M}_p$.

55.3.2
Material Designation

Thin-film coating material constants n and k are found in literature and databases, for instance in [55-11]. In many cases tables of λ, $n(\lambda)$ and $k(\lambda)$ are provided over a certain wavelength range. In this case, an interpolation has to be made for arbitrary wavelengths within the valid range.

For glass materials n is provided by databases and catalogues, for instance in [55-17], while k is assumed to be zero. In most cases constants for the *Sellmeier dispersion formula* (52-2) are given to enable the calculation of n for arbitrary wavelengths within a specified range (see section 52.2).

Alternatively to n and k, the optical properties can be represented as the complex dielectric function $\tilde{\varepsilon}$

$$\tilde{\varepsilon} = \varepsilon_1 - i\varepsilon_2 \tag{55-22}$$

(note: in physics the relation $\tilde{\varepsilon} = \varepsilon_1 + i\varepsilon_2$ is used)
which is related to the complex refractive index N by

$$\tilde{\varepsilon} = N^2 = n^2 - k^2 + i2nk \tag{55-23}$$

yielding the expressions

$$\varepsilon_1 = n^2 - k^2 \tag{55-24}$$

$$\varepsilon_2 = 2nk \tag{55-25}$$

Non-absorbing materials therefore have real refractive indices $N = n$ and real dielectric functions $\tilde{\varepsilon} = \varepsilon_1$.

For absorbing materials the extinction coefficient $k \neq 0$ is related to the absorption coefficient a in *Beer's Law*

$$I(x) = I(0)e^{-ax} = I(0)e^{-\frac{4\pi k}{\lambda}x} \tag{55-26}$$

in which the loss in irradiance $I(x)$ is described for light traveling in the material over a distance x.

In the data analysis of spectroscopic ellipsometry, the dielectric function is required. However, when the dielectric function of a sample is not known, it is necessary to model the dielectric function. There are many dielectric function models which are more or less appropriate to describe the physical properties of a material. Absorbing materials will often have a transparent wavelength region that can be modeled by the *Sellmeier dispersion formula*. However, the absorbing region must account for both real and imaginary optical constants. There are dispersion relationships using oscillator theory to describe absorption for various materials [55-12]. These include the *Lorentz, Harmonic*, and *Gaussian oscillators*, which share similar attributes, where the absorption features are described by an amplitude A, broadening B, and center energy E_c (related to the frequency of light). An offset $\varepsilon_{1,\text{offset}}$ to the real component is added to account for extra absorption outside the measured spectral region. As an example, the *Lorentz oscillator* will be described as

$$\tilde{\varepsilon} = \varepsilon_{1,\text{offset}} + \frac{AE_c}{E_c^2 - E^2 - iBE} \tag{55-27}$$

with the photon energy E given by

$$E = h\nu = \frac{hc}{\lambda} \tag{55-28}$$

where h is Planck's constant and c is the speed of light.

The photon energy is usually expressed in electronvolts eV and can be converted to the wavelength λ expressed in nm by

$$E(\text{eV}) = \frac{1240\text{eVnm}}{\lambda(\text{nm})} \tag{55-29}$$

When ellipsometric data is available, A, B, E_c and $\varepsilon_{1,\text{offset}}$ can be iteratively modeled, until the model corresponds with the measured data. The model can then be used to define a table of either λ, $n(\lambda)$, $k(\lambda)$ or λ, $\varepsilon_1(\lambda)$, $\varepsilon_2(\lambda)$.

55.3.3 Graded Interfaces

There are many applications in which the optical constants do not abruptly change between adjacent material layers, but rather vary gradually between the values for the pure materials on either side of the interface, mainly caused by diffusion processes or the individual production processes [55-13] and [55-14]. In the case of non-abrupt interfaces there is a resultant loss in specular reflectance, which can be approximated by multiplying the Fresnel reflection coefficients by an appropriate weighting factor or by introducing a diffuse interface describing the transition between n_i and n_j as well as between k_i and k_j (figure 55-9) [55-15].

For simulation purposes the transfer-matrix method can be applied by introducing a number N_g of thin constant layers describing a pseudo-smooth transition interface between the pure optical material constants. For the transition, an interface profile function must be defined, which best describes the material distribution.

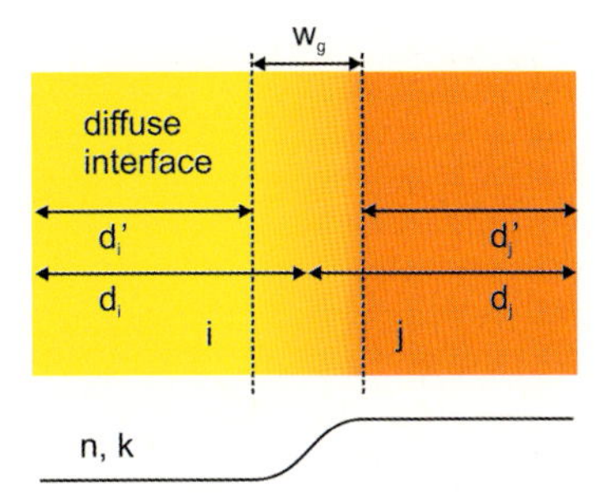

Figure 55-9: Diffuse interface of width w_g, in which n and k vary gradually between the pure material constants n_i, n_j and k_i, k_j.

In the following we describe:
a) a linear,
b) a sinusoidal; and
c) an exponential
transition between the layers i and j (figure 55-10).

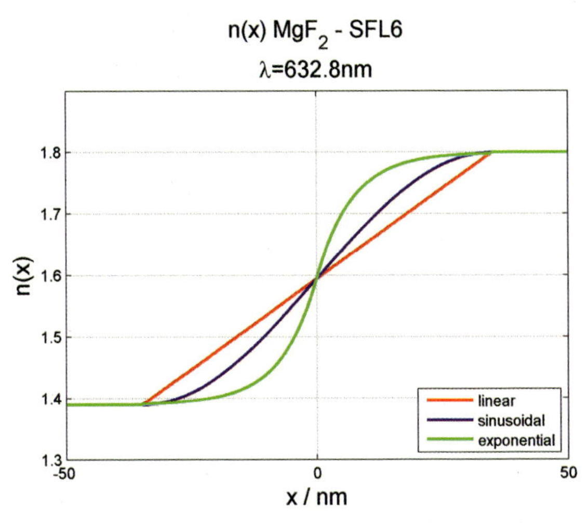

Figure 55-10: Transition of $n(x)$ between the refractive indices of MgF_2 and SFL6 for $\lambda = 632.8$ nm using a linear (red), sinusoidal (blue) and exponential (green) model.

The transition layers within the diffuse interface carry the number $l = 1 \ldots N_g$. The lth thin layer has the following optical constants [55-16] (p denotes either n or k):

a) Linear Transition

$$p_l = \frac{p_i + p_j}{2} + (p_i - p_j)\left(\frac{1}{2} - \frac{l}{N_g + 1}\right) \tag{55-30}$$

b) Sinusoidal Transition

$$p_l = \frac{p_i + p_j}{2} + \frac{p_i - p_j}{2} \sin\left[\pi\left(\frac{l}{N_g + 1} - \frac{1}{2}\right)\right] \quad (55\text{-}31)$$

c) Exponential Transition

$$p_l = p_i + \frac{p_j - p_i}{2} e^{4\sqrt{6}\left(\frac{l}{N_g+1} - \frac{1}{2}\right)} \quad \text{for} \quad l \le \frac{N_g}{2} \quad (55\text{-}32)$$

$$p_l = p_i + \frac{p_j - p_i}{2}\left(2 - e^{-4\sqrt{6}\left(\frac{l}{N_g+1} - \frac{1}{2}\right)}\right) \quad \text{for} \quad l > \frac{N_g}{2} \quad (55\text{-}33)$$

The thickness d_l of each transition layer is calculated by defining an interface or transition width $w_g \le d_i + d_j$ over which the smooth transition will occur:

$$d_l = \frac{w_g}{N_g} \quad (55\text{-}34)$$

In addition, a distribution factor X_g can be specified with $0 \le X_g \le 1$ describing the remaining thicknesses d_i' and d_j' of the pure material layers i and j:

$$d_i' = d_i - w_g(1 - X_g) \quad (55\text{-}35)$$

$$d_j' = d_j - w_g X_g \quad (55\text{-}36)$$

55.3.4
Surface Roughness

The roughness of an interface can be treated in the same way as a diffuse surface [55-12]. Since a light beam usually extends over a certain area in y and z, mean values for n and k will be effective. For the simulation of rough interfaces we can again introduce a number N_g of thin constant layers over an interface thickness w_g related to the roughness of the interface between layers i and j. The transitions of n and k must be modeled by a suitable transition function as described in the previous section.

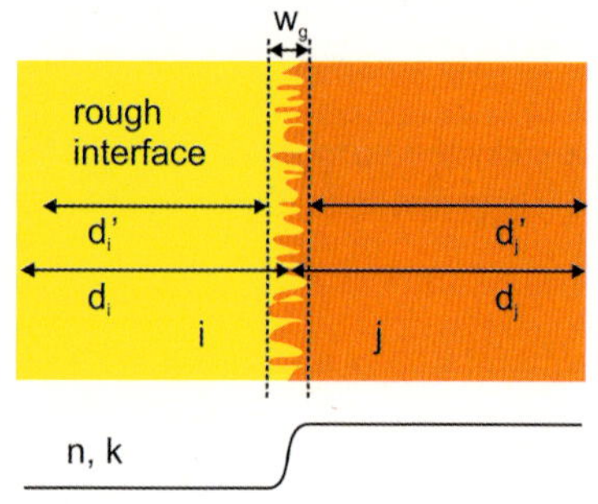

Figure 55-11: Model of a rough interface of width w_g, in which n and k vary gradually between the pure material constants n_i, n_j and k_i, k_j.

55.3.5 Data Analysis

In a coating production process it is necessary to control the final reflectance or transmittance of the components. Depending on the application and the requirement specifications, either every single part or some random samples are to be inspected to control the production process. It is also a common practice to inspect witness samples, which were coated along with the original parts.

In either case an appropriate *reflectometer* or *transmittometer* must be applied to measure reflectance $R(x,y,\lambda,\theta_i)$ or transmittance $T(x,y,\lambda,\theta_i)$ at one or several local positions x,y over a wavelength range $\lambda_1 \leq \lambda \leq \lambda_2$ and an incident angle range $\theta_i^{(1)} \leq \theta_i \leq \theta_i^{(2)}$.

If polarization matters, a *polarimeter* or an *ellipsometer* must be used to detect the s- and p-polarized reflectance components $R_s(x,y,\lambda,\theta_i)$ and $R_p(x,y,\lambda,\theta_i)$ or the s- and p-polarized transmittance components $T_s(x,y,\lambda,\theta_i)$ and $T_p(x,y,\lambda,\theta_i)$.

If polarization and phase matter, an *ellipsometer* must be used to detect the s- and p-polarized complex reflectance components $r_s(x,y,\lambda,\theta_i)$ and $r_p(x,y,\lambda,\theta_i)$ or the s- and p-polarized complex transmittance components $t_s(x,y,\lambda,\theta_i)$ and $t_p(x,y,\lambda,\theta_i)$.

Usual ellipsometers measure the full Jones matrix of a reflective sample. The sample is characterized by the parameters ψ and $\Delta\phi$ as a function of incident angle and wavelength, as explained in chapter 51.

The diattenuation ratio is denoted by $\tan\psi$

$$\tan\psi = \frac{|r_p|}{|r_s|} \text{ for reflection or } \tan\psi = \frac{|t_p|}{|t_s|} \text{ for transmission} \tag{55-37}$$

and $\Delta\phi$ denotes the retardance

$$\Delta\phi = \phi_s - \phi_p = \arctan\frac{\mathrm{Im}\{r_s\}}{\mathrm{Re}\{r_s\}} - \arctan\frac{\mathrm{Im}\{r_p\}}{\mathrm{Re}\{r_p\}} \text{ for reflection or}$$

$$\Delta\phi = \phi_s - \phi_p = \arctan\frac{\mathrm{Im}\{t_s\}}{\mathrm{Re}\{t_s\}} - \arctan\frac{\mathrm{Im}\{t_p\}}{\mathrm{Re}\{t_p\}} \text{ for transmission} \tag{55-38}$$

of the sample, where ψ and $\Delta\phi$ are functions of the wavelength, the incidence angle and the local position x,y on the surface under test.

As an example, figure 55-12 shows the reflectances R_s, R_p and their unpolarized average $R_u = \frac{R_s + R_p}{2}$ as well as the phases ϕ_s, ϕ_p and ψ and $\Delta\phi$ as a function of the incident angle θ_i and as a function of wavelength λ for an AR multi-layer coating on a BK7 substrate.

These curves will vary, if

- the thickness d_i,
- the refractive index $n_i(\lambda)$,
- the extinction coefficient $k_i(\lambda)$,

- the thickness w_g of the rough or diffuse interfaces, or
- the polarizing characteristics such as the depolarization, retardation, dichroism

of any layer i deviate from their expected magnitude.

For analysis purposes the measured curves can be used to estimate the real parameters of the multi-layer coating by *mathematical inversion*, in which the transfer-matrix method is used to iteratively alter the unknown layer parameters d_i, $n_i(\lambda)$, $k_i(\lambda)$ or the corresponding material model parameters (Sellmeier or oscillator coefficients, etc.) until the corresponding calculated ψ_{calc} and $\Delta\phi_{calc}$ values of the model best fit the measured values ψ_{meas} and $\Delta\phi_{meas}$ [55-12]. The technique is called *linear regression analysis*, for which a variety of different mathematical methods such as *Newton's method* or the *Levenberg–Marquardt method* can be employed [55-18] and [55-19].

Note that the optical model used in ellipsometry analysis merely represents an approximated sample structure and the results obtained are not necessarily correct even when the fit is sufficiently good. Thus, if the optical constants or thin-film structures are not well known, it is very likely that a model which deviates strongly from the real configuration will be obtained. The n- and k-values depend on the deposition process and the processing temperature. Different materials show different deposition growth characteristics such as columnar, porous or compact, producing unwanted depolarization effects. The model must then be justified by using different additional measurement methods such as scanning electron microscopy (SEM), transmission electron microscopy (TEM), atomic force microscopy (AFM) or X-ray reflectometry [55-12].

As an example, figure 55-13 (a) and (b) show the effect of a layer-thickness change by +10% in the AR coating from figure 55-12. ψ and $\Delta\phi$ are presented as a function of the incident angle θ_i and as a function of wavelength λ for an AR multi-layer coating. The solid curves represent the parameters with correct layer thicknesses d_i, while the dotted curves represent those with 1.1 d_i. The changes in ψ are very small compared with those of $\Delta\phi$. The latter seem to be sensitive to changes in wavelength.

Figure 55-13 (c) and (d) show the effect of layer-thickness change by +2% in the HR-coating from figure 55-1. ψ and $\Delta\phi$ react much more sensitively to thickness changes than the AR coating.

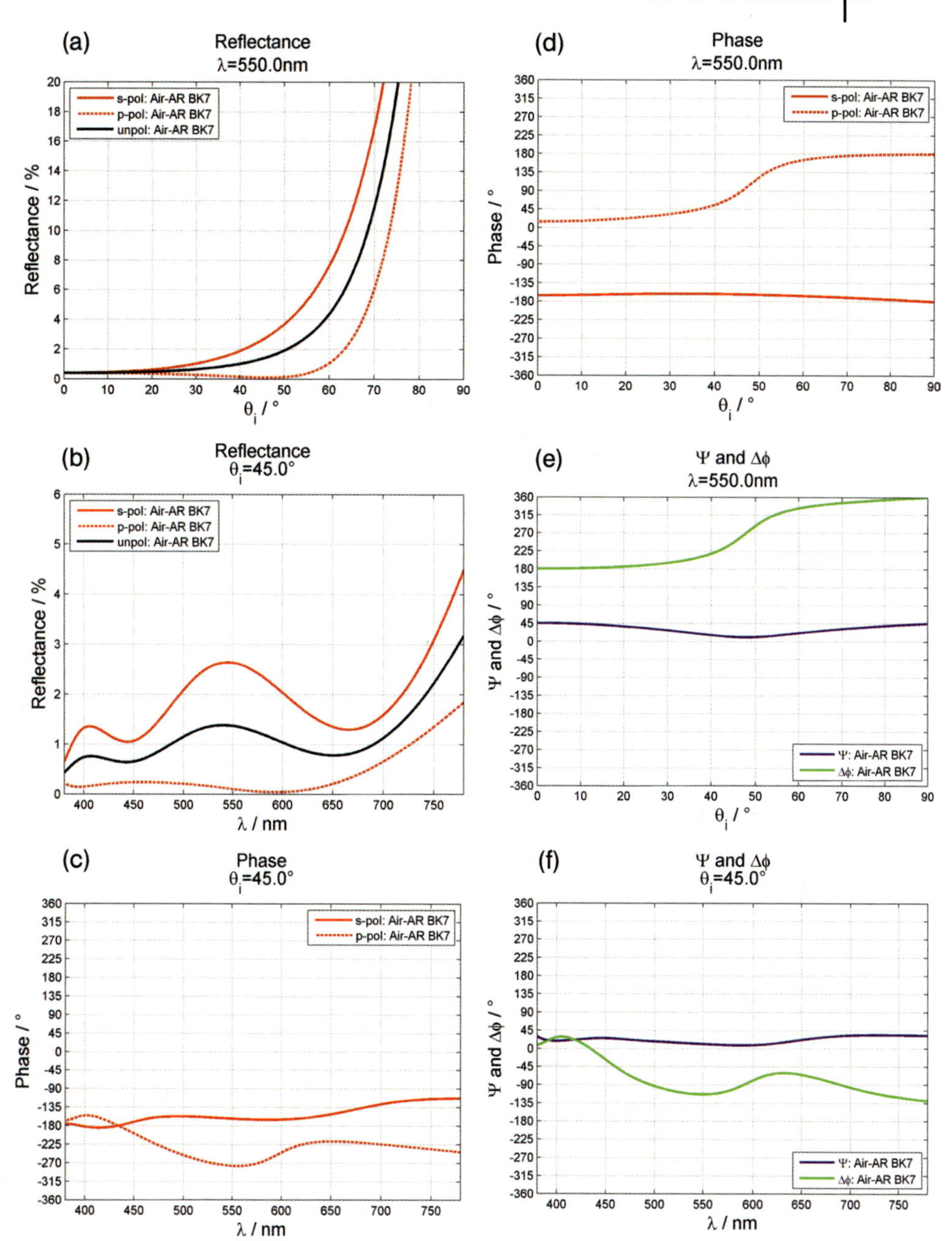

Figure 55-12: Coating characteristics of an AR multi-layer coating on a BK7 substrate: at $\lambda = 550$ nm as a function of incident angle θ_i: (a) reflectance; (b) phase; (c) ψ and $\Delta\phi$ at $\theta_i = 45°$; as a function of wavelength λ: (d) reflectance; (e) phase; (f) ψ and $\Delta\phi$.

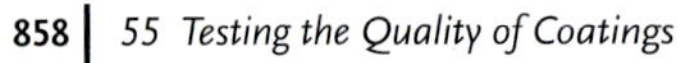

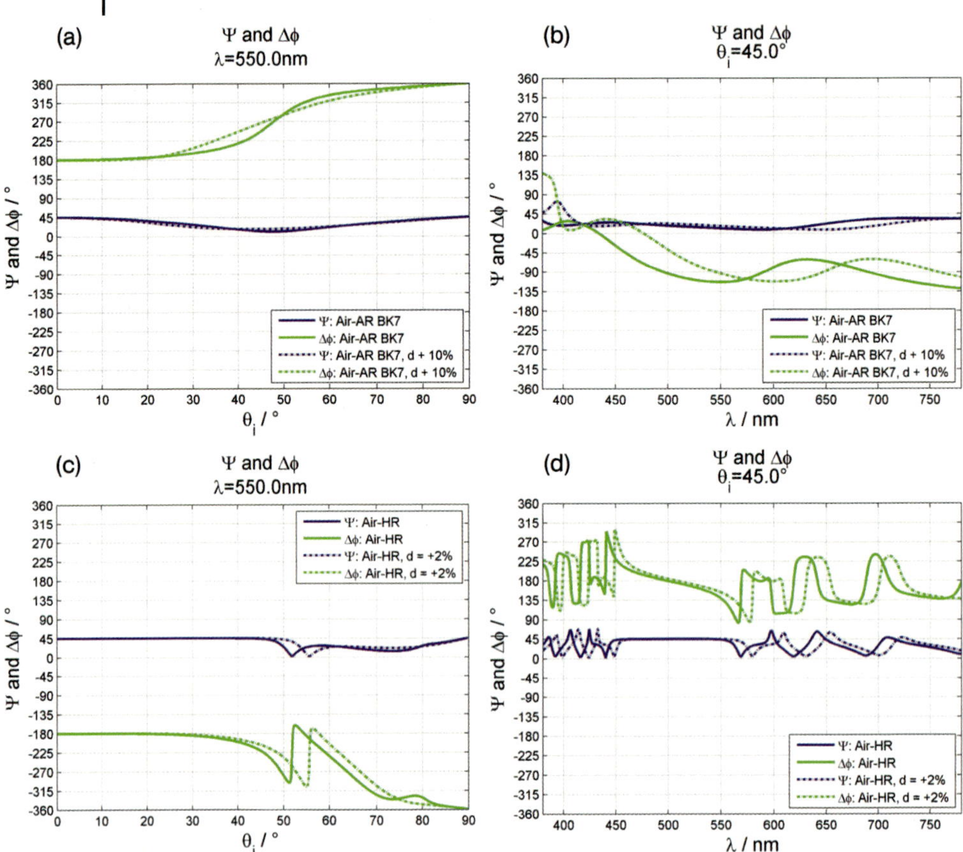

Figure 55-13: ψ and $\Delta\phi$ of an AR multi-layer coating ((a) and (b)) on a BK7 substrate and of an HR coating ((c) and (d)) with correct layer thicknesses d_i and thicknesses changed by +10% and +2%, respectively. (a) AR coating at $\lambda = 550$ nm as a function of incident angle θ_i; (b) AR coating at $\theta_i = 45°$ as a function of wavelength λ; (c) HR coating at $\lambda = 550$ nm as a function of incident angle θ_i; (d) HR coating at $\theta_i = 45°$ as a function of wavelength λ.

55.4 Coating Metrology

In this chapter we will shortly review the main photometer and ellipsometer configurations and explain their optical layouts. For further details concerning theory and evaluation techniques we refer to [55-20]–[55-25] and to chapter 51. We will discuss the metrology for spectral transmittance, spectral reflectance and spectral absorptance. For metrology concerning spectral scatter we refer to section 54.2.2.

55.4.1
Basics

Photometer and ellipsometer configurations generally consist of
a) a *sending channel* including

1. a light source,
2. a spectral filter or monochromator,
3. polarizing elements,
4. collimating optics,
5. a field stop,
6. an aperture stop

integrated in a common housing or unit
and
b) a *receiving channel* including

1. a detector.
2. a spectral filter or monochromator if not included in the sending channel,
3. polarizing elements if not included in the sending channel,
4. imaging optics,
5. an aperture stop

integrated in a common housing or unit.

The sample is placed between the sending and receiving units for transmittance inspection, or in front of the units for reflectance inspection. Depending on the application, both units are attached to positioning devices which enable tilt and lateral positioning. In the following examples, the sample is represented as a dashed line symbolizing either a reflective or a transmitting element. In principle all polarimetric configurations measuring a single spot on a sample can be arranged using this scheme.

Figure 55-14 shows a simple photometer setup used with monochromatic, unpolarized light. Light from a broadband light source is imaged onto the sample by means of collimators 1 and 2. The spot size on the sample can be adjusted by a field stop in front of the light source. The flux passing through the sending unit can be regulated by an aperture stop in front of collimator 1.

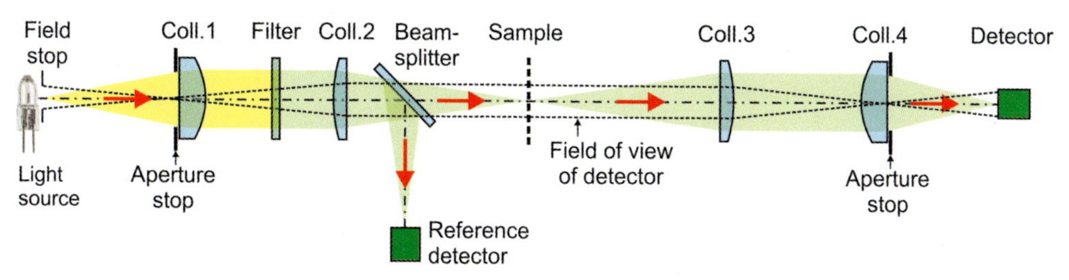

Figure 55-14: Monochromatic unpolarized photometer setup used as either a reflectometer or a transmittometer.

A color filter between the collimators provides the necessary spectral distribution. A reference detector at the exit of the sending unit receives its light via a beam-splitter. The signal from the reference detector is used to normalize the detector signal in the case of light source drifts and fluctuations. The beam-splitter must be designed so that it leaves the transmitted light unpolarized. Collimator 2 should have a low numerical aperture to provide only a small angle distribution for the test light arriving at the sample.

The receiving unit consists of collimators 3 and 4, which image the sample onto the detector. An additional aperture stop behind collimator 4 is used to adjust the received flux. The detector size must be large enough to provide enough space for the test light-spot image even in the case of small misalignments.

Figure 55-15 shows a polarized photometer, in which the color filter is replaced by a monochromator in order to select an arbitrary wavelength. Between collimators 1 and 2 a polarizer can be rotated to well-defined positions in order to select the required polarization plane. The exit slit of the monochromator is imaged onto the sample and then onto the detector by the optics in the receiving channel.

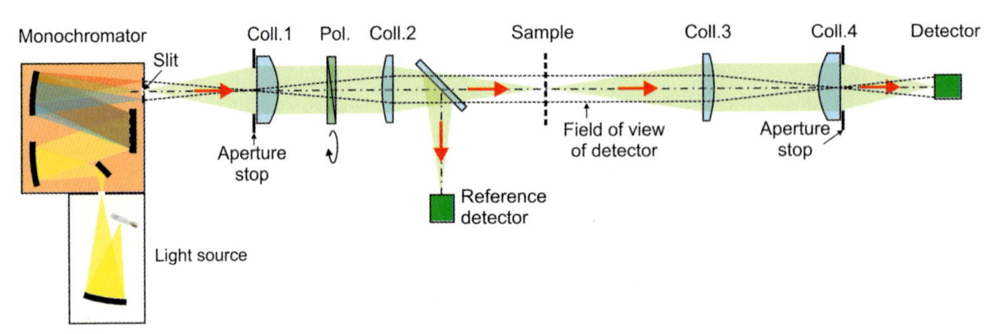

Figure 55-15: Spectral polarized photometer setup used as either a reflectometer or a transmittometer.

Figure 55-16 shows a spectral dual rotating-compensator polarimeter or ellipsometer in the PC_rSC_rA arrangement as described in section 51.4.5. It is used for spectral ellipsometry to determine all 16 elements of the Mueller matrix. A monochromator is used as a light source to select the required wavelength. A polarizer and a rotating compensator are placed between collimators 1 and 2. A rotating compensator and an analyzer are placed between collimators 3 and 4 of the receiving channel. The exit slit of the monochromator is imaged onto the sample and then onto the detector by the optics in the receiving channel.

Figure 55-17 shows a dual rotating-compensator spectroscopic polarimeter or ellipsometer in the PC_rSC_rA arrangement as described in section 51.4.5. With respect to figure 55-16 the monochromator is replaced by a broadband light source, whereas the detector in the receiving channel is now replaced by a multichannel spectrometer. The spectrometer evaluates the spectrum in real time using a fast line detector and appropriate signal processing. The instrument determines all 16 elements of

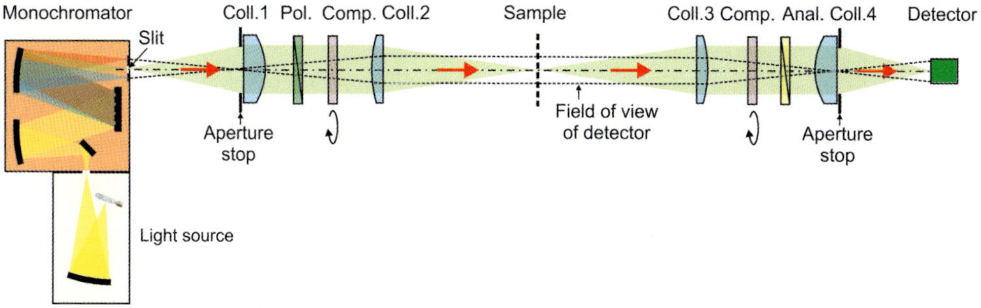

Figure 55-16: Configuration PC_rSC_rA of a dual rotating-compensator spectral polarimeter or ellipsometer using a monochromator to select arbitrary wavelengths.

the Mueller matrix. A polarizer and a rotating compensator are placed between collimators 1 and 2. A rotating compensator and an analyzer are placed between collimators 3 and 4 of the receiving channel. The exit slit of the light source is imaged onto the sample and then onto the entrance slit of the spectrometer by the optics in the receiving channel.

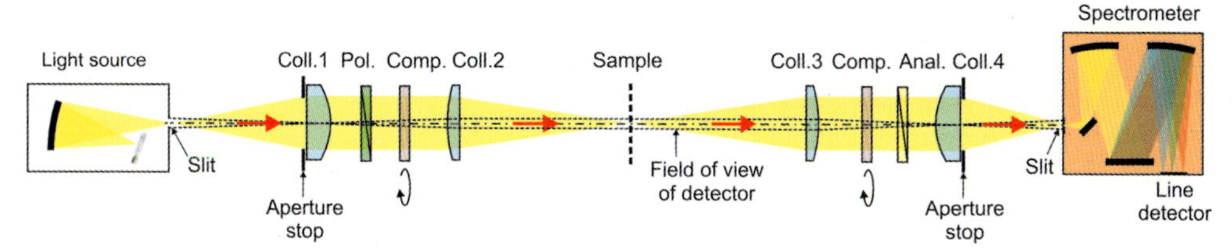

Figure 55-17: Configuration PC_rSC_rA of a dual rotating-compensator multichannel ellipsometer using a line detector spectrometer in the receiving channel.

55.4.2
Spectral Transmittance

Spectral photometers always integrate over a spectral range from λ_1 to λ_2. The measured value for transmittance is therefore

$$T_{\text{meas}}(\lambda) = \frac{\int_{\lambda_1}^{\lambda_2} T(\lambda)E(\lambda)S(\lambda)d\lambda}{\int_{\lambda_1}^{\lambda_2} E(\lambda)S(\lambda)d\lambda} \tag{55-39}$$

where

$E(\lambda)$ is the spectral emissivity of the light source and

$S(\lambda)$ is the spectral sensitivity of the detector system.

When $E(\lambda)$ and $S(\lambda)$ are constant over a spectral range from λ_1 to λ_2, (55-39) can be simplified to

$$T_{\text{meas}}(\lambda) = \frac{\int_{\lambda_1}^{\lambda_2} T(\lambda) d\lambda}{\lambda_2 - \lambda_1} \tag{55-40}$$

Transmittance of Plane-parallel Elements

When transmittance of a component is measured, the transmitted flux depends on the internal transmittance T_i of the component as well as on the reflection losses at both surfaces. We can calculate the expected transmittance using the transfer-matrix method including the substrate material and the coatings from both surfaces as shown in figure 55-18. We assume a stack of N layers on the front surface and a stack of M layers on the back surface.

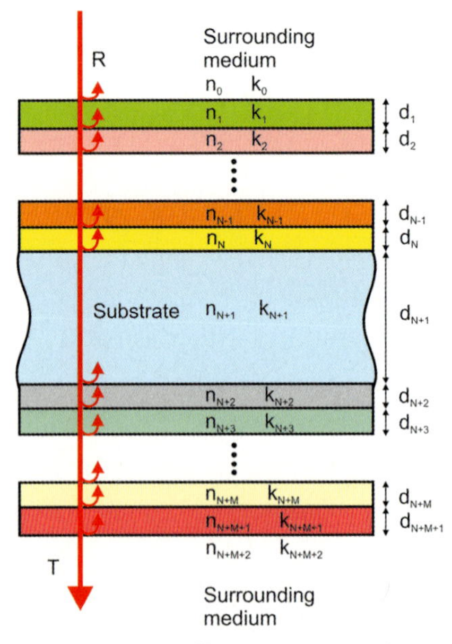

Figure 55-18: Thin-film coatings on a substrate consisting of N individual layers on the front surface and M layers on the back surface.

Since a plane-parallel substrate acts as a *Fabry–Perot resonator*, as described in section 46.2.14, the transmitted flux depends on the reflectivities of both surfaces and the thickness d_{N+1} of the substrate. Figure 55-19 shows the transmittances of a plane plate with $d_{N+1} = 1$ µm ((a) and (b)) and another plane plate with $d_{N+1} = 1$ mm ((c) and (d)), both made of BK7. The plane plates are uncoated (red), coated with a single-layer AR coating (green) and coated with a multi-layer AR-coating (blue). In normal incidence the transmittances vary strongly with wavelength λ. The frequency of the variation increases with the thickness d_{N+1} of the element. The amplitude of the variation increases with the reflectances of the surfaces. In order to measure mean transmittances, the spectral bandwidth from λ_1 to λ_2, as indicated in (55-39) and (55-40), should be large enough to average over several resonance periods in the spectrum. If possible, the element should carry a wedge, so that the measuring beam laterally extends over several periods.

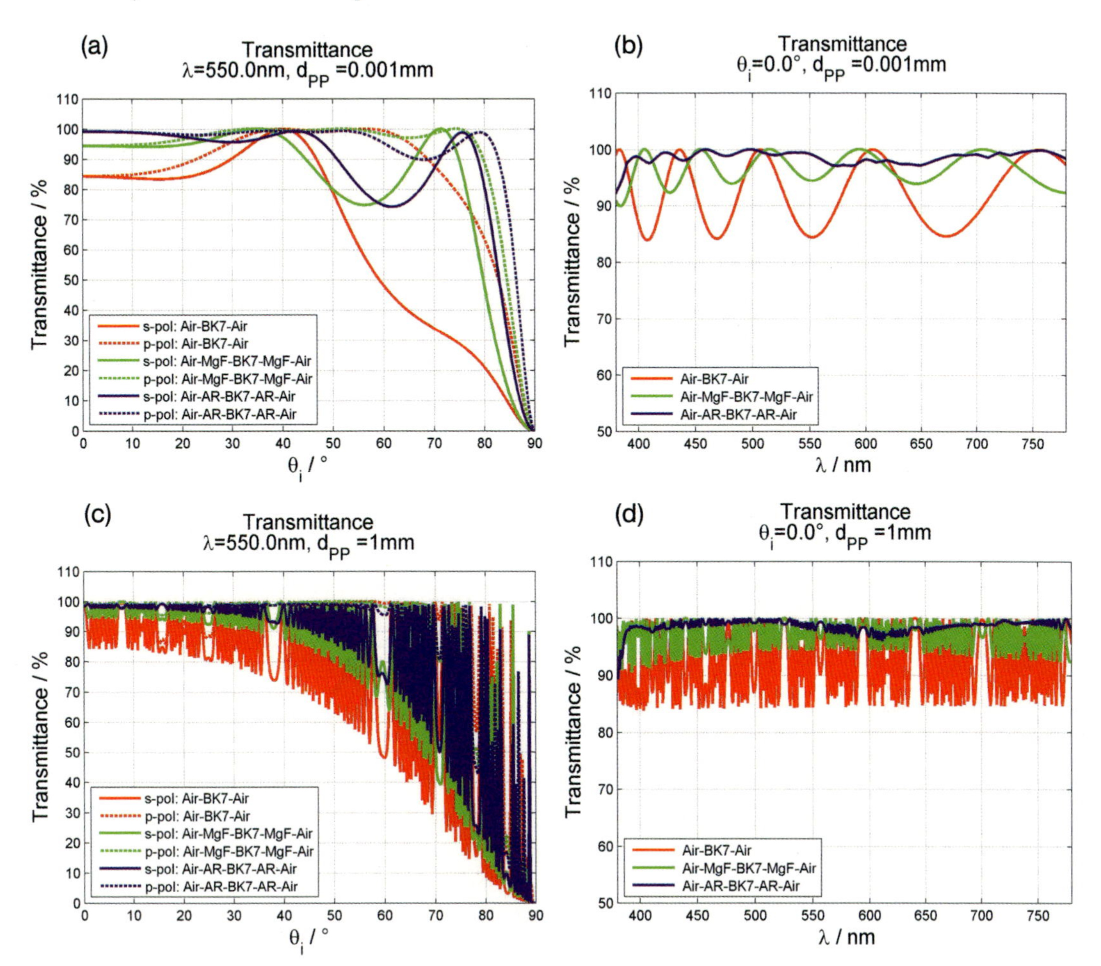

Figure 55-19: Transmittances of a plane plate made of BK7, uncoated (red), with single-layer AR-coating (green), with multi-layer AR-coating (blue): (a) thickness 1 μm, $\lambda = 550$ nm, T as function of incidence angle θ_i; (b) thickness 1 μm, normal incidence, T as function of λ; (c) thickness 1 mm, $\lambda = 550$ nm, T as function of incidence angle θ_i; (d) thickness 1 mm, normal incidence, T as function of λ.

Detector Surface Reflection

A further problem arising is the reflection at the detector surface. Depending on the detector almost 50% of the detected light might be reflected back to the sample. Since the sample and detector surfaces form conjugate images, it is very likely that most of the light is reflected at the sample to return to the detector. It is therefore necessary to tilt the detector to prevent the reflected light from reaching the sample, as shown in figure 55-20. Any detector sensitivity dependence upon the incidence angle must then be considered.

A receiving channel can also be provided by an integrating sphere, in which the entrance port represents a small hole, such that only a negligible small portion of the flux inside the sphere is reflected from the sample (figure 55-21) [55-26]–[55-29]. An integrating sphere should preferably be used to measure diffuse transmissivity for samples with rough surfaces or regular transmissivity and a sending channel which provides a broad range of incidence angles.

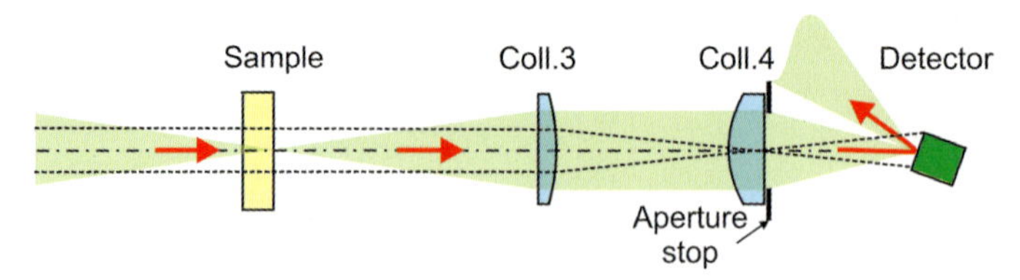

Figure 55-20: Receiving channel of a photometer with tilted detector to dispose of reflections from the detector surface.

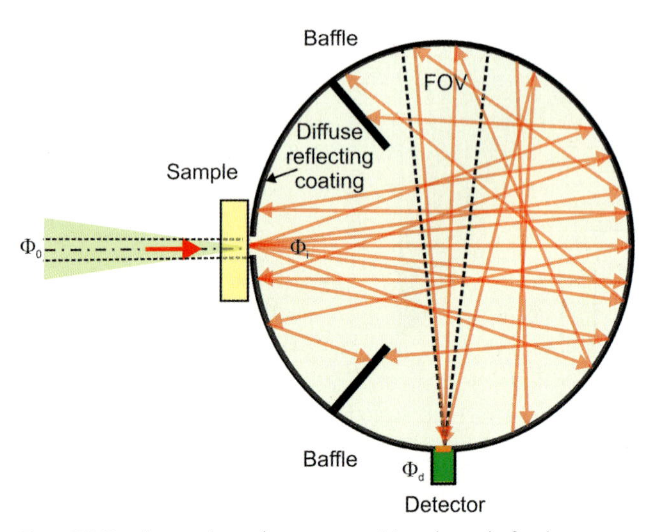

Figure 55-21: Integrating sphere as a receiving channel of a photometer.

Examples of Transmittance Metrology

The following examples show instuments providing well-defined positioning of sample and/or photometer units in order to generate transmissivity maps of transparent optical elements.

Figure 55-22 shows an example of a spectral photometer or polarimeter arrangement to measure the spectral transmittance of plane parallel optical elements. The sample is placed on an x/y-translation stage to select arbitrary points x,y for the measurement, while the sender and receiver units are fixed to the setup. The computer controls the x/y stage and the photometer or polarimeter units. It is possible to have fully automatic transmittance measurements to generate transmissivity maps of plane samples for different wavelengths. The lateral resolution is limited mainly by the spot size of the sender unit.

Figure 55-23 shows an example of a spectral photometer or polarimeter arrangement used to measure the spectral transmittance of optical lens elements. The element is placed on a rotary table, while the sender and receiver units are moved by linear translators in the x- and z-direction and are tilted by rotators in the ϑ-direction. The arbitrary points x,y can be selected for the measurement. The device is able to generate a transmissivity map of the sample with a lateral resolution limited mainly by the spot size of the sender unit.

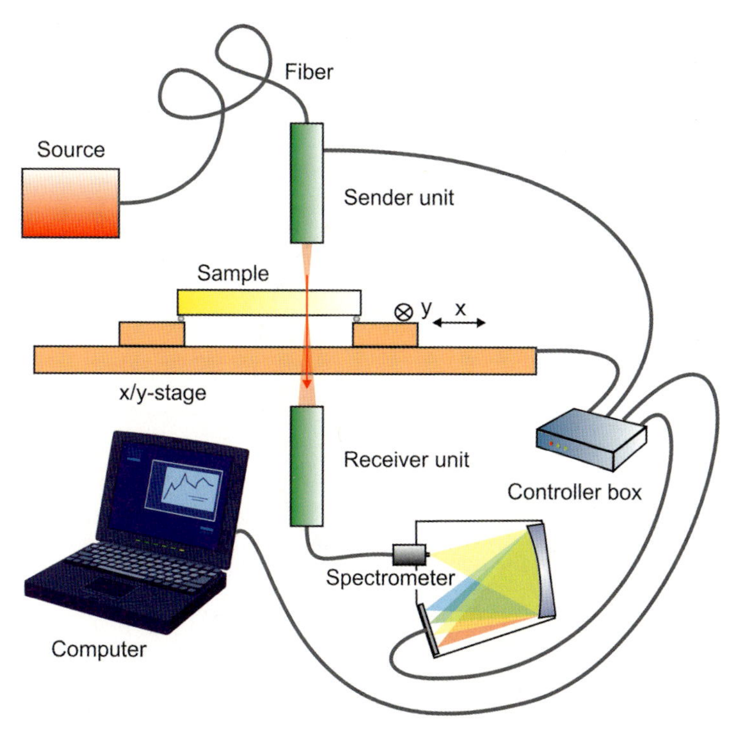

Figure 55-22: Spectral photometer or polarimeter arrangement used to measure the transmittance of plane optical elements at arbitrary points.

Typical specifications for photometer scanning devices are:

- Spot size 1 – 3 mm,
- Time per single measurement 1 – 3 s,
- Transmittance range 0 – 100 %,
- Repeatability $\Delta T/T$ for high transmissivities < 0.2%
- Accuracy $\Delta T/T$ for high transmissivities 0.2 – 1%.

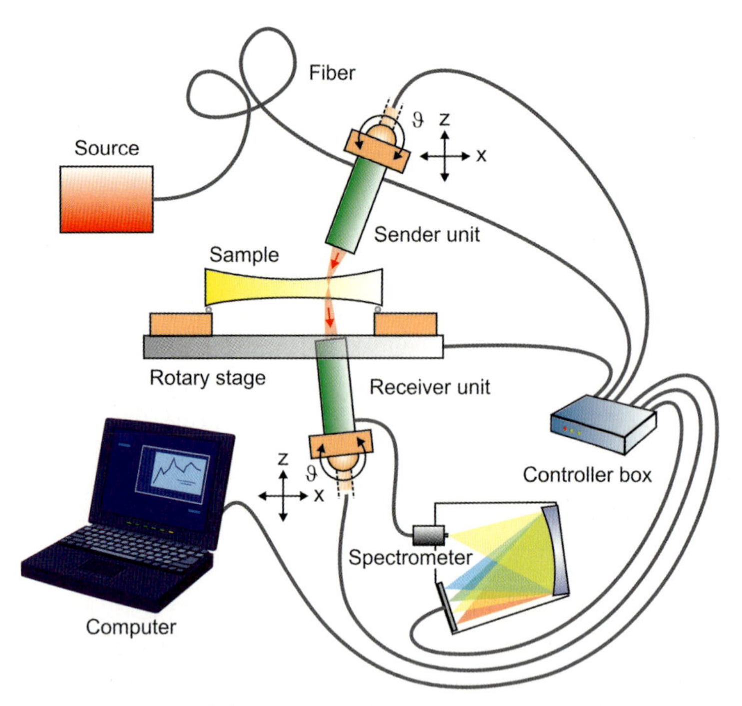

Figure 55-23: Spectral photometer or polarimeter arrangement used to measure the transmittance of optical lens elements at arbitrary points.

55.4.3
Spectral Reflectance

Due to the spectral bandwidth of the light source and the detector system, the measured value for the reflectance is

$$R_{\text{meas}}(\lambda) = \frac{\int\limits_{\lambda_1}^{\lambda_2} R(\lambda)E(\lambda)S(\lambda)d\lambda}{\int\limits_{\lambda_1}^{\lambda_2} E(\lambda)S(\lambda)d\lambda} \tag{55-41}$$

where
  $E(\lambda)$ is the spectral emissivity of the light source and
  $S(\lambda)$ is the spectral sensitivity of the detector system.

When $E(\lambda)$ and $S(\lambda)$ are constant over a spectral range from λ_1 to λ_2, (55-41) can be simplified to

$$R_{\text{meas}}(\lambda) = \frac{\int_{\lambda_1}^{\lambda_2} R(\lambda) d\lambda}{\lambda_2 - \lambda_1} \qquad (55\text{-}42)$$

Back-surface Reflection

When the reflectance of a component is measured, the reflected flux depends on the reflectances at both surfaces and also on the internal transmittance T_i of the component. If high reflectance coatings are measured, the transmitted flux can usually be neglected. We can calculate the expected reflectance using the transfer-matrix method. If necessary, we can consider the substrate material and the coatings from both surfaces as shown in figure 55-18.

By analogy with the previous section, the plane-parallel substrate acts as a *Fabry–Perot resonator*, and the reflected flux can be calculated accordingly. Figure 55-24 shows the reflectances of the same plane plates and the same coatings as shown in figure 55-19. For normal incidence the reflectances vary strongly with wavelength λ. The frequency of the variation increases with the thickness d_{N+1} of the element. The amplitude of the variation increases with the reflectances of the surfaces. In order to measure mean reflectances, the spectral bandwidth from λ_1 to λ_2, as indicated in (55-41) and (55-42), should be large enough to average over several resonance periods in the spectrum. If possible, the element should carry a wedge, so that the measuring beam laterally extends over several periods.

Back-surface Reflection Suppression

In most cases it is the reflectance of a single surface which is of interest. In the case of AR coatings or coatings with a considerable transmittance, the reflectance of the back surface must be suppressed. When the sample is thick enough, the back-side reflection can be prevented from entering the detector's field of view (FOV) by tilting the sample appropriately (figure 55-25). The FOV and thickness of the sample determine the minimum tilt angle necessary to avoid erroneous reflections from the back side.

Using thin witness samples for coating qualification does not allow for normal incidence or small tilt angles. In these cases the back surface must be roughened by a grinding process (sand-blasting, for example) see figure 55-26 (a). For critical cases, a rough surface must additionally be painted with a special black lacquer of similar refractive index to that of the substrate material (figure 55-26 (b)).

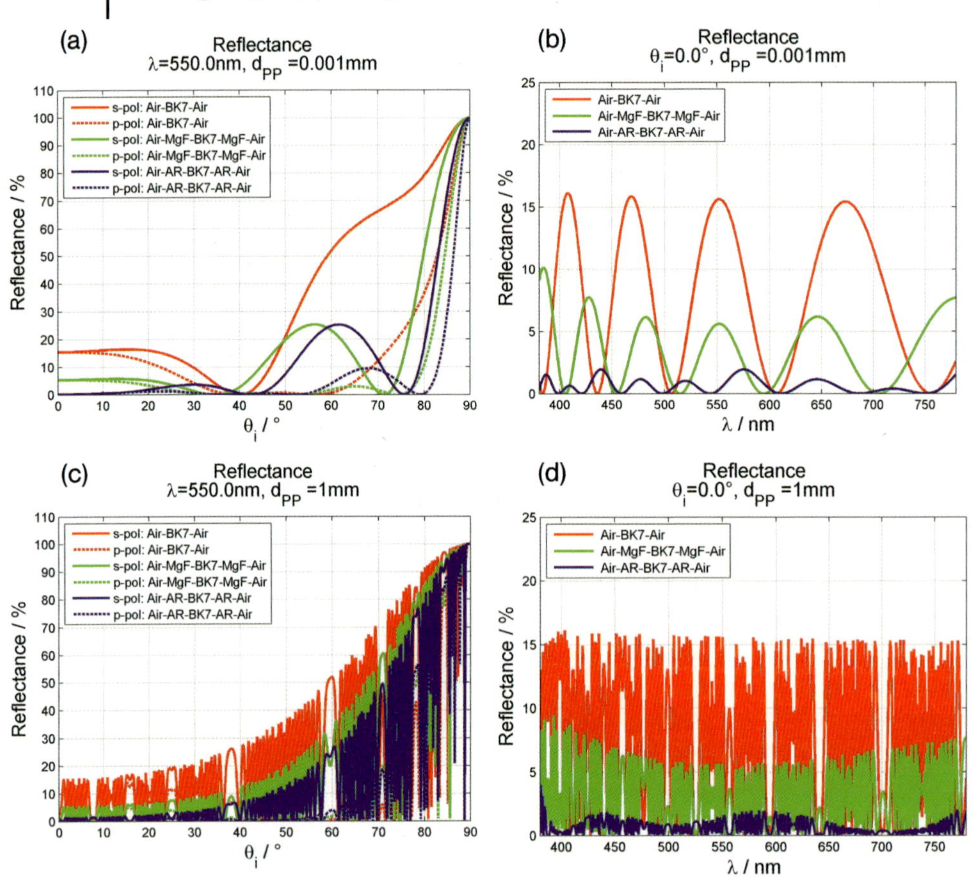

Figure 55-24: Reflectance of a plane plate made of BK7, uncoated (red), with single-layer AR coating (green), with multi-layer AR coating (blue): (a) thickness 1 μm, $\lambda = 550$ nm, R as a function of incidence angle θ_i; (b) thickness 1 μm, normal incidence, R as function of λ; (c) thickness 1 mm, $\lambda = 550$ nm, R as function of incidence angle θ_i; (d) thickness 1 mm, normal incidence, R as function of λ.

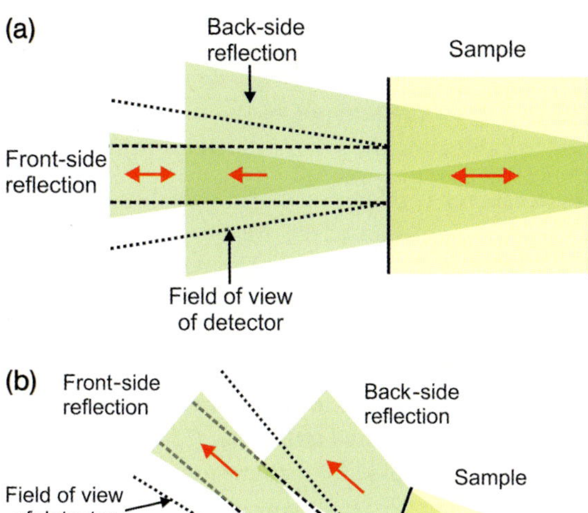

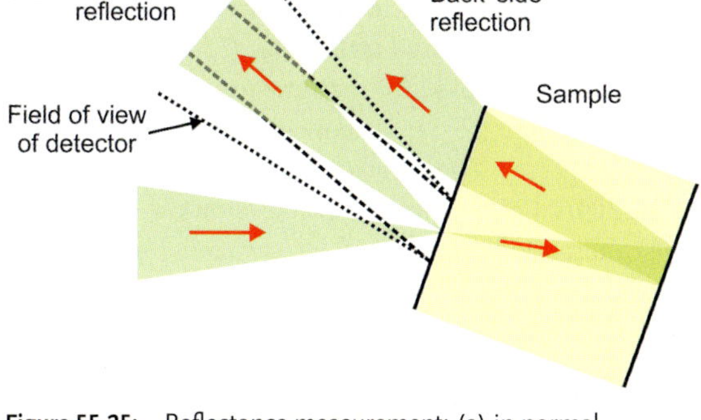

Figure 55-25: Reflectance measurement: (a) in normal incidence with the reflections from the front and back sides forming the signal; (b) in oblique incidence to keep the back-side reflection out of the detector's FOV.

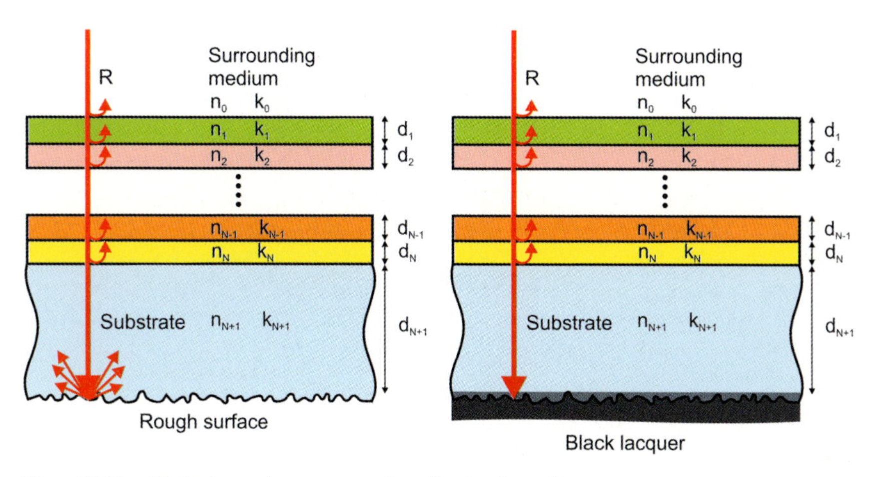

Figure 55-26: Methods used to suppress the reflection from the back side of the sample: (a) grinding to generate a rough surface; (b) covering the rough back surface with a black lacquer.

Examples of Reflectance Metrology

Reflectance can be measured

a) relative to a reference mirror of known reflectance, or
b) using the absolute method by calibrating the setup without the sample.

The relative method requires a calibrated reference mirror, which needs to be stable over a long time. Aging and contamination effects might change the reflectance, making recalibration necessary.

The absolute method avoids the need for a reference mirror; however, some components of the setup must be rearranged to enable a calibration without the sample in place. Two measurements then have to be carried out consecutively, one with, the other without, the sample. Provided that dark-signal correction has previously been carried out, the ratio of both results gives the required reflectance [55-30].

In the following, we give examples of setups used for the measurement of absolute reflectance. The well-known V-W arrangement [55-31] introduces two reflections at the sample, their location being laterally displaced. The displacement depends on the incidence angle and the geometric configuration. Figure 55-27 (a) shows a fixed-angle arrangement in V-mode, when the calibration is carried out. Figure 55-27 (b) shows the arrangement in W-mode, when the sample is measured. To change from V- to W-mode, the sample has to be introduced and the mirror M2 has to be brought to a new position.

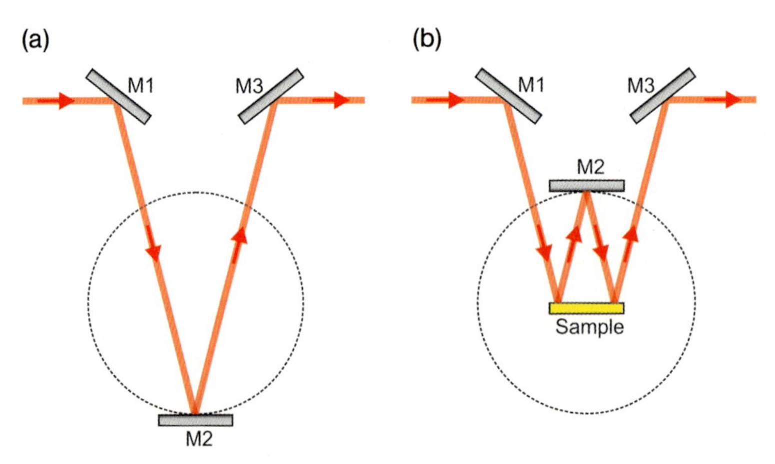

Figure 55-27: V-W setup to measure reflectance at a fixed angle of incidence: (a) V-mode for calibration, (b) W-mode for sample measurement.

The V-W setup can be modified to a variable angle system by placing the sample and the mirror M2 on separate rotary tables as shown in figure 55-28. For the calibration, the sample is removed and M2 is positioned accordingly.

The limitations of the V-W arrangement are as follows.

a) Samples of low reflectance generate a very low signal because of the two reflections. The arrangement is therefore suitable only for reflectance of approximately > 20%.

b) Small samples are unsuitable because of the reflection separations.
c) Measurements at normal incidence are not possible.
d) Realignment of the mirror M2 is critical. For basic adjustment an additional alignment laser should be used.

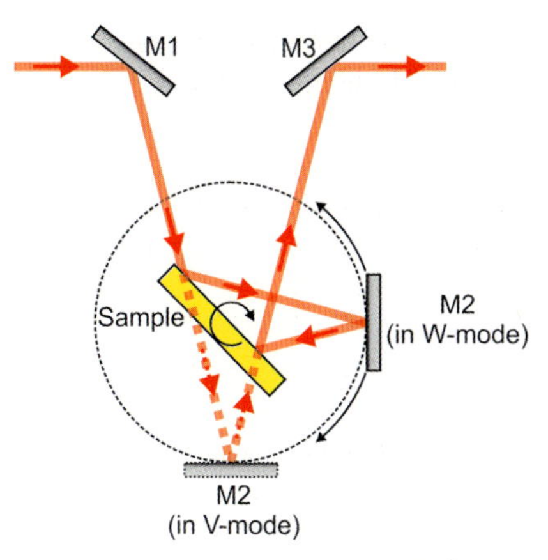

Figure 55-28: V-W setup to measure reflectance at variable angles of incidence.

To enable reflectance measurements for AR coatings, an arrangement with a single reflection at the sample must be found. The V-N setup shown in figure 55-29 is one example. For the calibration the mirror M2 is brought to the first position (a), for the sample measurement it is placed in the second position (b). Note that the output beam is inverted in position (b), so any misalignments or asymmetries in the components or detector sensitivities might lead to errors.

The V-N arrangement is suitable for all reflectance ranges and has the benefit of providing a small spot suitable for the measurement of small samples.

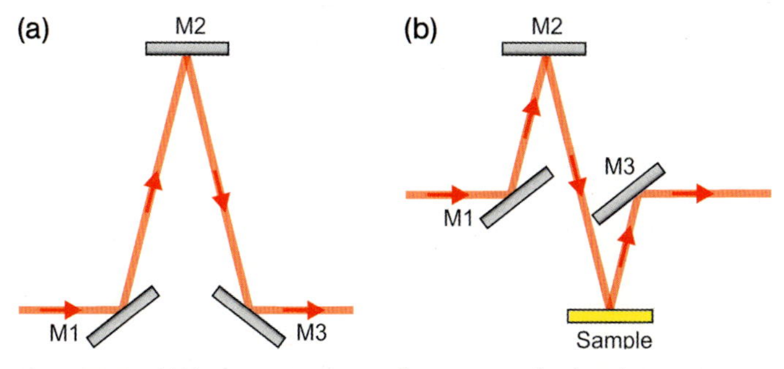

Figure 55-29: V-N setup to measure reflectance at a fixed angle of incidence: (a) V-mode for calibration, (b) N-mode for sample measurement.

A variable-angle V-N setup derivative is shown in figure 55-30. Switching between calibration (a) and sample testing (b) involves rotating the mirror M3 and translating the detector. Introducing different angles of incidence is achieved by the translation and rotation of M3 in order to direct the beam onto the sample. At the same time the detector is translated and rotated accordingly. Mirrors M1 and M2 form a unit for path-length compensation. If M3 is translated for different incidence angles then the unit is translated to maintain a constant optical path.

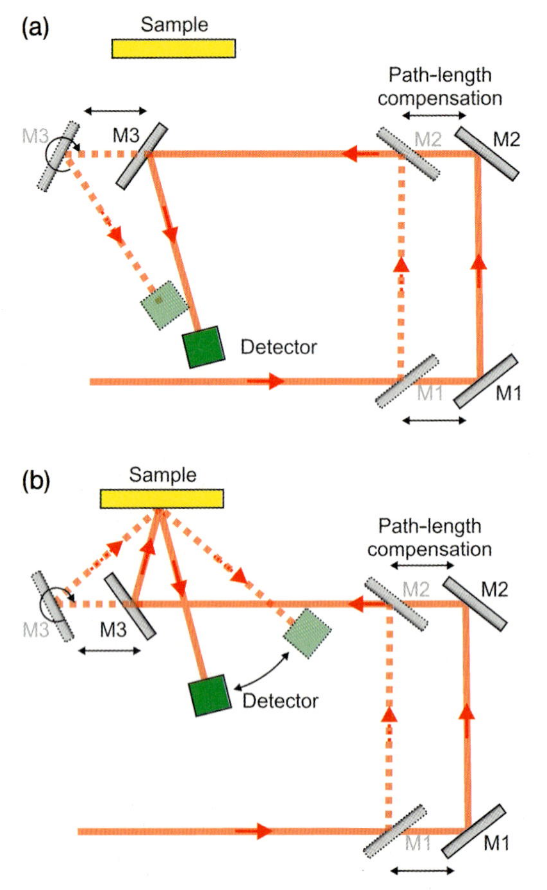

Figure 55-30: V-N setup derivative to measure reflectance at variable angles of incidence: (a) calibration mode; (b) sample-measurement mode.

A variable-angle setup with a single reflection is established by using an integrating sphere placed on a rotary table for correct positioning. The sample is rotated on a separate table to select the required incidence angle. Figure 55-31 shows the arrangement. For the calibration the sample is removed and the integrating sphere is positioned accordingly. The advantage of the integrating sphere is its larger tolerance to small misalignments and asymmetries in the illumination. To reduce systematic errors the sample can be measured at both the positive and its corresponding negative angle of incidence.

Figure 55-32 shows an example of a spectral photometer or ellipsometer arrangement used to measure the spectral reflectance of plane parallel optical elements. The sample is placed on an x/y-translation stage to select arbitrary points x,y for the measurement, while sender and receiver units are fixed to the setup. The angle of incidence is set at a well-defined small angle. Near-normal incidence is not possible in this arrangement, in fact it is avoided in order to get rid of the erroneous reflection from the back surface. The computer controls the x/y-stage and the photometer or ellipsometer units. Fully automatic reflectance measurements can be used to generate reflectance maps of plane surfaces for different wavelengths. The lateral resolution is limited mainly by the spot size of the sender unit.

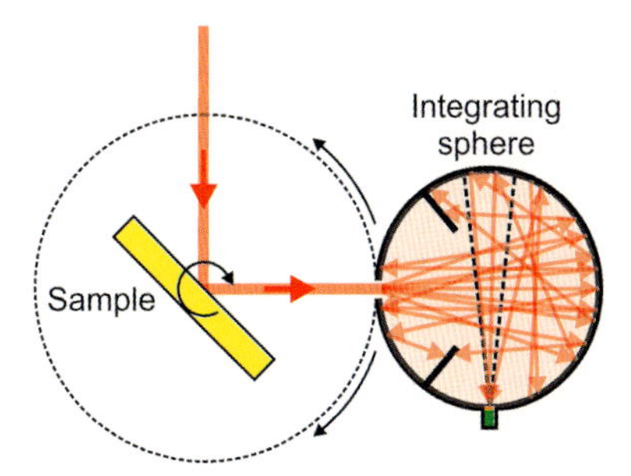

Figure 55-31: Integrating sphere setup to measure reflectance at variable angles of incidence.

Figure 55-33 shows an example of a spectral photometer or polarimeter arrangement used to measure the spectral reflectance of curved lens or mirror surfaces. The element is placed on a rotary table, while the sender and receiver units are moved by linear translators in the x- and z-direction and are tilted by rotators in ϑ-direction. Arbitrary points x,y and angles of incidence (except for near-normal incidence) can be selected for the measurement. The device is able to generate a reflectance map of the surface under test with a lateral resolution limited mainly by the spot size of the sender unit.

Typical specifications for photometer scanning devices are as follows.

- Spot size 1–3 mm,
- Time per single measurement 1–3 s,
- Reflectance range 0–100 %,
- Repeatability $\Delta R/R$ for AR coatings: < 2%
- Repeatability $\Delta R/R$ for HR coatings: < 0.2%
- Accuracy ΔR for AR-coatings 0.2–0.5%
- Accuracy ΔR for HR-coatings 0.3–0.6%.

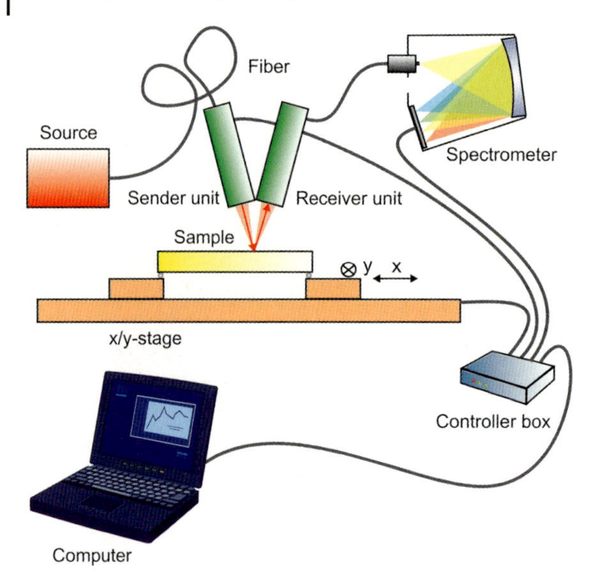

Figure 55-32: Spectral photometer or ellipsometer arrangement used to measure the reflectance of plane optical elements at arbitrary points and angles of incidence.

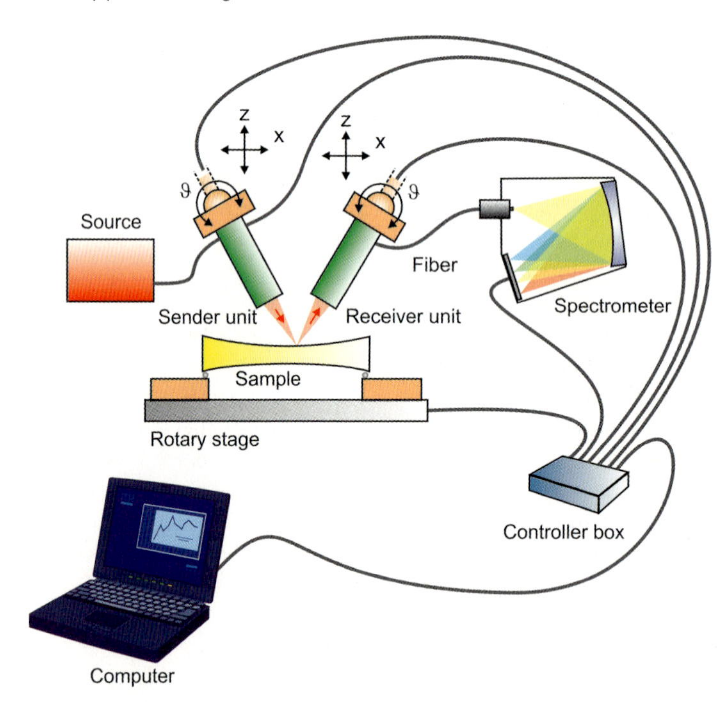

Figure 55-33: Spectral photometer or ellipsometer arrangement used to measure the reflectance of curved lens or mirror elements at arbitrary points and angles of incidence.

55.4.4
Spectral Absorptance

From (55-4) we see that the spectral absorptance of a sample is obtained by detecting its total spectral reflectance and transmittance and then calculating from

$$A(\lambda) = 1 - R(\lambda) - T(\lambda) \tag{55-43}$$

Note that stray light is contained in $T(\lambda)$ as the diffuse portion of the spectral transmittance (see (55-4)–(55-6)).

Since spectral reflectance and transmittance metrology always integrate over a spectral range from λ_1 to λ_2, the measured value for absorptance is therefore

$$A_{\text{meas}}(\lambda) = \frac{\int_{\lambda_1}^{\lambda_2} A(\lambda)E(\lambda)S(\lambda)d\lambda}{\int_{\lambda_1}^{\lambda_2} E(\lambda)S(\lambda)d\lambda} \tag{55-44}$$

where

$E(\lambda)$ is the spectral emissivity of the light source and
$S(\lambda)$ is the spectral sensitivity of the detector system.

When $E(\lambda)$ and $S(\lambda)$ are constant over a spectral range from λ_1 to λ_2, (55-44) can be simplified to

$$A_{\text{meas}}(\lambda) = \frac{\int_{\lambda_1}^{\lambda_2} A(\lambda)d\lambda}{\lambda_2 - \lambda_1} \tag{55-45}$$

Absorptance Metrology

The classical method used to determine the absorptance of a sample is to measure the sum of the total transmittance and total reflectance simultaneously by means of an integrating sphere [55-32]. Figure 55-34 shows an appropriate arrangement, in which the sample is placed within an integrating sphere outside the field of view of the detector. Care must be taken that any sample mounting parts are coated with the same diffuse reflecting coating as the walls of the sphere. The sample can be rotated to provide different angles of incidence. Calibration is carried out by making a separate measurement with the sample removed. After calibration the measurement result represents $R(\lambda) + T(\lambda)$, which leads to $A(\lambda)$ by using (55-43).

Laser calorimetry is a measurement procedure used to determine the absorption in laser optics [55-33]. It can characterize both optical materials and dielectric coatings [55-34]. This is achieved by high-precision temperature measurement on the sample of interest. A calibrated set of negative temperature coefficient resistors (NTCs) is applied in order to perform an absolute temperature measurement. An NTC is a resistor with an electrical resistance which decreases with temperature.

Figure 55-35 shows the arrangement for laser calorimetry. A laser (pulsed or cw) is attenuated to a defined level and is focused onto a beam dump after transmission through the sample. The sample and the NTCs are positioned within a thermal insulated environment.

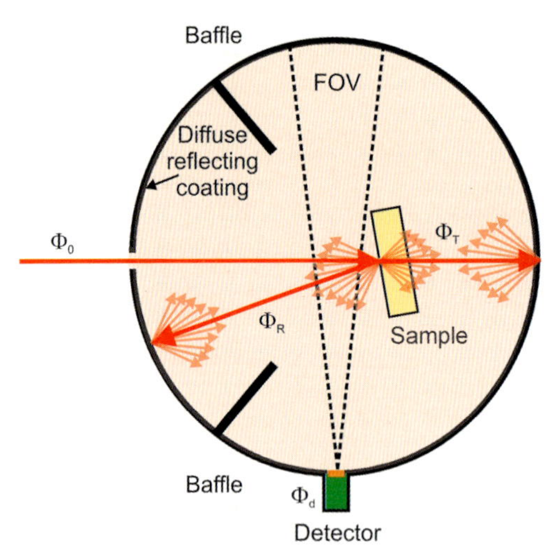

Figure 55-34: Integrating sphere arrangement to measure the total transmittance and total reflectance simultaneously.

In this configuration it is possible to detect temperature increases below 1 mK. When the temperature increase of an irradiated sample, the sample's effective heat capacity and also the applied laser power Φ_i have been measured, the absolute absorptance can be derived.

It must be taken into account that laser calorimetry depends on the sample geometry and the location of the temperature sensors on the surface of the sample. A specific disadvantage is the low temporal resolution, because a long irradiation time of between several seconds to several hundred seconds is needed for the accumulation of sufficient irradiation energy to create a measurable temperature rise.

Arrangements similar to figure 55-35 are able to detect absorptions below 0.1 ppm [55-35].

An alternative method to measure absorption is the so-called *photothermal deflection* (PTD) technique, which requires a temporal modulation of the temperature field induced by a short laser pulse or a periodic intensity modulation of continuous laser radiation. When laser radiation is applied to a sample, three effects can usually be observed (figure 55-36).

1. The sample's surface is deformed due to the thermal expansion of the substrate.
2. The refractive index of the substrate changes in the radiated area.
3. The refractive index of the adjacent coupling medium changes near the radiated area.

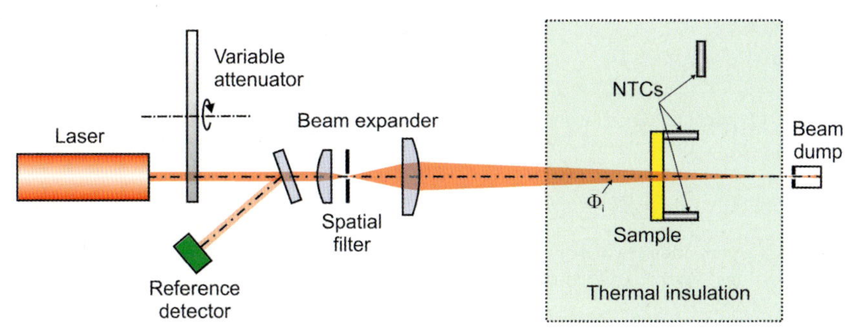

Figure 55-35: Laser calorimetry arrangement to measure the absorptance of optical elements.

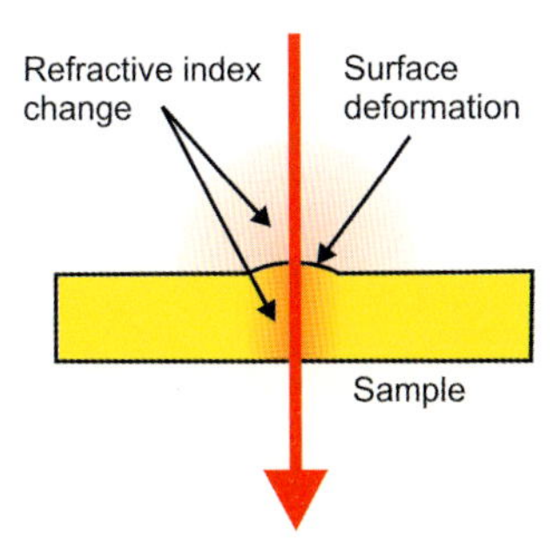

Figure 55-36: Photothermal effects at a radiated sample.

PTD techniques measure these thermal effects by applying fast and laterally resolving metrology detecting surface deformations or index changes. A probe beam propagating in a direction perpendicular or parallel to the sample surface will be deflected because of the modulated index gradient within the sample or in the adjacent medium.

PTD techniques are generally more sensitive compared with the calorimetric technique regarding the capability of measuring the absorption loss in a short period of time (in milliseconds) or with single-pulse irradiation. However, the PTD technique can measure only the relative value of the absorption loss. Calibration of the measured data is difficult because the recorded photothermal signal is a complicated function of the optical, thermal and thermoelastic properties of the investigated sample. Therefore, a combined technique using laser calorimetry and PTD with a single-pulse sensitivity and temporal resolution and the capability of measuring the absolute absorption loss is desirable [55-36].

55.5 Literature

55-1 ISO 9211-1 Optics and optical instruments – Optical coatings – Part 1: Definitions (2010).

55-2 ISO 9211-2 Optics and optical instruments – Optical coatings – Part 2: Optical properties (2010).

55-3 ISO 9211-3 Optics and optical instruments – Optical coatings – Part 3: Environmental durability (2008).

55-4 ISO 9211-4, Optics and optical instruments – Optical coatings – Part 4: Specific test methods, 2nd edition (2006).

55-5 ISO 80000-7, Quantities and units – Part 7: Light (2008).

55-6 ISO 6286, Molecular absorption spectrometry – Vocabulary – General – Apparatus (1982).

55-7 M. Born and E. Wolf, Principles of Optics, Pergamon Press, Oxford, 6th edition (1980).

55-8 O. S. Heavens, Optical Properties of Thin Films, Dover, New York (1965).

55-9 P. Yeh, Optical Waves in Layered Media, Wiley, New York (1988).

55-10 Z. Knittl, Optics of Thin Films, Wiley, London (1976).

55-11 E. D. Palik (Ed.), and G. Ghosh (Ed.), Handbook of Optical Constants of Solids, Five-volume Set: Handbook of Thermo-Optic Coefficients of Optical Materials with Applications, Academic Press, San Diego (1998).

55-12 H. Fujiwara, Spectroscopic Ellipsometry – Principles and Applications, John Wiley & Sons, Ltd, Chichester (2007).

55-13 C. L. Mitsas, and D. I. Siapkas, Generalized matrix method for analysis of coherent and incoherent reflectance and transmittance of multiplayer structures with rough surfaces, interfaces, and finite substrates, Appl. Opt. **34**, 1678–83 (1995).

55-14 C. C. Katsidis, and D. I. Siapkas, General transfer-matrix method for optical multilayer systems with coherent, partially coherent, and incoherent interference, Appl. Opt. **41**, 3978–87 (2002).

55-15 D. G. Stearns, The scattering of X-rays from non-ideal multi-layer structures, J. Appl. Phys. 65, 491–506 (1989).

55-16 D. L. Windt, IMD–Software for modeling the optical properties of multilayer films, Computers in Physics,Vol. **12**, 360–70 (1998).

55-17 Optical Glasses, Schott AG, Mainz, Germany (2009).

55-18 W. G. Oldham, Numerical techniques for the analysis of lossy films, Surf. Sci. **16**, 97–103 (1969).

55-19 W. H. Press, S. A. Teukolsky, W. T. Vetterling, and B. P. Flannery, Numerical Recipes in C++: The Art of Scientific Computing, 2nd edition, Cambridge University Press, Cambridge (2002).

55-20 R. M. A. Azzam, and N. M. Bashara, Ellipsometry and Polarized Light, Elsevier Science B.V., Amsterdam, The Netherlands (1987).

55-21 A. C. Boccara, C. Pickering, and J. Rivory (eds.), Spectroscopic Ellipsometry, Elsevier Publishing, Amsterdam (1993).

55-22 J.A. Woollam *et al.*, Overview of Variable Angle Spectroscopic Ellipsometry (VASE), Part I: Basic Theory and Typical Applications, Optical Metrology, vol. CR72, SPIE, Bellingham, Washington, pp 3–28 (1999).

55-23 B. Johs *et al.*, Overview of Variable Angle Spectroscopic Ellipsometry (VASE), Part II: Advanced Applications, Optical Metrology, vol. CR72, SPIE, Bellingham, Washington, pp 29–58 (1999).

55-24 J. A. Woollam, Ellipsometry, Variable Angle Spectroscopy in: J.G. Webster (ed.) Wiley Encyclopedia of Electrical and Electronics Engineering, John Wiley & Sons, New York, pp. 109–16 (2000).

55-25 H. G. Tompkins, and E. A. Irene (eds.), Handbook of Ellipsometry, William Andrew Publishing, New York (2005).

55-26 J. Kessel, Transmittance measurements in the integrating sphere, Appl. Opt. **25**, 2752–56 (1986).

55-27 F. Manoochehri, E. Ikonen, High-accuracy spectrometer for measurement of regular spectral transmittance, Appl. Opt. **34**, 3686–92 (1995).

55-28 L. Hanssen, Integrating-sphere system and method for absolute measurement of transmittance, reflectance, and absorptance of specular samples, Appl. Opt. **40**,. 3196–204 (2001).

55-29 J. Cheung, J. L. Gardner, A. Migdall, S. Polyakov, and M. Ware, High accuracy dual lens transmittance measurements, Appl. Opt. **46**, 5396–403 (2007).

55-30 I. Stemmler, Angular dependent specular reflectance in UV/Vis/NIR, Proceedings of the SPIE, Volume 5965, pp. 468–78 (2005).

55-31 J. Strong, Procedures in Experimental Physics, 1st ed., Prentice-Hall, Inc., New York, 376 (1938).

55-32 F. Grum, and R. J. Becherer, Optical Radiation Measurement, Vol. 1, Radiometry, Academic Press, New York, San Francisco, London (1979).

55-33 ISO 11551, Optics and optical instruments – Lasers and laser-related equipment – Test method for absorptance of optical laser components (2003).

55-34 U. Willamowski, D. Ristau, and E. Welsch, Measuring the absolute absorptance of optical laser components, Appl. Opt. **37**, 8362–70 (1998).

55-35 L. Jensen, I. Balasa, H. Blaschke, and D. Ristau, Novel technique for the determination of hydroxyl distributions in fused silica, Optics Express **17**, 17144–49 (2009).

55-36 B. Li, H. Blaschke, and D. Ristau, Combined laser calorimetry and photothermal technique for absorption measurement of optical coatings, Appl. Opt. **45**, p. 5827–31 (2006).

56
System Testing

Handbook of Optical Systems: Vol. 5. Metrology of Optical Components and Systems. First Edition.
Edited by Herbert Gross.

56.1 Introduction

56.1.1 System Measurement

Previously, all the basic methods and procedures used in the testing of various components within optical systems have been represented in detail. In the practical realization of low-cost consumer optics, these tolerances are mostly sufficient to guarantee an overall system performance within the desired requirements. Any additional testing stages of the complete system are costly and should be avoided, if possible. But there are also system types of higher quality, which cannot be completely specified by the individual tolerances. In these cases, the system must be tested and readjusted if necessary. The quality criteria are specific to the particular application and are comprehensively discussed in chapter 30. The corresponding measurement methodologies are described in this chapter. There is much literature on this subject [56-1]–[56-9].

56.1.2 Description of System Performance

A huge variety of possible specifications of a system's quality performance is available and so a suitable choice must be made for the particular application. It is impossibe to generalize or unify these recommendations. If the gathering of an image is performed by a digital sensor, resulting in a digital image being obtained from an optical instrument, it is also possible to define the quality by purely calculated measures of the discretized and quantized image. If, conversely, a human observer rates an image by direct observation, subjective psycho-physical aspects play a role and the performance description will be quite complicated. In this case statistics from tests taken by a huge number of observers with a direct rating will be necessary.

Here only the physical optical image formation with the corresponding performance data and measuring setups will be considered.

In accordance with chapter 30, the following types of quality characterization can be distinguished:

1. Geometrical optical aberrations such as transverse aberrations, spot diagrams, etc. These are extremely hard to measure and are therefore not recommended for testing purposes.
2. Wave aberrations, measured in the pupil of a system. These data can be measured very accurately with interferometric or other methods as described in chapters 46 and 47. The natural scaling on the wavelength provides good conditions for the comparison of different systems.
3. The point spread function, measured as a point object response in the image plane. This is a measure of the quality and is useful in highly corrected diffraction-limited systems such as microscopes.

4. Edge and line spread functions are useful criteria, especially in those cases where the object contains primarily corresponding linear structures.
5. Modulation transfer function, contrast and sine wave resolution are the most widespread methods and can be used very successfully.
6. Special aspects, which are not well covered by the methods above are:
   6.1 Distortion
   6.2 Chromatical aberrations
   6.3 Quality factor M^2 of laser beams
   6.4 Transmission
   6.5 Illumination uniformity
   6.6 False light
   6.7 Polarization.

In particular, for aspects 6.3, 6.5 and 6.7, note that the measurement of the light field data described here to characterize the optical system, assumes the illumination light to be free of errors or at least well known. Otherwise, the origin of measured data cannot be ascertained and cannot therefore be assigned to the system.

The testing of the different classical performance criteria 1-5 is discussed in sections 56.3 and 56.4 in detail. The special problems noted in 6 are considered in section 56.5.

56.1.3 Specifications

In the development of an optical system, one of the first steps is to establish a specification. The specification contains several data, which can be roughly divided into the following categories:

1. The functional specification
   This is related to the application and is valid for the complete system. This specification plays the major role of a basic agreement between the user or customer and the development company.
   Here two levels must be distinguished:
   1.1 Basic data
   1.2 Image quality.

2. The manufacturing specification
   It is usually necessary to state several points within the engineering approach of a system in order to be sure that the technological and business goals are clear. Here the following two aspects can be distinguished:
   2.1 Business specification, which will ensure economic development
   2.2 Engineering and technology specifications, particularly interface data.

3. Component specifications
   When the optical design is complete, it must be guaranteed that the theoretical results can be obtained in practice. Here the concept of tolerancing plays the major role, as described in chapter 35 in detail. To ensure the functional

specification of the entire system, all the complicated relationships and dependencies of the systems must be taken into account to calculate the necessary performance properties of the subsystems and components.

4. Assembly specification
   Almost all optical systems contain optical and non-optical parts and therefore the assembly and the interfaces between the mechanical and optical components and the procedure used to mount them must be fixed.
   From a more general point of view, signal and image processing software and the corresponding algorithms in digital recording systems must also be specified.

In principle, all specifications must be testable or measurable, otherwise verification will not make sense. Usually, a specification consists of a collection of numbers and drawings which give quantitative and unambiguous answers to all possible open questions.

There are a number of international standards for specification. These documents help to establish a common understanding of data without the need for interpretation. If components or subsystems are manufactured in different locations or companies, this unification is very important. Beneath the ISO-standard, the military norms of the US Defence Department are in widespread use. Some of the most important ISO standard documents are listed in table 56-1.

Table 56-1: ISO standard documents for optical system testing.

No	ISO-Number	Topic
1	9022	Environmental test methods
	9039	Measurement of distortion
	9334	OTF Definition and terms
	9335	OTF measurement
	11421	Accuracy of OTF measurements
	9358	Veiling glare
	13653	Measurement of relative irradiance in the image plane
	15529	Measurement of MTF of sampled imaging systems
	15795	Image degradation due to chromatic aberrations

In table 56-2 the most important specification data are listed. This collection of properties is not complete. Furthermore, the details of the specifications might be quite complicated. For example, the performance data can be fixed differently at various field positions for several wavelengths and different zoom positions.

Table 56-2: Specification properties of an optical system.

No	Category	Topic
1	Basic system parameter	Focal length Magnification f-number / numerical aperture / pupil diameter Wavelength interval Spectral weighting Field of view, horizontal / vertical aspect ratio Image distance Object distance Free working distance on object side Entrance pupil location Exit pupil location Zoom range, min/max values Sensor characteristics
2	Manufacturing specification	Overall size / total track Maximum diameter Number of lenses Cost of production Number of aspherics Material restriction (no plastic, toxity, etc.) Coatings Maximum weight Use of diffractive elements Use of currently available components Cosmetic properties Mechanical interface to mount the system Optical interface with connected systems Assembly requirements
3	Performance specification	Spectral transmission Vignetting Image quality (W_{rms}, Strehl, MTF, ...) Distortion Image field curvature Depth of focus Illumination uniformity Temperature range Vibration resistance Shock survival Ghost images Stray light level, veiling glare Telecentricity error Polarization preservation

If the environmental specifications are considered, one has to distinguish between those aspects which affect an operation due to bad handling or critical environmental conditions during the application. However, the survival of the system quality and function during delivery and storage of a system [56-10] is quite different. This last aspect mostly concerns vibration and shock, which can damage the system.

The correct choice of specification data is quite important in order to guarantee the function of an optical system while maintaining a reasonable cost of production. Both aspects are essential for the success of a product [56-11], [56-12].

56.2 Basic Parameters of Optical Systems

56.2.1 Focal Length

The focal length f of a system is one of the most important parameters of an optical system. It should be noted that the focal length of a complex system is defined with respect to the principal planes, the location of which may often be unknown [56-8].

In the literature, many different methods are proposed to measure the focal length of an optical system. In principle, many equations of optical properties containing f can be used to extract the focal length of a system. In fact, the most important methods can be divided into three general classes:

1. Procedures which are based on the image location
2. Methods using the focal length dependence of the magnification
3. Methods which determine the wavefront curvature of a focussed beam.

Several methods use special effects to enhance the accuracy of the focal length; for example, the Moire technique or the Talbot effect. It should be noted that not all of the proposed methods are able to measure negative focal lengths. Furthermore, very large and very short focal lengths create problems in most approaches.

The achievable accuracies mostly depend on the precision of focussing or adjusting a sharp image. Typical values of a relative accuracy in the range of 10^{-2}–10^{-4} can be obtained.

By definition, the focal length is the distance between the image-side principal plane and the image plane in the paraxial approximation. Therefore problems will arise for systems with large numerical apertures. In these cases, the refractive power as the reciprocal focal length is not invariant across the pupil and depends on the height of the marginal ray used.

Furthermore, it should be noted that the principal plane is a virtual plane, which is often located inside the system. From a practical viewpoint, the distance between the last surface vertex and the image plane is a real property. It is called the back focal length and has an identical value to the focal length but only in very special cases.

Measurement Setup with Collimator

The focal length of the optical system to be tested can be determined using a test pattern of a definite size y and a collimator with a focal length f'_c for collimated illumination, by measuring the size y' of the test pattern in the image plane [56-1], [56-5] and [56-6]. The following relationship

$$f' = -f'_c \frac{y'}{y} \tag{56-1}$$

can then be used to determine the unknown focal length. The corresponding setup is shown in figure 56-1.

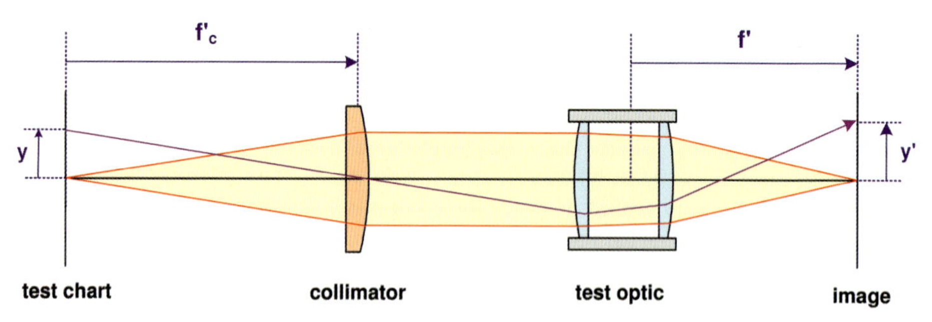

Figure 56-1: Measurement of the focal length using a collimator.

Method of Gauss

This approach assumes that the principal plane separation d_P of the optical system is either known or can be neglected [56-1], [56-5] and [56-6]. The determination of the focal length by the Gauss method proceeds in the following way as illustrated in figure 56-2.

An object and a receiving screen are placed at a distance L apart, which should be larger than the four-fold focal length. There are two z-positions of the test optics for which a sharp image is obtained, but which have different magnifications m. The separation D of these two positions can be experimentally determined. From the conjugate distance equation

$$\frac{1}{s'} - \frac{1}{s} = \frac{1}{f'} \tag{56-2}$$

and the relations

$$L = s' - s + d_P \tag{56-3}$$

for the two positions 1 and 2, one obtains with

$$s_2 - s_1 = D \tag{56-4}$$

the following expression for f'

$$f' = \frac{L - d_P}{4} - \frac{D^2}{4(L - d_P)} \qquad (56\text{-}5)$$

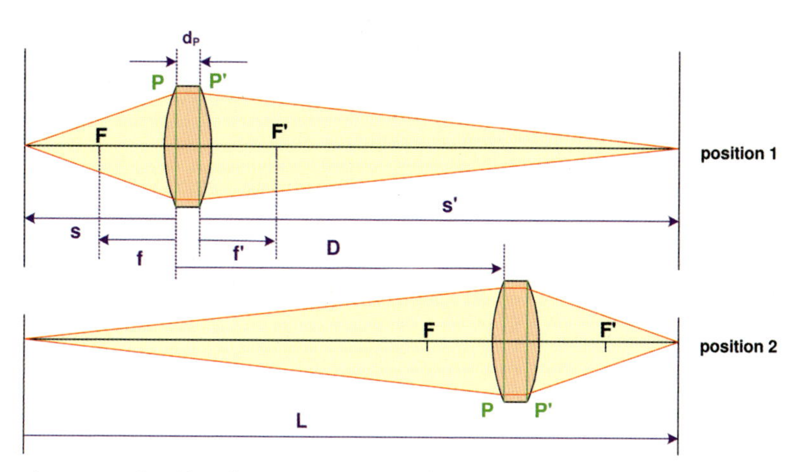

Figure 56-2: Focal length measurement using the Gauss method by shifting the whole optical system by D.

Focimeter

Focimeter are setups with a known afocal telescopic imaging of a test target [56-3]. This is the preferred method to measure systems with a long focal length. In the calibrated arrangement, the test target is imaged sharply. If the test lens is incorporated into the ray path at a distance $d{=}f_2$ from the second system lens, the image distance f_2 must be changed to x again to obtain a sharp image. From the lensmakers formula, the unknown focal length is determined by

$$\frac{1}{f} = \frac{1}{f_2} - \frac{x}{f_2^2} \qquad (56\text{-}6)$$

Figure 56-3 shows a focimeter arrangement.

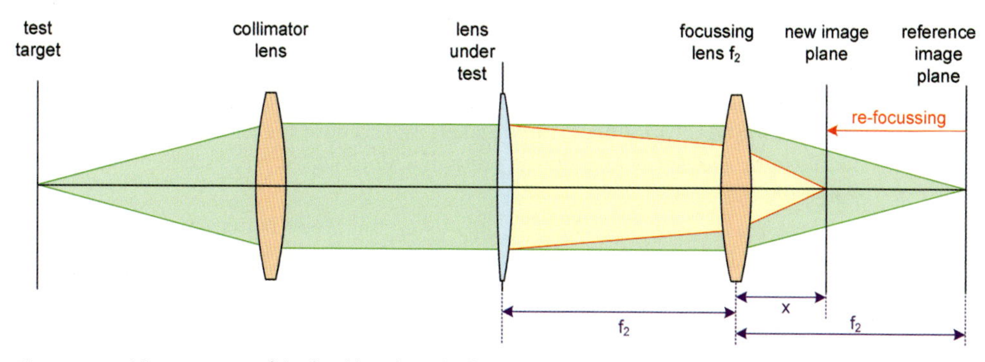

Figure 56-3: Measurement of the focal length in the focimeter.

Abbe Focometer

In the Abbe focometer [56-5], an auxiliary small telecentric microscopic system is used to observe the images of two test scales, which are arranged at a certain distance e apart. The microscope is moved along the optical axis direction in a constant distance y from the axis. According to the geometry shown in figure 56-4, the relationship

$$\tan u = \frac{y}{f} = \frac{y_2 - y_1}{e} \tag{56-7}$$

permits the determination of the focal length f.

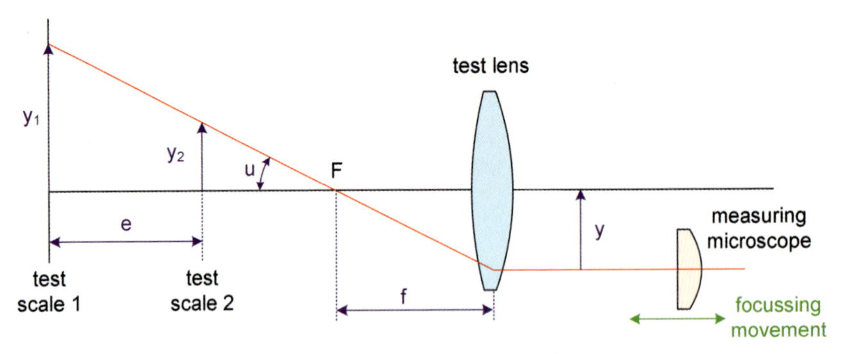

Figure 56-4: Measurement of the focal length using the Abbe focometer.

Moire Deflectometry

In this method, two Ronchi rulings are used in collimated light to measure the focal length of a lens by Moire deflectometry [56-13], [56-14]. If the light is no longer collimated due to the refractive power of the lens under test, a rotation of the Moire pattern by an angle α is obtained. With the pitch d of the gratings and the azimuthal inclination angle θ between the two gratings, the curvature of the beam is obtained from the formula

$$R = \frac{d}{\theta \tan \alpha} \tag{56-8}$$

If the lens under test is located quite near to the first grating, this value corresponds to the focal length of the lens. If there is a known distance between the principal plane and the grating, it can be taken into account for the determination of the focal length. Figure 56-5 shows the setup of a Moire-based measurement of the focal length.

There are also some modified setups using the Moire pattern of two gratings to evaluate the focal length of a system. Special setups are proposed, where the use of a collimator is not necessary. One of the advantages of the Moire-based approaches is the fact that a negative focal length can also be determined.

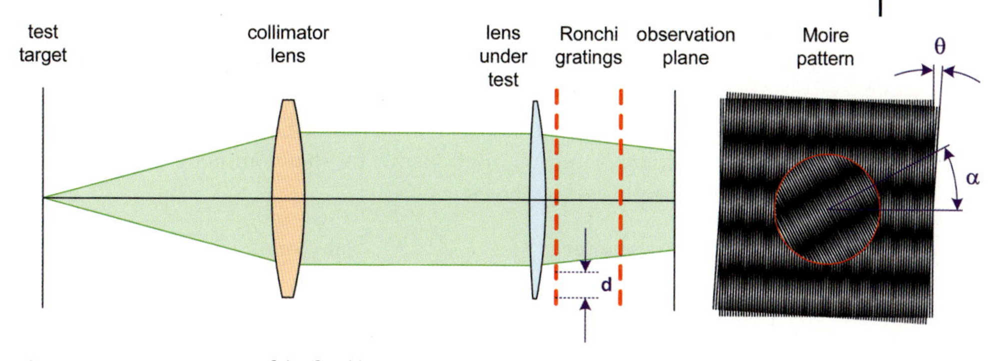

Figure 56-5: Measurement of the focal length using a Moire pattern.

Talbot Imaging

Another method uses the dependence of the Talbot images on the curvature of the illuminating beam [56-15]–[56-18]. A periodic Ronchi grating produces Talbot images in a coherent collimated light beam. These are used as objects for an imaging with the lens under test. Figure 56-6 shows the corresponding setup. A second identical Ronchi grating can be used to locate the images of the series of Talbot image planes. The object planes are equidistant and located at distances according to the Talbot effect. The corresponding images are given by the lens-makers' formula. In a reciprocal representation, there is a linear relationship for the positions, which can be analyzed with high precision. Due to the use of several focussing distances, the location of the principal plane is eliminated in the calculation of the focal length.

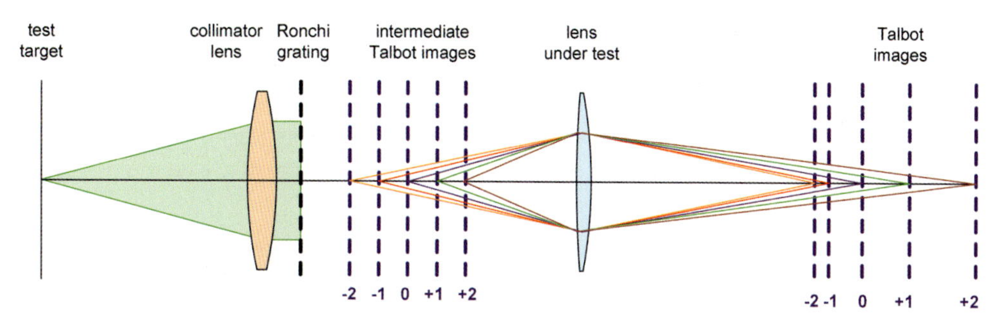

Figure 56-6: Measurement of the focal length using the Talbot imaging technique.

Interferometric Methods

If the location of the principal plane is well known, any interferometric method that determines the radius of curvature with high precision can be used to measure the focal length of a lens. For example, a Fizeau-type interferometer is proposed in the literature [56-19], or else a shearing interferometer [56-20], holographic setups [56-21] or digital holographic techniques [56-22].

Confocal Measurement Methods

An example of a confocal approach to measure the focal length of a lens is described in [56-23]. Here an autocollimation setup with a fiber is used to find the location of the best focus by maximizing the power of the re-coupled beam. The setup is shown in figure 56-7. A different method, used mainly for long focal length values based on the confocal principle, is described in [56-24].

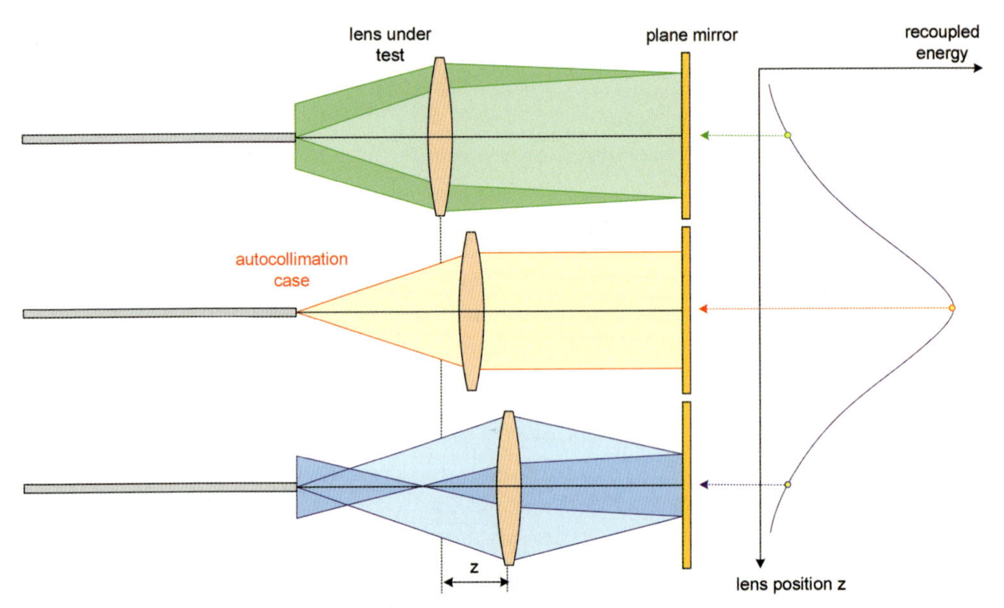

Figure 56-7: Measurement of the focal length by autocollimation and re-coupling of a fiber beam.

56.2.2
Focus and Image Location

The determination of the image location corresponds to the equivalent task of finding the best focus of the system. There are several criteria for the focus definition, which are not identical and generate slightly different image positions during an optimization procedure. Some of the most prominent criteria for focussing are:

1. The paraxial centre of curvature for the paraxial spherical wave of an on-axis object point.
2. The maximum of the Strehl ratio.
3. The smallest rms-value of the wave aberration.
4. The highest contrast for the modulation of an object feature with given spatial frequency.
5. The highest value of the slope of an edge.
6. The highest value of the entropy of the detected digital image.

In a practical experimental determination of the best focus location, a focus criterion should fulfil the following requirements:

1. A steep curve dependency in order to obtain high accuracy.
2. A robust definition to deliver a large dynamic range.
3. The suppression of side lobe effects to guarantee an unambiguous solution.
4. High-frequency pre-filtering to be noise insensitive.

In general, there are methods and criteria which are based on a local point-wise signal. This produces the risk of a focussing, which is not optimal in the complete field. On the other hand, methods based on digital images use the complete field of view and therefore it is easier to find the best overall focussing condition.

It should be mentioned, that the accuracy of a focussing method must always be considered in comparison to the wave optical depth of focus, which has the approximate size of the Rayleigh range

$$R_E = \frac{\lambda n}{NA^2} \tag{56-9}$$

The value of R_E is not an absolute lower bound for adjusting the focal plane, but gives an order of magnitude for z-differences, which cannot be distinguished in the image space.

There are significant differences in the performance of focal criteria if non-defocus aberrations are present [56-25]. In this realistic case, the simple focal criteria based on the energy are more robust and noise-insensitive than those which are based on Laplacian operators for digital image data or threshold criteria.

Another problem, in practice, is the sensitivity of the focal criterion on possible truncation or occlusion effects during the defocussing. A special approach, with an adapted support window, is described in [56-26]. Figure 56-8 illustrates the simple blur effect for defocussing.

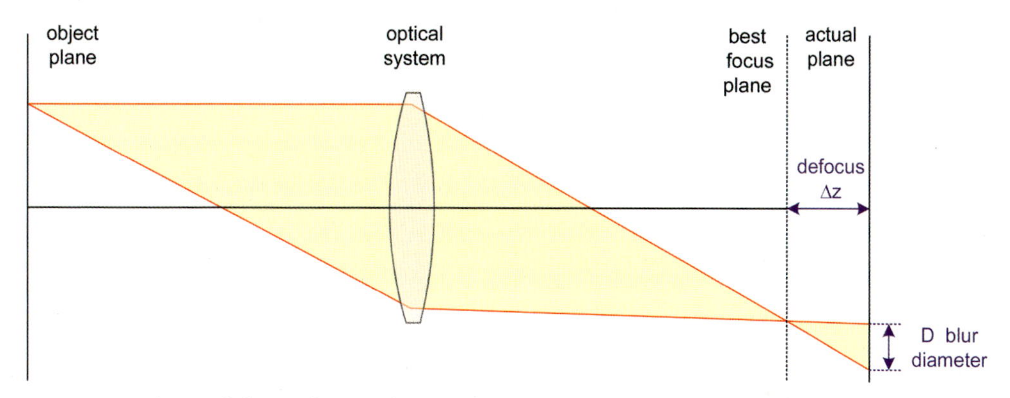

Figure 56-8: Blur in a defocussed image plane.

One quite simple method of determination of the paraxial image plane uses the relation between the normalized Zernike coefficient of defocussing c_4 with the linear focus shifted around the image plane Δz [56-27]. It is given by the formula

$$c_4 = -\frac{1}{4n\lambda}\,\Delta z\, NA^2 \tag{56-10}$$

Therefore, the measurement of the Zernike coefficients by means of a Ronchi ruling, with a varying grating location, can be used to find the paraxial image plane, according to figure 56-9. To evaluate the experiment, the dependence of the defocus Zernike as a function of the focus location z is analyzed. In the ideal case, this is a linear function but, in reality, a straight-line fit of the dependence can be used to find and interpolate the zero crossing point with high accuracy. It is recommended that the dependency of the spherical aberration coefficient c_9 should also be controlled and this should be independent of the focal position. If this condition is violated, the analysis concerning the data and assumptions will be critical as can be seen from figure 56-10.

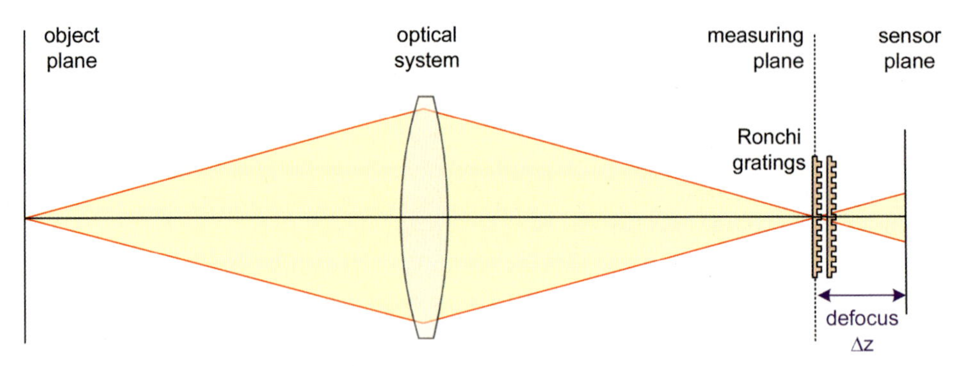

Figure 56-9: Detection of the image-plane location using a Ronchi ruling and determination of the Zernike coefficient c_4.

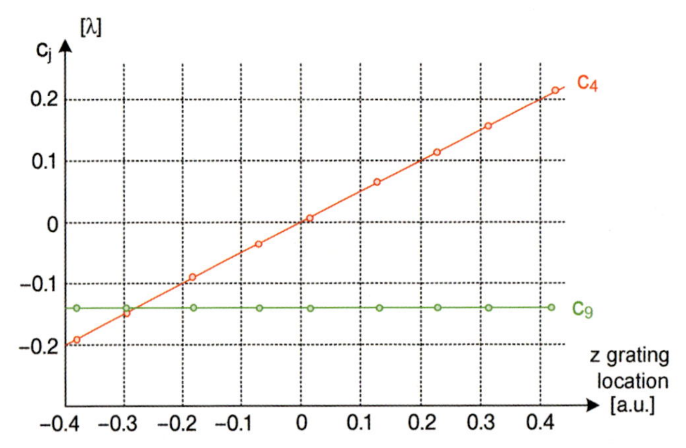

Figure 56-10: Change in the Zernike coefficients c_4 and c_9 as a function of the reference plane location.

If the image is gathered digitally, a software-based solution can be chosen to find the best focal position of an image. These methods mostly use a complete z-stack of images, which is analyzed. If the selection of z-planes is too coarse, a refinement can be performed during the iterative optimization.

In principle, there are several typical criteria-based features, which are used to find the best focus. If a reference image is available, a correlation technique can be used. If there is no known reference, the lateral gradient of the image can be used to find the z-position with best edge slopes. An even better insensitivity for noisy images can be obtained by methods which use curvature measures, for example, with the help of the Laplace operator.

If images with different defocus parameters in a non-telecentric system are compared, then the changes in magnification must be taken into account. The best way to proceed is to adjust an invariant mean value of the image brightness over all pixels.

The following definitions of a focussing criterion guarantee a good performance in practical cases [56-28]–[56-31].

Overall Image Energy

The total energy in an image takes its largest value in the image plane

$$P = \iint I(x,y)\,dx\,dy \tag{56-11}$$

Maximum of the Image Gradient

The magnitude of the image intensity gradient is given by

$$g = |\nabla I(x,y)| = \sqrt{\left(\frac{\partial I}{\partial x}\right)^2 + \left(\frac{\partial I}{\partial y}\right)^2} \tag{56-12}$$

A corresponding threshold criterion is known as the Tenengrad and is quite a good, robust criterion [56-31]. The best focus location using this approach determines the largest value of g.

Energy of the Image Gradient

$$G = \iint |\nabla I(x,y)|^2\,dx\,dy \tag{56-13}$$

which, with the help of the Parceval and the moment theorem of the Fourier transform, is identical to the second moment of the spectrum

$$G = \iint (u^2 + v^2)\,I(u,v)\,du\,dv \tag{56-14}$$

Figure 56-11 shows a corresponding image sequence with the rms value of the gradient as a function of the defocussing.

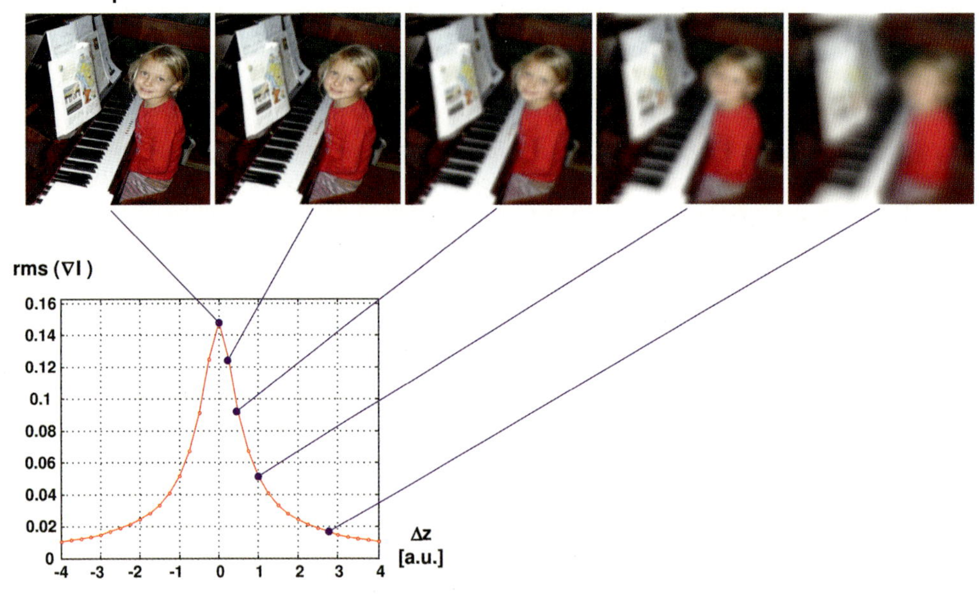

Figure 56-11: Focussing by analyzing the rms value of the image intensity gradients in an image z-stack.

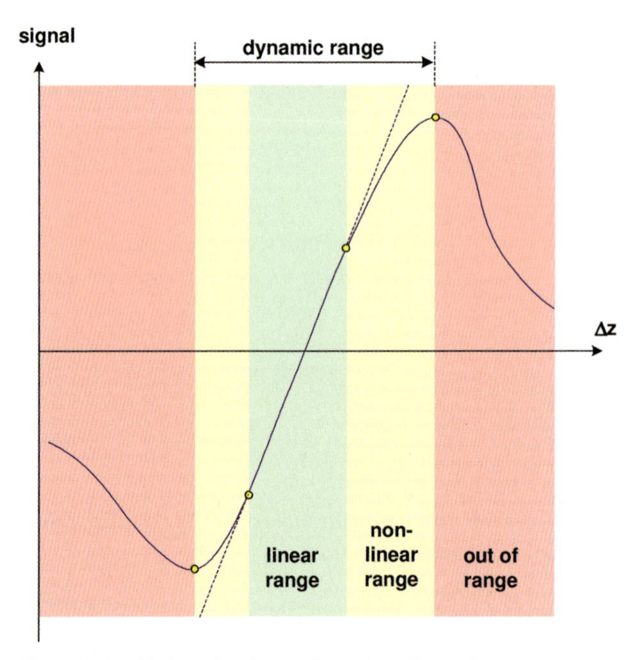

Figure 56-12: Typical signal curve for a focussing series.

The diagram in figure 56-11 exhibits the typical peaked shape of the focus criterion. In practice, it is hard to find the location of the maximum with high precision. Therefore, it is usual to calculate the first derivative of this curve by numerical methods. Then a typical curve, as shown in figure 56-12, is obtained. Here the dynamical range, being the largest z-interval that can be processed, is shown together with the linearity range, which allows a quite quick interpolation and iteration to the desired position with zero derivative.

Energy of the Laplacian

$$L = \iint \left|\nabla^2 I(x,y)\right|^2 dxdy \tag{56-15}$$

Entropy of the Image

The entropy of an image has its minimum when the image is perfectly focussed. Therefore the entropy, defined by [56-32]

$$E = -\sum_j w_j \log_2 w_j \tag{56-16}$$

describes the information content and the largest value is an indicator of the best image plane location. w_j is here the probability of the grey level with index j. The probability w_j can be replaced by a normalized intensity I_j and the criterion then corresponds to the minimum entropy of the intensity histogram of the image.

Contrast

One possible evaluation of the best image plane can be defined by using the modulation transfer function [56-33]. Usually the OTF is determined by calculating the Fourier transform of a measured point spread function intensity $I_{\rm psf}(x,y,z)$. The chosen spatial frequency for optimizing the contrast can be adapted to the dominant frequency content in the image. If this information is not available, a medium frequency, which is half of the cutoff frequency, is proposed. If the transfer function is not calculated explicitly, it is also possible to image an array of lines and the energy of the gradient can be evaluated directly from the slope regions in the image [56-34].

z-gradient of Intensity

According to the results of [56-35], which corresponds to the incoherent image formation in Fourier theory

$$I_{imag}(\vec{x},z) = \int I_{obj}(\vec{x}-\vec{x}')\, I_{\rm psf}(\vec{x}',z)\, d\vec{x}' \tag{56-17}$$

for all pixel locations x, the intensity of the point spread function has an extreme value according to the equation

$$\frac{\partial I_{psf}(\vec{x}, z)}{\partial z} = 0 \text{ for all } \vec{x} \tag{56-18}$$

This relation is also valid in the case of partial coherence and without telecentricity, which means that the pixel position is not constant during defocussing. Therefore, a focal function

$$R(z) = \sum_j \left| \frac{\partial I_{psf}(x_j, z)}{\partial z} \right| \tag{56-19}$$

can be defined and the condition for the best focal plane position z_0 corresponds to $R(z_0) = 0$. Note that the extremum can be a maximum or minimum for every individual pixel with index j.

If the best focus is defined by an image-related criterion, several aspects regarding practical use should be taken into account.

1. If an edge-based algorithm is used, a pre-processing involving searching for an optimal feature must be performed. Furthermore, it must be guaranteed that the selected feature stays in the process window during focussing.
2. A suitable choice for a region of interest is necessary. If it is too large, the calculations will take too long time, if it is too small, the selected sub-image may not be representative and the discretization error increases.
3. Most of the criteria are scalar quantities and therefore a linear optimization is necessary to find the optimal focal plane. The Fibonacci search methods are found to be the best performing algorithms in this case [56-31] considering the speed of convergence and final accuracy. Typically 10–15 iterations are necessary for the corresponding optimizations.

One special application of a high-accuracy focal measurement is the detection of the field curvature across the image field of an optical system. Here, all the above methods can be used in principle [56-8]. But it must be guaranteed in this case, that the focal position can be detected with a higher accuracy than the corresponding focal criterion and that the image field pattern is flat with a comparable quality.

56.2.3
Principal Planes

For simplicity, only systems in air will be considered here. In this case, the principal planes P and P′ coincide with the nodal planes N and N′, respectively. The easiest way to determine the location of the principal plane P′ is to illuminate the system with a collimated beam [56-3], [56-5]–[56-7]. The system under test must be movable along the z-direction and must be mounted on a rotatable stage. This testing device is known as a nodal-slide-bench [56-13]. A test pattern, which is imaged by a collimator and the test system, is observed in the detector plane. If the transverse position of the image is independent of the rotation of the system, the turning point coincides with the back principal plane P′. To avoid vignetting effects, small projection errors of relative size 1-cosφ and changes in the image due to residual aberrations for finite field locations, only small rotation angles in the range $\varphi = 0°–20°$ should be performed.

It can be shown by simple geometrical analysis of the arrangement [56-5], that a turning point at a distance t from the principal plane P′ causes a shift in the image position of size

$$\Delta y = t \sin\varphi \tag{56-20}$$

From this relationship, the accuracy of the method can be easily estimated. Figure 56-13 shows the setup for a measurement of the principal plane on the optical bench.

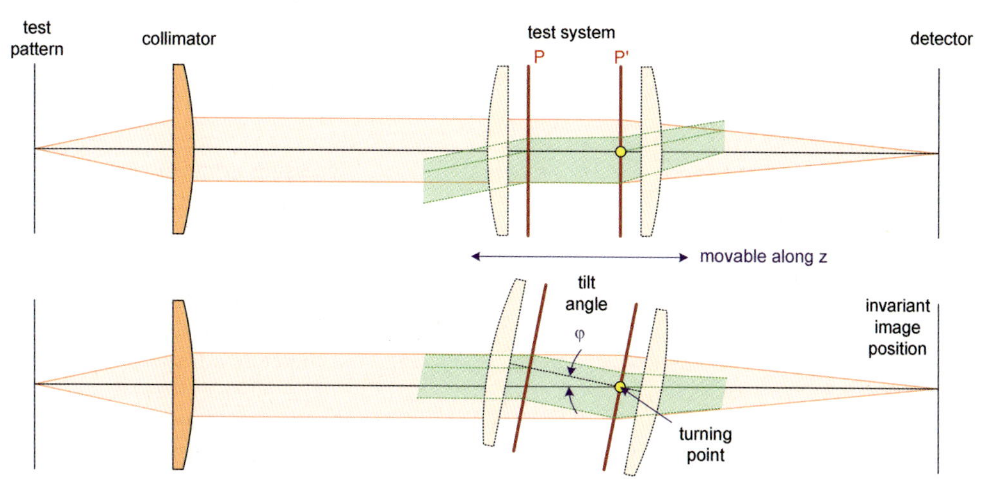

Figure 56-13: Measurement of the principal plane location by determination of the nodal points.

A more indirect method of evaluating the location of the principal planes is described in [56-36]. The simple lens-makers' formula is applied for a set of N measurements of a simple imaging setup with varying object distances s_j. The focal length and the distances of the object and image in the principal planes, respectively, can be extracted from the corresponding equations by linear regression to minimize measuring errors. Let Δ be the distance of the location of the front principal plane P from a fixed reference plane. Then the lens imaging equation for one imaging setup with index j reads

$$-\frac{1}{s_j} + \frac{1}{s'_j} = \frac{1}{f} \tag{56-21}$$

with the measured quantities

$$-\frac{1}{a_j + \Delta} + \frac{1}{a'_j - \Delta} = \frac{1}{f} \tag{56-22}$$

For a series of configurations the measuring error

$$\delta = f\left(a_j + a_j'\right) + \Delta\left(a_j - a_j'\right) - a_j a_j' + \Delta^2 \qquad (56\text{-}23)$$

should be minimized with respect to the unknown values of Δ and f. The corresponding quantities of this approach are indicated in figure 56-14.

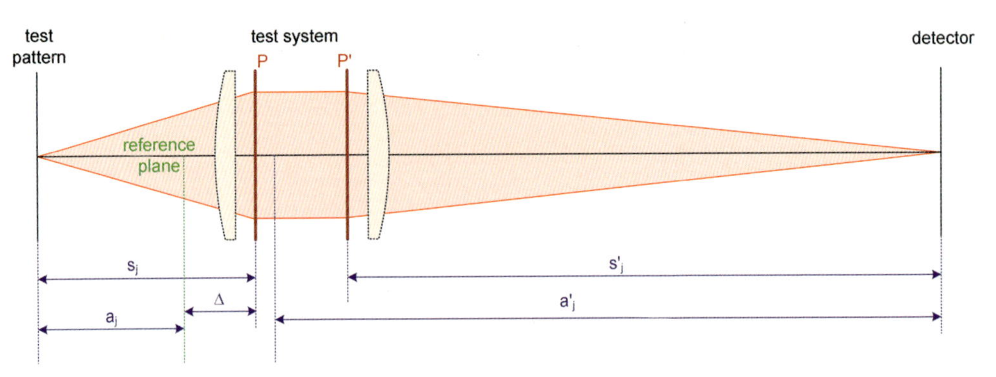

Figure 56-14: Measurement of the principal plane location by least-squares fit.

56.2.4 Magnification

At first sight, the measurement of the magnification appears to be quite an easy task. If calibrated rules are used to compare the size of an image with and without an optical instrument, the ratio of these two numbers gives the magnification of the system. However, in reality, there are several difficulties involved in defining and measuring a magnification with high precision, comparability and reproducibility.

First, several types of optical system must be distinguished. In relay systems with finite object and image distances, the magnification is given by the simple well-known formula

$$m = \frac{y'}{y} \qquad (56\text{-}24)$$

and is performed completely in the spatial domain. For afocal telescopic systems, only the magnification or a change in the field angle can be determined. The magnification is defined by

$$\Gamma = \frac{\tan w'}{\tan w} \qquad (56\text{-}25)$$

where the field angle is w. In the case of a loupe, or a collimator, the object side is finite and the image side is infinite. Here the conventional distance of $s_o = 250$ mm is used to define the magnification for visual use of the system.

$$\Gamma_{loupe} = s_o \frac{\tan w'}{y} \tag{56-26}$$

There are some further problems in the clear definition of magnification. The most important aspects are as follows.

1. If the system suffers from distortion, the absolute size of an image incorporates this effect.
2. A chromatic change in the magnification takes place when broad-band light is used for the measurement.
3. If the system is not telecentric, the definition and realization of the right image plane is essential and influences the measured magnification.
4. Usually the magnification depends on the absolute size of the object or image used in the measurement. If a power expansion with even powers is made in the form [56-37]

$$m = \sum_j c_j (D/D_{\max})^j \approx c_0 + c_2 (D/D_{\max})^2 \tag{56-27}$$

where D is the feature diameter and $D_{\max}$ the maximum field size, the quadratic term typically has a size $c_2 = 10^{-4}$.

Since the Lagrange invariant couples the field size with the numerical aperture, in many cases the evaluation of the system magnification can also be realized indirectly by measuring the diameter of the aperture angles in the object and the image space and using their inverse relationship with the aperture. In the case of a simple telescope of the Kepler or Galilean type, or for infinity-corrected microscopes with objective lens and tube lens, the magnification is also related directly by the ratio of the focal lengths of the two lens groups, respectively. Therefore, measurement of the individual focal lengths also allows the determination of the magnification.

For carefully prepared measurements, an accuracy of 10^{-4} relative error in the magnification can be achieved.

56.2.5
Pupil Location and Aperture Size

The numerical aperture and the location of the pupil are very important properties of an optical system. The measurement of the aperture is quite complex if vignetting occurs for outer field positions and if the shape of the pupil area or rim contour of the opened ray cone is complicated and cannot be described by a simple circle. Another critical case is a system with strong apodization across the pupil diameter. In this case, the true 'rim' of the pupil is not clearly defined.

The image-sided aperture angle can be measured, if an auxiliary lens with known focal length f is moved in the axis direction until the outgoing beam is collimated. A glass plate with a length scale in the collimated beam allows the measurement of the beam diameter D. This arrangement is shown in figure 56-15. Then the aperture angle is obtained by the simple relationship

$$\tan u = \frac{D}{2f} \tag{56-28}$$

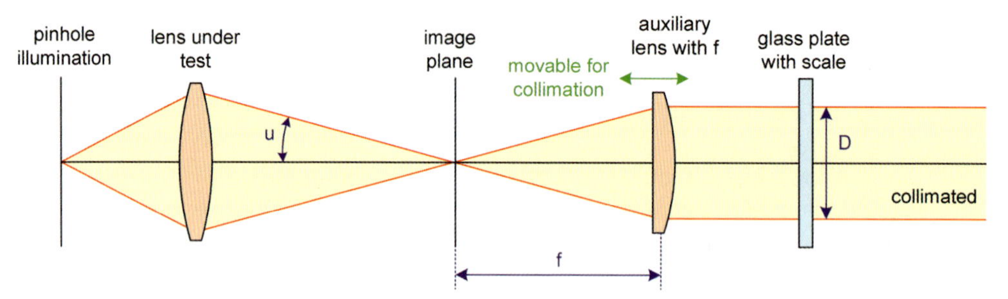

Figure 56-15: Measurement of the image-sided aperture angle.

If a system has a directly accessible pupil with a clearly defined boundary, the illumination distribution of the exit pupil can be observed by a telecentric measuring microscope, called a dynameter, which is very similar to the Abbe focometer shown in fig 56-16. This can, for example, be done in classical photographic lenses. It also allows the determination of the location of the pupil.

A more general approach uses the dynameter system to determine the exit pupil location and size of the system. In the special case of an afocal system, the principle is described in [56-38] and [56-39]. A collimator illuminates the test system with a parallel light beam, the auxiliary dynameter lens focusses the outgoing collimated light onto a CCD sensor. The dynameter lens must be moveable in the z-direction and in a lateral y-direction. First the rim of the exit pupil is sharply imaged at two diametric points, the calibrated lateral movement delivers the diameter of the exit pupil. In a second step the system is centered onto the axis and a reference point like the last surface apex is sharply focussed. The axial difference gives the location of the exit pupil. Figure 56-16 shows the setup for this method.

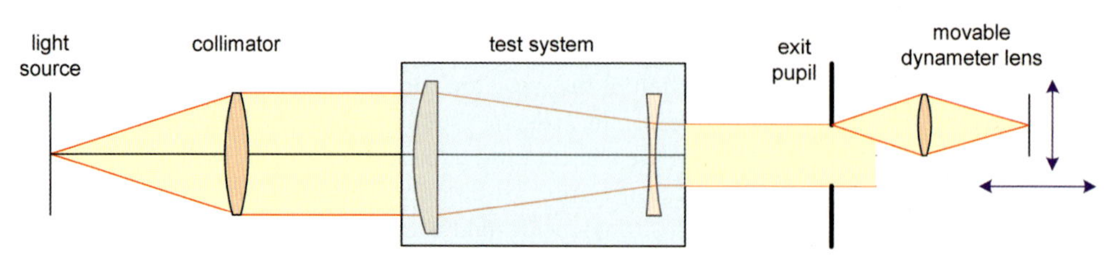

Figure 56-16: Measurement of the size and location of the exit pupil.

If a Ronchi grating with period g is located in the back focal plane of an objective lens, the numerical aperture on the image side can be measured using an interesting method (see [56-40] and also figure 56-17). In an extra-focal image plane, the diffraction orders have different propagation directions. This gives, for a certain distance z, a lateral offset of the orders, which can be measured by simple detection. If the diameter of the exit pupil is D_{ExP}, the diameter of the projection at a distance z from the image plane D, the relation

$$\frac{D_{ExP}}{f} = \frac{D}{z} \tag{56-29}$$

can be used. The grating equation delivers the condition for a diffraction angle of the first order

$$\sin \alpha = \frac{\lambda}{g} = \frac{\Delta x}{\sqrt{\Delta x^2 + z^2}} \tag{56-30}$$

Using the simple geometrical relationship

$$\sin \theta = \frac{D_{ExP}}{2\sqrt{\left(D_{ExP}/2\right)^2 + f^2}} \tag{56-31}$$

then for the determination of the numerical aperture, we finally get the following equation

$$\sin \theta = \frac{1}{\sqrt{1 + \left(\frac{2g \cdot \Delta x}{\lambda D}\right)^2 - 4\left(\frac{\Delta x}{D}\right)^2}} \tag{56-32}$$

It should be noted that the calculation of NA in the methods does not depend on the size of z, which can be chosen arbitrarily.

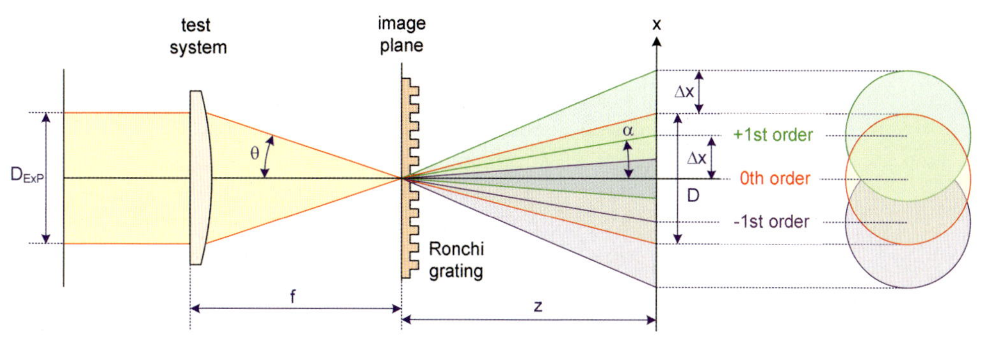

Figure 56-17: Measurement of the exit pupil size and location by means of a Ronchi ruling.

In microscopy, it is mostly the numerical aperture in the object space which is required, because it defines the resolution of imaging [56-42]. Since microscopic objective lenses are corrected for the sine condition, the relation

$$NA_{obj} = m \cdot \sin\theta_{image} \tag{56-33}$$

allows the determination of NA [56-41]. If the magnification m is known then the angle of the ray cone on the image side is evaluated by measuring the footprint diameter in the back focal plane of the lens and its distance s to the image plane by means of the formula

$$\sin\theta_{image} = \frac{D_{stop}}{2s} \tag{56-34}$$

A special setup, the so-called Abbe apertometer measures the numerical aperture for a known focal length of the objective lens by observing the back focal plane using an auxiliary lens [56-7]. The principle allows a conoscopic imaging and shows the boundary of the pupil, which is usually sharply imaged in microscopy. The evaluation gives the numerical aperture by

$$NA = \frac{D_{stop}}{2f} \tag{56-35}$$

A quite simple method of determining the numerical aperture uses the relation between the normalized Zernike coefficient of defocussing c_4 together with the linear focal shift around the image plane Δz [56-43]. It is given by the formula

$$c_4 = -\frac{1}{4n\lambda}\Delta z\, NA^2 \tag{56-36}$$

from which we find

$$NA = 2\sqrt{\frac{n\,\lambda\, c_4}{-\Delta z}} \tag{56-37}$$

The evaluation of the Zernike coefficients is performed for interferometric methods, for example, at several distances z near the focus location (see for instance section 46.3.2 or 46.3.6). The changes in the coefficients should be observed only in c_4. By calculating a regression line from the slope and the corresponding well-known changes in the distances z, the numerical aperture can be obtained as a factor of proportionality. In [56-43], a Ronchi grating is used to determine the Zernikes.

A special method is given in [56-44] for a very accurate measurement of large numerical apertures $NA > 1$ in microscopic systems. Fluorescence molecules in the sample directly behind the cover glass surface no longer radiate as perfect dipoles. But they emit light in a very large cone and show a discontinuity at total internal reflection (TIR) at the sample–glass interface. The angle of TIR is given by

$$\sin\theta_{TIR} = \frac{n_{sample}}{n_{glass}} \tag{56-38}$$

As can be seen in figure 56-18 [56-44], a sharp jump in the fluorescence intensity indicates this angle in the pupil illumination. If $r_{\max}$ and r_{TIR} are the corresponding radii in the pupil, the maximum numerical aperture is determined by

$$NA = \frac{r_{\max}}{r_{TIR}} \tag{56-39}$$

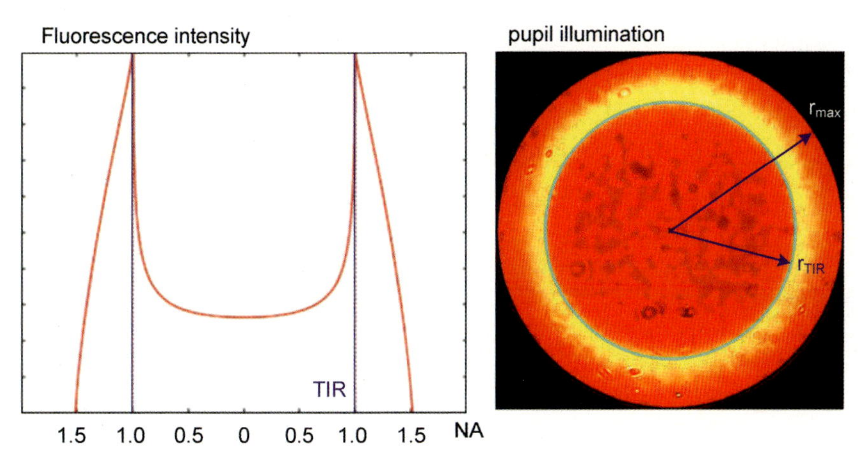

Figure 56-18: Measurement of the numerical aperture for high-NA microscopic lenses.

56.2.6 Telecentricity

There are several types of optical system, which are designed with a telecentric ray path. Particularly in microscopy, metrology and lithography, a telecentric system ensures an invariance in the magnification for moderate defocussed object distances. For the measurement of telecentricity, only a few proposals are found in the literature. One type of method uses changes in the magnification for several defocus positions to quantify any deviations from the exact telecentricity. Another type of approach detects the finite tilt angle of the wavefront in the corresponding pupil.

The simple definition of telecentricity is based on the chief ray, which must be parallel to the optical axis in the case of telecentricity in the object or the image domain. If the system suffers from vignetting for off-axis object points, the ray, which represents the energetic center of gravity, is a more suitable choice. Furthermore, if there is a special illumination pattern and a more complicated filling of the pupil illumination of the system, the definition of telecentricity by the center of gravity is mandantory. This is the case in applications of microscopy and lithography [56-45].

If the telecentricity in the object space is considered for simplicity, the violation of a perfect telecentricity can be described by a finite angle φ of the centroid ray [56-46] and [56-47]. If m is the magnification of the system and $\Delta x'$ the measured offset of the spot centroid in the image plane for a defocussing of the system in the range of the depth of focus, the error of telecentricity is given by (see figure 56-19)

$$\varphi_T = \frac{m\,\Delta x'}{\Delta z_F} \tag{56-40}$$

The pattern shift is proportional to the deviation error of the centroid ray from perfect telecentricity. In this definition, the depth of focus is given by Δz_F and indicates half of the interval.

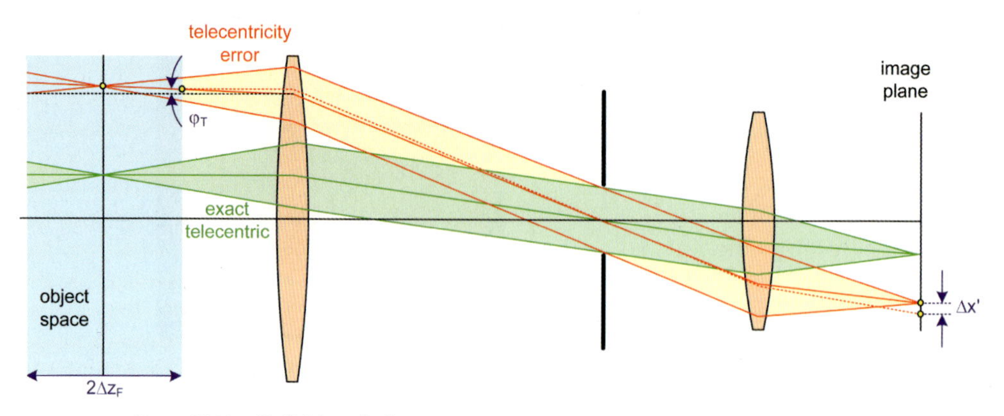

Figure 56-19: Definition of telecentricity error.

It should be noticed that the angle of telecentricity error can have both positive and negative values, corresponding to a centroid ray, which is inclined so that it is convergent or divergent with respect to the optical axis. In the practical measurement of the telecentricity error, a well-defined object, such as an edge or a cross, is imaged for several object distances in the range of the focal depth. The corresponding shift $\Delta x'$ in the image spot centroid is detected and plotted against the object distance. The averaged slope of the linear relationship $\Delta x'(z_O)$ gives the desired angle. If the reason for the telecentricity error is a residual error in the system aberration correction, the telecentricity error usually depends on the field size and therefore must be measured for several field points. As mentioned above, a uniform (for example, diffuse) illumination must be used to guarantee a complete filling of the pupil.

If the condition for ideal telecentricity is violated, the magnification will change. For a double telecentric system, the changes can be defined by non-perfect distances between the lens groups, the stop, the object and the image location.

If Δs_o is the difference in the correct object distance, Δs_a the deviation of the stop position relative to the front lens group, Δs_b the corresponding error behind the stop and Δs_i the difference in the optimal image location for a object-sided telecentric system, then the change in the magnification can be calculated by [56-48]

$$\frac{\Delta m}{m} = \frac{1 + \dfrac{1+m}{m - \Delta a/f_1}\,\dfrac{\Delta s_i}{s_i}}{1 - \dfrac{1+m}{m - \Delta s_a/f_1}\,\dfrac{\Delta s_o}{s_o}\,\dfrac{\Delta s_a}{f_1}} - 1 \tag{56-41}$$

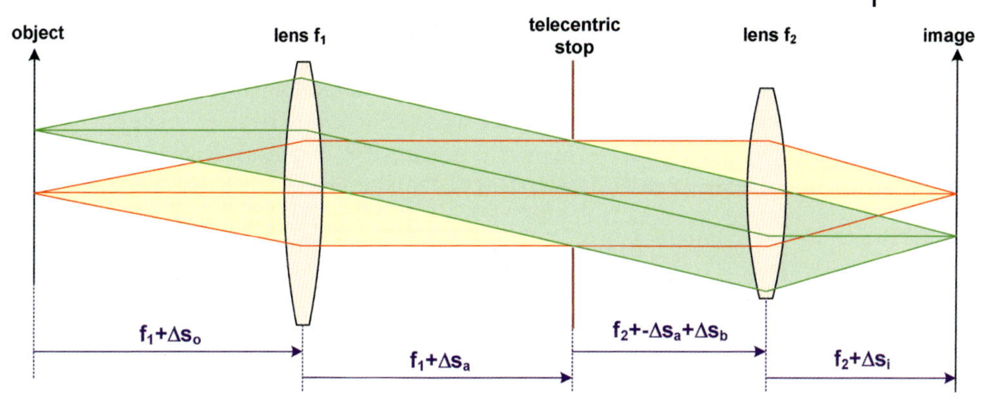

Figure 56-20: Changes in the magnification due to position errors in a double-sided telecentric 4-f system.

See figure 56-20 for the definition of the corresponding quantities. If the system is double-sided telecentric, we have the expression

$$\frac{\Delta m}{m}=\frac{1+\frac{\Delta s_a}{f_1}\left(1-\frac{s_o}{f_1}\right)}{1-\frac{\Delta s_a}{f_1}\left[\frac{s_o}{f_1}\left(1+\frac{\Delta s_0}{s_0}\right)-1\right] +\frac{\Delta s_b}{f_2}\left[\frac{\Delta s_b}{f_2}-m\left(1-\frac{s_o}{f_1}\right)-\frac{\Delta s_i}{s_i}\left\{\left(1-\frac{s_o}{f_1}\right)+1\right\}-\frac{1}{m}\frac{\Delta s_a}{f_1}\right]-\frac{\Delta s_i}{s_i}\frac{\Delta s_a}{f_1}\left(1-\frac{s_o}{f_1}+\frac{1}{m}\right)}-1 \tag{56-42}$$

It can be seen that the misadjustment errors influence the magnification to first order for single telecentricity, but to second order for double telecentric systems.

If the deviation from telecentricity is measured via the tilt coefficients $c_{2/3}$ of the wavefront Zernike representation [56-49], the conventional methods given in chapters 46 and 47 can be used to determine the system wavefront.

56.2.7 Lens Positions and Adjustment

The measurement of the distances between the lenses and components in a complete assembled setup for an optical system is very important in manufacturing systems. The precise control of nominal values and tolerances is complicated, especially in systems of large size and with strongly curved components. In principle, there are several possible ways to measure the positions of the lenses:

1. Mechanical contacting, this is mostly not applicable, since the housing does not allow direct contact with the components.
2. Illuminating the system from one side with a beam divergence that focusses exactly onto a vertex point of the desired surface. The conjugate image point

is detected and, if the imaging conditions through the transmitted surfaces are taken into account, the position of the surface can be determined. The problem with this method is the accurate identification of the reflected beam, especially in the case of well-coated surfaces with a rather small intensity in the reflected beam. This can be supported by kinematic dependences by moving either parts of the system or the illumination to be the nominal ghost image location. Another problem is the decreasing quality of imaging through the transmitted partial system in combination with extreme magnification changes.

3. The position is detected with optical coherence radar or OCT-based methods. In this case the axial coherence length is used to determine the axial distance. The above-mentioned problems with beam profiling and identification are the same. Furthermore, OCT can only be used with a broad-band illumination, which leads to a severe perturbation of the signal due to dispersion effects in the optical train used. On the other hand, the usable numerical aperture is not the parameter which limits the accuracy. The spectral width of the radiation used limits the precision to a value of approximately 1 μm, in practice.

In the following, only methods 2 and 3 are discussed in more detail: the concept of direct focussing onto a surface vertex and the observation of the reflected light.

In [56-50], a white light Mireau interferometer is proposed to measure the vertex positions of an assembled system contact-free. Due to the principal common path property of the Mireau interferometer, the aberrations induced in the transmitted front part of the test system are cancelled out in first approximation. For geometrical parameters like the microscopic objective lenses addressed in the [56-50], an accuracy of 0.2 μm can be achieved for the vertex locations in practice by this method.

In [56-51], the method outlined above is described in a modified arrangement. The basic methods of enhancing the performance of the method are as follows.

1. An annular beam is used for illumination. Due to the small ring width, the effects of spherical aberration in the transmitted part of the system considerably reduce the influence on the result.
2. Figure 56-21 shows the setup for the measuring system. There are two arms in the detection path of the reflected light, which contain a confocal pinhole to select the depth of the reflected light with high precision. The two positions of the confocal pinholes are shifted against one another. Finally, the signals from the two arms are subtracted. This causes a high depth discrimination and the resulting signal has the shape of a typical difference signal. This is the so called differential confocal principle of measurement, which is described in more detail in [56-52].

The overall accuracy is also determined by the numerical aperture angle of the illumination beam at the observed surface. A typical choice of parameters is a truncation ratio of the ring illumination of $D_{in}/D_{out} = 0.8$ and an aperture angle of 0.1. As indicated in [56-51] in detail, the change in the ray path through the transmitted surfaces must be taken into account, but can be easily calculated by simple raytracing.

The achievable accuracy of this method is in the range of a relative error in the surface positions in the order of 2×10^{-4}.

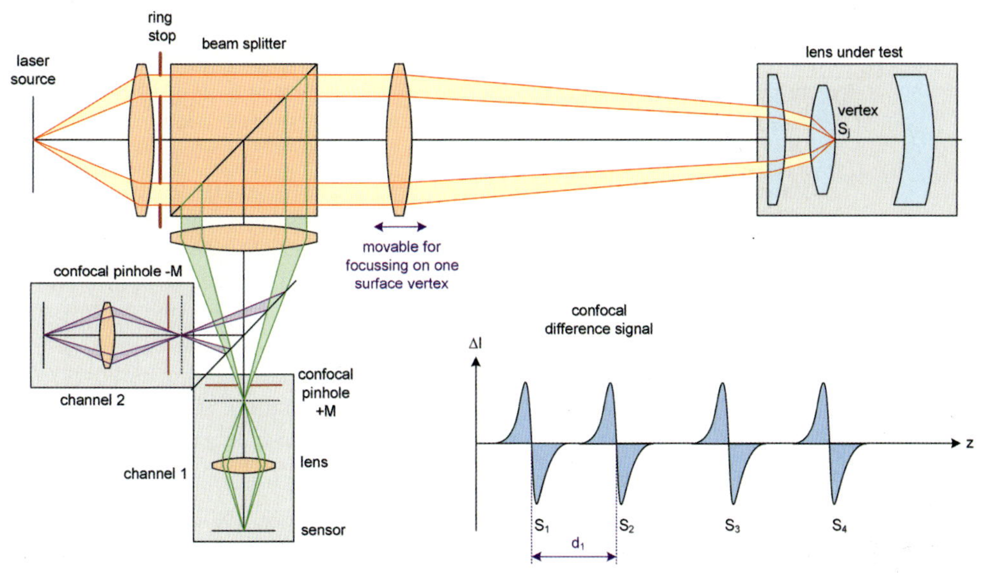

Figure 56-21: Measuring vertex positions in an assembled system by differential confocal methods.

The measurement of the vertex positions by low coherence interferometry is described in [56-53] and [56-54] (see section 53.3). Typically a near-infrared light source with a broad spectrum in the range of 50 nm–100 nm is used in a conventional interferometer setup. The accuracy of this method strongly depends on the knowledge of the wavelength dispersion in the glass materials, especially the group velocity dispersion of the light being used. Typically, the precision is in the range of 1 μm, but with special clever modifications using polarization and more sophisticated signal evaluation algorithms, the accuracy can be increased to 100 nm [56-54]. The focussing collimator must be adjusted to optimize the signal, if several interfaces are transmitted. It can be shown that focussing the light beam onto the vertex and concentric reflection at the surface delivers a good signal strength. In comparison to a geometric optical reflection measurement as described above, this method is preferred in the case of aspherical surfaces. Figure 56-22 shows the setup of this white-light interferometer. In particular, the relative measurement of the width of the distance between two lenses in air can be performed as a difference measurement and does not depend on an exact knowledge of the material dispersion.

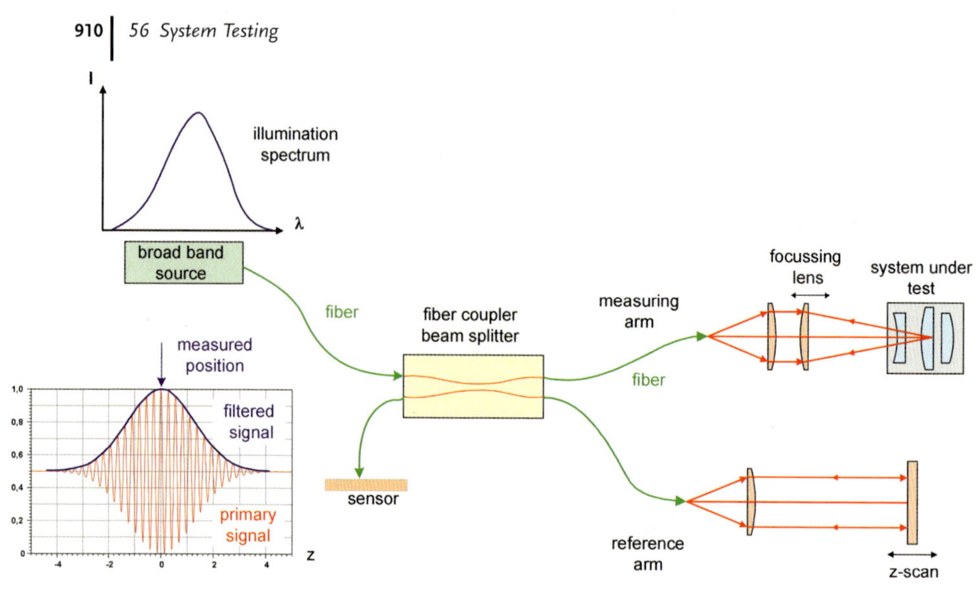

Figure 56-22: Measuring lens vertex positions using low-coherence interferometry.

56.2.8
Centering

An optical system generally contains more than one effective surface, each of which have to be aligned and centered with respect to each other. The centering of all surfaces and components belonging to a given system is one of the most complicated tasks involved in the adjustment and mounting of an optical device. The centering can be fixed, with sufficient positioning accuracy, simply using mechanical holders; or it can be adjusted by the provided degrees of freedom for various movements. In the latter case it is necessary to measure the exact centering state of all surfaces belonging to the optical system. The centering of individual optical components is described in detail in section 53.5.

Measuring Centering in Reflection

The centering or the evaluation of the centering state of a single optical surface typically proceeds by focussing a beam on the center of curvature using a zoom system so that the reflected image can be observed in autocollimation. By rotation of the surface or the lens around a mechanical reference axis and referencing the reflected image relative to a fixed reference, one can determine the inclination of the surface. As illustrated in figure 56-23, a deviation circle is seen in the detector plane, if the test surface has a residual centering error.

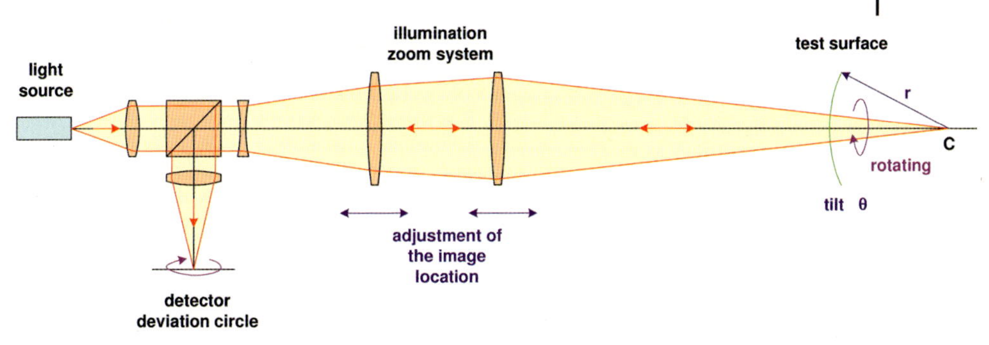

Figure 56-23: Centering measurement in autocollimation.

When the external surface of a lens with radius r is observed, with the image scale factor m from the center of curvature to the detector, one obtains the lateral off-set

$$v = m\,v_M = 2\,m\,r\,\theta \tag{56-43}$$

where θ denotes the inclination angle of the surface.

If the surface of an internal system is examined, the magnification is determined by the whole area of the system, which is penetrated by the beam. In addition, all surfaces traversed will affect the result by their imperfect centering state. The centering of the internal surface under consideration can be determined only in successive steps, eliminating the contribution of all other surfaces lying in the outward direction. In this case it is also difficult to adjust the correct position of the reflected image.

Measuring Centering in Transmission

Lenses can be centered in transmitted light by observing the lateral deviation of a narrow laser beam which is transmitted through the lens when the lens is rotated around a mechanical reference axis. In this case the contributions of both the front and the rear surfaces are simultaneously detected. In order to obtain an easily observable and analyzable small spot on the detector, it is usually necessary to use auxiliary optics to compensate for the focussing effect of the lens sample. Hence, this method is not adequate for lenses with short focal lengths. Moreover, the beam diameter on the lens should be within the paraxial range in order to avoid spurious spot broadening by aberration effects. The set-up of this measuring principle is shown in figure 56-24.

The achievable centering accuracy is of the order of 1′. It is limited by the multiple ghost reflections occurring in the lens, if there is no anti-reflecting coating.

In this general case it is not possible to determine the tilts of the separate surfaces unambiguously, since the exact location of the center of rotation is involved. For the same reason, the inclination of the entire lens cannot be observed.

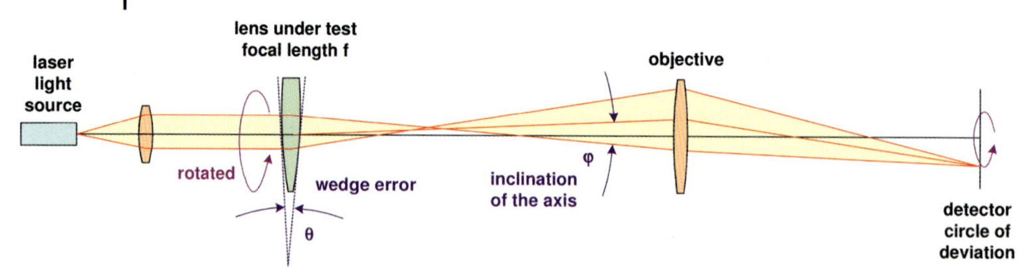

Figure 56-24: Lens centering by the transmission method.

In [56-55] and [56-56] a method is proposed to measure the centering error of all surfaces in a completely assembled objective lens with several components. The principle corresponds to the detection of a reflected beam as described above for a single lens. A variable shifted lens must be adjusted to focus the beam onto the vertex of the surface to be measured. As when measuring the vertex distances, the imaging of the transmitted beam through that part of the system between the test surface and the illumination must be taken into account. Therefore, the determination of the centering errors of the surfaces must be calculated in a sequence beginning with the outer surface. The knowledge of the tilt angles of all the transmitted surfaces will influence the outcome of the result for the inner surface.

56.3 Measurement of Image Quality

56.3.1 Wavefront Quality

One quite simple criterion used to evaluate the performance of an optical system is the wave aberration. A wave front is responsible for the point image quality for one field point and one wavelength. A complete characterization therefore consists of a larger number of values. To obtain a quantitative specification, there are several possible ways to describe the system performance as represented in chapter 30:

1. Peak–valley value

The simplest way to characterize a system quantitatively by wave aberrations is the computation of the peak-to-valley value W_{pv} of the wave surface in the exit pupil of the optical system. If x_p, y_p are the coordinates in the pupil, the pv value is defined as

$$W_{pv} = W_{\max}(x_p, y_p) - W_{\min}(x_p, y_p) \tag{56-44}$$

2. Rms value

The peak–valley value is a very conservative criterion for characterization of the wavefront and usually characterizes the wave aberration deviation as the worst case, but exists only in one single point of the complete pupil area. Therefore, in general, the rms value is used, which is comparable to the Gaussian second moment in the geometrical spot evaluation. It is defined by

$$W_{rms} = \sqrt{\langle W^2 \rangle - \langle W \rangle^2} = \sqrt{\frac{1}{A_{ExP}} \iint \left[W(x_p, y_p) - W_{mean}(x_p, y_p) \right]^2 dx_p \, dy_p} \tag{56-45}$$

with the average of the wave aberration over the pupil area being

$$W_{mean}(x_p, y_p) = \frac{1}{A_{ExP}} \iint W(x_p, y_p) \, dx_p \, dy_p \tag{56-46}$$

Here the integration extends over the area A_{ExP} of the exit pupil

$$A_{ExP} = \iint dx_p \, dy_p \tag{56-47}$$

If the system is well corrected and diffraction limited or near to this quality range, the Rayleigh or the Marechal criteria can be used to estimate a suitable size for acceptable wave aberrations.

If the description of the wave aberration needs to be more detailed, a decomposition of the wavefront surface into Zernike polynomials or other function systems can be used. See, for example, the representation in section 20.4.2. For high-performance optical systems, it is good practice to give upper limits to the first 36 Zernike coefficients to guarantee the system quality.

The measurement of the wave aberrations can be carried out by one of the various methods described in chapters 46 and 47. For further details see these chapters and the references cited there. So the most prominent metrology methods are as follows.

1. Interferometry
2. Shack–Hartmann sensor
3. Hartmann sensor
4. Point spread function retrieval.

It should be noticed that the system performance is not well described by the wave aberrations alone, if a strong apodization exists. In this case, the relative weighting of the pupil illumination plays a major role in the shape of the point spread function and image formation. Furthermore, if the system is illuminated by partially coherent light, a simple definition of the phase of the light is not possible. As noted in chapters 19 and 21, the image formation can only be described by incorporating the effect of the illumination and the modified interference capability of the radiation.

When quantifying the performance of the human eye, a special concept called the pupil fraction metric is proposed in the literature [56-57]. The basic idea is to define a wavefront criterion for a good performance W_{crit} (for example Marechal) and measure the fraction of the complete pupil area which fulfills this criterion. Usually the subaperture is chosen to be concentric and circular in shape.

$$P_{frac} = \frac{A_{W<Wcrit}}{A_{total}} \tag{56-48}$$

56.3.2
Point Spread Function PSF

If a single object point is considered, the so-called point spread function, which is the response of the optical system to a δ-excitation, gives important information on the performance of the system [56-58] and [56-59]. The definition and properties of the point spread function are described in chapters 20 and 30 in detail. A characterization of the PSF can be done in many ways and with several quantitative values. Some single- valued criteria of quality description and quantification of a point spread function are shown in figure 56-25 [56-60].

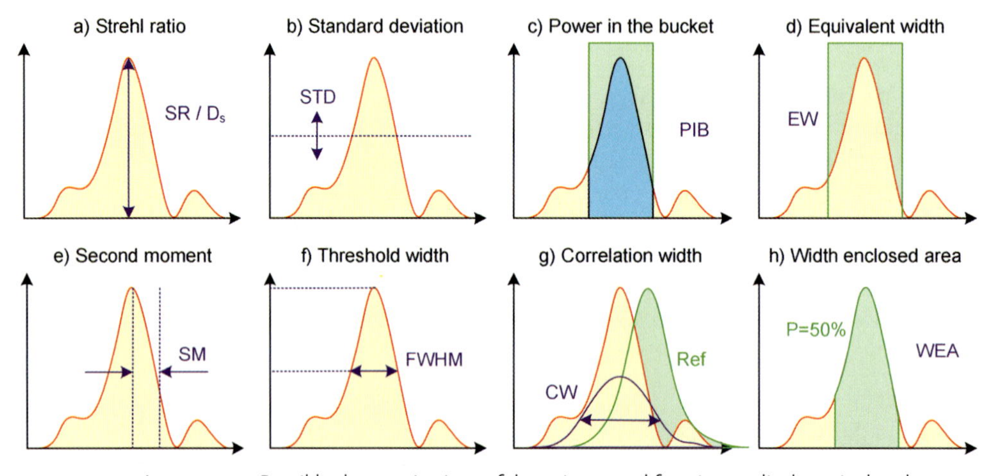

Figure 56-25: Possible characterizations of the point spread function quality by a single value.

In figure 56-25, the standard deviation is measured as the perturbation of the real PSF in comparison to the diffraction-limited PSF according to the formula

$$STD = \frac{\sqrt{\iint \left[I_{PSF}(x,y) - \bar{I}_{PSF}(x,y) \right]^2 dxdy}}{\sqrt{\iint \left[I_{PSF}^{(ideal)}(x,y) - \overline{I_{PSF}^{(ideal)}}(x,y) \right]^2 dxdy}} \tag{56-49}$$

The most important measures for characterizing the point spread function are briefly described in the following.

Intensity Moments

The second moments corresponds to the Gaussian moments and describe the spatial extent of the point spread function or intensity profile [56-61]. In one transverse dimension we have

$$M_{2,x} = \iint x^2 I(x,y)\, dx\, dy \tag{56-50}$$

The diameter of an intensity distribution, such as a beam section or a spot, can therefore be defined in every coordinate direction as

$$D_{2x} = 2\sqrt{\langle x^2 \rangle} = 2\sqrt{\frac{1}{P}\iint x^2\, I(x,y)\, dx\, dy} \tag{56-51}$$

with a corresponding equation for D_{2y}. The third-order moment can be used to describe the asymmetry of the intensity distribution. It is defined analogously, for example, in the x-section as

$$M_{3,x} = \iint x^3 I(x,y)\, dx\, dy \tag{56-52}$$

Strehl Ratio

The most obvious way to quantify the point spread function is by the indication of the Strehl ratio. Since the complete intensity distribution is generally a complicated function, a simple number is defined, which describes the behavior of the PSF by only one value. The definition used most often is the Strehl ratio, which is equal to the reduced height of the peaked point spread function intensity in the centroid of the energy for the real aberrated system referenced to the ideal aberration-free system. This is shown in figure 56-26.

$$D_S = \frac{I_{PSF}^{(real)}(0,0)}{I_{PSF}^{(ideal)}(0,0)} \tag{56-53}$$

is the expression for the Strehl ratio and is known as the Strehl definition.

If $A(x_p,y_p)$ is the amplitude in the pupil and $W(x_p,y_p\,)$ is the wave aberration, the definition of the Strehl ratio given by

$$D_S = \frac{\left|\iint A(x,y)\, e^{2\pi i\, W(x,y)}\, dx\, dy\right|^2}{\left|\iint A(x,y)\, dx\, dy\right|^2} \tag{56-54}$$

can be used, where A_{ExP} is the pupil area. As can be seen from this equation, every wave aberration W degrades the PSF and therefore lowers the Strehl number. Systems with apodization of the pupil or central obscuration require the full expression (56-54) for the calculation of the Strehl number.

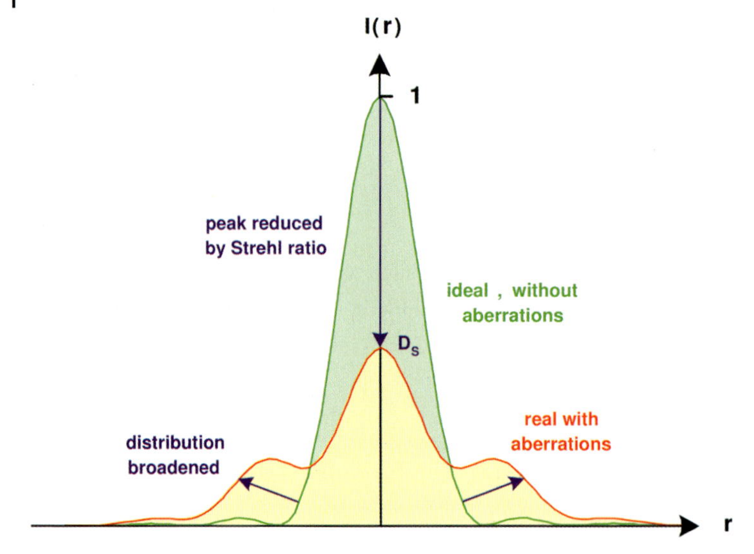

Figure 56-26: Deformation of the point spread function in the presence of aberrations.

The Strehl definition is a normalized aberration measure and can only take values between 0 and 1. A Strehl number of 1 corresponds to an ideal system. If the Strehl ratio is considered as a measure of quality, all the information in the fine structure of the intensity profile of the point spread function is reduced to only one number. Therefore, the Strehl ratio as a description of system quality only makes sense in systems with quite low wave aberrations, where the Strehl number is above approximately 0.6 and the complicated form of the PSF deviations from the ideal case, does not play an important role.

Most of the real systems show no significant difference between the centroid of the energy and the peak position of the intensity profile. But if there is a considerable amount of coma, the peak normally shows a lateral deviation from the centroid and the exact definition of the Strehl number is essential. But, in general, the Strehl ratio is considered as a reduction of the peak height, which is not exactly true in every case.

The Strehl definition is equal to the volume under the three-dimensional modulation transfer function surface. One possible definition of a polychromatic Strehl number is the use of the integral over the volume of the polychromatic MTF.

The Strehl ratio definition is not easily applicable to spectrally broad systems with polychromatic illumination. In this case, a generalization of D_S is complicated.

If the Fresnel number of a system is low or the numerical aperture is high and the polarization plays a major role, the reference in the formula of the Strehl ratio to the perfect case is much more complicated [56-62]. Therefore, the applicability of the Strehl number is limited to medium apertures and systems near the diffraction limit.

Power in the Bucket

The encircled energy is a quality criterion, which measures the energy of the intensity distribution of a spot or beam as the transmitted power through a stop of variable radius. In the case of rotational symmetry, with the help of the intensity $I(r)$, the energy $E(r)$ inside the radius r is given by the integral

$$E(r) = \int_0^r I(r)\, 2\pi\, r\, dr \tag{56-55}$$

The energy is a monotonously increasing function of the stop diameter $D = 2r$. If, in a practical application, a certain diameter is specified, the energy for this value is called the "power in the bucket", which also serves as a quality criterion. The beam must not necessarily be collimated and the principle is also valid in the case of convergent or divergent beams or bundles.

In two dimensions of arbitrary geometry; first, the centroid coordinates x_c, y_c must be determined. The energy function is then referenced to this point and the calculation is performed according to the equation

$$E(r) = \int_{y=-r}^{y=+r} \int_{x=-\sqrt{r^2-y^2}}^{x=+\sqrt{r^2-y^2}} I(x - x_c, y - y_c)\, dx\, dy \tag{56-56}$$

Measurement

Details of the measurement of the PSF which, when applied in astronomy and microscopy is called the Star test, are described in section 49.3 in detail.

A special problem in the PSF measurement is often the extreme variation in the intensity levels in the image plane and defocussed planes. Due to this fact, cameras with high dynamic ranges are necessary or special preprocessing steps for noise removal can be carried out. Highly sophisticated adaptations are also useful [56-63].

In the classical experimental procedure used to determine the point spread function, a very small quasi-point object is used corresponding to the definition of the PSF. If the system under test has a large numerical aperture or the wavelength is quite short in the ultraviolet, the size of this pinhole becomes quite small. The pinhole can be considered to be of negligible size, if it is not larger than approximately 1/3 of the Airy diameter (see section 47.7.3). As a result, the signal-to-noise ratio of the system is critical or the stability limits long exposure times. One possible solution to this problem is to use a pinhole of larger but well-known size and the calculate the true point spread function by deconvolution (see section 47.7.7).

Determination of the Point Spread Function via Edge Measurement

Another possible way to circumvent this problem is to measure the image of a straight knife edge in several different azimuthal positions and to determine the point spread function by the well-known Radon transform (see also section 56.4.3) [56-66].

Alternatively, it is recommended in the literature to measure the edge spread function for two perpendicular edge orientations, which has a significant better signal-to-noise ratio due to the integration and to calculate the point spread function digitally. The best performance for non-diffraction-limited systems are those methods in which a highly sophisticated edge model is fitted to the data. Furthermore, a clever algorithm can be used to reach sub-pixel resolution perpendicular to the edge direction [56-64] and [56-65]. This is explained in section 56.3.4 in more detail.

56.3.3
Axial Point Spread Function

Wave aberrations change the transverse shape of the point spread function as described in section 56.3.2. The intensity curve $I(0)$ of the complete caustic distribution $I(x,y,z)$ along the optical axis can additionally be analyzed to determine some aberrations types [56-67]. The superposition of all aberrations types for finite field points makes the separation of the influences on the axial distribution rather complicated [56-68]. Therefore, a reconstruction of the aberrations from the axial intensity behavior is mainly useful on-axis. As an example, figure 56-27 shows the changes in the function $I(0)$ due to primary spherical aberration, which is given by the Zernike coefficient c_{40}. If the paraxial image plane position is known, the corresponding Seidel coefficient of spherical aberration can also be deduced by analyzing the asymmetrical shift in the distribution relative to the image plane. It should be noted that, in cases of non-uniform pupil illuminations, the behavior of the axial intensity distribution becomes more complicated. If the system has an apodized pupil, which is not known, the analysis is critical. If simple apodization shapes, such as a ring or a gaussian distribution with known parameters are present, a highly accurate reconstruction of the aberrations is possible, as described in the references cited above.

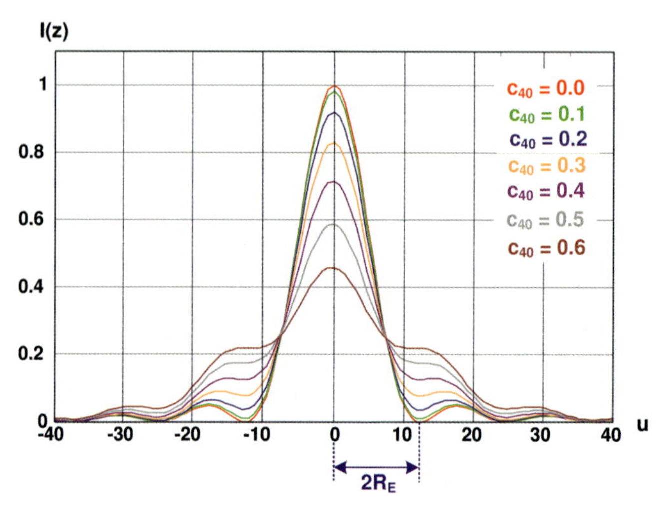

Figure 56-27: Axial intensity distribution of the point spread function in the presence of spherical aberration with Zernike coefficient c_{40} in λ.

56.3.4
Edge Spread Function ESF

A point spread function is obtained, if a quasi-point-like object is imaged. This automatically corresponds to a very low power level and degradation due to noise in the image. Therefore, well known objects with a simple geometry and a better power transmission are often preferred. One possible choice is a straight edge. The broadening of the ideal steep-edged object in the image is determined by diffraction and the quality of the system. The edge spread function therefore contains information about the system transfer properties similar to the point spread function. One major drawback is the asymmetrical geometry of the object arrangement. The orientation of the edge breaks the symmetry. Since the spatial frequency content along the edge is rather low, at least two perpendicular orientations of the edge must be used to characterize the system completely.

The edge spread function is related to the point spread function by the following equation

$$I_{ESF}(x_i) = \int_{-\infty}^{x_i} \int_{-\infty}^{\infty} I_{PSF}(x_o, y_i) dy_i dx_o \tag{56-57}$$

A direct relation exists between the edge and the line spread function via the following formulas

$$I_{ESF}(x_i) = \int_{-\infty}^{x_i} I_{LSF}(x_o) dx_o \tag{56-58}$$

$$I_{LSF}(x_i) = \frac{dI_{ESF}(x_i)}{dx_i} \tag{56-59}$$

The incoherent imaging of an ideal edge by the optic to be tested as shown in figure 56-28 offers one possible means of measuring the edge spread function. Two advantages of the measurement of the edge spread function in comparison to the point spread function are that an edge object can be manufactured much more easily than a pinhole as a quasi-point object and that the signal-to-noise ratio in the detection is significantly higher. This last property is particularly important in systems with high numerical aperture and if the wavelength used is short. In both cases, the pinhole size, which is scaled by the Airy diameter, must be manufactured to a very small size.

There are special image quality criteria, which are used in an evaluation of an edge spread function measurement. The most important terms are described in the following.

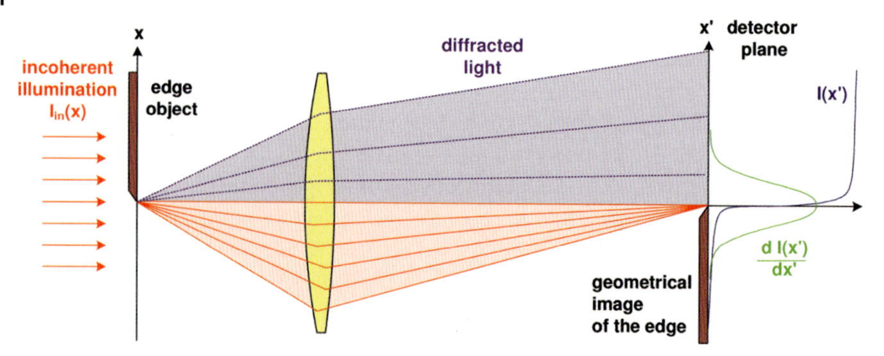

Figure 56-28: Measurement of the incoherent edge spread function.

Edge Width

According to Thomas, the quality criterion is defined as follows (see also chapter 30). A sketch of the various terms is illustrated in figure 56-29. As a reference, we use the lateral position of the 50 % intensity value of the maximum I_0. The left width, Δx_{left} of the edge image is defined by the extension of a rectangle with a height of $I = 0.5\ I_0$ and the same area as the integral under the image curve to the left of the reference position. A corresponding condition defines the right edge width Δx_{right}. As an ideal image, a corresponding perfect edge is assumed.

$$\Delta x_{left} = \frac{1}{I_0} \int_{-\infty}^{0} \left[I_{real}(x) - I_{ideal}(x) \right] dx \tag{56-60}$$

$$\Delta x_{right} = \frac{1}{I_0} \int_{0}^{\infty} \left[I_{real}(x) - I_{ideal}(x) \right] dx \tag{56-61}$$

The width of the edge image is then given according to Thomas by the mean value of these two measures

$$\Delta x_{width} = \frac{\Delta x_{left} + \Delta x_{right}}{2} \tag{56-62}$$

The difference between the left and right edge width measures the asymmetry of the image

$$\Delta x_{asym} = \Delta x_{left} - \Delta x_{right} \tag{56-63}$$

The sharpness of the edge image is given by the positive difference in the coordinates for the 25 % and the 75 % values of the intensity curve.

$$\Delta x_{sharp} = \left| \Delta x_{I=75\%} - \Delta x_{I=25\%} \right| \tag{56-64}$$

The definition of the edge width as defined above is a linear measure and not rms-based. There is the following relationship with the optical transfer function

$$\Delta x_{width} = \frac{1}{\pi^2} \int_0^\infty \frac{1 - \mathrm{Re}[H_{OTF}(\nu)]}{\nu^2} d\nu \tag{56-65}$$

The partitioning of (56-65) shows that this corresponds to a summation of a contrast and a phase- sensitive part.

$$1 - \mathrm{Re}[H_{OTF}(\nu)] = (1 - H_{MTF}(\nu)) + 2\, H_{MTF}(\nu)\, \sin^2\left(\frac{H_{PTF}(\nu)}{\nu}\right) \tag{56-66}$$

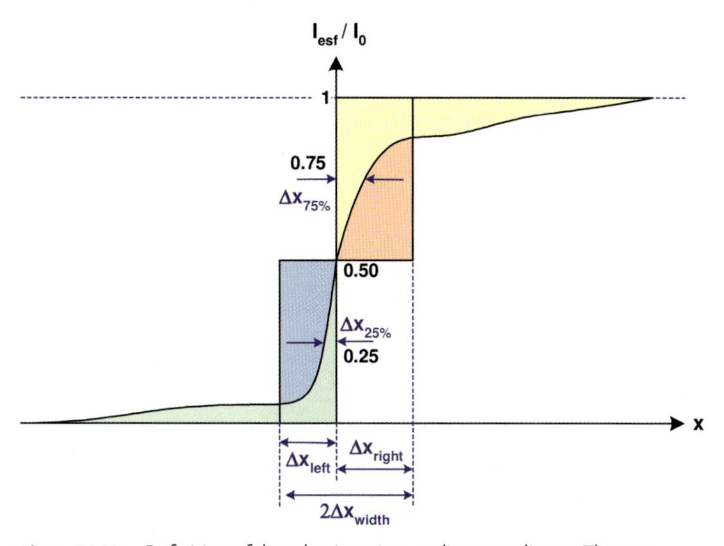

Figure 56-29: Definition of the edge imaging quality according to Thomas.

Edge Steepness

The maximum of the slope of an edge image is specified as the edge response. The relative edge response of a real curve in comparison with the ideal edge distribution is used as a quantitative figure of merit analogous to the Strehl number for the point spread function. The relative edge response can be calculated by integration of the one-dimensional transfer function

$$S_{rel} = \frac{\int_0^1 H_{MTF}(\nu) d\nu}{\int_0^1 H_{MTF}^{(ideal)}(\nu) d\nu} = \frac{3\pi}{4} \int_0^1 H_{MTF}(\nu) d\nu \tag{56-67}$$

In lithographic applications, the steepness of an edge is an important measure for the evaluation of the mask projection process. The exposure depends on the illumination in a logarithmic form, therefore the resolvable critical dimension of the object details,

$$\Delta x_{CD} = k \frac{\lambda}{n \sin u} \tag{56-68}$$

is used to define the steepness of the intensity gradient of the logarithm of an edge NILS, (normalized image log slope)

$$S_{NILS} = \Delta x_{CD} \left. \frac{d \ln I(x)}{dx} \right|_{thresh} = \frac{\Delta x_{CD}}{I_{thresh}} \left. \frac{dI(x)}{dx} \right|_{thresh} \tag{56-69}$$

Here I_{thresh} is an arbitrary threshold value of the intensity gradient. Typically, $I_{\text{thresh}} = I_{\max} / 2$ is used.

Acutance and Edge Defect

There are other common criteria for measuring the quality of an edge. The sharpness or acutance is defined by a normalized quadratic mean of the intensity gradient in the form

$$A_{acut} = \frac{\dfrac{1}{x_{\min} - x_{\max}} \displaystyle\int_{x_{\min}}^{x_{\max}} \left(\frac{dI_{ESF}(x)}{dx} \right)^2 dx}{I_{ESF}(x_{\max}) - I_{ESF}(x_{\min})} \tag{56-70}$$

This expression assumes finite limits $x_{\max}$ and $x_{\min}$ for the calculation of the gradient. For practical purposes, this expression can also be evaluated with the help of the line spread function. The advantage of this equation is the use of infinite limits in the calculation interval of the gradient

$$A_{acut} = \frac{\dfrac{1}{x_{\min} - x_{\max}} \displaystyle\int_{-\infty}^{\infty} I_{LSF}^2(x) dx}{I_{ESF}(x_{\max}) - I_{ESF}(x_{\min})} \tag{56-71}$$

Last, but not least, the following equation defines the so-called edge defect. It describes the broadening of an edge with the help of an rms criterion and is referenced to an ideal amplitude edge. The intensities are the values in the image.

$$K_{defect} = \sqrt{\left[\int \left(I_{ESF}^{(real)}(x) - I_{ESF}^{(ideal)}(x) \right) dx \right]^2} \tag{56-72}$$

In [56-69] and [56-70] a special technique is described to remove measurement errors and improve the spatial resolution when the edge and the sensor row is tilted against the scan direction at a small angle α. It is clear that the integration of the signals in a row assumes a perfect alignment with the edge. Otherwise, the relation of one detector stripe to one edge position is violated. It can be shown that, for a sensor length L along the edge, the relation

$$\alpha << \frac{1}{\nu L} \tag{56-73}$$

must be satisfied in order to give accurate data for the measurement of the spatial frequency ν of the MTF. If the numerical digital evaluation of the MTF from the rough edge spread function data is modified [56-70], this tilt effect can be used to gain a higher spatial sampling accuracy.

The basic principle is shown in figure 56-30. For an arbitrary rotated edge direction, the grid of the image-gathering device is usually not aligned with the scan positions. If an isoplanatic invariance of the edge spread function is assumed along the edge line, for the measured and further processed data, then at every intersection point of the scan direction with the lines of the grid of measured points, the next neighboring point is chosen. This delivers an accuracy of approximately 1/10 of a pixel width and gives the basis for high-precision digital processing of the data.

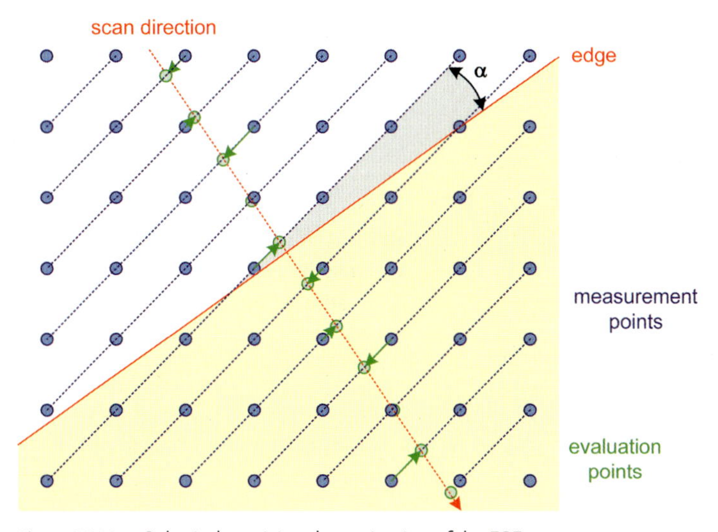

Figure 56-30: Sub-pixel precision determination of the ESF measurement points for a rotated-edge scan measurement.

56.3.5
Line Spread Function LSF

The line spread function can be used for image quality assessment in the same manner as the point spread function. Due to better conditions for the power and signal-to-noise ratio, the use of an extended line has some advantages. Furthermore, the line edge spread is quite well correlated to a visual subjective image quality impression [56-71]. The measurement setup for the determination of a line edge is quite similar to the point spread function. It should be noted that the slit integrates the intensity distribution in one direction and therefore specially oriented fine structures can be hidden by this method. Therefore, it is usual to scan the line across the field in various azimuthal directions.

The relation between the PSF and the LSF is given by a simple integration in one direction.

$$I_{LSF}(x) = \int I_{PSF}(x, y)\, dy \tag{56-74}$$

If the line is integrated in the other direction starting at a position x_o of the scan perpendicular to the slit direction, we obtain the edge spread function

$$I_{ESF}(x_i) = \int_{-\infty}^{x_i} I_{LSF}(x_o) dx_o \tag{56-75}$$

If the complete two-dimensional distribution $I(x,y)$ of the point spread function is required from slit measurements, then the azimuthal angle must be varied and a tomographic analysis of the rough data must be performed digitally [56-72]. If the rotation angle is θ and the coordinate along the slit is r, then the measurement of the power distribution p delivers the spatially resolved intensity by the inverse Radon transformation, which can be written as

$$I_{PSF}(x_i) = \int_0^{2\pi} d\theta \int_0^{\infty} r dr P(r, \theta) e^{-2\pi i\, r\, (\cos\theta + \sin\pi)} \tag{56-76}$$

The corresponding geometry and the principle is shown in figure 56-31.

An important aspect of this measurement is the number and sampling required to obtain highly accurate results. If a is the radius of the circular region to be covered and d_{slit} the finite width of the rotated slit, the sampling theorem requires the intervals [56-72]

$$\Delta r = d_{slit}\,,\ \Delta\theta = \frac{d_{slit}}{a} \tag{56-77a}$$

and the number of measurements is given by

$$N_r = \frac{2a}{\Delta r} = \frac{2a}{d_{slit}}\,,\ N_\theta = \frac{\pi}{\Delta\theta} = \frac{\pi a}{d_{slit}} \tag{56-77b}$$

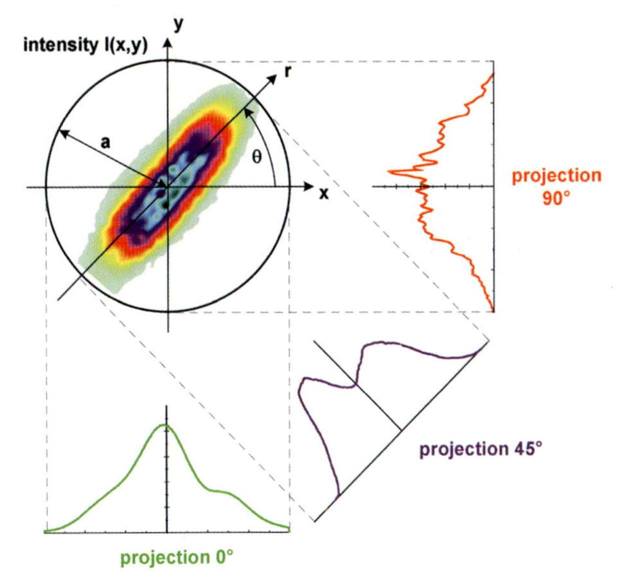

Figure 56-31: Principle of the inverse Radon transform by the tomographic measurement of projections.

For realistic values, these equations give quite a large number of angle positions in order to obtain appropriate results. But experience shows that the sampling Δr influences the accuracy far more sensitively than $\Delta\theta$. Also the slit width d_{slit} is not critical. Typically, 50–200 slit orientations must be considered. A critical aspect for this type of measurement is the alignment of the slit, the system axis and the rotation axis. Furthermore, numerical reconstruction delivers a stable algorithm such as, for example, the filtered back projection, if the data are noisy.

The method used to achieve sub-pixel accuracy in order to overcome the limitations of the sensor discretization, described in the previous section, can also be applied in the LSF measurement [56-70].

There are several special criteria defined in the literature in connection with a slit spread measurement, The most important quantities are described in the following.

Struve Ratio

Based on the line spread function LSF, in analogy to the Strehl ratio for the evaluation of a point image, according to Struve a quality measure for the image of a line can be defined by the equation

$$S_{LSF} = \frac{I_{LSF}^{(real)}(0)}{I_{LSF}^{(ideal)}(0)} \tag{56-78}$$

where the intensities are chosen in the middle of the line position at $x=0$. The value of the intensity in the real image is compared with the value for the ideal case.

Relative Ceiling

The normalized peak value of the line spread function is called the relative ceiling. It is a quality criterion very similar to the Strehl definition.

$$W = \frac{I_{LSF}^{(\max)}(x)}{I_{LSF}(0)} \tag{56-79}$$

If the relation between the line spread function and the modulation transfer function is inserted here, we get the formula

$$W = \int_{-\infty}^{\infty} H_{MTF}(v)\, dv \tag{56-80}$$

The relative ceiling, therefore, can be interpreted as the area under the MTF curve in one dimension, perpendicular to the orientation of the line.

56.3.6 Analysis of Image Degradations

There is another approach which can be taken when measuring the image quality of optical systems in very special cases. If object geometries are used, the determination of selected system performance criteria can be obtained by appropriate evaluation of the images. There are several different geometries of test plates used to detect special aberrations, which are proposed in the literature. These schemes show high sensitivity in selected cases. In practice, these measurement methods are mainly qualitative and only special cases allow for a quantitative evaluation of aberration values. The advantage, on the other hand, is the possibility of an *in situ* testing of systems and a very strong application-related detection of degradations. Since the image degradation is quite complicated in systems of low quality, measurements like this are mainly applicable in those optical systems which are highly corrected and are near to the diffraction limit.

In [56-73], for example, for microscopic objective lens inspection, the use of the following test pattern is discussed in detail.

1. Abbe test plate, parallel lines
2. Circle test plate
3. Crossed lines test plate
4. Star test plate with small pinholes.

Spherical aberrations can be seen to be quite sensitive, by means of the star test, when intra- and extrafocal positions are compared and the asymmetry is analyzed. Astigmatism is detected with the star test and circle test structures are the most sensitive. Field curvature over the complete field can be observed with the circle plate and the crossed lines by an appropriate inspection of the edge sharpness, depending on the field position. Distortion is, of course, best detected by a test with crossed lines. Coma can be observed very clearly using the star test, especially if the symmetry of the first diffraction ring is inspected. This is illustrated in figure 56-32 in detail.

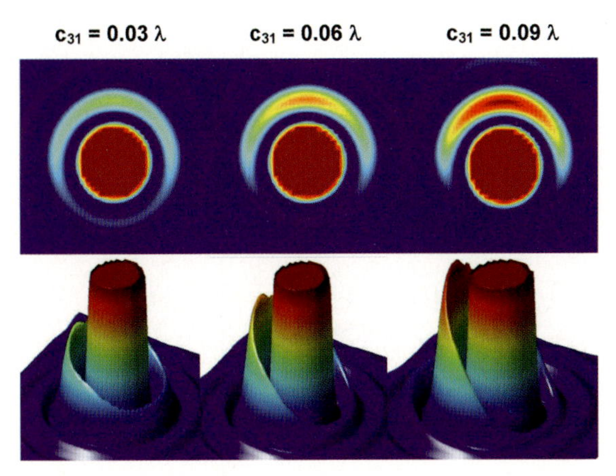

Figure 56-32: Measurement of coma aberration by an inspection of the start test. The coma is described by the Zernike coefficient c_{31}. Only those intensity levels below 10% of the peak are shown. The asymmetry of the first diffraction ring is very sensitive to the influence of coma.

In microlithography projection systems, the determination of the image degradation due to coma or astigmatism is of special interest. Furthermore, the object structures in lithography are mainly edges or lines. Special methods have been investigated and developed in recent years to qualify lithographic projection lenses on the basis of the image analysis for given regular geometrical structures. Since lateral displacements are of vital interest in lithography, coma is measured with high precision by the shift in the edge location [56-74]. An additional problem is the combination of the edge-shift effects with defocus in the case of a non-perfect telecentricity. Therefore, high-precision methods need both effects to be considered. Figure 56-33 shows an example of edge displacement as a function of coma and defocus from [56-74]. The basis for a quantitative evaluation is the sensitivity of a lateral shift Δx for the different Zernike coefficients defined as

$$S_j = \frac{\partial \Delta x(\Delta z, \sigma, NA)}{\partial c_j^{(zern)}} \qquad (56\text{-}81)$$

It should be noticed at this point, that the dependency in the figure shows only the third-order coma. Higher- order coma contributions have similar effects. Furthermore, the shift effects are influenced by the coherence, the alignment and the exact numerical aperture. This is indicated in a brief form in the formula above. Therefore, an accurate knowledge of these parameters of the system is needed. A special improvement in this approach is discussed in [56-75]. By using a mirror-symmetric bar pattern, the effects of odd and even aberrations can be separated and the accuracy increased. Alternatively, a phase mask object can be used to separate the odd and the even aberrations [56-76].

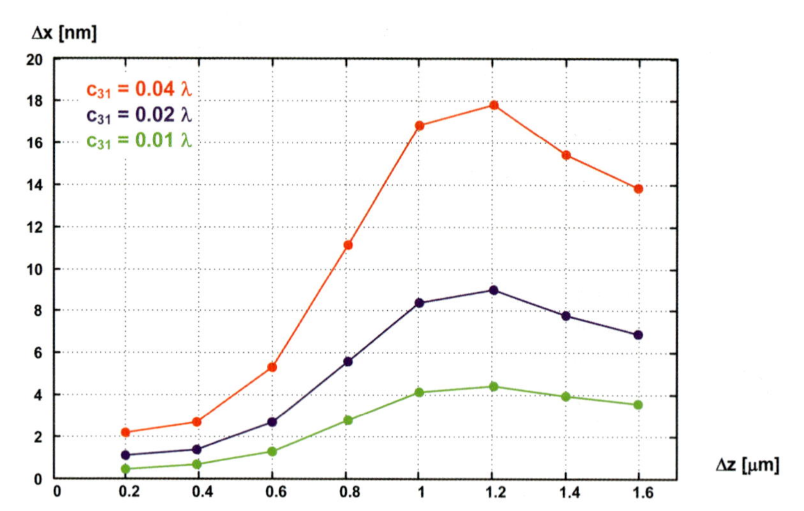

Figure 56-33: Measurement of third-order coma aberration given by the Zernike coefficient c_{31} by inspection of the edge displacement with defocus dependency.

Beneath the edge displacement, the changes in the line width data can also be used to detect coma [56-77]. By using corresponding analysis, the determination of trefoil aberrations within this method is also documented in the literature [56-78].

Other arrangements use special masks, which can be introduced inside the system to extract the aberrations in a type of intrinsic Hartmann sensor. If a movable hole mask is inserted near the pupil location and a pinhole object mask is observed, the lateral aberrations Δx, Δy can be detected as a function of the object position and the imaging subaperture in the pupil [56-79].

Recently, a more sophisticated mathematical analysis of the corresponding problem has been proposed in the literature [56-80]–[56-82]. Particularly for partially coherent illumination conditions, this allows a more accurate quantitative evaluation of image quality. If the image quality in the form of an aerial image intensity distribution is considered, a simulation of the characteristic image shapes and the correlation between a combination of elementary Zernike contributions can be used to perform a correlation analysis and also a principal component analysis of the system behavior. A mapping of the principal components space to the Zernike space allows a quantitative reconstruction of the Zernike coefficients from the degraded image. This can be done for arbitrary object patterns. This method assumes a linear response of the system to the Zernike coefficients and therefore it is only applicable in well-corrected systems.

56.3.7
Distortion

Most of the test procedures described in this chapter refer to the image quality for one object point. The complete system is characterized over the field by selecting discrete object points. The quality assessment for every single point is related to a blurred point spread function, which gives a degraded image.

The choice of the object points must be made carefully to represent the overall system performance accurately. Optical systems with circular or mirror symmetry can be inspected on a reduced area of the object due to the symmetry from a modeling point of view. But in reality, centering errors, local errors on the surfaces or striae in the bulk materials break the symmetry and the system must be evaluated and measured over the complete object field.

If the performance for single object points is characterized, the geometrical relation and shape of the point pattern is another important feature of the quality. This is the distortion of the system, in the general sense, which causes a deformed image. The resolution and the contrast may already be excellent, but the fidelity of the image and object is compromised.

The measurement of distortion, therefore, is quite different in comparison to all the wavefront or point spread function related measurement methods. It is described in section 49.5 in detail and the reader is referred to this chapter of the book.

56.3.8
Chromatical Aberrations

Most real systems are not used with monochromatic or quasi-monochromatic light. Therefore dispersion effects in the transparent media of components are an important aspect in the performance description of optical systems. In classical aberration theory, the chromatical effects are traditionally separated into the axial and the transverse effects.

The influence of the axial chromatical aberration is a difference in the image location of the system dependent on the wavelength. These focal shifts can be measured by all the methods described in sections 56.2.1 and 56.2.2 if the measurements are made at several illumination wavelengths. In reality, the situation is more complicated than this, since all aberrations occur simultaneously and the effect of the paraxial chromatical axial shift, separated from all the other influences affecting the best image location, is not trivial. Therefore, the literature proposes the calibration of the system by reverse raytracing to clearly separate the axial chromatical aberration [56-83]. Additional problems can also arise in the case of microlenses, as an example [56-84]. In this case, the diffraction effects of low Fresnel numbers cause an additional focal shift making it more complicated to extract the geometrical paraxial focal length effects.

The transverse chromatical aberrations are also known as the chromatical variation of the magnification. Large values of this aberration type produce colored fringes of edges in the image, which can be recognized very clearly in the image of the corresponding object structures. In principle, the measurement of the system magnification according to the methods described in section 56.2.4 allows the aberrations to be measured and quantified. Other approaches use the edge spread function with monochromatic light of different wavelengths to find the displacements accurately.

A completely different aspect is the overall color consistency across the field of view. In particular, in systems with strongly curved surfaces and high ray inclination angles the incidence and wavelength dependence of the coatings may be the reason for significant changes in the color performance depending on the location in the field.

56.4 Measurement of the Transfer Function

56.4.1 Introduction

If the image quality performance evaluation is completely switched from the spatial to the frequency domain according to the Fourier optical picture, a linear system theory approach describes the system by a transfer function. This step does not generate additional information, since the optical transfer function (OTF) is just the Fourier transform of the point spread intensity function

$$H_{OTF}(v_x, v_y) = \iint I_{PSF}(x', y') e^{2\pi i \left(v_x x' + v_y y'\right)} \, dx' \, dy'$$

The absolute value of the complex OTF is the modulation transfer function (MTF). It is used very often for quality assessment of optical systems. It gives the image contrast as a function of the spatial frequency v in the object and thus provides direct information on the minimal structural size for which good imaging can be expected. The accurate measurement of the modulation transfer function is relatively complex and will involve some approximation.

The systematic of assessment via the transfer function has some advantages and delivers several new insights [56-85]–[56-87].

With the help of the complex optical transfer function, the generation of an image by an optical system can be completely determined (see also chapters 20–23) and therefore allows a predictability of the image structure. In particular, in the case of digital pixelated sensors, the use of the transfer function description of optical systems has significant advantages and allows the corresponding discretization effects to be incorporated in the overall performance qualification.

As in the theory of the point spread function, the computation of the transfer function is quite simple in the limiting cases of completely incoherent or completely coherent illumination. In the regime of partial coherence in between these extreme cases, the conditions are complicated.

In the range of incoherent illuminated systems, the concept of the frequency response is a strictly linear model. This ensures the validity of a linear superposition of all frequency components as it is assumed in the Fourier theory of imaging.

Another very important point is the behavior and description of cascaded systems. If the systems are decoupled and worked in a linear information chain, the overall transfer function is given by the product of all single transfer functions. Inside optical systems, which consists of several subsystems, this concept is only applicable if, between the subsystems, ground glasses destroy every coherence across the beam area. If the phase is linearly transported and coherent interference is possible, the phases of the subsystems can compensate each other. This is the principle of optical correction in optical design and the multiplication of the incoherent optical transfer function is totally wrong.

Usually, the incoherent transfer function is used to qualify and test optical systems. Therefore, we will focus the measurement techniques described in this section on this case.

The theory of the transfer function is concerned with real imaging of structured objects with characteristic feature sizes. Therefore, in practical setups of measurement, auxiliary optical systems are necessary when afocal locations of the object or image are present. In these cases, the influence of the additional lenses must be taken into account. This means that the lenses of the measurement equipment must be characterized and their effect on the result should be removed by calibration or deconvolution in a digital post-processing step.

There are several approaches used to measure the transfer function of an optical system. A rough classification can be made as follows.

1. Imaging of special test structures and analyzing the corresponding image contrast behavior.
   1.1 If the structures are sine-grating structures, a single-frequency response is determined
   1.2 If the structures have a large frequency content such as points, lines, edges or bar patterns, a careful analysis of the higher frequency components and calculation of the OTF from the measurement data must be carried out.
2. Direct measuring of the autocorrelation function of the optical system pupil corresponding to the Duffieux-integral formulation of the transfer function.
3. Measurement of the point spread function and digital calculation of the transfer function by performing the Fourier transform.

The measurement of the modulation transfer function usually includes the imaging of a test structure with subsequent evaluation and backward computation. The following conditions have to be generally satisfied when measuring an incoherent modulation transfer function $H_{\mathrm{MTF}}(\nu)$.

1. An object is illuminated by incoherent light.
2. The object acquires, through its structures, all relevant spatial frequencies that have to be measured.
3. The object is imaged by the test system.

4. Spatial resolution is provided for the detection of the image intensity. As a rule this is achieved by an adjustable slit located in front of the detector. Alternatively, the slit can be fixed and scanning is achieved by the imaged grating or object structure.
5. The contrast is derived from the intensity distribution and analyzed as a function of the spatial frequency

It is very important, in particular, that the illumination of the test objects is incoherent since any coherent contribution could strongly affect the accuracy of the result.

In general, the measurements of the transfer function by imaging special target object geometries can be performed monochromatically or by using polychromatic light. The type of setup depends on the desired quality function.

Extensive literature on the transfer function and its measurement can be found in [56-4] and [56-10].

56.4.2 Test Targets

The object used plays an important role in the measurement of the transfer function. It is necessary to select an object, containing all those spatial frequencies which the measurement is required to resolve. There are some classical object geometries, which fulfill these conditions, at least in one direction. These are:

1. Point object
2. Edge object
3. Line object or slits
4. Bar pattern
5. Random transparencies
6. Sine gratings with one or several periods
7. Special test charts such as the Siemens star.

From a theoretical point of view, the transfer function properties are related to a sine-wave object, according to classical considerations in Fourier optics. In reality, a square-wave bar pattern with top-hat shape is much easier to realize and is often taken as a target. In this case, the changes in the detected spectrum must be corrected by the frequency characteristic due to the special shape of the object. This means that the system modulation transfer function is given [56-88] by

$$H_{MTF}(\nu) = \frac{H_{MTF-meas}(\nu)}{H_{MTF-\text{bartarget}}(\nu)} \tag{56-82}$$

The problem with the sine-wave gratings in real setups is the generation of continuously varying spatial frequency, which is hard to realize.

One of the drawbacks in the usual knife-edge-based method (see next section) is the scan across the edge. If a statistical target transparency with suitable properties is chosen, the scan of the knife-edge method can be avoided. According to the description given in [56-89], a white-noise target with a band-limited power spectrum $I_{PSD\text{-}in}(\nu)$ must be chosen. Figure 56-34 shows a typical transmission pattern and the corresponding power spectrum for illustration. In this case the measured output power spectrum is given by

$$I_{PSD-out}(\nu) = |H_{MTF}(\nu)|^2 I_{PSD-in}(\nu) \tag{56-83}$$

From this formula, the modulation transfer function can be reconstructed.

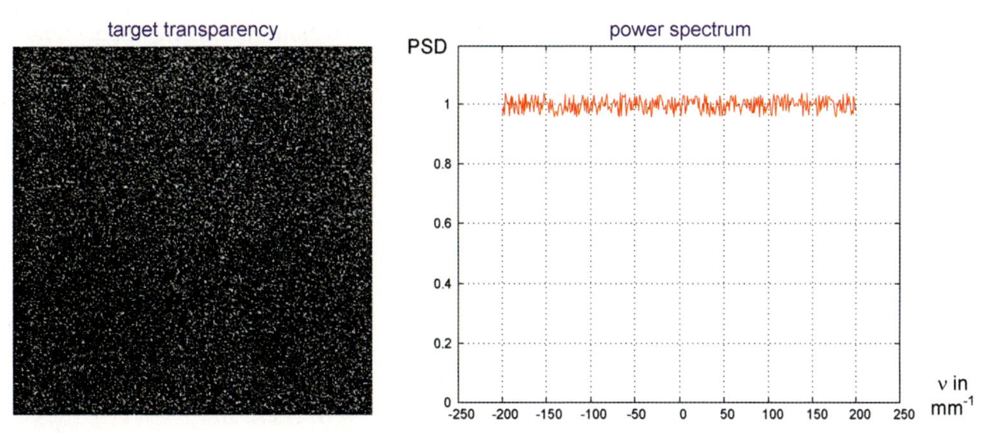

Figure 56-34: Statistical target transparency and corresponding band-limited power spectrum.

56.4.3
Measurement of the MTF via the Edge Spread Function

One of the most well-known methods used to determine the transfer function by measurement is the use of the edge spread function as a primary property and digital calculation of the MTF [56-90]–[56-93].

The line image equals the derivative of the edge image

$$I_{LSF}(x') = \frac{dI_{ESF}(x')}{dx'} \tag{56-84}$$

and vice versa

$$I_{ESF}(x') = \int_0^{x'} I_{LSF}(x)\,dx \tag{56-85}$$

Since the transfer function is equal to the Fourier transform of the line image function

$$H_{OTF}(\nu) = \hat{F}[I_{LSF}(x')] \tag{56-86}$$

it can be obtained from the measured edge image by differentiation and subsequent Fourier transformation. Figure 56-35 demonstrates the principle of such signal processing, also called the differentiation method.

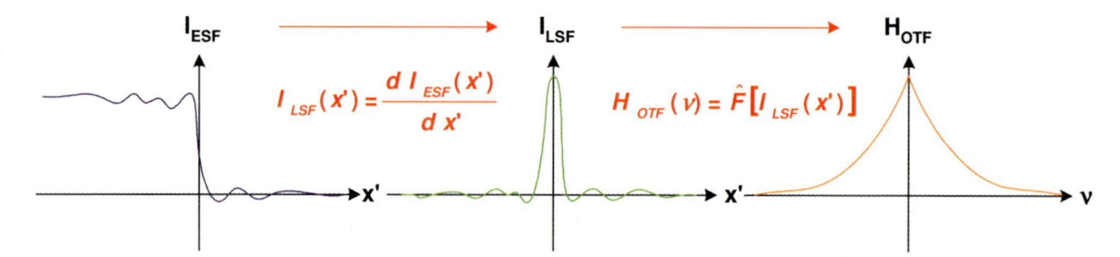

Figure 56-35: Principle of signal processing following MTF measurement by edge imaging

The inevitable measurement noise is very critical in the differentiation method and this kind of measurement is very sensitive and difficult to reproduce. Therefore, another type of evaluation is generally used.

The transfer function is separated in the equation governing the amplitude image

$$A'(\nu) = A(\nu)\, H_{OTF}(\nu) \tag{56-87}$$

Solving for it and taking only the absolute value, i.e., the MTF, one obtains

$$H_{MTF}(\nu) = \left|\frac{A'(\nu)}{A(\nu)}\right| \tag{56-88}$$

The spectrum of an ideal edge is

$$A(\nu) = \frac{\delta(\nu)}{2\pi} - \frac{i}{2\pi\,\nu} \tag{56-89}$$

and exhibits a singularity at the frequency $\nu = 0$. Therefore, this method is not appropriate for the center of the MTF at $\nu = 0$. For all other points having a finite frequency one obtains

$$H_{MTF}(\nu) = 2\pi\, i\, \nu\, A'(s) = 2\pi\, i\, \nu\, \hat{F}[I_{ESF}(x')] \tag{56-90}$$

which can be derived from the above formula by using the momentum theorem.

In reality, the edge image that has to be analyzed is recorded only over a distance L with a finite width around the edge. This means that the object is convolved with a rect-function

$$\bar{E}(x) = E(x) * rect\left(\frac{x - L/4}{L/2}\right) \tag{56-91}$$

or equivalently the Fourier spectrum is multiplied by the corresponding sinc-function

$$\bar{A}(\nu) = A(\nu) \cdot \sin c\left(\frac{\nu \cdot L}{2}\right) \tag{56-92}$$

As a consequence, the division of the spectra at the zeros of the sinc-function is impossible. That is why it is reasonable to adopt a better function for the window, which does not vanish. The Hanning filter whose spectrum

$$w(\nu) = \frac{1}{2}\mathrm{sinc}(L\nu) + \frac{1}{4}[\mathrm{sinc}(L\nu + 1) + \mathrm{sinc}(L\nu - 1)] \tag{56-93}$$

exhibits no zeros in the range from –2/L to +2/L, is one such possibility.

One special approach suggested in the literature to determine the transfer function from knife-edge measurements is to use a simple model function for the edge response, to fit the parameter of the functions to the measurement data and then to calculate the transfer function digitally. If the model function allows one to formulate the derivative analytically, the critical differentiation of the noise data can thus be avoided. In this sense, the smooth model functions work like a low pass filter and guarantee a stable analysis and computation of the data [56-89]–[56-91] and [56-94]. There are several analytical model functions proposed in the literature. The most promising candidates are the following.

1. The sum of three Fermi functions

$$I_{ESF}(x) = I_0 + \sum_{j=1}^{3} \frac{a_j}{e^{(x-b_j)/c_j} + 1} \tag{56-94}$$

2. The sum of gaussian profiles for the point spread function. By calculation of the derivative we get the line spread function, while integration delivers the edge spread function

$$I_{ESF}(x) = I_0 + \sum_{j=1}^{3} a_j \left[1 - erf\left(\frac{x - b_j}{c_j}\right)\right] \tag{56-95}$$

3. The sum of super-gaussian profiles for the point spread function. By a corresponding differentiation and integration the edge spread function is obtained. This calculation cannot be completely carried out analytically.

$$I_{PSF}(x) = \sum_{j=1}^{3} a_j \, e^{-\left(\frac{x-b_j}{c_j}\right)^{m_j}} \tag{56-96}$$

There is one problem in determining the transfer function by a knife-edge measurement, which should be noticed here [56-95]. The calculation scheme described above assumes ideal spatial incoherence of the edge illumination. If this condition is not fulfilled and a certain degree of coherence is observed, the result of the calculation becomes inaccurate. If the state of coherence of the illumination source and also the illumination system transfer function are known, the remaining error can be estimated. This error increases with the spatial frequency and therefore is not constant along the transfer function curve.

56.4.4
Measurement of the MTF via the Line Spread Function

Since the transfer function is equal to the Fourier transform of the line image function

$$H_{OTF}(\nu) = \hat{F}[I_{LSF}(x')] \tag{56-97}$$

as previously mentioned, it can be obtained from a measured edge image by a Fourier transform. In practice, the object and detection slits have finite widths a and a_D, respectively. That is why the measured MTF curve must be corrected by the Fourier transformed slit function

$$\bar{A}(\nu) = \frac{\sin(\pi\, a\, \nu)}{\pi\, a\, \nu} \tag{56-98}$$

as follows

$$H_{MTF}^{(corr)}(\nu) = H_{MTF}^{(mess)}(\nu) \frac{\pi\, a\, \nu}{\sin(\pi\, a\, \nu)} \frac{\pi\, a_D\, \nu}{\sin(\pi\, a_D\, \nu)} \tag{56-99}$$

where the spatial frequencies and the slit widths must have the same scale, i.e., both have to be calculated in the object (or image) space using the same scale ratio. Owing to the division by zero in the computation of the correction, the spatial frequency $\nu_{max} = 1 / a$ represents the maximum measurable frequency. This means that the slit width has to be matched to the frequencies that are measured. In practice, it is reasonable to have a slit width, which is substantially smaller than this limit; a value of $a = 0.25 / \nu_{max}$ is recommended.

A completely analogous correction should be performed when using digital detector arrays with a pixel width a_P. From the Nyquist theorem it follows that the maximum resolvable spatial frequency is given by

$$\nu_{max} = \frac{1}{2a_P} \tag{56-100}$$

56.4.5
Grating Imaging Measurement Setup

One quite direct method for the measurement of the modulation transfer function is based on the imaging of regular grating structures with selected frequencies and then evaluating the resulting contrast. The basic setup of this kind of measurement is quite easy and is shown in figure 56-36.

If the grating used is of sinusoidal shape, the transfer properties of the optical system can be deduced directly form the contrast measurement of the images. If a non-sinusoidal grating, like a bar pattern, is used, the effect of the higher harmonics must be removed by a digital post-processing step [56-87].

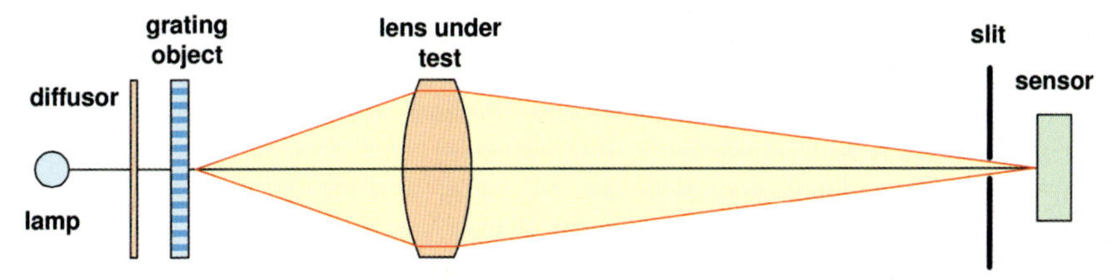

Figure 56-36: Setup for grating image-based transfer function measurement.

In the case of a sinusoidal grating, two types of structure can be distinguished:

1. Density type grating, where the sine wave is modeled by gray levels.
2. Area type gratings, where the sine wave is modeled by geometrical sine-shaped structures.

For the variation of the spatial frequency, step-by-step changes of grating or continuously varying structures are possible.

In order to achieve spatial resolution, the slit used for sampling must be moved in front of the detector, relative to the grating. In the simplest realization the grating is rotated. Several configurations are possible.

1. Rotating barrel with gratings of different spatial frequencies.

If these are discrete sine gratings, more of them are necessary in order to sample the spectrum with sufficient resolution. This type of object structure is depicted in Figure 56-37. Alternatively, one can use one rectangular grating with a spectrum containing all frequencies. In order to record the whole MTF curve, selection of the separate frequencies within the desired spectral window by an appropriate scan, is necessary.

Figure 56-37: MTF measurement by imaging sine-shaped intensity structures with different spatial frequencies.

2. Radial grating on a disk

A radial grating corresponds to a variable spatial frequency, which depends on the position. An additional sampling slit serves to illuminate only the desired grating area with a fixed spatial frequency. The rotation transforms the spatial frequencies into time frequencies on the sensor side. The brightness of the grating image on the detector exhibits a time dependence and this dependence can be analyzed. If the electronic amplifier is equipped with a variable band-pass filter, the separate narrow frequency regions can be isolated and other frequencies resulting, e.g., from the rectangular shape of the grating, can be excluded from the evaluation.

If the sampling slit is mounted in such a way that it can be rotated with respect to the radial grating, the spatial frequencies can be varied by the relative angle θ (see figure 56-38). In general, the detection slit in front of the sensor must be oriented perpendicular to the sampling slit so that the spreading of the radiation is measured in a direction normal to the grating. The effective spatial frequency for a rotation angle θ and a grating constant g is given by

$$\nu = \frac{\sin\theta}{g} \tag{56-101}$$

This projection effect is illustrated in figure 56-39.

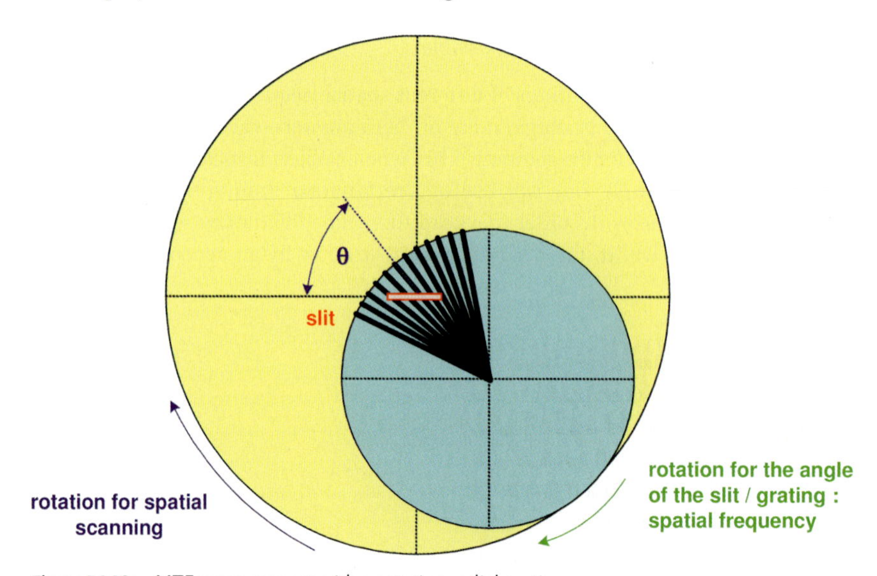

Figure 56-38: MTF measurement with a rotating radial grating.

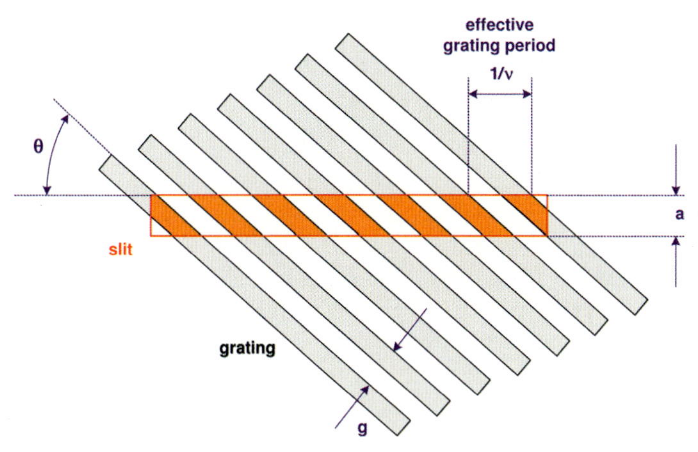

Figure 56-39: The effect of the relative angle on the frequency, registered in MTF measurements, with a rotating radial grating.

56.4.6
Measurement of the Pupil Autocorrelation Function

As is well known (see section 2.4.1), the optical transfer function can be written as the autocorrelation integral of the pupil function of an incoherent imaging system in the form of the so-called Duffieux representation.

$$H_{OTF}(v_x, v_y) = \frac{\int\limits_{-\infty}^{\infty}\int\limits_{-\infty}^{\infty} P(x_p + \frac{\lambda f' v_x}{2}, y_p + \frac{\lambda f' v_y}{2}) \cdot P^*(x_p - \frac{\lambda f' v_x}{2}, y_p - \frac{\lambda f' v_y}{2})\, dx_p\, dy_p}{\int\limits_{-\infty}^{\infty}\int\limits_{-\infty}^{\infty} \left|P(x_p, y_p)\right|^2 dx_p\, dy_p} \tag{56-102}$$

Therefore one special technique used to determine the transfer function is the measurement of the pupil autocorrelation [56-4]. The basic setup of this measurement is shown in figure 56-40. The arrangement is a modified Twyman–Green interferometer. Two roof-shaped mirrors in the two interferometer arms enable the performance of a lateral shear and readjustment of the longitudinal phase. The interference of the two split beams contains a superposition of the test lens pupil phase with two different lateral positions as defined in the Duffieux integral.

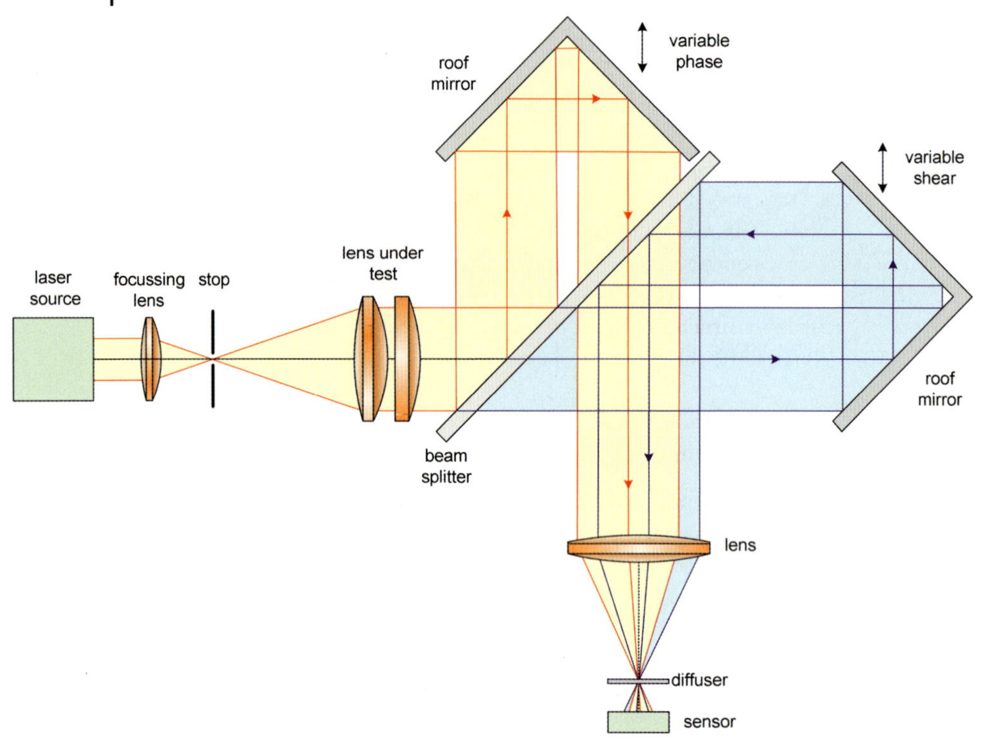

Figure 56-40: Basic setup for measuring the transfer function by the autocorrelation of the pupil function.

A modification of the pupil autocorrelation method is described in [56-96] for special application in microscopy. Here the autocorrelation of the complex focal field is detected by fluorescence. The interference of the focal field with a laterally-shifted identical field, shows an oscillation which is quite similar, but not identical, to the point spread function. In this sense this method strongly resembles a PSF characterization of the system rather than a transfer function assessment.

56.4.7
Special Measurement Aspects

There are some special aspects of the transfer function measurement techniques, which should be briefly mentioned here.

First, definition and measurement can also be possible at a large distance from the image plane in the Fresnel regime. This is described in more detail in [56-97].

Another special task is the determination of the transfer properties in a multichannel system, for example, an array-based multi-aperture system. If the system under test consists of several single imaging channels such as an array-based plenoptic camera, then in the determination of the transfer function the pixelation must be taken into account. According to [56-98], if the transfer function of the complete system is measured, the whole system is characterized by the transfer function

$$H_{MTF}(\nu) = \left| H_{MTF}^{(complete)}(\nu)\, H_{MTF}^{(channel)}(\nu) \right| \tag{56-103}$$

and the spatial frequency limiting the resolution cannot exceed the Nyquist frequency $1/x_{pitch}$ of the channel structure.

If measured values of contrast are compared with visual inspection it will be seen that the human visual system is based on a subjective quality metric. In particular, the contrast sensitivity function of the eye must be taken into account (see section 36.6.5), which defines the threshold of noticeable contrast depending on the spatial frequency.

In order to achieve high accuracy for a transfer function measurement several factors are involved. The most important influences are [56-4] as follows.

1. The mechanical tolerances of the movable parts of the setup, such as line scan, rotatable edges and alignment errors.
2. The application of precise correction factors for finite-sized slits.
3. The truncation errors of the finite-length structures of the object.
4. The calibration of the spatial frequency variable, in particular for finite fields of view with projection changes in the lengths and pattern widths.
5. Poorly known residual aberrations of the auxiliary optical components.
6. The use of incorrect spectral and coherence constraints of the illumination.
7. Shortcomings in the sensor performance.
8. Perturbing glare and stray light.

To obtain information on the accuracy of a measurement setup, the testing of a calibrated reference system is strongly recommended.

56.4.8
Image Quality Criteria Based on the Transfer Function

The optical transfer function and the modulation transfer function deliver a more or less complete description of the quality of an optical system. For a quantitative assessment of the system, it is necessary to extract a few numbers from the complicated MTF distribution function to provide a quality description. The most important quantities are listed in this section.

Hopkins Factor

There is a decreasing contrast in the image as a function of increasing spatial frequency, caused by the low- pass filter properties of optical systems, which occurs in ideally corrected systems without residual aberrations. The finite numerical aperture, and therefore the diffraction itself, causes this effect. To distinguish the effects of the aperture from the influence of the aberrations, it is usual to normalize the real modulation transfer function to the values of an ideal system with the same aperture.

The modulation transfer function is a quantitative measure for the behavior and performance of an optical system. But, in practice, for the optimization of an optical system with given numerical aperture and wavelength, it is only the influence of the residual system aberrations which are degrees of freedom and which can be influenced by suitable correction. Therefore, according to Hopkins, one defines the ratio of the real modulation to the modulation transfer function of an ideal system with the same aperture, as a figure of merit. The Hopkins factor is a function of the spatial frequency v

$$\Delta H_{MTF}(v) = \frac{H_{MTF}^{(real)}(v)}{H_{MTF}^{(ideal)}(v)} \tag{56-104}$$

Figure 56-41 illustrates this definition schematically. In practice, it is not the complete functional dependence, but the values of the Hopkins factor for distinct relevant frequencies, which are used to measure the performance of an optical system.

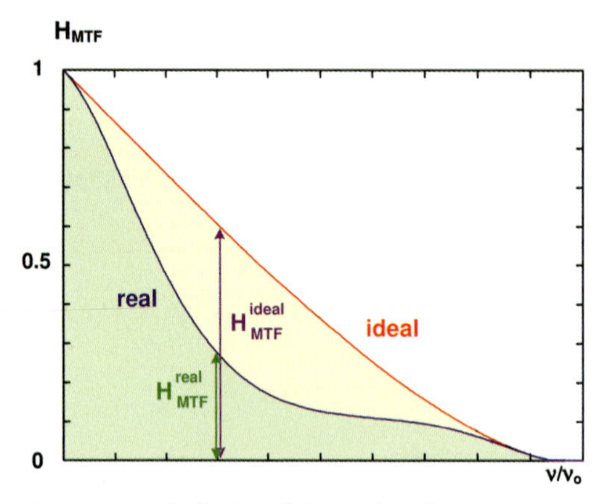

Figure 56-41: Definition of the Hopkins factor.

SQF Criterion

In practical applications, it is often only a special interval of the spatial frequencies which is relevant. Therefore, as a measure of performance, the area under the curve of the modulation transfer function is defined as a term which describes the quality. In only one dimension, the area definition of this integral measure is

$$A_{MTF} = K \int_{v_1 < v < v_2} H_{MTF}(v) dv \tag{56-105}$$

where K is a scaling factor. This is illustrated in figure 56-42.

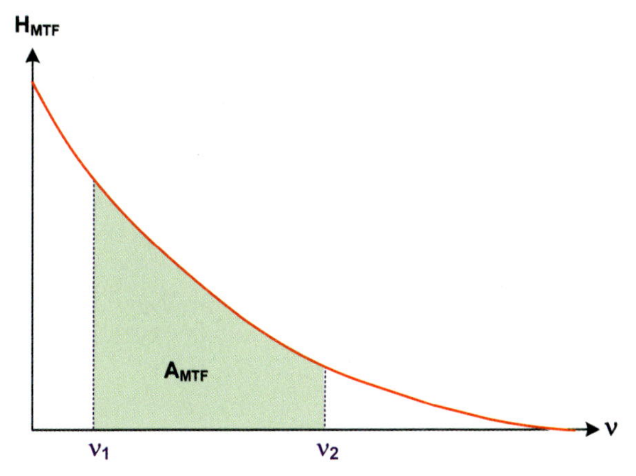

Figure 56-42: Definition of an area criterion under the modulation transfer curve.

In the more general case, when the transfer function depends on the azimuthal direction of the spatial frequency, the volume criterion under the surface of the modulation transfer function is defined as

$$V_{MTF} = K_\nu \iint\limits_{\nu_1^2<\nu_x^2+\nu_y^2<\nu_2^2} H_{MTF}(\nu_x, \nu_y) d\nu_x d\nu_y \tag{56-106}$$

This figure of merit is useful when there are strong azimuthal dependencies of the transfer behavior. See, for example, figure 56-43.

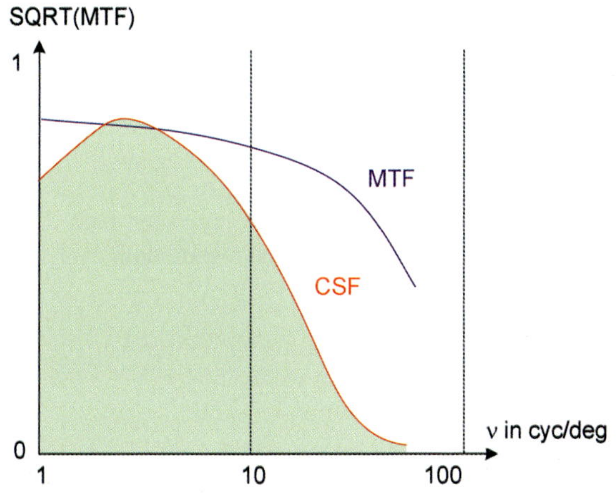

Figure 56-43: Dependence of the human contrast sensitivity function and modulation transfer function on the spatial frequency.

As a special test for this type of criterion, the so-called SQF (subjective quality factor) is defined for visual applications. The integration is performed between 10 and 40 cycles/mm on the retina on a logarithmic scale

$$M_{SQF} = K \int_{10cyc/mm}^{40cyc/mm} H_{MTF}(\nu)d(\log\nu) \tag{56-107}$$

Here the factor of normalization K is chosen in such a way that $M_{SQF} = 1$ for an ideal diffraction-limited system. The logarithmic weighting takes the lower frequencies more strongly into account.

SQRI Criterion

If the SQF function is multiplied by the human contrast sensitivity function (CSF), the resulting criterion is called the SQRI. According to Barton, this criterion nearly perfectly matches visual image quality assessment with a correlation factor $R = 0.99$.

Structural Content

The structural content is a normalized measure for the contrast capacity of the image. It is defined by

$$S = \frac{\iint I_{image}^2(x,y)\,dxdy}{\left[\iint I_{image}(x,y)dxdy\right]^2} = \frac{\iint \left|H_{OTF}(\nu_x,\nu_y)\right|^2 d\nu_x d\nu_y}{\left|\iint H_{OTF}(\nu_x,\nu_y)\,d\nu_x d\nu_y\right|^2} \tag{56-108}$$

With the help of the line spread function, the structural content in one dimension can be written as

$$S' = \int_{-\infty}^{\infty} I_{LSF}^2(x)\,dx = \int_{-\infty}^{\infty} \left|H_{OTF}(\nu)\right|^2 d\nu \tag{56-109}$$

where the normalization factor is omitted. It is also called the equivalent pass band and was first defined by Schade [56-99]. This one-dimensional formulation is of special interest for scanning systems.

56.5 Miscellaneous System Properties

56.5.1 Transmission

The transmission of an optical system is an important performance criterion. The transmission properties have an influence on the image contrast, the resolution, the signal-to-noise ratio, the color balance of the image and on the thermal behavior due to absorption in the system. In this discussion, several different consequences will be described.

1. For finite transmission, part of the signal light will be missed in the image, which usually decreases the performance.
2. Part of the light, which does not contribute to the regular image formation, can cause secondary effects on the system quality, such as thermal loading.
3. If some light does not follow the desired path of the imaging or signal train, it can reach the image plane at a different location. This again can decrease the overall performance; for example, as unwanted ghost light.

There are several possible reasons for loss of signal light in a system.

1. Absorption in the bulk material of the components.
2. Scattering in the bulk material by inclusions or finite scattering parameters.
3. Absorption in the coating of the surfaces.
4. Partial reflection or transmission by the coatings on transmissive or reflective surfaces.
5. Blocking of light via mechanical or diaphragm parts of the system due to vignetting.
6. Scattering of light by local surface imperfections or non-perfect polished surfaces.
7. Deflection of light by diffraction of the light at edges.
8. Deflection of light in unwanted higher orders of diffractive elements.

The finite transmission of an optical system depends on the following.

1. The field position.
2. The wavelength of light.
3. The pupil location used.
4. The polarization.

It is therefore a complicated task to characterize the transmission behavior in a complete manner. In practical cases, it is mainly the first two of these dependencies which are of interest and a separation of the transmission distributions for some major wavelengths are shown as a function of the field position. An initial, simple measure is an integrated overall transmission of the system.

One of the main problems of transmission measurements is an accurate reference, because the components of the measurement setups have a finite transmission and must be well known in order to be able to calibrate the properties of the system under test.

See also section 48.4 for the principles of photometric detection. The metrology for the transmittance of optical material and individual components is describe d in section 52.3, in relation with coatings in section 55.4. Figure 56-44 shows a simple typical setup for a transmission measurement [56-8]. For the calibration of the source power and the residual transmission loss of the setup components, such as collimator lenses, a highly reflecting mirror is used. The transmission of the system is then defined by the ratio

$$T = \frac{P_{out}}{P_{in}} \tag{56-110}$$

In critical cases, the reflectivity of the mirror can be taken into account. There are special setups proposed in the literature, where the mirror and an auxiliary lens are used in both setups and there is only quite a small difference resulting from using a small prism in the test lens measurement [56-100]. Depending on the exact setup, it is usually necessary to replace one or more components in the arrangement, which can induce inaccuracies in the measurement results due to misalignment errors. Therefore, more sophisticated configurations have been proposed in the literature [56-101], which use a double-pass technique for light passage through the system under test. Here simple concepts are possible, which require only a simple movement of one component in the switch between the test measurement and calibration.

One typical source of error in the results is due to spectral lines in the broadband lamp used for spectrally resolved measurements. If the setup is adjusted carefully, results with residual errors of 1 % are possible.

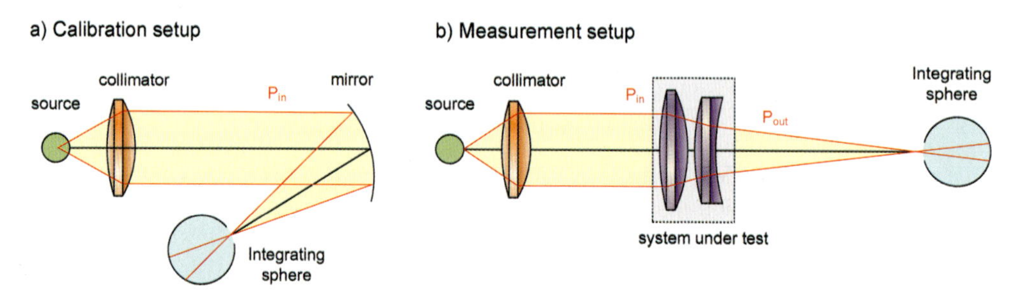

Figure 56-44: Setup for transmission measurements of optical systems.

A decrease in the transmission or a non-uniform distribution of the transmission over the field of view is called illumination falloff. For systems with circular symmetry, a decrease in the brightness due to a lowered transmission towards the field edges is often observed. As a result, the brightness of the image is not uniform. There are several reasons for this effect.

1. The natural vignetting due to projection effects of the pupils and the length of an inclined chief ray.
2. A reduced transmission caused by vignetting of light for strongly oblique bundles of light at the lens boundaries.
3. A change in the angles of incidence of the outer field bundles at strongly curved surfaces. The coatings are usually optimized for normal incidence. This effect can be especially strong for metallic coatings on mirrors [56-102].
4. An increasing inclination angle of the chief ray on the sensor, which shows an angle-dependent sensitivity.

The field dependence must be measured spatially resolved with small ray pencils. Depending on the system location and size of field and pupil, several transmission values are determined [56-103].

56.5.2 Spectral Transmission

Since the detection of color is important for image formation, the transmission information is often required for several wavelengths. Three different approaches are, in principle, applicable for color measurements [56-8] and [56-12].

1. The spectral method.

This photometric measurement with spectral resolution allows the so-called color valencies to be determined.

2. The trichromatic method.

This method uses three photo-detectors, which are calibrated with appropriate filters to match the sensitivity curve of the human eye.

3. The comparative method.

Various color tints are generated and visually compared.

The boundary conditions including illumination, angles of incidence and evaluation, brightness, spectral characteristics of the detectors, etc., are of particular importance in color measurement.

Spectral photometry offers the only possibility for an objective physical measurement of the color impression. The inspected radiation is detected photometrically by spectral resolution (see section 48.4). These spectrophotometers do not provide a color valency as a result of the measurement, but instead provide a spectral function, which depends on the light source and the sample filter function

$$\varphi(\lambda) = \beta(\lambda)\, S(\lambda) \tag{56-111}$$

Subsequently, the spectral response function of the eye $V(\lambda)$ is taken into account by computing the following integrals

$$S = k \int \varphi(\lambda)\, V_s(\lambda) d\lambda \tag{56-112}$$

$$M = k \int \varphi(\lambda)\; V_m(\lambda) d\lambda \tag{56-113}$$

$$L = k \int \varphi(\lambda)\; V_l(\lambda) d\lambda \tag{56-114}$$

k is a calibration factor which ensures the normalization of the color coordinates. In practice, the spectral dependence is recorded in discrete steps of 5 nm or 10 nm and the integrals are computed as a summation. The resulting color valency is given by

$$\vec{F} = S\,\vec{S} + M\,\vec{M} + L\,\vec{L} \tag{56-115}$$

The spectrophotometer for such an experiment includes a monochromator (see section 48.6), which is equipped with motor control for the spectral scan of the white-light source and a spectrally insensitive integrating sphere is used for the photometric measurement. Exact color measurement is a very complex task where the geometry of the arrangement, the calibration, the spectral resolution of the wavelength-selective components and the spectral effect of other components in the assembly, can play a critical role.

56.5.3
Illumination Distribution

In imaging optical systems and illumination systems, the generation of the desired intensity distribution is an important criterion of performance. This is the most important property, particularly for beam-shaping systems. There are several reasons for the degradation of a homogeneously illuminated image field of view. The most important effects are as follows.

1. Absorption in the component materials.
2. Absorption in the coatings.
3. Finite reflectivity of the coatings.
4. Vignetting of the aperture bundle for oblique chief rays.
5. Natural vignetting according to the $\cos^4$ law for oblique chief rays and projection effects of tilted planes.
6. False light from surrounding light sources, which reach the image plane.
7. Scattering of light at components of the system due to mechanical design.
8. False light due to ghost images or narcissus in infrared systems.

For example, figure 56-45 shows an image with a significant decrease in brightness in the corners due to the influence of points 4 and 5 above.

The measurement of intensity distributions is described in chapter 48 in detail and can also be found in classical handbooks or textbooks of radiometry and photometry [56-104]–[56-106]. These measurements are the same as in the previous section, where the transmission of systems is discussed. Here the influence of the

Figure 56-45: Demonstration of illumination falloff at the edges of a photographic image. In the upper corners, the homogeneous object brightness produces a radial gradient of intensity indicated by the yellow arrows.

optical system on the intensity distribution will be considered. Therefore, all the effects produced in the system by the light source or the illumination systems must either be well known or removed. The properties of the test setup detector will also influence the corresponding measurements, which are not of interest for system characterization.

There are some basic assumptions and special aspects, which must be taken into account when making radiometric measurements.

First, it is mostly assumed that the radiation is incoherent and follows the laws of geometrical optics. If this is not true, special care must be taken when diffraction or polarization effects occur or speckle structures perturb the homogeneous energy density. The exact speckle structure of a coherent illuminated area is not usually of interest and is not invariant on the global conditions of the system.

The characteristic of the test detection unit is particularly important in order to obtain objective measurement data. Noise, a spectral responsitivity, nonlinearity of the signal strength, incidence angle effects, discretization of the pixels and quantization of the detectable light levels are some of the points which should be considered.

In the majority of cases, only a relative measurement of the distribution of energy density is required, so a calibration of the various properties of the detector will not be necessary. The absolute power transfer capability of an optical system is considered in section 56.5.1, where the transmission is discussed. Nevertheless, a careful offset or dark-signal subtraction is always necessary to remove the effects of the electronic system or the environment.

Another important point, which should be considered in order to avoid wrong measurement results, is the specification of the spectral bandwidth used for the test procedure. Both monochromatic and broadband information on the intensity distribution are important. To ensure a repeatability of results, specification of the chosen spectral basis is always necessary.

56.5.4
Ghost Images and Veiling Glare

Veiling glare or flare is important when qualifying the systems properties concerning the generation of false light. This false light reduces the contrast of the image formation in the case of a broad and smooth distributed false light. Structured false light, such as ghost images, will produce unwanted artefacts in the image. The unwanted light can be signal light, which takes a wrong path through the system, or environmental light, which enters the detector plane through the system.

Unwanted light in the image can be very smoothly distributed and on a low level. The reasons for veiling glare of this type are: diffuse reflection at mechanical surfaces; non-perfect polished optical surfaces; or volume scattering in milky materials of components.

Structured false light is localized in specific areas of the image and arises from various sources: ghost light due to non-perfect coatings; unwanted diffraction orders of diffractive surfaces, e.g. from a regular structure of the surface topology due to diamond turning or edge diffraction generated at lens edges; stops or spiders.

In principle, there are two different types of veiling glare. If the object is surrounded by a scene of uniform radiance, an integral measurement with a black patch is recommended. A simple configuration used to measure veiling glare from surrounding light sources involves a black object field and measures the light in the field area of the image plane [56-8]. The light is gathered by a stop and a detector and the resulting irradiance may be I_{stray}. A second measurement uses a bright scene and delivers a level of irradiance I_{signal}. The ratio

$$i_{VG} = \frac{I_{stray}}{I_{stray} + I_{signal}} \tag{56-116}$$

is called the veiling glare index. The setup for the first part of the measurement is shown in figure 56-46 [56-8]. It is recommended that the intensity is measured by a detector of circular diameter 1/5 of the corresponding black area. In this case, it is also possible to take measurements at various field positions of the optical system.

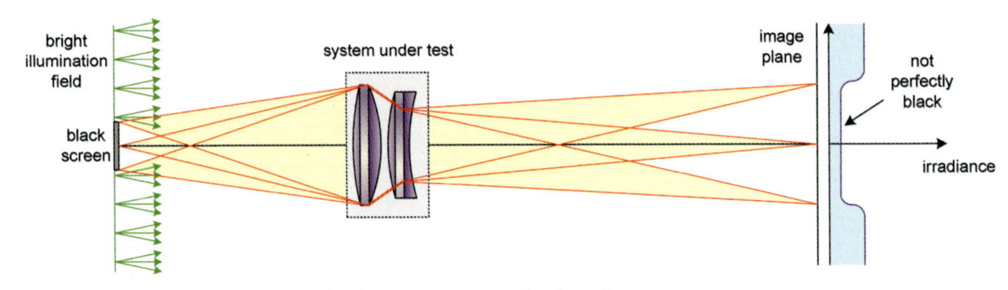

Figure 56-46: Setup for the measurement of veiling glare.

If, on the other hand, intense isolated light sources are present, the so-called glare spread function delivers a better description of the influence on the system performance [56-107] and [56-108]. In this technique, the point-like source is located on-axis in the object field and the intensity distribution in the image plane is recorded as a function of the transverse coordinate. Off-axis illumination points are also possible. The advantage of this method is a higher dynamic range.

Both described methods are primarily applied to photographic systems, which do not require auxiliary optical components [56-109]. However, the measurement of afocal telescope-type systems or very short focal lengths is more complicated. In microscopy particularly, veiling glare can be a great problem in applications, when dark-field techniques are used for contrasting the imaging of phase objects [56-110].

56.5.5 M^2 Beam Quality and Kurtosis

If the optical system under test does not have an imaging character but serves as a beam-guiding system, the performance measurement of the system determines the quality of the outgoing beam. This task requires the measurement of the spatial and angular extent of the beam. Primarily, the beam profile is one method used to describe the beam quality qualitatively. This is an intensity distribution and can be measured by simple camera techniques as outlined in section 48.2 or in classical textbooks on photometry and radiometry. The evaluation of the moment of the beam, for example, can be used to specify the beam quality by the well-known space–bandwidth criterion M^2. If this criterion alone is not sufficient to characterize the light distribution of the beam, higher-order intensity moments can be used, as described in section 30.6.3 and, in particular, the kurtosis parameter of section 30.6.4 which indicates the peaked nature of the profile.

The reasons for the use of special measurement techniques to characterize laser beam quality are as follows.

1. If the beam is coherent, there is no field of view and the spatial and angle performance are directly related.
2. In the case of multi-mode beams, the radiation is partially coherent. In the later case, statistical effects due to mode scrambling are present and therefore classical tests are not applicable.
3. The intensity distribution over the cross-section area can be complicated. Due to this large apodization, the measurement of the phase by interferometric techniques alone is not sufficient.
4. Many types of lasers are used in a pulse mode. This time-dependence requires special methods of measurement.
5. In the case of high-powered lasers, the energy density might be quite large. Here special care is necessary to prevent the camera from damage, to avoid non-linear effects, to operate in a non-saturating regime of the sensors and to control absorption and thermal problems. An attenuation of the beam without changing its spatial or angular properties is not easy.

There are several measurement techniques which can be used to determine the quality of a laser beam [56-111]–[56-114]. Since this topic is very specialized and more related to laser optics, the major aspect will only be mentioned briefly here. This topic is also discussed briefly in section 16.10. The definition of the M^2-measure can be found in section 30.6.5. The basic equation in one dimension reads

$$M_x^2 = \frac{\pi}{\lambda} \sqrt{\langle w_x^2 \rangle \langle \theta_x^2 \rangle - \langle w_x \theta_x \rangle^2} \qquad (56\text{-}117)$$

and shows that the second moment of the spatial and angular distributions of the beam must be determined. The second-order moments of the spatial and angular distributions are calculated by the integrals

$$\langle w_x^2 \rangle = \frac{4 \iint x^2 |E(x,y)|^2 \, dx \, dy}{\iint |E(x,y)|^2 \, dx \, dy} = \frac{4}{P} \cdot \langle x^2 \rangle \qquad (56\text{-}118)$$

$$\langle \theta_x^2 \rangle = \frac{4 \iint \theta_x^2 |E(\theta_x, \theta_y)|^2 \, d\theta_x \, d\theta_y}{\iint |E(\theta_x, \theta_y)|^2 \, d\theta_x \, d\theta_y} \qquad (56\text{-}119)$$

in the near and far-field, respectively. P is the total power of the beam in these formulas. In addition to these spatial and angle-oriented moments, a mixed moment with the definition

$$\langle w_x^m \theta_x^n \rangle = \frac{2^m \lambda^n}{(i\pi)^n} \cdot \frac{2 \iiint x^n E(x,y) \int \theta_x^m E^*(x',y) \, e^{ikx'\theta_x} \, dx' \, d\theta_x \, dx \, dy}{\lambda \iint |E(x,y)|^2 \, dx \, dy} + c.c. \qquad (56\text{-}120)$$

is also important. Using this quantity, the radius of curvature of the beam can be determined by the equation

$$R_{kx} = \frac{\langle w_x^2 \rangle}{\langle w_x \theta_x \rangle} \qquad (56\text{-}121)$$

In two lateral dimensions, with the mixed moment

$$M_{xy}^2 = \frac{\pi}{\lambda} \sqrt{\langle w_x \theta_y \rangle \langle w_y \theta_x \rangle - \langle w_x w_y \rangle \langle \theta_x \theta_y \rangle} \qquad (56\text{-}122)$$

the overall quality can be described by

$$M^2 = \sqrt{\frac{1}{2} \left(M_x^4 + M_y^4 \right) - M_{xy}^4} \qquad (56\text{-}123)$$

The beam quality measure M^2 has the following properties.

1. A gaussian beam has $M^2 = 1$, which is the best quality that can be achieved.
2. Real beams have $M^2 > 1$, large values correspond to a high quality degradation.
3. In paraxial systems, the value of M^2 remains invariant during propagation
4. Effects like phase aberration, decrease incoherence and energetic truncation, decrease the beam quality.

Laser beams are often almost collimated and the spatial beam profile can be measured in the near field. If a lens with long focal length is used to obtain the angular distribution in the focal plane, then the second term here can be evaluated. But note that the phase changes due to residual aberrations of the lens and also truncation effects can decrease the accuracy of the method. Furthermore, in practice, the profile measurement in the focal point of a lens can be critical concerning the spatial resolution of conventional CCD cameras.

The major topic in this context is the accurate measurement of the spatial beam profile of a laser beam. For this task, the following methods are proposed in the literature:

1. Scanning techniques
   1.1 Edge scanning (see also sections 16.10.2, 56.3.4 and [56-115]–[56-117]).
   1.2 Slit scanning (see also sections 16.10.3, 56.3.4 and [56-118]).
2. Camera-based, directly resolved, intensity recording.
3. Measuring the second moment with the help of a quadratic apodization mask [56-119].
4. Measurement with the phase space analyzer (see also section 47.4 and references therein)

If the spatial beam profile is detected, several computational steps are necessary to obtain the desired beam parameters. Specifically, these are:

1. Signal processing to reduce noise and to check for truncation effects.
2. Determination of the overall beam power.
3. Calculation of the centroid coordinates.
4. In the case of edge or slit-scanning techniques, reconstruction of the spatially resolved distributions, using the integrated power data, must be carried out.
5. Calculation of at least the second moment of the intensity profile.
6. Determination of main axes, beam-width ellipticity and gaussian profile fits are all possible analyses.

The calculation of the second moments is mandatory in order to determine the beam quality. Higher-order analysis can be used to determine peakedness, asymmetry, ellipticity or other properties. A particular problem is the evaluation of twisted beams, which does not allow measurement with the simple approach of separately decoupled x-y measurements [56-120] and [56-121]. This complicated scenario is not discussed further here.

One problem encountered in measuring the moments of a beam profile is the truncation of the beam. The theoretical definition of the moment is not given by a converging integral, if the beam is diffracted in the system at a hard edge. In practice, this is usually less critical, since the signal processing is necessary in any case, to remove low-level intensity pixels and noise and prevent them from influencing the result.

A more sophisticated analysis can be performed if the mode composition of a beam is reconstructed by caustic measurements [56-122] and [56-123].

One special quantity used to describe the shape of the beam profile with higher order moments is the so-called Kurtosis. It is defined in one dimension as

$$K_x = \frac{\langle x^4 \rangle}{3 \langle x^2 \rangle^2} \tag{56-124}$$

A value of $K=1$ corresponds to a gaussian intensity profile. Leptocurtic profiles with $K>1$ have a more peaked shape, platicurtic beams with $K<1$ appear more compact and resemble top-hat shapes. This behavior is illustrated in figure 56-47 for a supergaussian profile with exponent m

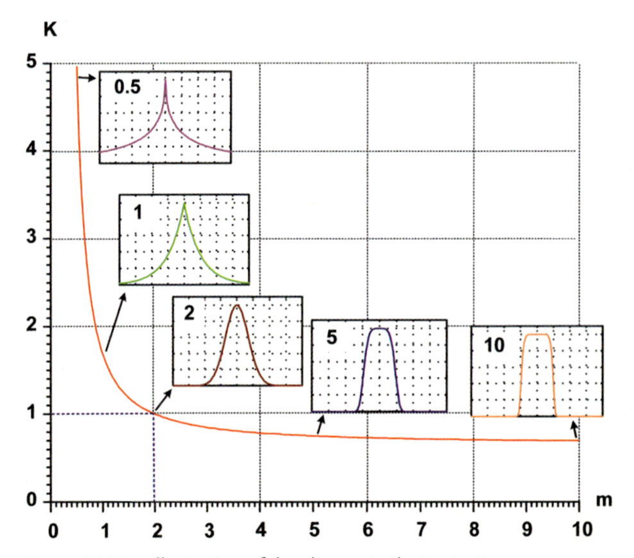

Figure 56-47: Illustration of the change in the Kurtosis parameter with the exponent m of a supergaussian profile.

56.5.6
Polarization Aberrations

The polarization-preserving properties of an optical system are important for the quality of performance in many applications. Residual polarization effects in optical systems reduce their image-forming potential. As can be seen in the following; both phase and amplitude changes are possible in this context, where the phase perturbations have a stronger impact on the system quality. In those systems where the polarization is used for signal generation, such as, for example, in polarization or differential interference microscopy, this is more obvious.

The measurement of the polarization properties of an optical system is based on polarimetry and ellipsometry. The methods used to measure the corresponding properties of a light field are described in chapter 51 in detail. The reader is referred to this chapter and to [56-124]–[56-126]. The determination of the polarization-changing properties of a system can only be detected by measuring the polarization behind a system for a well-known polarization input in front of the system. The performance quality description of polarization effects is described in chapter 28 and in [56-127]–[56-132].

There are several possible reasons for polarization changes in an optical system:

1. High numerical aperture effects due to a rotation of the field components [56-133] and [56-134].
2. Stress birefringence in the bulk materials during production or induced by the mountings.
3. Influence of coatings: this effect is mainly observed at the rim of strongly curved surfaces due to a large variation in the incidence angle [56-135] and [56-136]. Mirror coatings are particularly susceptible to these effects [56-137].
4. Birefringence in components made of crystals [56-138].

Figure 56-48 shows an example of a photographic image, which is recorded both with and without a polarization filter, respectively [56-139]. The influence of the polarization on the image properties can be clearly seen.

Figure 56-48: Influence of polarization on image formation.

Figure 56-49 shows a simple example for the distribution of the polarization in a beam of circular cross- section after being reflected by a corner cube prism. The changes in the polarization ellipse as a function of the position give a quite complete description of the outgoing polarization.

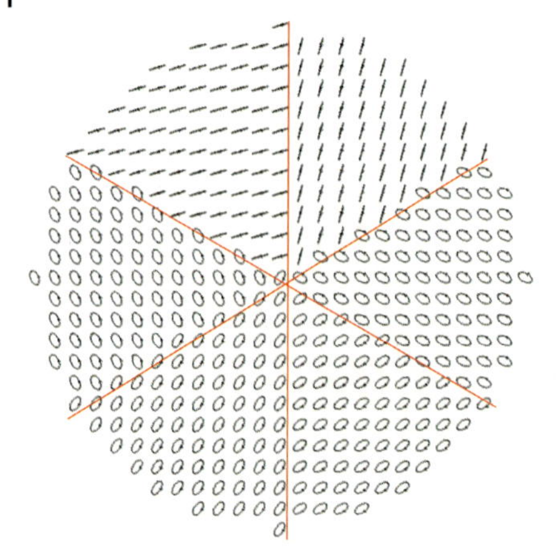

Figure 56-49: Representation of the polarization distribution behind a corner cube prism for input light, which is linearly polarized under 45° orientation.

Special schemes have been developed to refer the outgoing polarization onto the incoming one in order to select the effect of the system itself. For details, the reader is referred to the references. Only some of the most usual concepts are mentioned briefly here.

Jones Pupils

A system that influences the polarization of the transferred light can be described by the so-called Jones matrix pupil (see also chapter 28). This matrix changes the field vector of the incoming light in accordance with the equation

$$\begin{pmatrix} E_x^{(Exp)} \\ E_y^{(Exp)} \end{pmatrix} = \mathbf{J}_{pup} \begin{pmatrix} E_x^{(Enp)} \\ E_y^{(Enp)} \end{pmatrix} \tag{56-125}$$

between entrance and exit pupil. Here the Jones pupil matrix is defined as

$$\mathbf{J}_{pup} = \begin{pmatrix} J_{xx}(x_p, y_p) & J_{xy}(x_p, y_p) \\ J_{yx}(x_p, y_p) & J_{yy}(x_p, y_p) \end{pmatrix} \tag{56-126}$$

The Jones matrix elements are generally complex valued, therefore a matrix has eight degrees of freedom. The diagonal elements describe the preservation of the x- and y-polarized field components respectively. If the system has no polarization effect, J_{pup} will be the unit matrix. The off-axis elements describe the cross-talk or mixing between the orthogonal x- and y-field components. All the elements are two-dimensional matrices and can vary over the pupil area. As mentioned above, the matrix elements are generally complex valued and change the amplitude by corresponding damping and also the phase, by relative changes. Figure 56-50 shows, as an example, the four elements of a matrix for the amplitude and phase as a function of the pupil area [56-127].

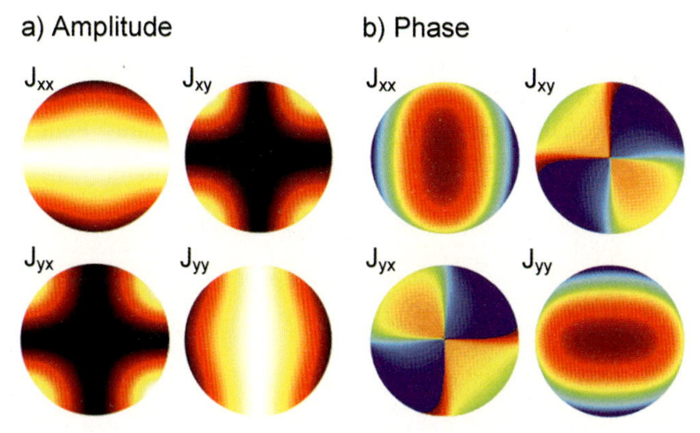

Figure 56-50: Example of a Jones pupil, with amplitude and phase components separated.

In particular, the cross-talk between the two orthogonal incoming polarization directions can be seen in practice, if the system is placed in between two perpendicular polarizers. Then the mixing element J_{xy} of the damping part can be seen directly. Due to larger angle effects in the diagonal directions, the so-called Maltese cross is often seen. This is illustrated in figure 56-51.

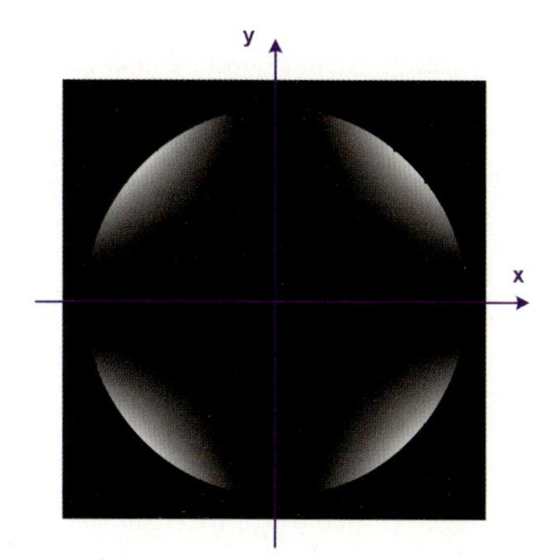

Figure 56-51: Typical Maltese cross as polarization cross-talk.

Retardance and Diattenuation Maps

It is well known from theory (see again chapter 28), that the Jones matrix can be written via an SVD decomposition as a product of a retardation and a diattenuation matrix. The corresponding mathematics reads as

$$\mathbf{J} = \mathbf{U} \cdot \mathbf{D} \cdot \mathbf{V}^{+}$$
$$\mathbf{J} = \mathbf{J}_{ret} \cdot \mathbf{J}_{dia}$$
$$\mathbf{J}_{ret} = \mathbf{U} \cdot \mathbf{V}^{+}, \quad \mathbf{J}_{dia} = \mathbf{U} \cdot \mathbf{D} \cdot \mathbf{U}^{+}$$

Consequently, the representation of these two matrices fully describes the relative changes in the system on an initial polarization. The retardance matrix describes a change in the relative phase and therefore in the polarization state. The diattenuation matrix describes an individual damping of the two polarization states. Figure 56-52 shows an example of two maps of this type.

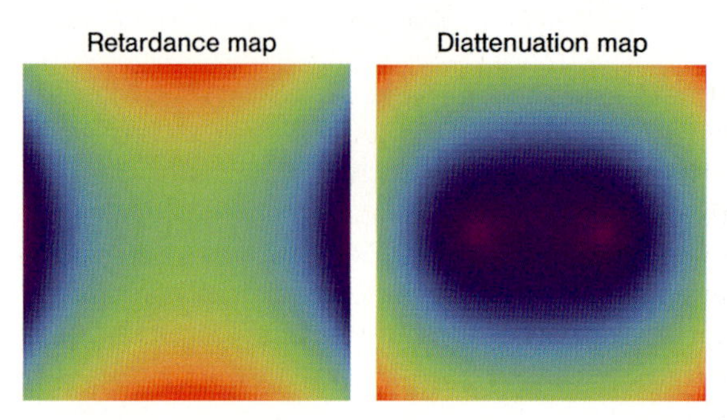

Figure 56-52: Representation of polarization aberrations as retardance and diattenuation maps.

Polarization Aberrations

The maps of the Jones pupil contain the integral effect of all the influences affecting the system and so analysis of the reasons could be complicated. Therefore, a decomposition of the overall distribution analog to a Zernike expansion of wave aberrations will be useful to achieve a better understanding. Figure 56-53 shows a simple example for the lowest orders [56-131] and [56-132] with different signs of the aberrations for illustration.

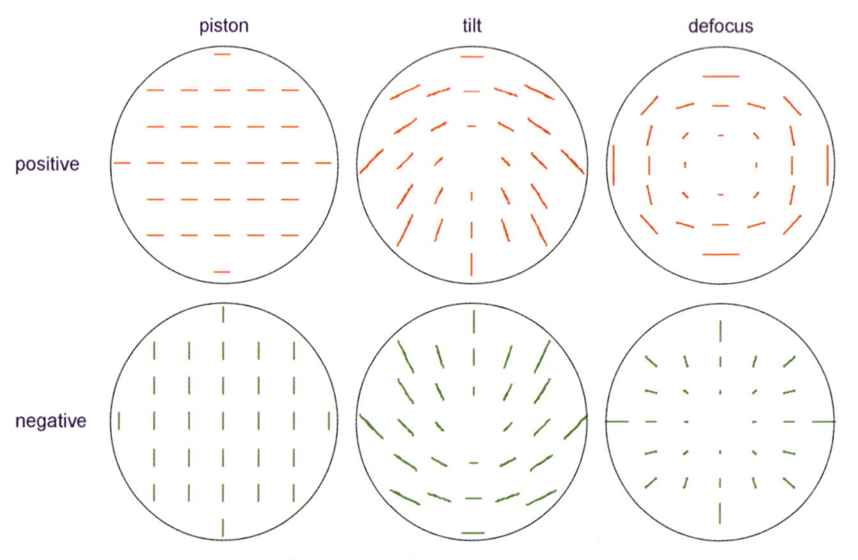

Figure 56-53: Representation of polarization aberrations as a decomposition of elementary low-order contributions.

Zernike Representations

Since the characteristic influence of the retardance and diattenuation maps on the image formation and system performance strongly depends on the shapes of the distributions, the spatially varying functions are expanded into a Zernike basis corresponding to the classical scalar aberrations. This approach is called the Jones–Zernike pupil representation. Details of this approach are described in section 28.13.2 or [56-127] and [56-128].

A more refined analysis of the distributions is described in [56-129] and [56-130]. The so-called orientation Zernike expansion is used to describe the polarization transmittance of a system. This generalization forms a systematic and complete basis for the effects of polarization in imaging. In comparison to the classical Zernike polynomials, which are well known from scalar aberrations theory, the orientation Zernikes are vectorial in nature and contain an additional third index for the orientation. These highly sophisticated approaches have been developed in recent times due to the fact that the polarization becomes one of the limiting factors in the imaging of modern lithographic systems.

Müller Matrix

Another possible way to visualize the polarization properties of an optical system is by means of a complete diagram of the Müller matrix, spatially resolved over the pupil area. The equation

$$\vec{S}_{ExP} = \mathbf{M}\,\vec{S}_{EnP} \tag{56-128}$$

describes the changes in the Stokes vectors for the transfer from the entrance to the exit pupil. The 4×4 Müller matrix M can be represented in an analogous way to the Jones pupil as a matrix of full-area maps. A typical example of this 4×4 matrix is shown in figure 56-54. Note that a system without polarizing effects has a perfect diagonal matrix; the incoming polarization state is preserved during the transfer through the system. For an example, see [56-132].

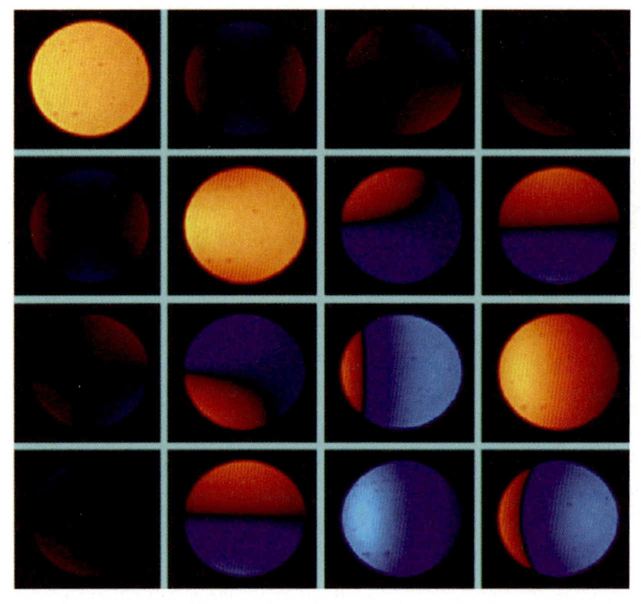

Figure 56-54: Representation of polarization aberrations as a Müller matrix visualization.

56.6 Literature

56-1 H. Naumann and G. Schröder, Bauelemente der Optik (Hanser Verlag, München, 1983).

56-2 K. J. Gasvik, Optical Metrology (Wiley, Chichester, 2002).

56-3 D. Malacara, Optical Shop Testing (Wiley, Hoboken, 2007).

56-4 T. L. Williams, The Optical Transfer Function of Imaging Systems (Inst. of Physics Publishing, Bristol, 1999).

56-5 J. Flügge, Einführung in die Messung der optischen Grundgrößen (Braun, Karlsruhe, 1954).

56-6 J. Picht, Mess- und Prüfmethoden der optischen Fertigung (Akademie Verlag, Berlin, 1955).

56-7 G. Schröder, Technische Optik (Vlg. Vogel, Würzburg, 1998).

56-8 J. Geary, Optical Testing (SPIE Press, Bellingham, 1993).

56-9 P. Mouroulis, Visual Instrumentation (McGraw Hill, New York, 1999).

56-10 R. Shannon, Optical Specifications, in M. Bass (Ed.), Handbook of Optics, Vol I, Ch. 35 (McGraw Hill, New York, 1995).

56-11 R. Ginsberg, Outline of tolerancing (from performance specification to toleranced drawings), Opt. Eng. **20**, 175 (1981).

56-12 R. Fisher and B. Tadic-Galeb, Optical System Design (McGraw Hill, New York, 2000) Chapter 1.

56-13 E. Keren, K. Kreske and O. Kafri, Universal method for determining the focal length of optical systems by Moire deflectrometry, Appl. Opt. **27**, 1383 (1988).

56-14 I. Glatt and O. Kafri, Determination of the focal length of nonparaxial lenses by Moire deflectometry, Appl. Opt. **26**, 2507 (1987).

56-15 D. Malacara-Doblado, D. Salas-Peimbert and G. Trujillo-Schiaffino, Measuring the effective focal length and the wavefront aberrations of a lens system, Opt. Eng. **49**, 053601 (2010)

56-16 P. Singh, M. Faridi, C. Shakher and R. Sirohi, Measurement of focal length with phase-shifting Talbot interferometry, Appl. Opt. **44**, 1572 (2005).

56-17 K. Sriram, M. Kothiyal and R. Sirohi, Talbot interferometry in noncollimated illumination for curvature and focal length measurement, Appl. Opt. **31**, 75 (1992).

56-18 J. Bhattacharya and A. Aggarwal, Measurement of the focal length of a collimating lens using the Talbot effect and the Moire technique, Appl. Opt. **30**, 4479 (1991).

56-19 Y. Kumar and S. Chatterjee, Techniques for the focal-length measurement of positive lenses using Fizeau interferometry, Appl. Opt. **48**, 730 (2009).

56-20 M. de Angelis, S. De Nicola, P. Ferraro, A. Finizio, G. Pierattini and T. Hessler, An interferometric method for measuring short focal length refractive lenses and diffractive lenses, Opt. Comm. **160**, 5 (1999).

56-21 A. Anand and V. Chhaniwal, Measurement of parameters of simple lenses using digital holographic interferometry and a synthetic reference wave, Appl. Opt. **46**, 2022 (2007).

56-22 B. DeBoo and J. Sasian, Precise focal-length measurement technique with a reflective Fresnel-zone hologram, Appl. Opt. **42**, 3903 (2003).

56-23 I. Ilev, Simple fiber-optic autocollimation method for determining the focal length of optical elements, Opt. Lett. **20**, 527 (1995).

56-24 W. Zhao, R. Sun, L. Qiu and D. Sha, Laser differential confocal ultra-long focal length measurement, Opt. Express **17**, 20051 (2009).

56-25 Y. Tian, K. Shieh and C. Wildsoet, Performance of focus measures in the presence of nondefocus aberrations, JOSA A **24**, B165, (2007).

56-26 T. Aydin and Y. Akgul, An occlusion insensitive adaptive focus measurement method, Opt. Express **18**, 14212 (2010).

56-27 J. Jeong, B. Lee and S. Lee, Determination of paraxial image plane location by using Ronchi test, Opt. Express **18**, 18249 (2010).

56-28 M. Subbarao, T. Choi and A. Nikzad, Focusing techniques, Opt. Eng. **32**, 2824 (1993).

56-29 F. Groen, I. Young and G. Ligthart, A comparison of different focus functions for use in autofocus algorithms, Cytometry **6**, 81 (1985).

56-30 A. Liu, Y. Ku and N. Smith, Through-focus algorithm to improve overlay tool performance, Proc. SPIE **5908**, 59081E (2005).

56-31 E. Krotkov, Focusing, Int. J. Comp. Vis. 1, **223** (1987).

56-32 L. Firestone, K. Cook, K. Culp, N. Talsania and K. Preston, Comparison of autofocus methods for automated microscopy, Cytometry **12**, 195 (1991).

56-33 F. Boddeke, L. van Vilet, H. Netten and I. Young, Autofocusing in microscopy based on the OTF and sampling, Bioimaging **2**, 193 (1994).

56-34 R. Attota, R. Silver, T. Gerner, M. Bishop, R. Larrabee, M. Stocker and L. Howard, Application of Through-focus Focus-metric Analysis in High Resolution Optical Metrology, Proc. SPIE **5754**, 1441 (2005).

56-35 G. Häusler and E. Körner, Simple Focusing Criterion, Appl. Opt. **23**, 2468 (1984).

56-36 B. Pernick and B. Hyman, Least-squares techniques for determining principal plane location and focal length, Appl. Opt. **26**, 2938 (1987).

56-37 D. Freeman, Measurement of Microscope Magnification, Appl. Opt. **3**, 1005 (1964).

56-38 F. Qisheng, X. Zuojiang, A. Zhiyong and S. Lixia, Research on testing techniques of exit pupil parameter of optical system based on CCD imaging theory, Proc. SPIE **5644**, 190 (2005).

56-39 L. Li, Z. An, R. Yang and X. Zhu, Research on the modern testing technology of optical parameters of sighting telescopes based on CCD; CLEO Pacific Rim, Shanghai (2009).

56-40 F. Lei and L. Dang, Measurement of the numerical aperture and f-number of a lens system by using a phase grating, Appl. Opt. **32**, 5689 (1993).

56-41 W. Lehmann and A. Wachtel, Numerical apertures of light microscope objectives, J. of Microscopy **169**, 89 (1993).

56-42 R. Juskaitis, Characterizing High Numerical Aperture Microscope Objective Lenses, in P. Török and F. Kao (Eds.), Optical Imaging and Microscopy (Springer, Berlin, 2007).

56-43 S. Lee, Direct determination of f-number by using Ronchi test, Opt. Express **17**, 5107 (2009).

56-44 L. Dai, I. Gregor, I. von der Hocht, T. Ruckstuhl and J. Enderlein, Measuring large numerical apertures by imaging the angular distribution of radiation of fluorescing molecules, Opt. Lett. **13**, 9409 (2005).

56-45 J. Shin, S. Lee, H. Kim, C. Hwang, S. KIm, S. Woo, H. Cho and J. Moon, Measurement technique of non-telecentricity of pupil-fill and its application to 60 nm NAND flash memory patterns, Proc. SPIE **5754**, 294 (2005).

56-46 N. Schuster and C. Maczeyzik, Messung und Kompensation des Telezentriefehlers, Photonik 60 (2005).

56-47 N. Schuster and S. Weber, Telezentriefehler und Telezentriebereich, Optik-Photonik, 57 (2008).

56-48 C. Berger, Design of telecentric imaging systems for noncontact velocity sensors, Opt. Eng. **41**, 2599 (2002).

56-49 A. Wong, Optical Imaging in Projection Microlithography (SPIE Press, Bellingham, 2005).

56-50 T. Sure, J. Heil and J. Wesner, Microscope objective production: On the way from the micrometer scale to the nanometer scale, Proc. SPIE **5180**, 283 (2003).

56-51 W. Zhao, R. Sun, L. Qiu, L. Shi and D. Sha, Lenses axial space ray tracing measurement, Opt. Express **18**, 3608 (2010).

56-52 W. Zhao, J. Tan and L. Qiu, Bipolar absolute differential confocal approach to higher spatial resolution, Opt. Express **12**, 5013 (2004).

56-53 J. Heinisch, P. Langenhanenberg and H. Pannhoff, Complete characterization of assembled optics with respect to centering error and lens distances, Proc. SPIE **8082**, 80821M (2011).

56-54 A. Courteville, R. Wilhelm, M. Delaveau, F. Garcia, F. de Vecchi, Contact-free on-axis metrology for the fabrication and testing of complex optical systems, Proc. SPIE **5965**, 596510 (2005).

56-55 J. Heinisch, P. Langenhanenberg and H. Pannhoff, Complete characterization of assembled optics with respect to centering error and lens distances, Proc. SPIE **8082**, 80821M (2011).

56-56 J. Heinisch, E. Dumitrescu and S. Krey, Novel technique for measurement of centration errors of complex, completely mounted multi-element objective lenses, Proc. SPIE **6288**, 628810 (2006).

56-57 L. Thibos, X. Hong, A. Bradley and R. Applegate, Metrics of optical quality of the eye, Proc. ARVO (2003).

56-58 R. Castaneda and J. Kross, PSF measurement using an Airy pattern as test object, Pure Appl. Opt. **3**, 259 (1994).

56-59 W. Welford, On the limiting sensitivity of the star test for optical instruments, JOSA **50**, 21 (1960).

56-60 L. Thibos, X. Hong, A. Bradley and R. Applegate, Metrics of optical quality of the eye, Proc. ARVO (2003).

56-61 N. Bareket, Second moment of the diffraction point spread function as an image quality criterion, JOSA **69**, 1311 (1979).

56-62 S. Szapiel, Marechal intensity formula for small Fresnel-number systems, Opt. Lett. **8**, 327 (1983).

56-63 C. Claxton and R. Staunton, Measurement of the point-spread function of a noisy imaging system, JOSA **A 25**, 159 (2008).

56-64 A. Tzannes and J. Mooney, Measurement of the modulation transfer function of infrared cameras, Opt. Eng. **34**, 1808 (1995).

56-65 S. Reichenbach, S. Park and R. Narayanswarmy, Characterizing digital acquisition devices, Opt. Eng. **30**, 170 (1991).

56-66 E. Marchand, From Line to Point Spread Function: The General Case, JOSA **55**, 352 (1965).

56-67 Q. Gong and S. Hsu, Aberration measurement using axial intensity, Opt. Eng. **33**, 1176 (1994).

56-68 J. Geary and P. Peterson, Spherical aberration: a possible new measurement technique, Opt. Eng. **25**, 286 (1986).

56-69 H.-S. Wong, Effects of knife-edge skew on modulation transfer function measurements of charge-coupled device imagers employing a scanning knife edge, Opt. Eng. **30**, 1394 (1991).

56-70 J. Olson, R. Espinola and E. Jacobs, Comparison of tilted slit and tilted edge superresolution modulation transfer function techniques, Opt. Eng. **46**, 016403 (2007).

56-71 N. Haig and T. Williams, Psychometrically appropriate assessment of afocal optics by measurement of the Strehl intensity ratio, Appl. Opt. **34**, 1728 (1995).

56-72 A. George and T. Milster, Spot distribution measurement using a scanning nanoslit, Appl. Opt. **50**, 4746 (2011).

56-73 M. Uhlig, Prüfung der einzelnen Abbildungsfehler von Mikroobjektiven an verschiedenen Testplatten, Mikroskopie **17**, 273 (1962).

56-74 M. Ma, X. Wang and F. Wang, A novel method to measure coma aberration of projection system, Optik **117**, 532 (2006).

56-75 D. Zhang, X. Wang, W. Shi and F. Wang, A novel method for measuring the coma of a lithographic projection system by use of mirror-symmetry marks, Opt. and Las. Tech. **39**, 922 (2007).

56-76 F. Wang, X. Wang, M. Ma, D. Zhang, W. Shi and J. Hu, Aberration measurement of projection optics in lithographic tools by use of an alternating phase-shifting mask, Appl. Opt. **45**, 281 (2006)

56-77 Z. Qiu, X. Wang, Q. Yuan and F. Wang, Coma measurement by use of an alternating phase-shifting mask mark with a specific phase width, Appl. Opt. **48**, 261 (2009).

56-78 Q. Yuan, X. Wang and Z. Qiu, Trefoil aberration measurement of lithographic projection optics based on linewidth asymmetry of the aerial image, Optik **121**, 1739 (2010).

56-79 N. Farrar, A. Smith, D. Busath and D. Taitano, In-situ measurement of lens aberrations, Proc. SPIE **4000**, 18 (2000).

56-80 L. Duan, X. Wang, A. Bourov, B. Peng and P. Bu, In situ aberration measurement technique based on principal component analysis of aerial image, Opt. Express **19**, 18080 (2011).

56-81 A. Burov, L. Li, Z. Yang, F. Wang and L. Duan, Aerial image model and application to aberration measurement, Proc. SPIE **7640**, 32 (2010).

56-82 K. Yamazoe and A. Neureuther, Aerial image calculation by eigenvalues and eigenfunctions of a matrix that includes source, pupil and mask, Proc. SPIE **7640**, N (2010).

56-83 K. Seong and J. Greivenkamp, Chromatic aberration measurement for transmission interferometric testing, Appl. Opt. **47**, 6508 (2008).

56-84 P. Ruffieux, T. Scharf, H.P. Herzig, R. Völkel and K. Weible, On the chromatic aberration of microlenses, Opt. Express **14**, 4687 (2006).

56-85 G. Boreman, Modulation Transfer Function in Optical and Electro-Optical Systems (SPIE Press, Bellingham, 2001).

56-86 T. Williams, The Optical Transfer Function of Imaging Systems (Inst. of Phys. Publ., Bristol, 1999).

56-87 C. Williams and O. Becklund, Introduction to the Optical Transfer Function (Wiley, New York, 1989).

56-88 G. Boreman and S. Yang, Modulation transfer function measurement using three- and four-bar targets, Appl. Opt. **34**, 8050 (1995).

56-89 A. Daniels, G. Boreman, A. Ducharme and E. Sapir, Random transparency targets for modulation transfer function measurement in the visible and infrared regions, Opt. Eng. **34**, 860 (1995).

56-90 A. Tzannes and J. Mooney, Measurement of the modulation transfer function of infrared cameras, Opt. Eng. **34**, 1808 (1995).

56-91 S. Reichenbach, S. Park and R. Narayanswarmy, Characterizing digital acquisition devices, Opt. Eng. **30**, 170 (1991).

56-92 B. Tatian, Method for Obtaining the Transfer Function form the Edge Response Function, JOSA **55**, 1014 (1965).

56-93 V. Nuzhin, S. Solk and A. Nuzhin, Measuring the modulation transfer functions of objectives by means of CCD array photodetectors, J. Opt. Techn. **75**, 111 (2008).

56-94 T. Li, H. Feng and Z. Xu, A new analytical edge spread function fitting model for modulation transfer function measurement, Chin. Opt. Lett. **9**, 031101 (2011).

56-95 M. Wernick and G. Morris, Effect of spatial coherence on knife-edge measurement of detector modulation transfer function, Appl. Opt. **33**, 5906 (1994).

56-96 M. Müller and G. Brakenhoff, Characterizing of high-numerical-aperture lenses by spatial autocorrelation of the focal field, Opt. Lett. **20**, 2159 (1995).

56-97 G. Waldman, J. Wootton and D. Holder, Imaging transfer function in the Fresnel approximation, Opt. Eng. **34**, 1818 (1995).

56-98 F. de la Barriere, G. Druart, N. Guerineau, J. Taboury, J. Primot and J. Deschamps, Modulation transfer function measurement of a multichannel optical system, Appl. Opt. **49**, 2879 (2010).

56-99 R. Kingslake, Applied Optics and Optical Engineering, Vol I, (Academic Press, 1965) pp. 172.

56-100 F. Liu and J. Geary, Novel method for measuring the spectral transmission of a photographic lens, Opt. Eng. **47**, 113602 (2008).

56-101 E. Mei and D. Gallinger, Optical system spectral transmission measurements in the visi-

ble and infrared, IEEE AutoTestCon, 242 (1989).
56-102 A. Wiebe and W. Courville, Measurement of the spectral transmission of an optical system employing four aluminized surfaces, Appl. Opt. **4**, 99 (1965).
56-103 Y. He, P. Li, G. Feng, L. Cheng, Y. Wang, H. Wu, Z. Liu, C. Zheng and D. Sha, Design and analysis of a sub-aperture scanning machine for the transmittance measurement of large-aperture optical systems, Proc. SPIE **7849**, 78491K (2010).
56-104 E. Zalevsky, Radiometry and Photometry, in M. Bass (Ed.), Handbook of Optics II (McGraw Hill, New York, 1995), Chap. 24.
56-105 V. Hodgkin, Radiometry - General, in R. Driggers (Ed.), Encyclopedia of Optical Engineering (Marcel Dekker, New York, 2003), Vol 3, p. 2287.
56-106 R. McCluney, Introduction to Radiometry and Photometry (Artech House, Boston, 1994).
56-107 ISO Standard 9358 (1994), Optics and Optical Instruments - Veiling Glare of Image-forming Systems - Definitions and methods of Measurement.
56-108 S. Martin, Glare characteristics of lenses and optical instruments in the visible region, Optica Acta **19**, 499 (1972).
56-109 S. Matsuda and T. Nitoh, Flare as applied to photographic lenses, Appl. Opt. **11**, 1850 (1972).
56-110 A. Grammatin, L. Agroskin and R. Larina, Field Glare in Reflecting Microscopes, Sov. Journ. Opt. Techn. **36**, 180 (1969).
56-111 C. Roundy, Current Technology of Beam Profile Measurements, in F. Dickey and S. Holswade (Eds.), Laser Beam Shaping (Marcel Dekker, New York, 2000).
56-112 N. Hodgson and H. Weber, Laser Resonators and Beam Propagation (Springer, Berlin, 2005), chapter 24.
56-113 M. Sasnett and T. Johnston, Beam characterization and measurement of propagation attributes, Proc. SPIE **1414**, 21 (1991).
56-114 N. Reng and B. Eppich, Definition and measurements of high-power laser beam parameters, Opt. and Quant. Elect. **24**, S973 (1992).
56-115 W. Plass, R. Maestle, K. Wittig, A. Voss and A. Giesen, High-resolution knife-edge laser beam profiling, Opt. Comm. **134**, 21 (1997).
56-116 A. Siegman, W. Sasnett and T. Johnston, Choice of clip levels for beam width measurements using knife-edge techniques, IEEE Jour. of Quant. Electr. **27**, 1098 (1991).
56-117 T. Johnston, Beam propagation (M^2) measurement made as easy as it gets: the four-cuts methods, Appl. Opt. **37**, 4840 (1998).
56-118 J. Zhang, S. Zhao, Q. Wang, X. Zhang and L. Chen, Measurement of beam quality factor (M^2) by slit-scanning method, Opt. and Las. Tech. **33**, 213 (2001).
56-119 Y. Champagne, C. Pare and P Belanger, Method for direct measurement of the variance of laser beams, Opt. Lett. **19**, 505 (1994).
56-120 M. Porras, Experimental investigation on aperture-diffracted laser beam characterization, Opt. Comm. **109**, 5 (1994).
56-121 B. Eppich, A. Friberg, C. Gao and H. Weber, Twist of coherent fields and beam quality, Proc. SPIE **2870**, 260 (1996).
56-122 M. Duparre, B. Lüdge and S. Schröter, Online characterization of Nd:YAG laser beam by means of modal decomposition using diffractive optical correlation filters, Proc. SPIE **5962**, 59622G (2005).
56-123 O. Schmidt, C. Schulze, D. Flamm, R. Brüning, T. Kaiser, S. Schröter and M. Duparre, Real-time determination of laser beam quality by modal decomposition, Opt. Express **19**, 6741 (2011).
56-124 A. Mahler and R. Chipman, Polarization state generator: a polarimeter calibration standard, Appl. Opt. **50**, 1726 (2011).
56-125 Q. Zhan and J. Leger, Imaging ellipsometry for high-spatial-resolution metrology, Proc. SPIE **4435**, 65 (2001).
56-126 M. Mijat and A. Dogariu, Real-time measurement of polarization transfer function, Appl. Opt. **40**, 34 (2001).
56-127 M. Totzeck, P. Gräupner, T. Heil, A. Göhnermeier, O. Dittmann, D. Krähmer, V. Kamenov, J. Ruoff and D. Flagello, How to describe polarization influence on imaging, Proc. SPIE **5754**, 23 (2005).
56-128 J. Ruoff, M. Totzeck, Using orientation Zernike polynomials to predict the imaging performance of optical systems with birefringent and partly polarizing components, Proc. SPIE **7652**, T (2010).
56-129 B. Geh, J. Ruoff, J. Zimmermann, P. Gräupner, M. Totzeck, M. Mengel, U. Hempelmann and E. Schmitt-Weaver,

The impact of projection lens polarization properties on lithographic process at hyper-NA, Proc. SPIE **6520**, 65200F (2007).

56-130 T. Heil, J. Ruoff, J. Neumann, M. Totzeck, D. Kraehmer, B. Geh and P. Gräupner, Orientation Zernike Polynomials – a systematic description of polarized imaging using high NA lithographic lenses, Proc. SPIE **7140**, 714018 (2008).

56-131 R. Chipman, Polarization analysis of optical systems, Opt. Eng. **28**, 90 (1989).

56-132 R. Chipman and L. Chipman, Polarization aberration diagrams, Opt. Eng. **28**, 100 (1989).

56-133 M. Shribak, S. Inoue and R. Oldenbourg, Polarization aberrations caused by differential transmission and phase shift in high-numerical aperture lenses: theory, measurement and rectification, Opt. Eng. **41**, 943 (2002).

56-134 B. Daugherty and R. Chipman, Low polarization microscope objectives, Proc. SPIE **7652**, S (2010).

56-135 D. Döring and K. Forcht, Coating-induced wave front aberrations, Proc. SPIE **7100**, 15 (2008).

56-136 Y. Li, W. Shen et al, Reduction of coating induced polarization aberrations by controlling the polarization state variation, J. Opt. **13**, 055701 (2011).

56-137 S. Edlou, L. Sun and C. Synborski, Coating induced phase aberrations in a Schwarzschild objective, Proc. SPIE **7067**, 09 (2008).

56-138 J. Lesso, A. Duncan, W. Sibbett and M. Padgett, Aberrations introduced by a lens made from a birefringent material, Appl. Opt. **39**, 592 (2000).

56-139 www.wikipedia.com

Index

Handbook of Optical Systems: Vol. 5. Metrology of Optical Components and Systems. First Edition.
Edited by Herbert Gross.

d

j

k

l